THE DESIGN
AND CONSTRUCTION OF
ENGINEERING FOUNDATIONS

THE DESIGN
AND CONSTRUCTION OF
ENGINEERING
FOUNDATIONS

EDITED BY

The late F.D.C. Henry

BSc (Eng), PhD, FICE, FAmSocCE
FIStructE, FIMunE, FGS

SECOND EDITION

London   New York

CHAPMAN AND HALL

First published 1956
by E. & F.N. Spon Ltd
Second edition 1986 published by
Chapman and Hall Ltd
11 New Fetter Lane, London EC4P 4EE
Published in the USA by
Chapman and Hall,
29 West 35th Street, New York NY 10001

© 1986 F.D.C. Henry

Printed in Great Britain by
J.W. Arrowsmith Ltd, Bristol

ISBN 0 412 12530 7

British Library Cataloguing in Publication Data

Henry, F.D.C.
 The design and construction of engineering
 foundations.
 1. Foundations
 I. Title
 624.1′5   TA775

 ISBN 0–412–12530–7

Library of Congress Cataloging in Publication Data

Henry, F.D.C., 1916–1985
 The design and construction of engineering
 foundations.

 Includes index.
 1. Foundations. I. Title.
 TA775.H38   1986   624.1/5   85–5708
 ISBN 0–412–12530–7 ✓

CONTENTS

v

CONTRIBUTORS

A.W. Astill, BScEng, CEng, MIStructE
Department of Civil Engineering and Construction
Aston University, UK

K. Elson DipCE, PhD, CEng, MICE
Department of Civil Engineering
Brighton Polytechnic, UK

W.G.K. Fleming BSc, PhD, MICE
Cementation Piling and Foundations Ltd
Hertfordshire, UK

D.A. Greenwood, BSc, PhD
Cementation Civil and Specialist Holdings Ltd
Hertfordshire, UK

I. Greeves, CEng, FICE, MICE
John Mowlem & Co. Plc
London, UK

F.D.C. Henry, BScEng, PhD, CEng, FICE, FASCE, FIStructE, FIMunE, FGS
Department of Civil Engineering
Brighton Polytechnic, UK

G.S. Littlejohn, BSc, PhD, CEng, FICE, MIStructE, FGS

Schools of Civil and Structural Engineering
University of Bradford, UK

J.A. Purkiss BScEng, PhD, CEng, MIStructE, MICE
Department of Civil Engineering and Construction
Aston University, UK

I.A. Rennie, BSc, PhD, MICE
Structural Engineering Division
Shell (UK), London

A.P.S. Selvadurai, MSc, PhD, DIC, AMAmeric, SCE
Department of Civil Engineering
Carleton University
Canada

K. Starzewski, BSc, PhD, CEng, MICE
Department of Civil Engineering and Construction
Aston University, UK

W.J. Walley, MSc, CEng, MICE, MIStructE
Department of Civil Engineering and Construction
Aston University, UK

PREFACE TO THE SECOND EDITION

The primary aim in this second edition, as it was in the first, published in 1956, is to link the design of foundations to geology, soil mechanics and structural analysis. Since 1956 advances in research, theory, design and construction have been made at accelerating rates and it has been necessary to enlist co-authors to present specialized subjects. It is perhaps surprising that in foundation construction, where for many years traditional practice was followed, ingenuity has led to the development of new equipment and techniques, particularly during the past two decades. On the other hand, some research carried out several decades ago stands in good stead, even though it is much older than many practising engineers, who may regard it as antiquated and hence of little account. Amongst such work can be mentioned Faber's experiments on bearing stress distribution (1933), Richart's experiments on pad foundations (1948) which led to some revision of Talbot's investigations (1913), and Tison's work on erosion at bridge piers (1940). The investigation at the Building Research Station by Glanville, Grime and Davies (1935) on driving stresses in piles is a classic study. There are, of course, many others, including numerous early investigations by Karl Terzaghi. Research may not lead directly to advances in analysis, design or construction because of the complexity of the problem but even if it reveals little more than a better understanding of behaviour, then something has been achieved.

Engineers are sometimes accused of lack of vision, however, the extent to which this trait can be remedied by eduction and training is a subject for debate. Nevertheless, because of the infinite variety of situations and requirements experienced in foundation engineering, a further aim of this edition, as it was in the first, is to encourage breadth of outlook. Somewhere there will be a place for the various analyses and designs presented herein, rare though some may be today, and perhaps they may be modified to advantage. Analyses vary, for example, from advanced techniques for continuous foundations to simplified methods for preliminary assessment of bank stability. Several types of independent foundations are treated and a number of different forms of retaining walls are illustrated. The topics of strength and deformation of deep beams and wall panels may seem rather remote from practical problems of underpinning, yet for many structures an understanding of relevant research is desirable. The list of references is extensive and provides an opportunity to gain a broad perspective on the subject.

In relation to site geology, breadth of vision is profoundly important, particularly in the interpretation of site investigations, a theme demonstrated many times in this edition. In the present state of engineering geology, a geologist can seldom reveal all site problems of the engineer mainly because of inadequate knowledge of engineering. Equally the engineer can fail to brief the geologist completely because of lack of experience in geology. There is no immediate remedy for this state of affairs, only close collaboration in the design office and in the field can produce the individuals required in engineering. The numerous references in chapter one to problems arising from site geology will, I hope encourage engineers to seek specialist advice and provide an adequate brief. In the USA engineering geologists have practised for about a century: some of their work is described in the Berkley Volume (Paige, 1950) published by the Geological Society of America, a manual worthy of careful study. Many engineers believe that deficiencies in effective site investigation have increased recently as a consequence of a decline in quality of sampling and testing, largely the result of keen competitive tendering. This must surely be a temporary lapse, since the work is vital to projects, and the situation must be rectified in the near future and, furthermore, improved by much closer collaboration between designers and investigators.

References and examples have been drawn from many parts of the world. Inevitably the text is influenced by the various authors' experiences but the lessons and principles are of universal application. The analyses and design methods presented herein are largely framed for manual computation since it is considered that this is the best way of demonstrating principles, and, in any case, many problems do not merit computer programs. Some of the manual computations can be translated into programs and, in certain cases, programs are available commercially.

The editor must express his gratitude to the co-authors and publishers for their cooperation and patience in the preparation of this edition.

F.D.C. Henry
Sompting, Sussex
May 1985

OBITUARY NOTE

Sadly, during the course of producing this book, the editor, Dr Henry and two of the contributors, Mr Astill and Dr Rennie, died.

Chapter 1

SITE INVESTIGATIONS AND PRELIMINARY CONSIDERATIONS

F.D.C. Henry

INTRODUCTION

The objects of a site investigation are to assess the general suitability of the site, to enable a design to be prepared, to predict possible difficulties in excavation and construction and to cater for future natural and artificial changes in conditions. Apart from the factors discussed in Sections 1.1–1.3, information should be sought on access to the site, availability of water, electricity and other services, local sources of materials and disposal of waste. British site investigation practice is treated in BS 5930 (BSI, 1981a) and American practice in Manual and Report No. 56 (ASCE, 1976).

The necessity for adequate site investigations requires no emphasis where major projects are concerned, but failures and difficulties encountered on minor works demonstrate that such studies can rarely be omitted on any project. For example, underground air-raid shelters were constructed in boulder clay at two sites in Middlesex in 1938 and running sand was encountered at both. A housing estate was under construction, principally on sand, in Lancashire in 1947 when a stratum of peat was discovered less than a metre beneath the surface. In both cases the difficulties encountered during construction could have been wholly or partly eliminated by minor alterations in the site layout or in design if preliminary site investigations of the most inexpensive character, such as hand-auger borings, had been made. The cost of site investigations varies with the type of project and its magnitude and with site conditions. It is commonly 1 or 2% of the total cost of the work but the range is broadly between 0.1 and 10%. For some projects it is usual to make preliminary site investigations, including borings, and these are followed by further investigations when the requirements can be specified in greater detail. On important projects it is now usual to make much more extensive investigations than formerly. Consideration should be given to

1

extraneous effects which may influence soils and foundations in the future, such as leakage of corrosive fluids from tanks or pipes. Thorough site investigations may well be repaid by relatively low tenders since contractors have then less risks to face. Nevertheless, in some cases site investigations may save little or no money but in others they may eliminate heavy expenditure on contingencies, indeed on many projects inadequate site investigations, including interpretation, are the most significant factors leading to overspending.

Unfortunately, cases still occur in which insufficient consideration has been given to methods of construction, site layout and preliminary works – at least broadly, if not in detail – in the design stage. Thus, for example, the influence of construction operations in disturbing sensitive clays or in setting up excess pore-water pressure sometimes has been ignored, with consequent adverse effects on excavation and construction operations. Experience shows that failures and difficulties which arise during construction are often the result of errors perpetrated by more than one of the parties concerned. For example, a designer may not be sufficiently specific in stating requirements for adequate site investigations. The site investigator may omit to make certain tests or recommendations which are desirable although perhaps unusual. Finally, the main contractors or the foundation sub-contractors may fail to take into account site conditions peculiar to the area but with which they might reasonably be expected to be familiar. The result is at least delay and additional expense. These remarks are not intended to cast aspersions in any particular direction but to call attention to the desirability of consultation between all concerned with design and construction. Ideally, consultation should commence in the early stages of design and, although this is rarely feasible, it has been effected on some large schemes. Experience demonstrates that it is desirable on many smaller projects. A further aspect of foundation engineering which requires early consideration is research into the behaviour of foundations and structures during construction and in service. When such investigations are considered worthwhile, it is essential to plan ahead in order that equipment can be fabricated or purchased and incorporated in the most appropriate locations as construction proceeds.

For some projects, investigations continue throughout construction either as records of geological and other details revealed as work proceeds in, for example, tunnels and dam foundations, or as site instrumentation to control construction processes in, for example, the formation of embankments over alluvium. Furthermore, in some cases investigation can be said to be never completed, since additional records may be made during service of a structure in connection with, for example, abnormal deformation of a tunnel lining or leakage beneath a dam.

Studies made by the Building Research Establishment in the United Kingdom indicate that the major factors leading to poor quality in the construction industry are unclear and missing information, the responsibility of the designer, and lack of care on the part of the contractors. These findings reinforce the Author's views expressed above that insufficient attention is paid to many aspects of design and construction, a failing tantamount to negligence. No doubt there are several inexcusable reasons for this state of affairs, such as undue haste for completion, lack of appreciation of extraneous factors which may influence the project and absence of pride in one's work. Further support is offered by Professor R.B. Peck in the Laurits Bjerrum Lecture of 1980 in which he ascribed most failures of dams to poorly understood, unquantifiable or neglected mechanisms and said that as long as the myth persists that only what can be calculated constitutes engineering, engineers will lack incentive or opportunity to apply the best judgement to the crucial problems that cannot be solved by calculation. With the advances in knowledge won over the last half-century there can be little or no excuse for many of the blunders suffered in the construction industry – occurrences which cost enormous sums every year. Perhaps this homily may seem out of place in a text book, yet one cannot be other than disturbed by events brought to notice, sometimes too late for effective remedy, and a plea is made to pause and consider, in design, whether all relevant factors have been taken into account and appropriate specialist advice has been sought and, in construction, whether the appropriate quality is being attained. Naturally, the Author is concerned mainly with inadequacy in geotechnical engineering and in the following pages reference is made repeatedly to this topic.

Cases of running sand at two sites in Middlesex were quoted above as examples of the merit of even small-scale site investigations. In fact the issues are much wider since running sand had been encountered in sewer headings and trenches and other excavations in the district. Regrettably, specialist geological information was not sought in those days over four decades ago. Yet study of 1/10 560 geological maps available at that time would have provided clear indications of problem sites and the extent of site investigations required. Much of the district was covered by up to 5 m of boulder clay overlying some 15 m of glacial sand and gravel, in turn overlying the London Clay. The moral today is that specialist information on geology, soils, hydrology and other factors, should be sought and may even be readily available on maps and in bulletins and memoirs.

1.1 GEOLOGICAL CONSIDERATIONS

In this Section the relationship between civil engineering and some of the branches of geology are studied. Recording and mapping of geological data are discussed in three reports (Anon, 1970b; Anon, 1972; Chaplow *et al.*, 1977) and in other publications such as those of Knill and Jones (1965) and Dearman and Fookes (1974). A series of papers dealing with a range of topics in engineering geology has recently been published (Chandler *et al.*, 1981). A review of problems in engineering geology has been presented by Henkel (1982).

1.1.1 Chemical and physical stability of rocks

The aggregate minerals of rocks may be altered by heat, stress and permeating fluids. Improvement in chemical stability by natural processes such as metamorphism is not generally a matter for concern in engineering, whereas deterioration in stability is often of considerable importance. Weathering comprises distinte-gration and decomposition and may lead to problems in excavation and construction: decomposition requires the presence of mobile water together with gases and solids in solution. Chemical weathering may involve the formation of new minerals by hydration, oxidation or carbonation and the removal of some of the constituents by leaching. The change of anhydrite to gypsum is an example of hydration, and in tunnels the floor has heaved and timbers have been crushed by the swelling which takes place when moisture gains access to the anhydrite as excavation proceeds. Quigley *et al.* (1973) and Penner *et al.* (1973) investigated heave of black pyritic shale: it is believed that this is caused by the production of sulphates, such as gypsum and jarosite, as a result of bacterial oxidation of ferrous sulphides. Preventative measures consist of coating newly exposed surfaces with gunite or bitumen to retard ingress of oxygen.

Chemical weathering proceeds very slowly in arid regions but in hot, humid climates it develops comparatively rapidly. It takes place principally in the surface layers of rocks, where the depth of decomposed rock may range from 1 or 2 m to 100 or 200 m. The ultimate product of surface weathering is a residual soil, sometimes containing residual boulders. In glaciated regions, however, the mantle of decomposed rock has been largely removed although it is often replaced by glacial deposits. Since faults, joints and other fissures form channels of access for water they often lead to zones of decomposed rock. Hamrol (1961) has invented parameters to express the degree of weathering of rocks and Fookes *et al.* (1971) presented a scheme of classification for weathering.

The following examples illustrate the influence of geochemical changes on engineering projects. The foundations for the main 207 m span of the Penrose Avenue Cantilever Bridge, Philadelphia, USA, were sunk through silt, sand and gravel and 6 m of decomposed rock to sound gneiss (Masters, 1951). The decomposed rock consisted partly of kaolinite, derived from the feldspar in the gneiss, and consolidation under heavy foundation loads would have taken place if the piers had been founded above the decomposed zone. Wash borings taken 22 years earlier failed to distinguish the softer rock from the intact gneiss. The piers for the lighter continuous girder approach spans were founded on piles driven to the sand and gravel. The United Engineering Center, New York, is founded on Manhattan Schist and the safe bearing capacity was assessed as $2000 \, kN \, m^{-2}$ on intact rock and $800 \, kN \, m^{-2}$ on decomposed rock (Bast, 1960). Davis Dam on the Colorado River is constructed partly on a fault belt and the fractures have encouraged decomposition of granite and rhyolite. The decomposed zones were excavated and back-filled with mass concrete (Bahmeier, 1950).

The properties of the decomposed granite of the batholith in the Hong Kong region have been discussed by Lumb (1962, 1965). In this case decomposition is ascribed to weathering whereas alteration in the granite batholith of Devon and Cornwall has been considered to be due to pneumatolysis – a late phenomenon of intrusion – although it has been suggested that it may be the result of groundwater movement in geologically more recent periods. In both cases the product is broadly a mass of kaolin, quartz and muscovite, interspersed with masses of partly decomposed granite, but in the case of weathering the effect of leaching is superposed on the alteration.

At Bombay Harbour, the basaltic Deccan Trap forms a stepped sea-bed due to degradation of successive lava flows with a westerly dip of 5° (White *et al.*, 1961). The rock varies considerably in hardness and natural channels incised in it are filled with disintegrated rock and conglomerate. Depth to sound rock is, therefore, very variable and in such situations close borings and probings are essential. The overlying material is an unconsolidated, viscous deposit varying in thickness up to 4 m. Since it was not possible to drive piles into the rock for the construction of a marine oil terminal, holes were drilled to accommodate the feet of precast columns which were grouted in position. Drilling was effected by a hammer-grab with two or three strong jaws which were locked open to serve as a chopping bit and unlocked to lift the broken material. Near Portland, Oregon, problems were presented at a site for a hospital by the presence of rubble-filled channels – collapsed tubes in basaltic lava, of structural not chemical origin – up to 12 m wide and 18 m deep.

At the Gibralter Dam on the Santa Ynez River, California, the foundation rock consisted of sound massive sandstone containing fossil shells which had provided the calcite for cementing the sand, but overnight the surface material slaked to become rock flakes, sand and chalky dust (Forbes, 1951). In San Francisco a fill consisted of weathered fragmented sandstone (Forbes, 1951). About 10 years after placing, a number of boreholes revealed that it had been reduced to a compact, relatively impervious sandy clay, in which kaolinization of feldspar had taken place.

Changes which occur in the properties of argillaceous deposits are commonly the result of physical instability. The geological history of such rocks plays an important part in determining the degree of stability. The Oahe Dam is constructed on relatively uncemented shales and marls with seams of bentonite (Johns *et al.*, 1963). These rocks disintegrate rapidly when exposed and, to prevent this, sealing layers of bitumen and cement mortar were placed shortly after excavation. In order to reduce differential heave due to the removal of overburden, foundation slabs were anchored by steel bolts penetrating at least 6 m into the rock. The process of slaking is illustrated by Mayer (1963) who refers to a sample of air-dried schist which literally exploded on immersion in water at Foum el Geiss Dam, Algeria. The phenomenon can be demonstrated by placing a piece of air-dried clay in water. Capillary forces draw water into the soil and air is trapped in the centre. The compressive forces imposed on the air are resisted by tensile forces in the soil structure which may cause rupture. The process is progressive and ultimately the clay becomes a soft mass. Weathering of exposed surfaces of shale has been prevented or retarded by spraying a water emulsion of a synthetic resin (Aerospray 52 – American Cyanamid Co.).

Squeezing and swelling are physical phenomena which occur in argillaceous rocks and decomposed igneous rocks (Baldovin and Santovito, 1973; Lee and Lo, 1976; Lindner, 1976). Both are most apparent in tunnelling (Terzaghi, 1945). Squeezing is generally due to a slow viscous flow of material towards the tunnel at almost constant water content and without perceptible volume increase and is a manifestation of creep. Swelling is accompanied by expansion of the clay and leads to greater stress on tunnel linings than does squeezing. The greatest swelling tendencies are found in clay and decomposed rock containing a high percentage of minerals of the montmorillonite group. A test for swelling clay is that a fresh sample increases in volume by more than 2% after immersion in water. Squeezing is due to plastic flow under stresses of magnitude less than the maximum resistance to shearing of the soil: a phenomenon which may have to be taken into account also in the assessment of forces on retaining walls. Swelling is the expansion, accompanied by increase in moisture content, which occurs when the stress on a specimen of clay is reduced. In tunnels the water required to permit the swelling is drawn from the clay behind the tunnel walls and not from the moist air in the tunnel. Swelling rocks are always at least moderately dense, having the consistency of stiff clay.

At one point, the Malgovert Hydro-Electric Intake Tunnel from the Tignes Dam, France, ran into swelling ground composed of badly crushed shales located near a fault (Pelletier, 1953). The shales were initially fairly dry, with the consistency of stiff clay and driving was easy, but they became wet a few days after excavating, high stresses were developed which buckled steel supports and at one time the track rose 1.2 m in a few hours. These conditions necessitated a diversion of the tunnel. In Israel, high stresses have been observed in pipes buried in expansive clays due to swelling arising from seasonal variations in moisture content of the soil and from irrigation (Kassiff and Zeitlen, 1962): expansive clays are discussed further in Section 1.4

Apart from causes such as chemical changes in clays or ingress of moisture, ground movements, including swelling, inevitably take place as a result of stress relief following excavation. Thus, for bridge abutments in Gault clay near Sevenoaks, Kent, the carriageway slab functioned as a strut between the contiguous bored pile walls, thus restraining lateral expansion, and the slab was designed with joints to cater for upward movement. The characters of swelling rocks and the problems presented by them are currently (1980) being studied by a commission of the International Society for Rock Mechanics.

Toxic and inflammable gases may be encountered in some rocks and additional ventilation or other measures may be required in confined excavations. Shale is often associated with coal-bearing strata from which methane may emanate. Small quantities of methane were encountered in excavations for the Forth Road Bridge and two explosions occurred (Anderson *et al.*, 1965). The ground-surface emission of methane through sandstones at Barnsley, Yorkshire, has been studied by Carden *et al.* (1983) and is considered to emanate from Coal Measures straka. At Bootle, Lancashire, the Rimrose Brook sewer tunnel passed near an old refuse tip and noxious gases were excluded from the tunnel by compressed air at $35\,kN\,m^{-2}$. When a 1 m diameter sewer tunnel was driven beneath an old tip near Dudley, two men died as a result of gases, probably CO and CO_2, entering the workings. Glossop (1954) has mentioned an unusual example of hazard in the use of compressed air at Dagenham, Essex, where it passed through peat, which was subjected to oxidation, and then emerged in a neighbouring shaft where the deficiency of oxygen led to fatalities. Methane and hydrogen sulphide, derived from Lower Carboniferous coals and oil-shales, were encountered in tunnels for a sewerage scheme in Edinburgh completed in 1976. Recent fatalities at Carsington, Derbyshire, have drawn attention to concentrations of carbon dioxide arising from reaction of calcium carbonate, in particular in limestones, with rainwater, the potentiality of which possibly being accentuated by humic acids derived from organic matter such as peat. The Mosul Tunnel, Iraq, passes through limestone, marl and clay. Gypsum occurs in the clay and sulphurous waters and free hydrogen sulphide are present in the limestone. Aluminous cement was employed to avoid corrosion of the concrete. The Mono Craters Tunnel, California – supplying water to Los Angeles – passes through a range of volcanic cones of Miocene to Pleistocene age. Carbon dioxide was encountered during driving but was dispersed by increasing the volume of fresh air supplied to the face.

The presence of gases arising from decomposition of fill is mentioned in Section 1.4.4 and consideration should be given to the possibility of gases entering buildings constructed on fill or on other suspect subsoil. For example, carbon dioxide seeped into houses built over old mine workings at Willington, Durham. Former industrial sites, such as gas works, may be contaminated with chemicals which could lead to corrosion and toxic and carcinogen emissions. The proceedings of a conference on reclamation of contaminated land have been published (SCI, 1980). Papers on similar topics have been presented at a colloquium (Smith *et al.*, 1982). At Hartford, Illinois, leakage from pipelines led to accumulation of petroleum products above groundwater which has been the source of noxious emissions and has caused fires: remedial measures comprised containment and floating recovery pumps (Mathes, 1982).

1.1.2 Structural geology in relation to engineering

The topics to be considered under this heading are joints, faults, natural and artificial slopes, superficial structures and disturbance of rocks under glacial and peri-glacial conditions. Joints commonly exhibit fairly uniform patterns and a prerequisite for important projects is a survey in which observations are made of the strike, dip and spacing of joints in the area. These data are plotted on polar diagrams, as described, for example, by Nevin (1949) and Terzaghi (1965), and the general characteristics of each set of joints can thus be determined. The interpretation of such data is discussed by Steffon *et al.* (1975). The application of terrestrial photogrammetry for mapping joints has been described by Moore (1974). Joints are important in engineering in several respects. The stability of strata exposed on hillsides or in cuttings depends partly on the frequency and orientation of the joints. In the process of excavating, including tunnelling and quarrying, advantage should be taken of the joint pattern to facilitate the extraction of rock by hand, machine or blasting. Zones of intensive jointing are likely to be very unstable in excavations and may be weathered to considerable depths. The flow of water in many rocks is controlled more by joints than by the proportions of the pores and an appreciation of joint width, frequency and pattern is essential to the control of the flow of water in relation to excavations involving, for example, pumping or cementation. Studies of the characteristics of fissures and shear zones in stiff overconsolidated clay were made at two sites in West Pakistan for design purposes, including the detection of possible loci of weakness in dam foundations and deep excavations (Fookes, 1965; Fookes and Wilson, 1966; Skempton, 1968). Joints and fissures in the London Clay have been investigated by Skempton *et al.* (1969) and fissures in Cretaceous sediments have been studied by Fookes and Denness (1969). The occurrence of joints in the field has been discussed by Doughty (1968) and Hancock (1968) and the mechanics of jointed rock by Lajtai (1969), Brown and Trollope (1970), Goodman *et al.* (1968), Hudson and Priest (1979), Hoek and Brown (1980) and Bandis *et al.* (1981). Seismic investigations have been employed to study fracturing in the Chalk (Grainger *et al.*, 1973).

Problems concerned with the stability of rocks can be approached by considering the mass as a continuum, in which case conventional soil mechanics analyses are commonly applicable, or as a discontinuum, in which case particular attention is paid to the shear strength of joints – a technique termed rock-mass analysis. Generally, this latter method is more appropriate and reference should be made to the publications cited immediately above and to those cited below in relation to stability of slopes.

A fault may appear as a fine crack, say, 2 mm wide, with almost plane interfaces or a zone of rock either is simply disturbed or is shattered to form fault breccia, mylonite (granulated rock) or gouge (rock flour). If the zone is very wide it is known as a shatter belt or shear zone and may be 30 m or more in width. Fault shattered rock is vulnerable to weathering. Sometimes faults, including brecciated zones, may be healed by autogenous welding or cemented by mineralization and may become as strong as intact rock. The presence of major faults is generally known from geological mapping but, although minor faults may be detected in boreholes, their presence is often revealed only as excavation proceeds. The extent of continuity in faults is demonstrated by the fact that zones of fracture, indicated at the surface by small valleys, were found in tunnels for Scottish Hydro-electric schemes at depths as great as 600 m. Features sometimes associated with faulting are intraformational shears, that is zones of shearing parallel to bedding (Salehy *et al.* 1977).

A fault can be classified as dead when no further movement is anticipated, dormant if it is considered liable to recurrent movements and active if the movements are perennial. For example, slight movement occurred, accompanied by a minor earthquake, at the Irwell Valley Fault, Lancashire, in 1889 and at the Craven Faults, Yorkshire, in 1944 – both can be classified as dormant. The San Andreas Fault, California, is active since there is a progressive creep in a north-westerly direction of the area lying to the south-west of the fault, and in 1906 a horizontal displacement of the order of 3 m was associated with the San Francisco earthquake. Earthquakes are due primarily to tectonic activity, sometimes accompanied by vulcanicity. In Great Britain, earthquakes of varying frequency and widespread distribution have been noted in historical times and are associated with sudden movements at faults but, since the maximum intensities seldom lead to more than fallen chimneys and cracked walls, no special precautions are taken to cater for seismic effects. Fault movements in Great Britain today probably never exceed a few millimetres at ground surface unless prompted by mining activity. Elsewhere, for example in the vicinity of the circum-Pacific and Euro-Asian mobile belts, displacements up to at least 15 m have been recorded. A tunnel beneath San Francisco Bay lies between the San Andreas and Haywards Faults. An initial seismic survey yielded details of the properties of the rocks and, subsequently, an earthquake detecting and recording system was installed, with geophones permanently located in boreholes. About 65 km north of Los Angeles, ground surface has risen by roughly 250 mm over a distance of nearly 200 km on a crescentic area trending east/west astride a section of the San Andreas fault zone. Movement at faults can be induced by extraneous factors, such as oceanic tides and changes in atmospheric pressure, in association, of course, with accumulated tectonic forces. In Great Britain, underground mining is often responsible for superficial movements, generally with a time lag on mining activity. In 1959 earthquakes were initiated by impounding water at the Hsinfengkiang Dam in China and during the last two decades extraction of groundwater in Arizona has caused progressive movement at a fault between Tucson and Phoenix. Dowrick (1981) has discussed earthquake risk and design for seismic effects in the United Kingdom, with particular reference to offshore regions.

The following is an example of the influence of a fault on a bridge project. The Kincardine-on-Forth, Scotland, road-bridge consists of a number of 15 to 30 m spans with a mid-stream swing-span 109 m long. Preliminary borings confirmed the existence of a fault near mid-stream. On the northern half of the river the piers were founded direct on sandstone bedrock, at depths less than 6 m below bed-level. On the southern half the throw of the fault placed bedrock at a depth such that the piers were carried on piles penetrating 9 to 18 m below bed-level and bearing on a stratum of gravel. The Kincardine bridge was completed about 50 years ago, but a recent example is afforded by the construction of a bridge on the Stonehaven bypass, Scotland, one foundation of which was located over a 3 m wide gouge-filled fault, a component of the Highland Boundary Fault zone. The proposed piled foundation was redesigned as a spread foundation.

Problems which may arise from the presence of a fault include movement following excavation, inflow of water to excavations, low bearing capacity of decomposed rock in a shatter belt, juxtaposition of rocks with different bearing capacities, earthquakes and possible horizontal and vertical displacements under tectonic forces. If the presence of a fault is known, it is not good practice to erect a structure over it without taking special precautions in design, unless it has been healed or mineralized and proved to be as strong as intact rock or unless it has been proved dead. It is often difficult to ascertain whether a fault is dead or dormant, but where considerable capital expenditure is involved, it is prudent to assume that it is dormant. The determination of the condition of a fault involves consideration of geological, historical and seismological evidence.

The removal of overburden, either by natural processes of degradation or by excavation for engineering purposes, leads to uplift, commonly termed rebound, and also to horizontal movements on the flanks of the denuded area. The nature of rebound and its influence on engineering works has been discussed by Nichols (1980). In order to minimize uplift, overburden load must be restored as soon as possible, either by dead load of construction or by rock bolts and other anchorages.

The stability of hillsides and cuttings can be considered according to the following classification. (1) (a) Rock creep, (b) talus creep, (c) soil creep. (2) (a) Mudflow, (b) slip or slump (c) debris slide, (d) rock slide. (3) (a) Debris fall, (b) rock fall. Some of these phenomena are illustrated in Fig. 1.1. Creep of rocks and soils occurs mainly at the surface and is known as terminal creep: it can be detected by studying the attitude of walls, posts and tree trunks (Fig. 1.1). At a site in Middlesex, it was necessary to underpin houses constructed on a clay slope of about 1 in 10 on account of soil creep, and a sewer about 1.5 m deep was relaid owing to opening of joints. Rock can also creep by movement on bedding planes

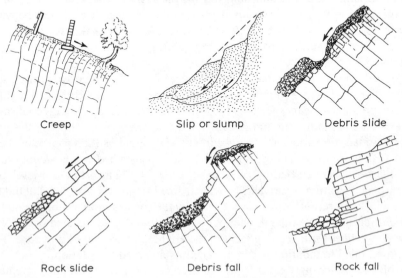

Creep   Slip or slump   Debris slide

Rock slide   Debris fall   Rock fall

Figure 1.1   Types of landslide and related phenomena.

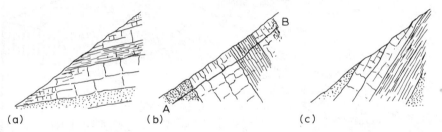

Figure 1.2 Stable and unstable strata. (a) Likely to be unstable; (b) stable (unless heavily jointed so that slip can occur on continuous joint plane (A–B); (c) stable.

(Fig. 1.2a). Talus creep is the intermittent and irregular movement of scree downhill. Mudflows are discussed by Vallejo (1979) but are not common in Great Britain, although streams of fluid peat, originating in bog bursts, occur occasionally (Clarke, 1961). Solifluction on slopes has been studied by McRoberts and Morgenstern (1974) and Harris (1977). Slip or slump has received much attention in soil mechanics and is discussed further in Section 3.2. Debris slides and falls involve the movement of disintegrated rocks whereas rock slides and falls are connected with movement of blocks. Descriptive terms for soil and rock slopes are illustrated in Figs. 1.3(a) and (b).

If the dip of strata is greater than the slope of a hillside, then the strata are relatively stable since there is no free outfall for the rock. If the beds dip into the hill, the strata should be stable unless the rock is markedly jointed. These features are shown in Fig. 1.2. Rock movements are most likely when the dip is steep but just less than the slope of the hillside; the danger being increased where some of the beds have low shear resistance. Although not on a hillside, shearing in a 3 mm seam of clay – undetected in initial borings – within a thin bed of shale was considered to be the cause of failure of a wall at Wheeler Dam, Tennessee (Anon, 1962). The stability of strata may be affected by the introduction of heavy foundation loads on or adjacent to slopes. Terzaghi (1962) deals at length with the stability of slopes on hard unweathered rock. The stability of slopes is discussed by Henkel (1961), Mencl (1966), John (1962; 1968), Chandler (1970), Eyre (1973), Hoek (1973), Hoek and Bray (1977), Hoek and Londe (1974), McGown *et al.* (1974), Hutchinson (1972), Franklin and Denton (1973), Jennings and Robertson (1969), Fookes and Sweeney (1976), Attewell (1977a) and Voight (1978, 1980). Stabilization of a rock face at Tremadog, Wales, has been described by Mercer (1982). The disintegration of rock surfaces in cuttings can be prevented by revetments of thin skins of brick, masonry or concrete tied to the rock (Fig. 1.3d). Stabilization of rocks on slopes can be effected by drilling into the face and restraining the rock by long bolts or cables grouted in position in conjunction with surface plates or walings or revetments. Rock bolting has been discussed by Thomas (1962), Leech and Pender (1961), Price and Knill (1967), Golder (1970), Littlejohn (1975), Littlejohn and Bruce (1975, 1976). At the Hanson Dam on Green River, Washington, USA, 25 mm diameter bolts up to 12 m long, with

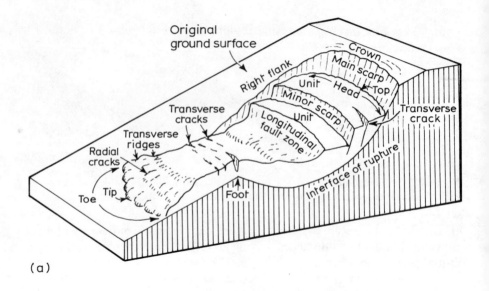

Original
ground surface

Crown

Main scarp

Right flank

Unit   Top

Head

Minor scarp

Transverse
cracks

Minor scarp

Unit

Longitudinal
fault zone

Transverse
crack

Transverse
ridges

Radial
cracks

Toe   Tip

Foot

Interface of rupture

(a)

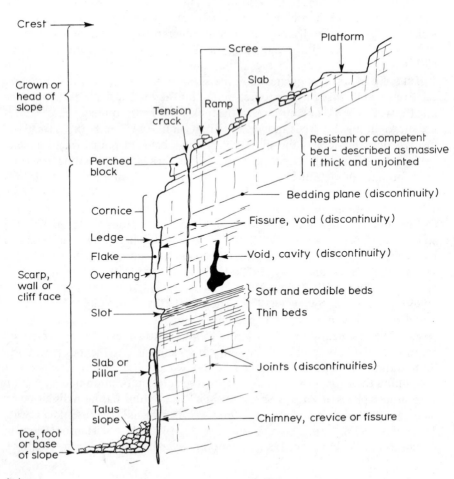

Crest

Platform

Scree

Crown or
head of
slope

Slab

Ramp

Tension
crack

Resistant or competent
bed – described as massive
if thick and unjointed

Perched
block

Bedding plane (discontinuity)

Cornice

Fissure, void (discontinuity)

Ledge

Flake

Void, cavity (discontinuity)

Overhang

Soft and erodible beds

Scarp,
wall or
cliff face

Slot

Thin beds

Slab or
pillar

Joints (discontinuities)

Talus
slope

Toe, foot
or base
of slope

Chimney, crevice or fissure

(b)

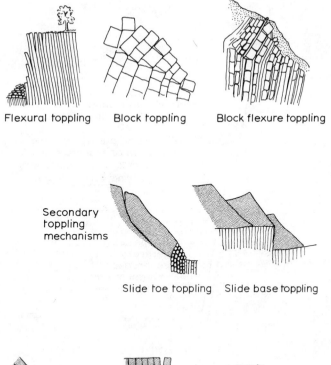

Flexural toppling   Block toppling   Block flexure toppling

Secondary
toppling
mechanisms

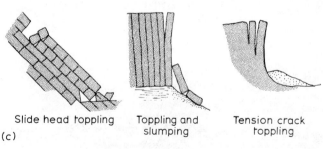

Slide toe toppling   Slide base toppling

Slide head toppling   Toppling and   Tension crack
                      slumping   toppling

(c)

Figure 1.3 (a) Descriptive terms for failures in soil slopes (after Eckel, 1958). (b) Descriptive terms for rock slopes (after Fookes and Sweeney, 1976). (c) Types of toppling failures (after Goodman and Bray, 1976).

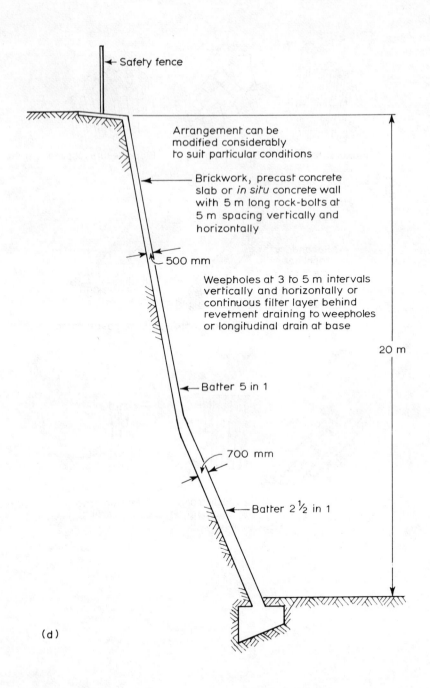

Safety fence

Arrangement can be
modified considerably
to suit particular conditions

Brickwork, precast concrete
slab or *in situ* concrete wall
with 5 m long rock-bolts at
5 m spacing vertically and
horizontally

500 mm

Weepholes at 3 to 5 m intervals
vertically and horizontally or
continuous filter layer behind
revetment draining to weepholes
or longitudinal drain at base

20 m

Batter 5 in 1

700 mm

Batter 2 ½ in 1

(d)

Figure 1.3 (d) Generalized arrangement of revetment to cutting in chalk in Sussex.

200 mm × 200 mm washers, were wedged and grouted in drill holes in the rock face (Christman, 1960). Clay cliffs near Los Angeles were stabilized by inserting reinforced concrete dowels ∼ 1.2 m diameter and 6 m long in holes bored 3 m into sandstone bedrock. The dowels thus projected 3 m into the clay above the surface of sliding.

The integrity of a rock mass can be severely damaged beyond the limits of excavation if powerful explosive charges are employed and the stability of faces may be best secured by using low power explosive, presplit blasting, hydraulic fracturing or other means, although this may retard progress and involve secondary fragmentation of large detached blocks. Excessive fracturing may reduce the capacity of rock bolts.

Near Northampton, relatively large-scale geological structures, unrelated to the regional structure, have been observed in beds of sandstone, limestone and ironstone overlying clay (Hollingworth *et al.*, 1944). These superficial structures include camber, gulls (widened joints partly filled with detritus) and valley bulges (Fig. 1.4). The processes leading to the formation of these structures include sub-surface erosion by the washing out of material at the spring line and plastic flow of underlying beds of clay towards the valley as a consequence of the removal of lateral support by degradation. In the valley bottom the clay is squeezed and intensely folded. These structures are relatively widespread in Great Britain, although little attention has been paid to them: much of the deformation probably occurred in the Pleistocene Period.

Gulls commonly occur in valley flanks where competent beds, such as limestones and sandstones, overlie incompetent beds, such as clays. Generally clays deform in a more or less ductile manner but the competent beds deform by increasing the gape of joints. In the Bristol area, houses have been founded on beams, instead of strip footings, spanning gulls in the Blue Lias which overlies Rhaetic sandstones. The gulls were conspicuous only below a depth of about 1.5 m and were packed with concrete before the foundations were placed. Indeed in the Bristol area and throughout the Cotswolds, the formation of cambers, gulls and valley bulges is widespread in the alternating sequences of limestones and argillaceous rocks of Jurassic age. Cambering sometimes leads to considerable displacement of intact blocks of rock between joints, often accompanied by

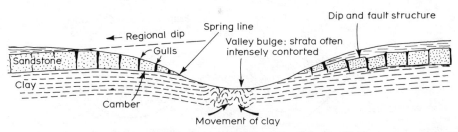

Figure 1.4 Superficial structures.

marked disorientation, as can be seen, for example, around Maidstone in the Hythe Beds sandstones which overlie the Atherfield Clay.

Reports of valley bulges appear to be scarce in engineering literature although they are relatively common features. In a valley in Sussex boreholes revealed fragmented sandstone which was interpreted as gravel. However, this may have been breccia formed in a valley bulge, particularly if the fragments were angular: valley bulges have been revealed in excavations elsewhere in the region. At Thorney, in the Colne valley west of London, a tunnel in Tertiary clays encountered a mass of upthrusted sandy beds bearing water under artesian head. This piercement structure displays a maximum uplift of the order 20 m and lies beneath a buried valley filled with Quaternary silts, sands and gravels. Although valley bulges are commonly produced by creep of valley flanks which induces buckling in the valley floor, they may be accentuated or produced solely by frost heave or by artesian water pressure when capped by relatively impermeable beds.

Various types of gravity collapse structures have been described by Harrison and Falcon (1936), as shown in Fig. 1.5: these appear to progress from stage 1 to stage 3 with time. During construction of the approach viaducts for the Medway Bridge, it was observed that fissures were pronounced in the chalk bedrock, mainly parallel with the river (Kerensky and Little, 1964). This suggests that

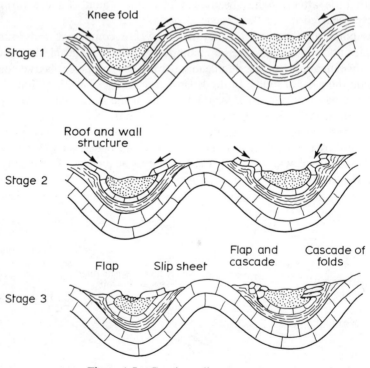

Figure 1.5   Gravity collapse structures.

either a tectonic structure governs the course of the river or minor cambering had occurred on the valley sides, possibly accentuated by frost action during the Pleistocene. Cement grout was injected to stabilize the rock.

Examples of the phenomenon of toppling have been given by de Freitas and Watters (1973). Four types of failure have been recognized, the fourth type being divided into five sub-types (Fig. 1.3c). This occurs where blocks of rock become unstable under gravitational force as a result of removal of underlying support and tilting. It will be seen in Fig. 1.1 that the near-surface blocks subjected to terminal creep will tend to topple, particularly if downhill lateral support is absent. The huge blocks of sandstone on the north flank of Otley Chevin, Yorkshire, have suffered toppling failure. The mechanics of toppling have been discussed by Goodman and Bray (1976), Wyllie (1980) and Evans (1981) and examples in Tasmania have been analysed by Caine (1982) Design charts have been prepared by Zanbak (1983).

Movements of viaduct and bridge piers towards a valley or a cutting have created problems which have been met by underpinning or the construction of inverted arches between the foundations. At Klosters, Switzerland, a reinforced concrete railway bridge was constructed in 1930 over the River Landquart (Haefeli *et al.*, 1953). The structure comprises an arch of 30 m span and 8 m rise with two 9 m approach spans on each side. In 1938 several cracks appeared in the bridge due to soil creep on one side of the valley; the other side was stable. A strut was suspended horizontally from the arch to take the thrust between the two abutments. A retaining wall, horsehoe-shape in plan, was constructed round the back of the foundations of the approach spans to protect them from lateral thrust. Measurements of the soil creep were made during the period 1938 to 1952 and it was found that the velocity of creep was practically constant. Two vertical shafts, about 11 m and 10 m deep, were sunk and measurements in these showed that the velocity of creep was practically constant throughout these depths. It was concluded that stable strata must lie at a considerable depth and that underpinning of the foundations was impracticable. This conclusion is supported by the fact that deformations had been observed in the railway tunnel beneath the hillside which connects with the approach spans. The greatest horizontal displacement on any one of the several observation stations which were established was about 40 mm per year at an altitude of about 52 m above the viaduct; the accompanying vertical displacement was about 13 mm. The average slope of the hillside was about 1 in 2.

During the Pleistocene, the thrust and drag of mobile ice caused deformation of bedrock and glacial drift. At a site on the Coal Measures in west Leeds about 2 m of thin-bedded sandstone overlies shale. At several places the lowest 1 m of sandstone is highly disturbed but the overlying sandstone and the underlying shale are little affected. It is possible that this is the result of ice drag – the upper 1 m of sandstone being frozen and hence more resistant to deformation than the lower 1 m which was free to move on its own bedding planes and on the shale, perhaps facilitated by excess pore pressure, under the drag of the Airedale glacier.

At Tickhill, about 20 km east of Sheffield, several superficial anticlinal structures have been observed in which Middle Permian Marl has penetrated Upper Magnesian Limestone to become exposed at ground surface. These superficial piercement or diapiric features are ascribed tentatively to periglacial activity. Sometimes glacial drift exhibits contortions which are inherited from ice deformed before it melted and deposited the drift. In some parts of Britain features occur which resemble those seen in permafrost regions and, for example, it is possible that disturbed zones in rocks in southern England may be the result of pingo formation: indeed several small bogs are known to occupy the sites of collapsed pingos. Problems caused by permafrost features near Wolverhampton have been described by Morgan (1971).

1.1.3 Considerations relating to sedimentary rocks and stratigraphy

Although igneous rocks form about 95% of the outer 15 km thickness of the Earth, sedimentary rocks outcrop over about 75% of the land area and igneous rocks cover the remaining 25%. These percentages include metamorphic rocks derived from sedimentary and igneous rocks. Of the principal sedimentary rocks, roughly 75% are shales and clays, 15% are sandstones and sands and 10% are limestones. In fact, about 70% of Great Britain is covered by sediments of Quaternary age although often these may be no more than 1 or 2 m thick. It will be appreciated that the engineer is more likely to encounter sedimentary rocks in foundation work than igneous rocks partly because of the relative frequency of outcrops and partly on account of other factors, such as the location of cities in low-lying terrain formed from the more easily degraded sedimentary rocks. Gravel, sand, silt, clay and peat are classified in geological nomenclature as sedimentary rocks and generally fall in the group of sediments known collectively as superficial deposits or drift, as opposed to bedrock or solid. Some sequences of rocks such as the Chalk (Section 1.4.1) possess more or less distinct engineering characteristics: Meigh (1976) has presented a study of Triassic rocks.

Perusal of engineering reports and other publications often reveals geological misnomers and these should not be perpetuated. Although the sequence known as the London Clay consists dominantly of over-consolidated silty clay, it includes also layers of septorian nodules and, along the Sussex coast, hard limestone beds. The term London clay is not recognized, it implies any clay occurring in or around London, and strictly one should refer to clay of the London Clay or limestone of the London Clay. Similarly the World Clay sequence comprises silty clay, thin fossiliferous limestones, thin argillocerus ironstones and lightly cemented sandstones: the term World clay is not recognized. The Millstone Grit includes, in addition to the characteristic coarse sandstones, micaceous sandstones known as flags and also shales. The sequence known as the Gault comprises clays and thin ironstones but it would be legitimate to refer to Gault clay. Major sequences are known, for example, as Jurassic or Eocene and these are sometimes divided into Series such as Upper,

Middle or Lower. These may be sub-divided into Stages such as Tournaisian and Visean, closely corresponding in Great Britain to the rock sequence known as Carboniferous Limestone, the Albian, closely corresponding to the Gault and Upper Greensand, and the Ypresian corresponding to the London Clay. This terminology cannot be treated further here: it is best understood by frequent usage, it is not suggested that engineers should learn it by rote, but they should be aware that such terms exist.

The following is an example of how a knowledge of local stratigraphy may assist in overcoming engineering problems. At a site in the south of England, investigations revealed coombe deposit overlying sand with some clay beneath. The coombe deposit was removed to reveal the sand, which was saturated and behaved as a quicksand. Work was halted and drainage was effected over a period of months by hand-dug ditches. Knowledge of local geology would have suggested that the sand was a raised beach deposit, that the clay was part of the Reading Beds, probably no more than about 5 m thick, and that this overlies the Chalk. Before excavation it would have been possible to sink vertical sand drains through the coombe deposit, sand and clay to the Chalk to permit drainage of the sand, thus avoiding the difficulties which were encountered.

The major factors contributing to the production of a sedimentary rock are provenance, transport and environment of deposition. The primary characteristics may be modified by diagenetic process, such as oxidation, reduction and cementation, taking place after deposition. Diagenesis is often complex and is considerably influenced by ionic and organic activity. The process of converting an incoherent mass of sediment into an indurated rock is known as lithification. This involves either consolidation or compaction and/or cementation. The lithification of argillaceous sediments is of particular interest in engineering. The natural water content, w, of the uppermost 200 mm or so of very recently deposited sediments is commonly close to the liquid limit and the liquidity index is, therefore, close to unity. A sample of silty clay from the shore of the estuary of the River Adur, Sussex, possessed $w_L = 50\%$, $w_P = 23\%$ and $w = 46\%$. The liquid limit of the more common argillaceous sediments ranges from about 30 for silty deposits to 80 for colloidal. Organic and highly colloidal clays possess values of liquid limit exceeding 100.

The liquid, w_L, and plastic, w_P, limits of clays consisting entirely of clay minerals provide an indication of mineral groups. Considering the three common groups, when $w_L > 150\%$ and $w_P > 60\%$, the minerals are most likely to be in the montmorillonite group: the values may rise respectively to about 700% and 100%, depending on the exchangeable ions present. When $w_L = 80$ to 120% and $w_P = 40$ to 60%, the mineral is likely to be in the illite group and when $w_L = 40$ to 60% and $w_P = 30$ to 40%, it is likely to be in the kaolinite. Test values of liquid and plastic limits for natural soils should be interpreted bearing in mind that soils may comprise more than one clay mineral, although one is frequently dominant, together with silt, sand and organic matter. The presence of silt and sand may reduce the values quoted above by as much as one-half or more. As a percentage,

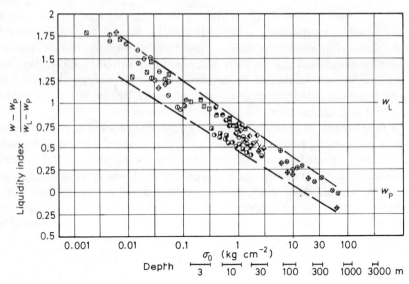

Figure 1.6 Relation between liquidity index and effective overburden stress for some named clays (after Skempton, 1970a).

clay content commonly is roughly mid-way between w_P and w_L and the plasticity index is commonly 0 to 15% less than the clay content but variations occur with different clay minerals and these broad generalizations may occasionally be erroneous. Clay mineralogy and the Atterburg limits have been studied by Seed *et al.* (1964a, b).

Although Fig. 1.6 shows that some clays possess values of liquidity index exceeding unity, values for many soft clays lie between 1.0 and 0.5, and for stiff clays, such as those of the Gault, Kimmeridge Clay and London Clay, commonly approach zero. Skempton (1944) states that for shales negative values may be attained. As consolidation of the sediment proceeds under the gradually increasing accumulation of overburden, so the natural water content is reduced, eventually approaching the plastic limit. However, sometimes it is found that, below the phreatic surface, the moisture content in a thick bed of normally consolidated homogenous clay is relatively constant throughout several metres depth. This may be due to thixotropy, which leads to an increase in strength with time after deposition, thus enhancing resistance to compression under the steadily increasing weight of overburden. The void ratio, therefore, remains sensibly constant for a given depth but, since there is a limit to the strength developed, eventually the weight of overburden becomes sufficient to cause further compression. Generalized relations are shown in Fig. 1.6 (Skempton, 1970). For depths down to about 900 m the effective overburden stress is equal to the submerged weight of overlying material. At values of effective stress exceeding the equivalent of 900 to 1500 m of overburden it seems likely that recrystallization takes place under natural conditions. It is probable that consolidation is the principal factor in lithification of argillaceous deposits down to at least 900 m with little, if any, cementation. Unless a cementing medium is introduced by mobile pore water, cementation does not occur until particles are brought into contact during consolidation, when partial recrystallization may take place. The microstructure of London Clay has been studied by Tchalenko (1968) and the pedological structure of the Keuper Marl by Davis (1968). The relationships between depth of overburden and engineering properties of the Oxford Clay have been discussed by Jackson and Fookes (1974). The classification of argillaceous soils and rocks has been studied by Morgenstern and Eigenbrod (1974). De Freitas *et al.* (1981) have presented a series of papers dealing with mudrocks, an omnibus term for primarily silty and clayey rocks, including impurities such as sand and calcareous, ferruginous and other matter which may be present.

Hydrostatic uplift in clays is influenced by the magnitude of the void ratio, e. When $e > 0.5$ it is probable that each clay particle is surrounded by a film of water capable of transmitting hydrostatic pressure. As e falls progressively below 0.5, water continues to be expelled until only layers of adsorbed water remain and these may not be capable of supplying buoyancy to particles. An indication of the low values of void ratio which can be attained is given by Jones (1944) who refers to tests on the Carboniferous Edale Shales of Derbyshire which revealed a value of 0.06. In addition to shales, other rocks, such as granite and compact limestone, undoubtedly possess low values of void ratio in the intact state. Hence in problems where buoyancy is to be taken into account, consideration must be given to the ability of the rock to transmit hydrostatic pressure through its pores. Each case must be considered on its merits but in doubtful cases it is prudent to assume the worst conditions, whether they obtain as a result of buoyancy or the absence of buoyancy. However, it is probable that pore pressure can be transmitted fully even when e is as low as 0.2. The value of e of every recently deposited CL clays may be about 0.35 and of CH clay about 0.45 but, as sampling depth beneath ground surface is taken to several hundred metres, the values fall, respectively, to about 0.25 and 0.35. For mudstones e is commonly between 0.10 and 0.25 but for shales the value may fall to that quoted above for the Edale Shales, although relief of overburden stress and weathering tend to increase the value. It should be observed that a mass buoyancy effect can arise where water occurs in joints and between bedding surfaces and, in this case, the greatest uplift can be expected where the contact area at the bedding surface is a minimum. Thus, for example, mass uplift may be relatively great in limestones in which bedding surfaces have become solution channels.

During rise and fall of river-water level with tides and floods there is a tendency for piers to rise and fall also. The nature of the response depends largely on the ease with which hydrostatic pressure is transmitted through the pores in the soil. If the effect of the additional water load on the river bed exceeds the effect of uplift on the pier, then the pier sinks as the water-level rises but returns as water-level falls. The return of the pier to normal level takes place only with long-established piers; some permanent settlement may occur with recently constructed piers. If the uplift effect exceeds water-load effect on the bed, then the pier rises with rise in water-level and sinks as it falls. If the groundwater-table level beneath a structure

Table 1.1 Till types and process (after McGown and Derbyshire, 1977)

Formative processes	Transportation processes	Depositional processes
Comminution till: produced by abrasion and interaction between particles in the basal zone of a glacier. It is a common element in most tills. Deformation till: produced by plucking, thrusting, folding and brecciation of the glacier bed.	Superglacial till: derived from frost riving of adjacent rocks or by differential melting out of glacier dirt bends. It may or may not become incorporated in the glacier and may suffer frost shattering and washing by meltwaters as it is transported on the top of the glacier. Emglacial till: derived from superglacial till subsequently buried by accumulating snow or entrained in shear zones. It is transported within the ice mass and is more abundant in Polar regions than in temperate zones. Basal till: derived from comminution products in the ice-rock contact zones particularly the lowermost regions of a glacier. It is generally transported in concentrated bands in the bottom metre or so of a glacier.	*Ablation till: accumulated by melting out on the surface of a glacier or as a coating on inert ice. *Melt-out till: accumulated as the ice of an ice–debris mixture melts out. Ice-inherited fabric may or may not be greatly changed during deposition. Lodgement till: accumulated subglacially by accretion from debris-rich basal ice. It may suffer varying degrees of shear and stress relief which may influence its fabric. Flow till: accumulated as a result of either lodgement or melt-out tills deforming by flow due to high pore-water pressures, slope instability or imposed stresses. Waterlain tills: accumulated on subaqueous surfaces under a variety of depositional processes and may thus show a wide variety of characters.

*The distinction between super glacial melt-out and ablation tills is based on the degree of disturbance of their ice-inherited fabric, including loss of fines. The term ablation till is thus best avoided.

rises, the weight of overburden and structure is reduced, in effect, by uplift, and the structure rises. The rise and fall of structures due to these effects is normally of the order of 5 mm but, with a seasonal change in water-level of, say, 3 m, a variation in foundation level of 10 to 15 mm may be experienced. A seismograph in County Hall, Westminster has responded to movement of the building with high and low water in the River Thames. Old Waterloo Bridge was known to rise at low tide and sink at high tide. Some of the circular section underground railway tunnels under the River Thames become slightly elliptical at high tide, regaining their normal shape as the tide falls.

The diagenetic process of mineral segregation leads to the formation of nodules and concretions, such as flints in chalk, doggers in sandstone and claystones in clay, and these occasionally cause difficulty during construction. When bored piles were put down for Runnymede Bridge over the River Thames, calcareous septarian nodules (claystones) up to about 0.3 m thick were encountered in the London Clay and these retarded the progress of the work.

Beds of boulders of primary sedimentary origin occur in a number of formations; for example, at the base of the marine Cambrian at Nuneaton, in the Precambrian tillite of Donegal and at the base of Pleistocene till in Saskatchewan where the boulders form a rude pavement when exposed.

Much of Great Britain is covered by Glacial and peri-Glacial deposits laid down during the Pleistocene Period. The properties of Glacial deposits have been discussed by a number of authors including Kazi and Knill (1969), McGown (1971), Anderson (1974), Boulton and Paul (1976), Marsland (1977), Hoole (1978) and Eyles and Sladen (1981). McGown and Derbyshire (1977) have published the classification for engineering purposes of till, commonly but less correctly known as boulder clay which may or may not contain boulders, shown in Table 1.1 and illustrated in Fig. 1.7. Post-Glacial deposits, laid down during the Recent or Holocene period, occur mainly along estuaries and in low-lying coastal areas. The glacial deposits, such as boulder clay and fluvio-glacial gravels, occur north of the Thames. Peri-Glacial deposits, including head and its variants coombe deposit and brickearth, are well-displayed mainly south of the Thames. The estuarine clays, silts, sands and peat are probably the most important of the late– and post–Glacial deposits. Sedimentation is taking place, of course, at many localities around the coast of Britain at the present time – the Humber estuary, the Wash and the creeks of west Sussex are examples.

Varve sediments comprise alternating very thin layers of, commonly, clayey and silty material. They are seasonally deposited in lakes, the material having been transported by glacial meltwater which is, of course, non-saline. Very little engineering research has been carried out on these deposits although their properties must be influenced markedly by orthotropy. Milligan *et al.* (1962) discuss experience with Canadian varve clays and Parsons (1976) discusses the varve silt and clay of New York.

Post-Glacial clays from a number of sites were investigated by Skempton (1948) and Skempton and Henkel (1953). At Shellhaven in the Thames estuary

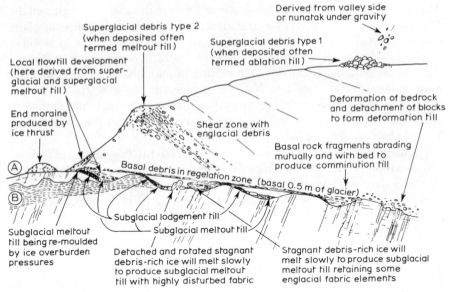

Derived from valley side
or nunatak under gravity

Superglacial debris type 2
(when deposited often
termed meltout till)

Superglacial debris type 1
(when deposited often
termed ablation till)

Local flowtill development
(here derived from super-
glacial and superglacial
meltout till)

Deformation of bedrock
and detachment of blocks
to form deformation till

End moraine
produced by
ice thrust

Shear zone with
englacial debris

Basal rock fragments abrading
mutually and with bed to
produce comminution till

(A)

Basal debris in regelation zone (basal 0.5 m of glacier)

(B)

Subglacial lodgement till

Subglacial meltout till

Subglacial meltout
till being re-moulded
by ice overburden
pressures

Detached and rotated stagnant
debris-rich ice will melt slowly
to produce subglacial meltout
till with highly disturbed fabric

Stagnant debris-rich ice will
melt slowly to produce subglacial
meltout till retaining some
englacial fabric elements

Figure 1.7 Acquisition, transportation and deposition of tills by a glacier (after McGown and Derbyshire, 1977).

about 14 m of post-Glacial clays with intercalated layers of peat were found to overlie about 2 m of late-Glacial clay overlying sandy gravel of the Flood Plain Terrace. Eight kilometres upstream at Tilbury a similar thickness of post-Glacial clays with peat was found. Tests revealed that these clays are normally consolidated except in the upper 3 m where desiccation has occurred. Below the zone of desiccation, the over-consolidation stress determined from tests was sensibly the same as the effective overburden stress. Shear strength was determined by *in situ* vane tests and by undrained triaxial compression tests on undisturbed samples, and the results of the two methods agreed throughout the full range of depth in accordance with the high activity of the clays of 1.3 to 1.4. The effect of disturbance during sampling is likely to be markedly apparent in strength tests when the activity is less than 0.75 and the liquidity index is high. The shear strength in a given bed of clay was found to increase at a sensibly linear rate with increase in depth except in the desiccation zone, where the relatively high strength near the surface falls rapidly with depth. Mineralogical analysis of the clay fraction revealed that the predominant mineral was illite, with some kaolinite but no montmorillonite. In view of the absence of this last mineral, it was considered that the high values of activity of these clays are due to the presence of organic colloids. The Shellhaven clays contained numerous microscopic fissures, although these were less evident in the Tilbury clays. It was found that the Shellhaven clays were appreciably more sensitive than the Tilbury clays and, since the salt content of the pore-water was practically the same, it was suggested that the difference in sensitivity was due to a higher salinity in the estuary at Shellhaven than upstream at Tilbury when the clays were deposited. The effect of leaching on a recent marine sediment on the coast of Nova Scotia has been studied by Moore *et al.* (1977).

Further studies of post-Glacial soils have been made by Kenney (1964) and Koutsoftas and Fischer (1976). The engineering characteristics of normally consolidated marine clays have been reported in a comprehensive study by Bjerrum (1967). Recent marine sediments have been studied by Keller (1967; 1969), Noorany and Gizienski (1970) and Buchan *et al.* (1972). Peat is often found interbedded with other Pleistocene and Holocene deposits and among the more recent studies dealing with this are publications by Weber (1969), Forrest and MacFarlane (1969), Berry and Poskitt (1972), Hollingshead and Raymond (1972), Amaryan *et al.* (1973) and Berry and Vickers (1975).

Peat is the most common unconsolidated cumulose deposit. It may occur on the surface in a completely unconsolidated state or below other soils in a partly consolidated state. In Great Britain there are two kinds; hill peat and fen peat. Both are highly organic, fibrous and spongy and contain up to 90% moisture. Oxygen is partly excluded whilst it is saturated in water, hence limited decomposition only of the vegetable matter takes place and the action of bacteria probably reduces it to its characteristic state. Hill peat consists principally of the remains of mosses and may enclose old tree trunks and branches. Fen peat consists principally of the remains of reeds, rushes, sedges and other aquatic plants; it may contain old buried forests and may be interbedded with sand, clay or other material deposited during temporary encroachment by water. Whereas hill peat has an acid character, fen peat tends to be alkaline. The maximum thickness of peat in Great Britain is about 15 m but it is more commonly less than 6 m thick. The consolidation characteristics of peat have been discussed by Berry (1983).

Fill is of little interest to the geologist but to the engineer this term describes an artificial deposit of natural or industrial material. Frequently it is deposited in depressions to raise the level of the terrain and consequently there may be soft and saturated natural soil beneath it leading to a considerabale total depth of unstable and very compressible material. The terms mud and ooze are sometimes used to describe soft deposits in the bed of an estuary or elsewhere. Neither word describes the soil type and, if possible, an alternative description should be found, such as unconsolidated organic silt.

A deposit may become slightly cemented, although not considered strictly as rock by the engineer, and moderate blasting may be necessary before it can be excavated. An example of this is Blackwall Rock, a slightly cemented gravel, which occurs in lenses about 1.5 m thick in the Thames estuary. This has led to difficulty in pile driving and caisson sinking. Provided the pile resistance is sufficient, it is often preferable to let the piles bear on such deposits than to drive through them with the assistance of blasting and provided also that no highly compressible or unstable sediments underlie the cemented deposit. A cemented

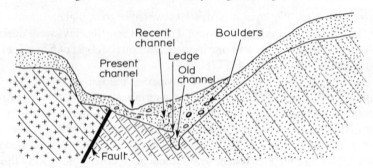

Figure 1.8 Typical formation of a buried channel.

layer in the flood plain gravels of the River Nene caused problems in sinking well-points for sewer construction near Peterborough, England.

Quicksand is a saturated sand which under certain conditions possesses a marked lack of shear strength: the conditions being caused either by an upward flow of water through the sand which reduces the effective stress between the grains or by disturbance of saturated sand when loosely packed. When due to disturbance, the phenomenon is referred to as spontaneous liquefaction. Running sand is a saturated sand which is carried into excavations by the lateral flow of ground water as excavation proceeds.

The following features may have a marked influence on foundation design and construction when they occur in Pleistocene and Holocene deposits:

(a) Rapid changes in thickness of beds in alluvial, glacial and other deposits.
(b) Soft and highly compressible beds of clay, silt or peat beneath firm sand and gravel.
(c) Running sand in alluvial deposits or in pockets in Glacial clay.
(d) Old buried valleys filled with more recent deposits (Fig. 1.8).
(e) Large boulders or a shelf of rock overhanging an old channel and mistaken for suitable foundation bedrock.

Correlation of strata in a line of boreholes in Pleistocene and Holocene deposits is often difficult because of discontinuities in the beds. This is illustrated by Fig 1.9(a). An example of a line of borehole logs is shown in Fig. 1.9(b) and it is instructive to attempt correlation along this section.

The Pleistocene was characterized by considerable fluctuations in sea-level arising from withdrawal of water from the oceans to form the ice masses and from isostatic depression of land masses under the weight of ice. Around the British Isles the maximum fall in sea-level below that obtaining today was of the order 100 m and the maximum rise 30 m: the corresponding coasts were respectively many kilometres off-shore of the present coast and generally no more than a few kilometres inland. Valleys were deepened during periods of low sea-level and buried as sea-level rose again. Raised beaches were formed during periods of high sea-level or depressed land-level. Buried valleys incised in bedrock occur adjacent

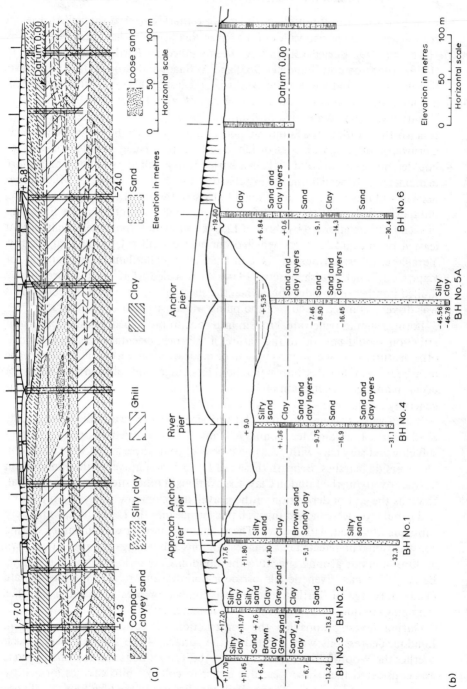

Figure 1.9 Site investigations for bridges across the River Tigris, Iraq; (a) Amara Bridge; (b) Kut Bridge (after Gelson and Plank, 1960).

to or beneath most, if not all, of the major estuaries around the British Isles. A section across the buried valley of the River Neath, Wales, has been presented by Harding (1949). Buried valleys have been studied by Parthasarathy and Blyth (1959), Anderson and Blundell (1965) and Al-Saadi and Brooks (1973). Sea-level fluctuations around the British Isles during the last 15 000 years are the subject of a comprehensive report edited by Greensmith and Tooley (1982).

Buried valleys may be filled with alluvial, glacial, estuarine or marine deposits. It is worth recording the failure in 1898 of a bridge over the St Lawrence, near Cornwall, Ontario, with loss of life, due to a pier being founded on a large boulder, mistaken for solid rock, in a buried valley. Boulders up to 10 m or more in diameter have been found during construction in glacial deposits. Much larger detached blocks and rafts are known to exist, although they are probably rare, but they provide relatively stable foundations on account of their size. Examples of such rafts are afforded by three of Liassic shales and ironstones, the largest at least 98 m long and 3 m thick, which occur in glacial till at Upgang, near Whitby, Yorkshire. Investigations for the construction of cofferdams in glacial drift for Cape Cod Canal, Massachusetts (1934–35), revealed what was interpreted as a limited number of boulders but, when the sheet piling for one of the cofferdams was driven, as much as 40% of the piling was obstructed by boulders.

Borings should penetrate 3 to 6 m into rock to prove bedrock and particular attention should be paid to the nature of the rock, orientation of bedding and other features so that comparisons can be made with known characteristics of the local bedrock. When it is known that boulders, ledges and solution cavities do not occur in an area, penetration of 2 m into bedrock, below any weathered zone, is generally adequate.

Buried valleys occur also on a relatively small scale. Site investigations for the Shell Centre, London (Measor and Williams, 1962), revealed fill overlying recent soft clay and silty clay with occasional beds of peat: these strata were formed in old riverside marshes. Beneath these are 1.5 to 5 m of gravel of the Flood Plain Terrace overlying the London Clay. Excavations at other sites have revealed that, towards the end of deposition of the gravel, deep, narrow valleys – both trunk and tributary – were incised through this terrace into the London Clay and were subsequently filled with gravel. These features may cause difficulties during construction if undetected in site investigations and borings have been put down at 30 m intervals around sites into the London Clay to locate possible valleys. Berry (1979) has given details of scour-hollows and related features of late Quaternary age in Tertiary rocks at a number of sites in London. Further examples are quoted by Wakeling and Jennings (1976).

During investigations for the construction of Battersea Power Station, London, depressions were located in the surface of the London Clay, which overlies the Woolwich and Reading Beds, the Thanet Beds and the Chalk, and it was considered that these were due either to collapse into cavities formed by transport of sand from the Thanet Beds into fissures in the Chalk or to collapse into solution cavities in the Chalk.

Buried cliffs of Pleistocene age may be encountered at various localities and at several levels in the littoral belt around the British Isles. At a site in Hove, Sussex, near the line of the ancient cliffs, the surface of the Chalk sloped downwards at about 1 in 3 in a south-westerly direction and was overlain by sandy clay of the Reading Beds topped with Coombe Deposit. This pre-Tertiary feature may represent either a degraded cliff or a valley side.

Maps which are useful for some projects are those showing bedrock surface contours and defining pre-Glacial drainage patterns. For the Hartlepool Nuclear Power Station contours were plotted of Triassic rockhead and of the top of glacial deposits, above which are estuarine sediments. Other maps can be produced to show isopachs – lines of equal thickness of a particular deposit or series of deposits. From site investigations for a sewage works near Erith, Kent, a map of isopachs for a stratum of gravel was prepared (Vick *et al.*, 1965). This stratum is overlain by silt and peat and underlain by the London Clay. Nearly all the structures were supported on piles and the map was of service in planning the layout since a minimum depth of 1.5 m of gravel below the toe of each pile was specified, provided the appropriate set was attained.

1.1.4 Rock mechanics

The subject of rock mechanics has developed considerably during the last two decades, largely in relation to dams, tunnels, shafts, underground chambers and mining. Rock mechanics can be applied also to problems concerned with excavations, natural and artificial slopes, quarrying and foundations for bridges and buildings. In practice, the prerequisites for the application of a particular theory are rarely satisfied completely and it may be necessary to employ analyses based on elastic and plastic theories together with photo-elastic and model studies in order to arrive at the best estimates of stresses in engineering design. Analyses are complicated by a number of factors such as absence of isotropy and lack of homogeneity arising from the presence of rocks of different types or zones of fractured or decomposed rock. The presence of joints renders the rock mass a discontinuum as opposed to the continuum which commonly obtains to a greater or lesser extent in soil mechanics. Although under neutral conditions horizontal stress should be less than about one-quarter of the present overburden stress, residual horizontal stress, induced by tectonic forces or earlier over-consolidation, may exceed the vertical stress.

The shear strength of many soils and rocks is influenced markedly by anisotropy or, perhaps more correctly in some cases, orthotropy. Because of the mechanism of sedimentation when soils are deposited in nature and because of the subsequent stress history, it is fairly certain that all soils are anisotropic to a greater or lesser extent. This applies to all modes of deposition, no matter whether it be in air or in water or whether the particles settle separately or as aggregates or whether the movement of the particles immediately prior to coming to rest is dominantly horizontal or vertical. Even residual soils derived from rocks possessing isotropy are likely to become anisotropic as a result of differential weathering. Disturbance of sediments after deposition by, for example, bioturbation or cryoturbation will lead to a difference in the anisotropic characteristics. Clearly sedimentary rocks will bear the characteristics of the unlithified state but these may be changed to some extent during lithification. In some igneous rocks the orientation of crystals is random and such rocks should display isotropy. On the other hand, in the whole or part of some igneous rock masses, anisotropy results, for example, from flow structure and marginal chilling. Similarly, some metamorphic rocks, such as some hornfels and marbles, can be expected to be isotropic whereas others, such as slates, schists and gneisses, which possess cleavage and band structures, display anisotropy.

The maximum compressive strength of a specimen possessing orthotropic strength characteristics is usually attained when the maximum principal stress is applied perpendicular to the plane of lowest shear strength – that is, say, the bedding plane or the cleavage plane. The lowest compressive strength is attained when the failure plane coincides with the plane of minimum strength – usually when the maximum principal stress is applied at an angle β between about 40° and 50° to the plane of minimum strength. A second maximum value of compressive strength is generally attained when the maximum principal stress is applied parallel to the plane of minimum strength, although this tends to be lower than the first maximum because of splitting along the weaker planes. These characteristics are illustrated in Fig. 1.10.

A full study of the behaviour of rock under load necessitates laboratory tests on

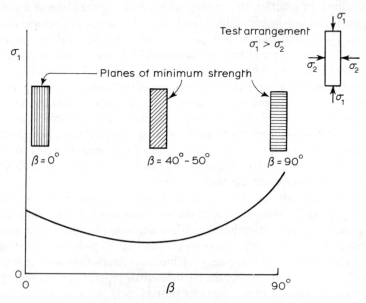

Figure 1.10 Influence of orthotropy on strength.

intact specimens to determine the intrinsic properties of the rock, together with field tests and other studies to assess mass properties in which the influence of joints and other structural features is taken into account. In connection with intrinsic properties, reference should be made to Sections 2.1 and 3.1 concerning the deformation and failure characteristics of rocks and soils.

The common laboratory tests on rock are as follows:

- Modulus of elasticity by static test on a column, resonance test on a beam or ultrasonic test on a block.
- Poisson's ratio by static test on a column.
- Modulus of rigidity by torsion test on a cylindrical specimen.
- Triaxial compression tests on cylindrical specimens with cell pressure sometimes up to $50\,000\,\mathrm{kN\,m^{-2}}$ or more for apparent cohesion and angle of shearing resistance.
- Tensile test on cylindrical specimen (tensile strength is commonly about 10% of the cube crushing strength, sometimes as low as 5%).
- Compression test on cubes.
- Direct double-shear test on cylindrical specimen (commonly much less than cube crushing strength) or shear-box test on prismatic specimens (Krsmanovic, 1967).
- Shearing resistance of joints (Paulding, 1970, observed that, although values of coefficient of friction of natural rock surfaces is sometimes between 0.5 and 1.0, the upper limit is often between 1 and 2).
- Permeability test.
- Compressibility and consolidation tests on soft rocks.

Tests for thermal conductivity and specific heat are required for projects involving high-temperature or refrigeration plant. Petrological analyses are needed for a variety of purposes and tests for chemical and physical stability may also be required.

The common site tests on rock are as follows:

- Plate bearing test for yield capacity (Section 1.2).
- Plate bearing test for modulus of subgrade reaction (Section 1.2).
- Shearing strength tests (Serafim and Lopes, 1961).
- Bolt or anchor resistance tests (ISRM, 1974; Littlejohn and Bruce, 1975, 1976).
- Tests to determine *in situ* stresses (Terzaghi, 1962; ASTM, 1967; BSI, 1981a).
- Tests relating to geotechnical processes.

Terzaghi (1946) mentions triaxial compression tests made by Richart on dry specimens of concrete protected by a watertight membrane from which was derived the empirical relationship

$$q_c = q_u + 4.1\sigma_c \tag{1.1}$$

where q_c is the confined or triaxial compressive strength, q_u is the unconfined or uniaxial compressive strength and σ_c is the confining stress. For saturated concrete the relationship is

$$q_c = q_u + 4.1(\sigma_c - u) \qquad (1.2)$$

where u is the pore-water pressure. This expression is valid for $(\sigma_c - u)$ up to about 2000 kN m^{-2} and probably for higher values. Tests on other materials, such as marble, give similar relationships and it is reasonable to expect that most intact rocks behave likewise.

The strength of rock can be estimated from the results of a point load test in which a specimen of any shape is split by loading between two conical rams (Broch and Franklin, 1972). The apparatus is portable and the load is applied by an hydraulic pump. The point load strength index

$$I_s = \frac{P}{d^2} \text{ MN m}^{-2} \qquad (1.3)$$

where P is the force required to split the specimen and d is the initial distance between the contact points.

Tests on cylindrical cores yield the most reproducible results but tests on irregular specimens show considerable scatter although this disadvantage can be overcome by testing a large number of specimens.

From a study of test results, Bieniawski (1974) determined the relationship

$$q_u = 24 I_s. \qquad (1.4)$$

Although this is derived from tests on 54 mm diameter cores, it is reasonable to assume that it is approximately true for specimens of other shapes. It is known, however, that the value of I_s increases as d decreases, the increase being of the order 20 to 30% for a 21.5 mm core compared with the value of I_s for a 54 mm core. Conversely the value of q_u will fall with larger specimens more representative of rock *in situ*. Clearly the determination of the strength of rock *in situ* requires much research and the factors influencing shear strength of soils which are discussed in Section 3.1 should be borne in mind. The effects of sample shape and size have been studied by Brook (1977) and Wijk (1979).

Another simple test employed to estimate strength and other engineering parameters of rock is based on the rebound of a piston from the surface of a specimen. The Shore scleroscope, employed to measure the surface hardness of metals, has been adapted for testing rock (Wijk and Hirengen, 1979). The measured rebound can be correlated with values of engineering properties. Where large crystals or grains of minerals are present in a specimen, the scleroscope may measure rebound from individual minerals. Hence a large number of test results are required to arrive at a reasonable mean value for rock mass. Furthermore, weathering affects surface hardness and values relating to intact rock must be determined from freshly exposed plane surfaces. Moisture content may also influence results. The Schmidt hammer, employed for concrete testing, is a type of scleroscope (Carter and Sneddon, 1977).

The Brazilian tensile test is another simple test although the disc specimens have to be prepared in the laboratory. Each thin circular disc is compressed to failure by two diametrically opposed point loads applied at the periphery. High values of compressive stress are generated very locally under the loading points but, beyond these zones, a sensibly uniform value of tensile stress is imposed normal to the diameter. The tensile stress at failure is given by $(2P/\pi Dt)$ where P is the load at failure, D is the diameter, and t the thickness of the disc. The specimen proportions are arranged so that the problem is one of plane strain but the possible influence of buckling cannot be discounted. Consequently some investigators believe that a short cylinder secures more consistent results. Tensile strength must be correlated with the desired engineering parameters. A number of authors have studied the Brazilian test including, for example, Fairhurst (1964) and Wijk (1979).

Bieniawski (1974) examined two empirical criteria for estimating the triaxial strength of rock proposed by Murrell (1965) and Hoek (1968), rewriting them in the forms given below and stating that the criteria are not contradictory and that the use of one in preference to the other is a matter of practical convenience. The Murrell criterion is

$$\frac{\sigma_1}{q_u} = k\frac{(\sigma_3)^A}{(q_u)} + 1, \tag{1.5}$$

where k and A are constants. If the reasonable assumption is made that the uniaxial tensile strength is equal to $0.1q_u$, the Hoek criterion is

$$\frac{\tau_f}{q_u} = B\frac{(\sigma_m)^C}{(q_u)} + 0.1 \tag{1.6}$$

where B and C are constants are σ_m is the mean normal stress. The tests examined by Bieniawski relate to norite, quartzite, sandstone, siltstone and mudstone and for these it was found that $k = 3.0$ to 5.0, $A = 0.75$, $B = 0.7$ to 0.8 and $C = 0.9$.

For any particular project, values of modulus of elasticity for rocks should be determined from laboratory tests or, preferably, field tests. Nevertheless, results of uniaxial and triaxial laboratory tests made by numerous investigators indicate that very broadly the range is commonly 10 to 100 GN m^{-2}. In some cases variation from the mean may be no more than 10% or 20% but in others it may be very much more. The value for Poisson's ratio commonly ranges between 0.1 and 0.3 but this also may be liable to marked variations.

Modulus of elasticity of a given rock determined in the laboratory by resonance methods is commonly up to about 20% greater than the value determined by static tests. Bruckshaw and Mahanta (1961) found that above 140 cycles s^{-1} the modulus is sensibly constant but it falls by about $2\frac{1}{2}$% from 120 to 40 cycles and it is likely that the value will decrease much more as the frequency is further reduced. If the rate of application of the fluctuating stress is sufficiently low it can be anticipated that the modulus will be equal to the static value.

The modulus of elasticity, E_r, of rock forming the foundations for an arch dam at Dukan Gorge in Northern Iraq was determined by field seismic methods (Brown and Robertshaw, 1953). The relationship between the velocity, V_p, of a longitudinal wave and E_r is

$$V_p = \sqrt{\left\{ \frac{E_r}{\gamma} \left[\frac{1-v}{(1+v)(1-2v)} \right] \right\}} \tag{1.7}$$

where γ is the weight and v Poisson's ratio for rock. It is found that E_r can be determined with sufficient accuracy for practical purposes direct from a graph, based on data given by Reich, without reference to γ or v. In practice the value of E_r determined by seismic methods may be much as 10 times the value derived from field plate bearing tests. The dynamic value is considered to be the least reliable for design purposes since the seismic waves follows paths which avoid open fissures. Furthermore, V_p is governed by the short range value of E_r, associated with instantaneous elastic strains, which has been discussed by Evans (1942). It is found from plate bearing tests that the effect of grouting a rock mass is generally to increase E_r to between two and five times the ungrouted value. Seismic determinations of E_r and v have been discussed also by Evison (1966) and shear wave measurements and prediction of settlement by Abbiss (1981, 1983).

The loads employed in field bearing and shear tests range between 500 and 50 000 kN or more, and groups of jacks are required for the higher values. It is desirable to allow the load at each increment to remain static for some time to study creep and the duration of a test at one location at least on a site may be several weeks. A complete programme of testing – from planning to final report – for an important project may take a year. To obtain statistically average values of rock properties necessitates tests at numerous locations on a site, but the cost of field testing generally prohibits this, although for some important projects exhaustive tests may be justified. The measurement of *in situ* stresses in rocks is a particularly delicate and expensive operation.

Figure 1.11 shows the results of a plate bearing test on gneiss in a tunnel entrance (Mayer, 1963). Each increment of load up to 8 on the graph produces deformation which is partly non-recoverable but, in the last cycle over a range of loading which does not exceed that of the previous cycle, an increment of permanent deformation is absent and the lower parts of curves 7–8 and 9–10 follow almost identical paths. The slope of the curves is related directly to E_r for which two values are obtained – one including permanent deformation and the other excluding it. In any particular project the selection of an appropriate value for E_r depends on whether initial or repeated service loading is being considered.

General texts on rock mechanics include Stagg and Zienkiewicz (1968), Coates (1970), Jaeger (1972) and Jaeger and Cook (1969). Many aspects of rock mechanics are discussed in the Proceedings of the First (1966), Second (1970) and Third (1974) Congresses of the International Society for Rock Mechanics, and in the periodicals *Engineering Geology, Rock Mechanics* and *International Journal of Rock Mechanics and Mining Science*. The shear strength of joints has been reviewed by Barton (1973) and, in a further publication, Barton (1981) discusses

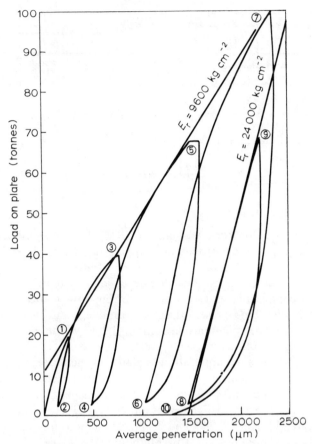

Figure 1.11 Plate bearing test on gneiss in a tunnel at St Jean du Gard.

the scale-effect leading to increased strength of laboratory specimens compared with field values. Trials of large scale excavation and compaction of rock have been reported by Williams and Stothard (1967). Laboratory and field testing techniques for rock mechanics are dealt with in two American publications (ASTM, 1966a, b) and in the proceedings of a conference (Anon, 1970). Other papers dealing with testing include, for example, Price (1958), Meigh and Greenland (1965), Rosenqvist (1965), Meigh (1968), Douglass and Voight (1969), Hobbs (1970a, b), Meigh *et al.* (1973), Bieniawski and van Heerden (1975), Raphael and Goodman (1979) and Marcuson and Curro (1981). In the absence of standard specifications for testing rocks, reference can be made to publications relating to concrete, such as BS1881, Parts 4 and 5 (BSI, 1970a), Sparks and Menzies (1973) and Sigvaldason (1966). A four-volume handbook on the mechanical properties of rocks has been produced by Lama and Vutukuri (1974, 1978) and a number of reports dealing with testing rocks have been issued by the

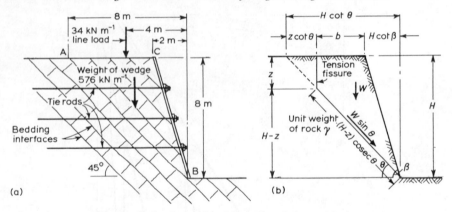

Figure 1.12 Analysis of stability of rock face with ties.

International Society for Rock Mechanics (ISRM, 1974, *et seq.*). The report of a conference on engineering properties of rocks has been edited by Attewell (1977b).

Because of the three-dimensional nature of joint systems, analysis of rock masses can be complicated and solution of problems is facilitated by reference to graphical and other data presented in, for example, some of the publications dealing with rock slopes mentioned in Section 1.1.2. Nevertheless, some problems concerning cuttings and natural cliffs can be treated as two-dimensional and Example 1.1 illustrates the application of statics in such cases. The stability of rock masses is closely connected with the shearing resistance of joints which is influenced by the roughness of the walls, the character of any detrital filling (Ludvig, 1981) and the degree of cementation. The strength of jointed rock masses has been treated by Hoek (1983) and graphical methods for the assessment of rock stability have been described by Matheson (1983).

EXAMPLE 1.1

Figure 1.12(a) is a section of a rock cliff which is to be stabilized by sets of three tie-rods at 4 m intervals horizontally. The unit weight of the rock is $24 \, \mathrm{kN \, m^{-3}}$ and the angle of shearing resistance between the bedding interfaces is $30°$. A masonry wall superposes a line load of $34 \, \mathrm{kN \, m^{-1}}$ run on top of the cliff.

Determine the total tensile force to be applied to each set of tie-rods to prevent sliding of wedge ABC down plane AB if shearing resistance in the rods is ignored and if the load factor to be applied to the disturbing force is 1.5.

Consider 4 m run of cliff.

Total vertical load $4(576 + 34) = 2440 \, \mathrm{kN}$.

Required resistance along BA $1.5 \times 2440/\sqrt{2} = 2588 \, \mathrm{kN}$.

Let total tension in tie-rods be P.

Resistance provided by friction under normal components of weight of wedge plus tie-rod tension and by tangential component of tie-rod tension, that is

$$\frac{2440 \times \tan 30°}{\sqrt{2}} + \frac{P \times \tan 30°}{\sqrt{2}} + \frac{P}{\sqrt{2}} = 2588$$

whence $P = 1427 \, \mathrm{kN}$.

It is useful to compare the resistance to be applied based on alternative data. For example, if the cliff face were vertical and the rock is treated as an elastic medium, calculate the force to be applied to each set of tie rods to restrain the face against outward displacement if Poisson's ratio for the rock is 0.2 and if the load from the wall is treated as a uniformly distributed surcharge on the surface of $8\,\text{kN}\,\text{m}^{-2}$.

From the concluding discussion in Section 3.7.2, $K_0 = v/(1 - v) = 0.25$. Horizontal stress at $z = 0$ is $0.25 \times 8 = 2\,\text{kN}\,\text{m}^{-2}$ and at $z = 8\,\text{m}$ is $0.25\,(8 + 8 \times 24) = 50\,\text{kN}\,\text{m}^{-2}$. Total thrust per $4\,\text{m}$ run is $4 \times 8\,(2 + 50)/2 = 832\,\text{kN}$. Applying a load factor of 1.5, required resistance $P = 1248\,\text{kN}$. This compares favourably with the resistance based on bedding interface slip, although given other conditions the values might be discordant. In fact the elastic analysis would be more appropriate with horizontal bedding and reasonably homogeneous and isotropic rock.

Hutchinson (1972) studied the stability of chalk cliffs in Kent, England. Referring to Fig. 1.12(b), assuming zero water pressure, the average shear stress in the inclined failure interface is

$$\tau = \frac{\gamma\{\tfrac{1}{2}H^2(\cos\theta - \cot\beta\sin\theta) - \tfrac{1}{2}z^2\cos\theta\}}{(H - z)\operatorname{cosec}\theta}$$

Treating θ as the independent variable, this is a maximum when $d\tau/d\theta = 0$ from which it will be found that

$$z = H\{1 - (\cot\beta\tan\theta)^{1/2}\}.$$

This is the value of z which leads to maximum shear stress for the given values of H, β and θ. The corresponding value of encroachment on the cliff top is

$$b = H\{(\cot\beta\cot\theta)^{1/2} - \cot\beta\}.$$

The maximum value of τ is

$$\tau_{\text{max}} = \gamma H \sin^2\theta\{\cot\theta - (\cot\beta\cot\theta)^{1/2}\}.$$

The required values of parameters to evaluate τ_{max} are derived from site investigations, with the exception of θ. The value of θ must be determined from actual failure surfaces, or assumed as one or more trial values, or taken as the dip of bedding.

In the problem posed in Fig. 1.12(a), $z = 0$ and, with $H = 8\,\text{m}$ and $\cot\beta = 0.25$, $\theta = 38°$: the greatest shear stress obtains when this is the dip of the strata. Since a dip steeper than $38°$ leads to an impossible negative value of z, a tension fissure will not be present when $\theta = 45°$. Based on the area of the wedge, the line load effectively increases the unit weight of the rock by $1.42\,\text{kN}\,\text{m}^{-2}$. With $\gamma = 25.42\,\text{kN}\,\text{m}^{-3}$ and a load factor of 1.5, calculation of τ reveals that the resistance to be provided by each set of tie-rods is $2588\,\text{kN}$, as calculated above. The statics of the two methods must, of course, be similar but the calculations have been performed by different routes.

1.2 SITE INVESTIGATIONS

The following are the main problems considered in site investigations:

(a) Vertical and horizontal variations in strata on the site and around its periphery and the engineering properties of the rocks and soils encountered.

(b) Groundwater conditions, including the occurrence of springs and seepages, on and adjacent to the site and the influence of changes in these or of other natural or artificial waters on the engineering properties of the rocks and soils.

(c) Liability of the site or adjacent land to flooding from river, lake or sea.
(d) Degree of stability of slopes on the site or on adjacent land.
(e) Liability of the site and peripheral land to subsidence arising from natural or artificial causes.
(f) Presence and condition of ditches, drains, wells or pits on and around the site.
(g) Presence and condition of mains, cable ducts and old backfilled trenches on and around the site: cable ducts often carry considerable quantities of water during rainstorms and old trenches may also permit the passage of water through backfilling.
(h) Presence of old foundations or other structures on the site.
(i) Liability of any existing fill materials to contain dangerous chemicals or to be subject to spontaneous combustion.
(j) Corrosion potential of soils, rocks or groundwater on site in relation to materials of construction.
(k) Location, species and age of trees and bushes on and peripheral to the site.
(l) Possible signs of distress in existing structures on and adjacent to the site and, where such signs are evident, the type and depth of the foundations together with recent geological and climatological history of the site.
(m) Collation of meteorological data over a period of time pertinent to the project.
(n) Potential hazards to foundations arising from use of proposed construction or other constructions around the site caused by, for example, spilt corrosive materials.

Not all of these items require investigation at all sites but occasionally it is necessary to consider problems not included in the list. During the past half-century increasing emphasis has been placed on the exploration of geological conditions at sites and the assessment of the engineering properties of the materials encountered. In many cases this forms the entire subject matter of the site investigation report and other relevant topics included in the above list are ignored with the result that clients or designers may believe that this is all that need be considered in relation to the site. The limitations of such reports should be made clear. In many cases, of course, clients or designers are better placed than drilling and testing firms to investigate, for example, buried or culverted streams, artificial drainage systems, wells, basements or old foundations on and around a site, often through the medium of archives such as old maps and plans. In connection with the redevelopment of a site, in the absence of early construction records, it may be useful to reconstruct by contours the natural geomorphology as a means of estimating depths of existing fill and other details, thereby supplementing borehole data: this was done for the proposed redevelopment of part of Brighton Station yards by East Sussex County Council.

An interesting, but now relatively old, discussion of site investigation methods was presented by Harding (1949). Tomlinson (1954) described exploration for maritime and river works, Stow (1962) described drilling techniques for water supply and McFarlane and Tomlinson (1974) have discussed investigations for structural foundations. Safety precautions relevant to inspection pits and boreholes are specified in BS 5573 (BSI, 1978), and Ward (1971) has discussed the merits of field observations in investigations for engineering projects. Current practice for site investigations in the United Kingdom is specified in BS 5930 (BSI, 1981a) and in the United States in, for example, Manual No. 56 (ASCE, 1976).

Tests for certain characteristics of soils and rocks are sometimes best made on the material *in situ*. These include the plate bearing test, the vane test and the penetration test. Other observations may be made at a site, including the measurement of pore-water pressure. Tests which form part of site investigations are discussed later in this Section. Instrumentation for construction control and long-term studies is treated, for example, a collected series of papers (ASTM 1975).

1.2.1 Field operations in site investigations

Site exploration is effected by excavating trial pits, trenches and headings, by boring or by geophysical surveying. Trial pits, trenches and headings generally give more information than borings but they are more expensive and are rarely used for deep exploration except in cases such as pilot tunnels where they can be incorporated in civil engineering works. Geological data can be recorded in detail by boring a hole about 0.75 m diameter and lowering observers in a protective cage into it. Following examination of surface outcrops and published details of the geology of a particular area, consideration may be given to the desirability of putting down inclined boreholes instead of, or in addition to, vertical, since these may clarify subsurface conditions.

The preparation of a programme for site investigations is facilitated by a study of geological maps. The conventional spacing of boreholes is at the intersections of a 15 m square grid on the site of a large structure, at the four corners and centre of a smaller structure or at 15 to 30 m intervals along the line of a narrow structure such as a retaining wall. These rules serve only as a guide to the intensity of investigation. A uniform spacing of boreholes is not necessarily desirable and local conditions may suggest other arrangements. Frequently only two or three boreholes are put down initially and when the logs and test data have been studied, the most advantageous positions of further boreholes can be decided. A marked disagreement between boreholes suggests that further investigations should be made. An additional borehole may be desirable at the location of a heavily loaded isolated foundation within or beyond the main structure. Consideration should be given to the possible existence of weak strata immediately beyond the location of a major structure since excavation through these in the future could lead to subsidence. The depth of borings is commonly $1\frac{1}{2}$ to 3 times the width of the foundation, with a minimum of 6 m but it should be determined principally by the desirability of founding on a resistant stratum or by the possible existence of a stratum of soft clay, silt or peat at a depth such that it may be overstressed or contribute appreciably to settlement. For a group of closely spaced buildings this depth is considerable and rough calculations should be made with assumed values of bearing stress to determine a depth beyond which the magnitudes of vertical and shear stresses are unimportant. If it is known from earlier investigations in the locality that soft layers are not present within this depth, then it may be sufficient to put down boreholes to 5 to 10 m only. For groups of foundations it is recommended in the *Earth Manual* of the US Bureau of Reclamation (1960) that, if B is the breadth of a footing (footings assumed square and the same size) and X is the distance between footings, the depth of borings, D, beneath founding level should be as follows:

$$X < 2B, D = 4\tfrac{1}{2}B : X = 2 \text{ to } 4B, D = 3B : X > 4B, D = 1\tfrac{1}{2}B.$$

With pile groups the depth is correspondingly greater and boreholes should penetrate into unlithified deposits to a depth below the pile cap equal to at least the pile-length plus $1\tfrac{1}{2}$ times the breadth of a pile group. Particular attention should be paid to the possible presence of compressible strata below pile founding level. For a bridge over a river, sufficient preliminary borings and other site investigations, such as geophysical exploration, are made to enable a soil profile (Fig. 1.9a) to be drawn and, when the positions of the piers have been established, not less than two further borings are usually made at each pier location. Disturbed samples are generally taken from borings at each major change in soil type and at 1 to 2 m intervals in an apparently homogeneous stratum. The whole of a diamond drill core of rock is usually preserved, although recovery may be considerably less than 100% and sometimes is less than 50% with soft rocks.

The importance of locating boreholes within the proposed plan area of major foundations, irrespective of administrative or physical problems, is illustrated by reference to road and bridge construction in a shallow col in Sussex which is described on p. 82. Because of protracted difficulties in giving access to the proposed site of one abutment and to avoid further delay in seeking legal right of entry, it was decided to put down a borehole a few metres outside the planned area on land where access was immediate. The abutment was designed to be founded at a relatively shallow depth on intact chalk in conformity with the borehole log. In fact excavation revealed that the planned area was located at the end of a pipe or swallow hole, roughly 30 m × 10 m and at least 10 m deep, filled with sand and clay, beneath which were several metres of soft fragmented chalk, and the abutment was redesigned to be carried on piles.

The following are the basic types of boring equipment.

(a) Post-hole auger. Usually 100 or 150 mm diameter, hand operated. Maximum depth about 5 m.
(b) Mechanical auger. Usually 150 to 300 mm diameter, although larger holes can be bored. Operated by two men. Maximum depth commonly 10 to 60 m depending on type of machine.

(c) Shell and auger gear. Usually 150 to 250 mm diameter. Hand or power operated. Maximum depth > 30 m.

(d*) Wash boring. Used in unlithified deposits. Casing about 75 mm diameter. Maximum depth about 30 m.

(e*) Percussion, cable-tool or churn drilling. 100 to 500 mm diameter, 300 m or more depth.

(f*) Mud flush rotary boring. 100 to 600 mm diameter, 300 m or more depth.

(g) Rotary chilled shot or calyx boring. 150 mm to 2 m diameter, 300 m or more depth. Core obtained.

(h) Rotary drilling with diamond or tungsten insert bits. 40 to 150 mm diameter, 300 m or more depth. Core obtained.

New and more versatile equipment is introduced from time to time. Serota (1972) has discussed drilling rigs in current use and Wilson (1975) has discussed investigation methods.

It is not always possible to allocate qualified personnel to watch boring metre by metre and to identify rocks and soils on the site. In the Institute of Geological Sciences Memoir 'Wells and Springs of Sussex', F. H. Edmunds gives the logs of two bores put down side by side. The two logs should be identical but in fact there is complete absence of correlation due to misidentification of rock and soil. Consequently cores or disturbed samples should be taken at intervals not exceeding 1 to 2 m, depending on the characteristics of the materials, and at all changes in strata for subsequent identification in the laboratory. However, it may be advantageous to take continuous samples in split 75 mm tubes in one or more boreholes and these are carefully examined and recorded so that the more critical strata can be identified and 250 mm diameter undisturbed samples are then taken from these horizons in additional boreholes. These large diameter samples provide more representative specimens for testing than conventional 100 mm tube samples. This technique has been discussed by Rowe (1972). A piston sampler is essential for retrieving good samples from soft clays.

Continuous coring certainly cannot be justified for every project, yet the following example illustrates the advantage derived by boring a hole in this way. A borehole was put down by calyx boring near Ilkley, Yorkshire, through shales of the Millstone Grit, anticipating that water would be met in coarse sandstone known to underlie the shales. In fact the sandstone was not penetrated at the expected depth. The core revealed that the borehole passed through a reverse fault, manifest in a brecciated zone about 3 m thick, which markedly increased the thickness of shale above the sandstone. Although the borehole could have been put down by other means without taking a core, the fault would have been undetected. The throw must have been considerable and could probably have been determined by careful examination of intact core above and below the fault, comparing lithology and fossil content.

---

*Not normally recommended since comminuted rock and soil is brought up in suspension.

Table 1.2 Soil sample quality

Quality class	Properties that can be reliably determined
1	Classification, moisture content, density, strength, deformation, consolidation characteristics
2	Classification, moisture content, density
3	Classification, moisture content
4	Classification
5	None, employed to determine sequence of strata only

A classification for quality of soil samples, evolved in Germany, is included in BS 5930 (BSI, 1981a) and is reproduced here (Table 1.2).

There appears to be little merit in this classification since appropriate equipment and technique would be selected when the required properties have been specified, without reference to class. The mass of soil required for various laboratory tests is tabulated in BS 5930 (BSI, 1981a) and ranges from 1 kg for classification tests on clay, silt and sand to 160 kg for comprehensive examination of coarse gravel used as a construction material. Although the specified values of mass serve as guides, it is preferable to consider carefully the tests required, together with procedure, increasing the quantities to allow for faulty sampling and preparation. Although British Standard procedures can be employed for many projects, it may be desirable to make modifications in certain cases.

Where observations for phreatic surface level are made in deposits of low permeability, several hours or days, or weeks in stiff clays, may be required for equilibrium level to be attained in a borehole since water must percolate into the void created by excavating the soil. Where the equilibrium level is not readily attained, a 50 mm diameter plastic standpipe with a permeable sleeve at the foot can be inserted in the borehole before the lining is withdrawn and the foot surrounded with sand or gravel so that observations can be maintained. In littoral belts, the time-lag on tidal high and low water before maximum and minimum groundwater level is reached, is of the order of $\frac{1}{2}$ to 2 h, and near the shore the range is usually about two-thirds to one-sixth that of the tide. The range is greater in gravels than sands. Range decreases with increase in distance from the shore and at 200 to 300 m may be sensibly zero, although in some localities tide response has been detected at distances up to about 1 km. The contact zone between bodies of fresh and salt water may be only about 0.3 m thick. During construction of the Mersey Road Tunnel, driven in sandstone, fresh water flowed into the top of the heading and salt water into the bottom. An example of observations of tide and water table levels is given in Fig. 1.13. After boring in water-bearing strata, particularly under artesian conditions, boreholes should be backfilled with well-compacted clay or concrete to prevent inflow of water during subsequent excavation.

Formerly a fall in the water table was more common than a rise and resulted from abstraction or leakage from aquifers: this led to problems such as general

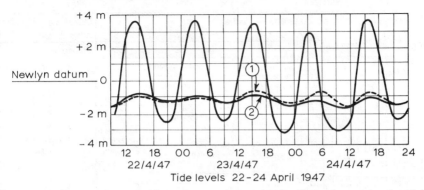

Figure 1.13 Observations of tide and groundwater levels. Groundwater levels: (1) in Blackheath beds, borehole No. 1, 113 m from river wall; (2) in Thames sands and gravels, borehole No. 2, 67 m from river wall.

settlement of an area and the rotting of timber piles. However, in some localities the water table has risen following a decrease in abstraction from boreholes caused by decline in industry and ready access to water supply. As a result, at Witton, Birmingham, factory basements have been flooded and in central Liverpool railway tunnels have been threatened. In the London Basin, water has been drawn from the Chalk through boreholes for nearly two centuries and this has led to a fall in groundwater level of the order 70 m in central London but the rate of abstraction has decreased during the last two decades with the result that the level is rising by about one metre per year in some places (Marsh and Davies, 1983).

Clearly, the greater the number of boreholes at any particular site, the greater the amount of data collected, although the additional information revealed by each additional borehole may be very little where lithology and structure are uniform. Careful selection of borehole locations is generally more advantageous than regular or random positioning. It is often desirable to observe water level over a period of at least several days and the changes recorded, together with the boring logs, should be carefully interpreted. The following example records the occurrence of a problem at a site where the number of boreholes was apparently reasonable. A 2.7 km long coastal sewer outfall tunnel 1.7 m diameter near Weymouth was commenced about 1980. The rocks comprise clays, sandstones and limestones and the sequence was determined in eight boreholes put down for design purposes. During construction the inflow of water from the 2 m thick Sandsfoot Grit Beds proved much greater than anticipated so that the tunnel diameter was increased to 2.1 m to accommodate a bolted segmental lining and compressed air working was introduced. Work was further impeded by inflow of sea-water. Despite the number of boreholes, the extent of the hazard was apparently not foreseen. Sudden copious inflow of water has been encountered many times in tunnelling but in this case it could reasonably be anticipated that the potential of the aquifer ought to have been revealed, for example, by changes in water level in at least some of the boreholes or by other evidence. However,

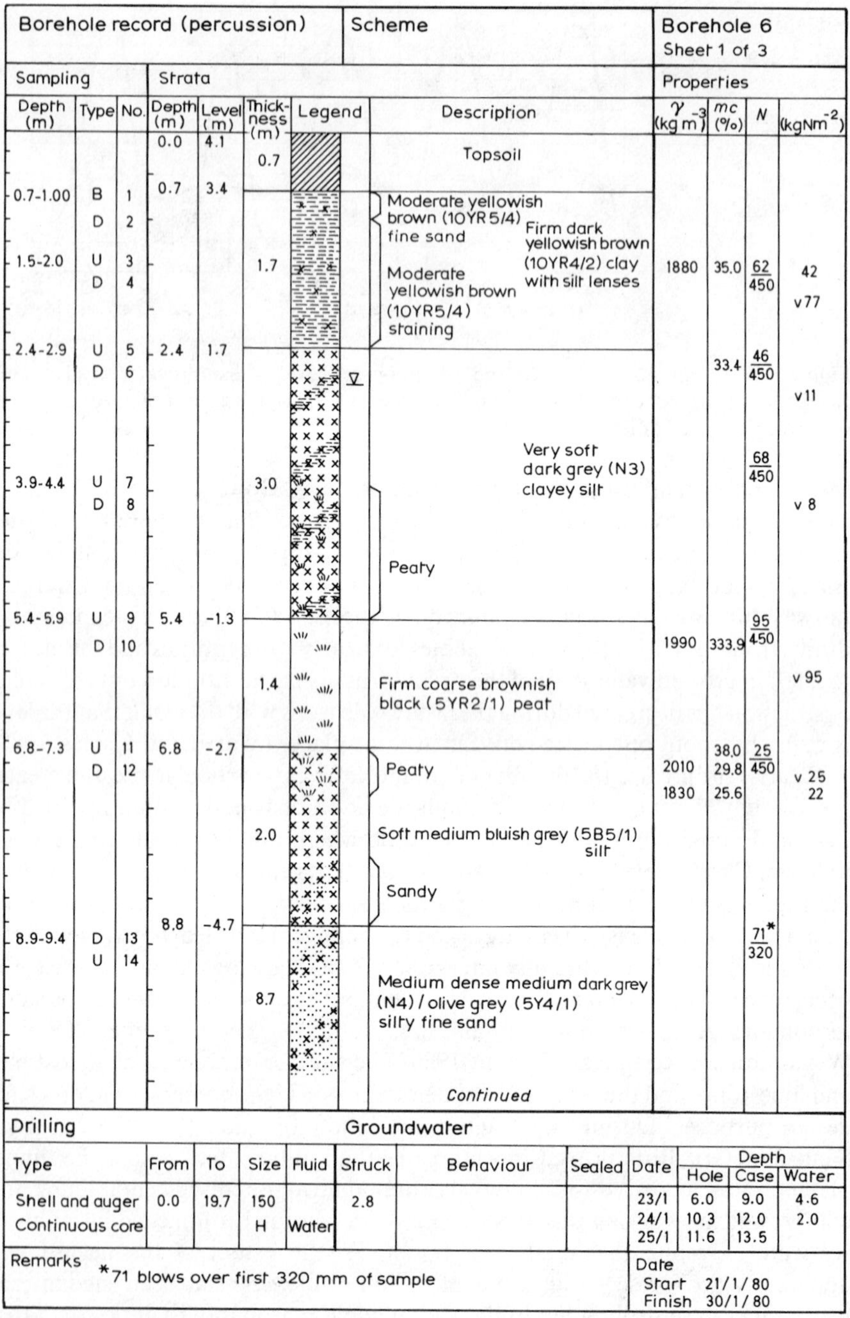

Borehole record (percussion)

Scheme	Borehole 6 Sheet 1 of 3

Sampling			Strata			Legend	Description	Properties			
Depth (m)	Type	No.	Depth (m)	Level (m)	Thickness (m)			γ (kg m^{-3})	mc (%)	N	(kgNm^{-2})
			0.0	4.1	0.7		Topsoil				
0.7–1.00	B	1	0.7	3.4			Moderate yellowish brown (10YR5/4) fine sand				
	D	2					Firm dark yellowish brown (10YR4/2) clay with silt lenses				
1.5–2.0	U	3			1.7		Moderate yellowish brown (10YR5/4) staining	1880	35.0	62/450	42
	D	4									v77
2.4–2.9	U	5	2.4	1.7					33.4	46/450	
	D	6									v11
							Very soft dark grey (N3) clayey silt			68/450	
3.9–4.4	U	7			3.0						v 8
	D	8					Peaty				
5.4–5.9	U	9	5.4	−1.3				1990	333.9	95/450	
	D	10									v 95
					1.4		Firm coarse brownish black (5YR2/1) peat				
6.8–7.3	U	11	6.8	−2.7			Peaty	2010	38.0	25/450	v 25
	D	12						1830	29.8		22
									25.6		
					2.0		Soft medium bluish grey (5B5/1) silt				
							Sandy				
8.9–9.4	D	13	8.8	−4.7						71*/320	
	U	14			8.7		Medium dense medium dark grey (N4) / olive grey (5Y4/1) silty fine sand				
							Continued				

Drilling

Type	From	To	Size	Fluid	Groundwater Struck	Behaviour	Sealed	Date	Depth Hole	Case	Water
Shell and auger	0.0	19.7	150		2.8			23/1	6.0	9.0	4·6
Continuous core				H Water				24/1	10.3	12.0	2.0
								25/1	11.6	13.5	

Remarks *71 blows over first 320 mm of sample

Date Start 21/1/80 Finish 30/1/80

Figure 1.14 Typical log for percussion drilled borehole.

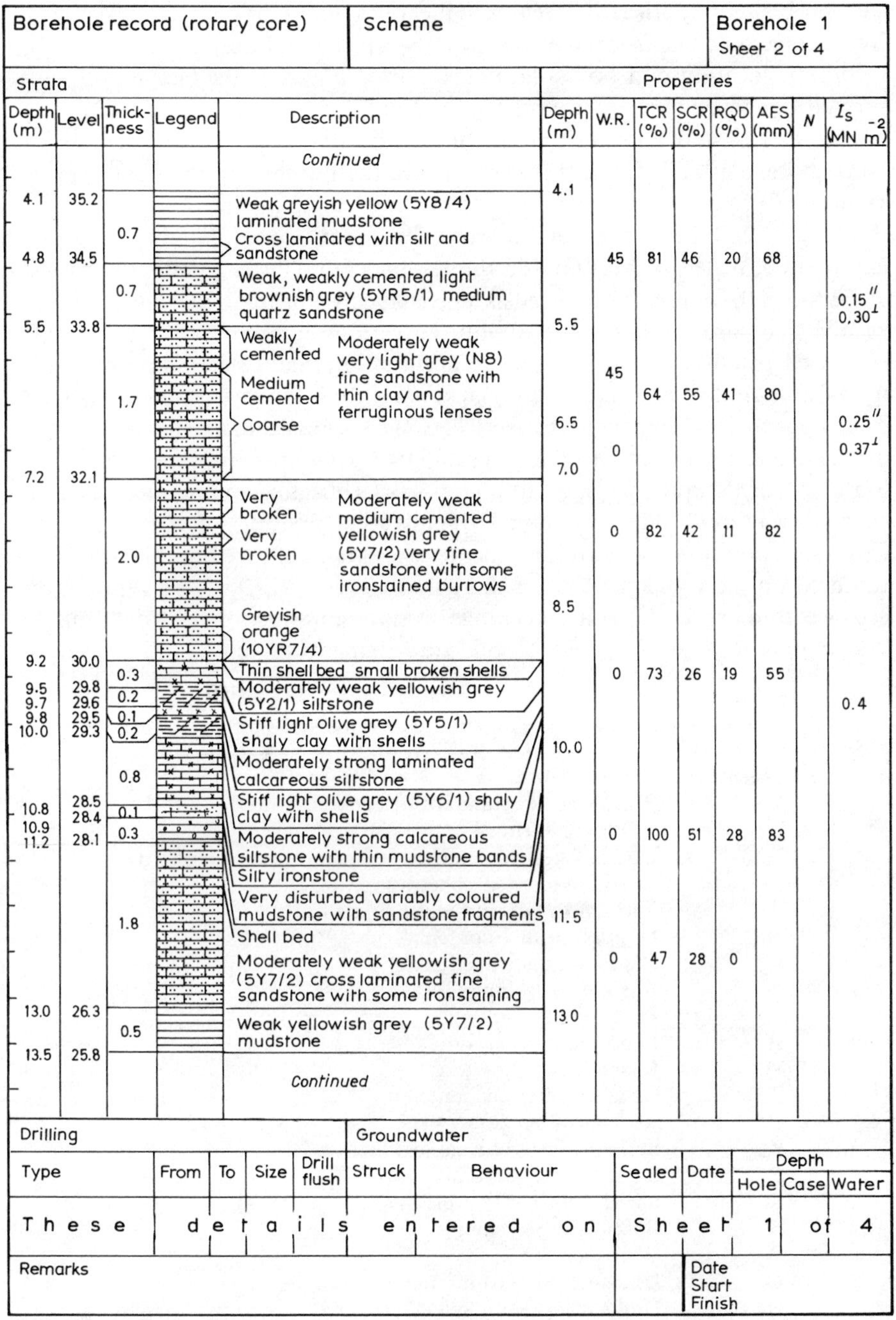

| Borehole record (rotary core) | Scheme | Borehole 1 |

Strata | **Properties**

Depth (m)	Level	Thickness	Legend	Description	Depth (m)	W.R.	TCR (%)	SCR (%)	RQD (%)	AFS (mm)	N	I_s -2 (MN m)
				Continued	4.1							
4.1	35.2	0.7		Weak greyish yellow (5Y8/4) laminated mudstone Cross laminated with silt and sandstone		45	81	46	20	68		
4.8	34.5	0.7		Weak, weakly cemented light brownish grey (5YR5/1) medium quartz sandstone								0.15" 0.30$^\perp$
5.5	33.8	1.7		Weakly cemented — Moderately weak very light grey (N8) fine sandstone with Medium cemented — thin clay and ferruginous lenses Coarse	5.5	45	64	55	41	80		
					6.5	0						0.25" 0.37$^\perp$
7.2	32.1	2.0		Very broken — Moderately weak medium cemented Very broken — yellowish grey (5Y7/2) very fine sandstone with some ironstained burrows Greyish orange (10YR7/4)	7.0	0	82	42	11	82		
9.2	30.0	0.3		Thin shell bed small broken shells	8.5	0	73	26	19	55		
9.5	29.8	0.2		Moderately weak yellowish grey (5Y2/1) siltstone								0.4
9.7	29.6	0.1		Stiff light olive grey (5Y5/1) shaly clay with shells								
9.8	29.5	0.2		Moderately strong laminated calcareous siltstone	10.0							
10.0	29.3	0.8										
10.8	28.5	0.1		Stiff light olive grey (5Y6/1) shaly clay with shells								
10.9	28.4	0.3		Moderately strong calcareous siltstone with thin mudstone bands Silty ironstone		0	100	51	28	83		
11.2	28.1	1.8		Very disturbed variably coloured mudstone with sandstone fragments Shell bed Moderately weak yellowish grey (5Y7/2) cross laminated fine sandstone with some ironstaining	11.5	0	47	28	0			
13.0	26.3	0.5		Weak yellowish grey (5Y7/2) mudstone	13.0							
13.5	25.8			*Continued*								

Drilling | **Groundwater**

Type	From	To	Size	Drill flush	Struck	Behaviour	Sealed	Date	Depth Hole	Case	Water

These details entered on Sheet 1 of 4

| Remarks | | Date Start Finish |

Figure 1.15 Typical log for rotary cored borehole.

such problems may arise from freak and inexplicable conditions, possibly created by local tectonic disturbance or open joints, which would be revelated only by fortuitous location of a borehole. Probe boring ahead of the face may enable precautions to be taken where such conditions are suspected.

Figures 1.14 and 1.15 are typical logs for lengths of percussion and rotary cored boreholes and Tables 1.3 and 1.4 give the relevant symbols and abbreviations. Colour names and classification are based on the Munsell system (Section 1.2.2). The properties recorded for rotary cored boreholes require explanation. Water recovery (WR) is the amount of drill flushing water discharged at the top of the borehole expressed as a percentage of the total used and is a crude relative measure of mass permeability at various horizons. It is commonly estimated by observation only: more accurate estimates necessitate metering inflow and outflow and this is rarely justified since *in situ* permeability tests are more appropriate. Total core recovery (TCR) is expressed as a percentage of the drill run, usually in core barrel lengths of 3 m. Losses arise from the presence of loose, soft or naturally fragmented material or artificially comminuted material carried away by flushing water or of highly compressible material or of open fissures. In addition, the barrel may fail to lift a stump of core but this should be retrieved on the subsequent run. Solid core recovery (SCR) is core with perfect cross-section expressed as a percentage of total core recovered. Rock quality designation (RQD) is the length of solid core segments exceeding 100 mm length

Table 1.3 Percussion borehole records. Symbols and abbreviations

N	Blows/penetration
	Bracketed no: standard or cone penetration test (Blows/300 mm)
	Unbracketed no: undisturbed sample (non-standard blows/recorded penetration)
	The suffix '*' denotes extrapolated value
mc	*In situ* moisture content
c	Cohesion (undrained state)
ϕ	Angle of shearing resistance
γ	Bulk density
lv	Laboratory shear vane test
v	*In situ* shear vane test
$\underline{\nabla}$	Ground water encountered
$\underline{\uparrow\nabla}$	Artesian water flow
$\blacktriangledown$	Standing level of ground water
P	Standpipe or piezometer tip
D	Small disturbed sample (jar)
B	Bulk disturbed sample
W	Water sample
U	Undisturbed 100 mm diameter sample
U̶	Undisturbed sample not recovered
S	Standard penetration test
C	Cone penetration test

Table 1.4 Rotary core borehole records. Symbols and abbreviations

WR	Estimated recovery of drill flushing water; expressed as a percentage of total used
TCR	Total core recovery; expressed as a percentage of the drill run (usually in core barrel lengths of 3 m)
SCR	Solid core recovery; expressed as a percentage of the total core recovered. Solid core must have a perfect unfractured cross-section
RQD	Rock quality designation; a percentage of the solid core greater than 100 mm in length against the total core recovered. Descriptive terms (BS 5930 (BSI, 1981a)).

RQD (%)			*Term*
0	to	25	Very poor
25	to	50	Poor
50	to	75	Fair
75	to	90	Good
90	to	100	Excellent

AFS	Average fracture spacing in mm
N	Blows/penetration Bracketed no.: standard or cone penetration test: blows/300 mm Unbracketed no.: blow count during driving for undisturbed sample. Blows/recorded penetration *denotes extrapolated value
I_s	Franklin point load strength ($MN\,m^{-2}$)

0–0.1	very weak
0.1–0.3	weak
0.3–1.0	moderately weak
1.0–3.0	moderately strong
3.0–6.0	strong
6–12	very strong
7–12	extremely strong
‖ measured parallel to bedding plane	
⊥ measured perpendicular to bedding plane	

▽	Ground water encountered
↑▽	Artesian water flow
▼	Standing level of ground water
P	Standpipe or piezometer tip

expressed as a percentage of the total core recovered. Average fracture spacing (AFS) is the average distance in mm between parallel or sub-parallel fractures; horizontal, vertical and inclined. The inverse of AFS is fracture frequency (FF), expressed as average number of fractures per metre.

It will be appreciated that all of these five parameters are related to some extent to the degree of fragmentation of the rock. Fractures which are obviously caused artificially during drilling or by desiccation or otherwise during storage should be omitted from fracture counts. It follows that some degree of correlation can be expected between these parameters and Deere (1968) presents a plot of RQD against FF for several rock types which can be expressed as $FF/m = (110 - RQD)/(5.5)$ within about $\pm 2\,m^{-1}$ which is valid between $RQD = 20$ to 100.

The borehole records shown in Figs 1.14 and 1.15 are based on investigations made by East Sussex Country Council. Other arrangements of records are employed by different organizations although the information presented is similar: standard forms are specified in BS 5930 (BSI, 1981a).

1.2.2 Description and classification of rocks and soils

Recommendations concerning the description of rocks and soils have been published by Engineering Group Working Parties of the Geological Society of London (Anon, 1970b; Anon, 1972a). The following extracts are of particular interest in site investigation.

The colour of soil can be specified according to the Munsell soil colour chart (1954) and of rock according to the colour chart based on the Munsell system and published by the Geological Society of America (1963). Colour is expressed in the Munsell system in terms of hue, a basic colour or mixture of colours, of chroma, the brilliance or intensity of colour, and of value, the lightness of colour. Unfortunately time can be wasted in determining colour precisely by reference to the charts, particularly as this commonly varies with moisture content. Dumbleton (1968) devised an abbreviated system. When colour charts are not available or when precision is not required, the following subjective scheme (Table 1.5) should prove adequate. The basic colour of soil or rock is selected from column 3 and is supplemented if necessary by a term from column 1 and/or column 2.

Table 1.5 Colour classification for soil and rock

1	2	3
Light	Pinkish	Pink
Dark	Reddish	Red
	Yellowish	Yellow
	Brownish	Brown
	Olive	Olive
	Greenish	Green
	Bluish	Blue
		White
	Greyish	Grey
		Black

Table 1.6 Classification of planar structures and discontinuities in rocks

Planar structures	Spacing	Discontinuities
Very thickly bedded	> 2 m	Very widely spaced
Thickly bedded	600 mm–2 m	Widely spaced
Medium bedded	200 mm–600 mm	Moderately widely spaced
Thinly bedded	60 mm–200 mm	Closely spaced
Very thinly bedded	20 mm– 60 mm	Very closely spaced
Laminated (sedimentary) Closely* (metamorphic and igneous)	6 mm–20mm| < 20 mm	Extremely closely spaced
Thinly laminated (sedimentary) Very closely* (metamorphic and igneous)	< 6 mm	

*Foliated, cleaved, flow-banded, etc., as apropriate.

The extent and characteristics of mottles can be expressed by the following terms, although these are not specified by the Working Parties.

- Contrast: faint, distinct, prominent.
- Abundance (percentage exposed surface mottled): few, < 2%; common, 2 to 20%; many, > 20%.
- Size (greatest dimension): extremely fine, < 1 mm; very fine, 1 to 2 mm; fine, 2 to 5 mm; medium, 5 to 15 mm; coarse, > 15 mm.

The terms listed in Table 1.6 can be employed to describe the spacing of planar structures such as bedding and lamination in sedimentary rocks, foliation in metamorphic rocks and flow-banding in igneous rocks. Discontinuities are fractures in rock, including joints, fissures, faults, cleavages and irregular shattering, and can be described by the terms in Table 1.6. Details should be specified also as to whether discontinuities are open or tight, healed, cemented or infilled, integral or incipient and whether the walls are plane, curved, irregular, smooth, rough or slickensided.

The spacings and orientations of joints are recorded along a scanline or transect parallel to the bedding over a distance of, say, 10 to 100 m as appropriate and the average is usually employed to give joint discontinuity spacing and orientation. Sometimes the readings are remarkably uniform but in other cases the range may be considerable and the information must be recorded. The reproducibility of joint spacing measurements has been studied by Ewan *et al.* (1981) and of joint orientation measurements by Ewan and West (1981).

Discontinuity spacing in three dimensions, used to describe the blocks forming the megastructure *in situ*, is specified in BS5930 (BSI, 1981a), as shown in Table 1.7.

The terms and grades listed in Table 1.8, can be employed to define degrees of weathering of rocks. Examples of their use have been given by Fookes *et al.*

Table 1.7 Discontinuity spacing in three dimensions

First term	Maximum dimension	Second term	Nature of block
Very large	> 2 m	Blocky	Equidimensional
Large	600 mm– 2 m	Tabular	Thickness much less
Medium	200 mm–600 mm		than length or width
Small	60 mm–200 mm	Columnar	Height (thickness)
Very small	< 60 mm		much greater than cross-section (length or width)

Table 1.8 Classification of weathering in rocks

Term	Grade symbol	Diagnostic features
Residual soil	W VI	Rock is discoloured and completely changed to a soil in which original rock fabric is completely destroyed. There is a large change in volume. *Genesis should be determined where possible.*
Completely weathered	W V	Rock is discoloured and changed to a soil but original fabric is mainly preserved. There may be occasional small corestones. The properties of the soil depend in part on the nature of the parent rock.
Highly weathered	W IV	Rock is discoloured; discontinuities may be open and have discoloured surfaces, and the original fabric of the rock near to the discontinuities may be altered; alteration penetrates deeply inwards, but corestones are still present. *The ratio of original rock to weathered rock should be estimated where possible.*
Moderately weathered	W III	Rock is discoloured; discontinuities may be open and will have discoloured surfaces with alteration starting to penetrate inwards; intact rock is noticeably weaker, as determined in the field, than the fresh rock. *The ratio of original rock to weathered rock should be estimated where possible.*
Slightly weathered	W II	Rock may be slightly discoloured, particularly adjacent to discontinuities, which may be open and will have slightly discoloured surfaces; the intact rock is not noticeably weaker than the fresh rock.
Fresh	W I	Parent rock showing no discolouration, loss of strength or any other weathering effects.

Table 1.9 Description of soil strength

Soil group	Term	Simple diagnostic character
Coarse grained soils	Indurated	Broken only with sharp pick blow, even when soaked. Makes hammer ring.
	Strongly cemented	Cannot be abraded with thumb or broken with hands.
	Weakly cemented	Pick removes soil in lumps, which can be abraded with thumb and broken with hands.
	Compact	Requires pick for excavation; 50 mm peg hard to drive more than 50 to 100 mm.
	Loose	Can be excavated with spade; 50 mm wooden peg easily driven.
Fine grained soils	Hard	Brittle or very tough.
	Stiff	Cannot be moulded with fingers.
	Firm	Moulded only by strong pressure of fingers.
	Soft	Easily moulded with fingers.
	Very soft	Exudes between fingers when squeezed.
	Friable	Non-plastic, crumbles in fingers.
Peat	Firm	Fibres compressed together.
	Spongy	Very compressible and open structure.
	Plastic	Can be moulded in hands and smeared between fingers.

Table 1.10 Description of soil structure

Soil group	Term	Character
Coarse Soils	Weathered	Particles are weakened, and may show concentric layering.
	Homogeneous	Material essentially of one type.
	Layered	Alternating layers of various types.
	Thinly layered	Stratified in thin layers.
Fine Soils	Aggregated	Strength decreases on working.
	Weathered	Usually exhibits crumb or columnar structure.
	Fissured	Breaks into polyhedral fragments.
	Intact	Not fissured.
	Homogeneous	Material essentially of one type.
	Layered	Alternating layers of various types.
	Thinly layered	Stratified in thin layers.
Peat	Fibrous, fine and coarse	Plant remains easily recognizable, retains structure and some of original strength; fine, diameter less than 1 mm; coarse, diameter greater than 1 mm.
	Amorphous–granular	Recognizable plant remains absent.

(1971). The significance of rock weathering in foundation engineering has been reviewed by Saunders and Fookes (1970) and a revised classification has been suggested by Dearman (1976).

The distinction between soil and rock can be defined for engineering purposes arbitrarily by stating that whereas rock is a natural aggregate of minerals connected by strong and permanent cohesive forces, soil is an aggregate of mineral grains that can be separated by such gentle means as agitation in water (Terzaghi and Peck, 1967). The boundary between soil and rock is inevitably broad and, in the list of terms (Table 1.9) describing strength of soils, indurated and strongly cemented soils can be classified also as rocks.

The structure of soils can be described according to the list of terms given in Table 1.10. Soil macrofabric has been discussed at length by McGown *et al.* (1980). Weathered conditions can be specified more precisely by reference to Table 1.11 and layered soils by reference to terms given in Table 1.6 for planar structures of rocks.

The terms given in Table 1.11 can be employed to describe the degree of weathering of soils: a distinction may have to be drawn between the state of individual particles of coarse soils and the state of the matrix.

The terms in Table 1.12 describe the strength of rock and is based on point load strength given by the portable field testing machine or on the uniaxial compressive strength.

In order to estimate construction costs, it is necessary to classify rock

Table 1.11 Degree of weathering of soils

Term	Grade Symbol	Diagnostic features
Completely weathered	W V	Soil discoloured and altered, with no trace of original structures.
Highly weathered	W IV	Soil mainly altered with occasional small lithorelics of original soil. Little or no trace of original structures. *The ratio of original soil to weathered soil should be estimated where possible.*
Moderately weathered	W III	Soil is composed of large discoloured lithorelics of original soil separated by altered material. Alteration penetrates inwards from the surfaces of discontinuities. *The ratio of original soil to weathered soil should be estimated where possible*
Slightly weathered	W II	Material is composed of angular blocks of fresh soil, which may or may not be discoloured. Some alteration starting to penetrate inwards from discontinuities separating blocks.
Fresh	W I	Parent soil showing no discolouration, loss of strength or any other defects due to weathering.

Table 1.12 Classification of rock strength

Term	Point load strength $(kN\,m^{-2})$	Uniaxial compressive strength (conforms also to BS5930 (BSI, 1981a) $(MN\,m^{-2})$
Extremely strong	> 12 000	> 200
Very strong	6000–12 000	100–200
Strong	3000– 6 000	50–100
Moderately strong	750– 3 000	12.5– 50
Moderately weak	300– 750	5– 12.5
Weak	75– 750	1.25– 5
Very weak	< 75	< 1.25

Table 1.13 Strength of cohesive soils and relative density of cohesionless soils

	Cohesive soils		Cohesionless soils
Term	Undrained shear strength $(kN\,m^{-2})$	Term	SPT N-values, blows/300 mm penetration
Very soft	< 20	Very loose	0– 4
Soft	20– 40	Loose	4–10
Firm	40– 75	Medium dense	10–30
Stiff	75–150	Dense	30–50
Very stiff or hard	> 150	Very dense	> 50

according to the degree of effort required in its excavation and, when the spoil is used as fill, the degree of effort required for its compaction. A simple classification system can be based on strength (point load, unconfined compressive or Brazilian tensile), a characteristic of intact rock, and on discontinuity spacing, a mass characteristic. Each classification group comprises a range of strength, say 12.5 to $50\,MN\,m^{-2}$, and a range of discontinuity spacing, say 200 to 600 mm. Classification groups commonly overlap either in strength or in spacing.

The strength of cohesive soils and relative density of cohesionless soils are specified in BS 5930 (BSI, 1981a) as shown in Table 1.13.

The following are examples of the use of the various terms listed above for rocks and soils.

- Dark olive brown, fine to medium-grained, massive, moderately widely spaced joints with majority of joints to 10 mm, slightly weathered, contact metamorphosed, *Dolerite*, strong, impermeable except along open joints.
- Dark grey, fine-grained, medium to thickly bedded and thinly laminated (within beds) closely to very closely jointed, fresh, *Shale*, strong, effectively impermeable, brittle.
- Light pinkish grey, coarse to very coarse-grained, massive, moderately widely

Table 1.14 Unified soil classification (after Wagner, 1957)

			Field identification procedures (excluding particles larger than 3 in and basing fractions on estimated weights)	Group symbols*	Typical names
Coarse-grained soils More than half of material is *larger than* No. 200 sieve size† (The No. 200 sieve size is about the smallest particle visible to naked eye)	*Gravels* More than half of coarse fraction is larger than No. 7 sieve size	Clean gravels (little or no fines)	Wide range in grain size and substantial amounts of all intermediate particle sizes	GW	Well graded gravels, gravel and mixtures, little or no fines
			Predominantly one size or a range of sizes with some intermediate sizes missing	GP	Poorly graded gravels, gravel–sand mixtures, little or no fines
		Gravels with fines (appreciable amount of fines)	Nonplastic fines (for identification procedures, see ML below)	GM	Silty gravels, poorly graded gravel–sand–silt mixtures
			Plastic fines (for identification procedures, see CL below)	GC	Clayey gravels, poorly graded gravel–sand–clay mixtures
	Sands More than half of coarse fraction is smaller than No. 7 sieve size (For visual classification, the ¼ in size may be used as equivalent to the No. 7 sieve size)	Clean sands (little or no fines)	Wide range in grain sizes and substantial amounts of all intermediate particle sizes	SW	Well graded sands, gravelly sands, little or no fines
			Predominantly one size or a range of sizes with some intermediate sizes missing	SP	Poorly graded sands, gravelly sands, little or no fines
		Sands with fines (appreciable amount of fines)	Nonplastic fines (for identification cedures, see ML below)	SM	Silty sands, graded sand–silt mixtures
			Plastic fines (for identification procedures, see CL below)	SC	Clayey sands, poorly graded sand–clay mixtures

Identification Procedures on Fraction Smaller than No. 40 Sieve Size

		Dry strength (crushing characteristics)	Dilatancy (reaction to shaking)	Toughness (consistency near plastic limit)		
Fine-grained soils More than half of material is *smaller* than No. 200 sieve size	Silts and clays liquid limit less than 50	None to slight	Quick to slow	None	ML	Inorganic silts and very fine sands, rock flour, silty or clayey fine sands with slight plasticity
		Medium to high	None to very slow	Medium	CL	Inorganic clays of low to medium plasticity, gravelly clays, sandy clays, silty clays, lean clays
		Slight to medium	Slow	Slight	OL	Organic silts and organic silt–clays of low plasticity
	Silts and clays liquid limit greater than 50	Silght to medium	Slow to none	Slight to medium	MH	Inorganic silts, micaceous or diatomaceous fine sandy or silty soils, elastic silts
		High to very high	None	High	CH	Inorganic clays of high plasticity, fat clays
		Medium to high	None to very slow	Slight to medium	OH	Organic clays of medium to high plasticity
Highly organic soils		Readily identified by colour, odour, spongy feel and frequently by fibrous texture			Pt	Peat and other highly organic soils

*Boundary classifications. Soils possessing characteristics of two groups are designated by combinations of group symbols. For example GW–GC, well graded gravel–sand mixture with clay binder.

†All sieve sizes on this chart are U.S. standard.

Field Identification procedure for Fine Grained Soils or Fractions

These procedures are to be performed on the minus No. 40 sieve size particles, approximately $\frac{1}{64}$ in. For field classification purposes, screening is not intended, simply remove by hand the coarse particles that interfere with the tests.

Dilatancy (reaction to shaking):

After removing particles larger than No. 40 sieve size, prepare a pat of moist soil with a volume of about one-half cubic inch. Add enough water if necessary to make the soil soft but not sticky.

Place the pat in the open palm of one hand and shake horizontally, striking vigorously against the other hand several times. A positive reaction consists of the appearance of water on the surface of the pat which changes to a livery consistency and becomes glossy. When the sample is squeezed between the fingers, the water and gloss disappear from the surface, the pat stiffens and finally it cracks or crumbles. The rapidity of appearance of water during shaking and of its disappearance during squeezing assist in identifying the character of the fines in a soil.

Very fine clean sands give the quickest and most distinct reaction whereas a plastic clay has no reaction. Inorganic silts, such as a typical rock flour, show a moderately quick reaction.

Dry Strength (crushing characteristics):

After removing particles larger than No. 40 sieve size, mould a pat of soil to the consistency of putty, adding water if necessary. Allow

Information required for describing soils	Laboratory classification criteria

Information required for describing soils

Give typical name; indicate
approximate percentages of sand
and gravel; maximum size;
angularity, surface condition, and
hardness of the coarse grains;
local or geologic name and other
pertinent descriptive information;
and symbols in parentheses

For undisturbed soils add
information on stratification,
degree of compactness,
cementation, moisture conditions
and drainage characteristics

Example:
Silty sand, gravelly; about 20%
hard, angular gravel particles
$\frac{1}{2}$ in. maximum size; rounded
and subangular sand grains
coarse to fine, about 15%
non-plastic fines with low dry
strength; well compacted and
moist in place; alluvial sand:
(SM)

(vertical text, center) Use grain size curve in identifying the fractions as given under field identification

(vertical text) Determine percentages of gravel and sand from grain size curve

(vertical text) Depending on percentage of fines (fraction smaller than No. 200 sieve size) coarse grained soils are classified as follows:
Less than 5% — GW, GP, SW, SP
More than 12% — GM, GC, SM, SC
5% to 12% — *Borderline* cases requiring use of dual symbols

Laboratory classification criteria

$$C_U = \frac{D_{60}}{D_{10}} \text{ greater than } 4$$

$$C_C = \frac{(D_{30})^2}{D_{10} \times D_{60}} \text{ between 1 and 3}$$

Not meeting all gradation requirements for GW

Atterberg limits below 'A' line, or PI less than 4	Above 'A' line with PI between 4 and 7 are *borderline* cases requiring use of dual symbols
Atterberg limits above 'A' line, with PI greater than 7	

$$C_U = \frac{D_{60}}{D_{10}} \text{ greater than } 6$$

$$C_C = \frac{(D_{30})^2}{D_{10} \times D_{60}} \text{ between 1 and 3}$$

Not meeting all gradation requirements for SW

Atterberg limits below 'A' line or PI less than 5	Above 'A' line with PI between 4 and 7 are *borderline* cases requiring use of dual symbols
Atterberg limits below 'A' line with PI greater than 7	

Give typical name; indicate degree
and character of plasticity,
amount and maximum size of
coarse grains; colour in wet
condition, odour if any, local or
geologic name, and other
pertinent descriptive information,
and symbol in parentheses

For undisturbed soils add
information on structure,
stratification, consistency in
undisturbed and remoulded states,
moisture and drainage conditions

Example:
Clayey silt, brown; slightly plastic;
small percentage of fine sand;
numerous vertical root holes;
firm and dry in place; loess;
(ML)

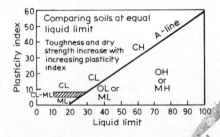

Plasticity chart for laboratory classification of fine grained soils

the pat to dry completely by oven, sun or air drying, and then test its strength by breaking and crumbling between the fingers. This
strength is a measure of the character and quantity of the colloidal fraction contained in the soil. The dry strength increases with
increasing plasticity.
High dry strength is characteristic for clays of the CH group. A typical inorganic silt possesses only very slight dry strength. Silty fine
sands and silts have about the same slight dry strength, but can be distinguished by the feel when powdering the dried specimen.
Fine sand feels gritty whereas a typical silt has the smooth feel of flour.
Toughness (consistency near plastic limit):
After removing particles larger than the No. 40 sieve size, a specimen of soil about one-half inch cube in size, is moulded to the
consistency of putty. If too dry, water must be added and if sticky, the specimen should be spread out in a thin layer and allowed to
lose some moisture by evaporation. Then the specimen is rolled out by hand on a smooth surface or between the palms into a
thread about $\frac{1}{8}$ in. diameter. The thread is then folded and re-rolled repeatedly. During this manipulation the moisture content is
gradually reduced and the specimen stiffens, finally loses its plasticity, and crumbles when the plastic limit is reached.
After the thread crumbles, the pieces should be lumped together and a slight kneading action continued until the lump crumbles.
The tougher the thread near the plastic limit and the stiffer the lump when it finally crumbles, the more potent is the colloidal clay
fraction in the soil. Weakness of the thread at the plastic limit and quick loss of coherence of the lump below the plastic limit
indicate either inorganic clay of low plasticity, or materials such as kaolin-type clays and organic clays which occur below the A-
line.
Highly organic clays have a very weak and spongy feel at the plastic limit.

Table 1.15 British soil classification system for engineering purposes

Soil groups (see 1)			Subgroups and laboratory identification			
GRAVEL and SAND may be qualified Sandy GRAVEL and Gravelly SAND, etc. where appropriate		Group symbol (see 2 and 3)	Subgroup symbol (see 2)	Fines (%) less than 0.06 mm)	Liquid limit (%)	Name
Slightly silty or clayey GRAVEL	G	GW	GW	0 to 5		Well graded GRAVEL
		GP	GPu GPg			Poorly graded/uniform/gap graded GRAVEL
Silty GRAVEL	G-F	G-M	GWM GPM	5 to 15		Well graded/poorly graded silty GRAVEL
Clayey GRAVEL		G-C	GWC GPC			Well graded/poorly graded clayey GRAVEL
Very silty GRAVEL	GF	GM	GML, etc.	15 to 35		Very silty GRAVEL; subdivide as for GC
Very clayey GRAVEL		GC	GCL GCI GCH GCV GCE			Very clayey GRAVEL (clay of low, intermediate, high, very high, extremely high plasticity)
Slightly silty or clayey SAND	S	SW	SW	0 to 5		Well graded SAND
		SP	SPu SPg			Poorly graded/uniform/gap graded SAND
Silty SAND	S-F	S-M	SWM SPM	5 to 15		Well graded/poorly graded silty SAND
Clayey SAND		S-C	SWC SPC			Well graded/poorly graded clayey SAND
Very silty SAND	SF	SM	SML, etc.	15 to 35		Very silty SAND; subdivided as for SC
Very clayey SAND		SC	SCL SCI SCH SCV SCE			Very clayey SAND (clay of low, intermediate, high, very high, extremely high plasticity)

COARSE SOILS less than 35% of the material is finer than 0.06 mm

GRAVELS More than 50% of coarse material is of gravel size (coarser than 2 mm)

SANDS more than 50% of coarse material is of sand size (finer than 2 mm)

FG	Gravelly SILT	MG	MLG, etc.		Gravelly SILT; subdivide as for CG
FG	Gravelly CLAY (see 4)	CG	CLG CIG CHG CVG CEG	<35 35 to 50 50 to 70 70 to 90 >90	Gravelly CLAY of low plasticity of intermediate plasticity of high plasticity of very high plasticity of extremely high plasticity
FS	Sandy SILT (see 4)	MS	MLS, etc.		Sandy SILT; subdivide as for CG
FS	Sandy CLAY	CS	CLS, etc.		Sandy CLAY; subdivide as for CG
F	SILT (M-SOIL)	M	ML, etc.		SILT; subdivide as for C
F	CLAY (see 5 and 6)	C	CL CI CH CV CE	<35 35 to 50 50 to 70 70 to 90 >90	CLAY of low plasticity of intermediate plasticity of high plasticity of very high plasticity of extremely high plasticity

FINE SOILS more than 35% of the material is finer than 0.06 mm

- SILTS and CLAYS 35% to 65% fines — Gravelly or sandy
- SILTS and CLAYS 65% to 100% fines

ORGANIC SOILS	Descriptive letter 'O' suffixed to any group or sub-group symbol.	Organic matter suspected to be a significant constituent. Example MHO: Organic SILT of high plasticity.
PEAT	Pt	Peat soils consist predominantly of plant remains which may be fibrous or amorphous.

1. The name of the soil group should always be given when describing soils, supplemented, if required, by the group symbol, although for some additional applications (e.g. longitudinal sections) it may be convenient to use the group symbol alone.
2. The group symbol or sub-group symbol should be placed in brackets if laboratory methods have not been used for identification, e.g. (GC).
3. The designation FINE SOIL or FINES, F, may be used in place of SILT, M, or CLAY, C, when it is not possible or not required to distinguish between them.
4. GRAVELLY if more than 50% coarse material is of gravel size. SANDY if more than 50% coarse material is of sand size.
5. SILT (M-SOIL), M, is material plotting below the A-line, and has a restricted plastic range in relation to its liquid limit, and relatively low cohesion. Fine soils of this type include clean silt-sized materials and rock flour, micaceous and diatomaceous soils, pumice, and volcanic soils, and soils containing halloysite. The alternative term 'M-soil' avoids confusion with materials of predominantly silt size, which form only a part of the group. Organic soils also usually plot below the A-line on the plasticity chart, when they are designated ORGANIC SILT, MO.
6. CLAY, C, is material plotting above the A-line, and is fully plastic in relation to its liquid limit.

Table 1.16 Field identification and description of soils

	Basic soil type	Particle size (mm)	Visual identification	Particle nature and plasticity	Composite soil types (mixtures of basic soil types)
Very coarse soils	BOULDERS		Only seen complete in pits or exposures	Particle shape:	Scale of secondary constituents with coarse soils
	COBBLES	200	Often difficult to recover from boreholes	angular, subangular,	Term — % of clay or silt
		60		subrounded,	slightly clayey / slightly silty — GRAVEL or SAND — under 5
	GRAVELS coarse		Easily visible to naked eye; particle shape can be described; grading can be described	rounded, flat, elongate,	
		20	Well graded: wide range of grain sizes, well distributed. Poorly graded: not well granded. (May be uniform: size of most particles lies between narrow limits; or gap graded: an intermediate size of particle is markedly under-represented)		— clayey / — silty — GRAVEL or SAND — 5 to 15
	medium	6			very clayey / very silty — GRAVEL or SAND — 15 to 35
	fine			Texture:	
Coarse soils (over 65% sand and gravel sizes)	SANDS coarse	2	Visible to naked eye; very little or no cohesion when dry; grading can be described	rough, smooth, polished,	Sandy GRAVEL / Gravelly SAND — Sand or gravel and important second constituent of the coarse fraction (see Table 1.17)
		0.6			
	medium		Well graded: wide range of grain sizes, well distributed. Poorly graded: not well graded. (May be uniform: size of most particles lies between narrow limits; or gap graded: an intermediate size of particle is markedly under-represented)		For composite types described as: clayey: fines are plastic, cohesive: silty: fines non-plastic or of low plasticity
		0.2			
	fine	0.06			
Fine soils (over 35% silt and clay sizes)	SILTS coarse		Only coarse silt barely visible to naked eye; exhibits little plasticity and marked dilatancy; slightly granular or silky to the touch. Disintegrates in water; lumps dry quickly; possess cohesion but can be powdered easily between fingers	Non-plastic or low plasticity	Scale of secondary, constituents with fine soils
		0.02			Term — % of sand or gravel
	medium	0.006			sandy / gravelly — CLAY or SILT — 35 to 65
	fine	0.002			— CLAY:SILT — under 35
	CLAYS		Dry lumps can be broken but not powdered between the fingers; they also disintegrate under water but more slowly than silt; smooth to the touch; exhibits plasticity but no dilatancy; sticks to the fingers and dries slowly; shrinks appreciably on drying usually showing cracks. Intermediate and high plasticity clays show these properties to a moderate and high degree, respectively	Intermediate plasticity (lean clay)	Examples of composite types
					(Indicating preferred order for description)
				High plasticity (fat clay)	Loose, brown, subangular very sandy, fine to coarse GRAVEL with small pockets of soft grey clay
Organic soils	ORGANIC CLAY, SILT or SAND	Varies	Contains substantial amounts of organic vegetable matter		Medium dense, light brown, clayey, fine and medium SAND Stiff, orange brown, fissured sandy CLAY
	PEATS	Varies	Predominantly plant remains usually dark brown or black in colour, often with distinctive small; low bulk density		Firm, brown, thinly laminated SILT and CLAY Plastic, brown, amorphous PEAT

Compactness/strength		Structure			Colour
Term	Field test	Term	Field identification	Intervals scales	
Loose	By inspection of voids and particle packing	Homogeneous	Deposit consists essentially of one type	Scale of bedding spacing	Red,
Dense		Interstratified	Alternating layers of varying types or with bands or lenses of other materals Interval scale for bedding spacing may be used	Term — Mean spacing (mm)	pink yellow brown olive green,
Loose	Can be excavated with a spade; 50 mm wooden peg can be easily driven	Heterogeneous	A mixture of types	Very thickly bedded — > 2000 Thickly bedded — 2000–600	blue, white, grey,
		Weathered	Particles may be weakened and may show concentric layering	Medium bedded — 600–200	black, etc.
Dense	Required pick for excavation; 50 mm wooden peg hard to drive			Thinly bedded — 200–60	
Slightly cemented	Visual examination; pick removes soil in lumps which can be abraded			Very thinly bedded — 60–20 Thickly laminated — 60–20 Thinly laminated — < 6	Supplemented as necessary with: light, dark, mottled, etc. and

Compactness/strength		Structure			Colour
Soft or loose	Easily moulded or crushed in the fingers	Fissured	Break into polyhedral fragments along fissures. Interval scale for spacing of discontinuities may be used		pinkish, reddish,
Firm or dense	Can be moulded or crushed by strong pressure in the fingers				yellowish, brownish,
Very soft	Exudes between fingers when squeezed in hand	Intact	No fissures		etc.
Soft	Moulded by light finger pressure	Homogeneous	Deposit consists essentially of one type	Scale of spacing of other discontinuities	
Firm	Can be moulded by strong finger pressure	Interstratified	Alternating layers of varying types. Interval scale for thickness of layers may be used	Term — Mean spacing (mm)	
Stiff	Cannot be moulded by fingers. Can be indented by thumb	Weathered	Usually has crumb or columnar structure	Very widely spaced — > 2000	
Very stiff	Can be indented by thumb nail			Widely spaced — 2000–600	
				Medium spaced — 600–200	
Firm	Fibres already compressed together			Closely spaced — 200–60	
Spongy	Very compressible and open structure	Fibrous	Plant remains recognizable and retain some strength	Very closely spaced — 60–20	
Plastic	Can be moulded in hand, and smears fingers	Amorphous	Recognizable plant remains absent	Extremely closely spaced — < 20	

spaced joints with occasional vertical joints open 5 mm, slightly to moderately weathered, porphyritic biotite *Granite*, very strong, slightly permeable.

- Reddish brown, compact, sub-angular, well graded, clean sandy *Gravel*, highly permeable.
- Dark grey, stiff, closely fissured, *Clay*, of high plasticity, slightly permeable, slakes slowly on exposure.

The geotechnical description of rock masses is specified in an international report (ISRM, 1974 *et seq.*) and is also the subject of a comprehensive report by Chaplow *et al.* (1977). Comparison of some of the tables in this latter publication with Tables 1.6, 1.8 and 1.9 above reveal some discrepancies in terminology which, in general, are not significant.

The best known system of soil classification was put forward by A. Casagrande in 1947 and subsequently this has been subjected to minor modifications by different authorities: the United States Unified Classification (Wagner, 1957) is given in Table 1.14. A major revision has been made by Dumbleton (1968, 1981) and this forms the basis of the classification set out in BS 5930 (BSI, 1981a) and reproduced here as Table 1.15. Table 1.16 provides a key to field identification and description. Tables 1.15 and 1.16 are taken from BS 5930 by permission of the British Standards Institution. Complete copies of the document can be obtained from BSI at Linford Wood, Milton Keynes, MK14 6LE. Figure 1.16 relates to the classification of fine soils and the finer part of coarse soils and Fig. 17

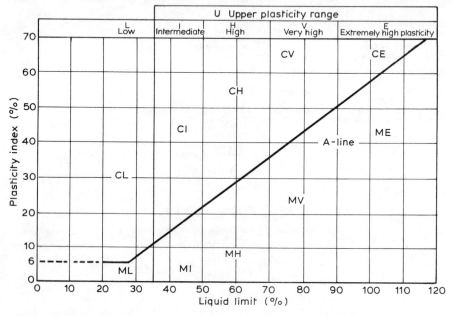

Figure 1.16 Plasticity chart for the classification of fine soils and the finer part of coarse soils (measurements made on material passing 425 μm sieve).

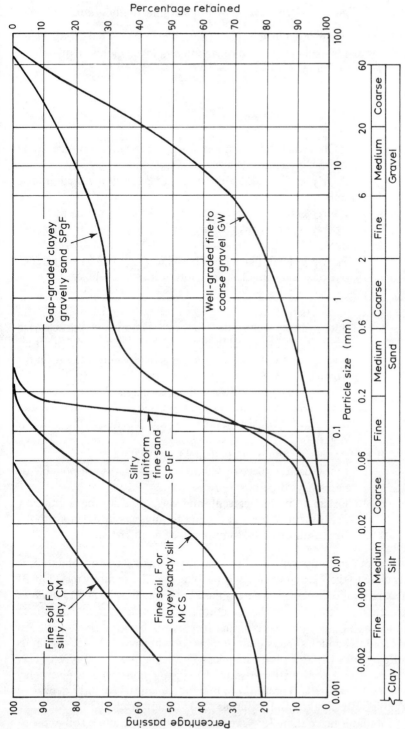

Figure 1.17 Grading for some soil types. The following parameters can be used to describe a particle size distribution curve, where d_{10}, d_{15}, d_{30} and d_{60} are the particle sizes corresponding respectively to 10, 15, 30 and 60% passing. Effective size d_{10}: sizes less than considered to influence behaviour of soil as much as remaining 90%, sometimes considered this should be d_{15}. Coefficient of uniformity $C_u = d_{60}/d_{10}$: measure of size range. Coefficient of curvature $C_c = d_{30}/d_{10} \times d_{60}$: broadly defines shape of curve. For GU or SU, C_u tends to unity. For GW or SW, C_u exceeds ~ 5 and C_c exceeds ~ 3. Example: SPuF, d_{10} = 0.07 mm, d_{30} = 0.12 mm, d_{60} = 0.14 mm, C_u = 2.0, C_c = 1.5.

shows grading curves for selected soil types. (Both figures are based on BS 5930.) The following is a brief explanation of the symbols employed in the classification. Note that $1 \mu m = 10^{-3} mm$.

Coarse fraction

G	Gravel; 2 to 60 mm.
S	Sand; 60 μm to 2 mm.
W	Well-graded; having no excess of material in any size range and no intermediate sizes lacking.
P	Poorly-graded; outside the grading limits for well-graded soils.
Pu	Poorly-graded; particles of relatively uniform size; particle size range is narrow, giving a large proportion of intermediate sizes.
Pg	Poorly-graded; particles of both large and small sizes present but relatively small proportion of intermediate sizes.

Fine fraction

M	Silt; primary definition relates to material in particle size range 2–60 μm; in relation to soil classification defined as a fine soil with consistency limits which plot below the A-line in Fig. 1.16.
C	Clay; primary definition relates to material in particle size range below 2 μm; in relation to soil classification defined as a fine soil with consistency limits which plot above the A-line in Fig. 1.16.
L, I, H, V, E, U	Ranges of liquid limits defined in Fig. 1.16.
F	Fines; used in place of M and/or C when it is not possible or necessary to be more precise.
O	Organic matter present and suspected to be significant constituent; added as a suffix to any classification symbol.
Pt	Peat; soil consists dominantly of plant remains.

The *in situ* moisture content of silts and clays, when well-consolidated, is commonly close to the PL and values approaching the LL are rarely met. The PI of a silt may be no more than 5% whereas the PI of a clay may be 50%. It follows that an increase of about 5% in the moisture content of a silt may cause it to behave almost as a liquid, whereas a similar increase in the moisture content of a clay would lead to a much smaller change in its consistency and hence also in its shear strength. Both LL and PI are influenced by the amount and type of clay minerals present in a soil. An increase in the organic content of a given clay increases both LL and PL. The *in situ* moisture content of recently deposited highly organic soils in lakes and estuaries tends to approach the LL.

A soil containing more than 35% silt and clay is termed a fine soil and one

Table 1.17

Term	Composition
Slightly sandy gravel	< 5% sand
Sandy gravel	5% to 20% sand
Very sandy gravel	> 20% sand
Gravel/sand	about equal proportions of gravel and sand
Very gravelly sand	> 20% gravel
Gravelly sand	20% to 5% gravel
Slightly gravelly sand	< 5% gravel

Table 1.18 Particle shape and character

Angularity	Form
Angular	Equidimensional
Subangular	Flat (alternatives are tabular or flaky)
Subrounded	Elongated
Rounded	Flat and elongated
	Irregular
Surface texture	
rough	More specific terms are etched, pitted,
smooth	honeycombed or polished

containing more than 65% sand and gravel is termed a coarse soil. The terminology applied to mixtures of sand and gravel specified in BS 5930 (BSI, 1981a) is shown in Table 1.17. Particle shape and character can be described by the terms specified in BS 5930 (BSI, 1981a) and shown in Table 1.18.

In relation to particle-size analysis, the results of which are illustrated by Fig. 1.17, it should be noted that the sedimentation process employed for silt and clay fractions is relatively expensive. Furthermore, the results are determined by the application of Stokes' Law which applies strictly to a single spherical particle sinking in a fluid of infinite extent. Inaccuracy is increased further by the effect of pre-treatment and by the degree of deflocculation. Again, the values of particle specific gravity, G_s for silt (commonly quartz, $G_s \simeq 2.65$) and clay (commonly clay minerals, $G_s = 2$ to 3) are likely to be slightly different: other minerals may be present, sometimes as coatings on particles. Although sedimentation analysis may be justified in a few cases, it will be appreciated that the results are of dubious value. It is more reasonable to determine particle size distribution by sieving to the finest possible size, usually slightly finer than 0.1 mm, and to extrapolate by eye as far as justifiable into the silt fraction. Commonly this enables the clay fraction to be estimated to within $\pm 10\%$ and a knowledge of the kind of clay minerals present, reflected in liquid and plastic limits, is of greater importance than accurate determination of clay fraction percentage. It should be observed that fractions coarser than clay may include aggregations of clay particles which, in service, can disintegrate by weathering or under repeated loads (Section 1.4.2). It should be observed also that wet sieving is generally more efficient than dry sieving and usually yields a few per cent more of finer material but it tends to be more expensive because of the necessity of removing large quantities of water. When it is desirable to minimize the influence of heat on the character of clay minerals, drying is best effected in a tunnel kiln with a fan since the temperature is usually less than 100°C. The results of analyses are expressed as percentages by mass but these are unlikely to correspond exactly with percentages by volume because of the slight differences in particle specific gravity. Values of particle specific gravity can usually be determined accurately to three significant figures, which is adequate for most purposes, but even this requires water bath temperature to be known and efficient removal of air by vacuum pump, and, because of this latter necessity, the determination of specific gravity of clay minerals and porous particles demands perseverance.

Figure 1.18 is one of several slightly different triangular diagrams produced by various authorities to provide names for soils, mainly for agricultural purposes. It is interesting to note the wide range of soil composition covered by the classification clay. In fact, of course, provided sufficient clay is present to just

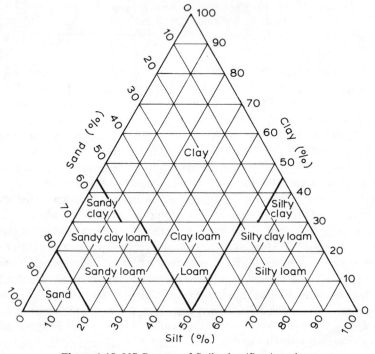

Figure 1.18 US Bureau of Soils classification chart.

more than fill voids in the coarser fractions, a soil will behave largely as clay. The classification illustrated has been employed by US Bureau of Soils and is sufficiently simple to be remembered. The diagram employed by Soil Survey of England and Wales is slightly more complex and, although the fractions vary by a few per cent, the nomenclature is very similar. The author believes that much time can be wasted in naming soils precisely, as in the case of colour identification mentioned above, since engineering behaviour depends not only on soil type but on degrees of compaction and saturation and other factors.

1.2.3 Site investigation plans and maps

The maps produced at scales of 1/50 000, 1/25 000 and 1/10 000, or their earlier equivalents, by the Institute of Geological Sciences for the United Kingdom are worthy of perusal at least in the planning stages of most projects. It is desirable to consult the Memoir relating to any particular map, if it is available, to make best use of the information. Within the last decade the practice of producing engineering geology plans and maps for specific projects has been developing. The principles and techniques involved in this have been specified in a report (Anon, 1972a) and applications have been described by Dearman and Fookes (1974), Burnett and Fookes (1974) and Dearman *et al.* (1977). The first part of a report on a symposium on engineering geological mapping at the University of Newcastle upon Tyne in 1977 appears in Bulletin No. 19 (1979) of the International Association of Engineering Geology and the second part in Vol. 12 (1979) of the *Quarterly Journal of Engineering Geology.*

During the last three decades, engineering soils maps have been produced for particular limited areas in a number of countries. Often the information has been linked to existing pedological maps produced for agricultural purposes. Generally these maps are of greatest service in regional planning and the preliminary stages of projects. Pedological maps are produced by the Soil Surveys of England and Wales and of Scotland and related memoirs or bulletins are usually available. The pedological classification of a soil is based on the characteristics of the entire soil profile (Fig. 1.19) and this rarely exceeds 1 or 2 m before the parent material is encountered. The engineering properties of the soil may vary from horizon to horizon within the profile and, because of the difficulty of presenting three-dimensional data on a map, it is often appropriate to sample from, say, 0.75 to 1.00 m only for engineering classification since this is reasonable depth for foundations of many structures. Pedological classification for mapping purposes includes thickness of profile, drainage conditions and other characteristics and an experienced interpreter should be able to predict engineering properties in broad terms solely by reference to pedological map and memoir.

The basic unit of pedological classification for mapping purposes is the soil series and this is defined as a group of soils possessing similar profiles and developed from lithologically similar parent material. A particular series is often named after the locality where the soil was first described or where it is extensively

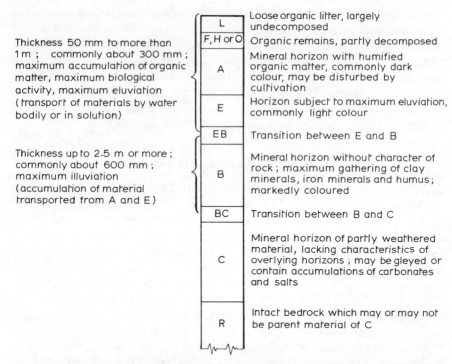

Thickness 50 mm to more than 1 m ; commonly about 300 mm ; maximum accumulation of organic matter, maximum biological activity, maximum eluviation (transport of materials by water bodily or in solution)

Thickness up to 2.5 m or more ; commonly about 600 mm ; maximum illuviation (accumulation of material transported from A and E)

L	Loose organic litter, largely undecomposed
F, H or O	Organic remains, partly decomposed
A	Mineral horizon with humified organic matter, commonly dark colour, may be disturbed by cultivation
E	Horizon subject to maximum eluviation, commonly light colour
EB	Transition between E and B
B	Mineral horizon without character of rock ; maximum gathering of clay minerals, iron minerals and humus ; markedly coloured
BC	Transition between B and C
C	Mineral horizon of partly weathered material, lacking characteristics of overlying horizons ; may be gleyed or contain accumulations of carbonates and salts
R	Intact bedrock which may or may not be parent material of C

Figure 1.19 Generalized soil profile possessing all principal horizons. The thickness of each horizon varies with soil series and may be zero. The notation given above is that employed by Soil Survey (Hodgson, 1974) and generally accords with recommendations of the International Society of Soil Science made in 1967. Suffixes in lower case are employed to express additional characteristics: for example, Ap implies that the A horizon has been cultivated and Bg that the B horizon has been subjected to gleying. Gleying is caused by periodic or continuous saturation by water, leading to reduction and segregation of iron and is characterized by reddish mottles or greenish and bluish colouring.

developed. Boundaries between soil series are seldom well defined and there is usually a transition, sometimes of appreciable width. In some localities the distribution of soils is so complicated, and the boundaries so intricate or indefinite, that delimitation of individual series is impossible at the scale employed for mapping: the soils in such localities are then distinguished as soil complexes. A soil series may be subdivided into phases based on differences in stone content, slope, depth to a lithologically contrasting material or degree of accelerated erosion. A soil association is a group of named taxonomic soil units which occur regularly in a well-defined pattern.

The areal redistribution of the constituents of decomposing rocks takes place by physical and chemical means. In physical redistribution, soil and weathered rock on high ground is transported downhill by creep, sheet-wash and groundwater and relatively coarse detritus is found on the upper steeper hillsides, followed by finer material on the lower slacker slopes and silt and clay in the valley. Chemical redistribution commonly involves a concentration of residual silica on high ground and leaching of other minerals which are transported downhill. Iron and aluminium compounds accumulate on the hillsides and calcium and magnesium minerals are transported into the valley where they accumulate if drainage is impeded or disperse via streams and rivers if drainage is free.

In a particular locality, this redistribution of minerals leads to the formation of a group – known as a soil catena – of contrasting but related soils derived from similar parent material. A common example of a catenary sequence of soils occurs in undulating country in the tropics where the ridges are covered with red loam overlying pan ironstone, the slopes are swathed with soil containing ferruginous concretions and the valleys have a veneer of greyish soils. Where clay occurs both on a hill and in a valley, it may be found that the dominant clay mineral on the hill is kaolinite and in the valley montmorillonite. In this case silica is leached and transported from the high ground to the valley and montmorillonite has a higher percentage of silica than kaolinite: the development of the former is favoured by an alkaline environment and the latter by acid conditions.

Several comprehensive engineering soil studies have been reported, including one by Thornburn *et al.* (1970) and another by Anon (1971). Thornburn and Larsen (1959) have investigated the number of samples required to characterize each soil type and Morse and Thornburn (1961) studied the reliability of soil map units. The kind of engineering information which can be derived from pedological surveys is illustrated in a study of soils in Berkshire, England (Jarvis *et al.*, 1979) and further details are given by Marsland *et al.* (1980).

Under certain climatic conditions, mainly tropical and sub-tropical, pedogenic processes lead eventually to the formation of an indurated bed, sometimes as much as 10 m or more thick, rich in particular minerals concentrated by seasonal downward and upward movement of water. Such beds are termed duricrusts and their areal extent may be tens or hundreds of square kilometres, when they are said to form carapaces or cuirasses, that is armour-plates. The minerals concentrated are commonly aluminous, ferruginous, siliceous or calcareous forming respectively bauxite, ferricrete, silcrete and calcrete. Bauxite are ferricrete are jointly known as laterite. Less frequently, duricrusts are formed of other minerals: for example, manganese. Duricrusts occur in many regions of the Earth: notable examples are the ferricrete of East and West Africa, the bauxite of Northern Australia, the silcrete of Western Australia and the calcrete of North Africa. The minerals concentrated are determined partly by the nature of the parent rock and partly by the climatic regime. It appears likely that many superficial duricrusts were formed in late Tertiary times on peneplains, where lateral movement of water would be minimal, but some have developed in the Quaternary. Duricrusts have also been formed during other geological periods and can be termed palaeoduricrusts. Subsequent to their formation, duricrusts may weather to boulders, gravel, sand and finer material. The processes and modes of occurrence described above are more complex and varied than may be inferred from this brief paragraph.

Within the last two decades methods of terrain evaluation have been developed for several purposes which are described by Mitchell (1973). Classification is based to a large extent on geomorphology and aerial photographs provide much of the data. Correlation can be established between terrain classifications and the engineering properties of soils, and the method has proved valuable for highway studies in developing countries (Dowling and Williams, 1964; Dowling, 1968; Dowling and Beaven, 1969; TRRL 1978). Terrain evaluation involves the recognition of distinctive patterns of landscape, termed land systems. These are determined by a number of interdependent factors such as geology, geomorphology, climate, soils and vegetation. It is considered that in different localities, where these factors operate together in the same manner, the form of the landscape will be essentially the same. The simplest unit of landscape is the land element, possessing uniform geology, geomorphology, soils and vegetation and is usually too small to be identified on aerial photographs at a scale of 1 in 20 000 or less. Land facets are formed from groups of genetically related land elements and can be readily identified on aerial photographs at scales between 1 in 10 000 and 1 in 80 000. They possess a reasonably high degree of homogeneity for most practical purposes and they are the most significant units of landscape in terms of engineering use. The recurrence of combinations of linked land facets characterize a particular land system. Thus, the terrain of a region is conceived as a series of land systems, each of which contains typical land facets which have similar features wherever they occur and which can be identified on aerial photographs. Within each land facet there are typical land elements with definable engineering characteristics. An essential part of terrain evaluation studies is the storage of information for subsequent use. Numerous references have been given by Mitchell (1973), including the valuable studies carried out by the Military Engineering Experimental Establishment at Christchurch, and a comprehensive study in Australia worthy of perusal has been reported by Grant (1970). Principles and applications of geomorphological mapping have been described by Brunsden *et al.* (1975) and Bromhead *et al.* (1983) and a report on land surface evaluation for engineering practice has recently been published (Edwards *et al.*, 1982).

Two survey techniques which can be employed to provide supplementary information in site investigations are photo-interpretation and geophysical exploration. Both techniques have considerable potential but require highly specialized and skilled interpretation and money can be wasted unless they are pursued with determination to make investigations as successful as possible. Aerial photographs may reveal features, such as old river courses and unstable slopes, which may not be detected in ground surveys or may be missed in the layout of boreholes. Photo-interpretation is commonly carried out with vertical stereo-pairs but oblique photographs may be revealing in some circumstances, and cliffs and excavation faces can be surveyed by terrestrial stereo-pairs. The plotting of major joints in the Oxford Clay by terrestrial photogrammetry has been described by Moore (1974). The use of aerial photographs in soil surveys has been described, for example, by Garner and Heptinstall (1974) and Carroll *et al.* (1977) and in terrain evaluation by Webster and Beckett (1970). The employment of geophysical surveys in site investigations has been assessed by the Working Party (Anon, 1972a) and by West and Dumbleton (1975). Methods for geophysical logging of boreholes are the subject of an international report (ISRM, 1974 *et seq.*) and geotechnical applications have been discussed by Crosby *et al.* (1981). Stewart and Beaven (1980) have described the use of seismic refraction surveys for highway engineering purposes.

The ancillary investigations discussed in this section can seldom replace the basic requirements of good topographical and geological maps and plans. Springs, ditches and other features are usually recorded on larger scale maps and contours are generally more significant to engineers than arrows indicating surface slope. It is difficult to represent three-dimensional sub-surface data on maps, although this is achieved on pedological maps for varying depths down to a metre or so by employing soil series and other mapping units, and engineers thus require geological sections produced as accurately as possible from maps and borehole data. The geology of an area is to some extent unique and the advice of a geologist versed by personal experience of the area in question should be sought when a project is first mooted. Much interpretation and advice depends on the minutiae of geology, such as subtle stratigraphic markers, variations in weathering and superficial structures.

1.2.4 Site plate bearing tests

The site plate bearing test is employed for determination of the settlement characteristics of rocks and cohesionless soils and, more particularly within the last decade, for studies of settlement characteristics and bearing capacity of cohesive soils. Since plate sizes are relatively small and observations relate only to soil characteristics immediately beneath the loaded area, laboratory consolidation and triaxial tests generally provide the most convenient means of estimating the settlement and bearing capacity of foundations on cohesive soils and site penetration tests are commonly employed for cohesionless soils. Plate bearing tests are more reliable for weak shales, limestones and sandstones and for clayey gravels, disturbed chalk and partly decomposed rocks such as weathered granite.

Deformation of soils and rocks under load is discussed in Section 2.1. Consolidation is negligible or absent in many rocks subjected to static or dynamic loads and in soils subjected to transient loads, as in the case of pavement subgrades. The term elastic implies that strains are completely recoverable on removal of load whereas plastic strains constitute permanent set. The value of modulus of subgrade reaction is commonly based on the secant modulus at the maximum working stress on the graph of bearing stress against settlement for a

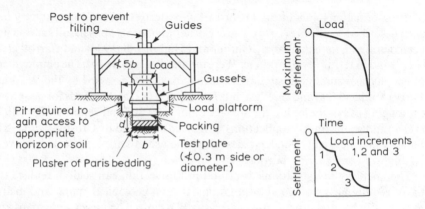

Figure 1.20 Plate bearing test.

plate bearing test. This value of the modulus is used for all settlement calculations between zero and the maximum working stress and implies the assumption of a linear stress/strain relationship.

A simple arrangement for a plate loading test employing kentledge is shown in Fig. 1.20. Relatively light loading can be provided conveniently by one or more rubbish skips filled with sand or other material. Where the required load is considerable, it is preferable to place the kentledge on a platform supported on piers standing clear of the test area and to apply the load by means of an hydraulic jack inserted between platform and test plate. Alternatively a jack can be inserted between the test plate and a yoke beam supported on blocks and anchored to rods or high-tensile steel wires grouted in boreholes in the rock. A short beam is desirable to avoid excessive deflection at high loads which may reduce the effective travel of the jack. Care should be taken to prevent interference between the stress field beneath the plate and the stress fields surrounding the anchorages. It is commonly considered that the distance between the anchorages and the plate centre should be not less than eight times the plate diameter. This distance can be reduced by grouting only the lower part of the anchorages but, in any particular case, stress patterns can be studied with a photo-elastic model. Settlement can be recorded with a geodetic level and invar staff or by dial gauges carried on a frame supported at points well clear of the test plate. Settlement observations should be made preferably at the quarter-points on the periphery, reading to 0.01, 0.1 or 1 mm as appropriate. Air temperature should be recorded at intervals during the observations to assist in accounting for inconsistent readings although it may not be possible to arrive at corrections for variations in temperature. The equipment should be protected from wind and sun when readings are taken to a high degree of accuracy. The pressure gauge on an hydraulic jack is not generally sufficiently accurate for the measurement of load and it is necessary to employ a proving ring or a dynamometer.

The confining stress of overburden influences the behaviour of soil and rock at depth and tests should be carried out in pits, if possible at estimated founding level, with plates which completely cover the pit bottom. This introduces complications in testing since plates are usually less than 1 m wide and, furthermore, plates of different sizes should be employed but such problem can be overcome, at least in part, by trimming the bottom metre of the pit to a taper, or by other means. Plate bearing tests can be successfully carried out in boreholes. Corrections based on assumed elastic conditions are discussed in Section 2.4 to take account of depth below ground surface and for the difference in proportions of plate and pit.

Plate bearing tests performed in boreholes require careful consideration since large-diameter boreholes are expensive. Apart from foundations for one- and two-storey buildings, most foundations are at least 1 m wide and, in order that tests should reasonably reflect the response of the ground to loading, plates should be as large as possible. Although plates seldom exceed 0.75 to 1.0 m, larger diameter load tests with water-filled tanks may be justified for very important structures, although such tests would not be executed in boreholes. All plate bearing tests should reflect the character of the ground in horizontal planes and, for example, a plate should cover a sufficient number of vertical fissures. The changing character of the ground with depth should also be reflected by the test. No test with a normal-size plate can be expected to satisfy these demands completely in relation to the behaviour of the ground subjected to loads from actual foundations. A further aspect of testing to be taken into account is that the ground surface should be cleared of loose material and trimmed carefully before bedding down the plate in order to obtain reliable results: the diameter of a borehole must be sufficient for a man to work in for efficient preparation, that is not less than 0.75 m. Nevertheless, it can be argued that in practice the ground surface is seldom trimmed carefully before placing concrete for foundations: this does not imply that slovenly workmanship in testing or construction can be condoned. Results may be influenced also by disturbance of joints in the floor and wall of a borehole and this effect may not be eliminated completely by careful trimming of the floor. However, concrete or mortar entering joints during installation of the test equipment may reduce the effect and this applies also to the installation of actual foundations. The smallest diameter of boreholes in which plate bearing tests have been carried out is probably 150 mm. Such small boreholes are usually specified on account of financial stringency, although some technical justification can be made in certain cases, the results can never be as reliable as those obtained in larger diameter holes.

Loads can be applied to a test plate in a borehole through a stiff mandrel or column in the borehole jacked down from a kentledge platform or a yoke spanning anchor piles of cables. Alternatively, load is applied hydraulically to a disposable jack or a flexible bag (5 MN m^{-2} maximum hydraulic pressure) at the bottom of the hole, reacting against concrete filling to the borehole. Settlement rods or other devices must be located on the test plate. It should be observed that settlement will be influenced to some extent by modification of the stress pattern in the ground around the jack arising from reaction of the jack against the concrete filling. Kay and Avalle (1982) and Kay and Parry (1982) have described the use of screw-plates in boreholes.

The zone of soil affected by a surface load is relatively shallow beneath a test plate, which rarely exceeds 0.75 m diameter, and the influence of vertical variations in strata on the behaviour of a full size footing will not necessarily be apparent from tests. Normally, plate tests can be employed as a guide to bearing capacity and settlement only where the soil is homogeneous and isotropic to a depth beyond which stress imposed by the full size footing will have little influence; that is, about twice the width of the footing. This applies only where footings are widely spaced otherwise the effect of overlapping of the stress bulbs must be taken into account. Where a plate test is employed on clay, the influence of consolidation will be included only partly in the results since it is not normally practicable to maintain loads for long periods.

The following procedure is commonly employed to determine bearing capacity by means of a plate bearing test. The load is applied in increments of 0.2 × estimated working stress. Settlement is observed for each increment at intervals of 1 h during a period of 6 h and then at 12 h intervals until no measurable settlement occurs during a period of 24 h. The next increment is then applied and the observations repeated. The maximum test load is either the ultimate load or twice the estimated working stress multiplied by the test plate area, whichever is the smaller. The ultimate load is reached when settlement continues without increase, or with very little increase, in stress. If failure is not conspicuous, then the ultimate load is taken as that at which the settlement is $0.2b$ where b is the plate width. It is desirable that the tests should be repeated with plates of two or three different sizes, say 300, 500 and 700 mm square or diameter. The ultimate stress for a full-size footing of width B on clay is taken equal to the ultimate test stress and on sand is assumed to be proportional to B. The allowable stress is the ultimate stress divided by a factor of safety.

When the test is employed to determine modulus of subgrade reaction, it is usual to apply the load in a series of cycles of loading and unloading up to at least the highest estimated working stress and observations are recorded at shorter intervals of time than in the test for settlement and bearing capacity. Subgrade reaction is the bearing stress q per unit area of the contact surface between a foundation and the subgrade on which it rests. The modulus or coefficient of subgrade reaction is $k_s = q/\rho$ where ρ is the settlement produced under the stress q. The value of k_s depends on the deformation characteristics of the subgrade and on the shape and dimensions of the contact area. These terms relate primarily to foundations which are horizontal or inclined. In the case of structures, such as sheet piling, which are vertical, the parameter is known as the coefficient of horizontal subgrade reaction k_h. It can be assumed that, for stiff clay, k_h has the same value at all points on the contact area but for cohesionless sands and gravels a better approximation is $k_h = m_h z$ where m_h is a parameter of constant magnitude in a particular case and z is depth below ground level or dredge line.

The following notes deal mainly with k_s but details concerning k_h, and k_s also, are given by Terzaghi (1955).

Where a foundation or bearing test plate is located on cohesionless soil, the ultimate settlement is attained immediately. The modulus of elasticity of such material increases with depth and the effective value of k_s increases with increase in width of foundation since the bulb of stress extends to greater depths. Investigations show that the value of k_s for a foundation of width B m on cohesionless soil can be expressed as

$$k_s = k_{sb}\left(\frac{B+b}{2B}\right)^2 \tag{1.8}$$

where k_{sb} is the value derived from a test on a plate of width $b = 0.3$ m (Terzaghi and Peck, 1967). As B increases, an asymptotic value of $0.25 k_{sb}$ is approached.

Where a foundation or test plate is located on stiff clay it can be assumed that settlement is proportional to the width of the contact area, in conformity with the theory of the behaviour of loaded elastic media, and hence

$$k_s = k_{sb}\frac{b}{B}. \tag{1.9}$$

The results of a comprehensive series of plate tests on rock for Karadj Dam, Iran, have been presented by Waldorf *et al.* (1963), with a discussion of the influence of joints, zones of decomposed rock and other factors. Plate bearing tests have been discussed also by Burmister (1962), Hansen (1961), Meigh and Nixon (1961), Meigh and Greenland (1965) and Anderson and Stenhamar (1982). The use of a pressure-meter for load tests on the wall of a small-diameter borehole has been described by Meigh and Greenland (1965). Equations expressing the results of plate bearing tests on chalk at Erith, Kent, have been quoted by Carey and Cumming (1961). Ward *et al.* (1965) found that plate bearing tests on London Clay are markedly influenced by the size of the plate in relation to the spacing of fissures and also by the length of time which elapses between completion of trial pit excavation and application of the test loads. Excessive initial settlement occurs with some rocks until the near-surface fissures close as the load is increased.

When a test is made on soft, brittle rocks such as chalk, it is sometimes found that preliminary failure occurs under comparatively small loads, due to rupture of individual blocks of rock, and this is followed firstly by a decrease in the rate of settlement as the load is increased, due to compaction of the disrupted rock, and finally by ultimate failure.

Another mode of behaviour is also observed with some grades of chalk. Burland and Lord (1969) found that the graph of stress against settlement could be divided into two phases distinguished by projecting the second part back to the stress axis giving, q_y, the nominal yield or threshold stress. For the grades tested, q_y lies between 400 and 600 kN m^{-2} in round figures and appears to be independent of depth beneath ground surface. Below q_y is q_e, the limit of approximate linearity.

There is a common tendency to load to unduly high values of stress, probably in an endeavour to ascertain if yield is evident. Since settlement is frequently the criterion of design and not bearing capacity, it is usually sufficient to load to 50% to 100% in excess of the anticipated working stress. In some tests the first increment of load has exceeded this upper limit and details of behaviour in the design range are therefore lacking. Nevertheless, account should be taken of possible increases in working stress if the soil or rock is found to be better than anticipated. In fact, of course, results should be studied as tests are performed and modifications in design can thus be anticipated and tests adjusted accordingly.

Immediate or instantaneous and time-dependent, delayed or creep strains are discussed in Section 2.1, together with elastic and plastic components. Elastic behaviour is exemplified by the absence of permanent set when stress is reduced to zero and the graph may be linear or curved and displaying hysteresis, thus demonstrating time-dependent behaviour. At relatively low stress, creep probably results mainly from closure of fissures and local fragmentation within the mass but, as stress is increased, it is probably related largely to displacements in

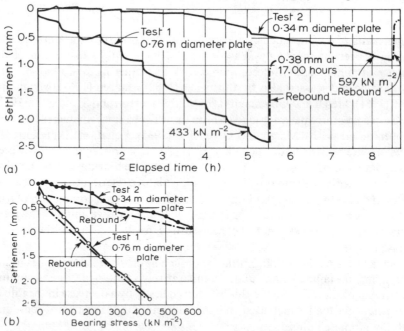

Figure 1.21 Plate bearing test observations. (a) Time–settlement curves for two plate bearing tests in stilling basin of Bayou Bodcau Dam reflect large initial movement of plate when load increment was first applied, followed by rather slow increase in settlement with increasing time. Settlement was measured at 1 minute intervals throughout loading period and during early stages of rebound period. (b) Load–settlement curves for two plate bearing tests of Bayou Bodcau Dam. Curves approximate straight lines, allowing for inconsistencies in seating and other possible test errors. Small permanent set on rebound. Greater settlement for test no. 1 reflects influence of larger loaded area. Curves were drawn by plotting final settlement under each load increment against bearing stress.

the micro-fabric. If the moisture content of soil or rock is sufficiently great, then some creep will be the result of consolidation.

Typical results of plate bearing tests are given in Fig. 1.21 (Shockley, 1950). The results of a test on Bunter Sandstone extending over a period of nearly three years have been described by Moore (1974) and tests on Keuper Marl have been reported by Marsland *et al.* (1983).

(a) Formulae from the theory of elasticity

The formulae given below relate to the settlement of loaded areas on the surface of an elastic half-space or semi-infinite elastic medium (Timoshenko and Goodier, 1951). A semi-infinite medium extends theoretically to infinity in all horizontal directions and vertically downwards. These formulae are useful for planning and interpreting tests for k_s and for foundation design. The total load on each area is Q, the average bearing stress is q and, for soil or rock, modulus of elasticity is E_r and Poisson's ratio v.

1. Circular loaded area, diameter B. Note that settlement is proportional to B.
   (a) Uniformly distributed, flexible loading.

   Settlement at centre of loaded area $\qquad \rho_o = \dfrac{1.00(1 - v^2)qB}{E_r}$ (1.10)

   Settlement at periphery of loaded area $\qquad \rho_d = \dfrac{0.64(1 - v^2)qB}{E_r}$ (1.11)

   Average settlement of loaded area $\qquad \rho_a = \dfrac{0.85(1 - v^2)qB}{E_r}$ (1.12)

   It will be seen that the difference between ρ_o and ρ_d is appreciable.
   (b) Load applied by rigid cylindrical block, load distribution non-uniform.

   Settlement of block $\qquad \rho_b = \dfrac{0.79(1 - v^2)qB}{E_r}$ (1.13)

   The average settlement under a flexible load is not every different from the settlement under a rigid load; $\rho_b = 0.92\rho_a$.
2. Square loaded area, width B, with uniformly distributed, flexible loading. Note that settlement is proportional to B.

   Settlement at centre of loaded area $\qquad \rho_o = \dfrac{1.12(1 - v^2)qB}{E_r}$ (1.14)

   Settlement at corners of loaded area $\qquad \rho_c = \dfrac{0.56(1 - v^2)qB}{E_r}$ (1.15)

   Average settlement of loaded area $\qquad \rho_a = \dfrac{0.95(1 - v^2)qB}{E_r}$ (1.16)

For this same value of q, the average settlement of a square area of width B is rather more than that of a circular area of diameter B.

Table 1.19 Influence factors for settlement of rectangular loaded areas

n	1.0	1.5	2.0	3.0	5.0	10.0	100.0
I_ρ	0.95	0.94	0.92	0.88	0.82	0.71	0.37

3. Rectangular loaded area, length L, breadth B, with uniformly distributed, flexible loading. $L/B = n$ and $LB = A$.

$$\text{Average settlement of loaded area} \qquad \rho_a = \frac{I_\rho(1 - v^2)q\sqrt{A}}{E_r} \qquad (1.17)$$

The value of the influence factor I_ρ depends on the ratio n (Table 1.19)

It should be noted that, for given values of q and A, and hence also total load Q, settlement decreases as n increases. This fact can be employed in foundation design to attain sensibly equal settlement of independent foundations of a structure.

(b) Evaluation of modulus of subgrade reaction from site or laboratory tests

Where the required test load is not large, a single plate can be employed on different parts of a site to give a mean value or a range of values. Where the required load is considerable, it is inconvenient to move heavy equipment and kentledge or to drill many anchorage holes, and it is usual to test on one or two parts of the site only, employing several plates of different sizes at each location. Each investigation must be treated on its merits, bearing in mind relative costs and the importance of the proposed structure. Terzaghi (1955) recommends that a test on clay should be carried out with a rigid plate, the thickness of which is at least equal to its width or diameter. Tests carried out with plates of different magnitudes of diameter yield a series of values of $k_s = q/\rho_b$, each related to the diameter of one plate, since $\rho_b \propto b$ (Equation 1.13 and Fig. 1.22).

Values for E_r and v can be determined from laboratory tests on rock specimens and employed in the relationship $C = E_r/(1 - v^2)$, which is independent of b. Then substituting, for example, for ρ_b from Equation 1.13.

$$k_s = \frac{E_r}{0.79(1 - v^2)b} = \frac{C}{0.79b}. \qquad (1.18)$$

From a site test on a single plate it is possible to calculate a value for the parameter C but a more reliable value should be obtained from a series of tests with plates of different diameter. A graph of $\log k_s$ against $\log b$ should yield a straight line from which values can reasonably be extrapolated for full-size foundations. The most satisfactory relationship should be given, however, by the application of the method of least squares to the observed values of k_s and b, applying weights where the relative reliability of each observation can be assessed (Example 1.2). The computed value of C can then be employed in any of the other equations to calculate settlement of foundations of different shapes and sizes. For

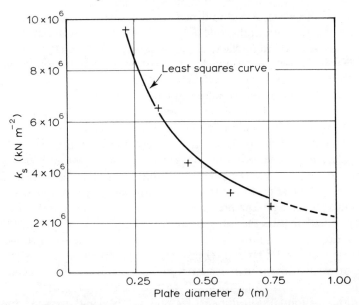

Figure 1.22 Site plate bearing test on sandstone of Lower Kinderscout Grit for modulus of subgrade reaction at Baitings Dam, Yorkshire.

example, the value of k_s applicable to the design of a uniformly loaded flexible rectangular foundation is found from Equation 1.17 in the form $k_s = C/(I_\rho \sqrt{A})$.

When test results are analysed, it should be borne in mind that test plates can seldom be regarded as completely rigid but, on the other hand, the condition of uniformly distributed loading is unlikely to be attained. Furthermore the location of the settlement dial gauges should be taken into account since they may not be located on the contour of average settlement. For each test on a plate, the value of k_s is the secant modulus of the graph of stress against settlement. In the case of tests for pavement design, k_s is based on an arbitrary standard settlement of 1.25 mm. For foundation design it is more appropriate to determine the secant modulus over the estimated working range of stress. Values of k_s can be derived from the graph obtained during loading or during unloading or the mean of the two can be employed for settlement calculations, whichever appears appropriate in a particular case. Although the equations derived from the theory of elasticity are applicable to stiff clays, Equation 1.8 should be employed for foundations less than, say, 3 to 6 m wide on cohesionless soils. Approximate values of modulus of subgrade reaction for sands and over-consolidated clays have been given by Terzaghi (1955).

The parameter $\lambda = {}^4\sqrt{[k_s/(EI)]}$ is employed in the analysis of flexible continuous footings, where E and I are the modulus of elasticity and second moment of area of the concrete foundation. Taking logarithms and differentiating yields $d\lambda/\lambda = dk_s/(4k_s)$. Thus a small error Δk_s in k_s leads to an error $\Delta\lambda$ of only $\Delta\lambda = \Delta k_s \lambda/(4k_s)$.

The process of design of pad and pier foundations from the results of plate bearing tests is almost inevitably one of successive trials. From design dead loads, and live loads also where applicable, a combination of foundation dimensions and net bearing stress is determined. The settlement of the test plate under the same magnitude of stress is read from the test graph. This value of settlement is converted to the settlement of a plate of equal size on the surface of the ground, which is assumed to be a homogeneous and isotropic elastic half-space. If the test has been carried out in a borehole, the appropriate depth correction factor is employed, taking into account whether or not the borehole was filled with concrete. Burland and Lord (1969) (Fig. 2.19) or Scherman (1969) (Fig. 2.20) factors are applicable when the borehole is unfilled but Fox (1948) (Fig. 2.18) or Janbu *et al.* (1956) (Fig. 2.21) factors are applicable when filled since the discontinuity arising from the presence of the borehole is effectively eliminated. This value of settlement is then employed to determine the settlement of the full-size foundation, assumed initially to be founded at ground level. Introduction of the Fox (Fig. 2.18) or Janbu *et al.* (Fig. 2.21) depth correction factors yields the estimate of settlement at the design founding depth. If this magnitude of settlement is not acceptable, then the calculations are repeated with other combinations of foundation dimensions and net bearing stress. Example 2.4 illustrates the procedure.

1.2.5 Field vane tests

The field vane test is employed primarily to determine *in situ* the shear strength of saturated soft to firm non-fissured cohesive soils. The apparatus consists of a cruciform vane on the lower end of a spindle (Fig. 1.23) which, when employed in

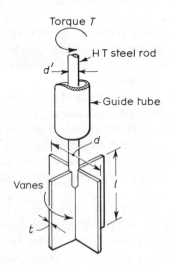

Figure 1.23 Vane test apparatus.

a borehole, is located in ball-bearing guides at 3 to 9 m intervals. The assembly is lowered into the borehole and the vane is forced into the soil at the bottom of the hole for a distance of not less than three times the diameter of the hole. The borehole is generally lined and the weight of the string of tools is carried by a thrust bearing fitted on the torque measuring apparatus at ground surface. The vane test has been employed at depths exceeding 30 m. The shear strength of the soil is measured by observing the torque required to rotate the vane by means of a torque head. The torque is equal to the moment developed by the shearing resistance acting over the surface of the cylinder of soil swept by the vanes. It is assumed that the clay is saturated and that $\phi_u = 0$. A vane test commonly gives a value of strength c_u within $\pm 10\%$ of the value derived from a laboratory undrained test, depending on the type of clay and its stress history and on sampling and testing techniques. Large discrepancies may be incurred with sensitive clays, when the vane test is likely to yield higher and more accurate results than laboratory tests, and also with fissured clays, when the swept cylinder is too small to take the full influence of fissures into account and laboratory tests on large specimens should yield lower and more reliable results. The sources of discrepancy in c_u derived from vane and laboratory tests have been discussed by Hansen and Gibson (1949) and by Bishop (1966).

The following is a brief specification for the proportions of the vanes.

For $c_u < 50 \, \text{kN m}^{-2}, d = 75 \, \text{mm}, l = 150 \, \text{m}$.

For $c_u = 50 \, \text{to} \, 75 \, \text{kN m}^{-2}, d = 50 \, \text{mm}, l = 100 \, \text{mm}$.

$$\text{Area ratio} = \left\{ \frac{8t(d - d') + \pi d'^2}{\pi d^2} \right\} 100 \ngtr 12\%$$

Assuming the shear stress is constant over the ends of the cylinder, integrating the moment of shearing resistance on annular elements at the ends and adding the moment of resistance on the cylindrical surface, it follows that the shear strength

$$c_u = \frac{T}{\pi d^2 \left(\dfrac{l}{2} + \dfrac{d}{6} \right)} \, \text{kN m}^{-2} \tag{1.19}$$

where T is the torque in kN m, l is the length of the vane and d is the swept diameter. For a given vane, a calibration graph can be drawn of this linear relationship to read c_u directly from measured values of T.

Mechanical friction in the apparatus and adhesion on that part of the spindle buried in the clay can be measured by carrying out a test under similar conditions with a dummy rod replacing the vane and the observed torque is deducted from the torque recorded when the vane is in position to give the corrected value of c_u. However, the buried part of the spindle is preferably enclosed in a sleeve to eliminate adhesion on it. The diameter of the swept volume of soil is slightly greater than the diameter of the vane and a factor of 1.05 can be applied to the vane diameter to give d if this is considered necessary. Some disturbance of the soil is inevitable even though the area ratio is restricted to 12%.

The vane test can be performed without a borehole by direct penetration from ground surface. The vane is protected with a shoe during driving and the rods are enclosed in a tube with bearing guides at 3 m intervals. On reaching the required depth, the vane is driven 0.5 m ahead of the shoe before measuring torque.

The rate of rotation of the vane should be about $0.1°$ to $0.2° s^{-1}$ and at this rate the shearing strength of the clay is fully mobilized in 3 to 10 min – a time of loading comparable with that adopted in laboratory tests. Carlson (1948) found that the torque increased as the rate of rotation was increased and, for example, one typical experiment showed that the torque at a rate of $1.0° s^{-1}$ was 26% greater than the torque at $0.1° s^{-1}$, the variation being roughly linear. In practice the rate of rotation may be less than $0.1° s^{-1}$ but even at very slow rates the torque is not likely to be more than 5% below the value at $0.1° s^{-1}$. Torque readings are generally taken at intervals of $5°$ rotation. The angle at maximum torque, corresponding to the peak strength, is also recorded and after this is attained a further five readings at $5°$ intervals are taken. On completion of this undisturbed test the vane is rotated through about $90°$ and the torque is measured at $5°$ intervals over a range of $25°$ to obtain the residual strength.

Applications of the vane test have been described by Skempton (1948), Gray (1957) and, in peat, by Northwood and Sangrey (1971). Experimental studies of the vane test have been made by Åas (1965), Flaate (1966) and Perlow and Richards (1977). Simple types of hand vane apparatus are available which enable shear strength to be determined at shallow depths. Torsion tests with special apparatus for the determination *in situ* of the shear strength of frictional soils have been described by Helenelund (1965). The vane test is specified in BS 1377 (BSI, 1975).

1.2.6 Penetration or sounding tests

The resistance to driving recorded in penetration and sounding tests can be correlated empirically with various properties, such as bulk density and shear strength, and the extent of a site investigation can thus be increased economically. Furthermore, marked local variations in the soil may be disclosed which might be overlooked if the investigation consisted only of a limited number of borings. The most effective correlation of engineering properties with resistance to penetration is obtained when the types of soil at given depths are known. The tests are best suited to sandy soils although some experience has been gained with clays and chalk. Heavy driving may be required in gravel and the presence of cobbles and boulders may invalidate results or halt driving. The tests most commonly employed are the Dutch static deep sounding test and the dynamic standard penetration test (SPT).

The Dutch deep sounding apparatus is fitted with a cone having a diameter of 35.6 mm, a projected base area of $10 cm^2$ and an apical angle of $60°$. The more powerful equipment is mechanically operated and a penetration force of up to 200 kN can be mobilized and depths of 30 to 40 m can be attained. When the cone resistance is measured mechanically it is commonly recorded at penetration intervals of 20 cm but when measured electrically a continuous record can be taken. A sleeve can be fitted above the cone and cone resistance and sleeve skin resistance can be recorded independently. The ratio of skin resistance to cone resistance facilitates identification of soil types: low values are associated with sands and gravels and high values with silty and clayey soils. Resistance can be plotted against depth at a succession of sounding locations along a section line and contours can be drawn in a vertical plane of equal resistance or, after correlation, values of other soil properties. The rate of testing is of the order 200 m per day but varies with local conditions. An inclinometer can be incorporated in the equipment so that the deviation of the instrument from the vertical can be recorded during driving.

Since the rate of penetration of the cone may be as much as 20 mm s^{-1}, the term static is misleading. A more satisfactory term would be constant rate of penetration to distinguish the test from the dynamic SPT, for which a more appropriate name is impulsive penetration test. Whereas some authorities believe that, with cohesionless soils, both cone resistance and skin resistance rise to maxima at certain depths, below which they remain roughly constant, others hold the view that both are related to lithology. In respect of cohesive soils, it is likely that better correlation between resistance and shear strength or bearing capacity is likely to be attained with unfissured clays than with fissured clays since the size of the cone is small in relation to the spacing of fissures. Applications of the static cone test have been discussed by Thomas (1965, 1968), Sanglerat (1972) de Ruiter (1971), Schmertmann (1970) and Marsland (1974). Investigations have been carried out, for example, by Mitchell and Durgunoglu (1973), Mitchell and Lunne (1978), Baligh *et al.* (1980) and Thorburn *et al.* (1981) in which correlations and comparisons with various soil properties have been made. A further development of continuous sounding is the Alimok BAT probe which distinguishes clays, sands and other soils by variations in the pore pressure generated as it penetrates these materials with differing values of permeability.

In the standard penetration test a split-barrel sampler is employed as the penetrometer. The split-barrel is 457 mm long and 50 mm external diameter and is fitted with a cutting shoe and a coupling or driving head with vent ports or ball valve. It is used in a borehole in conjunction with normal sampling. Loose soil is removed from the bottom of the borehole and the sampler is set on the soil at the end of a string of rods. Tests by Palmer and Stuart (1957) indicate that the weight of the rods is unlikely to influence resistance appreciably except in soft or loose soil. Steadies or, alternatively, very stiff rods are employed in boreholes deeper than 15 m. The sampler is driven an initial 150 mm with a 65 kg hammer falling freely through 760 mm and the number of blows recorded. It is then driven a further 300 mm and the number of blows required for each 75 mm penetration recorded. The penetration resistance, N, is the total number of blows required to penetrate 300 mm below the seating drive. Experiments indicate that a variation of up to $\pm$ 50 mm in a 760 mm drop of the hammer affects the measured resistance only slightly but precautions should be taken to ensure that energy is not lost by friction or otherwise. With very dense granular material the time required to drive 300 mm is excessively long and it is common practice to measure the penetration achieved by 50 blows and to determine N for 300 mm penetration in proportion. When driving in gravels the cutting shoe is replaced by a 60° cone. Artesian pressure creates upward seepage into a borehole and this may loosen sand and lead to a false reading for N. Under these conditions it is usual to test the soil at a greater depth below the bottom of the hole or alternatively the head can be balanced by water in the borehole. Saturated moderately dense fine or silty sand possesses an abnormally high resistance to penetration and it is necessary to apply an empirical correction (Terzaghi and Peck, 1967) to the observed value N' to obtain the working value $N = 15 + \frac{1}{2}(N' - 15)$. A representative sample of soil is taken from the core in the barrel.

Gibbs and Holtz (1957) carried out full scale laboratory tests with sands using a Raymond sampler with cutting shoe. It was confirmed that penetration resistance increases with increase in dry bulk density and also with increase in overburden stress. Furthermore, it was confirmed that the influence of the weight of the driving rods and sampler is appreciable only when the driving resistance is low. The tests were made with values of overburden stress of 0 and 276 kN m^{-2} – the latter is roughly equivalent to 15 m of overburden – and the results show that correlation of field observations should take the depth factor into account.

Information on the use of the SPT in silts is scanty but it is commonly accepted that, if $N < 10$, the silt is loose and if it exceeds this value it is medium to dense. If $N < 10$, the silt is likely to be highly compressible and of low bearing capacity unless it is treated by a geotechnical process or is pre-loaded. When used in sensitive clays, the SPT yields values of resistance which are low and unrepresentative of the resistance in the undisturbed state.

The SPT is standardized in BS 1377 (BSI, 1975) but the implications of nonstandardization are discussed by Ireland *et al.* (1970). Further studies have been reported by, for example, Rodin *et al.* (1974), Marcuson and Bieganousky (1977) and Schmertmann (1979). Parry (1978) comments that, if the SPT is carried out with care and with absolutely standard procedure, it is doubtful if any other tests, even apparently more sophisticated, will give better answers, with the possible exception of the plate bearing test when interpreted in relation to the character of strata in both horizontal and vertical directions.

1.2.7 Pore-water pressure measurement

Pore-water pressure is measured in preliminary site investigations by recording the levels of phreatic surfaces. Sometimes levels of more than one surface are observed. At a site near Worthing, Sussex, water was first encountered in silty clay at depths of 0.3 to 1.2 m beneath ground surface but subsequently a bed of gravel was met at 11 m and water under artesian pressure rose to ground level. Artesian water is frequently met in site investigations, particularly where gravel or fissured limestone, chalk or sandstone is overlain by deposits of relatively low permeability. Given a sufficiently long period of time, it is possible that an upper water table, when first encountered, would rise to the artesian level as water slowly permeates into the borehole until hydrostatic equilibrium is attained. Pore-water pressure is measured also for control during construction on certain projects such as cofferdams, earthdams, highway embankments and foundations for oil tanks. When a load is imposed on a saturated fine-grained soil, the excess pore pressure dissipates slowly, the soil consolidates and the shear strength increases. In some projects advantage can be taken of the increase in strength, and it is essential to observe pore pressure during construction to avoid overloading the soil before the appropriate strength has been attained.

The simplest apparatus for measuring pore-water pressure comprises a polythene standpipe about 25 mm diameter fitted at the foot with a permeable sleeve. The assembly is lowered into a borehole, the cylinder is surrounded by fine sand which is covered with clay to prevent leakage of water along the outside of the pipe. The hole is backfilled and the borehole casing withdrawn. The level of the water in the standpipe is measured preferably with an electric dip-stick. Where the level is likely to fluctuate relatively rapidly due to the influence of tides or floods, the standpipe can be connected to an automatic recorder. A disadvantage of the standpipe is that time is required for water to flow into or out of the pipe and the lag in response may be several hours or days. If the magnitude of the head can be estimated, the lag can be reduced by pouring water into the tube up to an appropriate level.

In closed systems, each piezometer cylinder or cell is connected by twin tubes to a central observation panel. Since air may be thrown out of solution and become trapped in a closed system, air relief values must be provided. Electrical transmission systems have been devised to eliminate response lag and air accumulation. Electric resistance strain gauges and vibrating wire gauges have been employed to measure the deformation of a diaphragm, incorporated in the permeable cylinder or cell, under changes in pressure. Gibson (1963) and Premchitt and Brand (1981) have investigated the problem of time lag and Vaughan (1969) has discussed sealing of piezometers in boreholes. The effect of tides on pore-water pressure has been studied by Margason *et al.* (1968).

1.2.8 Pressuremeters

During the last two or three decades various types of pressuremeter have been produced to measure soil properties *in situ*. The Menard pressuremeter comprises a cylindrical probe which is lowered into a borehole. At appropriate vertical intervals the probe is inflated and from the stress–strain relationship recorded at ground level predictions can be made of modulus of deformation and ultimate bearing capacity (Baguelin *et al.* 1978). The 80 mm diameter

Camkometer is a self-wash-boring pressuremeter covered by a rubber membrane and incorporating two cells for the measurement of pore-water pressure. The probe is inflated at suitable vertical intervals and the effective stress-strain relationship yields modulus of deformation, shear strength, pore-water pressure and lateral stress. This pressuremeter can be used in soils varying from soft clays to sands and gravels (Windle and Wroth, 1977; Dalton and Hawkins, 1982). Recently a spade-like flat dilatometer has been developed which is jacked into the ground (Marchetti, 1980; Tedd and Charles, 1981). Generally the pressuremeters described above can be used only in soils and weak rocks; other types are available for testing strong rocks in boreholes. Pressuremeter tests in sands have been described by Hughes *et al.* (1977) and in London Clay by Marsland and Randolph (1977).

EXAMPLE 1.2

Site tests on sandstone of the Lower Kinderscout Grit at Baitings Dam, Yorkshire, yielded the following results:

b	(m)	0.229	0.344	0.457	0.601	0.762
k_s	(kN m^{-3})	968×10^4	656×10^4	439×10^4	320×10^4	256×10^4

Employing the method of least squares, determine the value of C in kN m^{-2} from these results. The length of Baitings Dam is 472 m and the maximum width of the broad foundation is 44.5 m. For a rectangular foundation of these proportions the value of k_s can be taken as 0.348 of the value for a rigid circular foundation of diameter equal to the width of the rectangular foundation. Determine the appropriate value of k_s for Baitings Dam. $k_s = C/0.79b = C'b^{-1}$ where $C' = C/0.79$. Let C'_p be the most probable value of C' and e the residual in any observation equation $e = k_{so} - C'_p b_o^{-1}$ where k_{so} and b_o are the observed values of modulus of subgrade reaction and plate diameter. The least squares condition is that Σe^2 must be a minimum, or $\Sigma(k_{so}^2 - 2k_{so}C'_p b_o^{-1} + C'^2_p b_o^{-2})$ must be a minimum. Differentiating partially w.r.t. C'_p and dividing by 2,

$$\Sigma(C'_p b_o^{-2} - k_{so}b_o^{-1}) = 0.$$

The following quantities correspond to those given above.

b_o^{-2}	(m^{-2})	19.07	8.45	4.79	2.69	1.72	$\Sigma 36.72$
$k_{so}b_o^{-1}$	(kN m^{-4})	4227×10^4	1907×10^4	961×10^4	524×10^4	336×10^4	$\Sigma 7955 \times 10^4$

Then $36.72\, C'_p - 7955 \times 10^4 = 0$ whence $C'_p = 2166 \times 10^3\, \text{kN m}^{-2}$ and $C_p = 0.79 \times 2166 \times 10^3 = 1711 \times 10^3\, \text{kN m}^{-2}$ is the most probable value of C. The least squares graph of k_s against b is shown in Fig. 1.22. For a rigid circular foundation of diameter 44.5 m,

$$k_s = \frac{1711 \times 10^3}{0.79 \times 44.5} = 487 \times 10^2\, \text{kN m}^{-3}.$$

For a rectangular foundation of width 44.5 m,

$$k_s = 0.348 \times 487 \times 10^2 = 169 \times 10^2\, \text{kN m}^{-3}.$$

Laboratory tests on a sandstone column yielded $E_r = 8.3 \times 10^6\, \text{kN m}^{-2}$ and $v = 0.10$, whence $C = 8.4 \times 10^6\, \text{kN m}^{-2}$. Discrepancy between laboratory value of C and site value of $1.7 \times 10^6\, \text{kN m}^{-2}$ is probably due to the presence of microscopic fissures beneath plates, leading to greater compression than with rock in intact state. Calculations for settlement of a dam based on this analysis of site test data would be tempered by consideration of the characteristics of deeper strata, rigidity of dam and other factors. In particular, plate tests on underlying shales would be desirable.

1.3 POTENTIALLY CORROSIVE ENVIRONMENTS FOR FOUNDATION MATERIALS

The most important corrosive environments for concrete are associated with soft waters, sulphates and sulphides and saline waters (chiefly marine, estuarine and littoral). Artificial fills – particularly when ash or brick rubble with plaster or colliery shale is present – may contain a variety of potentially corrosive compounds. Consideration should be given to the environment generated in service, apart from the natural environment. For example, grillage beams encased in concrete have failed because of the corrosive action of water-quenched boiler ashes.

Soft water contains little or no dissolved calcium carbonate and free carbon dioxide in solution may be present. This aggressive carbon dioxide reacts with calcium carbonate in cement to form moderately soluble calcium bicarbonate which is leached out. Carbonic acid may also convert calcium hydroxide in cement to calcium carbonate which may, in turn, be converted to calcium bicarbonate. Fresh water with a pH value less than 7 can be considered potentially aggressive. These waters are associated with peat or with sandstones such as those of the Millstone Grit and the Coal Measures or sands such as those of the Folkestone Beds. The greater the strength of concrete – implying relatively high density and low permeability – the greater the resistance to attack by soft water. The best preventative is a carefully applied coat of bituminous paint which must be renewed periodically (Halstead, 1954).

Sulphates of calcium (gypsum), magnesium and sodium occur as crystals in soils and rocks or dissolved in groundwater. Naturally occurring sulphides, such as pyrite, are relatively insoluble and are not likely to lead directly to corrosion although oxidation converts these to sulphates. Gypsum and pyrite occur in varying amounts in the Coal Measures shales, Keuper Marl, Lias, Oxford Clay, Kimmeridge Clay, Weald Clay, Gault and London Clay. The top metre or so of a sulphate-bearing formation is generally free from salts because of leaching. The horizontal and vertical distribution of concentration in clay is likely to vary over a site and may be affected by changes in groundwater level. The sulphate content of groundwater may be fairly uniform in a restricted area but is liable to seasonal variations. Consideration should be given to the possibility of sulphates being imported into an area by groundwater flowing through disturbed ground such as trench backfill. In the Middle East and other regions, concrete laid on salt pan, formed in a shallow ephemerol lagoon known as a sabkah, is commonly attacked from beneath by gypsum and other evaporite minerals.

Sulphates tend to combine with lime and with hydrated calcium aluminates in concrete and new compounds are formed which have volumes greater than those of the original; internal stresses are set up and the concrete disintegrates.

Table 1.20 Requirements for concrete exposed to sulphate attack. Recommendations are for concrete in near-neutral groundwater – for acid conditions refer to **BRE** Current Paper CP 23/77 (Gutt and Harrison, 1977)

| Class | Concentrations of sulphates expressed as SO_3 | | | Type of cement | Requirements for dense fully compacted concrete made with aggregates meeting the requirements of BS882 or BS1047 | |
| | In soil | | In ground-water $(g\,l^{-1})$ | | Minimum cement content* $(kg\,m^{-3})$ | Maximum free water/cement* ratio |
	Total SO_3 (%)	SO_3 in 2:1 water:soil extract $(g\,l^{-1})$				
1	<0.2	<1.0	<0.3	Ordinary Portland Cement (OPC) Rapid Hardening Portland Cement (RHPC) –or combinations of either cement with slag† or pfa§ Portland Blastfurnace Cement (PBFC) — Plain concrete†	250	0.70
				Reinforced concrete	300	0.60
2	0.2–0.5	1.0–1.9	0.3–1.2	OPC or RHPC or combinations of either cement with slag or pfa PBFC	330	0.50
				OPC or RHPC combined with minimum 70% or maximum 90% slag¶ OPC or RHPC combined with minimum 25% or maximum 40% pfa**	310	0.55
				Sulphate Resisting Portland cement (SRPC)	290	0.55

3	0.5–1.0	1.9–3.1	1.2–2.5	OPC or RHPC combined with minimum 70% or maximum 90% slag	380	0.45
				OPC or RHPC combined with minimum 25% or maximum 40% pfa		
				SRPC	330	0.50
4	1.0–2.0	3.1–5.6	2.5–5.0	SRPC	370	0.45
5	>2	>5.6	>5.0	SRPC + protective coating††	370	0.45

* Inclusive of content of pfa or slag. These cement contents relate to 20 mm nominal maximum size aggregate. In order to maintain the cement content of the mortar fraction at similar values, the minimum cement contents given should be increased by 50 kg m⁻³ for 10 mm nominal maximum size aggregate and may be decreased by 40 kg m⁻³ for 40 mm nominal maximum size aggregate.

† When using strip foundations and trench fill for low-rise buildings in Class 1 sulphate conditions further relaxation in the cement content and water/cement ratio is permissible.

‡ Ground granulated blastfurnace slag. A new BS is in preparation.

§ Selected or classified pulverised-fuel ash to BS 3892. A new BS superseding BS 3892 is in preparation.

¶ % by weight of slag/cement mixture.

** % by weight of pfa/cement mixture.

†† See CP102 (BSI, 1973b). Protection of buildings against water from the ground.

Table 1.20 is taken from BRE Digest 250 by permission of the Controller, HMSO. Crown copyright.

Concrete attacked by sulphates shows cracking and spalling on the surface at first, more particularly at corners and edges, and the cement is reduced finally to a soft white paste. The more permeable the concrete, the faster is the attack – lean concrete may suffer considerably in a few months. The skin resistance of piles may be considerably reduced by deterioration of shaft concrete and, of course, the cross-section is also reduced. Where one side of a mass of concrete is in contact with sulphate-bearing water under hydrostatic pressure, the rate of attack is increased. Rounding of concrete sections is desirable to avoid attack on two faces at an arris. Concrete placed *in situ* is more vulnerable than precast units, since it is exposed while in a green state. The magnitude of concentration of the salts naturally has a marked effect on the rate of attack. Sodium and magnesium sulphates are much more soluble in water than calcium sulphate and are therefore potentially more dangerous. Concrete in dry ground containing sulphates is not attacked, since the presence of water is essential to the action. Recommended precautions are given in detail in BRE Digest 250 (BRE, 1981), from which Table 1.20 has been extracted. Sulphate concentration is recorded in terms of equivalent sulphur trioxide. In connection with the concrete mixes specified in Table 1.20, free water is the total weight of water less that absorbed by the aggregate.

A further corrosion problem is introduced where iron or steel is buried in soil containing both sulphates and sulphate-reducing anaerobic bacteria. Characteristically, corrosion from this source leads to deep pitting and a crust of black ferrous sulphide. The metal can be protected by carefully applied coats of coal-tar pitch or bitumen but it is believed that activity of the bacteria tends to be inhibited if the alkalinity of the soil can be raised above pH 9 by cathodic protection. Booth and Wormwell (1961) have discussed corrosion by sulphate-reducing bacteria.

Failures of reinforced concrete in saline environments are commonly due to inferior concrete with relatively high permeability which permits corrosion of steel and disintegration of concrete follows. The effects of adverse chemical reactions are aggravated by mechanical attrition and disruption by frost. Disintegration is often most marked in the tidal range, where alternate wetting and drying aid these processes. Magnesium chloride in sea water reacts with lime in cement; the chloride combining with the lime, leaving magnesia as a hydrate in its place. Magnesium sulphate in sea water causes the common sulphate attack. In some cases unstable felspar in coarse and fine aggregate has been considered to be responsible for disintegration of concrete when reactions with salts in sea water have occurred. For preventative measures an absolute minimum of 60 or 50 mm cover to reinforcement with Grades 40 and 50 concrete respectively is essential (Anon, 1972) and vibration of concrete is desirable. Particular attention must be paid to construction joints to prevent penetration of sea water. The durability of reinforced concrete in sea water is the subject of a report (H.M.S.O., 1960). Shalon and Raphael (1959) found that pH values of environments influence corrosion of reinforcement and, as a result, protection afforded by concrete is effective only when members are permanently immersed in sea water. Neville (1963) has presented a study of the deterioration of high-alumina cement concrete, including effects in marine environments. The mechanism of sea water attack on cement pastes has been discussed by Kalousek and Benton (1970) and the fatigue strength of concrete in sea water has been investigated by Patterson *et al.* (1982). It can be noted that in some parts of the World, such as the Persian Gulf, chlorides and sulphates occur not only in the ground and in water but in the atmosphere also and, aided by high ambient temperature, corrosion is rapid and widespread throughout many structures.

Various systems of cathodic protection of steelwork in saline environments have been employed within recent years. The rate of corrosion generally increases with rise in temperature and protective measures receive most attention in tropical and sub-tropical waters. Since steel corrosion is largely the result of electrolytic conduction, the attack can be reduced or prevented by applying an opposing potential difference in the threatened area. The steelwork is utilized as the cathode, sea water functions as the electrolyte and the anodes are specially provided. The two alternative techniques can be designated the natural or sacrificial method and the artificial or impressed method. In the natural technique, the anode is formed of a sacrificial mass of metal which is electropositive to the material to be protected. Zinc, aluminium or magnesium anodes are employed to protect steel. The anode is placed on the sea bed and is connected to the steelwork by an insulated cable, thus completing the circuit. The sacrificial mass is gradually reduced by electrolytic action and a protective coat is deposited on the steel. The life of anodes is usually between 2 and 20 years. Fittings of copper or brass must not be employed since these metals are electropositive to Zn, Al and Mg.

In the artificial method, the anode is formed of scrap steel, lead alloy, carbon or platinum-coated titanium and the current is induced by connecting the circuit to a low voltage d.c. supply. It may be necessary to adjust the current to seasonal variations in electrolytic activity arising, for example, from changes in temperature or salinity. The required current density is of the order 10 to $20\,\mathrm{mA\,m^{-2}}$ for coated steel and 50 to $100\,\mathrm{mA\,m^{-2}}$ for bare steel. Corrosion is restrained by cathodic protection below low water and up to about half-tide mark but in the upper half of the tidal range protection is only partial. The most effective protection is afforded with a coating, such as hot bitumen enamel, in addition to a cathodic system. Anodes must be renewed at intervals of a few years and the entire system must be carefully maintained, with inspections at intervals of 1 to 3 months. In addition to marine environment, corrosion may occur in littoral belts where sea spray and sand are transported by wind. Details of cathodic protection have been reported by Wright (1962), Anon (1963) and Milano (1961) and British practice is specified in CP 1021 (BSI, 1973).

In normal site investigations it is usual to check for sulphates and any other obvious source of corrosion. Special corrosion surveys can be made if necessary, as outlined by Skipp (1961). For corrosion of steel, full investigations include studies of soil and groundwater pH, electrical properties of soil, bacterial environment and activity, absorbed oxygen in groundwater and soil per-

meability. For corrosion of concrete, studies are made of soil and groundwater·
pH, aggressive carbon dioxide, temporary hardness, sulphates and sulphides.
Recommendations for sampling soil and groundwater for sulphates are given by
Bessey and Lea (1953). The number of borings under adverse conditions should
be two per $1000 \, m^2$, with a minimum of three on a small site, and five per km
along sewer lines. In boreholes samples should be taken at intervals of 1 to 2 m or
where changes in strata occur. It is essential in all types of sampling to prevent
contamination of soil and water obtained from any particular horizon. In transit
to the laboratory, precautions must be taken to prevent chemical and bacterial
changes. Procedures for the analysis of sulphate-bearing soils have been detailed
by Bowley (1979).

Various aspects of corrosion of metals have been discussed by Wormwell
(1969) and Ratcliffe (1983) and in a specialist publication (ICE, 1979). BS 5493
(BSI, 1977) is a guide to protection.

1.4 SPECIAL PROBLEMS RELATING TO PARTICULAR SITE CONDITIONS

This Section comprises a brief review of problems arising from particular site
conditions. The problems do not necessarily involve constructional hazards but
in some respects they are unique to particular sites. Certain conditions in Middle
East regions attain greater significance than in, say, north-west Europe: the
report of a conference dealing with relevant problems has been edited by
Chandler (1978) and geotechnical considerations for construction in Saudi
Arabia have been discussed by Oweis and Bowman (1981).

1.4.1 Foundations on limestone

The term limestone covers a wide range of rocks, all containing an appreciable
proportion of calcium carbonate. They include dense crystalline limestones,
characteristic of sequences in the British Devonian and Carboniferous suc-
cessions, dense calcareous mudstones, dolomitic, oolitic, pisolitic, shelly and
crinoidal limestones (sometimes notably porous), sandy limestones, chalk and
other varieties. Limestones are unique in engineering construction because of the
frequent occurrence of solution cavities, leading to problems additional to those
related to intrinsic mechanical properties such as strength and compressibility.
The classification of rocks containing carbonates has been discussed by Fookes
and Higginbottom (1975) and Burnett and Epps (1979).

Solution channels are pot holes, swallow holes, cavities, caves and enlarged
joints created by the solvent action of carbon dioxide in surface and underground
waters on limestone. Foundation construction on limestones may be complicated by the presence of numerous solution channels which divide the rock
into nearly vertical pillars. These features are typical of karst terrain and are well
displayed in a limestone pavement, such as the rock surface above Malham Cove,
Yorkshire, which is dissected by linear cavities known as grikes, leaving flat tops

on the pilars termed clints. Linear cavities in limestone may be a few centimetres or several metres in width and may be partly or wholly filled with detritus ranging from boulders to clay and may be water-bearing. Borings and probes are often useless in locating the bottom of cavities owing to obstruction by boulders. For important structures it is desirable during excavation to bore at least 3 m beneath each foundation to check for the presence of cavities. Even if solution cavities are not revealed in site investigations, it is usually prudent to anticipate that they may be encountered: this is particularly important where excavations are taken below groundwater level.

Rocks other than limestones which are fairly soluble in natural waters may also display voids and leaching of anhydrite and gypsum has led to the formation of cavities or to general subsidence. Hawkins (1979) has reported cavities formed as a result of diagenetic changes in Keuper rocks in the Bristol and Cardiff areas. The design of dam foundations on rocks containing soluble minerals has been discussed by James and Kirkpatrick (1980).

The types of foundations for a given structure and the proportions of cavities and pillars govern the treatment at any site on limestone but the following are possible solutions.

(a) Clean out cavities and backfill with concrete.
(b) Remove natural pillars by blasting or other means.
(c) Employ geotechnical processes to improve the quality of the natural cavity filling.
(d) Design the foundations to bridge the cavities.
(e) Arrange each footing so that it is located on a natural pillar which must be short and proved to be intact in depth.
(f) Piers can be constructed in cavities where these are very large and the pillars small.

The construction of power-station foundations on limestone has been described by Swiger and Estes (1959). Jennings *et al.* (1965) studied subsidence over solution cavities in limestone and Sowers (1975) has discussed a variety of failures in limestones. During construction of a 24 m span arch-bridge over the River Mole near Mickleham, Surrey, borings revealed swallow holes in chalk. Reinforced concrete domes were constructed over the holes founded on ledges cut in the chalk. The largest was 17 m diameter with 2.4 m rise. Each dome was furnished with an access shaft for maintenance inspection. Solution channels can create infiltration problems in cofferdam construction and may necessitate the application of geotechnical processes. Where soft or loose deposits overlie cavernous limestone and piles are to be located in bedrock, it is necessary first to drive a casing for each pile and take a small diameter core not less than 5 m long from the limestone to check the soundness of the rock. A hole is then bored to an appropriate level to accommodate the foot of each pile.

Problems associated with construction in Florida on Miami Limestone, a soft porous oolitic rock of Quaternary age, have been discussed by Kaderabek and Reynolds (1981). Geotechnical properties of limestones have been investigated by

McCann and Hobbs (1977) and Al-Jassar and Hawkins (1977). Modulus of elasticity for limestones appears to lie generally within the range (20 to 80) $\times$ 10^3 MN m^{-2} and Poisson's ratio 0.25 to 0.35. Soft limestones would undoubtedly yield values of modulus of elasticity below the bottom of the range quoted: compare these, for example, with the values for chalk given in Table 1.21. Apart from lithology, values depend on methods of testing and interpretation of results.

Chalk is a soft, fine-grained limestone with distinct properties and deserves further discussion. The Chalk formations of the British Isles contain a high proportion of $CaCO_3$ largely in the form of whole, and fragments of, microscopic shells of planktonic organisms, the remains including calcispheres, coccoliths, rhabdoliths and other forms, but present also in the remains of larger animals. The Lower Chalk of the British Isles sometimes contains up to about 50% clay in addition to calcite but, passing upwards, the calcite content increases markedly in the Middle Chalk and constitutes about 98% in the Upper Chalk. Nodular and tabular flint occurs as bands more or less parallel to the bedding and tabular flint as vertical or inclined veins. Flint occurs most frequently in the Upper Chalk, is less common in the Middle Chalk and is almost completely absent in the Lower Chalk. Marl bands occur in both the Upper and Middle Chalk and are generally between 100 and 200 mm thick. The unique characteristics of chalk appear to stem mainly from the fact that generally little or no recrystallization has taken place although this is seen in some beds and also locally due to tectonic stress. Because of this, many of the microscopic shells remain hollow and the void ratio is relatively high: the extreme ranges are 0.5 to 1.1 for the Middle and Upper Chalk and 0.2 to 0.7 for the Lower Chalk. The corresponding dry bulk density values are 1800 to 1300 kg m^{-3} and 2200 to 1600 kg m^{-3}. Frequently a well-compacted fill of chalk may have a dry density higher than that of the rock *in situ*.

The intensity of development of solution channels varies from zone to zone in the Chalk and, for example, the *Micraster cor-anguinum* Zone is noted for its abundant open fissures. Swallow holes frequently taper downwards in chalk and are known as pipes when they are partly or wholly filled with debris such as sand, silt, clay and flints. Pipes occur most frequently near the edges of outcrops of overlying Tertiary formations and their derivatives, such as the clay-with-flints, where clayey beds lead to the accumulation of large quantities of surface water during rainfall. The Thanet Sands are sometimes present at the base of the Tertiary formations and solution channels often occur in the Chalk beneath these beds. As the outcrop of the Tertiary formations gradually shrinks, solution channels formed in earlier geomorphological stages are eliminated by the normal processes of degradation. Studies of groundwater movement reveal that areas of high transmissibility in chalk correspond with river valleys and important coombes. This suggests that open fissures are likely to be well developed in such areas and indeed it is in these localities also that swallow holes may be found. Electrical resistivity methods of site investigation were partly successful in detecting filled fissures in the Carboniferous Limestone of Derbyshire (Early and

Dyer, 1964) and it is possible that these and other geophysical methods could be employed to locate buried pipes and cavities in the Chalk.

Disruption and disturbance is fairly common in chalk bedrock near ground surface and has been caused generally by frost action, mainly during the Pleistocene. The depth of frost-shattered chalk sometimes exceeds 6 m in the former peri-glacial region of southern England but north of the Thames it may approach 15 m beneath glacial deposits. Chalk which has been reduced to a highly comminuted form, either naturally or artificially, is known as putty chalk or chalk paste and generally possesses the characteristics of silt. Valley bulges (Fig. 1.4) are occasionally encountered in chalk terrain and Higginbottom and Fookes (1970) have observed one in Berkshire which is accompanied by thrusting upward of a plug of chalk. An isolated block of chalk, several cubic metres in volume, in the late-Quaternary Brickearth of the Sussex coastal plain may be an upthrust plug of underlying chalk, the remains of a Quaternary sea stack, a sea-ice transported boulder or may have some other origin. Major road and bridge construction was carried out in a shallow col between two valleys in the Chalk in Sussex. A thin veneer of Tertiary sands and clays overlies the site. The extent and frequency of pipes in the Chalk proved to be considerably greater than was anticipated from site investigations and problems were posed in connection with bridge foundations and stability of cuttings. The site was also characterized in places by a considerable thickness of soft, frost-shattered chalk.

Chalk is particularly susceptible to the effects of frost. Nevertheless, the construction thickness of major highways is sufficient, or nearly sufficient, to protect underlying chalk from freezing. The frost susceptibility of soils has been discussed by Croney and Jacobs (1968). Thrust developed in chalk as a result of frost heave behind retaining walls and revetments has received little attention in the British Isles although Toms (1966) stated that trouble was experienced with chalk fill behind railway platform walls. The best remedy is adequate drainage in the backing and a sufficient construction thickness for insulation.

Relatively undisturbed samples of chalk, particularly when frost-shattered, can be obtained only by careful rotary drilling and even this is often unsuccessful. The open drive sampler is employed to obtain material for identification and visual appraisal only.

The primary permeability of solid chalk is commonly less than 10^{-3} mm s^{-1}, typically about 10^{-5} mm s^{-1}; that is, comparable with the permeability of silt. However, the permeability of chalk *en masse*, which is governed largely by the proportions and spacing of fissures, is probably 10^{2} to 10^{4} times greater. Lewis and Croney (1966) determined values of primary permeability of 0.025×10^{-3} mm s^{-1} for soft chalk, with saturation moisture content 26%, and 0.005×10^{-3} mm s^{-1} for hard chalk, wtih saturation moisture content 10%.

Shear strength parameters for chalk are considerably influenced by disturbance in sampling, by the degree of saturation and by the methods of testing. In terms of effective stress, ϕ' is commonly 37° to 40° and c' is generally less than 50 kN m^{-2} and is sometimes zero. Undrained triaxial tests may yield c_u within the range 50 to 250 kN m^{-2} and ϕ_u may be as low as zero. Chalk possesses anisotropy in shear, a characteristic which may need further investigation.

The bearing capacity of chalk, even when frost-shattered, is commonly relatively high. Although possible bearing failure should be checked in design, the criterion of allowable bearing stress is usually settlement and this may well be no more than a few centimetres. It is probable that in some cases much of the settlement is related to closure of fissures and hair-cracks and, since chalk at normal founding levels is also often incompletely saturated, consolidation tests lead to meaningless or unreliable results. Samples are often subjected to disturbance during sampling, unless they are carefully cut from solid blocks, and this further detracts from the value of consolidation tests. Consequently, plate bearing tests, with a range of plate sizes, are advocated for foundation design, locating the plates in pits or boreholes at the anticipated founding depth. It may be found that settlement is fairly rapid as the load is applied initially, owing to closure of fissures and possibly also local crushing of the chalk, but settlement will subsequently decelerate when the fissures are closed and some additional compaction of the chalk has been achieved. Foundations are often located above groundwater level where the chalk is partly saturated and, in addition, hair-cracks and fissures provide ample passages for drainage as the rock is compressed under load. Consequently, consolidation is rarely significant and settlement of test plates can be interpreted in terms of elastic and plastic deformation. Burland and Davidson (1976) have recorded a case of additional settlement at the edges of a raft founded at a shallow depth, arising from yield in chalk beneath the edges, which led to fracture of columns supporting a silo. In fact, lateral yield of material beneath the edges of shallow foundations can be expected with any soil or rock possessing low cohesion since stress in the elastic state is greatest beneath the edges.

In some situations highly frost-shattered and comminuted chalk is saturated and under the influence of rapidly applied loads it may flow like a slurry. It is sometimes associated with solifluction of Pleistocene age and has been observed beneath disturbed clay-with-flints. Under continuous loading, excess pore pressure is soon dissipated and shearing resistance is increased.

For preliminary designs the allowable bearing stress of shallow foundations on frost-shattered, but not remoulded, chalk can be taken as 200 kN m^{-2} and on hard intact chalk 600 kN m^{-2}: with deeper foundations the value may rise to 1000 kN m^{-2} or more. Ward *et al.* (1968) classified chalk in the Middle Chalk at Mundford, Norfolk, as shown in Table 1.21. The basis of grading was three-fold, as follows:

(a) Visual description and classification *in situ* from inspections in a number of 0.75 m diameter boreholes, taking into account hardness, jointing and state of weathering.
(b) Each grade was quantified in terms of load deformation behaviour at three locations by 0.9 m diameter plate bearing tests.

(c) At one of these locations, load–deformation behaviour was studied in a large scale loading test, for which a water tank, 18.3 m diameter and 4500 tonnes maximum loaded weight, was employed. This test was required because of the very large size of the structure to be erected but could be dispensed with for most foundations. The *in situ* loading tests have been described by Burland and Lord (1970). Grainger *et al.* (1973) employed seismic refraction surveys at Mundford and found that seismic velocity increased with depth in increments which corresponded broadly with the grades defined in Table 1.21.

Standard penetration tests in chalk yield values of resistance which commonly range between 30 and 130 blows per metre, although higher values are sometimes recorded, but the results of such tests should be interpreted with caution since they may be influenced by the presence of fissures and the formation of chalk paste during driving. Comments on the correlation of SPT values and chalk grade have been made by Lord and Smith (1976). Recorded values for the ultimate skin resistance of piles in chalk have been summarized by Tomlinson (1976) and range between 8 and $500 \, kN \, m^{-2}$. The values are very variable and clearly depend on a number of unspecified factors. In the case of driven steel and concrete piles the value rarely rises above $50 \, kN \, m^{-2}$ and is largely independent of chalk grade (Table 1.21). The higher values are found with bored cast-in-place piles. Skin and base resistance is likely to improve during a period of a few days after driving as excess pore pressure, set up in chalk paste formed during driving, is dissipated.

Cliffs and excavations in chalk often stand to a vertical face but gradually degrade over a period of years, largely as a result of frost action. Vertical faces can be protected by brick or concrete revetments or screens of ivy draped from the top. However, in order that a reasonable growth of grass can be established, the slopes of cuttings and embankments should be 1 in $1\frac{1}{2}$ (about 34°). The natural angle of repose of chalk fragments is between 34° and 37° and slopes of embankments inclined at 1 in $1\frac{1}{2}$ or less should be stable. Generally there is a reduction in volume between cut in intact chalk and well compacted fill of the order 10% but, if balance of cut and fill is critical, preliminary laboratory and, in particular, full scale field studies are desirable. Heavy plant is generally required to effectively crush hard chalk during compaction but over-compaction of any grade of chalk may impede drainage and in the case of saturated soft chalk may generate positive excess pore pressure, although this is dissipated comparatively rapidly. When the *in situ* moisture content of chalk is close to the liquid limit of finely ground chalk, comminution by multiple handling may produce a slurry with negligible strength. An alternative method of constructing embankments is to preserve chalk in large fragments and place it as rock fill, thereby avoiding intensive compaction with its consequent disadvantages. In this case the volume of fill is likely to exceed that of cut. Excavation and fill of soft chalk have been discussed by Parsons (1967), Clarke (1977), Clayton (1977) and Ingoldby and Parsons (1977).

Table 1.21 Correlation between grade and the mechanical properties of chalk at Mundford

Grade*	Brief description	Approx. range of E_r $(kN\,m^{-2})$	Bearing stress causing 'yield' $(kN\,m^{-2})$	Creep properties	SPT[†] Value (N)
V	Structureless remoulded chalk containing lumps of intact chalk	$< 5 \times 10^5$	< 200	Exhibits significant creep	8–15
IV	Friable to rubbly chalk with open joints often infilled with soft remoulded chalk	5×10^5– 10×10^5	200–400	Exhibits significant creep	15–20
III	Medium to hard rubbly to blocky chalk with closely spaced, slightly open joints	10×10^5– 20×10^5	400–600	For stress not exceeding 400 $kN\,m^{-2}$ creep is small and terminates in a few months	20–25
II	Medium hard to hard chalk with widely spaced, tight joints	20×10^5– 50×10^5	> 1000	Negligible creep for stress of at least 400 $kN\,m^{-2}$	25–35
I	Hard brittle chalk with widely spaced, tight joints	$> 50 \times 10^5$	> 1000	Negligible creep for stress of at least 400$kN\,m^{-2}$	> 35

*To the grading defined above should be added grade VI, which is described as 'extremely soft structureless chalk, containing small lumps of intact chalk' and for which $N < 8$.
[†]Wakeling (1970)

The foregoing discussion of chalk outlines conventional ideas on engineering-related topics, although they do not always provide satisfactory answers to engineering problems. Thus, for example, it is often almost impossible to carry out effective compaction tests because of the difficulty of increasing the moisture content sufficiently to define the maximum dry bulk density and optimum moisture content (o.m.c.) – to do so may require marked perseverence. It follows that, if o.m.c. cannot be attained in the laboratory, it will be even more difficult to reach during construction. Again, the concept of uniformity in, say, the Upper Chalk is shattered when it is found that, for example, in one stratigraphic zone in Sussex, *Offaster pilula*, a face-shovel is required to excavate at one level whereas a scraper can be employed at another: the difference in character can, indeed, be seen in the styles of weathering on exposed faces in the two zones. Careful stratigraphic logging throughout the whole Chalk sequence reveals that lithology is very variable. Clearly engineering behaviour of each bed depends on its lithology and, of a series of beds operating *en masse*, on average values of parameters, weighted according to thickness of beds and other factors. The lithological character of a sample of chalk is revealed only by the high magnification of electron microscopy.

The terms 'hard' and 'soft' are often applied to rocks but, since these relate primarily to resistance to penetration, it is preferable to employ 'strong' and 'weak' for engineering descriptions, although the terms may, in practice, correlate. With sedimentary rocks, for a given type of cement, strength is dominantly a function of degree of compaction or consolidation and degree of cementation. It is sometimes found that beds of weak chalk occur between beds of strong chalk. The weak beds appear to have been deposited with high porosity, perhaps because of uniform grain size or as turbidites or slipped sub-aqueous masses. It is possible that slow dissipation of excess pore pressure in the high-porosity beds may be overtaken by cementation sufficient to resist subsequent overburden stress. When this kind of weak chalk is fractured and compacted by pile driving or other means, the reduction in porosity may lead to an effective increase in moisture content sufficient to develop excess pore pressure, thereby leading to much reduced strength, although this may be temporary only until excess pore pressure is dissipated.

Following extensive studies of the Chalk in field and laboratory, Mortimore (1979) identified five major types of lithology:

(a) Marl, comprising carbonates with 20–70% argillaceous matter.
(b) Griotte, chalk with irregular marly streaks.
(c) Nodular chalk.
(d) Laminate chalk, closely layered.
(e) Bioturbated chalk, disturbed by marine organisms, with burrows and related features.

Furthermore, engineering characteristics of the Chalk are influenced by geomorphological domains, classified as follows:

(a) Plateau, crest and valley flank.
  (i) Stress–relief effects, such as open joints and general instability, not present.
  (ii) Stress–relief effects present.
(b) Plateau, crest and col with pipes and swallow holes present: commonly found around margins of and beneath Tertiary formations and their derivatives, such as clay-with-flints.
(c) Dry valley, commonly displaying deep weathering, including frost shattering of chalk, with solution cavities: effects predominant towards open ends of valleys or coombes.

It will be apparent from earlier discussion that over the last two decades numerous attempts have been made to classify chalk and predict likely behaviour. Clearly more research is needed. This is true also of other sequences of rocks but chalk is unique and troublesome: a consequence of the dominance of calcium carbonate in its mineral composition, of its micro-fabric and its type of cementation. The simplest and most consistent and reliable guides to the quality of intact chalk are probably bulk density of intact specimens, oven-dried at, say,

60°C, and Brazilian crushing strength tests on cylinders instead of discs. For design and construction, rock mass data, such as fracture spacing, must be taken into account together with the character of intact rock.

The importance of recognizing geomorphological or engineering domains in the Chalk, and indeed in other formations, is illustrated by the following two examples. A sewer outfall was constructed seawards in tunnel in chalk and excessive inflow of water was experienced during driving. The tunnel followed, at least in part, the seawards extension of a dry valley or coombe. Undoubtedly the valley would have suffered deep weathering of the kind mentioned above when, during the Pleistocene, sea level was of the order 100 m below the present level. This may have been confirmed by the fact that the spuds of a drilling barge readily penetrated the chalk sea floor, although the spuds may have penetrated a weak high-porosity bed. Geophysical surveys and percussion drilling would be unlikely to reveal the true site conditions and cores would have been difficult to obtain by rotary drilling. The best predictions would have been obtained from a study of land geomorphology and of geological conditions revealed in the cliffs. Supplementary offshore investigations could then have been planned with the specific objective of, for example, defining the extent of the engineering domain. At another site a tunnel was driven on land through a spur of regularly fractured and massive chalk. Since this domain involved lateral stress relief, some movement of the hillside was considered possible during driving but the surface subsidence which occurred locally at a minor fault was perhaps rather greater and more rapid than anticipated.

1.4.2 Collapsing and aggregated soils

Certain soils exhibit an excessive degree of compaction when wetted under load and are known as collapsing or metastable soils. This is due primarily to a relatively high initial voids ratio. Sometimes the particles are cemented at their contacts without the matrix filling the voids and collapse may be sudden either as the load is applied or as the cement is softened by water. Generally the soils are of aeolian origin, such as loess and desert sands, or are residual, such as decomposed granite and felspathic sandstone where leaching has removed soluble and colloidal constituents. Where the thickness of a deposit can be measured in decametres, subsidence can often be measured in metres. Collapsing soils can be improved by appropriate geotechnical measures such as compaction, cementation, baking, pre-wetting and pre-loading.

Jennings and Knight (1957) have reported settlement under load arising from collapse of interparticle structure of ferruginous silty sands with high void ratio when saturated. Collapse of ferruginous aeolian sands was experienced at Nkana, Northern Rhodesia (Allaway, 1962) and of residual granite soils in Southern Africa (Brink and Kantey, 1961). Gibbs and Bara (1962) discuss collapse of fine-grained soils in California. Further details of collapsing soils have been given by Capps and Hejj (1968), Dudley (1970), Barden (1972), Clemence and Finbarr

(1981) and Knodel (1981). It is a matter of common experience that loose fill often collapses when wetted: an example is described by Penman and Godwin (1974).

Certain red clays occurring, for example, in Kenya comprise aggregations or clusters of clay particles cemented by iron oxides (Newill, 1961; Coleman *et al*, 1963). These soils initially appear to be coarse-grained but under repeated loading when soaked with water the structure gradually breaks down to produce a fine-grained soil. The heavily overconsolidated mudstones of the Keuper Marl contain considerable proportions of silt-sized aggregations of clay particles and their properties in the intact state may be considerably modified by weathering (Dumbleton and West, 1967; Davis, 1968; Chandler, 1969; Davis and Chandler, 1973). A classification of weathering of the Keuper Marl is shown in Table 1.22 (Chandler, 1969). Breakdown of soils can be studied in the laboratory by means of a grease-worker, the degree and effects of breakdown being plotted against the number of strokes of the apparatus.

Loess is a wind-borne deposit possessing typically a dominant silt fraction, a calcareous component, a porous structure, lack of stratification and a buff colour. The particles tend to be angular and consist principally of quartz with felspar, calcite, mica and clay minerals. Loess is penetrated by numerous vertical tubes, probably formed by roots of vegetation growing while the deposit was gradually built up. The material forming the vast areas of loess in China has been derived

Table 1.22 Weathering scheme for Keuper Marl (after Chandler, 1969)

Zone	Description	Notes
Fully weathered IVb	Matrix only	Can be confused with solifluction or drift deposits, but contains no pebbles. Plastic slightly silty clay. May be fissured
Partially weathered IVa	Matrix with occasional clay-stone pellets less than 3 mm diameter but more usually coarse sand size	Little or no trace of original (Zone I) structure, though clay may be fissured. Lower permeability than underlying layers
III	Matrix with frequent lithorelicts up to 25 mm. As weathering progresses lithorelicts become less angular	Water content of matrix greater than that of lithorelicts
II	Angular blocks of unweathered marl with virtually no matrix	Spheroidal weathering. Matrix starting to encroach along joints: first indications of chemical weathering
Unweathered I	Mudstone (often fissured)	Water content varies due to depositional variations

largely from the Gobi Desert. The loess deposits of Europe and North America occur in regions which were peri glacial during the Pleistocene and it is considered that the material was derived from glacial outwash and transported by winds blowing from the northern ice sheets to warmer southerly regions. Loess tends to become sandy towards the source areas.

Clevenger (1958) has described the engineering characteristics of the loess of the Missouri River Basin and these are broadly similar to the properties of most loessic soils, although marked variations occur locally. The Missouri River Basin loess consists primarily of silt particles bonded by small amounts of montmorillonite clay to form a porous structure. The *in situ* dry bulk density is generally between 1200 and 1500 kg m^{-3} and the natural moisture content is about 10%. When the moisture content is less than about 15 per cent, undisturbed loess has relatively high shear strength and low compressibility. The bearing capacity of low-density loess may be 500 kN m^{-2} when relatively dry but as low as 25 kN m^{-2} when wetted. However, when the dry bulk density exceeds about 1450 kg m^{-3}, the Missouri loess is capable of supporting the loadings normally imposed on it. Pre-consolidation of loess can be effected by applying a temporary load, such as fill, while the loess is maintained in a saturated condition. Where a deposit of loess is relatively thin and overlies a resistant stratum, it may be preferable to employ piles or piers founded on the lower bed.

Properties of loess have been discussed by Bolognesi and Moretto (1957) and Bally *et al.* (1969). Tests for index properties show that characteristically these soils are in the ML classification of Casagrande. Abelev and Askalonov (1957) describe stabilization processes applied to loess in Russia where heavy tampers and explosive charges have been used to effect compaction. *In situ* reinforced concrete friction piles suffered considerable settlement and were not considered satisfactory. Sodium silicate injections have also been employed, as also have thermal geotechnical processes. Karafiath (1957) has discussed deep compaction of loess in Hungary. Further details of foundation construction on loess have been given by Kassif (1957), Drashevska (1962), Larionov (1965), Varga (1965) and Audric and Bouquier (1976).

Brickearth is the name given to deposits of unstratified, porous, buff loam which occurs in southern England in thicknesses up to 5 m or more. Sometimes it lies directly on the Chalk and other solid formations but usually it overlies Coombe Deposit, often grading into it vertically and horizontally. Brickearth is of late- and post-Glacial age and is considered to be partly of aeolian origin but mainly to have been transported by sheetwash and is therefore more correctly of colluvial origin. Around Pegwell Bay, Kent, the Brickearth is probably a true loess and this material has been studied by Fookes and Best (1969).

The characteristic bearing failure on loess commonly involves punching of the foundation into the soil, probably connected with breaking through a crust. Rupture zone failure (see Section 3.3) cannot develop until considerable penetration has taken place. In the case of loess a rational method of calculating bearing capacity is that of W.S. Housel. The ultimate capacity of a foundation is

Table 1.23

Plate size B (m)	Yield load Q_f (kN)	Settlement at yield (mm)
0.30	39.9	6.1
0.91	139.5	5.2
1.52	199.3	4.3

$$Q_f = pA + sP \qquad (1.20)$$

where A is the area and P the perimeter of the foundation in plan, p is the compressive strength (kN m^{-2}) and s is the punching-shear strength (kN m^{-1}) of the soil. Clevenger (1958) reported site loading tests with square plates in pits 1.5 m deep on wetted loess with the results shown in Table 1.23.

Analysis of these results shows that approximately $p = 0$ and $s = 34.3 \text{ kN m}^{-1}$, indicating that yield in perimeter shear had occurred but resistance afforded by compression was negligible although this latter component would increase as settlement continued. After application of a factor of safety it will be seen that founding on loess with column loads exceeding about 100 kN is likely to be unsatisfactory unless the characteristics of the soil are improved by geotechnical process. In respect of the design of the foundation structure it should be observed that the upthrust from the soil is likely to be concentrated around the perimeter of the foundation. Long-period site plate loading tests may be desirable to ascertain whether or not creep in perimeter shear or compression is significant or whether there is a progressive breakdown of soil structure, particularly under saturated conditions.

The most satisfactory method of assessing bearing capacity and settlement on loess is to employ site plate loading tests, accompanied by borings and standard penetration tests to establish local variations in soil conditions. A factors of safety of 3 should be applied to the yield load and 2 to the load at which the settlement is 15 mm. The allowable load is taken as the lesser of the two values (Peck *et al.*, 1953).

1.4.3 Foundations on permafrost

Permafrost is soil or rock in which a temperature below freezing point has existed for a number of years and is a product of both Pleistocene and Holocene climates in high latitudes and/or at high altitudes. It attains thicknesses of as much as 600 m. The minimum ground-temperature recorded in Alaska is about $-10°C$ and occurs at depths of 30 to 60 m. The thickness of permafrost is governed mainly by air temperature, which may fall as low as $-60°$ C, and the flow of heat from the interior of the Earth but it is influenced also by vegetation and the movement of groundwater. The thermal conductivity of water near freezing-point is about 25 times that of air at the same temperature and the conductivity of ice is greater than that of water. All artificial constructions tend to disturb the natural thermal regime and locally permafrost is either destroyed or augmented. The surface layer of ground which freezes in winter and thaws in summer is known as the active layer. Ground below freezing point but devoid of ice is known as dry permafrost. The areal distribution of permafrost in any particular region may be either continuous, discontinuous or sporadic.

Frost heave involves the growth of ice lenses beneath the surface with a consequent upward displacement of the overlying soil. It occurs on a small scale during winter in the British Isles. On a large scale in permafrost regions it leads to the formation of isolated steep-sided hillocks, known as pingos, rising to as much as 100 m. The prerequisites for development of frost heave are an ample and free-flowing supply of water and the facility to attract it to a zone of ice accumulation. These conditions are related to at least medium permeability and relatively high capillarity. Silt and chalk possess both of these but other soils and rocks generally possess only one of these characteristics. Therefore silts and chalk are liable to suffer frost heave more than other soils and rocks. A contributory factor is the reduced vapour pressure of frozen water which encourages water in both liquid and gaseous phases to migrate from unfrozen soil to ice lenses. The growth of the ice is principally vertically, in the direction of heat-transfer, and hence upward heave is marked. On thawing, the liberated water tends to saturate the ground and soil softened in this way is said to be subjected to frost boil or, more correctly, frost melt. The low plasticity index of silts leads to a high degree of softening with a relatively small increase in moisture content. The slow movement of saturated soil and rock debris down a slope is known as solifluxion and, in permafrost regions, this takes place in summer on slopes as slack as 1° to 3°. The movement is generally amplified by the waxing and waning growth of ice lenses.

Vegetation in permafrost regions may give rise to a considerable thickness of cumulose deposits. Tundra is the name given to a wide variety of vegetation types found beyond the latitudinal and/or above the altitudinal limits of trees. Muskeg is tundra-like vegetation occupying swamps scattered in forest areas and includes stunted trees. The term muskeg is applied also to the cumulose deposit formed by this vegetation. Since permafrost constitutes an impermeable layer, it prevents infiltration of precipitation into the ground and tends to create concentrations of organic acids and mineral salts near the surface. The engineering characteristics of muskeg are discussed in a paper by Adams (1965) and in a publication edited by MacFarlane (1969).

Turning now to construction methods, two basic techniques are employed. In the passive method, site conditions are undisturbed except for excavation. Insulation between building and ground is provided so that the thermal regime is affected as little as possible. In the active method, either the permafrost is thawed before construction and subsequently maintained in this state or it is excavated and replaced by material not subject to the adverse effects of freezing. Russian practice favours the passive method in areas of continuous permafrost but in discontinuous or sporadic permafrost zones either the passive or the active method is employed, depending on local conditions.

For trial boreholes in permafrost, core drilling is more satisfactory than churn drilling but cooling is necessary to prevent thawing of the side of the hole and the sample. Test pits reveal site conditions in detail and are favoured by some engineers for depths down to about 6 m. Frozen ground stands without support although timbering is required near the surface. The ground surface at a permafrost site is often in a poor state and it is necessary to cover the surface with a layer of gravel or crushed rock to facilitate movement of plant.

Common practice in permafrost foundation engineering is to support structures on piles, piers or independent footings founded at a depth of not less than twice the thickness of the active layer. The object is to secure not only a firm support for the base of the foundation but also sufficient resistance by adfreezing of permafrost to the foundation to prevent uplift due to seasonal frost heave in the active layer. Generally, foundations are extended 1.0 to 1.5 m above ground surface to provide an open space beneath the structure, thereby reducing heat transfer from building to ground.

Pile placing necessitates either a hole drilled in the permafrost to accommodate each pile or the thawing of a cylindrical zone by a steam jet from a 25 mm diameter fluted pipe. The pile is inserted in the hole or thawed zone and hammered to refusal. There is a tendency for timber piles to float in the thawed zone so that temporary kentledge must be provided and each pile must be maintained in position by stays until the zone is again frozen. Complete refreezing may take 1 to 2 months. Boring reduces these difficulties to a minimum and also provides details of ground conditions. Another method of piling is to set the piles in an open trench, backfill with gravel and allow the backfilling to freeze before loading the piles. Adfreezing skin resistance increases with decrease in temperature: thus at $-1°C$ it may be $700 \, \mathrm{kN \, m^{-2}}$ and at $-5°C$ about $2000 \, \mathrm{kN \, m^{-2}}$. Because of creep, more particularly near freezing point, it is usual to apply a high factor of safety, sometimes as much as 10. Heat transfer through piles may be important. Loading tests on piles should be carried out in late summer when their resistance can be expected to be a minimum.

Piers were employed for the foundations of a building at Churchill, Canada, since piling would have been difficult owing to the presence of boulders (Dickens and Gray, 1960). The piers were founded at a relatively shallow depth and it was necessary to raise the level of the permafrost above founding level. This was achieved by placing about 1.5 m gravel filling over the ground surface after construction of the piers. The gravel was covered by about 0.6 m of moss.

The thaw-blast method of ground preparation was employed for the conventional spread footings of a power station in Alaska (Waterhouse and Sills, 1952). The site was underlain by permafrost of depth varying from 5 to 18 m beneath ground surface. The thermal characteristics of the building were such that permafrost could not be maintained above a depth of about 18 m and hence the frozen condition of some of the ground had to be eliminated. The process consisted of thawing to a depth of about 10 m by means of steam jetting pipes 20 mm diameter directed vertically into the ground at 2 m centres over the entire site. Tests indicated that subsequent thawing to a greater depth when the building was in service would not materially increase settlement. After thawing, explosive charges were placed at 7.5 m centres and at a depth of 7 m. These were followed by second and third charges, each offset 2 m in plan. The explosions were employed to compact the ground after the ice lenses had been thawed.

Construction programmes are governed to a considerable extent by climate and other natural phenomena in permafrost regions and a work feasibility chart is an essential preliminary to construction (Roberts, 1960). Details of permafrost and permafrost construction are given in numerous publications but the following should prove useful starting points for further reading, including references: Cass (1959), Pihlainen (1959), Crawford and Johnson (1971), Becker (1972), Robinsky and Bespflug (1973), Sykes *et al.* (1974), Mohan (1975), Rein *et al.* (1975), Johnston (1981), Linell and Ledrow (1981), Morgenstern (1981), McRoberts (1982) and Ladanyi (1983). A number of useful papers have been published in the *Canadian Geotechnical Journal* and in the proceedings of international conferences on permafrost. Problems arising from freezing are, of course, met on a small scale with cold stores, ice rinks and cryogenic gas tanks (Egan, 1973). Lightly loaded structures, such as ice rinks, are particularly at risk. Problems are also encountered when structures such as cold stores are demolished and thawing takes place. Construction in regions having cold winters should take into account the influence of single or multiple cycles of freezing and thawing. Thus, for example, fill may be frozen shortly after placing and may settle or fail suddenly on thawing.

1.4.4 Foundations on fill

Fill may be encountered in foundation engineering either as a deposit which has been placed at some time prior to the inception of a construction project or as a deposit placed under controlled conditions as a definite part of a scheme. If the fill has been placed under the former conditions a study should be made of the following:

(a) Depth of fill including variations over site.
(b) Nature of fill materials.
(c) Age of fill.
(d) Method of placing fill.
(e) Sequence and thickness of underlying strata.
(f) Nature of underlying strata.
(g) Extent of natural or artificial drainage.

Uneven depths of fill and fills of mixed composition or containing household or industrial refuse may lead to large differential settlements. Broken rock, concrete or brick, or gravel, coarse sand or fly ash generally produce the most satisfactory fills – especially if well-graded. Settlement of soft shale, clay or silt fills may continue over a period of many years. Fills which have been compacted at optimum moisture content in 150 to 300 mm layers are much superior to fills tipped at random. Deposits of waste natural materials, such as mine and quarry waste, are generally more satisfactory than domestic and industrial wastes. The latter are likely to give variable bearing capacity and may be subject to decomposition and combustion or lead to corrosion of foundation steel and concrete. Some fill material, such as over-consolidated clay and unburnt or partially burnt colliery shale, may absorb moisture and expand after placing, with consequent heave of floors and distress in other parts of structures. It is good practice to test fill for sulphate content except where it is known to be free from these minerals. In Wakefield, Yorkshire, calcium sulphate in hardcore fill has attacked concrete ground-floor slabs of houses and caused them to rise about 100 mm and to crack. Swelling of pyritic Jurassic shale fill has led to distress in houses in the Teesside area of England. Slag is sometimes used for fill but a distinction must be drawn between blast furnace slag and steel slag. The former is covered by British Standards and is used as fill provided sulphate contents are within certain limits, the latter tends to be variable in composition and contains free CaO and MgO. The lime and magnesia hydrate in association with water and expand: houses at Dudley, near Birmingham, have suffered considerable distress as a result of this.

Sanitary fill of domestic and industrial waste should be investigated on site and from records for the possible presence of hazardous or obnoxious materials, including toxic compounds, such as methane, which may be generated in the processes of decomposition. Gases have accumulated in buildings on fill – an explosion at an indoor bowling club at Christchurch, Dorset, in 1979 is one example of many such cases. Following an explosion in 1969 in a building on sanitary fill at Winston-Salem, North Carolina, wells and pumps were installed to remove methane. Explosions can occur when the methane content of air is as low as 5%. Placing an impermeable layer over an area through which noxious gases migrate may offer local relief but the gases are deflected laterally: complete protection may require also wells and pumps to pass the gases safely to waste or to a storage tank for use as fuel.

It may be possible to assess the allowable value of bearing stress on fill from site and/or laboratory tests, but possible variations over a site must be borne in mind and conservative values are often adopted, not exceeding say $50 \, kN m^{-2}$. Allowable values may be higher if fill has been carefully compacted during placing or if it is several decades old. Construction can often be commenced on completion of a fill which has been carefully compacted but should not normally take place until a minimum period of two years has elapsed if the fill compacts under its own weight. It may be advantageous to design structures in units capable of independent settlement. Raft, pier or pile foundations are generally most appropriate on fills. Rafts are best suited to light structures. Piers and piles may or may not penetrate beneath the base of the fill. Possible downward drag of consolidating fill on piers and piles should be taken into account in assessing allowable bearing stress. Short, bored piles may prove economical for fills of shallow depth. For domestic houses, reinforced concrete beam foundations may be suitable and, as a rough guide, can be designed to function both as simply-supported beams of span two-thirds of their length carrying a uniformly distributed load and as cantilevers one-third of their length. To eliminate differential settlement of some types of structures on fill, the provision of shallow foundations with facilities for jacking may prove cheaper than deep foundations. On sloping sites fill should be founded on horizontal steps excavated to soil or rock of appropriate bearing capacity and the stability against short- and long-term failure of the slope should be investigated.

The foundations for the columns of a four-storey factory on fill at Mansfield, Nottinghamshire, completed in 1927, consisted of continuous inverted reinforced concrete tee-beams. Similar foundations, on a 0.4 m thick bed of hoggin, were used for a stadium on fill at Minsk, Russia, in 1952. A cellular raft 21 m × 19 m × 2 m has been employed for the foundation of a three-storey office block at Stafford. The base slab is 0.3 m thick and the top slab forms the ground floor. The allowable bearing stress is $50 \, kN \, m^{-2}$.

The possibility of spontaneous combustion in some fills introduces unusual problems. Fires are raised by heat generated in chemical changes in the materials of the fill or by heat generated in the structure founded on the fill. Andrews (1944) states that a temperature as high as $830°C$ has been recorded in a burning fill. Prevention or quenching of a fire is best effected by injection of a material, such as cement or PFA grout, which seals the fill from fresh supplies of oxygen. For economy the area can be divided into bays separated by 6 m wide strips of injected material but each bay must be adequately sealed on all sides. No open excavations should be made in the fill since these increase the means of access for air. The application of water to reduce temperature or quench a fire enlarges air passages and conditions are aggravated. Combustion leads to a reduction in volume of the fill and cavities are formed which lead to large differential settlement. Temperature in fill can be measured by sinking 50 mm diameter tubes.

Compacted fill has been employed as a distinct part of the foundations for several heavy structures in the USA as an alternative to piles or caissons where rock lies at a considerable depth beneath soils of low bearing capacity. At San Diego, California (Smoots and Benton, 1961), the foundations for a power station were constructed in this manner. Soft soil was removed to a depth of 6 m and drainage ditches were formed round the floor of the excavation. The bottom 0.3 to 0.45 m thickness was excavated by a back-actor in order that the formation surface should not be disturbed by movement of plant. A 0.3 m layer of 20 to 40 mm crushed rock was spread on the bottom of the excavation and this was covered by 50 mm of pea gravel followed by 0.15 m of sand. The bulk of the fill consisted of fine sand obtained locally from littoral dunes. Compaction was effected in 0.3 m layers under saturated conditions with a minimum of three passes of a vibratory compactor. Within the fill, dikes with 1:1 slopes were formed of decomposed granite to serve as retaining walls for conduits in which water intake and discharge pipes could be accommodated. These pipes were concreted in position. The design bearing stress beneath the structural foundations varied from 135 to 175 kN m^{-2}. Compression in the fill under the foundation loads has been negligible and consolidation in the underlying soils has led to a maximum settlement at the surface of about 30 mm with differential settlement not exceeding 3 to 6 mm.

At Anacostia Naval Station, Washington, DC, the foundations of single-storey structures comprised reinforced concrete inverted T-beams on 2 m of fill ($w_L \not> 25$, $I_P \not> 8$, $\gamma_d \not< 18.4$ kN m^{-3}) and a 0.6 m basal layer of crushed stone, beneath which 2 m of recent silty sand fill overlies 18 m of soft silty clay above dense silty sand. The masonry walls were reinforced and provided with control joints. Arbitrarily, the T-beams were designed to be unsupported over a cantilever span of 3 m and a simply supported span of 6 m. Concentrated loads are suspended from the floor slabs which bear on fill and foundation beams. During five years from commencement of construction the maximum settlement reached nearly 0.5 m but no cracking or structural distress was observed. Calculations relating to the influence of a fill layer on settlement are demonstrated in Example 2.15: the additional settlement caused by the weight of fill should be included, although this should be relatively uniform. Where settlement is anticipated to be considerable and irregular, structures can be light, flexible and articulated or rigid and sufficiently strong to resist anticipated bending moments and shearing forces, as at Anacostia.

At London Airport, a 1.3 m bed of brickearth, overlying gravel, was removed and replaced by hoggin compacted in 0.2 layers to improve the capacity of foundations for a group of buildings (Edwards and Rigg, 1961). A large area of sand fill about 3 m thick was placed for a steel tube plant at Shotton, Flintshire (Brooks *et al.*, 1960). The fill was pumped from the River Dee. Before placing the fill, top soil and peat were removed and a new ditch was excavated around the site. Blocks of flats were constructed in Ayr over a quarry filled with 18 m of rubble and other material: *in situ* piles were put down through the fill into siltstone bedrock and the deep-beam effect of the load-bearing walls was utilized to reduce the number of piles (Orme and Thorburn, 1970).

Near Sevenoaks, Kent, it was found that undisturbed stiff clay of the Gault was weaker than the same clay in the remoulded state owing to the presence of fissures, a phenomenon mentioned in Section 3.1.5 in connection with values of sensitivity less than unity. Consequently, the slope adopted for cuttings was 1 in 6 and for embankments 1 in 4. Similar conditions were found in Gault clay near Cambridge and here cuttings were stabilized by excavating additional bank material and backfilling, the elimination of fissures by remoulding leading to an increase in strength. Since this practice imparts a new microstructure to clay, consideration should be given to long-term strength, although at first sight the effects appear to be generally beneficial. It should be noted that, if intact blocks of stiff clay are present in the fill, they will tend to continue to expand following stress relief from the *in situ* state before excavation and will absorb moisture over a period of weeks or months, thus leading to a reduction in strength.

Pulverized fuel ash (PFA), commonly known as fly ash, is being employed to an increasing extent for compacted fill (Raymond, 1961; Smith 1962, 1968; Sherwood and Ryley, 1966; Sutherland *et al.*, 1968; Di Gioia and Nuzzo, 1972; Sherwood, 1975b). The ash is incombustible residue and is derived mainly from clay minerals, calcite and pyrite. Some of the clay minerals are converted to complex silicates and the pyrite tends to form magnetite. The ash may also contain some coke (coal from which volatiles have been distilled). The range of particle sizes of the ash is from sand to silt but some ashes are comparable with uniform fine silts. Values of dry bulk density of 1000 to 1500 $kg\,m^{-3}$ are obtainable in the laboratory by compaction at optimum moisture contents ranging from 30 to 15% but the magnitude of density obtainable on site is commonly 80 to 90% of these values. The shear strength parameters of freshly compacted ash are generally $c_u = 30$ to $70\,kN\,m^{-2}$ and $\phi_u = 27°$ to 35°. Both parameters increase to some extent with time owing to natural cementing of the particles but sometimes the effect appears to be small. In some cases California bearing ratio (CBR) at 28 days may exceed twice the value obtained immediately after compaction. PFA possesses a low content of sulphates – mainly calcium sulphate – and foundation and floor slab concrete is commonly laid on polythene sheeting as a precautionary measure. Most ashes tend to be susceptible to frost heave but this can be prevented or reduced by covering the fill with a frost-resistant layer of PFA stabilized with 10% Portland cement and possessing a relative compaction of 85 to 90%. Stabilization with lime or cement is common for road bases: the addition of 6% to 15% increases strength and reduces permeability (Sutherland and Gaskin, 1970). In practice, the engineering characteristics of PFA vary with different generating stations and also with the source of the coal which, at any one station, may vary from time to time. Hence it is important to test the material before design and to make checks before and during construction.

Because compacted fill is currently employed as part of foundation structures, it may be useful to summarize some of the fundamental aspects of compaction – the process whereby soil particles are packed more closely through reduction in air voids. Soil is difficult to compress at very low values of moisture content (m.c.) and under these conditions values of dry bulk density (γ_d) are relatively low, and of air voids relatively high. Increasing the m.c. renders the soil more workable and higher values of density and less air voids are attained. Further increments in m.c. above the optimum (o.m.c.) hinder the escape of air with the result that γ_d becomes progressively less and air voids generally attain a sensibly constant value between 2 and 10% of the total volume of the specimen. A state of zero air voids is never reached in the compaction process. The peak of the γ_d against m.c. curve tends to be ill-defined when soil is uniformly graded and to be sharply defined when soil is well-graded. The o.m.c. of most soils lies between 10 and 20% but values outside this range are not uncommon. In fact, for silty and

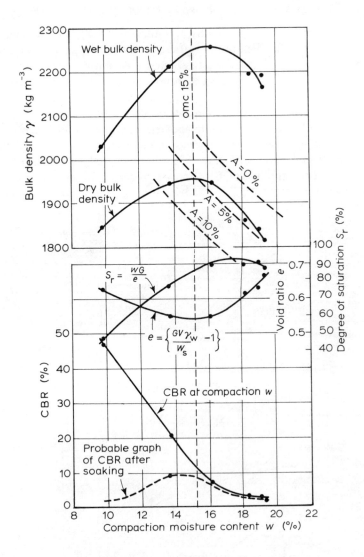

Figure 1.24 Compaction and CBR tests on lateritic concretionary gravel with fines from Nigeria. The CBR values are a measure of bearing capacity. The theoretical dry density lines (---) have been determined from the relationship $\gamma_d = (1 - A/100)/(1/G\gamma_w + m/100\gamma_w)$ where A is the air void expressed as a percentage of the total volume of the sample. Because of the ferruginous content, the effective value of G was assumed to be 3.0: this is probably a reasonable value since the $A = 5\%$ line passes close to the peak of the experimental dry density curve – a common characteristic with well-graded soils. Lower values of G displace the A lines to the left and the peak rarely occurs at values of A less than 5%. Only one value CBR was obtained after soaking in water: the compaction $m = 13.8\%$ was estimated to be near the optimum moisture content (o.m.c) and after soaking m had risen to 15.2%. With poorly graded soils, the peak of the dry bulk density curve may lie closer to the 10% A line.

clayey soils, o.m.c is usually slightly less than the plastic limit. Increase in compaction effort, either dynamic or static, on a given soil results in a decrease of up to about 20% in o.m.c., that is, up to about 4%, accompanied by an increase of up to about $250 \, \text{kg m}^{-3}$ in γ_d: the effects being greater with clayey soils than with sandy soils. However, the curves of γ_d against m.c., and also CBR against m.c., derived from tests with different compaction efforts (light to heavy) tend to run together at values of moisture content above the optima. At this stage, degree of compaction and also bearing capacity (or CBR) depend more on m.c. than on compaction effort. Figure 1.24 shows the complete analysis of the results of a compaction test together with CBR values. It is generally desirable to use fresh material each time the mould is filled in order to minimize the influence of

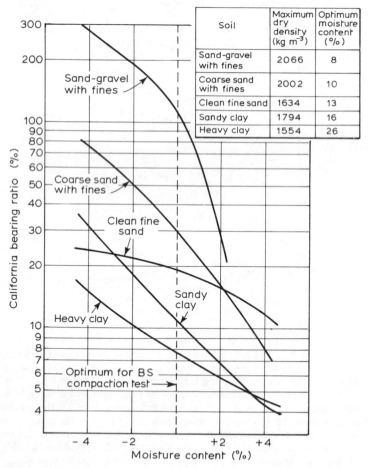

Soil	Maximum dry density (kg m^{-3})	Optimum moisture content (%)
Sand-gravel with fines	2066	8
Coarse sand with fines	2002	10
Clean fine sand	1634	13
Sandy clay	1794	16
Heavy clay	1554	26

Figure 1.25 Relationship between CBR and moisture content for soils at the maximum dry density given by the BS compaction test. Specimens statically compacted and tested unsoaked without surcharge.

comminution incurred during compaction. The variation of CBR value with moisture content for different soils is shown in Fig. 1.25. Black and Lister (1979) quote the empirical relationship $CBR = c_u/23$ for remoulded clays.

Occasionally it will be found that the graph of dry density against moisture content displays two peaks, the one at the lower moisture content generally corresponding to the lower value of peak γ_d. Excluding obvious experimental error in any particular case, the cause of double peaks must be related to the relative efficiencies of workability and air expulsion over particular ranges of moisture content. It is usual to base contract specifications on the higher value of optimum moisture content.

For the determination of *in situ* density of fill, it is good practice to make at each station one measurement with a sand bottle and another with a core cutter 100 mm diameter × 125 mm. Sand bottle measurements are generally regarded as more accurate than core cutter measurements, which may be higher or lower depending on whether the soil compacts or dilates as the cutter is inserted. In addition to this general effect, there are random variations arising from local differences in compaction effort and errors in sampling. The accuracy of relative density measurements has been investigated by Tavenas and La Rochelle (1972). It should be noted that, even with efficient compaction, there is a possibility of minor collapse of fill, more particularly sand or silt (Section 1.42), when inundated and this leads to differential settlement of magnitude which may be significant with some structures. Since nuclear *in situ* density testing is becoming more reliable, it may replace excavation techniques, which are lengthy and require laboratory work.

The degree of compaction achieved in tests or on site can be assessed by the parameters defined in Equations (1.21), (1.22) and (1.23).

$$\text{Relative compaction } \% = (\gamma_{di}/\gamma_{dm}) \, 100 \text{ (commonly 75\% to 100\%)} \qquad (1.21)$$

$$\text{Efficiency of compaction } \% = \left(\frac{\gamma_{di} - \gamma_{dl}}{\gamma_{dm} - \gamma_{dl}} \right) 100, \qquad (1.22)$$

where γ_{di} is the *in situ* dry bulk density, γ_{dm} is the maximum density achieved in tests, and γ_{dl} is the lowest density. Efficiency of compaction is zero in the loose state and 100% in the dense state.

The relative density of a soil in the field is defined as

$$D_r = \left(\frac{e_{max} - e}{e_{max} - e_{min}} \right) 100 = \left[\frac{(\gamma_{di} - \gamma_{dl}) \, \gamma_{dm}}{(\gamma_{dm} - \gamma_{dl}) \, \gamma_{di}} \right] 100, \qquad (1.23)$$

where the void ratio is e_{max} in the most loose state,
$\qquad\qquad\qquad\qquad e_{min}$ in the most dense,
$\qquad\qquad\qquad\qquad e$ in the field state.

If the field state is very dense, D_r tends to 100% and if it is very loose, D_r tends to zero. The field void ratio can be determined from the usual tests for *in situ* density. The minimum void ratio can be derived from the maximum γ_d obtained in a standard compaction test. The maximum void ratio is determined by pouring the soil in a dry state from a funnel into a standard compaction mould with a free fall of about 20 mm.

The respective values of e_{max} and e_{min} for uniform sands and silts are generally of the order 1.0 and 0.4 and for well-graded gravel, sand and silt 0.9 and 0.2. Lambe and Whitman (1969) describe relative density according to the nomenclature shown in Table 1.24.

Mixtures of material from different sources for mechanically stabilized fill can be proportioned to satisfy a given specification by graphical solution on a triangular diagram (Krynine, 1947). This is illustrated by Fig. 1.26 in which A, B and C represent the gradings of material from three sources and M represents the

Table 1.24 Density description

Relative density (%)	Descriptive term
0–15	Very loose
15–35	Loose
35–65	Medium
65–85	Dense
85–100	Very dense

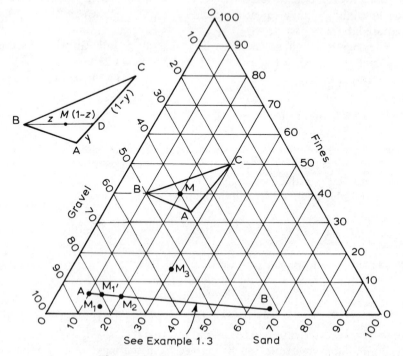

Figure 1.26 Proportions of soil mixtures.

specified grading of the mixture. BM is produced to interest AC in D. D represents a mixture of A and C in the proportion $y/(1 - y)$ where AC is assumed equal to unity: that is, y of C plus $(1 - y)$ of A equals unit volume of D. Similarly M represents a mixture of B and D in the proportion $z/(1 - z)$: that is, z of D plus $(1 - z)$ of B equals unit volume of M. Expressed as percentages, the final proportions are $100 (1 - y)z$ of A, $100 (1 - z)$ of B and $100\ yz$ of C. To obtain a reasonable degree of accuracy the sides of the triangular diagram should be about 10 cm.

A limiting condition obtains if M lies on AB, BC or CA, when the required proportion of one of the components is zero. If M lies outside ABC, then the mixture is unobtainable from A, B and C although the introduction of a suitable fourth component E in conjunction with one of the others, say C, could lead to a mixture F such that M would be enclosed within ABF.

The problem can be solved also by simultaneous equations in the following way. Let the percentage gradings be as follows:

	Gravel		Sand		Fines	
A	G_A	+	S_A	+	F_A	= 100
B	G_B	+	S_B	+	F_B	= 100
C	G_C	+	S_C	+	F_C	= 100
M	G_M	+	S_M	+	F_M	= 100

If the required percentages of A, B and C are a, b and c, then

$$aG_A + bG_B + cG_C = 100G_M$$
$$aS_A + bS_B + cS_C = 100S_M$$
$$aF_A + bF_B + cF_C = 100F_M$$

These equations should yield results identical with those derived by the graphical construction. The advantage of the graphical method is that the feasibility of the mixture or appropriate adjustments can readily be seen. If the mixture is unobtainable at least one of the unknowns will be negative.

An appropriate specification for a well-graded mixture of relatively high density can be evolved by application of Fuller's Rule for particle-size distribution of granular materials:

the percentage passing a given sieve

$$= 100\left(\frac{\text{aperture size of the sieve}}{\text{size of largest particles in mixture}}\right)^{1/2}$$

It is doubtful whether any method of determining particle-size distribution more refined than Fuller's Rule is justified for the production of fill and base material and often this needs to be satisfied only approximately. Indeed, in many cases, fill consists of a single available material, sometimes other than well-graded, and whatever value of strength that can be attained has to be accepted; often this proves to be more than adequate. For road bases it is commonly specified that, for material passing No. 36 sieve, $w_L \ngtr 25\%$ and $I_P \ngtr 6\%$.

The characteristics of compacted clays have been discussed by Seed and Chan (1959). The use of colliery shale has been discussed by Sherwood (1975a) and the properties of burnt shale have been investigated by Fraser and Lake (1967). Fills placed by hydraulic means have been discussed by Lambe (1969), Blight (1970), Whitman (1970) and Turnbull and Mansur (1973). Swelling of a blast furnace slag fill has been reported by Crawford and Burn (1969): the expansion appears to be of the order of 10% of the thickness of the fill. British laboratory and field practice for compaction of fill is dealt with in numerous reports published by the Transport and Road Research Laboratory, Crowthorne, Berkshire: these include TRRL (1974–1978). The proceedings of a conference on clay fill covers a wide field and includes many useful papers (Vaughan, 1979). Fill and hardcore for building construction are the subjects of digests published by BRE (1983). Earthworks practice is specified in BS 6031 (BSI, 1981b). The use of rock for fill has been reviewed by Penman (1971). A practical distinction between earthfill and rockfill is that excess pore pressure is not developed in the latter during compaction: satisfying this requirement may necessitate restriction of fines content although reduction in fines increases compressibility and reduces shear strength. Barton and Kjaernsli (1981) have made a fundamental study of shear strength of rock fill. Settlement rates of sanitary fill have been studied by Yen and Scanlon (1975). Field studies of the performance of compaction plant have been discussed by Lewis (1967) and of a heavy vibratory roller by Moorhouse and Baker (1969). A guide to compaction requirements of fill for highway construction is given in a government specification (HMSO, 1976). A form of fill which is now being used for embankments and retaining structures is reinforced earth, in which strips or meshes of material are embedded to resist tension. Reinforced earth has been described by Gedney and McKittrick (1975) and studied by Binquet and Lee (1975a, b), Romstad *et al.* (1976) and Shen *et al.* (1976).

1.4.5 Foundations on clays

Special factors may have to be taken into account in the design and construction of foundations on clays. These are frequently associated with changes in moisture content.

(a) Shrinkage of clays

Shrinkage in clay may be due to seasonal moisture changes or to absorption of water by plants. Both horizontal and vertical shrinkage may be experienced. Seasonal shrinkage effects have been detected at 2 m below ground surface although below 1 m the movement is likely to be small in Great Britain. The shallower the foundation the greater the settlement and vertical movements of 25 mm may occur at depths between 0.3 and 0.5 m (Cooling and Ward, 1948). In the Northern Hemisphere, the clay at the south face of a house can be expected to shrink more than at the north face and consequently south walls may settle more than north. Cracking of a structure arising from seasonal effects is usually cumulative over a period of years since the cracks close only partly during the wet season. Since any clay having a moisture content exceeding the shrinkage limit will shrink to some extent on drying, the term shrinkable is a relative one. The shrinkage limit (SL) of a soil is the moisture content below which further shrinkage is inappreciable and for most clays in the British Isles this is commonly between 12 and 14% although values between 10 and 15% may be encountered. The *in situ* moisture content of British clays is commonly just above the plastic limit and therefore is generally above these values, except in the uppermost 1 or 2 m subjected periodically to desiccation or excess saturation. The shrinkage ratio (SR) is probably a suitable criterion to express the potential shrinkage of a clay. This is normally defined as the ratio between the volume change (per cent of final dry volume of specimen) and the change in moisture content (per cent of dry weight of specimen). However, for the purpose of assessing the potential shrinkage of clay beneath foundations, the change in moisture content should cover the estimated or observed seasonal range of moisture content *in situ* and the volume change should be expressed preferably as a percentage of the volume at the lower moisture content. Much simpler than the test for SL is the test for linear shrinkage (LS), which is measured from the liquid limit to complete dryness and is effectively the same as drying to SL. It is likely that clays having values of LS exceeding about 10 per cent will display marked shrinkage and swelling during cycles of drying and wetting. Plasticity index is also a guide to potential shrinkage (BRE, 1980b): > 35% very high; 22 to 48%, high; 12 to 32%, medium; < 18% low. Since linear shrinkage is specified to be carried out on material passing BS sieve aperture 425 μm, the potential shrinkage or swelling of a given soil must be assessed in terms also of the percentage of material present larger than this, bearing in mind that the potential of such material is likely to be very low. Surface movements of clays in Canada have been studied by Baracos and Bozozuk (1957), Warkentin and Bozozuk (1961) and Hamilton (1963).

Remedial measures may necessitate underpinning an entire structure. Preventive measures essentially involve founding at a level below which seasonal shrinkage is negligible. One suitable type of foundation for houses comprises narrow walls about 1 m deep formed of concrete poured in 0.4 m wide trenches excavated by a chain-bucket trenching machine. Another type comprises 0.3 m or 0.35 m diameter bored concrete piles 2.5 to 3 m deep spanned by 0.3 m × 0.15 m reinforced concrete foundation-beams. From BRS Digest No. 96 (BRS 1957) it appears that the costs of these foundations were roughly the same although there may be some difference arising from current prices. The spacing of bored piles is commonly 2 to 2.5 m centres. A reinforced concrete foundation slab anchored to the piles can be used as an alternative to beams. One mechanical auger can drill eighty short pile holes per day under normal conditions and more than one hundred under ideal conditions.

No allowance should be made for support from the soil directly beneath beams carried on piles since the soil may shrink. The load on each span is, in practice, less than the weight of brickwork above the span owing to the influence of arching within the brickwork. Arching is mobilized as the beam deflects and tends to concentrate loads at the piles. In BRS Digest No. 42 (BRS, 1952) it is stated that, assuming elastic design, the bending moment can be taken as $WL/50$ when there are openings in the brickwork near the supports or $WL/100$ when the openings are in the centre of the span or when there are no openings, were W is the weight of the rectangle of brickwork plus the superimposed loads on the span of length L. The ratio of depth of beam to span must lie between $1/15$ and $1/20$ and the stress in the steel should not exceed $110 \, \text{N} \, \text{mm}^{-2}$. Additional piles are required for heavier parts of a building, such as chimney breasts. East (1951) describes the use of 0.35 m diameter bored piles, 2 m deep, penetrating through soft clay into stiff Gault clay. The safe bearing capacity per pile was 45 kN based on a factor of safety of two. Since the upper metre or so of over-consolidated clays commonly possesses relatively low values of shear strength, short bored piles can often be used to secure relatively high bearing capacity at, say, 4 m depth for structures such as three or four storey blocks of flats or offices.

Green (1961) loaded 0.3 and 0.35 m diameter bored piles, 3 m long, 1 year after casting *in situ* in clay and observed settlement for a period of 4 years. The rate of settlement was small after the first 3 months. Settlement at loads exceeding the design load were small compared with those of similar piles loaded shortly after casting. When a pile is cast, moisture diffuses into the clay in contact with the fresh concrete, thereby reducing the cohesion and adhesion of the clay. As the moisture is dispersed, cohesion and adhesion increase and the bearing capacity of the pile is enhanced. The improvement in these properties appears to be due in part also to the migration of chemical compounds from the concrete into the clay. Increase in bearing capacity arises also from the increase in shearing strength of the clay beneath the base of the pile brought about by consolidation under sustained load.

Shrinkage due to absorption of moisture by trees, particularly fast-growing varieties such as poplar, elm and willow, may be appreciable for depths down to 3 m. The root system of a tree commonly spreads to a radius slightly greater than the height of the tree. Differential movement of as much as 100 mm due to this effect has been measured in houses. Details of structural distress in a theatre at Stamford Hill, London, arising from clay shrinkage caused by poplar trees have been given by Skempton (1954). Observations of ground movements near elm trees in Ottawa have been reported by Bozozuk and Burn (1960), case histories of damage to structures in the Chicago area have been presented by Perpich *et al.* (1965) and shrinkage at a site in Michigan has been reported by Hammer and Thompson (1966). The reverse process of swelling often takes place, sometimes over periods of several years, when trees are removed and damage may be caused as a result of this: an example has been described by Samuels and Cheney (1974). The influence of vegetation on the swelling and shrinking of clays has been discussed at length in a symposium (Wakeling *et al.*, 1983).

Where shrinkage of clay is associated with trees, the remedy is to cut down immature trees or to underpin the structure. Mature trees can be left standing as they are unlikely to create further serious settlement. Preventive measures for new buildings are to construct beyond 15 to 30 m radius of fast-growing trees and to carry foundations 1 to 3 m below ground. More specific recommendations depend on local circumstances such as species and age of trees and depth to phreatic surface.

Exceptional climatic conditions were experienced in the British Isles during the spring and summer of 1976 when meagre rainfall was accompanied by increasing temperature which culminated in extremely hot weather. As a consequence many buildings – in particular two-storey houses – suffered distress sufficient to require remedial measures. It is likely that conditions were generally aggravated by the proximity of trees. As a result of this experience, some engineers have suggested that foundations on shrinkable clays should be not less than 1.5 m, even 2.5 m, deep. Nevertheless, partial relief of distress can be expected in some cases as the clay swells following a period of rainfall. For future construction in a particular area, the long-term behaviour of existing buildings in that area during and immediately following 1976 should provide a guide. Apart from potential shrinkage of a clay, the ability of any particular structure to accommodate differential settlement without distress or, alternatively, to resist the consequent forces should be borne in mind when deciding foundation depth. Low-rise buildings on shrinkable clay soils are treated in Digests 240, 241 and 242 (BRE, 1980).

Evidence that distress in a building is due to differential movement of foundations is generally afforded by cracks visible both internally and externally which extend downwards to the foundations and which become wider upwards. Where cracks are stable and not more than 5 to 10 mm wide, it is often possible to remedy the defects by cosmetic operations, employing plastic filler in the wider cracks. Cracks exceeding 10 mm width may be accompanied by other signs of distress, such as bulging walls, and where these are present the stability of the entire structure should be carefully assessed. Where the foundations are less than 0.6 m deep, it may be desirable to underpin the structure. In all cases of distress, the causes of the foundation movements should be determined and it may be desirable to initiate monitoring for a period up to 12 months to ascertain if movement is progressive.

(b) Swelling of clays

Fundamentally, shrinkage and swelling of clays are related phenomena and in the situations discussed above cycles of both these movements are generally responsible for distress in structures. However, swelling appears to be the dominant factor under certain conditions with some clays. These clays – termed expansive – occur in many regions including Burma, western USA, southern Africa, India, Israel, Peru and Spain. Distress is experienced in particular in light buildings. High capacity for shrinkage and expansion is generally associated with clays having a high montmorillonite content and the degree of movement depends on the proportion of the mineral present and on the kind and amount of exchangeable bases.

The physical chemistry of swelling is not wholly understood, but in some regions expansion appears to be due primarily to infiltration of moisture into the ground during the wet season. The development of cracks is progressive from season to season and the width of gaps is often several centimetres. Elsewhere in hot climates, swelling may be due primarily to thermo-osmotic transfer of moisture towards the zone of lower temperature beneath the centre of a building. The pattern of distortion is commonly dome-shaped, with the maximum movement in the centre of the building, although a wave-form is sometimes developed. In Burma, few cracks occur in buildings beneath which the bearing stress exceeds $100 \, \text{kN m}^{-2}$. Large volume changes can be expected in clay with a plasticity index exceeding about 30 and shrinkage limit less than about 12%. When these parameters are less than about 18 and more than 15 respectively, expansion is likely to be low. In broad terms, the shrinkage limit of pure montmorillonite clay is likely to be between 10% and 15%, of pure illite clay 15% to 18% and of pure kaolinite 25% to 30%: test values for natural clays depend on possible mixtures of clay minerals and organic matter content. The consolidation press can be used in conjunction with a proving ring to record swelling stress of expansive clays, but the deflection of the ring must be taken up as the stress of expansion is developed. Buried structures – such as pipes – in expansive clays may be subjected to high stress arising from differential vertical and horizontal swelling (Kassiff and Zeitlen, 1962). The following measures can be employed to prevent or reduce movement of structures on expansive clays.

(a) Design for a minimum bearing stress which is not less than the swelling stress – commonly of the order 150 to $200 \, \text{kN m}^{-2}$, although much higher stress may be developed under adverse conditions.

(b) Found structures at a level below the zone of varying moisture content. Under-reamed piles can be used for this purpose, designed to anchor the structure in this zone and resist the effects of upward swelling. Tests on *in situ* piles in expansive clays have been reported by Mohan and Chandra (1961).

(c) Reinforce the structure so that it functions as a rigid box which can tolerate the forces without distress. Rigby (1952) has described a method of design which resembles the method employed for the design of foundations liable to mining subsidence.

(d) Provide a flexible structure which will conform freely to the movements.

(e) Surround the building with a waterproof barrier by, for example, bituminous injections to prevent or retard movement of moisture.

(f) Stabilize the soil with lime or other additive.

(g) Excavate the expansive clay and replace with stable fill.

(h) Provide jacking pockets for re-levelling the structure at intervals.

(i) Construct the foundations in a honeycomb fashion so that the clay can expand laterally into the openings.

Many authors have discussed the problems of expansive clays: a review by Gromko (1974) includes a comprehensive list of references and other useful publications have been presented by Holtz and Gibbs (1956), Alpan (1957), Jennings (1961), Komornik and David (1969), Lytton and Meyer (1971), O'Neill and Ghazzaly (1977), Tucker and Poor (1978), ASCE (1980), O'Neill and Poormoayed (1980) and Schmertmann and Crapps (1980).

1.4.6 Structures in the open sea

Early examples of off-shore structures, apart from lighthouses, are the forts constructed at localities around the British Isles in the Second World War. Since then, the demand for natural gas and oil has led to development in marine situations for which specialized structures are required. Construction in the North Sea is particularly relevant at the present time and it is of interest to point out a few of the problems but, although the soil and structural mechanics principles are universal, analysis and design is highly specialized and restriction of space precludes more than brief mention in the present few paragraphs. Around the British Isles wave maxima occur on the western seaboard and the northern North Sea, with amelioration towards the centre of the Irish Sea and English Channel and Dutch coast.

Two kinds of structure are required: drilling rigs for exploration and platforms for exploitation. Where the depth of water is less than 90 m, floating drillships, with or without legs which can be extended into the sea bed, can be used for exploratory borings, but in deeper water, mobile floating rigs of the semi-submersible type, of variable draught, are employed. Instead of the conventional pre-fabricated framed production platform, pinned in position by long piles, a number of new types, relying mainly on gravity for stability, have been evolved to meet the needs in deep and rough waters. The total vertical load on the foundation of a platform may be of the order 2×10^6 kN and waves up to 30 m high can exert a lateral thrust approaching 8×10^5 kN and a downdrag of 4×10^4 kN. Large lateral forces require particular attention to be paid to stability against translation, involving sliding on the base or shear in the soil. Most platforms have a skirt of depth as much as 20 m around the foundation: this increases bearing capacity and resistance to translation and serves to protect against scour.

Bedrock in the North Sea is covered extensively with glacial, marine and estuarine deposits of varying thickness. At Ekofisk, for example, about 25 m of dense fine sand is underlain by clay, overconsolidated as a result of glaciation and commonly very stiff. Since the permeability of the fine sand is relatively low and the drainage path relatively long, the undrained strength of the sand is critical during the passage of a wave. Fortunately, because the sand is dense, shear stress tends to cause dilation, thereby inducing negative pore pressure with consequent increase in strength. However, if the sand is medium dense to loose, spontaneous liquefaction may be invoked with consequent loss of strength.

Under cyclic loads, normally consolidated clay tends to behave similarly to a loose sand but slightly over-consolidated clay tends to consolidate and gains in strength. If clay is heavily over-consolidated, the long-term effect involves reduction in pore pressure, accompanied by swelling and loss in strength. In some situations the fatigue strength of soil will need to be investigated. Oscillation of piles induced by vortex shedding is experienced in estuaries with currents of 5 knots and may present problems on some deep sea structures.

The order of mass structural displacement of production platforms is indicated by calculations for Ekofisk. Settlement under dead load is estimated to be 200 mm, of which 30 mm involves compression in the upper sand and 170 mm in the clay. For a wave occurring with frequency once in 100 years, the differential settlement across a diameter is expected to be ± 150 mm and the lateral translation about 150 mm.

The site investigation programme recommended by the Norwegian Geo-technical Institute involves a general survey of an area typically about 900 m $\times$ 600 m, followed by a detailed survey of an area about 200 m diameter. The general investigation includes, for example, a bathymetric survey, together with a seismic or other geophysical survey and two boreholes for stratigraphic correlation. The detailed survey includes sampling in about seven boreholes to a depth of 1 to 1.5 times the diameter of the structure and five to ten cone penetrometer tests to 5 to 30 m. In addition, a bathymetric survey, accurate to about 0.3 m, is carried out in conjunction with side scanning sonar and television camera studies and inspections by divers. Geophysical logging of boreholes can be carried out to supplement observations and facilitate cor-relation. Site investigations in the North Sea have been described by de Ruiter and Fox (1976) and drilling from a diving bell by van de Graaf and Smits (1983).

For a complete site investigation, about 1 month of good weather is required. A survey vessel may be anchored in position by moorings or it may be equipped with thrusters which enable it to maintain station dynamically while heading into wind and sea. Primary station location can be effected by one of several radio systems or by orbiting satellites to within ± 7 to 150 m, depending on the system and conditions. By laying on the sea bed battery-operated transponders with a life of 2 years, secondary location at the site can be made to within about ± 5 m.

Aspects of soil and structural mechanics in the open sea have been discussed by, for example, Bjerrum (1973), Hansen (1974), McClelland (1974), Lee and Focht (1975), George and Wood (1976), ICE (1977a, 1983), Toolan and Fox (1977), Madsen (1978), Ellis (1979), Dumas and Lee (1980), George and Sladden (1980), Parry (1980), Smith and Molenkamp (1980), Prevost *et al.* (1981a, b) and Wright (1983). Proceedings of a symposium on soils under cyclic and transient loading have been edited by Pande and Zienkiewicz (1980).

Wave and other hydrodynamic forces have been studied by Sainsbury and King (1971), Tickell *et al.* (1976), Hogben *et al.* (1977), Tickell (1977), Hewitt (1980) and Mynett and Mei (1982). The following publications will be found useful for the design of maritime structures: FIP (1977), Beeby (1978), BSI (1982a) and BSI (1978/9). Details of specific projects have been given by Leggatt and Bratchell (1973), Antonakis (1972), Marion and Mahfouz (1974), Derrington (1977), Duvivier and Henstock (1979) and Marr and Christian (1981).

In order to obtain data on environmental conditions and the response of marine structures and their foundations, some platforms are equipped with instruments to record, for example, information on meteorological conditions, currents, waves, structural static and dynamic strains, settlement and base translation, bearing stress and pore pressure. Monitoring the performance of marine structures has been discussed by Di Biagio (1982), Cuthbert and Poskitt (1983) and St John (1983).

EXAMPLE 1.3

The following details relate to tests to determine the field relative density of a sand (a) in the undisturbed state and (b) in a compacted state following pile driving. A standard compaction mould of internal volume $9.439 \times 10^{-4} \, m^3$ was used for the laboratory tests and a 152.4 mm internal diameter $\times$ 127.0 mm high core-cutter for the field tests. The specific gravity, G_s, of the soil particles was 2.656 and the density of water, γ_w, can be taken as $1000 \, kg \, m^{-3}$. The oven-dry values of mass of the sand in the mould and core-cutter were as follows.

Laboratory	*Field*
Most loose 1.249 kg	Undisturbed 3.264 kg
Most dense 1.621 kg	Compacted 3.819 kg

Calculate the field relative density in the undisturbed and compacted states. Volume of core-cutter $= 23.17 \times 10^{-4} m^3$

$$e = \left\{ \frac{G_s V \gamma_w}{W_s} - 1 \right\}$$

where V is the volume and W_s the mass of the dry sample; $e_{max} = 1.007$; $e_{min} = 0.547$; undisturbed $e = 0.885$; compacted $e = 0.611$.
Substituting in Equation (1.23):
   Undisturbed $D_r = 26.5\%$.   Compacted $D_r = 86.1\%$.
   According to Table 1.24 these states can be termed loose and very dense, respectively. The influence of pile driving in compacting sand below the phreatic surface is illustrated by details (Table 1.25) given by Nixon (1954). *In situ* dry density was obtained from undisturbed samples taken from boreholes by means of a compressed-air sampler. The borehole in the piled area was located in the centre of a group of four 0.45 × 0.45 m piles at 3 m centres both ways. The minimum and maximum values of density were obtained from

Table 1.25

Soil	Dry bulk density $(kN\,m^{-3})$			
	Unpiled area	*Piled area*	*Minimum*	*Maximum*
Medium sand	1360	1620	1310	1710
Fine sand	1280	1600	1310	1710

laboratory tests. The sands were obviously very loose *in situ* and in one instance the *in situ* density was less than the minimum laboratory density. This may be due to several layers of sand within the compass of the sampler having low values of the coefficient of uniformity and slightly different effective diameters whereas the minimum density is derived from a mixture of these layers, possibly having a high value of the coefficient of uniformity.

EXAMPLE 1.4

Investigate the production of fill for a road base using crude flint gravel and screened sand from Chichester, Sussex, having the following characteristics:

Fraction	A, Gravel (60 mm max.) (%)	B, Sand (10 mm max.) (%)
Gravel (> 2.00 mm)	84	32
Sand (0.06 to 2.00 mm)	10	67
Fines (< 0.06 mm)	6	1

From Fuller's Rule, the required percentage passing 2.00 mm is $100(2.00/60)^{1/2} = 18\%$ and passing 0.06 mm is $100(0.06/60)^{1/2} = 3\%$.

Hence the required proportions are: gravel 82%, sand 15% and fines 3% (M_1 in Fig. 1.25). Experience shows that for road bases it is usually desirable to increase the proportion of fines to improve cohesion and a better mixture might be secured with 5% to 15% fines and 20% to 30% sand (M_2 and M_3 in Fig. 1.26). Reference to Fig. 1.26 shows that the specified mix M_2 can be satisfied completely but, owing to the fact that only two components (A and B) are available, M_1 can be satisfied only approximately (M'_1) and M_3 cannot be secured. The proportions by volume of B to A should be 1 to 12.5 for M'_1 and 1 to 4.5 for M_2; respectively the percentages of gravel, sand and fines are 80, 14, 6 and 75, 20, 5. In fact, tests for CBR were made on specimens produced from a mixture of B and A in the proportions 1 to 3, yielding the percentages 71, 24, 5. CBR tests were also made on the crude gravel. The data shown in Table 1.26 were obtained in unsoaked CBR tests at maximum compaction in both cases, employing standard dynamic compaction. Surcharge in CBR tests 70 N.

CBR values of only 50% to 20% were recorded at lower values of density attained with moisture content below or above the optimum. The addition of the sand to the gravel leads to a marked increase in CBR. In fact, a well compacted mixture of gravel, sand and fines may possess a value of CBR considerably exceeding 100% and a very well compacted, relatively dry clay may possess a value approaching 100%. On the other hand, a clay compacted at a value of moisture content above the optimum may possess a value lower than 5%. Values of dry bulk density of about 2200 kg m^{-3} are sometimes attained with well-graded material. Material exceeding 20 mm is normally excluded when compaction is carried out in the standard 100 mm diameter mould but larger material can be included if the 150 mm diameter CBR mould is employed. The tests discussed above were carried out in CBR moulds. Although the presence of gravel tends to hinder the compaction of the

Table 1.26

Material	Optimum moisture content (%)	Max. dry bulk density (kg m^{-3})	CBR (%) Top	Bottom	Mean
Crude Gravel A	14.0	1886	52.5	59.7	56
Mixture B/A = 1/3	11.0	1954	110.5	99.4	105

matrix, the reduction in void ratio arising from the presence of the coarser displacers generally leads to an increase of, say, 5% or 10% in γ_d as the content of coarse material is increased. The CBR tests discussed above were carried out on the material without exclusion of particles exceeding 20 mm. However, consistent results in a wide range of tests on these particular materials suggested that, in fact, the CBR values so determined were valid.

The effects of cement stabilization of the crude Chichester gravel and of the 1/3 mixture were also investigated. As the cement content was increased from 5% to 10%, the optimum moisture content of the crude gravel increased from 13.7% to 14.8%, the dry bulk density increased from 1866 $kg\,m^{-3}$ to 1906 $kg\,m^{-3}$ and the 7 day compressive strength increased from 1655 $kN\,m^{-2}$ to 3034 $kN\,m^{-2}$. The optimum moisture content of the mixture decreased from 12.0% to 11.5% over the same range in cement content, the dry bulk density increased from 1903 $kg\,m^{-3}$ to 1928 $kg\,m^{-3}$ and the 7 day strength increased from 1793 $kN\,m^{-2}$ to 5309 $kN\,m^{-2}$.

When designing soil mixtures it may be desirable to ascertain whether or not the voids in the coarse fraction are filled with finer material. Assume that the following parameters have been determined: particle specific gravity of coarse fraction, G_s, minimum void ratio to which coarse fraction can be compacted, e_m, maximum dry bulk density of finer material, γ_{fm}. If the void ratio of the coarse fraction is any value, e, and the bulk density of water is γ_w(1000 $kg\,m^{-3}$), then the dry bulk density of the mixture is given by

$$\left\{\frac{1}{1+e}\right\}G_s\gamma_w + \left\{\frac{e}{1+e}\right\}\gamma_{fm}.$$

The maximum degree of compaction of the mixture obtains when $e = e_m$. Now assume that the ratio by dry mass of coarse fraction to finer material is A:B. The void ratio of the coarse fraction is given by equating masses. Thus

$$B\left\{\frac{1}{1+e}\right\}G_s\gamma_w = A\left\{\frac{e}{1+e}\right\}\gamma_{fm}$$

and, for the value of e so derived, the dry bulk density of the mixture is given by the above expression, provided $e > e_m$. If it is found that $e < e_m$, then the voids in the coarse fraction cannot be filled and the dry bulk density of the mixture is given by

$$\left\{\frac{1}{1+e_m}\right\}G_s\gamma_w + \frac{A}{B}\left\{\frac{1}{1+e_m}\right\}G_s\gamma_w.$$

In practice it will be found that the optimum moisture content of the mixture will differ from both that of the coarse fraction and that of the finer material and consequently calculations made as indicated above serve only as a guide to the characteristics of the mixture.

1.5 SELECTION OF FOUNDATION TYPES

The selection of foundation types for any particular project depends on three principal factors: (1) site conditions; (2) characteristics of superstructure; (3) methods of excavation and construction. Selection depends also on relative costs, which may have to be considered for the structure as a whole. Kany (1965) evolved a method for determining the most economical arrangement of groups of foundations. British practice for foundations is specified in CP2004 (BSI, 1972d) and for earthworks in BS6031 (BSI, 1981b). Factors to be taken into account in earthworks have been discussed by Horner (1980).

Table 1.27 Guide to selection of foundation types

Extreme structural type	*Deep firm bed*	*Firm bed overlying soft bed*	*Deep soft bed*	*Soft bed overlying firm bed*
		Site conditions		
Light, flexible structure	Pad or strip footings	Pad or strip footings or surface raft	Friction piles or surface raft	Bearing piles or piers
Heavy, rigid structure	Pad or strip footings	Buoyant raft or friction piles	Buoyant raft or friction piles	Bearing piles or piers

1.5.1 Site conditions

The selection of a foundation type to suit given site conditions involves mainly a study of horizontal and vertical variations in soils and rocks, including groundwater, and the determination of the compressibility and bearing capacity of different strata. Permeability of strata can be important where excavation and construction methods influence the selection of foundation types. It is unwise to be dogmatic, but certain types can be accepted as commonly suited to specific site conditions, as indicated in Table 1.27. A soft bed implies relatively high compressibility and low bearing capacity and a firm bed relatively low compressibility and high bearing capacity. However, the information given in Table 1.27 must be regarded as suggestions and each project should be treated on its merits. Bearing and settlement characteristics of soft or loose soils can sometimes be improved by geotechnical processes so that they become more like firm beds. Construction on marshes and other marginal land has been discussed by Rutledge (1970). A review of regional geology and construction in Bristol by Osborne (1970) provides an example of a regional study.

1.5.2 Characteristics of superstructure

The characteristics of the superstructure which are relevant to the selection of foundation types are principally the distribution and magnitudes of loads and the allowable settlement. Certain types of structures present particular requirements in respect of foundations and the following is a review – of necessity rather discursive – of some of these factors.

Some types of foundations have applications in a variety of structures. These include prestressed concrete foundations and ground anchors, which are discussed in papers presented at a conference (Maxwell-Cook, 1974). Post-tensioned foundations have been described by Cronin (1980). The technique of mounting structures on resilient bearings is being employed in some cases largely to ensure reasonable human comfort where sources of vibration are in close proximity. Thus a 24 000 tonne block of 96 flats in Ebury Street, London, is supported at first floor level on non-metallic reinforced polychloroprene bearings 103 mm thick; basement flats are independently isolated. For some structures it is necessary to study the interaction of superstructure, foundations and soil or rock, in some cases taking account of changes occurring with time; a report on interaction between soil and structure has been issued by the Institution of Structural Engineers (1978) and a review of methods of analysis of such problems has been presented by Hooper (1978). Although foundations for houses are commonly of very simple form, many failures have occurred throughout the British Isles and several publications have been issued within the last few years in an endeavour to rectify this adverse state of affairs (NHBC, 1977; Chapman *et al.*, 1978; Tomlinson *et al.*, 1978; Barnbrook, 1981). It must be said that some of the problems experienced with house foundations arise from ignorance or negligence. Attention may not be given to the consequences of building near the top or bottom of artificial or natural slopes. Soakaways are sometimes employed in relatively impermeable ground or located improperly: thus soakaways installed in gardens on clay slopes may cause slope failures, leading ultimately to destruction of house foundations. The influence of every excavation and construction on a site should always be taken into account.

(a) Tall buildings

The types of foundations employed for tall buildings – in which column loads may exceed 10^4 kN – include independent reinforced concrete footings, steel grillages and pedestals, rafts, piles and piers. Independent foundations commonly lead to maximum settlement beneath the centre of the structure, on account of overlap of stress bulbs, and, although the magnitude of differential settlement may be diminished by the monolithic character of the structure, it is desirable to attain as nearly uniform settlement as possible by adjusting the proportions and depths of the foundations.

A raft – more particularly when it is of box-form – increases the stiffness of the structure as a whole and thus tends to reduce differential settlement. Newspaper offices at Holborn Circus, London, afford an example of raft construction (Foot, 1962). Piles and piers are adopted generally to reach a resistant stratum. Sometimes a raft is employed in conjunction with piles or piers (Portland House, Stag Place, London: Mason and Frost, 1963; Millbank Tower Block, London: Davies, 1962; Shell Centre, London: Measor and Williams, 1962) and it is necessary to assess the relative proportions of load carried by each set of foundation components. Settlement observations on the Shell Centre tower suggest that about 40% of the gross load is carried on the base of the under-reamed piers and about 60% by the base of the raft and skin resistance around the pier shafts. Skin resistance may be relatively small since the presence of the raft base and the enlarged pier bases probably prevents full mobilization of this component. The weight of overburden removed is about 55% of the gross load and this corresponds reasonably with the estimated load supported on the base of the raft. Foundations for high buildings have been discussed by Skempton (1955). Further examples of tall buildings have been described by Frischmann *et al.* (1967), Williams and Rutter (1967), Dunican and Martin (1969), Green (1971), Frischmann *et al.* (1983) and Sainsbury and Shipp (1983). A tall building constructed over the Mersey Tunnel has been described by Bingham and Soane (1970). Deep basements are often associated with tall buildings and the design of these is discussed in a report (Dunican, 1975). Exclusion of groundwater from basements is treated in CP102 (BSI 1973b) and a guide to joint sealants has been prepared by Smith *et al.* (1970). An alternative material to bituminous and synthetic compounds is bentonite clay supplied in panels.

Wind loading is important on tall structures and some buildings are susceptible to the action of short, severe gust. In regions subject to hurricanes and typhoons the influence may be considerable. (Philcox, 1962; Thompson, 1966). Publications by Newberry and Eaton (1974) and Mayne and Cook (1978) are useful starting points for information on wind loading, including further references.

Tall buildings pose problems of setting-out, particularly at confined sites, and May (1964) describes a method of projecting vertically upwards, through holes formed in each floor as it is constructed, a base figure which is set-out on the ground-floor.

(b) Spacious buildings

The principal members of the superstructure of a single storey spacious building, such as a hangar, factory shed or exhibition hall, are commonly arch ribs or portal frames. Two important factors which must be taken into account in the design of foundations for these structures are lateral thrust and uplift. Lateral thrust arises from both vertical loads and wind loads and is generally catered for by abutments on rock, raking piles driven to a resistant stratum or by underfloor ties between opposite pairs of abutments. The elasticity of a tie must be taken into account in the analysis, as in the case of the 99 m span arch ribs for East Kilbride swimming pool (Haddow, 1967). A warehouse in Dublin was constructed with portal frames on vertical piles with a tie between pile caps (Lowe and Byrne, 1963). The superstructure of an exhibition hall in Turin consists of prestressed concrete ribs carried on raking columns hinged at top and bottom (Morandi, 1960). The bottom hinges are located on reinforced concrete pedestal foundations. At Ewell, Surrey, a 42 m diameter social centre is constructed of half-portal frames, each butting against a central compression ring at the crown and founded on piles: no underfloor ties are provided (Anon, 1970a). Uplift can be catered for by the means described below for towers, pylons and masts. The superstructure of single-storey spacious buildings is commonly relatively light and wind load may become very important, as in the case of a hangar at London Airport (Ward, 1953). Details have been given of foundations, having a factor of safety of 1.5 against uplift, for a hangar at Brize Norton and of the method of setting out the widely spaced steel bases (Haines *et al.*, 1968).

(c) Special structures

Certain special structures may introduce abnormal loads or impose restrictions on settlement or require unusual forms of foundations. Examples of such structures are the 180 m diameter radio telescope in West Virginia, USA (Tyrrell, 1959), the high-speed wind tunnel at Bedford (Greinig *et al.*, 1957), the gas turbine test houses at Pyestock, Hampshire (Hancock and Adams, 1962) and a rocket launching base in French Guiana (Freer, 1968).

(d) Electricity generating stations and large industrial structures

Special problems are introduced in the design of foundations for electricity generating stations. Loads on columns may be up to 10^4 kN or more. Very heavy loads may be imposed also by plant – the gross load at founding level beneath a nuclear reactor is commonly between 3×10^5 and 6×10^5 kN, leading to values of bearing stress approaching 300 kN m^{-2}. Beneath and around heat generating plant, soil is subjected to desiccation, and thermal stresses are induced in concrete. Since power stations are often located alongside rivers and coasts, basement walls must be designed as water-excluding structures (CP102, BSI, 1973b) with a safety margin for floods, and account must be taken of hydrostatic uplift. High degrees of accuracy in setting out and construction are commonly specified, although shrinkage and creep in large masses of concrete may eventually detract from initial tolerances. Differential settlement must be considered in this respect and, for example, at Berkeley Power Station the differential settlement of the reactor building was restricted to 40 mm with not more than 20 mm between reactor and gas blowers (Dick, 1959). Further details of reactor foundations have been given by O'Connor (1975). Where high standards of accuracy are to be attained on clay, it is generally desirable to spray bitumen on the soil after excavation to formation level and to cover this with a 100 mm blinding layer of concrete to afford protection from construction traffic. Unless the foundations can be located on bedrock, mass and reinforced concrete rafts are commonly adopted in power station design to reduce bearing stress and to reduce differential settlement. The considerations discussed above apply also to other large industrial structures, such as buildings in steelworks. Frequent reversals of stress occur in foundations of structures where overhead travelling cranes are provided but this factor seldom appreciably influences foundation design. On the other hand, limitation of differential settlement may be important in relation to alignment of crane tracks.

(e) Tall chimneys

The types of foundations which have been employed for tall chimneys are illustrated by the following examples. Chimneys up to about 300 m high have been constructed and both dead load and wind load may be considerable. The chimneys at Dunston Power Station, Newcastle upon Tyne, are each supported on a 13 m × 13 m × 2 m reinforced concrete block carried on 85 precast piles driven through fill, sand and silt, into sandy gravel overlying clay. Each chimney at Battersea Power Station, London, is supported on steel stanchions located on concrete piers, 2.5 m × 2.5 m, which are in turn carried by a reinforced concrete box, with a 2 m thick base slab, founded on stiff clay. At Valley Power Station, Bradford, the chimney is located above offices and the entire assembly is carried on a 13 m × 13 m × 2 m thick raft supported on 120 cast-in-place piles taken through clays and sandy gravels to sandy gravel overlying sandstone. The chimneys at Barton Power Station, Manchester, are each supported on four columns and these are located on independent pads on sandstone at the west end, and on a raft on sandstone at the east end. At the New Albany Power Station on the Ohio River, USA, the two chimneys are located above the water intakes and the foundations for each assembly were taken down to shale in cofferdams. The chimney at Drax Electricity Generating Station has been described by Judson and Morris (1974). The foundation of a chimney may crack as a result of a high thermal gradient imposed by hot flue gases: protection can be afforded by a layer of ventilated hollow tiles covered with fire-bricks. The oscillation of chimneys as a result of wind-excitation, including vortex-shedding, can introduce problems in foundation design. These aerodynamic phenomena have been discussed, for example, by Waller (1971), Tunstall (1973) and Jeary (1974).

(f) Bunkers, silos and water towers

Bunkers, silos and water towers – in contradistinction with many structures – are generally loaded in practice up to the full design load, although this may be intermittent. Furthermore, foundations of bunkers and silos may be subjected to considerable eccentricity of loading when the cells on one side only are loaded and this may be accentuated by wind load. Such structures are generally supported on independent foundations where there are a number of columns and these may be located on piles if necessary. Sometimes an annular strip foundation is provided to support a ring of columns or a circular load-bearing wall. Alternatively a raft is employed to carry a number of columns or a central shaft. A block of silos forms a stiff structure and, together with a raft foundation, functions as a monolithic unit in restricting differential settlement. Foundations of water-towers are illustrated by Ritchie (1957).

(g) Service reservoirs

Shallow underground service reservoirs are designed according to the require-ments for water-retaining structures. The roof is commonly supported on columns with independent foundations and the floor slab is cast in sections between the bases with flexible water-tight seals to cater for slight differential settlement. Particular attention should be paid to the protection of soil at formation level after excavation, as mentioned above in the case of power station structures. The design of service reservoirs has been discussed by Tattersall (1958) and Walmsley (1981). Concrete structures for storing water are the subject of a code of practice (CP 2007, BSI, 1970b).

(h) Towers, pylons and masts

Foundations for towers, pylons and guyed masts commonly have to be designed to cater for uplift and horizontal forces arising from wind load since these structures are often relatively light. Resistance to uplift is provided in soil by tension piles, mass concrete blocks, under-reamed cylindrical piers or buried grillages and in rock by bolts or cables grouted in drill holes. Foundations for towers, pylons and masts are discussed by Boscawen *et al.* (1962), Wild and Haslam (1962), Creasy *et al.* (1965a) and Mears and Charman (1966) and the soil mechanics aspects by Ramelot and Vandeperre (1950). The 190 m high Museum Radio Tower, London, is an example of a structure for which foundation studies are of particular importance (Creasy *et al.*, 1965b). The 11 m diameter shaft is supported on a cellular truncated pyramid with a 27 m square prestressed concrete base slab. Television towers have been treated generally by Leonhardt (1970) and the Emley Moor tower in Yorkshire has been described by Bartak and Shears (1972). Cooling towers at electricity generating stations have relatively large diameter ring bases and comprise a special group of structures akin to some water towers (Diver and Patterson, 1977).

(i) The influence of modern architectural concepts

Modern architectural concepts tend to influence the layout and proportions of foundations but the methods of foundation analysis and design are generally not affected. The tendency to mount buildings and other structures on a central stem commonly necessitates a single foundation, often with a plan area smaller than that of the building. A lecture theatre at the University of Edinburgh (Anon, 1960) and Simshill School, Glasgow (Anon, 1961), are typical examples. The school comprises three-storeys above ground-level and is 17 m wide and 17 m high but the central mat foundation is only 6.5 m wide. Other examples of compact foundations are piers for highway viaducts.

Reference must be made to modular co-ordination in relation to foundation design and construction (Edwards, 1969). Obviously, the centre to centre spacing of foundations is governed by this factor where the standard module has been adopted. The adoption of a 13.7 m (45 ft) module for a building at Dagenham, Essex, determined the foundation layout (Lax and Bunclark, 1960). The standardization of dimensions for individual foundations would undoubtedly facilitate the work of designer, quantity surveyor and various tradesmen and, since formerly the dimensions of foundations were commonly specified to the nearest 3 in., the standard module of 100 mm could readily be adopted. However, for various reasons, sub-module sizes will have to be accepted in some forms of construction for some years to come. Thus, for example, the diameter of power augers is generally a multiple of 6 in. (152 mm). In connection with tolerances in building construction in general, reference should be made to DD22 (BSI, 1972e) for definitions and fundamental principles and also to a paper by Akroyd (1966). The accuracy achieved in dimensions of reinforced concrete frame structures has been investigated by Thorogood (1975) and accuracy in setting-out is discussed in Digest 234 (BRE, 1980a).

Another aspect of modern construction is industrialized building (McMeekin, 1964; Anon, 1966). This may involve special joints between precast columns and precast or *in situ* foundations. Precast foudations can be bedded on mortar on a blinding layer. Where the jack-block system of construction is employed, the foundations may have to be designed to cater for forces imposed by jacks during construction and, where prestressed floors are adopted, prestressing equipment (Adler, 1964).

1.5.3 Methods of excavation and construction

Methods of excavation and construction are generally arranged to meet design requirements and these operations rarely influence the selection of foundation types except in matters of detail. Thus, for example, the construction of new pile foundations beneath an existing building necessitates the use of excavation and pile-driving machinery designed to operate in limited headroom and the piles or casing would be designed to be placed in sections spliced together. Consideration should always be given in the design stage to the methods of construction, yet problems arise from time to time due to lack of forethought. Samuels (1958) illustrates the difficulty involved in placing concrete where the entire reinforcement for a turbogenerator foundation is fixed before pouring is commenced. In these circumstances it may be advantageous to place the aggregate dry and inject the cement by pressure grouting: this technique is the subject of a report (Houghton *et al.*, 1969). For economical and speedy construction it is desirable to employ a minimum of different items of plant on a project. The tower crane has revolutionized construction yet it may not have the capacity to lift heavy loads at large operating radii. A grillage for a stanchion of a tall building may have a mass of 20 tonnes and, in the design stage, it may be desirable to arrange to assemble the steelwork in two or more parts and thus avoid the use of a mobile crane of high capacity in addition to a tower crane.

During and after construction, problems are caused by thermal expansion and contraction of concrete poured in large masses. Traditionally, such problems have been overcome by employing low content of cement, preferably of low-heat type, limiting pour volume and lift height and introducing construction joints. Although this is suitable for unreinforced concrete, an alternative technique facilitates the continuous pouring of large volumes for reinforced concrete (FitzGibbon, 1976). Internal thermal strain is controlled by preventing the temperature of any part of the mass from becoming more than 20° C cooler than the hottest part: this is achieved by insulating exposed surfaces of the mass. The maximum temperature attained by hydrating concrete is the temperature of the pour at placement plus 12° C for each 100 kg of cement per cubic metre of concrete: this applies to mixes between 300 and 600 kg m^{-3} where the dimensions of the mass exceed 2 m. Where reinforcement is continuous across construction joints, it is important to avoid locking-in expansion stresses which later become contraction stresses: this is achieved by not casting between opposing or adjacent faces of hardened concrete. If possible, it is desirable to place each mass in one continuous pour. This may lead to imposition of large forces on shuttering which must be carefully designed to resist them. Stresses generated by drying shrinkage are superposed on stresses caused by thermal contraction. Techniques for relieving locked-in stresses in welded steel bases of columns have been in use for many years: these are based on heat treatment.

All aspects of safety are of paramount importance in both design and construction and these are covered extensively in national regulations and codes. Safety in excavation has received considerable attention because of recurrent fatalities in this work. In addition to safety of persons, the risk of partial or complete collapse of structures must be considered. Mistakes have been made in the past but it is obvious that checks should be made of, say, the adequacy of falsework or formwork during placement of concrete in high lifts, in retaining walls, or of temporary and semipermanent fixings for column bases. Problems arising from vibrations generated by construction processes have been reviewed by Wiss (1981). A code of practice for falsework has recently been published (BSI, 1982b).

1.5.4 Example of the selection of foundation types

The following is a typical example of the train of considerations leading to the selection of foundation types for a specific project. The site conditions at Ringsend Power Station, Dublin (MacDonald, 1952), consist of sand, gravel, cobbles and silt from ground-level at $+7$ m above datum to -10 m below datum, estuarine silt and clay from -10 m to -30 m, boulder clay, gravel and cobbles from -30 m to -34 m, beyond which limestone bedrock was proved by boring to -40 m. Consideration was given to a foundation of piles driven to sand and gravel at -5 m but this scheme was rejected because the consequent heavy loading on the underlying clay would lead to excessive settlement. A 4.5 m deep basement supported on a raft or piles was unacceptable on the grounds of expense and of consequent alterations to mechanical and electrical plant already ordered. Piles about 37 m long driven to the boulder clay necessitated splicing and difficult driving would be experienced in sand and gravel between -5 m and -10 m. Eventually it was decided to sink 1 m diameter steel cylinders to bedrock.

The load per cylinder was about 4500 kN, including the weight of cylinder and concrete filling and the downdrag derived by skin resistance induced by settlement of the surrounding ground. The cylinders were sunk by the Benoto system under their own weight accompanied by slight rotation, and excavation was effected by grab.

1.6 PHILOSOPHICAL OBSERVATIONS ON FOUNDATION ENGINEERING

Philosophy can be defined as the pursuit of wisdom, an attribute much needed in foundation engineering, and this comprises experience and knowledge coupled with the power of judging rightly. Recasting the words of R.B. Peck (1962), soil and rock mechanics are engineering sciences but foundation engineering is an art. The former enable qualitative and quantitative analyses to be made of engineering problems whereas the latter involves a synthesis of all aspects of a project and, finally, the ability of judging rightly. Armitage (1981) states that engineering judgment depends on innate common sense, that is knowing what is important, on the acceptance of personal responsibility, that is professionalism, on competence, including awareness, and on willingness and ability to test the validity of common-sense conclusions and to learn from experience.

Again recasting the comments of Peck, the attributes necessary for the successful practice of foundation engineering are experience, which can be obtained by reading precedents and by personal observation, profound knowledge of soil and rock mechanics and familiarity with geology. This specification is perhaps rather more demanding than that pronounced by Peck but it represents the peak of attainment which is likely to be reached by a few with many years of experience. Although a working knowledge of the principles and practice of geology may be adequate for much foundation work, the greatest benefits can be derived from familiarity with local or regional conditions which enables adjustments in the development of designs to be made to advantage. Peck believes that the most important attribute is precedent or experience: certainly the greatest practitioner will endeavour to learn a lesson from each of his experiences.

The great works of engineers from Roman times to the eighteenth and nineteenth centuries were erected successfully without the advantage of modern knowledge in engineering and geological sciences but solely on wise decisions based on experience. On reflection it is, therefore, disturbing to encounter difficulties in foundation construction to-day which were predictable and should have been anticipated. Nevertheless, it must be admitted that failure of completed foundations is comparatively rare. In the experience of the author, the primary cause of these difficulties is generally the omission to seek local geological knowledge. This can be provided by specialists serving in the Institute of Geological Sciences, in universities and polytechnics or in industry. Frequently, study of geological maps and memoirs and site investigation reports is insufficient. In some cases only a knowledge of the geology of an area enables difficulties to be anticipated at a particular site and advice on this should be sought initially before site investigations are carried out. Less commonly the cause of difficulties or failures stems from a lack of appreciation of the principles of soil and rock mechanics.

Turning now to the subjects of sampling and testing, it can be said that the primary object of these operations is to provide quantitative data which will enable the foundations of a structure to be constructed and used with a low risk of failure or of excessive settlement. Clearly the acceptable degree of risk for a high head dam will be less than that for a single storey building. Sometimes it is difficult to eliminate a relatively high degree of uncertainty in the results of site investigations because, for example, of difficulty in securing samples representative of the material *in situ*. Since every engineering project entails risk to a greater or lesser extent, it is reasonable to expect that the degree of risk attached to any particular site investigation, or part thereof, be specified in the report as high, medium or low.

Sundry aspects of site investigation are discussed earlier in the present chapter, in the introductions to Chapters 2 and 3 and in Sections 2.1 and 3.1. Although uniform arrangement of boreholes and frequency of sampling may suit certain sites on relatively uniform strata, it would be unsuited to a site over a buried valley filled with a variety of deposits. In some cases the most efficacious investigation comprises a single continuous core examined visually in detail, followed by sampling at selected horizons in further boreholes. Some routine tests, such as consistency limits, on large numbers of samples may be significant for certain projects, such as highways, but may be of little value for deep foundations in, say, London Clay, yet money has been wasted in this way in the past. Obviously unnecessary testing should be avoided but an excess of samples is preferable to too few, in order to provide a reserve should some yield defective specimens or should some results prove suspect. Consideration should be given to testing a relatively small number of large specimens instead of a larger number of smaller specimens since the former generally yield more reliable results.

From test results it is necessary to select appropriate values of parameters for design purposes. Generally it is reasonable to accept mean values for any bed or horizon and it is excessively conservative to base design on, say the lowest value of shear strength or the highest value of coefficient of compressibility. However, such conservative practice may be justified in some cases but consideration must be given to the probable reliability of test results, the validity of design assumptions, the prescribed values of load and safety factor, the type of structure to be erected and other circumstances. In fact, if a particular foundation of a framed or stiffly cladded structure is founded at a point where locally the soil is weak or compressible, excessive penetration or settlement will be accompanied by some relief of loading which is automatically transferred elsewhere and the adoption of ultra-conservative values of parameters would not necessarily be justified. Clearly each case must be considered on its merits. Consideration

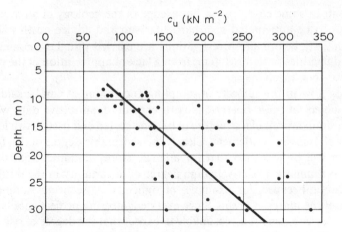

Figure 1.27 Variation of undrained shear strength with depth in London Clay at a site in Victoria Street, London.

should be given to the use of a probability relationship such as Equation (1.26) which yields a slightly lower value of design strength than the mean value of the test results. Figure 1.27 shows the results of undrained triaxial tests on specimens of London Clay at a site in Victoria Street, London (Hodgson and Bryan, 1975). The scatter of results for this relatively homogeneous stiff clay is probably due to the presence of fissures and variations in the degree of sampling disturbance and possibly also to variations in the sedimentological character of the deposit which have not yet been studied. The statistical analysis of test results is discussed by Wilun and Starzewski (1972).

The three principal methods of ensuring the safety of a construction are (1) the application of a factor of safety to strength of a material, (2) the application of a load factor to the applied load, and (3) the application of partial safety factors to both strength and load. Although partial safety factors are based on general probability theory, this latter can be carried further in design. In addition, design can be based on limits placed on acceptable deformation. The applications of probability, risk, limit state and related concepts and theories to general engineering and geotechnical problems have been discussed by Casagrande (1965), Lumb (1970), Hoeg and Murarka (1974), Wu (1974), Cochrane and Montgomery (1980), Bolton (1981), Lambe *et al.* (1981), Semple (1981), Simpson *et al.* (1981), Smith (1981), Tang (1981), Wu and Wong (1981), Thoft-Christensen (1982), Bowles and Ko (1984) and Whitman (1984). Meyerhof (1970) and Danilevsky (1982) have discussed a factor of safety in soil mechanics. The use of probability-based partial safety factors is ostensibly a more refined technique for design than a single overall factor of safety but much depends on the viewpoint of the designer and some authorities believe that the use of the Gaussian continuous probability distribution can be misleading. Indeed, in the field of geotechnics the character of a design may depend finally on sound judgment in which probability and risk may be largely intuitive.

Although the design of structural elements is now commonly based on the concept of limit states, the determination of permissible bearing stress beneath foundations has not been standardized on a similar basis. Generally the important limit states are the ultimate or collapse state and the serviceability or deformation state, which includes deflection, local damage, such as cracking, and vibration. Serviceability limit states are analysed for working loads by the application of elastic theories whereas the ultimate limit state is analysed by either elastic theory with redistribution or ultimate strength theory in conjunction with suitable partial safety factors. In soil mechanics the ultimate state of bearing failure is analysed by plastic theories and the serviceability state of settlement by elastic theories. With the latter are employed values of modulus of elasticity and coefficient of compressibility which generally include both elastic and plastic components of deformation, a practice which is followed also in reinforced concrete design. Broadly, the analysis of deformation for serviceability states is based on similar concepts for soil, rock, concrete and steel but analysis for the collapse state for soil and rock is different from that for concrete and steel since partial safety factors are not employed for the former. The principles of limit state design are fully described in CP 110 (BSI, 1972c) but it is useful to examine a few of the details concerning assessment of loads and of materials strength pertinent to soil and rock mechanics. Foundation design in relation to CP 110 has been discussed by Astill (1974). The strength of *in situ* concrete has been investigated by Maynard and Davis (1974).

Design is based primarily on characteristic loads which preferably should be obtained from a statistical survey of the loads for different service conditions but commonly are based on tabulated values specified in appropriate codes of practice. Partial safety factors are applied to characteristic loads to establish the design load: higher factors are adopted for imposed loads than for dead loads in view of the greater uncertainty in the prediction of the former, and the values depend also on which combination of dead, imposed and wind loads are to be considered. For buildings up to about four storeys high, wind load is generally not significant in design and the ultimate design load F is given by

$$F = 1.4G_k + 1.6Q_k \qquad (1.24)$$

where G_k is the characteristic dead load, Q_k is the characteristic imposed load and 1.4 and 1.6 are the corresponding partial safety factors.

By analogy with structural concrete, the design strength of soil or rock can be assessed as

$$f^* = \frac{f_k}{\gamma_m}, \qquad (1.25)$$

where f_k is the characteristic strength of the material and γ_m is the corresponding partial safety factor.

$$f_k = f_m - 1.64\sigma \tag{1.26}$$

where f_m is the mean strength of the specimens tested and σ is the standard deviation relating to the test results. The number of specimens should be sufficient to render this statistical relationship reasonably valid and the coefficient 1.64 ensures that not more than 5% of the test results will fall below f_k. The value of the partial factor of safety for steel is commonly taken as 1.15 and for concrete 1.5: the higher value being related to the greater variability to be expected in concrete. The variability of many soils and rocks over a site may be greater than that revealed in tests and the value of factor of safety for such materials can reasonably be expected to be about 2. Reference to Equation 1.24 indicates that F is likely to be about 1.5 times the total load. The overall factor of safety given by comparing (ultimate bearing capacity based on f_m)/2 with F is likely to be of the order of 3, i.e. equal to the factor of safety commonly employed in current practice and given by (ultimate bearing capacity based on f_m)/(total characteristic load). It follows that adoption of ultimate load limit state for determination of bearing areas should not lead to designs markedly different from those based on conventional practice which, in general, are known to be safe. However, the application of limit state design to foundation engineering needs to be thoroughly investigated before a recommended practice can be laid down. Even in structural engineering, where concepts of collapse theory and limit state have been propounded and used for several decades, opinions are divided on the validity and relative merits of various approaches to design. Confusion is sometimes experienced but, it should be noted, the limit state concept requires only that service requirements are restricted to certain limits: the application of partial safety factors is not inherent in it and the results can be achieved by any other rational means.

A sequel to this discussion on limit state applied to the bearing capacity of soil and rock is to consider the design of the foundation structure on the assumption that failure of structure and soil or rock is concomitant, a condition to which the author has previously drawn attention (Henry, 1955). The greatest difficulty to be overcome in solving this problem is the prediction of bearing stress distribution at failure. Logical though this procedure may be, very little advance has been made in this direction, but the universal acceptance to-day of limit state design for structures may encourage further progress. Some studies on reinforced concrete pad footings have been reported by Meyerhof and Rao (1974): Fig. 1.28 illustrates a few of the results for structural failure although failure in the underlying material does not appear to have been complete, particularly when this is steel. Flexural failure only was considered and it was assumed that diagonal tension capacity was adequate. In the case of axial loading on clay the distribution was sensibly uniform. The tests on steel show that with rigid materials, including hard rocks, the marginal area of a foundation is ineffective unless the foundation has sufficient thickness for the column load to be dispersed over the full plan area of the base or, expressed in another way, unless

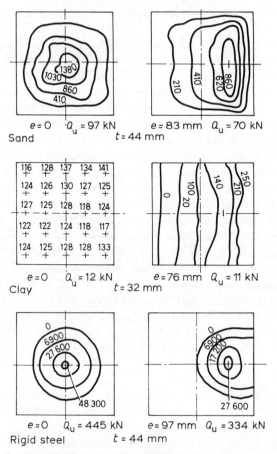

Figure 1.28 Distribution of bearing stress at failure of footings supported on different media. Footing size 305 mm × 305 mm × thickness *t*. Column size 76 mm × 76 mm. Ultimate load Q_u. Eccentricity *e*. Stress in kN m^{-2}.

the foundation has a stiffness comparable with that of the underlying material. It should be noted that, as a foundation penetrates into soil or rock during concomitant failure, forces will be developed by adhesion and/or friction around the periphery which will considerably influence bending moment and shearing force at failure. Investigations related to these philosophical aspects of design have been reported by Muspratt (1969) and Jain *et al.* (1979). Ingold (1982) has investigated the use of partial load factors for retaining wall design.

In the case of continuous footings and rafts, it may prove that the concomitant collapse limit state involves local failure of soil and base slab over, say, circular areas centred on each column and of radius equal to one-quarter of the distance between columns: the actual area is governed, of course, by relative stiffness and strength of soil and foundation in addition to column spacing. An extension of yield line and other theories of slab failure may prove a satisfactory approach to these problems.

Turning now to serviceability limit states, it will be observed from the discussion in Section 2.6 on the stiffening influence of cladding on framed structures that this leads to reduction in differential settlement and, by restraining deformation of members, to reduction in flexural stresses. In the case of continuous beam bridges, differential settlement of 25 mm between piers is unlikely to lead to changes in maximum bending moment of more than 20% when the spans exceed 10 m. On the other hand, calculations based on elastic theory indicate that 25 mm differential vertical or horizontal displacement of the abutments of 10 m span fixed ended arches or portal frames may lead to undesirably high values of stress and 1° rotation at the supports may lead theoretically to disaster. Therefore, it has always been regarded as the best practice to found arches and portals on rock if possible. Nevertheless, there are many such structures showing no sign of instability although founded on clay, sand or gravel, some on piles and most without underfloor ties. Undoubtedly several factors contribute to this unexpectedly satisfactory behaviour. These include, for example, the stiffening influence of spandrel filling when present, local yield in the soil and the fact that many such structures are seldom loaded to full capacity. It is possible that the most important factor is that rotation and differential displacements of foundations scarcely influence the magnitude of collapse loading. From this brief discussion it will be appreciated that the interaction of soil and structure is at present largely a matter of speculation and a subject overdue for research.

Undoubtedly the digital computer can be of great service in foundation engineering, particularly for routine, repetitive or highly theoretical studies including, for example, the statistical analysis of test results, the stability of slopes and the design of raft foundations. However, judgement is usually of major importance in foundation engineering and the use of a computer may lead to loss in appreciation by the elimination of visible intermediate steps and the sheer mechanization of the process. Furthermore, in some cases each foundation beneath a structure may require some measure of individual treatment. Consequently the problems in foundation engineering for which computers are justified should be carefully selected.

On the subject of costs, it is often possible to evolve several alternative schemes for foundations but it may be found that the advantage of cheapness is outweighed by the greater probability of satisfactory service. Again, the economics of usage may be more important than the economics of construction. For example, in the case of improvement of an existing highway over an appreciable thickness of alluvial and estuarine deposits, the cost of a viaduct on piles may be several times the cost of an embankment but the former can be put into service shortly after construction whereas, in the interest of stability and settlement, a high embankment may have to be constructed in stages over a considerable period followed by several years of maintenance before settlement is sensibly complete. From a national standpoint, the delay to traffic during construction and maintenance of the embankment, taking account of fuel, wear and tear, personal time and other factors, may be more expensive than the additional cost, including loan charges, of the viaduct. It is difficult to make reliable comparisons in such cases but these wider issues are worthy of consideration and, although the final decision may be economic, the basic problem is geotechnical. Some aspects of minimum cost design have been investigated by Lane and Harriman (1975).

Finally, it may be useful to discuss a few issues concerning construction. The erection of structures has always been subjected to some measure of control, commencing with setting out and the securing of tolerances followed by, for example, the use of a plumb bob or a theodolite to maintain perpendicularity. Within the last few decades, increasing attention has been paid to monitoring soil pore pressure together with settlement and lateral deformation of soil and foundation structures, for which purpose various types of equipment are installed in the ground. Site instrumentation is employed for the following purposes.

(a) For observations on trial structures, such as embankments and piles, to facilitate prediction of the behaviour of projected structures.
(b) For observations during construction to avoid unstable conditions, for example, in a bed of soft clay or to restrict deformation of, say, piles under lateral stress from an embankment.
(c) For research and development in analysis and design by short-term and long-term observations.

The cost of site instrumentation is usually a small proportion of the cost of a project and, although the provision of large numbers of expensive items is rarely justified, sufficient should be provided to ensure that as much useful information is obtained as possible even if a few items become unserviceable. In this respect it is imperative to avoid damage by construction operations, a goal which necessitates very determined effort by geotechnical staff. A wide range of information on site instrumentation is given in the proceedings of a symposium (Anon, 1973).

Because relatively large numbers of individuals of different trades are involved in construction, coupled with inevitable looseness of control, and because of other factors inherent in the nature of construction, more failures can be expected to arise from construction operations than from defective design. Nevertheless, construction defects are likely to lead generally to local failures whereas defective design may well lead to major failures although, for example, collapse of formwork can be a catastrophic event and even in this case the fundamental cause may be in design. Investigations of the strength of concrete in structures illustrate the nature of problems arising from constructions operations (Drysdale, 1973; Maynard and Davis, 1974). It is almost inevitable that the *in situ* strength of concrete is likely to be lower than the cube strength, generally it is in the range 70 to 100% of the latter. This is catered for in design by a factor of safety of 1.5

applied to strength. However, because of the nature of foundation construction, it is reasonable to expect that the quality of foundation concrete may fall below this range. Whereas design is normally based on 28 day strength, there is a further gain in strength of 30% or more in a period of 6 to 12 months and this effectively increases the factor of safety, provided the full service load is not applied for several months. Since foundations are poured early in construction, it is reasonable to expect foundation concrete to benefit in this way, thus to some extent compensating for intrinsic poorer quality.

The strength of concrete in columns is found characteristically to vary from a maximum at the base to a value between 5 and 20% less at the top. This is ascribed to greater consolidation at the base under self-weight and migration of water upwards. The characteristic is observed also in other deep structures such as concrete walls. The phenomenon can be expected to be manifest in cast-in-place piles and piers but added to this is the possibility of malformation or of loss of concrete in fissures in rock. In suspect cases it may be necessary to take a full length core in the centre of the pile to prove continuity. In passing, it should be noted that, as a result of possible deterioration of high alumina cement, it is considered undesirable to employ this material in foundations where a suitable alternative is available. Furthermore, attention should be paid to a possible alkali/cement reaction, a phenomenon known for several decades, although the process is not yet fully understood. It appears that a critical amount of reactive silica must be present and sufficient moisture must be available to support the reaction. In order to prevent the problem, the alkali content of cement should not exceed about 0.6%, although not all cements conform to this requirement. Deterioration arising from alkali/cement reaction has occurred on bridges completed about 1970 near Plymouth, Devon, on the A38 road.

These philosophical observations are rather discursive and the discussion is far from complete. Nevertheless, the object of this section is to draw attention to the need for wide vision in foundation engineering and to draw away from the very restricted approach to design inherited from the past.

Chapter 2

DEFORMATION AND GROUNDWATER PROBLEMS IN FOUNDATION ENGINEERING

F.D.C. Henry

INTRODUCTION

Broadly speaking, the roots of the subject of soil mechanics extend back to the eighteenth and nineteenth centuries, in which such names as Coulomb and Rankine have a distinguished place. Today an enormous volume of new theories and experimental data pours forth from a multiplicity of sources but, in spite of this, the application of investigations and theories to practical problems achieves the desired conclusion only in part. Two factors which tend to damp the enthusiastic application of theories are that soil is the most variable material with which the engineer deals, involving changes in characteristics in both vertical and horizontal directions at a site, and is rarely isotropic, and that observations of laboratory tests do not necessarily represent the behaviour of soil *in situ*, even if relatively undisturbed samples can be secured. In particular, it is sometimes difficult to take into account the influence of pore-water on the behaviour of soils. Theoretical and experimental soil mechanics investigations are fundamental to the design of all structures and therefore such limitations as obtain today should serve as a spur to seek the truth, in particular through the medium of full-scale field observations. In fact it has been suggested that a realistic approach to the provision of support to excavations is, in some cases, to take observations with measuring instruments as the work proceeds and to modify the design of temporary work or the techniques of excavation and construction as necessary (Bjerrum *et al.*, 1965; Peck, 1969). This principle could be applied also to other aspects of foundation design and construction although it should never be employed as an alternative to thorough preliminary site investigations.

Many engineering problems, including those in soil and rock mechanics, are concerned with two aspects of material behaviour: that of deformation and that

141

of failure. In this chapter are treated problems related to deformation and to groundwater flow. Inevitably soils and rocks behave according to the same laws which govern the behaviour of other materials but, with different materials, the components of a generalized stress/strain relationship assume different degrees of importance. Thus, for example, although some pores are probably present in all materials, their influence is commonly negligible, except in the case of soils and many rocks where the effects, particularly under saturated conditions, can be dominant. Furthermore, both soils and rocks are commonly heterogeneous and anisotropic *en masse.*

It is probable that no physical property of any material extends over such an enormous range of magnitude as does permeability (see Tables 2.17, 2.18). Consequently, discrepancies incurred in calculating discharges to excavations by simple formulae or by inexact flow nets are generally small in relation to the errors incurred in determining appropriate values of permeability. It follows that the required capacity of pumps can rarely be estimated precisely and the validity of calculations must be assessed in the light of practical experience and related to the acceptable degree of risk. Calculations for discharge and for stability of excavations are commonly made assuming media are isotropic but such a condition rarely obtains in nature. In any particular case the influence of anisotropy should be reviewed at least qualitatively in relation to calculations.

Although refined techniques of analysis are available for some problems, they are often subject to limitations, and in Chapters 2 and 3 relatively simple analyses, generally applicable to a wide range of problems, are presented, but it should be borne in mind that some are based on broad assumptions. For important structures, such as the piers and abutments of long span bridges and the foundations of tall buildings, more refined techniques are desirable and, where possible, references are quoted which will provide at least an approach to such problems. However, it should be observed that refined analyses must be supported by test data of adequate reliability and this may necessitate considerable effort in site investigation and laboratory testing.

In spite of the large magnitude of errors which may be involved in values of parameters or in simplifying assumptions, remarkably good correlation is often achieved between predicted and observed behaviour in geotechnical engineering. This must surely be due to the fact that the errors are largely compensating, although it must be recognized that occasionally correlation can be expected to be poor. However, it must be acknowledged that advances in field and laboratory testing and in the analysis of problems and in computation enable predictions of behaviour to be made in some cases with greater confidence. A comprehensive review of settlement of structures has been presented by Burland *et al.* (1977).

2.1 DEFORMATION CHARACTERISTICS OF SOILS AND ROCKS

Tests on specimens of soils and rocks commonly reveal a delayed component of elastic strain response to stress, manifest in elastic hysteresis, and also plastic

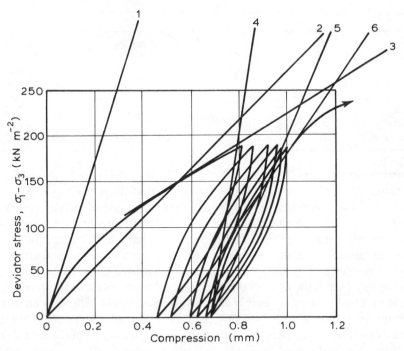

Figure 2.1 Modes of modulus of elasticity determined from repeated cycles triaxial test.

deformation, manifest in permanent set. Much of the plastic deformation is delayed response to stress but it is possible that there is also an instantaneous component, although tests on concrete show that when the rate of loading is very fast, permanent set is very small and this suggests that it may become zero in the limiting case. Instantaneous permanent set can be distinctly imposed on other materials such as copper and lead. Delayed elastic and plastic strains lead to the phenomenon of creep, which is discussed below: the elastic component is, of course, recoverable on removal of the stress. The elastic constants commonly required for studies of deformation of soils and rocks are modulus of elasticity and Poisson's ratio.

Modulus of elasticity can be measured in several modes, according to the requirements for a particular problem. It can be measured as the initial tangent to the stress/strain curve, the tangent at some particular stress or the secant modulus defined by a chord to the curve between the limits of a given range of stress. These forms of modulus of elasticity are illustrated in Fig. 2.1, which shows the results of repeated loading and unloading cycles on a 76 mm long clay specimen in an undrained triaxial test. Values obtained from the first and last loading curves are as shown in Table 2.1.

In the triaxial test, compression is measured over the full length of the specimen and, owing to the influence of lateral restraint at the plattens, the measured value

Table 2.1

	First loading $(kN\,m^{-2})$		Last loading $(kN\,m^{-2})$	
Initial tangent modulus	(1)	66 100	(4)	126 600
Secant modulus in deviator stress range 0 to $150\,kN\,m^{-2}$	(2)	20 800	(5)	50 700
Tangent modulus at working deviator stress of $150\,kN\,m^{-2}$	(3)	13 100	(6)	30 400

of modulus of elasticity should be reduced by 2% when Poisson's ratio is 0.25, and 8% when Poisson's ratio approaches 0.50 (Girijavallabhan, 1970).

Much of the permanent set displayed in the first cycle of loading shown in Fig. 2.1 is probably due to closure of fissures and bedding interfaces which have been opened during sampling, and this state is not characteristic of the soil *in situ*. Whereas the set generally decreases with each additional cycle, there is an increase in the third cycle, probably caused by sudden closure of a fissure or bedding interface. If the number of cycles were increased, permanent set per cycle would eventually become sensibly zero and the soil would behave as an elastic medium in response to relatively rapid loading and unloading, although hysteresis would still be present. After the first cycle, all modes of the modulus of elasticity show only slight increases and with a larger number of cycles would become sensibly constant. The mode of modulus of elasticity for a particular analysis should be selected on the basis of similarity between conditions in service and in tests. The first cycle is unlikely to be employed for evaluation of modulus of elasticity because of the influence of disturbance in sampling. For some purposes it may be necessary to employ values derived from the unloading curve of any one cycle.

Since the modulus employed in calculations is rarely solely elastic and almost invariably contains a plastic component, the term modulus of deformation is often employed instead of modulus of elasticity.

Commonly in homogeneous clays both modulus of elasticity and undrained strength (Fig. 3.3) increase with depth, and experience shows that this characteristic can be expressed approximately by expressions of the form $E_{rz} = E_{r0} + C_1 z$ and $c_{uz} = c_{u0} + C_2 z$ where E_{rz} and c_{uz} are the values at depth z, E_{r0} and c_{u0} are the values at the base of the weathered layer and C_1 and C_2 are constants.

Poisson's ratio varies with magnitude of stress but commonly accepted values are 0.1 to 0.2 for sandstone, 0.3 for limestone and granite and 0.2 to 0.3 for dense soils at working stresses. Precise values depend on several factors and differences frequently occur between values determined on laboratory specimens and values derived from site tests: disturbance in sampling influences laboratory values and the frequency and width of fissures and variations in the degree of weathering influence site values. If dilatancy occurs in a specimen as failure is approached, the value may exceed 0.5. It is assumed in some problems that the soil or rock is

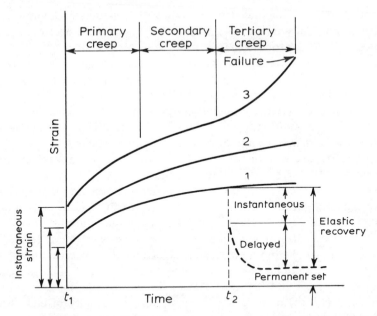

Figure 2.2 Creep of soils and rocks under constant deviator stress.

saturated and that water is incompressible and, providing no volume change occurs under these conditions, Poisson's ratio is 0.5. Ward *et al.* (1965) determined a value of 0.4 for London Clay from undrained compression tests, and Bozozuk (1963) obtained correlation between field measurements for modulus of elasticity, assuming a value of 0.4 for Poisson's ratio, and the initial tangent modulus from unconfined compression tests.

The creep behaviour of soils and rocks under a constant deviator stress ($\sigma_1 - \sigma_3$) is illustrated in Fig. 2.2. When the deviator stress is relatively low – say less than about 30 or 40% of the stress at failure – creep strain is commonly small and ceases after a period of time (curve 1). If the stress is relieved completely at time t_2, the strain components can be identified as shown by the dotted curve. At stresses up to, say, 60 or 70% of the stress at failure, creep strains may continue indefinitely (curve 2). The stress above which this occurs is known as the creep strength. At stresses above, say, 70% of the stress at failure the rate of strain may eventually increase rapidly, and rupture finally takes place (curve 3). Creep is sometimes divided into phases: primary creep may be largely elastic, secondary largely plastic and tertiary probably involves progressive local rupture which terminates in complete rupture. A detailed study of creep in an organic silty clay has been reported by Arulanandan *et al.* (1971). The general stress–strain–time function for soils has been studied by Singh and Mitchell (1968) and time-dependent deformation by Kavazanjian and Mitchell (1980). Further studies of creep have been reported by Bishop and Lovenbury (1969), Singh and Mitchell (1969), Finn and Shead (1973), Hyde and Brown (1976) and Nelson and Thompson (1977). Stress relaxation in soils has been investigated by Lacerda and Houston (1973). Keedwell (1971) has proposed a theory for the rheological behaviour of clays and Naylor (1978) has reviewed the stress–strain laws for soils. The influences of content and structure of clay minerals on deformation have been discussed by Olson and Mesri (1970) and Barden (1972).

In the case of concrete, consideration has been given in recent years to creep Poisson's ratio in addition to elastic Poisson's ratio. In fact these two terms are incompatible; either creep and instantaneous parameters are implied or plastic and elastic. In the case of soils and rocks, very little attention has apparently been given to the two parameters and the precise distinction has not become important. Nevertheless, multiaxial stress systems need to be studied in soil and rock mechanics and the appropriate parameters will have to be selected and investigated in order to solve certain problems in the future.

Studies of clay from the London Clay by Ward *et al.* (1965) revealed that specimens prepared from blocks carefully extracted from excavations yielded values of secant modulus of elasticity many times greater than those yielded by specimens cut from 100 mm tube samples.

Marsland (1973b) investigated the modulus of elasticity of clay from the London Clay by laboratory and *in situ* tests, and drew the following conclusions:

Values of modulus determined from laboratory triaxial tests on 38 and 98 mm diameter specimens are very variable and difficult to interpret;

The highest and most reproducible values were obtained from 865 mm diameter plate tests for which the surface was prepared by hand excavation of 50 to 70 mm from the floor of the borehole before bedding the plate on plaster. With present techniques, tests on smaller diameter plates made in boreholes too small to permit a man to descend and prepare the base give lower, and probably less reliable, values;

The average values of modulus determined from tests on 865 mm plates in well prepared boreholes were 1.8 to 4.8 times those obtained from undrained triaxial tests on 38 and 98 mm diameter specimens. The values obtained from the plate tests were more compatible with the observations of the movements of the ground around excavations and foundations.

The values determined by *in situ* plate tests are affected by such factors as the disturbance during drilling, the lapse of time between excavation and loading the plate, the size of the plate and of the borehole, and the degree of restraint applied to the sides of the borehole in the vicinity of the plate. It is therefore essential, when quoting values of moduli, to give details of all the relevant aspects.

Similar factors affect values of modulus determined from laboratory tests. In the case of tests on specimens prepared from block samples, the time between sampling and testing is probably the most important single factor.

Even the best *in situ* tests give values of modulus lower than the *in situ* values before excavation. It should be noted, however, that relevant values for design purposes are not necessarily the undisturbed pre-excavation *in situ* values. In many cases it is appropriate to employ values corresponding to those after excavation and, in some cases, additional allowances must be made to cater for the effects of the methods of construction.

Ajaz and Parry (1976) have described a laboratory bending test to investigate the behaviour of soils in tension. A review of the dynamic properties of soils and their relation to engineering problems has been given by Richart (1975) and includes an extensive list of references.

2.2 STRESSES IN SOILS

2.2.1 Stresses beneath a point load

The results of analyses by Boussinesq, published in 1885, and by later investigators are commonly employed for problems in soil and rock mechanics involving a state of elastic equilibrium. It is assumed that the elastic medium is homogeneous and isotropic and is a half-space bounded by an upper horizontal surface and extending to infinity in all horizontal directions and vertically downwards. In Fig. 2.3(a) it is assumed that the point load Q operates on the surface of the medium and the vertical stress σ_v at any point with radial co-ordinates (z, r) below the surface is given by the Boussinesq equation quoted in the diagram. Since modulus of elasticity and coefficient of compressibility of soils and rocks both vary with intensity of stress, the application of this equation, and stress distributions based on it, to the determination of strains – for example, in prediction of settlement – involves the assumption of a mean value over a given stress range. Furthermore, it is generally considered desirable that stresses should not exceed about one-third of the magnitude of stresses which would create a state of plastic equilibrium.

When vertical stresses are required to be calculated at a depth z exceeding three times the breadth B of a square footing, it is generally sufficiently accurate to treat the surface load as a point load, and values of influence factor I_σ can be read

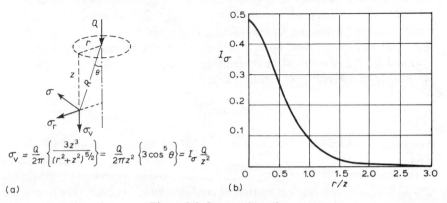

$$\sigma_v = \frac{Q}{2\pi}\left\{\frac{3z^3}{(r^2+z^2)^{5/2}}\right\} = \frac{Q}{2\pi z^2}\left\{3\cos^5\theta\right\} = I_\sigma \frac{Q}{z^2}$$

(a)

(b)

Figure 2.3 Stresses in soils.

from Fig. 2.3(b) and substituted into the equation for σ_v. Where $z < 3B$, the foundation plan is subdivided into square compartments as in Example 2.1 below. Subdivision of the area is useful also where the foundation plan is rectangular or irregular or where the bearing stress distribution is not uniform (see Section 2.3).

Tests by Turnbull *et al.* (1961) with surface loads on sand and clayey silt showed that measured stresses on vertical and horizontal planes were generally in good agreement with those predicted by the theory of elasticity, but correspondence of displacements tended to be less reliable at shallow depths. The stress/strain characteristics of the sand proved to be nearly linear and of the clayey silt roughly linear.

The direct application of the Boussinesq equation to the case of buried foundations may lead to considerable errors in calculated stresses (Geddes, 1966), more particularly in and near the zone vertically beneath the foundation, although with shallow footings the over-estimates of stresses may not exceed 10 to 30% in practical problems. Transfer of part of the load by skin resistance around a foundation shaft influences magnitudes and distribution of stress. Charts and/or tables for the determination of stresses beneath foundations have been presented by Jurgenson (1934), Sneddon (1946), Geddes (1966, 1969), Poulos (1967), Huang (1969), Mirata (1969), Milović (1970, 1973), Milović *et al.* (1970), Jumikis (1971), Carrier and Christian (1973), Poulos and Davis (1974) and Bhushan and Haley (1976). The publication by Poulos and Davis (1974) covers a wide range of problems in elasticity, including piles subjected to axial and lateral loading. A computer program for the calculation of stresses and displacements in a multilayer elastic medium has been presented by Thrower (1968, 1971). Geddes (1966) determined values of influence factors for stresses in a medium subjected to a point load beneath the surface and also to a transfer of load by skin resistance around the shaft of a foundation. It was assumed in the analysis that the medium was capable of resisting tensile stress and the values of influence factor for vertical stress beneath founding level given in Table 2.2 for Poisson's ratio $v = 0.5$ probably lead to underestimates of stress. If the presence of that part of the medium above founding level is ignored, the Boussinesq solution is applicable for a point load, but this probably leads to an overestimate of stress. On this basis Geddes (1969) determined further values of influence factors for point load and skin resistance load. Values of vertical stress generated by combinations of point load and skin resistance can be determined by superposition. The magnitude of vertical stress derived from Geddes' analyses is

$$\sigma_v = \frac{I_\sigma Q}{D^2} \tag{2.1}$$

where I_σ is the Geddes' influence factor and D is founding depth below ground level.

The influence on stresses of anisotropy in an elastic half-space, in particular the special case of orthotropy, is worthy of consideration. Terzaghi (1943) refers to investigations by K. Wolf and by D.L. Hall and presents the information shown in

Table 2.2 Influence factors for vertical stress caused by a vertical point load acting below the surface at depth D; Poisson's ratio $= 0.50$ (after Geddes, 1969)

z/D	\multicolumn{9}{c}{r/D}
	0	0.2	0.4	0.6	1.0	1.4	2.0	2.5	3.0
1.0	—	0.115	0.103	0.086	0.051	0.026	0.008	0.003	0.001
1.2	6.067	0.150	0.194	0.094	0.050	0.027	0.010	0.004	0.001
1.4	1.574	0.934	0.338	0.144	0.055	0.029	0.011	0.005	0.002
1.6	0.732	0.577	0.328	0.174	0.065	0.033	0.013	0.006	0.003
1.8	0.431	0.378	0.268	0.172	0.073	0.037	0.015	0.007	0.004
2.0	0.289	0.266	0.212	0.154	0.076	0.040	0.017	0.009	0.004
2.2	0.209	0.197	0.168	0.133	0.075	0.042	0.019	0.010	0.005
2.4	0.160	0.153	0.136	0.114	0.071	0.043	0.020	0.011	0.006
2.6	0.126	0.123	0.112	0.097	0.066	0.042	0.021	0.012	0.007
2.8	0.103	0.101	0.093	0.083	0.061	0.041	0.022	0.013	0.008
3.0	0.086	0.084	0.079	0.072	0.055	0.039	0.022	0.014	0.008

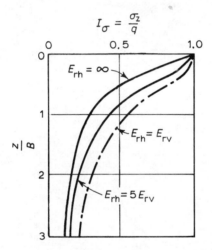

$$I_\sigma = \frac{\sigma_z}{q}$$

Figure 2.4 Influence of orthotropy on stresses in an elastic medium when $L = \infty$.

Fig. 2.4 for a strip unit loading q of width B and infinite length on the surface of an elastic medium possessing orthotropy, for which the value of modulus of elasticity E_{rh} in a horizontal direction is greater than the value E_{rv} in a vertical direction. It will be seen that, for example, when $E_{rh} = 5E_{rv}$, the values of stress on the centre-line beneath the strip generally are 25 to 30% less than the values which obtain when the medium is isotropic. Hence estimates of settlement based on the latter as an assumption are likely to exceed observed values by roughly 30 to 35% where such orthotropic conditions in fact occur.

2.2.2 Calculation of stresses beneath distributed loads

The calculation of stresses imposed by non-uniform loads or by a number of independent foundations is facilitated by reference to charts, such as those devised by Newmark (1942). Figure 2.5 shows an influence chart for vertical stress σ_v at any specified depth z generated by a surface load. The geometric pattern of

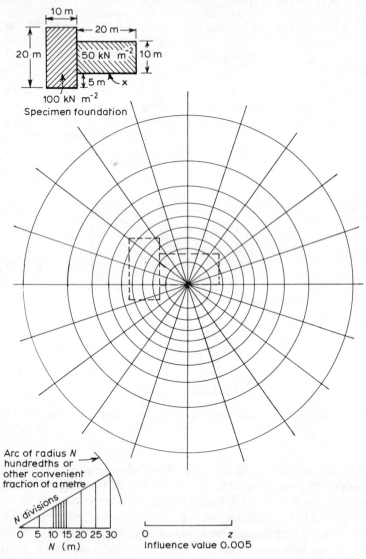

Figure 2.5 Influence chart for vertical stress σ_v (after Newmark, 1942, and Terzaghi and Peck, 1967).

influence charts for all values of z is the same but the scale is different. It is generally convenient to employ a single chart and to vary the scale of the foundation plan to obtain values for stress at different depths. The scale of the plan is such that $0z$ in Fig. 2.5 is equal to the depth z of the specified point below the underside of the foundation. The distance $0z$ can be readily divided into any number, N, of divisions by the construction shown in Fig. 2.5. The foundation plan is drawn to the same scale on tracing paper and placed on the influence chart so that the plan position of each specified point at one depth z is located in turn at the centre of the chart. In the particular chart shown in Fig. 2.4 each compartment bounded by two adjacent radii and two adjacent arcs has an influence value of 0.005. The influence value of the original charts of Newmark is 0.001. The number, M, of compartments, including fractions, covered by the plan is counted and the vertical stress in $kN\,m^{-2}$ at the specified point is given by $0.005\,Mq$ where q is the foundation bearing stress in $kN\,m^{-2}$ within each compartment. A specimen foundation plan is shown in dotted lines in Fig. 2.5 and is drawn to a scale corresponding with a depth to a specified point X of 30 m. If the bearing stress distribution is as shown, at X $\sigma_v = (0.005 \times 11 \times 100 + 0.005 \times 17 \times 50)$ $= 9.75\,kN\,m^{-2}$.

2.2.3 Stress bulbs

Typical patterns of iso-stress lines generated by various load distributions on the surface of an elastic half-space are shown in Fig. 2.6. The general pattern of these stress distributions has led to the concept of a stress bulb – this is the zone within which the half-space is stressed appreciably by the surface load. It is convenient to assume that the limit of the bulb is defined by the iso-stress line which is equal to, say, 5 or 10% of the applied loading. Theoretically, the stresses extend to infinity in all horizontal directions and vertically downwards but outside the bulb the stresses are negligible for practical purposes. The stress patterns are based on the assumptions that the half-space has no weight and that the modulus of elasticity, E_r, is constant throughout the mass. Furthermore, some distributions depend on a specific and constant value of Poisson's ratio v although with soils and rocks this varies with stress intensity.

The stress bulbs most commonly required for practical problems are those for vertical stress and shear stress. The iso-shear-stress lines relate to the absolute maximum shear stress at a given point and the direction of the stress is unspecified; the orientation of these maxima varying from point to point. It is generally assumed that the surface loading is uniformly distributed and is flexible so that it can be maintained uniform as the surface deforms to the settlement pattern. Such loading rarely obtains in practice and the distributions discussed in Section 2.3 commonly apply, in which cases stress distributions based on stress bulbs are approximations. Vertical stresses determined from two bulbs for different loadings can sometimes be superposed but shear-stress bulbs cannot be so employed because of variations in stress orientation.

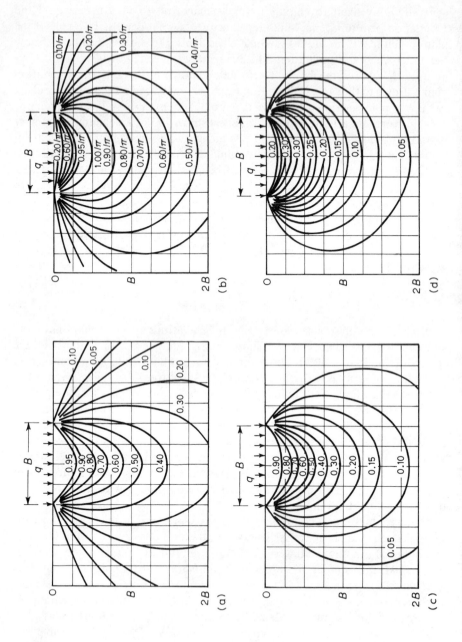

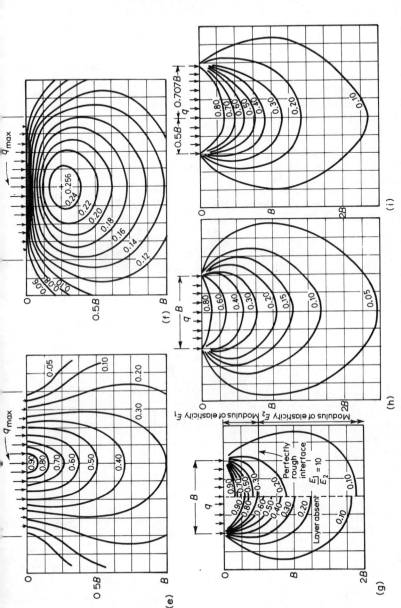

Figure 2.6 Bulbs of stress in an elastic half-space with various surface loadings of intensity q/unit area. The values relating to iso-stress lines are influence factors I_σ and I_τ for vertical stress σ_v and absolute maximum shearing stress τ_m. (a) Uniform strip loading, $\sigma_v = I_\sigma q$. (b) Uniform strip loading, $\tau_m = I_\tau q$. (c) Uniform circular loading, $\sigma_v = I_\sigma q$. (d) Uniform circular loading, Poisson's ratio $\nu = 0.45$, $\tau_m = I_\tau q$. (e) Linearly varying strip loading, $\sigma_v = I_\sigma q_{max}$. (f) Linearly varying strip loading, $q_m = I_\tau q_{max}$. (g) Uniform circular loading. Effect of vertical stress distribution of stiff upper layer of thickness $0.5B$, $\sigma_v = I_\sigma q$ (after Fox, 1948). (h) Uniform square loading. Vertical stress along axes, based on Westergaard's theory, $\sigma_v = I_\sigma q$ (after Beggren, 1961). (i) Uniform square loading. Vertical stress along axes and diagonals, based on Boussinesq's theory, $\sigma_v = I_\sigma q$.

EXAMPLE 2.1

The centres of two columns A and B are 4 m apart. Column A is supported by a footing 2 m × 2 m in plan and founded 5 m below ground level. Column B is supported by a footing 1.5 m × 1.5 m in plan and founded 2 m below ground level. The net bearing stress beneath A is 400 kN m^{-2} and beneath B is 200 kN m^{-2}. Assuming the soil beneath the footings behaves as an elastic medium, determine the increment in vertical stress created by these loaded footings on a horizontal plane 10 m below ground level (a) vertically beneath the centre of A and (b) mid-way between the verticals through A and B.

The following values of influence factors have been taken from Terzaghi (1943) although values of accuracy sufficient for many practical problems can be read from Fig. 2.3. For intermediate values of r/z, determine I_σ by linear interpolation.

r/z	0	0.1	0.2	0.3	0.4	0.5
I_σ	0.478	0.466	0.433	0.385	0.329	0.273

Refer to Fig. 2.7. Since the horizontal plane is less than $3B$ beneath footing A, divide A into four compartments. From Fig. 2.3 $\sigma_v = I_\sigma Q/Z^2$.

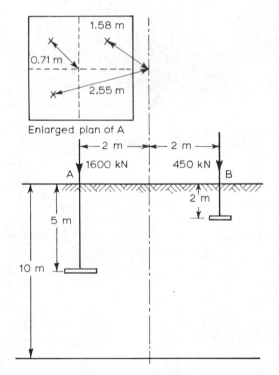

Figure 2.7 Calculation of stress beneath foundations.

Vertically beneath A:

$$\text{Derived from A with } \frac{r}{z} = \frac{0.71}{5} = 0.14. \qquad \sigma_v = \frac{4 \times 0.453 \times 400}{5^2} = 29.00$$

$$\text{Derived from B with } \frac{r}{z} = \frac{4}{8} = 0.5. \qquad \sigma_v = \frac{0.273 \times 450}{8^2} = 1.92$$

$$\underline{30.92\,\text{kN m}^{-2}}$$

Mid-way between A and B:

$$\text{Derived from A with } \frac{r}{z} = \frac{1.58}{5} = 0.316. \qquad \sigma_v = \frac{2 \times 0.376 \times 400}{5^2} = 12.02$$

$$\text{and with } \frac{r}{z} = \frac{2.55}{5} = 0.510. \qquad \sigma_v = \frac{2 \times 0.267 \times 400}{5^2} = 8.55$$

$$\text{Derived from B with } \frac{r}{z} = \frac{2}{8} = 0.250. \qquad \sigma_v = \frac{0.409 \times 450}{8^2} = 2.88$$

$$\underline{23.45\,\text{kN m}^{-2}}$$

Since the Boussinesq formula relates to surface loading, the values of stress calculated above are somewhat excessive. The estimates of stress calculated below are based on Equation 2.1 and Geddes' values of influence factor and are probably rather low. The most correct estimates will lie between these extreme values.

Refer to Table 2.2. Then $\sigma_v = I_\sigma Q/D^2$
Vertically beneath A:

$$\text{Derived from A with } \frac{r}{D_A} = \frac{0.71}{5} = 0.14 \text{ and } \frac{z}{D_A} = \frac{10}{5} = 2.0$$

$$\sigma_v = \frac{4 \times 0.277 \times 400}{5^2} = 17.73$$

Derived from B with $\frac{r}{D_B} = 2.0$ and $\frac{z}{D_B} = 5.0$. I_σ in this case, and also below

is beyond the range of Table 2.2 and is estimated from the trend of values.

$$\sigma_v = \frac{0.01 \times 450}{2^2} = 1.13$$

$$\underline{18.86\,\text{kN m}^{-2}}$$

Mid-way between A and B:

$$\text{Derived from A with } \frac{r}{D_A} = \frac{1.58}{5} = 0.316 \text{ and } \frac{z}{D_A} = 2.0.$$

$$\sigma_v = \frac{2 \times 0.236 \times 400}{5^2} = 7.55$$

$$\text{and with } \frac{r}{D_A} = \frac{2.55}{5} = 0.510 \text{ and } \frac{z}{D_A} = 2.0$$

$$\sigma_v = \frac{2 \times 0.179 \times 400}{5^2} = 5.73$$

$$\text{Derived from B with } \frac{r}{D_B} = 1.0 \text{ and } \frac{z}{D_B} = 5.0. \; \sigma_v = \frac{0.01 \times 450}{2^2} = 1.13$$

$$\underline{14.41\,\text{kN m}^{-2}}$$

2.3 DISTRIBUTION OF BEARING STRESS

2.3.1 Observed and theoretical distributions of bearing stress

It is commonly assumed in the design of foundation structures that the bearing stress distribution at the soil/foundation contact surface is uniform when the structural load is applied axially and varies linearly when the load is applied eccentrically, as shown in Fig. 2.8 (a) and (b). In fact, for a rigid foundation, the bearing stress distribution reaches a maximum beneath the centre when the soil is cohesionless and a minimum beneath the centre when the soil is cohesive. These distributions conform roughly to parabolic and elliptic figures respectively (Fig. 2.8a and b). The fall in stress at the edges in cohesionless soil is attributed to lateral movement of particles which is made possible by the absence of shearing resistance at the edges resulting from zero overburden stress. This influence is more pronounced when the foundation is located at ground level than when located beneath the surface owing to the greater degree of confinement of particles with increase in depth. When a rigid foundation is located beneath ground level and is surrounded on all sides by overburden, the stress distribution on both cohesive and cohesionless soils tends to become more uniform. Scanty evidence suggests that at depth in well-compacted cohesionless soil the bearing stress may even be greatest around the edges. It is known that in cohesive soil bearing stress distribution varies slightly with time and is probably associated with the progressive development of consolidation and secondary settlement: the magnitude of the variation appears to be of the order ± 10 to 20% about the mean value at any point beneath a foundation. Site distributions are influenced also by factors such as variation in soil properties and seasonal changes in soil-moisture content.

Faber (1933) carried out laboratory experiments with a 0.30 m diameter model footing loaded axially and located on Leighton Buzzard sand and on undisturbed blue clay from the London Clay (Fig. 2.9). The soil was contained in a box about 0.75 m diameter and 0.45 m deep. In relation to the size of the footing these dimensions are probably insufficient to eliminate completely the effects of confinement on strains in the soil. The tests were arranged to simulate conditions for a founding level at ground surface and, by imposing a surcharge on the soil around the foundation, at a depth of about 9 m.

Observations on rigid foundations on sand reported by Rodstein in 1959 and cited by Burmister (1962) revealed stress distributions which were roughly elliptic beneath the centre, rising to peaks about one-sixth of the width from the edge and falling to zero at the edge. Repeated cycle tests on a rigid circular foundation on the surface of dry sand by Ho and Lopes (1969) showed a distribution which was initially parabolic but as the load was increased the central value became depressed, thus leading to a peak at about one-quarter of the radius from the centre: this latter pattern persisted in subsequent cycles irrespective of the magnitude of the load. When the foundation was embedded at a depth equal to twice its diameter, the form of the distribution was a flat parabola with depressed

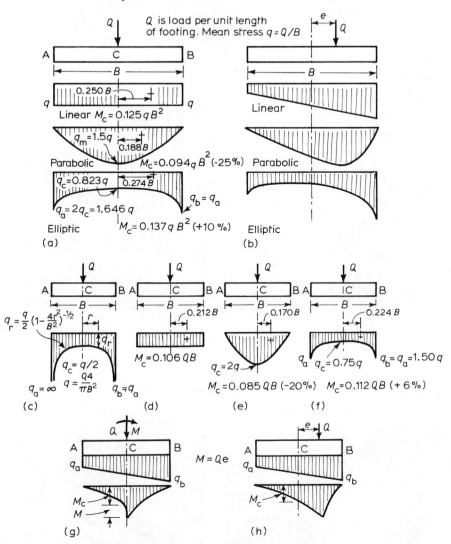

Figure 2.8 Distribution of bearing stress beneath foundations. Centres of stress marked +. Percentage increase or decrease in M_c compared with M_c for linear distribution given in parentheses. (a) Strip footing, axial load; (b) strip footing, eccentric load; (c) elastic; (d) linear; (e) parabolic; (f) elliptic; (c) to (f) circular footing, axial load; (g) and (h) bearing stress distributions and bending moment diagrams for eccentric loading on a rectangular footing.

points at the centre and at one-half the radius from the centre and peaks at one-third and two-thirds of the radius. The stress distribution beneath a vibrating foundation on sand was studied by Chae *et al.* (1965). At low frequencies the distribution was parabolic but as the frequency increased there was a tendency for the stress at the edge to increase sharply. Other site and laboratory

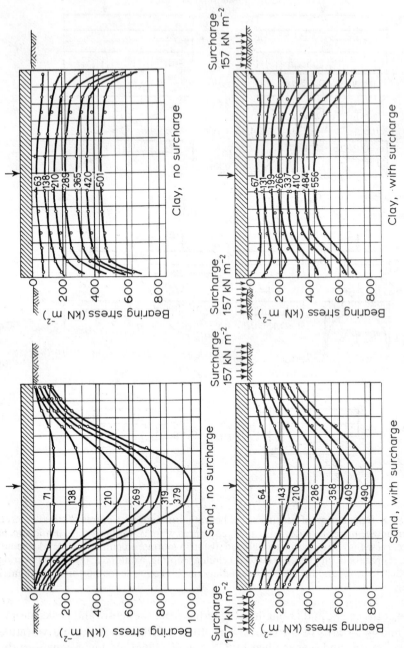

Figure 2.9 Typical stress distributions obtained by Faber (1933) in tests on model footings. Average stress in $kN\,m^{-2}$ quoted for each distribution.

investigations have been reported by Schultze (1961), Sutherland and Lindsay (1961), Lee (1965), Zaretsky and Tsytovich (1965) and Muspratt (1969).

The theoretical distribution of stress at the contact between a rigid circular die and the surface of an elastic half-space is shown in Fig. 2.8(c). The edge stress is infinite but, in practice, plastic flow in the half-space and deformation of the base of the die reduce this to a finite value not exceeding the yield stress of the half-space. In order to provide the necessary total upthrust, the stress beneath the centre would then slightly exceed $\sigma/2$ and the modified distribution would be roughly elliptical.

The above discussion has been concerned essentially with rigid foundations which do not deflect under load. Very few foundations can be considered completely rigid, with the exception of piers and deep blocks of concrete. If a foundation is very flexible, when axially loaded the stress distribution is a maximum beneath the load on all types of soil. The influence of parabolic and elliptic distributions on the design of footings is discussed in the following paragraphs.

(a) Strip footings

Axial loading (Fig. 2.8a)

Consider 1 m run of a strip footing of width B founded at ground surface. With the parabolic distribution, the bending moment M_c at the centre is 25% less than that given by a uniform distribution. Shearing force, S_c, at the centre is the same with both distributions although the parabolic distribution leads to a reduction in shearing force towards the footing edge. The influence of an elliptic distribution on values of bending moment and shearing force depends on the relative magnitudes of the minimum and maximum values of bearing stress. From Fig. 2.9 it appears that the stress at the edge will be of the order twice that at the centre and with this distribution beneath a strip footing M_c is 10% greater than with a uniform distribution.

Eccentric loading (Fig. 2.8b)

In this case values of bending moment and shearing force are influenced by the nature of the applied load – that is, whether it is a truly eccentric load or an axial load plus a moment at the column foot which is treated as an equivalent eccentric load (Fig. 2.8g and h). Compared with a linear distribution, the parabolic distribution for a truly eccentric load leads to a reduction in M_c of 25% when $e = 0$ and a reduction of zero when $e = B/6$, with intermediate values between these limits. Similarly the elliptic distribution leads to values of M_c which exceed the values based on a linear distribution by amounts which depend on e. Consideration of the diagrams reveals that shearing forces will follow similar trends.

(b) Pad footings

Axial loading

Typical patterns of bearing stress distribution beneath circular pad footings are illustrated in Fig. 2.9 and are expressed in a general form in Fig. 2.8(e) and (f). The distributions beneath square footings are likely to be roughly similar to these patterns.

Eccentric loading

No information appears to be available on the distribution of bearing stress beneath eccentrically loaded circular and square pad footings but the general form must involve a combination of Figs 2.8(b) and (e) or (f).

(c) General conclusions relating to design

The main conclusion which can be drawn from this discussion is that linear distributions of stress can generally be assumed for design purposes. In the worst case the maximum bending moment is not likely to exceed 10% of the value based on a linear distribution. On sandy soils it may be less than the value calculated for a linear distribution. In view of the variable nature of soil and the fact that the precise distribution of bearing pressure cannot be predicted there is no good reason for departing in general from current methods of design. However, consideration of the diagrams will suggest certain features connected with individual problems which may be catered for if designers see fit. For example, the outer beams of a grillage on clay and centre beams on sand are likely to be the most heavily loaded. Diagonal tension is likely to be higher for a reinforced concrete footing on clay than the value based on a linear distribution. Marshall (1944) carried out tests on 0.20 m square reinforced concrete model foundations 40 mm thick. Well graded pit sand and moist, yellow clay from the London Clay were placed in a box to serve as the foundation soils. Most failures were in diagonal tension. The average failing load when supported on clay was about 0.6 of that on sand and, although no comment was made in the original publication, this confirms that the stress at the edge of a foundation on clay is higher than that on sand.

It is possible that partial support may be afforded by adhesion and friction between the soil and the sides of a buried footing, thus reducing the total upthrust on the base. Some field observations appear to confirm this, although the evidence is not conclusive. Side resistance representing 10 or 20% of the column load would materially influence values of bending moment and, to a lesser extent, shearing force. The magnitude of such resistance depends on soil properties, depth beneath the surface, area of the footing sides and the method of construction.

2.3.2 Linear distributions of bearing stress

Expressions for maximum and minimum values σ_a and σ_b which obtain with linear distributions of bearing stress are readily derived. Fig. 2.10 relates to a footing of length L and breadth B. For a strip footing B is taken as 1 m. In each case expressions for stress are derived by equating total bearing upthrust to downward load ($\Sigma V = 0$) and equating the moments of these forces about the centre line ($\Sigma M_c = 0$), in conformity with requirements for static equilibrium. In case 4 of Fig. 2.10 the unknowns are σ_b and L'. If the eccentric load consists of an axial load Q plus a moment M, the equivalent eccentricity can be determined from $e = Q/W$. A general expression which covers cases 1, 2 and 3 is σ_a or σ_b $= Q/(LB) \pm 6 \, We/(BL^2)$ and follows from the common formula for stress in short columns $\sigma = W/A \pm M/Z$. Data for the determination of stress beneath foundations of various shapes are given in Fig. 2.11.

The concent of the core of a section – employed in column analysis – is a useful adjunct in the solution of problems concerning bearing stress. Consider the rectangular footing shown in Fig. 2.12(a). If a load is applied eccentrically on one or other of the principal axes of the footing, then it follows from the data given in

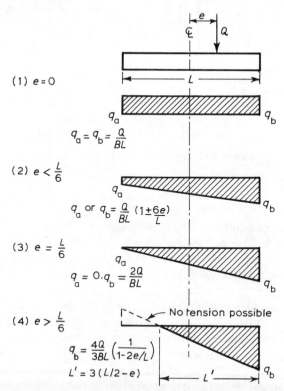

Figure 2.10 Minimum and maximum value of bearing stress beneath foundations.

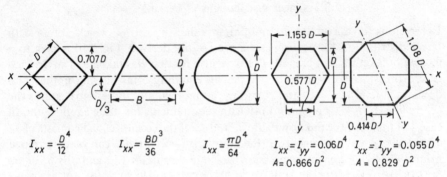

Figure 2.11 Area and second moment of area of various shapes of foundation plan.

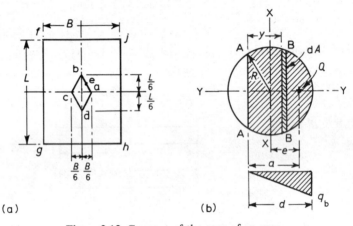

Figure 2.12 Concept of the core of an area.

Fig. 2.10 that zero stress exists at one side of the footing when the eccentricity is equal to or exceeds $L/6$ or $B/6$. If the load is not applied on a principal axis, zero stress obtains at one corner of the footing when the point of application of the load lies on the side of, or outside, the rhombus abcd. Consider a load Q applied at any point e on ab. This load can be resolved into two parallel components Q_a and Q_b acting at a and b respectively. The line of zero stress for Q_a lies at the edge fg of the footing and for Q_b lies at the edge of gh. Since a triangular distribution of stress is developed along each of the axes under the loads Q_a and Q_b, it follows that, when the two distributions are superposed, zero stress exists only at the corner g. From the symmetry of the rhombus, zero stress cannot exist beneath the footing if the point of application of the load lies within the rhombus abcd.

The core of a circular footing of radius R is obviously a concentric circle of some smaller radius r. When zero stress exists at one point on the periphery of the footing $\sigma_a = Q/(\pi R^2) - QeR64/[\pi(2R)4] = 0$ and hence $e = r = R/4$.

Now consider a circular footing where the eccentricity of loading lies outside the core (Fig. 2.12b). The bearing stress beneath any strip BB of area δA parallel to the line of zero stress AA is $\sigma_y = y\sigma_b/d$. For static equilibrium the sum of the vertical forces must be zero and hence $Q - \int(y\sigma_b dA/d) = 0$ and the sum of the moments of the vertical forces about AA must also be zero and hence $Q_a - \int(y^2\sigma_b dA/d) = 0$.

Then $a = I_{AA}/A_{AA}$ where $I_{AA}(= \int y^2 dA)$ and $A_{AA}(= \int y dA)$ are the second and first moments respectively of the bearing area about AA. Thus for any position of AA the distance a can be calculated and the problem is usually one of trial and error in order to determine a position of AA for which a point located on the axis YY by the distance a coincides with the point of application of the load. When coincidence is obtained, the maximum bearing stress σ_b can be calculated from $\sigma_b = Qd/A_{AA}$. This method can be employed for a footing of any shape provided the load is applied on one of the principal axes and is useful for the analysis of foundations subjected to differential settlement due to mining subsidence.

EXAMPLE 2.2

Figure 2.13 shows an ideal parabolic distribution of bearing stress beneath a strip footing subjected to a continuous eccentric line load. Show that

$$b_1 = \left(\frac{B}{2} + 4e\right) \quad \text{and} \quad b_2 = \left(\frac{B}{2} - 4e\right).$$

Explain how the bearing stress distribution changes between $e = 0$ and $e = B/6$.

Consider 1 m run of footing. With a parabolic distribution of stress, $q_m = \frac{3}{2}q$ irrespective of the magnitude of e.

Taking moments about a

$$Q\left(\frac{B}{2} + e\right) - \frac{10}{24}q_m b_1^2 - \frac{2}{3}q_m b_2\left(b_1 + \frac{3}{8}b_2\right) = 0.$$

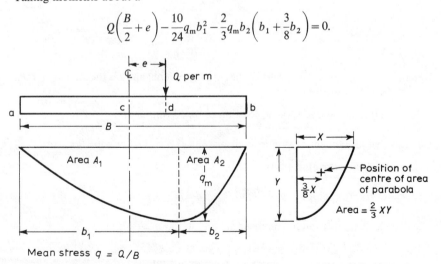

Mean stress $q = Q/B$

Figure 2.13 Parabolic distribution of bearing stress beneath eccentrically loaded foundation.

Substituting $(B - b_1)$ for b_2 and $\frac{2}{3}q_m B$ for Q yields

$$b_1 = \left(\frac{B}{2} + 4e\right) \quad \text{and} \quad b_2 = \left(\frac{B}{2} - 4e\right).$$

The centre of stress must always lie vertically beneath Q. When $e = B/8$, $b_1 = B$ and $b_2 = 0$: this is the limit for parabolic distribution with q_m at the vertex. As e passes beyond $B/8$, the curvature of the distribution diagram decreases and the vertex lies to the right of the footing and is imaginary. The maximum stress is now at b and rises from $q_m = \frac{3}{2}q$ when $e = B/8$ to $2q$ when $e = B/6$ and the distribution becomes triangular.

Although this problem is rather academic, it does give an idea of the changes in distribution which will occur as e increases from zero, even though the forms are less geometrically perfect in practice.

EXAMPLE 2.3

It is required to investigate soil/structure interaction for the design of an aircraft hangar. The external columns are H m high from the top of the pad footings to the seats of the roof truss. The pad footings are square, of side B m, and are founded at a shallow depth so that resistance at the sides of the footings does not enter the present problem. Each column can be assumed to act as a propped cantilever, fixed at the footing and propped by the roof truss. The wall sheeting is attached directly to the columns and the wind can be assumed to impose a uniformly distributed horizontal load w kN m^{-1} on the entire height H of each column. The pitch of the roof is small and wind load on it is negligible. The truss seats are displaced δ m under wind load. The axial load from the roof truss is W kN: eccentricity of loading arising from δ can be ignored. The footings can be assumed stiff and their flexibility disregarded. Bearing stress distribution beneath the footings can be assumed linear and each footing rotates through an angle θ under wind load. The modulus of subgrade reaction of the soil is k_s kN m^{-3}. Show that the moment at the column foot is

$$M = - \; \frac{\dfrac{3\delta EI}{H^2} + \dfrac{wH^2}{8}}{1 - \dfrac{36EI}{Hk_s B^4}}$$

where E is the modulus of elasticity of the column material and I is the second moment of area of the column section.

Fixed end moment due to wind load on propped cantilever $\qquad -\dfrac{wH^2}{8}$

Changes in end moment arising from displacement δ and rotation θ:

due to displacement of prop $\qquad\qquad -\dfrac{3EI\delta}{H^2}$

due to rotation of footing $\qquad\qquad +\dfrac{3EI\theta}{H}$

Slope–deflection equation for moment at column foot

$$M = \frac{2EI}{H}\left(1.5\theta - \frac{1.5\delta}{H}\right) - \frac{wH^2}{8}. \tag{2.2}$$

Bearing stress distribution is shown in Fig. 2.14 and, since $y_2 = q_1/k_s$ and $y_2 = q_2/k_s$, $\theta = (q_2 - q_1)/k_s B$.

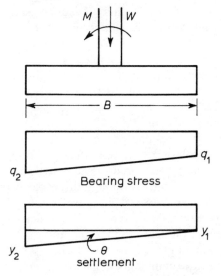

Figure 2.14 Rotation of footing.

Now
$$q_1 = \frac{W}{B^2}\left(1 - \frac{6e}{B}\right) \quad \text{and} \quad q_2 = \frac{W}{B^2}\left(1 + \frac{6e}{B}\right).$$

Hence
$$\theta = \frac{\left(\dfrac{6eW}{B^3} + \dfrac{6eW}{B^3}\right)}{k_s B} = \frac{12We}{k_s B^4}.$$

If e is equivalent eccentricity of loading, $M = We$ and then

$$\theta = \frac{12M}{k_s B^4}. \tag{2.3}$$

Substituting for θ in Equation 2.2

$$M = \frac{2EI}{H}\left(\frac{18M}{k_s B^4} - \frac{1.5\delta}{H}\right) - \frac{wH^2}{8}$$

or

$$\frac{M\left(1 - \dfrac{36EI}{Hk_s B^4}\right) = -\dfrac{3\delta EI}{H^2} - \dfrac{wH^2}{8}}{8}$$

whence follows the required equation.

2.4 SETTLEMENT OF FOUNDATIONS ON COHESIVE SOILS

2.4.1 Review of the problem of settlement and related laboratory tests

The normal settlement of a foundation arises solely through compression of soil and rock under the load imposed by the foundation. Abnormal settlement may arise from a number of causes discussed briefly in Section 2.6 but is excluded from the present considerations. Conventionally, it is assumed that settlement involves three components – immediate, consolidation and secondary – in relative proportions which vary with soil types. Cohesive soils commonly display these components more distinctly than cohesionless soils. Generally, each component is calculated independently and the sum is taken as the total settlement, but this process of superposition is not strictly correct since the stress/strain relationships for soils are non-linear. However, for many practical problems it is sufficient to know the order of magnitude and rate of settlement. More precise estimates are required for certain structures and particular site conditions, for which more refined techniques of analysis and computation can be employed.

In respect of the loading considered to cause settlement, both dead and live load lead to immediate settlement, although this is partly recovered on removal of live load. Loads which are imposed infrequently or for short periods generally are unimportant in consolidation and secondary settlement on clays, but should be taken into account in the case of sands and gravels. Wind load is important when it constitutes a large part of the total load.

The following terms are commonly employed in foundation design.

- Total overburden stress, σ_0, is the stress on any horizon due to the weight of soil solids and water above that horizon.
- Effective overburden stress on any horizon, σ_0', is the total overburden stress minus the pore water pressure at that horizon.
- Gross loading intensity or bearing stress q is the stress at the base of a foundation due to all loads above that level.
- Net loading intensity or bearing stress $q_n = (q - \sigma_0)$ when σ_0 is determined at founding level.
- Ultimate bearing capacity q_f is the net bearing stress at which the soil fails in shear.
- Safe bearing capacity q_s is the maximum bearing stress which the soil will support without risk of shear failure and is determined by applying a factor of safety to q_f.
- Allowable bearing stress, q_a, is the maximum permissible net bearing stress, taking into account safe bearing capacity, the estimated magnitude and rate of settlement and the ability of the structure to accommodate the settlement.
- Presumed bearing stress is employed in preliminary design and is the net bearing stress considered appropriate to the particular type of ground under consideration: it is usually determined from general or local experience.

The estimation of settlement on a cohesive soil generally involves determination of compression characteristics of the soil by testing in a consolidation press. For a particular group of foundations, it may be possible to estimate for a given bed of soil the overburden stress, σ_0, and the maximum net stress, σ_n, induced by the foundations and in this case it may be sufficient to test over the single stress increment σ_0 to $(\sigma_0 + \sigma_n)$. However, it is usual to test over a series of stress increments in order to permit a wide range of adjustment in the

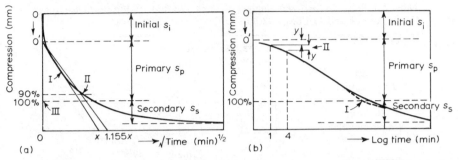

Figure 2.15 Analysis of observations for single pressure increment in consolidation test. (a) I, Tangent to straight part of curve locates zero $0'$ for primary and also distance x mm. II, Intersection of line $0'$ to $1.155x$ with curve locates 90% primary. III, Add 1/9 to $0'$ to 90% to locate 100% primary. (b) I, Intersection of projections of straight parts of curve locates 100% primary. II Assume initial part of natural curve is parabola, then compression at, say, 4 min is twice compression at 1 min, and zero $0'$ for primary located by equal distances y.

proportions and dispositions of foundations. Two methods of plotting and analysing the observations for a single stress increment are shown in Fig. 2.15. The techniques are semi-empirical and, since the results depend on personal interpretation of the curves, close correspondence in the magnitudes of the three components derived by the two methods cannot always be expected.

Initial compression, s_i, is attributed largely to compression of gas, closure of fissures, disturbance of structure during sampling and bedding down of the porous plates. This compression is considered to be completed almost instantaneously but, because of the nature and variability of the components, it is not employed in settlement calculations. Furthermore, immediate settlement involves lateral and vertical strains whereas no lateral strains can be developed in the consolidation test. Primary compression, s_p, involves expulsion of water from the pores and is the hydrodynamic process which is assumed to be governed by the laws of the mathematical theory of consolidation. Theoretically it is completed only in an infinite period of time. Dissipation of pore pressure is accompanied by increase in inter-particle stress. Secondary compression, s_s, is considered here to involve plastic yield which is accompanied by rearrangement of the inter-particle structure. If secondary compression is dominant, it may be necessary to apply each stress increment in the consolidation test for a period of a week or more instead of the usual 2 days. The rate of development of secondary compression is controlled partly by the plastic characteristics of the soil and partly by the rate of transfer of stress during the dissipation of pore pressure.

The parameters derived from a consolidation test are as follows: strictly they should be derived from primary compression as shown in Fig. 2.15.

Coefficient of compression

$$m_v = \frac{\delta h}{h.\delta\sigma'} = \frac{\delta e}{(1+e)\delta\sigma'}\mathrm{m^2\,kN^{-1}} \tag{2.4}$$

Table 2.3 Relation between degree of consolidation and time factor (after Cappe, Cassie and Geddes, 1980)

	Types of stress distribution (one-way drainage)		
Top (permeable)	(1)*	(2)	(3)
Bottom (impermeable)	▢	◺	◹

	Time factor, T_v		
Degree of consolidation	*Stress distribution* (1)	*Stress distribution* (2)	*Stress distribution* (3)
0.1	0.008	0.047	0.003
0.2	0.031	0.100	0.009
0.3	0.071	0.158	0.024
0.4	0.126	0.221	0.048
0.5	0.197	0.294	0.092
0.6	0.287	0.383	0.160
0.7	0.403	0.500	0.271
0.8	0.567	0.665	0.440
0.9	0.848	0.940	0.720

* For two-way drainage condition (1) is used for all linear distributions of stress.

where h is the initial thickness of the specimen, δh is the compression of the specimen under an effective stress increment $\delta \sigma'$, e is the void ratio corresponding to h and δe is the decrement in void ratio corresponding to δh.

Coefficient of consolidation

$$c_v = \frac{T_v d^2}{t} \, \mathrm{m^2 \, year^{-1}} \qquad (2.5)$$

where T_v is the time factor corresponding to the appropriate degree of consolidation and stress distribution (Table 2.3), t is the time corresponding to the degree of consolidation and d is the length of the drainage path ($d = h$ for a half-closed layer and $d = h/2$ for an open layer).

$$\text{Permeability } k = m_v c_v \gamma_w \qquad (2.6)$$

where γ_w is the unit weight of water.

The rates of development of both primary and secondary compression are influenced by temperature changes. The coefficient of consolidation can be reduced to the arbitrary standard field temperature of 10°C by the application of a coefficient K to the value c_{vt} derived from laboratory tests at temperature t°C. Thus $c_{v10} = K c_{vt}$ where K is taken from Table 2.4.

Table 2.4 Factors for the correction of coefficient of consolidation to standard temperature

$t(°C)$	5	10	15	20	25
K	1.17	1.00	0.87	0.77	0.69

Strictly, the coefficient of compressibility, m_v, coefficient of consolidation, c_v, and permeability, k, should be determined from the primary compression but the rearrangement of particles involved in secondary compression will influence the values of k and c_v to some extent. The following are typical ranges of m_v in $m^2 kN^{-1}$: unweathered overconsolidated clays $< 10^{-4}$, normally consolidated and weathered overconsolidated clays 10^{-4} to 10^{-3}, approaching 10^{-2} when the organic content is high. If a sample of overconsolidated clay is disturbed in sampling, the compressibility determined by the consolidation test is too great and it is necessary to make an adjustment (Terzaghi and Peck, 1967). The true compressibility characteristics of sensitive clays with $S_t > 8$ can be obtained only from undisturbed samples (Terzaghi and Peck, 1967). Although both m_v and k may decrease with depth in a given bed of clay, c_v tend to remain sensibly constant: this is confirmed by inspection of Equation 2.6.

Both c_v and k tend to increase with decrease in liquid limit w_L and, very broadly, c_v can be expected to be between 0.1 and 1.0 m^2 year^{-1} at $w_L = 100$ and between 1.0 and 50 m^2 year^{-1} at $w_L = 20$.

Contemplation of the conventional analysis of consolidation test results and the prediction of settlement of foundations reveals that the techniques employed are indeed arbitrary. The three components of compression must, in fact, be combined to comprise a single continuous process. Many investigators have evolved a variety of rheological models for representing the behaviour of soils under compression. For example, if instantaneous elastic compression is ignored and consolidation only is considered, then the mechanism can be represented for the one-dimensional case by the model proposed by Barden (1968) and shown in Fig. 2.16. It must be remembered that consolidation involves the expulsion of water from voids and that normally 100% saturation is assumed in theoretical analyses. Barden proposed that the rheological behaviour of the clay skeleton, relating to primary and secondary consolidation, can be represented by a non-linear spring in parallel with a non-linear dashpot. It appears that, in this model, strain is completely recoverable on removal of the applied load and in order to cater for plastic strain, a dashpot should be inserted in series with the parallel assembly. Other investigations concerning secondary compression have been reported by Christie (1965, 1966), Hansen and Inan (1969), Poulos et al. (1976) and Sridharan and Rao (1982). Field observations of settlement, however, indicate that there is frequently, perhaps invariably, an instantaneous strain response to loading of the soil which is interpreted as elastic in nature. This presumably arises largely from incomplete saturation of the soil, resulting in compression of gas accompanied by deformation of the soil skeleton, and from

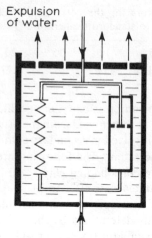

Figure 2.16 Rheological model for clay under compression (after Barden, 1968).

partial or complete closure of fissures. Because some of this deformation is not recoverable, a part of it must be plastic in nature. Most models are one-dimensional, whereas the field case is three-dimensional, although when appropriate corrections, such as those proposed by Skempton and Bjerrum (1957), are made, good correspondence is frequently obtained between predicted and observed settlement. Predicted magnitude of settlement on normally consolidated clay is usually reasonably good except when the imposed net stresses are relatively low, since the influence of disturbance on diagenetic changes brought about by desiccation, cementation and other processes is most conspicuous under these conditions, with the result that settlement is over-estimated. Settlement on overconsolidated clay may also be over-estimated because of both disturbance and the release of *in situ* horizontal stresses.

Commonly the rate of settlement cannot be predicted very accurately, although a rough estimate may be adequate for many purposes. Tests on large diameter specimens with vertical and radial drainage often yield values of c_v many times greater than those derived from tests in a conventional one-dimensional drainage consolidation press. This is related to the facts that drainage is tri-axial and that permeability parallel to the bedding is often greater than permeability normal to the bedding (Section 2.7.1), although this may be reversed by biological activity. Consequently, rate of settlement is related to the thickness and nature of the consolidating layers of soil and to the areal extent of the foundation and such conditions complicate calculations.

At present, the total process of compression of soils cannot be expressed perfectly in quantitative terms, even if the mechanisms are reasonably well understood qualitatively. Many studies of consolidation and settlement have been reported and it is difficult to select a few for further reference.

Comprehensive studies have been published by Lowe (1974), Lambe (1964) and Murray (1978) and a review of the deformation characteristics of soils, including an extensive list of references, has been presented by Scott and Ko (1969). Janbu (1969) has examined rate of settlement in terms of strain instead of pore pressure. Studies on the behaviour of stratified soils have been reported by Rowe (1959, 1964), Horne (1964), Rowe and Shields (1965) and Davis and Lee (1969). The settlement of embankments on soft sediments has been investigated by Murray (1971, 1973), Dallard (1971) and Thorburn and Beevers (1981) and Naylor and Jones (1973) have analysed compression within an embankment and give further references. Proceedings of a symposium on settlement and stability of embankments have been published by TRRL (1978).

2.4.2 Immediate settlement

Immediate settlement takes place without dissipation of pore pressure and is treated as an elastic phenomenon. Equations 1.10 to 1.17 can be employed to calculate immediate settlement of a loaded area on the surface of a deep bed of clay. The value of E_r is commonly taken as the secant modulus to the maximum working stress on the stress/strain curve derived from an undrained triaxial test (Fig. 2.1). Experience indicates that, in respect of the determination of E_r, all clays are sensitive to disturbance to some extent. For example, the secant modulus of a specimen carefully prepared from an undisturbed block of London Clay is much greater than that of a specimen prepared from the same clay obtained by an open drive sampler. It is known also that values of E_r determined from site plate loading tests are influenced by the size of the plate in relation to the frequency of fissures and by the time delay between completion of excavation of the trial shaft and testing. Values of E_r in $kN\,m^{-2}$ commonly range between 10^3 for soft clays and 10^5 for stiff clays.

At the low values of stress developed at depths exceeding about $4B$ beneath a foundation, the effective value of E_r is the initial tangent modulus and this exceeds the secant modulus. Furthermore, E_r commonly increases with depth. Therefore, strains below a depth of $4B$ contribute very little to immediate settlement at the surface and, ignoring these displacements, Skempton (1951) determined the values of I_ρ given in Table 2.5 for estimating the mean settlement of a uniformly loaded area from

$$\rho_a = \frac{I_\rho(1 - v^2)qB}{E_r} \qquad (2.7)$$

Table 2.5 Influence factors for average immediate settlement of uniformly loaded rectangular and circular areas on the surface of an elastic medium

L/B	1	2	5	10	Circle (diameter B)
I_ρ	0.82	1.00	1.22	1.26	0.73

The relationship

$$\rho_i = \frac{I_\rho(1 - v^2)qB}{E_r} \tag{2.8}$$

gives the settlement ρ_i beneath the corner of a flexible, uniformly loaded rectangular area on the surface of a semi-infinite elastic medium or elastic half-space and the influence factor I_ρ was evaluated by F. Schleicher in terms of the ratio L/B. Steinbrenner extended this analysis to cater for a medium of finite depth, Z, supported on a rigid base. Terzaghi (1943) transforms Equation 2.8 into $\rho_i = qBI'_\rho/E_r$, where $I'_\rho = (1 - v^2)F_1 + (1 - v - 2v^2)F_2$ and gives a chart of F_1 and F_2 values. However, when $v = 0.5$ the second term on the r.h.s. is equal to zero and $F_1 = I_\rho$ in Equation 2.8, which can be expressed as

$$\rho_i = \frac{0.75qBI_\rho}{E_r}. \tag{2.9}$$

The values given in Table 2.6 relate to I_ρ in Equation 2.9.

Settlement of any other point within the loaded area can be calculated by superposition. Thus at X in Fig. 2.17

$$\rho_i = \frac{0.75q}{E_r}\{I_{\rho 1}B_1 + I_{\rho 2}B_2 + I_{\rho 3}B_3 + I_{\rho 4}B_4\}. \tag{2.10}$$

This technique of superposition can be employed also when the load on each subarea or compartment is different. Settlement of the surface beyond the perimeter of the loaded area can be determined by treating appropriate terms in Equation 2.10 as negative (Fig. 2.17).

Corrections are applied to calculated values of immediate settlement to cater for foundations located beneath ground surface, for the presence of a basal rigid

Table 2.6 Influence factor I_ρ for settlement beneath the corner of a flexible, uniformly loaded rectangular area $(L \times B)$ on the surface of an elastic medium of depth Z on a rigid base when $v = 0.5$

Z/B	L/B			
	1	2	5	10
0.20	0.008	0.006	0.006	0.006
0.50	0.05	0.04	0.04	0.04
1	0.15	0.12	0.11	0.11
2	0.29	0.29	0.27	0.26
5	0.44	0.53	0.56	0.53
10	0.48	0.64	0.76	0.77
∞	0.56	0.77	1.05	1.27

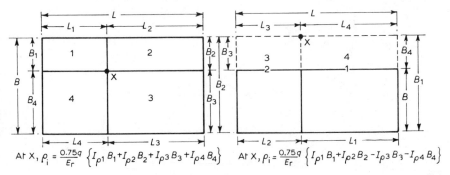

At X, $\rho_i = \dfrac{0.75q}{E_r}\left\{I_{\rho 1}B_1+I_{\rho 2}B_2+I_{\rho 3}B_3+I_{\rho 4}B_4\right\}$ At X, $\rho_i = \dfrac{0.75q}{E_r}\left\{I_{\rho 1}B_1+I_{\rho 2}B_2-I_{\rho 3}B_3-I_{\rho 4}B_4\right\}$

Figure 2.17 Calculation by superposition of settlement at any point X induced by flexible uniformly loaded area on the surface of an elastic medium of depth Z to a basal rigid boundary.

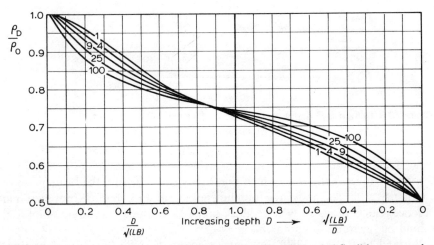

Figure 2.18 Ratio ρ_D/ρ_0 of mean settlement ρ_D of uniformly loaded flexible rectangular foundation at depth D to that ρ_0 of similar foundation on surface of elastic medium. The numbers on the curves denote values of the ratio L/B where L is the length and B the breadth of the foundation. Poisson's ratio $v = 0.5$ for elastic medium (after Fox, 1948).

boundary (this correction is catered for in the values given in Table 2.5) and for the presence of a relatively stiff layer overlying the main elastic medium. The ratio of mean settlement of a uniformly loaded flexible footing at a depth D to settlement of a footing at the surface is shown in Fig. 2.18 for different values of L, B and D assuming Poisson's ratio to be 0.5 (Fox, 1948). The abscissae in this diagram change from $D/\sqrt{(LB)}$ to $\sqrt{(LB)}/D$ at unit abscissa in order that the infinite range of depth can be covered in a finite range of plotting. Furthermore the ratio $D/\sqrt{(LB)}$ and its reciprocal have been chosen rather than D/L or D/B to avoid crowding the curves for different values of L/B. It will be seen that the settlement of a footing founded at some depth D is less than a similar footing

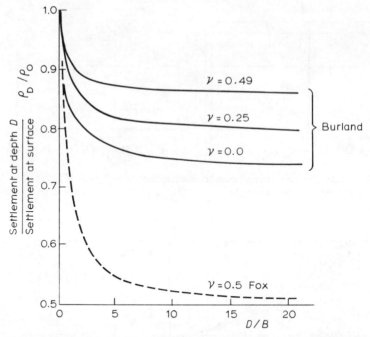

Figure 2.19 Influence of depth on settlement of bearing plate in unlined borehole (after Burland and Lord, 1969).

located at the surface where the net stress is the same in both cases. The theoretical investigations on which Fig. 2.18 is based involve the assumption of continuity of the soil above the founding level and, since this does not obtain in practice, calculations in which values taken from these curves are employed should be regarded as approximations.

Burland and Lord (1969) took into account the influence of discontinuity caused by an unlined borehole on the settlement of a circular plate located at the bottom of the hole (Fig. 2.19). It will be seen that the influence is considerable and the effect would be greater if it is assumed that the soil or rock is incapable of resisting tension. Scherman (1969) broadly confirmed these conclusions and showed that the value of the correction for a rigid plate is practically the same as that for, presumably, the mean settlement of a uniformly distributed load. It should be observed that the parameter B/B' is employed in Fig. 2.20, where B is the plate diameter and B' is the borehole diameter, and the case of a plate on the surface of an elastic medium obtains when $B' = \infty$. The curves shown in Fig. 2.21(b) were derived by Janbu *et al.* (1956).

The presence of a rigid stratum beneath an elastic medium influences settlement. Figure 2.22 illustrates the influence of the ratio Z/B on the settlement of a flexible uniformly distributed load on a circular area of diameter B when $v = 0.5$. Note that as Z/B approaches 1/3, the settlement becomes greatest at a distance of

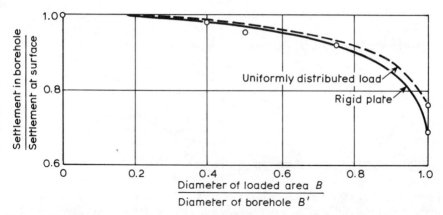

Figure 2.20 Influence on settlement of ratio of plate diameter to borehole diameter (after Scherman, 1969).

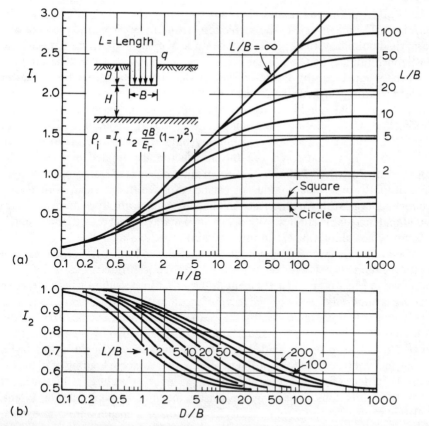

Figure 2.21 (a) Plan shape and rigid boundary factor I_1; (b) depth factor I_2 (After Janbu *et al.*, 1956.)

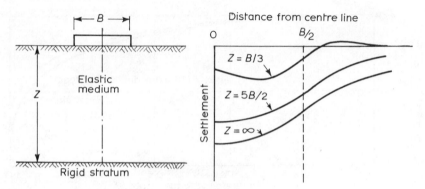

Figure 2.22 Influence of a rigid stratum beneath an elastic medium on the settlement of a uniformly distributed load on a circular area of diameter B for $v = 0.5$ (after Terzaghi, 1943).

about $B/3$ from the centre and also that the free surface rises just beyond the periphery. The magnitude of the correction factor, arising from the presence of a basal rigid boundary, to be applied to the value of ρ_i calculated for a semi-infinite elastic medium can be taken from Fig. 2.21(a) (Janbu *et al.*, 1956). The problem has been studied also by Poulos (1967). It should be noted that the values of influence factor given in Table 2.5 cater for the presence of a basal rigid boundary. Furthermore, that the Janbu factors are applied to the term $\{qB(1 - v^2)/E_r\}$ irrespective of foundation plan shape, since this is taken into account in Fig. 2.21(a) and it is necessary to employ data from this graph even when $H \to \infty$. The Fox factor and the Burland factor are applied directly to the appropriate Equations 1.10 to 1.17. Brown and Gibson (1972) studied settlement of a deep stratum, the modulus of elasticity of which increases with depth. Following large-scale loading tests on chalk at Mundford, Burland *et al.* (1973) found that increase in stiffness with depth has a marked effect on settlement and is a factor which should be taken into account in refined analysis.

When a load is imposed on a relatively stiff elastic layer overlying a less stiff elastic medium, there is a tendency for stress to be concentrated in the stiffer layer, as shown in Fig. 2.6(g), and settlement at the surface is less than where the same loading is imposed directly on the less stiff medium. Clearly, values of stress and displacement will be influenced by the ratio of modulus of elasticity E_{r1} of the upper layer to that E_{r2} of the underlying medium and also by the ratios of the length and breadth of the loaded area to the thickness H of the upper layer. Burmister (1943) analysed the problem of a circular uniform loading of diameter B on the surface of a thin relatively stiff elastic layer, extending to infinity in all horizontal directions, overlying an elastic medium of semi-infinite extent. Different ratios of modulus of elasticity in the two elastic media were employed in the analysis and Poisson's ratio was taken as 0.5. Figure 2.23 gives values of the

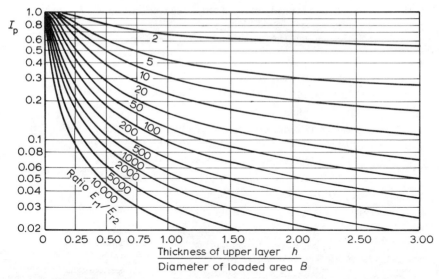

Figure 2.23 Influence factor I_p for settlement of centre of uniformly loaded flexible area on surface of layer with modulus of elasticity E_{r1} overlying deep elastic medium with modulus of elasticity E_{r2}.

influence factor I_p for settlement at the centre of the loaded area. Then

$$\rho_i = \frac{0.75qBI_p}{E_{r2}}. \tag{2.11}$$

The data given in Fig. 2.23 are used freely in Example 2.15 and the calculations inevitably lack rigour but estimates of settlement are essential for design purposes and data, however scant, must be utilized to the best advantage. The resulting estimates will certainly give the order of magnitude of settlement and in many problems this is all that is required. Further limited data are given in *Soil mechanics for road engineers* (H.M.S.O., 1952). A stiff layer overlying a less stiff soil reduces settlement by concentrating stress in the upper layer and not by distributing the load, as will be seen by reference to Fig. 2.6(g). It should be noted that both stress distribution and settlement will be modified to some extent if it is assumed that the overlying layer is incapable of resisting tension.

2.4.3 Consolidation settlement

The pore pressure parameters A and B (Skempton, 1954) were taken into account by Skempton and Bjerrum (1957) in the theory of consolidation. B depends mainly on the degree of saturation of the soil and is unity when $S_r = 100\%$. A depends mainly on the stress history of the soil (Table 2.7) and on the magnitude of the applied stress. Although lateral strains are developed during consolidation

Table 2.7 Typical values of A for the working range of stress beneath foundations

Soil	A
Very sensitive soft clays	>1
Normally consolidated clays	$\frac{1}{2}-1$
Over-consolidated clays	$\frac{1}{4}-\frac{1}{2}$
Heavily over-consolidated sandy clays	$0-\frac{1}{4}$

Table 2.8 Variation of the parameter α with thickness Z of the consolidating stratum

$\dfrac{Z}{B}$	α	
	Circular foundation, diameter B	Strip foundation, breadth B
0	1.00	1.00
0.25	0.67	0.80
0.5	0.50	0.63
1	0.38	0.53
2	0.30	0.45
4	0.28	0.40
10	0.26	0.36
∞	0.25	0.25

beneath a foundation, they cannot be developed in the consolidation test and it is necessary to introduce a parameter α to take these into account. If ρ'_p is the settlement beneath a foundation given by the direct application of test results for primary compression, then the corrected settlement is

$$\rho_p = \mu\rho'_p \qquad (2.12)$$

where $\mu = A + \alpha(1 - A)$, provided m_v and A are constant with depth beneath the foundation, and α (Table 2.8) depends only on the geometry of the problem, provided $S_r = 100\%$. For approximate calculations μ can be taken as 1.0 for normally consolidated clays and 0.5 for over-consolidated clays. In connection with the evaluation of α, Wood (1959) pointed out that, although in the case of a circular foundation the lateral stress increments $\Delta\sigma_2$ and $\Delta\sigma_3$ are equal, in the case of a strip foundation they are unequal and, taking this into account, determined the approximate values given in the third column in Table 2.8.

The distribution of vertical stress on horizons beneath a foundation can be derived from iso-stress diagrams or from tables and charts given, for example, by Terzaghi (1943). Button (1961) has computed influence factors which enable consolidation settlement beneath the centre and corners of the loaded area to be readily determined. Since many foundations are sensibly rigid, the bearing stress distribution is not uniformly distributed and therefore the loading for which most tables, charts and diagrams are applicable does not obtain. Consequently, the following method based on an assumed 2 in 1 dispersal of load beneath a foundation is adequate for many cases. The vertical stress on any horizon at a depth z beneath a foundation of length L and breadth B carrying a net load Q_n is taken as

$$\sigma_z = Q_n/(L+z)(B+z). \tag{2.13}$$

The degree of reliability of this method can be assessed approximately by comparing the stress at any depth z/B calculated from Equation 2.13 with the mean stress at the same depth derived from iso-stress diagrams. Generally it will be found that values of stress are within $\pm 20\%$ and sometimes correspond closely. Furthermore, consolidation settlement calculated by this method is commonly tolerably close to observed settlement. The settlement of a rectangular foundation due to consolidation of a stratum of clay between depths z_1 and z_2 is

$$\rho_p = \int_{z1}^{z2} m_v \sigma_z dz = m_v Q_n \int_{z1}^{z2} \frac{d_z}{(L+z)(B+z)}. \tag{2.14}$$

The solution to Equation 2.14 is

$$\rho_p = \frac{2.303 m_v Q_n}{(L-B)} \log_{10}\left[\left(\frac{z_2+B}{z_2+L}\right)\left(\frac{z_1+L}{z_1+L}\right)\right]. \tag{2.15}$$

For a square foundation with breadth B

$$\rho_p = m_v Q_n \left[\frac{z_2-z_1}{(z_2+B)(z_1+B)}\right]. \tag{2.16}$$

For a circular foundation with diameter B

$$\rho_p = \frac{4}{\pi} m_v Q_n \left[\frac{z_2-z_1}{(z_2+B)(z_1+B)}\right]. \tag{2.17}$$

If the value of m_v varies with depth, it can be taken into account in the integration, provided the variation can be simply expressed in relation to z. Alternatively, a tabular form of computation can be employed. In this method the greatest accuracy is attained by employing relatively thin sub-divisions of the consolidating strata. Since the 2 in 1 dispersal method of estimating consolidation settlement is not based on theoretically established stress distributions, it must be regarded as approximate and justification for its use lies principally in the fact that it leads to reasonable results. Where it is considered that part of the load is transferred to the ground by skin resistance, as in the case of some piles, it is commonly assumed that the load is dispersed at 2 in 1 from a level one-third of the buried length of the shaft above the base or from some other level appropriate to a particular problem.

If the distribution of stress σ_z cannot be expressed as a simple function of z for substitution in the integral

$$\rho_p = \int_{z1}^{z2} m_v \sigma_z \mathrm{d}z,$$

then a graph can be plotted of $m_v \sigma_z$ against z and the area of the graph will give ρ_p. If m_v is constant, the area of the graph σ_z against z can be multiplied by m_v to give ρ_p.

For the calculation of rate of consolidation, the value of T_v is selected from Table 2.3 for the stress distribution which fits most closely the actual distribution. The length of the drainage path d is established after the relative values of permeability of the materials at the upper and lower boundaries of the consolidating layer of clay have been assessed. Some classes of concrete have a permeability greater than many clays but steel and bitumen are regarded as impermeable. If the consolidating layer is underlain by another bed of clay, the approximate values of permeability of the two beds should be compared and a decision made to treat the lower bed as either permeable or impermeable or alternatively to calculate rates of consolidation on the assumption of both open and half-closed layers and to treat the periods so derived as lower and upper limits. In order that any bed can be regarded as a drainage layer it must be relatively continuous laterally. The rate of consolidation of a bed of clay may be appreciably influenced by intercalated continuous seams of sand or silt. A pocket of sand or gravel confined in clay does not constitute a drainage layer except when it outcrops at some point at ground surface or can drain freely elsewhere. Settlement rate can be increased artificially by the provision of sand drains.

The value of T_v employed for the calculation of settlement rate is based on assumed one-dimensional drainage and this applies most aptly when the dimensions of the loaded area are large compared with the thickness of the underlying bed of clay. Rate of settlement is increased by components of flow in horizontal directions and can be augmented further by the fact that in some sediments the permeability parallel to the bedding is greater than the permeability perpendicular to the bedding.

It is reasonable to apply the depth correction factor to both consolidation and secondary settlement, since both depend at least in part on elastic stress distribution. The rigid stratum correction is not employed since integration is to a lower boundary and, furthermore, values of the parameter α are determined for the same lower limit. Unquestionably, the application of various correction factors extends the degree of empiricism of calculations but it is justified on the grounds that influences are known to exist and it is desirable to cater for them in a reasonable manner.

It will be seen that a discrepancy is revealed between Equations 2.15 and 2.16 when $L = B$ is substituted in the former. However, if the expressions are expanded in logarithmic and binomial series, respectively, immediately after integration, it will be found that the first three terms, at least, of the series

$$\frac{z_2 - z_1}{B^2}\left[1 - \frac{(z_2 + z_1)}{B} + \frac{(z_2^2 + z_1 z_2 + z_1^2)}{B^2} \cdots \right] \qquad (2.18)$$

obtain in both cases.

2.4.4 Secondary settlement

Secondary compression is considered to involve plastic deformation but it should be observed that some plastic deformation probably occurs during immediate settlement. The coefficient of secondary compression is

$$m_s = \frac{\delta h}{h \delta \sigma'}, \qquad (2.19)$$

where $\delta h = S_s$ is the secondary compression determined from Fig. 2.15. It is assumed that the magnitude of compression of a layer of clay under a given sustained stress is proportional to the thickness of the layer. The secondary compression/stress relationship can be regarded as linear over small ranges of stress only, as in the case of primary compression. The magnitude of secondary settlement can be calculated by employing any of the techniques applicable to the calculation of consolidation settlement, including the use of Equations 2.15, 2.16 and 2.17, substituting m_s for m_v. It can be observed that frequently primary and secondary compression are not separated in laboratory tests and m_v is quoted for the combination. Fortuitously, the magnitude of combined consolidation and secondary settlement is predicted correctly according to the methods presented above. It is reasonable to apply the depth correction factor and the Skempton/Bjerrum factor to calculated values of secondary settlement.

In the case of concrete, the rate of creep – involving plastic flow and consolidation – has been expressed by a variety of empirical equations including exponential and hyperbolic functions. For practical purposes it is often sufficient to assume that at time t secondary settlement ρ_s can be expressed in the form

$$\rho_s = \rho_{su}(1 - e^{-Kt}) \qquad (2.20)$$

where ρ_{su} is the ultimate secondary settlement and K is a constant governing the rate of progress. Values of δh and K can be determined from the graphical analysis of consolidation test results, as demonstrated in Example 2.17. Magnitude and rate of secondary settlement predicted by this method must be regarded as approximate.

2.4.5 Settlement progression

A graph showing the estimated progress of settlement with time should comprise immediate, consolidation and secondary components and take into account the influence of load changes during the construction period. Where the depth to founding level exceeds a few metres, it may be necessary to estimate the magnitude of heave in the bottom of the excavation which will be incurred following the removal of overburden stress. This heave will comprise immediate and reversed consolidation components and the latter can be estimated from an expansion test over an appropriate stress range carried out as part of the consolidation test on the soil. The coefficient of expansion m_v and the coefficient of swelling c_v are required to estimate the magnitude and rate of reversed consolidation. It is desirable to reimpose, by means of structural weight, a stress equal to that of the overburden as soon as possible in order to reduce heave, and consequently settlement also, and to avoid deterioration in shearing strength of the soil caused by the increase in moisture content associated with swelling.

Reversed consolidation may become appreciable if the bottom of an excavation is exposed for more than a few weeks and may not necessarily be terminated by the weight of the foundations alone. For example, because of financial problems, construction was halted for several months on a block of flats in Sussex after the foundations had been placed at a depth of about 3 m in clay. The foundations were disrupted by swelling of the clay – facilitated in this instance by heavy rainfall. Heave of up to 35 mm was experienced in clay in most of the cofferdams for the new Waterloo Bridge, London, but at pier no. 4 the heave was 75 mm (Buckton and Cuerel, 1943). It was pointed out that these values were about double those experienced at Chelsea Bridge further upstream and this was explained by a combination of harder driving of sheet piles and a higher degree of lamination in the clay, with the result that the laminations admitted water which led to swelling of the clay. It was noticed that skin resistance on the sheet piling exerted a considerable check on swelling in its immediate vicinity. Skempton and Ward (1953) have given details of investigations concerning a deep cofferdam in clay. An exceptional case of heave in shale was observed during excavation for Oahe Dam, Dakota, where differential movements occurred due to the presence of a complex system of faults (Underwood *et al.*, 1964). Observations of heave in excavations have been reported by Serota and Jennings (1959) and by Peterson and Peters (1963).

It is often sufficient to assume that, during the construction period, the rate of increase of load on the foundations is linear. The rate of increase of immediate settlement is commonly assumed to be also linear (Fig. 2.24). Divergence from a linear rate of loading necessitates special consideration and may involve rather arbitrary adjustments to the settlement curves. The sudden increases in loads caused by rapidly filling a silo or a water tank or hoisting a bridge superstructure on to abutments are examples of the need for adjustments in the curves for all components of settlement. The slight curvature in the graph of heave and counter-heave in Fig. 2.24 is drawn arbitrarily to cater for hysteresis. Settlement progression is complicated if several structures forming a connected group are constructed at different times.

The consolidation settlement curve is constructed in the following way. Initially the basic settlement curve $O_c AB$ for full net stress σ_{max} applied

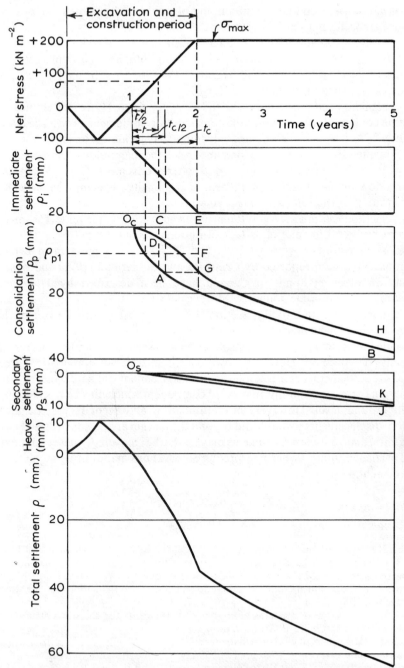

Figure 2.24 Construction of graphs of settlement progression.

instantaneously (Fig. 2.24) is drawn. At any time t less than the period t_c the net stress is σ. It is assumed that at t the amount of consolidation is the same as if the stress σ had been imposed for a period $t/2$. Thus the settlement at t is $\rho_{p1}\sigma/\sigma_{max}$ $= \rho_{p1}\,CD/EF = CD$ since $EF = \rho_{p1}$. The construction is repeated for different values of $t < t_c$. The settlement curve GH following completion of construction is a replica of the basic curve from A. Terzaghi (1943) extends the method to cater for non-linear rates of loading.

The secondary settlement curve is derived from $\rho_s = \rho_{su}\,(1 - e^{-Kt})$, employing the same construction as used for the consolidation settlement curve. The curve O_sJ (Fig. 2.24), based on instantaneous loading, may be almost linear for the first few years after application of the load and, since the magnitude of settlement at this stage may be relatively small, it is often convenient to carry out the construction on an enlarged settlement scale, finally reducing the scale for incorporation in the total settlement curve.

This method of predicting settlement progression is adequate for many practical problems in spite of its lack of rigour.

Settlement records covering a period of 65 years for a building in British Columbia have been reported by Crawford and Sutherland (1971) and observations on other buildings in Canada have been described by Trow and Bradstock (1972) and De Jong and Morgenstern (1973).

As a concluding note to this section it is of interest to discuss briefly the sequence of displacements involved in progressive settlement. Beneath the centre of a loaded area the immediate, consolidation and secondary settlements are all vertically downwards. However, towards the margin of the loaded area, immediate settlement has a small horizontal component outwards and beyond the margin there is commonly an upward component which may be manifest in a rise in ground level. However, as consolidation and secondary settlements develop, displacements towards and beyond the margin are downwards and part of the immediate outward movements may be also be recovered. These comments are of particular interest when interpreting observations made during monitoring of settlement.

EXAMPLE 2.4

Figure 2.25 shows the load/settlement record for a plate bearing test on chalk in a pit 1.0 m deep. The test plate was 0.3 m square. The bottom of the pit was well above the phreatic surface and it is assumed that no consolidation settlement will be incurred.

(a) Determine the secant modulus of subgrade reaction at ground surface for the stress range 0 to $1000\,kN\,m^{-2}$.

(b) Calculate the value of modulus of elasticity for the chalk for the stress range 0 to $1000\,kN\,m^{-2}$, assuming Poisson's ratio is 0.2.

(c) Estimate the settlement of a 0.9 m diameter concrete pier founded at a depth of 1.0 m in the chalk and carrying a gross load of 750 kN.

Burland depth factor (Fig. 2.19) for plate with $v = 0.2$ and $D/B = 3.3$ is 0.83. Since this applies primarily to circular plate, it is approximate only for square plate.

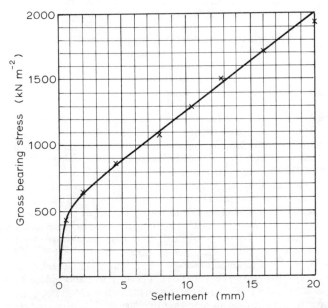

Figure 2.25 Load/settlement record for plate bearing test on chalk.

At $1000 \, \text{kN m}^{-2}$, $\rho_i = 6.5 \, \text{mm}$.

Hence at ground surface $k_s = \dfrac{0.83 \times 1000}{6.5 \times 10^{-3}} = 128 \times 10^3 \, \text{kN m}^{-3}$

From Equation 1.16, intrinsic value of $E_r = 128 \times 10^3 \times 0.95(1 - 0.2^2)0.3$
$$= 35 \times 10^3 \, \text{kN m}^{-2}$$

Bearing stress beneath 0.9 m diameter pier $q = \dfrac{750}{\pi \times 0.45^2} = 1179 \, \text{kN m}^{-2}$

Settlement of 0.3 m plate at $1179 \, \text{kN m}^{-2}$ is 9.0 mm.
Fox depth factor (Fig. 2.18) for pier assuming value for square foundation of equivalent area applicable and with $B/D = 0.9$ is 0.68.

Then from Equations 1.13 and 1.16 settlement of pier $\rho_i = \dfrac{0.68 \times 0.79 \times 0.9 \times 9.0}{0.83 \times 0.95 \times 0.3}$
$$= 18.4 \, \text{mm}.$$

Depth factor should be applied to settlement based on net load for both plate and pier but in this case overburden stress is small in relation to gross load and is ignored. Strictly, the appropriate range on the stress/settlement graph is from overburden stress to gross stress. The assumed value of v for the chalk is 0.2 whereas Fig. 2.18 is based on $v = 0.5$ and it is likely that the settlement of the pier will be less than the calculated value, perhaps by about 10%. The magnitude of settlement will probably be reduced also by transfer of part of the load to the chalk by skin resistance around the shaft although this will be partly offset by the greater compressibility of the chalk near ground surface. If the pier is relatively long, settlement at the column foot should include compression of the concrete.

EXAMPLE 2.5

Figure 2.26 shows records of constant rate of penetration tests to failure on two cast-*in-situ* piles founded in the Beaumont Clay of Pleistocene age at Houston, Texas (Reese and O'Neill, 1969; O'Neill and Reese, 1972). From ground surface to a depth of 9 m the soil is a stiff clay with average values of plastic limit and liquid limit about 25% and 60%, leading to the Casagrande Classification of CH: the *in situ* moisture content was generally within ± 5% of the plastic limit.

Pile 1 was 0.75 m diameter and 7 m deep and pile 2 has a shaft 0.75 m diameter and 5.5 m deep with an under-reamed base expanding to 2.30 m diameter at a depth of 7 m. The cycles of loading were completed within 1 to 2 hours and, since the clay was probably nearly completely saturated, the value of Poisson's ratio applicable in these tests can be taken as 0.4. Base settlement can be determined by deducting compression of pile from cap settlement. Pile compression can be calculated with sufficient accuracy from the average shaft load and Pile 2 can be assumed to be of uniform diameter 0.75 m for its entire length. Compression due to self weight of the pile can be ignored since this was completed before tests commenced. Furthermore, applied loads can be assumed to be net loads since the unit weight of pile concrete will be only about 10% more than the unit weight of overburden. The modulus of elasticity of the pile concrete was recorded as $40 \, \text{kN mm}^{-2}$.

Determine the values of modulus of subgrade reaction at base level for both piles when the cap loads were 750 kN on Pile 1 and 3115 kN on Pile 2 and reduce these values to ground surface standard if Fox's depth correction factor is 0.53 for Pile 1 and 0.58 for Pile 2. Strain gauges within the piles showed that the base loads corresponding to these cap loads were about 65 and 2400 kN: the difference being transferred to the soil by skin resistance. Also determine the values of modulus of elasticity of the soil at ground surface from both results.

Aide mémoire (Equation 1.13): $\rho_i = \dfrac{0.79(1 - v^2)Bq}{E_r}$.

Pile 1

Average load $\dfrac{750 + 65}{2} = 407.5 \, \text{kN}$.

Shaft compression $= \dfrac{407.5 \times 7 \times 1000}{\pi 0.375^2 \times 40 \times 1000^2} = 0.161 \, \text{mm} \simeq 0.2 \, \text{mm}$.

From Fig. 2.26(a), settlement of cap = 1.5 mm.
Hence settlement of base = 1.3 mm.

Bearing stress $= \dfrac{65}{\pi 0.375^2} = 147.1 \, \text{kN m}^{-2}$.

Modulus of subgrade reaction $k_s = \dfrac{147.1 \times 1000}{1.3} = 113 \, 154 \, \text{kN m}^{-3}$.

Reduced to ground surface $k_s = 0.53 \times 113 \, 154 = 59 \, 972 \, \text{kN m}^{-3}$.
Modulus of elasticity for soil $E_r = k_s 0.79(1 - v^2)B$
$$= 59 \, 972 \times 0.79 \times 0.84 \times 0.75 = 29 \, 848 \, \text{kN m}^{-2}.$$

Pile 2

Average load $\dfrac{3115 + 2400}{2} = 2757.5 \, \text{kN}$.

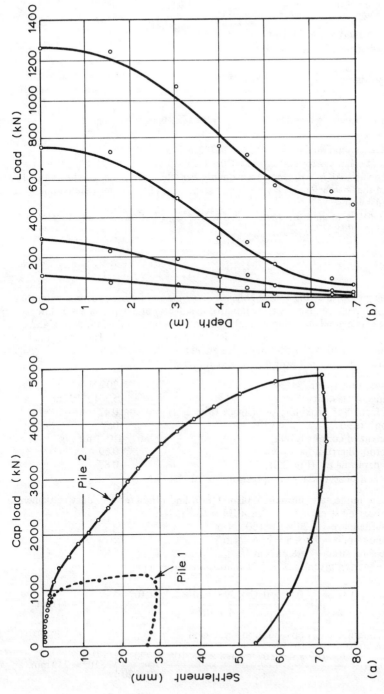

Figure 2.26 Records of load tests on cast-*in-situ* piles (after Reese and O'Neill, 1969). (a) Load against settlement for pile 1 and pile 2. (b) Load distribution for pile 1.

Shaft compression $= \dfrac{2757.5 \times 7 \times 1000}{\pi 0.375^2 \times 40 \times 1000^2} = 1.092\,\text{mm} \simeq 1.1\,\text{mm}.$

From Fig. 2.26(a), settlement of cap $= 25.0\,\text{mm}$.
Hence settlement of base $= 23.9\,\text{mm}$.

Bearing stress $= \dfrac{2400}{\pi 1.15^2} = 577.7\,\text{kN}\,\text{m}^{-2}.$

Modulus of subgrade reaction $k_s = \dfrac{577.7}{23.9} = 24\,172\,\text{kN}\,\text{m}^{-3}.$

Reduced to ground surface $k_s = 0.58 \times 24\,172 = 14\,020\,\text{kN}\,\text{m}^{-3}$.
Modulus of elasticity for soil $E_r = 14\,020 \times 0.79 \times 0.84 \times 2.30 = 21\,398\,\text{kN}\,\text{m}^{-2}$.
    The two values of E_r correspond reasonably well: the discrepancy can be ascribed partly to the fact that below 9 m is clayey silt and settlement of the under-reamed pile would be influenced by this.
    Fig. 2.26(b) is of interest since it shows the distribution of load transfer by shaft resistance for Pile 1.

EXAMPLE 2.6

The pivot pier for a swing bridge consists of a rigid reinforced concrete cylindrical box 20 m diameter externally. It is to be founded in stiff clay at a depth of 8 m below river bed level. The clay extends to a depth of 20 m beneath the base of the foundation and overlies bedrock. The mean stage of the river is 8 m above bed level. The gross load on the foundation is 2×10^5 kN including the weight of the cylinder. Values of various properties of the clay and other relevant data are as follows:

Saturated unit weight	$20\,\text{kN}\,\text{m}^{-3}$
Modulus of elasticity	$5 \times 10^4\,\text{kN}\,\text{m}^{-2}$
Janbu factor for proximity of bedrock (Fig. 2.21)	0.42
Poisson's ratio	0.5
Coefficient of compressibility	$10^{-4}\,\text{m}^2\,\text{kN}^{-1}$
Skempton–Bjerrum factor μ	0.69
Janbu depth factor (Fig. 2.21)	0.87
Coefficient of secondary compression $5 \times 10^{-5}\,\text{m}^2\,\text{kN}^{-1}$	

Calculate immediate, ultimate consolidation and ultimate secondary settlements.
Gross bearing stress $= 2 \times 10^5/\pi \times 10^2 = 636.6\,\text{kN}\,\text{m}^{-2}$
Overburden stress $= 20 \times 8 = 160\,\text{kN}\,\text{m}^{-2}$
Buoyancy uplift $= 9.8 \times 8 = 78.4\,\text{kN}\,\text{m}^{-2}$
Net bearing stress $= 398.2\,\text{kN}\,\text{m}^{-2}$
From equation given in Fig. 2.21 immediate settlement

$$\rho_i = \dfrac{0.87 \times 0.42 \times 398 \times 20 \times 0.75 \times 10^3}{5 \times 10^4} = 44\,\text{mm}.$$

From equation 2.17 consolidation settlement

$$\rho_p = \dfrac{0.87 \times 0.69 \times 4 \times 10^{-4} \times 398 \times \pi \times 10^2}{\pi} \left\{ \dfrac{20}{40 \times 20} \right\} \times 10^3 = 239\,\text{mm}.$$

From Equation 2.17, substituting m_s for m_v, secondary settlement

$$\rho_s = \frac{0.87 \times 0.69 \times 4 \times 5 \times 10^{-5} \times 398 \times \pi \times 10^2}{\pi} \left\{\frac{20}{40 \times 20}\right\} \times 10^3 = 120\,\text{mm}.$$

It should be observed that for the purpose of calculating settlement it is assumed that all the load is transferred to the soil at the base contact surface and the influence of transfer by skin resistance is ignored. It can be taken into account arbitrarily by an assumed enlarged base area defined, for example, by lines inclined at 4 in 1 from the top of the adhesion zone and intersecting the plane of the base to give a larger diameter. In this example, if it is assumed that skin resistance operates from bed level, equivalent enlarged diameter at base level is 24 m and $\rho_i = 26$ mm, $\rho_p = 141$ mm and $\rho_s = 71$ mm. The adoption of this concept reduces the total settlement by about 40% but, of course, skin resistance would not be mobilized until some penetration had taken place and it is possible that the actual reduction might not be as great as this.

EXAMPLE 2.7

The following details relate to one of the rectangular piers of Waterloo Bridge on the River Thames (Skempton *et al.*, 1955). It is founded in the London Clay which is underlain by the stiff clays and dense sands of the Woolwich and Reading Beds.

Length $L = 36$ m, breadth $B = 8$ m.
Net load on base $= 78 \times 10^3$ kN.
Founding depth below river bed level $D = 7$ m.
Thickness of the London Clay between base of pier and
   top of Woolwich and Reading Beds $= 24$ m.
Properties of pre-consolidated clay of London Clay:
$m_v = 84 \times 10^{-6}\,\text{m}^2\,\text{kN}^{-1}$. $c_v = 0.33\,\text{m}^2\,\text{year}^{-1}$. $E_r = 10^5\,\text{kN}\,\text{m}^{-2}$.
Poisson's ratio $v = 0.5$.

Calculate immediate and consolidation settlement. Also calculate the time required for 60% consolidation settlement to be developed assuming that the Woolwich and Reading Beds are (a) impermeable and (b) permeable if the time factors for these conditions are respectively 0.200 and 0.287.
   Janbu depth factor (Fig. 2.21) with $L/B = 4.5$ and $D/B = 0.9$ is 0.85. Janbu factor for rigid basal boundary (Fig. 2.21) with $L/B = 4.5$ and $H/B = 3.0$ is 0.95 and this takes account also of plan shape. London Clay is heavily over-consolidated, hence assume pore pressure coefficient A is 0.25. With $Z/B = 3.0$, interpolation in Table 2.7 for rectangular foundation gives $\alpha \simeq 0.35$. Then $\mu = 0.25 + 0.35\,(1 - 0.25) = 0.51$.
Net bearing stress $q = 78 \times 10^3/36 \times 8 = 271\,\text{kN}\,\text{m}^{-2}$
From equation given in Fig. 2.21 immediate settlement

$$\rho_i = \frac{0.85 \times 0.95 \times 271 \times 8 \times 0.75 \times 10^3}{10^5} = 13\,\text{mm}$$

From Equation 2.15 consolidation settlement

$$\rho_p = \frac{0.85 \times 0.51 \times 2.303 \times 84 \times 10^{-6} \times 78 \times 10^3}{28}\,\log_{10}\left\{\frac{32 \times 36}{60 \times 8}\right\} \times 10^3 = 89\,\text{mm}.$$

From Equation 2.5:
  half-closed layer $t_{60} = 0.20 \times 24^2/0.33 = 349$ years.
  open layer $\quad\quad t_{60} = 0.287 \times 12^2/0.33 = 125$ years.

Observations on four piers yielded the following average settlement: at the end of the construction period of almost 3 years, 28 mm; at the end of a period of about 16 years, 88 mm. Settlement at the end of construction comprises immediate settlement plus a small proportion of consolidation settlement. Values of E_r employed in earlier calculations were 34×10^3 and $37 \times 10^3\,\mathrm{kN\,m^{-2}}$ and these lead to about 37 mm for immediate settlement. More recent studies suggest that the *in situ* value can be expected to be of the order $10^5\,\mathrm{kN\,m^{-2}}$ although there are no recent experimental results relating specifically to Waterloo Bridge. The relatively large settlement at 16 years suggests that either the calculated ultimate consolidation is too small or the estimated rate of settlement is slow. Since drainage beneath these piers is three-dimensional, the discrepancy possibly arises mainly from the rate of settlement. Although it has been assumed that the Woolwich and Reading Beds are comparatively rigid, some further settlement will be incurred due to compression within this formation but the magnitude is likely to be relatively small.

Settlement of the foundations of new London Bridge over the River Thames, completed in 1973, has been discussed by Simpson (1976).

EXAMPLE 2.8

The foundation for the Auditorium Tower, Chicago, is $30.5\,\mathrm{m} \times 20.4\,\mathrm{m}$ in plan and is founded at a depth of 5.2 m in normally consolidated soft clay which extends 13 m below the foundation (Skempton *et al.* 1955). Beneath are 3.4 m of stiff clay followed by sands and very stiff clays. The phreatic surface is 3.0 m below ground surface. The value of m_v at founding level is $0.0007\,\mathrm{m^2\,kN^{-1}}$ and at the base of the soft clay $0.0003\,\mathrm{m^2\,kN^{-1}}$: the variation can be assumed linear with depth. The average value of c_v for the soft clay is $4.2\,\mathrm{m^2\,year^{-1}}$. The time factor for 90% consolidation for the prevailing vertical distribution of vertical stresses is approximately 0.7. The net bearing stress is $122\,\mathrm{kN\,m^{-2}}$. The value of the Skempton–Bjerrum correction factor is 0.91 and of the Fox depth correction factor 0.94.

Assume that the net load is dispersed within a zone defined by planes inclined at 2 in 1 from the base of the foundation, divide the 13 m bed of clay into five layers and calculate the ultimate magnitude of settlement due to consolidation. Also calculate the time required for 90% of this settlement to develop if the underlying stiff clay is assumed relatively impermeable.

Calculate the immediate settlement if the modulus of elasticity of the soft clay is $3750\,\mathrm{kN\,m^{-2}}$ and Poisson's ratio is 0.5. From Fig. 2.21, the Janbu depth factor $I_2 = 0.96$ and the shape and rigid basal boundary factor $I_1 = 0.4$ if $H = 13\,\mathrm{m}$ or $I_1 = 0.8$ if the stiff clays and sands are assumed to be elastic, having a value of modulus of elasticity equal to that of the soft clay, so that H tends to ∞.

Total net load $= 122 \times 30.5 \times 20.4 = 75\,908\,\mathrm{kN}$.

Net stress on plane in middle of each layer $= 75\,908/A$.

Area $A = (L + z)(B + z)$. Compression of each layer $\delta H = m_v H \delta \sigma$ where $H = 13/5 = 2.6\,\mathrm{m}$

z (m)	$L+z$ (m)	$B+z$ (m)	A (m²)	Net stress $\delta\sigma(\mathrm{kN\,m^{-2}})$	m_v (m² kN⁻¹)	δH (m)
1.3	31.8	21.7	690	110.0	0.0007	0.200
3.9	34.4	24.3	836	90.8	0.0006	0.142
6.5	37.0	26.9	996	76.2	0.0005	0.099
9.1	39.6	29.5	1168	65.0	0.0004	0.068
11.7	42.2	32.1	1353	56.1	0.0003	0.044
						0.553

Ultimate magnitude of consolidation settlement $\rho_p = 0.91 \times 0.94 \times 0.553 = 0.473$ m.
Time required for 90% consolidation to develop

$$t_{90} = \frac{0.7 \times 13^2}{4.2} = 28.17 \text{ or say 28 years.}$$

If the stiff clays and sands are assumed to be permeable $t_{90} = 28/4 = 7$ years.
Immediate settlement (Fig. 2.21)

$$\rho_i = I_i I_2 q \frac{B}{E_r}(1 - v^2)$$

If $H = 13$ m, $\rho_i = 0.4 \times 0.96 \times 122 \times 20.4 \times 0.75/3750 = 0.191$ m.
If $H = \infty$ $\rho_i = 0.8 \times 0.96 \times 122 \times 20.4 \times 0.75/3750 = 0.382$ m.

The calculated immediate plus ultimate consolidation settlement is therefore $(0.473 + 0.191) = 0.664$ m or $(0.473 + 0.382) = 0.855$ m. In fact, both these values would be augmented by consolidation in the stiff clays and sands and by secondary compression in the complete sequence of strata: no information appears to be available on secondary compression. In the absence of complete details, a reasonable assessment of total ultimate settlement is probably about 0.8 m.

The observed settlement at the end of construction, which took about 2.5 years, was 0.23 m and this includes immediate settlement plus a small proportion of consolidation and secondary settlement. It appears that the stiff clays and sands functioned as a rigid medium. The maximum observed settlement which has been published is about 0.61 m: this includes immediate, consolidation and secondary components. The observations on this structure have been maintained for at least 68 years and 90% consolidation settlement

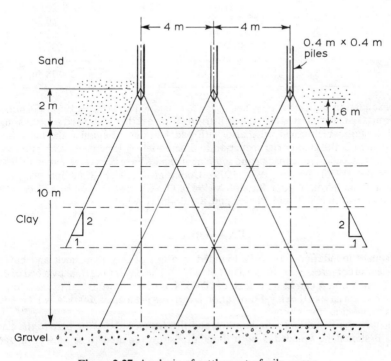

Figure 2.27 Analysis of settlement of pile group.

was considered to have been attained about 20 years after completion. It appears that little consolidation took place in the stiff clays and sands and that secondary compression was negligible in all strata. Furthermore, the stiff clays were relatively impermeable. These conclusions are drawn on the assumption that the theories employed in the analyses and the values of compression parameters are all reasonably correct.

EXAMPLE 2.9

The 0.4 m square piles in a large group supporting a reinforced concrete mat are arranged in plan on a square grid at 4 m centres. The load at the foot of each pile is 500 kN. The foot of each pile is located in a stratum of dense sand at a distance of 1.6 m above the base of the stratum. Beneath this sand is a 10 m thick stratum of soft clay overlying gravel. The coefficient of compressibility of the clay is $0.00050 \, m^2 \, kN^{-1}$.

Calculate the settlement of the inner piles of the group due to consolidation in the clay assuming a dispersal of load on areas enclosed by planes inclined at 2 in 1 from the foot of each pile.

The stratum of clay is divided into five layers each $H = 2 \, m$ thick (Fig. 2.27). Overlap of stressed zones on vertical axis of piles is taken into account.

Layer	Bearing area at centre of layer beneath each pile (m^2)	Stress beneath single pile $(kN \, m^{-2})$	Number of piles contributing	Stress beneath each pile taking account of overlap $(kN \, m^{-2})$
1	9	55.56	1	55.56
2	25	20.00	1	20.00
3	49	10.20	1	10.20
4	81	6.17	9	55.53
5	121	4.13	9	37.17
				Σ 178.46

Consolidation settlement $\rho_p = \Sigma m_v \sigma_v H = 5 \times 10^{-4} \times 178.46 \times 2 \times 10^3 = 178 \, mm$

The marked increase in stress at depth arising from overlap of stressed zones, should be noted, a state which could arise also with pad footings. The stiffer the mat and the superstructure, the greater the reduction in differential settlement between piles.

Various aspects of the capacity and settlement of piles have been investigated by Booker and Poulos (1976), Poulos (1977, 1979), Cooke *et al.* (1976), Randolph *et al.* (1979), Randolph and Wroth (1982), Randolph (1983), O'Neill *et al.* (1982a, b), O'Neill (1983), Potts and Marlin (1982) and Poulos and Randolph (1983).

EXAMPLE 2.10

Two square foundations are to be founded in a bed of clay 18 m thick and having a coefficient of compressibility $m_v = 0.0001 \, m^2 \, kN^{-1}$. The clay is underlain by a bed of dense sand. Foundation A carries a net load of 1000 kN, is of unknown breadth and is founded 15 m above the base of the clay. Foundation B carries a net load of 2000 kN, is 1.2 m square and is founded at an unknown depth.

Calculate the breadth of foundation A and the distance above the base of the clay at which B should be founded if both foundations are required to settle 0.05 m due to consolidation of the clay, both calculated to the nearest safe 0.1 m. Assume a dispersal of load beneath each foundation within a zone defined by planes inclined at 2 in 1 from the edges of the foundation. The value of the Skempton–Bjerrum correction factor is estimated to be 0.63 for A and 0.64 for B and the Fox depth correction factor is 0.56 for A and 0.53 for B.

Also calculate the settlement of foundation A in 10 years if open layer conditions are assumed and the coefficient of consolidation $c_v = 2.5\,\text{m}^2\,\text{year}^{-1}$. For the distribution of pressure which obtains beneath the foundation, the time factors are 0.287, 0.403 and 0.567 respectively at 0.6, 0.7 and 0.8 degrees of consolidation.

From Equation 2.16

$$A: 0.05 = 0.56 \times 0.63 \times 0.0001 \times 1000 \left\{ \frac{15 - 0}{(15 + B_A)(0 + B_A)} \right\}, \text{ hence}$$

$$B_A^2 + 15B_A - 10.58 = 0$$

Trial substitution for B_A yields $B_A = 0.68\,\text{m}$ or say 0.7 m.

$$B: 0.05 = 0.53 \times 0.64 \times 0.0001 \times 2000 \left\{ \frac{z_2 - 0}{(z_2 + 1.2)(0 + 1.2)} \right\}, \text{ hence}$$

$$0.157z_2 - 1.44 = 0 \text{ or } z_2 = 9.17\,\text{m or say 9.1 m}$$

From Equation 2.5

$$T_v = \frac{2.5 \times 10}{7.5^2} = 0.444$$

By linear interpolation, degree of consolidation for $T_v = 0.444$ is 0.725. Then settlement at 10 years is $0.725 \times 0.05 = 0.036\,\text{m}$.

EXAMPLE 2.11

It is desired to determine the reduction in consolidation settlement of a square foundation on the surface of the ground carrying a given net load Q which can be achieved by reducing the bearing stress from q to $q/2$ and $q/4$. If the breadths producing these values of bearing stress are respectively B_A, B_B and B_C, determine the ratio of settlement of B_B and B_C in terms of the settlement of B_A when consolidation takes place,

(a) between $z_1 = 0$ and $z_2 = 4B_A$ in all cases, and
(b) between $z_1 = 0$ and $z_2 = 10B_A$, $10B_B$ and $10B_C$ for the respective foundations.

Equation 2.16:

$$\rho_p = m_v Q \left[\frac{z_2 - z_1}{\{z_2 + B\}\{z_1 + B\}} \right]$$

$$B_B = 1.41B_A \text{ and } B_C = 2B_A$$

(a) $$\frac{\rho_{pB}}{\rho_{pA}} = \frac{(4B_A + B_A)(B_A)(4B_A)}{(4B_A + 1.41B_A)(1.41B_A)(4B_A)} = 0.66$$

$$\frac{\rho_{pC}}{\rho_{pA}} = \frac{(4B_A + B_A)(B_A)(4B_A)}{(4B_A + 2B_A)(2B_A)(4B_A)} = 0.42$$

(b) $\dfrac{\rho_{pB}}{\rho_{pA}} = \dfrac{(10B_A + B_A)(B_A)(14.1B_A)}{(14.1B_A + 1.41B_A)(1.41B_A)(10B_A)} = 0.71$

$\dfrac{\rho_{pC}}{\rho_{pA}} = \dfrac{(10B_A + B_A)(B_A)(20B_A)}{(20B_A + 2B_A)(2B_A)(10B_A)} = 0.50$

These calculations show that, in order to reduce settlement appreciably, foundation size must be considerably increased and this may be physically or economically undesirable. These conclusions apply to consolidation within deep homogeneous beds of soil but, if a bed is thin and close to the base of the foundation, consolidation within it will be appreciably reduced by increasing breadth. The analysis is complicated if immediate, consolidation and secondary components are taken into account, together with various correction factors although the conclusions are still broadly valid.

EXAMPLE 2.12

A flexible foundation 20 m × 20 m in plan is founded on the surface of a deep homogeneous mass of soil having a coefficient of compressibility which varies linearly from $0.0001\ m^2\ kN^{-1}$ at ground surface to $0.000\,02\ m^2\ kN^{-1}$ at a depth of 40 m. The uniformly distributed bearing stress is $100\ kN\,m^{-2}$. Divide the foundation area into sixteen equal compartments and calculate the vertical normal stress at depths of 10, 20, 30 and 40 m beneath the centre of the foundation. Estimate the settlement of the centre of the foundation due to consolidation of the mass of soil between ground surface and a depth of 40 m.

The Boussinesq equation (Fig. 2.3) is $\sigma_z = I_\sigma Q/z^2$ where the influence factor I_σ is determined from Table 2.9, when necessary by linear interpolation.

Table 2.9

r/z	0.00	0.10	0.20	0.30	0.40	0.50	0.60	0.70	0.80	0.90	1.00	1.10	1.20
I_σ	0.478	0.466	0.433	0.385	0.329	0.273	0.221	0.176	0.139	0.108	0.084	0.066	0.051

Divide plan area into sixteen compartments (Fig. 2.28), each 5 m square: then $Q = 2500\ kN$.
Divide 40 m depth into four layers, each $H = 10\ m$ thick.

$z(m)$	Compartment	Multiplier	$r(m)$	r/z	I_σ	$\sigma_v(kN\ m^{-2})$	$\Sigma\sigma_v(kN\ m^{-2})$
10	1	4	10.61	1.06	0.073	7.3	
	2	4	3.54	0.35	0.358	35.8	
	3	8	7.91	0.79	0.142	28.4	71.5
20	1	4	10.61	0.53	0.257	6.4	
	2	4	3.54	0.18	0.441	11.0	
	3	8	7.91	0.40	0.329	16.5	33.9
30	1	4	10.61	0.35	0.358	3.9	
	2	4	3.54	0.12	0.461	5.1	
	3	8	7.91	0.26	0.405	9.0	18.0
40	1	4	10.61	0.27	0.400	2.5	
	2	4	3.54	0.09	0.468	2.9	
	3	8	7.91	0.20	0.433	5.4	10.8

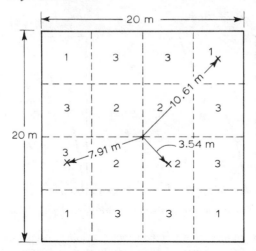

Figure 2.28 Division of foundation into compartments for stress calculations.

Calculate mean stress $\bar{\sigma}_v$ in each layer and also mean value $\bar{m}_v$.
Total consolidation settlement $= \Sigma \bar{m}_v \bar{\sigma}_v H$

$$= 10 \times 10^{-5}(9 \times 85.8 + 7 \times 52.7 + 5 \times 26.0 + 3 \times 14.4)10^3 = 131 \, \text{mm}$$

Instead of calculating mean stress, value of stress at centre of each layer can be determined at $z = 5, 15, 25$ and $35 \, \text{m}$: discrepancy between values commonly $+$ or $-$ a few per cent.

EXAMPLE 2.13

It is proposed to support an oil tank, 20 m diameter, on a flexible base on the surface of the ground and the bearing stress distribution will thus be uniform. The bearing stress q is $50 \, \text{kN m}^{-2}$. The influence factors for vertical stress σ_v on horizontal planes beneath uniformly loaded circular areas are taken from Fig. 2.5(c).

Table 2.10

			Centre			Periphery		
$z(m)$	z/B	$m_v(m^2kN^{-1})$ $\times 10^{-5}$	$I\sigma$	$\sigma_v(kN \, m^{-2})$	$m_v\sigma_v$	$I\sigma$	$\sigma_v(kN \, m^{-2})$	$m_v\sigma_v$
0	0	80	1.00	50.0	0.040	1.00	50.0	0.040
5	0.25	75	0.91	45.5	0.034	0.40	20.0	0.015
10	0.50	70	0.65	32.5	0.023	0.31	15.5	0.011
15	0.75	65	0.43	21.5	0.014	0.25	12.5	0.008
20	1.00	60	0.28	14.0	0.008	0.19	8.5	0.005
25	1.25	55	0.20	10.0	0.006	0.15	7.5	0.004
30	1.50	50	0.14	7.0	0.004	0.12	6.0	0.003
35	1.75	45	0.11	5.5	0.003	0.09	4.5	0.002
40	2.00	40	0.09	4.5	0.002	0.07	3.5	0.001

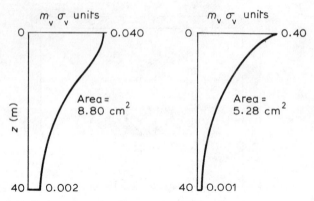

Figure 2.29 Calculation of settlement from area of graphs of $m_v\sigma_v$ against z.

The coefficient of compressibility of the soft clay forming a deep bed on which the tank is to be founded varies linearly from $0.0008\,\mathrm{m^2\,kN^{-1}}$ beneath the tank base to $0.0004\,\mathrm{m^2\,kN^{-1}}$ at a depth of 40 m.

Determine the settlement beneath the centre of the tank and also beneath the periphery arising from consolidation of the clay to a depth of 40 m.

Stress $\sigma_v = qI_\sigma$.

From the tabulated computations (Table 2.10) graphs are drawn of $m_v\sigma_v$ against z. Original scales employed for Fig. 2.29:

$m_v\sigma_v$ 0.0125 unit cm^{-1}, z 5 m cm^{-1}, hence area scale is $5 \times 0.0125 = 0.0625\,\mathrm{m\,cm^{-2}}$.

Consolidation settlement:

centre $\rho_p = 8.80 + 0.0625 \times 10^3 = 550\,\mathrm{mm}$
periphery $\rho_p = 5.28 \times 0.0625 \times 10^3 = 330\,\mathrm{mm}$

EXAMPLE 2.14

A long length of fill is to be constructed with a base-width of 180 m, a top-width of 140 m, a height of 10 m and side slopes of 1 in 2. The unit weight of the compacted fill will be $20\,\mathrm{kN\,m^{-3}}$. The fill is to be placed on the surface of a stratum of normally consolidated clay 30 m thick overlying bedrock and the coefficient of compressibility of the clay is $4 \times 10^{-5}\,\mathrm{m^2\,kN^{-1}}$. The phreatic surface is at ground surface.

Calculate the settlement beneath the centre, beneath the edge of the top and beneath the edge of the base of the fill due to consolidation of the clay.

The loading of the fill is produced by superposing a triangular distribution of negative loading with $B = 140$ m on a triangular distribution of positive loading with $B = 180$ m as shown in Fig. 2.30, and the values of influence factor are determined from Fig. 2.5(e), employing the appropriate ratio z/B in each case. The distributions of resultant stress are shown in Fig. 2.30.

From the tabulated computations (Table 2.11) graphs are drawn of resultant stress σ_v against z. Original scales employed for Fig. 2.30:

σ_v 200 kN m^{-2} cm^{-1}, z 12.5 m cm^{-1},   hence   area   scale   is $200 \times 12.5 = 2500\,\mathrm{kN\,m^{-1}\,cm^{-2}}$

Consolidation settlement:

centre $\rho_p = 4 \times 10^{-5} \times 2.36 \times 2500 \times 10^3 = 236\,\mathrm{mm}$
edge of top $\rho_p = 4 \times 10^{-5} \times 1.96 \times 2500 \times 10^3 = 196\,\mathrm{mm}$
edge of base $\rho_p = 4 \times 10^{-5} \times 0.32 \times 2500 \times 10^3 = 32\,\mathrm{mm}$.

Table 2.11

z(m)	z/180	z/140	Beneath centre					Beneath edge of top					Beneath edge of base					
			+		−		Res.	+		−		Res.	+		−		Res.
			I_σ	σ_v	I_σ	σ_v	σ_v	I_σ	σ_v	I_σ	σ_v	σ_v	I_p	σ_v	I_σ	σ_v	σ_v
0	0	0	1.00	900	1.00	700	200	0.22	200	0	0	200	0	0	0	0	0
7.5	0.042	0.054	0.94	846	0.92	644	202	0.22	200	0.05	35	165	0.04	36	0.01	7	29
15.0	0.083	0.107	0.89	801	0.86	602	199	0.23	207	0.07	49	158	0.05	45	0.02	14	31
22.5	0.125	0.161	0.83	747	0.79	553	194	0.24	216	0.08	56	160	0.07	63	0.03	21	42
30.0	0.167	0.214	0.79	702	0.74	518	184	0.25	225	0.10	70	155	0.08	72	0.04	28	44

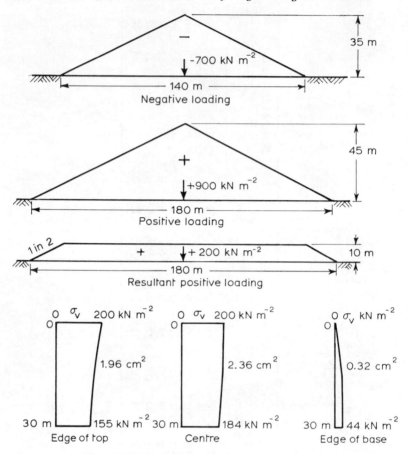

Figure 2.30 Vertical distribution of vertical stress.

EXAMPLE 2.15

At Newhaven, Sussex, 1 m compacted PFA fill was placed over 3 m stiff to soft clay which overlies at least 9.5 m silty sand. Relevant test data are as follows:

Fill.	Modulus of elasticity	$11 \times 10^4 \, \text{kN m}^{-2}$
Stiff to soft clay.	Average modulus of elasticity	$11 \times 10^3 \, \text{kN m}^{-2}$
	Average coefficient of compressibility	$5 \times 10^{-4} \, \text{m}^2 \text{kN}^{-1}$
	Average coefficient of consolidation	$10 \, \text{m}^2 \text{year}^{-1}$

Assume that a foundation, 3 m × 2 m in plan, is to be constructed imposing a uniform gross bearing stress of $100 \, \text{kN m}^{-2}$.
Two conditions are to be examined:

(a) Foundation constructed on surface of fill.
   (i) Calculate immediate settlement due to compression of the fill and the underlying soil.
   (ii) Calculate settlement due to consolidation of the 3 m thick layer of clay.

(iii) Calculate the time required for 90% consolidation to be attained if open layer conditions are assumed and if the relevant value of the time factor is 0.848.

(b) Fill omitted and foundation constructed on surface of clay.

  (i) Calculate immediate settlement.

  (ii) Calculate settlement due to consolidation of the 3 m thick layer of clay assuming a 2 in 1 dispersal of load.

  (iii) Calculate the time required for 90% consolidation to be attained if half-closed layer conditions are assumed and if the relevant value of the time factor is 0.848.

The clay is saturated and the value of Poisson's ratio can be taken as 0.5 but the value for the PFA fill is likely to be of the order 0.2. The Burmister factor (Fig. 2.23) relates to $v = 0.5$ for both the stiff layer and the deep underlying medium and to settlement at the centre of a uniformly loaded circular area. Hence the use of the factor in this case is an arbitrary and approximate means of taking into account an influence which is undoubtedly present. The equivalent diameter of the 3 m × 2 m foundation is $B = 2.76$ m: hence $h/B = 0.36$ and with $E_{r1}/E_{r2} = 10$, the value of the Burmister factor is 0.6.

(a) (i) From Equation 1.17, with shape factor of 0.94 for $L/B = 1.5$,

$$\rho_i = \frac{0.6 \times 0.94(1 - 0.2^2)100 \sqrt{6 \times 10^3}}{11 \times 10^3} = 12 \,\text{mm}$$

From the equation given in Fig. 2.21, with Janbu shape factor of 0.9 for $L/B = 1.5$ and $H/B \to \infty$,

$$\rho_i = \frac{0.6 \times 0.9(1 - 0.2^2)100 \times 2 \times 10^3}{11 \times 10^3} = 9 \,\text{mm}.$$

(ii) Although Fig. 2.23 is not strictly applicable in this case because the ratio h/B does not correspond and because v is not 0.5 for the stiff layer, study of the iso-stress lines suggests that the stress increment at the middle of the clay bed will be of the order $0.20q = 20 \,\text{kN m}^{-2}$. The Skempton–Bjerrum factor μ is taken as unity for this normally consolidated clay. Then

$$\rho_p = 5 \times 10^{-4} \times 20 \times 3 \times 10^3 = 30 \,\text{mm}$$

(iii) Time for 90% consolidation to be completed (Equation 2.5)

$$t_{90} = \frac{0.848 \times 1.5^2}{10} = 0.19 \,\text{year}.$$

(b) (i) As for (a)(i), from Equation 1.17

$$\rho_i = \frac{0.94(1 - 0.5^2)100 \sqrt{6 \times 10^3}}{11 \times 10^3} = 16 \,\text{mm}$$

and from the equation given in Fig. 2.21

$$\rho_i = \frac{0.9(1 - 0.5^2)100 \times 2 \times 10^3}{11 \times 10^3} = 12 \,\text{mm}$$

(ii) From Equation 2.15, with $\mu = 1$,

$$\rho_p = \frac{2.303 \times 5 \times 10^{-4} \times 100 \times 6}{1} \log_{10} \left\{ \frac{5 \times 3}{6 \times 2} \right\} \times 10^3 = 67 \,\text{mm}.$$

(iii) From equation 2.5 $t_{90} = \frac{0.848 \times 3^2}{10} = 0.76 \,\text{year}.$

It should be observed that settlement arising from the weight of fill is additional to settlement calculated above. Settlement due to compression of the silty sand has not been specifically taken into account although it is included arbitrarily in immediate settlement since this stratum has not been treated as a rigid basal medium below the clay. In fact stress values will be relatively low beneath the base of the clay and these are unlikely to lead to settlement exceeding 10 or 20% of the total. Clearly calculations for a problem of the nature treated above can yield only the order of magnitude of settlement.

EXAMPLE 2.16

Figure 2.31 shows the load–settlement records for plate-bearing tests at two locations A and B on compacted PFA fill at Newhaven, Sussex. The test plate was 0.3 m × 0.3 m and the tests were made in the bottom of pits 0.75 m × 0.75 m × 2 m deep. The fill is deep and the water table is below its base. No consolidation settlement is likely to be incurred and the test plate data can be assumed to be representative of the settlement characteristics of the fill.

(a) Comment on the discordance between the two graphs shown in Fig. 2.31.
(b) Determine the secant modulus of subgrade reaction at ground surface for the bearing stress range 0 to 200 kN m^{-2} if the depth factor is 0.98.
(c) Calculate the value of modulus of elasticity for the PFA for the bearing stress range 0 to 200 kN m^{-2}, assuming Poisson's ratio is 0.2.
(d) Estimate the settlement of a 1.0 m diameter concrete pier founded at a depth of 2.0 m in the fill and carrying a gross load of 175 kN if the depth factor is 0.75.

Aide mémoire:

$$\rho_i = \frac{I_p(1 - v^2)Bq}{E_r},$$

where I_p is 0.95 for the average settlement of a square plate and 0.79 for the settlement of a rigid circular foundation.

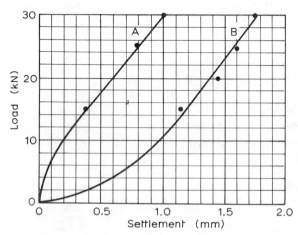

Figure 2.31 Load/settlement record for plate bearing tests on PFA fill.

(a) Excessive initial settlement for B probably due to bedding down since graphs are parallel when load exceeds 10 kN. Hence A can be assumed representative of load–settlement relationship at both locations.

(b) At 200 kN m^{-2} load on plate is 18 kN and corresponding settlement is 0.52 mm. Hence at ground surface $k_s = (0.98 \times 200)/(0.52 \times 10^{-3}) = 377 \times 10^3$ kN m^{-3}.

(c) Intrinsic value of $E_r = 377 \times 10^3 \times 0.95(1 - 0.2^2)0.3 = 103 \times 10^3$ kN m^{-2}.

(d) Gross bearing stress beneath 1.0 m diameter pier $q = 175/(\pi \times 0.5^2) = 222.8$ kN m^{-2}. At this stress, load on plate is $222.8 \times 0.3^2 = 20$ kN and corresponding settlement is 0.60 mm.
Settlement of foundation

$$\rho_i = \frac{0.60 \times 0.75 \times 0.79 \times 1.0}{0.98 \times 0.95 \times 0.3} = 1.27 \text{ mm, say } 1.3 \text{ mm.}$$

EXAMPLE 2.17

The following compression dial readings were observed at the times stated during a consolidation test on a specimen of clay initially 20.0 mm thick under a stress increment from 400 to 600 kN m^{-2}.

Time after primary compression completed (min)	0	60	120
Compression dial reading (mm)	6.867	6.814	6.781

Calculate the magnitude of the ultimate secondary compression ρ_{su} and the value of the rate factor K in the Equation 2.20

$$\rho_s = \rho_{su} (1 - e^{-Kt})$$

Also calculate the coefficient of secondary compression m_s for this stress range.

Data of the kind tabulated above are obtained most conveniently from a graph of compression against log t (Fig. 2.15b) since the secondary compression part of the curve can be inspected for the desired linear characteristic. If the compression readings are ρ_{s1}, ρ_{s2} and ρ_{s3} at times t_1, t_2, and t_3 respectively, where

$$(t_2 - t_1) = (t_3 - t_2) = \Delta t, \text{ then}$$

$$\rho_{s1} = \rho_{su} - \rho_{su} e^{-Kt_1} \qquad (2.21)$$
$$\rho_{s2} = \rho_{su} - \rho_{su} e^{-(Kt_1 + K\Delta t)} \qquad (2.22)$$
$$\rho_{s3} = \rho_{su} - \rho_{su} e^{-(Kt_1 + K2\Delta t)} \qquad (2.23)$$

From Equations 2.21 and 2.22 and from Equations 2.22 and 2.23 whence

$$\rho_{s2} - \rho_{s1} = -\rho_{su} e^{-Kt_1}(e^{-K\Delta t} - 1)$$
$$\rho_{s3} - \rho_{s2} = -\rho_{su} e^{-Kt_1} e^{-K\Delta t}(e^{-K\Delta t} - 1)$$

$$\frac{\rho_{s3} - \rho_{s2}}{\rho_{s2} - \rho_{s1}} = e^{-K\Delta t} \qquad (2.24)$$

Substituting in Equation 2.24

$$e^{-K60} = \frac{0.033}{0.053} = 0.623 = e^{-0.473}$$

whence $K = 0.0079$ min^{-1}.

The units of K must be correct for subsequent use in calculations for secondary settlement and may be more conveniently expressed, for example, as year^{-1}. Substituting in Equation 2.22

$$\rho_{su} = \frac{\rho_{s2}}{(1 - e^{-Kt_2})} = \frac{0.053}{(1 - e^{-0.473})} = 0.141 \text{ mm}$$

$$m_s = \frac{0.141}{20 \times 200} = 3.53 \times 10^{-5} \text{ m}^2 \text{ kN}^{-1}$$

EXAMPLE 2.18

From the following details relating to estimates of heave and settlement for a certain structure, construct graphs showing the progress of settlement for a period of 5 years from the commencement of excavation, giving relevant calculations in tabular form.

Excavation period: $\frac{1}{2}$ year. Construction period: $1\frac{1}{2}$ years. Heave at end of excavation 10 mm, completely eliminated when net stress attains zero during construction. Ultimate settlement components under maximum net stress: Immediate 20 mm, Consolidation 100 mm, Secondary 60 mm.

Net stress at founding level -100 kN m^{-2} when excavation completed and $+200 \text{ kN m}^{-2}$ when construction completed. The rate of unloading and loading can be assumed linear and no additional load will be imposed after completion. It has been estimated that 90% consolidation will be attained in 30 years if the maximum net stress is applied instantaneously. For secondary settlement the rate factor K is 0.0460 year^{-1}.

Immediate settlement is proportional to the applied load and in this case the graph of ρ_i against time is linear (Fig. 2.24). Reference to Table 2.3 enables Table 2.12 to be constructed since, from $c_v = \dfrac{T_v d^2}{t}$, $t \propto T_v$ or $t = T_v$ (30.0/0.848) when d and c_v are constant.

The basic graph $O_c AB$ of consolidation settlement is drawn for these values of ρ_p and time and the modified graph $O_c GH$ is constructed as outlined above (Fig. 2.24).

The basic graph of secondary settlement is derived from Equation 2.20 by substituting the data as follows:

$$\rho_s = 60(1 - e^{-0.0460t})$$

Reference is made to any standard tables of exponential functions or to an electronic calculator in order to produce the results shown in Table 2.13.

The basic graph $O_s J$ of ρ_s against time is drawn and the construction employed for consolidation settlement is repeated to define the modified graph $O_s K$ of secondary settlement. The graphs of ρ_s in Fig. 2.24 were drawn originally with ρ_s enlarged five times.

Table 2.12

Consolidation settlement ρ_p (mm)	Degree of consolidation U (%)	T_v	Time t (years)
90	90	0.848	30.00
40	40	0.126	4.45
30	30	0.071	2.51
25	25	0.049	1.73
20	20	0.031	1.10
15	15	0.018	0.64
10	10	0.008	0.28
5	5	0.002	0.07

Table 2.13

Time t (years)	Kt	e^{-Kt}	$(1 - e^{-Kt})$	Secondary settlement ρ_s (mm)
$\frac{1}{2}$	0.0230	0.9773	0.0227	1.362
1	0.0460	0.9550	0.0450	2.700
2	0.0920	0.9121	0.0879	5.274
3	0.1380	0.8711	0.1289	7.734
4	0.1840	0.8319	0.1681	10.086

2.5 SETTLEMENT OF FOUNDATIONS ON COHESIONLESS SOILS

2.5.1 The use of plate bearing tests for predicting settlement

Except where a foundation is less than about 1 m wide, the criterion of design on sand or gravel is settlement since, in order to limit this, the plan area must be of dimensions which lead to a factor of safety against bearing failure exceeding 3. Reliable estimates of settlement on cohesionless soils cannot be made from laboratory tests, because of difficulties in sampling and testing, and site tests are usually employed for this purpose. From theoretical, experimental and site data Terzaghi and Peck (1967) derived the equation

$$\rho = \rho_b \left(\frac{2B}{B+b} \right)^2 \tag{2.25}$$

where ρ is the settlement of a square footing of width B m under a given bearing stress q on an unsubmerged deep homogeneous bed of sand and ρ_b is the settlement of a test plate $b = 0.3$ m $\times$ 0.3 m under the same stress. When B is large, ρ tends to 4 ρ_b. Equation 2.25 caters for a normal increase in E_r with depth. Approximately, it is applicable also to continuous footings since the restraint afforded by a long footing to longitudinal movement of sand compensates for the compression of the sand at greater depths induced by the greater length of a continuous footing. Theory and experiment indicate that settlement on sand which is submerged to the underside of a footing located at or near the surface is approximately twice the settlement on unsubmerged sand. Care is required in the interpretation of results given by Equation 2.25, since the effect of variations in compressibility of deeper strata will not be apparent in the test observations owing to the small width of the loaded area. Because of the relatively high permeability of sands and gravels, water is readily expelled from the pores in these soils and settlement is generally finished when the structure is completed and loaded. The site test for cohesionless soils caters for settlement arising from elastic, plastic and consolidation deformations. Settlement of a foundation on sand can be expected normally to decrease with increase in founding depth. Excessive settlement may develop on sand due to vibrations.

Since the design width of a footing cannot be decided definitely until the amount of settlement associated with it has been estimated, the application of Equation 2.25 to a practical problem is a process of trial and error. In a site test, observations of settlement are made with a 0.3 m × 0.3 m plate covering a range of stress from zero up to, say, twice the estimated working stress and a graph is plotted of ρ_b against q. An initial design width B_1 is chosen arbitrarily and the net bearing stress q_1 under the given column load is calculated. The settlement ρ_b of the test plate under the stress q_1 is read on the graph and this value is substituted together with B_1 in Equation 2.25. If the settlement ρ is too great or too small, a second width B_2 is chosen, the stress q_2 is determined and the new value of ρ_b is substituted together with B_2 in Equation 2.25 to give a new value of ρ. If this value is not acceptable, the process is repeated until a suitable combination of footing width and settlement is found. Parry (1978) investigated settlement in sand estimated from plate-bearing tests.

Tschebotarioff (1951) records an average observed settlement of about 90 mm for a structure 30 m × 27 m in plan, founded on 20 m sand overlying very stiff clay, under a bearing stress of $350\,\mathrm{kN\,m^{-2}}$. The settlement of a 0.6 m × 0.6 m test footing was 9 mm under the same stress. The test and founding levels were both about 3 m below ground level and the corresponding Fox depth factors are 0.55 and 0.98. The estimated settlement ρ is derived from Equation 2.25 in the following way

$$\frac{9}{0.55}\frac{(0.6+0.3)^2}{(1.2)} = \rho_b = \frac{\rho}{0.98}\frac{(27+0.3)^2}{(54)} \quad \text{where } \rho = 35\,\mathrm{mm}.$$

Since the water table was about 6 m beneath the foundation, it is reasonable to expect the settlement to be of the order 65 mm. The discrepancy of 25 mm between estimated and observed values may be accounted for partly or wholly by the fact that the resistance recorded in a penetration test was much lower at a depth of 6 m than between 3 and 5 m although at greater depths the resistance increased steadily. Consequently, the test footing probably did not yield representative data for estimating settlement of the structure. Obviously, site plate tests should be accompanied by penetration tests in order that estimated settlement can be adjusted for variations in characteristics of the soil. Prediction of settlement on sand has been studied by Jorden (1977) and Arnold (1980).

2.5.2 The use of penetration tests for predicting settlement

The settlement of shallow foundations in sand and gravel is commonly predicted by the modified Terzaghi–Buisman semi-empirical formula

$$\rho = \sum \frac{H\sigma_0'}{E_r}\log_e\frac{\sigma_0'+\Delta\sigma'}{\sigma_0'} \tag{2.26}$$

where H is the thickness of any given layer of soil, σ_0' is the effective overburden stress at the middle of the layer, $\Delta\sigma'$ is the increment in effective stress at the middle of the layer generated by the foundation load, and E_r is the equivalent modulus of elasticity of the soil.

Since settlement of a foundation on sand or gravel is sensibly completed when the working load is first applied, the value of E_r should cater for elastic, plastic and consolidation deformations. The equivalent value of E_r is derived empirically from a static cone penetration test which yields

$$E_r = 2q_c \qquad (2.27)$$

where q_c is the static cone resistance in $\mathrm{kgf\,cm^{-2}}$. The coefficient 2 was derived by Schmertmann but earlier Meyerhof suggested the value should be 1.9 and the original value of Buisman was 1.5. Equation 2.27 can be employed to calculate settlement on silts for which the coefficient is taken as 1.5 and the results are treated with caution.

If the standard penetration test is employed in a site investigation instead of a static penetration test, then further correlation is required to arrive at a value for E_r. This appears to depend largely on particle size distribution and, from consideration of the results of several investigations, Thorburn (1970) produced a graph from which are extracted the coefficients given in Table 2.14: then $q_c = C_N N$. It is axiomatic that empirical correlations should be used with caution and is particularly true in this case because of the kind and extent of the correlations.

If standard penetration tests indicate that a deep bed of cohesionless soil is relatively homogeneous, then the allowable bearing stress q_a can be determined by reference to Fig. 2.32, which was constructed by Terzaghi and Peck (1967) largely on empirical evidence. For a given penetration test value N and a given width of foundation B, the corresponding allowable bearing stress q_a should lead to a settlement of about 25 mm. If the water table is at or above the base of the foundation, the value of q_a derived from Fig. 2.32 should be reduced by 50%. If the water table is at a depth $z_w < B$ beneath the foundation, the chart value of q_a should be reduced by $[(B - z_w/B)] \times 50\%$. Formerly the penetration test did not yield reliable measures of the degree of compaction of gravels and Terzaghi and

Table 2.14 Ratio of static cone resistance to standard penetration resistance

Soil fraction	Coefficient C_N
Medium silt (part extrapolated)	1.8 to 2.6
Coarse silt	2.6 to 3.3
Find sand	3.3 to 4.0
Medium sand	4.0 to 4.7
Coarse sand	4.7 to 5.5
Fine gravel	5.5 to 6.2
Medium gravel (extrapolated)	6.2 to 7.0
Coarse gravel (extrapolated)	7.0 to 7.7

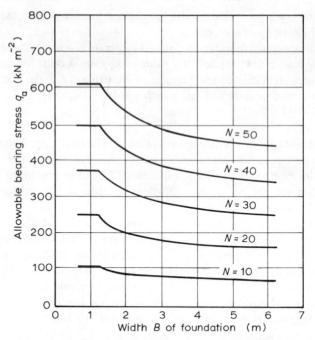

Figure 2.32 Correlation between standard penetration resistance N blows/0.3 m, foundation width and allowable bearing stress (after Terzaghi and Peck, 1967).

Peck suggested that the *in situ* bulk density should be determined in trial pits and the value of q_a taken equal to the value for a sand possessing the same degree of compaction. However, more reliable values of N are considered to be attainable in gravels today when the solid cone is employed in the SPT.

Since the values of N recorded in a test are influenced by stresses induced by the weight of overburden, Sutherland (1963) suggested the following procedure for estimating q_a.

(a) Calculate the effective overburden stress σ_0' corresponding to the depth at which the standard penetration test was made.
(b) On Fig. 2.33(a) locate the intersection a of the observed penetration value N_0 with σ_0'. Draw a vertical line through a to intersect the Terzaghi and Peck curve at b.
(c) Draw a horizontal line through b to give the equivalent penetration value N_e.
(d) Assume a trial value for B and determine q_a on Fig. 2.32 corresponding to N_e. If the load which can be carried on a foundation of these proportions combined with the derived value of q_a does not correspond with the required load, further trial values of B must be assumed until this condition is satisfied.

Whereas Sutherland employed effective overburden stress as a parameter, Thorburn (1963) constructed Fig. 2.33(b) with depth of overburden as a

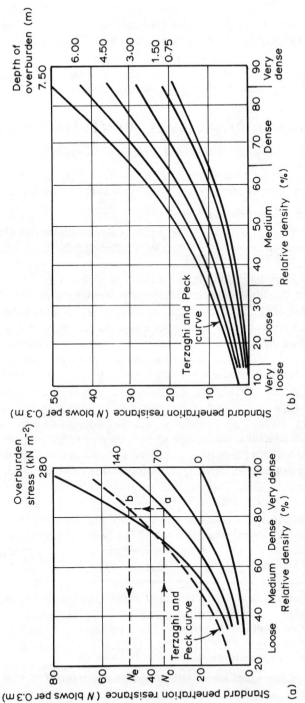

Figure 2.33 Relationships between standard penetration resistance, relative density, depth of overburden and effective overburden stress. (a) After Sutherland (1963); (b) after Thorburn (1963).

parameter. In this chart it is assumed that the Terzaghi and Peck curve corresponds with an effective overburden stress of $144\,\mathrm{kN\,m^{-2}}$. The procedure suggested by Sutherland can be followed also with Fig. 2.33(b).

Rodin *et al.* (1974) reviewed the use of penetration tests for predicting settlement and collected an extensive list of references. The compressibility of rock fill has been studied by Sowers *et al.* (1965).

2.6 THE EFFECT OF SETTLEMENT ON SUPERSTRUCTURES

2.6.1 The general effects of settlement on superstructures

Settlement may be due to a number of causes other than those associated with normal compression of soil or rock. They include the following.

(a) Insufficient support or loss of ground during excavation in the vicinity of an existing structure. For minor underground excavations the technique of pipe-jacking has been employed as a means of reducing subsidence (Clarkson and Ropkins, 1977).
(b) Overloading of the ground in the vicinity of a structure.
(c) Subsidence due to mining or other underground excavation. Subsidence arising from coal mining is treated in Chapter 9. Winning other materials, such as metalliferous ores, evaporites and stone for masonry and aggregate, can also lead to subsidence. Large ore bodies are commonly irregular in shape and mining these has sometimes led to sudden, irregular subsidence of several metres. Alternatively, mining veins may lead to no more than minor local subsidence. In Cheshire, England, mining shallow, thick beds of rock salt has caused violent and erratic subsidence, sometimes as much as 10 m. Mining was succeeded by pumping natural brine, occasionally drawn over distances of several kilometres, and then by controlled pumping in which water is forced into the bottom of the bed and drawn off from the top. When pumping natural brine, the magnitude and character of subsidence depend on the proximity of adjacent brine runs and their depth. Salt is won in a number of localities in North America by mining and by brine pumping but the thick beds are at greater depths than in Cheshire and subsidence is less. Other evaporites, such as anhydrite in County Durham and gypsum in Sussex, are often worked by mining and the characteristics of subsidence are controlled by the factors which govern subsidence caused by coal mining. Mining stone for masonry is an industry which commenced in times at least as ancient as the Pyramids of Giza, Egypt. If mining of any kind is beneath open country, then subsidence may occur almost unnoticed, but beneath urban areas the effects on buildings call attention to the phenomenon. Extensive mining for gypsum of Eocene age has led to subsidence in Montmartre in the centre of Paris.
(d) Lowering of the water table. The progressive settlement of London is due partly to abstracting water, mainly from the Chalk beneath the London Clay,

through boreholes within the city. Little more than a century ago, artesian water flowed from boreholes in the centre of the London Basin but today piezometric levels are some 100 m or more beneath ground surface. The considerable settlement in Mexico City is due partly to withdrawal of water from sands interbedded with highly compressible clays. Serious settlement caused by pumping from aquifers is encountered in California, Nevada, Texas and other locations in the USA. Where the areas concerned cover tens or hundreds of square kilometres, subsidence may attain 4 m or more and may be accompanied by faulting involving displacements of as much as 0.5 m or more. The resulting problems include flooding, reduction of freeboard of dams, harbours, coastal defences, rivers and canals, reduction in headroom of bridges over water and change of gradient in pipelines. Flooding leads not only to health hazards and loss of amenity but to other problems such as hydrostatic uplift and flotation of basements and tanks. Horizontal compression induced in the surface beds of the sagging ground may lead to high passive earth forces on buried structures.

Remedial measures include the palliative of reducing the rate of abstraction, artificial recharge to raise the water table and installation of float-controlled drainage systems around structures to alleviate the effects of flooding.

As a result of lowering the water table, timber piles may rot, thus leading to settlement of superstructures. Drainage of peat leads to marked shrinkage and sometimes also decomposition. Cessation of compressed-air working in tunnelling may allow water to drain towards the tunnel, thereby leading to reduced piezometric levels.

Analogous with the fall in pore-pressure arising from lowering of the water table is the reduction caused by the abstraction of oil and gas. Settlement of 1 to 4 m or more can be anticipated over a period of 10 years. Notable examples of oil-field settlement are Long Beach Harbour, California, and Galveston–Houston, Texas. Normally, groundwater from the periphery of the field infiltrates to replace the oil and gas but, if the movement is restricted by faults, dykes or low-permeability beds, the consequent reduction in pore-pressure leads to settlement. The problems caused by oil-field settlement, and also remedial measures, are similar to those caused by abstraction of water.

(e) Vibrations in the vicinity; more particularly with loose sands. These may be created by road and rail traffic and by excavation and construction plant, including tunnelling equipment.

(f) Earthquakes; either general lowering of an area or local settlement.

(g) Failure of the foundation structure; either from faulty design or construction or deterioration of materials.

(h) Alteration of properties of the ground, in particular fill, due to natural or artificial causes.

(i) Instability of hillside on which structure is founded. Marked settlement may occur at any time following construction, sometimes after many years, and often it may be initiated by relatively minor excavations or constructions in the vicinity.

The purpose of this section is to consider in more detail the effects of normal settlement on the superstructure. If the settlement of a building is completely uniform then distress will be experienced only in the service pipes and cables connected from the external mains to the building. Even so, the relative displacements may not necessarily be abrupt since the ground surrounding the building is generally brought down slightly by the process of settlement. If differential settlement develops between different parts of the structure, then architectural or structural damage may be experienced. Differential settlement arises from variations in the load distribution to the foundations of a structure and from the characteristic basin-form of settlement associated with a raft or independent foundations with overlapping bulbs of stress. It may arise also from variations in strata beneath a site, from the presence of old foundations or from pre-consolidation of soil by earlier structures. In the case of underground structures, such as car parks, the gross foundation load may be less than the weight of overburden excavated and the resulting negative net stress may lead to heave of several centimetres. The settlement of a foundation may be increased by the downward drag of fill placed against it compacting within itself. Fill should be well compacted when placed to reduce this effect.

It is often instructive to estimate the effect on bending moments and shearing forces of a differential vertical settlement of 25 mm between adjacent columns of framed structures or between supports of continuous structures. In addition, for portals or arches, the influence of a horizontal displacement of 25 mm at an abutment can be investigated and, if the structure is fixed-ended, the effect of $1°$ rotation at an abutment. Arvidsson (1975), Coull (1974), Coull and Chantaksinopas (1974), Leonards (1949), Meyerhof (1947), Morris (1966), Majid and Cunnell (1976) and Majid and Rahman (1982) have presented methods of analysis to estimate the influence of settlement on structures: these problems constitute part of the wider field of study known as soil/structure interaction.

The ability of a steel structure to accommodate itself to differential settlement is due to its high ductility, whereas the limit of extensibility of concrete in tension is of the order 0.01% and in compression 0.15%. If it is impossible to predict the relative directions of movement of footings subject to differential settlement, then the directions of moments in the members of the frame due to this cause are not known. Under these circumstances a reinforced concrete design catering for settlement moments requires reinforcement in both faces of members. The aim should be to secure yield of the steel in tension as opposed to crushing of the concrete, thereby creating a plastic hinge which relieves moments. The steel in the compression zone assists slightly in preventing crushing of the concrete. Studies by Wood (1952, 1958), Thomas (1953), Holmes (1961, 1963), Mainstone (1972, 1974) and Smolira (1973), confirm that a framed structure is considerably strengthened and stiffened by panel filling, even though it may be pierced by door and window openings, and flexural stresses may be only 30% of the stresses computed for the bare frame. If a building frame is designed with a factor of safety of three, it follows that cladding will effectively increase this factor to, say, six. New approaches to design, which take into account the composite nature of buildings, are at present being evolved and when eventually such methods are employed with a specified load factor there will be no excess of safety. Consequently, designs based on these analyses must be accompanied by thorough site investigations and refined methods of estimating bearing capacity and settlement.

Although not commonly regarded as normal settlement, it is of interest to examine briefly the general settlement of an area arising from changes in groundwater level. General settlement can lead, for example, to inundation during floods and, when irregular, to the creation of backfalls in sewers. Consider first a bed of sand 6 m thick overlying a bed of clay 10 m thick overlying bedrock, with the water table initially 1 m below ground surface but subsequently falling to a level 5 m below the surface. Assume that in both cases the hydrostatic conditions in the ground are stabilized. If the unit weight of moist sand is $18 \, \text{kN m}^{-3}$ and of saturated sand is $20 \, \text{kN m}^{-3}$ and assuming the unit weight of water is $10 \, \text{kN m}^{-3}$, the effective stress on the top of the clay is initially $[18 \times 1 + (20 - 10)5]$ $= 68 \, \text{kN m}^{-2}$ but becomes $[18 \times 5 + (20 - 10)1] = 100 \, \text{kN m}^{-2}$ when the level falls. Assuming that the clay is of alluvial or estuarine origin and normally consolidated, a reasonable value for m_v is $0.0005 \, \text{m}^2 \, \text{kN}^{-1}$. The increase in effective stress would lead to consolidation settlement of $(0.0005 \times 32 \times 10)$ $= 0.16 \, \text{m}$.

Now consider the case of a bed of saturated clay 15 m thick overlying bedrock and assume that the piezometric level is initially 1 m below ground surface but falls to 5 m. It is assumed that, because of capillary rise and gravitational water, the clay above the piezometric level is always saturated at practically constant water content. If the unit weight of saturated clay is $20 \, \text{kN m}^{-3}$, the initial effective stress at a depth of $(20 \times 1 + (20 - 10)4) = 60 \, \text{kN m}^{-2}$ and becomes subsequently $(20 \times 5) = 100 \, \text{kN m}^{-2}$. With $m_v = 0.0005 \, \text{m}^2 \, \text{kN}^{-1}$, the increase in effective stress leads to consolidation settlement of $(0.0005 \times 40 \times 10) = 0.20 \, \text{m}$. Consolidation would occur also between 1 and 5 m depth, leading to settlement of $(0.0005 \times 20 \times 4) = 0.04 \, \text{m}$. The total settlement would thus be $0.24 \, \text{m}$.

These simple calculations give some idea of the effects which can be experienced. Where changes in water level are comparatively rapid – for example, as in tides and floods – the permeability of clay is insufficient to permit the immediate stabilization of hydrostatic conditions and the resulting increases and decreases in settlement are more complex.

2.6.2 Classification of settlement effects

Several studies have been made of the tolerance of structures to deformation arising from differential settlement. The results of one investigation were published by Skempton and MacDonald (1956) who divide damage in frame buildings arising from settlement into three categories:

(a) Structural: involving only the frame (beams, stanchions) and likely to occur when the angular distortion $\delta/L > 1/150$ where L is the span and δ the differential settlement between adjacent columns.
(b) Architectural: involving only the panel walls or floors and likely to occur when $\delta/L > 1/300$.
(c) Combined structural and architectural.

Table 2.15 summarizes the limits suggested by Skempton and MacDonald. In view of uncertainties in the determination of settlement contours, it is suggested that a factor of safety of not less than 1.5 should be applied to angular distortion, giving a limit of about 1/500 and this may be reduced to 1/1000 under exceptional conditions. For differential settlements a factor of safety of 1.25 may be sufficient, giving limits of 38 and 25 mm on clay and sand, respectively. Terzaghi and Peck (1967) state that most ordinary structures, such as office buildings, flats or factories, can stand a differential settlement between adjacent columns of 18 mm, and this corresponds closely with $\delta/L = 1/300$ when $L = 6$ m, a common spacing for columns. It is considered that on sand, differential settlement of 18 mm will not be exceeded, through local variations in compressibility of the soil, if the settlement of the largest footing is limited to 25 mm, even if the footings are of different sizes and at different depths.

The values for maximum settlement in Table 2.15 do not apply to buildings where the settlement arises essentially from the consolidation of deep-seated clay layers, nor to a building which is erected adjacent to an existing structure to which it has to be bonded or aligned, nor where settlement is due largely to general filling and not to the building itself. The allowable maximum settlement for a box-foundation or a chimney or water-tower foundation on clay can exceed 102 mm, provided that such settlement does not encourage tilting or other undesirable consequences. The presence of an overhead travelling crane gantry may impose a rigorous limit on differential settlement although a maximum

Table 2.15 Damage limits for load-bearing walls or for panels in traditional-type frame buildings.

Criterion	Isolated foundations		Rafts
Angular distortion		1/300	
Greatest differential settlement	Clays	44 mm	
	Sands	32 mm	
Maximum settlement	Clays	76 mm	76–127 mm
	Sands	51 mm	51–76 mm

inclination of 0.003 for the rails is acceptable in the USSR. Lightly clad factory sheds can generally withstand greater differential settlement than, say, an office block where cracking of panel filling would be objectionable. The assessment of damage to walls and ground-floor slabs of low-rise buildings is treated in Building Research Establishment Digest 251 (BRE,1981).

In the discussion on the paper by Skempton and MacDonald, K.Terzaghi stated that the conclusions were sweeping and implied that the problems were over-simplified. Furthermore, W.H. Ward suggested that ordinary records of settlement were inadequate for determination of criteria of distress in buildings and that measurements of curvature were more logical than angular distortion, but Grant *et al.* (1974) consider that the advantage is marginal. Nevertheless, data published by Polshin and Tokar (1957), relating to construction in the USSR, are not very different from those of Skempton and MacDonald although they are more detailed. Field observations of settlement have been reported by Feld (1965) and McKinley (1965). Grant *et al.* (1974) investigated the behaviour of a large number of buildings, extending and broadly confirming the findings of Skempton and MacDonald (1956).

2.6.3 Methods of reducing differential settlement

Although theoretically a structure can be designed to resist stresses imposed by irregular displacements, the prediction of differential settlement is often hazardous and it may be preferable to endeavour to reduce differential settlement or to attain sensibly uniform settlement of all parts of the structure. The following methods can be employed to reduce differential settlement:

(a) Adjustment of design loads on foundations: not often practicable. Load control devices for varying loads in service have been discussed by Clark *et al.* (1973).
(b) Adjustment of proportions or depths of individual foundations: generally effected by trial and error.
(c) Transfer of loads to deeper, less compressible strata by piers or piles.
(d) Provision of a rigid raft foundation: the raft must be sufficiently strong to resist stresses imposed by non-uniform distribution of bearing stress.
(e) Increase depths of foundations so that net stress is reduced by increased overburden stress: this reduces total, and hence also differential, settlement. Generally this requires a hollow foundation. In addition, founding depth can be varied beneath parts of a structure carrying considerably different loads in order to attain sensibly equal net loads. The component blocks of the Shell Building, London, were founded at different depths to reduce differential settlement (Williams,1957).
(f) Provision of facilities for jacking superstructures. This was arranged on New Waterloo Bridge, London: the average 25 mm heave which developed during excavation was eliminated as subsequent settlement (Buckton and

Cuerel, 1943). At some sites, where considerable settlement has been anticipated with shallow foundations, provision has been made for jacking to maintain levels and this has proved cheaper than founding at greater depths. In the case of relatively light structures, such as tanks, mud jacking can be employed – in this process pulverized fuel ash grout is injected immediately beneath the structure in stages. Peynircioglu (1965) describes provisions for jacking and injection or mortar beneath a foundation liable to excessive settlement and tilting.

(g) Adding kentledge or fill to lightly loaded areas in or immediately adjoining the structure.

(h) Precompression of strata (see below).

(i) Excavating highly compressible material and replacing it with well-compacted granular fill.

A conference was held at Evanston, Illinois, in 1964 on the design of foundations for control of settlement and has been reported in *Journal of the Soil Mechanics and Foundations Division (ASCE)*, **90** (1964, Sept.) and **91** (1965, March). In order to reduce settlement of oil tanks in service and to permit relevelling if necessary, the tanks can be preloaded with water for periods ranging between 1 and 12 months (Darragh, 1965; Penman and Watson, 1965). The principle of precompression of strata has been applied also to highway and runway construction and to building sites by placing excess fill in depths of 0.3 to 3 m for periods of 3 months to 4 years, after which it is removed (Aldrich, 1965: Jonas, 1964: Kleiman, 1964: Sowers, 1964). Halton *et al.* (1965) describe the pre-consolidation of soil at Philadelphia by pumping to lower the phreatic surface, followed by the application of a vacuum. The provision of sand drains accelerates consolidation settlement, although in some cases the technique shows little advantage. A symposium on vertical drains has been presented by Wood (1981). Pore pressure and settlement observations during the preloading period are essential for accurate settlement predictions (Roberts and Darragh, 1962).

2.6.4 Observations of settlement

The datum to which settlement of a structure is referred should be absolutely static and this condition is not readily secured at many sites. Ideally, the datum should be established on rock beyond the influence of the load from the structure or other disturbing factors such as adjacent recently constructed buildings, tidal effects and mining subsidence. For the Shell Centre, London, a datum was located at a depth of 75 m on the Chalk (Williams, 1957). Boreholes were sunk to basement floor level prior to excavation, in order to record both heave and settlement, and level points were established on column bases and, later, in the ground floor. The establishment of deep bench marks in clays and permafrost has been discussed by Bozozuk *et al.* (1962). In order to facilitate observation of differential settlement, consideration should be given to installing a permanent water-tube levelling system within a structure. High degrees of accuracy can be attained with this type of equipment over distances exceeding 30 m.

Many comparisons between observed and calculated settlement have been reported: useful earlier studies are those of Cooling and Gibson (1955), MacDonald and Skempton (1955), Skempton *et al.* (1955) and Deere and Davisson (1961), and more recent investigations have been published by Hooper (1973), Cooke *et al.* (1981) and Foott and Ladd (1981). Commonly, the theoretical settlement beneath a large building is dish-shaped but observations indicate that the iso-settlement lines are often displaced laterally and that the maximum settlement may not be near the centre. This appears to be due partly to the rigidity of the structure and partly to variations in conditions over the site. Discrepancies between calculated and observed settlement may be due also to disturbance in sampling, to differences between assumed and actual load distributions on the structure, to the fact that stresses imposed on the soil do not conform to those derived from the theory of elasticity and to the fact that drainage may be largely horizontal instead of vertical. In addition to settlement, significant horizontal movements have been recorded by Hardy and Ripley (1961). The measurement of very slow movements in large structures – in particular the Tower of London and Rochester Bridge – has been discussed by Cox and Mitchell (1952).

Whereas formerly, attention has been concentrated on vertical movements beneath foundations, during the last decade predictions and observations have been made of horizontal displacements in and around excavations and in the vicinity of structures. Standard solutions based on the theory of elasticity, assuming simplified conditions, may be adequate for many predictions but it is usual to employ powerful methods of computation for large and important projects. These methods include finite element and boundary element techniques (Burland, 1978; Butterfield and Banerjee, 1981). Techniques and examples of observation of stresses and deformations have been discussed by several authors (Penman and Charles, 1971; Cole and Burland, 1972; Burland and Moore, 1973; Lambe, 1973; Ward and Burland, 1973; Burland and Hancock, 1977; Burland *et al.*, 1977; Simpson *et al.*, 1979; Furley and Curtis, 1981; Geddes, 1981; O'Rourke, 1981; Toombs *et al.*, 1982; Chard and Symons, 1982; Frischmann *et al.*, 1983; MacLeod and Paul, 1984). Studies of movement around excavations retained by a system of earth reinforcement have been reported by Shen *et al.* (1981a, b).

2.7 FLOW OF WATER TO EXCAVATIONS AND BOREHOLES

2.7.1 Permeability

Turbulent flow rarely occurs in water moving through pores in soil or rock, except occasionally in gravels close to a borehole or adjacent to sheet piling. Calculations for movement of ground-water are normally based on the assumption of laminar flow, to which Darcy's Law is applicable, and the

Table 2.16 Variation of kinematic viscosity of water with temperature

$t°C$	0	5	10	15	20	25	30
$\mu\,g\,mm^{-1}\,s^{-1}$	0.00179	0.00152	0.00131	0.00114	0.00101	0.00089	0.00080

Table 2.17 Permeability of soils

Soil	Permeability $k\ (mm\,s^{-1})$
Gravel	$10^3–10$
Sand, sand-gravel mixture	$10–10^{-2}$
Fine sand, silt, stratified clay, sand-silt-clay mixture	$10^{-2}–10^{-6}$
Unweathered homogeneous clay	$10^{-6}–10^{-8}$

discharge q through a cross-sectional area A of soil or rock is given by $q = kAi$ where k is the permeability of the medium in relation to water and i is the hydraulic gradient. Permeability can be expressed as a generalized coefficient $K = k\,\mu/\gamma_w$ which depends solely on the medium and is independent of the fluid, where μ is the kinematic viscosity and γ_w is the density of water.

Over the working range of temperature for water, γ_w can be considered constant but variation in μ may cause 10 to 20% change in k. A laboratory value k_t at $t°C$ should be reduced to normal standard field temperature of 10°C, thus $k_{10} = k_t\mu_t/\mu_{10}$, where values of μ are taken from Table 2.16.

The void ratio of a remoulded sample of cohesionless soil can be gradually reduced in a constant head permeameter by initially compacting loosely and gently tapping the permeameter after each set of readings. A sensibly linear graph of e against $\log_{10}k_{10}$ can then be plotted. If the field value of e is calculated from *in situ* bulk density tests and a value of specific gravity G of solid particles is estimated or known, the field value of k can be derived from the graph. However, disturbance during sampling is likely to alter the shape and arrangement of pores even when the soil is re-compacted at the field value of e. Gas trapped in the voids reduces the magnitude of k: a rise in temperature or decrease in pressure causes some gas to come out of solution.

It is necessary to distinguish between primary permeability, involving flow of water through pores in soil and rock, and secondary permeability, involving flow through fissures. Very little research has been carried out on flow in fissures and this section is concerned mainly with primary permeability. Typical ranges of value of primary permeability of soils are given in Table 2.17 and of mass permeability of rocks, taking account of both pores and fissures, in Table 2.18 (Anon, 1972). The permeability of a given sand in a loose state can be expected to be 2 to 4 times the value in a dense state. Studies of permeability have been

Table 2.18 Permeability of rock masses

Rock mass description	Permeability	
	Term	$k\,(mm\,s^{-1})$
Very closely to extremely closely spaced joints	Highly permeable	10^3–10
Closely to moderately widely spaced joints	Moderately permeable	10–10^{-2}
Widely to very widely spaced joints	Slightly permeable	10^{-2}–10^{-6}
Unjointed, solid	Effectively impermeable	$< 10^{-6}$

reported by Loudon (1952) and Amer and Awad (1974) and various aspects of turbulent and laminar flow through porous media have been investigated by Anandakrishnan and Varadarajulu (1963), Ward (1964), Rumer and Drinker (1966) and Arbhabhirama and Dinoy (1973). Field tests for permeability have been discussed by Golder and Gass (1962), Mansur and Dietrich (1965), Gibson (1966, 1970) and Wilkinson (1967, 1968). The results of field tests in chalk have been reported by Haswell (1969). False measurements of permeability arising from hydraulic fracturing of soils and rocks under high pressure in field testing have been investigated by Bjerrum *et al.* (1972). Improvements in the Lugeon or Packer test have been described by Pearson and Money (1977).

Terzaghi (1943) and Evans (1962) show that the permeability of a group of beds, individually homogeneous and isotropic, is always greater parallel (k_p) to the bedding than normal (k_n) to it (Example 2.24). Since all sediments bear a measure of stratification, isotropic permeability probably does not exist in nature, although for simplicity in computation this condition is commonly assumed. In some sediments the difference between k_p and k_n may not be great. However, the presence of a seam of clay 1 mm thick in a bed of sand may be sufficient to make the permeability ratio k_p/k_n large. The permeability of stratified soils has been discussed by Kenney (1963) and Chan and Kenney (1973). Nixon (1954) states that the most reliable estimate of ground-water lowering conditions in laminated soils is formed by detailed examination of a continuous series of undisturbed samples and gives illustrations of specimens prepared in the manner of a soil monolith.

In some alluvial and estuarine deposits the value of k_n is found to be greater than that of k_p, in spite of well defined stratification. Careful examination of the sediments reveals root holes and animal burrows, often filled with relatively coarse material, and these features influence mass permeability characteristics. In some cases the original fine stratification may be largely destroyed by bioturbation. The permeability of other sediments may be influenced by the presence of fissures.

Owing to variations in conditions over a site and to discrepancies arising from the factors discussed above, computations based on values of permeability derived from laboratory or field tests may be considerably in error and the accuracy may be further reduced by broad assumptions made in analyses of practical problems. Frequently, quantitative estimates can be made only in this way but the degree of accuracy attained must be carefully assessed.

2.7.2 Flow nets

Examples of flow nets are shown in Fig. 2.34. The construction of a flow net for two-dimensional flow involves the solution of the Laplace equation for two-dimensional steady conditions. Computation can be effected, for example, by application of relaxation or finite element techniques. If a number of similar problems are to be studied consecutively, flow patterns can be studied in the laboratory with models in a seepage tank. Dixon (1967) has investigated the use of a silicone to eliminate capillary rise in seepage models. For the electrical analogy, electrolytic tanks have been replaced to some extent by the introduction of special conducting paper.

The graphical technique of trial and error is adequate for many problems although skill is required in the construction of the flow nets. Since permeability may very considerably over a site, more accurate constructions are rarely

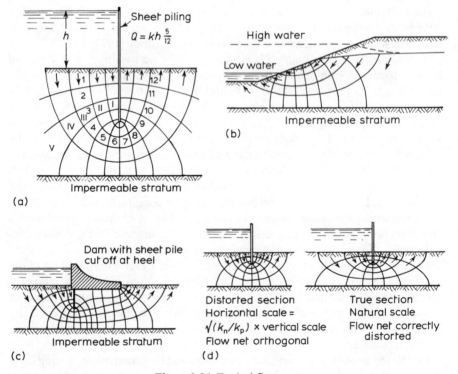

Figure 2.34 Typical flow nets.

justified. The hydraulic boundary conditions are generally as follows. Horizontal ground surfaces on each side of a structure often constitute equipotential surfaces. The base and buried sides of a structure are flow lines and the contact between an aquifer and an aquiclude is also a flow line. The sloping face of a bank (Fig. 2.34 b) and the vertical drainage layer behind a retaining wall are neither flow lines nor equipotential lines. Since water flows in the direction of the steepest hydraulic gradient, the flow lines always intersect the equipotential lines at right angles. The profile of a curved free water surface, for example, in an earth dam, can be established precisely by lengthy relaxation computations but it can be defined approximately by the L. Casagrande construction, which is described in most text books on soil mechanics. Problems involving a free water surface have been treated by Taylor and Brown (1967).

Where the soil is anisotropic, a section is drawn with the horizontal scale $\sqrt{(k_n/k_p)}$ of the vertical scale and the flow net is constructed by treating the soil as isotropic (Fig. 2.34 d). The net is then transferred to a section drawn to a natural scale in which the horizontal dimensions of the net, taken from the distorted section, are multiplied by $\sqrt{(k_p/k_n)}$. The total seepage, Q, is

$$Q = h \frac{N_f}{N_e} \sqrt{(k_p k_n)}$$

where N_f is the number of flow channels and N_e is the number of fields enclosed in one entire channel.

The construction of flow nets is facilitated by electrical analogue plotters used in conjunction with special conducting paper. A cross-section of the structure and the strata is drawn on the paper, the relatively impermeable zones, such as an aquiclude or concrete foundations or sheet piling, are cut out and the paper is pinned to a drawing board. Two fine-wire electrodes are fixed to the paper by metallic paint along two equipotential lines – commonly boundary lines. The plotter comprises basically a potentiometer. It is set to record any simple fraction of the full potential difference across the electrodes and the equipotential line conforming to this fraction is drawn on the paper by a single probe carrying a 2H pencil lead. In this way a series of equipotential lines can be drawn and the flow lines can then be sketched to produce an orthogonal net. Alternatively, the flow lines can be traced using the plotter in an electrical field imposed by electrodes fixed along any two known flow lines, such as the base or side of the structure and the interface between strata. The equipotential lines can then be sketched or both equipotential and flow lines can be traced independently by the plotter and the results superposed. Errors should be lightly marked in pencil since erasure damages the paper. It is desirable to provide wide marginal areas to satisfy the theoretical assumption of a medium extending to infinity in one or more directions and thus avoid excessive distortion in plotting. Since the resistivity in one direction on the paper may be as much as 25% different from the resistivity in a direction normal to it, this method of plotting flow nets is not as accurate as the electrolytic tank but it is satisfactory for many problems. The resistivity of the paper can be increased by piercing holes in it with a needle or punch and flow nets for multi-layer problems can thus be studied. The required density of the holes must be found by experiment and the maximum increase in resistivity which can reasonably be attained is about ten times.

2.7.3 Stability of the floor of an excavation under the influence of hydrodynamic pressure

Instability of the floor of an excavation may be local or general. Piping is a form of local instability. The smallest square at exit on the flow net locates the zone in which piping is most likely to develop and the exit hydraulic gradient is determined to ascertain whether or not the critical gradient is exceeded. A critical condition is generally found adjacent to sheeting and, should it develop in the field, collapse of the sheeting may follow owing to the removal of lateral support.

However, Terzaghi (1943) states that model tests demonstrate more extensive, but still local, uplift may occur – which can conveniently be termed a blow – when a mass of soil abcd (Fig. 2.35) tends to rise. A critical condition obtains when the upward force U_e of excess hydrodynamic pressure on the base of the prism abcd is just equal to the submerged weight W_b of the soil particles in the prism. Since the vertical effective stress is then zero, the horizontal effective stress is also zero and, theoretically, frictional resistance on the sides of the prism should

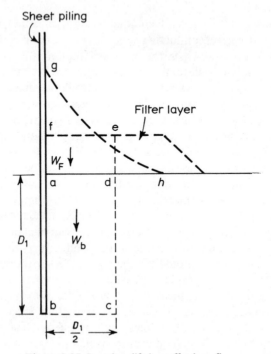

Figure 2.35 Local uplift in cofferdam floor.

be nil. The excess hydrodynamic pressures at b and c are determined from the flow net and the mean is assumed to operate vertically on the base of the prism. The factor of safety, E, against uplift can be increased by placing a filter layer of relatively high permeability over the top of the prism (Fig. 2.35). The flow pattern is barely influenced by the permeability of this layer. Then $F = (W_b + W_f)/U_e$ where W_F is the weight (buoyant if submerged) of the prism adef. For the case of a single row of sheet piling (Fig. 2.35), the critical position of the base of the prism is close to the toe of the piling. For other structures, such as a double row of sheet piling, bc is likely to be above the toe and the critical position must be found by trial and error. Theoretically, in order to balance excess hydrodynamic pressure effectively, the filter layer should be formed to the profile gh.

Since the spacing of flow lines in a cofferdam floor is proportional to the width of the excavation, values of hydraulic gradient at exit are greater in a narrow cofferdam than in a wide cofferdam. It should be observed also that a drainage sump or ditch in the floor of an excavation causes flow lines to be concentrated around it, with consequent danger of local instability.

Material for a filter layer should be sufficiently coarse to permit relatively free flow of water yet fine enough to prevent transport of fine particles from the underlying stratum. The empirical rules for the requirements of a filter devised by K. Terzaghi were modified by the US Corps of Engineers at Vicksburg to read as follows:

$$\frac{d_{15} \text{ filter}}{d_{85} \text{ soil}} < 5$$

$$4 < \frac{d_{15} \text{ filter}}{d_{15} \text{ soil}} < 20$$

$$\frac{d_{50} \text{ filter}}{d_{50} \text{ soil}} < 25$$

where d_{15}, d_{50} and d_{85} are the particle sizes of the 15, 50 and 85% passing in particle size analyses of the filter material and the soil to be protected. Where a filter is of necessity thick, it may be convenient to construct it in a series of layers, each graded to serve as a filter to the layer immediately beneath it. Ward (1948) reported the use of a filter of beach sand to stabilize a seam of sand in Tertiary clays which caused degradation of the upper part of cliffs at Newhaven, Sussex, by internal erosion. Fig. 2.36 gives details of the seam sand and the beach sand: the requirements of the three parameters quoted above are satisfied by the values 2.8, 7.3 and 10.8 respectively. The use of geotextiles as filters has been discussed by Hoare (1984).

Piping and blowing are most likely to occur in sands but general heave is likely to occur if a bed of clay in the floor of an excavation is subjected to an upward excess hydrostatic pressure developed in a bed of sand, gravel or fissured rock. Since the clay particles will not be in suspension, owing mainly to cohesion, a mass of clay plus interstitial water will be lifted by the excess pressure. Following

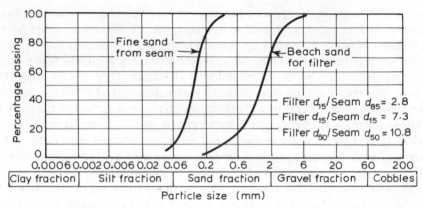

Figure 2.36 Particle size analyses for design of a filter layer (after Ward, 1948).

heave in clay in the floor of an excavation for a pump house at Cowes, arising from artesian pressure in underlying sand, a reinforced concrete foundation slab was cast under water and anchored in a bed of limestone by prestressed cables located in steel box piles. Construction continued when the cofferdam was dewatered (Coates and Slade, 1958). Anchors were employed in part of the floor of a dry dock at Belfast where boulder clay overlies sandstone (Ross *et al.*, 1972). During construction of locks at Shoreham Harbour, the stability of the excavation floor was threatened by artesian pressure in the Chalk which underlies clay of the Reading Beds (Ridehalgh, 1958).

The floor of an excavation can be stabilized either by increasing the load on the floor or by reducing the excess pressure. The latter can be achieved by excavating and concreting under water or by sinking relief wells through the clay floor to the underlying artesian aquifer (Ward, 1957) or, in the case of land excavations, by installation of well-point or deep well drainage outside the excavation. The use of boreholes in an excavation floor has been described also by Huder (1969). Increasing the length of the drainage path, for example, by driving sheeting deeper, may also increase stability. In general, any feature, such as a sump, which leads to a concentration of flow should be avoided, except where the soil is not susceptible to piping or where the velocity of flow is low or where the possibility of piping is of no consequence.

The possibility of encountering artesian pressure can often be predicted, before boreholes are put down, from historical evidence or from known geology and geomorphology. For example, fresh water springs have long been known to emerge from the Chalk forming the bed of the English Channel and also from the bed of the Persian Gulf. Artesian pressure can be anticipated where an aquifer, dipping towards the site, is confined beneath an aquiclude. The problem at Shoreham Harbour involved the main Chalk aquifer but a few kilometres west the Tertiary clays overlying the Chalk are, in turn, overlain by Quaternary gravel and silty clay, the gravel holding artesian water which rises to ground level and

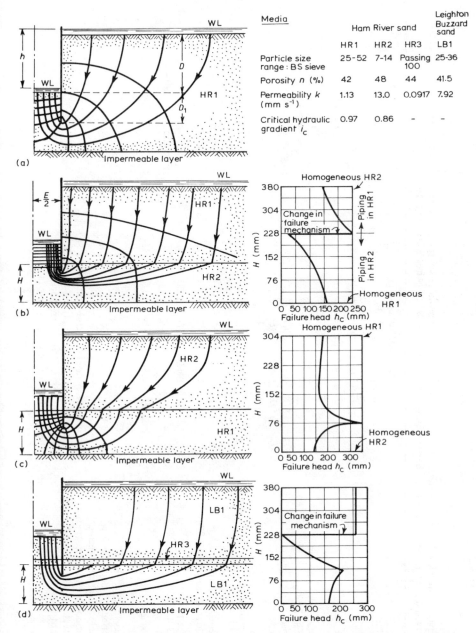

The following data table appears within figure (a):

Media	Ham River sand			Leighton Buzzard sand
	HR1	HR2	HR3	LB1
Particle size range : BS sieve	25-52	7-14	Passing 100	25-36
Porosity n (%)	42	48	44	41.5
Permeability k (mm s^{-1})	1.13	13.0	0.0917	7.92
Critical hydraulic gradient i_c	0.97	0.86	–	–

Figure 2.37 Typical results obtained in experiments on model cofferdams (Marsland, 1953). (a) Homogeneous medium; (b) fine medium over coarse medium; (c) coarse medium over fine medium; (d) fine seam in coarse medium.

Table 2.19 Typical values for minimum penetration of sheeting for excavations based on the available mathematical and experimental data using a factor of safety of 1.5 applied to i_c for dense sands and U_c for loose sands. (After Marsland, 1953).

Site conditions	Minimum penetration D_1 Width of excavation E (D = depth of excavation)						Remarks
	8D	4D	2D	D	0.5D	0.25D	
(a) Homogeneous with infinite depth (Fig. 2.37a):							Mathematical analysis
(i) Loose	0.7D	0.8D	0.9D	1.0D	1.2D	1.4D	
(ii) Dense	0.4D	0.5D	0.6D	1.8D	1.0D	1.3D	
$\dfrac{H_1}{D}$ = infinity							
(b) Homogeneous with impervious layer below (dense) (Fig. 2.37a):							Obtained by electrical analogy and mathematical analysis
(i) $\dfrac{H_1}{D} = 1$	—	0.4D	0.4D	0.6D	0.8D	0.9D	
(ii) $\dfrac{H_1}{D} = 2$	—	0.4D	0.5D	0.8D	1.1D	1.3D	
(c) Coarse layer under fine (Fig. 2.37b): (i) interface at depth > E below foot of sheeting	As in (a)						
(ii) interface between excavation level and a depth < E below foot of sheeting	This condition needs careful study since head required to cause piping may reach very low value. Reduce pressure until head difference between coarse layer and excavation level = 0.66 H_4.						

(iii) coarse layer reaches above excavation level	Homogeneous values are safe but at some stage before interface is reached the fine material will become completely unstable	Dangerous conditions will arise at some time before excavation reaches coarse layer: probably most economical to excavate under water
		With thin coarse layer, heave occurs in centre before occurring at sides
(d) Coarse layer above fine (Fig. 2.37c):	Generally safer than homogeneous. If coarse layer is all above excavation level, neglect it	
(e) Fine thin layer in a homogeneous soil mass (Fig. 2.37d): (i) at a depth $> E$ below foot of sheeting	Use homogeneous values with impervious base at upper interface of layer.	As thin layer approaches foot of sheeting, failure is partly heave and partly wedge adjacent to sheeting. Well points can be used where the quantities to be pumped are low, otherwise deep submersible pumps
(ii) Lying below excavation base but above position (i)	Very dangerous. Pressure relief required under fine layer such that excess pressure does not exceed height of soil above base of fine layer: compare with (c) (ii) above	
(iii) Layer above excavation base	Complete excavation safer than homogeneous case but dangerous condition may arise during construction as fine layer is approached, as in (ii) above	Temporary pressure relief required until excavation is below fine layer (this may be provided by previous part of excavation if of trench type)

presumably is derived from access to the Chalk at some location. Artesian water was encountered in the sands of the Bracklesham Beds at the site of a new bridge over the River Itchen, Southampton: the source is not immediately apparent.

The question of whether uplift operates on buoyant weight or saturated weight of a soil or rock mass may require careful consideration. Such factors as degree of continuity between pores and the time required to attain hydrodynamic equilibrium should be taken into account. In the case of blocks of rock it may be reasonable to assume that uplift cannot obtain at points of contact between interfaces.

Although this section is concerned largely with problems encountered during construction, permanent solutions may have to be provided for some structures in water-bearing ground. Thus anchorages or surcharge loads may have to be provided for underground structures, such as tanks at petrol stations, to avoid flotation. Coastal and riverine sites are obvious locations for such problems and consideration must be given to the influence of fluctuations in water-level arising from tides and floods. Flint and Neill (1977) describe uplift problems catered for in the National Theatre, London, and Morris *et al.* (1971) discuss the influence of groundwater on construction around the City of London Guildhall. The effect of uplift on reactors founded on a sequence of mudstones and limestones has been described by Haydon and Hobbs (1977).

Seepage into excavations has been treated by McNamee (1951), Marsland (1953), Gray and Nair (1967), King (1967), King and Cockroft (1972) and Rao and Rao (1973). The experimental work of Marsland is of particular interest. Figure 2.37 illustrates some of the experiments and Table 2.19 summarizes information for design purposes. In tests with loose homogeneous sand it was found that for narrow excavations ($D > E$, where D is the depth and E the width of the excavation), failure was preceded by scattered eruptions over the entire bottom and completed by general heaving in which the soil particles moved more or less independently of each other. With wide excavations ($D < E$), initial eruptions occurred adjacent to the sheeting and failure was completed by a triangular wedge of soil, base uppermost, blowing alongside the sheeting. The head h required to cause failure in this case was about 10% more than that required with narrow cofferdams. The criterion for instability occurs when the pressure head loss between the foot of the sheeting and the excavation bottom is the critical value U_c, equal to the buoyant weight of the soil adjacent to the sheeting, whence $U_c = D_1 (G - 1)/(1 + e)$. If frictional forces develop, this value may be exceeded and instability may not obtain until the exit gradient is equal to the critical value i_c. Failure occurred in the experiments when conditions were between the critical stages defined by U_c and i_c.

With dense homogeneous sand, failure generally occurred when the exit gradient became equal to or slightly exceeded i_c, probably on account of inter-particle and sheeting-face friction which had to be eliminated before movement of the particles could take place. Failure in a wide cofferdam is likely to occur in the form of a wedge adjacent to the sheeting and in a narrow cofferdam is likely to become manifest in numerous eruptions over the excavation bottom. The measured exit gradients varied by 10 to 15% with loose homogeneous sand and by 5% with dense homogeneous sand due to non-uniformity of compaction.

Further tests were made with loosely compacted sands in various layered arrangements, as shown in Figs 2.37(b), (c) and (d). A third type of failure was recognized in the case shown in Fig 2.37(d): when the thin layer of fine material was at or just above the level of the foot of the sheeting, heave was manifest by the rising of a plug, consisting of the fine layer and the coarse material above it, across the full width of the excavation.

It was found that the stability of an excavation can be improved if fine material is excavated under water inside a cofferdam and replaced by coarse material before dewatering. The improved stability is due to a change in the flow pattern leading to the pressure at the foot of the sheeting becoming a smaller fraction of the total head and to attenuation of the upward seepage drag.

When excavating in a dewatered cofferdam, the exit gradient of the seepage water should not exceed a value beyond which a man cannot move without planks underfoot. Experiments show that a man can stand on sand when the gradient does not exceed about 0.5 and this implies a factor of safety of about 2 applied to i_c. If there is adequate shallow drainage within the excavation, the factor can be reduced to 1.5 since the depth of the bearing failure zone beneath a man's foot is only about 0.1 m. Shallow drainage is particularly important in silty or clayey sands because movement on the surface may produce liquefaction. A 0.15 m layer of coarse sand or gravel placed over the excavation floor may eliminate the need for shallow drainage and serve as a drainage layer to reduce uplift pressures after the foundation concrete has been placed and relief wells provided. Observations on heave of excavation floors have been reported by Milligan and Lo (1970) and by Moore and Longworth (1979).

2.7.4 Calculation of discharge of water to excavations and boreholes

(a) Equilibrium or steady flow conditions

Calculations for the discharge of water to the sides of excavations and to boreholes can be made by simple analyses provided the following conditions are satisfied.

(a) Aquifer is homogeneous and isotropic and extends to infinity in all horizontal directions.
(b) Excavation or borehole completely penetrates aquifer and is unlined: this latter condition is reasonably satisfied by a perforated lining.
(c) Flow is through pores in the aquifer and obeys Darcy's Law.
(d) Rate of pumping is constant and equilibrium is attained.
(e) Phreatic surface is horizontal, except where depressed, and recharge by rainfall does not take place.
(f) Interfaces of aquifer and aquicludes are horizontal.

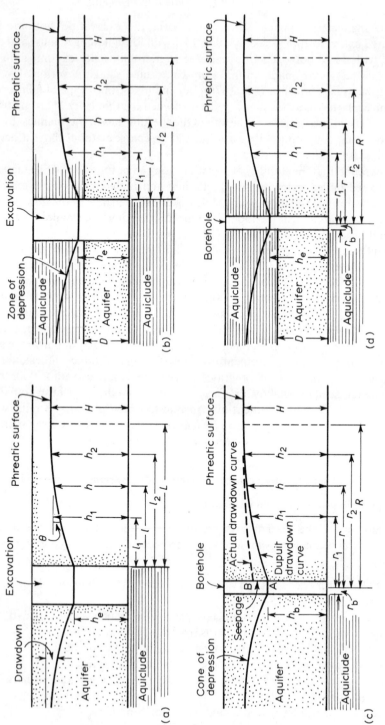

Figure 2.38 Discharge of water to excavations and boreholes.

Excavation

Aquiclude beneath aquifer only (Fig. 2.38a).

Discharge through 1 m length of one side of excavation is

$$q = kih = kh\,dh/dl \text{ where } dh/dl \simeq i \text{ (strictly } i = \sin\theta).$$

Integrating yields $\qquad (h_2^2 - h_1^2) = \dfrac{2q}{k}(l_2 - l_1)$ $\qquad\qquad$ (2.28)

Aquiclude above and beneath aquifer (Fig. 2.38b).

Discharge through 1 m length of one side of excavation is

$$q = kD\,dh/dl$$

Integrating yields $\qquad (h_2 - h_1) = \dfrac{q}{kD}(l_2 - l_1)$ $\qquad\qquad$ (2.29)

Borehole

Aquiclude beneath aquifer only (Fig. 2.38c)

Discharge through cylindrical surface of radius r is

$$q = kh2\pi r\,dh/dr$$

Integrating yields $\quad (h_2^2 - h_1^2) = \dfrac{q}{\pi k}\log_e\dfrac{r_2}{r_1} = 0.734\dfrac{q}{k}\log_{10}\dfrac{r_2}{r_1}$ $\qquad$ (2.30)

Aquiclude above and beneath aquifer (Fig. 2.38d)

Discharge through cylindrical surface of radius r is

$$q = kD2\pi r\,dh/dr$$

Integrating yields $\quad (h_2 - h_1) = \dfrac{q}{2\pi Dk}\log_e\dfrac{r_2}{r_1} = 0.367\dfrac{q}{Dk}\log_{10}\dfrac{r_2}{r_1}$ $\quad$ (2.31)

Equations 2.28 to 2.31 can be solved for any one unknown quantity. Thus if h_1 and h_2 are recorded in two observation boreholes at radii r_1 and r_2 respectively from the centre of a discharging borehole, and if q is measured, then k can be calculated from 2.30 or 2.31 as appropriate. This is the basis of one form of field test for measuring permeability. For a borehole, the radius of influence R is the radius at which drawdown is zero. Similarly, for an excavation, there is a distance of influence L. Provided q and k are known, L can be determined, for example, by substituting $h_1 = h_e$, $h_2 = H$, $l_1 = 0$ and $l_2 = L$ and R by substituting $h_1 = h_b$, $h_2 = H$, $r_1 = r_b$ and $r_2 = R$. It can be observed that a fourfold increase, say, in the diameter of a borehole above about 0.15 m can be expected to lead to no more than 30 to 40% increase in discharge. Since the logarithm of the ratio R/r_b appears in Equations 2.30 and 2.31, any reasonable assumed value of R will yield a reasonable value of any unkown if R is not known exactly. Experience shows that, for practical purposes, the value of R can usually be taken between 100 and 300 m, although careful observation of water table may reveal minor drawdown far beyond these distances.

In the above analyses it is assumed that flow through any vertical section is horizontal but, since the free surface is curved in the zone of depression, there is also a small vertical component. Furthermore, the real phreatic surface intersects the excavation side or borehole at some point B above the water level A and seepage occurs in the face between A and B. In spite of these discrepancies it can be concluded from model studies and relaxation computations by Boulton (1951) that Equation 2.30 gives discharge with good accuracy for the assumed conditions and also represents the profile of the phreatic surface with reasonable accuracy when r exceeds $1.5h$. It follows, incidentally, that it is desirable in field tests to locate observation boreholes at radii exceeding $1.5h$ but not approaching R, where the drawdown is small and difficult to measure accurately. For such tests, h and R must be estimated initially. Further studies of equilibrium flow to boreholes have been reported by Kashef (1965) and Murray and Monkmeyer (1973). Pumping tests may be expensive if they are lengthy or if the discharge must be conveyed in a long pipeline to carry it clear of the cone of depression.

It is useful here to examine briefly the problem of recharge of aquifers through boreholes, either unlined or with perforated lining. In these cases the water level in the borehole is above the original phreatic surface and an inverted cone of impression is formed. The head falls away from the borehole and dh becomes negative: as a result, the signs of the h_1 and h_2 terms in Equations 2.30 and 2.31 are reversed, as will be seen in Example 2.21.

(b) Discharge from partially penetrating boreholes

Where a borehole does not penetrate to the base of an aquifer, the pattern of flow in the vicinity of the borehole diverges from the assumed radial flow to become three-dimensional – markedly so when the penetration is relatively short.

For the case of a borehole penetrating a distance h_p below the top of a confined aquifer, G.J. de Glee determined in 1930 that the discharge q_p is given by

$$q_p = \frac{4\pi k(h_{2D} - h_e)}{\left[\dfrac{2}{h_p}\log_e\dfrac{\pi h_p}{2r_b} + \dfrac{0.20}{D}\right]} \tag{2.32}$$

where heads to the phreatic surface are measured from the base of the aquifer and h_{2D} is the head measured at a radius 2D. The equation is valid for $1.3h_p \leqslant D$ and $h_p/2r_b \geqslant 5$. To a close approximation the drawdown $(H - h_e)$ at the borehole is given by

$$(H - h_e) = \frac{q_p}{2\pi k}\left[\frac{1}{h_p}\log_e\frac{\pi h_p}{2r_b} + \frac{0.10}{D} - \frac{1}{D}\log_e\frac{R}{2D}\right]. \tag{2.33}$$

The ratio $K = q_p/q$, where q is the discharge from a fully penetrating borehole having the same drawdown $(H - h_e)$ as the partially penetrating borehole, depends on the penetration fraction h_p/D and the borehole slimness $h_p/2r_b$. Figure 2.39(a) gives values of K for the case when $R/r_b = 1000$.

For the case of a borehole penetrating a distance h_p below the original water

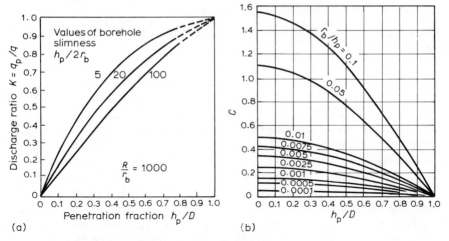

Figure 2.39 Factors relating to partially penetrating boreholes.

table in an unconfined aquifer, the discharge is given approximately by

$$q_p = \frac{4\pi k(h_{2H} - h_e)}{\left(\dfrac{2}{h_p}\log_e\dfrac{\pi h_p}{2r_b} + \dfrac{0.20}{H}\right)} \tag{2.34}$$

providing the drawdown is small in relation to H, where h_{2H} is the head measured at a radius $2H$. The similarity between Equations (2.32) and (2.34) enable Fig. 2.39(a) to be employed also for the case of an unconfined aquifer, when D is replaced by H.

N.K. Girinsky derived in 1950 the following equation for the discharge from a borehole penetrating a distance h_p below the base of an aquiclude into a semi-infinite aquifer where the phreatic surface is wholly within the aquiclude.

$$q_p = \frac{2\pi k h_p \Delta_b}{\log_e\left(\dfrac{1.6 h_p}{r_b}\right)} \tag{2.35}$$

where Δ_b is the drawdown at the borehole.

For the case of a borehole penetrating a distance h_p below the top of a confined aquifer, J. Kozeny determined in 1927 that the discharge is given by

$$q_p = q'(1 + C) \tag{2.36}$$

where q' is the discharge from the borehole assuming if fully penetrates an aquifer of thickness h_p and C is a correction factor taken from Fig. 2.39(b). The factor $C = 7(r_b/2h_p)^{1/2}\cos(\pi h_p/2D)$ accounts for the upward flow to the borehole where the aquifer is of thickness $D(>h_p)$.

(c) **Discharge of boreholes in fissured strata**

Whereas groundwater flow in sandstone, such as Bunter or Keuper, is principally through pores, flow in limestone, such as the Chalk, is principally through fissures. The Dupuit equations were derived fundamentally for porous media and, for a given value of drawdown of the water-level in a borehole, the yield increases as the diameter of the borehole increases. Although increase in diameter in a fissured aquifer may lead to no increase in yield if no additional fissures are intercepted, a statistical analysis by Ineson (1959a, b) indicates that in general the influence of diameter on yield is greater in fissured aquifers than in porous aquifers. Ineson evolved the following relative values of yield for equal drawdown shown in Table 2.20, taking the yield of a borehole 0.46 m diameter as unity. At some sites it may be preferable to sink a number of small diameter boreholes instead of one large one. The terms fissured and porous are employed here in a relative sense to indicate the principal channels of flow in an aquifer. It can be added that, for a given value of drawdown, the influence of borehole diameter on yield is reduced as the radius of influence increases in different aquifers. For unconfined conditions the radius of influence may not exceed 600 m, but for confined aquifers it may be as much as 3000 m. It can be noted also that 75% penetration of an aquifer theoretically results in a yield of at least 85% of that obtained by complete penetration.

The majority of boreholes in any one aquifer conform to a system of type curves (Fig. 2.40) when yield is plotted against depression or drawdown of the

Table 2.20 Comparison of yield of boreholes in fissured and porous aquifers.

Borehole diameter (m)	0.10	0.20	0.30	0.41	0.46	0.61	0.81	0.01
Fissured non-porous aquifer	0.30	0.62	0.80	0.94	1.00	1.14	1.28	1.33
Homogeneous porous aquifer	0.56	0.81	0.90	0.98	1.00	1.07	1.13	1.15

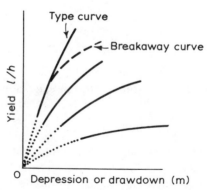

Figure 2.40 Typical yield/depression curves for both confined and unconfined aquifers.

water level in a borehole. The yield for any given depression is a function of transmissibility of the aquifer, diameter of the borehole and additional head losses due to screens or to turbulence. Some boreholes do not follow the type curves entirely but deviate from them at relatively high yields. Fundamentally, these breakaway curves develop due to excessive head losses at high velocities. It is logical to conclude that all curves become breakaway curves if sufficiently high velocities can be attained. During initial pumping tests, discontinuities in yield–depression curves may arise as sediment is removed from fissures by the flow of water, thereby increasing yield. The characteristics of these curves may be altered also by shot-firing or by surging or by any other development process.

The figures given in Table 2.20 are significant in relation to the interpretation of borehole permeability tests. Graphical extrapolation of the figures indicates that sensibly constant values of the factors of 1.5 and 1.3 are reached, respectively, for non-porous and porous aquifers when the borehole diameter is about 3 m. Since $k \propto q$, provided other parameters remain constant, it follows that the intrinsic large-scale permeability of fissured strata will be greater than the value derived from borehole tests. The factor by which the calculated value should be multiplied can be determined by comparing the figure in Table 2.20 relating to the borehole diameter with 1.5 or 1.3. These observations relate primarily to lithified rocks where fissures are present but, in the case of homogeneous and isotropic sands and gravels in which fissures are absent, it seems likely that the range of values between borehole diameters of 0.1 and 3 m should be less–arbitrary figures of 0.85 and 1.05 appear reasonable.

(d) Influence of a body of open water on an adjacent single borehole with aquiclude below aquifer

The method of images can be employed for the case of a single borehole located near a body of open water, i.e. when the body is closer than the radius of influence R of the pumping borehole. Figure 2.41(a) shows a real discharge borehole P adjacent to a river or lake. An imaginary recharge borehole I is considered to be established at a distance IC from the boundary of the free body of water equal to the distance PC where ICP is straight and normal to the boundary ACB. Both boreholes have a radius r_b. It is assumed that the characteristics of the aquifer are the same for both boreholes, that the rates of discharge and recharge are equal and of magnitude q, and that equilibrium conditions obtain. Along the boundary water line ACB the excess recharge head or drawup equals the discharge drawdown and the effective drawdown is thus zero. From Equation 2.30 the relationships for any observation borehole X and P and for X and I are, respectively,

$$(H^2 - h^2{}_{XP}) = \frac{0.734q}{k} \log_{10}\left(\frac{R}{r_P}\right) \tag{2.37}$$

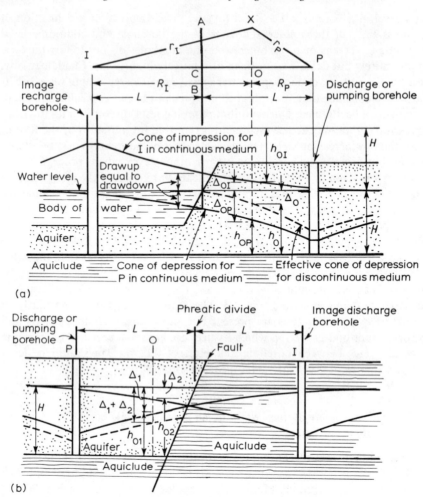

Figure 2.41 Analysis of discharge to boreholes in discontinuous aquifers.

and

$$(H^2 - h^2{}_{\text{XI}}) = \frac{0.734q}{k} \log_{10}\left(\frac{R}{r_\text{I}}\right). \tag{2.38}$$

The effective cone of depression for P, taking into account the influence of the boyd of open water, is derived by deducting the drawup for I from the drawdown for P. Generally it is required to establish the effective cone of depression along the line ICP. Assuming that the radius of influence for a given discharge q is known, Equations 2.37 and 2.38 can be solved for h_{OP} and h_{OI} at any point O. Then the drawdown at that point is

$$\Delta_O = (H - h_O) = (\Delta_{OP} - \Delta_{OI}) = (H - h_{OP}) - (H - h_{OI}) = (h_{OI} - h_{OP}). \quad (2.39)$$

If the radius of influence is not known, the following method can be employed to calculate Δ_O approximately by assuming that

$$(H + h_{OP}) \simeq (H + h_O) \simeq (H + h_{OI}).$$

Multiplying appropriate terms in Equation 2.39 by $(H + h_O)$, $(H + h_{OP})$ and $(h + h_{OI})$

$$(H - h_O)(H + h_O) \simeq (H - h_{OP})(H + h_{OP}) - (H - h_{OI})(H + h_{OI})$$

or

$$(H^2 - h_O^2) \simeq (H^2 - h_{OP}^2)(H^2 - h_{OI}^2)$$

whence

$$(H^2 - h_O^2) \simeq \frac{0.734q}{k} \log_{10}\left(\frac{r_I}{r_p}\right), \quad (2.40)$$

from which h_O, and hence Δ_O, can be calculated. Alternatively, any other single unknown, such as k, can be determined from field observations.

(e) Influence of an upstanding impermeable barrier on an adjacent single borehole with aquiclude below aquifer

The principle of images can be employed also where a borehole is adjacent to an upstanding impermeable barrier, such as a sheet pile cut-off or an aquiclude brought into juxtaposition with an aquifer by a fault (Fig. 2.41(b)). In this case both pumping and image boreholes are discharge boreholes and the phreatic divide between their cones of depression is arranged to coincide with the boundary. It is assumed that the boundary is straight in plan and that the line joining the real borehole P and the image borehole I is normal to it. It will be observed that, unless the boundary is vertical, the position of the phreatic divide must first be estimated and then checked at the end of computations to ascertain that it is located within tolerable limits, making an adjustment and re-calculating if necessary. Equations similar to Equations 2.37 and 2.38 are used to calculate drawdown but in this case the total effective drawdown is

$$(H - h_{O1}) + (H - h_{O2}) = (2H - h_{O1} - h_{O2}). \quad (2.41)$$

(f) Unsteady flow conditions

When pumping is commenced and maintained at a constant rate q from a borehole penetrating an extensive aquifer, the radius of the cone of depression increases with time at a diminishing rate (Fig. 2.42). The discharge leads to a reduction in storage associated with the compressibility of the aquifer. The rate of decline of head decreases with time. The differential equation governing unsteady

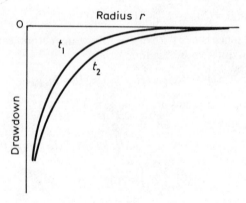

Figure 2.42 Cone of depression at times t_1 and t_2 with non-equilibrium conditions.

flow to a borehole in a confined aquifer (aquiclude above and beneath) can be solved by analogy with flow of heat and the resulting exponential integral can be expanded as a convergent series. Provided r is relatively small and t is relatively large, the later series terms become negligible and the following approximate solution for drawdown $(H - h)$ at radius r, according to C.E. Jacob, is satisfactory in many cases.

$$H - h = \frac{2.303q}{4\pi T} \log_{10}\left(\frac{2.246Tt}{r^2S}\right) \qquad (2.42)$$

where T is the transmissibility of the aquifer and is equal to kD where D is the thickness of the aquifer, S is the storage coefficient and is a measure of the compressibility of a confined aquifer, h is head above datum radius r from the centre of the borehole, H is head above datum beyond the radius of influence of the cone of depression and t is time after the commencement of pumping. Thus a graph of $(H - h)$ against $\log_{10} t$ is a straight line (Fig. 2.43) where $(H - h)$ is the drawdown in an observation well at time t after commencement of pumping. This linear relationship should not be assumed when $u = r^2S/(4Tt)$ exceeds about 0.01, in order to avoid large errors.

The difference in drawdown Δh metres per log cycle of time is read on the graph, together with the time intercept t_0 for zero drawdown. In fact, Δh relates to the difference between head h_2 at time t_2 and h_1 at t_1, and since $t_2/t_1 = 10$,

$$\Delta h = \frac{2.303q}{4\pi T}$$

when Equation 2.42 is written for h_2, t_2, and h_1, t_1 and one equation is deducted from the other. Then

$$T = \frac{0.1832q}{\Delta h}. \qquad (2.43)$$

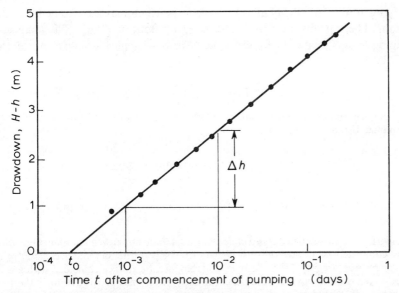

Figure 2.43 Analysis of observation borehole data with non-equilibrium conditions.

When the drawdown is zero, Equation 2.42 is zero and, since $q > 0$, this condition is satisfied only when

$$\log_{10}\left(\frac{2.246\,Tt_o}{r^2 S}\right) = 0,$$

that is, when

$$\frac{2.246\,Tt_o}{r^2 S} = 1.$$

Then

$$S = \frac{2.246\,Tt_o}{r^2}. \tag{2.44}$$

The units of the various quantities in Equations 2.43 and 2.44 are as follows: T, $m^3\,day^{-1}\,m^{-1}$; q, $m^3\,day^{-1}$; $t_o\,day$; S, dimensionless; Δh, m; r, m.

Although the non-equilibrium analysis relates primarily to a confined aquifer, it can be applied also to an unconfined aquifer (aquiclude beneath only) when the drawdown is small compared with H. The analysis can be modified to cover the case of recovery after pumping is ceased. Unsteady flow to boreholes is treated at length by, for example, Todd (1980). The problem was investigated theoretically and experimentally by Boulton (1954).

For water supply purposes, extensive tests on boreholes are employed in order to determine the characteristics of the aquifer and the yield and efficiency of the borehole. For the former, a constant discharge is adopted and for the latter a step-drawdown test at different discharge rates. The influence of leakage from an overlying aquifer and delay in yield are factors which may have to be taken into account. These problems were investigated by Boulton (1963). The analysis of pumping tests is treated in, for example, *Manual of applied geology for engineers* (ICE, 1976).

EXAMPLE 2.19

The geological succession at the cooling water pumphouse for the Shellhaven Refinery in the Thames Estuary is as follows:

Ground level	+ 2.1 m o.d.	
		Clay with impersistent peat seams
	− 14.1 m o.d.	
		Sand and gravel
	− 25.7 m o.d.	
		Clay

Construction was carried out in a cofferdam surrounded by eleven pumping wells which almost completely penetrated the sand and gravel. The perimeter of the area enclosed by the well system was roughly 300 m. At high tide the phreatic surface coincided with ground level, the steady pumping rate from the entire excavation was 6810 litres min^{-1}, the water level in the wells was about − 10.7 m o.d. and the level in an observation well 41 m from the perimeter was − 2.4 m o.d.

Ignoring the proximity of the river and treating the well system as the side of an excavation, calculate the average permeability in mm s^{-1} of the bed of sand and gravel. Also calculate the distance from the well perimeter to the point at which the drawdown is just zero.

If under these pumping conditions the maximum artesian head in the centre of the cofferdam stood at − 4.0 m o.d., calculate the maximum depth to which excavation could be taken without artesian pressure heave taking place in the bottom, assuming the unit weight of saturated clay is 18 kN m^{-3} and the unit weight of water is 9.8 kN m^{-3}. Also calculate the maximum depth to which the excavation could be taken if ground water lowering is not employed.

$$\text{Discharge per m of perimeter} = \frac{6810}{300 \times 10^3 \times 60} = 0.000378 \text{ m}^3 \text{ s}^{-1}.$$

$$\text{From Equation 2.29} \qquad k = \frac{0.000378 \times 41 \times 10^3}{8.3 \times 11.6} = 0.161 \text{ mm s}^{-1}.$$

$$\text{From Equation 2.29} \qquad L = \frac{12.8 \times 0.000161 \times 11.6}{0.000378} = 63.2 \text{ m}.$$

Maximum artesian head above top of sand and gravel is 10.1 m. Minimum thickness of clay to resist this head = $10.1 \times 9.8/18 = 5.5$ m. Hence maximum possible depth of excavation is 10.7 m. Maximum possible depth of excavation without groundwater lowering is

$$\left(16.2 - \frac{16.2 \times 9.8}{18}\right) = 7.4 \text{ m}.$$

The calculated value of k is reasonable for a sand and gravel mixture (see Table 2.11). This example is based broadly on data published some years ago by Soil Mechanics Ltd and shows that a very simplified analysis can often yield results of practical value, bearing in mind the remarks made in the Introduction to this chapter. In any case it would be desirable to install standpipes and instruments to monitor piezometric heads and ground movements during excavation.

EXAMPLE 2.20

Field observations for permeability gave drawdown values of 0.832 m and 0.466 m in observation wells at radii of 15 and 45 m, respectively, from a pumping well when the discharge was 1700 litre m^{-1} under equilibrium conditions. The aquifer is horizontal and has an average thickness of 13 m and is confined between upper and lower aquicludes. The pumping well completely penetrated the aquifer and the water table was initially horizontal.

Calculate (a) The permeability of the aquifer in mm s^{-1}.
　　　　　(b) The drawdown at a radius of 30 m from the pumping well.
　　　　　(c) The radius of influence of the cone of depression.
　　If, in fact, the borehole penetrated only 3.25 m below the top of the aquifer and if the borehole diameter is 0.30 m, calculate
　　　　　(d) Discharge employing Fig. 2.39(a), checking that the graph can reasonably be employed.
　　　　　(e) Discharge employing Fig. 2.39(b).

Using Equation (2.31).

(a) Permeability $k = \dfrac{(2.303 \times 1700 \times 1000 \log_{10} 45/15)10}{[2\pi(0.832 - 0.466)60 \times 13 \times 100 \times 100]} = 1.04$ mm s^{-1}.

(b) Drawdown at 30 m $= 0.832 - (0.832 - 0.466)\dfrac{\log_{10} 30/15}{\log_{10} 45/15} = 0.601$ m.

(c) Radius of influence R given by $\log_{10} R/15 = \dfrac{(0.832 - 0)}{(0.832 - 0.466)}\log_{10} 45/15$, whence

$R = 182$ m.

(d) Penetration fraction $h_{\mathrm{p}}/D = 3.25/13 = 0.25$.
　　$R/r_{\mathrm{b}} = 182/0.15 = 1213$: this is probably sufficiently close to 1000 to render Fig. 2.39(a) valid although the value $R = 182$ m is calculated for the fully penetrating condition.
　　Borehole slimness $h_{\mathrm{p}}/2r_{\mathrm{b}} = 3.25/0.30 = 10.8$.
　　From Fig. 2.39(a), $K \simeq 0.43$, hence $q_{\mathrm{p}} = 0.43 \times 1700 = 731$ litre min^{-1}.

(e) $r_{\mathrm{b}}/h_{\mathrm{p}} = 0.15/3.25 = 0.046$ and $h_{\mathrm{p}}/D = 0.25$.
　　From Fig. 2.39(b), $C \simeq 0.98$.
Discharge from borehole fully penetrating aquifer of thickness 3.25 m,

$$q' = (3.25/13)1700 = 425 \text{ litre min}^{-1}.$$

Hence $q_{\mathrm{p}} = q'(1 + C) = 425 \times 1.98 = 842$ litre min^{-1}.

EXAMPLE 2.21

The stratigraphic sequence at two sites is shown in Fig. 2.44. The sands of the Folkestone Beds can be treated as homogeneous and isotropic with a permeability of 5000 litre day^{-1} m^{-2}. The Gault and Sandgate Beds can be assumed impermeable for the purpose of this study. At both sites the bedding and the phreatic surface are horizontal and the latter is 10 m beneath ground level.
　　It is intended to sink 0.5 m diameter boreholes to the base of the Folkestone Beds at both

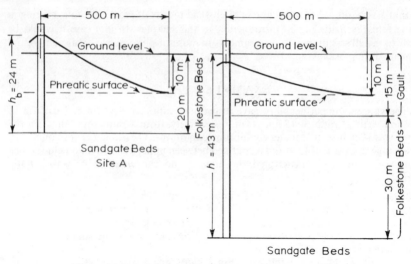

Figure 2.44 Recharge of aquifer through boreholes.

sites and to inject recharge water at the rate of 1×10^6 litre day^{-1}. The radius of the cone of impression at which the drawup is zero can be taken as 500 m.

Calculate the head of water in the boreholes above the base of the Folkestone Beds required to meet these steady-state conditions.

Equation 2.30 modified (see end of Section 2.7.4(a)).

$$(h_b^2 - H^2) = 0.734 \frac{q}{k} \log_{10} \frac{R}{r_b}.$$

Then
$$h_b^2 = \frac{0.734 \times 10^6}{5 \times 10^3} \log_{10} \frac{500}{0.25} + 10^2.$$

whence
$$h_b = 24 \text{ m}.$$

Equation 2.31 modified (see end of Section 2.7.4(a)).

$$(h_b - H) = 0.367 \frac{q}{Dk} \log_{10} \frac{R}{r_b}.$$

Then
$$h_b = \frac{0.367 \times 10^6}{30 \times 5 \times 10^3} \log_{10} \frac{500}{0.25} + 35.$$

whence
$$h_b = 43 \text{ m}.$$

EXAMPLE 2.22

Field observations for permeability have been made at a site on the flood plain of a river. Boreholes have been put down on a line normal to the river bank. The pumping borehole is 85 m from the bank and observation boreholes A, B and C are 70, 55 and 40 m respectively from the bank. The average depth of the base of the aquifer below the phreatic surface, which is coincident with the water level in the river, is 13.00 m and the permeability of the

aquifer is 62.83 litre min^{-1} m^{-2}. The radius of influence of the cone of depression is 180 m.
   Assuming equilibrium conditions, calculate the drawdown at the three observation wells when the pumping rate is 1700 litre min^{-1}.
   The following table is a convenient form of computation for Equations 2.37 and 2.38.

$$\frac{0.734q}{k} = \frac{0.734 \times 1700}{62.83} = 19.85 \text{ m}^2$$

Observation borehole	r_p(m)	$\dfrac{180}{r_p}$	**Pumping borehole** $\log_{10}\dfrac{180}{r_p}$ $= Y$	$19.85\,Y$ $= X$	$h_{OP}^2 =$ $169 - X$	h_{OP}(m)
A	15	12.00	1.0792	21.41	147.6	12.13
B	30	6.00	0.7782	15.45	153.5	12.38
C	45	4.00	0.6021	11.93	157.1	12.53

Observation borehole	r_i(m)	$\dfrac{180}{r_i}$	**Image borehole** $\log_{10}\dfrac{180}{r_i}$ $= Y$	$19.85\,Y$ $= X$	$h_{OI}^2 =$ $169 - X$	h_{OI}(m)	$\Delta_0 =$ $(h_{OI} - h_{OP})$ (m)
A	155	1.161	0.0648	1.288	167.7	12.93	0.80
B	140	1.286	0.1092	2.165	166.8	12.91	0.53
C	125	1.439	0.1581	3.140	165.9	12.88	0.35

The following table is a convenient form of computation for the approximate Equation 2.40.

$$\frac{0.734q}{k} = 19.85 \text{ m}^2$$

Observation borehole	r_P(m)	r_i(m)	$\dfrac{r_I}{r_P}$	$\log_{10}\dfrac{r_I}{r_P}$ $= Y$	$19.85\,Y$ $= X$	$h_0^2 =$ $169 - X$	h_0(m)	$\Delta_0 =$ $(H - h_0)$ (m)
A	15	155	10.33	1.0141	20.13	148.87	12.20	0.80
B	30	140	4.67	0.6693	13.28	155.72	12.47	0.53
C	45	125	2.78	0.4440	8.81	160.19	12.65	0.35

In this example, Equation 2.40 yields the same values of drawdown as Equations 2.37 and 2.38.

EXAMPLE 2.23

Figure 2.45 shows one-half of a flow net constructed for a long cofferdam 12 m wide internally. Draw to scale a diagram showing the horizontal water pressure in $kN\,m^{-2}$ on the inside and the outside of the sheeting, taking into account losses due to seepage. The unit weight of water should be taken as $9.8\,kN\,m^{-3}$.

Estimate the following:

(a) The rate of seepage in $m^3\,h^{-1}\,m^{-1}$ of sheeting if the soil is homogeneous and isotropic with a permeability of $10^{-4}\,mm\,s^{-1}$.
(b) The maximum exit hydraulic gradient in the floor of the excavation.
(c) The water level in the cofferdam below which, on pumping the water out, piping is likely to commence.
(d) The factor of safety against uplift of prism abcd, assuming the saturated unit weight of the soil is $20\,kN\,m^{-3}$.

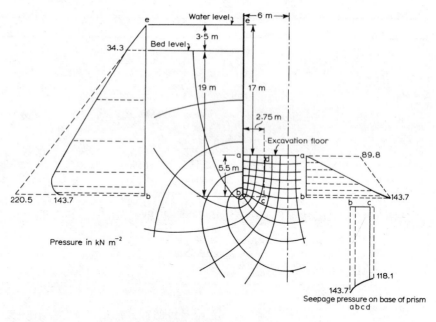

Figure 2.45 Flow net for seepage into cofferdam.

Hydrostatic pressure outside sheeting without seepage loss: At 3.5 m, $34.3\,kN\,m^{-2}$. At 22.5 m, $220.5\,kN\,m^{-2}$.
Total seepage loss in pressure $17 \times 9.8 = 166.5\,kN\,m^{-2}$.
Number of equipotential fields $N_e = 13$.
Loss in pressure per field $= 166.5/13 = 12.82\,kN\,m^{-2}$.

Progressive loss outside sheeting involves six decrements of $12.82\,\text{kN}\,\text{m}^{-2}$ leading to total of $76.8\,\text{kN}\,\text{m}^{-2}$ and inside sheeting involves seven decrements of $12.82\,\text{kN}\,\text{m}^{-2}$ leading to total of $89.7\,\text{kN}\,\text{m}^{-2}$.

(a) Number of flow channels $N_f = 6$

$$\text{Total seepage } Q = k\frac{h}{N_e}N_f = \frac{10^{-4}}{10^3} \times \frac{17}{13} \times 6 \times 60^2 = 2.82 \times 10^{-3}\,\text{m}^3/\text{h/m run.}$$

(b) Smallest field at exit is adjacent to sheeting and its length is 0.9 m.

$$\text{Hence maximum value of } i = \frac{17}{13 \times 0.9} = 1.45.$$

Since the submerged unit weight of the solid particles is $\gamma_w(G-1)/(1+e)$, suspension will just commence when $i\gamma_w = \gamma_w(G-1)/(1+e)$ and the critical hydraulic gradient is $i_c = (G-1)/(1+e)$. Substituting average values of $G = 2.65$ (quartz sand) and $e = 0.6$ yields $i_c \simeq$ unity. The value of 1.45 is clearly very unsafe.

(c) Piping is likely to commence when $i = 1 = \dfrac{h}{13 \times 0.9}$, that is when $h = 11.7$ m or when water level is 5.3 m above floor level.

(d) Submerged unit weight of soil is $(20 - 9.8) = 10.2\,\text{kN}\,\text{m}^{-3}$
Mean excess pressure on base of prism abcd is approximately $130\,\text{kN}\,\text{m}^{-2}$

$$\text{Hence } F = \frac{W_b}{U_e} = \frac{5.5 \times 2.75 \times 10.2}{130 \times 2.75} = 0.43.$$

EXAMPLE 2.24

To examine values of permeability, parallel and normal to the bedding in parallel strata, which may be horizontal or dipping. Consider the strata shown in Fig. 2.46: it is assumed that isotropic conditions prevail in relation to permeability in each bed.

When the flow is parallel to the bedding planes, the hydraulic gradient in each bed must be the same. Hence the total discharge per metre width of flow is

$$q = k_p i H_1 = (k_1 iH + k_2 iH_2 + k_3 iH_3)$$

where k_p is the effective mean permeability in the three beds. Then

$$k_p = \frac{1}{H}(k_1 H_1 + k_2 H_2 + k_3 H_3) \tag{2.45}$$

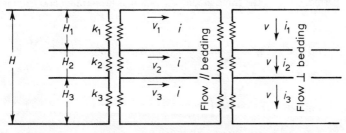

Figure 2.46 Permeability parallel and normal to bedding.

When the flow is normal to the bedding planes, continuity of flow demands that the velocity is the same in each bed. Hence

$$v = k_1 i = k_2 i_2 = k_3 i_3 = \frac{h}{H} k_n$$

where h is the head lost in the distance H and k_n is the effective mean permeability in the three beds. Also

$$h = i_1 H_1 + i_2 H_2 + i_3 H_3$$

Substituting

$$\frac{h k_n}{H k_1} = i_1, \quad \frac{h k_n}{H k_2} = i_2 \quad \text{and} \quad \frac{h k_n}{H k_3} = i_3$$

$$h = \frac{h k_n}{H} \left(\frac{H_1}{k_1} + \frac{H_2}{k_2} + \frac{H_3}{k_3} \right)$$

whence

$$k_n = H \bigg/ \left(\frac{H_1}{k_1} + \frac{H_2}{k_2} + \frac{H_3}{k_3} \right) \tag{2.46}$$

Now consider two layers, each of thickness $H/2$. From Equations 2.45 and 2.46

$$k_p = \tfrac{1}{2}(k_1 + k_2) \quad \text{and} \quad k_n = 2 k_1 k_2 / (k_1 + k_2)$$

Then

$$\frac{k_n}{k_p} = \frac{4 k_1 k_2}{(k_1 + k_2)^2} = \frac{k_1^2 + 2 k_1 k_2 + k_2^2 - k_1^2 + 2 k_1 k_2 - k_2^2}{(k_1 + k_2)^2}$$

or

$$\frac{k_n}{k_p} = \frac{(k_1 + k_2)^2 - (k_1 - k_2)^2}{(k_1 + k_2)^2} = 1 - \left(\frac{k_1 - k_2}{k_1 + k_2} \right)^2 \tag{2.47}$$

For all practical values of k_1 and k_2 the term in parenthesis is less than unity and, when squared, is always positive. Hence k_n/k_p must always be less than unity and therefore k_n is always less than k_p. Although only two beds are considered in this analysis, the same result holds for three or more beds. The mass permeability of a sequence of beds is thus orthotropic, although within each bed permeability conditions are isotropic.

Consider two beds of sand, each 3 m thick, separated by a seam of silty clay 3 mm thick. The permeability of the sand is 10^{-1} mm s^{-1} and of the silty clay 10^{-5} mm s^{-1}. Calculate the mass permeability of the three media when flow is (a) parallel and (b) normal to the bedding.

From Equation 2.45　$k_p = \dfrac{1}{6.003}(10^{-1} \times 3 + 10^{-5} \times 0.003 + 10^{-1} \times 3)$

$$= 0.995 \times 10^{-1} \text{ mm s}^{-1}.$$

From Equation 2.46　$k_n = 6.003 \left(\dfrac{1}{3/10^{-1} + 0.003/10^{-5} + 3/10^{-1}} \right)$

$$= 0.167 \times 10^{-1} \text{ mm s}^{-1}.$$

Thus the influence of the silty clay parting on k_p is negligible whereas k_n is only one-sixth of k_p. In the medium/coarse sands of the Folkestone Beds in southern England, seams or partings of silty clay follow the main and current bedding interfaces. These seams markedly reduce permeability normal to the interfaces compared with that parallel to the interfaces. Since the seams are impersistent, particularly those following current bedding, the effect would be experienced mainly in excavations but to a lesser extent in regional hydrology; in Example 2.24 isotropic conditions are assumed.

Chapter 3

STABILITY PROBLEMS IN FOUNDATION ENGINEERING

F.D.C. Henry

INTRODUCTION

Since failures in foundation engineering are often catastrophic in nature, they have been the main subjects of field studies. In contrast with this, although indeed a number of investigations of settlement have been made, observations of deformation have received insufficient attention. Thus, for example, there is a paucity of field studies concerning deformation of sheet pile walls and of pile groups subjected to vertical and horizontal loads. On the other hand, there have been many investigations of failures of banks and hillsides during the past three decades.

Frequently the factor of safety of a failure is found by relatively simple analysis to be close to unity. Nevertheless, the satisfaction afforded by this becomes somewhat tarnished as the advancing frontiers of knowledge reveal factors which were not taken into account in the analysis. As mentioned in the Introduction to Chapter 2, apparently good correlation is often achieved because some of the ignored factors are partly compensating.

Where details are known of groundwater static and seepage pressure, effective stress analysis generally affords the most reliable design method, except for the case of saturated clay in the undrained state. Desirable refinements include, for example, the use of pore pressure parameters and the concept of critical state. The testing of models in a centrifuge is an analytical innovation outside Russia where it has been employed for about four decades.

Study of the proceedings of the symposium on the stress–strain behaviour of soils (Parry, 1972) reveals a great variety of analytical techniques and refinements but, in practice, it is rarely impossible to ignore completely the fact that soil is the most variable material with which the engineer deals. The overall geology of the site must be borne in mind in the application and interpretation of any analysis. At the present time it is difficult to take account of a multiplicity of variations in strata, although it seems possible that finite and boundary element analyses will enable this to be done but whether or not such refinements will be justified

246

economically will certainly be debatable. The following examples illustrate the heterogeneous and anisotropic nature of rocks and soils.

The Upper Chalk is commonly regarded as a formation of relatively uniform lithology, apart from the obvious intercalated thin bands of marl and flint. Detailed examination reveals evidence of many cycles of sedimentation, with basal hardgrounds and a variety of other types of lithology brought about, for example, by current activity and bioturbation. Some characteristics of chalk have been discussed in Section 1.4.1. The influence of fossil borings and burrows on the permeability of late- and post-Glacial sediments has been mentioned in Section 2.7: these effects are superposed on those resulting from stratification. The density of these features varies from bed to bed.

It is well known that many clays show marked orthotropy in shear, brought about by stratification or by stress or both. Where rocks have been subjected to tectonic forces, the orientation of shear planes can be seen to vary from bed to bed, thus reflecting the influence of composition of each bed on shear strength. Until these and other diverse characteristics can be taken into account, it cannot be said that any technique of analysis is completely adequate.

The argument can be taken one stage further when the influence of the manner and rate of construction operations is considered. Thus, for example, the magnitude of stress imposed on tunnel linings and on excavation supports is often governed to a marked extent by the length of time soil or rock remains unsupported. The coefficient of earth stress at rest is particularly difficult to assess because of this where the backing is undisturbed or because of variable degrees of compaction where the backing is fill, even though laboratory techniques can be employed to determine this parameter accurately for prescribed conditions. Unfortunately, it is often impossible to predict the conditions reliably and the value of the coefficient must be estimated from the knowledge that it lies between the values of the active and passive coefficients, using judgment to decide where it lies between these limits. Whether or not laboratory tests will assist in resolving the issue depends on the circumstances of each case. In the case of driven piles, the bearing capacity of soil around the shaft is attained for the full buried length during the process of driving and the coefficient of earth stress must be of the order of the passive state value initially, although subsequent relaxation may cause this to fall. Clearly both theoretical and empirical approaches to soil and rock mechanics have a long haul before they meet on common ground.

3.1 FAILURE CHARACTERISTICS OF SOILS AND ROCKS

3.1.1 General observations concerning failure of soils and rocks

This section is a brief review of some of the aspects of shear failure of soils and rocks. When dense sand is subjected to a shear box or triaxial test it initially compacts very slightly while the shearing load is increased but thereafter dilates until failure occurs, when the shearing resistance drops and the volume may

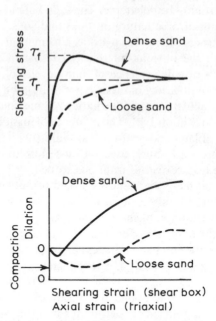

Figure 3.1 Characteristics of shear tests on cohesionless soils.

decrease slightly (Fig. 3.1). On the other hand loose sand compacts markedly under the shearing action but its ultimate resistance is approximately the same as that of dense sand after failure. The maximum resistance is known as the peak strength and the ultimate resistance as the residual strength.

The research of Hvorslev from 1934 to 1937 revealed that the true cohesion in saturated clay depends solely on water content and furthermore that the angle of internal friction influences the shear strength of every soil including clays. The results of tests on soil specimens prepared to determine shear strength are commonly analysed on the assumption that they conform to the Navier–Coulomb equation

$$\tau_f = c' + \sigma' \tan \phi' \qquad (3.1)$$

However, the parameters c' and ϕ' so obtained represent not intrinsic properties of the soil but instead quantitites which depend to a large extent on the method of testing. These empirical quantities are usually termed the apparent cohesion, or cohesion intercept, and the angle of shearing resistance of the soil, both determined here with respect to effective stress.

The deformation characteristics of many materials under progressively increasing stress can be represented reasonably well by elastic deformation up to the yield point, above which plastic deformation takes place, manifest in continuing yield at constant stress. With some materials, yielding is accompanied by work-hardening of the material and the stress increases as plastic deformation takes place until the ultimate strength is reached. However, in the case of shearing in soils, with a few exceptions such as loose sand under drained conditions, yielding is accompanied by work-softening and beyond the peak strength τ_f the stress decreases until the ultimate or residual strength τ_r is attained.

A brittle material is one which displays little or no plastic deformation before fracture occurs, whereas a ductile material displays appreciable plastic deformation before rupture. Arbitrarily, rock is termed brittle when the strain to fracture is less than 3% and ductile when the strain exceeds 5%. When the strain is between 3 and 5% the rock is termed semi-brittle. Characteristically, brittle clays display marked work-softening and Bishop (1967) introduced the parameter

$$\text{brittleness index } I_B = \frac{\tau_f - \tau_r}{\tau_f}. \tag{3.2}$$

However, different brittle clays may possess similar values of I_B although the strain to peak strength may be different and the best criterion of brittleness is undoubtedly the overall proportions of the stress–strain curves. Clay from the London Clay commonly possesses a value of I_B of about 0.7 whereas boulder clay is markedly ductile, with a value approaching zero. Sands commonly yield values of I_B between 0.3 and 0.4 although values outside this range have been recorded and loose sand may yield a value of zero.

The theoretical value of tensile strength of any particular material can be calculated from the atomic forces inherent in single crystals but this is found to be much greater than the value attained in a test on a specimen of the material. In 1921, Griffith suggested that the discrepancy is caused by minute cracks present throughout the mass of the material and demonstrated their existence and the general validity of relationships for strength based on this hypothesis. At the end of some cracks there is a concentration of tensile stress and, as the stresses on a specimen are increased, local failure occurs, the cracks are extended and general failure follows. The theory is of particular importance in the case of brittle materials. However, some authorities believe that progressive failure is accompanied by an increase in number of cracks rather than extension of earlier ones, and furthermore, that concentrations of compressive stress at flaws contribute to failure.

It is useful to consider briefly research on concrete in relation to the behaviour of brittle materials. Uppal and Kemp (1973) analysed the problem of prismatic columns of concrete under uniaxial compression, assuming that failure occurs by buckling of a macro-structure of longitudinal strips which presumably develop as the lateral stress reaches the yield point in tension in the vicinity of Griffith cracks. The phenomenon of splitting in uniaxial compression tests on concrete has been known for many years and is probably characteristic of most brittle materials. The theoretical results of Uppal and Kemp broadly confirm experimental results reported by Newman and Lachance (1964). The ultimate compressive stress is constant when the ratio of length to breadth of a column is between 2.5 and 3.0

but when the ratio is 2.0 the observed compressive stress is about 8% greater.

Since the particles in a sand or silt uncontaminated with clay are roughly spherical, they do not become preferentially oriented in engineering shear failures, although this does not necessarily apply where intense natural tectonic forces are involved. A parallel arrangement of clay minerals invokes a relatively low angle of shearing resistance, of the order 10°, but the presence of sand or silt in a clay inhibits to some extent a parallel re-orientation of clay particles and contributes to the intrinsically higher angle of shearing resistance. It is of interest to note that Bishop (1966) has observed, under high stresses, marked comminution of grains during shearing of sand and to a lesser extent during consolidation of sand.

Observation and statistical theory indicate that the width of rupture zones is of the order ten times the particle diameter, presumably roughly the mean diameter where soil or rock comprises particles of different sizes. Examinations of rupture zones in failures of clay slopes reveal widths varying from about 5 mm to 2 m but these comprise numerous thinner bands of reoriented particles or are complicated by fissures. Faults and thrusts of tectonic origin in rocks, involving displacements measured in metres or even kilometres, may have rupture zones no more than a few centimetres or decimetres in width, though they may extend to hundreds of metres. Clearly width depends not only on the micro-structure of the material but also on other factors.

Further details of the Griffith criterion of brittle failure are given, for example, by Jaeger (1969) and Price (1966). The strength of soils has been discussed at length by Bishop (1966) and the proceedings of a symposium on the stress–strain behaviour of soils has been edited by Parry (1972). Developments in sampling and testing procedures have been discussed by Broms (1980) and Bishop *et al.* (1973). Interest in the effects of earthquakes has led to studies of liquefaction of soils and two useful introductions to this topic are Martin *et al.* (1975) and Valera and Donovan (1977). The construction of offshore structures has prompted a number of studies of the behaviour of soils under dynamic, including cyclic, loading (Van Eekelen and Potts, 1978; Andersen *et al.*, 1980; Houston and Herrmann, 1980; Koutsoftas and Fischer, 1980; Martin *et al.*, 1980; Prakash and Puri, 1981).

Within recent years it has become possible to detect incipient failure in the laboratory and in the field through acoustic emissions as the microfabric of rocks and soils is disturbed and eventually slowly disintegrates. The technique is clearly of use in laboratory and field research but its greatest value is probably in monitoring the behaviour of constructions threatened by failure: some details are given, for example, by Koerner *et al.* (1978) and by Drnevich and Gray (1981).

3.1.2 Failure in cohesionless soils

A cohesionless soil can be defined as one in which interparticle forces or bonds make a negligible contribution to the mechanical behaviour of the soil. The strength of a cohesionless soil depends to a considerable extent on the magnitude of stresses normal to the potential shear plane and, at low stresses, shearing is accompanied by dilatancy in all but the loosest specimens. Because of the difficulty of obtaining undisturbed specimens of cohesionless soils, the vast majority of laboratory tests on these soils are performed on remoulded specimens and the consequent rearrangement of the particles is bound to influence engineering properties to some extent.

The influence of various factors on angle of shearing resistance is illustrated by the following equation derived by Koerner (1970) which represents the recommended design value for the angle of shearing resistance of cohesionless single mineral soils, based mainly upon test results of saturated quartz soils.

$$\phi_f = 36° + \Delta\phi_1 + \Delta\phi_2 + \Delta\phi_3 + \Delta\phi_4 + \Delta\phi_5$$

in which ϕ_f is the angle of shearing resistance with dilatation removed.

$\Delta\phi_1$ is the correction for particle shape; $\Delta\phi_1 = -6°$ for high sphericity and subrounded shape and $\Delta\phi_1 = +2°$ for low sphericity and angular shape.

$\Delta\phi_2$ is the correction for particle size (effective size, d_{10}); $\Delta\phi_2 = -11°$ for $d_{10} > 2.0$ mm (gravel), $\Delta\phi_2 = -9°$ for $2.0 > d_{10} > 0.6$ (coarse sand), $\Delta\phi_2 = -4°$ for $0.6 > d_{10} > 0.2$ (medium sand) and $\Delta\phi_2 = 0$ for $0.2 > d_{10} > 0.06$ (fine sand).

$\Delta\phi_3$ is the correction for gradation (coefficient of uniformity, CU); $\Delta\phi_3 = -2°$ for CU > 2.0 (well graded), $\Delta\phi_3 = -1°$ for CU = 2.0 (medium graded) and $\Delta\phi_3 = 0$ for CU < 2.0 (poorly graded).

$\Delta\phi_4$ is the correction for relative dansity (D_R); $\Delta\phi_4 = -1°$ for $0 < D_R < 50\%$ (loose), $\Delta\phi_4 = 0$ for $50 < D_R < 75$ (intermediate) and $\Delta\phi_4 = +4°$ for $75 < D_R < 100$ (dense).

$\Delta\phi_5$ is the correction for type of mineral; $\Delta\phi_5 = 0$ for quartz, $\Delta\phi_5 = +4°$ for feldspar, calcite, chlorite and $\Delta\phi_5 = +6°$ for mica.

Experiments reveal that, of the three most important failure criteria, the extended Tresca and the extended von Mises criteria give better agreement than the Mohr–Coulomb $\{(\sigma_1' - \sigma_3') = \sin\phi'(\sigma_1' + \sigma_3')\}$ over a limited range only, and the latter is of more universal application, particularly when compared with results obtained from plane strain tests. The Mohr envelope for dry or saturated dense cohesionless soils is usually slightly curved and should pass through ($\tau_f = 0 = \sigma$), since cohesion is clearly absent. However, the best straight line passing through the origin is drawn to determine the angle of shearing resistance. Slight cohesion may be induced in partly saturated sand by surface tension between particles. Undrained tests on saturated cohesionless soils are relevant to studies of liquefaction in relation, in particular, to catastrophic flows of such materials. Bishop and Edlin (1950) found that, with respect to changes in total stress applied under undrained conditions, saturated sand behaves as a purely cohesive material with an angle of shearing resistance $\phi_u = 0$ and a Mohr–Coulomb cohesion intercept $c_u = \frac{1}{2}(\sigma_1 - \sigma_3)$, except for dilatant samples in the low stress range where cavitation sets a lower limit to pore pressure change during shearing.

Laboratory studies of sands show that values of ϕ' calculated from the Mohr–

Coulomb criterion are usually higher in plane strain than in triaxial compression ($\sigma'_2 = \sigma'_3$); the difference varying from sensibly zero for loose sands to about $5°$ for dense sands. The shear box test yields values a few degrees lower than plane strain tests over the full range of porosities (Skempton and Hutchinson, 1969).

3.1.3 Failure in cohesive soils

A cohesive soil can be defined as one in which interparticle forces or bonds make a significant contribution to the mechanical behaviour of the soil. Cohesive soils possess both cohesive and frictional components of strength. A marked characteristic of many cohesive soils is the fact that both remoulding effected by a disorderly stress system, as employed in kneading, and rupture induced by an orderly stress system lead to re-arrangement of the particles and destruction of the original cohesive bond. Furthermore, pore pressure is generally of greater importance than it is in the case of cohesionless soils. The effective stress σ' in any given direction in an element of saturated soil is

$$\sigma' = \sigma - u_w \tag{3.3}$$

where σ is the total stress acting in that direction and u_w is the pore-water pressure. If the soil is partially saturated then

$$\sigma' = \sigma - [u_w + (1 - \chi)(u_a - u_w)] \tag{3.4}$$

where u_a is the pressure of the air in the pores and is somewhat greater than the pressure u_w of the water in the pores, and where χ is an experimental coefficient. When the soil is fully saturated, $\chi = 1$ and when the degree of saturation $S_r = 0$, $\chi = 0$. When $S_r \geqslant 90\%$, χ tends to unity and $(u_a - u_w)$ is small and hence the product term can be neglected. In partially saturated soil, a hydraulic piezometer measures u_w whereas a diaphragm piezometer measures u_a. When S_r approaches 100%, both types give approximately the same readings.

The difference between peak and residual strengths of silty clays is generally small but, as the clay content increases, the difference becomes greater owing to re-orientation of clay particles in the rupture zone. As the stiffness of the clay increases, the difference increases further owing to dilatancy, superposed on re-orientation, which is accompanied by an increase in moisture content. Experiments suggest that, before a constant value of residual strength is reached, shear displacements of the order of 1 m may be required and that the residual strength may be appreciably lower than in reversal shear box tests (Skempton and Hutchinson, 1969).

The zone of preferred orientation is a thin band in which the laminar clay particles assume parallelism with the direction of shearing. This has been observed in both laboratory and field specimens. The thickness of the primary band comprising the main rupture zone has been found in field failures to be between 0.01 and 0.02 mm. Associated with this are several secondary bands displaying parallel and sub-parallel orientation in a zone between 30 and 50 mm thick but beyond this scarcely any re-orientation is observed (Skempton, 1964) (see also Section 3.1.1).

For the purpose of design it is usually assumed that the shear strength parameters of a clay are governed only by particular effective stress conditions in laboratory tests which are deliberately related to the likely conditions of failure in the field. Indeed, this assumption is a prerequisite for simplicity of analysis but research shows that shear strength characteristics may vary considerably with a number of other factors. The determination of peak strength is influenced by the rate of shearing, by orientation with respect to bedding, an effect related to anisotropy, and, in the case of fissured clay, by the size of the specimen. The influence of anisotropy on strength is probably greatest at peak value but the effect may be almost lost at residual value owing to the re-orientation of particles.

In terms of effective stresses, the failure criterion is sometimes known as the Coulomb–Terzaghi relation

$$\tau_f = c' + \sigma' \tan \phi'. \tag{3.5}$$

When c' and ϕ' are written without subscripts it is assumed that they relate to peak strength, and the residual values are distinguished as c'_r and ϕ'_r. Very limited tests on clays, referred to by Skempton and Hutchinson (1969), suggest that the plane strain strength expressed by the Coulomb–Terzaghi criterion is 5 to 12% higher than the triaxial strength. Some of the factors which influence shear strength of clays and which lead to discrepancies between values recorded in laboratory tests and values developed in the field are discussed below.

Skempton (1964) plotted values of ϕ'_r against clay fraction for normally and over-consolidated clays (Fig. 3.2). Broadly, there is little difference in results for

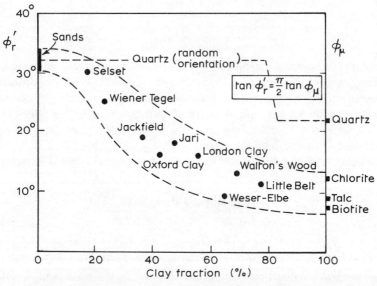

Figure 3.2 Variation of ϕ' with clay content.

clays in either condition and the fall in ϕ'_r with increasing clay content attains an ultimate range of values comparable with that of other layer-lattice minerals, including biotite, chlorite and talc, in a saturated state. The range is commonly from 20° to 10° for clays with low and high clay fractions respectively. Drained residual strength of cohesive soils is the subject of a comprehensive study by Lupini *et al.* (1981). Voight (1973) correlated Atterberg limits with residual shear strength. The application of the method of least squares to defining failure envelopes has been outlined by Bland (1983). In relation to sampling stress-relief, Kirkpatrick and Khan (1984) found from tests on specimens of manufactured kaolin and illite clays that the unconsolidated undrained test is misleading for predicting *in situ* undrained behaviour where stress relief is appreciable.

3.1.4 Anisotropy

In situ vane tests by Åas (1965) on three normally consolidated sensitive clays showed that the ratio of undrained strengths on vertical and horizontal surfaces was between 1:2 and 2:3. In this case the surfaces presumably correspond closely to surfaces normal and parallel to the bedding planes. On the other hand (Bishop, 1966), tests on heavily over-consolidated London Clay indicated that the strength parameters parallel to the bedding are less than those normal to the bedding but these relationships may apparently be reversed under the influence of pore pressures. Test results on anisotropic soils commonly show considerable scatter. The shear strength of laminated clays has been investigated by Lo and Milligan (1967): in such soils, each layer may be anisotropic within itself. The strength of clays when anisotropically compacted, as in the case of embankments, has been studied by Lee and Morrison (1970).

3.1.5 Sensitivity

Thixotropic hardening is exhibited by most clays to a greater or lesser extent, becoming soft when kneaded and slowly returning to a stiffer condition when allowed to stand. Thixotropy is defined as the property possessed by certain gels of liquifying when shaken and of returning to their characteristic jelly-like solid states on standing. No change in moisture content is required for this decrease and increase in strength to take place.

Many clays lose a proportion of their strength when remoulded. The degree of loss is expressed as the sensitivity

$$S_t = \frac{\text{undisturbed shear strength}}{\text{remoulded shear strength}}$$

where the strengths are measured in terms of cohesion intercept in undrained tests at the same moisture content. High degrees of sensitivity are found chiefly in Pleistocene soft clays deposited in brackish water or sea-water and in soft clays derived from volcanic ash. Some Scandinavian clays have sensitivities as high as

Table 3.1 Degrees of sensitivity

S_t	Type
~1	Insensitive
1–2	Slightly sensitive
2–4	Medium sensitive
4–8	Very sensitive
8–16	Slightly quick
16–32	Medium quick
32–64	Very quick
> 64	Extra quick

1000. Rosenqvist (1953) amended the classification of Skempton and Northey (1952) to include groups of quick clay (Table 3.1)

Laboratory and field evidence shows that high sensitivity is displayed by estuarine and marine clays in which the original concentration of salt in the pore-water has been reduced by leaching. Incidentally, artificial sedimentation tests (Bjerrum and Rosenqvist, 1956) show that clay deposited in fresh water possesses a lower water content and a higher shear strength than clay deposited in salt water and that both clays display the same sensitivity. Extensive landslides occur in Scandinavian quick clays which were deposited in a marine environment and subsequently overlain by ice sheets during the Pleistocene. It is probable that sensitivity up to about 3 is accounted for by thixotropy alone since this is capable of increasing the strength of some remoulded clays by up to about 200%. Sometimes clays exhibit a small increase in strength in laboratory tests on remoulding. This may be due to loss in moisture content and to reduction in void ratio by excessive consolidation during remoulding or it may be due to the presence in the undisturbed state of fissures which are eliminated by remoulding or it may be due to some undefined physico-chemical phenomenon. Table 3.2 is a summary of causes of sensitivity given by Mitchell and Houston (1969).

It is sometimes found that the value of sensitivity determined from *in situ* vane tests is higher than the value determined from laboratory tests. This could be attributed to partial loss of strength caused by disturbance during sampling. However, although the undisturbed strengths determined by laboratory tests and *in situ* vane tests on a late-Glacial normally consolidated silty clay from a site on the Sussex coastal plain were practically the same, the remoulded laboratory strength was about twice the so-called remoulded *in situ* strength. Examination of blocks of soil taken from strata in which vane tests have been performed show that the sheared zone is bounded by a fine and almost perfectly cylindrical rupture surface. In this case the original interparticle structure is replaced by one in which the orientations accord in a very thin zone with the rotary shearing movements. It is normal practice to base remoulded strength on the resistance developed around the surface after failure in the field vane test but in fact it is the residual strength which is measured. On the other hand, during remoulding in the

Table 3.2 Summary of causes of sensitivity development in clays (after Mitchell and Houston, 1969)

Mechanism	Type of reaction	Limit of sensitivity	Predominant soil types affected
Metastable particle arrangements	Physical	Slightly quick (8–16)	All clays
Silt skeleton-bond clay	Physical	Very sensitive (4–8)	Clay–silt–sand mixtures
Cementation	Chemical	Slightly quick (?) (8–16)	All soils containing potential cementing compounds
Ion exchange	Physico-chemical	Slightly quick (?) (8–16)	Leached and weathered clays
Leaching of salt	Physico-chemical	Extra quick (> 64)	Glacial and post glacial marine clays
Weathering	Chemical	< 1.0 to medium sensitive (1–4)	All soils – magnitude of effect depends on mineralogy
Thixotropic hardening	Physico-chemical	Medium sensitive* to slightly quick (2–16)	Clays
Dispersing agent addition	Physico-chemical	Extra quick (> 64)	Clays – particularly organic bearing or organic deposit associated

*Pertains to samples hardening starting from present composition and water content. Role of thixotropy in causing sensitivity of clays *in situ* is indeterminate.

laboratory a new structure is imparted to the soil which is not the same as the structure around the periphery of the sheared zone in the vane test. It would be preferable, therefore, to determine the *in situ* remoulded strength after remoulding the mass of soil in and around the vane by working the apparatus up and down in the borehole but this operation is generally not easily effected.

A different nature of sensitiveness to disturbance is displayed by some stiff clays. In this case the process of taking a sample, followed by extrusion and preparation in the laboratory, leads to opening of laminations and fissures. Even careful cutting and extraction of a block of clay from an excavation does not prevent a certain amount of disturbance. However, driving a standard 100 mm diameter sampling tube creates more disturbance than careful extraction of a block and specimens of clay from the London Clay obtained by tube sampling have possessed shear strengths about 75% of those possessed by specimens taken from blocks. This form of sensitivity involves disturbance of macrostructure whereas the sensitivity of soft clays discussed above involves disturbance of microstructure. Further investigations concerning London Clay have been reported by Bishop *et al.* (1965) and Hooper and Butler (1966).

Ward *et al.* (1965) found that rotary-coring methods lead to a minimum of disturbance when sampling London Clay. In the case of normally consolidated clays, provided that a saturated sample can be obtained without mechanical disturbance and despite the non-elastic behaviour of such clay and the very considerable changes in stress which take place even during perfect sampling, the undrained strength obtained in laboratory tests is virtually the same as the corresponding strength of the clay in the ground (Skempton and Sowa, 1963).

3.1.6 Organic content

The organic content of a soil may vary from one of dominance, as in the case of peat, to a few per cent, as in the case of very slightly organic sands, silts and clays. Clearly the strength of soil varies with both the amount and type of organic matter present. The strength of peat has been studied by McFarlane (1969) and Hanrahan *et al.* (1967). Peat has a relatively low shear strength but sometimes the value is markedly greater than the strength of silts and clays at the same site. Enhancement of the strength of peat is thought to be due to the tensile strength of plant fibres but in some types of peat these are largely destroyed by organic decomposition.

Little appears to be known about sands, silts and clays with an organic content of, say, less than 20%, and these often introduce uncertainties in analysis and design, particularly in preliminary studies when the results of strength tests may not be available. Characteristically these soils are of late- and post-Glacial age and were formed in lacustrine, deltaic and estuarine (including mud-flat) environments where both animal and vegetable remains contribute to organic content. Clearly the influence of this organic material on shear strength depends on the degree of decomposition but frequently the material can be expected to be in colloidal form, thus supplementing the colloidal mineral content, and may be combined in a variety of ways with different mineral constituents of soil. Generally, the presence of organic matter can be expected to have an adverse effect on shear strength but Franklin *et al.* (1973) concluded that, for the soils which they studied, the influence is much less than that of minor mineralogical or structural differences. Determination of organic content has been investigated by Al-Khafaji and Andersland (1981).

3.1.7 Specimen size

Clearly a test specimen should be of size sufficient to include a representative proportion of large particles, such as gravel, and of discontinuities, such as fissures, although in practice this requirement cannot often be satisfied. Large scale *in situ* tests and large diameter triaxial tests on stiff, fissured clay of the London Clay suggest that the *in situ* value of undrained shear strength is 65 to 80% of values derived from 75 mm × 38 mm diameter specimens in the triaxial test. Investigations by Marsland and Butler (1968) on stiff, fissured clay from the

Barton Beds reveal an undrained strength for 200 mm × 100 mm diameter triaxial specimens of 65 to 85% of the strength of 75 mm × 38 mm diameter specimens. It follows that the *in situ* strength of fissured clays is likely to be roughly 70% of the strength derived from 75 mm × 38 mm specimens but, in the case of intact clays, the *in situ* strength is likely to be sensibly the same as the strength derived from 75 mm × 38 mm specimens, ignoring, of course, the influence of sample disturbance. Furthermore, it is likely that the *in situ* strength of a fissured clay will be reasonably represented by the strength of triaxial specimens 200 mm × 100 mm or 300 mm × 150 mm.

3.1.8 Rate of shearing

The rheological influence of rate of shearing leads to differences in strength between relatively rapid failures in laboratory tests and slower failures in the field. The following rates are generalized from a discussion by Skempton and Hutchinson (1969).

In a drained test, peak strength in usually attained in about 24 h and, in the case of overconsolidated clays, the *in situ* strength would be about 90% of the laboratory value if the time to failure in the field were 3 years. Hence the strength in long term failures may be 10 to 15% less then the laboratory value. The reduction in residual strength with rate of shearing is less than this. Some normally consolidated clays show much greater reductions in peak strength.

In undrained tests, peak strength is normally attained in about 15 min. If the time to failure in the field is 1 day, the peak strength of many clays can be expected to be 10 to 15% less then the laboratory value although some normally consolidated clays may experience a much greater reduction.

If it is desired to investigate the influence of rate of shear on strength, then laboratory tests can be made at, say, five different rates ranging from 15 min to 1 week to peak undrained strength or 1 day to 1 month to peak drained strength. The resulting graph of strength against log time can be extrapolated for investigation of longer term field failures.

3.1.9 Progressive failure

In a mass of clay subjected to disturbing forces, progressive failure develops along a potential rupture surface as the strains corresponding to peak strength are exceeded locally and the consequent local reduction in shearing resistance imposes excessive stresses on other parts of the potential rupture surface where the sequence is repeated. Failure may occur before the residual strength is reached along the entire length of the rupture surface. The strength at fissures is likely to be less than the peak value and stress concentrations at these discontinuities are likely to encourage the early development of strength approaching the residual value. Consequently, the average strength along a potential rupture surface in a fissured clay can be expected to be less than the peak value. Failure is encouraged by a relatively high level of the phreatic surface since this would lead to relatively low values of σ' and hence also residual strength. The average reduction in shear strength along a rupture surface can be expressed as a residual factor $R = (\tau_f - \bar{\tau})/(\tau_f - \tau_r)$ where $\bar{\tau}$ is the average shear strength at failure (Skempton, 1964). In effect, R is the proportion of the total length of rupture surface along which the strength has fallen to the residual value. Skempton and La Rochelle (1965) found that R varied between 0.2 and 0.45 in short term failures in London Clay at Bradwell, Essex, whereas in long term failures, occurring 20 to 50 years after excavation, R varies between 0.55 and 0.8. The operational strength of fissured clays has been studied by Lo (1970) and the application of residual strength to the design of cuttings in overconsolidated fissured clays has been discussed by Symons (1968).

Closely connected with long-term progressive failure of clays is the phenomenon of softening. If the lateral support to a mass of soil is removed by excavation or by erosion, there is a tendency for the soil to expand laterally, with consequent increase in void ratio accompanied by an increase in moisture content. The removal of overburden, either artificially or naturally, also leads to softening although the effect may extend to a depth of a few metres only since lateral restraint restricts opening of vertical fissures. Softening involves a reduction in shear strength of clays although the process may extend over many years before equilibrium is attained owing to the low permeability of clays. The process is facilitated by the presence of vertical fissures which tend to open as lateral support is removed. The shear strength of the clay adjacent to the fissure walls decreases first and, in some situations, failure may develop along the fissures even though the strength of the soil in the blocks bounded by the fissures shows little deterioration. In the field, softening is hastened by prolonged spells of wet weather.

Progressive failure has been studied by Lo (1972) and by Lo and Lee (1973).

3.1.10 Variation of shear strength with depth

The shear strength of a thick homogeneous bed of clay commonly varies with depth (Skempton and Bishop, 1950). Three zones are observed in the soft clays of late- or post-Glacial age (Skempton and Northey, 1952). In zone 1 the clay is normally consolidated under the effective overburden stress and the shear strength increases linearly (Fig. 3.3a). In zone 2 the strength is relatively constant and the increase above the normally consolidated value may be due to intermittent desiccation. The strength in zone 3 increases rapidly towards the ground surface, provided this is above the phreatic surface, and although this is commonly considered to be due to desiccation, Moum and Rosenqvist (1957) consider that with Scandinavian Pleistocene marine clays it is due to mineralogical changes associated with weathering. The rate of increase of shear strength with depth, expressed as the ratio c_u/σ_o where σ_o is the effective overburden stress, is sometimes not as great as would be expected from consolidated undrained tests

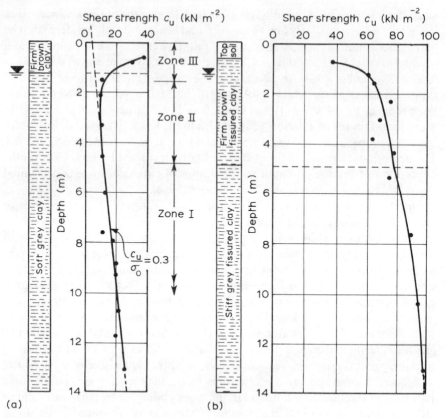

Figure 3.3 Variation of strength with depth for two clays. (a) Recent estuarine clay normally consolidated below 5 m, $W_L = 80\%$, $W_P = 25\%$, $W = 40$ to 65%. (b) Heavily over-consolidated Eocene clay, $W_L = 75\%$, $W_P = 24\%$, $W = 26$ to 30%.

on undisturbed samples. This may be due to discrepancies between field and laboratory values of the ratio σ_1/σ_3. For a normally consolidated clay the ratio of the undrained shear strength to the vertical effective stress under which it was consolidated *in situ* shows a close correlation with the plasticity index, I_p which can be expressed as follows (Skempton, 1957)

$$\frac{c_u}{\sigma_o} = 0.11 + 0.0037\, I_p. \tag{3.6}$$

Stiff clays generally have been pre-consolidated by a weight of overburden which now has been removed. The strength is relatively great but in the upper 1.5 to 5 m it is commonly less than the pre-consolidated value owing to softening (Fig. 3.3b).

3.1.11 Further discussion of the properties of stiff fissured clays

The properties of stiff fissured clays have been discussed generally in the foregoing notes but it is desirable to extend the details in the light of further research. Marsland and Butler (1968) have described a typical stiff fissured clay from the Barton Beds, Hampshire, as follows.

The detailed structure of the clay varies with both areal position and depth. In the top 3 m, the greater part is composed of hard irregular shaped lumps of clay which generally have dimensions between 5 and 100 mm but locally up to about 200 mm. The horizontal dimensions of the lumps are usually greater than the vertical, the maximum horizontal dimension being 1 to 3 times the vertical. The lumps fit closely together when the clay is *in situ* but fall apart when the clay is excavated. Fissures separating the lumps rarely form continuous planes over more than one or two lumps. The dimensions of the lumps generally increase with depth and attain 400 mm at depth of 10 m. The more sandy clays do not contain fissures.

The shear strength characteristics of stiff fissured clays have been summarized by Marsland (1971) who draws the following conclusions.

For a particular clay the strength measured in field and laboratory tests depends on the extent of fissuring; the size of the element of soil tested in relation to the spacing of the fissures; the inclination, orientation, shape, and surface roughness of the fissures; the stress changes which occur during boring, excavation, sampling and testing of the clay, and the length of time that the clay remains under each stress condition. Different clays have different basic properties and previous stress histories which also vary with location and depth. These factors affect the properties of the intact clay and the nature of the fissures.

The strength applicable to a particular engineering problem depends on some or all of these factors. For clays containing fissures which are not too widely spaced in relation to the volume of soil tested, and provided that the fissures occur with a range of inclinations, laboratory and field tests of a reasonable size can give reliable measures of the large-scale strength. It should be noted that these strengths are applicable to the clay only at the time of sampling and testing and the possibility of further deterioration must be considered in the case of long-term stability problems.

In clays where the discontinuities are more widely spaced the determination of the appropriate strength for a particular application becomes more difficult except in special cases where extensive joint planes occur at critical positions and inclinations. It has yet to be shown whether or not the range of parameters, extending from those corresponding to shear along discontinuities to those corresponding to the intact clay between discontinuities, can be significantly reduced by tests on large specimens. In spite of obvious limitations, Marsland considers that tests on large specimens in which long periods are allowed for equilibrium to be reached will reduce the variation to more tolerable limits. Wherever possible these studies should be made in conjunction with measurements to determine the stability or instability in the field. Studies of the fissures are obviously an important part of all such investigations.

Since fissured clays are complex materials it is very unlikely that a single, simple criterion will be found for the estimation of their full-scale strength. Caution should be exercised in transferring the principles obtained from measurements on one fissured clay to another. At the present time it is essential that each type of fissured clay be regarded as a new material and studied by all the means available. In this way the uses and limitations of laboratory tests, *in situ* field tests and other methods of field investigation will be established.

The relief of *in situ* stress following excavation has a marked effect on shear strength. Tunnelling in stiff fissured clays has revealed that fissures open fairly rapidly during and following excavation of the face. Limited data from 152 mm diameter plate tests in a shaft in London Clay (Ward *et al.*, 1965) indicated that strength measured 4 to 8 h and 2.5 days after excavation were approximately 85 and 75% respectively of those measured 0.5 h after excavation. Laboratory tests on 38 mm diameter specimens cut from blocks extracted from other shafts in London Clay yielded values of strength after 150 days storage of about 75% of the values measured after 5 days. It is considered that this time-dependent effect is possibly due to progressive extension of fissures and microcracks within the specimens. It may be due also to softening of the clay in the walls of fissures brought about by expansion of the clay as stress is relieved in the fissures, leading to an increase in void ratio accompanied by an increase in moisture content, the moisture being drawn from the interior of the intact block.

Typical results of triaxial tests on 38 mm diameter specimens cut from a block of fissured clay from the London Clay are shown in Fig. 3.4. These illustrate the influence of the frequency, continuity as opposed to complexity, shape, roughness and orientation of fissures on strength. In order to assess a realistic value of strength it is necessary to examine the composition and structure of each specimen. The best estimates of strength are generally derived from 865 mm and 292 mm diameter plate bearing tests, which incidentally may require penetration up to 50% of the plate diameter, and from *in situ* tests with a shear box 610 mm square.

The highest and most reproducible results were obtained from 865 mm diameter plate tests for which the surface was prepared by hand excavation of 50 to 70 mm from the floor of the borehole before bedding the plate on plaster. With present techniques, tests on smaller diameter plates made in boreholes too small to permit a man to descend and prepare the base give lower, and probably less reliable, values.

In connection with bearing capacity and shear strength tests with plates in boreholes and pits, an investigation by Marsland (1972) is particularly relevant. The conclusions apply strictly to unfissured soft clays and may be subject to modification in relation to stiff fissured clays, although the observations for London Clay in Fig. 3.5 are reasonably consistent. Tests on a plate having a diameter equal to the diameter of the borehole showed that the maximum

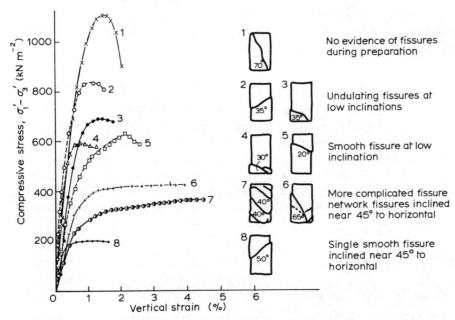

Figure 3.4 Relation of strength measured in triaxial tests of 38 mm diameter specimens to the fissures in the specimens.

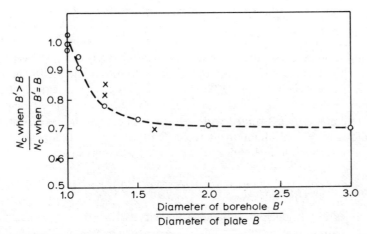

Figure 3.5 Effect of ratio of borehole diameter/plate diameter on bearing capacity of soil, where ○ represents model tests in soft clay and $_*$ represents large diameter *in situ* tests in stiff fissured London Clay (after Marsland, 1972).

bearing stress was the same for unlined boreholes, lined boreholes in which the liner remained stationary and lined boreholes in which the liner moved down at the same speed as the plate. Tests on plates in boreholes with a diameter larger than that of the plate showed that the bearing capacity dropped rapidly as the ratio of the hole to plate diameter was increased from 1.0 to 1.5, beyond which it became asymptotic to the capacity of a plate located at ground surface (Fig. 3.5).

3.2 ANALYSIS OF THE STABILITY OF HILLSIDES AND BANKS

Evidence of instability of hillsides is seen in landslides; these are manifestations of geomorphological processes of degradation. Unstable banks may be encountered in artificial excavations and fills. For the analysis of any particular case it is necessary to recognize and classify the type of movement which has occurred or which may occur in the future, to determine the properties of the materials involved in the movement and to select the appropriate techniques of analysis. The discussion which follows is concerned largely with hillsides and banks of cohesive soil.

3.2.1 The classification of failures in hillsides and banks of cohesive soils

Falls (Fig. 3.6(a))

These occur mainly in sea and river cliffs and in steep sided excavations. Failure is commonly preceded by bulging at the toe and the formation of a tension fissure at the rear, which sometimes develops at a joint.

Rotational slide, slip or slump (Fig. 3.6(b))

These movements occur characteristically in moderately steep slopes of fairly uniform clay. Generally the ratio D/L of maximum thickness D to maximum length L lies between 0.15 and 0.33 when the slides are deep-seated. The more uniform the clay, the more likely the slide will be nearly circular. Non-circular slides are commonly associated with stiff clays where the degree of weathering varies along the potential sliding interface. Shallow rotational slides are common on slopes of moderate inclination in weathered or colluvial clays.

Compound slides (Fig. 3.6(c))

These slides are associated with heterogeneity within the slope, such as the presence of a resistant stratum at moderate depth. The failure interface involves both curved and planar elements.

Translational slides (Fig. 3.6(d))

These require the presence of a relatively resistant layer at a shallow depth beneath the slope. This condition is met, for example, when a more or less uniform thickness of weathered material overlies relatively unweathered material. At one extreme, more or less intact blocks may slide independently, and at the other, a

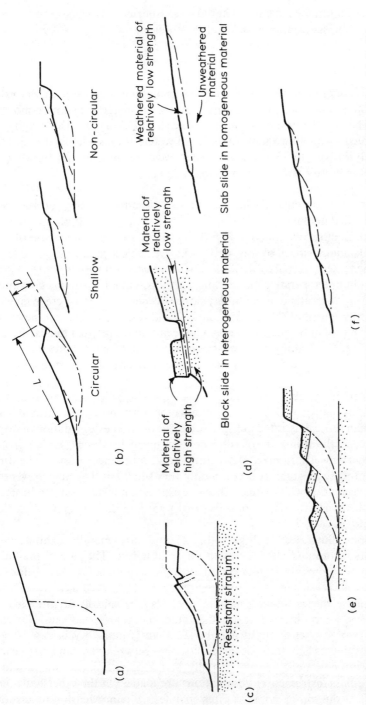

Figure 3.6 Classification of slope failures. (a) Falls; (b) rotational slides; (c) compound slide; (d) translational slides; (e) multiple slides; (f) successive slides.

more or less continuous mass of material is transported and, when excess water is present, the mechanism approaches that of a flow.

Flows

These commonly involve the movement of material at relatively high moisture content, often with more or less intact blocks of material at lower moisture content. The rate of flow is commonly between 5 and 25 m year^{-1} although when accompanied by an appreciable quantity of water, as in a bog burst, the rate is more likely to be 5 to 25 m min^{-1}. Under cycles of freezing and thawing, the movement is more correctly termed solifluction.

Various combinations of the mechanisms described above, or successive movements in different parts of a slope, lead to a variety of multiple and complex slides, such as those illustrated in Figs 3.6(e) and (f). In addition to these obvious movements, many slopes of soils and rocks are subject to creep.

This phenomenon is partly seasonal, involving only surface material to a depth of about 1 m in temperate climates, and partly perennial, involving continuous slow shearing at greater depth. On clay slopes steeper than 1 in 10, creep may well lead to displacement of shallow foundations and opening of joints in pipes. The rate of creep probably seldom exceeds a few centimetres per year and is often less than 1 mm year^{-1}. A study of the various processes involved in the degradation of cliffs on the Hampshire coast, England, has been described by Barton and Coles (1984).

Associated with the movements discussed above will be found superficial structures, as shown in Fig. 1.4. A number of other processes, generally less obvious than those illustrated in Fig. 3.6, contribute to the degradation of slopes. These include the transport of soil and rock particles down sloping ground surfaces by rainwash, rainbeat, flow in rills and other mechanisms. The direct influence of these on stability is commonly very small but they may prevent the growth of vegetation which could otherwise exert a minor stabilizing influence by reducing the moisture content of soils or by retarding the transport of particles down slopes.

The slides illustrated in Figs 3.6(b), (c) and (d) generally exhibit minor movements before and after the main phase of failure. The rate of pre-failure movement is commonly several centimetres per day for a few days before the main failure, increasing to several decimetres on the day preceding failure. During the main phase, the rate of movement in quick clays is typically 0.5 to 1.5 m s^{-1} – about 10 m s^{-1} was observed at the Rissa landslide, near Trondheim, Norway – but in clays of low sensitivity it may be less than 10 mm s^{-1}. The rate of post-failure movements has been observed to vary between zero and 6 m year^{-1}. Overstressing commonly occurs initially at the toe of slopes but is seldom apparent unless instrumental observations are made. On the other hand, later movement at the top of slopes is often manifest in minor tension fissures and scarps.

Comments on the classification of slope failure

The mechanisms illustrated in Fig. 3.6 can be detected in many failures, such as the Sevenoaks, Kent, slip in the railway cutting in Weald Clay in 1939, the repeated failures in the Folkestone Warren cliffs, Kent, the planar slide in Coal Measures shales at Jackfield, Shropshire, in 1951 and others reported from many regions of the Earth. Most of these failures are large scale but occasionally the mechanisms are displayed on a small scale, such as a 3.4 m radius circular arc slip in a 2.0 m high bank in the Weald Clay at Ewhurst, Surrey, which occurred in 1962. Nevertheless, Fig. 3.6 may give a false impression of facility with which slope failures can be analysed: there are many failures which cannot be related directly to these mechanisms. The Tertiary beds above the Chalk cliffs at Newhaven, Sussex, are dominantly clays interbedded with seams of sand. About 1943 water seeping through a sand seam towards the cliffs eroded the sand, the overlying beds collapsed and formed a dam behind which water accumulated, eventually bursting through the dam carrying slurry and blocks of clay over the cliffs. Erosion of the sand was prevented by placing a filter layer of beach sand over the outcrop (Section 2.7.3 and Fig. 2.36). Similar modes of failure can be seen on the wholly Tertiary cliffs of the Hampshire and Isle of Wight coasts, although here marine erosion of the toe and the influence of surface water flowing over the top of the cliffs are further contributory factors. The boulder clay cliffs of the Yorkshire coast often contain seams and lenses of sand and gravel and similar failures can be seen here. In 1948, deep excavations were made in the Coal Measures shales and sandstones near Burnley, Lancashire, and some of the excavated material was dumped on peat on the moor top. Peat was displaced and formed a dam across a stream, behind which accumulated a large quantity of water. Eventually the dam burst and thousands of cubic metres of peat and excavated material flowed down the gully in the escarpment, across a road and into fields beyond. The Hastings Beds of south-east England broadly comprise thick sequences of shaly clays and weak sandstones, both with a marked silt content. Here also seepage of water leads to failures of complex kinds. Solifluction lobes and minor slips of Quaternary age are widespread in the British Isles and commonly form hummocky ground: the interface of rupture must be very irregular but over it the strength of the soil must be at the residual value. Most of the failures discussed in this paragraph are of relatively minor importance, although occasionally large-scale failures occur in the geological formations mentioned which display the mechanisms shown in Fig. 3.6.

Two classic natural slopes in the British Isles are Herne Bay cliffs, Kent, in London Clay and Folkestone Warren cliffs, Kent, where the Chalk overlies the clayey Gault which, in turn, overlies the sandy Folkestone Beds. These have been the subjects of numerous papers: recent publications giving earlier references are Bromhead (1979) and Hutchinson et al. (1980), the former deals also with transitions between various types of movements. The form of some rupture interfaces can be represented fairly closely by a series of straight lines and the soil or rock involved in the sliding mass can be divided into triangular and trapezoidal masses. Examples of such analyses have been reported, for example, by Wood (1956) and Lambe *et al.* (1981).

Although many failures are due to natural causes, others are of artificial origin. Seepage of water may emanate from fractured sewers and water mains or from cable ducts or along poorly backfilled trenches. Of course, it is possible that in some cases sewers and mains are fractured as a result of preliminary movement associated with incipient slope failure. Furthermore, there is no question that heavier and denser traffic on roads in also contributing directly and indirectly to slope failures. Again there are numerous instances of failures caused by forming banks to unsafe inclinations simply to increase the level area above or below the bank for development purposes. Inevitably problems arise in apportioning responsibility in cases discussed in this paragraph, particularly when at least one of the parties involved has gone into liquidation.

3.2.2 Stability analyses

(a) General remarks on stability analyses

Occasionally it is difficult to arrive at a factor of safety of unity, within $\pm$ a few per cent, when analysing a slope failure: discrepancies in any or all of the elements in the problem may contribute to this in varying degrees. Clearly the geometry of the slope before failure and of the rupture interface must be reasonably accurate. The choice of rational values of soil parameters may be in question, particularly shear strength, since field conditions at failure may not be simulated sufficiently closely in laboratory tests. The extent of the influence of weathering or of sensitivity or of the reduction to residual strength may be impossible to assess accurately. The most likely source of error will probably be found in ground-water conditions and consequent values of pore pressure since these are often difficult to establish accurately after failure. Prolonged heavy rainfall over days, weeks or months commonly precipitates failures. However, it is often difficult to decide whether or not equilibrium conditions have been attained and, therefore, also the position of the phreatic surface. Minor slope movements are commonly associated with incipient failure and these may lead to fracture of buried pipelines or opening of pipe joints. This will at least increase the supply of water to the soil and, if under pumping pressure, will cause a marked local increase in pore water pressure. The pressure immediately outside the pipe will probably be appreciably less than the pumping pressure and is likely to be inappreciable beyond a few metres. Nevertheless, the influence on slope stability may be conspicuous. Relatively high pressure in gravity sewers may be incurred during intense rainfall which can lead to backing up in manholes, thereby increasing static head. Errors in analyses performed by manual computation can usually be detected more readily than when performed by computer: in the latter case the program must be studied very carefully to ensure that data conform to the prerequisites.

During the last decade the technique of testing models in a centrifuge has provided a further method of slope stability analysis: some papers giving further references are Hird *et al.* (1978), Fragaszy and Cheney (1981), Davies and Parry (1983) and Padfield and Schofield (1983). A comprehensive review of the long-term stability of slopes has been presented by Simons and Menzies (1978) and of the stability of embankments on soft ground by Menzies and Simons (1978). Random interface generation is the subject of a study by Siegel *et al.* (1981) and Karal (1977a, b) has given details of the energy method for stability analyses. Charles and Soares (1984) have published charts to facilitate the analysis of rockfill slopes, taking account of marked curvature of the Mohr failure envelope at low and medium stresses which is characteristic of the behaviour of rockfill.

(b) Short-term and long-term conditions

In current practice, limit equilibrium analyses are made to assess the degree of stability of a slope. It is postulated that incipient failure obtains along a continuous interface of known or assumed shape. It is usual to treat the problem as two-dimensional and the factor of safety so determined is unlikely to exceed 10% less than that determined by taking into account the three-dimensional nature of the real problem (Skempton and Hutchinson, 1969): recent studies have been published by Baligh and Azzouz (1975) and Hovland (1977).

One of the stress changes brought about by forming an excavation in saturated soil is a reduction in pore pressure in the soil adjacent to a newly exposed face. In the case of sands and gravels, adjustment of the pore pressure to new equilibrium conditions is rapidly achieved but the low permeability of silts and clays considerably retards the completion of this process for periods of several months or years. Thus in the case of silts and clays it is often necessary to consider short-term conditions before equilibrium is re-established and also long-term conditions when equilibrium is attained. In the case of many geological faults and thrusts, movement is undoubtedly facilitated by reduction in shearing resistance invoked by high pore pressure generated by frictional heat and the phenomenon has been investigated in relation to landslides by, for example, Voight and Faust (1982).

In terms of effective stress the shear strength mobilized under conditions of limiting equilibrium when a factor of safety F is applied is

$$\frac{\tau_f}{F} = \frac{c'}{F} + (\sigma - u)\frac{\tan \phi'}{F}. \tag{3.7}$$

Only rarely is the total normal stress on the actual or potential shear surface statically determinate but it is usually possible to determine an approximate value of sufficient accuracy for practical purposes. The value of u at the shear interface is either measured by field observations or estimated from a flow net. The analysis of the long-term stability of a slope is usually made in terms of effective stress.

In terms of total stress, assuming $\phi_u = 0$, the shear strength mobilized under conditions of limiting equilibrium when a factor of safety F is applied is

$$\frac{\tau_f}{F} = \frac{c_u}{F}.$$ (3.8)

In such terms analyses are considerably simplified since neither the total normal stress nor the pore pressure are required for the determination of shear strength. However, it is necessary that in laboratory undrained tests, the condition $\phi_u = 0$ be satisfied in order that the results can be applied in total stress analyses. The analysis of the short-term stability of a slope is usually made in terms of total stress.

An interesting note relating to the principle of the mid-slope circle in a $\phi_u = 0$ analysis has been presented by Brand and Shen (1984).

(c) Falls

From experiments on a small gelatine model of a vertical cut of height H, with a horizontal surface to the backing, Terzaghi (1943) concluded that the maximum tensile stress at the surface of the backing is attained at a distance of about $H/2$ from the face and hence a tension crack is most likely to develop here. The critical height of a vertical face can be estimated by the Rankine–Bell equation for active stress on retaining walls. If the relief of stress due to tension in the soil is taken into account, the total horizontal force is

$$P_a = \frac{\gamma H^2}{2N_\phi} - \frac{2c'H}{\sqrt{N_\phi}}.$$ (3.9)

The critical height H_c for a vertical bank of laterally unsupported soil is that at which P_a is zero; that is, when

$$H_c = \frac{4c'}{\gamma}\sqrt{N_\phi}.$$

For clay with $\phi_u = 0$, $H_c = 4c_u/\gamma$. It must be assumed for this equation to be satisfied that the soil is competent to resist the greatest tensile stress to which it is subjected. In practice, cohesive soil is able to resist a certain amount of tensile stress, but not the full amount required by this assumption, and consequently tensile cracks will develop to a depth $z_c < 2c_u/\gamma$. Then it follows that, for $\phi_u = 0$,

$$P_a = \left(\frac{\gamma H^2}{2} - \frac{\gamma z_c^2}{2}\right) - \left(2c_u H - 2c_u z_c\right)$$ (3.10)

and when $P_a = 0$, $H = H_c = 4c_u/\gamma - z_c$.

R.M. Wynne-Edwards made tests in 1938 to determine the critical height for a vertical bank of remoulded clay from the London Clay. The clay was filled to different levels in a box 3 m × 2 m × 3 m deep and side was rapidly removed each time until the level at which the bank failed was found. The failure was repeated

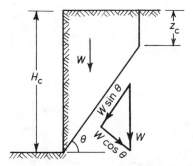

Figure 3.7 Analysis of a fall.

three times and the mean experimental value of H_c was found to be 2.4 m. Tests gave $c_u = 13.4 \, \text{kN} \, \text{m}^{-2}$ and $\gamma = 1.77 \, \text{kN} \, \text{m}^{-3}$ and hence $4c_u/\gamma = 3.0 \, \text{m}$. Whereas $2c_u/\gamma = 1.5 \, \text{m}$, the mean depth z_c of the tension crack was observed to be 0.5 m. Hence the calculated value of

$$H_c = \left(\frac{4c_u}{\gamma} - z_c \right) = 2.5 \, \text{m}$$

and this compares favourably with the observed value. The mean angle of inclination of the slip plane was 51° whereas theoretically, with $\phi_u = 0$, it should have been 45°.

Where a failure has occurred and the values of H_c, θ and z_c (Fig. 3.7) have been recorded, the problem can be analysed by equating the disturbing force on the slip interface to the resisting force. Thus

$$W \sin \theta = c'(H_c - z_c) \operatorname{cosec} \theta + W \cos \theta \tan \phi' \tag{3.11}$$

where $W = (\gamma/2)(H_c^2 - z_c^2) \cot \theta$. The values of γ, c' and ϕ' can be determined from tests: the relevant shear strength parameters may be either the peak or the residual values. It is possible that the strength may not have fallen to the residual value along the entire length of the slip plane and then the residual factor would be less than unity but greater than zero.

(d) Circular slip interfaces

The effective stress analysis commonly employed for circular slip interfaces is the Bishop (1955) Simplified Method – a modification of the method of slices developed by Fellenius in 1927 – which is sufficiently accurate for most purposes. Consider the slice shown in Fig. 3.8. It is found, in practice, that for all the slices in the soil mass above the slip interface, the algebraic sum of the terms involving the shear forces X and the normal forces E on the sides of the slices can be neglected without serious loss of accuracy. The factor of safety is then found by equating the sum of the disturbing moments about 0, that is, $R \sum W \sin \alpha$, with the sum of the

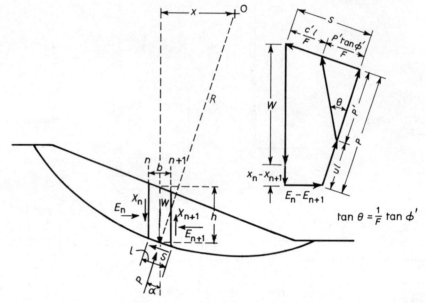

Figure 3.8 Forces in the slices method.

moments about 0 of the potential shear strength around the basal arc of each slice. The simplest form of this relationship was published by Krey in 1926 and is

$$F = \frac{R\sum[c'l + (W\cos\alpha - ul)\tan\phi']}{R\sum W\sin\alpha} \tag{3.12}$$

and, of course, R cancels. This form of analysis is known as the conventional method but it may lead to an underestimate of the factor of safety of over 20%, largely as a result of underestimating the normal effective stresses acting on the more steeply inclined parts of the slip interface. Bishop (1955) has shown that a more accurate form of the relationship is

$$F = \frac{\sum\left[\{c'b + (W - ub)\tan\phi'\}\dfrac{1}{m_\alpha}\right]}{\sum W\sin\alpha} \tag{3.13}$$

where

$$m_\alpha = \cos\alpha\left(1 + \frac{\tan\alpha\tan\phi'}{F}\right). \tag{3.14}$$

Since F appears on both sides of Equation 3.13, the solution is affected by successive approximation. Values of m_α are given on a chart (Fig 3.9) for any assumed value of F. The following is a convenient tabular form of computation based on Equation (3.12).

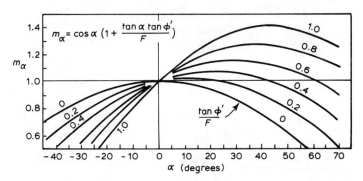

Figure 3.9 Values of factor m_α (after Janbu *et al.*, 1956).

Slice	Stratum	Area A	γ	$W = \gamma A$	$W\sin\alpha$	$W\cos\alpha$	u	l	ul	ϕ'	$\tan\phi'$	$(W\cos\alpha - ul)$ $\tan\phi'$	c'	$c'l$
				$\sum W\sin\alpha$		$\sum(W\cos\alpha - ul)\tan\phi'$								$\sum c'l$

In terms of total stresses, with $\phi_u = 0$, the factor of safety for the method of slices is given by

$$F = \frac{R\sum c_u l}{R\sum W\sin\alpha}. \tag{3.15}$$

Where line loads are superposed on a slope by, for example, retaining walls, crane tracks, buildings (Fig. 3.10), a more convenient form is

$$F = \frac{(c_w L_1 + c_{u2} L_2)R}{W_w x_w + W_L x_L + W_{c_1} x_{c_1} + W_{c_2} x_{c_2}}. \tag{3.16}$$

The following is a convenient tabular form of computation based on Equation (3.16)

Stratum	Area A	γ	Weight $W = \gamma A$	Moment arm x	Wx	Length of arc L	c_u	$c_u L$
					$\sum Wx$			$\sum c_u L$

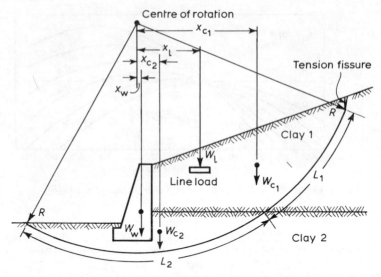

Figure 3.10 Total stress analysis for bank with superposed load. Shear strength of clays: c_{u1} for 1, c_{u2} for 2.

(e) Slip interfaces other than circular and planar

A generalized solution for slip interfaces of any shape, based on the method of slices, was developed by Morgenstern and Price (1965, 1967). In this method the shape of the slip interface must be known or assumed and, in order to make the problem statically determinate, an assumption must be made regarding the distribution of internal forces, although the factor of safety does not appear to be very sensitive to this latter prerequisite. The analysis requires the use of a computer to avoid an immense amount of manual calculation.

An alternative method proposed by Janbu *et al.* (1956) yields factors of safety slightly less accurate than the method of Morgenstern and Price but the discrepancies appear to be no more than ± a few per cent. In terms of effective stress

$$F = \frac{f_0 \sum \left[\dfrac{c'b + (W - ub)\tan\phi'}{n_\alpha} \right]}{\sum W \tan\alpha} \tag{3.17}$$

where

$$n_\alpha = \cos^2\alpha \frac{(1 + \tan\alpha\tan\phi')}{F} = m_\alpha\cos\alpha \quad \text{and} \quad m_\alpha$$

is given by Equation 3.14. The factor f_0 (Fig. 3.11) depends on the shear parameters and the form of the slip and takes account of the influence on the factor of safety of the vertical shear forces between the slices.

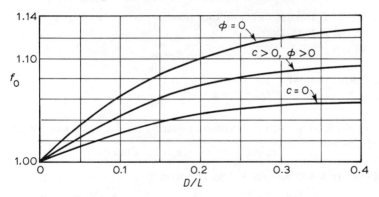

Figure 3.11 Values of factor f_0 for different proportions of slip (after Janbu *et al.*, 1956).

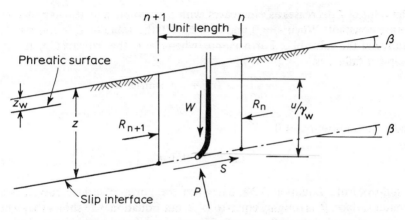

Figure 3.12 Forces in the infinite slope analysis.

Analyses in terms of total stress, with $\phi_u = 0$, can be made by a semi-empirical slices method, based on the method of Janbu *et al.* (1956). The factor of safety is given by

$$F = f_0 \frac{\sum \dfrac{c_u l}{\cos \alpha}}{\sum W \tan \alpha}. \tag{3.18}$$

Tabular forms of computation for Equations 3.13, 3.17 and 3.18 can be modelled on the tables shown above.

(f) Planar slides in infinite slopes

Consider the prism shown in Fig. 3.12. It is assumed that the soil is fully saturated in the zone of depth z_w to the phreatic surface. The lateral forces on the sides of the prism are equal and opposite and thus need not be taken into account. The shear stress on the slip interface is

$$\tau = \gamma \sin \beta \, z \cos \beta. \tag{3.19}$$

The effective normal stress on the slip surface is

$$\sigma' = (\gamma'z + \gamma_w z_w)\cos^2 \beta. \tag{3.20}$$

In terms of effective stress the ultimate shearing resistance on the slip interface is

$$\tau_f = c' + (\gamma'z + \gamma_w z_w)\cos^2 \beta \tan \phi'. \tag{3.21}$$

The factor of safety against failure is, therefore,

$$F = \frac{c' + (\gamma'z + \gamma_w z_w)\cos^2 \beta \tan \phi'}{\gamma'z \sin \beta \cos \beta}. \tag{3.22}$$

The value of F decreases as z increases, with z_w constant, and also as z_w decreases, with z constant. When $c' = 0$ and also $z_w = 0$, the value of F is constant irrespective of the value of z. Furthermore, when $c' = 0$, the critical slope $\tan \beta_c$ for incipient failure is

$$\tan \beta_c = \frac{(\gamma'z + \gamma_w z_w)\tan \phi'}{\gamma z} \tag{3.23}$$

and if, in addition, $z_w = 0$

$$\tan \beta_c = \frac{\gamma'}{\gamma}\tan \phi'. \tag{3.24}$$

It follows from Equation 3.24 that when the phreatic surface coincides with ground surface, β_c is roughly equal to $\phi'/2$, but Equation 3.23 shows that when it is below ground surface, β_c exceeds this value, attaining a maximum of ϕ' when $z_w = z$.

In terms of total stress, assuming $\phi_u = 0$,

$$F = \frac{c_u}{\gamma z \sin \beta \cos \beta}. \tag{3.25}$$

3.2.3 Assessment of the stability of slopes by stability number and stability coefficients

Taylor (1937) evolved a method of analysis of homogeneous and isotropic banks based on total stress and introduced the dimensionless parameter known as stability number N_s, graphs relating to which are shown in Fig. 3.14. Bishop and Morgenstern (1960) evolved a method of analysis based on effective stress and introduced two stability coefficients m and n and a pore pressure ratio $r_u = u/\gamma z$ (Fig. 3.13). The general solutions are based on the assumption that r_u is constant throughout the section. Approximations to cater for variable pore-pressure distributions, such as those relating to the steady seepage case, are given by Bishop and Morgenstern. Charts relating to the rapid drawdown condition have been presented by Morgenstern (1963) and tables relating to sidehill benches by Huang (1977). For a given value of stability number $N_s = c'/\gamma H$, the factor of safety depends on the geometry of the section and on r_u and ϕ'. Computations were made for various combinations of $N_s = 0, 0.025$ and 0.050, $r_u = 0, 0.3$ and 0.7, $\cot \beta = 2:1$ to $5:1$, $\phi' = 10°$ to $40°$ and depth factor $D = 1.0, 1.25$ and 1.50. Linear extrapolation can be employed for the range of stability number 0.05 to 0.10. A sensibly linear relationship obtains for the factor of safety

$$F = m - nr_u. \tag{3.26}$$

Charts and tables of m and n for various values of specified parameters are given by Bishop and Morgenstern: the use of these is illustrated in Example 3.1. Stability number can be derived most simply by employing dimensional analysis. The elements governing the stability of a bank with horizontal top surface and base are given in Fig. 3.13. The rupture interface is assumed to be cylindrical and the possible presence of a tension fissure is ignored. The geometry of the section is defined by a, α, β and θ and the scale parameter is H. The moment about the centre of rotation of the disturbing force is a function

$$F'[\gamma, a, \alpha, \beta, \theta, H, r_u], \tag{3.27}$$

and the moment about the centre of rotation of the resisting force is a function

$$F''[c', \phi', \gamma, a, \alpha, \beta, \theta, H, r_u]. \tag{3.28}$$

At incipient failure the factor of safety F is unity, the value of H is the critical

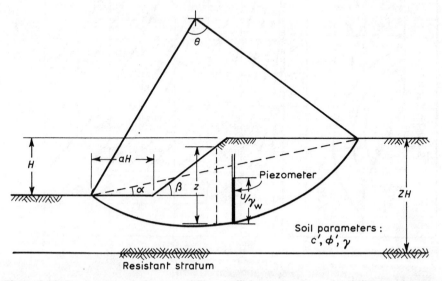

Figure 3.13 Parameters relating to stability coefficients m and n and stability number N_s.

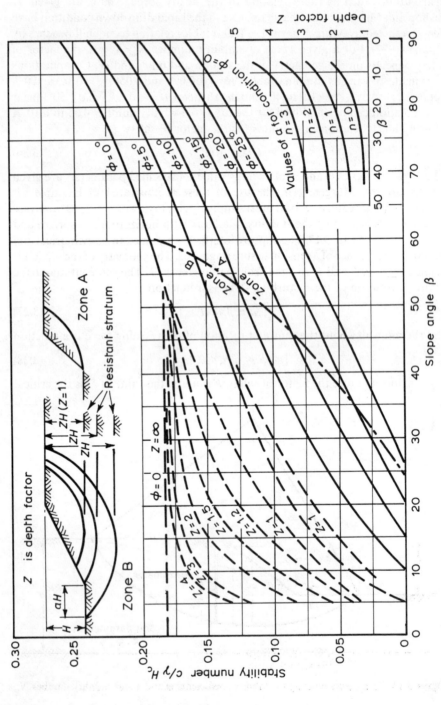

Figure 3.14 Values of stability number.

height H_c and the functions 3.27 and 3.28 are equal. The following arrangement secures dimensional homogeneity.

$$\gamma^2 H_c^2 f'[a, \alpha, \beta, \theta, r_u] = c' \gamma H_c f''[\phi', a, \alpha, \beta, \theta, r_u]$$

whence the dimensionless stability number

$$N_s = \frac{c'}{\gamma H_c} = f[\phi', a, \alpha, \beta, \theta, r_u]. \tag{3.29}$$

Taylor's stability number can be employed as a rapid check on the stability of a bank, bearing in mind the limitations of the original total stress analysis. In addition the curves shown in Fig. 3.14 have considerable value in demonstrating generally the modes of failure of slopes and the factors which control stability. In zone A the rupture interface passes through the toe of the slope and the lowest horizon intercepted is at the toe. In zone B, the lowest horizon is below toe-level. Where the lines are continuous the interface passes through the toe but where they are shown by dashes it passes below the toe. If the value of ϕ exceeds about $6°$ or $7°$ the interface can be assumed to pass through the toe. As the value of ϕ falls below about $10°$ and tends to zero, the stability of slopes at moderate and low values of β is influenced to an increasing extent by the presence of a resistant stratum.

The factor of safety against failure of a bank can be based either on height H or on shearing strength. When $\phi \neq 0$, the two values so obtained in a particular case are different although if one exceeds unity, the other does also. When $\phi = 0$ the factors of safety with respect to H and to c are the same. The safe height H for the case $\phi = 0$ is determined from

$$H = \frac{H_c}{F} = \frac{c}{\gamma N_s F} \qquad \text{w.r.t. height, and from}$$

$$H = \frac{c/F}{\gamma N_s} = \frac{c}{\gamma N_s F} \qquad \text{w.r.t shear strength.}$$

For a c–ϕ soil

$$H = \frac{H_c}{F} = \frac{c}{\gamma N_s F} \qquad \text{w.r.t. height and}$$

$$H = \frac{c/F}{\gamma N_s'} = \frac{c}{\gamma N_s' F} \qquad \text{w.r.t. shear strength}$$

where N_s' is related to a diminished value of angle of shearing resistance equal to $\tan^{-1}(\tan \phi / F)$, bearing in mind the Coulomb equation, written in the form $\tau_f/F = c/F + \sigma \tan \phi / F$. These relationships can be employed to find F for an existing bank where the actual height H is substituted for safe height. For the case of a $c - \phi$ soil, the calculation of F with respect to shear strength is effected by successive approximations, commencing with a value of N_s' based on ϕ. This yields F which is used to give a diminished value of ϕ and hence N_s', yielding a second value of F. The process achieved in the values of F on both sides of $F = c/\gamma N_s' H$.

The problems which occur in practice fall into three categories

(1) To check F for an existing bank: c, ϕ, γ, β and H known.
(2) To determine β for a projected bank: c, ϕ, α, H and F known.
(3) To determine c and ϕ following failure of a bank: γ, β, H and $F = 1$ known (check against laboratory tests for c and ϕ).

3.2.4 Stability of banks of cohesionless soil

Provided the angle of slope β of a bank of clean dry sand or gravel is less than the angle of shearing resistance of the soil, the bank will be stable irrespective of its height. The factor of safety is

$$F = \frac{\tan \phi}{\tan \beta}. \tag{3.30}$$

It is desirable to prevent water flowing through a bank by a cut-off filter drain along the top and a filter drain beneath the toe or by other drainage systems which may be more appropriate. Where water flows through a bank, with or without drainage, and is likely to threaten safety, then a flow net must be constructed and the stability of trial wedges investigated, taking into account seepage pressure on the rupture interface.

3.2.5 The application of analytical techniques to the solution of practical problems

The study of case records of failures of banks and hillsides, in which field conditions are compared with stability analyses based on laboratory tests, is invaluable in advancing knowledge of the factors governing the stability of slopes. Skempton and Hutchinson (1969) summarized a number of case records and grouped the failures under the headings:

(a) First-time, short-term slides.
(b) First-time, long-term slides.
(c) Repeated slides on existing slip interfaces.

First-time, short-term slides generally occur in artificial excavations during, or shortly after, formation. The cases examined by Skempton and Hutchinson confirmed the validity of total stress analysis for such failures, provided account is taken of all the significant factors which influence shear strength.

First-time, long-term slides occur in natural banks or in cuttings, years (often fifty or more) after formation. Effective stress analyses prove valid for such failures but, apart from factors governing shear strength, careful consideration must be given to pore-water pressure. These slides are commonly progressive and the average shear strength operating at failure lies between peak and residual strength values.

Repeated slides on existing slip interfaces commonly occur in natural cliffs and hillsides only, since remedial measures are normally executed when a slide occurs for the first time in an artificial bank. The cases examined by Skempton and Hutchinson confirmed the validity of effective stress analyses for repeated slides, employing residual shear strength parameters.

The inclination at which a slope in nature finally becomes stable against any form of landslide is known as its angle of ultimate stability. This angle depends upon the properties of the soil, groundwater conditions and climate. Other processes of degradation, such as rain beat, sheet wash and creep, continue after a state of stability against landslides has been attained. Initial downcutting by streams is accompanied by relatively shallow landslides, which are commonly translational, but, as the depth of cutting increases, deep rotational slides occur. Skempton and Hutchinson record the value of the angle of ultimate stability as 22° to 26° in boulder clay at Peterlee, County Durham. For slopes in London Clay, the angle of ultimate stability against first-time slides is about 9°: stable slopes of 10° are probably associated with a lower maximum level of the phreatic surface. Instability on slopes as slack as 8° probably involves renewed movements on solifluction slip interfaces of Pleistocene age and stable slopes of 7° or less can be observed.

It is generally accepted that the Bishop Simplified Method, expressed by Equation 3.13 and, when $\phi_u = 0$, by Equation 3.15, is both reliable and economical in time for the majority of investigations involving circular slip interfaces. Little and Price (1958) have adapted this method for use with a computer and programs are now available commercially. For problems involving slip interfaces other than circular and planar, the most satisfactory approach is to employ the Janbu Method, expressed by Equation 3.17, to locate the critical interface approximately and to check the factor of safety by the Morgenstern and Price (1965, 1967) Method, although the former alone may be adequate for many problems. The methods discussed above are those which are perhaps best known and have proved satisfactory but limitations of space preclude discussion of others which have been advanced.

Stability number and stability coefficients, discussed in Section 3.2.3 above, can be employed respectively for $\phi_u = 0$ and $c'-\phi'$ cases in the design of unimportant banks and the initial design of others. The application of both methods is restricted by the requirements of homogeneous and isotropic material together with simple geometry of the bank and generally broad assumptions must be made in order to achieve these.

The value of factor of safety for permanent fill or cutting should be normally at least 1.5 but for temporary works this can be reduced to not less than 1.3. Natural slopes can be considered to be reasonably safe if the value lies between 1.3 and 1.5, but historical evidence and observations of adjacent similar sites by, for example, walking and study of aerial photographs should be taken into account. Any reduction in the lower limits of factor of safety should normally be supported by refined analyses based on the results of accurate field and laboratory tests and a review of potential life and property hazards.

First-time slides in over-consolidated clays have been discussed further by Skempton (1970). Progressive failure in clay slopes has been studied by Lo and Lee (1973) and by Kenney and Uddin (1974). Chandler (1974) and Chandler and Skempton (1973) have investigated the long-term stability and design of cuttings in stiff fissured clays.

The following two examples of analyses of the stability of slopes illustrate graphical means of presenting data concerning factor of safety for design purposes. It will be appreciated that the amount of work involved in analyses is often considerable but it may be possible to define certain limits in a given problem. For example, it is unlikely that a slip interface will penetrate below the top of a resistant stratum nor is it likely that the toe of a slip will extend into the opposite face of an excavation. On the other hand, it is possible that a slip interface will follow a particularly weak seam, thus producing a plane interface at the base of the slip. Now that computer programs are available for most methods of analysis, defining limiting conditions is not as important as it was formerly, when between 100 and 200 trial interfaces were analysed for important projects. When a program is employed, it is necessary to analyse manually one or two interfaces only to check that data have been presented correctly to the computer.

Eau Brink Cut failure

Figure 3.15 gives details of a slip which occurred in 1943 in the west bank of the Eau Brink Cut, near King's Lynn, Norfolk. The cut was excavated about 1820. The actual slip interface is indicated by the circular arc of radius 16.8 m and the values of factor of safety for this are 1.30 based on total stress and 1.15 based on effective stress.

A comprehensive series of analyses was made during the investigation (Skempton, 1945) which serves to illustrate the extent of studies required for design purposes in a similar situation. The sand and gravel is a resistant stratum and it is most unlikely that any slip will penetrate below the top of this bed.

In both total stress and effective stress analyses, three horizons were selected, at − 3.7, − 4.9 and − 6.1 m, and a number of circular arcs having different centres were constructed tangential to each horizon. The values of factor of safety derived from total stress analyses of one family tangential to each horizon were plotted at the corresponding centres of rotation and iso-FS lines were drawn by interpolation, as shown in Fig. 3.15 for the total stress analyses with slip interfaces tangential to the top of the sand and gravel. In the total stress analyses it was found that families of interfaces tangential to horizons above − 6.1 m gave higher values of factor of safety. On the other hand, in the effective stress analyses the absolute minimum value of factor of safety of 1.02 was found by interpolation to be given by an interface tangential to the − 4.3 m horizon.

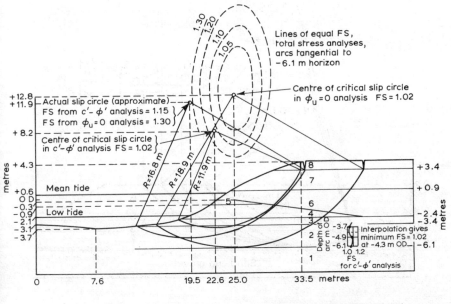

Stratum no.	Soil	$\phi_u = 0$	$c'-\phi'$		γ (kN m^{-3})
		c_u (kN m^{-2})	c' (kN m^{-2})	ϕ'	
8	Made ground	—	—	—	18.5
7	Brown silt and brown silty clay	21.6	6.7	30°	18.5
6	Grey silt	23.9	6.7	30°	17.9
5	Peat	—	—	—	11.0
4	Buttery clay	22.5	11.0	20°	16.0
3	Grey silty clay	18.2	6.7	21°	17.6
2	Buttery clay	22.5	11.0	20°	15.7
1	Sand and gravel	—	—	—	—

Figure 3.15 Failure in Eau Brink Cut (after Skempton, 1945).

These analyses and resulting values of factor of safety may be subject to minor modification in the light of advances made since 1943 in the determination of strength parameters and in stability analyses.

Design of channel at Wiggenhall St Peter

Figure 3.16 shows trial profiles for a flood relief channel on the River Great Ouse near Wiggenhall St Peter, Norfolk (Skempton, 1946). Berm widths of 6.1, 9.1, 12.2 and 15.2 m were investigated with various radii of slip interfaces which are assumed to pass through the toe and to be tangential to the top of the Lower Peat which is not penetrated on account of its relatively high strength. The analyses

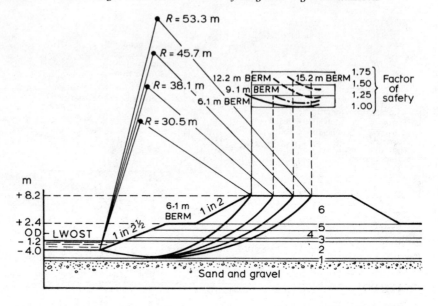

Figure 3.16 Design of a bank at Wiggenhall St Peter, Norfolk (after Skempton, 1946).

Stratum no.	Soil	Level (m OD)	c_u (kN m^{-2})	γ (kN m^{-3})
6	Fill	— +8.2 —	12.0	17.3
5	Brown silt	— +2.4 —	23.9	18.5
4	Grey silt	— +0.9 —	22.7	18.1
3	Upper peat	— -0.6 —	33.5	11.0
2	Buttery clay	— -1.2 —	22.7	16.0
1	Lower peat	— -5.5 —	33.5	11.0
		— -6.1 —		

were based on total stress with $\phi_u = 0$, presumably because it was anticipated that failure would be most likely to occur during, or immediately after, construction. No tension crack was taken into account in the analysis but its omission makes little difference to the values of factor of safety since the depth is only about 1.5 m and the increase in disturbing moment is compensated by the increase in resisting moment. A berm width of 12.2 m was employed in the final design and the factor of safety was 1.2.

3.2.6 Preventative and remedial measures

Preventative and remedial measures for instability of slopes are briefly as follows:

(1) Remove part of load tending to cause failure. If this consists solely of the weight of soil, the disturbing load is reduced either by reducing the slope or by

forming berms at one or more levels in the face. If excess load is due to a structure which cannot be resited, then it must be carried on piles, the feet of which should be driven well below the critical slip interface.

(2) Provide external support. This may consist of either a retaining wall or sheet piling at the toe, extending below the slip interface, or a heavy vertical load on the toe. Slopes have been stabilized by placing vertical piles in the threatened mass of soil: this problem has been studied by Wang and Yen (1974).

(3) Increase the strength of the soil. This is usually effected by sub-surface drainage in the form of a longitudinal interceptor drain along the top of the embankment and counterfort drains laid in the direction of the slope, each collecting from a series of herringbone or chevron drains and discharging into a longitudinal collector drain at the toe. The stabilization of the clay cliffs at Herne Bay, Kent, by drainage has been described by Duvivier (1939).

(4) Plant grasses and shrubs on the slope thereby reducing moisture content, reinforcing the surface layers by a network of roots, reducing the erosive effects of rain beat and sheet wash and providing a blanket which restricts the effects of frost. Moran (1949) and Toms (1949) have discussed the use of vegetation in stabilizing slopes. In relation to deforestation on slopes, Brown and Sheu (1975) found that removal of trees, with consequent reduction in overburden and wind loading, leads to reduction in creep rate and increase in stability but that both reduction in soil tenacity, arising from decay of root systems, and rise in water table, occasioned by decrease in evapotranspiration, lead to increase in creep rate and decrease in stability. Increase in moisture content arising from removal of trees and other vegetation appears to have been a major factor in some slope failures in the British Isles.

In the event of a slip occurring or becoming imminent, the most easily applied remedial measure is generally to place a surcharge on the toe, provided this is accessible and provided the surcharge is not so great as to cause local failure. A desirable immediate measure when failure is threatened is to insert at intervals down the slope several flexible polythene tubes, 20 to 50 mm diameter, through the estimated location of the slip interface by means of a steel mandrel which is withdrawn after driving. The tubes are deformed as failure develops and the position of the slip interface can be located by measuring the depths of the deformities. In addition, surface markers should be located on the slope and related to stable reference points beyond the slope in order to record the progress of displacements.

Rates of movement of landslides have been discussed earlier in this chapter. If observations are maintained over a period of several months or years, some landslides display a rate of movement which increases with time and can be represented reasonably well by an exponential growth curve. The progress of movement is illustrated in Fig. 3.17, although this relates to the failure of a retaining wall where active and passive earth thrusts were probably involved instead of a rotational landslide. Broadbent and Ko (1972) record details of a

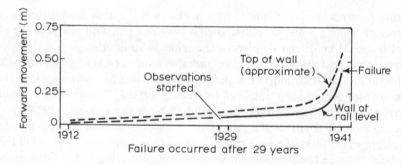

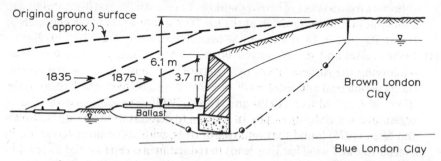

Figure 3.17 Failure at Kensal Green cutting (after Skempton, 1964).

landslide for which observed displacements were fitted to the equation

$$\delta = Ce^{kt} \tag{3.31}$$

where δ is the total displacement at time t and C and k are constants. By extrapolation, the approximate date of failure was predicted, based on an estimated displacement at failure. The displacement of some landslides takes place in a series of steps of increasing magnitude, with a halt between each increment, but the general progress may well be represented by a growth curve.

EXAMPLE 3.1

An excavation 30 m deep is to be made in boulder clay overlying bedrock. The depth to bedrock is defined by the depth factor $D = 1.5$ (Fig. 3.14). The slope of the banks is $\tan \beta = 1/3$. Data for the clay are $c' = 24\,\mathrm{kN\,m^{-2}}$, $\phi' = 30°$ and $\gamma = 20\,\mathrm{kN\,m^{-3}}$. Investigations at existing banks in the clay indicate $r_u \simeq 0.5$. Table 3.3, giving related parameters for $c'/\gamma H = 0.025$ and 0.05 is based on values given by Bishop and Morgenstern (1960). It has been found that when $c'/\gamma H = 0.025$, $D = 1.50$ is seldom more critical than $D = 1.25$. Since the actual value of $c'/\gamma H = 0.04$ lies between the tabulated values, the factor of safety is found ultimately by interpolation.

For a given set of parameters β, ϕ' and N_s, there is a value of pore-pressure ratio r_{ue} for which F when $D = 1.00$ is the same as F when $D = 1.25$. This value can be expressed as

$$r_{ue} = \frac{m_{1.25} - m_{1.00}}{n_{1.25} - n_{1.00}} \tag{3.32}$$

Table 3.3

				Stability coefficients	
$c^\circ/\gamma H$	D	ϕ'	$Cost\,\beta$	m	n
0.025	1.00	30°	$3:1$	2.235	2.078
0.025	1.25	30°	$3:1$	2.431	2.342
0.05	1.00	30°	$3:1$	2.574	2.157
0.05	1.25	30°	$3:1$	2.645	2.342
0.05	1.50	30°	$3:1$	2.964	2.696

If the design value of the pore-pressure ratio is higher than r_{ue}, then F for $D = 1.25$ is less than F for $D = 1.00$. A similar rule obtains in relation to $D = 1.50$ and $D = 1.25$ based on

$$r_{ue} = \frac{m_{1.50} - m_{1.25}}{n_{1.50} - n_{1.25}}. \tag{3.33}$$

When $c'/\gamma H = 0.025$, Equation 3.32 gives $r_{ue} = 0.742$ and hence F for $D = 1.25$ is greater than F for $D = 1.00$. The most critical level is therefore $D = 1.00$, and from Equation 3.26

$$F = 1.196.$$

When $c'/\gamma H = 0.05$, Equation 3.32 gives $r_{ue} = 0.384$ and Equation 3.33 gives $r_{ue} = 0.902$. Hence F for $D = 1.25$ is less than F for $D = 1.00$ and F for $D = 1.50$ is greater than F for $D = 1.25$. The most critical level is therefore $D = 1.25$ and from Equation 3.26

$$F = 1.474.$$

Interpolating linearly for $c'/\gamma H = 0.04$

$$F = 1.196 + 0.6 \times 0.278 = 1.36.$$

3.3 BEARING CAPACITY OF SOILS

3.3.1 Characteristics of bearing failure of soils

The bearing capacity of a soil is assessed from assumed failure conditions involving either rupture zones or a rupture interface in the soil. In the rupture zone hypothesis, for the practical case of a rough base beneath a strip foundation (Fig. 3.18a), the tendency for the soil immediately beneath the base to spread laterally is resisted by friction and adhesion at the contact between soil and base and the soil in Zone I remains in a state of elastic equilibrium. In Zones II and III the soil is in a state of complete or partial passive plastic equilibrium. Terzaghi (1943) assumed that when the founding depth is no more than shallow (D < B), the overburden at the sides of the foundation can be treated as a surcharge. Equating the sum of vertical forces at the base of the foundation

$$Q_f + \gamma \frac{B^2}{4} \tan\phi - 2P_p - Bc\tan\phi = 0. \tag{3.34}$$

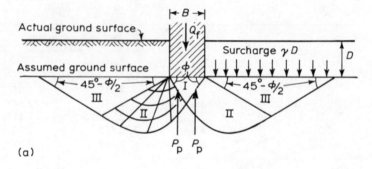

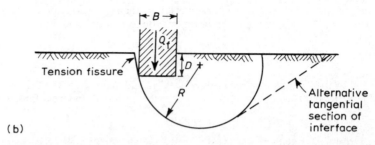

Figure 3.18 Bases of bearing capacity theory. (a) Rupture zone hypothesis; (b) rupture interface hypothesis.

From this is evolved the relationship for ultimate gross base bearing capacity of a strip foundation

$$q_f = cN_c + \gamma D N_q + \tfrac{1}{2}\gamma B N_\gamma \tag{3.35}$$

Where N_c, N_q and N_γ are bearing capacity factors which depend only on the value of ϕ. Although ultimate bearing capacity is defined in CP2004 (BSI, 1972d) as the net stress at which soil beneath a foundation fails in shear, it has been usual in theoretical investigations to apply this term to gross stress and it is convenient in this Section to employ it in this sense unless otherwise specified.

From a study of experimental data, Terzaghi concluded that the ultimate gross bearing capacity of a circular footing of diameter B is given by

$$q_f = 1.3cN_c + \gamma D N_q + 0.3\gamma B N_\gamma \tag{3.36}$$

and of a square footing of side B by

$$q_f = 1.3cN_c + \gamma D N_q + 0.4\gamma B N_\gamma. \tag{3.37}$$

It should be noted that when $\phi = 0$, $N_q = 1$ and $N_\gamma = 0$ and then q_f is independent of B: this applies to cohesive soils in the undrained state. For all other cases, which include cohesionless as well as c–ϕ soils, q_f is a function of B.

If soil is soft or loose, considerable penetration of a footing occurs before the state of general shear failure is reached and, in order to cater for this in design, Terzaghi introduced the concept of local shear failure for which arbitrary reductions are made in the values for ultimate gross bearing capacity. However, the concept does not appear to be necessary and may lead to confusion; it is sufficient to regard the reduced values of bearing capacity as being necessary by virtue of the large amount of penetration. The magnitude of penetration occurring during bearing failure has received little attention and it may be a more critical factor in design than ultimate bearing capacity. The relationship between compressibility and bearing capacity has been discussed by Ismael and Vesić (1981).

In the rupture interface hypothesis it is assumed that failure occurs in the soil towards one side only of the foundation and the mass of soil plus the foundation rotates about some point (Fig. 3.18b). Kinematically the rupture interface must be cylindrical but the form has been modified by various investigators. Wilson (1941) derived the following equation for the ultimate net bearing capacity of a strip foundation based on this hypothesis.

$$q_f = 5.5c_u(1 + 0.38D/B) \qquad (3.38)$$

when $\phi_u = 0$ and $D \leqslant 1.5B$. The safe gross bearing capacity is

$$q_s = q_f/F + \gamma D. \qquad (3.39)$$

Button (1953) employed the rupture interface hypothesis to determine the bearing capacity of soil in layers of different strength. This hypothesis was adopted also by Raymond (1967) for determination of the bearing capacity of clay supporting large footings and embankments (see Section 3.3.8).

The characteristics of bearing failure have been observed in the laboratory and in the field. Both theory and experiment show that shearing failure in the soil can be expected to develop initially in zones near the edges of a foundation and this does not constitute a stress condition in the rupture interface hypothesis. Furthermore, a cylindrical rupture interface would lead to maximum heave at the intersection of ground and rupture interface and this rarely, if ever, appears to be attained. In fact, the mode of failure is satisfied better by the rupture zone hypothesis than by the rupture interface. These arguments are required solely for shallow foundations since the rupture zone hypothesis alone satisfies conditions at failure when foundations are located at greater depths.

In the case of axially loaded model strip foundations, failure is often asymmetrical, with greater disturbance of soil on one side than on the other, but beneath axially loaded square footings, the rupture zones tend to develop more or less symmetrically on all four sides. Asymmetrical failure may appear to support the rupture surface hypothesis but, in fact, it can be explained by slight unintentional eccentricity of loading and non-uniformity of soil. In the case of a surface strip footing on cohesionless soil a difference of only $\frac{1}{2}°$ in the values of ϕ on each side of the footing leads to a difference in bearing capacity of the order of 10%. If the footing is unrestrained by beams or walls, it tends to drift towards the side with the lower capacity as the load approaches the ultimate value. On the other hand, if the footing is restrained it penetrates vertically downwards and the ultimate load is slightly greater than when unrestrained. Normally, this difference in capacity is not significant in practical problems. Eastwood (1955) found that partial restraint of axially and eccentrically loaded strip footings 150 mm wide on cohesionless soil led to bearing capacities about 6% more than those obtained when the footings were unrestrained.

Model footings, loaded axially or eccentrically, often rotate through a few degrees as the ultimate load is approached but bearing capacity is not influenced substantially by slight tilting. This tilting is not connected with a rotational rupture interface failure: in fact, the direction of tilt is usually opposite to that which would be associated with such a failure. The direction of application of load to a footing deviates from the vertical when lateral movement of the footing takes place but, since this is unlikely to exceed a few degrees before failure occurs, it cannot have any significant effect on bearing capacity. If a footing tilts, a vertical load can be resolved into normal and tangential components. If the angle of tilt is about 6°, the tangential component is about one-tenth of the applied load but the normal component Q_n is practically equal to the applied load and, since the shearing resistance between cohesionless soil and a cast *in situ* concrete base is equal to $Q_n \tan \phi$, sliding due to tilting is unlikely to occur. However, lateral movement due to tilting may occur on cohesive soils.

In spite of the arguments in favour of the rupture zone hypothesis, there remains a suspicion that a rupture interface failure may occur where lateral restraint is completely absent. This condition is satisfied by structures such as a battery of silos or tanks founded at no more than shallow depths (see Transcona silos, Section 3.7.3).

It can be observed that, since some penetration is required in order to mobilize resistance fully, the capacity of a foundation, in particular when founded on the ground surface, is augmented by the influence of penetration depth. In practice, failure may appear to occur at a value of bearing stress less than the ultimate given by the rupture zone equations. This is often due to the fact that, as penetration develops, the load on the foundation is relieved either by automatic transfer to other members of the structure or by deliberately unloading the structure. The magnitude of penetration of model footings at failure is generally not less than $0.1B$ and this is seldom attained with foundations in practice. However, the penetration at a working load of one-third of the ultimate value is commonly very small and is probably represented in practice by the immediate settlement of a foundation.

In the evaluation of the bearing capacity factors for shallow foundations by Terzaghi (1943), shearing resistance of the soil above founding level was ignored but this contribution to bearing capacity should be taken into account. At shallow depths the rupture zone extends to ground level but at greater depths it is entirely below ground level (Fig. 3.19).

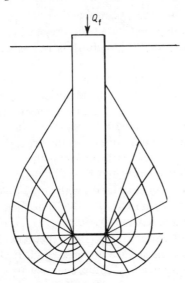

Figure 3.19 Rupture zones for a deep foundation.

3.3.2 Estimation of bearing capacity based on the rupture zone hypothesis

Many theoretical studies of bearing capacity have been propounded but generally there is a paucity of supporting evidence from observations on model and full-scale foundations, although a number of investigations on model pile foundations have been reported. One factor which discourages the testing of models is that of scale, which is difficult to overcome with cohesionless soils although corrections can be made in the case of cohesive soils in the undrained state. Furthermore, it is difficult to reproduce the same degree of compaction in cohesionless soils and $1°$ difference in the angle of shearing resistance can make an appreciable difference in bearing capacity. Amongst the important theoretical investigations, those of Terzaghi (1943), Meyerhof (1951), Sokolovsky (1960), Hansen (1961, 1966) and Balla (1962) are perhaps best known but there are others of considerable interest. The basic theory of limiting equilibrium has been treated by Harr (1966). Vesić (1973) has taken soil compressibility into account in the determination of bearing capacity. A review of bearing capacity theories, with particular reference to foundations on clays, has been presented by de Mello (1969). The special case of the capacity of cohesive soils in the undrained state has been investigated by Skempton (1951) and is discussed later. Uncertainty in the bearing capacity of sands has been investigated by Ingra and Baecher (1983).

The general form of bearing capacity equations derived by most investigators is similar to that presented by Terzaghi for a strip foundation (Equation 3.35) but for design purposes it is necessary to assess the reliability of the values of bearing capacity factors published by these authors. In fact these values cover fairly wide ranges and clearly estimated capacity is sensitive to assumptions made in the analyses. Although the method of construction of a particular foundation will influence the magnitude of penetration to failure, it will probably have little effect on ultimate base resistance, although skin resistance on the shaft of the foundation will often be affected.

Milović (1965) compared experimental values of ultimate bearing capacity with values calculated by the methods advanced by several investigators and concluded that, with cohesionless and slightly cohesive soils, the best correlation was secured employing the Balla (1962) theory but it should be noted that the experimental founding depth did not exceed the breadth of the foundation. For soils with appreciable cohesion, Milović found that the Hansen (1961) theory gave the closest correlation with experimental results. Abdul-Baki and Beik (1970) secured generally good correlation with cohesionless soils employing the theories of Meyerhof (1951), Hansen (1961,1966) and their own, for ratios of depth to breadth between 0 and 8. Ko and Davidson (1973) reported good correlation between experimental and Sokolovsky (1960) theoretical values for the case of smooth strip footings on the surface of dense sand. It will be appreciated that these conclusions are insufficiently comprehensive to serve as a guide to the validity of the theories investigated and, furthermore, perusal of the discussions concerning the relevant literature reveals some contradictory conclusions. Bearing capacity factors have been computed by Griffiths (1982) using finit element analyses.

Values of bearing capacity factors presented by Meyerhof (1951) and Hansen (1961) for strip footings on the surface of $c - \phi$ soils are practically the same; in fact, similar relationships appear to have been evaluated in both cases. Meyerhof (1951) also presented values for shallow and deep foundations, but if founding depth is very shallow the use of these overestimates bearing capacity. Subsequently, Meyerhof (1963) published depth factors for shallow foundations where $D \not> B$ and these are applied to corresponding terms in the bearing capacity equation in which values of N_c, N_q and N_γ (Fig. 3.20) for surface foundations are employed. The magnitudes of these depth factors are not markedly different from those published by Hansen (1961, 1966). Values of N_c, N_q and N_γ for rectangular foundations can be estimated by interpolation between the graphs for square and strip foundations: it is reasonable to assume that a rectangular foundation can be treated as a strip foundation when L is equal to $5B$ or more. The Meyerhof depth factors are as follows.

$$D_c = 1 + 0.2\frac{D}{B}\sqrt{N_\phi} \tag{3.40}$$

$$D_q = D_\gamma = 1 \quad \text{when } \phi = 0 \tag{3.41}$$

$$D_q = D_\gamma = 1 + 0.1\frac{D}{B}\sqrt{N_\phi} \quad \text{when } \phi > 10° \tag{3.42}$$

where $N_\phi = \tan^2(45° + \phi/2) = (1 + \sin \phi)/(1 - \sin \phi)$. \tag{3.43}

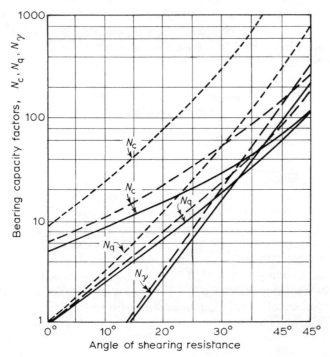

Figure 3.20 Bearing capacity factors for foundations, where --- represents strip and -- represents square and circular surface foundations, and --- represents piles (after Meyerhof, 1963).

Table 3.4 Values of bearing capacity factors for deep foundations

ϕ	10°	15°	20°	25°	30°	35°	40°	45°
N_c	20	32	57	100	200	420	1200	3500
N_q	4	9	20	48	120	300	820	3000
N_γ	2	6	18	47	130	470	1800	7000

The graphs of N_c and N_q for piles in Fig. 3.20 relate to driven square or circular piles with 60° points, provided the pile foot is located at a depth in the bearing stratum not less than about $4B \sqrt{N_\phi}$. Values of N_γ are not given because the term in the bearing capacity equation involving this is commonly small, since B is small, in relation to the other terms. A review of bearing capacity and settlement of piles has been given by Meyerhof (1976).

For deep foundations on c–ϕ soils where D exceeds, say, $4B$ to $10B$, values of bearing capacity factors given in Table 3.4 can be employed (Meyerhof, 1951). If ϕ exceeds 25° to 30°, it is likely that cohesion will be small so that the soil can be treated as cohesionless, and values of N_c for $\phi > 25°$ are rarely required. The magnitude of bearing capacity predicted by the application of these values is sometimes considered to be excessively high, no field evidence appears to have been recorded to confirm this. Nevertheless, it will generally be found that the predicted safe bearing capacity of deep foundations in cohesionless soils is greater than the required value, even when a conservative load factor of 5 is applied. It is, of course, recognized that the criterion of design in such cases is likely to be settlement. Alternative calculations can be made for strip foundations by employing the relationship

$$q_f = 0.5\gamma B N_{\gamma q} \qquad (3.44)$$

given by Abdul-Baki and Beik (1970), where the factor $N_{\gamma q}$ is selected from Fig. 3.21. The form of Equation 3.44 was earlier employed by Meyerhof (1951). It will be found that when D is less then about $5B$, the magnitude of bearing capacity given by Equation 3.44 may exceed that given by Equation 3.35. According to Equation 3.53 the bearing capacity of square and rectangular foundations will be somewhat greater than the value determined from Equation 3.44 for a strip foundation, probably of the order 20 to 30% in many cases, but it is prudent to assume that the shape factor is unity. This reasonably applies also to the capacity of square and rectangular foundations calculated by employing the values given in Table 3.3, although Meyerhof (1963) has given approximate relationships for shape factors.

Adhesion and friction between soil and the shaft of a foundation influence the shape of the rupture zone to some extent but in addition they contribute to bearing capacity. Frictional resistance on a shaft is calculated from the friction angle between soil and shaft and the distribution of horizontal stress on the shaft surface, as demonstrated in Example 3.7. The estimation of lateral stress necessitates assessment of the coefficient of earth stress which could vary from the

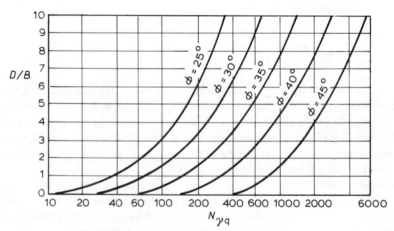

Figure 3.21 Relationships between $N_{\gamma q}$, D/B, and ϕ.

active value to the elastic and passive values, depending on the method of installation of the foundation. If the total frictional resistance on the shaft is F_w, the safe total gross bearing capacity of a foundation on cohesionless soil based, for example, on Equation 3.44 is

$$Q_s = \left(\frac{0.5 \gamma B N_{\gamma q} A + F_w - \gamma D A}{F} \right) + \gamma D A \tag{3.45}$$

where A is the base area of the foundation and F is the load factor.

The case of saturated cohesive soils in an undrained state requires special consideration. Skempton (1951) tested model foundations on clays. The load at a given penetration represented the ultimate bearing capacity at that depth. The tests were commenced with the foundations on the surface and also buried at depths of 0.5, 1.5 and 4.0 B (Fig. 3.22a). In each case the shearing resistance of the soil was fully mobilized when the penetration was roughly 0.5 B. When $D \geqslant 4$ to 5 B, penetration continued under a constant net stress and the rupture zones did not reach ground surface when D exceeded this value. The observations were corrected for increase in shear strength brought about by consolidation during testing, arising from the short drainage path associated with small models, and for difference in rate of strain with the model tests compared with the rate employed in shear strength tests (Meigh, 1950). Design curves based on this work are given in Fig 3.22b and can by employed for a foundation at any depth in saturated cohesive soil ($\phi_u = 0$). The value of N_c includes the influence of shearing resistance of the soil above founding level. The ultimate gross base bearing capacity of the soil is therefore.

$$q_f = c_u N_c + \gamma D. \tag{3.46}$$

For the case of deep circular foundations in clay, Gibson (1950) extended an analysis originated by Bishop *et al.* (1945) for metals and found that

$$N_c = \frac{4}{3}\left(log_e \frac{E_r}{c_u} + 1 \right) + 1 \tag{3.47}$$

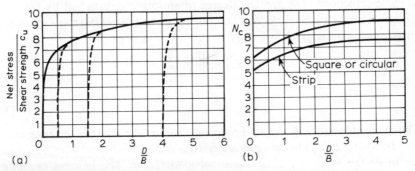

Figure 3.22 (a) Dimensionless graphs of load/penetration for 20 mm diameter model foundation on clay. (b) Design graphs for bearing capacity of foundations on clays.

where E_r is the modulus of elasticity of the clay, reasonably taken in practice as the secant modulus at one-half the yield stress. This demonstrates that N_c depends on E_r/c_u and, for the majority of undisturbed clays, the ratio varies between 50 and 200: the corresponding values of N_c are 7.6 and 9.4. the Skempton value is 9.0 for $D/B \geqslant 4$.

If the total adhesion resistance on the perimeter of the shaft of the foundation is C_w, the safe total gross bearing capacity based on Equation 3.46 is

$$Q_s = \left(\frac{c_u N_c A + C_w}{F}\right) + \gamma D A, \qquad (3.48)$$

where A is the base area of the foundation and F is the load factor. A poorly compacted backfill may detract from bearing capacity by downward drag as the fill settles and C_w may become zero or negative. When C_w is negative it should be omitted from Equation 3.48 and should be treated as part of the applied load, as shown in Example 3.7.

The load factor is not applied to the overburden stress γD in Equations 3.45 and 3.48. Before excavation the soil is in a neutral state of stress, but the process of excavation imposes a form of prestressing on the soil. Some of the excavated overburden may be backfilled over parts of the foundation but the load imposed by this is included in the gross applied foundation stress which must not exceed q_s. If a foundation is not closely surrounded by overburden then γD must be omitted in Equations 3.45 and 3.48 and the value of $N_{\gamma q}$ or N_c should be that corresponding to $D = 0$.

The bearing capacity of saturated anisotropic cohesive soil in the undrained state has been investigated by Davis and Christian (1971). The conventional method of loading in triaxial tests is with the axis of the specimens roughly normal to the bedding, assuming the strata are approximately horizontal *in situ*. The shear strength c_{un} of anisotropic soil derived in this way is different from the value c_{up} given when the specimen axis is roughly parallel to the bedding. The quantities c_{un} and c_{up} are not the values of cohesion normal and parallel to the bedding but the values of cohesion yielded by applying the major principal stress in those directions. The greatest degree of anisotropy examined by Davis and Christian was $c_{up}/c_{cn} = 0.73$ and for this condition the net bearing capacity of a strip foundation is given by

$$q_f = 0.86 N_c c_{un} \qquad (3.49)$$

When $c_{up}/c_{un} = 1.56$,

$$q_f = 1.13 N_c c_{un}. \qquad (3.50)$$

The range of broadly $\pm 15\%$ in the discrepancy between the capacity of isotropic and anisotropic clays in the cases analysed is not excessive and would be catered for in design by the normal values of load factor. The bearing capacity of anisotropic soils has been studied also by Reddy and Srinivasan (1970). Brown and Meyerhof (1969) have reported studies of the bearing capacity of stiff clay overlying soft clay and of soft clay overlying stiff clay. Other studies of two layers of soil have been reported by Meyerhof (1974) and Purushothamaraj *et al.* (1974).

3.3.3 Bearing capacity of square and ractangular foundations

The investigations of Skempton (1951) show that the unit bearing capacity of a square foundation on the surface of clay is about 20% more than the capacity of a strip foundation on the same soil. Whereas a strip foundation can be considered to have an infinite length, a square or rectangular foundation has a finite length and the influence of the ends on bearing capacity must be taken into account. The total ultimate capacity of a rectangular foundation on the surface of clay can be considered to comprise the total capacity of a square foundation ($B \times B$) plus the total capacity of a portion of a strip foundation $(L - B)B$, that is

$$Q_f = 1.2c_u N_c B^2 + c_u N_c (L - B)B,$$

where N_c is the factor for a strip foundation. The ultimate unit capacity is therefore

$$q_f = \frac{Q_f}{BL} = c_u N_c \frac{(n + 0.2)}{n} = C_R c_u N_c, \tag{3.51}$$

where $n = L/B$. In the case of foundations on clay, it is reasonable to assume that a rectangular foundation can be treated as continuous when $C_R = (n + 0.2)/n$ is less than about 1.05: that is when $n \geqslant 4$.

For shallow foundations on soils for which $\phi > 0$ the bearing capacity factors can be estimated from Fig. 3.20 by interpolating between the factors for a strip foundation ($B/L = 0$) and a square foundation ($B/L = 1$) in proportion to the ratio B/L. Meyerhof (1963) gives the following relationships for shape factors for rectangular and square foundations which are applied to corresponding terms in Equation 3.35 for strip foundations.

$$S_c = 1 + 0.2 \frac{B}{L} N_\phi \tag{3.52}$$

$$S_q = S_\gamma = 1 \text{ when } \phi = 0 \tag{3.53}$$

$$S_q = S_\gamma = 1 + 0.1 \frac{B}{L} N_\phi \text{ when } \phi > 10° \tag{3.54}$$

where N_ϕ is given by Equation 3.43. It can be observed that the increment yielded by Equation 3.54 does not accord with the decrement involved when comparing Equations 3.35 and 3.37.

Whereas the problem of bearing capacity of a strip foundation can be treated as one in plane strain, that of a rectangular or square foundation is one in triaxial strain. Since the angle of shearing resistance in plane strain compression tests is roughly 10% greater than in triaxial compression, Meyerhof (1963) suggests that the angle of shearing resistance for a rectangular foundation can be taken as approximately $(1.1 - 0.1\, B/L)\phi$ when ϕ is determined from a triaxial test.

3.3.4 Bearing capacity of eccentrically loaded foundations

Bearing failure beneath an eccentrically loaded surface or shallow foundation usually occurs with soil on the side of the eccentricity passing into a state of plastic equilibrium whereas soil on the opposite side may show no external evidence of failure. Since the mechanics of the problem are ill-defined (involving both elastic and plastic states) and probably vary with the amount of eccentricity, the following analysis for a strip foundation on the surface of cohesionless soil must be regarded as an approximate interpretation. The two forces P_a and P_b in Fig. 3.23 replace the two equal forces P_p in the case of axial loading.

Equating vertical forces

$$Q_{fe} + \frac{\gamma B^2}{4} \tan \phi - P_a - P_b = 0. \tag{3.55}$$

Taking moments about c

$$Q_{fe}e + \frac{P_a B}{6} - \frac{P_b B}{6} = 0. \tag{3.56}$$

Multiplying Equation 3.56 by $6/B$ and adding to Equation 3.55

$$Q_{fe}\left(1 + \frac{6e}{B}\right) - \left(-\frac{\gamma B^2}{4}\tan \phi + 2P_b\right) = 0. \tag{3.57}$$

At failure P_b equals the value P_p for an axially loaded foundation and thus the last two terms in Equation 3.57 represent the load Q_f at failure of an axially loaded

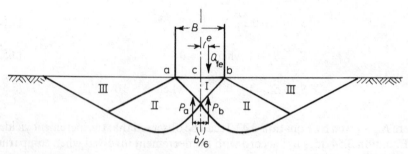

Figure 3.23 Eccentrically loaded strip foundation with a rough base on the surface of cohesionless soil.

foundation. Hence

$$Q_{fe} = \frac{Q_f}{(1 + 6e/B)} = C_e Q_f, \tag{3.58}$$

where $C_e = 1/(1 + 6e/B)$. It has been assumed above that P_a is located at a distance $B/6$ from the axis of the foundation, consistent with a state of plastic equilibrium in the soil on that side of the foundation. This is probably true when e/B is small but when $e/B = 1/6$, $P_a = 0$, and hence the soil on that side must be in a state of elastic equilibrium. If it is assumed that P_a is located at $B/4$ from the axis and a similar analysis is carried out, it will be found that approximately

$$Q_{fe} = \frac{\frac{5}{6}Q_f}{(1 + 4e/B)} = C_e' Q_f \tag{3.59}$$

When $e/B = 1/6$, $C_e = 0.5 = C_e'$. Therefore, even when an elastic state is most likely to prevail on one side, the appropriate value of the reduction factor is given by C_e. Consequently it is reasonable to employ C_e as the factor within the practical range $E/B = 0$ to $1/6$ for strip foundations on the surface of cohesionless soil.

Eastwood (1955) tested model rectangular foundations on the surface of sand; the proportions of some of these tended towards those of strip foundations. Under eccentric loading the observed load at failure was between 1 and 9% in excess of the load calculated from Equation 3.58. Ramelot and Vandeperre (1950) tested model square foundations on the surface of sand. The ultimate loads calculated from Equation 3.58 correspond closely with the observed loads. It follows that the bearing capacity of eccentrically loaded strip, rectangular and square foundations on the surface of cohesionless soils can be calculated from Equation 3.58 with reasonable accuracy. From the scanty information available, it appears that Equation 3.58 can be employed also for foundations on cohesive soils. For the estimation of bearing capacity of eccentrically loaded foundations, Meyerhof (1953) introduced the concept of a reduced base width $B' = (B - 2e)$. The bearing capacity of eccentrically loaded foundations has been studied also by Prakash and Saran (1971) and by Purkayastha and Char (1977).

Apart from considerations of bearing capacity, it is desirable to limit the maximum bearing stress beneath a foundation in order to reduce excessive settlement on one side. Some tilting of a foundation carrying an eccentric load appears inevitable and a rotation of less than $1°$ is usually sufficient theoretically to relieve a footing of the bending moment at the column foot, thereby reducing the effective eccentricity, although there will be a transfer of moment to other parts of the structure. Consequently, the application of the factor C_e can be regarded as a rational method of good design to be applied to all eccentrically loaded foundations on all types of soils. An important exception to these considerations is a retaining wall since rotation commonly leads to an increase in eccentricity. In concluding this discussion on eccentrically loaded foundations it should be observed that when foundations are buried the problem is modified by the influence of lateral soil stress.

3.3.5 Bearing capacity of foundations subjected to inclined loads

Investigations by Meyerhof (1953, 1963) show that the vertical component q_{fv} of bearing capacity of a foundation subjected to a load inclined at an angle $\alpha°$ to the vertical can be calculated approximately by the application of the inclination factors

$$I_c = I_q = (1 - \alpha°/90°)^2 \qquad (3.60)$$

and

$$I_\gamma = (1 - \alpha°/\phi°)^2. \qquad (3.61)$$

In the general case

$$q_{fv} = D_c I_c c N_c + D_q I_q \gamma D N_q + \tfrac{1}{2} D_\gamma I_\gamma \gamma B' N_\gamma \qquad (3.62)$$

where $B' = (B - 2e)$ caters for eccentricity e of loading. Factors to cater for inclination of load have been given also by Hansen (1961). The horizontal component of load can be assumed to be resisted by friction and adhesion at the interface between soil and base plus passive resistance of the soil at the side of the foundation. Since the lateral displacement required to mobilize full passive resistance may be appreciable, a factor of safety of three or more may be desirable to restrict translation of the structure. For a given inclination of the load, an inclined foundation with a base normal to the resultant load gives a value of bearing capacity greater than that of a horizontal foundation (Meyerhof, 1953). The bearing capacity relationship for inclined foundations is similar to that for foundations located adjacent to the sloping face of a bank (Meyerhof, 1957). Studies of the behaviour of shallow foundations subjected to inclined loads have been reported by Muhs and Weiss (1973).

3.3.6 The influence of groundwater on bearing capacity

In the case of a cohesive soil, variations in groundwater conditions lead to changes in cohesion which, in turn, influence the bearing capacity of the soil. The shearing resistance of partially saturated clay above the phreatic surface may be enhanced by a frictional component which increases bearing capacity, although this is not normally taken into account owing to a possible rise in phreatic surface. Certain clays carrying lightly loaded foundations swell when the phreatic surface rises.

The bearing capacity beneath both surface and buried foundations on cohesionless soil is influenced by changes in buoyancy effects on account of changes in the level of the phreatic surface. Meyerhof (1955) deals with this problem in the following way. For a foundation on the surface of cohesionless soil

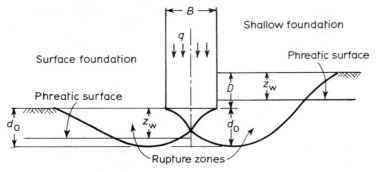

Figure 3.24 Groundwater conditions beneath a strip foundation.

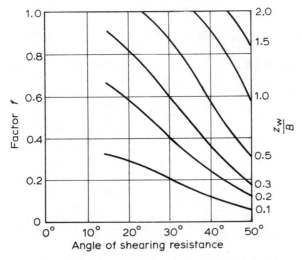

Figure 3.25 Relation between weight factor and angle of shearing resistance.

and phreatic surface at depth z_w (Fig. 3.24), the ultimate bearing capacity for a strip foundation is

$$q_f = [\gamma' + f(\gamma - \gamma')]\frac{B}{2}N_\gamma \qquad (3.63)$$

where f is a factor varying between zero when $z_w = 0$ and unity when $z_w \geq d_o$. Values of f for a strip foundation with a rough base are given in Fig. 3.25. Commonly, $\gamma' \simeq \gamma/2$ and therefore q_f for fully submerged conditions is roughly one-half of q_f for the unsubmerged case.

For a circular or square foundation on the surface of cohesionless soil it has been found that d_o does not exceed B and a linear increase in f with z_w leads to the approximate relationship.

$$q_f = [\gamma' B + (\gamma - \gamma')z_w]\frac{N_\gamma}{2}. \qquad (3.64)$$

If a strip foundation is located at a shallow depth $(D/B \not> 1)$ in cohesionless soil (Fig. 3.24) and if $z_w \not> D$, the capacity is approximately

$$q_f = \frac{\gamma'B}{2}N_\gamma + [\gamma'D + (\gamma - \gamma')z_w]N_q + \gamma_w(D - z_w), \tag{3.65}$$

and if $D < z_w < (D + d_o)$

$$q_f = [\gamma' + f(\gamma - \gamma')]\frac{B}{2}N_\gamma + \gamma D N_q, \tag{3.66}$$

where f is given in Fig. 3.25 by substituting $(z_w - D)$ for z_w. When $D/B > 1$ in cohesionless soil

$$q_f = \left[\gamma' + \frac{(\gamma - \gamma')z_w}{D + d_o}\right]\frac{B}{2}N_{\gamma q} + \gamma_w(D - z_w), \tag{3.67}$$

where d_o can be estimated from the value of z_w/B for $f = 1$ in Fig. 3.25, taking $d_o = z_w$.

The general ultimate gross bearing capacity Equation 3.35 can be expressed in terms of effective stress in the form

$$q_f' = cN_c + \sigma_o'N_q + 0.5B\gamma_B N_\gamma \tag{3.68}$$

where σ_o' is the effective overburden stress at founding level and γ_B is the vertical component of the effective body force per unit volume of soil, taking into account buoyancy and seepage pressure: If seepage is absent, $\gamma_B = \gamma$ above the phreatic surface and $\gamma_B = (\gamma - \gamma_w) = \gamma'$ below the phreatic surface.

The allowable gross bearing stress in terms of total stress is

$$q_a = \left[\frac{q_f - \sigma_o}{F}\right] + \sigma_o. \tag{3.69}$$

but

$$q_f' - \sigma_o' = (q_f - u) - (\sigma_o - u) = q_f - \sigma_o,$$

and hence in terms of effective stress

$$q_a = \left[\frac{q_f' - \sigma_o'}{F}\right] + \sigma_o. \tag{3.70}$$

Generally, conditions immediately after the service load is applied are the most critical in predicting bearing capacity and hence total stress analysis using Equation 3.35 is appropriate. Consolidation of the ground beneath a foundation during and after construction leads to some improvement in shear strength although soil at the sides may scarcely be affected and appropriate values of shear strength parameters should be employed in effective stress analysis.

If water percolates upwards through the soil it can be shown that for a continuous foundation in fully submerged soil

$$q_f = (\gamma' - \gamma_w i)\frac{B}{2}N_{\gamma q} + \gamma_w D \tag{3.71}$$

where i is the average vertical hydraulic gradient in the rupture zone. Flow net analysis shows that the bearing capacity of the soil is less than the value derived from the average hydraulic gradient on account of piping around the foundation edges and a greater seepage pressure on the theoretical failure surface.

3.3.7 Additional comments on bearing capacity

The criterion for determination of allowable bearing capacity for a foundation on cohesive soil may be either a risk of shear failure or an unacceptable magnitude of settlement. In the case of cohesionless soil has criterion is settlement unless the foundation is less than about 1 m wide. If cohesionless soil is loose, it may be desirable to employ some means of compaction to reduce settlement and improve bearing capacity.

It is convenient to mention here that the weight of a foundation enters calculations for stability but, together with the soil reaction due to the weight, is generally omitted for convenience in the forces and moments for design of the foundation structure. Soil conditions must be examined for all combinations of loading on a structure. The factor of safety for shear failure should normally be less than three but on unimportant structures, such as domestic houses or single-storey factories or where maximum loads are only occasional, a value of two is sometimes adopted. In the case of sensitive soils it may be desirable to increase the factor of safety by 0.5 or 1.0 in order to reduce the possibility of local shearing leading to a progressive reduction in bearing capacity. In CP 2004 (BSI, 1972d) it is suggested that, where the bearing stress due to wind exceeds 25% of that due to dead and live loads, foundations should be so proportioned that the combined stress does not exceed the allowable bearing stress by more than 25%. Where the proportion is less than 25%, stress due to wind is neglected.

The bearing capacity of foundations on sand can be determined approximately from the magnitude of ϕ estimated from the results of site standard penetration tests. The correlation given in Table 3.5 is taken from a chart given by Peck *et al.* (1953): these figures provide only a rough guide since N is governed by factors other than angle of shearing resistance. The ultimate resistance of driven piles in cohesionless soils in relation to the static cone penetration test (Section 1.2.6.) has been studied by Thorburn (1976).

For relatively unimportant structures and for preliminary designs for major projects, it is often possible to assess the safe bearing capacity from tabulated values. Table 3.6 is based on CP 2004 (BSI 1972d) and gives values of presumed bearing capacity. If the phreatic surface is at a depth less than about B from the

Table 3.5 Approximate correlation between angle of shearing resistance of sand and standard penetration test results

N	10	20	30	46	56	70
ϕ	30°	33°	36°	40°	42°	44°

Table 3.6 Presumed bearing values under vertical static loading

Group	Class	Types of soils	Presumed bearing value $kN\ m^{-2}$	Remarks
II	9	Compact gravel, or compact sand and gravel	> 600	Width of foundation (B) not less than 1 m
Non-cohesive soils	10	Medium dense gravel, or medium dense sand and gravel	200–600	Groundwater water level assumed to be a
	11	Loose gravel, or loose sand and gravel	< 200	depth not less than B below the
	12	Compact sand	> 300	base of the
	13	Medium dense sand	100–300	foundation
	14	Loose sand	< 100	
III	15	Very stiff boulder clays and hard clays	300–600	Group III is susceptible to
	16	Stiff clays	150–300	long-term
Cohesive soils	17	Firm clays	75–150	consolidation
	18	Soft clays and silts	< 75	settlement
	19	Very soft clays and silts	Not applicable	

These values are for preliminary design purposes only, and may need alteration upwards or downwards. No addition has been made for the depth of embedment of the foundation. Reference should be made to other parts of the code when using this table. Group I is rock: see Table 3.7. This extract from CP 2004 is reproduced by permission of the British Standards Institution. Complete copies of the document can be obtained from BSI at Linford Wood, Milton Keynes, MK 14 6LE.

base of a footing located near the surface, the values for cohesionless soils should be reduced; by 50% when the phreatic surface is at the footing base. If the surface of the site is sloping or if a bed of soft soil is present beneath the foundation stratum it may be necessary to reduce the tabulated values. Bearing stress should be limited to a value which produces a net loading intensity at the level of the soft stratum not exceeding its safe bearing capacity. On cohesionless soil, bearing capacity increases approximately in direct proportion to the width of the footing, but settlement also increases with increase in width and this is usually limited to an acceptable value.

Theoretically the bearing capacity of cohesive soils does not vary with B. Settlement of many foundations on firm and stiff clays does not exceed 50 mm since the safe bearing capacity, based on a factor of safety of three, seldom exceeds the pre-consolidation stress, but on normally consolidated soft clay, low values of bearing stress may be necessary to limit settlement. However, it should be noted that settlement on a deep bed of clay is governed largely by total net load and is not markedly influenced by changes in foundation width (Example 2.11).

The California Bearing Ratio test is a penetration test and the correlation between CBR and bearing capacity has been investigated by Black (1961). For a circular foundation on the surface of a saturated clay the bearing capacity in

$kN\,m^{-2}$ is approximately $q_f = 70$ (CBR) where the CBR is expressed as a percentage. Black (1962) investigated also the correlation of CBR with plasticity data.

Stuart (1962) has studied the effect on bearing capacity of coalescence of rupture zones of independent foundations at close spacing. Under these conditions the bearing capacity of footings on the surface of sand tends to increase with decrease in spacing and can rise to as much as two or three times the value for an isolated footing. Mandel (1963) confirms these conclusions in general although the increase with cohesive soil ($\phi_u = 0$) is insignificant. Further studies of this problem have also been reported by Stuart and Hanna (1961), Mandel (1965) and West and Stuart (1965).

The bearing capacity of foundations located on soil slopes has been investigated by Meyerhof (1957) who found that the capacity of foundations on the face or near the upper edge of a bank decreases with increase in inclination of the slope, especially in the case of cohesionless soils. It is, of course, generally undesirable to found on or near a bank because of the influence on stability of the slope but foundations are often located on gently sloping ground.

L'Herminier *et al.* (1965) confirmed, as might be expected, that, where the ground surface is at different levels on each side of a foundation, the bearing capacity is equal to the value corresponding to the smaller depth of burial. The estimation of bearing capacity and settlement from pressuremeter tests has been discussed by Ménard (1965). Vesić *et al.* (1965) studied the bearing capacity of footings on sand under dynamic loads. Studies of the bearing capacity of reinforced earth have been reported by Binquet and Lee (1975a, b) and by Akinmusuru and Akinbolade (1981).

3.3.8 The bearing capacity of cohesive soil in the undrained state beneath an embankment

The rupture interface hypothesis was employed by Raymond (1967) to determine the bearing capacity of cohesive soil beneath an embankment founded at ground level. It was assumed that, with $\phi_u = 0$, cohesion increases linearly with depth, and at any depth z is

$$c_u = c_{uo} + \lambda z \qquad (3.72)$$

where c_{uo} is the cohesion at ground level and λ is a constant.

In order to predict the value of the load factor against bearing failure, the embankment loading is converted to an equivalent uniformly distributed loading of magnitude q_e, as shown in Fig. 3.26(a). Several trial values of e are assumed and values of q_e and a (Fig. 3.26a) are determined from Fig. 3.26(b), which relates to the two-dimensional case of a long embankment. With a slide rule having a reciprocal scale, $8\lambda a/q_e$ is set on the D-scale opposite q_e/c_{uo} on the reciprocal scale. The position of the cursor is adjusted until $8\lambda a/q_f$ on the reciprocal scale and q_f/c_{uo} on the D-scale gives a point on the graph shown in Fig. 3.26(c) and thus

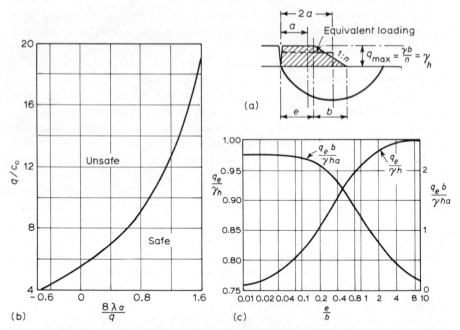

Figure 3.26 Data for the analysis of the stability of soils beneath embankments. (a) Details of embankment and equivalent loading; (b) failure criteria; (c) conversion of embankment loading to equivalent loading.

q_f is determined. The load factor is then q_f/q_e. Failure does not occur within the side slope but when e is equal to or exceeds zero. If the value of cohesion is, or can reasonably be assumed to be, constant with depth, then q_e is determined from Fig. 3.26(b) for trial values of e and, from Equation 3.38,

$$q_f = 5.5c_u. \tag{3.73}$$

The influence of increasing strength with depth on the bearing capacity of clays has been studied also by Davis and Booker (1973).

A relatively common example of the engineering problem for which the solution presented above is relevant is that of a highway embankment founded on normally consolidated soils. In such cases, the height of the embankment is likely to be small in relation to the radius of the rupture interface. Furthermore, if the embankment fill is cohesive, a tension fissure should be assumed to be present, and if it is cohesionless, the shearing resistance developed along that part of the rupture interface within the fill is likely to be relatively small. Consequently, the magnitude of the load factor is unlikely to be underestimated by more than 10% if shearing resistance within the fill is ignored, as it is in the analysis of Raymond. Another factor which is difficult to take into account in a standard solution is the influence of higher strength in a desiccated zone near ground surface, as shown in Fig. 3.3. A satisfactory analysis can probably be effected by first locating the critical rupture interface, ignoring the influence of the desiccated zone. Then a number of trial rupture interfaces should be selected, with mean values of shear strength assigned to short lengths of arc and values of load factor determined by tabular computation to establish the minimum value. With complex conditions it would, however, be preferable to employ a computer program.

Kenney (1964) has drawn attention to the influence of anisotropic permeability in layered soils where migration of water predominantly in a horizontal direction beneath an embankment can lead to increases in pore pressure beyond the embankment which would not be suspected if isotropic permeability were assumed. The influence on stability of embankments of thin layers of peat in alluvium has been examined by Thorburn and Beevers (1981). Proceedings of a symposium on settlement and stability of embankments have been published by TRRL (1978).

EXAMPLE 3.2

The gross axial load on the base of a square footing is 5000 kN, including the weight of the footing and backfilling. The footing is to be founded at a depth of 3 m below ground surface in a soil having a saturated unit weight of $16 \, \text{kN} \, \text{m}^{-3}$ and shear strength parameters $c_u = 50 \, \text{kN} \, \text{m}^{-2}$, $\phi_u = 0$ and $c' = 20 \, \text{kN} \, \text{m}^{-2}$, $\phi' = 20°$.

Determine the plan proportions of the footing to the nearest 0.1 m, applying a load factor of 3, with respect to both total and effective stress for the two conditions in which

(a) the phreatic surface is at a depth exceeding $(3 + B)$ m below ground surface
(b) the phreatic surface is at ground surface.

The Skempton value for a square footing is estimated to be (Fig. 3.22b) $N_c = 7.2$. For (a) it is assumed that soil above the phreatic surface is saturated by capillarity. The Meyerhof values for a square footing are estimated to be (Fig. 3.20) $N_c = 21$, $N_q = 8$ and $N_\gamma = 3$ and $D_c = 1.2$ and $D_q = D_\gamma = 1.1$: the depth factors are calculated from an estimated value of $B = 4 \, \text{m}$ (Equations 3.40 and 3.42). Alternatively, the depth factors can be included in Equation 3.35, with the unknown B in Equations 3.40 and 3.42. Skin resistance on the sides of the foundation is ignored.

Total stress: Equation 3.48

$$\text{(a) and (b)} \quad q_a = \frac{5000}{B^2} = \frac{7.2 \times 50}{3} + 16.3 \text{ whence } B = 5.5 \, \text{m}.$$

In (b) the effects of buoyancy on foundation and on overburden balance.

Effective stress: Equation 3.35

$$\text{(a)} \quad q_a = \frac{5000}{B^2} = \left\{ \frac{20 \times 21 \times 1.2 + 16 \times 3 \times 8 \times 1.1 + 0.5 \times B \times 16 \times 3 \times 1.1 - 16 \times 3}{3} \right\}$$

$$+ 16 \times 3$$

or $8.8B^3 + 340.8B^2 - 5000 = 0$ whence trial substitution yields $B = 3.7 \, \text{m}$.

$$\text{(b)} \quad q_a = \frac{5000}{B^2} = \left\{ \frac{20 \times 21 \times 1.2 + 6 \times 3 \times 8 \times 1.1 + 0.5 \times B \times 6 \times 3 \times 1.1 - 6 \times 3}{3} \right\}$$

$$+ 16 \times 3$$

or $3.3B^3 + 262.8B^2 - 5000 = 0$ whence trial substitution yields $B = 4.3 \, \text{m}$.

EXAMPLE 3.3

Extrapolation of results of loading tests (East, 1951) on 0.3 m diameter bored piles at Folkestone indicate that the ultimate capacity of a pile penetrating 1.83 m in the ground was about 90 kN and penetrating 2.44 m was about 100 kN. Assuming frictional components of resistance were negligible, calculate the cohesion c_u of the soil and adhesion c_w between pile and soil if the latter is effective only below a depth of 1.0 m from ground surface. The value of the bearing capacity factor N_c can be taken as 9 and the unit weight of the soil as 20 kN m^{-3}.

From Equation 3.48

$$1.83 \text{ m pile} \qquad 90 = (9c_u + 20 \times 1.83)\pi \times 0.15^2 + \pi \times 0.30 \times 0.83c_w$$

$$2.44 \text{ m pile} \qquad 100 = (9c_u + 20 \times 2.44)\pi \times 0.15^2 + \pi \times 0.30 \times 1.44c_w$$

Solving these two equations yields $c_u = 118$ kN m^{-2} and $c_w = 16$ kN m^{-2}. $\alpha = c_w/c_u = 0.14$: this value is low in relation to the normal range of 0.25 to 0.7. Both c_u and c_w can be expected to vary markedly in the weathered zone (Fig. 3.3) of the stiff fissured Gault clay in which the piles were installed: this is confirmed by the much higher resistance afforded by a pile penetrating 3.05 m. Additional and more accurate details of tests of this kind are required in order to achieve reliable analyses of the results.

EXAMPLE 3.4

At a certain site the undrained shear strength in terms of apparent cohesion c_u of a deep bed of saturated overconsolidated clay increases according to the relationship

$$c_u = [100 + 2(z - 3)] \text{kN m}^{-2}$$

where z m is the depth below ground surface and is greater than 3 m. The relative merits of bored cylindrical piers at depths of 6 and 15 m below ground surface are to be investigated. The axial load to be applied to the top of each pier at ground surface is 2000 kN. The unit weight of the clay is 20 kN m^{-3} and of the concrete 22 kN m^{-3}.

The Skempton bearing capacity factor can be taken as 9 for both piers and the load factor is to be 3. Adhesion between pier and clay in the uppermost 3 m is to be ignored but below this it is to be taken as one-half of the cohesion.

Calculate to a close approximation the theoretical diameter of each pier.

$$\text{Gross safe load} \quad Q_s = 2000 + \frac{\pi}{4}B^2 z 22$$

$$\text{From Equation 3.48} \quad Q_s = \left\{ \frac{c_u N_c \pi B^2/4 + (z - 3)\pi B c_w}{3} \right\} + \gamma z \pi B^2/4$$

At $z = 6$ m, at base $c_u = 106$ kN m^{-2} and mean $c_w = 51.5$ kN m^{-2}.

Then

$$2000 + 103.673B^2 = 249.757B^2 + 161.792B + 94.248B^2$$

Collecting terms and substituting trial values for B yields remainder of 3 when $B = 2.57$ m
At $z = 15$ m, at base $c_u = 124$ kN m^{-2} and mean $c_w = 56$ kN m^{-2}.

Then

$$2000 + 259.185B^2 = 292.168B^2 + 703.717B + 235.619B^2$$

Trial substitution for B yields remainder of 5 when $B = 1.72$ m.

Calculations of this nature could be employed for preliminary studies in a project but final design of bored piles would be based on the methods treated in Section 3.5.

EXAMPLE 3.5

A cylindrical foundation, 1 m diameter, is to be constructed by boring in clay. The vertical load on the foundation at ground level is 120 kN. The unit weight of the foundation concrete is 22 kN m^{-3} and of the overburden is 19 kN m^{-3}. The undrained strength of the clay is 40 kN m^{-2} and the angle of shearing resistance can be taken as zero. Adhesion of 20 kN m^{-2} operates on the shaft below a depth of 1 m. A load factor of 3 is to be applied to the ultimate net bearing capacity of the foundation. Determine the minimum safe founding depth to the nearest 0.1 m, given the following values of the Skempton bearing capacity factor N_c for a circular foundation.

D/B	1.5	2.0	3.0
N_c	8.1	8.4	8.8

Gross load $= 120 + \pi \times 0.5^2 \times 22D$ kN
Total gross ultimate capacity $= 40 \times \pi \times 0.5^2 N_c + \pi \times 1 \times 20(D-1) + \pi \times 0.5^2 \times 19D$

Then $\quad 120 + 17.28D = \dfrac{31.42N_c + 62.83(D-1)}{3} + 14.92D$

or $140.94 - 18.59D - 10.47N_c = 0$

For trial values of D, remainder R is calculated and graph plotted of R against D (Fig. 3.27).

$D(m)$	1.5	2.0	3.0
D/B	1.5	2.0	3.0
N_c	8.1	8.4	8.8
R	28.24	15.80	-6.97

When $R = 0$, $D = 2.5$ m.

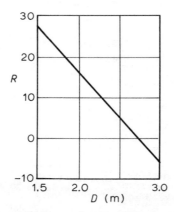

Figure 3.27 Determination of D when $R = 0$.

EXAMPLE 3.6

The pivot pier for a swing bridge consists of a rigid reinforced concrete cylindrical box 20 m diameter externally. It is to be founded in stiff clay at a depth of 8 m below river bed level. The clay extends to a depth of 20 m beneath the base of the foundation and overlies bedrock. Mean low water and mean high water are 8 m and 16 m above bed level respectively. The clay has a saturated unit weight of $20 \, kN \, m^{-3}$ and undrained shear strength parameters $c_u = 96 \, kN \, m^{-2}$ and $\phi_u = 0$.

Adhesion between clay and concrete is $48 \, kN \, m^{-2}$ and is assumed to operate on the lower 6 m of cylinder. The gross load on the foundation is 2×10^5 kN including the weight of the cylinder. The unit weight of water is taken as $9.8 \, kN \, m^{-3}$.

Determine the load factors against bearing failure at both high and low water. For a circular foundation with a ratio of depth/diameter of 0.4 the value of the Skempton bearing capacity factor N_c is 7.0.

Total weight of saturated overburden $= 20 \times 8 \times \pi \times 10^2 = 50\,265 \, kN$
Additional uplift at low water $= 8 \times 9.8 \times \pi \times 10^2 = 24\,630 \, kN$
Additional uplift at high water $= 16 \times 9.8 \times \pi \times 10^2 = 49\,260 \, kN$
Net load at low water $= 200\,000 - 50\,265 - 24\,630 = 125\,105 \, kN$
Net load at high water $- 200\,000 - 50\,265 - 49\,260 = 100\,475 \, kN$
Total adhesion resistance $= \pi \times 20 \times 6 \times 48 = 18\,095 \, kN$
Ultimate total net bearing capacity $= 7.0 \times 96 \times \pi \times 10^2 + 18\,095 = 229\,210 \, kN$
Hence at low water $F = 229\,210/125\,105 = 1.83$.
and at high water $F = 229\,210/100\,475 = 2.28$.

EXAMPLE 3.7

A $0.3 \, m \times 0.3 \, m$ reinforced concrete pile 12 m long is to by driven through 6 m of soft clay fill and 6 m into an underlying bed of sand. The phreatic surface is well below the foot of the pile.

For the sand, the apparent cohesion is zero, the angle of shearing resistance is $30°$ and the corresponding value of N_q is 50 (Fig. 3.20). The friction angle between the sand and the pile is $20°$ and the adhesion between the fill and the pile is $15 \, kN \, m^{-2}$. The unit weight of the fill and of the sand is $20 \, kN \, m^{-3}$. The coefficient of lateral earth stress of the sand can be taken as 0.5.

Calculate the safe load on the head of the pile if a load factor of 3 is employed. The total weight of the pile is 24 kN. Deduct for downward drag of the fill due to compaction.

Horizontal stress:
at $z = 6 \, m$, $\sigma_h = 0.5 \times 20 \times 6 = 60 \, kN \, m^{-2}$
at $z = 12 \, m$, $\sigma_h = 0.5 \times 20 \times 12 = 120 \, kN \, m^{-2}$
Total frictional resistance on pile in sand

$$F_w = 4 \times 0.3 \times 6 \left(\frac{60 + 120}{2} \right) \tan 20° = 236 \, kN$$

Safe gross capacity of pile

$$Q_s = \left(\frac{\gamma D N_q A + F_w - \gamma D A}{F} \right) + \gamma D A \tag{3.74}$$

$$= \left(\frac{20 \times 12 \times 50 \times 0.3^2 + 236 - 20 \times 12 \times 0.3^2}{3} \right) + 20 \times 12 \times 0.3^2 = 453 \, kN$$

Total downward drag of fill is $4 \times 0.3 \times 6 \times 15 = 108 \, kN$

Safe load on head of pile is $453 - 108 - 24 = 321\,\mathrm{kN}$, say $320\,\mathrm{kN}$.

This is the normal order of capacity of a 0.3 m square pile, perhaps rather lower than usual, but a more reliable estimate would be given by tests on a pile *in situ*.

The shaft capacity of driven piles has been investigated by Randolph and Wroth (1981). Bearing capacity of pile foundations has been discussed at length by Meyerhof (1976) and also by Cooke (1978). Tests of full-scale piles in sand have been analysed by Coyle and Costello (1981).

Downdrag depends on a number of factors, some of which can rarely be assessed with reasonable accuracy. Thus the lateral stress imposed on the shaft of a displacement pile driven in cohesionless soil depends on the degree of displacement, which varies with pile section, on void ratio of the soil and on type of driving equipment. Since downdrag is difficult to estimate and, in any case, reduces the available bearing capacity of a pile, it is generally desirable to eliminate it or, at least, reduce its magnitude by a slip coating of bitumen or bentonite around the shaft. Several authors have reported studies of skin resistance, downdrag and preventative measures, including Bjerrum *et al.* (1969), Endo *et al.* (1969), Fellenius and Broms (1969), Poulos and Mattes (1969), Bozozuk (1972), Fellenius (1972), Heijnen and Lubking (1973), Walker and Darvall (1973), Claessen and Horvat (1974), Poulos and Davis (1975), Parry and Swain (1977), Kraft *et al.* (1981), Chandler and Martins (1982), Poulos (1982) and Potts and Martins (1982). Load transfer from pile to soil and related deformation of a pile have been investigated by Randolph and Wroth (1978).

EXAMPLE 3.8

Field tests have been carried out on two 0.3 m × 0.3 m section reinforced concrete piles driven entirely in sand: the penetration of one pile is 5 m and of the other 10 m. The load at failure of the former was 559 kN and of the latter 1337 kN: the difference between weight of displaced soil and weight of concrete can be ignored. The unit weight of the sand was $20\,\mathrm{kN\,m^{-3}}$.

In addition to the field tests, a laboratory shear box test has been carried out in which the bottom half of the box contains a block of concrete with surface finish similar to that of the piles and the top half is filled with sand from the site. The results of this test are as follows.

Normal stress ($\mathrm{kN\,m^{-2}}$)	100	200	300
Shear stress ($\mathrm{kN\,m^{-2}}$)	36.4	72.8	109.2

Determine the unkown values of the parameters relating to these bearing capacity tests.

From Equation 3.37, $q_f = \gamma D N_q$, the term $0.4\gamma BN\gamma$ is neglected since B is small.

From plot of laboratory test results, angle of friction δ between sand and concrete given by $\tan\delta = 109.2/300 = 0.364$ whence $\delta = 20°$.

$$Q_{f1} = \gamma D_1 N_q A + 4B\tfrac{1}{2}K_0\gamma D_1^2 \tan\delta$$
$$= 20 \times 5 \times N_q \times 0.3^2 + 2 \times 0.3 \times K_0 \times 20 \times 5^2 \tan\delta$$

whence $9N_q + 109K_0 = 559$.   (1)

$$Q_{f2} = \gamma D_2 N_q A + 4B\tfrac{1}{2}K_0\gamma D_2^2 \tan\delta$$
$$= 20 \times 10 \times N_q \times 0.3^2 + 2 \times 0.3 \times K_0 \times 20 \times 10^2 \tan\delta$$

whence $18N_q + 437K_0 = 1337$.   (2)

Multiply (1) by 2 and subtract from (2)

$$18N_q + 437K_0 = 1337$$
$$\underline{18N_q + 218K_0 = 1118}$$
$$219K_0 = 219 \quad \text{whence } K_0 = 1.0.$$

Substituting for K_0 in (1)

$$9N_q + 109 = 559 \text{ or } 9N_q = 450 \quad \text{whence } N_q = 50.$$

By reference to appropriate tables or charts the angle of shearing resistance can be determined from N_q relating to deep square-section piles with, say, $60°$ points at toe: from Fig. 3.20, $\phi \simeq 30°$.

EXAMPLE 3.9

The perimeter of a group of six piles encloses a plan area $2.7\,\text{m} \times 1.5\,\text{m}$. The piles are $0.3\,\text{m} \times 0.3\,\text{m} \times 12\,\text{m}$ long and are spaced at $1.2\,\text{m}$ centres. They are driven through $6\,\text{m}$ of soft clay fill and $6\,\text{m}$ into an underlying bed of sand. The phreatic surface is more than $3\,\text{m}$ below the feet of the piles.

The angle of shearing resistance of the sand is $30°$ and the corresponding Meyerhof values of N_q and N_y for a deep foundation are 120 and 130 respectively (Table 3.3). The shape factor can be taken as unity. The friction angle between sand and pile is $20°$ and both cohesion in the fill and adhesion between fill and pile are $15\,\text{kN}\,\text{m}^{-2}$. The unit weight of the fill and the sand is $20\,\text{kN}\,\text{m}^{-3}$.

Assuming that the piles and the enclosed sand function as a single unit, calculate the safe load on the head of the group if a load factor of 3 is employed. The total weight of each pile is $24\,\text{kN}$. Deduct for downward drag of the fill around each pile due to compaction. The coefficient of lateral earth stress in the sand can be taken as 0.5.

Horizontal stress: see Example 3.7:

at $z = 6\,\text{m}$, $\sigma_h = 60\,\text{kN}\,\text{m}^{-2}$.   At $z = 12\,\text{m}$, $\sigma_h = 120\,\text{kN}\,\text{m}^{-2}$

Total frictional resistance on perimeter in sand

$$F_w = (4 \times 0.6 + 2 \times 0.3)6 \times \tfrac{1}{2}(60 + 120) \tan 20° + 6 \times 0.9 \times 6 \times \tfrac{1}{2}(60 + 120) \tan 30°$$
$$= 2274\,\text{kN}$$

Safe gross capacity of assembly

$$Q_s = \left\{ \frac{(20 \times 12 \times 120 + 0.5 \times 20 \times 1.5 \times 130 - 20 \times 12)2.7 \times 1.5 + 2{,}274}{3} \right\}$$
$$+ 20 \times 12 \times 2.7 \times 1.5$$

$$= 42\,919\,\text{kN}$$

Weight of soil within perimeter is $(2.7 \times 1.5 - 6 \times 0.3^2)12 \times 20 = 842\,\text{kN}$
Weight of 6 piles is $\qquad\qquad\qquad\qquad\qquad 6 \times 24 = 144\,\text{kN}$
Downdrag on outer faces of piles and on soil between piles is
$$2(2.7 + 1.5)6 \times 15 = 756\,\text{kN}$$

$$\overline{1742\,\text{kN}}$$

Safe load on head of group is $42\,919 - 1742 = 41\,177\,\text{kN}$.

This very high capacity is greater than any load likely to be applied in practice, even if a conservative load factor of 5 is employed. The design criterion will clearly be settlement. Note the discordance between the values of bearing capacity factors used above and the value for a single pile in Example 3.7.

EXAMPLE 3.10

A stiff rectangular foundation is supported on a group of 20 reinforced concrete piles, $10\,\text{m}$ long, as shown in Fig. 3.28. The inner rows consist of $300\,\text{mm} \times 300\,\text{mm}$ piles and the outer rows of $400\,\text{mm} \times 400\,\text{mm}$ piles: the cross-sectional areas in equivalent concrete units are

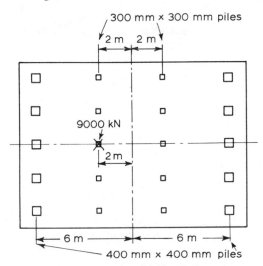

Figure 3.28 Pile layout for example 3.10.

$105\,000\,\text{mm}^2$ and $185\,000\,\text{mm}^2$, and the modulus of elasticity of concrete can be taken as $15\,\text{kN mm}^{-2}$. The foundation is subjected to an eccentric load of $9000\,\text{kN}$, including the weight of the foundation, which is applied on the longitudinal axis and $2\,\text{m}$ from the transverse axis.

The piles are driven through $7\,\text{m}$ of soft clay and $3\,\text{m}$ into an underlying bed of sand. The phreatic surface is several metres below the feet of the piles. The angle of shearing resistance of the sand is $30°$ and the corresponding Meyerhof value of N_q is 50 for piles with $60°$ points (Fig. 3.20). The apparent cohesion of the sand is zero. The friction angle between the sand and the pile is $20°$ and the coefficient of lateral earth stress in the sand can be taken as 0.5. The bulk density of the clay and of the sand is $20\,\text{kN m}^{-3}$.

Determine the lowest factor of safety against bearing failure of the $400\,\text{mm}$ piles assuming that no load is transmitted to the clay and that the spacing of the piles is too great for the assembly to be treated as an equivalent pier. The weight of a $400\,\text{mm}$ pile is $35\,\text{kN}$.

The load Q on any given pile at a distance x from the neutral axis in an eccentrically loaded group of N vertical piles is given by

$$Q = \frac{\sum Q}{N} \pm \frac{ex\sum Q}{\sum x^2} \tag{3.75}$$

where $\sum Q$ is the total vertical load and e is the eccentricity of loading. Where the group comprises two or more types of piles differing in cross-sectional area A, length L or modulus of elasticity E, analysis of the group can be effected in the following way. It is assumed here that the load on the head of the pile is transmitted wholly to the foot, driven to an unyielding stratum, and transfer to the peripheral soil is ignored. Let the compressibility of any pile be $K = AE/L$. Consider, say, two pile types B and C in a group comprising N_B of B and N_C of C. The compression δ_A due to an axial load $\sum Q$ on the group is the same for all piles and hence $\delta_A = Q_B/K_B = Q_C/K_C$ where subscripts B and C relate to the two pile types. Then $\sum Q = N_B Q_B + N_C Q_C = N_B \delta_A K_B + N_C \delta_A K_C$ or

$$\delta_A = \frac{\sum Q}{N_B K_B + N_C K_C}. \tag{3.76}$$

The compression or extension due to a moment is $\delta_E = Q/K$ where $Q \propto x$ if a linear variation in δ_E is assumed. Let the maximum positive and negative values of δ_E and x be Δ_E and X. The resisting moment afforded by the B piles is

$$\sum Q_B x_B = K_B \sum \delta_E x_B = K_B \frac{\Delta_E}{X} \sum x_B^2$$

and by the C piles is

$$\sum Q_C x_C = K_C \frac{\Delta_E}{X} \sum x_C^2.$$

The total resisting moment is therefore

$$\frac{\Delta_E}{X} \left\{ K_B \sum x_B^2 + K_C \sum x_C^2 \right\} = e \sum Q.$$

For any given pile $\delta_E/x = \Delta_E/X$ and therefore numerically

$$\delta_E = \frac{ex \sum Q}{K_B \sum x_B^2 + K_C \sum x_C^2}. \tag{3.77}$$

The compression or extension in any given pile arising from axial load plus moment is

$$\delta = \delta_A \pm \delta_E = \frac{\sum Q}{N_B K_B + N_C K_C} \pm \frac{ex \sum Q}{K_B \sum x_B^2 + K_C \sum x_C^2} \tag{3.78}$$

This method of analysis was employed by McNulty and O'Brian (1961) for the extension and underpinning with concrete-filled cylindrical piles of an existing foundation supported on timber piles.

Compressibility of piles:

$$300 \text{ mm } K = \frac{AE}{L} = \frac{105\,000 \times 15}{10\,000} = 157.5 \text{ kN mm}^{-1}$$

$400 \text{ mm } K = 277.5 \text{ kN mm}^{-1}$

Compression of 400 mm piles:

$$\delta = \frac{9000}{10 \times 157.5 + 10 \times 277.5} + \frac{9000 \times 2000 \times 6000}{157.5 \times 10 \times 2000^2 + 277.5 \times 10 \times 6000^2}$$

$= 2.07 \pm 1.02$ whence maximum compression is 3.09 mm.

Thus maximum load on any 400 mm pile is $3.09 \times 277.5 = 857 \text{ kN}$

Lateral stress on pile at 7 m is $0.5 \times 20 \times 7 = 70 \text{ kN m}^{-2}$ and at 10 m is 100 kN m^{-2}

Total frictional resistance on 400 mm pile is $4 \times 0.4 \times 3 \times \left(\dfrac{70 + 100}{2} \right) \tan 20° = 148 \text{ kN}$

Safe load on 400 mm pile

$$Q_s = 857 + 35 = \left(\frac{20 \times 10 \times 50 \times 0.4^2 + 148 - 20 \times 10 \times 0.4^2}{F} \right) + 20 \times 10 \times 0.4^2$$

Hence $F = 2.0$.

For the 300 mm piles it is found that $F = 2.6$.

The behaviour of pile groups has been studied by many investigators including Vesić (1969), Butterfield and Banerjee (1971a, b), Poulos (1979), Randolph and Wroth (1979) and Cooke *et al.* (1980).

3.4 LOCAL OVERSTRESSING BENEATH FOUNDATIONS
AND THE BEARING CAPACITY OF FOUNDATIONS ON
A THIN LAYER OF COHESIVE SOIL

3.4.1 Overstressing in a thick bed of cohesive soil

The conditions which may lead to an overstressed zone beneath a foundation on a thick bed of clay can arise in the following way. The maximum shearing stress in the soil beneath, for example, a uniformly distributed strip load q per unit area on the surface of the ground is q/π and extends to a depth $B/2$ beneath the centre of the loaded area of width B (Fig. 2.6b). Thus in cohesive soil, with shear strength $\tau_f = c_u$, local shear failure can occur when $q = 3.14c_u$, whereas general bearing failure does not occur until q is equal to at least $5c_u$. Hence the cross-sectional area of the overstressed zone increases as q is increased from $3.14c_u$ to $5c_u$. Under these conditions excessive settlement may occur whilst the soil accommodates itself to the overstress by plastic flow. Since the overstressed zone is contained within a mass of soil in elastic equilibrium, it is possible that changes in differential settlement would be the only effects observed in some cases. However, increased total settlement may result from disruption of particle structure by shearing within the overstressed zone.

Although shear strength may vary with depth owing to variations in the composition of the soil, in homogeneous beds of normally consolidated clay it is sometimes found that down to 1 or 2 m below the surface the clay has relatively high strength (developed by desiccation) and may be as much as $100 \, kN \, m^{-2}$ but below that depth the strength may fall to values of the order $20 \, kN \, m^{-2}$ (Fig. 3.3a). Where a strip foundation is supported on the surface of a bed of clay with a fairly uniform shear strength c_u the safe bearing capacity will be less than $2c_u$ and local overstress need not be anticipated. On the other hand, where a foundation is supported on clay with a relatively high shear strength c_{u1} in a zone near the surface but a lower value c_{u2} beneath that zone, then it may exceed 3.14 c_{u2} even though the safe bearing capacity does not exceed $2c_{u1}$. The maximum shearing stress beneath the centre of a circular area of diameter B on the surface of an elastic medium and loaded with a uniformly distributed load q has a magnitude of about $q/3$ and occurs at a depth of about $B/3$. The possibility of overstressing beneath a circular or square footing can be checked roughly by comparing this stress with the shear strength at the same depth.

Cooling (1942) cites the case of a building 38 m long and 19 m wide supported on a concrete raft founded 1 m below ground level. Continuous gradual settlement amounted to about 250 mm in 12 years. The gross bearing stress was $40 \, kN \, m^{-2}$ and the net stress $27 \, kN \, m^{-2}$. The topmost stratum of soil was firm clay with an average shear strength of about $14 \, kN \, m^{-2}$ but beneath this was softer material down to a depth of 12 m, with a minimum shear strength of $7 \, kN \, m^{-2}$, beyond which the strength increased to above $11 \, kN \, m^{-2}$. The 250 mm settlement could not be accounted for by consolidation alone. However,

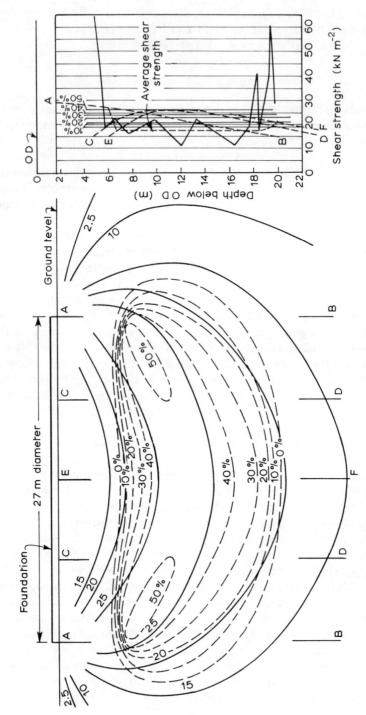

Figure 3.29 Determination of extent of overstressing in soil beneath a foundation, where ——— represents iso-stress lines: shear stresses (kN m^{-2}) and - - - represents iso-overstress lines: percentage overstress.

at a depth of 9 to 12 m the shear stress was as much as $9 \, \mathrm{kN \, m^{-2}}$ and the maximum overstress in this zone was therefore $2 \, \mathrm{kN \, m^{-2}}$. Consequently plastic yield occurred, leading to excessive settlement. This mode of overstressing is illustrated also by conditions beneath a group of four tanks (Glossop, 1945). The gross bearing stress when the tanks were full was $86 \, \mathrm{kN \, m^{-2}}$. The ultimate consolidation settlement was estimated to be 340 mm. After a period of 2 years following completion, the observed settlement was about 150 mm although the estimated value was only about 60 mm. The excess settlement was attributed to overstressing of a zone beneath the centre of the group.

Analysis of a particular problem can be effected by superposing graphs of vertical distributions of shear stress, derived from the stress bulb or otherwise, on the graph of shear strength, as shown in Fig. 3.29. The mean strength is increased by various percentages in order that the extent of a particular degree of overstressing can be assessed. Since the presence of a zone in a state of plastic equilibrium within the soil mass will influence the pattern of iso-stress lines, the results of an analysis will be approximate. The degree of overstressing shown in Fig. 3.29 has been arranged to be large in order to illustrate the principles of analysis: in fact the elastic stress distribution would not obtain and general shear failure (Section 3.3.1) is imminent. Some distributions of shear stress depend on Poisson's ratio and an appropriate value must be selected for any particular problem; 0.5 is commonly employed for saturated clay when the load is applied relatively rapidly. This type of analysis indicates only the relative degree of overstressing and whether or not preventive or remedial measures are likely to be effective. Assuming that a stratum having an adequate load-bearing capacity cannot be reached by normal economical types of foundation, consideration can be given, as a preventive measure, to increasing the plan area of the foundation on the surface in order to reduce shearing stress. Remedial measures vary widely but in some cases, such as that of an oil tank, it is possible that little can be done economically to save the structure.

Another example quoted by Cooling (1942) is uncommon and concerns a single-storey building situated at the foot of the bank of a cutting. The floor of that part of the building near the toe of the bank cracked and rose as much as 90 mm following heavy rain. The movement was stopped by cutting back the bank. Site investigations showed that the bank and the soil immediately beneath the structure consisted of stiff clay but a stratum of very soft clay about 1 m thick extended from the toe of the bank beneath the structure at a depth of roughly 1.5 to 3 m. Overstressing of the soft stratum was caused primarily by the load of the bank. Golder and Palmer (1955) investigated overstressing beneath a bank on the Isle of Sheppey.

3.4.2 The bearing capacity of a foundation on a thin layer of cohesive soil

Where a foundation is located on a relatively thin layer of cohesive soil overlying a comparatively rigid stratum, the mode of failure discussed in Section 3.3.2 does

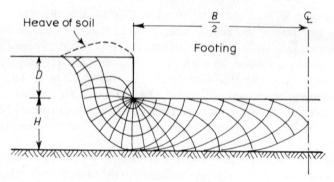

Figure 3.30 Plastic zone in thin layer of cohesive soil involved in bearing failure beneath footing of width B with a rough base and overlying a rigid stratum with a rough surface.

not obtain. Under a vertical load, the soil in the layer moves horizontally outwards from beneath the centre of the foundation. Analysis of the case of a block of material compressed vertically between two rigid platens was extended by Meyerhof and Chaplin (1953) to take account of the horizontal continuity of the soil layer. For the two-dimensional case of a long block, relatively wide in relation to its thickness, which is compressed between two rough-faced platens larger in area than the block, the yield stress is

$$q_f = c_u \left[\frac{B}{2H} + \frac{\pi}{2} \right]. \tag{3.79}$$

The pattern of plastic failure in a thin layer of clay is illustrated in Fig. 3.30. The bearing stress at failure increases with increase in adhesion between the cohesive soil and the foundation and the rigid stratum. It increases also with decrease in thickness H of the layer beneath the foundation. For a strip foundation of width B

$$q_f = c_u \left(\frac{B}{2H} + N_c - 1 \right) + \gamma D \tag{3.80}$$

and for a circular foundation of diameter B

$$q_f = c_u \left(\frac{B}{3H} + N_c - 1 \right) + \gamma D \tag{3.81}$$

where N_c is the bearing capacity factor appropriate to a deep layer of cohesive soil. When the cohesive layer is relatively thick (strip foundation $B/H < 2$, circular foundation $B/H < 6$), the methods discussed in Section 3.3.2 should be employed to evaluate bearing capacity. Since local overstressing occurs in thin layers with lower values of bearing stress than in thick layers, a check should always be made for local shear failure. The maximum shearing in a thin cohesive layer beneath both strip and circular foundations is approximately equal to one-half of the net bearing stress. The bearing capacity of thin layers has been studied by Suklje and Vidmar (1973).

Meyerhof (1952) describes a foundation failure in Essex which occurred beneath a 10 m high steel tank supported at ground level on a 38 m diameter reinforced concrete raft 150 mm thick with a ring beam round the perimeter. Site conditions consist of a layer of post-Glacial soft brown clay varying in thickness from about 0.1 m at one side of the raft to about 2 m at the other. Beneath this is a bed of Glacial sand and gravel, including a 3 m layer of stiff blue clay, extending to a depth of about 17 m beneath ground surface and overlying about 60 m of London Clay. The settlement after 1 year was about 10 mm at one side of the raft and 100 mm at the other. In addition to tilting, the raft was fractured along a line where the thickness of the brown clay layer changed rapidly. The net bearing stress beneath the raft was $6.4 \, \text{kN m}^{-2}$ when the tank was empty and $100 \, \text{kN m}^{-2}$ when full. This latter stress was much less than that required to cause shear failure in any underlying soils except the brown clay, the shear strength of which varied from about $72 \, \text{kN m}^{-2}$ near the surface to about $24 \, \text{kN m}^{-2}$ at 1 m depth, below which it was constant at this value. Serious overstressing of the brown clay occurred below 1 m depth and plastic flow probably commenced when the tank was not quite half-full, with the bearing stress about $50 \, \text{kN m}^{-2}$. This led to settlement exceeding normal consolidation settlement. Another aspect of this type of failure is related to movement of the clay radially outwards when in a state of plastic flow beneath a foundation which induces tensile stress in the foundation by adhesion. In the case described above, the maximum tensile stress in the raft was about $1000 \, \text{kN m}^{-2}$ but the ultimate tensile strength of $1:2:4$ mix concrete can be expected to be about $1400 \, \text{kN m}^{-2}$.

EXAMPLE 3.11

A stiff raft foundation is 20 m diameter and the gross bearing stress is $50 \, \text{kN m}^{-2}$. It is founded at a depth of 1 m in a stratum of clay, beneath which is a stratum of dense sand. The top of the sand is 3 m below founding level. The following values relate to various properties of the clay:

Unit weight	$20 \, \text{kN m}^{-3}$	
Apparent cohesion	$24 \, \text{kN m}^{-2}$	} undrained test
Angle of shearing resistance	zero	
Coefficient of compressibility	$0.001 \, \text{m}^2 \, \text{kN}^{-1}$	
Coefficient of consolidation	$1.0 \, \text{m}^2 \, \text{year}^{-1}$.	

Determine the following:
(a) The factor of safety against shear failure if the bearing capacity factor N_c is 6.2.
(b) The approximate factor of safety against local overstressing.
(c) The approximate magnitude of ultimate settlement arising from consolidation in the clay.
(d) The time for 90% consolidation settlement to develop if the appropriate value of the time factor is 0.848 and the clay layer is treated as open.

Since $B/H = 6.7 (\nless 6)$, treat as thin layer.

(a) From Equation 3.81, $50 = 24/F(20/(3 \times 3) + 6.2 - 1) + 20 \times 1$, whence $F = 5.9$
(b) Net bearing stress $= 30 \, \text{kN m}^{-2}$

Hence maximum shearing stress in thin layer is approximately $30/2 = 15\,\mathrm{kN\,m^{-2}}$
Then $F = 24/15 = 1.6$.

(c) Assuming 2 in 1 dispersal of load, at middle of 3 m layer of clay increment in net stress
$= 20^2 \times 30/21.5^2 = 26\,\mathrm{kN\,m^{-2}}$.
Hence consolidation settlement $\rho_p = 0.001 \times 26 \times 3000 = 78\,\mathrm{mm}$.

(d) Time for 90% consolidation settlement $t_{90} = 0.848 \times 1.5^2/1.0 = 1.9$ years.

3.5 THE DESIGN OF BORED PIERS AND PILES

The bearing capacity of foundations is discussed in general terms in Section 3.3 but it is useful to consider further the capacity of bored piers and piles. Whereas bearing capacity and settlement can generally be considered independently for most foundations, in the case of deep bored piers and piles in clay the two factors are commonly intimately related and are considered jointly in this Section. The total ultimate capacity Q_f of a pier or pile comprises the bearing resistance developed beneath the base plus the weight of overburden excavated plus the skin resistance developed around the shaft and can be expressed as

$$Q_f = \frac{\pi B^2}{4}(N_c c_u + \gamma D) + \pi B D' c_w \qquad (3.82)$$

where B is the diameter of the pier of pile (assumed to be of uniform section), N_c is the Skempton bearing capacity factor, c_u is the shearing strength of the soil in the undrained saturated state in the zone surrounding the base, γ is the average unit weight of the soil within the depth D, D is the depth of the base below ground surface, D' is the length of the shaft over which skin resistance is effective and c_w is the skin resistance expressed as unit adhesion between clay and concrete.

Skempton (1959) examined the records of loading tests for a number of 0.25 to 0.9 m diameter bored piles, 2.5 to 27 m long, founded in London Clay. When $D \geqslant 4B$ to $5B$, N_c is a constant value of 9 and this condition obtains for most bored piers and piles. If $\bar{c}_u$ is the average shearing strength at natural moisture content, over the length D', $c_w = \alpha \bar{c}_u$, where α is a factor less than unity. Evidence suggests that c_w is about 80% of the shearing strength of the clay adjacent to the concrete. The water content of the London Clay within a zone about 75 mm thick adjacent to the shaft has been observed to be about 4% greater than the natural water content. The primary factor leading to this moisture increase is expansion of the soil at the walls of the borehole on removal of lateral support by excavation and the additional moisture may be drawn from the clay mass beyond the expanding zone, from fissures, from water in the borehole or finally from the concrete. Since an increase of 1% in water content of the London Clay leads to a reduction of 20% in α, it is desirable to take all measures which reduce the increase in moisture content. Reduction in the period during which the clay is unsupported reduces the amount of expansion and therefore excavation should be followed immediately by concreting. Skempton considers that generally α lies within the range 0.3 to 0.6 and that the value is unlikely to be less than 0.25 or more than 0.7. As a general rule, α can be taken as 0.45 but c_w should not exceed $100\,\mathrm{kN\,m^{-2}}$.

However, there are grounds for believing that shaft adhesion is related to the effective stress parameters of the soil (Burland, 1973) and on this basis c_w for London Clay can be taken as roughly equal to $8z$ kN m^{-2} where z(m) is depth below ground surface down to at least 12 m.

The allowable bearing capacity is

$$Q_a = \frac{\left(\beta Q_f - \frac{\pi B^2}{4}\gamma D\right)}{F} + \frac{\pi B^2 \gamma D}{4} \qquad (3.83)$$

where F is a load factor and β is a factor which takes into account the modus operandi of a group of piles. Skempton recommends that generally a load factor of 2.5 should be employed except that for piles of large diameter a higher value may be necessary to reduce the amount of penetration settlement, since at ultimate load this may be about $0.1B$ although the load/penetration graph is not linear, and under working load the penetration is very much less. The penetration

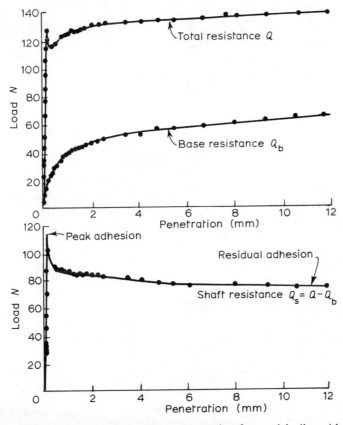

Figure 3.31 Typical graphs of load against penetration for model piles with enlarged bases in remoulded clay. (After Cooke and Whitaker, 1961.)

required to fully mobilize the skin resistance is, however, less than 0.01B. More recently Burland and Cooke (1974) have stated that, for straight shafted single piles, it is common to employ a load factor of 2 and even lower values if capacity is based on pile loading tests but, of course, acceptable penetration must be taken into account. It is reasonable to ignore adhesion in the first 1.5 to 3 m of clay measured down from the ground surface on account of seasonal shrinkage. Cooke and Whitaker (1961) tested model piles with enlarged bases in very soft remoulded clay. It was found that the shaft skin resistance was mobilized at a penetration of about 0.5% of the shaft diameter whereas a penetration of 10 to 15% of the base diameter was required to mobilize the ultimate bearing capacity of the base (Fig. 3.31). It is possible that with stiff clays the penetration required to develop full base resistance would be less than this value. The results suggest that the minimum penetration required to attain a given capacity would be secured with the longest possible shaft and a small diameter base. The factor β is unity for an isolated pile but Whitaker (1957) found, for example, that when piles are spaced at centres 3B apart, β lies between 0.7 and 0.8 for soft clays. For piles in London Clay, Skempton (1959) considers that β will be greater than these values. The reduction in capacity of a group arises from the fact that the soil between the piles has to support loads transferred from each pile by adhesion.

Tests on full-scale bored piles in London Clay were made by Whitaker and Cooke (1965): Fig. 3.32 shows the results of one of the tests. It will be seen that for

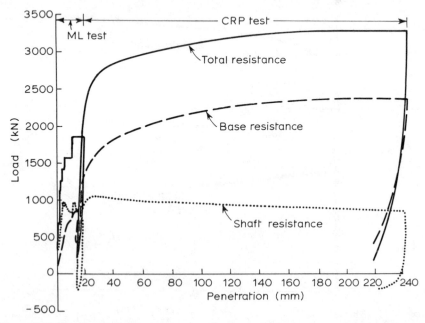

Figure 3.32 Load against penetration graphs for typical pile with enlarged base (after Whitaker and Cooke, 1965).

penetration up to nearly 15 mm the greater proportion of the load is carried by skin resistance on the shaft. The corresponding penetration in other tests has been as much as 50 mm and this is the maximum acceptable settlement for some structures. Under constant load there is generally a gradual reduction in skin resistance with time and a corresponding increase in load carried by the base and this is accompanied by an increase in penetration. On removing the applied load, the direction of the skin resistance is reversed and a residual load in excess of the weight of the shaft is imposed on the soil beneath the base. The distinction between 'maintained load' (ML) tests and 'constant rate of penetration' (CRP) tests is that in the former the load is applied in increments, each of which is maintained for a period of time, whereas in the latter the load is steadily increased to secure a constant rate of penetration (Whitaker and Cooke, 1961). Observations of shaft adhesion on a test bored pile have been reported by O'Riordan (1982).

Since the ultimate values of capacity of the shaft and of the base are reached at different penetrations, Whitaker and Cooke (1965) suggest that, to achieve compatibility at working load, the allowable load on a bored pier or pile should be taken as $[(\text{ultimate shaft resistance}/F_s) + (\text{ultimate base resistance}/F_b)]$ where F_s and F_b are load factors which vary with penetration. The curves shown in Fig. 3.33(a) can be used in design in the following way. The dimensions of the pile are selected and the ultimate shaft resistance and base capacity are estimated. It is assumed initially that S_c and S_b are both equal to the allowable settlement and a value of F_s and an approximate value of F_b are derived from Fig. 3.33(a). Working values of shaft resistance and base load are then calculated, the compression of the shaft is determined and a more closely approximate value for the settlement of the base is thus found. This leads to an improved estimate of F_b from Fig. 3.33(a). A more accurate assessment of working load can be obtained by continued successive approximations. Figure 3.33(a) shows empirical relation-

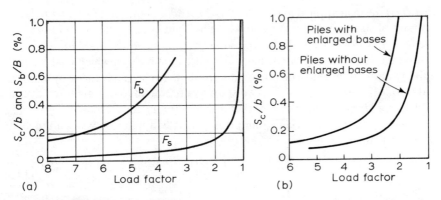

Figure 3.33 Graphs of settlement against load factor for piles in London Clay, where b = shaft diameter, B = base diameter, S_c = settlement of pile cap, S_b = settlement of pile base, F_s = shaft load factor and F_b = base load factor (after Whitaker and Cooke, 1965).

ships and should be used with caution as a guide to the design of piles in London Clay only: Example 3.12 illustrates the application of the method. The influence of an enlarged base on penetration and load factor is illustrated in Fig. 3.33(b), although these graphs cannot be used directly for design.

Further details relating to bearing capacity of bored piers and piles are given in the discussion concerning the construction of the Shell Centre, London, (Williams, 1957; Measor and Williams, 1962), in the reports on two symposia (Tomlinson, *et al.*, 1961; ICE; 1966) and in a paper by Hobbs (1963). Opinions differ on the degree of skin resistance developed at the shaft of an under-reamed pile but it seems likely that the resistance is ineffective only in the vicinity of the enlarged base.

Normal settlement of individual bored piers and piles supporting load mainly on their bases is calculated by the methods discussed in Section 2.4. The settlement of structures supported on groups of closely spaced adhesion piles in clays has been observed to be greater than that of a single pile. Furthermore, the ultimate capacity of a group of piles is less than the ultimate capacity of a single pile multiplied by the number of piles in the group. Both effects are more pronounced as the pile spacing is reduced. The ultimate capacity of a group can be predicted either by applying an efficiency factor to (capacity of single pile x number of piles) or by calculating the capacity of the block of piles and soil when treated as a pier (Example 3.9). Cooke (1975) studied the settlement of groups of piles in clays and produced the design chart shown in Fig. 3.34 which enables the settlement of a group of a given arrangement to be predicted approximately by multiplying the settlement of a single pile by the settlement ratio R_s. The settlement ρ of a single pile is best determined by a loading test but in the absence of this it can be estimated approximately (Cooke, 1975) from

$$\rho = 1.8 \frac{Q}{DE_r} \qquad (3.84)$$

where Q is the load applied to the pile, D is the length of the pile shaft, and E_r is the modulus of elasticity of the soil, and where no rigid basal layer is present near the pile foot and Poisson's ratio for the soil is 0.5.

A method of design of bored piers and piles in stiff clays, which enables both stability and settlement to be taken into account, has been advanced by Burland and Cooke (1974). The following two stability criteria impose a minimum overall load factor F with the proviso that the minimum load factor F_b for base resistance must not be exceeded and this normally applies only when the shaft adhesion is fully mobilized.

$$Q_s \leqslant \frac{Q_{fb} + Q_{fs}}{F} \qquad (3.85)$$

$$Q_s \leqslant Q_{fs} + \frac{Q_{fb}}{F_b}, \qquad (3.86)$$

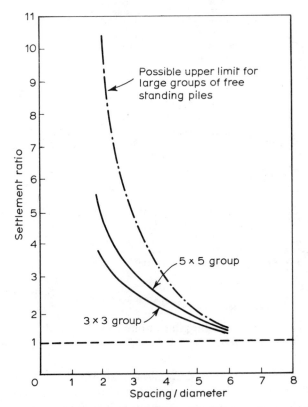

Figure 3.34 Settlement ratio against pile spacing for small square pile groups and for large groups in any configuration (after Cooke, 1975).

where Q_s is the maximum safe working load, $Q_{fb} = (\pi B^2/4)N_c c_u$ is the ultimate capacity of the base and $Q_{fs} = \pi BD' c_w$ is the ultimate capacity of the shaft as defined in Equation 3.82.

It is known that, even with careful supervision, the settlement of bored piles may vary appreciably and, in this respect, under-reamed piles are particularly sensitive to installation conditions. Consequently complex and sophisticated analyses are seldom justified. Fig. 3.35 shows schematically load/penetration graphs for straight-sided and under-reamed piles when subjected to CRP tests. The graphs for pile shaft resistance are sensibly linear up to full mobilization, which is attained at a relatively small penetration of about 0.5% of the shaft diameter, and it is penetration or settlement of the base which is of greater importance. Burland *et al.* (1966) found that by plotting the results of plate loading tests in the non-dimensional form ρ/B against q/q_f a unique relationship was obtained, irrespective of depth, for any given site. This relationship is sensibly

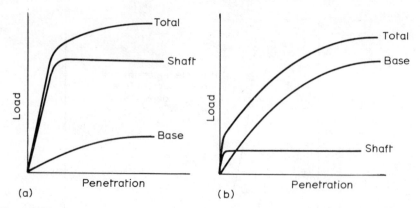

Figure 3.35 Characteristics of load against penetration graphs for bored piles. (a) Straight-shafted pile; (b) under-reamed pile.

linear up to $q/q_f = 1/3$ and for this range

$$\frac{\rho}{B} = \kappa \frac{q}{q_f} \qquad (3.87)$$

where ρ is the penetration of the plate or pile base, B is the diameter of the plate or base, q is the bearing stress corresponding to ρ, q_f is the ultimate bearing stress, and κ is a constant.

Plate loading tests in London Clay have yielded values of κ between 0.01 and 0.02 and it is reasonable to employ 0.02 for design purposes, knowing that with careful installation the settlement might be only one-half of the calculated value. Burland and Cooke (1974) recommend that for a large project a set of design charts, which facilitate achievement of a balanced design, should be prepared particular to the site.

The behaviour of bored piles in sands has been studied by Touma and Reese (1974), in stiff fissured clays by Hammond *et al.* (1980) and in both sands and clays by Reese (1978) and Wright and Reese (1979). A specification for drilled piles has been prepared by the ACI (1978a). In the case of laterally loaded piles, displacement at the head is generally more important than ultimate bearing capacity: the behaviour of short rigid piers has been investigated by Vallabhan and Alikhanlov (1982).

EXAMPLE 3.12

A 2 m diameter bored pile is to be sunk 20 m into the London Clay. Values of the shear strength parameters of the clay are $c_u = 500 \text{ kN m}^{-2}$, $\phi_u = 0$ and $c_w = 200 \text{ kN m}^{-2}$ and the unit weight is $\gamma = 20 \text{ kN m}^{-3}$. The modulus of elasticity and the unit weight for concrete are $E_c = 15 \text{ kN mm}^{-2}$ and $\gamma_c = 22 \text{ kN m}^{-3}$.

The allowable settlement is 10 mm and adhesion is assumed to be ineffective over the top 4 m of the shaft.

The value of the Skempton bearing capacity factor is $N_c = 9$. Determine the allowable load on the pile cap, referring to Fig 3.33 for values of load factor.

$S_c/b = S_b/B = 10/2000 = 0.5\%$: corresponding $F_s = 1.1$ and $F_b = 4.3$.

Safe total capacity $Q_s = \dfrac{\pi \times 1^2 \times 9 \times 500}{F_b} + \dfrac{\pi \times 2 \times 16 \times 200}{F_s} + \pi \times 1^2 \times 20 \times 20$

$$= \dfrac{14\,137}{F_b} + \dfrac{20\,106}{F_s} + 1257$$

That part of the load supported by skin resistance causes compression in the top 4 m of shaft and can be assumed to operate over one-half of the remaining length. Hence the compression due to this is

$$\dfrac{20\,106 \times 12}{1.1 \times \pi \times 1^2 \times 15 \times 10^6} = 0.004\,65 \text{ m or } 4.65 \text{ mm}$$

Compression caused by base load is

$$\dfrac{14\,137 \times 20}{4.3 \times \pi \times 1^2 \times 15 \times 10^6} = 0.001\,40 \text{ m or } 1.40 \text{ mm}$$

Total compression of shaft is therefore 6.05 mm and allowable settlement of base is thus 3.95 mm

$S_b/B = 3.95/2000 = 0.2\%$: corresponding $F_s = 1.6$ and $F_b = 7.0$. The calculations are repeated with these and subsequent values of load factors until sensibly constant values of F_s and F_b are attained. In this case final values are $F_s = 1.3$ and $F_b = 6.0$ when compression due to skin resistance load is 3.94 mm and due to base load is 1.00 mm, and allowable settlement of base is 5.06 mm, whence $S_b/B = 0.25\%$. Employing these values of load factors

$$Q_s = 2356 + 15\,466 + 1257 = 19\,079 \text{ kN}$$

From this is deducted the weight of concrete in the shaft, which is $\pi \times 1^2 \times 20 \times 22 = 1382$ kN. Hence the safe load on the pile cap is 17 697 kN. It has been assumed for simplicity in these calculations that the weight of soil excavated and the weight of concrete are approximately equal and the net effect on settlement and on load factors is negligible.

It is instructive to examine this problem in the light of the design recommendations of Burland and Cooke (1974). Assume that the load on the pile cap is to be $Q_s = 17\,700$ kN. Substituting in Equations 3.85 and 3.86

$$17\,700 \leqslant \dfrac{14\,137 + 20\,106}{F} \text{ whence } F = 1.9.$$

Commonly the value of F would be specified as not less than 2 and of F_b not less than 3.

$$17\,700 \leqslant 20\,106 + \dfrac{14\,137}{F_b}.$$

In this case Q_s is more than provided by shaft resistance, in effect with $F_s = 1$. Ultimate shaft resistance is usually attained at a penetration of about 0.5% B. Hence Q_s will be attained at less than $17\,700 \times 0.005 \times 2000/20\,106 = 9$ mm: probably about 8 mm since the shaft resistance against penetration graph is not linear near the ultimate value. Furthermore, during penetration, some base resistance will be invoked and the magnitude required to provide the specified capacity by both shaft and base resistance would probably be about 5 mm. The previous analysis has shown that compression in the shaft would be about 5 mm, and, therefore, the limit of 10 mm settlement of the pile cap is likely to be just satisfied. If the permissible settlement were greater, the base capacity would be increased as penetration increased and it would be possible to calculate base penetration by Equation 3.84.

3.6 BEARING CAPACITY OF ROCK

The compressibility of soft, decomposed or disrupted rock may be sufficient to lead to appreciable settlement under heavy loads and this may be the criterion for allowable bearing stress. Compressibility can be studied either by laboratory consolidation tests or, generally more satisfactorily, by field plate bearing tests. However, allowable bearing stress for rock is frequently assessed in relation to ultimate strength. If a rock mass can be treated as an isotropic continuum, bearing failure is similar to the rupture zone failure of soil mechanics and triaxial compression tests can be employed to determine the shear strength parameters of the rock. Unfortunately, within, say, 30 m of ground surface the joints are commonly open and relatively closely spaced and the rock mass comprises a discontinuum of more or less independent blocks. Where joints are vertical and open, each block can be expected to behave as a short column when covered completely or almost completely by the whole or part of a foundation, and bearing capacity can be based on laboratory unconfined compression tests, provided the rock in each block can be guaranteed intact: failure in this case may involve numerous vertical cracks induced by lateral tension. If the foundation area is less than about 1/25 of the plan area of the block on which it is sited centrally, then a roughly pyramidal zone of rock beneath the foundation is likely to be retained in elastic equilibrium by friction at the rock/foundation interface as failure is approached and this tends to split the block vertically: in this case only a few cracks are formed. It is unlikely that a block will ever function as a long column since wide and deep joints are necessary to permit lateral deflection and such joints should be cleaned out and filled with concrete in the normal course of foundation preparation. The bearing capacity of concrete and rock, mainly in relation to isolated blocks of these materials, has been studied by Meyerhof (1953), Shelson (1957), Au and Baird (1960), Chen and Drucker (1969), Coates (1970), Stagg and Treharne (1970) and Oehlers and Johnson (1981).

Before bearing capacity of rock can be assessed rationally, much experimental and theoretical work is still required and field studies are necessary to present joint data in a suitable form. Features such as stratification and the preferred orientation of crystals, which lead to anisotropy, will have to be taken into account. The possible influence of slaking or repeated freeze/thaw cycles on rock after construction should also be considered. Geotechnical processes may alter the characteristics of rocks, e.g. cementation may convert, for practical purposes, a discontinuum to a continuum. Assuming that a particular mass of rock functions as an elastic medium, high values of bearing stress may be developed under the edges of a foundation, but plastic flow in rock and foundation will mitigate this condition. Although some rocks will sustain very high bearing stress, the allowable values are limited by the strength of the foundation structure. Furthermore, the area of a foundation is sometimes governed by the plan area of the structure and, if this is large, then stress values must perforce be relatively low. In practice, these (and other) considerations often lead to values of stress having factors of safety of 10 or more based on cube crushing strength. Unless site conditions justify a theoretical approach to the determination of bearing capacity in any particular case, the assessment of safe values is commonly based on experience coupled with cube crushing tests.

Presumed bearing values for horizontal foundations on unweathered rock under vertical static loading, as specified in CP 2004 (BSI, 1972d), are shown in Table 3.7. Lower values should be adopted where beds dip steeply. The presumed bearing value of thinly bedded limestones and sandstones, occurring in beds several centimetres thick and often separated by clays or soft shales, and of heavily shattered rocks can be assessed only after inspection. Blasting should not be permitted within 1 or 2 m of founding level.

The following details relating to cube crushing tests may be of interest since no code of practice is available for testing rocks, although BS 1881, Parts 4 and 5 (BSI, 1970a) provides a guide for crushing strength and modulus of elasticity. Generally, tests loads are applied in the same direction relative to bedding, cleavage or other structure as will obtain under service loading at the site. Cube crushing strength is influenced by the size of specimens: the unit strength of a

Table 3.7 Presumed bearing values under vertical static loading (taken from BSI, 1972d, with permission)

Group	Class	Types of rocks	Presumed bearing value $(kN\,m^{-2})$	Remarks
I	1	Hard igneous and gneissic rocks in sound condition	10 000	These values are based on
Rocks	2	Hard limestones and hard sandstones	4 000	the assumption that the
	3	Schists and slates	3 000	foundations
	4	Hard shales, hard mudstones and soft sandstones	2 000	are carried down to
	5	Soft shales and soft mudstones	600–1 000	unweathered
	6*	Hard sound chalk, soft limestone	600	rock
	7	Thinly bedded limestones, sandstones, shales	To be assessed after inspection	
	8	Heavily shattered rocks		

These values are for preliminary design purposes only, and may need alteration upwards or downwards. No addition has been made for the depth of embedment of the foundation. Reference should be made to other parts of the code when using this table. Groups II and III soils: see Table 3.6. *Requires pick at least for excavation and does not include disturbed chalk.

50 mm cube may approach 50% in excess of that obtained with 75 or 100 mm cubes. The crushing strength of cubes loaded, in turn, in three perpendicular directions may vary by as 50% owing to the presence of structures such as bedding or foliation. The common mode of failure of a cube of rock possessing isotropy comprises two pyramids of relatively intact rock (retained in a state of elastic equilibrium by frictional restraint at the rock/platen interfaces) surrounded on the four exposed sides by rock which has yielded. If 3 mm thick sheets of cardboard are inserted between rock and platens, the mode of failure generally comprises a series of vertical plates which are formed by splitting under the action of lateral tension induced by the vertical compressive forces.

A suitable number of specimens to be submitted for test is not less than six, the mean strength of which generally should give a value within $\pm 10\%$ of the mean of a larger number of specimens. It is suggested that the sample from which a 100 mm cube is sawn should be not less than 300 mm cube, should show no sign of incipient fractures due to blasting and should be split carefully from larger pieces by plugs and feathers. Contact faces should be ground plane and parallel. Cracking load is noted at first visible crack and crushing load at yield. The strength of rock is influenced by desiccation, which commonly leads to an increase in strength, and it is desirable to prevent this by protection of cores and specimens immediately after extraction from the mass. Subsequent soaking of a dried specimen does not necessarily yield the *in situ* strength, although it is sometimes specified that, before testing, the specimens should be soaked in water at 15 to 20°C for 24 h. The ranges of crushing strength which may be expected are 30 to 110×10^3 kN m^{-2} for sandstone, 30 to 140×10^3 kN m^{-2} for limestone and 110 to 230×10^3 kN m^{-2} for granite. A 100 mm cube may require a testing machine capacity of 300 tonnes or more.

Duncan and Hancock (1966) represent the *in situ* strength q_m of a rock mass in the form

$$q_m = C(q_u + d_q) \tag{3.88}$$

where C is a joint contact factor, which takes account of the influence of fissures and never exceeds unity, q_u is the laboratory unconfined compressive strength of the rock and d_q is a confinement increment, the magnitude of which depends on the particular conditions of confinement. The influence of anisotropy on the bearing capacity of shales has been investigated by Dvořák (1966).

3.7 THE INVESTIGATION OF FAILURES

Failures of earthworks and foundations offer unrivalled opportunities to assess the validity of sampling, testing and analytical techniques. The studies outlined below indicate how the principles of soil mechanics can be applied in such investigations. The examples have been simplified for presentation here but more refined and comprehensive studies are usually justified in practice and all contributory factors, such as those which influence shear strength (Section 3.1), should be taken into account. Other field investigations serve the same purpose in connection with deformation of soils and structures, such as those relating to anchored bulkheads or settlement of foundations: Examples 2.7 and 2.8 are concerned with the latter. Investigation of failures has been reviewed by Leonards (1982).

3.7.1 Failure of banks and hillsides

Skempton and Hutchinson (1969) reviewed many failures of banks and hillsides and gave references to the original more complete reports. The Eau Brink Cut bank failure has been discussed briefly in Section 3.2.5 and the following is an example of failure on a hillside at Jackfield, Shropshire (Henkel and Skempton, 1955; Skempton, 1964).

The strata outcropping on the hillside at Jackfield comprise very stiff clays and mudstones alternating with marl-breccia and occasional coal seams, with a slack dip roughly in the direction of the slope of the hillside. The plane of rupture was at an average depth of 5.5 m and parallel to ground surface, which was inclined at $10°$. The average depth to the phreatic surface was 0.5 m. The failure occurred within a zone of weathered, fissured clay extending to a depth of 6 to 7.5 m and having the following values of drained shear strength parameters.

$$\text{Peak } c' = 10.5 \, \text{kN m}^{-2}. \quad \phi' = 25°$$
$$\text{Residual } c'_r = 0. \quad \phi'_r = 19°$$

The unit weight of the clay when fully saturated was $20.5 \, \text{kN m}^{-3}$ and the unit weight of water is $9.8 \, \text{kN m}^{-3}$.
From Equation 3.22, based on peak strength

$$F = \frac{10.5 + (10.7 \times 5.5 + 9.8 \times 0.5)\cos^2 10° \tan 25°}{20.5 \times 5.5 \sin 10° \cos 10°} = 2.04$$

and based on residual strength

$$F = \frac{(10.7 \times 5.5 + 9.8 \times 0.5)\cos^2 10° \tan 19°}{20.5 \times 5.5 \sin 10° \cos 10°}. \quad = 1.10.$$

Clearly the strength along the rupture interface had fallen completely to residual value and, in fact, it is likely that ϕ'_r was slightly less than $19°$. The landslide developed progressively during the period 1950 to 1951. The length of the sliding mass measured up the slope was 170 m and the downhill displacement exceeded 18 m.

3.7.2 Failure of retaining walls

A retaining wall failure at Uxbridge affords an example of the method of analysis. Strata at the site comprised up to $1\frac{1}{2}$ m topsoil and glacial gravel overlying the London Clay, consisting of an upper layer 3 to 5 m thick of brown weathered clay

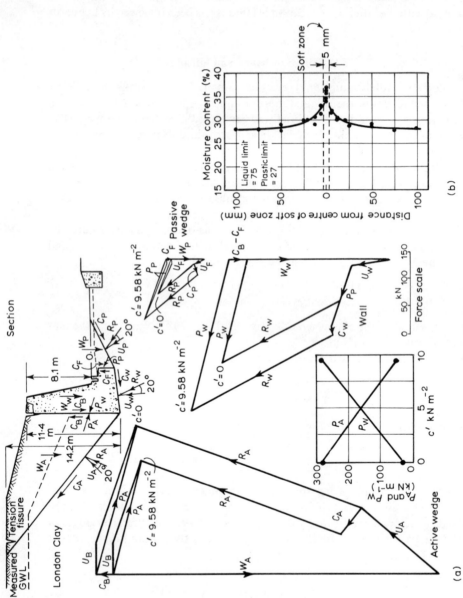

Figure 3.36 Failure of retaining wall at Uxbridge (after Henkel, 1956). (a) Effective stress analysis of wall failure, $\phi' = 20°$; (b) moisture content in vicinity of thin soft failure zone in blue London Clay 9 m below surface.

below which is blue clay. A railway cutting was excavated in 1902 and slips were observed in the north bank in 1914. The track was widened and a retaining wall, about 9 m high above track level, constructed on the south side in 1937. Slips were experienced in excavations during the execution of this work. In 1954 a slip occurred in which the wall moved forward a maximum distance of about 0.4 m. Various immediate remedial measures were taken, including loading the ground in front of the toe with ballast, thus raising the track by about 1 m. Borings and a trial trench revealed a failure zone of maximum thickness about 2 m in which a number of impersistent slip interfaces were located. In the blue clay the *in situ* moisture content increased markedly in a very narrow band, as shown in Fig. 3.36(b). Undrained triaxial tests yielded shear strengths of 78 to 139 (average 107) $kN m^{-2}$ for clay outside the rupture zone and 11 to 76 (average 46) $kN m^{-2}$ for clay within the rupture zone, although the zone of maximum softening was only 5 mm thick. Drained triaxial tests yielded $c_d = 10$ to $14 kN m^{-2}$ and $\phi_d = 20°$ to 20.5°. Fig. 3.36(a) shows an analysis of the problem (Henkel, 1956) based on effective stress in which ϕ' was assumed to be 20°. Active thrust and passive thrust plus base resistance were determined for trial values of $c' = 0$ and $c' = 9.58 kN m^{-2}$. These thrusts are equal at the intersection of the lines in Fig. 3.36(a) and c' was thus found to be 5.0 $kN m^{-2}$. Observations for ground-water enabled pore pressure to be taken into account in the analysis.

Another example is afforded by a retaining wall constructed in the London Clay at Mill Lane, Cricklewood, London, in 1902: Figure 3.37 shows simplified details. The wall was constructed in trench and the soil behind and in front of the wall was undisturbed. Slight movement of the wall appears to have occurred at intervals, culminating in a rapid movement forward of about 1 m in 1943. The approximate positions of the active and passive rupture interfaces (assumed plane) were located and active and passive thrusts were analysed by a technique similar to that employed for the Uxbridge wall but it was assumed that $\phi_u = 0 = \delta$.

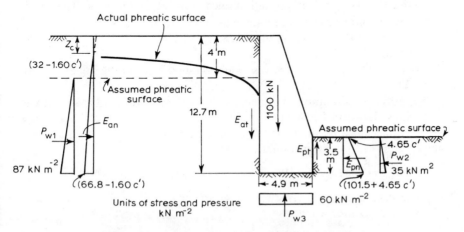

Figure 3.37 Failure of retaining wall at Cricklewood.

The active thrust and the passive thrust plus shearing resistance beneath the base were calculated for three trial values of c_u and it was found that when these forces were equal $c_u = 30.5\,\text{kN}\,\text{m}^{-2}$. Undrained tests on samples taken from near the rupture interfaces yielded values of c_u ranging from 24 to 38 $\text{kN}\,\text{m}^{-2}$. The original undrained strength of the clay was probably about 110 $\text{kN}\,\text{m}^{-2}$. This failure can be analysed approximately by employing the Rankine–Bell relationships.

For the case of vertical wall back and horizontal ground surface at back and front of the wall, the Rankine–Bell relationships for horizontal effective stress at depth z beneath ground surface are in the active state

$$\sigma'_{an} = K_A \sigma'_v - K_{AC} c' \tag{3.89}$$

and in the passive state

$$\sigma'_{pn} = K_p \sigma'_v + K_{pc} c', \tag{3.90}$$

where σ'_v is the vertical effective stress at depth z and is γz when the soil is homogeneous and above the phreatic surface. The values of the earth stress coefficients K_A, K_{AC}, K_p and K_{pc} were determined by Packshaw (1946) for various values of c_w and δ and are reproduced in CP2 (ISE, 1951). When it can be assumed that $c_w = 0 = \delta$, $K_A = 1/N_\phi$, $K_{AC} = 2/\sqrt{N_\phi}$, $K_p = N_\phi$ and $K_{pc} = 2\sqrt{N_\phi}$ where $N_\phi = (1 + \sin\phi)/(1 - \sin\phi) = \tan^2(45° + \phi/2)$. The depth of a tension fissure behind a wall is $z_c = (2c'\sqrt{N_\phi})/\gamma$. If the back of the wall or the ground surfaces are inclined then reference can be made to the tables given, for example, by Caquot and Kerisel (1948).

The Crickewood failure was a long term failure and should be analysed in terms of effective stress: as a first approximation it is reasonable to assume that $\phi' = 20° = \delta$. The problem will be analysed to determine the corresponding value of c': it is assumed that $c_w = c'$. The unit weight of water is taken as $10\,\text{kN}\,\text{m}^{-3}$ and of the saturated soil $20\,\text{kN}\,\text{m}^{-3}$. The weight of the wall is about $1100\,\text{kN/metre}$ run. The mean level of the phreatic surface is taken at a depth of 4 m behind the wall and at ground surface in front of the wall. The mean hydrostatic pressure beneath the base is assumed to be $60\,\text{kN}\,\text{m}^{-2}$.

For the given values of ϕ', δ and c_w, $K_A = 0.40$, $K_{AC} = 1.60$, $K_p = 2.90$ and $K_{pc} = 4.65$. $\tan 20° = 0.364$.

The theoretical depth of the tension fissure is found by equating Equation 3.89 to zero. Then $z_c = K_{AC}c'/K_A = 0.20c'$.

Substitution into Equations 3.89 and 3.90 leads to the active and passive stress distributions shown in Fig. 3.37. The effect of buoyancy on the soil below the phreatic surface must be taken into account to determine effective stress and, for example, at $z = 12.7\,\text{m}$, from Equation 3.89

$$\sigma_{an} = 0.40(20 \times 4 + 10 \times 8.7) - 1.60c' = (66.8 - 1.60c')\,\text{kN}\,\text{m}^{-2}$$

The active and passive normal thrusts per metre run of wall are given by the areas of the stress diagrams, whence

$$E_{an} = (493.78 - 20.32c' + 0.16c'^2)\,\text{kN}$$

and

$$E_{pn} = (177.63 + 16.28c')\,\text{kN}.$$

The corresponding tangential forces are

$$E_{pt} = E_{pn}\tan 20° + 3.5c_w = (64.66 + 9.43c')\,\text{kN}.$$

The hydrostatic thrusts are

$$P_{w1} = 378.45\,\text{kN},\ P_{w2} = 61.25\,\text{kN and } P_{w3} = 294\,\text{kN}.$$

The total vertical force on the base

$$V_b = 1100 + E_{at} - P_{w3} - E_{pt} = (921.08 - 4.13c' - 0.14c'^2)\,\text{kN}.$$

The base resistance to translation is

$$H_b = V_b\tan 20° + 4.9c_w = (335.27 + 3.40c' - 0.052c'^2)\,\text{kN}.$$

At failure the sum of the horizontal forces is zero, then

$$E_{an} + P_{w1} - H_b - E_{pn} - P_{w2} = 298.08 - 40c' + 1.212c'^2 = 0.$$

Solving yields $c' = 7.77\,\text{kN m}^{-2}$.

This is the order of average magnitude of cohesion available when the residual strength is approached in clay in the London Clay. In fact the assumed value of $\phi = 20°$ is probably slightly high and $\phi' = 17°$ might be more appropriate: in this case the magnitude of c' can be expected to be slightly greater.

In the analysis of retaining wall failures, it should be borne in mind that the lateral displacement required to evoke the full active state is of the order of $H/1000$ for dense sand, $H/500$ for loose sand, $H/100$ for firm clay and $H/20$ for soft clay, where H is the height of soil in contact with the wall. The lateral displacement required to evoke the full passive state is of the order of ten times the value quoted for each soil in the active state. These values are broad generalizations.

Although this section is concerned with failure of retaining walls involving active or passive states, it is convenient to comment briefly here on the intermediate state at rest, mentioned in the *Introduction* to this chapter. According to the theory of elasticity, the coefficient of earth stress at rest is

$$K_0 = \frac{v}{1-v}$$

where v is Poisson's ratio. Since v is 0.5 or less, it follows that K_0 should not exceed unity. In 1944, J. Jacky proposed the relationship

$$K_0 = 1 - \sin\phi'$$

where ϕ' is the effective stress angle of shearing resistance. Other investigators have corroborated this. Again, K_0 should not exceed unity. Values for soils reported in a number of publications generally like between 0.35 and 0.75.

Nevertheless, observations on retaining and abutment walls reveal that higher values may obtain, depending on the compaction effort applied to backfilling. Furthermore, the *in situ* value in the undrained state of an over-consolidated clay may be as much as 3 or more, although excavation before construction of a wall would lead to lateral stress relief. The following are a few of the publications relevant to this topic: D'Appolonia *et al.* (1970), Andrawes and El-Sohby (1973), Massarsch and Broms (1976), Carder *et al.* (1977, 1980), Ingold (1979a, b) and Mayne and Kulhawy (1982) (see also Section 6.1.1).

Because of increases in magnitude and frequency of traffic loading, leading to excessive campaction of fill, many abutments, wing walls, arches and spandrel walls of highway bridges are now severely distressed. In future designs it may be prudent to cater for the resulting high lateral forces or alternatively reinforcing or cementing the fill to function with the main structure as a composite unit.

3.7.3 Bearing failures beneath foundations

Most of the bearing failures which have been reported appear to relate to saturated clayey soils in the undrained state, with $\phi_u = 0$. Generally the calculated capacity at failure is rather greater than the observed capacity and this discrepancy is due to some extent to insufficient penetration to mobilize full shearing resistance in the rupture zones. In fact the calculated capacity would be enhanced slightly if the observed penetration is added to the founding depth in the calculations which follow.

(a) Kippen, near Stirling (Skempton, 1942)

2.74 m × 2.44 m footing founded at 1.68 m in clay. Near surface $c_u = 48\,\text{kN}\,\text{m}^{-2}$. Below footing $c_u = 17\,\text{kN}\,\text{m}^{-2}$. $\gamma = 18\,\text{kN}\,\text{m}^{-3}$. Settlement during failure is about 0.25 m. Gross load at failure is about 1000 kN, thus observed $q_f = 150\,\text{kN}\,\text{m}^{-2}$. Assume a mean value of $c_u = 30\,\text{kN}\,\text{m}^{-2}$ in the rupture zone and that adhesion $c_w = 20\,\text{kN}\,\text{m}^{-2}$ operates over a depth of 0.6 m of footing. From Equation 3.46 and Fig. 3.22(b) (Skempton): $D/B = 0.69$, $N_c = 7.2$ since $B \simeq L$.

$$\text{Gross } q_f = 7.2 \times 30 + 1.68 \times 18 + \frac{2(2.74 + 2.44)0.6 \times 20}{2.74 \times 2.44}$$

$$= 265\,\text{kN}\,\text{m}^{-2}$$

From Equation 3.38 (Wilson):

$$\text{Net } q_f = 5.5 \times 30\,(1 + 0.38 \times 0.69)$$
$$= 208\,\text{kN}\,\text{m}^{-2}$$

Add, say, 20% for square footing and 18 × 1.68 for overburden displaced, then gross $q_f \simeq 280\,\text{kN}\,\text{m}^{-2}$. Both estimated values are higher than the observed value, probably due to the fact that failure was incomplete since the penetration was only about $0.1B$ below $D = 0.69B$. Referring to Fig. 3.22(a), it is likely that only about 75% of the load required to cause failure (Fig. 3.22b) had been attained at this penetration. Hence the stress at complete failure would have been about $150 \times 4/3 = 200\,\mathrm{kN\,m^{-2}}$.

(b) Transcona silos, Winnipeg (Peck and Bryant, 1953)
$59.44\,\mathrm{m} \times 23.47\,\mathrm{m}$ foundation. $D = 3.66\,\mathrm{m}$. 9 to 15 m of clay overlying limestone. $\gamma = 19\,\mathrm{kN\,m^{-3}}$. Immediately below foundation $c_u = 54\,\mathrm{kN\,m^{-2}}$; at greater depths $c_u = 31\,\mathrm{kN\,m^{-2}}$; weighted average $c_u = 45\,\mathrm{kN\,m^{-2}}$. Observed gross stress at failure $q_f = 293\,\mathrm{kN\,m^{-2}}$. From Equation 3.46 and Fig. 3.22(b) (Skempton): $D/B = 0.16$, $N_c = 5.2$, $L/B = 2.53$, $C_R = 1.08$ from Equation 3.51.

$$q_f = 1.08 \times 5.2 \times 45 + 19 \times 3.66 = 322\,\mathrm{kN\,m^{-2}}.$$

Adhesion on the sides of the foundation would increase this probably to about $330\,\mathrm{kN\,m^{-2}}$. From Equation 3.38 (Wilson):

$$\begin{aligned} q_f &= 1.08[5.5 \times 45(1 + 0.38 \times 0.16)] + 19 \times 3.66 \\ &= 353\,\mathrm{kN\,m^{-2}}. \end{aligned}$$

The discrepancy between observed and estimated values may be due partly to a progressive reduction in strength from peak to residual value and partly to sensitivity $S_t = 2$. Furthermore, the entire structure tilted as failure developed and it is doubtful whether sufficient penetration was attained to develop full capacity. Heave of soil occurred adjacent to the structure, thus conforming to the concept of rupture zone failure.

(c) Battery of cement silos, Alsen, New York State (Tschebotarioff, 1951)
$68.58\,\mathrm{m} \times 14.94\,\mathrm{m}$ foundation. $D = 3.20\,\mathrm{m}$ on side X and 0.91 m on side Y. About 21 m of varve clay followed by 1.5 m of granular soil overlying bedrock. Weighted average $c_u = 53\,\mathrm{kN\,m^{-2}}$. Observed gross stress at failure $q_f = 292\,\mathrm{kN\,m^{-2}}$. $\gamma \simeq 19\,\mathrm{kN\,m^{-3}}$.

From Equation 3.46 and Fig. 3.22(b) (Skemption): $D/B = 0.06$ (side Y), $N_c = 5.2$. $L/B = 4.59$, $C_R = 1.04$ from Equation 3.51

$$\begin{aligned} \text{Gross } q_f &= 1.04 \times 5.2 \times 53 + 19 \times 0.91 \\ &= 304\,\mathrm{kN\,m^{-2}} \text{ (side Y)}. \end{aligned}$$

Adhesion on sides probably small on account of relatively shallow depths. From Equation 3.38 (Wilson):

$$\begin{aligned} \text{Gross } q_f &= 1.04[5.5 \times 53(1 + 0.38 \times 0.06)] + 19 \times 0.91 \\ &= 327\,\mathrm{kN\,m^{-2}}. \end{aligned}$$

Progressive reduction in c_u would arise from the change from peak to residual strength and from S_t varying from 2 at founding level to 5 at greater depths. Some degree of local overstressing may have occurred beneath the foundation. Heave of soil occurred adjacent to the structure on side, Y, thus conforming to the concept of rupture zone failure. Collapse finally occurred after a period of about 8 months, during which filling of the silos was intermittent. Consequently effective stress analysis might be more rational in this case.

(d) Oil tank, Thames Estuary (Nixon, 1949)

6.71 m diameter × 9.14 m tank on chalk mat 1 m thick. $D = 0.14$ m of soft clay, with lenses of peat, overlying sandy gravel. The value of c_u falls from 34 kN m^{-2} near the surface to 10 kN m^{-2} between 1.5 and 3 m, below which it increases to 20 kN m^{-2} at 9 m: the average value in the rupture zones is probably about 20 kN m^{-2}. Observed gross stress at failure $q_f = 79$ kN m^{-2}. From Equation 3.46 and Fig. 3.22(b) (Skempton):

$$\text{Gross } q_f = 6.2 \times 20 = 124 \text{ kN m}^{-2}.$$

Discrepancy between observed and estimated values probably due to insufficient penetration which was only about 0.15 m when rapid failure took place and the tank overturned. In the case of a circular foundation, rupture interface failure is, in theory, impossible kinematically.

(e) Oil tank, Forth Estuary (Saurin, 1949)

35.36 m diameter × 9.14 m tank on burnt shale mat 0.6 m thick. $D = 0$. 1.5 m of stiff clay above 30 m of soft, silty clay overlying boulder clay. The value of c_u falls from 40 kN m^{-2} at the surface to 10 kN m^{-2} at 4.5 m, below which it remains constant to 15 m: the average value is probably about 17 kN m^{-2}. Observed gross stress at failure $q_f = 99$ kN m^{-2}. From Equation 3.46 and Fig. 3.22(b) (Skempton):

$$\text{Gross } q_f = 6.2 \times 17 = 105 \text{ kN m}^{-2}.$$

The correspondence in this case is good: the settlement at failure was 0.6 m but ultimate failure can be expected to occur when settlement is greater than this and the observed bearing capacity would then be higher.

Studies of bearing failures beneath oil tanks have been reported also by Bjerrum and Overland (1957) and by Brown and Peterson (1964). The collapse of a block of silos in North Dakota, USA, has been reported by Norlund and Deere (1970): in this case also maximum heave occurred close to the side of the structure. Failure of a raft supporting a silo has been studied by Butting and Wood (1982). Further studies have been reported by Abbs and Sinclair (1979) and Bell and Iwakiri (1980).

EXAMPLE 3.13

When observations are made in the field or laboratory, it is sometimes useful to treat them statistically or empirically in order to establish generalized relationships. The following illustrate applications of these techniques.

(a) *In situ* plate bearing tests were made in chalk for the design of a jetty at Erith, Kent (Carey and Cumming, 1961), and from these were evolved the following empirical formulae.

Ultimate bearing stress in kN m^{-2} $q_f = q_D + \dfrac{350}{B}$

and settlement in mm $\rho_i = \dfrac{qB}{55}$

where q_D is a parameter which varies with founding depth, having values of 3700, 7700 and 13 100 kN m^{-2} respectively at depths of 2.5, 5.0 and 7.5 m, B is the diameter of the circular test plate in m, and q is any value of stress in kN m^{-2} on the plate within the elastic range. All figures in SI units are approximate conversions from the original values. Discuss the rationality of these formulae in relation to plastic and elastic theory and the known behaviour of fissured soils and rocks.

For circular loaded areas,

$$\text{from Equation 3.36} \quad q_f = 1.3cN_c + \gamma DN_q + 0.3\gamma BN_\gamma$$

$$\text{and from Equation 1.12} \quad \rho_i \text{ average} = \frac{0.85(1 - v^2)qB}{E_r}$$

If it is assumed that shearing resistance of the chalk is a function of friction only, the first term in Equation 3.36 is zero. Then $q_f = \gamma DN_q + 0.3\gamma BN_\gamma$. In this equation γ is assumed constant for particular beds of chalk at a particular site. The first term increases with D and the first term in the empirical formula conforms with this. Furthermore, N_q is known to increase with D, probably at an increasing rate, and this conforms with the first term in the empirical formula.

The second term in the empirical formula is inversely proportional to B, whereas the second term in Equation 3.36 is proportional to B. The latter caters for the behaviour of granular soil whereas the former takes account of the influence of fissures in soil and rock, leading to lower values of strength with large specimens than with small specimens. However, N_γ increases with depth and on balance it might be anticipated that the second term in the empirical formula underestimates bearing capacity at depth. Nevertheless, it is possible that the two terms in the two equations do not necessarily correspond and, in fact, the influence of depth in both terms of Equation 3.36 may be catered for wholly in the first term of the empirical formula. It can be concluded that the empirical formula is a rational expression of bearing capacity at the site.

The empirical formula for ρ conforms to the theoretical relationship Equation 1.12 for elastic settlement, assuming v and E_r are sensibly constant at the site. It is of interest to calculate the modulus of subgrade reaction in kN m^{-3} for the 140, 290 and 445 mm diameter plates employed in the tests.

$$k_s = \frac{q}{\rho_i} = \frac{55}{B}$$

whence for 140, 290 and 445 mm diameter plates k_s is respectively 392 860, 189 660 and 123 600 kN m^{-2}. These values are consistent with closely jointed friable chalk.

(b) In an investigation concerning anisotropy in shear in a saturated clay, the following values of maximum torque were obtained with the five vanes employed. The axis of the vane was normal to the bedding in each test.

Vane length	l mm	150	150	150	150	150
Vane diameter	d mm	50	75	100	125	150
Torque	T N m	30	63	120	173	285

Calculate the most likely values of the apparent cohesion c_{un} normal and c_{up} parallel to the bedding.

The total torque T comprises the part due to shearing resistance around the cylindrical surface $(\pi l d^2 c_{un}/2)$ and the part due to shearing resistance across the two ends of the swept cylinder $(2 \times 2\pi c_{up} \int_0^{d/2} r^2 \delta r)$.

Thus
$$T = \frac{\pi}{2} l d^2 c_{un} + 2\frac{\pi}{12}d^3 c_{up} = A c_{un} + B c_{up}$$

Let n and p be the most probable values of c_{un} and c_{up}.[K] Observation equations are five in number and of the form

$$An + Bp - T = \text{residual } R$$

Squares of residuals are of the form

$$(An + Bn - T)^2$$

and ΣR^2 should be a minimum.

Then $\dfrac{\partial \Sigma R^2}{\partial n} = 0$ or $2A_1(A_1 n + B_1 p - T_1) + \ldots + 2A_5(A_5 n + B_5 p - T_5) = 0$

or $(A_1^2 + \ldots + A_5^2)n + (A_1 B_1 + \ldots + A_5 B_5)p - (A_1 T_1 + \ldots + A_5 T_5) = 0$

and $\dfrac{\partial \Sigma R^2}{\partial p} = 0$ or $2B_1(A_1 n + B_1 p - T_1) + \ldots + 2B_5(A_5 p + B_5 p - T_5) = 0$

or $(A_1 B_1 + \ldots + A_5 B_5)n + (B_1^2 + \ldots + B_5^2)p - (B_1 T_1 + \ldots + B_5 T_5) = 0$

l(mm)	d(mm)	T Nm	$ld^2/2$	$d^3/6$	$A = \pi L d^2/2$	$B = \pi d^3/6$	A^2 ($\times 10^{-10}$)	B^2 ($\times 10^{-10}$)	AB ($\times 10^{-10}$)	AT^* ($\times 10^{-10}$)	BT^* ($\times 10^{-10}$)
150	50	30.0	187 500	20 830	589 050	65 440	35	0.4	3.8	1.8	0.20
150	75	63.0	421 900	70 310	1 325 440	220 890	176	4.9	29.3	8.4	1.39
150	100	120.0	750 000	166 670	2 356 190	523 610	555	28.4	123.4	28.3	6.28
150	125	173.0	1 171 900	325 500	3 681 630	1 002 590	1355	104.6	376.5	53.7	17.69
150	150	285.0	1 687 500	562 500	5 301.430	1 767 140	2811	312.3	936.8	151.1	50.36
							4932	450.6	1469.8	253.3	75.92

*T is N mm

Normal equations

$$4932n + 1470p - 253.3 = 0$$
$$1470n + 450.6p - 75.92 = 0$$

Solving yields $n = 0.041\ 16\ \text{N mm}^{-2} = 41\ \text{kN}^{-2}$ and $p = 0.03421\ \text{N mm}^{-2} = 34\ \text{kN}^{-2}$.

(c) The data given below relate to undrained triaxial tests on saturated specimens of remoulded clay from the Reading Beds in Sussex. Plot the results on one or more types of graph paper and hence determine an empirical relationship between shear strength, c_u, and moisture content, w.

[†] $\pi/2$ could be incorporated in n and $\pi/6$ in p, thus simplifying calculations, and dividing resulting values of n and p by these constants.

$\dfrac{w}{\%}$	$\dfrac{c_u}{\mathrm{kN\,m^{-2}}}$
18.1	126.4
19.8	85.5
21.8	57.9
23.7	40.5
25.8	25.5
28.0	17.7
29.4	13.7
31.0	10.6
33.5	8.8
35.1	6.2
37.3	5.1
39.6	3.1

A plot on double linear graph paper (Fig. 3.38) suggests a relationship of the form $c_u = AW^{-n}$

whence $$\log c_u = \log A - n \log w.$$

A plot double logarithmic graph paper yields a straight line for which $n = -4.583$ and $A = 77.97 \times 10^6$ (from $w = 1$)

hence $$c_u = 77.97 \times 10^6 w^{-4.583}.$$

A least squares analysis can be effected as follows, with tabular computations as in example 2 above. For the relationship $\ln_u = \ln A - n \ln w$, the observations are of the form $R = \ln A - n \ln w - \ln c_u$ where A and n are the most likely values of those constants.

Then $R^2 = (\ln A)^2 + n^2(\ln w)^2 + (\ln c_u)^2 - 2n \ln A \ln w - 2 \ln A \ln c_u + 2n \ln w \ln c_u.$

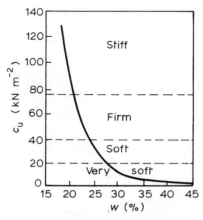

Figure 3.38 Variation of shear strength (c_u) with moisture content ($w\%$) of a clay.

Let the number of observations be N, then

$$\Sigma R^2 = N(\ln A)^2 + N^2\Sigma(\ln w)^2 + \Sigma(\ln c_u)^2 - 2n \ln A\Sigma \ln w - 2 \ln A\Sigma \ln c_u$$
$$+ 2n\Sigma \ln w \ln c_u$$

Normal equations

$$\frac{\partial \Sigma R^2}{\partial A} = 2N\frac{\ln A}{A} - \frac{2n}{A}\Sigma \ln w - \frac{2}{A}\Sigma \ln c_u = 0$$

$$\frac{\partial \Sigma R^2}{\partial n} = 2n\Sigma(\ln w)^2 - 2 \ln A\Sigma \ln w + 2\Sigma \ln w \ln c_u = 0$$

The plastic limit w_P of the clay tested was 24% and the liquid limit w_L 51% the Casagrande classification is therefore CI/CH. At values of w below w_p, c_u against w is almost linear and the loss in strength is roughly $15\,\text{kN}\,\text{m}^{-2}$ per one unit of % increase in w. At values of w above 40%, the graph becomes asymptotic and $c_u \simeq 2\,\text{kN}\,\text{m}^{-2}$ at $w_L = 51\%$. Within the range of w employed in the tests, consistency ranges from 'stiff' to 'very soft' (Table 1.13 and Fig. 3.38).

Chapter 4

INDEPENDENT FOUNDATIONS

A.W. Astill

4.1 GENERAL NOTES

Independent foundations are designed in a wide variety of types. The variety has been increased by the current tendency to depart from traditional patterns and experiment with new. Examples of this are afforded by the steel pedestal base, produced as an alternative to the grillage, and by the casting of stanchions directly in concrete without slab or gusseted bases. The development of new types will grow and should be encouraged, with the objective being the more economical use of materials and labour. It may be found that advantages apart from economy may accrue from newly evolved patterns. It is the intention that the discussion and examples presented in this chapter shall merely point the way and that innovation shall be encouraged whilst the principles of sound design are retained.

In this chapter the term footing is used to describe the part of the foundation which is designed to spread the loads from columns or walls over the ground. In America this is common usage but in Great Britain the word base in often used. This causes confusion particularly in steel construction where the term base is also used, as it is in this chapter, to describe the enlarged part of the steel column which spreads the column load over the concrete or beams of the footing.

The relative advantages of different types of independent foundations are difficult to assess in general. Each case must be considered on its merits but the choice depends to some extent on the individual designer's preferences and experience. Reinforced concrete footings are generally used where the framework of the building is in reinforced concrete since the connection of reinforced concrete columns to steel grillages produces unnecessary complications. Plain concrete footings may be used with reinforced concrete superstructures but the inclusion of column starter bars or dowels tends to cause the plain concrete to be priced at the higher rates associated with reinforced concrete. Plain concrete footings are usually employed with steel frameworks or where adequate ground bearing capacity can be found only at a depth such that a great deal of building up is required. Steel grillage footings are used much less than in the past, their principal advantage is to cater for high shearing forces.

Both steel column bases and pedestal bases spread the load from the column to the footing and are discussed in Sections 4.9 and 4.10. The pedestal base can be made to cover a larger area more economically than is probably the case with a more conventional slab or gusseted base so that it can be used to help reduce shearing stresses in reinforced concrete footings.

Steel and reinforced concrete foundations may both be designed by working load or limit state design methods. The current British code for design of structural concrete is CP 110 (BSI, 1972c) which is based on the principles of limit state design. The ACI (1977) code for reinforced concrete is also based on limit state design. BS 449 (BSI, 1969a) for steelwork is generally based on working load design but plastic design methods are permitted which is also the case with the AISC code. The publication of a new British code for steelwork, which is based on limit state design principles, is imminent.

4.2 TRANSFER OF LOAD FROM COLUMN TO FOOTING

4.2.1 Reinforced concrete columns

Column design is frequently based on ultimate strength methods although it may be coded as working load design. The load is carried by the steel and concrete in the column and the stresses in each can be determined by strain compatibility. If the column is axially loaded to failure the steel will reach its yield stress before the concrete fails, so that the ultimate axial load N is given by the usual addition formula.

$$N = A_{sc} \gamma_{ms} f_y + A_c \gamma_{mc} f_{cu} \qquad (4.1)$$

where, for the longitudinal reinforcement, A_{sc} is the area, γ_{ms} is the partial factor of safety for strength and f_y is the characteristic strength, and, for the concrete, A_c is the area, γ_{mc} is the partial factor of safety for strength and f_{cu} is the characteristic cube strength. If a working stress design is employed then $\gamma_{ms} f_y$ and $\gamma_{mc} f_{cu}$ are replaced by the appropriate permissible stresses.

In transferring the load from the column to the foundation, the stress in the concrete is transferred direct but the stress in the steel must be transferred by surface shear (bond) stresses from the steel to the foundation concrete. The foundation thickness may therefore be determined by the bond length needed to transfer the load from the column or starter bars (dowels) to the base. This bond length cannot be reduced by the use of hooks. Increasing the cross-sectional area of the starter bars above the area of the main bars in the column is also likely to have little effect since both concrete and steel stresses are reduced. It can be shown, using strain compatibility, that the reducton in steel stress is small. There is some evidence, (Pfister and Mattock, 1963), to show that where starter bars are used, a part of the load in the main column bars is transferred to the base by end bearing of the main bars on the base concrete. It has also been shown by Astill and Al-Sajir (1980) that the joint capacity is related to the ratio (footing area)/(column area) and to the quantity of footing reinforcement. The design of footings is discussed further in Section 4.4.

Where the column transmits a moment as well as an axial force, some of the column reinforcement may be in tension. Proper anchorage must be provided for the tension bars. In general the tensile stresses are low and the effect on the design of the foundation is minimal but where the axial force is low and the moments high, as in a single storey portal frame, anchorage of the tension bars may cause problems. In this case the use of hooks reduces the anchorage length. Increasing the area of tension bars reduces the tensile stresses, but doubling the area does not reduce the stress by half. Horizontal forces are transmitted to the footing by shear in the normal way.

4.2.2 Steel columns

(a) Axially loaded columns

The load in an axially loaded steel column is usually assumed to be transmitted to the concrete foundation by a uniform bearing stress over the whole area of the steel base. The allowable bearing stress on the concrete determines the plan area of the base. In the past the magnitude of the allowable bearing stress has been similar to that of the compressive stress allowed in an axially loaded reinforced concrete column except that, due to the containment of the loaded concrete by the unloaded concrete, increased bearing stresses have been allowed.

Hawkins (1968a) and others have studied the problem of bearing stresses for loads applied to a part of the gross area of the top of a concrete block. The effect of applying loads through rigid plates both concentrically with the area of the top surface of the block and at varying eccentricities until the edge of the loaded area coincided with the edge of the block was considered and the following formula was proposed.

$$\frac{q}{f_c'} = 1 + \frac{K}{\sqrt{f_c'}}(\sqrt{R} - 1), \tag{4.2}$$

where q is the bearing stress on the loaded area, f_c' the cylinder strength of concrete, and K is a factor which depends on the tensile strength of the concrete and the angle of internal friction. These are difficult to evaluate but experiments suggests that K may be taken as 50 in most cases.

$$R = \frac{\text{effective unloaded area}}{\text{loaded area}} < 40.$$

The effective unloaded area must be assumed concentric with the loaded area and if it is rectangular the ratio of the longer side to the shorter side should not exceed 4. The ratio of q/f_c' is independent of the depth of the block provided that the formation of the failure cone or pyramid under the loaded area is not restricted by the proximity of the base of the block.

Hawkins (1968b) also tested blocks loaded by a rigid punch through flexible steel plates. Starting from the bearing capacity for the rigid punch alone, he showed that the bearing capacity increased in direct proportion to the thickness of the plate whilst the steel of the plate yielded before failure of the concrete. As the plate thickness approaches that of a rigid plate, there is a rapid increase in the bearing capacity of the block.

The ACI (1977) code recommendations are based in part on this work. For cases where the bearing area is equal to the gross area of the block the design ultimate bearing stress is given by $0.85 \ \phi \ f'_c$ where $\phi = 0.70$. Where the load is applied to a part only of the block the design ultimate bearing stress is given by

$$\left(\frac{A'_b}{A_b}\right)^{1/2} \times 0.85 \ \phi \ f'_c,$$

where A_b is the loaded area and A'_b is the area of the base of a right frustrum of a cone or pyramid which is wholly contained within the block. The side slope of the cone or pyramid should not be steeper than 1 vertical to 2 horizontal nor should the ratio $(A'_b/A_b)^{1/2}$ exceed 2. This side slope should not be used to determine the block thickness, such a flat slope is given to ensure that there is sufficient concrete surrounding the bearing area to ensure a proper triaxial state of stress. It should be noted that all Hawkins' test pieces had flat tops. Additionally, the bases should be thick enough to allow formation of the inverted cone or pyramid below the loaded area. Apex angles of 35° to 40° are quoted which indicate a base thickness of at least 1.7 times the larger base plate dimension. See Section 4.4 for the design of reinforced concrete footings and Section 4.5 for the design of plain concrete footings.

(b) Eccentrically loaded columns

The foregoing considerations of concrete stress still apply to eccentrically loaded columns no matter what the eccentricity may be. If the eccentricity of the load is less than one-sixth of the relevant base dimension the stress on the concrete foundation block is usually calculated by treating the base as a short column of homogeneous material in which no tension is developed. Nominal holding down bolts are usually provided. It becomes necessary to determine the forces in the holding down bolts as tension develops. The bolts are assumed to act in a similar manner to the reinforcement in a reinforced concrete column; the analysis is discussed further in Section 4.11. The strength of the bolts is calculated on the cross-sectional area at the root of the thread, for which an allowable tensile stress of $130 \ \text{N} \, \text{mm}^{-2}$ is given in BS 449 (BSI, 1969a). This is less than the normal allowable tensile stress for mild steel and is reduced to allow for the increase in stresses induced by the notch effect of the threads and the pretensioning caused by tightening the nuts during erection.

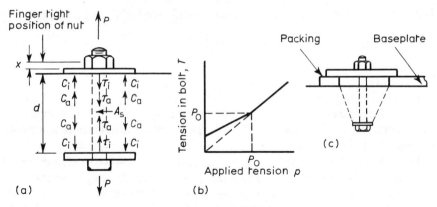

Figure 4.1 Stresses in foundation bolts.

The following gives some indication of methods which may be used to calculate the stresses in holding down bolts due to pretightening. In order to simplify the problem consider a single bolt with top and bottom plates arranged on a block of concrete as shown in Fig. 4.1(a). The plates are considered to be stiff but thin compared with the depth of the block in order that the distance d can be taken as both the length of the bolt and the depth of the block. Assume that no bond or friction exists between the barrel of the bolt and the concrete. Let the plan area of the plates and of the block be A_c and the cross-sectional area of the bolt be A_s. Now assume that the nut is tightened and that the travel of the nut is a distance x from the finger-tight position. The tension in the bolt is then T_i and the compression in the block C_i.

The extension of the bolt plus the contraction of the concrete must equal the travel of the nut, or

$$\frac{T_i d}{A_s E_s} + \frac{C_i d}{A_c E_c} = x. \tag{4.3}$$

where the modulus of elasticity of the steel is E_s and of the concrete E_c. For static equilibrium $T_i = C_i$ and substituting for C_i in Equation 4.3

$$T_i = \frac{A_s E_s x}{d[1 + (A_s E_s)/(A_c E_c)]}. \tag{4.4}$$

Now let an additional tensile force P be applied externally to the ends of the bolt and let the corresponding increase in tension in the bolt be T_a and the decrease in compression in the concrete be C_a. The increase in extension of the bolt must equal the decrease in compression of the concrete or

$$\frac{T_a d}{A_s E_s} = \frac{C_a d}{A_c E_c}. \tag{4.5}$$

For static equilibrium $T_a + C_a = P$, and substituting for C_a in Equation 4.5

$$T_a = \frac{A_s E_s P}{A_c E_c + A_s E_s}.$$ (4.6)

The total tension in the bolt is then $(T_i + T_a)$.

The nut is lifted clear when $C_a = C_i$, that is when $P = T_a + T_i$. Substituting for T_a and T_i from Equations 4.6 and 4.4 it follows that when the nut is just lifted clear the applied tension P_0 is given by $P_0 = A_s E_s x/d$. These considerations are shown graphically in Fig. 4.1(b).

The values of A_s and d are known and E_s is commonly known to a fair degree of accuracy. On the other hand, A_c and E_c are not known accurately in practice. It is convenient to consider two limiting cases. Firstly assume either A_c or E_c or both, are very large; say approaching infinity. Then $T_i = A_s E_s x/d$, thus showing that the strain in the bolt is equal to the travel of the nut and the strain in the block is zero. Also T_a tends to zero under these conditions and hence no additional tension is developed in the bolt by the external tensile force until P just exceeds T_i. As the external force is increased, the nut is lifted clear of the plate and the total tension in the bolt is P. If now E_c approaches zero, T_i tends to zero and T_a is nearly equal to P. In practice conditions lie between these two extremes; it is common to accept $E_s = 15E_c$ and it is reasonable to assume that A_c is commonly about 30 to 60 times A_s. Then $A_c E_c$ equals 2 to 4 times $A_s E_s$, and

$$T_i = \frac{2}{3} \text{ to } \frac{4}{5} \times \frac{A_s E_s x}{d} \quad \text{and} \quad T_a = \frac{1}{3} \text{ to } \frac{1}{5} \times P.$$

These values are, of course, very broad generalizations. In practice these idealized conditions are modified by several factors:

Bedding down of plates

The irreversible displacement of unknown magnitude due to the plates bedding down on the concrete represents local crushing and compaction of concrete, dust and rust at the contact faces of plates and concrete.

Concrete shrinkage

Shrinkage of the concrete has a considerable influence on stresses and is generally of the order 4×10^{-4} per unit length over a period of about 12 months, roughly three-quarters of this occurring in the first 3 months. The extension of mild steel under a stress of $10 \, \text{N mm}^{-2}$ is about 5×10^{-5} per unit length. Thus if the initial stress is $80 \, \text{N mm}^{-2}$ the shrinkage of the concrete is about equal to the extension of the mild steel bolt and much of the initial tension will be lost due to shrinkage.

Concrete creep

In addition to shrinkage, there is creep of the concrete under the initial compression, leading to an indefinite reduction in initial tension. Even if anchor bolts are not subjected to any initial tension until the concrete is 1 or 2 months old, it is highly probable that this will become practically zero due to the combined effect of shrinkage and creep. Shrinkage and creep are unlikely to affect resistance to external load but if the combined effect is large the bearing plate may lift clear of the concrete.

Bond

Undoubtedly stress distribution in the bolt and the block under external tension will be affected by bond between steel and concrete but this does not affect the maximum tension in the bolt which, until the concrete cracks, is probably developed only between the bottom of the nut and the concrete. Part of the external force is transferred to the concrete at the bolt head and part is transferred by bond. When cracks develop at the limit of extensibility of the concrete, that is about 0.01%, the maximum tension is developed over most of the length of the barrel. Until cracks occur, the concrete contributes to the resistance to an externally applied tension, a factor which was ignored in deriving a value for P_0 above. Where the undesirable practice of anchoring bolts in position by grouting alone has to be adopted, particular attention should be paid to the bond between grout and concrete, and the hole should be of conical form in order to increase, by wedge action, the resistance of the grout plug to pulling out.

Flexibility of plate

The flexibility of the bearing plate affects both the magnitude of the tension in the bolts and also the stress distribution in the concrete, and obviously complicates the assessment of the area of the plate which is to be included in calculations based on Fig. 4.1.

Length of bolt above contact surface

It was assumed that the nuts are located close to the contact surface but on many bases the nuts bear on stiffeners or other components some distance above the base plate. The increased extension of the bolt leads to increased flexibility of the assembly.

Much of the above discussion also relates to the anchroage of bolts in rock but in that case the problem is further complicated by the presence of fissures.

The analysis of the single bolt in a block of concrete discussed above is applicable to any foundation subjected to direct tension or compression. The problem of a foundation, length h and breadth b, such as that shown in Fig. 4.2, subjected to both axial load and moment can now be investigated.

Assume that there are four anchor bolts, one near each corner of the base plate. The combined cross-sectional area of each pair of bolts is A_s. If the nuts are tightened equally on each bolt, the total tension developed in each pair is $T_i/2$ and the total compression in the concrete is C_i. The magnitude of T_i and C_i is

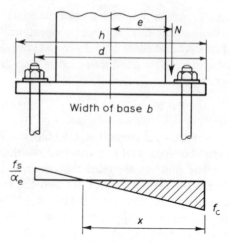

Figure 4.2 Analysis of a steel column base.

determined from Equation 4.4 in which $2A_s$ is substituted for A_s. If an axial load N and a moment $M(=Ne)$ about the centre of the plate are now applied, where e is the equivalent eccentricity of load N, the induced additional tension in one pair of bolts is T_a. Because of bond it can be assumed that the other pair of bolts contribute a resistance to compression C_a. Assume that the base plate is lifted clear at one pair of bolts under the action of the moment and that the stress distribution is linear, as in Fig. 4.2 Hence, if α_e is the modular ratio,

$$f_s = \frac{\alpha_e f_c(d-x)}{x} = \frac{T_i + T_a}{A_s}$$

and

$$C_a = \frac{\alpha_e f_c(x-h+d)A_s}{x}.$$

The sum of the vertical forces must be zero and therefore

$$C_i - T_i + \frac{f_c x b}{2} - T_a + C_a - N = 0, \qquad (4.7)$$

and the sum of the moments of the vertical forces about the centre line of the base plate must be zero; the moments of T_i and C_i are zero and hence

$$\frac{f_c x b}{2}\left(\frac{h}{2} - \frac{x}{3}\right) + T_a\left(d - \frac{h}{2}\right) + C_a\left(d - \frac{h}{2}\right) - Ne = 0. \qquad (4.8)$$

Substitute $\quad (T_i + T_a) = A_s \alpha_e f_c \dfrac{(d-x)}{x} \quad$ in Equation 4.7

$$C_a = \frac{\alpha_e f_c(x-h)A_s}{x} \qquad \text{in Equations 4.7 and 4.8}$$

and

$$T_a = \frac{A_s \alpha_e f_c (d - x)}{x} - T_i \qquad \text{in Equation 4.8.}$$

Now T_i, C_i, N, e and α_e are known, h, d, b and A_s are assumed and f_c and x are unknown. These equations can be solved by substituting an assumed value for x in Equation 4.7 to give a value for f_c. This value of f_c, together with the assumed value of x, are substituted into Equation 4.8 and if there is a remainder the process is repeated with different values for x until consistent values of x and f_c are found, as indicated previously. If the base plate is not lifted clear on the tension side under the action of the moment, then the whole area of concrete beneath the base plate is in compression and this analysis must be modified accordingly.

All these considerations are so varied that it is not prudent to specify a precise form of calculation or of site practice. However, these notes should serve as a guide to a rational solution of this problem and designers can formulate rules of procedure to conform to site erection methods practised by their firms. Anchor bolts should be free from grease, loose rust and scale before placing. They should extend a proportion (say 20 bolt diameters) of full bond distance into the concrete when provided with 100 mm × 100 mm × 10 mm washers at the head, although this is not necessary when they are anchored to the underside of a grillage. The practice of anchoring a base plate temporarily by means of the device shown in Fig. 4.1 (c) is to be deprecated owing to the tendency for the concrete within the dotted lines to be torn out of the block, possibly leading to accidents during erection. Undoubtedly a final check of anchor bolts should be made with a spanner set to give a specified torque when a structure is completed and, possibly also, at the end of the contract maintenance period.

Theoretical studies by Salmon *et al.* (1957) of the moment rotation characteristics of column bases are of interest in connection with the semi-rigid and plastic design of structures. A method of analysis of the forces in the anchor bolts of unstayed slender towers, such as distillation columns with annular base rings, has been presented by Boyd (1958).

Very little information is available concerning the anchor resistance of a bolt acting in conjunction with the concrete in which it is embedded. Somerville (1955) investigated this problem in the following way. Two holes 65 mm diameter and 320 mm deep were drilled in a concrete base. A 25 mm diameter bolt with a 50 mm diameter washer was inserted in each hole. The bolts were grouted with a 2 to 1 sand–cement grout of creamy consistency to within 65 mm of the top. After working to remove entrapped air, the holes were filled to the top with well-compacted thick mortar. Wet curing was maintained for 7 days. The bolts were tested by jacking under a yoke. The maximum load applied was 111 kN per bolt. No rupture was evident under this load and the bond stress between grout and concrete was then 1.83 N mm^{-2}. This load represented a stress of 310 N mm^{-2} at the root of the thread and it seems likely that yield of the steel had just commenced. No progressive creep in the filling was detected when loads were maintained for periods of 30 min. but very slight permanent set may have occurred under loads exceeding 70 kN per bolt. Under maximum load the movement of the embedded bolt was about 0.1 mm. Conard (1969) also tested ordinary hexagon headed bolts grouted into holes drilled in a concrete block. Three different grouts were used and the bolts were tested in direct tension and in shear. The best results were obtained using a pre-mixed nonshrink grout with water as the mixing liquid, next was a sand–cement grout with a polymer resin as the liquid and last was a sand–cement grout with water as the mixing liquid. Conard also compared his results with previously published tests on patent anchor bolts which clearly showed that the patent anchor bolts were better. Further information concerning anchor bolts is provided by Constrado (1980), the Agrément Board (1981), Cannon *et al.* (1981), Elfgren *et al.* (1980, 1981). Horizontal forces are transmitted to a footing partly by friction between the steel and concrete and partly by shear on the holding down bolts.

4.3 TRANSFER OF LOAD FROM FOOTING TO SOIL

4.3.1 General details

For the purpose of designing footings it is usual to assume that the soil bearing stress under the footing varies linearly. There is, however, nothing to prevent the designer from using any suitable soil stress distribution, as discussed in Chapter 2. The footing should be proportioned to achieve a stress distribution as near to uniform as possible. Where the axial load on a footing is large compared with the moment on the column foot, as for example, in multi-storey structures, the equivalent eccentricity of loading is small and the effect on the footing dimensions is not great. On the other hand, if the axial load is small compared with the column foot moment, say, in a single storey portal frame structure, then equivalent eccentricity is large and the effect on footing dimensions is considerable.

If eccentric loads of approximately equal magnitude and displacement occur on any one of the four sides of a column independently, e.g. through wind loading, then the plan shape of the footing should be square. On the other hand, conditions may arise where eccentric loads occur independently on two opposite sides of a column only and it may be an advantage to adopt a footing rectangular in plan. Increasing the length of the footing in the direction of the applied moment and decreasing its width reduces the ratio of equivalent eccentricity to length of footing and the resulting maximum and minimum soil bearing stresses are more nearly equal. In broad terms, economical construction is obtained if the length of the footing is about one and a half times the width but this obviously depends on the equivalent eccentricity of the load. An asymmetrical footing, similar to that illustrated in Fig. 4.3, may be worth considering in certain circumstances as discussed below.

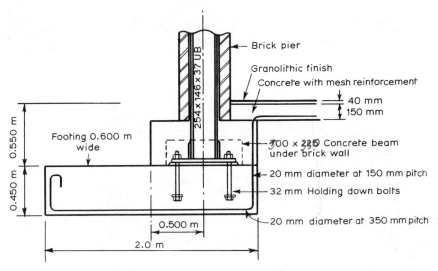

Figure 4.3 Foundation for fixed-base welded portal frame.

4.3.2 Eccentricity about one axis

Suitable dimensions for eccentrically loaded footings can be determined by trial and error or by fixing the ratio of breadth to length and substituting in a formula. If a trial and error method is employed, then an area some 10 to 25% greater than that required for axial load only may be a suitable first approximation. The following formula is usually used:

$$\frac{N}{BL} + \frac{6M}{BL^2} = \sigma \not> \sigma_a, \tag{4.9}$$

where N is the axial force in the column plus the weight of the footing, M is the moment at the column base, B is the breadth of the footing, L is the length of the footing, σ is the calculated maximum soil stress under the footing and σ_a is the allowable safe soil stress. If $\sigma > \sigma_a$ or if σ is significantly less than σ_a, adjustment is made to L and/or B and the calculation is repeated.

If the base size is to be determined for a fixed ratio $B/L = K$, then substitution in Equation 4.9 gives

$$NL + 6M = \sigma_a KL^3$$

which can be solved for L and then $B = KL$. In the above σ_a is the allowable gross bearing stress and σ is the actual gross bearing stress. For the purpose of design of the footing slab, the stress due to overburden and weight of concrete is deducted, since it does not contribute to shearing forces or bending moments on the slab.

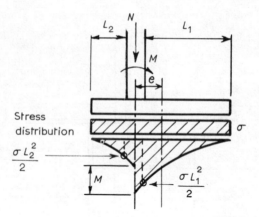

Figure 4.4 Eccentrically loaded column with footing giving uniform stress distribution.

Savran (1962) has presented charts for the direct determination of foundation proportions.

When the eccentric load occurs on one side only of a column or when the range of eccentricities is significantly greater on one side than the other, the most economical construction should be obtained by offsetting the footing relative to the column. In this way the footing centre lies within the range of the points of application of the equivalent eccentric loads and the soil stress is as near uniform as possible. The resulting bending moment diagram for such a footing is shown is Fig. 4.4. Figure 4.3 shows an asymmetrical footing for a fixed-base portal frame for a warehouse where the vertical force is small relative to the moment at the base.

Where columns are subjected to lateral force it may be desirable to check the foundations for horizontal movement. The resisting forces will be contributed by base friction and adhesion and passive resistance as for retaining walls (see Chapter 6). This is normally critical only on light, sheeted, single storey buildings. Although it may be sufficient to treat all the foundations of a structure as one in resisting lateral forces, conditions may arise where uplift due to wind at windward columns reduces the frictional resistance of the base and leads to possible movement at individual bases. This effect may be increased by the flexibility of the column and its connections. The factors of safety against sliding and against overturning can each be expressed in different forms which lead to different values of factor of safety. In the case of sliding of a foundation it is reasonable to express the resisting force in terms of vertical force N and coefficient of friction $\tan \delta$ and adhesion c_w between soil and concrete. Then

$$\text{Factor of safety} = \frac{c_w LB + N \tan \delta}{H},$$

where H is the applied horizontal force. The passive resistance developed in the soil at the side of the foundation may also be included in the resisting force if account is also taken of the active thrust on the opposite side. However, considerable horizontal movement is required to develop the passive thrust. The factor of safety chosen should therefore be higher if passive resistance is included than if it is ignored but obviously there should be some gain by including it.

In the case of overturning, the factor of safety can be defined as

$$\text{Factor of safety} = \frac{\text{resisting moment}}{\text{overturning moment}}.$$

The application of a moment to a foundation inevitably causes a small amount of rotation of the foundation which may relieve the moment. Even if the rotation is less than $1°$, the relief may be sufficiently large to appreciably influence the design of the structure. Theoretical and field studies of rotation of foundations have been reported by Weissman and White (1961).

4.3.3 Footings loaded eccentrically in two directions

If a footing is loaded with axial load N plus two moments M_1 and M_2 acting concurrently at right angles to each other, the soil stress can be obtained as indicated below. If a single moment acts on the footing in some direction other than along the axes, it can be resolved into two components and treated in the same way. It is assumed that no uplift exists under the footing; it is not good practice to permit uplift unless special arrangements, such as tension piles, are adopted to cater for it. Referring to Fig. 4.5 the soil stresses at the corners are

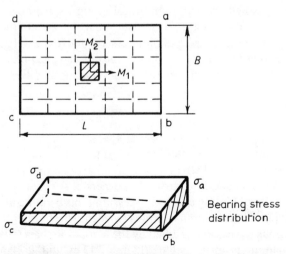

Figure 4.5 Footing loaded eccentrically in two directions.

$$\sigma_a = \frac{N}{BL} + \frac{6M_1}{BL^2} + \frac{6M_2}{LB^2}; \quad \sigma_b = \frac{N}{BL} - \frac{6M_1}{BL^2} - \frac{6M_2}{LB^2}$$

$$\sigma_c = \frac{N}{BL} + \frac{6M_1}{BL^2} - \frac{6M_2}{LB^2}; \quad \sigma_d = \frac{N}{BL} - \frac{6M_1}{BL^2} + \frac{6M_2}{LB^2}$$

(4.10)

Examination of the stress distribution diagram reveals that design must involve approximations in order to limit calculations. It is probably better to be generous in fixing footing dimensions rather than spend too much time in over-precise calculations, particularly as little is known of the behaviour of eccentrically loaded footings. For the purpose of design it may be advantageous to divide the footing into rectangles and assume uniform stress over each rectangle. In the case of a reinforced concrete footing the rectangles could be fairly small but, for a grillage, strips parallel to the beams may be more suitable. In both cases the pitch of the bars or beams should remain reasonably uniform or error in construction is liable to occur.

4.4 DESIGN OF REINFORCED CONCRETE COLUMN FOOTINGS

4.4.1 Axial loading

For many years the design of reinforced concrete column footings was based on the work of Talbot (1913) who carried out tests on footings of 1.5 m × 1.5 m with effective depths of 250 to 300 mm. The load was applied through central square columns of 300 mm side. The footings were supported on a test bed of helical springs. More recent investigations by Richart (1948) and Hognestad (1953) have given a better understanding of the mechanism of failure which has led to some modifications in design procedure. In the following discussion it is assumed that the stress applied to the soil is uniform over the whole area of the footing. For a footing of uniform thickness part of the stress σ is due to the weight of the footing and the earth above it; the rest, σ_c, which is the most significant part, is due to the column load. It is σ_c which causes the shear, bending and bond stresses in the foundation.

(a) Bending

Talbot (1913) concluded that the force producing bending at the face of the column represented the soil stress acting on an area contained between lines at 45° to the column face radiating from the corners of the columns. He concluded also that the width of footing effective in resisting this bending moment could be taken as the width of the column face plus twice the effective depth of the footing. These conclusions have been modified by the work of Richart (1948) who carried out tests on footings, principally 2.13 m × 2.13 m and 2.74 m × 1.83 m and 3.05 m × 1.53 m. Effective depths varied between 200 and 400 mm and the columns were 350 mm × 350 mm. Both plain and deformed bars were used as reinforcement and the footing bed consisted of coil springs. The maximum bending moment is theoretically beneath the column centre but measurements showed no greater stresses here than at the column faces. Hence the use of the column face as the critical plane for bending moment calculations is justified. As the load producing yield in the tension reinforcement was approached, the measured strains corresponded to the bending moment derived from the entire rectangle on one side of the critical plane. Additionally, the strains in the outer bars of square footings were similar in magnitude to those in the centre bars, indicating that the full width of the footing is effective in resisting bending moment. The central transverse bars of rectangular footings showed a greater concentration of stress than the outer bars. This fact justifies the design procedure whereby lateral bars are concentrated in the centre part of the footing.

In the ACI (1977) code and the British code CP 110 (BSI, 1972c) the critical section for bending is taken at the face of a rectangular column. For a circular column the ACI code suggests the face of a square column of equivalent area. For a metallic column base the ACI code recommends that the critical section be taken on a line midway between the column face and the edge of the base. The bending moment at the critical section, which extends across the complete width of the base, is that moment due to all forces on one side of the section. The longitudinal reinforcement so calculated is distributed uniformly across the full width of the base. The transverse reinforcement is divided into two parts, the greater part being concentrated in a centre band of width equal to the width of the footing. The quantity of reinforcement in the centre band is given by

$$\frac{2}{\beta_1 + 1} \times \text{total area of transverse reinforcement},$$

where β_1 is the ratio of the longer to the shorter side of the base. This gives 80% of the transverse reinforcement in the centre band when $\beta_1 = 1.5$ and 66.7% when $\beta_1 = 2.0$.

(b) Bond

Tests indicate that bond is an important factor in footing design. Flexural (shear or local) bond stresses can be high and must always be checked. The critical section for flexural bond is the same as for bending, that is, at the face of the column or at any change of section. The anchorage bond stresses must also be checked to ensure a proper build up of stress in the tension bars. If the anchorage bond stresses are low (less than 0.8 of the allowable stress), flexural bond need not be checked.

Where the footing is associated with a reinforced concrete column provision must be made for the transfer of the force in the column bars into the footing. Thus the thickness of the footing is often dictated by the bond length required, which is prescribed in detail in the ACI (1977) code but not in CP 110 (BSI, 1972c). There is evidence (Astill and Al-Sajir, 1980) that the permissible bond stresses given in CP 110 are too low and consequently lead to excessively thick footings or to the use of a plinth.

(c) Shear (diagonal tension)

Talbot (1913) found that diagonal tension failures occurred on a plane approximately 45° to the horizontal, extending from the column face at the top of the footing to a point at the level of the reinforcement in the base of the footing. The critical plane for diagonal tension was therefore assumed to be a vertical plane distance d from the column face.

When a footing failed in diagonal tension in tests by Richart (1948), the failure was sudden and the column, together with a pyramidal block of concrete, broke through the footing. Diagonal tension stresses at failure were higher on thin footings than on thick, varying from $0.09f_c'$ to less than $0.05f_c'$ when calculated for a critical plane at distance d from the column face. As the ratio of length to width of rectangular footings is increased, so the diagonal tension failure changes from the pyramidal type to a failure on two opposite faces of the column, indicating that the footing then functions as a double cantilever. About 75% of the pad footings tested failed primarily in shear after local yielding of the tension reinforcement but before the ultimate load in flexure was reached.

Hognestad (1953) reviewed the work of Richart (1948) and suggested that shearing stresses on vertical planes at the column faces were a better measure of shearing strength. The ultimate strength of slabs failing in shear appears to depend mainly on the following factors:

(a) The quality of the concrete expressed as the cylinder (or cube) crushing strength.
(b) The amount, type and strength of tension, compression and shear reinforcement.
(c) The size and shape of the loaded area in relation to slab thickness.
(d) The span, support conditions and degree of edge restraint of the slab.

The results of tests indicate that shear failure generally occurs after initial flexural cracking of the concrete, either just before or whilst some of the tension reinforcement is beginning to yield. For this range Hognestad (1953) proposed the following empirical relationship for the ultimate shearing strength, v, in $N\,mm^{-2}$ at the face of a square column of side h:

$$v = \frac{V}{4h7d/8} = \left(0.035 + \frac{0.07}{\phi_0}\right)f_c' + 0.9, \qquad (4.11)$$

where $4h$ is the length of the perimeter of the square column, $7d/8$ is an approximate value for the lever arm, V is the shearing force round the column perimeter and $\phi_0 = V_s/V_f$ where V_s is the ultimate shearing capacity and V_f the ultimate flexural capacity of the slab. Equation 4.10 is valid up to a value of

$d/h = 1.14$ and possibly also above this value. Furthermore, Equation 4.10 was based on tests in which $f'_c = 14$ to $35\,\mathrm{N\,mm^{-2}}$ but for $f'_c < 12\,\mathrm{N\,mm^{-2}}$ it is probably unsafe. Such a low concrete strength is, however, unlikely to be used for reinforced concrete work.

Whitney (1957) studied the shear strength of footings, primarily in connection with the load factor method of design and concluded that the then current method of analysis, based on the relationship $v = V/bz$, where z is the lever arm, was inadequate because it took no account of the quantity of tension steel. Designs based on these methods are too conservative where the flexural reinforcement is heavy and relatively unsafe where it is light. Whitney (1957) also suggested that the most consistent values of shear stress were obtained when the critical section was located at $d/2$ from the column face.

Moe (1961) also carried out work on slabs and footings and proposed the formula

$$v = V/bd = [1.25(1 - 0.075h/d) - 0.44\phi_0]\sqrt{f'_c}\,\mathrm{N\,mm^{-2}}. \qquad (4.12)$$

This relationship correlates well with test results. It should be noted that if the formula were to be used for practical design, the shear capacity should be at least equal to the flexural capacity and that ϕ_0 should therefore be at least equal to one.

All this work was reviewed by a joint ACI–ASCE (1962) committee which found that the proposal by Whitney (1957) of taking the critical section at $d/2$ from the column face gave very similar results to Moe's formula and has the advantage of ease of use. The report of this committee, and subsequent work, led to the recommendations of the ACI (1977) code in which the footing is treated either as a beam with a shear crack forming across the entire section at distance d from the face of the column or punching shear is considered, with the critical section taken at a distance $d/2$ from the face of the column so that the periphery b_0 is a minimum. The shear stress may be taken as $v_u = V_u/b_w d$ for beam-type shear failure or $v_u = V_u/b_0 d$ for punching shear, where b_w is the web width and b_0 is the periphery of the critical section for the footing.

If the footing is treated as a beam the allowable shear stress is given by

$$v_c = 0.85\,(0.16\sqrt{f'_c} + 17.2\rho V_u d/M_u) \qquad (4.13)$$

where ρ is $A_s/b\,d$ and A_s is the area of tension reinforcement. The ratio V_{ud}/M_u must not be taken as greater than 1.0 nor must v_c exceed $0.85 \times 0.29\sqrt{f'_c}$. If the punching shear at $d/2$ from the column face is considered, the allowable shear stress is $0.85 \times 0.083(2 + 4/\beta_c)\sqrt{f'_c}$ but not more than $0.85 \times 0.33\sqrt{f'_c}$, where β_c is defined as the ratio of the longest overall dimension of the effective loaded area to the shortest overall dimension. Shear reinforcement may be provided but it must be designed to carry all shear stress in excess of $0.85 \times 0.17\sqrt{f'_c}$. A 50% increase in shear capacity is the maximum permitted. In cases where shear is a problem, load distributors may be fabricated from steel I or channel shapes. In this case the maximum permitted increase in shear capacity is 75%.

CP 110 (BSI 1972c) specifies that the critical section for shear should be at a distance 1.5d from the column face but the allowable stresses, given in a table, allow empirically for both the strength of the concrete and the quantity of flexural steel present. These allowable stresses are the same as those used for beams. The shear stress is still calculated from $v = V/bd$, where b is either the breadth of the base measured at a distance 1.5d from the face of the column or the circumference of an area, the boundary of which is 1.5h from the column face. In making this decision the British Code committee were no doubt influenced by the results shown in Fig. 4.6 which has been taken from the Handbook on the Unified Code (CP 110, BSI, 1972c); the original version of which appeared in *The Shear Strength of Reinforced Concrete Beams* (1969). The results shown on the graph are for beams and corbels with very short shear spans. It can be seen that the shear strength rises rapidly for loads applied near the supports.

Shear reinforcement causes complications in fixing steel and in concreting and, unless there are compelling reasons for its introduction, it is probably best avoided by increasing the footing thickness.

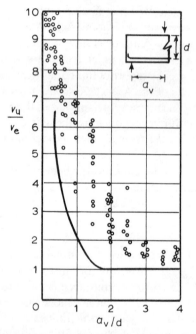

Figure 4.6 Ultimate shear stresses for beams loaded close to supports with v_c taken from CP 110 (BSI, 1972c), where o is experiment and—is CP 110 line. Reproduced from the *Handbook on the Unified Code for Structural Concrete*, by permission of the Cement and Concrete Association.

4.4.2 Eccentric Loading

The considerations described for axially loaded footings apply in all cases to eccentrically loaded footings but there are some additional points to consider.

(a) Bending

The critical section is the same as for the axially loaded footing. When calculating the bending moments for the longitudinal reinforcement the trapezoidal or other distribution of the earth stress must be taken into consideration. For the transverse reinforcement, the average earth stress may be used to compute the bending moments. The reinforcement so obtained must be divided into two parts, as described in Section 4.4.1 part(a), and the greater part concentrated under the column. Some designers consider that the non-uniform earth stress leads to higher transverse bending moment where the earth stress is greater than average. The conservative designer must make his own provision either by judgement or by calculating the bending moments due to an earth stress greater than the average.

(b) Bond

Critical sections for bond in the footing reinforcement are as indicated in Section 4.4.1 part(b), but bond for the column bars must be considered carefully. The compression reinforcement in the column can be dealt with as before but, if the eccentricity is such that large tensile stresses are induced in the column reinforcement, proper anchorage must be provided. The use of hooks to provide anchorage for column bars in tension helps to reduce footing thickness.

(c) Shear

When shear stress on a transverse section is considered then calculations using the trapezoidal earth stress distribution are simple but when the shear on the circumference of an area around the column is investigated, the average shear stress can be readily determined as for an axially loaded column but the maximum shear stress will be more difficult to calculate and the calculations may not be reliable. The bending moment is transferred into the footing by a combination of shear, torsion and bending. Neither in CP 110 (BSI, 1972c) nor in the ACI (1977) code is there any information concerning this problem but CP 110 does include some guidance in its section on flat slabs. For cases where the stability of the structure is provided by shear walls or bracing and the ratio of adjacent spans does not exceed 1.25, the allowable shear stresses should be reduced by 20%. For other cases the shear force should be increased by a factor $(1.0 + 12.5M/VL)$, where M is the moment transmitted, V is the shear force and L

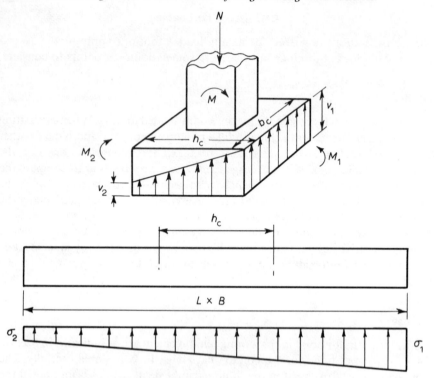

Figure 4.7 Shear in an eccentrically loaded footing.

is the shorter of the two spans in the direction in which bending is being considered. The derivation of the factor is given by Regan (1974). The method is not directly applicable to footing slabs and an alternative method is to consider the forces to be transferred at the face of a rectangular critical section as shown in Fig. 4.7.

$$\sigma_1 = N/BL + 6M/BL^2$$
$$\sigma_2 = N/BL - 6M/BL^2$$
$$M_1 = \{(L - h_c)^2[\sigma_1(5L - h_c) + \sigma_2(L + h_c)]\}b_c/48L$$
$$M_2 = \{(L - h_c)^2[\sigma_1(L + h_c) + \sigma_2(5L - h_c)]\}b_c/48L.$$

Consider the equilibrium of the block within the critical section.
Resolving vertically

$$N = (v_1 + v_2)b_c d + 2(v_1 + v_2)h_c d/2 + (\sigma_1 + \sigma_2)h_c b_c/2. \tag{4.14}$$

Taking moments

$$M = M_1 - M_2 + (v_1 - v_2)b_c d\frac{h_c}{2} + \frac{2(v_1 - v_2)}{2}\frac{h_c^2 d}{6} + \frac{(\sigma_1 - \sigma_2)}{2}\frac{h_c^3 b_c}{6L}, \tag{4.15}$$

hence

$$v = \frac{2N - (\sigma_1 + \sigma_2)h_c b_c}{4d(h_c + b_c)} + \frac{12L(M - M_1 + M_2) - (\sigma_1 - \sigma_2)h_c^3 b_c}{4Ldh_c(3b_c + h_c)}. \tag{4.16}$$

The ACI–ASCE (1962) committee used the following formula and found good correlation with such few tests results as were available:

$$v = \left[\frac{N}{A_c} + \frac{kM}{J_c} \left(\frac{h_c}{2} \right) \right], \tag{4.17}$$

where

$$A_c = 2(h_c + b_c)d \tag{4.17a}$$

$$J_c = \frac{2dh_c^3}{12} + \frac{2h_c d^3}{12} + 2b_c d \left(\frac{hc}{2} \right)^2. \tag{4.17b}$$

A_c and J_c can be increased to allow for the dowel action of the tension steel by multiplying individual terms in A_c and J_c by $[1 + (\alpha_e - 1)\rho]$. A mean value for $k = 0.20$.

In Equation 4.16 the terms containing σ_1 and σ_2 deal with part of the column load transferred to ground within the shear perimeter. If these terms are removed, Equations 4.16 and 4.17 are of similar form, the actual differences being that kM replaces $(M - M_1 + M_2)$ and that the second term on the right-hand side of Equation 4.17b is not present in Equation 4.16. Equation 4.17b implies a torsional transfer of moment across the face of length h_c, whereas in Equation 4.16 a linear distribution of shear stress is assumed.

EXAMPLE 4.1 REINFORCED CONCRETE FOOTING DESIGN

Column data

Axial load	$= 450\,\text{kN}$
Bending moment at column foot	$= 90\,\text{kN m}$
Column section	$= 300\,\text{mm} \times 300\,\text{mm}$
Concrete cube strength f_{cu}	$= 25\,\text{N mm}^{-2}$
Column reinforcement	$= 4\text{–}25\,\text{mm diameter}$
Reinforcement strength f_y	$= 410\,\text{N mm}^{-2}$

Foundation data

Design ultimate bearing capacity $= 160\,\text{kN m}^{-2}$
This bearing capacity is obtained from the considerations of Sections 2.4, 2.5 and 3.3 and corresponds to the ultimate limit state.
Concrete cube strength $f_{cu} = 25\,\text{N mm}^{-2}$.
This is the lowest strength permitted (Table 19, CP 110, BSI, 1972c) for underground work. Minimum cover to reinforcement is 40 mm.
Steel strength, $f_y = 410\,\text{N mm}^{-2}$

Soil bearing stress

Try a footing 2.5 m long × 1.8 m wide.
Ignoring footing and overburden weight

$$\sigma_c = \frac{450}{1.8 \times 2.5} \pm \frac{90 \times 6}{1.8 \times 2.5^2} = 52 \text{ or } 148 \text{ kN m}^{-2}.$$

This is lower than the permissible but assume that the remainder will be taken up by footing and overburden weight. The soil stress at the face of the column is

$$\sigma_f = 52 + 96 \times \frac{1.4}{2.5} = 108 \text{ kN m}^{-2}$$

and at $1.5d$ ($= 0.45$ m) from the face of the column

$$\sigma_{0.45} = 52 + 96 \times \frac{1.85}{2.5} = 123 \text{ kN m}^{-2}$$

Longitudinal reinforcement

Bending moment at column face, i.e. on line ab due to force on area abcd (Fig. 4.8)

$$M_u = \frac{108}{2} \times 1.8 \times \frac{1.1^2}{3} + \frac{148}{2} \times 1.8 \times \frac{2 \times 1.1^2}{3} = 146.7 \text{ kN m}$$

for a base 350 mm thick

$$d = 350 - 40 - 10 = 300 \text{ mm}$$

$$\frac{M_u}{bd^2} = \frac{146.7 \times 10^6}{1800 \times 300^2} = 0.91 \text{ N mm}^{-2}.$$

From Graph 2, CP 110 Part II (1972)

$$\frac{100 A_s}{bd} = 0.27$$

$$A_s = \frac{0.27 \times 1800 \times 300}{100} = 1460 \text{ mm}^2$$

8 bars 16 mm diameter = 1608 mm^2

8 bars give a spacing of $\dfrac{1800}{8} = 225$ mm.

To control tension cracks maximum spacing permitted = 185 mm (Table 24, CP 110). Hence use 13 bars 12 mm diameter = 1471 mm^2 and then

$$\text{spacing} = 139 \text{ mm}.$$

Local bond (flexural or shear bond)

Shear force at face of column due to force on area abcd

$$V = \frac{(108 + 148)}{2} 1.8 \times 1.1 = 254 \text{ kN}$$

$$f_{bs} = \frac{254 \times 10^3}{490.1 \times 300} = 1.72 \text{ N mm}^{-2} < 2.5 \text{ (Table 21, CP 110)}.$$

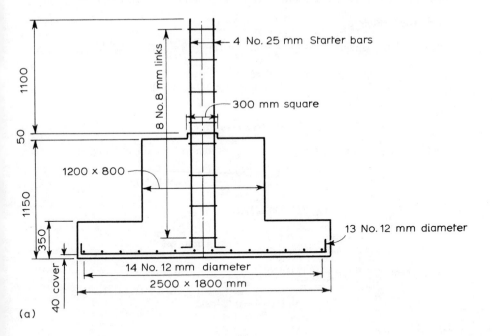

4 No. 25 mm Starter bars

8 No. 8 mm links

300 mm square

1200 × 800

13 No. 12 mm diameter

14 No. 12 mm diameter

2500 × 1800 mm

1100

50

1150

350

40 cover

(a)

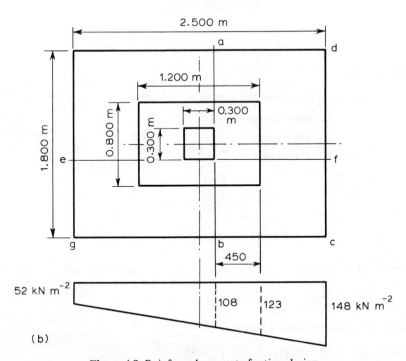

2.500 m

a

d

1.200 m

0.300 m

0.800 m

0.300 m

1.800 m

e

f

g

b

c

450

52 kN m⁻²

108   123

148 kN m⁻²

(b)

Figure 4.8 Reinforced concrete footing design.

Anchorage bond

$$\text{Length of bar available} = \frac{2.5 - 0.3}{2} - 0.1 = 1.0\,\text{m}$$

$$\text{Bond stress} = \frac{0.87 \times 410 \times 12}{4 \times 10^3} = 1.07\,\text{N mm}^{-2} < 1.9\ (\text{Table 22, CP 110})$$

Transverse reinforcement

Bending moment at column face, i.e. on line ef due to force on area efcg

$$M_\text{u} = 100 \times 2.5 \times \frac{(0.9 - 0.15)^2}{2} = 70.4\,\text{kN m}$$

$$d = 300 - 20 = 280\,\text{mm}$$

$$\frac{M_\text{u}}{bd^2} = \frac{70.4 \times 10^6}{2500 \times 280^2} = 0.36\,\text{N mm}^{-2}.$$

From Graph 2, CP 110 Part II (1972c)

$$100\frac{A_\text{s}}{bd} = 0.11 < 0.15.$$

Use 0.15%, then

$$A_\text{s} = \frac{0.15 \times 2500 \times 280}{100} = 1050\,\text{mm}^2.$$

$$\text{10 bars 12 mm diameter} = 1131\,\text{mm}^2$$

Local bond

Shear force at face of column due to force on area efcg

$$V = 100 \times 2.5 \times 0.75 = 187.5\,\text{kN}$$

$$f_\text{bs} = \frac{187.5 \times 10^3}{377.0 \times 280} = 1.78\,\text{N mm}^{-2} < 2.5\ (\text{Table 21, CP 110}).$$

Steel area spread over centre 1.8 m strip =

$$\frac{2 \times 1050}{2.5/1.8 + 1} = 880\,\text{mm}^2$$

$$\text{8 bars 12 mm diameter} = 905\,\text{mm}^2$$
$$\text{Pitch} = 225\,\text{mm}.$$

Provide 10 bars to reduce pitch to 180 mm and two bars at each end, i.e. 14 bars total.

Transverse shear

Take the critical section at $1.5d$ from the face of the column

$$V = 0.65 \times 1.8 \times \frac{(123 + 148)}{2} = 159\,\text{kN}$$

$$v = \frac{159 \times 10^3}{1800 \times 300} = 0.295 \, \text{N mm}^{-2} < 0.35 \; (\text{Table 5, CP 110})$$

Punching shear

The critical section is taken at $1.5d$ from the column face

$$V = 450 \left[1 - \frac{(\pi \times 450^2 + 1.200^2 - 4 \times 0.450^2)}{2500 \times 1.800} \right] = 288 \, \text{kN}$$

$$v = \frac{288 \times 10^3}{300 \times (300 \times 4 + \pi \times 900)} = 0.238 \, \text{N mm}^{-2}.$$

The above makes no allowance for eccentricity and a further check may be made using Equation 4.16.

$$h_c = b_c = 0.3 + (2 \times 0.45) = 1.20 \, \text{m}$$

$$M_1 = \frac{1.2}{48 \times 2.5} \{(2.5 - 1.2)^2 \, [148(5 \times 2.5 + 1.2) + 52(2.5 + 1.2)]\}$$

$$= 31.52 \, \text{kN m}$$

$$M_2 = \frac{1.2}{48 \times 2.5} \{(2.5 - 1.2)^2 \, [148(2.5 + 1.2) + 52(5 \times 2.5 - 1.2)]\}$$

$$= 19.18 \, \text{kN m}$$

$$v = \frac{[2 \times 450 - (148 + 52)1.2^2]10^3}{4 \times 0.3(1.2 + 1.2) \times 10^6}$$

$$\pm \frac{12 \times 2.5(90 - 31.52 + 19.18)10^9 - (148 - 52)1.2^4 \times 10^9}{4 \times 2.5 \times 0.3 \times 1.2(3 \times 1.2 + 1.2) \times 10^{12}}$$

$$= 0.2125 \pm 0.1233 = 0.336 \; \text{or} \; 0.089 \, \text{N mm}^{-2}$$

Anchorage length for column starter bars

To transmit the load from the column bars to the base, a suitable transfer length is required.

$$\text{Anchorage length} \; \frac{410 \times 25}{4 \times 2.4} = 1070 \, \text{mm} \quad (\text{Table 22, CP 110}).$$

The base is obviously not thick enough to provide this. Hooks cannot be expected to work in compression, particularly near to the concrete face. There are two alternatives, either to increase the base thickness to accommodate the 1070 mm bond length or to provide a pedestal so that the stress in the column starter bars is reduced.
Depth of pedestal $= 1070 - (300 - 1.5 \times 16) = 794 \, \text{mm}$.
For a maximum 45° angle of spread (cl. 3.6.1.3, CP 110) the pedestal must fall within a circle of maximum radius of

$$\sqrt{2} \times 150 + 794 + 40 = 1046 \, \text{mm}.$$

Assuming that the pedestal is unreinforced, then to avoid tension the eccentricity of load

must not exceed 1/6 of the pedestal length.
Eccentricity of loading $= 90\,000/450 = 200$ mm.
Length of pedestal in direction of bending moment $=$

$$6 \times 200 = 1200 \text{ mm}.$$

Make pedestal 1200 mm × 800 mm.
Actual maximum radius $= (400^2 + 600^2)^{0.5} = 720$ mm.

$$\text{Maximum bending stress } = \frac{2 \times 450 \times 10^3}{800 \times 1200} = 0.94 \text{ N mm}^{-2}$$

This stress is very low and the pedestal could be reduced in cross-section provided reinforcement is introduced to resist the tensile stresses which would be induced. The pedestal would probably be constructed as a truncated pyramid if visible above the floor level but since it is likely to be concealed a rectangular block is easier to cast. The one deficiency of the rectangular block is that the re-entrant angles lead to relatively high local stresses but in this case stresses are generally low so that overstressing is unlikely to arise. The completed design is shown in Fig. 4.8.

Alternative designs

The principal alternatives are related primarily to the bond length for the column starter bars. In the design outlined above it could be argued that the bending moments and shearing forces for the foundation slab should be calculated from the face of the pedestal, whereas in fact the column face was used as a datum. In this case the bending moments would be reduced considerably. The longitudinal steel would be reduced to the minimum of 0.15% instead of the 0.27% provided but the transverse steel of 0.15% would be unchanged. The critical sections for shear would be near the edge of the pad so that shear stresses would obviously be low and need not be calculated.

If the foundation pad were 1150 mm thick there would be sufficient room to accommodate the full bond length for the column bars but the slab reinforcement would be increased since the 0.15% minimum would still apply.

4.5 PLAIN CONCRETE FOUNDATION BLOCKS

Plain concrete foundation blocks are commonly used to distribute the load from steel slab bases or small-area grillages to the soil where the eccentricity of loading is small. The contact area between base and block can be calculated on an allowable bearing stress of 5 N mm^{-2} for concrete of 20 N mm^{-2} cube strength. The plan area of the block is based on the allowable soil stress. The ACI (1977) code specifies the minimum thickness of the edge of the block to be not less than 200 mm.

In the United Kingdom the design of plain concrete foundation blocks is usually based solely on diagonal tensile stress. Bending stresses are usually very low, as a quick check will show, and shear stresses are covered by consideration of diagonal tensile stress. To design for diagonal tension the practice is to ensure that a line drawn at 45° to the vertical from the edge of the steel base or grillage intersects the side not less than 50 to 100 mm above the underside of the block, as shown in Fig. 4.9. Failure by diagonal tension becomes possible if the line

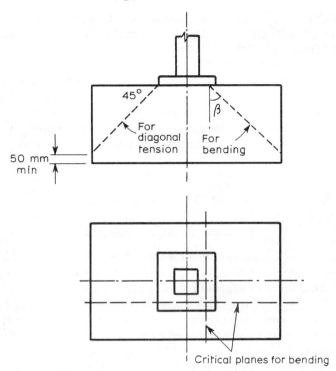

Figure 4.9 Analysis of plain concrete foundation blocks.

intersects the underside. British practice does not usually permit the line to fall within the base. Hamilton (1945) suggested that as an alternative a line at angle $\beta = \arctan 0.09\,(f_{cu}/\sigma)^{1/2}$ should intersect the side, as shown in Fig. 4.9 and that β should not exceed 45°. This is based on a maximum permissible tensile stress of $0.03\,f_{cu}$ for a working load design.

In the USA the design is based on the flexural tensile stress in the concrete of the block which is limited to $0.27\sqrt{f_c'}$ for an ultimate load design. The shear stresses are checked in the same way as for reinforced concrete footings.

The plan area of the block is decided by the allowable soil stress, some allowance having been made for the weight of the block. The block is usually made square, or nearly so, and its thickness is slightly greater than half the greatest difference between base dimension and block dimension, both measured parallel to the same side. A check on soil stress may necessitate minor adjustments.

It is sometimes economical to place a plain concrete pedestal between a column and a foundation block or footing. The width of the pedestal is usually about one-quarter or one-third of the width of the block. The minimum cross-section is determined from the applied load and the concrete strength. The height of the pedestal should not exceed three times its least width and should be not less than twice the projection of the pedestal beyond the column base. The compressive stress on the gross area of an axially loaded pedestal should not exceed $0.25 f'_c$ unless it is reinforced and treated as a reinforced concrete column.

The work of Hawkins (1968a, b) on bearing stress, discussed in Section 4.2.2. part (a), has some relevance to the design of plain concrete footings since his tests were made on plain concrete blocks which failed by tensile splitting from the loaded side. It might by expected that plain concrete footings would fail in the same way but Hawkins' blocks were supported on rigid supports which are unlikely to be achieved in real foundations where normal bending failure is more likely.

4.6 STEEL GRILLAGE DESIGN

4.6.1 Concrete encased grillages

Grillages consisting of one, two or three tiers of beams may be used to spread column loads to the soil or to a plain concrete block, which in turn transmits the load to the soil. Allowable stresses in grillage beams are specified in BS 449 (BSI, 1969a). Briefly, if the beams (not including hollow compound girders) are encased in a dense concrete with a works cube strength of not less than 21 N mm^{-2} at 28 days*, working stresses may be increased by 33% over those for uncased beams, provided that (a) the distance between the flanges is not less than 75 mm, (b) the concrete cover is not less than 100 mm and (c) the concrete is solidly tamped round the beams.

For Grade 43 (mild) steel to BS 4360 (BSI, 1972b) the maximum allowable stress in bending for uncased rolled I beams and channels is 165 N mm^{-2} and the allowable average shear stress in unstiffened webs of I beams is 100 N mm^{-2} for webs not greater than 40 mm thick and 90 N mm^{-2} for webs over 40 mm thickness. Reference should be made to BS 449 (BSI 1969a) for full details of allowable stresses. It is commonly accepted that the clear distance between the flanges of grillage beams should be not more than $1\frac{1}{2}$ to 2 times the flange width, with a maximum of 300 mm in order that the concrete and steel should act together.

4.6.2 Uncased grillages

During erection, and when grillages are used in temporary works, it sometimes happens that considerable loads may be carried by an uncased grillage. For this situation, normal allowable stresses should be used although some increase in

---

* It should be noted here that CP 110 (BSI, 1972c) does not permit the use of concrete strengths lower than 25 N mm^{-2} in foundation work but the minimum concrete cover to the reinforcement is only 40 mm. The ACI (1977) code calls for a minimum cover to the steel of 75 mm where the concrete is adjacent to earth.

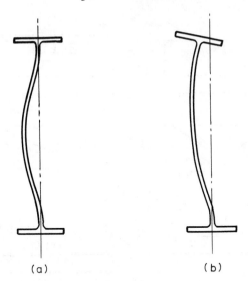

(a)   (b)

Figure 4.10 Buckling of webs of rolled beams.

stress is usually permitted for temporary works. The main difference arises when checking web stress. In the encased grillage, web buckling cannot occur, but in uncased grillages there is no concrete to prevent buckling failure of the webs. In recent years several bridges in America and Europe have collapsed during construction and failure has been attributed at least in part to buckling of the webs of unstiffened grillage beams.

In most codes web buckling is checked by treating a part of the web as a strut. In BS 449 (BSI, 1969a) and BS 153 (BSI, 1972a) the length of web assumed to be acting is taken as the length of stiff bearing plus a length, equal to half the beam depth, on each side of the bearing if available. The effective height of the strut is taken as half the beam depth. The form of buckling failure envisaged is shown in Fig. 4.10(a) but the absence of stiffeners or diaphragms may make it possible for a sway type failure to occur, as shown in Fig. 4.10(b). For this condition the effective height is not less than the beam depth and the allowable stresses are at least halved and the effects of horizontal or eccentric forces are magnified. In situations where the uncased grillage is likely to carry considerable load, the safe procedure is to provide sufficient stiffeners or diaphragms, or both, to ensure that sway type buckling failure of the webs cannot occur.

4.6.3 Grillage loaded axially

Various approaches to the calculation of bending moments in grillages are shown in Fig. 4.11. Method (d) is probably the most rational and is common practice. Method (a) is also frequently used to reduce work. For both methods a degree of flexibility is assumed in both the base plate and the grillage beams. In order to

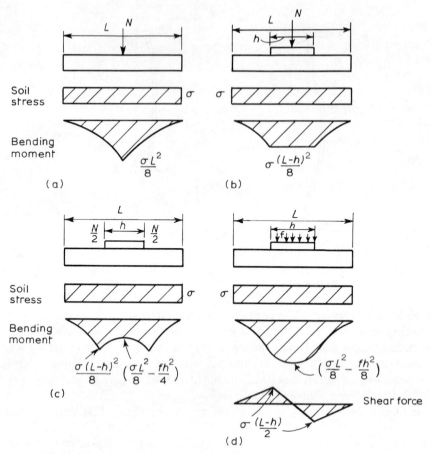

Figure 4.11 Distribution of bending moments and shearing forces on centrally loaded grillage beam. (a) Knife edge loading; (b) beams rigid over length h; (c) ends of rigid column base ride upon flexible beams; (d) column applies uniformly distributed load on beams. N is total column load, n is number of beams in tier, σ is soil bearing stress per metre of beam, $\sigma = N/nL$, f is column base stress per metre of beam, $f = N/nh$.

limit the deflection of the ends of the beams, their depth should be not less than about one-seventh of their projection from the edge of the base plate or the tier above.

4.6.4 Grillages loaded eccentrically along one axis

An eccentrically loaded column can be considered as carrying either an axial load plus a bending moment or a load offset some distance from the column axis. In the design of grillages there is a tendency to confusion if the latter is adopted. Once the soil stress has been determined, it is more convenient to work in terms of axial load plus moment. Figure 4.12 illustrates the approach to this problem. If a

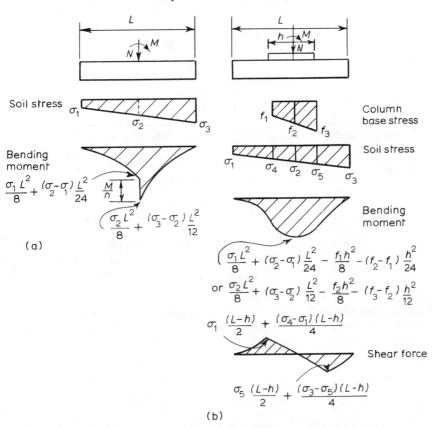

Figure 4.12 Distribution of bending moments and shearing forces on eccentrically loaded grillage beams. (a) Knife edge loading; (b) distributed load from column. N is total column load, n is number of beams in tier, f_1, etc. are column base stresses per metre of beam, σ_1, etc. are soil bearing stresses per metre of beam.

knife-edge load on the grillage can be assumed, as in Fig. 4.12(a), then the bending moment diagram takes the form indicated. On the other hand, if a bearing plate is provided, the bending moment diagram takes the form shown in Fig. 4.12(b) with a maximum at the centre. In this case the moment from the column is applied through the medium of the base plate, thus producing a continuous bending moment diagram, whereas in the previous case it must be considered as being applied by an external couple at the knife edge.

The top tier and, in a three tier grillage, the bottom tier can be designed from the diagrams shown in Fig. 4.12. The beams in each tier are uniform throughout, although tiers will differ from each other so that the beams for the second tier will be designed for the load occurring on the beam at the end where the soil stress is highest.

The approach indicated above deals with a foundation where the column is

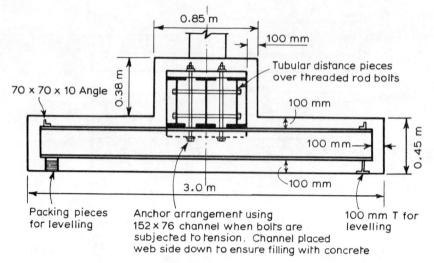

Figure 4.13 Typical grillage.

placed centrally on the footing and any moment is resisted by non-uniform soil stress. For an applied moment, equilibrium can be obtained also by placing the column eccentrically on the footing so that a uniform soil stress is obtained.

Figure 4.13 shows details of a grillage.

EXAMPLE 4.2 STEEL GRILLAGE DESIGN

Axial load in column 2000 kN
Column base 650 mm × 650 mm
Allowable soil bearing stress 250 kN m^{-2}
Assume the overall base size to be 3.0 m × 3.0 m.

Top tier

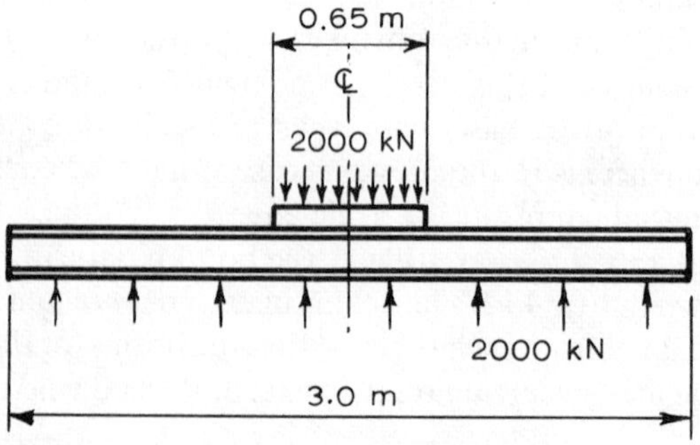

0.65 m

₵

2000 kN

2000 kN

3.0 m

Figure 4.14 Loading for design of top tier of grillage.

BM at centre $= 1000 \times 0.75 - 1000 \times 0.65/4 = 587.5\,\text{kN}\,\text{m}$.
For Grade 43 (mild) steel, allowable bending stress $= 165 \times 133/100\,\text{N}\,\text{mm}^{-2}(\text{MPa})$.

$$\text{Section modulus required} = \frac{587.5 \times 10^6}{165 \times 1.33} = 2.67 \times 10^6\,\text{mm}^3.$$

Try $3 \times (457 \times 152 \times 52\,\text{UB})$.
Tabulated gross second moment of area $= 21\,345 \times 10^4\,\text{mm}^4$.
Tabulated net second moment of area $= 19\,035 \times 10^4\,\text{mm}^4$.
The tension flange is drilled so that top and bottom layers can be bolted together. There are usually two bolts at each intersection, the bolts being placed diagonally opposite to each other, and thus at any section allowance should be made for one bolt hole. The tabulated values of net second moment of area allow for one bolt in each flange, therefore as an approximation, use the mean of gross and net second moment of area to allow for only one hole.

$$\text{Section modulus (3 beams)} = \frac{3(21\,345 + 19\,035) \times 10^4 \times 2\,\text{mm}^3}{2 \times 449.8}$$

$$= 2693 \times 10^3\,\text{mm}^3$$

$$f_\text{b} = \frac{587.5 \times 10^6}{2693 \times 10^3} = 218\,\text{N}\,\text{mm}^{-2} < 165 \times 1.33$$

Shear

Shear force $= 1000\,\text{kN}$.

$$\text{Average shear stress} = \frac{1000 \times 10^3}{3 \times 449.8 \times 7.6} = 97.5\,\text{N}\,\text{mm}^{-2} < 100 \times 1.33.$$

Web buckling need not be checked for grillage beams solidly encased in concrete. Hence use $3 \times (457 \times 152 \times 52\,\text{UB})$.

Bottom tier

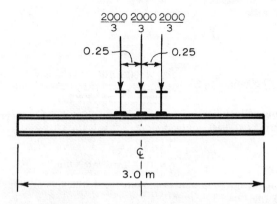

Figure 4.15 Loading for design of bottom tier of grillage.

$$\text{BM at centre} = \left(1000 \times \frac{3}{4}\right) - \left(\frac{2000}{3} \times 0.250\right) = 583.3 \,\text{kN m}.$$

$$\text{Section modulus required} = \frac{583.3 \times 10^6}{165 \times 1.33} = 2.66 \times 10^6 \,\text{mm}^3.$$

Try $8 \times (254 \times 146 \times 31 \,\text{UB})$ section modulus $= 8 \times 353.1 \times 10^3$

$$= 2824.8 \times 10^3 \,\text{mm}^3.$$

Bolt holes will be in the compression flange, therefore the gross modulus of section can be used.

Shear

$$\text{Average shear stress} = \frac{1000 \times 10^3}{8 \times 251.5 \times 6.1} = 81.7 \,\text{N mm}^{-2} < 100 \times 1.33.$$

Hence use $8 \times (254 \times 146 \times 31 \,\text{UB})$ for bottom tier.

The mass of the footing based on the encasing concrete is

$$3.0 \times 3.0 \times 0.45 \times 2400 + 0.850 \times 0.38 \times 3.0 \times 2400 = 12\,000 \,\text{kg}.$$

$$\text{Soil bearing stress} = \frac{2000 \times 10^3 + 12\,000 \times 9.8}{3.0 \times 3.0 \times 10^3} = 235 \,\text{kN m}^{-2}.$$

Some designers may wish to revise the design to achieve a soil bearing stress nearer to the allowable but in this case the economy in materials would be small. In addition the weight of backfill has not been included in soil bearing stress.

The grillage beams are connected by the bolts between top and bottom tiers but it is common practice to use additional spacers. These may be special cast iron spacers or short pieces of joist or channel which act as diaphragms or may be simply bolts with tube spacers. For the bottom tier an angle bolted to each top flange near the end is frequently considered sufficient. In view of the remarks in Section 4.7.2, consideration should be given to the provision of stiffeners and diaphragms to prevent web buckling before the grillage is cased.

If in the same example a load of 1500 kN is applied to the uncased grillage, then buckling can be investigated as indicated in BS 449 (BSI, 1969a).

The slenderness ratio of the unstiffened web

$$\frac{l}{r} = \frac{d_3}{t}\sqrt{3} = \frac{407.7\sqrt{3}}{7.6} = 92.9$$

$$p_a = 87 \,\text{N mm}^{-2} \text{ and } f_a = \frac{1500 \times 10^3}{3(650 + 449.8) \times 7.6} = 59.8 \,\text{N mm}^{-2}.$$

Hence web buckling will not occur if sidesway is prevented. If sidesway is not prevented, the slenderness ratio is at least doubled.

For
$$l/r = 2 \times 92.9 = 185.8,$$
$$p_a = 27 \,\text{N mm}^{-2} \quad f_a = 59.8 \,\text{N mm}^{-2}.$$

Hence web buckling in sidesway mode is possible and stiffeners or diaphragms must be provided, as shown in Fig. 4.16.

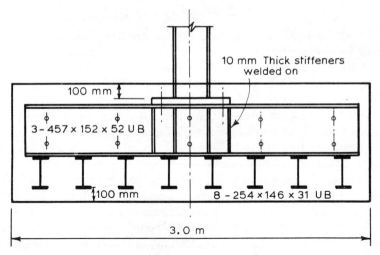

Figure 4.16 Completed grillage design.

4.7 CONSTRUCTION OF FOOTINGS

4.7.1 General notes

These notes are generally applicable to individual and combined footings and rafts. In the United Kingdom, soil formation level for a foundation on silt, chalk and fine sand should be not less than 0.5 m beneath ground level in order to avoid frost-heave and on clay not less than 1 m to avoid the volume changes in the upper layers due to seasonal variations in moisture content. These effects may be artificially induced, for example, by a refrigerating plant or a large boiler in a basement and special consideration must be given to foundations subjected to these conditions. Where necessary, adequate drainage of the soil at ground level or below should be provided.

When depths of foundations are being decided, a point worth remembering is that for hand excavation 1.5 m is the normal throw, and therefore it may, for example, be preferable to make the formation level 1.5 m or 3 m deep and not, say, 2 m. Currently machine excavation is more commonly used with a certain amount of trimming by hand. For individual pits the hydraulic back acter is mostly used. Back acters are able to produce reasonably flat bottoms and vertical sides, requiring a minimum of hand trimming, and some can excavate to depths exceeding 6 m. When these machines are fitted with an actuated grab, they can excavate to slightly greater depth but more trimming is necessary.

4.7.2 Reinforced concrete

Reinforced concrete footings may be flat, stepped or sloped on the top surface. Stepped footings are undesirable, as high local stresses are introduced at the re-entrant angles and sudden changes in bond stress are likely to be produced. Sloped footings are better and, if the slope is limited to 1 in 2, no special calculations are required for bending but calculations for diagonal tension and bond should be made at intermediate sections to check that permissible stresses are not exceeded. Also if the slope exceeds 1 in 2, formwork may be necessary for the top-surface; generally 1 in 4 is a convenient slope. Sloped and stepped footings should be cast as one unit.

Bottom reinforcement can be supported on steel or plastic chairs in order to obtain the necessary cover under the bars. Concrete or mortar blocks wired to the underside of the reinforcement are perhaps better than steel chairs, which may corrode, spall the concrete and eventually permit corrosion of reinforcement. Concrete walkway blocks of depth 50 mm greater than the height of the reinforcement are useful either as individual stepping stones or to carry scaffold boards, the blocks being of plan size to fit easily between the reinforcement mesh.

The ACI (1977) code specifies that the minimum thickness of a footing above the reinforcement should be not less than 150 mm for footings supported on soil and 300 mm on piles. Footings circular or octagonal in plan lead to the most even bearing stress distribution but these shapes complicate the arrangement of reinforcement so that rectangular shapes are more common. Where steel slab bases, pedestals or bridge bearings carry loads over about 500 kN, it is good practice to provide a mat of 10 mm diameter bars at 150 mm centres both ways 75 mm beneath the surface of the concrete under the base.

The maximum size of aggregate is normally 40 mm for reinforced concrete footings, 50 or 75 mm for large plain concrete footings and up to 150 mm for large mass blocks. Appropriate slumps are 50 to 120 mm for reinforced footings and 25 to 80 mm for mass construction. If the concrete is vibrated the slump can be reduced by about 50%.

4.7.3 Steel grillages

The beams in each tier of a grillage can be separated as described in Example 4.2. A disadvantge of separators and distance plates is that continuity of the concrete is broken but both assist in spreading isolated loads from one beam to the next, although concrete should adequately perform this function in encased grillages. Separators, plates, or tubular pieces are normally placed under concentrated loads and not further apart than five times the depth of the girder, with a maximum of 1.2 to 1.5 m. Pairs of tubular pieces should be used for web depths exceeding 200 m. In Fig. 4.16 the base plate is shown bolted to the flanges of the top tier in addition to the fastening afforded by the holding down bolts and channels. It is not essential to provide both and, in practice, one alone should be sufficient.

Grillages can be erected by setting the bottom tier on 100 mm × 100 mm tees, which are carefully levelled with shims and grouted under. Upper tiers and base plates can be levelled by inserting thin shims if necessary. Sometimes four sets of 100 mm × 60 mm or 100 mm × 100 mm packing plates are used to level the bottom tier; one set near each corner. The sets are made up of 25 mm, 12 mm, 6 mm, etc., plate as may be necessary. The end angles of the bottom tier span from the two supported beams and serve to hold up the remaining beams. Alternatively, the bottom tier can be levelled by precast concrete spacer blocks and shims set on the blinding layer. The grillage is normally set about 3 mm proud of the specified level to allow for bedding down under load. Figure 4.13 shows details of a grillage.

4.8 CANTILEVER FOUNDATIONS

A cantilever foundation is a useful form of construction where it is impossible to place a foundation directly beneath a column or other load because of limitations imposed by adjacent buildings or where extreme eccentric loading produces inadmissible soil stress. In Fig. 4.17 the load W_1 from the outer column is balanced by the load W_2 from the inner column acting about X as fulcrum. $W_2 l_2$ should exceed $W_1 l_1$ by not less than 50% under all conditions of loading on the columns, giving a factor of safety of at least 1.5 against rotation. If no column load W_2 is available, anchorage must be effected by a mass concrete block or tension piles giving the same factor of safety. The load on base 2 is that part of W_2 not required to balance W_1. The foundation blocks 1 and 2 are designed as concentrically loaded foundations. The beam is designed as a simply supported beam with an overhanging end. Excessive deflection in the beam should be avoided since it would affect the uniform distribution of bearing stress beneath the foundation blocks and, together with differential foundation settlement might affect superstructure moments and forces.

In order that central loading of the foundation blocks may be reasonably

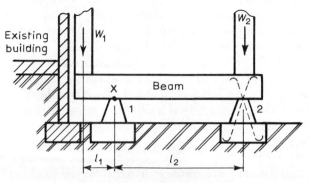

Figure 4.17 Cantilever foundation.

attained in practice, cast steel or built-up mild steel pedestals can be placed under steel beams, or the top surface of a reinforced concrete block may be slightly haunched, terminating in a level rectangular surface to carry either steel or reinforced concrete beams. It is desirable to provide restraint to lateral movement at the seats of the cantilever beam in the form of foundation bolts or reinforcement, as indicated in foundation block 2 in Fig. 4.17. On a narrow building site between two existing buildings a simply supported beam with two overhanging ends may be adopted to transmit column loads to the two foundation blocks. If either of the columns is fixed at the cantilever or if unequal settlement of the foundation blocks is anticipated, an estimate should be made of any bending moments which may be developed and the beam designed to cater for them. Design for such conditions should be to generous proportions. The overhanging length of cantilever is usually short and consequently adequate anchorage for tensile reinforcement must be provided. Diagonal tension and bond stresses may be high at the support and govern the depth of the beam at that point.

The cantilever foundation is really a steelwork design which has been translated into reinforced concrete in many cases. Where reinforced concrete foundations are being constructed, it is often better to design a combined footing for two or more columns, the column at the edge of the site being, of course, on the extreme end of the footing. It may be necessary to use combined footings which are not rectangular in plan to achieve a reasonably uniform soil stress.

4.9 STEEL COLUMN BASES

4.9.1 General notes

Bases for steel columns are normally either gusseted or slab. However, on the Port Talbot, Wales, steelworks (Atkins, 1950), pockets with corrugated sides were left in the concrete foundation blocks. The lower sections of steel stanchions were temporarily supported and positioned by short transverse beams and side packings and then cast in the pockets. Short lateral plates or channels were welded to the lower sections to distribute the load to the concrete filling. It is claimed that this type of base gives greater continuity between superstructure and foundation, is cheaper in steel than normal type bases and allows greater tolerance in setting out. On the other hand, some difficulty may be experienced in packing concrete under lateral plates and channels.

A variety of hinged and fixed bases for steel portal frames are illustrated in BCSA Publication No. 9 (BCSA 1955) and Husband and Best (1953) give details of hinged bases for the steel portal frames of a large industrial building. The design of these can sometimes be based on details relating to bridge bearings given in Chapter 10.

In the following discussion relating to gusseted and slab bases, direct compression and bending stresses are analysed. It may be necessary to investigate horizontal shearing forces even though, in many cases, they are adequately resisted by friction at the contact surface between the base plate and the concrete foundation. Such horizontal forces may arise from lateral loads on the structure or they may be the result of vertical loads on rigidly jointed portal frame type structures.

4.9.2 Gusseted bases

Gusseted bases consist of gusset plates, cleats, stiffeners and base plate welded or riveted together; the assembly being sufficiently stiff to transmit the column load uniformly to the foundation proper. The base plate is normally 12 to 25 mm thick or up to 50 mm for very heavy loads. For riveted construction the end of the column shaft is often machined after the cleats and gussets have been attached to ensure that a proper bearing is obtained on the machined top face of the base plate. If this is the case then the gussets, cleats, and fastenings are usually designed to transmit the whole of the bending moment plus 60% of the axial load. If machined surfaces are not provided for bearing, the whole of the forces and the moments must be transmitted through the fastenings. Where machining is carried out after attachment of cleats, allowance must be made for the reduction in thickness of the outstanding legs of the cleats; with care in assembly this should not exceed 2 mm. For welded construction the gussets and welds must be designed to transmit the whole of the forces and moments in the column shaft to the base plate.

It is usual practice to provide holding down bolts for all column bases axcept where the concrete is cast directly round the column shaft. If the eccentricity of load is small the bolt size is chosen to be in proportion with the rest of the base. For great eccentricities the bolts are designed to resist the tensile stresses developed. Where stresses are caused by wind loads the allowable stresses on the plates and bolts may be increased by 25%, provided that normal stresses are not exceeded when the wind forces are not acting.

The transfer of load from column base to foundation is discussed in Section 4.2 and the anchorage of bolts in tension is discussed in some detail. Nevertheless, the slightly simpler method of analysis usually adopted for the determination of the size of base plate and holding down bolts is based on the elastic theory for a reinforced concrete member subjected to direct compression and bending. The analogy is not strictlty correct because the initial tension in the bolts is not zero, the base plate does not remain plane and the apparent elastic modulus of the concrete is changed due to the confining effect of the unloaded concrete surrounding the base plate area.

Referring to Fig. 4.2, the stress distribution is assumed to be linear and hence

$$x = \frac{\alpha_e f_c d}{f_s + \alpha_e f_c}$$

then

$$f_s = \frac{\alpha_e f_c (d - x)}{x}.$$

For static equilibrium the algebraic sum of the vertical forces must be zero and the algebraic sum of the moments of forces must be zero.

Resolving vertically

$$\frac{f_c x b}{2} - A_s f_s - N = 0$$

or

$$\frac{f_c x b}{2} - A_s \alpha_e \frac{f_c (d - x)}{x} - N = 0, \tag{4.18}$$

and taking moments about the centre line

$$\frac{f_c x b}{2}\left(\frac{h}{2} - \frac{x}{3}\right) + \frac{A_s \alpha_e f_c}{x}(d - x)\left(d - \frac{h}{2}\right) - Ne = 0. \tag{4.19}$$

In Equations 4.18 and 4.19 N, e and α_e are known and values of h, d, b, and A_s are assumed. There remain two unknowns, f_c and x. From Equations 4.18 and 4.19 it is possible to derive a cubic equation for x. It is probably simpler, however, to assume a value for x (generally $0.4d$ to $0.6d$) which is substituted into Equation 4.18 to give a value for f_c. This value of f_c and the assumed value of x are substituted into Equation 4.19. If the remainder in this expression is not zero the value of x should be changed, usually by increments of $\pm 0.05d$, and the whole process repeated. If a positive and a negative value for the remainder are found, a very close approximation to x can be obtained by interpolation.

4.9.3 Slab bases

Slab or bloom bases vary in thickness from about 40 mm for a column load of 1000 kN to 120 mm for 10 000 kN, but much thicker slabs can be provided if necessary. The thicknesses and loads given depend on the relative plan sizes of the column and slab. Slabs up to about 50 mm can be hydraulically flattened and do not need to be machined. Slabs above this thickness, and many below it, are planed to ensure a proper bearing for the machined end of the column shaft. The machined end of the column shaft bears directly on the slab base and is connected to it by light cleats which are sufficiently strong to resist the stresses of handling and transportation and also serve to locate the column in the correct position on the slab.

It is claimed, for slab bases, that the labour required for fabrication is considerably less than for gusseted bases but they require more steel. For heavy columns, slab bases are probably more economical but they are unlikely to be used where bending moments are other than very small because the connections approach those required for gusseted bases.

The formula given in BS 449 (BSI, 1969a) for the thickness of rectangular slabs under axially loaded rectangular section columns is

$$t = \left[\frac{3w}{p_{bct}}\left(A^2 - \frac{B^2}{4}\right)\right]^{1/2} \tag{4.20}$$

where t is the slab thickness (mm), A the greater projection of the plate beyond the stanchion (mm), B the lesser projection of the plate beyond the stanchion (mm), w the uniform stress on the underside of the base (N mm^{-2}), and p_{bct} the permissible bending stress in the steel (185 N mm^{-2} for all steels).

EXAMPLE 4.3 THE DESIGN OF A WELDED STEEL COLUMN BASE

The working axial load on the column is 400 kN and the bending moment at the column base is 250 kN m. The characteristic cube strength of the concrete in the reinforced concrete footing is 25 N mm^{-2}. From considerations of concrete strength and ratio of loaded area to unloaded area it can be assumed that the allowable bearing stress on the concrete is 5.6 N mm^{-2} for a working load design. Assume that the modular ratio is 15. Try a base 1000 mm $\times$ 500 mm with 4 $\times$ 30 mm diameter bolts.

$$d = 950 \text{ mm}, \quad b = 500 \text{ mm}, \quad A_s = 1122 \text{ mm}^2 \text{ for 2 bolts.}$$

If $x = 400$ mm, from Equation 4.18 $f_c = 5.22$ N mm^{-2},
and from Equation 4.19 remainder $= -5 \times 10^6$.
If $x = 380$ mm, $f_c = 5.72$ N mm^{-2} and remainder $= 17 \times 10^6$.
  Interpolating, $x = 395$ mm, $f_c = 5.33$ N mm^{-2} and remainder $= 0.58 \times 10^6$.
  The remainder appears to be rather large but, if x is increased by 1 mm to 396 mm, $f_c = 5.30$ N mm^{-2} and remainder $= -0.70 \times 10^6$. Therefore x falls between 395 and 396 mm.

When $\qquad x = 395$ mm, $f_s = 15 \times 5.33 \dfrac{(950 - 395)}{395} = 112$ N mm^{-2}.

Both steel and concrete stresses are below the allowable values. Some designers may wish to change the base dimensions to achieve slightly more economical proportions.
Tensile force in one bolt $= 112 \times 561 \times 10^{-3} = 62.8$ kN.
Base plate thickness can be determined by considering tension in bolts assuming 20 mm thick gusset.
BM due to bolt $= 62.8 \times (65 - 20) = 2826$ N m.
Lawton (1951) suggests that the width of plate resisting this bending moment may be taken as the width between lines at 60$°$ on each side of the bolt.
Then width of plate $= 2 \times (65 - 20)\sqrt{3} = 155.9$ mm, but only 50 mm is available on one side, therefore assume 100 mm.

$$\text{Plate thickness} = \left(\frac{2826 + 10^3 \times 6}{165 \times 100} \right)^{1/2} = 32.06 \text{ mm}$$

Hence use 35 mm plate.
Alternatively, considering stress on concrete

$$\text{BM} = \frac{5.33 \times 85^3}{10^3 \times 2} = 19.25 \text{ N m per mm width.}$$

This is less than that due to the tension in the bolts.
Assume all forces and moments to be transferred from column to gussets through welds.
Maximum force in weld connecting column and gusset

$$= \frac{400}{4} + \frac{250 \times 10^3}{2 \times 432} = 389.35 \text{ kN.}$$

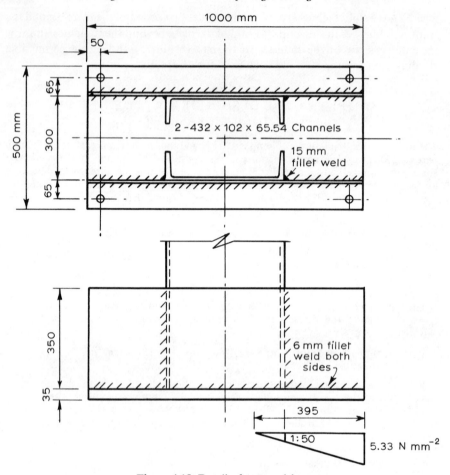

Figure 4.18 Detail of gusseted base.

Length of 15 mm fillet weld $= \dfrac{389.35}{1.21} = 322$ mm.

Use 350 mm deep gusset plate.

The thickness of the gusset plates must be determined and the criterion for the design is the buckling of the unstiffened part of the gusset. In Fig. 4.18 the unstiffened gusset extension is shown as a rectangular extension but no tests of such gussets have been reported although Salmon *et al.* (1964) and Martin (1979) have reported tests of triangular gussets. Martin indicates that material added to the free edge of a triangular gusset does increase its strength but only by a small margin; the rectangular gusset shown in Fig. 4.18 may therefore by safely treated as a triangular gusset. Martin gives an approximate formula for the thickness of a triangular gusset where $l/r \not> 185$

$$t = \frac{2(P_{\mathrm{u}} - M_{\mathrm{p}}/s)s[(L/H)^2 + 1]}{f_y L^2} + \frac{L\sqrt{3}}{138.75[(L/H)^2 + 1]^{1/2}}$$

where t is the gusset plate thickness, P_u the ultimate load on the gusset, M_p the plastic moment of resistance of the base plate, s the distance of the resultant force P_u from the enclosed corner of the gusset, L the length of the loaded side of the gusset, H the height of the gusset, and f_y the yield strength of the material.

For checking that Martin's approximate formula is applicable

$$l/r = \frac{2\sqrt{3}}{[(L/H^2) + 1]^{1/2}} \left(\frac{L}{t}\right). \tag{4.22}$$

For checking cases where $l/r > 185$, Matin (1979) gives a more accurate formula. Later tests by Martin and Robinson (1980) have shown that buckling of the gusset plate occurs before the plastic hinge in the base plate can develop and the gusset thickness should be obtained from the revised formula.

$$t = \frac{2P_u s[(L/H)^2 + 1]}{f_y L^2} + \frac{L\sqrt{3}}{138.75[(L/H)^2 + 1]^{1/2}}. \tag{4.23}$$

In the present example, the ultimate load may be taken as

$$P_u = \frac{(1.5 + 5.33)}{2} \times \frac{284 \times 250 \times 1.7}{1000} = 412.2 \, \text{kN},$$

where 1.7 is the global safety factor and

$$s = \frac{(1.5 \times 284^2/6 + 5.33 \times 284^2 \times 2/6}{[(1.5 + 5.33)/2] \times 284} = 168.5 \, \text{mm}$$

$$t = \frac{2 \times 412.2 \times 10^3 \times 168.5[(284/350)^2 + 1]}{284^2 \times 250} + \frac{284\sqrt{3}}{138.75[(284/350)^2 + 1]^{1/2}}$$

$$= 11.42 + 2.75 = 14.17 \, \text{mm}.$$

Hence use 15 mm plate.

The simplified method given above is applicable only for slenderness ratio less than 185 when calculated according to the Equation 4.22

$$l/r = \frac{2\sqrt{3}}{[(284/350)^2 + 1]^{1/2}} \left(\frac{284}{14.17}\right) = 53.9.$$

Hence the simplified method given above may be used.

It is also of interest that Martin (1979) shows that the greatest economy is obtained when L and H are approximately equal.

4.10 STEEL PEDESTAL BASES

Steel pedestal bases can be used as an alternative to grillages for spreading column loads. Their design, fabrication and erection are described by Bartlett and Danks (1941) in a paper which includes dimensions of pedestals to carry up to 17 700 kN. Figure 4.19 illustrates a base to carry 6000 kN and is sketched from details given by Bartlett and Danks. A base of this type contains about half the amount of steel in a grillage but requires fabrication. It is stiffer than a grillage and should lead to a more uniform stress distribution on the foundation concrete. A pedestal is easier to position and level than a grillage, particularly as four or more levelling screws can be threaded through the base plate.

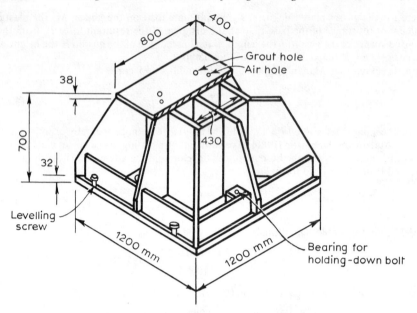

Figure 4.19 Steel pedestal to carry 6000 kN.

The top bearing plate is machined on both surfaces and the bottom plate on the top surface only. Core and rib plates are machined on all bearing edges. If the pedestal is to be encased in concrete the permissible stresses are increased to $33\frac{1}{3}\%$ above normal, with the exception of the core where normal stresses only are permitted. Diaphragm plates are assumed to act as stiffeners and not to carry direct load.

Care is required to prevent warping during fabrication. The core is first welded, end milled to length and welded to the base. The ribs are then welded to the base and the core. The top plate, stiffeners and diaphragm plates are then welded to complete the assembly. Finally the top plate upper surface is machined.

During erection a 75 mm space is left between base and concrete which is packed with fine concete after levelling. The centre pockets are grouted and the concrete surrounding the pedestal is brought up to floor level, with a minimum cover of 100 mm to the steel.

4.11 DEEP FOUNDATIONS

Deep foundations are specified in CP 2004 (BSI, 1972) as foundations exceeding 3 m in depth. Nevertheless it is recognized that certain small foundations of less than 3 m in depth may be treated as deep foundations. Deep foundations may be divided into two main types, depending on how stability is maintained. Basement or hollow boxes in general carry the loads from a number of columns and overturning is unlikely because the horizontal dimensions are significantly greater than the vertical and bearing stress under the basement is virtually the sole criterion. Such foundations are designed as rafts (Chapter 5) but have features in common with retaining walls (Chapter 6). Other foundations have high depth to breadth ratios and their stability depends partly on lateral support from the soil. The commonest type of deep foundation is the pile which may be driven or bored (Chapter 8); large diameter bored piles may be up to about 2 m in diameter and function more in the nature of piers. If the ground is bad or waterlogged, cofferdams, or caissons may be necessary: these are discussed in Chapter 7.

4.12 FOUNDATIONS SUBJECTED TO OVERTURNING MOMENTS AND UPLIFT

The stability of pad foundations subjected to moments and to horizontal forces has been treated in Section 4.3. The following discussion is concerned mainly with foundations subjected to uplift and to foundations of pier-type subjected to overturning moments.

Consider the pylon structure illustrated in Fig. 4.20, square in plan and supported on four pad foundations. Let W be the weight of each foundation plus one-quarter of the vertical dead load from the superstructure and let P be one-half of the total horizontal force on the superstructure. The vertical force on each foundation due to the horizontal force is $V = \pm Ph_1/l$ and the factor of safety against overturning about foundation 2 is Wl/Ph_1.

Assuming that the horizontal thrusts at foundations 1 and 2 are equal and of magnitude $H = P/2$, the maximum potential resistance to horizontal force at

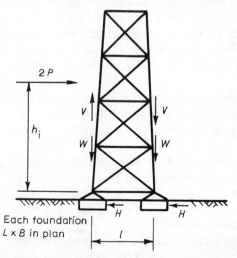

Figure 4.20 Foundations subjected to uplift.

foundation 1 is

$$\{P_p + c_w LB + (W - V)\tan\delta\},$$

where P_p is the total passive resistance of the soil on the side of the foundation, and at foundation 2 is

$$\{P_p + c_w LB + (W + V)\tan\delta\}.$$

The factor of safety against sliding of the pylon is

$$(2P_p + 2c_w LB + 2W\tan\delta)/2H.$$

The horizontal displacement required to invoke the full resistance may be several centimetres and a factor of safety of 3 or more may be necessary to limit movement under working loads. If each foundation is free to rotate about one of its sides, it may be necessary to investigate the stability of each foundation against overturning, as indicated in Section 4.3

The factor of safety of foundation 1 against uplift is W/V. In practice the resistance to uplift may involve several components in addition to W. The following are three possible cases.

(1) Cables or piles bonded to bedrock. The major part of the anchorage resistance is provided by the bond to bedrock with contributions, in the case of a pile, from the weight of the pile and the skin resistance developed in the overburden. Consideration should be given to possible rupture of an inverted cone of intact rock or of a prismatic block of rock where joints are well developed; however, the mechanics of this problem has received little attention. An estimate of the capacity of this type of anchorage is best made from the results of field tests. Moore (1964) carried out tests on a 50 cm diameter concrete cylinder 2.6 m long cast in bedrock comprising fractured and weathered shale and sandstone in San Francisco; reaction to the test load was provided by skin resistance only and under a load of 5000 kN the total displacement was 32 mm, of which 13 mm was elastic compression in the concrete. No sign of failure was observed. The permanent set was 16 mm. The skin resistance developed under this load was 1000 kN m^{-2}.

(2) Piers or piles of constant diameter founded in soil and extending to ground surface. Anchorage resistance comprises the weight of the pier or pile and skin resistance. The latter can be estimated from field tests or from calculations based on experimental values of adhesion and angle of friction.

(3) An under-reamed pier or a horizontal slab or encased grillage founded in soil and extended to the surface as a pier with a cable or rod. The pull-out resistance of such anchorages depends on the relative depth D/B of the anchor or under-reamed portion, the slope and size of the anchor, and the soil properties. Such anchors may be considered to be shallow or deep, depending on whether the surface is involved in any failure or not. In shallow anchors the anchorage resistance is provided by the weight of the anchor plus the weight of the soil above

it and any shear resistance of the soil which may be developed; the shape of the shear surface may be more cylindrical or more conical depending on the soil properties. In deep anchors the anchorage resistance is provided by the weight of the anchor and the bearing resistance of the soil above it.

Theoretical solutions for the pull-out resistance of anchors have been proposed by a number of writers including Balla (1961), Vesić and Barksdale (1963), Matsu (1967) and Meyerhof and Adams (1968) for shallow anchors and Vesić and Barksdale (1963), Mariupol'skii (1965) and Meyerhof and Adams (1968) for deep foundations. Davie and Sutherland (1977), reporting model tests by Davie (1973), compared the shallow anchor theories of Balla (1961), Vesić (1971), Matsu (1967) and Meyerhof and Adams (1968) showing that the theories of Vesić, Matsu and Meyerhof and Adams gave broadly similar values of uplift resistance factor, whilst Balla's theory gave uplift resistance factors which were significantly higher than the others. Davie and Sutherland (1977) also compared Vesić's theories for shallow and deep foundations with the model test results by Davie and with the field test results by Meyerhof and Adams. For shallow anchors the model test results were of the order of 50% of the theoretical results whilst the field test results were lower still. However, in reporting their results, Meyerhof and Adams noted that the field tests were carried out in stiff clays which were brittle and fissured and that the laboratory soil strengths would probably be greater than the true strength. For deep anchors, Vesić's theory appears to give reasonable results when soil-compressibility is taken into account. The comparison of theoretical and test results by Davie and Sutherland is shown in Fig. 4.21. For shallow anchors Vesić (1971) proposed the formula

$$q_0 = cF_c + \gamma D F_q,$$

where q_0 is the soil stress on the upper surface of the anchor and the breakout factors F_c and F_q depend on the shape and relative depth of the anchor and on the angle of shearing resistance of the soil. Values of F_c and F_q for $D/B < 5$ are tabulated by Vesić (1971).

For deep anchors a similar formula

$$f_u = cF_c' + \gamma D F_q'$$

where f_u is the stress required to expand a cavity in the soil and F_c' and F_q' depend on the rigidity index I_r for the soil. Tables for I_r are given by Vesić (1971).

The uplift resistance of piers in cohesionless soils has been studied by Kulhawy *et al.* (1979) and Ismael and Klym (1979).

Foundations for electricity transmission pylons and towers have been described by Turner (1962), Wild and Haslam (1962) and Boscawen *et al.* (1963). Sutherland (1965) investigated the problem of jacking a shaft upwards through cohesionless soils from a tunnel by analogy with foundations subjected to uplift. In connection with uplift it is specified in BS 449 (BSI, 1969a) that the anchorage should be designed so that the least restoring moment, including anchorage, shall

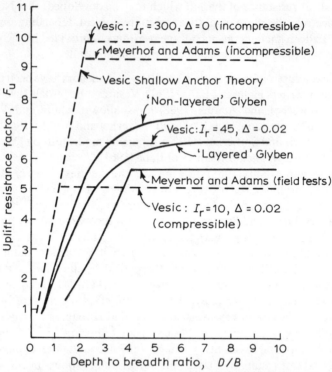

Figure 4.21 Comparison of theoretical and test results. Reproduced from Davie and Sutherland (1977) with permission of the American Society of Civil Engineers.

be not less than the sum of 1.2 times the maximum overturning moment due to dead loads and 1.4 times the maximum overturning moment due to imposed loads. Wind loading shall be treated as imposed loading. To ensure stability at all times account should be taken of probable variations in dead load during construction, repair or other temporary measures.

When a foundation of pier-type is subjected to an overturning moment, as shown in Fig. 4.22, it tends to rotate about some point and resistance to overturning is afforded by passive resistance in front of and behind the pier. The problem is analogous to that of the cantilever sheet pile wall and can be solved by considering active and passive stresses, equating to zero both horizontal forces and moments of those forces about the base, as in Fig. 4.22(a). The depth of pier required to resist the overturning moment is thus found. Altenatively, a depth can be selected and a stress distribution can be assumed, as shown in Fig. 4.22(b). The maximum stresses required to resist the overturning moment are calculated and a check is made to ascertain that these do not exceed the values which can be invoked in the soil. In these analyses it is assumed that the problem is two-

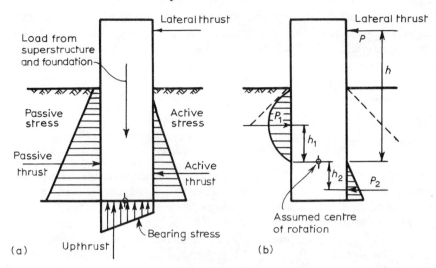

Figure 4.22 Pier subjected to overturning force. (a) Active and passive stress distribution; (b) assumed parabolic and triangular stress distribution. $P = P_1 + P_2$; $P_h = P_1 h_1 + P_2 h_2$. Solve for P_1 and P_2 and hence determine the maximum value of parabolic and triangular stress distributions. The maximum potential passive resistance less the active stress is shown dotted. In this case it is nowhere exceeded by the assumed pressure. Vertical loads are omitted for simplicity.

dimensional and, although this is justified in the case of sheet piling, it can be no more than a crude representation in the case of a pier.

Biarez and Capelle (1961) found that the shafts of foundations subjected to a horizontal force above ground level tend to rotate about a point at a height above the base equal to approximately one-quarter of the embedded depth, the point being displaced slightly in the direction in which the force is acting. A pear-shaped zone of soil adjacent to the shaft rotates with the shaft. There is little doubt that the resistance of cohesionless soil surrounding a pier less than say 1 m long or wide, would be much reduced by the lateral escape of particles. On the other hand, the resistance would be augmented by skin resistance developed at the sides of the pier in all types of soil. In the case of some clays there may be a tendency to progressive overturning if the creep strength is relatively low. The plan shape of the pier may also influence the resistance although little is known concerning this factor. The rotation required to develop the full resistance of the soil may be appreciable and a factor of safety of three or more may be desirable to reduce this to an acceptable value. In fact, analysis based on the modulus of horizontal subgrade reaction may provide a better means of calculating the required depth of the pier. Similar problems are presented by piles and bridge piers subjected to lateral loads. It should be observed that the contributions to overturning and resisting moments arising from the weight of the foundation and superstructure have been ignored in this discussion. In the case of cohesionless soil, the load from the superstructure and the substructure may invoke sufficient frictional resistance at the base to restrict translation at this level and consequently the centre of rotation may be practically in the plane of the base. Research on the behaviour of pier-type foundations under lateral loads is needed before reliable recommendations for analysis can be made.

4.13 MACHINE FOUNDATIONS

4.13.1 General notes

Vibrations in machines occur because of lack of balance of the machine parts, as in rotating or reciprocating machinery, or due to a sudden release of energy, as in presses or when starting machines. Whereas the generation of vibrations by some types of machinery, such as forging hammers is unavoidable, theoretically no unbalanced forces should occur in rotating machinery, such as turbines, but factors of design, manufacture and maintenance usually lead to some lack of balance. The out of balance forces are created by lack of coincidence of centre of gravity and centre of rotation and arise, for example, from inaccuracies in manufacture or from damage, corrosion or wear and gravitational deflection of shafts. In reciprocating machinery, such as engines and compressors, a reasonable balance can be attained by the use of fly wheels and balance weights but some vibration is unavoidable.

The operation of any machine on an unsuitable foundation can lead to large amplitudes of oscillation which may, in an extreme case, lead to failure of the machine or its foundation. This condition involves resonance, in which a primary oscillator agitates a body at or near to its natural frequency of oscillation. The nearer the frequency of the primary oscillator is to the natural frequency of the agitated body, the more likely it is that resonance will occur. Resonance in undamped bodies leads to very high amplitudes, even though the disturbing force is small; the mathematical formulae are given in Section 4.13.4 and the magnification effect is illustrated in Fig. 4.26.

In the special case of foundations, although an apparent resonance occurs and it is possible to treat the phenomenon as resonance when performing calculations, the soil does not have a unique natural frequency. The apparent natural frequency depends on the mass of the machine and its foundation block, the area of the block in contact with the soil, the existing force, and the properties of the soil itself.

4.13.2 Criteria for design

The vibrations due to machinery must be limited to prevent damage to the machine and to avoid annoyance to persons in the vicinity. When specifying the permissible magnitude of vibration, a restriction may be placed on the amplitude, the velocity, the acceleration or combinations of these three.

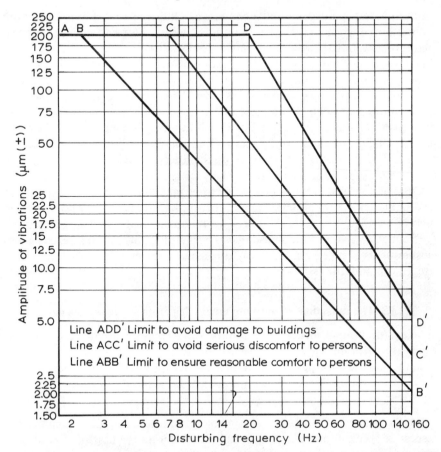

Figure 4.23 Amplitude limits of foundation block. Reproduced from CP 2012 (BSI, 1974b) with permission from the British Standards Institution.

Richart (1962) assembled data from several sources and produced a diagram giving recommendations for such limits. CP 2012 (BSI, 1974b) includes a simpler diagram, part of which is reproduced as Fig. 4.23, which may be used as a guide when specifying allowable limits of vibration. It is also stated in CP 2012 that the limits given include no safety factor and there should be a considerable margin between the calculated and recommended amplitudes. It should be noted that the effect of vibrations on persons is likely to be more significant that the effect on machines. The psychological effects must be taken into account, the physical limits given in Fig. 4.23 may be considered to be reasonable by the owner of the machine and its operator, whilst the same vibration extending into neighbouring property could be described by the neighbour as troublesome. Occasionally foundations must be designed for machines which can withstand only very small accelerations, magnitudes of $10^{-4}g$ are quoted for electron microscopes and calibration test stands and similar equipment. Often these must be isolated from ground-borne vibrations despite the frequent necessity for setting them near to a source of vibration. Building and human response to vibrations is discussed in a digest (BRE, 1983). A guide to the evaluation of human exposure to vibrations is to be found in BSI (1984).

Although the principal source of concern with machine foundations is the effect of vibrations on machines and personnel, the effects of vibrations on the ground must also be considered. The effect of vibrations will usually be to increase, periodically, the static bearing stress and may have a compacting effect on the ground, thus increasing settlement above that due to static loads. An illustration of this effect is the use of vibratory rollers for the compaction of soils.

Costs are always an important factor to be taken into account in any design and in this field, perhaps more than any other, the initial costs of the foundation must be weighed against the costs of down-time and repairs. Since large machines are frequently involved the shut-down of a single key machine may even lead to closure of the whole plant, whereas other machines may be halted with little overall effect on operations.

4.13.3 The reduction of vibration emanating from machine foundations

Whereas preventive measures to reduce vibrations are usually relatively inexpensive when incorporated in the original design, remedial measures applied after the machinery is put in operation are likely to be costly. The following are some of the means by which vibrations can be reduced. Sometimes they are employed in combination.

(1) Adjustment of machine either in respect of balance of moving parts or in respect of speed of operation. Husband (1947) quotes the case of a motor and compressor mounted on a concrete block which created serious vibrations in a cinema 100 m away. The unit was then mounted on a heavy steel frame supported on spring pads but, when in operation, alarming vibrations developed in the unit. The trouble was overcome by improved balancing in the unit carried out by the manufacturers.

(2) Provision of a suitable mass foundation for the machine. In some cases a larger mass in the block will be advantageous, in others it may be better to use a lighter block. Piled foundations may be advisable in certain circumstances, obviously if the soil bearing stress or the settlement or the amplitude of vibration is excessive then the use of piles is indicated. The use of a piled foundation may be indicated if ground water or a rigid stratum is present within a depth of about twice the breadth of the foundation block. Piles may also be used to minimize the disturbance transmitted to other foundations in the vicinity. In cases where a machine causes vibrations in several modes and it is not possible to avoid resonance in one mode when using a simple foundation block, a combination of vertical and raking (batter) piles make it possible to adjust the foundation

characteristics as required. Singh *et al.* (1977) have presented a simplified method of design, with an example of the design of a combined foundation for a pair of compressors.

(3) Provision of steel springs or layers of flexible material, such as cork or rubber, either between the machine and the foundation or between the foundation and a base slab overlying the soil. If the foundation is large, it can be installed in a pit with an air-gap between the foundation and the sides of the pit and with strips of flexible material to cover the top of the gap. Although a flexible mounting frequently reduces the transmitted vibrations, the machine may be subjected to vibrations of increased amplitude which can lead to unsatisfactory operation or damage to the machine.

(4) Site machinery at an appropriate distance from other buildings so that vibrations reaching those buildings are tolerable.

The common vibration isolators are leaf, helical and disc springs and cork and rubber mats, pads and connectors. Where static deflections exceed about 5 mm, metal springs are appropriate. Neither helical nor disc springs possess the property of self-damping but leaf springs normally possess a damping factor, arising from friction between the leaves, of about 0.15 which can be increased to about 0.70 by the manufacturer. Disc springs are produced in standard diameters from 25 to 400 mm and are useful where heavy loads are to be supported with relatively small deflections. They are also very adaptable and the capacity of a group for a given deflection can be increased by increasing the number in parallel, whereas the deflection under a given load can be increased by increasing the number in series.

The compressibility and resilience of cork are directly related to its cellular structure, with air contained in the cells. The dynamic modulus of elasticity of cork is higher than the static modulus of elasticity; values in the range 25 to 40 N mm^{-2} have been recorded. Cork can be treated to resist deterioration in the presence of oil and water.

Rubber possesses resilience but is not durable in the presence of oil. Both cork and rubber possess the property of self-damping, the damping characteristics of both are mentioned by Steffens (1964). A considerable proportion of the energy is reflected at the interface when a shock wave passes from a cork or rubber mat into concrete. A wide variety of rubber connectors and mountings are produced for example by André Rubber Co. Ltd, Dunlop Rubber Co. Ltd, and Metalastik Ltd. Rubber can be employed in tension, compression, shear and torsion but the greatest flexibility over a range of deflections is obtained by using it in shear. Bearings of neoprene are produced by the Du Pont Co. (UK) Ltd and it is claimed that these are resistant to oils, soil chemicals, ozone and severe weather. Other types of vibration isolators have been employed and, for example, Crocket and Hammond (1947) report a machine of 110 tonne mass mounted on compressed-air springs with aperiodic hydraulic damping.

The following are examples of foundations designed to cater for the influence of

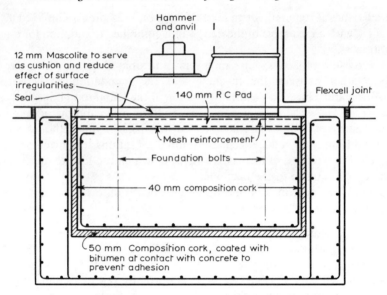

Figure 4.24 Foundation for forging hammer at Sheffield.

vibrations. Figure 4.24 shows a foundation for a forging hammer at Sheffield. The hammer operates at 2 to 4 blows per second and the natural frequency of the foundation was arranged to be well above the exciting frequency. Rubber layers have been used in a similar type of foundation for a 425 kN roll grinding machine in a steelworks in Wales. The mass of the foundation is 46 tonne. Klein and Crockett (1953a, b) described the foundation for an 80 kN forging hammer at Huntington, West Virginia, in which spring-mounted inertia blocks of pre-stressed concrete were employed. A review of forging hammer foundations has been given by Crockett (1958). The mass of foundation and backfilling for a forging hammer may be as much as 100 tonne per 10 kN of hammer. Hammond (1959) described hydraulic dampers fitted as a remedial measure to a 40 kN drop stamp at Saltley, Birmingham. Crockett and O'Neil (1959) described a pre-stressed foundation for a heavy-gun testing-hammer which is subjected to horizontal impact forces. Sometimes vibrations can be reduced by altering the vibration characteristics of the ground. Tschebotarioff and Ward (1948) record a case of exessive vibration at a pumping station where the compressor foundation blocks were carried on timber friction piles, 18 m long, driven in a deep deposit of silt and clay with water level less than 0.3 m below ground surface. Resonance in the compressor–foundation–soil system occurred at speeds of about 30 rpm. Remedial measures consisted of lowering the ground water level by about 0.75 m by continuous pumping from a drainage system installed around the structure, thus increasing the effective weight of the foundation. Thornton (1951) cites the case of a turbo-generator which, after operating for some years without excessive vibration, was more or less suddenly subjected to a serious state or resonance.

This phenomenon was found to be due to the lowering of the level of the phreatic surface. The use of injections to alter the natural frequency of foundation soils has been discussed by Gnaedinger (1961). A study of the effects of soil stabilization techniques on the dynamic properties of soils has been presented by Au and Chae (1980). Driven piles unconnected to the foundation block may be used to compact the soil beneath the foundation, thus improving bearing capacity and elastic properties. The case of machines founded on a firm crust of soil overlying softer deposits has been investigated by Gazetas (1981).

The problem of vibration reduction sometimes occurs in reverse when an structure or a machine needs protecting from vibrations emanating from a external source. This involves passive isolation as opposed to the active isolation of an exciting machine. Crockett and Hammond (1947) have reported a small building housing delicate machinery which was constructed in the form of a box resting on a number of cork pads in order to reduce the amplitude of vibrations generated by an external source. Precision grinding machines in proximity to a railway at Colchester are mounted on rubber mats to isolate them from vibrations generated by passing trains. The use of piles as isolation barriers has been suggested by Richart *et al.* (1970) and some preliminary investigations have been made by Woods *et al.* (1974) and Liao and Sangrey (1978). Rocking vibrations of footings have been studied by Sreekantiah (1982). Foundations for equipment and machinery are treated in Special Publication 78 (ACI, 1982).

4.13.4 Theory of vibrations

Most of the methods of analysis of foundations reduce to the basic mechanical system of a mass supported on a spring with damping provided by a dashpot, as shown in Fig. 4.25. The effect of the machine is to impose a forced vibration on the system which is modified by the natural vibrations. The theory of such vibrations is fully treated in a number of text books; the basic formulae for a single degree of freedom are given below without derivation.

For such a system with zero damping, if the spring constant is k N m^{-1} and the

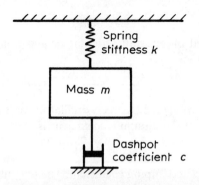

Figure 4.25 Mass with spring and dashpot assembly.

oscillating mass is m kg, the undamped natural circular frequency is

$$\omega_n = \left(\frac{k}{m}\right)^{1/2} \text{ rad sec}^{-1},$$

and the undamped frequency is

$$f_n = \frac{1}{2\pi}\left(\frac{k}{m}\right)^{1/2} \text{ Hz(cycle sec}^{-1}).$$

If the system is operated on by a cyclical force,

$$Q = Q_0 \sin \omega t,$$

where Q_0 is the maximum value of the force and

$$\omega = 2\pi f,$$

where f is the frequency of the applied force in Hz. The amplitude of the vibrations is given by

$$a = \frac{Q_0}{m(\omega_n^2 - \omega^2)}.$$

Resonance occurs when $\omega_n = \omega$, under which condition the amplitude will theoretically become infinite if allowed to continue although breakdown is more likely to occur.

Damping can be used to control the amplitude of the oscillations. Assuming viscous damping, in which the damping force is proportional to the velocity, the usual assumption for foundations, then the amplitude becomes

$$a = \frac{Q_0}{[c^2\omega^2 + m^2(\omega_n^2 - \omega^2)^2]^{0.5}},$$

where c is the damping coefficient expressed as a force per unit velocity (N sec m^{-1}).

If a static force Q_0 is applied to the spring, the displacement is

$$z = Q_0/k.$$

The ratio a/z is called the magnification factor and is given by

$$M = \frac{k}{[c^2\omega^2 + m^2(\omega_n^2 - \omega^2)^2]^{0.5}}.$$

If viscous damping is applied to the freely oscillating mass it can be shown that no oscillation will occur if the damping coefficient is

$$c_c = 2(km)^{1/2}.$$

This is called the critical damping coefficient. The magnification factor can now be expressed solely in terms of the frequency ratio ω/ω_n and the damping ratio

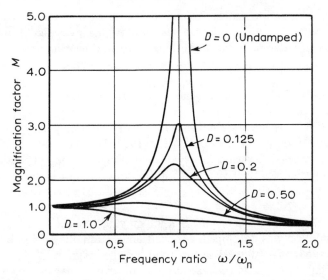

Figure 4.26 Variation of magnification factor with frequency ratio and damping ratio.

$D = c/c_c$:

$$M = \frac{1}{\left\{\left(\frac{2D\omega}{\omega_n}\right)^2 + \left[1 - \left(\frac{\omega}{\omega_n}\right)^2\right]^2\right\}^{1/2}}.$$

In Fig. 4.26 the value of magnification factor is plotted against the frequency ratio ω/ω_n for several values of the damping ratio $D = c/c_c$. These curves show that the magnification is not large for frequency ratios between 0.5 and 1.5 and that one of the effects of damping is to reduce the frequency ratio at which maximum amplitude occurs to less than 1.0.

4.13.5 Analysis of oscillating systems

A freely supported body has six degrees of freedom and the corresponding modes of vibration involve translational and rotational oscillations in relation to the three major axes. Frequently two or more modes of vibration occur at the same time but it is sometimes possible to distinguish one mode as being more significant than the others. Even with one mode of oscillation it may be found that there are several combinations of frequency and amplitude operating at the same time which may or may not be in harmonic relationship. Nevertheless it is common practice to reduce the forced vibrations to simple harmonic motion.

The machine is usually bolted to a foundation block which serves the dual purpose of spreading the loads on the ground and increasing the mass of the machine. The foundation may be varied in overall size to change the stress on the soil or to increase the total oscillating mass or it may have cavities formed within it to produce a better relationship between mass and bearing area on the soil.

The soil also forms part of the vibrating system and its characteristics contribute to the natural frequency of that system. The method by which the soil properties are included in the system is the principal difference between the approaches of different authorities.

The DEGEBO organization in Germany carried out many tests in the period 1928 to 1936 (Hertwig *et al.*, 1933; Lorenz, 1934; Hertwig and Lorenz, 1935). Their results satisfied the basic theory of oscillations outlined in Section 4.13.4 but it was not possible to derive a single spring constant for a given soil since the resonant frequency varied with changes in base plate area and changes in exciting force.

Subsequently, the concept of a mass of soil moving in phase with the machines and its foundation was developed. Crockett and Hammond (1948), Rao (1961) and Pauw (1953) have all used this method but the magnitude of the in-phase mass is difficult to estimate because of the number of variables which affect it. Richart (1975) has presented a summary of work on the effects of soil properties on soil-structure interaction.

The Modulus of Dynamic Subgrade reaction method depends on the use of a modulus k'_z (N m^{-3}) such that the spring constant used in calculating the natural frequency is $k_z = k'_z A$, where A is the area of foundation. This modulus can be determined by *in situ* tests either by using a vibrator or by repeated static loading and unloading tests. If the static tests are employed, the repetition of loading and unloading is essential to enable separation of the elastic deformation from other effects (Barkan, 1962). When interpreting the results of such tests, allowance must be made for the relative area of the real or prototype and the test or model foundation. Thus

$$k'_2 = k'_1 \left(\frac{A_1}{A_2}\right)^{1/2}$$

where k'_1 and A_1 relate to the model and k'_2 and A_2 relate to the prototype. This is reasonably reliable if A_1 and A_2 are fairly similar but, if A_2 is much larger than A_1, k'_2 will be underestimated. Barkan (1962) has published a table in which k'_2 for base areas of $10 \, \text{m}^2$ is related to the safe stress on the soil and these data can be used if more reliable information is not available.

The use of the elastic half-space is undoubtedly the most elegant procedure; Reissner (1936) developed the analytical solution which has since been improved and extended by others.

Lysmer and Richart (1969) introduced a modified dimensionless mass ratio

$$B_z = \frac{(1-v)}{4} \frac{m}{\rho r_0^3},$$

and if the spring constant is taken equal to the static value,

$$k_z = \frac{4Gr_0}{1-v}.$$

The best fit for the damping coefficient is then given by

$$C_z = \frac{3.4r_0^2}{1-v}(\rho G)^{1/2},$$

and the damping ratio is thus

$$D = \frac{C_z}{C_0} = \frac{3.4r_0^2(\rho G)^{1/2}}{1-v} \cdot \frac{1}{2}\left(\frac{1-v}{4Gr_0 m}\right)^{1/2} = \frac{0.425}{B_z},$$

which gives good answers for $B_z \gtrsim 1$.

The use of these equations is illustrated in Example 4.4. It should be noted that an equivalent circular base is used, the radius being chosen to give the same value of A, I or J as the rectangular base considered.

EXAMPLE 4.4 DESIGN OF A MACHINE FOUNDATION

Data
Mass of machine, 5 tonne.
Minimum plan dimensions of foundation, 2.4 m × 1.2 m.
Concrete density, $2.4 \, \text{kg m}^{-3}$.
Top of foundation block to be 600 mm above ground level.
Maximum vertical exciting force 2.5 kN at 600 cycles per min, simple harmonic motion.
Allowable single amplitude = 0.04 mm.
Allowable net bearing stress on soil, $100 \, \text{kN m}^{-2}$.
Modulus of dynamic subgrade reaction, $k_z' = 20 \, \text{MN m}^{-3}$ for a base area of $10 \, \text{m}^2$.
Soil density, $\rho = 1800 \, \text{kg m}^{-3}$.
Shear modulus of soil, $G = 17.0 \, \text{MN m}^{-2}$.
Poisson's ratio of soil, $v = 0.4$.
Assume dimensions of foundation block = 2.4 m × 1.2 m × 1.2 m.
Mass of block = 2.4 × 1.2 × 1.2 × 2.4 = 8.29 tonne.
Mass of excavation = 2.4 × 1.2 × 0.6 × 1.8 = 3.11 tonne.
Mass of machine and foundation = 5.0 + 8.29 = 13.29 tonne.

Gross static stress on soil $= \dfrac{13.29 \times 9.81}{2.4 \times 1.2} = 45.27 \, \text{kN m}^{-2}$.

Net static stress on soil $= \dfrac{(13.29 - 3.11) \times 9.81}{2.4 \times 1.2} = 34.68 \, \text{kN m}^{-2}$.

The influence of dynamic load on settlement may be greater than that of static load and, when calculating soil stress, the part due to dynamic load should be increased by a suitable factor.
Additional soil stress due to exciting force

$$= \frac{2.5}{2.4 \times 1.2} = 0.87 \, \text{kN m}^{-2}.$$

Three times additional soil stress plus net static soil stress

$$= 34.68 + 3 \times 0.87 = 37.29 \, \text{kN} \, \text{m}^{-2},$$

which is less than the allowable $100 \, \text{kN} \, \text{m}^{-2}$.

The dynamic calculations will be carried out by two methods:

The Modulus of Dynamic Subgrade reaction method

Adjust k'_z for new base area

$$k'_z = 20 \left(\frac{10}{2.88} \right)^{1/2} = 37.26 \, \text{MN} \, \text{m}^{-3}$$

$$k_z = 37.26 \times 2.4 \times 1.2 \, \text{MN} \, \text{m}^{-1}$$

$$\omega_n = \left(\frac{37.26 \times 2.4 \times 1.2 \times 10^3}{13.29} \right)^{1/2} = 89.9 \, \text{rad} \, \text{sec}^{-1}$$

$$f_n = \frac{89.9}{2\pi} = 14.3 \, \text{Hz} = 858 \, \text{cycles} \, \text{min}^{-1}.$$

Static deflection due to exciting force

$$z = \frac{2.5 \times 10^3}{37.26 \times 2.88 \times 10^6} = 0.023 \times 10^{-3} \, \text{m}.$$

For this method there is no damping factor and thus

$$\frac{\omega}{\omega_n} = \frac{10}{14.3} = 0.699$$

$$M = \frac{1}{1 - (0.699)^2} = 1.96.$$

Amplitude $a = 1.96 \times 0.023 = 0.045 \, \text{mm}$.
Reduce the mass of the base block by introducing 30% voids.

$$\text{Mass of block} = 8.29 \times 0.7 = 5.80 \, \text{tonne}$$

$$\omega_n = \left(\frac{37.26 \times 2.4 \times 1.2 \times 10^3}{5.80 + 5.0} \right)^{1/2} = 99.7 \, \text{rad} \, \text{sec}^{-1}$$

$$f_n = \frac{99.7}{2\pi} = 15.87 \, \text{Hz} = 952 \, \text{cycle} \, \text{min}^{-1}$$

$$\frac{\omega}{\omega_n} = \frac{10}{15.87} = 0.63$$

$$M = \frac{1}{1 - (0.630)^2} = 1.66$$

Amplitude $a = 1.66 \times 0.023 = 0.038 \, \text{mm}$.

The elastic half-space method

The radius of a circular foundation with the same area as the rectangular foundation is

$$r_0 = \left(\frac{2.4 \times 1.2}{\pi}\right)^{1/2} = 0.96 \, \text{m}$$

$$k_z = \frac{4Gr_0}{1-v}$$

$$= \frac{4 \times 17.0 \times 10^6 \times 0.96}{1 - 0.4} = 108.8 \times 10^6 \, \text{N m}^{-1}.$$

$$\omega_n = \left(\frac{108.8 \times 10^6}{13.29 \times 10^3}\right)^{1/2} = 90.5 \, \text{rad sec}^{-1}$$

$$f_n = \frac{90.5}{2\pi} = 14.4 \, \text{Hz} = 864 \, \text{cycles min}^{-1}.$$

The dimensionless mass ratio

$$B_z = \frac{(1-v)}{4} \frac{m}{\rho r_0^3}$$

$$= \frac{(1 - 0.4) \times 13.29 \times 10^3}{4 \times 1800 \times 0.96^3} = 1.25$$

$$D = \frac{0.425}{\sqrt{B_z}} = \frac{0.425}{\sqrt{1.25}} = 0.38$$

$$\frac{\omega}{\omega_n} = \frac{10}{14.4} = 0.694$$

$$M = \frac{1}{[(2 \times 0.38 \times 0.694)^2 + (1 - 0.694^2)^2]^{1/2}} = 1.35.$$

Static deflection due to exciting force

$$z = \frac{2.5 \times 10^3}{108.8 \times 10^6} = 0.023 \times 10^{-3} \, \text{m}.$$

Amplitude $a = 0.023 \times 1.35 = 0.031$ mm.
For foundation block reduced by 30% in weight

$$\omega_n = \left(\frac{108.8 \times 10^6}{10.8 \times 10^3}\right)^{1/2} = 100.4 \, \text{rad sec}^{-1}$$

$$f_n = \frac{100.4}{2\pi} = 15.98 \, \text{Hz} = 959 \, \text{cycle min}^{-1}$$

$$B_z = \frac{(1 - 0.4) \times 10.8 \times 10^3}{4 \times 1800 \times 0.96^3} = 1.02$$

$$D = \frac{0.425}{\sqrt{1.02}} = 0.42$$

$$\frac{\omega}{\omega_n} = \frac{10}{15.98} = 0.63$$

$$M = \frac{1}{[(2 \times 0.42 \times 0.63)^2 + (1 - 0.63^2)^2]^{1/2}} = 1.24.$$

Amplitude $a = 0.023 \times 1.24 = 0.29$ mm.

This example shows the effect on the amplitude of vibrations of the natural soil damping which is not included in the Dynamic Subgrade reaction method. It also shows how in this case reducing the weight of the block increases the natural frequency, thus making it less likely that resonance will occur.

Chapter 5

COMBINED FOUNDATIONS

A.P.S. Selvadurai

INTRODUCTION

The term *combined foundations* covers a variety of structural foundations, including continuous footings, grids, mats and rafts. The conventional designs of these foundations are based on semi-empirical methods which are not wholly satisfactory. A rational design approach should take into consideration the influence of factors such as flexibility of the foundation, flexibility of the superstructure or the manner in which this flexibility is transmitted to the foundation, time independent and time dependent deformation characteristics of the soil medium, construction sequences of the foundation and the superstructure and the nature of the external loading. However, there are as yet no efficient techniques for the design of combined foundations which take into consideration the influence of all the above factors. The design procedure recommended by the ACI Committee 436 (1966) is perhaps the most significant advance in this direction.

This chapter presents a brief introduction to the theoretical knowledge that has been advanced, in recent years, in the field of soil–foundation interaction. In particular, attention is restricted to the elastic behaviour of the foundation and the supporting soil medium. To effect an analysis of the soil–foundation interaction problem it is convenient to idealize the deformation characteristics of the soil. Generally these idealized models of elastic soil behaviour can be classified in two groups: (i) idealized behaviour of the soil mass (e.g. mathematical or mechanical models such as the Winkler model, the two parameter model, etc.) and (ii) idealized behaviour of the soil (e.g. continuum behaviour of the soil such as the isotropic elastic half-space, etc.). In the former type, the parameters describing the idealized soil model are not intrinsic material properties of the soil. However, in particular circumstances they may be related to the intrinsic material properties of the soil or soil configuration. Vlazov and Leontiev's (1966) interpretation of the constants describing the two parameter model in terms of the elastic constants and thickness of a constrained isotropic elastic layer and Gibson's (1967) interpretation of the modulus of subgrade reaction in terms of the linear variation of the shear modulus of an incompressible non-homogeneous

405

elastic half-space are two such typical examples. It must be emphasized that these idealizations of soil behaviour are, by definition, only approximations to the behaviour of real soils and soil masses, valid under limited conditions of operation. None the less they are particularly helpful in reducing the analytical rigour expended in the treatment of soil–foundation interaction problems and at the same time retaining essential features of the soil behaviour.

When flexibility of the foundation and the idealized deformation characteristics of the soil medium are taken into consideration, the analysis of the interaction problem essentially reduces to the determination of the contact stresses at the soil–foundation interface for any given external loading condition. In this chapter is presented a brief exposition of particular analytical techniques that can be adopted for the elastic analysis of soil–foundation interaction of beams and plates of infinite and finite extent, drawing attention wherever possible to significant developments. Limitations of space prevented the inclusion of a variety of examples. The final section in this chapter is devoted to the design and construction techniques adopted for mat, raft and tank foundations.

5.1 GENERAL CONSIDERATIONS IN THE ANALYSIS OF STABILITY AND SETTLEMENT OF COMBINED FOUNDATIONS

When structures such as raft foundations and combined footings are founded on soil media, it is important that preliminary investigations should be made to assess the factor of safety against a bearing failure of the soil and the amount of settlement to be expected. The former is a function of the strength parameters of the soil and the latter depends upon the stress–deformation–time characteristics of the soil medium. In order to carry out such an investigation it is necessary to establish the manner in which external loads are transmitted to the soil–foundation interface as contact stresses. The exact distribution of these contact stresses, however, cannot be determined without solving the complete soil–foundation interaction problem, taking into account the flexibility of the foundation. The theoretical results given by Borowicka (1936) for the contact stress distribution beneath uniformly loaded circular and strip flexible foundations resting on an elastic medium are shown in Fig. 5.1. It is clear that the relative flexibility of the foundation–elastic medium system considerably alters the contact stress distribution. If, for example, the soil possesses definite yield characteristics of an ideal elastic-plastic material (Hill, 1950) then failure will occur at the zones of high concentration of stress at the boundary of the foundation, even though there may not be any danger of a complete bearing failure.

Schultze (1961) has examined the development of plastic zones beneath a rigid foundation by combining the contact stress distributions obtained for the elastic and plastic cases. It is found that the contact stress distribution varies with the factor of safety against bearing failure. When the factor of safety is unity, the contact stress has a parabolic shape. For values of the factor of safety greater than

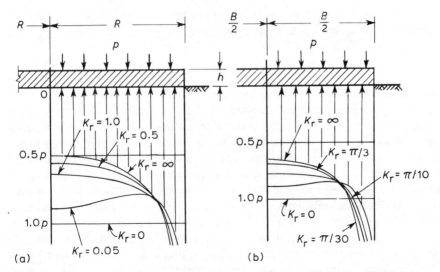

Figure 5.1 Stress distribution under footings of varying rigidity. (a) Circular footing, (b) infinite strip footing. (After Borowicka, 1936, 1938.)

e.g.
$$K_r = \left(\frac{1 - v_s^2}{1 = v_f^2}\right)\left(\frac{E_f}{6E_s}\right)\left(\frac{h}{R}\right)^3$$

where subscripts f and s denote foundations and soil, respectively.

three, the stress distribution is saddle-shaped, similar to that obtained for the elastic case (Fig. 5.1). Measurements of contact stresses under rigid foundations have confirmed these theoretical findings, except for the case of shallow foundations in cohesionless soils. Also, at conditions of complete plastic failure, the contact stress distribution at the base of rigid footings is generally non-uniform (Terzaghi, 1943; Sokolovski, 1965; Larkin, 1968; Chen, 1975).

The mode of bearing failure and the magnitude of ultimate settlements in a combined foundation can be significantly influenced by the flexibility of the foundation. There are, however, no simple ways of incorporating the flexibility of the foundation in conventional bearing capacity or settlement calculations. Therefore, from the point of view of a preliminary assessment of bearing capacity and settlement, it is prudent to consider the foundation to be either completely flexible or completely rigid. Continuous footings can be considered as being perfectly rigid with a uniform contact stress distribution across their width. In raft foundations the contact stress distribution can vary within wide limits depending upon the magnitude and distribution of external loading, relative stiffness of the raft to that of the superstructure and the size of the raft in relation to the thickness of underlying compressible strata. Some general comments on the various aspects of the analysis of bearing capacity and settlement of combined foundations are presented in the following notes. The methods of estimating settlement and bearing capacity discussed in Chapters 2 and 3 are assumed to be generally applicable to combined foundations.

5.1.1 Bearing capacity

An assessment of the bearing capacity of combined foundations can be approached via two principal methods. The first examines, on the basis of the classical theory of elasticity, the state of stress in the soil medium due to the external loads and compares the resulting maximum shear stress with the shearing strength of the soil medium. The ultimate value of the external loading is then regarded as that load which causes a shear stress equal to the shear strength of the soil at the most unfavourable location. An improvement on this particular method is based on consideration of the plastic zones developed in the soil due to overstressing (Frohlich, 1934). Techniques of this type suffer from the drawback that the behaviour of the soil when stressed close to its yield value cannot be adequately described by a state of stress obtained on the basis of an elastic analysis; thus the theoretical boundary between the elastic and plastic zones is somewhat fictitious. Also, the stresses within both elastic and plastic zones are calculated on the basis of the theory of elasticity. None the less, this method serves as a useful preliminary check for assessing the bearing capacity of the soil medium.

An extension of this type of analysis assumes that owing to the yielding of the soil, according to a specified yield criterion, plastic regions are developed in the soil and these regions satisfy stress–strain relations generally different from those applicable to the elastic regions. Such elastic–plastic methods of stress analysis have received considerable attention owing to the development of numerical methods of stress analysis, such as relaxation and finite element techniques. Examples of the application of relaxation methods to elastic–plastic problems are given by Allen and Southwell (1950), Jacobs (1950) and Spencer (1965) and those related to finite element techniques are summarized in Zienkiewicz and Holister (1965), Zienkiewicz (1971), Gudehus (1977) and Desai and Christian (1977). The development of plastic regions in an elastic–plastic half-plane has been studied, particularly in relation to bearing capacity problems, by Ang and Harper (1964), Christian (1968) and Hoeg et al. (1968). Further discussions of the occurrence of elastic–plastic zones under rigid strip foundations are given by Terzaghi (1943) and Schultze (1961), Sedykh (1964), Gorbunov-Posadov (1965), Brown (1968) and Ho and Lopes (1969).

The second approach employs the mathematical theory for ideal plastic solids (Hill, 1950). If the soil behaves as an ideal compressible plastic medium, by loading beyond the elastic limit, the medium passes into a partly or entirely plastic state. As a result slip interfaces occur in the medium leading to contained or complete plastic flow. The external load required to produce complete shear failure is usually referred to as the bearing capacity of the soil. There exist several analytical results from which the ultimate bearing capacity of a foundation resting on soil can be calculated. However, from a design point of view, the following factors should be taken into account:

(a) Foundation type (rigid or flexible).
(b) Foundation shape (strip, rectangular or circular).
(c) Foundation location (surface or embedded).
(d) Foundation interface (rough, smooth, or bonded).
(e) Character of the loading (inclined or vertical, static or dynamic).
(f) Type of soil and yield criteria (granular, cohesive or granular–cohesive).
(g) Weight of soil (saturated or dry).

To produce a generalized analytical solution for the ultimate bearing capacity of a foundation embodying all these factors is, of course, a futile task. Most calculations for the assessment of the ultimate bearing capacity are usually based on the theory of plasticity combined with some simplifying assumptions. The shape of the rupture or sliding surface is first assumed and by examining the forces acting in the limit state of equilibrium the ultimate load can be determined. In this method, the bearing capacity is usually expressed in terms of bearing capacity factors which indicate the effect of the width and the depth of the combined foundation together with the effects of cohesive and frictional properties and the weight of the soil medium. Extensive treatment of the subject of the ultimate bearing capacity of foundations is contained in the works of Terzaghi (1943), Meyerhof (1951, 1963), Sokolovski (1965) and Harr (1966).

5.1.2 Settlement

The estimation of settlement is also an important problem in design of combined foundations. The movements which can be tolerated by different structural systems depend upon many factors, including the type, size, location and intended use of the structure and the pattern, rate, cause and source of settlement. A discussion of the settlements that can be tolerated by various structures and materials form the subject of the papers by Skempton *et al.* (1955), Skempton and Macdonald (1956) and Skempton and Bjerrum (1957). References to further work are also given by Sowers (1962), Bjerrum (1963), D'Appolonia *et al.* (1968, 1971), Davis and Poulos (1968, 1972), Kerisel and Quatre (1968*a,b*) and in the review articles by Meyerhof (1965) and Seed (1965).

Settlement of foundations in sands and gravels not underlain by clayey soils are usually complete or nearly complete at the end of the construction period. This immediate settlement in granular media can be assessed on the basis of either (a) the theory of elasticity or (b) empirical methods. The use of the theory of elasticity is still the most universal despite the fact that at working loads granular soils exhibit a marked departure from perfectly elastic continua. Various modifications, to the classical isotropic elastic continuum, such as (a) anisotropy, (b) non-homogeneity, (c) uniform and non-uniform layering, (d) physical non-linearity, etc., can be employed to further refine the settlement analysis (Feda,

1963). The common empirical method employs results obtained for the relative density of the granular medium from standard penetration tests. The correlation of the driving resistance with settlement of plate loading tests and of large scale structures forms the basis for the empirical settlement computations. Details of these empirical methods are found in Peck *et al.* (1953), Meyerhof (1965), D'Appolonia *et al.* (1968), Terzaghi and Peck (1969), Schmertmann (1970), Sutherland (1975) and Burland (1977).

It is generally accepted that the settlement of a foundation in clay is composed of three different components, namely: the immediate settlement, the primary consolidation settlement and the secondary settlement. The immediate settlement is commonly assumed to occur without any change in volume of the soil and hence it is mainly due to shear distortion of the clay medium. This settlement is estimated by using the theory of elasticity. The consolidation settlement is the gradual expulsion of the pore-water from the clay. The magnitude of the consolidation settlement is generally calculated on the basis of a one-dimensional oedometer test. In the case of foundations constructed on thick clay strata, the simple one-dimensional approach may not accurately represent the consolidation process. Here, it is advisable to estimate the consolidation settlement by using a complete three-dimensional analysis of the consolidation problem (Gibson and McNamee, 1957; McNamee and Gibson, 1960). In secondary compression, the settlement is assumed to be due to creep processes in the structure of the clay medium. One-dimensional viscoelastic models are generally used to explain secondary compression phenomena (Taylor, 1948; Gibson and Lo, 1961; Schiffman, 1963; Schiffman *et al.*, 1964). Estimates for secondary settlement can be obtained, to a reasonable degree of accuracy, by employing the elementary rheological model approach. Certain unsaturated soils and soft rocks may also exhibit creep effects in the absence of a process similar to consolidation. The theory of viscoelasticity can be effectively adopted for the analysis of such creep effects (see, for example, Kravtchenko and Sirieys, 1966; Jaeger and Cook, 1976; Selvadurai, 1978).

5.2 IDEALIZED SOIL RESPONSE FOR THE INTERACTION ANALYSIS OF COMBINED FOUNDATIONS

The evaluation of the response of soil media to external loads constitutes a factor of fundamental importance to the analysis of soil–foundation interaction problems. Such an evaluation can be made only from a knowledge of the complete stress–strain characteristics of the soil. The stress–strain relations are the mathematical description of the mechanical properties of the soil – its constitutive equations. A complete stress–strain relationship for a soil will furnish, at least in theory, the stresses and strains in a soil medium at any particular time under any given loading condition. Owing to the variety of soils and soil conditions that can be encountered in engineering practice, it seems unlikely that generalized stress–strain relations will be developed to fulfill the requirements of every type of soil behaviour, especially in relation to foundation design. This limitation is clearly illustrated by the fact that in the history of development of soil mechanics, elastic, consolidation, creep and failure processes in soils have been analysed by completely separate theories of material behaviour. Any generalized stress–strain relationships that are developed, although useful from a research point of view, can extend their applicability to only very simple states of stress encountered in uniaxial, plane-strain or triaxial test conditions (see, for example, Gudehus, 1977). The analytical or numerical solution of practical boundary value problems using these generalized stress–strain relations presents formidable difficulties.

The inherent complexity in the behaviour of real soils has led to the development of many idealized models of soil behaviour, especially for the analysis of soil–foundation interaction problems. The classical theories of elasticity and plasticity are two such idealizations commonly employed in the analysis of problems in soil mechanics. The idealized foundation models or, for that matter, the generalized stress–strain relations for soils, are not exact descriptions of even the gross physical properties of real soil media. The best that can be said of any such model of soil response or material behaviour is that it provides a useful description of certain features of soil-like materials under limited conditions of operation. The mathematical or physical idealization of soil behaviour is particularly instrumental in reducing the analytical rigour expended in the solution of many complex boundary value problems in soil mechanics.

Idealized soil models prove to be particularly useful in the analysis of soil–foundation interaction problems. The relevant choice of a particular form of idealized behaviour of the soil for a soil–foundation interaction problem depends primarily on the type of soil and soil conditions, the type of foundation and the nature of external loading. In addition to these, due consideration should be given to factors such as the method of construction, the life span and purpose of the structure, economical considerations and engineering judgement.

In this section is presented a brief exposition of some idealized models which, from the point of view of soil–foundation interaction problems, take into account the time-independent behaviour of soil media. The response of each idealized model is typified by the surface deflection it experiences under the application of an external system of forces. These surface deflections in general represent the displacement characteristics of the upper boundary of the soil which is in contact with the foundation, or the soil–foundation interface, and forms a vital part of the information necessary in soil–foundation interaction analysis. It must be emphasized that these idealized models are primarily intended to model the response of soil media and not the response of elements within the soil medium.

5.2.1 The Winkler model

For the idealized model of soil media proposed by Winkler (1867) it is assumed that the deflection (w) at any point on the surface of the soil medium is directly

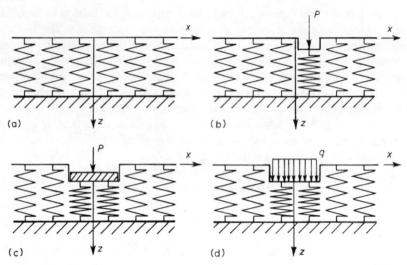

Figure 5.2 Deformations of the Winkler model, shown unloaded in (a), due to a concentrated force (b), rigid load (c) and uniform flexible load (d).

proportional to the contact stress pressure (q) at that point and independent of contact stress at other points; that is

$$q(x, y) = k_s w(x, y), \tag{5.1}$$

where k_s is the modulus of subgrade reaction with units of stress per unit length. There are indications in the literature that this assumption is already to be found in the works of Euler, Fuss, Bubnov and Zimmermann (Korenev, 1960; Hetényi, 1966; Selvadurai, 1979). Equation 5.1 is usually termed the response function or the kernel function for the Winkler model. Physically Winkler's idealization of the soil medium consists of a system of mutually independent spring elements. One important feature of this soil model is that the displacement occurs immediately under the loaded area and outside this region the displacements are zero (Fig. 5.2). Also it can be seen that for this particular model the displacements of a loaded region will be constant whether the soil is subjected to an infinitely rigid load or a uniform flexible load. Apart from soil–foundation interaction, there are many engineering problems for which this model represents an accurate idealization of actual operating conditions. In problems associated with floating bridges and floating ice sheets in bending, the relationship, $q = k_s w$, represents a simple consequence of Archimede's principle. Also it can be useful in the analysis of grillage type structures.

5.2.2 The elastic solid model

The inability of the Winkler model to deform outside the loaded area restricts its applicability to soil media which possess the slightest amount of cohesion or transmissibility of applied forces. It is common experience that, in the case of soil media, surface deflections occur not only immediately under the loaded region but also within certain limited zones outside the loaded region. In attempts to account for this behaviour, soil media have often been idealized as three-dimensional continuous elastic solids or elastic continua. Generally, the displacements and stresses in such media remain continuous under the action of external force systems. The initial impetus for the continuum representation of soil media stems from the work of Boussinesq (1885), who analysed the problem of a semi-infinite homogeneous isotropic linear elastic solid subjected to a concentrated force which acts normal to the plane boundary. This basic solution can be used to obtain the response function for the three-dimensional elastic soil medium. For example, in the case of a three-dimensional plate resting on an elastic half-space the Boussinesq solution for the concentrated load on the surface of the elastic half-space serves to obtain a singular integro-differential equation for the deflection of the plate. A similar technique can be employed for the analysis of the two-dimensional beam or plate probem by making use of Flamant's solution (Timoshenko and Goodier, 1970) for a line load acting on the surface of a half-plane.

In general, the application of the continuum theory of classical elasticity to soil–foundation interaction presents a complex mathematical problem. A number of solutions to boundary value problems of particular interest to soil–foundation interaction and references to further work are given by Gorbunov-Posadov (1941, 1949), Korenev (1954, 1960), Galin (1961), Lur'e (1964), Vlazov and Leontiev (1966), Harr (1966), Hetényi (1966) and Selvadurai (1979). In recent years the elastic half-space and the half-plane models have undergone various generalizations and modifications, including variations of material properties such as non-homogeneity (Korenev, 1957; Olszak, 1959; Zaretsky and Tsytovich, 1965; Gibson, 1967; Golecki and Knops, 1969; Carrier and Christian, 1973), anisotropy (Savin, 1940; Lekhnitskii, 1960; Barden, 1963b; Selvadurai and Moutafis, 1975) and other spatial variations of properties such as layered, structured or porous media (Westergaard, 1938; Burmister, 1945; Lemcoe, 1960; Król, 1964; Harr, 1966; Selvadurai, 1973). The relevance of elastic anisotropy and non-homogeneity to soil mechanics problems is further discussed by Poulos and Davis (1974) and Selvadurai (1979).

5.2.3 Two-parameter models

The inherent deficiency of the Winkler model in depicting the continuous behaviour of real soils and the mathematical complexities of the elastic continuum models has led to the development of many other simple soil response models. These models possess the characteristic features of a continuous elastic solid (Kerr, 1964; Hetényi, 1966). The term 'two-parameter' signifies the fact that the model is defined by two independent elastic constants. The development of these two-parameter soil models has been approached along two distinct lines.

The first proceeds from the discontinuous Winkler model and eliminates its discontinuous behaviour by providing interaction between the individual spring elements. Such physical models of soil behaviour have been proposed by Filonenko-Borodich (1940), Hetényi (1946), Pasternak (1954) and Kerr (1964) where interaction between the spring elements is provided by either elastic membranes, elastic beams or elastic layers capable of purely shearing deformation. The second approach starts from the elastic continuum model and introduces constraints or simplifying assumptions with respect to the distribution of displacements and stresses. The soil models proposed by Reissner (1958) and Vlazov and Leontiev (1966) take into consideration such simplifications.

(a) Filonenko-Borodich model

The model proposed by Filonenko-Borodich (1940) acquires continuity between the individual spring elements in the Winkler model by connecting them to a thin membrane under a constant tension T(Fig. 5.3). By considering the equilibrium of the membrane–spring system it can be shown that for three-dimensional problems (e.g. rectangular or circular foundations) the surface deflection of the soil medium due to a normal stress q is given by

$$q(x, y) = k_s w(x, y) - T\nabla^2 w(x, y), \qquad (5.2)$$

where
$$\nabla^2 = \frac{\partial^2}{\partial x^2} + \frac{\partial^2}{\partial y^2},$$

is Laplace's differential operator in rectangular cartesian co-ordinates. In the case

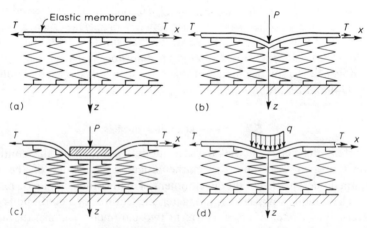

Figure 5.3 Deformations of the Filonenko–Borodich model (a), due to a concentrated force (b), rigid load (c), and uniform flexible load (d).

of two-dimensional problems (e.g. a strip foundation) Equation 5.2 reduces to

$$q(x) = k_s w(x) - T\frac{\mathrm{d}^2 w}{\mathrm{d}x^2}.$$ (5.3)

The two elastic constants necessary to characterize the soil model are k_s and T. Typical examples of surface deflection profiles of this particular model due to concentrated, flexible and rigid external loads are shown in Fig. 5.3.

(b) Hetényi model

In the model proposed by Hetényi (1946) interaction between the independent spring elements incorporates an elastic plate in the case of three-dimensional problems or an elastic beam in the case of two-dimensional problems. The response function for this model is given by

$$q(x, y) = k_s w(x, y) - D\nabla^4 w(x, y),$$ (5.4)

where $D[=E^*h^3/12(1 - v^{*2})]$, is the flexural rigidity of the plate. For two-dimensional problems, Equation 5.4 reduces to

$$q(x) = k_s w(x) - \frac{E^* h^3}{12}\frac{\mathrm{d}^4 w(x)}{\mathrm{d}x^4}.$$ (5.5)

(c) Pasternak model

For the model proposed by Pasternak (1954) it is assumed that there is shear interaction between the spring elements. This can be accomplished by connecting the spring elements to a layer of incompressible vertical elements which deform in transverse shear only (Fig. 5.4). The deformations and forces maintaining equilibrium in the shear layer are shown in Fig. 5.4. By assuming that the shear layer is isotropic in the x, y plane, with shear moduli $G_x = G_y = G_p$,

$$\tau_{xz} = G_p\gamma_{xz} = G_p\frac{\partial w}{\partial x}; \tau_{yz} = G_p\gamma_{yz} = G_p\frac{\partial w}{\partial y}.$$ (5.6)

The total shear forces per unit length of the shear layer are

$$N_x = \int_0^1 \tau_{xz}\,\mathrm{d}z = G_p\frac{\partial w}{\partial x}; N_y = \int_0^1 \tau_{yz}\,\mathrm{d}z = G_p\frac{\partial w}{\partial y}.$$ (5.7)

For force equilibrium in the z direction:

$$\frac{\partial N_x}{\partial x} + \frac{\partial N_y}{\partial y} + q - r = 0.$$ (5.8)

Using the condition $r = k_s w$, and introducing Equation 5.6 into Equation 5.8:

$$q(x, y) = k_s w(x, y) - G_p\nabla^2 w(x, y).$$ (5.9)

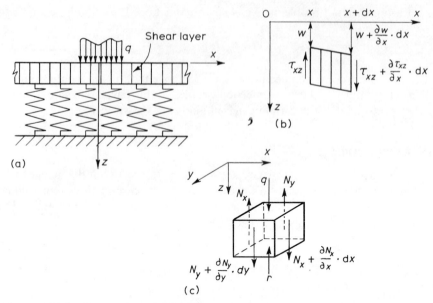

Figure 5.4 The Pasternak model. (b) Stresses in the shear layer; (c) forces acting on the shear layer.

It can be seen that Equation 5.9 is identical to Equation 5.2 if T is replaced by G_p. Thus the surface deflection profiles for this model are very similar to those obtained for the Filonenko-Borodich model. It is of interest to note that, with the two-parameter models considered so far, the Winkler case can be recovered as a limiting case, as T, D and G_p tend to zero.

(d) Vlazov model

The model of soil response proposed by Vlazov (1949) presents an example of the second type of two-parameter elastic model which is derived by introducing displacement constraints that simplify the basic equations for a linear elastic isotropic continuum. Vlazov's approach to the formulation of the soil model is based on the application of a variational method. By imposing certain restrictions upon the possible distribution of displacements in an elastic layer he was able to obtain a soil response function similar in character to Equations 5.2 and 5.9. The details of the general variational method of analysis, together with the solutions to many practical problems, are given by Vlazov and Leontiev (1966) but in view of the particular importance of this model, especially in relation to the analysis of soil–foundation interaction problems, a brief exposition of the method is presented here.

Consider firstly the state of plane strain in the elastic layer (Fig. 5.5) in the x–z plane. The state of strain in the foundation layer is assumed to be such that the

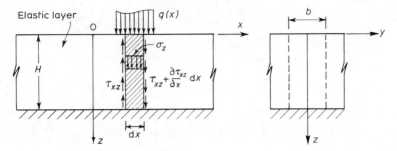

Figure 5.5 The Vlazov model; stresses in the elastic layer.

displacement components are

$$u(x, z) = 0; w(x, z) = w(x)h(z). \tag{5.10}$$

The function $h(z)$ describes the variation of displacement $w(x, z)$ in the z direction. Several such variations have been proposed by Vlazov and Leontiev (1966), including the linear and exponential variations

$$h(z) = (1 - \eta); h(z) = \frac{\sinh\left[\gamma(H - z)/L\right]}{\sinh\left[\gamma H/L\right]}, \tag{5.11}$$

where $\eta = z/H$ and γ and L are constants.

Using the stress–strain relations for plane-strain conditions (Timoshenko and Goodier, 1970):

$$[\sigma_{xx}; \sigma_{zz}] = \frac{E_0}{(1 - v_0^2)} w(x) \frac{dh(z)}{dz} [v_0; 1],$$

$$\tau_{xz} = \frac{E_0}{2(1 + v_0)} \frac{dw(x)}{dx} h(z), \tag{5.12}$$

where $E_0 = E_s/(1 - v_s^2)$, $v_0 = v_s/(1 - v_s)$ and E_s and v_s are, respectively, the elastic modulus and Poisson's ratio for the elastic material. The equation of equilibrium in the z direction is obtained by Lagrange's principle of virtual work; i.e. by equating to zero the total work of all internal and external forces on an element (Fig. 5.5) over any arbitrary virtual displacement.

The virtual work contribution from external forces

$$U_e = bq(x)h(0)dx + b \int_0^H \frac{\partial \tau_{xz}}{\partial x} h(z) dx dz. \tag{5.13}$$

The virtual work contribution from internal forces

$$U_i = -b \int_0^H \sigma_{zz} \frac{dh(z)}{dz} dz dx. \tag{5.14}$$

Using the condition $U_e + U_i = 0$, and Equations 5.12 to 5.14, the response function is:

$$q(x) = k_s w(x) - 2t \frac{d^2 w(x)}{dx^2}, \tag{5.15}$$

where

$$k_s = \frac{E_0}{(1 - v_0^2)} \int_0^H \left(\frac{dh}{dz} \right)^2 dz; t = \frac{E_0}{4(1 + v_0)} \int_0^H (h)^2 dz. \tag{5.16}$$

By comparing Equation 5.15 with Equations 5.3 and 5.9 also, when expressed for the two-dimensional case, it is apparent that the shear modulus G_p, the membrane tension T and k_s are directly related to the elastic constant E_s and v_s of the soil layer. Here, therefore, is a physical interpretation of the modulus of subgrade reaction k_s. An alternative derivation of the relationship between the Vlazov and Pasternak type models is given by Ting (1973). The constant k_s is a measure of the capacity of the soil medium to deform under applied compressive stress and t is a measure of the transmissibility of an applied force to neighbouring elements or the load spreading capacity. It is clear that the response of various soil media can be modelled by assigning suitable expressions for the function $h(z)$. For the linear variation of $h(z)$ given by Equation 5.11, the stresses in the soil layer are

$$\sigma_{zz} = -\frac{E_0}{H(1 - v_0^2)} \frac{dw}{dz}, \tau_{xz} = \frac{E_0}{2(1 + v_0)} \left(1 - \frac{z}{H} \right) \frac{dw}{dx}, \tag{5.17}$$

and

$$k_s = \frac{E_0}{H(1 - v_0^2)}, t = \frac{E_0 H}{12(1 + v_0)}. \tag{5.18}$$

Results similar to Equation 5.18, for the exponential variation of $h(z)$ (Equation 5.11), are given by Vlazov and Leontiev (1966).

Consider the variation of the vertical stress along a vertical through the elastic layer (Fig. 5.5): it can be shown (Vlazov and Leontiev, 1966) that the normal stresses are non-zero at points on the foundation surface which are located outside the loaded region. This is a direct consequence of employing a variational procedure for the problem. In such a technique the equations of equilibrium take an integral form, which is applicable to a region, rather than the differential form, which is applicable to every single point within the medium. As mentioned earlier, the intention here is to model the soil medium in such a way that foundations resting on soil media can be successfully analysed. It is therefore unreasonable to expect, from these idealized soil models, an accurate description of the state of stress within the soil medium. It should be noted that, in contrast to the elastic continuum model, the displacements and stresses in the elastic layer model due to a concentrated force are finite and bounded.

(e) Reissner model

The model proposed by Reissner (1958) is also derived by introducing displacement and stress constraints that simplify the basic equations for a linear elastic isotropic continuum. By assuming that the in-plane stresses (in the x, y plane) throughout a soil layer of thickness H are negligibly small ($\sigma_{xx} = \sigma_{yy} = \tau_{xy} = 0$) and that the displacement components u, v and w in the rectangular cartesian co-ordinate directions x, y, z respectively satisfy conditions

$$u = v = w = 0 \text{ on } z = H; u = v = 0 \text{ on } z = 0, \tag{5.19}$$

it can be shown that the response function for the soil model is given by

$$c_1 w - c_2 \nabla^2 w = q - \frac{c_2}{4c_1} \nabla^2 q, \tag{5.20}$$

where w is the vertical displacement of the surface of the elastic layer, $z = 0$, and q is the external load. The constants c_1 and c_2 characterizing the soil response (Equation 5.20) are related to E_s and v_s by $c_1 = E_s/H$ and $c_2 = HG_s/3$ where E_s and G_s are the elastic modulus and shear modulus, respectively, of the soil layer. It should be noted that, for a constant or linearly varying stress, after re-defining $c_1 = k_s$ and $c_2 = G_p$, Equation 5.20 is identical to Equations 5.9 or 5.15. Here again, as a consequence of assuming that the in-plane stresses σ_{xx}, σ_{yy} and τ_{xy} are zero, the shear stresses τ_{xz} and τ_{yz} are independent of z. These stresses are constant throughout the depth of the elastic layer for a given location x, y. Such an assumption is particularly unrealistic for a thick soil layer. An alternative interpretation of Reissner's model, together with a discussion of its applicability to clayey soils and compacted granular soils, is reported by Kerr (1965). The experimental work of Faber (1933) and Siemonsen (1948) is also used by Kerr (1965) to substantiate the theoretical results.

5.2.4 Inelastic and time-dependent soil models

The elastic soil models described in the previous section assume that the stresses in the soil medium nowhere exceed its limiting strength. It was noted above that assumptions of elastic behaviour of the soil medium are usually satisfied when the factor of safety against bearing failure exceeds three (Terzaghi, 1943; Schultze, 1961; Selvadurai and Kempthorne, 1980). Normally the relatively large dimensions of raft and mat foundations ensure a high factor of safety against a bearing failure. On the other hand, in continuous footings, owing to the relatively narrow width, the factor of safety at working loads may be less than three. In this case it is necessary to account for plastic deformations which occur at the perimeter of the footing due to high concentration of stress. Here, again, the analysis of the interaction problem can be performed on the basis of a modified elastic-plastic continuum approach or the two-parameter models can be modified to include finite strength (Kerr, 1965; Rhines, 1969; Selvadurai, 1976).

The methods of analysis are, however, beyond the scope of this chapter. As far as the design of foundations is concerned, the effects of plastic flow of the soil medium can be taken into account by considering the contact stress distribution at ultimate load conditions (Terzaghi, 1943). In the case of continuous footings located at or near the surface of the soil medium, the contact stress distribution is sensibly uniform for cohesive soils and parabolic for granular soils. For a granular–cohesive soil, the contact stress distribution can be considered to be a combination of the uniform and parabolic cases.

In addition to plasticity effects, it is necessary to consider the variation of the contact stress distribution with time owing to consolidation or creep. The contact stress distribution beneath foundations resting on granular soils is not likely to change substantially with time, except for the occurrence of plastic flow. With cohesive soils, however, the contact stress and displacement distributions could have a marked variation (Fig. 5.6) with time (Biot and Clingan, 1942; Teng, 1949; Marvin, 1972). The estimation of the effects of consolidation and creep on the contact stress distribution is a complex analytical problem. Analyses based on consolidation, viscoelasticity and creep theories are referred to by Korenev (1960), Arutiunian (1966), Hetényi (1966), Chiarella and Booker (1975), Agbezuge and Deresiewicz (1975), Desai and Christian (1977), Gudehus (1977) and Scott (1978). As a design procedure it is generally advisable to estimate the final contact stress distribution by using the elastic soil models in which the elastic constants, E_s and v_s (and hence k_s, G_p etc.) are determined from drained tests. This procedure

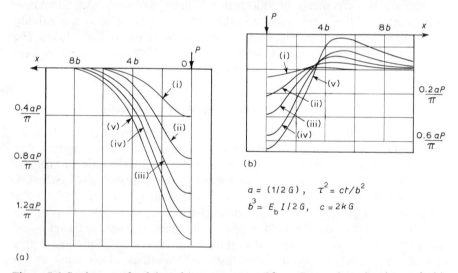

$$a = (1/2\,G), \quad \tau^2 = ct/b^2$$
$$b^3 = E_b\,I/2\,G, \quad c = 2kG$$

Figure 5.6 Settlement of a slab under a concentrated force P at various time intervals. (a) Pervious slab, (b) impervious slab. (After Biot and Clingan, 1942.) Curves: (i) $\sqrt{\tau} = 0.5$; (ii) $\sqrt{\tau} = 1.0$; (iii) $\sqrt{\tau} = 1.5$; (iv) $\sqrt{\tau} = 2.0$; (v) $\sqrt{\tau} = 2.5$. G is the shear modulus for the completely consolidated clay, k the coefficient of permeability, t the time, $E_b\,I$ the flexural rigidity of the slab.

is somewhat analogous to the technique used by Skempton and Bjerrum (1957) to calculate consolidation settlements.

5.2.5 Determination of constants describing the elastic soil models

The basic disadvantage of idealizing the response of the soil medium as a Winkler or a two-parameter model is that the constants describing these models are no longer intrinsic properties of the soil. They may, of course, be unique for a particular foundation problem. With the elastic or elastic-plastic continuum idealizations, however, the constants describing the models, namely the elastic moduli, Poisson's ratio, yield criteria etc., are characteristic and unique properties of the material. These parameters can usually be determined, to a reasonable degree of accuracy, from laboratory or field tests (see, for example, Baguelin *et al.*, 1977; Feda, 1978; Selvadurai, 1979; Selvadurai and Nicholas, 1979; and Selvadurai *et al.*, 1980). The determination of the constants describing the Winkler model (k_s) or the two-parameter soil models (k_s and G_p) can, therefore, be approached in the following ways:

(a) the constants of the Winkler and two-parameter soil models can be determined from *in situ* tests (e.g. plate loading tests);
(b) the constants can be related to the elastic constants E_s and v_s of the soil (or soils) which may be determined from laboratory tests;
(c) the constants can be related to the elastic constants (E_s and v_s) of the soil by comparing solutions to a particular soil–foundation interaction problem.

Comprehensive accounts of the evaluation of the modulus of subgrade reaction are given by Terzaghi (1955) and Selvadurai (1979). It is observed that k_s is not a unique property of the soil medium and the following factors affect the determination of k_s from plate loading tests.

(i) *Size of plate*

On the basis of experimental results, Terzaghi (1955) has shown that the value of k_s decreases with increase in width (B) of the rigid plate. For relatively long plates

$$k_s = k_{sl} \left[\frac{B + 0.305}{2B} \right]^2, \quad \text{and} \quad k = \frac{k_{sl}}{B}, \tag{5.21}$$

for granular and cohesive soils respectively, where k_s is the modulus of subgrade reaction for a long plate of width B(m) and k_{sl} is the modulus of subgrade reaction for a long plate of width 0.305 m.

(ii) *Shape of plate*

For plates having the same width B under the same uniformly distributed load, the value of k_s decreases with increasing length (L) of the plate according to

$$k_s = \frac{2k_{ss}}{3}\left[1 + \frac{B}{L}\right],\qquad(5.22)$$

where k_{ss} is the modulus of subgrade reaction for a rigid square plate.

(iii) Depth of the plate

The modulus of subgrade reaction k_s' measured from a rigid plate located at a depth D from ground level is related to the modulus of subgrade reaction (k_s) of a plate located at the surface of the soil medium by

$$k_s' = k_s\left[1 + 2\frac{D}{B}\right].\qquad(5.23)$$

Combining the depth and size effects, for a square plate on a granular medium,

$$k_s = k_{s1}\left[\frac{B + 0.305}{2B}\right]^2\left[1 + 2\frac{D}{B}\right],\qquad(5.24)$$

but not greater than $2k_{s1}[(B + 0.305)/B]$. In the case of cohesive soils k_s can be assumed to be independent of the depth. For a partly cohesive and partly granular medium, the expression equivalent to Equation 5.24 is

$$k_s = k_{sa}\left[\frac{B + 0.305}{2B}\right]^2\left[1 + 2\frac{D}{B}\right] + \frac{k_{sb}}{B},\qquad(5.25)$$

where k_{sa} and k_{sb} should be evaluated by performing at least two tests using two different sizes of plates. Further discussions of the measurement of k_s are given by Brebner and Wright (1952), Little (1961), Teng (1962), Weissmann (1965, 1972), Tomlinson (1975) and Wilun and Starzewski (1972). To obtain reliable data from plate bearing tests Weissmann and White (1961) suggest the following procedure.

(a) The area of the test plate should be about 10 to 15% of the area of the foundation provided the foundation has a geometrical shape with a centre of symmetry (circles, squares etc.). For foundations having a different geometrical shape the size of the plate has to be determined by theoretical considerations based on the analysis of a rigid plate resting on an elastic half-space (Weissmann, 1972).
(b) The plate should be located at a depth equal to that of the foundation and if possible at the exact location of the proposed foundation.
(c) The plate should be loaded so that the average contact stress between the plate and the soil medium is equal to the maximum estimated contact stress of the foundation. This maximum contact stress can be evaluated on the basis of a planar contact stress distribution. The load should be maintained until the creep of the plate becomes insignificant.

(d) The coefficient of subgrade reaction should be determined for an alternating stress superimposed on an average stress equal to the average calculated contact stress between the foundation and subsoil. General aspects of plate bearing tests performed on granular soils, cohesive soils and rocks are further discussed by Coates and Gyenge (1965), Ward *et al.* (1965), Zienkiewicz and Stagg (1965), Harrison and Richardson (1967), D'Appolonia *et al.* (1968), Hendron *et al.* (1970), Rocha (1970), Tsytovich *et al.* (1970), Marsland (1971*a,b*) and Hanna (1973).

The methods outlined for the determination of k_s for the Winkler model are, of course, applicable for the determination of k_s in the two-parameter models. To determine the value of G_p it is necessary to observe not only the settlement of the loaded plate but also the deflections of the soil surface in the immediate vicinity of the plate. This settlement profile is a measure of the constant G_p. For example, in the case of the two-parameter foundation, the surface deflection outside the rigid plate (in the plane-strain case) takes the form

$$w(x) = \frac{Pe^{-\alpha(x-B/2)}}{(k_s B + 2\alpha G_p)}, \qquad (5.26)$$

where P is the load per unit length of the long plate and $\alpha = \sqrt{(k_s/G_p)}$. Similarly, in the case of a circular rigid plate of radius a subjected to an axially symmetric loading condition, the surface deflection profile takes the form

$$w(r) = \frac{\alpha P^* K_0(\alpha r)}{\pi a k_s [\alpha a K_0(\alpha a) + 2 K_1(\alpha a)]}, \qquad (5.27)$$

where P^* is the axial load and K_0 and K_1 are the 0th and 1st order Bessel functions of the second kind, respectively. Similar results can be obtained also for square or rectangular rigid plates. It is therefore possible to estimate the value of G_p from the surface deflection profile of the soil outside the loaded rigid plate. Equations 5.26 and 5.27 indicate that, unlike the case of the elastic continuum model (where the surface deflections corresponding to Equations 5.26 and 5.27 are proportional to $\ln x$ and $1/r$ respectively), the surface deflections for the two-parameter soil foundations are restricted to a zone very close to the loaded area. In practice, of course, it is difficult to obtain surface deflection profiles of the soil from any routine plate bearing tests. Measurement of surface deflections in model tests are reported by Kögler and Scheidig (1938), Bond (1961), Eggestad (1963), Brown and Pell (1967) and Ueshita and Meyerhof (1967). Experimental techniques adopted for the field measurement of surface deflection profiles are discussed by Waldorf *et al.* (1963), Bergdahl and Broms (1967), Ward *et al.* (1968), Forrest and MacFarlane (1969), Mitchell and Gardner (1971), Cooke and Price (1973a), Eisenstein and Morrison (1973) and Ward and Burland (1973). The experimental data indicate that, in general, the deflections of the soil surface are restricted to the close vicinity of the loaded area.

Using the second approach, k_s and G_p can be related to the elastic constants E_s and v_s. It has been shown (Equation 5.18) that, for a single layer with a linear variation of normal stresses, k_s and G_p are given by

$$k_s = \frac{E_s}{H(1 + v_s)(1 - 2v_s)}, \quad G_p = \frac{E_s H}{6(1 + v_s)}. \tag{5.28}$$

The values of E_s and v_s can be determined from triaxial tests. The Vlazov type of soil model considered in Section 5.2.3 consists of a single elastic layer. Expressions equivalent to Equation 5.28 can be developed for multi-layered soil media with non-linear variations of normal stresses and non-zero distributions of lateral displacements u and w. The resulting expressions for k_s and G_p are, however, much more complicated. Details of such analyses are given by Vlazov and Leontiev (1966) and Rao *et al.* (1971).

Biot (1937) has expressed the modulus of subgrade reaction k_s in terms of the elastic constants of the soil medium (E_s and v_s) by a comparison of solutions for a particular soil–foundation interaction problem, using both Winkler and elastic continuum models. In particular, the correlation between k_s and E_s and v_s is obtained by comparing the maximum bending moment in an infinite beam subjected to a concentrated force (P) using both soil models. Consider the case of a beam of flexural rigidity $E_b I$ and width B resting on an elastic half-space. The bending moment at any point x of the beam is given by [see, for example, Selvadurai, 1979]

$$M(x) = \frac{P_c}{\pi} \int_0^\infty \frac{\alpha \cos(\alpha x/c)}{[\alpha^3 + \Psi(\beta)]} d\alpha, \tag{5.29}$$

where α, $\beta (= B\alpha/2c)$ are dimensionless parameters; c denotes $[C(1 - v_s^2)E_b I/E_s]^{1/3}$, the characteristic length of the beam; C is a function of β ($C = 1.00$, if the bearing stress distribution across the width of the beam is uniform and $1.00 < C < 1.13$ if the deflection across the width of the beam is uniform). The function $\Psi(\beta)$ is tabulated for $\beta > 0.1$ and given as an asymptotic expression (Biot, 1937) for $\beta < 0.1$. The maximum value of Equation 5.29 can be written as

$$M_{max} = 0.166 \, PB \left[16C(1 - v_s^2) \frac{E_b I}{E_s B^4} \right]^{0.277}. \tag{5.30}$$

By comparing Equation 5.30 with the maximum bending moment ($M = 0.176 PB[16E_b I/k_s B^5]^{1/4}$) for the Winkler case, Biot obtained

$$k_s = \frac{1.23 E_s}{(1 - v_s^2)B} \left[\frac{E_s B^4}{16C(1 - v_s^2)E_b I} \right]^{0.11}. \tag{5.31}$$

Equation 5.31 is, therefore, a measure of k_s in terms of E_s, v_s and the properties of the infinite beam. Biot has also shown that for the infinite beam problem, the much simpler Winkler model can be used with the modified value of k_s (Equation 5.31) to obtain reasonable distributions of bending moment and deflection. It must, however, be pointed out that correlation between k_s and E_s and v_s is obtained on the basis of (i) the analysis of the infinite beam, (ii) a comparison of maximum bending moment and (iii) a special case of external loading. Therefore, expressions similar to Equation 5.31 can be obtained by varying these three factors (e.g. by comparing the deflection in a finite beam for a uniformly distributed external load). Techniques of this nature, substantiated by experimental work, have been used by many investigators (Drapkin, 1955; Vesić, 1961a, b; Barden, 1962a, b, 1963a; Vesić and Johnson, 1963) to obtain expressions similar to Equation 5.31 for k_s. Galin (1961), Vesić (1961a, b)), Barden (1962a, b; 1963a, b) and Selvadurai (1979) have suggested the following expressions:

$$k_s = \frac{0.65 E_s}{B(1 - v_s^2)} \left[\frac{E_s B^4}{E_b I} \right]^{1/12}, \tag{5.32}$$

and

$$k_s = \frac{0.65 E_s}{B(1 - v_s^2)}, \quad k_s = \frac{\pi E_s}{2B(1 - v_s^2) \log_e (L/B)}, \tag{5.33}$$

respectively, for relatively long beams ($L/B > 10$, where L is the length of the beam).

Another interesting interpretation of the modulus of subgrade reaction k_s was proposed by Gibson (1967, 1974). The stresses and displacements in an incompressible non-homogeneous elastic half-space, whose shear modulus $G(z)$ increases according to

$$G(z) = G(0) + mz, \tag{5.34}$$

were obtained by Gibson (1967). It is found that when $G(0) = 0$, the surface deflection $w(0)$ is $q/2m$ within the loaded area zero outside the loaded area, where q is the stress intensity of the uniform external load. This result is applicable to any loaded area of any shape to which a uniform contact stress is applied. This interpretation of $k_s(= 2m)$, based on a continuum model, is applicable to deep beds of soil such as normally consolidated clay, chalk, etc. (Ward *et al.*, 1968; Marsland, 1971a, b; Ward and Burland, 1973) strained under conditions of no volume change.

5.3 CONTINUOUS FOOTINGS

An analysis of the soil–foundation interaction problem which takes into account the flexibility of the foundation is the rational approach to design of continuous footings. It can be argued that the effort expended on such analyses may not be justified by the validity of the basic assumptions regarding the response of the soil or the accuracy of field or laboratory testing techniques available for determining the various parameters. To a certain extent this is a justifiable criticism. On the other hand, the designer must appreciate the fact that traditional methods have little or no scientific basis. The analysis of combined foundations based on infinitely rigid foundation behaviour with assumed planar or other geometrically simple distribution of contact stresses completely disregards the deformation characteristics of the soil and the flexural properties of the foundation. In this Section the prime concern is with the methods of elastic analysis of soil–foundation interaction which can be conveniently adopted in the design of engineering foundations. It is not intended that elastic methods of analysis should completely replace the traditional methods of analysis but they enable the designer to refine his approach to more complex practical problems. It is probably more satisfactory for the designer to concentrate his attention on one of the theoretical elastic methods so that facility is gained in its application and, of course, the drudgery of analysis can be avoided by employing computer-aided design.

5.3.1 Analysis of infinitely long beams

Consider first the *plane-strain* problem of an infinitely long elastic beam resting on a linearly deformable elastic soil medium. From a practical point of view plane strain conditions can occur in certain regions of large rafts and plates which are subjected to concentrated line loads or uniformly distributed loads (Fig. 5.7).

The analysis presented here is restricted to the class of slender beams whose bending response is governed by the classical Bernoulli–Euler theory of beams. This particular beam theory is based on the fundamental assumption that cross-sections of the beam remain plane and normal to the axis of bending. It is implied that the strains and rotations of the beam are small compared to unity. For classical elastic behaviour of the beam material, the stresses (σ) and curvatures (κ)

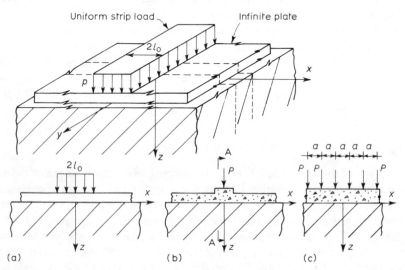

Figure 5.7 Plane strain conditions in an infinite plate. (a) Uniform strip load; (b) line load; (c) section A–A.

in the beam take the forms

$$\sigma(x, \bar{z}) = \frac{\bar{z}M(x)}{I(x)}, \quad \kappa(x) = \frac{M(x)}{E_b I(x)}, \tag{5.35}$$

where $M(x)$ is the bending moment $E_b I(x)$ is the flexural rigidity at the location x, and $\bar{z}$ is the distance from the neutral axis of the particular fibre under consideration. In addition, the shearing strains are assumed to have a negligible effect on the axial strains. The cross-sectional dimensions are such that the deflections due to shearing stresses can be neglected. The beams are assumed to be straight and prismatic with proportions that will prevent failure by twisting, lateral collapse or local wrinkling. Normally, for conditions of plane strain, the width of the beam is assumed to be unity. In the formulation which follows the beam is assumed to be of width B. In the case of thick beams, effects of shear deformations (Essenburg, 1962) have to be incorporated into the Bernoulli–Euler beam theory, or the beam may be treated as an elastic layer (Timoshenko and Goodier, 1970).

(a) Infinite beam subjected to a uniform load

The generalized analysis of a beam of infinite length presented here is applicable to beams resting on a Winkler, two-parameter, or elastic solid type of soil medium. This generalized formulation is facilitated by the applicability of the principle of superposition to the elastic soil models. Consider the plane-strain problem of an infinite beam resting on an idealized soil medium. The contact between the soil and beam is assumed to be smooth. The beam is loaded by a uniform strip load of stress intensity p and width $2l_0$ (Fig. 5.8). Consider firstly the problem in which the surface of the soil medium is subjected to the sinusoidal normal stress

$$q(x, 0) = q(x) = q_0 \cos(mx/a), \tag{5.36}$$

where q_0, m are constants and a is a typical length parameter. By virtue of the symmetry of the external stress system and the linearity of the governing response functions, the surface deflection of the soil medium can be written in the form

$$w(x, 0) = w(x) = \frac{aq_0}{K} \cos(mx/a). \tag{5.37}$$

The constant K depends upon the material properties of the particular soil medium. For example, in the case of the Winkler model, $K = ak_s$, where k_s is the modulus of subgrade reaction defined by Equation 5.1. Now consider the flexure of the infinite beam. It is assumed that the beam is subjected to an external sinusoidal stress distribution

$$p(x) = p_0 \cos(mx/a), \tag{5.38}$$

and that the contact stress distribution at the interface is $q(x)$. Assuming that there

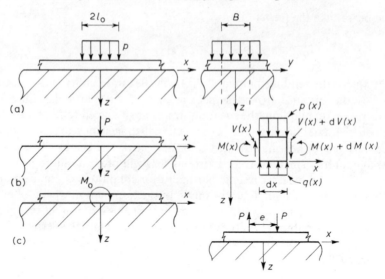

Figure 5.8 Plane strain problem of the infinite beam. (a) Uniform load of finite width; (b) concentrated line load; (c) concentrated moment.

is no loss of contact at the beam–soil interface, the differential equation for the deflection of the beam is

$$E_b I \frac{d^4 w}{dx^4} + Bq(x) = Bp(x),$$ (5.39)

where B is the width considered and $E_b I$ is the flexural rigidity of the beam. From Equations 5.36 and 5.37 note that $w(x) = aq(x)/K$. Using this relation in Equation 5.39

$$w(x) = \frac{Bp_0 a^4 \cos(mx/a)}{E_b I [m^4 + Ka^4/E_b I]}.$$ (5.40)

By superposing sinusoidal stresses of the form of Equation 5.38, any arbitrary stress distribution $p(x)$ symmetrical about the z-axis can be represented in the form of a Fourier integral

$$p(x) = \int_0^\infty \bar{p}(\xi) \cos(mx/a) d\xi,$$ (5.41)

where

$$\bar{p}(\xi) = \int_0^\infty p(\zeta) \cos(m\zeta/a) d\zeta.$$ (5.42)

In the case of a uniform strip load of stress intensity p and width $2l_0$ (Fig. 5.8),

Equation 5.41 reduces to

$$p(x) = \frac{2p}{\pi} \int_0^\infty \frac{1}{m} \sin(ml_0/a) \cos(mx/a) dm. \tag{5.43}$$

From Equations 5.40 and 5.43 it can be concluded that the deflection of the infinite beam due to the uniform strip load is given by

$$w(x) = \frac{2pa^4 B}{\pi E_b I} \int_0^\infty \frac{\sin(ml_0/a)\cos(mx/a)}{m[m^4 + \Omega]} dm, \tag{5.44}$$

where

$$\Omega = \frac{Ka^3 B}{E_b I}. \tag{5.45}$$

(Note that the length parameter a can be set equal to the width of the beam B or to $\sqrt{[E_b I/K]}$.)

Similarly, expressions for the bending moment $M(x)$, shearing force $V(x)$ and contact stress $q(x)$ can be written in the concise form:

$$[M(x); V(x); q(x)] = \frac{2pB}{\pi} \int_0^\infty \frac{\sin(ml_0/a)\cos(mx/a)}{[m^4 + \Omega]}$$
$$\cdot \left[-ma^2; m^2 a \tan(mx/a); \frac{\Omega}{B} \right] dm. \tag{5.46}$$

Expressions for the non-dimensional parameter Ω for various soil models are as follows.

(a) Winkler model: $\quad \Omega = \dfrac{k_s a^4 B}{E_b I}. \tag{5.47a}$

(b) Pasternak model: $\quad \Omega = \dfrac{a^2 B}{E_b I} [m^2 G_p + k_s a^2]. \tag{5.47b}$

(c) Reissner model: $\quad \Omega = \dfrac{4 E_s \eta a^2 B}{E_b I} \left[\dfrac{3\eta^2 + m^2 \chi}{12\eta^2 + m^2 \chi} \right], \tag{5.47c}$

where $\eta = a/H$ and $\chi = G_s/E_s$.

Note that as G_p (or χ) tends to zero, both Equations 5.47b and 5.47c reduce to the Winkler case if E_s in Equation 5.47c is replaced by $k_s H$. The results for the two parameter Vlazov foundation can be obtained by substituting values of k_s and $2t$ defined by Equation 5.16.

(d) For an isotropic elastic continuum with shear modulus G_s and Poisson's ratio v_s

$$\Omega = \frac{G_s m a^3 B}{(1 - v_s^2) E_b I}. \tag{5.47d}$$

Expressions for Ω, similar to Equations 5.47a to d, for an orthotropic elastic medium and a non-homogeneous elastic medium where G_s varies linearly with depth, can also be obtained from Lekhnitskii (1960) and Gibson (1967; 1974).

(b) Infinite beam loaded by a concentrated force

The solution to the problem of an infinite elastic beam loaded by a concentrated line force of intensity P (load per unit length) at the origin (Fig. 5.8) can be obtained as a limiting case, when the width of the strip load $2l_0 \to 0$, and the total load per unit length $2pl_0 \to P$. Considering that

$$\underset{ml_0/a \to 0}{\text{Lt}} \frac{\sin(ml_0/a)}{ml_0/a} = 1, \tag{5.48}$$

Equations 5.44 and 5.46 reduce to

$$[w(x); M(x); V(x); q(x)] = \frac{PB}{\pi} \int_0^\infty \frac{\cos(mx/a)}{[m^4 + \Omega]} \left[\frac{a^3}{E_b I}; -am^2; m^3 \tan(mx/a); \frac{\Omega}{aB} \right] dm. \tag{5.49}$$

(c) Infinite beam loaded by a concentrated couple

A couple M_0 per unit length acting on the infinite beam can be represented as a limiting case of a combination of two concentrated forces acting at a small distance e apart, as shown in Fig. 5.8. It is assumed that as $e \to 0$, Pe approaches the value of M_0. Using Equation 5.49, the equation of the deflected shape can be written as

$$w(x) = -\frac{Pa^3 B}{\pi E_b I}[\Phi(x+e) - \Phi(x)], \text{ for } x > 0, \tag{5.50}$$

where

$$\Phi(x+e) = \int_0^\infty \frac{\cos[m(x+e)/a]}{[m^4 + \Omega]} dm. \tag{5.51}$$

Using the condition that $Pe \to M_0$ as $e \to 0$, Equation 5.50 reduces to

$$w(x) = M_0 \frac{a^3 B}{\pi E_b I} \frac{d\Phi}{dx}, \tag{5.52}$$

or

$$w(x) = -\frac{M_0 a^2 B}{\pi E_b I} \int_0^\infty \frac{m \sin(mx/a)}{[m^4 + \Omega]} dm, \tag{5.53}$$

and the expression for the bending moment, shearing force and contact stress is

$$[M(x); V(x); q(x)] = \frac{M_0 B}{\pi} \int_0^\infty \frac{\sin(mx/a)}{[m^4 + \Omega]} \left[m^3; \frac{m^4}{a} \cot(mx/a); \frac{m\Omega}{a^2 B} \right] dm. \tag{5.54}$$

In the solutions for the infinite beam presented in this section, it is explicitly assumed that there is no loss of contact at the soil–beam interface and that the interface is capable of sustaining tensile stresses. Normally, the weight of the foundation imposes sufficient compressive stresses at the interface to prevent loss of contact. Alternatively, a uniform load can be imposed throughout the length of the infinite beam. By virtue of its symmetry, a normal load of this type gives rise to a constant vertical displacement on the surface of the soil medium. The infinite beam is then subjected to a rigid body translation in the z direction and the flexure of the beam is thus unaffected. This constant stress is simply added to the contact stress distribution.

Techniques for evaluating infinite integrals of the type in Equation 5.44 are outlined by Biot (1935, 1937), Drapkin (1955), Vesić (1961a,b) and Selvadurai (1979).

5.3.2 Three-dimensional effects in the infinite beam problem

In Section 5.3.1 the problem of an infinite plate, loaded in a manner such as to permit an analysis of the plate on the basis of two-dimensional plane strain, has been presented. In most foundation problems, such as long strip footings, combined footings, rail tracks, where the plan dimension of the foundation in any one direction is small compared to that in the other, it becomes necessary to consider the three-dimensional nature of the soil–foundation interaction problem. An analysis which takes into account the flexibility of such strip foundations as a complete plate is rigorous and complex. (Referring to Fig. 5.9,

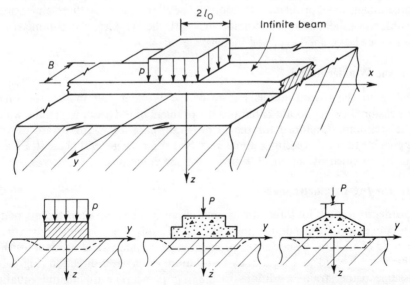

Figure 5.9 Three-dimensional effects: infinitely long beam of finite width.

this implies flexure of the foundation in both the x, z and y, z planes.) The three-dimensional analysis of the infinitely long strip foundation can, however, be greatly simplified by assuming that the flexural rigidity of the foundation in the transverse plane (y, z) is much greater than that in the longitudinal plane (x, z). This assumption is reasonably satisfied in most strip foundation bases with a ribfooting or a haunched-footing (Fig. 5.9). In these cases the vertical displacement across the width of the foundation is assumed to be constant. Alternatively, it can be assumed that the contact stress distribution across the width of the foundation is uniform (i.e. the flexural rigidity of the foundation in the y, z plane is zero and the external loading is uniform across the width of the beam). Such an assumption is not usually realized in practice, although it leads to a simple analysis of the three-dimensional problem. It is also possible to carry out the analysis of the three-dimensional problem by taking into account average values of displacement $(\bar{w})$ and contact stresses $(\bar{q})$ across the width of the foundation, defined by

$$[\bar{w}; \bar{q}] = \frac{1}{B} \int_{-B/2}^{B/2} [w(y); q(y)] \mathrm{d}y. \tag{5.55}$$

Only the case of uniform displacement across the width of the foundation will be considered here. The analysis of the three-dimensional problem of the infinite beam differs from that of the two-dimensional problem only in one respect: the contact stress distribution at the interface $q(x)$ should now account for the three-dimensional nature of the interaction problem. Alternatively, K (see Equation 5.37) has to be modified to incorporate three-dimensional effects. The expressions developed for the plane strain problem are, therefore, directly applicable to the three-dimensional problem. Attention will be restricted to the three-dimensional problems of the infinite beam associated with the Winkler, two-parameter and isotropic elastic solid type soil models.

(a) The Winkler model

Owing to the discontinuous nature of the Winkler soil model, the three-dimensional effects do not enter into the problem of the beam with finite width. The solutions developed for the plane strain case (of width B) are directly applicable to the three-dimensional case. Note that the general plane strain case can be obtained by setting B to unity in these equations.

(b) The two-parameter model

In order to account for three-dimensional effects in the two-parameter soil model, it is convenient, from a design point of view, to adopt the following approximate method of analysis. Consider the problem of an infinitely rigid beam of width B which is subjected to a rigid body translation w^*, in the z direction (Fig. 5.10). Assume plane strain conditions in the y, z plane. The differential equation governing the deflected shape of the surface has the same form as Equation 5.3 or

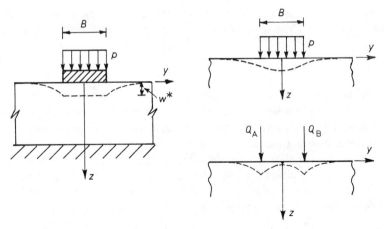

Figure 5.10 Three-dimensional effects in the infinite beam: deformations of the Vlazov model.

5.9 where the x co-ordinate is replaced by the y co-ordinate. Thus

$$q(y) = k_s w(y) - G_p \frac{d^2 w}{d y^2}(y). \tag{5.56}$$

Under the rigid foundation AB (Fig. 5.10) the displacement is constant; therefore, from Equation 5.56, the contact stresses under the foundation are also constant and of magnitude $q(y) = k_s w^*$. The shear traction t_z on any plane $y =$ constant beneath the beam is zero. Beyond the edges of the rigid beam, however, the traction is non-zero. It implies that the traction t_z ($t_z = \int_0^H \tau_{xz} h(z) dz = G(dw/dy)$, Vlazov and Leontiev (1966)) is discontinuous at the edges $y = \pm B/2$. Hence, concentrated reactions Q_A and Q_B arise at the rigid beam edges due to the stresses in the elastic region beyond these edges. The proof of their existence is further established by the fact that a uniform stress $k_s w^*$ gives rise to a dish-shaped surface deflection profile instead of the assumed uniform deflection. It can be shown that the surface deflection beyond the foundation is

$$w(y) = w^* e^{-\alpha(y - B/2)}, \tag{5.57}$$

where $\alpha = \sqrt{(k_s/G_p)}$. Taking into account the shear traction (t_z), $Q_A = Q_B = G_p \alpha w^*$. The relationship between the reactive force per unit length $Bq(x)$ and w^* can be obtained by considering the equilibrium of forces acting on the rigid beam AB per unit length in the x direction. Then

$$Bq(x) = Bk_s w^* \left[1 + \frac{1}{B} \sqrt{\left(\frac{G_p}{k_s} \right)} \right]. \tag{5.58}$$

Since w^* was taken as an arbitrary deflection, the form of Equation 5.58 is assumed to be valid throughout the length of the infinite beam. Therefore, the result for the infinite beam of finite width resting on a two-parameter foundation is obtained by replacing K (Equation 5.37) by a modified value K^* where

$$K^* = k_s a \left[1 + \frac{1}{B} \sqrt{\left(\frac{G_P}{k_s} \right)} \right]. \qquad (5.59)$$

Note that as $G_P \to 0$, the approximate three-dimensional solution converges to the Winkler case. The solution to the three-dimensional problem can be obtained by replacing Ω by Ω^*, where $\Omega^* = K^* a^3 B / E_b I$, in the various integral expressions for deflection, bending moment, etc.

The assumption of plane-strain conditions in the y, z plane imposes restrictions on the preceding analysis. In order to investigate the true three-dimensional nature of the problem it is necessary to consider the surface deflection of the soil model produced by a two-dimensional normal stress distribution

$$q(x, y) = q_0 \cos (mx/a) \cos (ny/a). \qquad (5.60)$$

It was noted above that, in the case of a uniform surface displacement, the contact stresses consist of a uniform distribution of stress together with edge reactions. Keeping x constant, the imposed stress (Equation 5.60) can be integrated, using the Fourier integral theorem, to give either a uniform stress of finite width or a concentrated force. Avoiding algebraic details, it can be shown that the complete three-dimensional problem of an infinite beam resting on a two-parameter soil model is obtained by replacing Ω by Ω^* where

$$\Omega^* = \frac{a^4 b k_s}{E_b I} \left[1 + \frac{1}{B \left[\dfrac{m^2}{a^2} + \dfrac{k_s}{G_P} \right]^{1/2}} \right]. \qquad (5.61)$$

Again as $G_P \to 0$, Equation 5.61 converges to the Winkler case.

(c) The isotropic elastic solid model

As an approximate method of taking into account the three-dimensional effects, it can be assumed that the distribution of contact stress across the width of the foundation is the same as that obtained on the basis of restricting the deformations in the y, z plane to a state of plane strain. The contact stress distribution $q(y)$ under a rigid circular foundation subjected to a central load was obtained by Sadowsky (1928). The contact stresses are given by

$$q(y) = \frac{2P}{\pi \sqrt{(B^2 - 4y^2)}}, \qquad (5.62)$$

and the resulting uniform displacement w^* of the rigid foundation is $P(1 - v_s) \log_e 2 / \pi G_s$. Using Equations 5.37, 5.45 and the equilibrium condition, $P = Bq(x)$, it can be shown that an approximate solution for the three-dimensional infinite beam problem is obtained by employing

$$\Omega^* = \frac{\pi G_s a^4}{(1 - v_s)E_b I \log_e 2}. \tag{5.63}$$

The complete three-dimensional problem of a beam resting on an isotropic elastic solid medium has been analysed by Biot (1937), Rvachev (1956, 1958), Lekkerkerker (1960), and Selvadurai (1979). In these analyses the surface deflection of the half-space due to the two-dimensional surface stress (Equation 5.60) is examined. The stress (Equation 5.60) can be integrated to obtain a uniform displacement profile across the width of the beam. The modified value of Ω in this case can be shown (Biot, 1937; Lekkerkerker, 1960; Selvadurai, 1979) to be equal to

$$\Omega^* = \frac{2G_s a^3 B\beta}{C(1 - v_s)E_b I}\Psi(\beta), \tag{5.64}$$

where $C = 1.13$ and the function $\Psi(\beta)$ (where $\beta = Bm/2a$) is numerically calculated. The details of these calculations are given by Biot (1937). For example

$$\begin{array}{ccccccc} \beta = 0.1 & 0.5 & 1 & 3 & 8 & \infty \\ \psi(\beta) = 4.80 & 1.90 & 1.42 & 1.13 & 1.04 & 1.0 \end{array}$$

and for $\beta < 0.1$; $\Psi(\beta) \approx 2/\pi\beta \{\log (1/\beta) + 0.923\}$.

5.3.3 Analysis of finite beams

Foundation beams and continuous footings encountered in engineering practice possess finite dimensions. Such structural foundations should in general be analysed on the basis of finite beams which not only satisfy continuity conditions at the soil–foundation interface but also boundary conditions at the ends of the beam. In this section are considered some special cases of analysis of finite beams resting on a Winkler, a two-parameter or an isotropic elastic solid type soil medium. It is assumed that the stiffness of the finite beam in the y–z plane is large enough to prevent flexure in that plane and, as before, that the soil–foundation interface is capable of sustaining tensile stresses. The analysis of finite beams do not, in principle, present any difficulties except for certain computational obstacles encountered in satisfying the boundary conditions of the problem; for example, in fulfilling continuity conditions at regions or points where uniform external load or its derivatives become discontinuous and at points of application of concentrated forces and moments. There exist several methods of analysis which overcome these difficulties but attention is restricted here, where relevant, to *the method of superposition* proposed by Hetényi (1936), *the strain energy method* proposed by Christopherson (1956) and the approximate numerical techniques of Ohde (1942) and Zemochkin and Sinitsyn (1947).

Numerous alternative solutions have been advanced for the analysis of finite beams on elastic media; references to these can be found in the review articles by Korenev (1960), Hetényi (1966) and Selvadurai (1979). Some of these references of particular interest to this section will be mentioned here. For the Winkler model, the method of initial parameters (sometimes referred to as the Pusyrevskii–Krylov method) has been proposed by Umansky (1933); Levinton (1947) has represented the contact stress distribution as a series of redundant reactions; Wright (1952) has employed a relaxation technique; Thoms (1960) gives experimental results for a beam, supported on helical springs, which has been designed by a dimensional analysis technique; Popov (1951) has applied graphical successive approximations; Pipes (1943) and Iwinski (1967) have demonstrated the application of integral transform techniques; Malter (1958) has proposed a step-by-step integration process and a method of finite differences; influence-line methods have been employed by Ray (1958) and Iyengar and Anantharamu (1965); matrix methods have been employed by Iyengar (1965) and Mozingo (1967); dimensionless influence functions have been developed by Dodge (1964) and beam-column problems have been studied by Lee *et al.* (1961), Iyengar and Anantharamu (1963) and Matlock and Wayne (1963). Several other investigators, including Volterra and Chung (1955) and Miranda and Nair (1966), have contributed to the literature on Winkler type soil–foundation interaction problems. From a design point of view, Kramrisch and Rogers (1961) have employed a semi-empirical method applicable to combined footings with a specific form of periodic column loading. For a particular column spacing and relative flexibility of the soil–foundation system, the contact stress distribution comprises a series of linear functions with maxima beneath columns and minima at mid-span. Examples of the applicability of the method to mat and grid foundations are also given. Other design techniques include the soil line method proposed by Baker (1957) and Szava-Kovats (1967).

The analysis of finite beams resting on a two-parameter medium has received only limited attention. The works of Filonenko-Borodich (1940), Pasternak (1954) and Vlazov and Leontiev (1966) contain comprehensive treatment of the general methods of analysis. Further modifications to Vlazov and Leontiev's technique have been proposed by Rao *et al.* (1971). Examples of analysis of finite beams, together with specimen computer programs, are given by Harr *et al.* (1969).

The general analysis of flexible beams resting on elastic continua requires the use of complex mathematical techniques. As such most of the available literature deals with approximate methods of analysis. Here again, the reader is referred to the works of Ohde (1942), Zemochkin and Sinitsyn (1947) and Gorbunov-Posadov (1949) for further information. It is not uncommon to find that the approximate methods of analysis developed by these authors, and others, are accepted as conventional methods for the design of structural foundations, especially in the Eastern European countries. The theoretical analysis of Vesić (1961*a*, *b*) is concerned with an investigation of the similarity of results for the infinite beam obtained by using the Winkler and elastic half-space models. Ohde's method for analysis of finite beams has been empolyed by Vesić and Johnson (1963) to verify results obtained for model studies of beams on silt subgrades. Zemochkin's method has been used by Barden (1962*a*,*b*), who extends the analysis to the domain of transversely isotropic elasticity. Similar results have been presented by Król (1964), who takes into account the effects of the rigidity of the superstructure. Experimental results obtained from photoelastic models and from granular subgrades have also been reported by Barden (1962*a*,*b*, 1963*a*) Zemochkin's method has been used by Glassman (1972) to examine the experimental results obtained for crossed beams resting on granular subgrades. Krsmanović (1961) has employed a system of free discontinuous beams with two supports as a means of analysis for a continuous beam.

Finite element methods of finite beam problems have been considered by Just *et al.* (1971), Zienkiewicz (1971) and Bowles (1974). The non-linear effects in soil–foundation interaction problems due to loss of contact have been considered by Tsai and Westmann (1967), Cheung and Nag (1968), Weitsman (1970), Lin *et al.* (1971), Farshad and Shahinpoor (1972) and Gladwell (1975, 1976).

5.3.4 Finite beams on a Winkler medium

The general differential equation governing the deflected shape of a beam resting on a Winkler medium is

$$E_b I \frac{d^4 w}{dx^4} + k_s B w = B p(x). \tag{5.65}$$

This has a homogeneous solution

$$w(x) = e^{\lambda x}[C_1 \cos \lambda x + C_2 \sin \lambda x] + e^{-\lambda x}[C_3 \cos \lambda x + C_4 \sin \lambda x], \tag{5.66}$$

where $\lambda^4 = k_s B/4E_b I$, and $C_1 \ldots C_4$ can be determined by making use of boundary conditions at the ends of the beam for any particular loading condition $p(x)$. The parameter λ includes the flexural rigidity of the beam as well as the elastic characteristics of the supporting medium. From Equation 5.66 it is evident that λ influences the deflected shape of the beam. For this reason λ^{-1} is usually termed the characteristic of the system and (with dimensions of length) is regarded as the characteristic length; consequently λx is a dimensionless parameter. Equation 5.66 represents the general solution for the deflected shape of a beam when $p(x)$ is zero and an additional term is necessary to account for distributed loads. The constants $C_1 \ldots C_4$, which depend on $p(x)$, have constant values along each portion of the beam where $w(x)$ and its derivatives are continuous. Their values can be obtained from the boundary conditions at the two ends of each continuous portion. Of the four quantities, the deflection (w), slope (θ), bending moment (M) and shearing force (V), two are usually known at each end. When discontinuous loadings are encountered the beam must be resolved into continuous portions, and at intermediate points continuity requirements for the beam furnish necessary data for the determination of integration constants. It is therefore evident that, as the number of continuous portions increases, so do the computational difficulties. The method of superposition, which tends to alleviate this difficulty, is discussed below.

(a) The method of superposition

The solution of the appropriate infinite beam problem forms the basis of the method of superposition proposed by Hetényi (1936). The finite beam is then considered as occupying a portion of the infinite beam. To satisfy boundary conditions relevant to the finite beam, certain unknown *end conditioning forces* (bending moments and shearing forces) must be applied at the points corresponding to the ends of the finite beam. The correct values of these end conditioning forces are determined from four simultaneous equations representing simultaneous fulfillment of boundary conditions at the ends of the finite beam. This technique possesses remarkable clarity and no ambiguity arises from the singular character of external loading. Before attempting a detailed treatment of Hetényi's method, closed form solutions to the problem of the infinite beam subjected, separately, to a concentrated force, a uniformly distributed load of finite width and a concentrated couple are considered. These results can be obtained also from the integral representations derived in Section 5.3.1.

(i) Infinite beam subjected to a concentrated force

The deflected shape of an infinite beam loaded by a concentrated force (P) applied at the origin (Fig. 5.11) is given by (Hetényi, 1946)

$$w(x) = \frac{P\lambda e^{-\lambda x}}{2Bk_s}[\cos \lambda x + \sin \lambda x] = \frac{P\lambda}{2Bk_s}A(\lambda x). \qquad (5.67)$$

The slope (θ), bending moment (M) and shearing force (V) are related to the

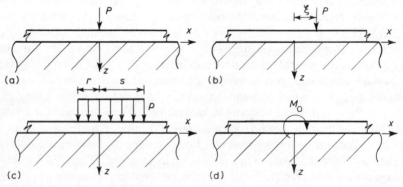

Figure 5.11 Infinite beam on a Winkler medium. (a), (b) Concentrated force problems; (c) uniform load of finite width; (d) concentrated moment problem.

deflection of the beam by

$$\theta = \frac{dw}{dx}; \quad M = -E_b I \frac{d^2 w}{dx^2}; \quad V = -E_b I \frac{d^3 w}{dx^3}. \tag{5.68}$$

Using Equation 5.67 in Equation 5.68:

$$\theta(x) = -\frac{P\lambda^2}{Bk_s} e^{-\lambda x} \sin \lambda x = -\frac{P\lambda^2}{Bk_s} B(\lambda x), \tag{5.69a}$$

$$M(x) = \frac{P}{4\lambda} e^{-\lambda x} [\cos \lambda x - \sin \lambda x] = \frac{P}{4\lambda} C(\lambda x), \tag{5.69b}$$

$$V(x) = -\frac{P}{2} e^{-\lambda x} \cos \lambda x = -\frac{P}{2} D(\lambda x). \tag{5.69c}$$

From Equation 5.69c it is evident that the shearing force at the origin is $P/2$. The curves represented by the functions $A(\lambda x) \dots D(\lambda x)$ are plotted in Fig. 5.12. It will be seen that they have the characteristics of damped waves with gradually reduced amplitude. For this reason λ is sometimes referred to as the damping factor. Note that, for example, when $\lambda x > 3\pi/2$ the value of any of the four

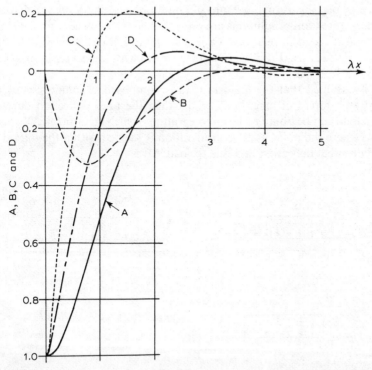

Figure 5.12 Curves for functions A,B,C, and D. (See also Table 5.3.)

functions is less than 0.01. This implies that for all practical purposes the manner in which the beam is supported at a distance $x > 3\pi/2\lambda$ from the point of application of the load will have only minor effects on the deflected shape. It can be shown that a finite beam of length $l \approx 2\pi/\lambda$ loaded by a concentrated force will have roughly the same deflected shape as that for an infinite beam loaded in a similar manner. The sign convention adopted in this method should be carefully noted. When the load acts in the positive z direction, $w(x)$ is nominally positive both to the left and to the right of the load, but the real sign is determined by whether $A(\lambda x)$ is positive or negative. The slope $\theta(x)$ is nominally positive to the left of the load and negative to the right; the real sign being influenced by the sign of $B(\lambda x)$. The bending moment $M(x)$ is nominally positive both to the left and to the right of the load but is influenced by the sign of $C(\lambda x)$. Shearing force is nominally positive to the left and negative to the right but influenced by the sign of $D(\lambda x)$. Loads which act in the positive z direction are taken as positive. The value of x is positive both to the left and to the right of the load.

(ii) *Infinite beam subjected to uniform load of finite width*

Consider the case of the infinite beam subjected to a uniform load of stress intensity p and width $(r + s)$ (Fig. 5.13). The solution to this problem can be obtained by making use of the results of Equations 5.67 and 5.69a to c derived for the concentrated force. Solutions are presented here by considering the location of the uniform load in relation to a specific point (J) which may lie outside or inside the loaded area (Fig. 5.13). The location of J with respect to the origin of co-ordinates is not specified.

If it is assumed that the uniform load is composed of infinitesimally small concentrated forces of magnitude $bp\,\mathrm{d}\xi$, then the total deflection due to the uniform load can be obtained by an integration of such concentrated forces. For the case where the point under consideration is located within the loaded area, the infinitesimal deflection at J due to load AB is

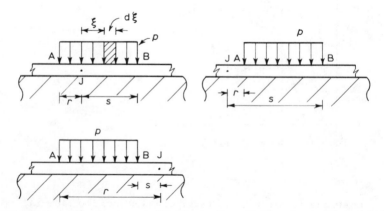

Figure 5.13 Infinite beam on a Winkler medium. Uniform load of finite width.

$$dw_J = \frac{p\lambda d\xi}{2k_s} e^{-\lambda\xi}[\cos(\lambda\xi) + \sin(\lambda\xi)]. \qquad (5.70)$$

The total deflection at J due to load AB is

$$w_J = \frac{p\lambda}{2k_s}\left[\int_0^r e^{-\lambda\xi}\{\cos(\lambda\xi) + \sin(\lambda\xi)\}d\xi + \int_0^s e^{-\lambda\xi}\{\cos(\lambda\xi) + \sin(\lambda\xi)\}d\xi\right].$$
$$(5.71)$$

Equation 5.71, in which the distance r to the left can be regarded as negative, can be reduced to the form

$$w_J = \frac{p}{2k_s}[2 - D(\lambda r) - D(\lambda s)]. \qquad (5.72)$$

Thus, if r and s are known, the deflection at any point within the loaded region AB can be determined by using Equation 5.72. Similarly, expressions for θ_J, M_J and V_J are obtained

$$[\theta_J; M_J; V_J] = \frac{pB}{4\lambda}\left[\frac{2\lambda^2}{k_s B}\{A(\lambda r) - A(\lambda s)\}; \frac{1}{\lambda}\{B(\lambda r) + B(\lambda s)\}; \{C(\lambda r) - C(\lambda s)\}\right].$$
$$(5.73)$$

Note that if r and s are large $D(\lambda r)$ and $D(\lambda s)$ will be small and $w_J \approx p/k_s$. Hence, under the centre portion of a long uniformly distributed load, the bending of the beam can be neglected and p is transmitted directly to the supporting medium. When the point J lies to the left/right of the loading (Fig. 5.13),

$$w_J = \pm\frac{p}{2k_s}[D(\lambda r) - D(\lambda s)]; \; \theta_J = \frac{+}{-}\frac{p\lambda}{2k_s}[A(\lambda r) - A(\lambda s)],$$
$$(5.74)$$
$$M_J = \mp\frac{pB}{4\lambda^2}[B(\lambda r) - B(\lambda s)]; \; V_J = \frac{pB}{4\lambda}[C(\lambda r) - C(\lambda s)].$$

(iii) Infinite beam subjected to a concentrated moment

A further standard solution useful for design purposes is that of the infinite beam which is subjected to a concentrated moment M_0 at the origin (Fig. 5.11). As outlined in Section 5.3.1, two concentrated forces can be combined to yield a concentrated moment. It can be shown that for the portion of the infinite beam to the right of the point of application of the moment

$$[w(x); \theta(x); M(x); V(x)] = \frac{M_0}{B}\left[\frac{\lambda^2}{k_s}B(\lambda x); \frac{\lambda^3}{k_s}C(\lambda x); \frac{B}{2}D(\lambda x); -\frac{\lambda B}{2}A(\lambda x)\right]. \quad (5.75)$$

In this case $w(x)$ is nominally positive to the right of the point of application of the clockwise moment and negative to the left; $\theta(x)$ is nominally positive to both the left and right of the origin; $M(x)$ is nominally positive to the right and negative to the left and $V(x)$ is negative to both the left and right. The signs are reversed if M_0 acts in the anticlockwise direction. All these are subject to the sign of $A(\lambda x), \ldots, D(\lambda x)$.

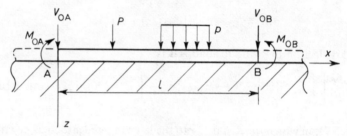

Figure 5.14 Method of superposition: notation for a footing of finite length.

The essential features of the method of superposition proposed by Hetényi (1936) are now examined. As an example of such an analysis, consider a beam of finite length l (Fig. 5.14) subjected to the arbitrary external loads P and p. The ends A and B are assumed to be free. Firstly, this finite beam is assumed to occupy a portion of an infinite beam with the same external loading. In this case the origin of co-ordinates coincides with one end of the beam. The bending moments and shearing forces at the locations A and B (M_A, V_A and M_B, V_B) in the infinite beam are calculated. The boundary conditions at the ends are satisfied by introducing the end conditioning moments (M_{OA}, M_{OB}) and shearing forces (V_{OA}, V_{OB}) which reduce the calculated bending moments and shearing forces at the ends of the finite beam to zero simultaneously. This is effected by making use of the results derived for the infinite beam under the action of an external concentrated force and couple. The forces V_{OA} and V_{OB} give rise to moments $V_{OA}\{C(\lambda 0)/4\lambda\}$ and $V_{OB}\{C(\lambda l)/4\lambda\}$ respectively at A. Similarly, the moments M_{OA} and M_{OB} produce moments $M_{OA}\{D(\lambda 0)/2\}$ and $M_{OB}\{D(\lambda l)/2\}$ respectively at A. Therefore, the zero moment condition at the free end A requires that the equation

$$\frac{V_{OA}}{4\lambda}C(\lambda 0) + \frac{V_{OB}}{4\lambda}C(\lambda l) + \frac{M_{OA}}{2}D(\lambda 0) + \frac{M_{OB}}{2}D(\lambda l) = -M_A \quad (5.76)$$

be satisfied. Conditions similar to Equation 5.76 can be derived for the remaining three boundary conditions. As a result, four simultaneous equations are obtained for the unknowns M_{OA}, M_{OB}, V_{OA} and V_{OB}, which can be represented in the matrix form

$$
\begin{bmatrix}
\dfrac{C(\lambda 0)}{4\lambda} & \dfrac{C(\lambda l)}{4\lambda} & \dfrac{D(\lambda 0)}{2} & \dfrac{D(\lambda l)}{2} \\[2mm]
\dfrac{C(\lambda l)}{4\lambda} & \dfrac{C(\lambda 0)}{4\lambda} & \dfrac{D(\lambda l)}{2} & \dfrac{D(\lambda 0)}{2} \\[2mm]
-\dfrac{D(\lambda 0)}{2} & \dfrac{D(\lambda l)}{2} & -\dfrac{\lambda A(\lambda 0)}{2} & \dfrac{\lambda A(\lambda l)}{2} \\[2mm]
\dfrac{D(\lambda l)}{2} & \dfrac{D(\lambda 0)}{2} & -\dfrac{\lambda A(\lambda l)}{2} & \dfrac{\lambda A(\lambda 0)}{2}
\end{bmatrix}
\begin{bmatrix}
V_{OA} \\[2mm] V_{OB} \\[2mm] M_{OA} \\[2mm] M_{OB}
\end{bmatrix}
+
\begin{bmatrix}
M_A \\[2mm] M_B \\[2mm] V_A \\[2mm] V_B
\end{bmatrix}
= 0. \quad (5.77)
$$

For conciseness write Equation 5.77 as

$$[C][R] = [F], \tag{5.78}$$

where $[C]$ is the coefficients matrix (note that $A(\lambda 0)$, $C(\lambda 0)$ and $D(\lambda 0)$ are unity), $[R]$ is the unknown end conditioning force matrix and $[F]$ is the external force matrix. By inverting the matrix relationship (Equation 5.78)

$$[R] = [C]^{-1}[F], \tag{5.79}$$

where $[C]^{-1}$ is the inverse of the coefficients matrix. It can be shown that

$$\left. \begin{array}{c} V_{0A} \\ V_{0B} \end{array} \right\} = P_0' \pm P_0''; \qquad \left. \begin{array}{c} M_{0A} \\ M_{0B} \end{array} \right\} = M_0' \pm M_0'', \tag{5.80a}$$

where

$$P_0' = 4E_1[S_A'(1 + D(\lambda l)) + \lambda M_A'(1 - A(\lambda l))];$$
$$P_0'' = 4E_2[S_A''(1 - D(\lambda l)) + \lambda M_A''(1 + A(\lambda l))], \tag{5.80b}$$

$$M_0' = -\frac{2E_1}{\lambda}[S_A'(1 + C(\lambda l)) + 2\lambda M_A'(1 - D(\lambda l))];$$

$$M_0'' = -\frac{2E_2}{\lambda}[S_A''(1 - C(\lambda l)) + 2\lambda M_A''(1 + D(\lambda l))],$$

and

$$\left. \begin{array}{c} S_A' \\ S_A'' \end{array} \right\} = \tfrac{1}{2}(V_A \pm V_B); \qquad \left. \begin{array}{c} M_A' \\ M_A'' \end{array} \right\} = \tfrac{1}{2}(M_A \pm M_B), \tag{5.80c}$$

$$\left. \begin{array}{c} E_1 \\ E_2 \end{array} \right\} = [2(1 + D(\lambda l))(1 - D(\lambda l)) - (1 \mp A(\lambda l))(1 \pm C(\lambda l))]^{-1}. \tag{5.80d}$$

Equations 5.80a to d are identical to those derived by Hetényi (1946), from first principles, by specifying that M_A', S_A' and M_A'', S_A'' are respectively the bending moments and shearing forces at the ends of the footing for the symmetrical and asymmetrical loading conditions (Fig. 5.15). The symbols adopted by Hetényi have, therefore, been retained in this section for purposes of comparison. Explicit values for the unknowns V_{0A}, V_{0B}, M_{0A} and M_{0B} for any external loading condition can be determined using the results derived here. These results, of course, apply only for the particular case of a finite beam with free ends. Similar results can easily be developed for other boundary conditions. The matrix inversion procedure is ideally suited for evaluation by computer.

From the design point of view, if the cantilever end of a footing extends sufficiently far beyond the end load or column (Fig. 5.16), then the values of the end conditioning forces become small and the application of end conditioning is not necessary.

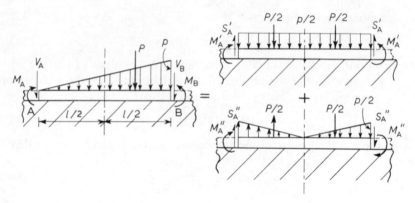

Figure 5.15 Symmetrical and asymmetrical external loading. Hetényi's notation for continuous footings of finite length.

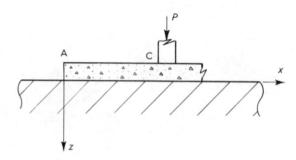

Figure 5.16 Projection of footings beyond column loads.

If $\lambda x > 3.0$, then M_A and M_B are less than 10% of the moments under the column loads. Very approximately, for this situation to exist (with $E_b \approx 13.8 \times 10^3$ MN m^{-2}, $k_s = 54.3$ MN m^{-3}), the cantilever projection AC (Fig. 5.16) should exceed 10 to 30 times the depth of the footing. Such long cantilever ends are uncommon in practice. Therefore, any preliminary design of a footing should be approached on the basis of the analysis of a beam of finite length.

(b) The strain energy method

The strain energy method described in the following paragraphs was developed by Hetényi (1946) and Christopherson (1956). The variational approach based on energy principles is applied to the problem of a finite beam resting on a Winkler medium. This method employs Fourier series approximations for the variables encountered in the problem and the final solutions are presented in series form. The accuracy of the method therefore depends upon the number of terms included in the calculations. Consider the problem of a finite beam of length l and

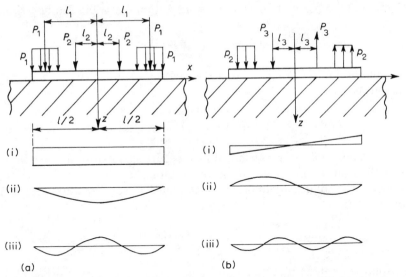

Figure 5.17 Strain energy method of analysis of finite footings on a Winkler medium. (a) Symmetrical loading: (i) $w(x) = w_o$, (ii) $w(x) = a_1 \cos (\pi x/l)$, (iii) $w(x) = a_3 \cos (3\pi x/l)$. (b) Asymmetrical loading: (i) $w(x) = -2w_1 x/l$, (ii) $w(x) = a_1^* \sin (2\pi x/l)$, (iii) $w(x) = a_2^* \sin (4\pi x/l)$.

rigidity $E_b I$ resting on a Winkler foundation (Fig. 5.17). For any arbitrary external loading, the displacement of the beam can be regarded as consisting of three parts (Fig. 5.17): (i) a rigid body translation w_0 along its entire length; (ii) a rigid body rotation about its centre of symmetry leading to a displacement $2xw_1/l$ at any point; (iii) a deformation due to the flexure of the beam which is composed of wave forms of the type $a_n \cos (n\pi x/l)$ (in the symmetric case) such that the total deflection is

$$w(x) = \sum_{n=1,3,5}^{\infty} a_n \cos\left(\frac{n\pi x}{l}\right). \tag{5.81}$$

The individual terms of Equation 5.81 are represented graphically in Fig. 5.17. The series $n = 1, 3, 5$, is chosen so that the displacements at the ends A and B are zero. The strain energy method leads to the evaluation of w_0 and the Fourier coefficients a_n; these values, combined with the relation $q = k_s w$, determine the contact stress distribution. The contact stress distribution can, in turn, be employed to evaluate the deflections, slopes, etc., in the finite beam.

(i) Symmetrical loading

Consider firstly any symmetrical loading as shown in Fig. 5.17. By definition, the rotational displacement is zero over the entire length of the beam and the deflection of the beam at any point distant x from the origin can be written as

$$w(x) = w_0 + \sum_{n=1,3}^{\infty} a_n \cos\left(\frac{n\pi x}{l}\right). \tag{5.82}$$

For equilibrium, the total external load (W) must be equal to the total soil reaction; hence

$$W = \int p\,dx + \sum_{r=1}^{t} P_r = Bk_s\left[lw_0 + \int_{-l/2}^{l/2} \left\{ \sum_{n=1,3}^{\infty} a_n \cos\left(\frac{n\pi x}{l}\right)\right\} dx \right], \tag{5.83}$$

where p is any uniformly distributed load occupying an arbitrary region of the beam; the limits of the integral for p are, therefore, kept indefinite. The term ΣP_r represents the total magnitude of concentrated forces. Equation 5.83 can be reduced to the form

$$\frac{W}{Bk_s l} = w_0 + \sum_{n=1,3}^{\infty} (-1)^{(n-1)/2} \frac{2a_n}{n\pi}. \tag{5.84}$$

The total work done by external loads is

$$U_W = \frac{Ww_0}{2} + \frac{1}{2}\int_{-l/2}^{l/2} \left\{ \sum_{n=1,3}^{\infty} pa_n \cos\left(\frac{n\pi x}{l}\right)\right\} dx + \frac{1}{2}\sum_{n=1,3}^{\infty}\sum_{r=1}^{t} P_r a_n \cos\left(\frac{n\pi\xi}{l}\right), \tag{5.85}$$

where ξ is the distance from the origin to each concentrated load and the last summation of Equation 5.85 includes all the concentrated loads for all values of n and r. The work done by all the loads when a displacement $\cos(n\pi x/l)$ is imposed on the beam is

$$U_U = \frac{1}{2}\int_{-l/2}^{l/2} \left\{ p\cos\left(\frac{n\pi x}{l}\right)\right\} dx + \frac{1}{2}\sum_{r=1}^{t} P_r \cos\left(\frac{n\pi\xi}{l}\right). \tag{5.86}$$

Using Equation 5.86, Equation 5.85 can be reduced to the form

$$2U_W = Ww_0 + 2\sum_{n=1,3}^{\infty} U_U a_n. \tag{5.87}$$

The strain energy stored in the Winkler medium is

$$2U_s = \int_{-l/2}^{l/2} Bk_s\{w(x)\}^2 dx, \tag{5.88}$$

and substituting for $w(x)$, defined by Equation 5.82,

$$2U_s = Bk_s \int_{-l/2}^{l/2} \left\{ w_0^2 + 2w_0 \sum_{n=1,3}^{\infty} a_n \cos\left(\frac{n\pi x}{l}\right) + \sum_{n=1,3}^{\infty} a_n^2 \cos^2\left(\frac{n\pi x}{l}\right) \right.$$

$$\left. + 2\sum_{n=1,3 \cdot m=1,3}^{\infty}\sum_{m=1,3}^{\infty} a_n a_m \cos\left(\frac{n\pi x}{l}\right)\cos\left(\frac{m\pi x}{l}\right)\right\} dx. \tag{5.89}$$

Using the orthogonal properties of the double Fourier series, the last term of the integral in Equation 5.89 is always zero for all values of n and m where $n \neq m$. Then

$$2U_s = Bk_s l \left\{ w_0^2 + 2w_0 \sum_{n=1,3}^{\infty} (-1)^{(n-1)/2} \frac{2a_n}{n\pi} + \frac{1}{2} \sum_{n=1,3}^{\infty} a_n^2 \right\}. \qquad (5.90)$$

Similarly it can be shown that the strain energy in the beam due to flexure

$$U_F = \int_{-l/2}^{l/2} \frac{E_b I}{2} \left(\frac{d^2 w}{dx^2} \right)^2 dx, \qquad (5.91)$$

can be written in the form (neglecting the product term in the integral as in Equation 5.89)

$$2U_F = \frac{E_b I}{2} \sum_{n=1,3}^{\infty} \frac{a_n^2 n^4 \pi^4}{l^3}. \qquad (5.92)$$

Combining Equations 5.90 and 5.92,

$$2(U_S + U_F) = Bk_s l \left\{ w_0^2 + 2w_0 \sum_{n=1,3}^{\infty} \left[(-1)^{(n-1)/2} \frac{2a_n}{n\pi} + \frac{a_n^2}{2} \right] \right\} + \frac{E_b I}{2} \sum_{n=1,3}^{\infty} \frac{a_n^2 n^4 \pi^4}{l^3}. \qquad (5.93)$$

The principle of minimum strain energy is now applied to this problem. Assume that a single value of a_n is varied, keeping w_0 and all other values of a_n constant, thus prescribing displacements in the elastic system. If this prescribed displacement is such that the total strain energy in the foundation and the beam is a minimum, then the variation of strain energy $[\delta(U_S + U_F)]$ becomes completely identical with the variation in potential energy (δU_W) such that

$$\delta U_W = \delta(U_S + U_F). \qquad (5.94)$$

Therefore

$$\frac{\partial U_W}{\partial a_n} = \frac{\partial}{\partial a_n}(U_S + U_F). \qquad (5.95)$$

Using Equations 5.85, 5.93 and 5.95,

$$U_u = \frac{Bk_s l w_0}{n\pi}(-1)^{(n-1)/2} + \frac{Bk_s l a_n}{4} + \frac{E_b I n^4 \pi^4 a_n}{4l^3}, \qquad (5.96)$$

from which

$$a_n = \frac{4Wn^2\pi^2}{Bk_s l C(n)} \left[\frac{U_U}{W} - \frac{Bk_s l w_0 (-1)^{(n-1)/2}}{Wn\pi} \right], \qquad (5.97)$$

where

$$C(n) = n^2 \pi^2 \left\{ 1 + \frac{n^4 \pi^4 E_b I}{Bk_s l^4} \right\}. \qquad (5.98)$$

The coefficients a_n are determined from Equation 5.97 in terms of known quantities and the parameter w_0. This parameter is obtained by substituting Equation 5.97 into Equation 5.84. Then

$$w_0 = \frac{W}{Bk_s l}\left[1 - \sum_{n=1,3}^{\infty}\frac{(-1)^{(n-1)/2}}{C(n)}8n\pi\frac{U_U}{W}\right]\left[1 - \sum_{n=1,3}^{\infty}\frac{8}{C(n)}\right]^{-1}. \qquad (5.99)$$

(ii) Asymmetrical loading

The case of a purely asymmetrical loading (Fig. 5.18) can also be analysed by adopting a similar technique to that developed for symmetrical loading. Therefore, only the relevant results are presented here. The deformation of the beam takes the form

$$w(x) = \frac{2xw_1}{l} + \sum_{n=1,2}^{\infty}a_n^* \sin\left(\frac{2n\pi x}{l}\right). \qquad (5.100)$$

The superscript * refers to the case of asymmetrical loading. It can be shown that the moment of the external forces about the axis of symmetry is

$$M_W = Bk_s l^2\left[\frac{w_1}{6} + \sum_{n=1,2}^{\infty}a_n^* \sin\left(\frac{2n\pi x}{l}\right)\right], \qquad (5.101)$$

and the total work done by the loads is

$$U_W^* = w_1\frac{M_W}{l} + \frac{1}{2}\int_{-l/2}^{l/2}\left\{\sum_{n=1,2}^{\infty}p^*a_n^* \sin\left(\frac{2n\pi x}{l}\right)\right\}dx$$
$$+ \frac{1}{2}\sum_{n=1,2}^{\infty}\sum_{r=1,2}^{t}P_r^*a_n^* \sin\left(\frac{2n\pi\xi}{l}\right). \qquad (5.102)$$

As before, the work done by all the loads when a displacement $\sin(2n\pi x/l)$ is

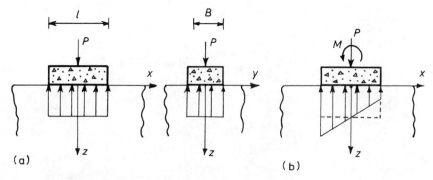

(a)   (b)

Figure 5.18 Rigid footing resting on a Winkler medium. (a) Symmetrical loading, $q(x) = P/Bl$; (b) asymmetrical loading, $q(x) = P/Bl - 12Mx/Bl^3$.

imposed on the footing is

$$U_U^* = \frac{1}{2} \int_{-l/2}^{l/2} \left\{ p^* \sin\left(\frac{2n\pi x}{l}\right) \right\} dx + \frac{1}{2} \sum_{r=1,2}^{t} P_r^* \sin\left(\frac{2n\pi\xi}{l}\right). \quad (5.103)$$

Then

$$U_W^* = \frac{w_1 M_W}{l} + \sum_{n=1,2}^{\infty} a_n U_U^*. \quad (5.104)$$

The strain energy in bending (U_F^*) and in compression of the soil medium (U_S^*) can be shown to be equal to

$$2(U_S^* + U_F^*) = \frac{Bk_s l w_1^2}{3} + 4Bk_s l w_1 \sum_{n=1,2}^{\infty} (-1)^{n-1} \frac{a_n^*}{2n\pi}$$

$$+ \frac{Bk_s l}{2} \sum_{n=1,2}^{\infty} (a_n^*)^2 + \frac{E_b I}{2} \sum_{n=1,2}^{\infty} \frac{16 n^4 \pi^4 (a_n^*)^2}{l^3}. \quad (5.105)$$

Again, using the variational condition

$$\frac{\partial U_W^*}{\partial a_n^*} = \frac{\partial}{\partial a_n^*}(U_S^* + U_F^*) \quad (5.106)$$

it follows that

$$a_n^* = \frac{4 M_W n^2 \pi}{Bk_s l^2 D(n)} \left[\frac{U_U^* l}{M_W} - \frac{Bk_s l^2 w_1 (-1)^{n-1}}{2 M_W n\pi} \right], \quad (5.107)$$

where

$$D(n) = n^2 \pi^2 \left\{ 1 + \frac{16 n^4 \pi^4 E_b I}{Bk_s l^4} \right\}. \quad (5.108)$$

The displacement w_1 required to determine the coefficients a_n^* for the case of asymmetrical loading is given by

$$w_1 = \frac{6 M_W}{Bk_s l^2} \left[1 - \sum_{n=1,2}^{\infty} \frac{(-1)^{n-1} 2n\pi l U_W^*}{D(n) M_W} \right] \left[1 - \sum_{n=1,2}^{\infty} \frac{6}{D(n)} \right]^{-1}. \quad (5.109)$$

The displacements of the footing are now completely defined and the contact stress distribution can be determined. The shear force and bending moment at any point due to the external loads and the contact stresses can be found by integration between appropriate limits. If the magnitude of the second moment of area cannot be assessed from previous designs, it can be estimated from consideration of shearing forces at the columns assuming, say, a linear contact stress distribution obtained on the basis of a 'rigid beam analysis' as mentioned earlier. This method can also be extended to take into account variation in $E_b I$ and/or k_s along the length of the beam. In such cases the product term integrals of

Equations 5.89 and 5.102 are not, in fact, zero. When these terms are retained, a set of simultaneous equations are obtained for a_n and a_n^*. As pointed out in Section 5.3.2, owing to the discontinuous nature of the Winkler medium, three-dimensional effects do not affect the analysis of the finite beam.

(c) Classification of finite beams on a Winkler medium in relation to their stiffness

At this point it is pertinent to include a classification of the response of finite beams resting on a Winkler medium in relation to their stiffness. It was noted earlier that, in the analysis of soil–foundation interaction based on the Winkler assumption, the quantity λl, where λ is the characteristic length and l is the length of the beam, indicates the stiffness of the beam in relation to the supporting soil medium. Thus λl determines the magnitude of the curvature of the deflected shape and defines the rate at which the effects of external loads diminish along the length of the beam. The non-dimensional parameter λl can, therefore, be employed to classify beams according to their length. Such a classification (Hetényi, 1946) is of considerable importance to the design of beams since it offers the possibility of using approximations to the general analysis of beams of finite length by neglecting certain quantities (flexibility of beam, length of beam, etc.) in particular instances.

Group I: Short beams $\lambda l < \pi/4$
Group II: Beams of medium length $\pi/4 < \lambda l < \pi$
Group III: Long beams $\lambda l > \pi$.

Group I includes foundation beams whose dimensions are such that flexural deformations can be neglected altogether. In fact, such beams identically satisfy the conditions that are embodied, somewhat unwittingly, in the conventional design of footings and the short beam can be treated as rigid. Under the action of a symmetrical load the contact stress distribution under this type of foundation is uniform (Fig. 5.18). The magnitude of the contact stresses can be obtained by a direct appeal to the equations of equilibrium. For asymmetrical loading the contact stress distribution is linear (Fig. 5.18).

A common error among designers is to assess new dimensions for the footing or beam, based on the contact stress distributions thus obtained, in order to reduce either material or construction costs. There is, of course, no guarantee that the new dimensions resulting from such a calculation will satisfy conditions requisite for the treatment of the foundation as a rigid footing. It is, therefore, advisable that the final design is checked to verify that the condition $\lambda l < \pi/4$ is satisfied.

Beams which satisfy the requirements of Group II can be classified as finite flexible beams where an accurate analysis is necessary. The method of super-position (or any other technique) can be used for the purposes of analysis. In Group III, the length of the beam is large enough to neglect effects of end conditioning forces. Note that as λl becomes large (> 3.0), the values of

$A(\lambda l) \ldots D(\lambda l)$ are negligible and the resulting computations are greatly simplified. The analytical techniques developed for the infinite beam (Section 5.3.1) are directly applicable to this particular class of problems.

5.3.5 Finite beams on a two-parameter elastic medium

The superposition technique outlined in Section 5.3.4 can be extended also to the analysis of finite beams resting on a two-parameter elastic soil medium. Here, the finite beam can be considered as occupying a portion of an infinitely long strip foundation (Fig. 5.19), in which case the problem is treated on the basis of two-dimensional plane strain or is considered as a beam with finite plan dimensions (Fig. 5.19) which is analysed as a complete three-dimensional beam problem.

In the analysis of the latter class of problems it is convenient to assume that the cross-section of the finite beam remains rigid so that the displacements across the width of the beam are uniform. Also, the form of the contact stress distribution across any such rigid cross-section is assumed to be the same as that obtained on the basis of plane strain conditions in the y, z plane. The basic differential equation governing the flexure of a beam resting on a two-parameter elastic medium can be written in the form

$$E_b I \frac{\mathrm{d}^4 w}{\mathrm{d}x^4} - G_p B^* \frac{\mathrm{d}^2 w}{\mathrm{d}x^2} + k_s B^* w = B p(x), \qquad (5.110)$$

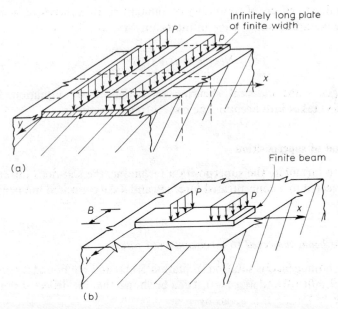

Figure 5.19 Finite plate on a two-parameter elastic medium. (a) Plane strain problem of the finite beam; (b) three-dimensional problem of the finite beam.

where $B^* = B$ in the case of the plane strain problem and $B^* = B[1 + \sqrt{(G_p/B^2 k_s)}]$ in the case of a beam of finite width. The general homogeneous solution of Equation 5.110 is

$$w(x) = [A_1 e^{\mu\lambda x} + A_2 e^{-\mu\lambda x}]\cos\beta\lambda x + [A_3 e^{\mu\lambda x} + A_4 e^{-\mu\lambda x}]\sin\beta\lambda x, \qquad (5.111)$$

where

$$\left.\begin{matrix}\mu\\\beta\end{matrix}\right\} = \sqrt{\left(1 \pm \frac{G_p\lambda^2}{k_s}\right)}, \qquad (5.112)$$

and $\lambda^4 = k_s B^*/4E_b I$. Note that as $G_p \to 0$; $\mu \to \beta$ and Equation 5.111 becomes equal to Equation 5.66. In obtaining Equation 5.111 it has been assumed that $(G_p\lambda^2/k_s) < 1$. From Equation 5.18 note that, by adopting Vlazov's interpretation of k_s and G_p; since $E_s > 0$ and $v_s < \frac{1}{2}$, k_s and G_p are always positive. It is possible that $(G_p\lambda^2/k_s) \geqslant 1$; in which case the form of Equation 5.111 changes. When $(G_p\lambda^2/k_s) = 1$, β is zero and Equation 5.111 can be directly transformed. When $(G_p\lambda^2/k_s) > 1$, however, the proper homogeneous solution of Equation 5.110 is obtained by replacing β in Equation 5.111 by $i\bar{\beta}$ where $i^2 = -1$. Furthermore, note that the inequality $(G_p\lambda^2/k_s) < 1$, which can be rewritten as $G_p < 2\sqrt{(k_s E_b I/B^*)}$, is satisfied in a majority of foundation problems, since k_s is usually greater than G_p. The homogeneous solution (Equation 5.111) is, therefore, adopted for the development of the method of superposition.

In discussing three-dimensional effects associated with two-parameter soil models, it was observed that concentrated reactions are developed at the ends of the beam due to effects of continuity of foundation. It is, therefore, necessary to define a generalized shear force in the beam $N(x)$ as

$$N(x) = V(x) + G_p B^* \frac{dw}{dx}, \qquad (5.113)$$

where $V(x)$ is the classical shearing force defined by Equation 5.68 and $GB^*(dw/dx)$ takes into account the concentrated reactions.

(a) Method of superposition

In order to formulate the superposition technique, the solutions for an infinite beam subjected to a concentrated force (P) and a concentrated moment (M) are required.

(i) Infinite beam subjected to a concentrated force

Using the homogeneous solution (Equation 5.111) and the boundary conditions $N(0) = P/2$, $\theta(0) = 0$ and $w(\infty) = 0$, it can be shown that the deflected shape of the infinite beam for $x > 0$ is given by

$$w(x) = \frac{P\lambda}{2B^* k_s} e^{-\mu\lambda x}\left[\frac{1}{\mu}\cos\beta\lambda x + \frac{1}{\beta}\sin\beta\lambda x\right] = \frac{P\lambda}{2B^* k_s} A^*(\mu, \beta). \qquad (5.114a)$$

Using this solution in Equations 5.68 and 5.113,

$$\theta(x) = -\frac{P\lambda^2 e^{-\mu\lambda x}}{B^* k_s \mu \beta} \sin \beta\lambda x = -\frac{P\lambda^2}{B^* k_s} B^*(\mu, \beta), \qquad (5.114b)$$

$$M(x) = \frac{P}{4\lambda} e^{-\mu\lambda x}\left[\frac{1}{\mu}\cos \beta\lambda x - \frac{1}{\beta}\sin \beta\lambda x\right] = \frac{P}{4\lambda} C^*(\mu, \beta), \qquad (5.114c)$$

$$N(x) = \frac{P}{2} e^{-\mu\lambda x}\left[\cos \beta\lambda x + \frac{(\mu^2 - \beta^2)}{2\mu\beta}\sin \beta\lambda x\right] = -\frac{P}{2} D^*(\mu, \beta). \quad (5.114d)$$

Note that the form of Equations 5.114 a to d is very similar to that obtained for the Winkler case in Section 5.3.4. The functions A* ... D* have to be evaluated for values of the two variables μ and β. Tables of these functions are given by Vlazov and Leontiev (1966).

(ii) Infinite beam subjected to a concentrated moment

In the case of a clockwise moment M acting at the origin, the deflection, slope etc., in the infinite beam (for $x > 0$) are given by

$$[w(x); \theta(x); M(x); N(x)]$$
$$= \frac{M}{B^*}\left[\frac{\lambda^2}{k_s}B^*(\mu, \beta); \frac{\lambda^3}{k_s}C^*(\mu, \beta); \frac{B^*}{2}\bar{D}^*(\mu, \beta); -\frac{\lambda B^*}{2}A^*(\mu, \beta)\right], \qquad (5.115)$$

where

$$\bar{D}^* = e^{-\mu\lambda x}\left[\cos \beta\lambda x - \frac{(\mu^2 - \beta^2)}{2\mu\beta}\sin \beta\lambda x\right]. \qquad (5.116)$$

Except for a slight change of signs, Equation 5.115 is very similar to Equation 5.75. The procedure adopted for the analysis of a finite beam on a two-parameter elastic medium is exactly the same as that outlined for the Winkler case. The solution of the corresponding infinite beam problem furnishes M_A^*, M_B^*, N_A^* and N_B^*. These correspond to the external force matrix $[F^*]$. The end conditioning forces M_{0A}^*, M_{0B}^*, N_{0A}^* and N_{0B}^* (or $[R^*]$) are obtained from a matrix relation of the type

$$[R^*] = [C^*]^{-1}[F^*], \qquad (5.117)$$

where $[C^*]$ is the influence coefficients matrix. For the case of a finite beam with free ends, the matrices $[F^*]$ and $[C^*]$ take the form

$$[F^*] = \begin{bmatrix} -M_A^* \\ -M_B^* \\ -N_A^* + \alpha B^* w_A^* G_P \\ -N_B^* - \alpha B^* w_B^* G_P \end{bmatrix} \qquad (5.118a)$$

$$
[C^*] = \begin{bmatrix}
\dfrac{C^*(0)}{4\lambda} & \dfrac{C^*(l)}{4\lambda} & \dfrac{\dot{D}^*(0)}{2} & \dfrac{\dot{D}^*(l)}{2} \\[2ex]
\dfrac{C^*(l)}{4\lambda} & \dfrac{C^*(0)}{4\lambda} & \dfrac{\dot{D}^*(l)}{2} & \dfrac{\dot{D}^*(0)}{2} \\[2ex]
-\dfrac{D^*(0)}{2} - \dfrac{\lambda}{2\alpha}A^*(0) & \dfrac{D^*(l)}{2} - \dfrac{\lambda}{2\alpha}A^*(l) & -\dfrac{\lambda}{2}A^*(0) - \dfrac{\lambda^2 B^*(0)}{\alpha} & \dfrac{\lambda}{2}A^*(l) - \dfrac{\lambda^2}{\alpha}B^*(l) \\[2ex]
-\dfrac{D^*(l)}{2} + \dfrac{\lambda}{2\alpha}A^*(l) & \dfrac{D^*(0)}{2} + \dfrac{\lambda}{2\alpha}A^*(0) & -\dfrac{\lambda}{2}A^*(l) + \dfrac{\lambda^2 B^*(l)}{\alpha} & \dfrac{\lambda}{2}A^*(0) + \dfrac{\lambda^2}{\alpha}B^*(0)
\end{bmatrix}
$$

(b) The strain energy method

The strain energy method proposed by Hetényi (1946) and Christopherson (1956) can also be adopted for the analysis of finite beams resting on a two-parameter elastic medium. However, the resulting expressions tend to be rather lengthy and unwieldy for the purposes of design calculations.

(c) Classification of finite beams resting on a two-parameter medium in relation to their stiffness

To date there is very little information available on the classification of finite beams on two-parameter media in relation to their stiffness. In view of the fact that two constants are necessary to describe the response of the soil medium, it becomes difficult to extract from the homogeneous solution (Equation 5.111) a single characteristic parameter. Both $\alpha\lambda$ and $\beta\lambda$ influence the flexural deformations of the beam. The only information available is that due to Vlazov and Leontiev (1966). They employ the non-dimensional parameter $\lambda^* l$ where λ^* can be reduced to the form

$$(\lambda^*)^3 = \frac{(1-v_s)G_s B^*}{8(1-2v_s)E_b I}, \tag{5.119}$$

and k_s and G_p (or $2t$) are related to v_s and G_s by Equation 5.16. Results for the deflections, bending moments etc, in a finite beam have been developed for the exponential variation of $h(z)$ (Equation 5.11) for $\gamma = 1.5$, $v_s = 0.23$ and $H \to \infty$. For this very restricted case of the two-parameter model, and for the case of a concentrated loading, the following classification holds.

Group I: short beams $\lambda^* l < 0.86$
Group II: beams of intermediate length $0.86 < \lambda^* l < 1.85$
Group III: long beams $\lambda^* l > 1.85$

From the design point of view, owing to the absence of further information on the stiffness classification, it is advisable to adhere to the classification for finite beams on an elastic solid medium which is presented in Section 5.3.6.

5.3.6 Finite beams on an elastic solid medium

The analysis of finite beams resting on elastic solid media presents a much more complex problem than the two cases described earlier. In the case of the Winkler and two-parameter models, analysis is facilitated by the fact that their respective response functions are of a particularly simple form. As a result, the basic solutions for the infinite beam subjected to a concentrated force or a moment can be solved in closed form in terms of elementary functions. Techniques such as the method of superposition can then be effectively employed in the solution of finite beam problems. In the case of finite beams resting on elastic solid media, solutions for deflection or contact stresses cannot be obtained in closed form. To illustrate this, consider the problem of a finite beam of narrow width in which flexure takes place only in the x–z plane (Fig. 5.20). The distribution of contact stress in the transverse direction is assumed to be the same as that obtained by Sadowsky (1928) on the basis of the plane strain problem. It can then be shown that the flexural deflection of a finite beam resting on an isotropic elastic medium is given by

$$E_b I \frac{\mathrm{d}^4 w}{\mathrm{d}x^4} + \tilde{q}(x) = Bp(x). \tag{5.120}$$

In Equation 5.120 the contact stress per unit length of the beam $\tilde{q}(x)$ is given by

$$w(x) = \frac{2(1 - v_s^2)}{\pi^2 E_s} \int_{-1/2}^{1/2} \frac{\tilde{q}(\xi)}{|x - \xi|} F\left[\frac{\pi}{2}; \frac{iB}{2(x - \xi)}\right] \mathrm{d}\xi, \tag{5.121}$$

where $F[\pi/2; iB/2(x - \xi)]$ is the complete elliptic integral of the first kind. Clearly the solution of Equations 5.120 and 5.121 can be obtained only in an approximate form. Also, since the basic solutions for the infinite beam subjected to a concentrated force or a moment can be obtained only in integral form, it is not always convenient to adopt the method of superposition. The superposition technique has been applied by Drapkin (1955) to obtain solutions to the problem of the centrally loaded finite beam.

The approximate methods of analysis of finite beam problems proposed by Ohde (1942) and Zemochkin and Sinitsyn (1947) are considered below.

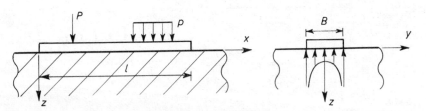

Figure 5.20 Finite beam resting on elastic solid medium; flexure of the beam takes place in the x–z plane.

(a) Zemochkin and Sinitsyn's method

The method of analysis of finite beams proposed by Zemochkin and Sinitsyn (1947) is based on the idea of approximating the contact stress distribution at the interface. It is assumed that contact between the beam and the soil exists only at discrete points. The finite beam is first divided into a number of equal segments. Continuity of displacement at the interface is achieved by a series of hinged rigid bar connections (Fig. 5.21) located at the centres of these segments. The reactive forces in these bars, $X_1, X_2 \ldots X_n$, constitute the basic unknowns of the problem. To simplify the analysis, a cantilever beam is employed as the basic structure (Fig. 5.21). The actual displacement (w_0) and rotation (θ_0) at the free end then constitute additional unknowns of the problem.

By considering the continuity of displacement between the beam and the elastic half-space at the rigid bar locations the n equations are obtained:

$$\sum_{i=1}^{n} X_i(v_{mi} + w_{mi}) - w_0 - a_m\theta_0 + \Delta_{m0} = 0, \ (m = 1, 2 \ldots n). \qquad (5.122)$$

In Equation 5.122, v_{mi} represents the deflection in the cantilever at m due to a unit load at i (Fig. 5.22); w_{mi} represents the surface deflection of the half-space at m due to a unit load uniformly distributed over the area aB, at point i (Fig. 5.22), and Δ_{m0} is the deflection in the cantilever at the point m due to the external loads P and p. The displacements v_{mi} and Δ_{m0} can be easily evaluated by using the classical area-moment theorems of structural analysis (Timoshenko and Young, 1965). The deflection w_{mi} can be calculated by integrating Boussinesq's solution for the surface deflection of an isotropic elastic half-space subjected to a concentrated

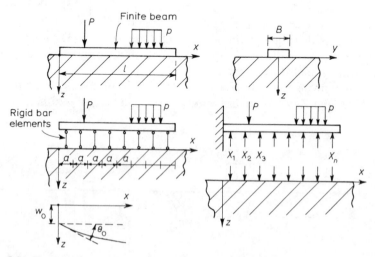

Figure 5.21 Zemochkin's method for the analysis of finite beams resting on an elastic solid medium.

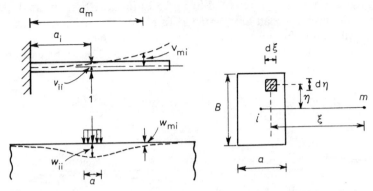

Figure 5.22 Influence coefficients for Zemochkin's method of analysis.

force (P).

$$w(r) = \frac{P(1 - v_s^2)}{\pi E_s r} \tag{5.123}$$

where $r(=x)$ is the radial co-ordinate.

Referring to Fig. 5.22, the deflection at m due to a concentrated elemental load $d\eta d\xi/aB$ (stress intensity of the uniform load is $1/aB$) is

$$dw_{mi} = \frac{(1 - v_s^2)}{aB\pi E_s} \frac{d\eta d\xi}{[\eta^2 + \xi^2]^{1/2}}. \tag{5.124}$$

Integrating Equation 5.124, w_{mi} can be written

$$w_{mi} = \frac{(1 - v_s^2)}{\pi E_s a} F_{mi}, \tag{5.125}$$

where

$$
\begin{aligned}
F_{mi} = \frac{a}{B} &\left\{ 2\ln\frac{B}{a} - \ln\left[\left(\frac{2x}{a}\right)^2 - 1 \right] - \frac{2x}{a}\ln\left[\frac{2x+a}{2x-a} \right] \right. \\
&+ \frac{B}{a}\ln\left[\frac{\left(\frac{2x}{a} + \frac{a}{B}\right) + \left\{\left(\frac{2x}{a} + \frac{a}{B}\right)^2 + 1\right\}^{1/2}}{\left(\frac{2x}{a} - \frac{a}{B}\right) + \left\{\left(\frac{2x}{a} - \frac{a}{B}\right)^2 + 1\right\}^{1/2}} \right] \\
&+ \frac{2x}{a}\ln\left[\frac{1 + \left\{\left(\frac{2x}{B} + \frac{a}{B}\right)^2 + 1\right\}^{1/2}}{1 + \left\{\left(\frac{2x}{B} - \frac{a}{B}\right)^2 + 1\right\}^{1/2}} \right] \\
&+ \left. \ln\left[1 + \left\{\left(\frac{2x}{B} + \frac{a}{B}\right)^2 + 1\right\}^{1/2} \right]\left[1 + \left\{\left(\frac{2x}{B} - \frac{a}{B}\right)^2 + 1\right\}^{1/2} \right] \right\}.
\end{aligned} \tag{5.126}
$$

Typical variations of F_{mi} presented in Fig. 5.23 indicate that effects due to the shape of the uniform load are restricted to the neighbourhood of the applied stress. Values for the coefficient F_{mi} for various values of x/a and B/a are presented in Table 5.1. The n equations (Equation 5.122) can be written in the form

$$\sum_{i=1}^{n} X_i \left\{ v_{mi} + \frac{(1 - v_s^2)}{\pi E_s a} F_{mi} \right\} - w_0 - a_m \theta_0 + \Delta_{m0} = 0, \quad (m = 1, 2 \ldots n). \quad (5.127)$$

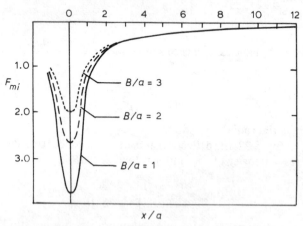

Figure 5.23 Influence functions F_{mi} for surface deflection of an isotropic elastic medium subjected to a uniform rectangular load.

Table 5.1 Influence coefficients for surface deflections (Zemochkin and Sinitsyn's method)

					F_{mi}				
$\dfrac{x}{a}$	$\dfrac{a}{x}$	$\dfrac{B}{a} = \dfrac{2}{3}$	1	2	3	4	5	$\dfrac{x}{a}$	F_{mi}
0	∞	4.265	3.525	2.406	1.867	1.542	1.322	11	0.091
1	1	1.069	1.038	0.929	0.829	1.746	0.678	12	0.083
2	0.500	0.508	0.505	0.490	0.469	0.446	0.424	13	0.077
3	0.333	0.336	0.335	0.330	0.323	0.315	0.305	14	0.071
4	0.250	0.251	0.251	0.249	0.246	0.242	0.237	15	0.067
5	0.200	0.200	0.200	0.199	0.197	0.196	0.193	16	0.063
6	0.167	0.167	0.167	0.166	0.165	0.164	0.163	17	0.059
7	0.143	0.143	0.143	0.143	0.142	0.141	0.140	18	0.056
8	0.125	0.125	0.125	0.125	0.124	0.124	0.123	19	0.053
9	0.111	0.111	0.111	0.111	0.111	0.111	0.110	20	0.050
10	0.100	0.100	0.100	0.100	0.100	0.100	0.099		

These n equations, together with the equations for force and moment equilibrium

$$\sum_{i=1}^{n} X_i - \sum P = 0; \quad \sum_{i=1}^{n} X_i a_i - \sum M = 0, \tag{5.128}$$

where $\sum P$ and $\sum M$ denote, respectively, the sum of all the external forces and external moments about the origin, constitute the $(n+2)$ equations for the unknowns X_n, w_0 and θ_0. The general character of Equations 5.127 and 5.128 remains unchanged for different choices of the basic structure and supporting soil media. Król (1964) has employed a simply supported beam as the basic structure. A number of interesting practical applications of the Zemochkin and Sinitsyn technique to finite beams resting on non-uniform layered media are presented by Król (1964). Barden (1962a) has extended this method to the analysis of finite beams resting on a transversely isotropic elastic half-space. References to further applications involving non-homogeneous elastic media are given by Korenev (1960) and Selvadurai (1979).

(b) Ohde's method

In the method of analysis proposed by Ohde (1942) it is also assumed that the contact stress distribution can be approximated by a series of uniform blocks of stress (Fig. 5.24). These stresses (q_n) are selected as the unknowns of the problem.

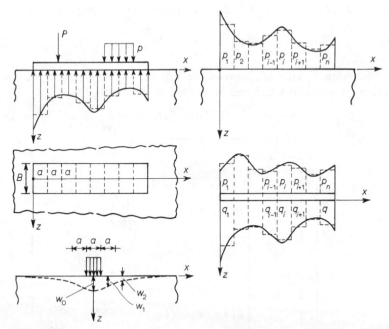

Figure 5.24 Ohde's method of analysis of a finite beam resting on an elastic solid medium.

In Ohde's method it is assumed that continuity of *slope* between the beam and the elastic half-space exists at discrete points. Consider the finite beam shown in Fig. 5.24. The surface deflection, relative to the x-axis, (w_i) at the centre of the elemental stress (q_n) can be written as

$$w_i = \frac{a(1 - v_s^2)}{E_s} \sum_{j=1}^{n} c_{i-j} q_j, \tag{5.129}$$

where c_{i-j} are the influence coefficients for the surface deflection of point i due to a unit load uniformly distributed over the elemental area (aB) at j. These coefficients are also obtained by using Boussinesq's basic solution (Equation 5.123). By comparing Equations 5.125 and 5.129, note that $c_{i-j} = F_{ij}/\pi a^2$.

The moment of external forces at the ith section can be written as

$$M_i = M_L + a^2 B \sum_{j=1}^{i} (q_j - p_j)(i - j), \tag{5.130}$$

where M_L is the clockwise external moment at the end of the beam and $p(= P_j/aB)$ and P_j are, respectively, the external stress and external force acting on the jth section. For overall equilibrium of the foundation

$$aB \sum_{i=1}^{n} q_i = \sum_{i=1}^{n} P_i. \tag{5.131}$$

Using Clapeyron's three moment equation, the deflections w_i can be related to the moments M_i. For continuity of the elastic curve at the ith point

$$-w_{i-1} + 2w_i - w_{i+1} = \frac{a^2}{6E_b I} \{M_{i-1} + 4M_i + M_{i+1}\}. \tag{5.132}$$

By eliminating w_i and M_i between the Equations 5.129 and 5.130, a set of equations is obtained which can be written in the matrix form

$$[C][q] = [p], \tag{5.133}$$

where $[q]$ and $[p]$ are the column matrices

$$[q] = \begin{bmatrix} q_1 \\ q_2 \\ q_3 \\ \vdots \\ q_n \end{bmatrix}, [p] = \begin{bmatrix} \beta\left(p_1 + \dfrac{p_2}{6} - \dfrac{M_L}{a^2 B}\right) \\ \beta\left(2p_1 + p_2 + \dfrac{p_3}{6} - \dfrac{M_L}{a^2 B}\right) \\ \beta\left(3p_1 + 2p_2 + p_3 + \dfrac{p_4}{6} - \dfrac{M_L}{a^2 B}\right) \\ \vdots \\ \beta\left(np_1 + (n-1)p_2 + \cdots + \dfrac{p_{n+1}}{6} - \dfrac{M_L}{a^2 B}\right) \end{bmatrix} \tag{5.134}$$

and

$$[\mathbf{C}] = \begin{bmatrix}
(C_1 + \beta) & \left(C_0 + \dfrac{\beta}{6}\right) & C_1 & C_2 & C_3 & \cdots & C_n \\[2ex]
(C_2 + 2\beta) & (C_1 + \beta) & \left(C_0 + \dfrac{\beta}{6}\right) & C_1 & C_2 & \cdots & C_{n-1} \\[2ex]
(C_3 + 3\beta) & (C_2 + 2\beta) & (C_1 + \beta) & \left(C_0 + \dfrac{\beta}{6}\right) & C_1 & \cdots & C_{n-2} \\[2ex]
\cdot & \cdot & \cdot & \cdot & \cdot & \cdots & \cdot \\[1ex]
\cdot & \cdot & \cdot & \cdot & \cdot & \cdots & \cdot \\[1ex]
(C_n + n\beta) & (C_{n-1} + (n-1)\beta) & \cdot & \cdot & \cdot & \cdots & C_1
\end{bmatrix}$$

$$(5.135)$$

In Equation 5.135, $\beta = a^3 BE_s/(1 - v_s^2)E_b I$ is a non-dimensional parameter and the constants C_i are related to c_i by the equation

$$C_i = c_{i-1} - 2c_i + c_{i+1}. \qquad (5.136)$$

Equation 5.133, together with the equations of force and moment equilibrium, can then be used to obtain the unknown soil reaction q_n for any arbitrary external loading condition. In the case of purely symmetrical external loading and, say, for $n = 10$, $p_1 = p_{10}$, $p_2 = p_9 \ldots$, etc., and $q_1 = q_{10}$, $q_2 = q_9 \ldots$, etc. Here only the equation of force equilibrium is required. Similarly, for purely asymmetric loading, $p_1 = -p_{10} \ldots$ and $q_1 = -q_{10} \ldots$ and only the equation of moment equilibrium is required. Ohde (1942) has applied this technique to the analysis of beams of variable stiffness. An extension of Ohde's method to non-homogeneous and anisotropic media has been presented by Kany (1959). Sommer (1965) has also applied this method to investigate the influence of the rigidity of the superstructure on the flexural deformations in foundations.

(c) Classification of finite beams on an elastic solid medium in relation to their stiffness

The stiffness classification of finite beams resting on an isotropic elastic medium presented here is due to Gorbunov-Posadov and Serebrjanyi (1961). The non-dimensional parameter characterizing the relative stiffness of the soil–foundation system is $\lambda_e l$ where

$$\lambda_e^4 = \frac{\pi E_s B}{32(1 - v_s^2)E_b I l}. \qquad (5.137)$$

Gorbunov-Posadov and Serebrjanyi (1961) examined the displacements, contact stresses etc., in finite beams subjected to a variety of uniformly distributed and localized loading (localized loading is usually assumed to be remote from the ends of the finite beam) and for various ratios of plan dimensions (varying from 10 to

10^2). Results obtained by Gorbunov-Posadov indicate the following classification.

Group I: Short beams $\lambda_e l < 0.84$
Group II: Beams of medium length $0.84 < \lambda_e l < 1.78$
Group III: Long beams $\lambda_e l > 1.78$

In Group I, the results for the finite flexible beam practically coincide with those obtained for the infinitely rigid beam with $\lambda_e l \to 0$. The contact stresses in this type of foundation can be estimated on the basis of a 'rigid stamp' type of analysis (Galin, 1961). In Group III the results for the finite beam are very similar to those obtained for the infinitely long beam. In Group II, the flexibility of the beam should be taken into account, using a finite beam analysis.

Investigations carried out by Vesić (1961*a*,*b*) on the applicability of the Winkler analysis to beams resting on elastic media suggest the following classification.

Group I: Short beams $\lambda l < 0.80$
Group II: Beams of medium length $0.80 < \lambda l < 2.25$
Group III: Moderately long beams $2.25 < \lambda l < 5.00$
Group IV: Long beams $\lambda l > 5.00$

where $\lambda^4 = k_s B/4E_b I$, and k_s is given by Equation 5.32.

Results given by Vesić (1961*a*,*b*) indicate that for short beams resting on elastic continuous media, an analysis based on the Winkler assumption leads to inaccurate results. However, a rough estimate of the contact stress distribution beneath these short beams can be made by assuming absolute rigidity of the beam and a linear distribution of stresses. In the case of long beams or moderately long beams on an elastic medium, an analysis based on the Winkler theory gives reasonable results for centrally located loads. Vesić (1961*a*) has also shown that the same criterion ($\lambda l > 2.25$) can be used when several loads are present on the beam with the total free length outside the extreme loaded regions being greater than *l*.

5.3.7 Design of continuous footings

In the preliminary design of continuous footings it is generally assumed that the foundation is infinitely rigid and that the contact stress distribution is planar. Therefore, to achieve a desirable uniform contact stress distribution beneath the continuous footing it is necessary to arrange the centre of area of the foundation directly beneath the centre of gravity of the external loads. This may necessitate trapezoidal or irregular-shaped footings in plan. If equal column loads are symmetrically disposed about the centre of the footing, then the contact stress distribution is uniform. For asymmetrical loading the contact stress distribution is planar but non-uniform. Using the contact stress distribution and the external loads it is possible to evaluate the flexural moments and shearing forces in the footing.

The footing is designed as a conventional reinforced concrete rectangular or T-beam. Its depth is commonly constant although it need not be if economics dictate otherwise. Transverse reinforcement is required to cater for bending moments due to cantilever action, more particularly in T-beams.

Once the overall dimensions of the continuous footing are determined, an interaction analysis can be carried out to ascertain the influence of the flexibility of the foundation and the deformations of the soil medium. In the following example, the flexural moments, deflections, etc., in a footing of finite length obtained from a preliminary 'rigid footing analysis' are compared with results obtained by assuming Winkler behaviour of the supporting soil medium. It is, of course, possible to extend the analysis to other types of soil models, such as the two-parameter or elastic continuum models, but owing to space limitations such analyses will not be attempted here. Several such examples are contained in the literature already cited. The design of a continuous footing supporting five concentrated loads has been described by Panak *et al.* (1972). The ultimate flexural strength of combined footings has been studied by Nathan (1978) employing yield line analysis.

EXAMPLE 5.1

A continuous reinforced concrete footing of plan dimensions $15.25\,\text{m} \times 1.525\,\text{m}$ is subjected to column loads as shown in Fig. 5.25. The footing rests on a saturated clay medium ($v_s = 0.5$) with undrained elastic modulus $E_s = 95.5\ \text{MN}\,\text{m}^{-2}$. The second moment of area of the rectangular footing section is $0.0288\,\text{m}^4$ (concrete units): it is assumed that there is no cracking of the cross-section due to flexure of the footing. The weight of the footing is neglected. The modulus of elasticity of the concrete is $E_b = 13.8 \times 10\ \text{MN}\,\text{m}^{-2}$.

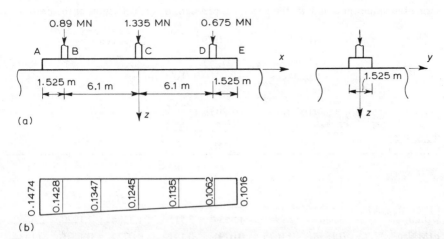

Figure 5.25 Continuous footing; (b) contact stress distribution, infinitely rigid footing.

(a) Rigid footing method

The calculations for this method will not be detailed here; it can be shown that the eccentricity of loads is 0.4667 m to the left of C (Fig. 5.25(a)). The contact stress distribution is shown in Fig. 5.25(b). For future reference, note that at C the shear force $V_C = 0.69$ MN and that the bending moment $M_C = 0.76$ MNm. The effective depth required to resist the maximum bending moment is 609.6 mm. The maximum shear stress is 0.883 MN m^{-2}. If this value exceeds the permissible shear stress, then the depth of the footing can be increased or extra reinforcement has to be provided.

(b) Strain energy method

The modulus of subgrade reaction can be evaluated approximately by using Equation 5.33

$$k_s = \frac{0.65 E_s}{B(1 - v_s^2)} \text{ whence } k_s = 54.33 \text{ MN m}^{-3}.$$

To apply the strain energy method, divide the external loading into its symmetrical and asymmetrical components (Fig. 5.18). To simplify calculation take $P = 0.445$. For the symmetrical case $W = 6.5P$. Evaluating Equation 5.86,

$$U_1 = 2.041P; U_3 = 0.084P; U_5 = 3.250P; U_7 = 0.084P.$$

Using these values of U and noting that $(E_b I \pi^4 / B k_s l^4) = 0.008\,66$, then, from Equation 5.99

$$w_0 = 2.918 \text{ mm}.$$

Evaluating Equation 5.97,

$$a_1 = 0.828 \text{ mm}; a_3 = 0.795 \text{ mm}; a_5 = 0.599 \text{ mm}; a_7 = 0.0304 \text{ mm}.$$

For the asymmetrical case, $M_W = Pl/5$, and evaluating Equation 5.103,

$$U_1^* = -0.1469P; U_2^* = 0.2378P; U_3^* = -0.2378P.$$

Evaluating Equation 5.109, $w_1 = 0.462$ mm. Using the values of U^* and the value of w_1 in Equation 5.107,

$$a_1^* = 0.076 \text{ mm}; a_2^* = 0.058 \text{ mm}; a_3^* = -0.0203 \text{ mm}.$$

Using the values of a and a^*, the general expression for the deflection of the beam is

$$w(x) = 2.918 - 0.828 \cos\frac{\pi x}{l} + 0.795 \cos\frac{3\pi x}{l} + 0.599 \cos\frac{5\pi x}{l}$$

$$+ 0.0304 \cos\frac{7\pi x}{l} - 0.924\frac{x}{l} + 0.076 \sin\frac{2\pi x}{l}$$

$$+ 0.058 \sin\frac{4\pi x}{l} - 0.0203 \sin\frac{6\pi x}{l} + \cdots.$$

Table 5.2

	Location						
	A	B	B–C	C	C–D	D	E
w (mm)	3.3807	2.9947	1.4554	3.5154	1.3284	2.1920	2.4562
q (MN m^{-2})	0.1838	0.1625	0.0791	0.1912	0.0722	0.1191	0.1334

The displacements (w) and corresponding contact stresses (q) at various points under the footing are shown in Table 5.2. Using this contact stress distribution and the external loads it is possible to obtain bending moments and shearing forces in the footing. The bending moment at C obtained from the strain energy method is $M_C = 0.6632$ MNm.

(c) Method of superposition

For the Winkler model, the characteristic length is obtained from the relation

$$\lambda^4 = \frac{Bk_s}{4E_bI}.$$

Evaluating this expression, $\lambda = 0.477\,\text{m}^{-1}$. Hence for the overall length of footing $\lambda l = 7.284$. Therefore, if all the external loads were located near the centre of the beam, since $\lambda l > \pi$, it would be sufficient to use the results for the infinite beam. However, two external loads are located near the ends of the footing and it is pertinent to proceed with the finite beam analysis. By considering the expressions for the bending moment and shearing forces in the infinite beam due to a concentrated force (Equations 5.69b and c), it is possible to obtain the bending moments and shearing forces M_A, M_E, V_A and V_E at points which correspond to the free ends of the finite beam. Values for $C(\lambda x)$, $D(\lambda x)$..., etc., are determined by reference to Table 5.3.

The load at B produces at A ($x = 1.525$ m; $\lambda x = 0.7284$) and E ($x = 13.72$ m; $\lambda x = 6.556$) the following:

$$M_A^B = 0.0188 \text{ MNm}, \qquad V_A^B = 0.1606 \text{ MN},$$
$$M_E^B = 5.12 \times 10^{-4} \text{ MNm}, \quad V_E^B = -6.23 \times 10^{-4} \text{ MN}.$$

The load at C produces at A ($x = 7.62$ m; $\lambda x = 3.642$) and E ($x = 7.62$ m; $\lambda x = 3.642$) the following:

$$M_A^C = -73.33 \times 10^{-4} \text{ MNm}, \quad V_A^C = -153.52 \times 10^{-4} \text{ MN},$$
$$M_E^C = -73.33 \times 10^{-4} \text{ MNm}, \quad V_E^C = 153.52 \times 10^{-4} \text{ MN}.$$

The load at D produces at A ($x = 13.72$ m; $\lambda x = 0.7284$) and E ($x = 1.524$ m; $\lambda x = 0.7284$) the following:

$$M_A^D = 3.84 \times 10^{-4} \text{ MNm}, \qquad V_A^D = 4.67 \times 10^{-4} \text{ MN},$$
$$M_E^D = 140.68 \times 10^{-4} \text{ MNm}, \quad V_E^D = -0.1205 \text{ MN}.$$

Combining the values of M and V produced by loads at B, C and D,

$$M_A = 118.08 \times 10^{-4} \text{ MNm}, \quad V_A = 0.1457 \text{ MN},$$
$$M_E = 72.47 \times 10^{-4} \text{ MNm}, \quad V_E = 0.1057 \text{ MN}.$$

The column matrix [F] of Equation 5.78 is now known. The elements of the coefficients matrix [C] can be determined by using the values of λ and λl. Solving for the end conditioning forces M_{0A}, M_{0E}, V_{0A} and V_{0E}

$$M_{0A} = -0.6568 \text{ MNm}; \qquad V_{0A} = 0.1457 \text{ MN},$$
$$M_{0E} = -0.4714 \text{ MNm}; \qquad V_{0E} = 0.4368 \text{ MN}.$$

Alternatively, the end conditioning forces can be obtained from Equations 5.80a to d. The deflected shape of the finite beam is calculated by combining the deflections due to both the external loads and the end conditioning forces. These are shown in Table 5.4.

The shearing forces and bending moments in the continuous footing can be calculated in

Table 5.3 Functions A, B, C, D, E_1 and E_2 for use with the Winkler model

λx	A	B	C	D	E_1	E_2
0	1.0000	0	1.0000	1.0000	∞	∞
0.1	0.9907	0.0903	0.8100	0.9003	2.7634	1492.537
0.2	0.9651	0.1627	0.6398	0.8024	1.5265	233.645
0.3	0.9267	0.2189	0.4888	0.7077	1.1249	74.6826
0.4	0.8784	0.2610	0.3564	0.6174	0.9323	34.8797
0.5	0.8231	0.2908	0.2415	0.5323	0.8239	19.7941
0.6	0.7628	0.3099	0.1431	0.4530	0.7584	12.6566
0.7	0.6997	0.3199	0.0599	0.3798	0.7178	8.8028
0.8	0.6354	0.3223	−0.0093	0.3131	0.6931	6.5147
0.9	0.5712	0.3185	−0.0657	0.2527	0.6795	5.0582
1.0	0.5083	0.3096	−0.1108	0.1988	0.6739	4.0740
1.1	0.4476	0.2967	−0.1457	0.1510	0.6745	3.3807
1.2	0.3899	0.2807	−0.1716	0.1091	0.6800	2.8746
1.3	0.3355	0.2626	−0.1897	0.0729	0.6892	2.4967
1.4	0.2849	0.2430	−0.2011	0.0419	0.7017	2.2066
1.5	0.2384	0.2226	−0.2068	0.0158	0.7166	1.9802
1.6	0.1959	0.2018	−0.2077	−0.0059	0.7338	1.7997
1.7	0.1576	0.1812	−0.2047	−0.0235	0.7524	1.6550
1.8	0.1234	0.1610	−0.1985	−0.0376	0.7724	1.5369
1.9	0.0932	0.1415	−0.1899	−0.0484	0.7933	1.4396
2.0	0.0667	0.1230	−0.1794	−0.0563	0.8145	1.3593
2.1	0.0439	0.1057	−0.1675	−0.0618	0.8358	1.2927
2.2	0.0244	0.0895	−0.1548	−0.0652	0.8571	1.2365
2.3	0.0080	0.0748	−0.1416	−0.0668	0.8775	1.1900
3.6	−0.0366	−0.0121	−0.0124	−0.0245	1.0256	0.9771
3.7	−0.0341	−0.0131	−0.0079	−0.0210	1.0276	0.9751
3.8	−0.0314	−0.0137	−0.0040	−0.0177	1.0287	0.9738
3.9	−0.0286	−0.0140	−0.0008	−0.0147	1.0292	0.9732
4.0	−0.0258	−0.0139	0.0019	−0.0120	1.0290	0.9733
4.1	−0.0231	−0.0136	0.0040	−0.0095	1.0282	0.9738
4.2	−0.0204	−0.0131	0.0057	−0.0074	1.0271	0.9747
4.3	−0.0179	−0.0125	0.0070	−0.0054	1.0258	0.9758
4.4	−0.0155	−0.0117	0.0079	−0.0038	1.0241	0.9773
4.5	−0.0132	−0.0108	0.0085	−0.0023	1.0222	0.9790
4.6	−0.0111	−0.0100	0.0089	−0.0011	1.0205	0.9805
4.7	−0.0092	−0.0091	0.0090	+0.0001	1.0186	0.9822
4.8	−0.0075	−0.0082	0.0089	0.0007	1.0167	0.9839
4.9	−0.0059	−0.0073	0.0087	0.0014	1.0149	0.9857
5.0	−0.0046	−0.0065	0.0084	0.0019	1.0132	0.9872
5.1	−0.0033	−0.0057	0.0080	0.0023	1.0115	0.9889
5.2	−0.0023	−0.0049	0.0075	0.0026	1.0099	0.9903
5.3	−0.0014	−0.0042	0.0069	0.0028	1.0084	0.9918
5.4	−0.0006	−0.0035	0.0064	0.0029	1.0071	0.9931
5.5	0.0000	−0.0029	0.0058	0.0029	1.0059	0.9943
5.6	0.0005	−0.0023	0.0052	0.0029	1.0047	0.9953
5.7	0.0010	−0.0018	0.0046	0.0028	1.0036	0.9964
5.8	0.0013	−0.0014	0.0041	0.0027	1.0028	0.9972
5.9	0.0015	−0.0010	0.0036	0.0026	1.0021	0.9979

2.4	−0.0056	0.0613	−0.1282	−0.0669	0.8974
2.5	−0.0166	0.0492	−0.1149	−0.0658	0.9160
2.6	−0.0254	0.0383	−0.1019	−0.0636	0.9336
2.7	−0.0320	0.0287	−0.0895	−0.0608	0.9498
2.8	−0.0369	0.0204	−0.0777	−0.0573	0.9642
2.9	−0.0403	0.0132	−0.0666	−0.0534	0.9772
3.0	−0.0423	0.0070	−0.0563	−0.0493	0.9883
3.1	−0.0431	0.0019	−0.0469	−0.0450	0.9982
3.2	−0.0431	−0.0024	−0.0383	−0.0407	1.0065
3.3	−0.0422	−0.0058	−0.0306	−0.0364	1.0131
3.4	−0.0408	−0.0085	−0.0237	−0.0323	1.0184
3.5	−0.0389	−0.0106	−0.0177	−0.0283	1.0226

6.0	0.0017	−0.0007	0.0031	0.0024	1.0014	0.9986
6.1	0.0018	−0.0004	0.0026	0.0022	1.0008	0.9992
6.2	0.0019	−0.0002	0.0022	0.0020	1.0003	0.9997
6.3	0.0019	+0.0001	0.0018	0.0018	0.9999	1.0001
6.4	0.0018	0.0003	0.0015	0.0017	0.9997	1.0003
6.5	0.0018	0.0004	0.0012	0.0015	0.9994	1.0005
6.6	0.0017	0.0005	0.0009	0.0013	0.9992	1.0008
6.7	0.0016	0.0006	0.0006	0.0011	0.9990	1.0010
6.8	0.0015	0.0006	0.0004	0.0010	0.9989	1.011
6.9	0.0014	0.0006	0.0002	0.0008	0.9988	1.0012
7.0	0.0013	0.0006	0.0001	0.0007	0.9988	1.0012

Table 5.4

Load or moment	A	B	B–C	C	C–D	D	E
Infinite footing							
B	1.7518	2.5705	0.6645	− 0.1044	− 0.0414	0.0036	0.0043
C	− 0.1372	− 0.1567	0.9962	0.8557	0.9962	− 0.1567	− 0.1372
D	0.0033	0.0028	− 0.0310	− 0.0782	0.4983	1.9294	1.3134
M_{OA}	0	− 0.5812	− 0.1671	0.0226	0.0104	− 0.0011	− 0.0010
P_{OA}	1.7496	1.1913	0.0480	− 0.0622	− 0.0058	0.0030	0.0018
M_{OE}	− 0.0008	− 0.0080	0.0074	− 0.1199	− 0.1199	− 0.4168	0
P_{OE}	0.0013	0.0020	− 0.0041	0.0450	0.0345	0.8590	1.2616
Finite footing	3.3680	3.0279	1.5137	3.5586	1.3723	2.2204	2.4429

a similar manner. For comparison with the results given by methods (a) and (b), $M_C = 0.6528$ MNm. The effective depth required to resist this bending moment is 560 mm. Recalculation involving a revised value of I is not worthwhile in this example but a smaller value leads to a slight reduction in M_C. The value of M_C obtained by taking into account the flexibility of the footing and the deformability of the soil (on the basis of the Winkler model) is 8% less than the value given by the conventional analysis. Therefore, the dimensions for the combined footing obtained from the preliminary design can be retained.

5.4 MAT AND RAFT FOUNDATIONS

Mat and raft foundations are most commonly used where the soils encountered have low bearing capacity or high compressibility. A mat foundation usually consists of a concrete slab with constant thickness throughout its plan area. Such foundations are usually subjected to relatively small column or distributed loads. In these circumstances the occurrence of local shear or punching failure at the column loads is prevented by extra reinforcement rather than by an increase in the thickness of the slab. The theories of elastic plates offer considerable possibilities for the analysis and design of this type of mat or slab foundation (see, for example, Selvadurai, 1979). As the magnitudes of the local column loads increase, it becomes necessary to adopt a beam and slab type of construction. Foundations of this type essentially consist of a network of beams integrated with a base slab. The analysis of rafts by elastic methods is fundamentally similar to the analysis of continuous footings but is naturally more complex. Rafts are commonly designed on the assumption that the structure is rigid and that the contact stresses at the raft–soil interface are uniform. Within recent years, numerous papers have dealt with theoretical and experimental research on plain slabs and composite structures of the beam and slab type. Although many of these papers are concerned solely with floor systems, interesting aspects of advanced studies in raft design are often brought forward. For example, arching action in the form of membrane compression may have to be taken into account as the stress analysis of raft foundations becomes refined (Ockleston, 1958; Christiansen, 1963; Park, 1965). In some cases membrane tension may be important (Park, 1964; Basu and Chapman, 1966). Composite action between beams and slabs has been treated by Allen and Severn (1961, 1962, 1964). Some of these investigations are concerned with plastic yield-line theories (Johansen, 1962) which could provide an alternative to the current elastic theories for raft design. Shearing in slabs has been treated by Yitzhaki (1966) and Long and Bond (1967). References to papers dealing specifically with raft foundations are given by Pengelley *et al.* (1955), Parkes (1956), de Beer (1957), Allen and Severn (1960, 1961, 1963), Korenev (1960), Meyerhof (1962), Severn (1962, 1966), Ranganatham and Hendry (1963), Barden (1965b), Scriven and Pilgrim (1965), Hetényi (1966), Sawko (1972), Szilard (1974) and Selvadurai (1979). Raft foundations can also be analysed as grillage structures or as slabs with orthotropic elastic properties, depending upon the relative dimensions of the slab and foundation-beam network (Bares and Massonnet, 1968). Raft foundations commonly encountered in engineering practice can be considered under the following broad classification.

(a) Stiff raft supporting stiff superstructure

In the case of exceptionally large loads similar to those imposed by tall buildings, it becomes convenient to adopt cellular raft construction. Generally, this type of raft functions as a rigid body and the distribution of settlement and contact stresses will be sensibly linear and the rigid method of analysis can be employed. Here again, possibility of concentration of contact stresses at the centre, when the raft is founded on sand, and at the perimeter, when the raft is founded on clay, should be taken into account. The stiffness of the entire assembly should be employed in the analysis by an elastic method (Meyerhof, 1953; Chamecki, 1956; Sommer, 1965; Heil, 1969; Lee and Harrison, 1970; Brown and Lee, 1972; King and Chandrasekaran, 1974). Elastic behaviour of the soil medium is an appropriate approximation when the soil or rock beneath the raft foundation is relatively stiff. Grasshoff (1957) found that the higher the degree of fixity at the connections between superstructure and foundation and the stiffer the soil, the more favourable is the distribution of bending moments in the foundation. A complete and rigorous analysis of interaction problems associated with cellular rafts can be attempted by employing numerical methods of stress analysis such as the finite difference or finite element techniques.

There are, of course, alternative continuum theories of plate behaviour (Woźniak, 1970; Selvadurai, 1973) which could be successfully adopted for the analysis of cellular type structures. The latter techniques merit further investigation owing to their adaptability in the general analysis of cellular raft foundations as elastic plates endowed with microstructure.

(b) Flexible raft supporting stiff superstructure

An example of this type of foundation is a battery of hoppers or silos supported on columns on a flexible raft. The assembly functions similar to (a) but with an additional factor arising from the flexibility of the raft, namely, local concentrations of contact stresses beneath columns. Generally the flexible raft should be analysed on the basis of an elastic interaction problem. The validity of an analysis based on the assumption that the raft is completely rigid depends upon the distribution and spacing of columns.

(c) Stiff raft supporting flexible superstructure

In this case the raft foundation supports heavy machinery or concentrated loading and can be expected to function similar to (a), except that the stiffness of the superstructure is negligible; the raft may tend to hog or sag thus leading to a broad redistribution of contact stresses.

(d) Flexible raft supporting flexible superstructure

This type of raft will be subjected to varying concentrations of contact stresses beneath columns and to either hogging or sagging. A method of analysis based on elastic response of the soil media is generally desirable, although it may not be justified by the degree of importance of the building or by relatively light static and dynamic loads. Depending upon the relative stiffness of the soil–raft system and the column spacing the conventional rigid method may also be applied.

5.4.1 Plate theories

It is instructive to consider first the various theories that can be adopted for analysing the flexural behaviour of elastic plates. Elastic theories for the flexure of plates can be classified as follows: (a) thick slab or thick plate; (b) thin plates; (c) thin plates with large deformations; (d) membranes.

A slab or plate can be called a *thick plate* if the strain energy of the transverse shearing stresses is large enough to prevent linear or nearly linear distributions of the normal bending stresses through the thickness at each point. A comprehensive account of these various thick plate theories is given in a review article by Naghdi (1972). Formulations of thick plate theories, with a special reference to soil–foundation interaction problems, are given by Reissner (1945, 1955), Naghdi and Rowley (1953), Frederick (1957), Pister (1961), Pister and Westmann (1962) and Selvadurai and Adjeleian (1977). Bibliographies of such works are also given by Timoshenko and Woinowsky-Krieger (1959) and Selvadurai (1979). The thick plate theory is ordinarily formulated as a three-dimensional problem in the theory of elasticity, whose solution exactly satisfies boundary conditions on the bounding surfaces. As regards the conditions on the edges, these are satisfied approximately so that by virtue of St Venant's principle, reasonably exact solutions are obtained in many loading cases for points remote from the edges. Such thick plate theories are, of course, analytically rigorous. The thick plate theory proposed by Reissner (1945) incorporates effects of transverse normal and shearing stresses by using energy principles. Pister and Westmann (1962) modified Reissner's plate theory by assuming a more precise weighted average of the deflections across the thickness of the plate. The non-linear variations of bending stresses appear to be particularly significant in thick plate problems involving highly localized surface loads. In the particular case of concentrated loads, even the thick plate theories may not give an accurate description of the state of stress in the vicinity of the applied loads. In these circumstances the three-dimensional theory of elasticity can be employed to analyse the local state of stress (Fig. 5.26).

A slab or a plate can be regarded as a *thin plate* if the only strain energy that need be considered is that of bending and twisting by stresses that are proportional to the distance from the middle plane of the plate. The classical Poisson–Kirchoff theory for the flexure of thin elastic plates embodies the following basic assumptions.

(a) The thickness of the plate is assumed to be small in comparison to a characteristic length of the plate.
(b) The lateral displacements (w) of the middle surface of the plate are small compared to the thickness (h) of the plate, such that second and higher order terms of w/h can be neglected.
(c) The component of stress normal to the middle surface is assumed to be small compared to the other components of stress and hence can be neglected in the stress–strain relations.
(d) Plane cross-sections normal to the undeformed middle surface are assumed to remain normal to the middle surface when deformed.

The first assumption defines the term *thin plate* and the second is essential in the classical theory of small strains and small displacements. In classical elasticity the latter condition ensures the linearity of the governing differential equations.

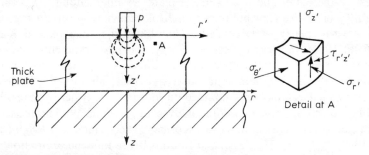

Figure 5.26 Three-dimensional effects in a thick elastic plate or a thick mat foundation due to localized loading.

The last two assumptions emphasize the distinction between thin and thick plates. The thin plate theory has been widely applied in the interaction analysis of slabs and flexible mats resting on soil media. The analysis of infinite plates resting on Winkler media has been considered by Hertz (1884), Westergaard (1943, 1948), Pickett *et al.* (1951) and Timoshenko and Woinowsky-Krieger (1959). The analysis of plates of finite extent has been considered by many including Reissman (1954), Reissner (1955), Leonards and Harr (1959), Richart and Zia (1962), Hudson and Matlock (1966) and Kerr (1966). Solution of finite plate problems using relaxation techniques and finite element methods have been considered by Allen and Severn (1960, 1961, 1963), Severn (1966), Hooper (1978) and Selvadurai (1979). Infinite and finite plate problems associated with two-parameter elastic models have been considered by Pasternak (1956), Reissner (1958), Galletly (1959), Pister and Williams (1960) and Vlazov and Leontiev (1966). Infinite plate problems related to elastic solid supporting media have been considered by Hogg (1938), Holl (1938), Sneddon *et al.* (1975), Selvadurai (1977a, b, 1979) and Gladwell (1980). Corresponding finite plate problems have been treated by Borowicka (1936, 1943), Bergstrom (1946), Volterra (1947), Gorbunov-Posadov (1949, 1951), Pickett *et al.* (1951), Pickett and McCormick (1951), Pickett and Janes (1953), Gorbunov-Posadov and Serebrjanyi (1961), Barden (1965), Brown (1969a, b) and Selvadurai (1980a, b). Numerical methods of analysis of finite plates are further discussed by Habel (1937), Hsieh (1960), Cheung and Zienkiewicz (1965), Chakravorty and Ghosh (1975) and Chattopadhyay and Ghosh (1976). The methods of Zemochkin and Sinitsyn (1947) and Ohde (1942), described in Section 5.3.6, can also be adopted in the analysis of flexible plates resting on elastic solid media (see, for example Selvadurai, 1979). In these cases the plate is represented by a system of crossed beams. References to further work can be found in Timoshenko and Woinowsky-Krieger (1959), Korenev (1960), Hetényi (1966) and Selvadurai (1979).

When thin plates are subjected to large lateral displacements, appreciable stresses occur in the middle plane of the plate making it necessary to take into account not only the energy of bending and twisting but also that of stretching. In this case the lateral deflections of the plate are large in comparison with the thickness of the plate. The strains in the plate are small enough to permit the applicability of Hooke's law but the strain displacement relations are non-linear. A well known theory of large deflection of plates is due to von Karman (Mansfield, 1964). In large deflection theory, the basic differential equations governing the flexure of the plate are non-linear. Large deflection theory is particularly relevant to the interaction analysis of thin slabs or highly flexible mat foundations, such as tank bases, founded on soils that are either compressible or susceptible to creep. Large deflection analysis of plates on Winkler media have been considered by many authors (see Korenev, 1960; Hetényi, 1966) including Sinha (1963), Nash and Ho (1960) and Bolton (1972). Influence of friction at the interface of the raft foundation on its flexural response has been examined by Hooper (1975, 1976).

In membranes, only energy due to stretching is considered and bending energy is neglected. It seems unlikely that the large deflection plate theory or the membrane theory will be of any real importance to the analysis and design of conventional slab foundations. The indications are that these theories will acquire considerable importance in structural foundation problems associated with highly flexible synthetic materials. From the design point of view it is sufficient to treat flexible slab foundations as thin elastic plates satisfying the Poisson–Kirchoff theory. Localized loadings may be analysed on the basis of the theory of elasticity.

5.4.2 The analysis of the infinite plate

Consider the flexure of a plate of infinite extent resting on a linearly deformable soil medium. The external loading is assumed to be such that the state of strain in the plate and in the soil medium is symmetrical about the z-axis (Fig. 5.27). The plate is subjected to a circular load of uniform stress intensity (p). This problem can be regarded as the three-dimensional analogue of the infinite beam problem. The analysis of an infinite plate is not only of purely academic interest; many structural foundations which are subjected to highly localized loading such as circular or square column loads, flexible pavements subjected to aircraft or traffic

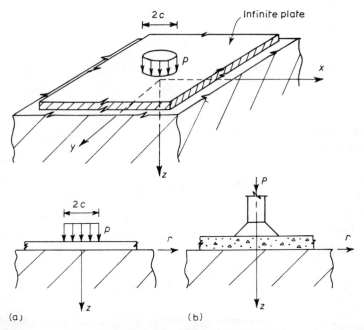

Figure 5.27 Axisymmetrical loading of an infinite plate resting on an elastic medium. (a) Uniform flexible load; (b) isolated column load.

loads, etc., can be adequately represented by the infinite plate problem. Owing to the assumed linearity of the soil response models, the solution to the infinite plate problem can be developed in a completely generalized fashion.

Assume that the surface deflection of the soil medium when subjected to an external stress

$$q(r) = q_0 J_0(\xi r/a),\tag{5.138}$$

can be written in the form

$$w(r, 0) = w(r) = a\frac{q_0}{C} J_0(\xi r/a),\tag{5.139}$$

where C is a constant which depends upon the material properties of the particular soil model, a is a typical length parameter and $J_0(\xi r/a)$ is the zeroth-order Bessel function of the first kind (Watson, 1944). For axially symmetric deformations, the differential equation governing flexure of the plate is

$$D\nabla^2\nabla^2 w(r) + q(r) = p(r),\tag{5.140}$$

where $D = E_p h^3/12(1 - v_p^2)$ is the flexural rigidity of the plate, $q(r)$ and $p(r)$ are the contact stress and external stress respectively, E_p and v_p are the elastic constants of the plate material and h is the thickness of the plate. Also, the Laplace operator ∇^2 is given by

$$\nabla^2 = \frac{d^2}{dr^2} + \frac{1}{r}\frac{d}{dr}.\tag{5.141}$$

In the particular case where the plate is subjected to the external surface stress distribution,

$$p(r) = p_0 J_0(\xi r/a),\tag{5.142}$$

it can be shown that, provided there is no loss of contact at the interface, Equations 5.139 to 5.142 give

$$w(r) = \frac{p_0 a^4 J_0(\xi r/a)}{D[\xi^4 + Ca^3/D]}.\tag{5.143}$$

An external load distribution of uniform intensity p_0 acting over an area of radius c, and of zero intensity for all $r > c$, can be represented in the form of a Fourier–Bessel integral

$$p(r) = \frac{p_0 c}{a}\int_0^\infty J_0(\xi r/a)J_1(\xi c/a)\,d\xi,\tag{5.144}$$

where $J_1(\xi c/a)$ is the first-order Bessel function of the first kind. Equation 5.144 is developed from the orthogonality of Bessel functions in a manner similar to the development of the Fourier integral (Equation 5.43). Using Equations 5.143 and

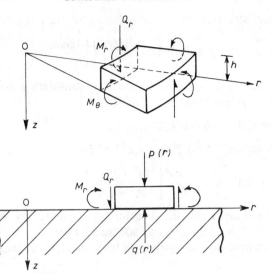

Figure 5.28 Flexural moments and shear forces in a thin plate subjected to an axisymmetrical state of stress.

5.144, the deflection of the plate due to the uniform load is given by

$$w(r) = \frac{p_0 c a^3}{D} \int_0^\infty \frac{J_0(\xi r/a) J_1(\xi c/a)}{[\xi^4 + \Delta]}, \qquad (5.145)$$

where $\Delta = Ca^3/D$. The flexural moments and shearing forces per unit length, M_r, M_θ and Q_r (Fig. 5.28) are given by:

$$M_r = -D\left\{\frac{d^2w}{dr^2} + \frac{v_p}{r}\frac{dw}{dr}\right\}; \quad M_\theta = -D\left\{v_p\frac{d^2w}{dr^2} + \frac{1}{r}\frac{dw}{dr}\right\},$$

$$Q_r = -D\frac{d}{dr}\{\nabla^2 w\}, \qquad (5.146)$$

and owing to the axial symmetry of the problem the twisting moment $M_{r\theta}$ and the shearing forces Q_θ are zero. From Equations 5.145 and 5.146

$$[M_r; M_\theta; Q_r] = p_0 ac \int_0^\infty \frac{J_0(\xi r/a) J_1(\xi c/a)}{[\xi^4 + \Delta]} [m_r; m_\theta; q_r] d\xi, \qquad (5.147)$$

where

$$m_r = \xi^2 J_0(\xi r/a) - (1 - v)(\xi a/r) J_1(\xi r/a),$$
$$m_\theta = v\xi^2 J_0(\xi r/a) + (1 + v)(\xi a/r) J_1(\xi r/a), \qquad (5.148)$$
$$q_r = -(\xi^3/a) J_1(\xi r/a).$$

By substituting Equations 5.144 and 5.145 into Equation 5.140 it can be shown that the contact stress at the soil–foundation interface is given by

$$q(r) = \frac{p_0 c}{a} \int_0^\infty \frac{\Delta J_0(\xi r/a) J_1(\xi c/a)}{[\xi^4 + \Delta]} d\xi. \tag{5.149}$$

Different expressions for Δ can be obtained by considering the response of each soil model to a surface load $q(r) = J_0(\xi r/a)$.

(a) Winkler model $\Delta = a^4 k_s/D$ (5.150a)

(b) Pasternak model $\Delta = [\xi^2 G_p + k_s a^2](a^2/D)$ (5.150b)

(c) Reissner model $\Delta = \dfrac{4E_s \eta a^3}{D} \left[\dfrac{3\eta^2 + \xi^2 \chi}{12\eta^2 + \xi^2 \chi} \right],$ (5.150c)

where $\eta = a/H$ and $\chi = G_s/E_s$. As before, the result for the Winkler model can be recovered from Equations 5.150b and 5.150c.

(d) For an isotropic elastic continuum with shear modulus G_s and Poisson's ratio v_s

$$\Delta = \frac{G_s \xi a^3}{(1 - v_s)D}. \tag{5.150d}$$

Again, expressions similar to Equations 5.150a to d can be obtained for orthotropic, layered or non-homogeneous elastic media.

Using the results given here it is possible to evaluate the stresses in the plate at any location. Techniques for evaluating integrals of the type in Equation 5.144, in terms of Fourier–Bessel series expansions, are presented by Biot (1935), Reissner (1955), Hoskin and Lee (1959) and Selvadurai (1977b, 1979).

The solution to the problem of an infinite plate subjected to a concentrated force (P) can be obtained as a limiting case of the uniform load solutions, as $\pi p_0 c^2 \to P$ and as

$$\operatorname*{Lt}_{\zeta \to 0} \frac{J_1(\zeta)}{\zeta} = \frac{1}{2}, \tag{5.151}$$

the deflection of the infinite plate is given by

$$w(r) = \frac{Pa^2}{2\pi D} \int_0^\infty \frac{\xi J_0(\xi r/a)}{[\xi^4 + \Delta]} d\xi. \tag{5.152}$$

Flexural moments, shear forces and contact stresses can be obtained by using Equations 5.146 and 5.140.

5.4.3 The analysis of finite plates

Only a brief exposition of the finite plate problem is presented here. Because of the variety of external loading conditions, boundary conditions, shapes of plates, types of supporting media etc., the subject of finite plates is so vast that to include more is beyond the scope of this chapter.

The analysis of the finite plate problem involves the solution of the differential equation governing flexure of the plate

$$DV^4w(x, y) + q(x, y) = p(x, y), \tag{5.153}$$

subject to various boundary conditions. In Equation 5.153, V^2 is Laplace's operator given by

$$V^2 = \frac{\partial^2}{\partial x^2} + \frac{\partial^2}{\partial y^2}, \tag{5.154}$$

and $p(x, y)$ is the external loading. The contact stresses $q(x, y)$ depend on the character of soil response. For example, in the case of

(a) the Winkler model, $q(x, y) = k_s w(x, y)$ (5.155a)
(b) the Pasternak model, $q(x, y) = k_s w(x, y) - G_p V^2 w(x, y)$ (5.155b)
(c) the isotropic elastic halfspace model, the integral relationship

$$w(x, y) = \frac{(1 - v_s)}{2\pi G_s} \int_{-l/2}^{l/2} \int_{-b/2}^{b/2} \frac{q(\xi, \zeta)\mathrm{d}\xi\mathrm{d}\zeta}{\{(x - \xi)^2 + (y - \zeta)^2\}^{1/2}} \tag{5.155c}$$

is required, where l and b are the length and width of, say, a rectangular plate. The bending moments and twisting moment per unit length M_x, M_y and M_{xy} are given by

$$M_x = - D\left[\frac{\partial^2 w}{\partial x^2} + v_p\frac{\partial^2 w}{\partial y^2}\right]; \ M_y = - D\left[\frac{\partial^2 w}{\partial y^2} + v_p\frac{\partial^2 w}{\partial x^2}\right],$$

$$M_{xy} = - D(1 - v_p)\frac{\partial^2 w}{\partial x \partial y}, \tag{5.156}$$

and the shear forces per unit length Q_x and Q_y are

$$Q_x = - D\frac{\partial}{\partial x}(V^2 w), \ Q_y = - D\frac{\partial}{\partial y}(V^2 w). \tag{5.157}$$

The boundary conditions at the edges of the plate can be any of the following.

(i) *Simply supported boundary.* Consider the case where the edge of the plate, $x = x_0$, is supported in such a way that there is no lateral displacement. In this case the boundary conditions are

$$\text{(a) } w(x_0, y) = 0, \text{ (b) } M_x(x_0, y) = 0. \tag{5.158}$$

(ii) *Clamped boundary.* In the case where the edge of the plate $x = x_0$ is clamped

$$\text{(a) } w(x_0, y) = 0, \text{ (b) } \left(\frac{\partial w}{\partial x}\right)_{x = x_0} = 0. \tag{5.159}$$

(iii) *Free boundary.* Consider the case where the boundary of the plate $x = x_0$ is free of external traction. In general the boundary conditions for the free end are

$$\text{(a) } M_x(x_0, y) = 0, \text{ (b) } M_{xy}(x_0, y) = 0, \text{ (c) } Q_x(x_0, y) = 0. \qquad (5.160)$$

These boundary conditions were originally proposed by Poisson (Timoshenko and Woinowsky-Krieger, 1959). By using a variational method, Kirchoff (1850) (see Naghdi, 1972) proved that it would, in general, be impossible to satisfy all three boundary conditions (Equation 5.160) in the small deflection thin plate theory. For axially symmetrical deformations, this difficulty does not arise since $M_{r\theta}$ is identically equal to zero and Equation 5.160 reduces, effectively, to two boundary conditions $M(r_0) = 0$ and $Q_r(r_0) = 0$. It can be shown that, by invoking St Venant's principle, the last two of Poisson's boundary conditions (Equations 5.160b and c) can be replaced by a single one. The two boundary conditions for a free end $x = x_0$ are, therefore,

$$\text{(a) } M_x(x_0, y) = 0, \text{ (b) } \left(Q_x - \frac{\partial M_{xy}}{\partial y} \right)_{x = x_0} = 0. \qquad (5.161)$$

These boundary conditions are applicable only to plates supported by Winkler or elastic solid media. In the case of the two-parameter medium, allowance has to be made for the concentrated edge reactions that occur at the ends of the plate due to the continuity of the supporting medium. For example, the free end boundary conditions for the edge $x = x_0$ are

$$\text{(a) } M_x(x_0, y) = 0, \text{ (b) } \left(Q_x - \frac{\partial M_{xy}}{\partial y} + Q_x^* \right)_{x = x_0} = 0 \qquad (5.162)$$

where

$$Q_x^* = G_p \left[\alpha w + \frac{\partial w}{\partial x} - \frac{1}{2\alpha} \frac{\partial^2 w}{\partial y^2} \right]$$

and $\alpha = \sqrt{(G_p/k_s)}$. In addition, the concentrated edge reactions at the corners of the plate (R^*) can be expressed in terms of the displacements at the corners (w_c) as $R^* = 3G_p w_c/4$.

Generally, the following methods have been employed in the analysis of the soil–plate interaction problem: (a) exact solutions; (b) power series approximations; (c) trigonometric series approximations; (d) finite difference techniques; (e) other discrete approximations, such as Zemochkin and Sinitsyn's (1947) and Ohde's (1942) techniques.

Solutions to several problems of engineering interest are available in the literature and references to these are found in the papers cited in Section 5.4.1. Reference should also be made to the finite element technique (Zienkiewicz, 1971; Svec and Gladwell, 1973; Hooper, 1974, 1978; Selvadurai, 1979) which could be successfully employed in the analysis of mat foundations, especially where complicated shapes are involved.

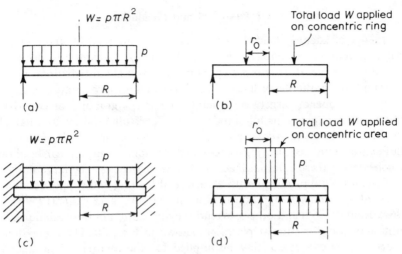

Figure 5.29 Flat plate formulae.

(a) Max. at centre: $M_r = M_\theta = \dfrac{W}{16\pi}(3 + v_p)$;

   max. deflection: $w(0) = \dfrac{W(1 - v_p)(5 + v_p)R^2}{64\pi E_p I_p}$.

(b) Max. at centre: $M_r = M_\theta$
$$= \frac{W}{4\pi}\left[\frac{1}{2}(1 - v_p) + (1 + v_p)\log\left(\frac{R}{r_0}\right) - (1 - v_p)\frac{r_0^2}{2R^2}\right],$$

   max. deflection: $w(0) = \dfrac{W(1 - v_p^2)}{8\pi E_p h^3}\left[\dfrac{(3 + v_p)R^2 - (1 - v_p)r_0^2}{2(1 + v_p)} - r_0^2\left\{\log\left(\dfrac{R}{r_0}\right) + 1\right\}\right]$.

(c) Max. at edge: $M_r = M_\theta/v_p = \dfrac{W}{8\pi}$;

   at centre: $M_r = M_\theta = \dfrac{W}{16\pi}(1 + v_p)$

   max. deflection: $w(0) = \dfrac{W(1 - v_p^2)R^2}{64\pi E_p h^3}$

(d) Max. at centre: $M_r = M_\theta$
$$= \frac{W}{4\pi}\left[(1 + v_p)\log\left(\frac{R}{r_0}\right) + \frac{1}{4}(1 - v_p)\left(1 - \frac{r_0^2}{R^2}\right)\right].$$

   max. deflection:
$$w(0) = \frac{W(1 - v_p^2)}{64\pi E_p h^3}\left[4r_0^2\log\left(\frac{R}{r}\right) + 2r_0^2\left(\frac{3 + v_p}{1 + v_p}\right) + \frac{r_0^4}{R^2} + R^2\left(\frac{7 + 3v_p}{1 + v_p}\right) + \frac{(R^2 - r_0^2)}{R^2}r_0^2\right]$$

M_r, M_θ: bending moments per unit length in the radial and tangential directions E_p, v_p: elastic constants of the plate material. $w(0)$: vertical deflection. h: thickness of plate. I_p: second moment of area of a strip of the plate of unit width.

5.4.4 Design of mat foundations

The theory of laterally loaded thin plates can be adopted for the preliminary design of reinforced concrete slab foundations. In preliminary designs, the contact stresses are usually represented by approximate distributions which are either uniform, parabolic or linear. The choice of the approximate contact stress distribution depends largely on whether the soil is granular or cohesive. By superposition, expressions for standard cases of loading on circular flat plates (Fig. 5.29) can be made to cover a variety of loadings on circular slab foundations. The flexural moments, shearing forces and deflections in the plate can be obtained by considering these external loads and contact stresses. The large number of solutions for plate problems with a variety of loading and support conditions, presented by Nadai (1925), Timoshenko and Woinowsky-Krieger (1959), Flügge (1962), Roark (1965) and others, extend the application of this technique. Some additional cases for circular plates are shown in Fig. 5.30. The expressions for circular plates can reasonably be applied for the analysis of hexagonal or octagonal plates, provided the external loading corresponds to an axisymmetrical state. Formulae for rectangular plates are not readily expressed in simple

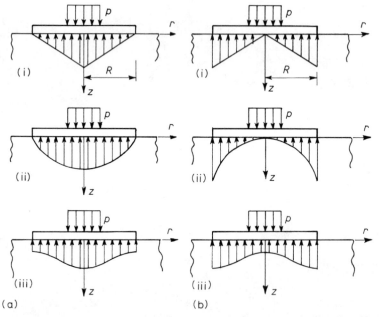

Figure 5.30 Non-uniform contact stress distributions beneath flat plates, axisymmetrical case. (a) cohesionless soils (i) $q(r) = q_s(1 - r/R)$, (ii) $q(r) = q_s(1 - r^2/R^2)$, (iii) $q(r) = q_s(1 + \alpha \cos \pi r/R)$; (b) cohesive soils (i) $q(r) = q_s r/R$, (ii) $q(r) = q_s r^2/R^2$, (iii) $q(r) = q_s(1 - \alpha \cos \pi r/R)$. The values of q_s, which are different for each case, can be evaluated by considering vertical equilibrium of external forces and contact stresses. In each case limiting values of contact stresses at the perimeter are governed by plastic flow conditions (Schultze, 1961).

form, as account must be taken of the ratio of plate length to width and of the ratio length to width of the loaded area.

In design, the maximum deflection of the foundation should be calculated to ascertain whether or not it conforms to the basic assumptions of the theory for a thin plate with small deflection. For this purpose the elastic constants of the concrete can be assessed from uniaxial compression or torsion tests. The modulus of elasticity of concrete varies with the elastic properties of the mix constituents, their volume fractions and the time elapsed after initial hardening. Some typical values for the elastic constants are given by Evans and Wood (1937a, b), Anson (1964), Anson and Newman (1966), Newman (1966) and Neville (1971). The various theories for multiphase materials can also be employed to obtain reasonably accurate values for the elastic constants (Newman, 1966).

It is assumed that, up to a deflection of about one-half of the thickness, the middle surface of the plate of homogeneous material remains unstressed but with greater deflections it becomes appreciably stressed. Such stresses enable the plate to carry part of the load as a diaphragm in direct tension. This tension is balanced by radial tensile stress or circumferential compression stress, depending on whether the edges are clamped or free (Fig. 5.31). Absolute restraint at the edges appears never to be achieved in practice and some rotation occurs. The thin plate theory neglects deflection due to shear but, if the foundation has a large opening, shear effects may contribute appreciably to the total deflection. Once the basic dimensions of the foundation have been estimated it is, of course, essential to carry out, wherever possible, a complete interaction analysis of the problem. The interaction analysis is necessary not only to arrive at the efficient design but also to ensure that the assumptions of the preliminary design are not radically violated.

As an example, consider a circular or octagonal flat slab foundation for a water tank supported on a large diameter shaft concentric with the slab. Assume, as a first approximation, a uniform contact stress distribution. The maximum radial bending moment at the centre is obtained by subtracting M_r for case (b) (Fig. 5.29) from M_r for case (a). This technique can, of course, be extended to other forms of

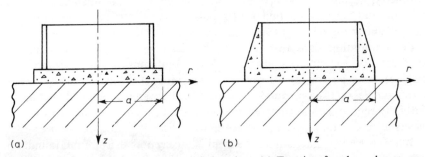

Figure 5.31 Boundary conditions for a finite plate. (a) Traction free boundary, $r = a$. (b) restrained rotation at $r = a$.

contact stress distribution, such as those of the parabolic or triangular type. Reinforcement in the form of a square mesh is normally provided in this type of foundation, although three layers mutually inclined at 60° have been used. Radial stresses diverge from the centre so that a square mesh does not cater directly for the radial and tangential stresses but resists components of these stresses. The ideal reinforcement consists of bars radiating from the centre together with bars bent into concentric circles to cater for tangential stresses. The latter are relatively easily erected but difficulty arises at the intersection of the radial bars at the centre. Consideration may be given to the introduction of a flat steel plate, say 1.00 m diameter, of thickness equal to the diameter of the radial bars to which the radial bars are site welded. To assist in maintaining continuity in the concrete, either holes can be punched in the plate or dowels can be welded to the plate. Some difficulty may be experienced in concreting beneath the plate. The increase in radial bar spacing with increase in radius from the centre automatically caters for the reduced radial stresses towards the edge.

Additional reinforcement is required, of course, where the bending moment is reversed due to fixed or continuous conditions. Whether or not this idealized layout of reinforcement is worthy of development is a matter of conjecture but a practical design based on it should be considered satisfactory (Taylor *et al.* 1966). The foundation for the district heating accumulator at Pimlico, London, consisted of a reinforced concrete raft 12.2 m diameter and 1.52 m thick. The accumulator was 9.14 m diameter and the raft was designed by combining loading for two cases which led, in fact, to the case represented by Fig. 5.29(d). In foundations where localized loading occurs, for example, a shaft of a chimney or water tower, calculations should be made to assess the shearing stresses, which are likely to be most critical around the periphery of the shaft. The design of slab foundations for tall chimneys has been discussed by Smith and Zar (1964) and Chu and Afandi (1966).

The design of rectangular slabs can also be approached along the lines described earlier. The following is a semi-empirical elastic method of analysis for a rectangular slab supporting an array of columns at more or less uniform spacing. The slab is divided into strips in the direction of the two axes of symmetry, basing the width of each strip on the mid-span distances between columns. Each strip in the two orthogonal directions is treated as a beam on an elastic supporting medium employing the full column loads for each strip.

For foundation slabs with an irregular shape in plan, when asymmetrical loadings are encountered, it is prudent to obtain the flexural moments, displacements, etc., on the basis of a complete interaction analysis using perhaps a very simple soil response model. The empirical design techniques do not always faithfully reproduce the complex states of stress encountered due to torsional effects at re-entrant corners or at zones with local stiffening or at openings.

Where excessive differential settlement is unacceptable, a stiff raft foundation or a mat foundation supported on piles is employed. In the latter case piles are generally located on a more or less square grid at say 2.00 m centres. The spacing

L can be determined from the loaded area contributing to the load on each pile and from the safe bearing capacity Q_s of each pile. Thus $L \approx \sqrt{[(Q_s - W)/p]}$, where p is the uniformly distributed load including the weight of the mat and W is the weight of the pile. The piles around the periphery cannot conform strictly to the grid and commonly they do not carry a full share of the uniform load, but they may have to carry additional load supported by a ring beam. In this case it is reasonable to design the mat as a flat floor slab supported on columns (see BSI (1972), and ACI Code, 1963) with or without drops and with or without caps on the piles, although a slab of uniform thickness (not less than 0.3 m) with local stiffening for concentrated loads is generally the most convenient. Tests on a slab supporting a number of columns have been reported by Jirsa *et al.* (1966). The slab-drop panel type and the slab-perimeter beam type mat foundations described by Gifford and Butler (1966) and Green (1957) are also of interest. Since the spans are usually short, shearing forces may be the critical factor in design. Account should also be taken in design of the possible misplacement of adjacent piles. An interaction analysis based on a Winkler type soil response (Terzaghi, 1943) is suitable depending upon the flexibility of the pile and the spacing of piles in the grid. For closely spaced piles, group action becomes much more important and the elastic half-space approach is more appropriate.

5.4.5 Design of raft foundations

In general, the preliminary design of raft foundations is based on the assumption that the contact stress distribution at the raft–soil interface is either uniform or otherwise planar. The raft is assumed to function as an inverted reinforced concrete floor consisting of a slab or slab with beams. Since the monolithic action of a beam and slab network usually results in a relatively rigid raft, there is always the possibility, at least on theoretical grounds, that the contact stress distribution is non-uniform and tends towards a parabolic or elliptic shape. For example, when rigid rafts are founded at or near the surface of a cohesionless soil, the contact stress distribution across a section of the raft tends to be of a parabolic shape (Fig. 5.32a) and as the depth of the founding level increases the pressure distribution tends to be more uniform (Schultze, 1961). For relatively large rigid rafts, with which the superstructure contributes further to the rigidity of the entire assemblage, non-uniform contact stresses may occur only in the vicinity of the perimeter of the raft (Faber, 1933; Smoltczyk, 1967). Rigid rafts constructed on stiff clay soils, however, tend to exhibit non-uniform contact stress distributions (Fig. 5.32b) irrespective of whether the founding level is at the surface or at a finite depth (Faber, 1933; Schultze, 1961; Vesić and Johnson, 1963; Selvadurai, 1979; Selvadurai and Kempthorne, 1980). It is therefore prudent to consider, even at the preliminary stages of the design, effects that are introduced by these non-uniform contact stress distributions. The magnitudes of such contact stresses may be assessed by treating the foundation as infinitely rigid and by selecting an appropriate model of the particular soil conditions.

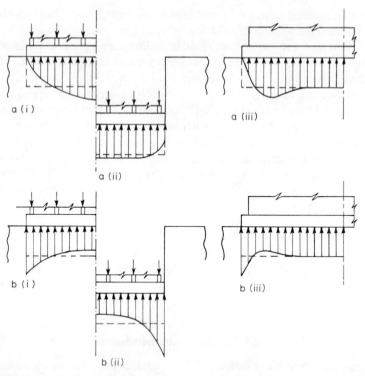

Figure 5.32 Typical contact stress distributions along the cross-section of rigid found-ations resting on (a) cohesionless and (b) cohesive soil media, due to symmetrical loading. (i) Surface or shallow foundations of small width, (ii) deep foundations of small width, (iii) surface or shallow foundations of large width.

The plan of the raft is usually arranged in a way such that the centre of gravity of the external loads coincides with the centre of area of the raft; if this cannot be achieved the eccentricity of loads should be reduced to a minimum. The following notes give a typical specification of dimensions for a raft; the total slab thickness is 0.20 to 0.25 m for a raft carrying a number of columns loaded with 500 to 2000 kN or about 0.60 m for up to 4000 kN; generally rather more than the calculated thickness of the slab may be necessary to provide an adequate flange for the T-beams and requisite cover to the reinforcement. Where the contact stress varies owing to eccentric loading, the slab thickness can be varied over the area of the raft. Transverse ribs work out at about 0.5 m overall depth and longitudinal ribs 1.2 to 1.8 m. Shear reinforcement is required in both. The details of reinforcement required in the slab and the beam network are calculated according to accepted principles of reinforced concrete design (Baker, 1949; Reynolds, 1964; CP 110, BSI, 1972c). Beams may be haunched. If large bending moments occur in the ribs beneath the columns, a cellular raft may be suitable;

the top or deck slab forming an upper flange. The beams should be cast at the same time as the slab in beam-slab type rafts.

To illustrate the preliminary design procedure, specific examples of raft foundations subjected to both symmetrical and asymmetrical loading conditions are considered in the following sections.

(a) Symmetrical loading (Fig. 5.33)

The distance between transverse ribs is commonly one-third of the distance between columns. In this example the contact stress is assumed to be uniform and of intensity (q_0). For the cantilever sides of the slab $M = q_0 l_1^2/2$. The interior sections of the slab are designed as continuous and supported on four sides or as a slab spanning the transverse beam. The appropriate choice of methods is very much in the hands of the designer who should make assumptions that are consistent with given conditions.

For the transverse ribs it is assumed that the sum of the centre reactions from the transverse ribs equals the total load from the central columns. The end ribs r_1 carry only part of the load imposed on ribs r_2 and hence the centre reaction R_1 for r_1 is assumed to be nR_2, where n is a factor based on comparative soil contact areas. The central reactions $5R_2 + 2nR_2 = 2W_1 + W_2$ and hence R_1 and R_2 can be evaluated. It is then assumed that one-half of the upthrust from the contact stress per rib not resisted by R_1 and R_2 represents the end reactions R_1' and R_2' of the ribs. The bending moments and shearing force diagrams are then constructed and the ribs designed as T-beams.

The loading on the longitudinal ribs consists of the reactions from the transverse ribs and the column loads. The outer longitudinal ribs carry, in addition, the upthrust from the cantilever portion of the slab. These ribs are also

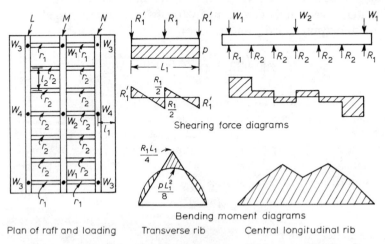

Shearing force diagrams

Bending moment diagrams

| Plan of raft and loading | Transverse rib | Central longitudinal rib |

Figure 5.33 Analysis of a symmetrically loaded beam and slab raft.

designed as T-beams. The forms of the shearing force and bending moment diagrams shown in Fig. 5.33 are typical but vary in shape with individual examples.

The manner in which the external loads are distributed within the monolithic beam and slab network, as presented here, is of course not meant to be unique; there are numerous other interpretations (Baker, 1957; Jumikis, 1971; Manning, 1972; Reynolds, 1972) each worthy of consideration. For example the beams of the raft are designed to resist the bending moments obtained by taking the algebraic sum of the static moments of the column loads to the left or right of any point of the beam, plus the moment of the uniform contact stresses tributary to the beam taken about the same point. Where the raft beam is long, and where it can be shown by reference to reliable soil data and proper analysis that contact stresses greater than average (developed over the entire raft area) will occur under heavily loaded columns or areas, without exceeding the permissible deflection for the raft beam or of the superstructure, then the bending moment may be adjusted in accordance with such analysis.

In this example, it has been assumed that the contact stress distribution at the raft–soil interface is uniform. This technique can be extended to include other non-uniform distributions of contact stress which are symmetrical about the axes of the raft. Finally, it is pertinent to carry out an interaction analysis either by assuming monolithic behaviour of the beam network, without slab action, or by considering the raft as a slab foundation with variable rigidity. The choice of the relevant idealization of the raft depends upon the relative dimensions of the beams and the slab.

(b) Asymmetrical loading (Fig. 5.34a)

Assume that the contact stress at the raft–soil interface has a non-uniform planar distribution. In general the contact stress distribution beneath a rigid raft with an arbitrary non-symmetrical plan shape (Fig. 5.34b) is given by

$$q(x, y) = \frac{W}{A} + \left[\frac{\tilde{M}_y I_x - \tilde{M}_x I_{xy}}{I_x I_y - I_{xy}^2} \right] x + \left[\frac{\tilde{M}_x I_y - \tilde{M}_y I_{xy}}{I_x I_y - I_{xy}^2} \right] y, \qquad (5.163)$$

where W is the total load on the raft; A is the plan area; $\tilde{M}_x = We_x$, $\tilde{M}_y = We_y$ and e_x and e_y are the eccentricities measured from the centroidal axes x and y respectively; I_x and I_y are the second moments of area of the raft about the x- and y-axes respectively; I_{xy} is the product of inertia. For a raft of rectangular plan shape, there are two axes of symmetry and $I_{xy} = 0$. Therefore, the contact stress distribution (Equation 5.163) reduces to

$$q(x, y) = \frac{W}{A} + \frac{We_y}{I_y} x + \frac{We_x}{I_x} y. \qquad (5.164)$$

The contact stresses are thus completely defined in terms of the load and the

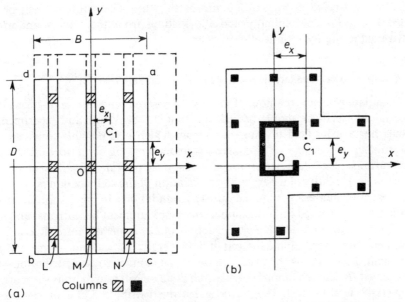

Figure 5.34 Analysis of an asymmetrically loaded raft. C_1 is the centre of gravity of external loads; O is the centre of area of the raft.

dimensions of the loaded area. The convention of signs used in applying Equations 5.163 or 5.164 is that a compressive load W is positive, and the values of x and y in the upper right-hand quadrant (Fig. 5.34) are also positive (i.e. M_y is positive if e_y is positive).

For the eccentric loading case, assume that the raft consists of a slab and three beams L, M and N (Fig. 5.34a). The slab is designed by considering a 1 m wide strip along the edges *da* and *bc*, both of which are subjected to uniformly varying contact stresses. Shearing forces and bending moments can be calculated by treating the strip as a continuous footing. Intermediate 1 m wide strips can also be considered. For economy, it may be desirable to vary the slab thickness transversely, or longitudinally, or both. Nominal longitudinal reinforcement is required in addition to the transverse reinforcement. The reactions on the slab strips at the beams can be determined by dividing the sum of the column loads on any beam in proportion to the contact stress variation beneath it. On any beam the sum of the reactions on all 1 m wide strips must equal the sum of the column loads. The longitudinal T-beams are designed from this distribution of slab reactions and the known column loads; shearing forces and bending moments being calculated as for a continuous footing.

In order to achieve a theoretically uniform contact stress distribution, the slab and beams can be extended so that the centre of area of the raft coincides with the centre of gravity of the external load system, as shown by the dotted lines in

Fig. 5.36a, provided conditions on site permit this. The slab can then be of uniform section longitudinally and the loading from the slab is uniformly distributed along the length of each beam.

(c) Contact stress measurements beneath mat and raft foundations

To formulate efficient techniques for the design of mat and raft foundations it is essential that recourse should be made not only to theoretical and experimental studies but also to field investigations from which it is possible to assess the performance of mat and raft foundations under working conditions. Such field observations of contact stress are necessary to provide an estimate of the accuracy and validity of a particular idealization of the soil medium adopted in the interaction analysis of mat and raft type foundations. In this respect there are only a few documented cases of contact stress measurement beneath mat and raft foundations, during and after construction of the superstructure. Records of settlement are more abundant but it is generally difficult to estimate from settlement distributions the exact nature of the contact stress distribution.

Amongst the limited amount of research done on raft foundations the work of Teng (1949) is of interest. Teng investigated the distribution of contact stresses beneath a 55 m × 21 m raft constructed on the volcanic clay deposits of Mexico City. The raft is of composite construction consisting of a 0.3 m thick reinforced concrete mat to which column loads are applied through a series of steel trusses. An important conclusion is that the contact stress distribution depends largely on variations in the compressibility of the soil and that for design purposes it is desirable to assume limiting conditions. It is of interest to note that this stiff raft, which supports a flexible superstructure, was successful in reducing differential settlements to amounts much less than is experienced for most buildings in Mexico City. While the settlement took place over a number of years with repeated application of dead and live loads, it was noted that the contact stress distribution varied to some extent.

The measurement of contact stresses beneath a nuclear reactor foundation, resting on dense alluvium underlain by marl, at Marcoule, France is reported by l'Herminier *et al.* (1957). The raft was designed to resist contact stresses which were either uniform, sinusoidal or of forms obtained on the basis of an elastic continuum analysis. The measured contact stresses indicate a distribution which varies between the uniform and the elastic continuum cases.

Lazebnik (1970) measured the contact stress distribution beneath a mat foundation, 106 m × 13 m × 1 m thick, located at a depth of 5 m on a layer of fill mainly consisting of sand and loam and also beneath a ribbed raft foundation, 38 m × 15.5 m in plan with a 0.8 m thick mat and 1.5 m ribs, located at a depth of 11 m on a thick layer of undisturbed marly loam. Contact stresses were measured along transverse sections in both structures during and after the period of construction. In the case of the first structure, contact stresses due to the weight of the mat alone indicate that, at first, higher contact stresses developed at the edges of the mat and lower stresses at the centre. As construction progressed, the column loads at the centre tended to increase the contact stresses at the centre but concentrations of stress at the edges of the mat still persisted. In the case of the ribbed raft, the contact stresses during the initial stages of construction appear to have been roughly uniform. As construction progressed the central loads caused an increase in contact stresses at the centre. Here reduced stresses were observed at the edges; thus most of the external loads were supported by the central portion of the raft. Preliminary calculations indicate that the rafts can be considered to be flexible in the transverse direction. Similar field observations for an annular foundation are also discussed.

Gerrard *et al.* (1971) have observed the contact stress distribution beneath two raft foundations, with sizes ranging from 26.8 m × 28.6 m × 1.2 m thick to 30.2 m × 31.1 m × 1.5 m thick, sited on an alluvial layer approximately 9.8 m thick, comprising interbedded layers of grey and brown silty clay and sand, underlain by dark grey shale. The rafts are in contact with a sand layer. Measurements indicate that in the case of the larger raft, the contact stress distribution attained a maximum at the edges, resembling that associated with a rigid foundation resting on an elastic half-space. In the case of the second raft, the shape of the contact stress distribution appears to have been more uniform.

Contact stress measurements beneath a raft, supporting a stiff fifteen-storey building, resting on a 20 m thick layer of clay are given by Eden *et al.* (1973). It was observed that contact stresses near the edge of the raft were up to three times those near the centre and this condition persisted over a period of 2 years. The non-uniformity of the contact stress distribution is discussed in relation to the stiffness of the superstructure. Also, the measured contact stress distribution appears to be markedly different from that obtained on the basis of the design procedure suggested by ACI Committee 436 (1966).

The results of contact stress measurements beneath several rigid structures, which were summarized by Schultze (1961), clearly indicate the existence of non-uniform saddle-shaped variations associated with cohesive soils. Similar observations have been reported by Sikso and Johnson (1964) for a powerhouse stilling basin at Garrison Dam, USA, founded on clay shale (the structure resembles a raft located at a large depth or a basement) and by Kaufman and Sherman (1964) for a rigid U-shaped structure at Port Allen Lock, USA, founded on silt underlain by sand. These contact stresses have been re-assessed by Duncan and Clough (1971) in the light of the constructional sequence adopted in placing of backfill. Contact stress measurements beneath small scale models have also been investigated by Muhs (1965), Leussink *et al.* (1966) and Tetior and Litvinenko (1971).

Crowser *et al.* (1974) deduced the contact stress distribution beneath a raft, supporting a silo group on glacial till, from measured settlements. The finite element method of analysis of an isotropic elastic half-plane has been employed to evaluate the contact stresses and rather large contact stresses were predicted around the perimeter of the raft. It should be mentioned that in elastic analysis, the state of contact stress is highly sensitive to the displacement field prescribed at the edges. As observed by Galin (1961), Goodier and Loutzenheiser (1965) and others, the contact stress distribution beneath a foundation can be markedly altered by changing the profile of the foundation at its perimeter. In reality, it seems unlikely that such large concentrations of contact stress occur, as the state of stress within the soil is governed by the yield criterion for the particular soil medium. Furthermore, the occurrence of excessive plastic deformations, physical non-linearity of material behaviour, creep effects, etc., can lead to contact stresses around the perimeter of the foundation much lower than in the ideal elastic case (Shtaerman, 1956; Schultze, 1961; Sedykh, 1964; Brown, 1968; Shirkov *et al.*, 1971; Wood and Larnach, 1974: Selvadurai and Kempthorne, 1980).

The performance of two mat foundations, 66.8 m × 18.3 m and 85.1 m × 20 m, founded on clay, is reported by De Simone and Gould (1972). Here again the measured settlement profiles indicate the possible occurrence of non-uniform contact stresses and that a concentration of contact stresses can be sustained at the edges of the foundation. A summary of these contact stress distributions is also given by Selvadurai (1979).

The measurements of contact stresses that have been recorded so far are insufficient to form any quantitative judgements as to the contact stress distribution beneath raft type foundations. However, they illustrate trends which are consistent with certain theoretical assumptions. It is apparent that in addition to the behaviour of the soil, the influence of factors such as (a) flexibility of the raft, (b) flexibility of the superstructure or the manner in which flexibility of the superstructure is transmitted to the raft by, for example, column or wall connections, (c) time effects including consolidation or creep of soil, and (d) constructional sequences (excavation, dewatering, or constructional stages of the superstructure) should be given consideration in the design of mat and raft foundations.

In the remainder of this section some typical cases of raft foundations that have been adopted in foundation design are briefly described. Further examples are recorded by Peck *et al.* (1953), Little (1961), Dunham (1962), Teng (1962), Tomlinson (1975) and Manning (1972). The application of theory to design involving variable and yielding soils has been treated by Hooper (1983a, b) and by Hooper and West (1983). A typical layout of reinforcement for a 13 m × 12 m beam and slab raft supporting a total load of 11 000 kN was constructed at Millwall, London (Anon, 1947). The weight of the raft together with the earth filling is an additional 3400 kN. The raft is supported on clay. The maximum eccentricity under any system of loading is about 0.6 m and the maximum and minimum contact stresses are about 100 kN m^{-2} and 50 kN m^{-2}. The slab is 0.3 m thick. The ribs ranged in height from about 0.5 to about 0.9 m, and are 0.3 and 0.4 m wide. The raft carried twelve vats and four reaction vessels at a lead works. The floor is brought up to the upper level of the haunched beams by filling the cells between the beams.

The foundation beneath a guyed flare stack at Grangemouth, Scotland (Pike and Saurin, 1952) consists of a 0.3 m reinforced concrete slab, octagonal in plan and 9 m between opposite sides, with eight radial inverted T-beams, each 0.4 m wide by 0.8 m deep at the centre and tapering to 0.4 m at the periphery.

The foundation for the main building of a power station at Morro Bay, California (Thon and Coltrin, 1958), is a 6 m deep cellular raft, 81 m × 57 m in plan, and comprises a 1.5 m thick base slab, piers, perimeter walls and a deck slab. The superstructure columns are supported on piers. The turbo-generator pedestals are separated from the remainder of the structure and are carried direct on the base slab. The gross contact stress beneath the foundation varies between 120 and 134 kN m^{-2} but removal of overburden and hydrostatic uplift leads to a net contact stress of 24 to 38 kN m^{-2}.

Large and deep rafts are best constructed in sections, breaking joint vertically or horizontally, in order to reduce shrinkage stresses and facilitate construction. The size of each section may depend partly on the amount of concrete which can be poured in one day. In certain cases where thick slabs and beams are involved, the heat generated during hydration of the concrete could be a significant factor. Construction joints should be located at sections where the shearing stresses are relatively low. If different parts of a raft carry loads of markedly different intensity, it is generally preferable to divide the raft into sections with joints between each in order to avoid excessive stresses due to differential settlement. Generally, the overall dimensions of a raft can be changed very little and, therefore, the net contact stresses and settlements can be reduced only by increasing the overburden depth.

(d) Buoyant, floating or compensated foundations

Buoyant foundations are commonly designed as cellular rafts although they are often constructed in the manner of shallow open caissons. The principle of a buoyant foundation is that the weight of the soil excavated to accommodate the foundation should be equal to the total load imposed by superstructure and foundation on the soil. The net contact stress is, therefore, zero and within reasonable limits the settlement should also be zero. However, a certain amount of heave may be experienced during excavation and construction but this normally leads only to minor settlement as the loads are increased to their working values. Furthermore, the stresses in the soil are assumed to return to the neutral state when the working load is attained and therefore the factor of safety against bearing failure is theoretically infinite. In practice there is usually a certain amount of positive net contact stress and some settlement is incurred. Occasionally, with relatively light structures, heave of the entire structure may take place. Buoyant foundations can be employed, therefore, either to reduce the magnitudes of maximum and differential settlements or to reduce the risk of bearing failure. Buoyant foundations are relatively expensive to construct but the cost may be justified by the reduction in settlement and enhanced stability. Furthermore it may be possible to use the cellular foundation as basement accommodation. A general review of buoyant foundations has been given by Golder (1965). The term 'buoyant' suggests flotation in a liquid medium and the term 'compensated' is a better alternative (Girault, 1965). When the net contact stress is zero, the foundation is said to be fully compensated and when it has a positive value the foundation is said to be partly compensated. The following major systems of loading must be catered for in the design of a compensated foundation.

(i) Gross contact stress, including any hydrostatic uplift pressure beneath the foundation. In the case of the fully compensated foundation, the contact stress distribution beneath a stiff foundation is theoretically uniform provided the centre of gravity of external loads coincides with the centre of area of the foundation; careful arrangement of load disposition and the plan of the raft may be required to achieve this. Tilting of the foundation may occur if the loading is eccentric. If the foundation is to be sunk to founding level, a rectangular plan is desirable, although not essential, for ease of sinking. Loads which are located just outside the rectangular plan of the foundation can sometimes be supported on cantilever extensions of the deck. For a foundation which is partly compensated some variation from linear distribution will occur.

(ii) Loading on the deck from superstructure, machinery, plant and other superimposed loads. Generally the deck and base slab are designed to function in combination with the longitudinal and transverse walls to resist bending moments and shearing forces imposed by this loading acting in opposition to the contact stresses.

(iii) Lateral earth stresses and water pressure on the external walls. External and internal walls must be designed to resist bending moments, shearing forces and direct thrusts imposed by this loading. If the cells are square, the bending moments in the internal walls are, theoretically, equal to zero.

(iv) Launching or sinking loads. These loads often lead to more severe conditions than those which are imposed by the service loads since the structure lacks the stiffening influence of the deck and base slabs. Generally, arbitrary distributions of launching and sinking loads have to be assumed. For the Grangemouth foundations (Pike and Saurin, 1952) the following cases were considered: (a) The assembly was assumed to be supported solely by resistance between the outer walls and the soil. (b) The assembly was assumed to be supported solely by the soil beneath walls of a few cells at the centre.

Other conditions may be appropriate in different situations and, for example, the assembly could be assumed to be supported solely by soil beneath the walls of cells in two diagonally opposite corners.

The following details of compensated foundations at Grangemouth (Pike and Saurin, 1952) on the estuary of the River Forth illustrate typical methods of construction. Each raft comprised 3 m × 3 m cells and the depths varied from 4 to 6.5 m. The internal walls were formed of precast reinforced concrete panels, 0.15 m thick, connected at the intersections by *in situ* concrete. Continuity of reinforcement was secured by welding. The 0.3 m thick reinforced concrete external walls were formed *in situ* after erection of the internal walls. Sinking was commenced about 1 week after completion of the raft construction and was effected in 0.6 m stages by grab, commencing each stage at the corner cells and working sideways and inwards. As sinking progressed, the excavation in the centre was maintained 0.6 to 1.2 m above the excavation in the peripheral cells for ease of control. The lowest metre of excavation was executed by hand to reduce disturbance of soil at founding level. The downward movement of the raft was arrested in stages by plugging first the bottoms of the perimeter cells with plain rapid-hardening cement concrete. The cells were then plugged successively towards the centre and finally the centre cells were plugged when founding level was reached. Reinforced concrete seals were then formed above the plugs, continuity of reinforcement being secured by splice-bars inserted in holes through the bottom of each precast wall panel. On occasions some of the plugs failed before the seals were formed and had to be replaced. In other situations, geotechnical processes may facilitate the placing of plugs. A deck was constructed on top of each raft and consisted of a 0.3 m thick reinforced concrete slab laid on 78 mm thick prestressed precast concrete planks. The planks served as shuttering and the assembly was designed to function as a unit. Beams and plinths were employed to spread the concentrated loads over the deck slab. It can be noted that, if tilting cannot be eliminated completely during sinking, the effect can be taken into account to some extent by varying the thickness of the deck slab in order to secure a level floor.

Compensated rafts in British Guiana (Little, 1961) were constructed in prestressed concrete on the ground surface and sunk to founding level. Cellular compensated rafts at Syracuse, New York State, were constructed in open excavations (Hahn, 1964). Compensated foundations for the Shell Centre Tower, London, are described by Williams (1957). In Mexico City, inverted barrel vaults (Fig. 5.35) have been employed in compensated foundations (Enriquez and Fierro, 1963). Generally the axes of the shells run in the direction of the lesser plan dimension to reduce the length of the edge girders. Although, theoretically, bending moments may be induced in the shells by a uniform contact stress, experience indicates that the stress redistribution is sufficient to make these moments negligible. Consequently, shell thickness is governed by diagonal tensile stress. There appears to be a need for further investigations on the distribution of contact stresses beneath shells employed in this way. This form of construction leads to a foundation which is relatively stiff and of low weight, thus reducing both material cost and contact stresses but increasing the cost of construction.

The City Hall, Havana, comprises a 24-storey tower block including three basement storeys, with wing blocks connected to the tower block by hinged beams and slabs (Martin and Ruiz, 1959). The foundation of the tower block is a folded plate raft (Fig. 5.36). The axes of the folded plates span the shorter plan dimensions of the raft. The inclined formation surface of the ground was shaped by a motor grader and no formwork was required in placing the concrete for the

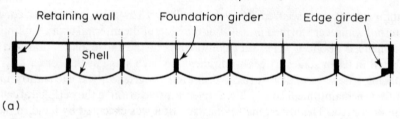

(a)

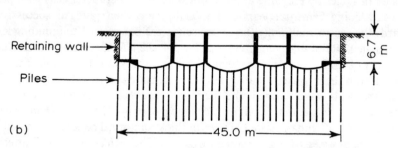

(b)

Figure 5.35 Compensated foundations, barrel vault type. (a) Typical section of compensated foundation employing barrel vaults, Mexico City; (b) longitudinal section through compensated foundation employing barrel vaults and piles for Nonoalco Tower office block, Mexico City.

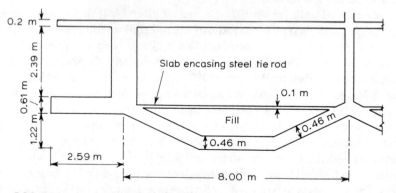

Figure 5.36 Compensated foundations, folded plate type. Cantilever and part span section of folded plate raft for City Hall, Havana, Cuba.

folded plates. Less reinforcement is required in the folded plate, which functions as a polygonal arch, than in a flat plate. The unbalanced horizontal thrust at the end spans is resisted by steel tie rods encased in a concrete slab. The cost of the folded plate raft was estimated to be about 30% less than the cost of a flat plate raft.

5.5 TANK FOUNDATIONS

The unique problems that are encountered in the design of tank foundations are illustrated by the following example described by Terzaghi (1935). The case study refers to a steel tank, 30 m diameter and 10 m high, constructed at Angern, Germany, for containing molasses. It is founded at a depth of 2 m below ground surface in a bed of soft clay, with streaks of fine sand and silt, and 4 m beneath the tank is a stratum of very stiff clay. The load on the bottom of the tank is 162 kN m^{-2} and, since this is a fluid pressure, the distribution of bearing stress is theoretically uniform provided the foundation is flexible. The form of settlement under this type of loading on a clay medium is usually dish-shaped. The maximum settlement was estimated to be 0.1 m and it was decided to support the tank on a 0.1 m thick concrete slab covered with a layer of bituminous material, since the differential settlement, even if it was equal to the maximum settlement, would not cause distress in a 30 m diameter slab of this thickness. Settlement observations confirmed the dish-shaped deformation, although irregularities occurred due to local variations in the compressibility of the soil.

A tank was also constructed at a neighbouring site with similar soil conditions and here it was supported on a reinforced concrete slab 0.5 m thick and with 1 m × 0.5 m ribs. In this case, although the load from the tank was uniformly distributed, the settlement tended to be more uniform because of the stiffness of the raft and the contact stress distribution was consequently a minimum at the centre and a maximum at the edges. Flexural moments and shearing forces were thus induced in the raft which failed because it was insufficiently strong, the rivets in the bottom of the tank were sheared and the contents flowed out. The additional expenditure in constructing the rigid slab thus led to failure, which could have been averted (once the decision was made to adopt this method of construction) only by making the raft stronger and hence more rigid. The greater rigidity aggravates conditions so that the effects are cumulative although, of course, a raft can be designed with sufficient strength to cater for bending moments and shearing forces induced by its rigidity.

Another type of failure of tank foundations results from the bearing failure of the supporting soil medium. Since tank foundations are relatively flexible, the loading imposed on the soil is known fairly accurately. The factor of safety against bearing failure is, therefore, commonly accepted close to unity; taking into account the improvement in shearing resistance which may result from the compaction or consolidation of the soil medium due to the external loading. Owing to this low factor of safety, bearing failures of tank foundations are not uncommon. Such failures may be caused either by local shear failure or complete bearing failure. The presence of soft or variable strata is commonly responsible for bearing failure during the loading or precompression stages. The mode of failure of tank foundations is largely governed by the flexibility of the foundation base. The failure of an oil tank in Fredirikstad, Norway, was reported by Bjerrum and Øverland (1957). In this case the failure of the tank foundation was caused by overstressing of a zone of soft clay. The failure of a large oil tank, 79 m diameter and 19.5 m high, constructed on a mat foundation, 0.3 m thick, supported on driven cast *in situ* piles at Fawley, England, has also been recorded. Neville (1966) describes the failure of a circular prestressed concrete reservoir, constructed full-depth in an excavation, which collapsed due to eccentric lateral loading imposed by irregular placing of backfill around the periphery, augmented by loading from an earth-moving machine. Tanks are sometimes constructed wholly or partly above ground and surrounded by an earth bank, but in all cases backfill should be placed uniformly in stages around such structures and construction plant should be carefully supervised to ensure that excessive loading is not imposed locally on the walls or roof. Settlement arising from artificial and natural compaction of backfill leads to a downward drag, through the medium of skin resistance, on the walls of the structure and this loading should be taken into account in assessing contact stresses.

Although most tank foundations are designed to function as flexible structures, an accurate assessment of the short- and long-term settlement due to the external loads is perhaps more important than a comprehensive interaction analysis (see, for example, Penman 1977). Also, in view of the low factors of safety against bearing failure, the possibility of short-term settlements in tank foundations being of a predominantly plastic nature should not be ignored. In poor soil conditions it is necessary to employ a relatively thick mat as the foundation. In this case it is generally advisable to carry out an interaction analysis to assess the flexural moments in the mat. The thin plate theories described in Section 5.4.1 are directly applicable to the thick mat type of tank foundation. The behaviour of square, rectangular and circular reinforced concrete tanks has been described by Ghali (1958), Davies (1962*a,b*, 1963*a,b*, 1964) and Hanna and Dawood (1965).

Where a circular tank, for example, is founded on a thin reinforced concrete slab, theoretically no flexural stresses or shearing stresses should be imposed on the slab by the liquid loading and the resulting contact stresses but owing to their relative flexibility, tank foundations usually experience large displacements. Therefore, flexural stresses will be imposed on the slab as a result of bending brought about by differential settlement. Before examining a specific example in which the bending stresses induced due to differential settlement are estimated, consider the problem of a thin circular plate of radius a which is subjected to a uniform peripheral radial moment $M_r = M_0$. This particular plate flexure problem represents, to a fair degree of accuracy, conditions in a region $(r < a)$ of a flexible tank foundation which is loaded by fluid pressure (Fig. 5.37).

It is assumed that the contact stresses are equal to the applied fluid pressure and that there is no loss of contact at the interface; thus the deflection of the soil surface corresponds to the dishing of the circular plate area. It should be noted that the surface deflection of the soil medium does not always correspond to this assumption and furthermore that dishing of the tank foundation may not be spherical. Restricting attention to the small deflection theory, it can be shown that in this particular axisymmetric spherical bending case, the deflection of the

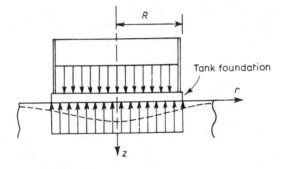

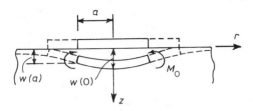

Figure 5.37 Flexible tank foundation loaded by fluid pressure.

centre of the plate relative to the deflection at $r = a$ is given by

$$\delta = w(0) - w(a) = \frac{M_0 a^2}{2D(1 + v_{\mathrm{p}})},$$ (5.165)

where $D = E_{\mathrm{p}} h^3/12(1 - v_{\mathrm{p}}^2)$, is the flexural rigidity of the plate, h is the thickness of the plate and E_{p}, v_{p} are the elastic constants of the plate material. It is assumed that M_0 is the only bending moment acting at $r = a$. The radial and tangential bending moments M_{r} and M_{θ} are both equal to M_0. As noted in Section 5.4.1, when the deflections of the plate are of the same order as the plate thickness, the small deflection theory is inadequate. Large deflections are usually accompanied by membrane action and may cause tensile fracture in brittle materials such as concrete. To illustrate this point further consider the following example.

EXAMPLE 5.2

It is proposed to support an oil tank, circular in plan, on a flexible base on the surface of the ground and the contact stresses are assumed to be sensibly uniform. The diameter of the base of the tank is 12.2 m and the external loading is equivalent to a fluid pressure of 53.6 kN m^{-2}. The influence factors for vertical stress on horizontal planes beneath a uniformly loaded circular area on the surface of the ground are given in Table 5.5, where B is the diameter of the circle and z is the depth below the base of the area. The coefficient of compressibility of the deep bed of soft clay on which the tank is to be founded varies linearly from 0.000 76 m^2 kN^{-1} beneath the tank base to 0.000 38 m^2 kN^{-1} at a depth of 24.5 m.

Table 5.5 Influence coefficients

I_σ	z/B								
	0.00	0.25	0.50	0.75	1.00	1.25	1.50	1.75	2.00
centre	1.00	0.91	0.65	0.43	0.28	0.20	0.14	0.11	0.09
periphery	1.00	0.40	0.31	0.25	0.19	0.15	0.12	0.09	0.07

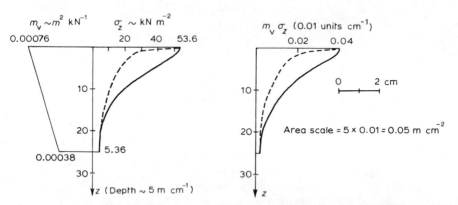

Figure 5.38 Variation of normal stress and compressibility beneath the flexible tank foundation. ——— section through centre; - - - - section through periphery.

It is assumed that settlement is induced in the tank foundation as a result only of consolidation of the clay medium. The surface settlement of the clay medium is computed on the basis of the one-dimensional theory of consolidation. In this case, the settlement at a point on the ground surface is assumed to be due to the compression of the entire soil medium directly beneath this point. In practice it would be desirable to introduce the Skempton and Bjerrum (1957) correction or to employ a refined analysis which estimates settlement on the basis of the consolidation of a semi-infinite clay stratum (De Josselin de Jong, 1957; McNamee and Gibson, 1960) or a clay layer (Gibson *et al.* 1970; Chiarella and Booker, 1975; Selvadurai, 1979). Since the present example is concerned mainly with illustrating principles, it is convenient to adopt the one-dimensional analysis. The vertical stress σ_z at any depth z beneath the base of the loading is $I_\sigma q$ where q (kN m^{-2}) is the uniform contact stress at the surface. The surface settlement at any point ($\rho(r)$), due to consolidation of the clay between depths z_1 and z_2, is

$$\rho(r) = \int_{z_1}^{z_2} m_v(z)\sigma_z(r, z)\,\mathrm{d}z$$

and this is the area of the dimensionless quantity $m_v\sigma_z$ against z, (Fig. 5.38). The relative settlement between the centre of the loaded area and the location $r = a$ is given by

$$\delta = \rho(0) - \rho(a) = \int_{z_1}^{z_2} m_v(z)\{\sigma_z(0, z) - \sigma_z(a, z)\}\,\mathrm{d}z.$$

In Fig. 5.38 the quantity $m_v\sigma_z$ was plotted originally at the scale of 0.01 units cm^{-1} and the depth at $5\,\mathrm{m\,cm^{-1}}$. Considering the areas for $m_v\sigma_z$ against z, it can be shown that the differential settlement between the centre of the loaded area and its perimeter is $0.147\,\mathrm{m}$.

Assume that the tank is to be seated on an unreinforced concrete slab $0.10\,\mathrm{m}$ thick, with elastic properties $E_p = 13.8 \times 10^6\,\mathrm{kN\,m^{-2}}$ and $v_p = 0.15$. From Equation 5.140 the flexural rigidity of the slab is

$$D = \frac{(13.8)(10^6)(0.1)^3}{12(1-0.15^2)} = 1176.5\,\mathrm{kNm},$$

The bending moment induced by the relative settlement is given by Equation 5.165

$$M_0 = \frac{2(0.147)(1+0.15)D}{(6.1)^2} = 10.7\,\mathrm{kNm\,m^{-1}}.$$

The maximum stress in a metre wide strip

$$\sigma = \frac{12M_0(0.1)}{2(0.1)^3} = 6420\,\mathrm{kN\,m^{-2}}.$$

This stress considerably exceeds the tensile strength of the concrete and cracking would be widespread in the slab. The differential settlement exceeds one-half of the thickness of the slab and other factors such as diaphragm action and friction would influence the stresses. The radius of curvature $\bar{r}$ of the slab is

$$\bar{r} \approx \frac{(d/z)^2}{2\delta} = 126.56\,\mathrm{m},$$

and the slab thickness does not exceed 1% of $\bar{r}$. The slab can be reinforced to cater for the high tensile stresses. The advantage of a concrete slab is that it forms an even surface on which to erect the tank but the slab may be omitted if a good surface can be achieved with, say, $0.3\,\mathrm{m}$ or $0.6\,\mathrm{m}$ of granular fill or fly ash.

Tank foundations are generally designed to be flexible and, since tank farms are often located along estuaries or coasts on relatively soft or loose soils, differential settlements of 0.3 to $0.8\,\mathrm{m}$ may develop. Fig. 5.39 illustrates settlements which developed beneath three tanks, $45.7\,\mathrm{m}$ diameter and $14.6\,\mathrm{m}$ high, at a site in Venezuela during test loading with water (Carlson and Fricano, 1961). The water was pumped into each tank at a uniform rate over a period of about 1 month, during which time some degree of consolidation took place, thereby leading to an increase in the shearing resistance of the soil. In spite of this, excessive settlement occurred in one quadrant beneath tank No. 39 and it is probable that this was the result of local shear failure. A reduction in differential settlement can be achieved by precompression (Aldrich, 1965). This process leads also to improved shearing resistance and hence increases the factor of safety against bearing failure. In order that the factor of safety should be not less than unity at any time, the precompression load should be applied slowly. Excessive and irregular differential settlement may set up high stresses in the cylindrical shell, cause the shell to buckle, cause the bottom to tear from the shell or, when a floating roof is provided, cause the roof to bind with the shell.

A tank foundation of asphalt on compacted fill is illustrated in Fig. 5.40(a). The provision of a ring beam (Fig. 5.40b), with or without a flexible slab, is advocated

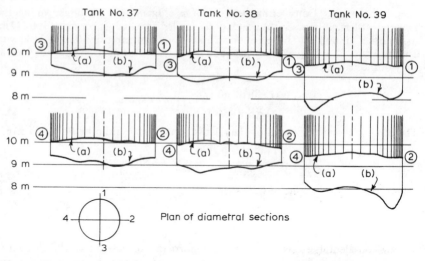

Figure 5.39 Settlement of interiors of tanks, Venezuela (after Carlson and Fricano, 1961). (a) Original and (b) final elevation of tank bottom.

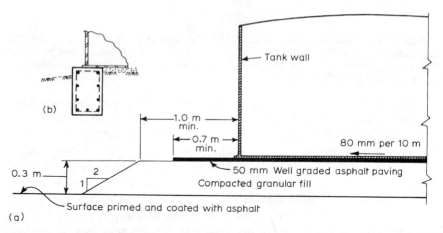

Figure 5.40 Details of tank foundations. (b) Detail of ring beam.

by some authorities, although others claim that it is not generally justified. There is, however, a marked stress concentration beneath the cylindrical shell, arising from the weight of the shell and frequently also the weight of the roof. The shell tends to expand and contract radially under changes in liquid loading and temperature (Hetényi, 1946; Olander, 1970) whereas the tank bottom tends to be restrained by friction. Consequently, flexing at the junction of the shell and tank bottom, coupled with stress concentration, sometimes leads to wear in the foundation surface around the periphery of the base. This can be overcome by

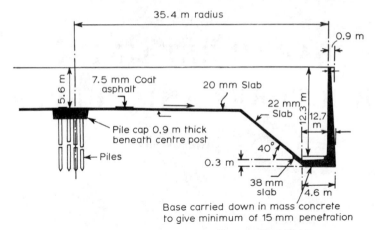

Figure 5.41 Foundation for gas holder.

using resistant surfacing. A reinforced concrete ring beam caters for shearing stresses beneath the shell. Depending on the diameter and height of the tank, ring beams generally vary between 0.45 and 1.0 m in depth and 0.45 and 0.6 m in width. The width is determined by making the contact stresses on the base of the ring, derived from structure and fluid, approximately equal to the contact stresses beneath the centre of the tank, derived from the liquid. The depth of the ring beam for an oil tank foundation was determined by Braswell (1958) from soil mechanics considerations of a particular site. Since the remainder of the foundation is flexible, it is reasonable to provide a flexible beam and, therefore, the depth of the beam should be relatively small. The beam will be subjected to hoop tension developed from lateral 'at rest' stress of the earth inside the ring subjected to the vertical fluid surcharge. Passive resistance of the soil outside the ring should be ignored.

General reviews of foundations for tanks have been presented by Roberts (1961) and Harris (1976) and performance criteria for settlement have been treated by Marr *et al.* (1982). Observations on tanks constructed at Teesmouth, England, have been reported by Penman and Watson (1967). The foundation for a tank on the Somerset levels has been described by Hirch and Martin (1976). The Durley dome type of oil tank foundation is discussed by Harding and Glossop (1951). This comprises a lightly reinforced concrete dome with a rise to span ratio of about 1/10, supported at the edge by a heavily reinforced ring beam. Granular fill is placed above the dome to provide a level surface for the tank bottom. Only minor differential settlements of the tank bottom should be experienced with this type of foundation. Complex foundations are required for gas holders. Fig. 5.41 shows details of a foundation for a gas holder at Bristol (Hughes and Turner, 1953). The welded holder is water sealed and of the four-lift spiral-guided type.

Chapter 6

EARTH RETAINING
STRUCTURES AND CULVERTS

K. Starzewski

INTRODUCTION

The major problem in the design of earth retaining structures and culverts is the determination of the magnitude and distribution of the forces that the soils exert or transfer to them. These forces are either induced by the sole action of the gravity effects (the body forces) or are resultants of the combined action of the body forces, applied loading and other effects such as temperature or volume changes or dynamic loading, e.g. earthquakes (Section 6.1). The knowledge of these forces, or their statistical estimates, enables the engineer to design individual components of these structures and to check their overall stability and hence their safety against all possible modes of failure (Sections 6.2 to 6.4).

It will be seen from the sections of this chapter that the estimation of the forces acting on the retaining structures and culverts, together with the analysis of stresses within them, present the engineer with complex problems which, in practice, are usually solved on the basis of simplified procedures.

With the above in mind, and taking into consideration the fact that in most cases the solutions are based on rather crude soil investigations supplemented by some routine tests which most frequently are carried out on more or less disturbed samples, it must be admitted that the design of this type of structure is still generally based on *simple empirical methods*; it is therefore prudent to design them with generous proportions and an adequate margin of safety.

6.1 EVALUATION OF LOADS ON EARTH RETAINING
STRUCTURES

Great progress has been made over the last decade in obtaining complete and correct solutions to the boundary value problems of which retaining structures are a special case. This was due partly to the introduction of the finite element method of analysis into the field of soil mechanics and partly to some meticulous laboratory and field research which has enabled the formulation of realistic stress–strain relationships for soils necessary in this type of analysis. For reviews of progress in this field the reader is referred to papers by Morgenstern and Eisenstein (1970) and by Wroth (1972).

While it is anticipated that, with more field and laboratory data and observations of the type presented by Cole and Burland (1972), these new methods will become more widely used, it must be appreciated that for a long time to come their use will be limited to major projects for which the high quality of site investigation justifies the use of the sophistications of these analyses. Thus there is still a need for practical methods of design of retaining structures based on consideration of the 'at rest' condition or the 'active' and 'passive' states of limiting equilibrium (Section 6.1.1) and on semi-empirical knowledge of stress distributions associated with different types of structures. These can, of course, be improved by additional field observations, particularly with regard to the stress distributions, but can never be extended to take directly into consideration the deformations of the retained soils which, as in the case of retaining structures situated in the vicinity of existing buildings, may be the most important design consideration.

In the following sections the total thrust on retaining structures will be considered, in general, to comprise the following components:

(a) thrust due to the self weight of soil (effective body forces);
(b) resultant of pore water pressure (hydrostatic and hydrodynamic);
(c) thrust due to surcharge loads;
(d) special effects such as compaction, temperature or volume changes, creep and dynamic loading.

In the case of cohesive soils the first two components of the thrust will be considered together in the so-called *total stress analysis*.

6.1.1 The state of stress 'at rest' and the concept of 'active' and 'passive' states of limiting equilibrium

The stresses which exist in the soil due to its self weight are referred to as overburden stresses. In the case of a horizontal ground surface the vertical and horizontal components of the overburden stresses are also principal stresses and therefore the state of stress at any point M (Fig. 6.1a) within the soil mass can be simply defined by stress circle PQ.

Unless external loading (surcharge) is applied to the soil or the soil is allowed to deform under the action of internal (body) forces, the stresses within it will remain equal to the overburden stresses. This means that, for example, if the soil to the left of the vertical line AM (Fig. 6.1a) is replaced with a retaining structure and the soil to the right of that line is not allowed to deform during or after its construction, then the lateral stresses on the structure remain equal to the lateral overburden stresses σ_{0x}. Because these stresses are associated with zero deformation of the soil they are referred to as the *lateral stresses at rest*.

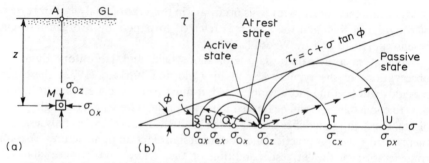

Figure 6.1 Graphical representation of different states of stress at point M.

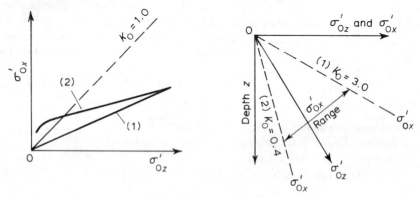

Figure 6.2 Effect of over-consolidation on K: (1) normally consolidated soil; (2) over-consolidated soil.

Introducing the concept of the coefficient of lateral stress at rest, K_0, the lateral overburden stress can be expressed in terms of the vertical stress component, σ_{0z}:

$$\sigma_{0x} = \sigma_{0y} = K_0 \sigma_{0z}. \tag{6.1}$$

The coefficient, K_0, refers to the effective overburden stresses and its value depends on the type of soil and on its past geological history. For normally consolidated soils, K_0 varies between 0.4 and 0.5 for sands and between 0.5 and 0.7 for silts and silty clays; for most practical purposes its value can be estimated from the simplified Jaky expression:

$$K_0 = 1 - \sin \phi', \tag{6.2}$$

where ϕ' is the angle of internal shearing resistance obtained with respect to effective stresses.

For over-consolidated soils, K_0 is usually greater than 1.0 (Fig. 6.2) and can be as high as 2.5 or even 3.0. The determination of K_0 for these soils is very difficult but with considerable research being done in this field the situation may improve.

For a comprehensive review on the subject of K_0, the reader is referred to work by Wroth (1972).

Now consider a case in which, during or after the construction of a smooth retaining structure to the left of line AM, the soil at M is allowed to expand laterally under the action of the body forces. The lateral expansion of the soil against a smooth wall does not affect the vertical stress at M, but leads to a decrease in the lateral stress σ_{0x} to some new value σ_{ex}; the new state of stress at M is now defined by stress circle PR. If the expansion is allowed to proceed further, a state of stress is reached eventually which is represented by a stress circle tangential to the Coulomb–Mohr envelope (circle PS). This implies that the soil at M is now in the state of limiting equilibrium. Because this stress is associated with continuous (active) deformation of the soil, it is referred to as the *active lateral stress*; it represents the minimum possible lateral stress on the wall at M. From the theoretical expression for the Coulomb–Mohr failure criterion:

$$\sigma_{ax} = \sigma_{0z}\tan^2(45° - \phi/2) - 2c\tan(45° - \phi/2) \qquad (6.3.)$$

where ϕ and c are soil strength parameters as defined in Fig. 6.1(b).

Introducing the coefficient of lateral active stress $K_a = \tan^2(45° - \phi/2)$,

$$\sigma_{ax} = \sigma_{0z}K_a - 2c\sqrt{K_a}. \qquad (6.3a)$$

In the case of cohesionless soils ($c = 0$) the last term in the above expression is equal to zero.

If now the line AM is considered to represent the back of a smooth wall which, under the action of external horizontal force, is pushed into the soil (i.e. to the right of line AM), then the soil at M is compressed laterally. Again, the vertical stress at M is not affected by this compression but the lateral stress σ_{0x} increases to some new value σ_{cx} and the new state of stress is defined by stress circle PT (Fig. 6.1b). As before, if the compression proceeds further a state of stress is eventually reached which is represented by circle PU tangential to the failure envelope; the soil at M is again in the state of limiting equilibrium and further lateral compression takes place at a constant lateral stress σ_{px}. Because this stress is associated with the passive resistance of the soil it is referred to as the *passive lateral stress*; it represents the maximum resistance of the soil that the wall will encounter at M. Again, utilizing the theoretical expression for the Coulomb–Mohr failure criterion:

$$\sigma_{px} = \sigma_{0z}\tan^2(45° + \phi/2) + 2c\tan(45° + \phi/2) \qquad (6.4)$$

and introducing the coefficient of lateral passive stress $K_p = \tan^2(45° + \phi/2)$

$$\sigma_{px} = \sigma_{0z}K_p + 2c\sqrt{K_p}. \qquad (6.4a)$$

As before, in the case of cohesionless soils the last term in the above equation is equal to zero.

In the above simple cases the coefficients of lateral active and passive stresses are functions of the angle of internal shearing resistance, ϕ, only (Table 6.3) and

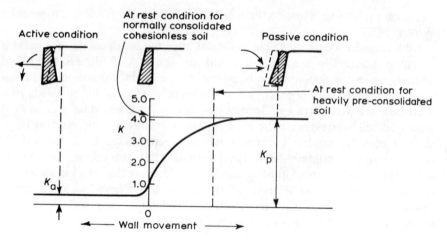

Figure 6.3 Transition from the 'at rest' state to fully active and passive conditions.

are inter-related as follows:

$$K_a = \frac{1}{K_p}. \tag{6.5}$$

Using Equation 6.2 it can be shown that for normally consolidated soils

$$K_p > K_0 > K_a.$$

The gradual transition from the 'at rest' state to the fully 'active' and 'passive' conditions is illustrated in Fig. 6.3. Experimental and field evidence indicates that in the case of cohesionless soils the active condition is attained when the top of the wall moves outwards by less than 0.5% of the wall height. On the other hand, to obtain the total passive condition, inward lateral displacement of the order of 5% for dense soils and up to 30% for loose soils has to be attained (Rowe and Peaker, 1965). In the case of heavily preconsolidated cohesive soils, the situation is almost reversed, i.e. the lateral stresses at rest are closer to the passive values than to the active values and large outward lateral displacement is necessary to attain the active state (Fig. 6.3); theoretical analysis by Di Biagio (1966) of an unbraced excavation indicates that the magnitude of the lateral displacement of the sides is proportional to K_0. Practical implications of the above results are discussed in more detail in Section 6.1.3.

6.1.2 Methods of evaluating loads on earth retaining structures due to the self weight of soil

The methods most commonly used for the determination of loads on earth retaining structures can be divided into two main groups:

(a) Methods based on the solution of the plane-strain boundary problems of the theory of limiting equilibrium to accord with the Coulomb–Mohr failure criterion; these include Sokolovsky's general solution, Rankine's method and other 'plastic stress field' solutions.

(b) Methods based on the well-known Coulomb wedge hypothesis with either plane or curved slip surfaces; these include the wedge, slip circle, logarithmic spiral and ϕ-circle methods.

According to the theory of plasticity of soils, originally developed by Drucker and Prager (1952), the solutions in group (a) can be classified as lower bounds and those in group (b) as upper bounds to the solution of the general problem of limiting equilibrium. If unrealistic collapse mechanisms are used in the upper bound methods of solution then errors on the unsafe side will be introduced as, for example, in the passive resistance of cohesive soils determined by means of the Coulomb wedge method.

(a) Limiting equilibrium methods

Rigorous and approximate solutions of a great variety of the limiting equilibrium problems which take into account, for example, wall friction and surcharge effects, were presented by Sokolovsky (1960); these included simple cases of the active and passive states of plastic equilibrium solved in the middle of 19th century by Rankine (Section 6.1.1). The tables and figures given below summarize the results most frequently used in practice.

General case of active and passive thrust in cohesionless soil

Using theoretical slip surface patterns (Figure 6.4), Sokolovsky (1960) has obtained approximate values of coefficients of active and passive lateral stress for the case of a rough inclined wall with horizontal ground surface (Tables 6.1 and 6.2, respectively). Note that in the case of a smooth ($\delta = 0°$) wall with a vertical

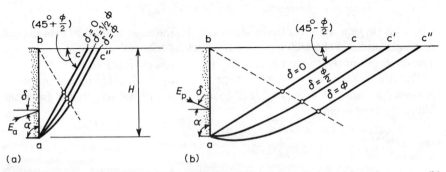

Figure 6.4 Slip surfaces for Sokolovsky's approximate solution: (a) active state; (b) passive state.

Table 6.1 Values of coefficient of active stress K_a for rough inclined wall with horizontal ground surface (Wilan and Starzewski, 1975)

ϕ		10°			20°			30°			40°		
α	δ	0°	5°	10°	0°	10°	20°	0°	15°	30°	0°	20°	40°
120°		0.49	0.45	0.44	0.27	0.24	0.23	0.13	0.12	0.12	0.06	0.05	0.05
110°		0.58	0.54	0.52	0.35	0.32	0.30	0.20	0.18	0.17	0.11	0.10	0.09
100°		0.65	0.61	0.59	0.42	0.39	0.37	0.26	0.24	0.24	0.16	0.14	0.15
90°	K_a	0.70	0.66	0.65	0.49	0.45	0.44	0.33	0.30	0.31	0.22	0.20	0.22
80°		0.72	0.70	0.68	0.54	0.51	0.50	0.40	0.37	0.38	0.29	0.27	0.28
70°		0.73	0.70	0.70	0.57	0.54	0.54	0.46	0.44	0.45	0.35	0.34	0.38
60°		0.72	0.69	0.69	0.60	0.57	0.56	0.50	0.48	0.50	0.42	0.41	0.47

Table 6.2 Values of coefficient of passive stress K_p for rough inclined wall with horizontal ground surface (Wilan and Starzewski, 1975)

ϕ		10°			20°			30°			40°		
α	δ	0°	5°	10°	0°	10°	20°	0°	15°	30°	0°	20°	40°
120°		1.52	1.71	1.91	2.76	3.67	4.51	5.28	9.07	13.5	11.3	28.4	56.6
110°		1.53	1.69	1.83	2.53	3.31	4.04	4.42	7.38	10.8	8.34	19.5	39.0
100°		1.49	1.64	1.77	2.30	2.93	3.53	3.65	5.83	8.43	6.16	13.8	26.6
90°	K_p	1.42	1.55	1.66	2.04	2.55	3.04	3.00	4.62	6.56	4.60	9.69	18.2
80°		1.31	1.43	1.52	1.77	2.19	2.57	2.39	3.62	5.02	3.37	6.77	12.3
70°		1.18	1.28	1.35	1.51	1.83	2.13	1.90	2.80	3.80	2.50	4.70	8.22
60°		1.04	1.10	1.17	1.26	1.48	1.72	1.49	2.08	2.79	1.86	3.17	5.43

back ($\alpha = 90°$) the values of K_a and K_p are the same as obtained from Rankine's solution (Table 6.3). Then the active thrust E_a is given by:

$$E_a = \frac{1}{2} K_a \frac{\gamma H^2}{\sin^2 \alpha} \tag{6.6}$$

and the passive thrust E_p by

$$E_p = \frac{1}{2} K_p \frac{\gamma H^2}{\sin^2 \alpha}, \tag{6.7}$$

where γ is the unit weight of the retained soil, δ is the angle of friction between soil and wall, and α is the inclination of the back of the wall as shown in Fig. 6.4. The direction of the thrusts E_a and E_p are as indicated in Fig. 6.4.

The distribution of stress in both cases is linear with depth and the stress components normal and tangential to the wall can be determined from the following expressions:

$$\sigma_n = K \gamma z \frac{\cos \delta}{\sin \alpha} \tag{6.8}$$

$$\tau_n = \sigma_n \tan \delta, \tag{6.9}$$

where K is the coefficient K_a or K_p from Tables 6.1 and 6.2.

Table 6.3 Values of coefficients of active and passive lateral stress for the case of smooth vertical wall with horizontal ground

ϕ (degrees)	K_a $= tan^2 \left(45 - \dfrac{\phi}{2} \right)$	$\sqrt{K_a}$ $= tan \left(45 - \dfrac{\phi}{2} \right)$	K_p $= tan^2 \left(45 + \dfrac{\phi}{2} \right)$	$\sqrt{K_p}$ $= tan \left(45 + \dfrac{\phi}{2} \right)$
5	0.8396	0.9163	1.1900	1.0913
10	0.7041	0.8391	1.4204	1.1918
15	0.5887	0.7673	1.6983	1.3032
20	0.4903	0.7002	2.0395	1.4281
25	0.4059	0.6371	2.4640	1.5697
30	0.3333	0.5774	3.0000	1.7321
35	0.2710	0.5206	3.6902	1.9210
40	0.2174	0.4663	4.5989	2.1445
45	0.1716	0.4142	5.8274	2.4142

Modified Rankine's solution for cohesive soils

It can be seen from Equation 6.3a, which defines the active lateral stress on a smooth vertical wall with horizontal ground surface, that down to a certain depth the lateral stresses are negative. This is due to the presence of cohesion and suction in the pore-water, i.e. due to the existence of compressive forces between the soil particles, which enable the soil to resist a certain amount of tension. To find the depth z_0 of the tension zone, within which the soil is obviously able to stand unsupported, the left-hand side of Equation 6.3a is equated to zero:

$$0 = K_a \sigma_{0z} - 2c \sqrt{K_a}$$

or

$$\sigma_{0z} = \frac{2c}{\sqrt{K_a}}.$$

In homogeneous soils with no surcharge on the surface, $\sigma_{0z} = \gamma z$ and hence

$$z_0 = \frac{2c}{\gamma \sqrt{K_2}}. \tag{6.10}$$

Where a uniformly distributed surcharge q is present on the ground surface, z_0 must be measured from a height $h_s = q/\gamma$ above it. If the value of h_s exceeds z_0, then the tension zone has been suppressed by the surcharge and the soil exerts compressive stresses over the entire height of the wall. Values of K_a are obtained from Table 6.3.

When calculating the total active thrust on the wall the tensile stresses within the tension zone are ignored. The distribution of lateral active stresses on the wall is obtained from Equation 6.3a and the magnitude of the total thrust is given by

$$E_a = \tfrac{1}{2} K_a \gamma H^2 - 2cH \sqrt{K_a} + 2c^2/\gamma. \tag{6.11}$$

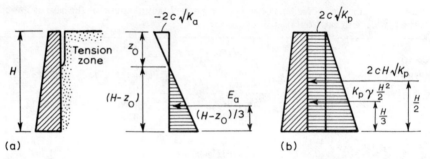

Figure 6.5 Active and passive lateral stresses. Rankine's solution: (a) active case; (b) passive case.

The last term in Equation 6.11 allows for the neglected tensile stresses. The point of application of E_a is as shown in Fig. 6.5(a).

Water pressure on the wall in the tension zone should be taken into consideration regardless of the actual ground-water condition because rain-water can accumulate in it.

The passive resistance of cohesive soil in the case of a smooth vertical wall with horizontal ground surface is obtained directly from integration of Equation 6.4a:

$$E_p = \tfrac{1}{2}K_p\gamma H^2 + 2cH\sqrt{K_p}. \tag{6.12}$$

Values of K_p are obtained from Table 6.3 and the point of application of E_p is obtained from consideration of the stress diagram in Fig. 6.5(b).

Approximate allowance for soil–wall friction in Rankine's solutions

Rankine's solutions for both cohesionless and cohesive soils can be modified to allow for the friction between the soil and wall simply by substitution in appropriate equations of coefficients from Tables 6.1 and 6.2 or 6.5 and 6.8 corresponding to the given value of δ. Because of the approximate nature of this approach, the last term in Equation 6.11 is usually neglected. Some authors (e.g. Rowe, 1952) recommend the use of coefficients obtained from the Coulomb wedge analysis, substituting in Equations 6.14 and 6.17.

This approximate method is particularly useful when dealing with layered soils. Note that the resultant thrusts E_a and E_p are now inclined at an angle δ to a line normal to the wall i.e. to the horizontal; in some cases this is ignored, thereby leading to an increase in the safety factor.

(b) Coulomb's wedge methods

These methods are based on finding a plane or curved slip surface which corresponds to the maximum active thrust or minimum passive thrust on a retaining wall. According to the Civil Engineering CP 2 (ISE, 1951), Coulomb's

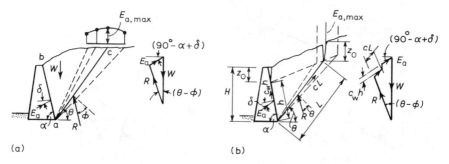

Figure 6.6 Graphical determination of active thrust (a) for cohesionless soils, (b) for cohesive soils.

plane slip surface wedge analysis is generally acceptable for the determination of active thrusts in all cases of gravity and cantilever retaining walls and passive thrusts for cohesionless soils in cases where $\delta \leqslant \phi/3$; in other cases, solutions based on curved slip surfaces give more realistic results.

General case of active thrust – simple wedge method

In the general case of a rough retaining wall, taking into account friction and/or adhesion between soil and wall, and irregular ground surface, the active thrust is found by the graphical procedure shown in Fig. 6.6.

In the case of cohesionless soils, a plane slip surface is chosen as shown in Fig. 6.6(a) and the area of the wedge is determined. The magnitude of the vertical force W can then be evaluated from the knowledge of the density of soil within the wedge. Directions of the thrust on the wall E_a and resultant of the shearing resistance R are determined from the geometry of the problem and from knowledge of the angle of friction between the soil and wall, δ, and angle of internal shearing resistance of the soil, ϕ. The magnitude of the thrust on the wall is then determined from a triangle of forces. The procedure is repeated with other slip planes until sufficient values have been obtained to enable the maximum thrust to be found by graphical interpolation; not less than three – preferably five – slip surfaces should be used.

In the case of cohesive soils the active thrust is found by a graphical procedure similar to that above but two additional factors are taken into consideration: (a) the existence of the tension zone of depth z_0, and (b) the existence of adhesion between the soil and wall, c_w (Fig. 6.6b). The depth of the tension zone is calculated from Equation 6.10.

In both cases the point of application of E_a can be taken on the point of intersection with the back of the wall of a line drawn through the centre of gravity of the wedge parallel to the slip plane of the wedge.

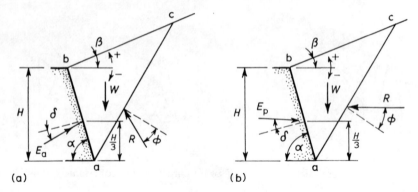

Figure 6.7 Notation for analytical determination of active and passive thrusts for cohesionless soils: (a) active case; (b) passive case.

Special cases of active thrust in cohesionless soils – simple wedge method

In the case of a rough retaining wall with an inclined or horizontal ground surface (Fig. 6.7a) the maximum active thrust E_a can be calculated from the following equations:

$$E_a = \frac{1}{2} \frac{K_a \gamma H^2}{\sin \alpha \cos \delta} \tag{6.13}$$

where

$$K_a = \frac{\sin^2(\alpha + \phi) \cos \delta}{\sin \alpha \sin(\alpha - \delta)\left\{ 1 + \sqrt{\left[\dfrac{\sin(\phi + \delta) \sin(\phi - \beta)}{\sin(\alpha - \delta) \sin(\alpha + \beta)} \right]} \right\}^2} \tag{6.14}$$

For specific values of ϕ, δ, α and β, the values of K_a can be obtained from Table 6.4.

Special cases of active thrust in cohesive soils – simple wedge method

The active thrust in cohesive soils can be calculated for the case of a wall with a vertical back and with horizontal ground surface from the following equation (Packshaw, 1946):

$$E_a \cos \delta = \tfrac{1}{2} K_a \gamma (H^2 - z_0^2) - c(H - z_0) K_{ac} \tag{6.15}$$

where the values of the coefficients K_a and K_{ac} depend upon ϕ, δ, and c_w/c and are given in Table 6.5. The actual stress distribution is assumed as shown in Fig. 6.8; water pressure in the tension crack is always taken into consideration as rain-water may accumulate in it.

General case of passive resistance – simple wedge method

For cohesionless soils the graphical procedure for determination of the minimum passive thrust by the simple wedge method is basically similar to that used to

Table 6.4 Values of coefficient K_a given by Equation 6.14

ϕ (deg)	α (deg)	$-20°$		$-10°$		$0°$		$10°$		$20°$		$30°$	
		δ		δ		δ		δ		δ		δ	
		0	$2/3\phi$	0	$2/3\phi$	0	$2/3\phi$	0	$2/3\phi$	0	$2/3\phi$	0	$2/3\phi$
15	110			0.421	0.350	0.467	0.395	0.558	0.489				
	100			0.476	0.406	0.531	0.462	0.634	0.572				
	90			0.525	0.457	0.589	0.525	0.704	0.653				
	80			0.567	0.502	0.642	0.585	0.772	0.736				
	70			0.602	0.542	0.691	0.643	0.840	0.803				
20	110	0.302	0.240	0.325	0.261	0.357	0.292	0.412	0.344	0.664	0.612		
	100	0.355	0.290	0.385	0.319	0.426	0.359	0.492	0.427	0.773	0.743		
	90	0.401	0.336	0.440	0.375	0.490	0.426	0.569	0.510	0.883	0.883		
	80	0.440	0.376	0.489	0.426	0.551	0.492	0.644	0.597	1.000	1.044		
	70	0.470	0.408	0.532	0.474	0.608	0.559	0.721	0.689	1.132	1.239		
30	110	0.174	0.132	0.185	0.142	0.199	0.154	0.220	0.172	0.258	0.206	0.468	0.413
	100	0.228	0.180	0.245	0.195	0.266	0.214	0.296	0.242	0.348	0.292	0.605	0.569
	90	0.279	0.228	0.304	0.251	0.333	0.279	0.374	0.320	0.441	0.389	0.750	0.750
	80	0.326	0.274	0.360	0.308	0.400	0.349	0.454	0.405	0.539	0.500	0.910	0.973
	70	0.367	0.316	0.414	0.365	0.468	0.423	0.537	0.501	0.646	0.631	1.098	1.066
40	110	0.090	0.067	0.095	0.070	0.100	0.074	0.107	0.080	0.117	0.089	0.138	0.106
	100	0.138	0.106	0.146	0.113	0.156	0.122	0.169	0.133	0.187	0.150	0.222	0.182
	90	0.187	0.151	0.201	0.164	0.217	0.179	0.238	0.198	0.267	0.226	0.318	0.278
	80	0.236	0.198	0.259	0.219	0.283	0.244	0.313	0.275	0.355	0.319	0.426	0.398
	70	0.283	0.246	0.316	0.281	0.352	0.320	0.396	0.368	0.454	0.435	0.551	0.554

Table 6.5 Values of coefficients K_a and K_{ac} for cohesive soils (After the Civil Engineering CP 2 (ISE, 1951); by permission of Council of the Institution of Structural Engineers.)

		K_a		K_{ac}			
c_w/c		*All values*		0	1.0	0.5	1.0
ϕ	δ	0	ϕ	0	0	ϕ	ϕ
0°		1.00	1.00	2.00	2.83	2.45	2.83
5°		0.85	0.78	1.83	2.60	2.10	2.47
10°		0.70	0.64	1.68	2.38	1.82	2.13
15°		0.59	0.50	1.54	2.16	1.55	1.85
20°		0.48	0.40	1.40	1.96	1.32	1.59
25°		0.40	0.32	1.29	1.76	1.15	1.41

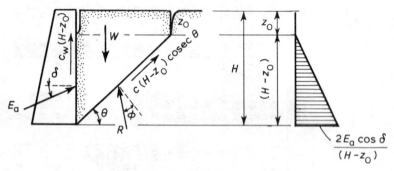

Figure 6.8 Active thrust in cohesive soil.

determine the maximum active thrust but the directions of the reactive forces are changed (Fig. 6.7b). While for practical purposes the accuracy of the simple Coulomb wedge analysis is sufficient in the active case, the passive resistance of soils can be significantly overestimated by this method. This is particularly true for soils with ϕ greater than about 20° and δ greater than $\phi/3$. These differences are due to the divergence of the assumed plane slip surface from the shape of the actual slip surface which is curved. The use of the Coulomb wedge method is not recommended for the determination of passive resistance of cohesive soils. The actual slip surface in these soils is markedly curved and cannot be approximated by a plane surface.

Special cases of passive resistance of cohesionless soils – simple wedge method

In the case of rough wall with inclined or horizontal regular surface (Fig. 6.7b) the minimum passive thrust E_p can be calculated from

$$E_p = \frac{1}{2}\frac{K_p\gamma H^2}{\sin\alpha\cos\delta},$$

(6.16)

where

$$K_p = \frac{\sin^2(\alpha - \phi)\cos\delta}{\sin\alpha \sin(\alpha + \delta)\left\{1 - \sqrt{\left[\dfrac{\sin(\phi + \delta)\sin(\phi - \beta)}{\sin(\alpha + \delta)\sin(\alpha + \beta)}\right]}\right\}^2}. \tag{6.17}$$

For $\beta = 0$ and $\alpha = 90°$, the values of K_p can be obtained from Table 6.6 but it should be noted that the values in brackets are high, for many purposes on the unsafe side, and in these cases the values in Table 6.2 are recommended.

Other methods of determination of passive resistance

For more accurate determination of the passive resistance of both cohesionless and cohesive soils by the wedge method, it is necessary to introduce a curved slip surface which should, as closely as possible, correspond to the actual surface of failure. The two most common of such methods are (a) the logarithmic spiral method and (b) friction or ϕ-circle method; both involve laborious graphical determination of the minimum passive resistance and their use is seldom justifiable.

For determination of the passive resistance, E_p of cohesive soils the special case of the wall with a vertical back and with horizontal ground surface the Civil

Table 6.6 Values of coefficient K_p for $\alpha = 90°$ and $\beta = 0°$

ϕ (deg)	$\delta = 0°$	$\delta = \frac{1}{3}\phi$	$\delta = \frac{2}{3}\phi$	$\delta = \phi$
10°	1.42	1.52	1.61	1.70
20°	2.04	2.40	2.81	(3.31)
30°	3.00	4.08	(5.74)	(8.74)
40°	4.60	7.93	(16.73)	(70.93)

Note: values in brackets are high on the unsafe side and in these cases values in Table 6.2 are recommended.

Table 6.7 Values of coefficients K_p and K_{pc} for cohesive soils (After the Civil Engineering CP2 (ISE, 1951); by permission of Council of the Institution of Structural Engineers.)

		K_p				K_{pc}		
c_w/c		All values		0	0.5	1.0	0.5	1.0
ϕ	δ	0	ϕ	0	0	0	ϕ	ϕ
0°		1.0	1.0	2.0	2.4	2.6	2.4	2.6
5°		1.2	1.3	2.2	2.6	2.9	2.8	2.9
10°		1.4	1.6	2.4	2.9	3.2	3.3	3.4
15°		1.7	2.2	2.6	3.2	3.6	3.8	3.9
20°		2.1	2.9	2.8	3.5	4.0	4.5	4.7
25°		2.5	3.9	3.1	3.8	4.4	5.5	5.7

Engineering CP 2 (ISE, 1951) suggests the following equation based on the ϕ-circle method (Packshaw, 1946):

$$E_p \cos \delta = 1/2 K_p \gamma H^2 + cHK_{pc} \tag{6.18}$$

where the values of the coefficients K_p and K_{pc} depend upon ϕ, δ, and c_w/c and are given in Table 6.7. The point of application of E_p is obtained from consideration of the stress diagram which is similar to that in Fig. 6.5(b).

(c) Layered soils and uniform surcharge loading

In the general case, the determination of active and passive thrusts in layered soils can only be satisfactorily carried out using the simple Coulomb wedge method. However, in the most important case from the practical point of view, that is the case of a vertical wall with horizontal ground surface, Rankine's solutions, with or without wall friction and adhesion, can conveniently be used. The presence of uniform surcharge on the ground surface does not present difficulties and in the Coulomb method is simply allowed for in the weight of the wedge.

In Rankine's general solution (Equations 6.3a and 6.4a) the surcharge q is added to the overburden stress thus

$$\sigma_{0z} = q + \gamma z.$$

For calculation of active or passive stresses in an individual soil layer the weight of the overlying soil layers is included in the surcharge thus

$$q = q_0 + \sum \gamma_i h_i \tag{6.19}$$

where q_0 is the surcharge acting on the ground surface, γ_i and h_i are the unit weight and thickness of soil in any layer i.

The basic stress Equations 6.3a and 6.4a can be rewritten in the form:

$$\sigma_{ax} = \gamma(h_s + z)K_a - 2c\sqrt{K_a}, \tag{6.20}$$

and

$$\sigma_{px} = \gamma(h_s + z)K_p + 2c\sqrt{K_p} \tag{6.21}$$

where z is the depth measured from the surface of the layer under consideration, h_s is the equivalent height of surcharge of unit weight γ above each layer, i.e. $h_s = q/\gamma$, K_a and K_p are coefficients, the values of which are derived from Tables 6.1 and 6.2 or Equations 6.14 and 6.17, if wall friction is considered and from Table 6.3 if it is neglected.

The resultant thrust or reaction from an individual layer is obtained by integration of the appropriate stress diagram and if $\delta \neq 0$ it is assumed to act at an angle δ to the horizontal. As mentioned in the preceding section, the presence of a surcharge affects the depth of the tension zone in all the underlying layers of cohesive soils.

6.1.3 Effects of movement and deformation of retaining structures on total thrust and stress distribution

In order to develop completely the state of limiting equilibrium or the Coulomb type of failure mechanism in the retained soil, the movement or deformation of the wall must have certain characteristics and must be of sufficient magnitude. For example, to develop Rankine's state of stress or the Coulomb type of failure it is theoretically necessary for the wall to rotate bodily so that its displacement increases linearly from the bottom (Fig. 6.9a) and exceeds a certain minimum value at the top (Fig. 6.3).

If the displacement and rotation or deformations of the wall are inadequate, the soil mass remains wholly or partly in the state of elastic equilibrium, and the distribution and magnitude of stresses on the wall are substantially different from those predicted by consideration of the soil in the state of limiting equilibrium; three such cases are illustrated in Fig. 6.9(b), (c) and (d) for a cohesionless soil.

A series of large-scale tests by Terzaghi (1934) and many subsequent tests and *in situ* measurements by him and other researchers (Kjaernsli, 1958; Di Biagio and Kjaernsli, 1961; Terzaghi and Peck, 1967; Lambe 1970; Bjerrum *et al.*, 1972) have clearly demonstrated the importance of wall movement on the stress distribution and have helped in the development of empirical rules which facilitate the extension of both the limiting equilibrium and Coulomb wedge methods to a wide range of engineering problems for which theoretical solutions do not exist at present or are cumbersome.

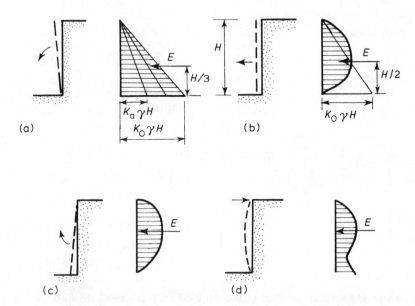

Figure 6.9 Effect of wall deformation on stress distribution.

(a) Gravity and cantilever retaining walls

For these walls the movements are generally of the type necessary to develop the state of limiting equilibrium or the Coulomb failure mechanism in the retained soil but their magnitude may not be sufficient to develop the fully active condition. For design purposes, the stress distributions are usually taken as predicted by one of the above theories but the magnitude of the total thrust is taken to be between the at rest and the fully active state, depending on the wall support conditions. Thus for granular fill behind gravity walls founded on rock or on raking and vertical piles, the at rest condition is usually assumed, whereas for cantilever walls founded on soil or on vertical end-bearing piles only the active thrust is taken (see also Sections 6.2.3 and 6.2.4).

(b) Strutted excavations

The nature of the deformation in a strutted excavation as shown in Fig. 6.10 is an excellent example of limited movements which do not permit the development of the fully active condition but at the same time disturb the 'at rest' state of stress. Construction begins with the driving of the sheet piles. Usually the first row of struts is placed near the top of the excavation immediately after a small cut, i.e. before any appreciable movement occurs (Fig. 6.10a). The subsequent rows of struts however, are placed after a considerable amount of soil has been removed and, therefore, some deformation does take place in the lower part of the excavation (Fig. 6.10b). The final deflected shape of the sheeting and the redistribution of stresses depend on the type of soil in which the excavation is carried out, particularly on its deformation characteristics, and on the geometry of the excavation but they are influenced also by the method of construction. The effect of the type of soil on the redistribution of stresses is illustrated by two extreme cases shown in Fig. 6.10(c) and (d) (Bjerrum *et al.*, 1972); the deflections of the sheeting in a strutted excavation in dense sand are small and the sum of the strut loads is nearly equal to the theoretical active thrust, whereas in soft clay the deflections are relatively large and therefore additional loads are thrown onto the

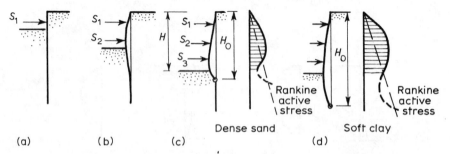

Figure 6.10 Deformation of sheet piles in strutted excavation and redistribution of the active stresses.

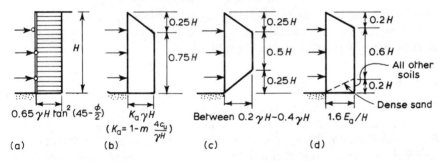

Figure 6.11 Design stress distributions for strutted excavations: (a) in sands; (b) on soft to medium clays; (c) in stiff-fissured clays; (d) for all types of soils.

struts so that the sum of the strut loads exceeds the theoretical active thrust as computed for the depth of the excavation H. To allow for these effects in the design of the struts, a number of semi-empirical methods have been developed, the best known of which are described below.

The semi-empirical design rules published by Terzaghi and Peck (1967) are based on evaluation of the total active thrust down to the excavation level using the Rankine method ($\delta = c_w = 0$) (Sections 6.1.2a and 6.1.2c) and on empirical modifications of this thrust and of stress distributions (based on field measurements) to allow for the above-mentioned differences due to the type of soil in which the excavation is carried out. For that purpose the soils are divided into three categories: (a) sands, in which a rectangular stress distribution is assumed (Fig. 6.11a), (b) soft to medium clays in which trapezoidal stress distribution is assumed, as shown in Fig. 6.11(b), and (c) stiff-fissured clays in which a different trapezoidal stress distribution is taken (Fig. 6.11c). The value of m for the soft to medium clays is taken as 1.0 unless the excavation is in truly normally consolidated clay (very rare) for which m can be as low as 0.4. For the stiff-fissured clays, lower stress intensity is used only when deformations can be kept to minima and construction time is short. For excavations in sands, water pressure must also be taken into consideration as discussed in Section 6.1.4.

The above method does not cover such commonly occurring soils as sandy clays, clayey sands, cohesive silts and layered or non-homogeneous soils. Until more observations are available it is suggested that for excavations in these soils to a depth of less than about 6 m the recommendations given is Civil Engineering CP2 (ISE, 1951) can be used. In this method the design thrust is taken as 44% greater than the active thrust obtained from the Rankine's solution ($\delta = c_w = 0$) and is assumed to be distributed as shown in Fig. 6.11(d). For excavations deeper than about 6 m the suggestions put forward by Bjerrum *et al.* (1972) can be followed. This method takes into consideration the deformation of the soil beneath the excavation level and requires knowledge of depth H_0 below which the deformations of the sheeting are insignificant (see Fig. 6.10c and d); the value of H_0 is estimated either from the consideration of the particular geological conditions or on the basis of experience from other jobs. Design forces in struts and bending moments in sheet piles can be computed assuming a statically determinate system as shown in Fig. 6.11(a).

(c) Cantilever sheet pile walls

The resistance to movement of cantilever sheet piles is provided by the passive resistance of the soil mobilized against the embedded portion of the piles (see Potts and Burland, 1983). The movement of these piles (Fig. 6.12a) is not appreciably different from that required to mobilize states of active and passive plastic equilibrium and therefore straightforward stress distributions as obtained from the Rankine or Coulomb wedge analysis are acceptable (Fig. 6.12b). Because of the approximate nature of the problem, the simple Rankine solution is generally used and wall friction and adhesion are ignored ($\delta = c_w = 0$).

In the determination of the required depth of penetration of the piles it is assumed that the moment of failure the sheet pile wall rotates about a point O. The resulting system of forces for cohesionless soils is as shown in Fig. 6.12(b). For design purposes this is simplified (Fig. 6.12c) by assuming $(E_{p2} - E_{a2})$ to be a point load with its line of action passing through O. The theoretical depth of penetration d is then found by considering the equilibrium of moments about O. In order to provide a suitable factor of safety and to limit rotation of the wall necessary to mobilize the passive resistance of the soil only one-half of the passive thrust E_p, is taken into account and the actual depth of penetration D is taken as $1.2d$. Thus

$$0.5E_p \times d/3 = E_a(H+d)/3, \tag{6.22}$$

and utilizing Rankine's solution for cohesionless soils

$$0.5(K_p \gamma d^2/2) \times d/3 = [K_a \gamma (H+d)^2/2](H+d)/3, \tag{6.23}$$

which on substitution of $K_p = 1/K_a$ results in a cubic equation

$$d^3 = 2K_a^2(H+d)^3. \tag{6.24}$$

In cohesive soils, the presence of the tension zone must be considered, but otherwise the procedure is the same. In the special case of $\phi = 0$, $K_a = K_p = 1.0$ and unless $c/\gamma H$ is greater than 0.25, the equilibrium conditions cannot be satisfied.

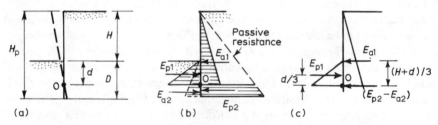

Figure 6.12 Deformation and forces on cantilever sheet pile wall.

(d) Anchored sheet pile walls

The height of sheet pile cantilever walls is generally limited by the large bending moments set up in the wall and excessive depth of penetration which may be necessary to mobilize the required fixity in the soil. It is therefore common in deep excavations to introduce horizontal walings and tie bars near the top of the wall so that the sheet piling acts in a manner similar to that of a propped cantilever. The tie bars are anchored to concrete blocks, short length of sheet piling or pile caps carried on raking piles which are positioned sufficiently far from the retaining wall so that their stability is not affected by its movement (Tsinker, 1983; Anderson *et al.*, 1983).

As in the case of strutted excavations, the deformation of anchored sheet pile walls diverges appreciably from that assumed in both Rankine and Coulomb theories. In the case of cohesionless soils the actual stress distribution is of the form shown in Fig. 6.13(a). The redistribution of the contact stresses is due to the deflection of the wall: high stresses are generated in zones H (Fig. 6.13c) due to restricted 'active' deflections or due to large 'passive' deflections whereas in zone L large 'active' deflections result in low contact stresses (Bjerrum *et al.*, 1972).

The increased stresses on the lower part of the back of the wall and on the front, just below the dredge level, induce a 'fixing' moment which reduces the maximum 'span' moment in the wall (Fig. 6.13b); this effect was described by Rowe as 'flexure below dredge level'. The increased stresses on the back of the wall above the anchor have a similar effect which was referred to by Rowe as 'flexure above anchor level'. Finally, the relatively low stresses on the back of the wall between anchor and dredge levels also decrease the maximum 'span' moment; this effect is generally referred to as 'arching'.

The most significant effect, from the practical point of view, is 'flexure below dredge level'. The degree of fixity provided by the soil at the bottom of the wall depends on the depth of penetration of the wall D, on its flexural stiffness EI, and on the deformation resistance characteristics (relative density) of the penetrated

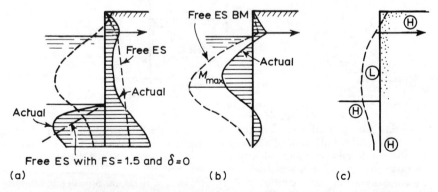

Figure 6.13 Anchored sheet pile walls, effects of deformation on (a) stress distribution, (b) bending moments, (c) zones of stress intensity.

soil; the latter two properties are dominant and are discussed further in Section 6.2.8. Over-simplifying the problem, at this stage, one can say that if D is small (or EI large or relative density low), the pile can mobilize passive resistance only on the excavation side (Fig. 6.14a to d) and, therefore, it can be assumed that no fixity is provided; this case is referred to as the 'free earth support' (free ES) condition. If D is large (or EI small or relative density high) then passive resistance will be mobilized on both sides of the pile wall (Figs. 6.13c and 6.14e) and partial or complete fixity can be assumed; the latter case is known as the 'fixed earth

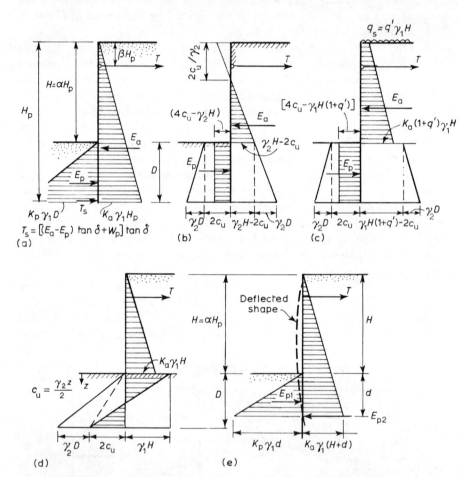

Figure 6.14 Design earth stress distributions. (a) Free ES, pile driven into and retaining cohesionless soil or fully drained over-consolidated cohesive soil with $c' = 0$. (b) Free ES, pile driven into and retaining cohesive soil, $\phi_u = 0$. (c) Free ES, pile driven into over-consolidated cohesive soil, ($\phi_u = 0$) and retaining cohesionless soil. (d) Free ES, pile driven into normally consolidated cohesive soil (c_u increasing with depth, $\phi_u = 0$) and retaining cohesionless soil. (e) Fixed ES, pile driven into and retaining cohesionless soil.

support' (fixed ES) condition. The two simplified systems of forces used in determination of D for different soil conditions are shown in Fig. 6.14a to d; note that in evaluation of the total active thrust, pore-water pressure must be taken into consideration. For practical comments on the choice of the design conditions see Section 6.2.8.

For practical design purposes, the stress distributions in both active and passive zones are taken as shown in Fig. 6.14. The methods used in evaluation of the active and passive thrusts are generally based on either the simple Rankine or Coulomb wedge solution. The effects of wall friction on the 'active' side of the sheet pile wall are included only in the evaluation of the coefficient of active lateral stress, K_a, and the resultant thrust is assumed to act in a horizontal plane. The same applies to the effects of wall friction on the 'passive' sides, although in some methods of design these effects are completely ignored, i.e. δ is assumed equal to zero. In the case of the 'immediate' ($\phi_u = 0$) analysis of sheet piles in cohesive soils, the wall adhesion c_w may be taken into consideration. For piles driven into cohesionless soils toe frictional resistance T_s may be taken into consideration in the manner suggested by Rowe (1952) and shown in Fig. 6.14(a) (see also example 6.3, Section 6.2.8, part b).

To allow for large strains necessary to mobilize the passive resistance of soils a factor of safety of at least 1.5 is usually introduced, particularly when the soils are loose or normally consolidated. Terzaghi (1954) has recommended that the factor of safety in clean sands or silty sands should be between 2 and 3, depending on the accuracy with which the active thrust can be estimated; for silts and clays, values between 1.5 and 2 (applicable to K_{pc} only, see Equation 6.18) are recommended.

(e) Rigid earth-retaining structures

With rigid self-supporting earth-retaining structures, such as monolithic concrete culverts, portal frame bridges, and rigid basements strutted by means of a heavy raft and floor slabs or beams, the deformation is usually insufficient to mobilize active state of stress and stresses will approximate to the earth stresses at rest; in heavily overconsolidated clays this may mean stresses close to the passive condition (Fig. 6.3). A comprehensive discussion of lateral stresses on rigid walls, with case histories, was presented by Gould (1970). Broms and Ingelson (1972) have presented interesting results of a full-scale test on a bridge abutment of a multi-span portal frame bridge: considerable variation of lateral earth stresses with seasonal variation in temperature and shrinkage was observed. The highest stresses, equivalent to $K = 0.4$, were measured during the summers (maximum expansion of the deck) and the lowest during the winter; the highest stresses were approximately equal to these immediately after compaction of the sandy gravel and medium sand backfill. Further studies of stresses on walls have been published by Jones and Sims (1975), Coyle and Bartoskewitz (1976), Sherif and Mackey (1977), Jones (1979), Ingold (1980) and Symons (1983).

6.1.4 Pore-water pressure and seepage effects (hydrostatic and hydrodynamic pressures)

The calculations in the preceding sections did not include the effects of the pore-water pressure or seepage. Nor did they distinguish between total stresses and effective stresses. To take the water forces into consideration, it is generally necessary to make calculations in terms of effective stresses; this, of course, involves the use of effective strength parameters c' and ϕ'. The principles outlined in previous sections remain unchanged, except that the water forces are now included in the calculations.

(a) Effects of water pressure in limiting equilibrium methods

The general theory of plastic equilibrium of soils is limited to homogeneous materials and any inclusion of the effects of pore-water pressure or seepage can be considered only as approximate. In the case of the methods discussed in Section 6.1.2a, only the effects of a horizontal ground water table can be taken into account.

In Sokolovsky's approximate method and in the Rankine solution for cohesionless soils ($\delta = 0$) the effective stress components normal and tangential to the wall are computed from Equations 6.8 and 6.9. The hydrostatic water pressure on the wall is then determined and added vectorially to the effective stresses to obtain the total stresses and hence the total thrust on the wall.

EXAMPLE 6.1

Using the given data determine the distribution of stresses and total active thrust per lineal metre of the wall if ground water table is present at 2 m below the ground surface (Fig. 6.15).

Data: $\gamma = 16.0\,\mathrm{kN\,m^{-3}}$; $\gamma_{sat} = 19.5\,\mathrm{kN\,m^{-3}}$; $\phi' = 30°$; $\delta = 20°$; $\alpha = 80°$; $\gamma_w = 9.8\,\mathrm{kN\,m^{-3}}$.

From Table 6.1, $K_a = 0.38$. Effective stress components normal to the wall are obtained from Equations 6.8 and 6.9:

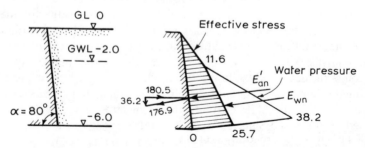

Figure 6.15 Determination of active thrust with ground water present.

at -2.0 level:

$$\sigma'_{an} = 0.38 \times 16.0 \times 2.0 \times \frac{\cos 20°}{\sin 80°} = \frac{0.38 \times 16 \times 2 \times 0.940}{0.985} = 11.6\,kN\,m^{-2}$$

at -6.0 level:

$$\sigma'_{an} = \frac{0.38[16.0 \times 2.0 + (19.5 - 9.8) \times 4] \times 0.940}{0.985} = 25.7\,kN\,m^{-2}$$

$$E'_{an} = \frac{11.6 \times 2.0}{0.985} + \frac{11.6 + 25.7}{2} \times \frac{4.0}{0.985} = 23.6 + 75.8 = 99.4\,kN\,m^{-1}$$

$$\tau'_{an} = E'_{an}\tan\delta = 99.4 \times 0.364 = 36.2\,kN\,m^{-1}.$$

Water pressure at -6.0 level:

$$\sigma_w = 9.8 \times 4 = 38.2\,kN\,m^{-2}$$

$$E_{wn} = \frac{38.2 \times 4.0}{2 \times 0.985} = 77.5\,kN\,m^{-1}.$$

Total active thrust normal to the wall:

$$E_{an} = E'_{an} + E_{wn} = 176.9\,kN\,m^{-1}.$$

Resultant total active thrust:

$$E_a = \sqrt{[(36.2)^2 + (176.9)^2]} = 180.5\,kN\,m^{-1}.$$

In the modified Rankine solution for active thrust in cohesive soils, two cases must be considered: (a) when the ground water table is below the tension zone, and (b) when it is at or above the ground level. In the first case the procedure is similar to that used above for the Rankine solution for cohesionless soils, but additional pressure due to rain-water in the tension zone is taken into consideration.

In the second case the depth of the tension zone is determined using the submerged unit weight of soil $(\gamma' = \gamma_{sat} - \gamma_w)$ in Equation 6.10, but otherwise the procedure is the same as for cohesionless soils, i.e. the total thrust is obtained by summation of the effective thrust and the resultant of the pore-water pressure.

(b) Effects of water pressure and seepage in the Coulomb wedge method

When these additional effects are taken into account, the principles of the wedge analysis outlined in Section 6.1.2b remain unchanged except that water forces are now included in consideration of the equilibrium of forces. The method illustrated in Fig. 6.16 involves the use of the total weight of the wedge, together with boundary water pressures and effective stresses. One can also use an exactly equivalent system of forces involving buoyant weight of the wedge, seepage force, and boundary effective stresses.

(c) Effects of unbalanced water pressure and seepage in analysis of sheet piles

When the water table in the retained soil is above the free water level in front of a sheet pile wall, then seepage takes place downwards through the retained soil,

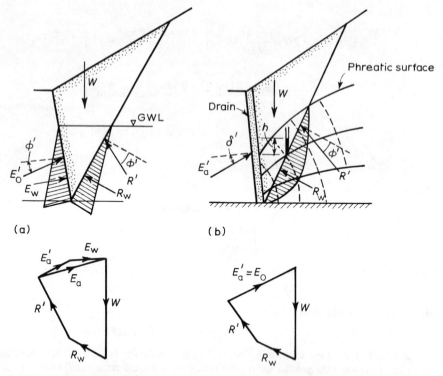

Figure 6.16 Effects of water pressure and seepage in Coulomb's wedge analysis.

around the lower edge of the sheet piles and upwards beyond the outer face, as shown in Fig. 6.17(a); this is a common situation in the water-front retaining structures when the water table behind the wall lags behind a receding tide or a rapidly receding flood water in a river. Converse conditions are not normally critical in water-front structures although they are in coffer dams. The unbalanced water pressure exerts an additional thrust on the wall and the seepage pressure exerted by the rising ground water reduces the effective unit weight of the soil in contact with the outer face of the wall and hence reduces its passive resistance. Both these effects must be taken into consideration in the design of cantilever or anchored sheet pile walls.

The distribution of the unbalanced water pressure on both sides of the sheet piling can be determined by means of a flow net (Fig. 6.17a), provided the excess hydraulic head ΔH and values of coefficient of permeability of the soils in contact with the wall are known. For simple cases of piles driven into homogeneous soils or into some layered deposits, simplified distributions of water pressure can be assumed in the design as shown in Fig. 6.17(b) and (c) (Terzaghi 1954); a range of flow nets for anchored sheet pile walls in homogeneous sand was published by McNamee (1949). To minimize the effects of the unbalanced water pressure, weep

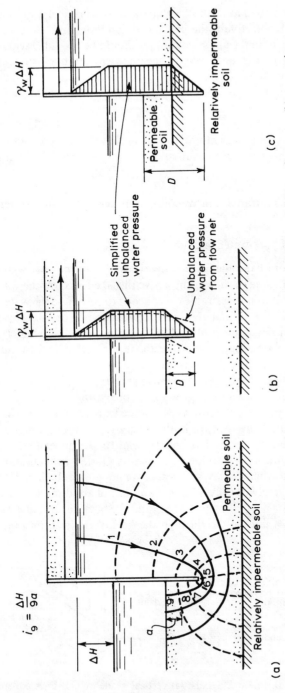

$i_g = \dfrac{\Delta H}{9a}$

Permeable soil

Relatively impermeable soil

(a)

Simplified unbalanced water pressure

Unbalanced water pressure from flow net

$\gamma_w \Delta H$

D

(b)

$\gamma_w \Delta H$

Permeable soil

Relatively impermeable soil

D

(c)

Figure 6.17 Determination of unbalanced water pressure on sheet pile retaining walls: (a) flow net; (b) approximate distribution of unbalanced water pressure in homogeneous soil, and (c) in layered deposits.

holes must be provided at the bottom of the exposed face of a sheet pile but these can only be fully effective if the soil immediately behind the wall is of a very permeable type; in tidal waters, weep holes should be provided with hinged flaps.

The decrease in the unit weight of soil with depth, arising from seepage pressure, is calculated from knowledge of the variation of hydraulic gradient i in the soil adjoining the outer face of the wall. Thus at any point

$$\Delta\gamma' = i\gamma_w. \tag{6.25}$$

Hence the effective unit weight γ' of the soil at that point in contact with the outer face of the wall to be used in evaluation of the passive resistance, is:

$$\gamma' = (\gamma_{sat} - \gamma_w) - \Delta\gamma' = \gamma_{sat} - \gamma_w(1 + i) \tag{6.26}$$

where γ_{sat} is the saturated unit weight of soil and γ_w the unit weight of water.

(d) Reduction of pore-water pressure by drainage

It can be seen from Example 6.1 that the thrust due to the pore water pressure is almost of the some magnitude as the active thrust exerted on the wall by the soil. Therefore, if possible, it is desirable to reduce or to eliminate the water pressure by provision of drainage. Other reasons for provision of drainage are to prevent softening and subsequent loss of strength of cohesive backing material, to reduce the swelling following ingress of water through fissures formed during hot dry spells and to prevent frost heave during cold weather which, in silty materials, may lead to considerable increases in the lateral stresses and to subsequent failure of the wall. Drainage can be achieved by the provision of a permeable backfill or installation of special drainage layers placed at an angle in the backfill or immediately behind the wall; in both cases weep holes or longitudinal drains must be provided along the base of the wall to permit exit of the water (see Fig. 6.18). A free-draining backfilling, such as graded broken stone, gravel or sand, is desirable, whereas clay, silt or peat are not good: chalk needs to be placed so that it drains freely. If only a limited amount of good material is available, it should be used for backfilling to just beyond the back surface of a wedge formed by the wall and a line drawn through the heel of the wall and inclined to the

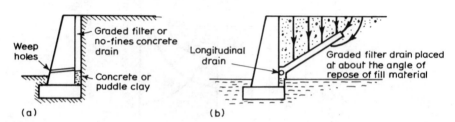

Figure 6.18 Typical drainage details: (a) vertical graded or no fines concrete drain; (b) inclined graded filter drain.

horizontal at the angle of shearing resistance of the poorer material; the poorer material can be used outside this zone. Pervious drainage layers for slow-draining backing should be about 300 mm thick; according to Terzaghi and Peck (1967) there is no need for this material to be filter graded.

Weep holes have the disadvantage that they discharge water over the toe of the wall which may lead to deterioration of the soil where bearing stresses are highest, unless an impervious surface is provided; on the other hand, longitudinal drains are more difficult to maintain. Weep holes, should be 100 mm diameter and of stoneware or asbestos since metal pipes lead to staining on the wall. They should be spaced at about 3 m centres horizontally and not more than 1.5 m vertically. Any water-carrying pipe, such as water main, in or near the backfilling should be surrounded by a gravel drain with conspicuous outlets, so that leakage due to fracture is evident and the stability of the wall not immediately endangered.

(e) Total stress analysis

In the determination of active or passive stresses in cohesive soils immediately after completion of construction and before any change of water content has taken place, the calculations can be carried out in terms of total stresses which automatically take into consideration the effects of ground water. In such cases, of course, undrained strength parameters c_u and ϕ_u must be used; in saturated clays ϕ_u is equal to zero and this leads to simplified calculations ($K_a = K_p = 1.0$).

6.1.5 Effects of surcharge loads

(a) Uniformly distributed loading

The introduction of a distributed surcharge load on the ground surface does not present any problems and can easily be allowed for in both limiting equilibrium and Coulomb wedge method, see Section 6.1.2b and 6.1.2c.

(b) Concentrated surcharge line load

The problem of a concentrated surcharge line load q per unit length cannot be solved by any of the limiting equilibrium methods. If the load is small and its influence on the slip surface can be ignored, then the Coulomb method can be used to calculate the resultant lateral thrust. In practice, however, this procedure is cumbersome and the following semi-empirical approach suggested by Terzaghi (1954), based on Boussinesq's theory, is recommended.

The lateral stress on the wall at depth (Fig. 6.19a) $z = nH$ due to a line load q situated at a distance $x = mH$ from the back face of the wall, where H is its height, can be obtained from the following expression:

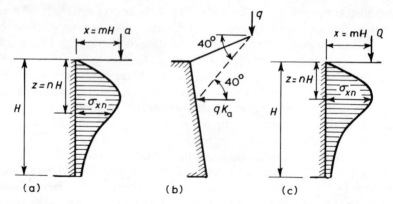

Figure 6.19 Lateral stresses due to surcharge loads: (a) and (b) surcharge line load; (c) surcharge point load.

for $m > 0.4$

$$\sigma_{xn} = \frac{q}{H} \frac{4}{\pi} \frac{m^2 n}{(m^2 + n^2)^2} = \frac{q}{H} K_L,$$
(6.27a)

and for $m \leqslant 0.4$

$$\sigma_{xn} = \frac{q}{H} \frac{4}{\pi} \frac{0.16 n}{(0.16 + n^2)^2} = \frac{q}{H} K_L.$$
(6.27b)

Values of coefficient K_L can be taken from Table 6.8.

The Civil Engineering CP 2 (ISE, 1951) suggests the use of a simple empirical procedure for determination of the lateral thrust due to a surcharge line load. The line load q per unit length is considered to exert a horizontal force $K_a q$ per unit length, where K_a is the appropriate coefficient of active lateral stress. The point of application of this force is taken as the intersection of the back face of the wall with a line drawn from the point of application of the load at an angle 40° to the horizontal (Fig. 6.19b).

(c) Concentrated surcharge point load

Since none of the existing theories can adequately account for the stress distributions on the back of the wall due to a concentrated surcharge load Q the use of semi-empirical expressions developed from experimental results by Terzaghi (1954) is recommended (Fig. 6.19c):

for $m > 0.4$

$$\sigma_{xn} = \frac{Q}{H^2} \frac{1.77 m^2 n^2}{(m^2 + n^2)^3} = \frac{Q}{H^2} K_c$$
(6.28a)

Table 6.8 Values of coefficients K_L and K_c for evaluation of lateral stresses due to surcharge loads

	n	\multicolumn{5}{c}{m}				
		0.4	0.6	0.8	1.0	1.2
K_L	0.1	0.705	0.335	0.193	0.125	0.087
	0.2	1.020	0.573	0.353	0.235	0.167
	0.3	0.977	0.680	0.459	0.322	0.235
	0.4	0.796	0.678	0.510	0.379	0.287
	0.5	0.606	0.616	0.515	0.408	0.321
	0.6	0.452	0.531	0.490	0.413	0.340
	0.7	0.338	0.444	0.447	0.401	0.345
	0.8	0.255	0.367	0.399	0.379	0.339
	0.9	0.195	0.302	0.349	0.350	0.326
	1.0	0.152	0.248	0.315	0.319	0.308
K_c	0.1	0.576	0.126	0.041	0.018	0.008
	0.2	1.415	0.399	0.144	0.063	0.031
	0.3	1.630	0.629	0.262	0.123	0.064
	0.4	1.382	0.723	0.353	0.181	0.100
	0.5	1.027	0.701	0.401	0.227	0.132
	0.6	0.723	0.615	0.407	0.253	0.157
	0.7	0.505	0.508	0.385	0.262	0.174
	0.8	0.354	0.407	0.345	0.257	0.181
	0.9	0.252	0.322	0.301	0.242	0.181
	1.0	0.181	0.253	0.256	0.221	0.175

and for $m \leqslant 0.4$

$$\sigma_{xn} = \frac{Q}{H^2} \frac{1.77 \times 0.16n^2}{(0.16 + n^2)^3} = \frac{Q}{H^2} K_c. \qquad (6.28b)$$

Values of coefficient K_c can be taken from Table 6.8.

EXAMPLE 6.2

Determine the additional lateral thrust on a 6 m high sheet pile wall shown in Fig. 6.20 due to a surcharge line load of intensity $30\,\mathrm{kN\,m^{-1}}$ at a distance of 2.0 m from its back face.

$$m = \frac{x}{H} = \frac{2.0}{6.0} = 0.33 < 0.40$$

Because $m < 0.4$, Equation 6.27b is used in evaluation of lateral stresses σ_{xn}:

$$\sigma_{xn} = \frac{30}{6} K_L = 5K_L.$$

Values of K_L are obtained from Table 6.8 for $m = 0.4$ and the distribution of stress is shown in Fig. 6.20a.

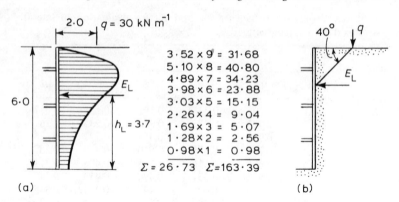

Figure 6.20 Lateral stresses and resultant thrust due to surcharge line load: (a) Terzaghi's semi-empirical method; (b) Civil Engineering Code of Practice empirical method.

The resultant thrust can be taken as approximately equal to

$$E_L = \frac{H}{10} \times (\sigma_{x,0.1} + \sigma_{x,0.2} + \ldots \sigma_{x,0.9})$$

$$= \frac{6.0}{10} \times 26.7 = 16.0\,\text{kN}\,\text{m}^{-1}.$$

(6.29)

Its point of application from the bottom of the wall is given approximately by the following expression:

$$h_L = \frac{H}{10} \times \frac{(9 \times \sigma_{x,0.1} + 8 \times \sigma_{x,0.2} + \ldots 1 \times \sigma_{x,0.9})}{(\sigma_{x,0.1} + \sigma_{x,0.2} + \ldots \sigma_{x,0.9})}$$

$$= \frac{6.0}{10} \times \frac{163.4}{26.7} = 3.7\,\text{m}$$

(6.30)

Using the empirical procedure suggested by the Civil Engineering CP 2 and taking K_a from Table 6.8 for the uppermost layer of fine sand ($\phi = 30°$), the resultant lateral thrust is

$$E_L = 0.33 \times 30 = 10\,\text{kN}\,\text{m}^{-1},$$

and its point of application from the bottom of the wall (Fig. 6.20b):

$$h_L = 6.0 - 2 \times \tan 40° \approx 4.3\,\text{m}.$$

Note that the moment of the resultant about the toe of the wall is $59.2\,\text{kN}\,\text{mm}^{-1}$ in the case of the semi-empirical Terzaghi's approach and $43\,\text{kN}\,\text{mm}^{-1}$ in the case of the empirical method suggested by the Code; the use of the latter method is recommended only in the case of cantilever retaining walls.

6.1.6 Influence of special effects

A discussion of loading on earth-retaining structures would be incomplete without mention of several effects associated with the natural complex properties of soils which may significantly alter the magnitude and distribution of lateral earth stresses. Experience has shown that lack of consideration for the effects of, for example, compaction of backfill or its volume changes and complete reliance on calculated theoretical values in the design of earth-retaining structures may result in severe damage and even failure. Therefore, although it is difficult to quantify these effects, it is necessary to accept their existence and, in some practical manner, allow for them in the design.

(a) Compaction of backfill and its effects

Granular materials are generally used as backfilling because it is easier to control their compaction and because they are not susceptible to frost heave or swelling (see also Section 6.1.4, part d). Although it is desirable to achieve maximum compaction in the backfill, heavy compaction can induce lateral stresses on the wall well in excess of the active stresses usually used in the design (Sowers *et al.*, 1957). Stresses induced by heavy compaction of a cohesive backfill may, in combination with stresses induced by swelling or frost action (Sandergen *et al.*, 1972), lead to failure of retaining structures designed only to resist active stresses. In a full scale test carried out on a 2.5 m high basement wall, Rehnman and Broms (1972) observed that for loosely placed granular soil, without compaction, the distribution of lateral stresses was approximately linear and corresponded to $K = 0.35$ for a gravelly sand and $K = 0.31$ for a silty fine sand; in the latter soil, redistribution of stresses took place with time. For the compacted backfill of gravelly sand the stress distribution was approximately rectangular whereas for the silty fine sand it was considerably higher close to the surface than at the bottom. Furthermore, there was a considerable increase in the lateral stresses as the silty fine sand was allowed to freeze. In another full scale test on a bridge abutment, Broms and Ingelson (1972) recorded approximately linear lateral stresses (equivalent to $K = 0.4$) for a sandy gravel and medium sand backfill compacted in layers by a 3.0 tonne vibratory roller and by a 140 kg vibratory plate compactor. Broms (1971) has proposed a method of analysis for cohesionless soils which takes into consideration the effect of compaction on the lateral earth stresses.

Backfilling should be compacted in 100 to 150 mm layers when hand-punned or about 300 mm layers when compacted by power rammer. The layers should not slope towards the wall to prevent sliding of one layer on another. Care is needed during backfilling to prevent damage to any waterproof membrane on the wall back or, in the case of sheet pile walls or U-abutments, to tie rods. The filling in front of the toe and in front of anchorages must be adequately compacted before backfilling. If possible, backfilling should be carried out at optimum moisture content; the presence of excessive quantities of water is generally undesirable.

(b) Effects of temperature

Broms and Ingelson (1972, 1979) have shown that in the case of self-supporting earth-retaining structures such as, for example, the abutments of a multi-span portal frame bridge, the seasonal and even daily changes in temperature have a considerable effect on the magnitude of the lateral earth stresses through the thermal strains within the structure; these effects are insignificant in the case of free-standing retaining walls.

(c) Effects of volume changes in the backfill

Although the use of cohesive soils as a backfill material is not generally recommended, there are cases when it cannot be avoided. If the total protection against infiltration of water from the surface or capillary rise from a high ground water table cannot be provided, there is always a danger of swelling of the soil which will induce additional lateral stresses on the retaining structure; a granular filter layer between the wall and a cut face will usually prevent the development of excessive lateral swelling stresses. A method of estimation of the lateral swelling stresses has been put forward by Kézdi (1972). Volume changes in the backfill behind retaining walls may be due also to frost heave if penetration of frost is deep enough and the backfill consists of frost-susceptible soils. The lateral stresses induced by frost heave can considerably exceed the active lateral earth stresses and may result in tilting and cracking of retaining structures. In a similar manner, frost heave can take place behind trench timbering left exposed during the winter; on thawing of the soil, timbering may fall down and be followed by collapse of the sides of excavation. To prevent this, sheeting, walings and struts must be properly connected together and kept in contact by wedges driven between them.

(d) Effects of creep

If a retaining structure can yield at the same rate as the soil creeps, then increases of lateral earth stresses on it, due to the creep phenomenon, will be only minimal. On the other hand, if the movement of the structure is stopped, or if it is a rigid structure, then the loading on it will gradually increase and approach the 'at rest' stress condition. A comprehensive treatment of this subject can be found in the works of Šuklje (1969). Creep effects are most pronounced when the backing is cohesive. It should be remembered that in the case of the heavily overcon-solidated clays, the lateral stresses 'at rest' are greater than the vertical overburden stresses – see, for example, Sills *et al.* (1977).

(e) Dynamic effects

A comprehensive review of this problem can be found in the paper by Seed and Whitman (1970). Dynamic effects are generally only considered in areas where the probability of earthquakes is high.

6.1.7 Physical and mechanical properties of soils

The choice of correct values of physical and mechanical properties for retained materials is of prime importance in the design of retaining structures. Some typical values of densities and strength parameters of soils are given in Civil Engineering CP 2 (ISE, 1951), and Terzaghi (1954) presents a table of further typical values of soil properties for the use in the design of anchored steel pile walls. Such typical values can be used in the preliminary designs of all walls and in the final design of small or medium height retaining walls but, for the final design of walls exceeding about 3.0 m in height, it is recommended that all the properties of the natural soils to be retained or of backfill materials should be determined experimentally. The design of small or medium height walls can be based also on the empirical 'equivalent fluid method' suggested by Terzaghi and Peck (1967) in which the backfill materials are classified into five different types to which different 'fluid' properties are ascribed.

The strength parameters c and ϕ must be determined under the conditions corresponding to those under which the soil will be stressed *in situ*. Thus, for short term excavations or for immediate (undrained) thrust calculations in cohesive soils, c_u and ϕ_u are obtained from undrained tests and for evaluation of the long-term conditions (in both cohesive and cohesionless soils), c' and ϕ' are obtained from drained tests or undrained tests with pore pressure measurements. In general, the long-term conditions are critical and control the factor of safety of a slope or the thrust for which a retaining structure must be designed, both are evaluated by effective stress analysis using pore-water pressures determined by the natural ground water conditions. In the case of a temporary excavation which is to remain open for a time that is short compared with that required for the dissipation of excess pore-water pressures induced by the excavation, its stability can be analysed in terms of total stresses using undrained strength parameters.

In the case of stability of retaining structures supporting natural slopes in localities where previous mass movements have taken place or where the soils have been deformed and folded by tectonic activities or ice advances, pre-existing slip planes may be present along which the shear strength parameters for cohesive soils are reduced to the residual values of $c'_r = 0$ and ϕ'_r ranging between 6° and 18°.

Long-term stability of artificial slopes in stiff fissured clays should be investigated in terms of drained strength parameters obtained with due allowance for softening of the soil which occurs as the pore-water suction induced by the mobilized shear stresses (due to removal of excavated material) dissipates. The dissipation process may be very slow (Vaughan and Walbancke, 1973) and may delay the release of the high locked-in horizontal stresses which are usually present in the weathered zones of heavily over-consolidated deposits such as Keuper Marl or London Clay. As these stresses are gradually released, large strains take place which may eventually result in the damage of structures situated above the excavation and sometimes culminate in the failure by lateral translation of the retaining structure.

Particularly dangerous in the sides of excavations are the zones of bedding planes between clay or silty clay and saturated sand or silt; the excavation results in the decrease of stresses in the soil and with a readily available supply of water the cohesive soil can swell and soften fairly rapidly. If the dip of the bedding plane is steep, a slip may develop in a very short time after excavation.

If, for construction purposes, a large space is provided behind a retaining structure which is subsequently backfilled with imported soil or if a wall is specifically designed to retain fill, then the strength parameters of the fill material should be determined under conditions corresponding to its *in situ* state of compaction. A cohesionless soil in its loosest state has an angle of internal friction equal to the critical angle of friction; nominal compaction by any crawler track vehicle will generally increase its state of compaction to medium dense and hence substantially increase its angle of internal friction.

Inclusion of wall friction (defined by angle of friction δ) does not only reduce the active thrust on a wall (which is often ignored) but also reduces its moment about the toe of the base. Observed values of δ according to Packshaw (1946) vary between $\frac{1}{2}\phi$ to $\frac{3}{4}\phi$ (many authorities recommend the use of $\frac{2}{3}\phi$); in no case can δ exceed ϕ. Terzaghi (1954) quotes values of δ for use in the design of the anchored sheet pile retaining wall and relates them to the possibility of relative movement between the soil and wall which is necessary to mobilize wall friction.

The presence of wall adhesion in cohesive soils, in addition to wall friction, has the same beneficial effects. In soft clays the adhesion c_w can be taken as equal to the cohesion c_u; in stiffer clays c_w is less than c_u. According to the Civil Engineering CP 2 (ISE, 1951), c_w should not be taken as greater than $50\,\mathrm{kN\,m}^{-2}$.

In general, the presence of wall friction or adhesion increases the calculated passive resistance of the soil. Full advantage, however, should be taken into consideration only when a wall can settle appreciably in relation to the ground which is providing the support. When such settlement is not certain, values of δ or c_w should be reduced to one-half of the full values or their presence should be completely ignored. Because of the large deformations usually required to mobilize full passive resistance of soils, not more than one-half of the available passive thrust should be considered in stability calculations.

6.2 DESIGN AND CONSTRUCTION OF RETAINING WALLS

The information presented in the preceding sections enables one to determine all the forces (other than external loads) that may act on a retaining structure (Sections 6.1.2 to 6.1.5) and to allow for other effects which cannot easily be quantified but which, in practice, can either be eliminated or minimized (Section 6.1.6). With the knowledge of the forces (and external loads) it is now possible to design the structure and its components so that adequate factors of safety against failure are ensured and possibility of excessive local damage or excessive deflection is eliminated.

In relation to the overall stability of retaining structures, the following modes of failure must be considered: (a) bearing failure; (b) failure by forward movement (sliding); (c) failure by rotation: about the top, in the case of anchored sheet pile walls and about the toe, in the case of retaining walls; (d) failure by slip of a mass of soil, including the wall. In order to eliminate the possibility of excessive local damage and deflection the total and differential settlement of the retaining structures must be kept within the acceptable, practical limits.

All the existing practical methods of stability analysis, regardless of their outer theoretical appearance, are basically semi-empirical design rules and, as such, must be used within the bounds of established practice. It would be unwise, for example, to change the values of the established factors of safety used with these methods without detailed reconsideration of the entire problem in terms of basic principles of statics and in the light of up-to-date practical experience and theoretical knowledge.

In the design of individual components of the retaining structures the calculated forces should be treated as characteristic loads; the individual components should be designed in accordance with up-to-date codes of practice relevant to the materials used.

6.2.1 Stability and serviceability considerations

(a) Bearing failure

Since retaining walls are subject to both vertical and horizontal loads, the soil beneath their foundations has to resist the combined action of these forces; the displacement of soil from underneath the foundation is only counteracted by the passive resistance of the soil within the area BCD'E'F' (Fig. 6.21) on which, in addition to the vertical load Q_z, a horizontal load Q_x is acting. This considerably decreases the bearing capacity of the soil in relation to that evaluated from consideration of the vertical forces only. The bearing capacity depends now on the ratio of Q_x/Q_z (expressed in terms of angle δ' as shown in Fig. 6.21 as well as on the shear strength parameters of the soil c and ϕ;

$$q_{ult} = cN'_c + \gamma_0 DN'_q + \gamma_1(B/2)N'_\gamma \qquad (6.31)$$

where N'_c, N'_q and N'_γ are bearing capacity coefficients dependent on ϕ and δ' (Fig. 6.22), γ_0 is the unit weight of soil above foundation level, and γ_1 the unit weight of soil below foundation level.

Equation 6.31 is directly applicable to strip foundations but its use can be

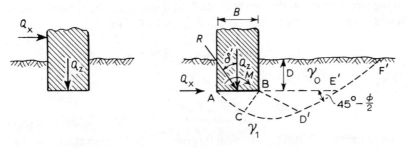

Figure 6.21 Combined vertical and horizontal loading.

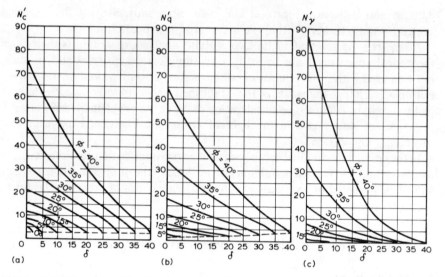

Figure 6.22 Nomogram for determination of bearing capacity coefficients N'_c, N'_q and N'_γ (after Sokolovsky – see Harr, 1966).

extended to cover rectangular foundation of any length to width ratio ($L:B$) by empirical modification of the bearing capacity factors N'_c and N'_γ; this is done by multiplication of these factors by $(1 + 0.3B/L)$ and $(1 - 0.2B/L)$, respectively. The ultimate loading evaluated from Equation 6.31 should be treated as an average loading:

$$q_{ult} = Q_{z\,ult}/B.$$

The safe bearing capacity is then evaluated from the following equation:

$$q_{safe} = \frac{q_{ult}}{F_b} + \gamma_0 D \qquad (6.32)$$

where F_b is the factor of safety against failure which should be not less than 2.0.

If the moment about the centre of the base is such that the ratio of the extreme stresses beneath the foundation ($\sigma_{max}/\sigma_{min}$) is less than 1.20, then.

$$\sigma_{av} = (\sigma_{max} + \sigma_{min})/2 \leqslant q_{safe}. \qquad (6.33)$$

If this ratio is equal to or greater than 1.20, then

$$\sigma_{max} \leqslant 1.10 q_{safe}. \qquad (6.34)$$

To avoid excessive tilting of the wall it is recommended that the ratio of σ_{max} to σ_{min} should never be greater than 3.

(b) Failure by sliding

Both gravity and reinforced concrete walls may fail by forward sliding if the base friction and adhesion and passive resistance of the ground in front of the wall are insufficient; if there is any possibility of removal of the ground from the front of the wall then only the base resistance should be included in the calculations.

For *in situ* concrete foundations on cohesionless soils the angle of friction beneath the base can be taken as equal to the angle of shearing resistance ϕ' of the soil; when the base is precast it should be taken as equal to δ, the angle of wall friction. For foundations on cohesive soils, if the base is cast *in situ* on a porous (no fines) concrete blinding layer, the angle of friction can be taken as equal to ϕ' or full undrained shearing resistance can be considered; for saturated clays $\phi_u = 0$ and, therefore, the undrained shear strength is equal to c_u. For precast bases and for *in situ* bases with no porous blinding layer on cohesive soils, the base resistance should be taken as equal to the undrained wall adhesion c_w.

If the base resistance only is considered, the factor of safety against forward sliding should be not less than 1.5, but if passive resistance of the ground in front of the base is also taken into account then it should be not less than 2.0. If the amount of forward translation of the wall necessary to induce full passive resistance cannot be tolerated then only a proportion of the full value should be included. This resistance against forward sliding can be increased by an inclined underside to the base (Fig. 6.23a) or by inclusion of a rib beneath the base (Fig. 6.23b, c and d). In the case of bases founded on soils, the angle of inclination of the underside of the base θ should not exceed about 10°, but for bases on rock it can be selected so that the resultant of all the forces is normal to it. For walls retaining not more than about 2 m of backing, construction of a rib beneath the base may be more economical.

If the rib is positioned below the toe of the base (Fig. 6.23b) then passive resistance over the increased thickness of the base h_2 can be taken into account. For the position of the rib shown dotted in Fig. 6.23(b), additional passive resistance across the face BC is mobilized and can be taken into account. When a rib is located under the heel of the base, forward movement of the wall occurs along plane AC which joins the underside of the rib with the toe of the base; to avoid active thrust over the back face of the rib (DE) it should be positioned in front of a line through the heel of the base and inclined to the horizontal at the angle of shearing resistance of the soil ϕ (Fig. 6.23d).

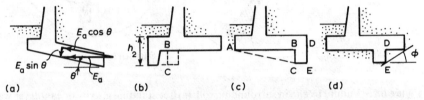

Figure 6.23 Different methods of increasing sliding resistance of the base.

(c) Failure by rotation

Gravity and reinforced concrete retaining walls may fail by rotation about the toe as shown in Fig. 6.24(a) and (b) but, with the exception of walls founded on rock, this type of failure is usually preceded by bearing failure. The Civil Engineering CP 2 (ISE, 1951) recommends a factor of safety against this mode of failure of at least 3 for gravity walls, i.e. the resultant must be within the middle third of the base, and of at least 2 for all other types of wall. In the case of cantilever and counterfort walls, the active thrust E_a is assumed to act at the virtual back of the wall, as defined in Fig. 6.24(b).

Failure by rotation of anchored sheet pile walls (Fig. 6.24c) is discussed in detail in Section 6.2.8. The factor of safety against this type of failure, in the case of the 'free earth support' condition, should be taken as equal to or greater than 2, when the passive resistance in front of the wall is evaluated with the effects of wall friction, and as equal to or greater than 1.5 when the wall friction is neglected.

(d) Failure by slip in surrounding soil

In cohesive soils, a slope stability failure may occur (Fig. 6.25). The factor of safety against this type of failure should be not less than 1.25 if the strength properties of the soil have been obtained by analysis of a previous failure in the same strata or if

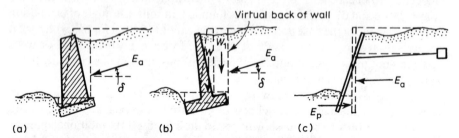

Figure 6.24 Failure of retaining wall by rotation: (a) gravity wall; (b) cantilever or counterforted wall; (c) anchored sheet pile wall.

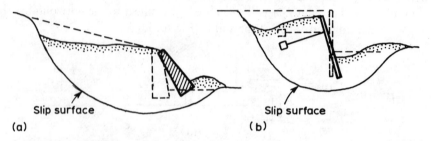

Figure 6.25 Failure by slip in surrounding soil: (a) free standing gravity or cantilever wall; (b) anchored wall.

the residual strength parameters are used, and it should be not less than 1.5 if the strength properties, other than residual, have been obtained from laboratory tests. The stability of banks is treated in Section 3.2.

(e) Total and differential settlement

Settlement of retaining structures is not usually taken into consideration in design but in practice it often accounts for the local damage or even overall failure of these structures – forward tilting is the most common manifestation of the differential settlement of retaining walls. As with other types of structures some retaining walls are more sensitive to total and differential movements than others and this should be taken into account in the selection of the type of retaining structure for the particular soil conditions on a given site. The following simple rules should be remembered: (a) the greater the total settlement of the structure, the greater may be the differences between settlements of individual points within it, and (b) the greater the variation in the contact stress beneath a wall foundation, the greater will be its tilt.

Overall final settlements of retaining walls (with ratio of the length to width of the base greater than 10) can be checked using the following equation

$$S = \frac{2.12 B(1 - v^2)}{E_v} \sigma_{av} \tag{6.35}$$

where B is the width foundation, v is Poisson's ratio of the soil ($\simeq 0.3$ for final settlement), E_v is the deformation modulus of the soil, σ_{av} is the average contact stress. The above equation is based on the assumptions that the soil beneath the wall can be treated as uniform to a considerable depth and that deformation modulus of the soil is known; for practical purposes, values of E_v, as given by Wilun and Starzewski (1972), can be used.

Differential settlement or angle of tilt β of the base can be estimated from the equation derived by Jegorov (Vasiljev 1955) which is subject to the same limitations:

$$\tan \beta = \frac{5.1 (1 - v^2)}{E_v B^2} M_o \tag{6.36}$$

where M_o is the moment of the resultant thrust about the centre of the base per metre run of the wall, and B, v and E_v are as defined above.

To avoid tilting of the wall it is desirable to have a uniform distribution of contact stress beneath the base. This is not as difficult to achieve as is commonly believed (see Section 6.2.4) but, in any case, if for practical reasons uniform distribution of the contact stress cannot be achieved, the ratio of σ_{max} to σ_{min} should not be allowed to exceed 3 or preferably 1.5. For free-standing walls of up to 3.0 m in height the calculated tilt should not exceed 1 in 100 and it should be smaller for higher walls. The permissible tilt may also be limited by other considerations, as for example in bridge abutments (see Chapter 10). In order to eliminate the objectionable appearance of forward tilting retaining structures, the front face of walls is usually given a batter of approximately 1:25; brick and masonry walls are provided with a coping but reinforced concrete walls should be provided with a corbel overhanging 70 to 150 mm.

To minimize the effects of differential settlements (other than forward tilting) in long retaining structures of variable heights, or founded on variable soils, the walls are divided into suitable lengths and movement joints are introduced between them. Longitudinal reinforcement is also introduced in the base and near the top of the wall to enable it to act as a deep beam in the direction of its length.

(f) Other factors affecting serviceability of retaining structures

In Great Britain thermal movement (max/min range about 40°C) is approximately equal to the magnitude of shrinkage of concrete: that is about 10 mm per 10 m wall length. To avoid damage due to these effects, it is necessary to introduce expansion joints at intervals of not more than 30 m and preferably about 20 m; additional construction joints may be desirable, depending on the type of construction adopted.

To prevent percolation of groundwater through construction or expansion joints, a proprietory plastic strip (water bar) is usually incorporated in them; for effective waterproofing of these joints it is necessary to make them simple and easy to construct.

6.2.2 Mass (gravity) retaining walls

The stabilizing force of a mass retaining wall is provided by its dead weight. Simplified profiles of mass walls are shown in Fig. 6.26. Their economic height is usually not more than 2.0 to 3.0 m, although in certain circumstances these limits are exceeded, particularly where deleterious salts are present in the soil or ground water, or in coastal works, and mass concrete is used in preference to reinforced concrete; a mass concrete riverside wall at Lambeth, London was built to an overall height of nearly 13.5 m (Anon, 1949).

Walls of rectangular profile (Fig. 6.24a) are used only up to a height of about

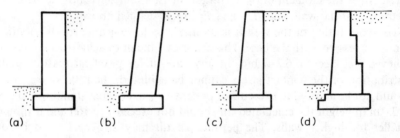

(a)   (b)   (c)   (d)

Figure 6.26 Simplified profiles of mass concrete walls.

2 m. For greater heights, walls of parallelogram, trapezoidal or stepped profiles are used (Fig. 6.26b and d); the first needs care to ensure stability during construction, i.e. before backfilling. For economic reasons (cost of formwork) faces of mass concrete walls are usually made flat although considerable savings of material could be achieved with curved profiles which follow the line of thrust. The range of heights of mass concrete walls can be extended by the use of relieving platforms (see Section 6.2.5).

6.2.3 Reinforced concrete cantilever walls

Cantilever retaining walls consist of two basic elements: a vertical (or almost vertical) stem and a horizontal base slab which are monolithically joined together. The stabilizing force of a cantilever wall is provided by the weight of the soil above the heel of the base together with the weight of the concrete.

The stem is designed as a cantilever carrying the lateral thrusts and its main (vertical) reinforcement is located in the back face adjacent to the retained soil. Stem thickness at the top is seldom less than 200 mm and usually increases towards the bottom where bending moment attains a maximum value. When the stem thickness exceeds 150 mm, longitudinal temperature and shrinkage steel is provided in the front face, normally 0.1 to 0.2% of the cross-sectional area, and is supported on vertical 16 mm diameter bars at 500 mm centres. Longitudinal distribution steel in the back is usually about 0.2%.

The base comprises toe and heel slabs which are designed as cantilevers carrying vertical upwards and downwards loads respectively; the weight of soil over the toe is commonly ignored.

The main problem in the design of cantilever retaining walls is selection of the base proportions: the width and the ratio of the length of the heel to that of the toe. Preliminary design for a wall retaining cohesionless soil can be based on the following simplified assumptions: (i) because of the relatively small cross-sectional area of these walls it can be assumed that the unit weight of concrete is the same as that of soil and hence the vertical force G (Fig. 6.27a) can be taken as equal to $\gamma H l$ and as acting half way along the length of the heel, and (ii) the active thrust is assumed to consist of two horizontal components acting at the virtual back of the wall, $E_1 = 0.5\,\gamma H^2 K_a$ and $E_2 = qHK_a$, with the points of application being at a distance of $H/3$ and $H/2$ above the underside of the base, respectively.

Consider first the possibility of failure of the wall by forward sliding. The total active thrust is

$$E_t = E_1 + E_2. \tag{6.37}$$

This is opposed by the base resistance and passive resistance of the ground in front of the wall which initially will be neglected. If the coefficient of friction between the soil and the underside of the base is taken as $\tan \delta = \tan \phi$ (for *in situ* base), then the factor of safety against forward sliding, F_1, is

$$F_1 = \frac{G\tan\phi}{E_t} = \frac{\gamma l \tan\phi}{K_a(0.5\gamma H + q)}. \tag{6.38}$$

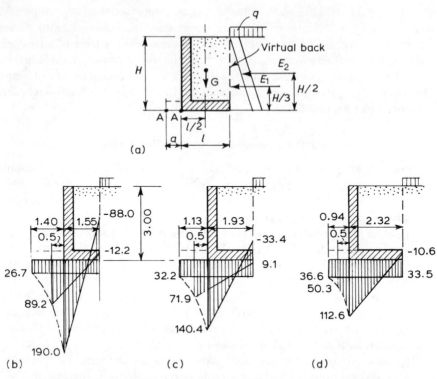

Figure 6.27 Preliminary design of a cantilever retaining wall: (a) general details and notation; (b) contact stresses and base details for $F_1 = 1.0$; (c) for $F_1 = 1.25$; (d) for $F_1 = 1.50$.

For given height of the wall H and surcharge q, the length of the heel l is the only variable that can be modified to change the value of F_1.

Consider now overturning about the toe of the wall (point A, Fig. 6.27a). If the wall is not provided with a projecting toe, the factor of safety against overturning, F_2, is

$$F_2 = \frac{M_{SA}}{M_{OA}} = \frac{G \times l/2}{E_1 \times H/3 + E_2 \times H/2} = \frac{\gamma l^2}{2E_1/3 + E_2}. \qquad (6.39)$$

If the length of the heel l was determined from Equation 6.38, F_2 is fixed. If it is smaller than 2, a toe must be added so that the stabilizing moment M_{SA} is increased. Then

$$F_2 = \frac{\gamma l^2 (1 + 2a/l)}{2E_1/3 + E_2}. \qquad (6.40)$$

The remaining and most important condition from the point of view of serviceability of the wall is the distribution of contact stresses beneath the base: the ratio of σ_{max} to σ_{min} should be less than 3.0 and, if possible, less than 1.5. This can be achieved by appropriate selection of the heel length l and toe length a. To determine the stress distribution, the moment of all the forces about the centre of the base is considered:

$$M_0 = M_{0A} - G \times a/2. \tag{6.41}$$

The extreme stresses at the edges of the base are

$$\sigma = \frac{G}{A_b} \pm \frac{M_0}{W_b}, \tag{6.42}$$

where $A_b = 1 \times (l + a) =$ area of the base, $W_b = l \times (l + a)^2/6 =$ section modulus of the base.

The following example illustrates a trial and error procedure for selection of dimensions and proportions of a 3.0 m high wall.

EXAMPLE 6.3

Data $\qquad H = 3.0\,\text{m}, \ \gamma = 17\,\text{kN}\,\text{m}^{-3}, \ q = 20\,\text{kN}\,\text{m}^{-2}$
$\qquad\qquad \phi = 30°, \ K_a = 0.33, \ \tan\phi = 0.577$

Step 1. Find l for $F_1 = 1.0; 1.25; 1.5$. In the first two cases, base friction is insufficient, and the passive resistance of soil in front of the wall will eventually need to be taken into account, and a rib beneath the base may have to be provided to give a combined factor of safety against sliding of 2.

$$l = \frac{F_1 K_a (0.5\gamma H + q)}{\gamma \tan\phi} = 1.55 F_1. \tag{6.43}$$

For clarity, the results are tabulated (Table 6.9) and plotted in Fig. 6.27(b), (c) and (d). The active thrust components are

$$\begin{aligned} E_1 &= 0.5\gamma H^2 K_a = 25.5\,\text{kN}\,\text{m}^{-1} \\ E_2 &= qHK_a \qquad = 20.0\,\text{kN}\,\text{m}^{-1}. \end{aligned}$$

The factor of safety against overturning for $a = 0$

$$F_2 = \frac{\gamma l^2}{2E_1/3 + E_2} = 0.46 l^2. \tag{6.47}$$

Moment about the centre of the base for $a = 0$

$$M_o = E_1 \times H/3 + E_2 \times H/2 = 55.5\,\text{kNm}\,\text{m}^{-1}.$$

Hence the extreme stresses for $a = 0$

$$\sigma = \frac{G}{1 \times l} \pm \frac{M_o \times 6}{1 \times l^2} = 51 \pm \frac{333}{l^2}. \tag{6.48}$$

As can be seen from the tabulated results it is necessary to introduce a toe to improve F_2 in the first two cases and to improve the distribution of contact stresses (σ_{min} is negative in all cases).

Table 6.9

	Factor of safety F_1		
	1.0	1.25	1.50
$a = 0$ m			
l (m)	1.55	1.93	2.32
l^2 (m^2)	2.40	3.73	5.40
F_2	1.10	1.72	2.48
σ_{max} (kN m^{-2})	190.0	140.4	112.6
σ_{min}	-88.0	-38.4	-10.6
$a = 0.5$ m			
a/l	0.323	0.259	0.216
F_2	1.81	2.61	3.55
σ_{max} (kN m^{-2})	89.2	71.9	50.3
σ_{min}	-12.2	9.1	33.5
$\sigma_{max} = \sigma_{min}$			
a (m)	1.40	1.13	0.94
a/l	0.906	0.585	0.405
F_2	3.09	3.73	4.49
$\sigma_{max} = \sigma_{min}$	26.7	32.2	36.3

Step 2. Introduce a toe of length, say, $a = 0.5$ m and proceed as before.

$$F_2 = 0.46 l^2 (1 + 2a/l) \tag{6.49}$$

$$M_0 = 55.5 - 25.5 la \tag{6.50}$$

$$\sigma = \frac{51}{(1 + a/l)} \pm \frac{6 M_0}{l^2 (1 + a/l)^2}. \tag{6.51}$$

For the actual values see Table 6.9. As one would expect, the values of F_2 increase and the stress distribution becomes more uniform.

Step 3. Finally, the case of uniform distribution of contact stress is considered: this is satisfied when $M_0 = 0$. The necessary length of the toe is then given by

$$a = 2.8/l. \tag{6.52}$$

The plotted results (Fig. 6.27b, c and d) show clearly the effects of length of the toe on the magnitude of $\sigma_{max}/\sigma_{min}$ ratio and F_2; it can be seen that as the tension beneath the heel is eliminated, F_2 becomes greater than 2. For uniform distribution of stress, the overall length of the base is approximately equal to the height. Rejman (1955) has devised a method of determination of the base proportions for both cantilever and counterfort walls which leads to an overall economic design; and earlier work by Jones (1943) also deals with this subject and gives charts to facilitate calculations.

Reinforced concrete cantilever walls are economic up to a height of about 8.0 m, beyond which the counterfort type of construction takes over. Prestressed cantilever walls have been constructed, for example, at the Shell Centre London (Measor and Williams, 1962) but are expensive in both material and labour.

6.2.4 Reinforced concrete counterfort walls

Counterfort walls (Fig. 6.28) become economic when the retained height of soil exceeds about 7.0 m; the most expensive items in construction of these walls are the formwork and fixing of reinforcement.

The structural principles of the counterfort construction are as follows: the

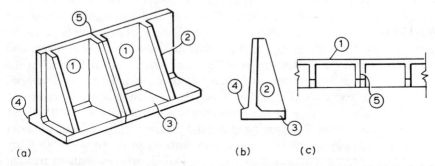

Figure 6.28 Counterfort retaining wall. (a) General view from the back: (1) vertical slab, (2) counterforts, (3) heel and (4) toe of the base slab; (5) expansion joint; (b) transverse section, (c) horizontal section.

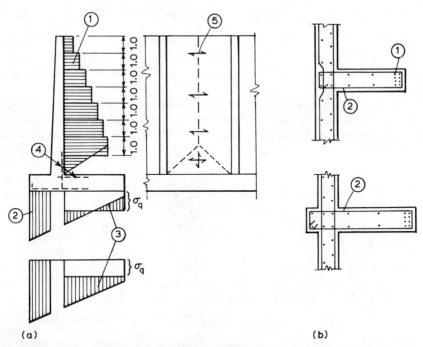

Figure 6.29 Loading and construction details of counterfort walls. (a) Loading details: (1) stem slab loading and hence counterfort loading, (2) toe loading, (3) alternative loadings of the heel of the base slab, (4) reinforcement to cater for fixed moments, (5) directions of main reinforcement in stem slab; (b) details of counterfort–stem slab connection: (1) main counterfort reinforcement, (2) links transmitting reactions from stem slab to counterfort.

earth is retained by a vertical slab that acts as a continuous beam spanning (in plan) between vertical counterforts spaced at 2.5 to 3.5 m centres; the counterforts act as cantilevers and are usually triangular in shape. The heel part of the base slab acts also as a continuous beam spanning between the counterforts whereas the toe forms a simple cantilever. The intensity of loading on the vertical slab increases with depth (Fig. 6.29a) and this is accommodated by increasing either the amount of reinforcement or thickness of the slab towards the base; the minimum thickness of the stem slab is of the order of 250 mm and the front face usually has a batter of 1:25. Because of the monolithic connection between the stem slab and base, fixing moments are generated in both these elements and must be allowed for in the design (Fig. 6.29a). Care is required also in detailing the junctions between counterforts and stem and base since tension is developed between these elements (Fig. 6.29b). Base slab design loading is shown in Fig. 6.29(a). The thickness of the counterforts is normally about 0.05 of the spacing between them. The spacing of the end counterforts is reduced to about 70% of the spacing of intermediate counterforts in order to maintain fairly constant maximum bending moments in the stem slab; twin counterforts are used at expansion joints (Fig. 6.28c). Vertical steel in the face of the stem slab is about 0.1% and a similar amount of the horizontal steel must be continuous and acts as temperature and shrinkage steel.

6.2.5 Reinforced concrete and mass retaining walls with relieving slabs

In order to shift the position of the resultant of the forces acting on a retaining structure away from the toe of the base, and hence reduce its width and also bending moments in the stem, relieving slabs or platforms can be introduced as shown in Fig. 6.30.

Detailed analysis of stability of a mass concrete retaining wall with a relieving slab is given in the following example. Further uses of relieving platforms are illustrated in Section 6.2.6.

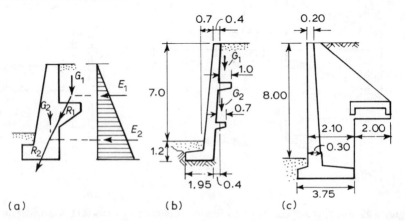

Figure 6.30 Details of relieving slabs: (a) mass concrete wall with reinforced concrete relieving slab; (b) reinforced concrete cantilever wall; (c) counterfort wall.

EXAMPLE 6.4

Check the stability of the mass concrete retaining wall shown in Fig. 6.31. Note that the base of the wall has a reinforced concrete relieving slab to improve the distribution of contact stresses and stability of the wall against rotation about the toe.

For stability analysis, the active thrust is evaluated at the virtual back of the wall using equation 6.6 and $\delta = \phi = 30°$.

$$E_a = 0.5 \times 0.31 \times 16.0 \times 5.7^2 + 0.31 \times 2.0 \times 5.7 = 85.7 + 35.3 = 121.0 \, \mathrm{kN \, m^{-1}}.$$

The horizontal component of the active thrust is

$$E_{ah} = E_a \cos \delta = 121.0 \times 0.866 = 104.5 \, \mathrm{kN \, m^{-1}}.$$

Net moment about the toe

$$M_t = 511.5 - 227.0 = 284.5 \, \mathrm{kN \, mm^{-1}}.$$

Eccentricity of the resultant about the centre of the base is

$$e_c = \frac{B}{2} - \frac{M_t}{\Sigma G} = 1.0 - 0.975 = 0.025 \, \mathrm{m} \leqslant \frac{B}{6} = 0.33 \, \mathrm{m}.$$

Hence the resultant falls within the middle third. The resultant forces and contact stresses

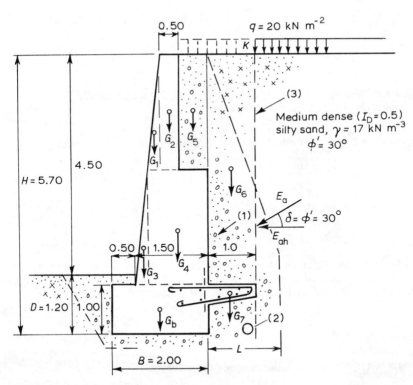

Figure 6.31 Cross-section of a mass concrete retaining wall. (1) Gravel drainage layer, (2) longitudinal drain, (3) virtual back of the wall.

Table 6.10 Moments about the toe of the base

Force $(kN\,m^{-1})$	Lever arm (m)	Moment $(kNm\,m^{-1})$	Moment $(kNm\,m^{-1})$
E_{ah} = 104.5	2.17		−227.0
G_1 = 6.5	0.92	6.0	
G_2 = 20.6	1.20	24.7	
G_3 = 6.5	0.67	4.3	
G_4 = 64.5	1.37	88.4	
G_5 = 37.4	1.70	63.6	
G_6 = 104.8	2.50	262.0	
G_7 = 7.5	2.47	18.5	
G_b = 44.0	1.00	44.0	
$\sum G = 291.8$		511.5	−227.0

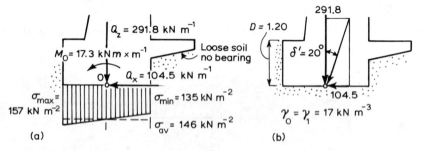

Figure 6.32 Resultant forces at foundation level. (a) Contact stresses; (b) assumed forces for evaluation of the ultimate bearing capacity.

are shown in Fig. 6.32a. Contact stresses are evaluated in the usual manner:

$$\sigma_{max} = \frac{Q_z}{A} + \frac{M_0}{W} = \frac{291.8}{1.0 \times 2.0} + \frac{6 \times 7.3}{1.0 \times 2.0^2} = 146 + 11 = 157\,kN\,m^{-2}$$

$$\sigma_{min} = \frac{Q_z}{A} - \frac{M_0}{W} = 146 - 11 = 135\,kN\,m^{-2}$$

$$\sigma_{max}/\sigma_{min} = 157/135 = 1.16 < 1.25.$$

(i) To determine ultimate bearing capacity of the soil, the vertical and horizontal forces only are taken into consideration, as shown in Fig. 6.32(b). The bearing capacity coefficients for $\phi = 30°$ and $\delta' = 20°$ are found from Fig. 6.22: $N_q' = 8.5$ $N_\gamma' = 3.5$ and, therefore,

$$q_{ult} = \gamma_q N_q' + \gamma_1(B/2)N_\gamma' = 17 \times 1.25 \times 8.5 + 17 \times 2 \times 3.5/2 \times 181 + 59 = 240\,kN\,m^{-2}$$

Using a factor of safety of 2, the bearing capacity is

$$q_{safe} = \frac{q_{ult}}{F_b} + \gamma_0^D = \frac{240}{2} + 1.20 \times 17 = 120 + 20 = 140\,kN\,m^{-2}.$$

This is only 5% smaller than σ_{av} and can be accepted. Note that if only the vertical force is taken into account then $\delta' = 0$ and $N'_q = 18.0$ and $N'_y = 15.0$. Hence

$$q_{ult} = 17 \times 1.25 \times 18.0 + 17 \times 2 \times 15/2 = 382 + 255 = 647 \, \text{kN} \, \text{m}^{-2},$$

and

$$q_{safe} = 647/2 + 20 = 343 \, \text{kN} \, \text{m}^{-2} > \sigma_{max}.$$

(ii) Factor of safety against sliding is determined for two conditions: firstly for base friction only, and secondly for base friction and passive resistance of soil in front of the wall:

(a) base friction only (*in situ* concrete, hence $\delta = \phi = 30°$)

$$F_s = \frac{\sum G \tan \delta}{E_{ah}} = \frac{291.8 \times 0.577}{104.5} = 1.61 > 1.5,$$

(b) base friction and passive resistance of soil (from Table 6.2 for $\delta_w = \frac{2}{3}\phi = 20°$, $K_p = 5.6$)

$$F_s = \frac{\sum G \tan \delta + 0.5 \times \gamma D^2 K_p}{E_{ah}} = \frac{291.8 \times 0.577 + 0.5 \times 17 \times 1.2^2 \times 5.6}{1091}$$

$$= \frac{168.2 + 68.5}{104.5} = \frac{236.7}{104.5} = 2.26 > 2.0.$$

Note that from the design point of view the first condition represents the worst possible case and, therefore, if $F_s \geqslant 1.5$, there is no need to consider the second condition.

(iii) Factor of safety against rotation about the toe of the base is obtained directly from the tabulated results

$$F_R = \frac{\text{Resisting moment}}{\text{Overturning moment}} = \frac{511.5}{227.0} = 2.25 > 2.0.$$

It is interesting to note that for a retaining wall with a relieving slab, an almost uniform distribution of contact stresses is not indicative of a high factor of safety against overturning, as is the case with the straight-forward mass and cantilever retaining structures.

(iv) With the existing soil conditions there is no need to check the factor of safety against possible slip failure in the surrounding soil.

(v) Settlement and tilting of the wall can be checked using Equations 6.35 and 6.36. Unless test results are available, values of E_v and v can be taken after Wilun and Starzewski (1975) provided the type of soil and its consistency or relative compaction are properly defined.

For silty sand of medium density ($I_D = 0.5$), $E_v = 30 \, \text{MN} \, \text{m}^{-2}$ and $v = 0.30$. Average settlement of the wall under a uniform contact stress of $146 \, \text{kN} \, \text{m}^{-2}$ is evaluated from Equation 6.35:

$$S = \frac{2.12\sigma_{0y}(1 - v^2)B}{E_v} = \frac{2.12 \times 146 \times 2.0 \times 0.91}{30} = 19 \, \text{mm}.$$

Tilt of the wall due to moment M_0 is evaluated from Equation 6.36:

$$\tan \beta = \frac{16 M_0(1 - v^2)}{\pi E_v B^2} = \frac{16 \times 7.3 \times 0.91}{\pi \times 30 \times 10^3 \times 2.0^2} = \frac{1}{350} < \frac{1}{100}.$$

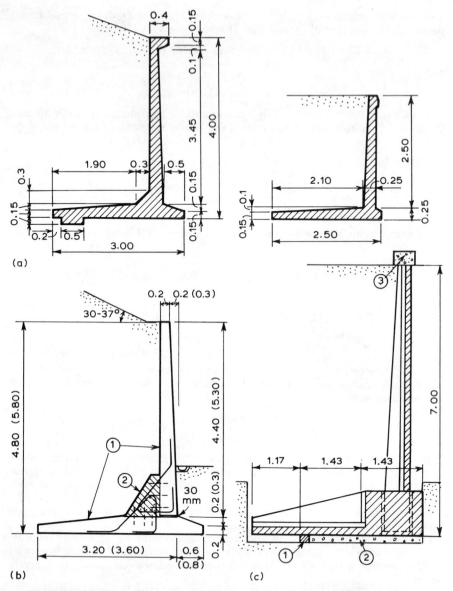

Figure 6.33 Precast cantilever retaining walls: (a) simple cantilever units; (b) precast walls for erection at below 0°C, (1) precast stem and base slabs, (2) *in situ* joint; (c) wall consisting of ribbed stem and base units, (1) precast concrete blocks, (2) bedding layer on concrete, (3) *in situ* capping beams.

Both the above figures are within acceptable practical limits for massive rigid structures (Wilun and Starzewski, 1975).

6.2.6 Precast reinforced concrete walls

The advantages and limitations of the production of any type of precast concrete elements under controlled factory conditions are well known and need not to be considered here. The two major advantages which particularly apply to the use of precast concrete elements in the construction of retaining structures are: (i) the relatively short erection time and (ii) an almost complete independence of the construction on weather conditions. With the present day highly mechanized construction industry, both these advantages may result in considerable reductions of the contract period and hence of the costs.

However, construction of precast retaining walls similar in detail to the traditional *in situ* mass concrete or reinforced concrete structures, encounters considerable practical difficulties: the precast elements are usually large and heavy and satisfactory joints between them are difficult to achieve. Their use, therefore, is generally limited to walls of up to 4.0 m in height, whereas for higher walls new types of structures have been developed. These consist of smaller elements and utilize the weight of the retained soil or rock fill to obtain the required degree of stability.

(a) Cantilever walls

Most precast concrete manufacturers produce a range of simple free-standing reinforced concrete units which act as vertical cantilevers with an L or inverted T-section as shown in Fig. 6.33(a); the height of these units ranges between 1.0 and 4.0 m and the mass between 600 and 2000 kg. Two precast cantilever walls of greater heights are illustrated in Figs. 6.33(b) (Bulgakov, 1961) and (c) (Stepanov, 1972).

(b) Crib walls

Retaining walls fabricated with individual units that form a series of cells into which fill is placed to form an integral part of the retaining structure are known as crib walls (Fig. 6.34a); precast reinforced concrete elements are particularly suitable for formation of this type of structure. The face or stretcher units are anchored back by ties or header units to a second face (Fig. 6.34b) or else special Y-shaped ties are used. The cells so formed are usually filled with cohesionless soil or crushed rock and the whole acts as a retaining structure (Fig. 6.34a). For higher walls, the structure is usually tilted backwards to increase its stability (Fig. 6.34c) or its width is doubled or trebled by addition of extra rows of interlocking cells, or both (Fig. 6.34d). Tschebotarioff (1965) points out that wide crib walls are very sensitive to transverse differential settlements and cells and fill

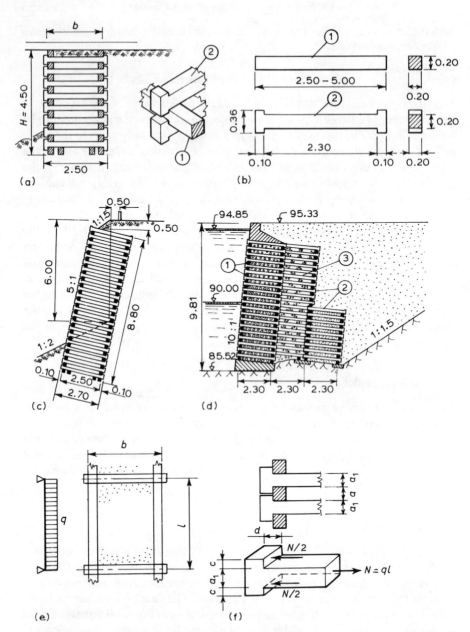

Figure 6.34 Details of crib wall: (a) vertical face, (1) face or stretcher units, (2) ties; (b) details of face and tie units; (c) tilted wall; (d) tilted 'treble' crib wall, (3) filter-graded backfill; (e) internal loading on stretchers; (f) tie loading.

may not be monolithic unless the fill is compacted with special care; in a multi-cell structure, settlement of the central longitudinal cells should not be greater than that of the front and back cells because this may result in reversal of the direction of the wall friction and thereby decrease its resistance to forward sliding.

Design of crib walls involves checking of their overall stability against overturning and forward sliding and includes the design of the individual component stretchers and ties. For stability calculation the weight of the wall G can be taken as

$$G = 1.06\gamma bH \qquad (6.53)$$

where γ is the unit weight of the cell filling, and b and H are the width and height of the wall (Fig. 6.34a). The active earth thrust is usually calculated taking $\delta \leqslant \frac{2}{3}\phi$.

It is assumed in evaluation of loads on the individual components of a crib wall that, when the crib cells are filled prior to placing backfilling, the crib skeleton behaves like a silo and the soil or rock filling the cells exerts lateral stresses on the two faces: the stretchers are therefore subject to horizontal loading which produces bending and the ties provide the horizontal reactions.

For cohesionless soils the intensity of the horizontal loading can be evaluated approximately from the theory developed for the determination of lateral stresses in silos or bunkers:

$$\sigma_x = \frac{\gamma bl}{2(b+l)\tan\phi}. \qquad (6.54)$$

where γ is the unit weight of filling in the crib cells, ϕ is the angle of shearing resistance of the soil, and b and l are the plan dimensions of the crib cells (Fig. 6.34e).

The horizontal uniformly distributed loading acting on an individual stretcher (Fig. 6.34e) is taken as

$$q = (a + a_1)\sigma_x. \qquad (6.55)$$

The tensile force transmitted to individual tie units is equal to

$$N = ql = (a + a_1)\sigma_x l. \qquad (6.56)$$

The cross-sectional dimensions of the tie units are normally the same as those of the stretchers and therefore their design reduces to the determination of the reinforcement necessary to keep the tension cracks within the prescribed limits and selection of dimensions c and d (Fig. 6.34f).

(c) Other types of precast or partly precast walls

One of the simpler forms of partly precast construction involves the use of precast reinforced concrete panels supported by structural steel columns (H-piles) tied

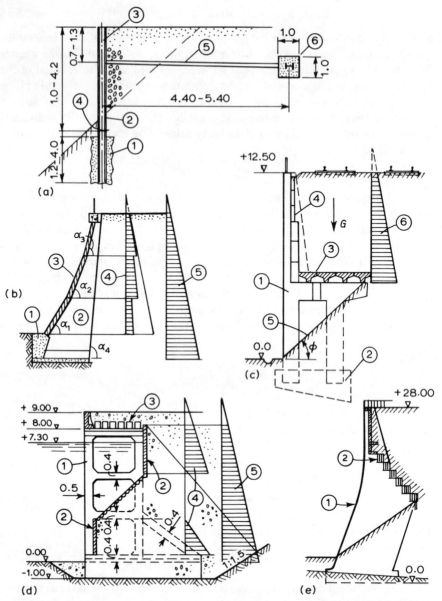

Figure 6.35 Composite precast concrete retaining walls: (a) anchored composite wall, (1) lean concrete base, (2) H-Pile column, (3) precast wall panels, (4) steel angle supporting panels, (5) 38 to 100 mm × 10 mm steel tie bar, (6) lean concrete encasement; (b) wall with *in situ* sloping panels, (1) foundation block, (2) counterfort, (3) retaining panels, (4) lateral stress distribution on retaining panels, (5) loading on counterforts; (c) wall with relieving platform, (1) counterfort frame at 5.5 m centres, (2) reinforced concrete foundation, (3) precast T-beams, (4) precast retaining slabs, (5) spill-through soil, (6) lateral reduced thrust between counterforts; (d) composite construction wall, (1) counterfort, (2) prestressed concrete retaining slabs, (3) presetressed concrete ribbed beams, (4) lateral stress distribution on retaining panels, (5) loading on counterforts; (e) 28 m high wall in Tibilisi, USSR, (1) *in situ* counterfort at 6 m centres, (2) retaining panels in groups of 2, 3 or 4.

back to either steel or concrete anchorages; details of such a wall constructed near Pasadena (USA) for widening an existing road on a slope are shown in Fig. 6.35(a).

A range of more complex retaining structures is shown in Fig. 6.35(b) to (e). These consist of rather substantial, precast or *in situ*, counterforts with precast retaining panels spanning between them; the backfill at the lower levels is, wherever possible, allowed to spill through to reduce the total lateral thrust on the wall (Fig. 6.35c to e). The retaining panels either slope backwards to reduce the lateral stresses on them (Fig. 6.35b), or are arranged so that the weight of some of the backfill is utilized in increasing the stability of the wall (Fig. 6.35c and e); distributions of the lateral stresses used in the design of the precast retaining panels and counterforts are shown in the same figures. A detailed treatment of such complex structures has been presented by Rossyiskii (1961).

(d) Reinforced earth

An ingeneous method of supporting fill by reinforcing it with metal strips or rods, attached to light metal or precast reinforced concrete fascia panels, was introduced by Vidal (1966) and is generally known as 'reinforced earth', the use of this method is still under license held by the inventor. The basic principles of the 'reinforced earth' are the same as those of the laminated rubber bridge bearings; the presence of the horizontal laminar reinforcement transforms rubber into a strongly anisotropic material capable of carrying much greater vertical loads than the rubber alone, and at the same time exhibiting considerably reduced deformations. In the case of the 'reinforced earth', the closely spaced horizontal metal strips transform a granular soil into a coherent material capable of supporting its own weight as well as applied vertical loading. The retaining structure so formed does not require any special foundations, is very flexible and thus can withstand large differential movements. Fascia units are necessary to retain the soil and protect it against erosion.

The results of research on reinforced earth have been published by, for example, Lee *et al.* (1973), Romstad *et al.* (1976), Shen *et al.* (1976), Chang and Forsyth (1977a, b), Schlosser (1978), McKittrick (1979), Jones and Edwards (1980), Murray (1980), Ingold (1981, 1983), Murray and Irwin (1981) and Bolton and Pang (1982). The behaviour of this new, anisotropic material was found to be complex and hence, for practical purposes, simple semi-empirical design rules were developed using the basic methods described in Section 6.1. The proceedings of a symposium on reinforced earth have been published by TRRL (1979).

6.2.7 Strutted excavations

The evaluation of forces necessary for the design of struts, walings and sheet piles or timbering used in supporting the sides of narrow deep trenches is discussed in Section 6.1.3(b). However, provision of adequate bracing to deep excavations

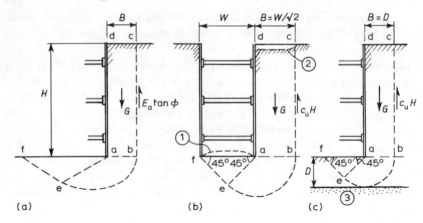

Figure 6.36 Stability of the floor excavations: (a) excavation in sand; (b) and (c) excavations in clay, (1) heave in floor, (2) settlement of ground surface, (3) resistant stratum.

does not eliminate the possibility of failure in the bottom of the excavation, particularly in cohesive soils of relatively low strength. Failure is manifested in heave of the floor of the excavation, accompanied by settlement of the adjoining ground surface and is caused by the weight of retained soil on one or both sides of the excavation which induces a form of bearing failure of the soil in the vicinity of the bottom. This type of failure is discussed below.

(a) Excavations in sand

The following analyses are considered as two-dimensional problems in which a 1.0 m long portion of an excavation is studied. Figure 6.36(a) shows a section across an excavation in clean sand which extends to a depth not less than about 0.5H below the excavation floor, with the ground water table not rising above that level. The weight of overburden in a zone such as abcd of some width B on each side of the excavation tends to cause shear failure in a zone aef in the bottom with consequent heave of the floor between a and f. It is assumed that the timbering is inserted in stages as excavation progresses and that it does not bear on the floor; experience suggests that danger of failure is increased if timbering is insufficient to prevent horizontal displacements. The failure is assumed to be similar to that which develops beneath a foundation of width 2B on the surface of the ground. The mass of soil involved is abef and the resistance developed is approximately one-half the value developed beneath a foundation of width 2B. The load from block of soil abcd on ab is

$$Q = G - E_a \tan \phi = \gamma H (B - 0.5 H K_a \tan \phi), \qquad (6.57)$$

where K_a is the coefficient of active lateral stress for the sand; frictional resistance cannot develop along ad because the timbering is free to move downward with the block of soil. At failure, a state of limiting plastic equilibrium is induced in zone abef. From analogy with a strip foundation of width $2B$ the ultimate bearing capacity of soil beneath ab is given approximately by

$$Q_{ult} = 0.5 \times 0.5\gamma(2B)^2 N_\gamma = \gamma B^2 N_\gamma. \tag{6.58}$$

The factor of safety against failure is

$$F = Q_{ult}/Q = \frac{\gamma B^2 N_\gamma}{\gamma H(B - 0.5HK_a \tan \phi)}. \tag{6.59}$$

Terzaghi (1943) has shown that F is approximately a minimum when

$$B = H \tan^2 (45 - \phi/2) \tan \phi = HK_a \tan \phi. \tag{6.60}$$

For values of $\phi = 30°$ and $40°$ the values of B are $0.19H$ and $0.18H$ and of F approximately 8 and 50, respectively. Thus shear failure in the bottom of an excavation in clean sand, well above the ground water table, is theoretically impossible. When the ground water table coincides with the floor of the excavation, the above F values are approximately halved but are still adequate. For higher positions of the ground water table, particularly when the excavation is maintained dry by pumping from a sump in the floor, seepage is set up which in the extreme may lead to piping in the bottom. This condition should preferably be eliminated by a geotechnical process such as external drainage by well points behind the timbering.

(b) Excavations in clay

Now consider the case of an excavation in saturated clay ($\phi_u = 0$) which extends to a depth not less than $W/\sqrt{2}$ below the floor (Fig. 6.36b). The total vertical load on ab is the weight of the block abcd less cohesive resistance along bc

$$Q = G - c_u H = H(\gamma B - c_u). \tag{6.61}$$

Following the same arguments as in the case of excavation in sand, the ultimate bearing capacity of soil beneath ab is given approximately by

$$Q_{ult} = 0.5(2B)c_u N_c = c_u B N_c. \tag{6.62}$$

The factor of safety against failure is then given by

$$F = \frac{c_u B N_c}{H(\gamma B - c_u)} = \frac{c_u N_c}{H(\gamma - c_u/B)}. \tag{6.63}$$

Thus F decreases as B increases, but a limit is placed on the value of B by the width of the excavation W:

$$B = W/\sqrt{2}. \tag{6.64}$$

Hence the minimum value of F is

$$F = \frac{c_u N_c}{H(\gamma - \sqrt{2}(c_u/W))}. \tag{6.65}$$

The critical depth of excavation H_c for incipient failure can be obtained approximately by assuming that W is large and that $N_c = 5.0$:

$$H_c = \frac{5c_u}{\gamma}. \tag{6.66}$$

If the excavation is carried down close to the top of a resistant stratum where D is less than $W/\sqrt{2}$ (Fig. 6.36c), then $B = D$ and

$$F = \frac{c_u N_c}{H(\gamma - c_u/D)}. \tag{6.67}$$

It will be seen that the value of F increases as D decreases and that a minimum value is attained again when $B = D = W/\sqrt{2}$.

Bjerrum and Eide (1956) reported a number of failures in clay which confirm the general mechanics of the problem – heave of the floor being accompanied by settlement of the ground surface above zone abcd. For shallow excavations with H/W of the order of 1 to 2, the stability analysis outlined above was found to be adequate. However, this has proved unreliable for deep excavations and shafts in which H/W exceeds about 4 since the shear strength of the soil in the upper part of zone abcd is not fully mobilized. For such cases, Bjerrum and Eide suggest that the problem is analogous with a bearing capacity failure of a deep foundation subject to a negative load equal to the relief of load by the excavation. For a deep foundation on saturated clay the ultimate unit bearing capacity can be taken as

$$q_{ult} = c_u N_c + \gamma H, \tag{6.68}$$

where the Skempton bearing capacity factor N_c is 7.5 for a deep strip foundation and 9.0 for a deep circular or square foundation. Neglecting the overburden stress on the right-hand side of the equation, and assuming that at failure the negative stress on the floor of the excavation is

$$q_{ult} = \gamma H_c, \tag{6.70}$$

it follows that

$$H_c = \frac{c_u N_c}{\gamma}. \tag{6.71}$$

The critical depths for excavations in clay can be summarized as follows:

(i) clay extends to depth not less than $W/\sqrt{2}$ below floor;

long, narrow excavation, $H/W > 4$ $H_c = 7.5c_u/\gamma$
circular or square shafts, $H/W > 4$ $H_c = 9.0c_u/\gamma$
long, wide excavation, $H/W = 1$
    for the general case $H_c = 5.0c_u/(\gamma - \sqrt{2(c_u/W)})$
    for small values of $\sqrt{2(c_u/W)}$ $H_c = 5.0c_u/\gamma$

(ii) top of resistant stratum at depth less than $W/\sqrt{2}$ below floor:
long, wide excavation, $H/W = 1$ $H_c = 5.0c_u/(\gamma - c_u/D)$.

Intermediate values of H_c can be selected for other proportions of length, width and depth of the excavation. If the ratio of length to depth of an excavation is relatively low then the resistance to shearing on the ends of the zone dcbef should be taken into account and this will increase the factor of safety against heave in the bottom. In fact this increase in stability is utilized when the processes of excavating and concreting are effected in a series of short lengths. Alternatively, advantage can be taken of the increased stability afforded by the reduction in width of a deep excavation when a dumpling of soil is retained temporarily in the centre of the area to be excavated. It should be remarked that possible deterioration in shear strength of a clay should be taken into account if an excavation is to remain open for more than a relatively short time; the relief of stress in the bottom of an excavation permits steady expansion of clay accompanied by an increase in water content.

An example of the prediction and observation of performance of a braced excavation has been discussed at length by Golder *et al.* (1970) and Lambe *et al.* (1970).

6.2.8 Cantilever and anchored sheet pile walls

In the past, the use of sheet pile walls, also known as bulkheads, was limited almost exclusively to waterfront retaining structures but with the introduction of ground anchors extensive use is now made of anchored sheet pile walls for the temporary support of the sides of deep excavations to obtain an unobstructed working space for the highly mechanized earth moving operations and foundation construction work. In waterfront construction, where the ground is excavated in front of the sheeting after driving, the bulkhead is known as a dredge bulkhead, and where filling is placed behind the sheeting after driving, it is known as a fill bulkhead.

Sheet pile walls are of two broad types: (a) cantilever, relying for support solely on fixity in the soil at the foot (see also Section 6.1.3c), and (b) anchored, with one or more rows of ties in addition to fixity in the soil (see also Section 6.1.3d). Both these types of the sheet pile walls can be constructed in timber, steel, precast reinforced or prestressed concrete and cast *in situ* reinforced concrete. Typical cross-sections of piles in these materials are shown in Fig. 6.37 and properties of some of the most commonly used steel piling are given in Table 6.11.

The problem of selection of the type of piles is complex and involves the following factors: (a) character of the proposed structure (temporary, semi-permanent or permanent), (b) required length of piles, (c) availability of materials, (d) type of loading, (e) ground conditions and factors causing deterioration, (f) maintenance problems, (g) economic considerations (initial cost, life expectancy, cost of maintenance and availability of funds). The reader may find the following general remarks useful in the selection of the type of piles.

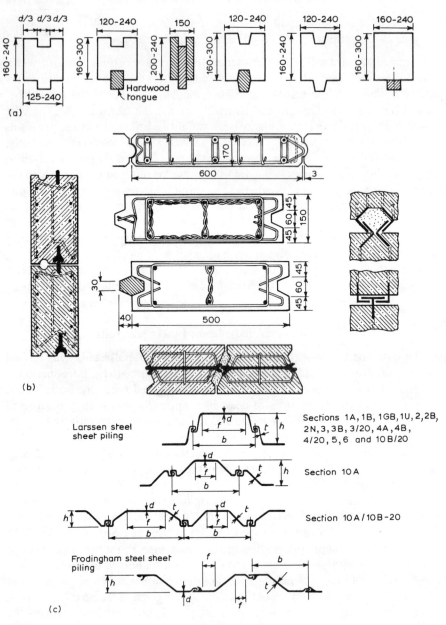

Figure 6.37 Typical cross-sections of sheet piling in different materials: (a) timber; (b) reinforced concrete; (c) steel.

Table 6.11 Steel sheet piling – dimensions, properties and flexibility numbers

Section	O/A Depth of section h (mm)	Second moment of area $I \times 10^{-6}$ ($mm^4\,m^{-1}$)	Section modulus $Z \times 10^{-3}$ ($mm^3\,m^{-1}$)	Mass per m^2 of wall (kg)	Weight per m^2 of wall (kN)	Approx max length (m)	Maximum working BM ($kN\,m\,m^{-1}$) for recommended working stress ($N\,mm^{-2}$)			Flexibility number $\rho \times 10^3$ for $E = 0.20\ MN\,mm^{-2}$ $[\rho = H_p^4/EI\ (m^4\ MN^{-1}\ mm^{-2}\ m^{-1})]$ Pile length H_p (m)							
							125	160	170	6	8	10	12	14	16	18	20
Frodingham Sheet Piling – Normal Profiles*																	
1A	146	41.10	563	89.1	0.87	6	70.4	90.1	95.7	0.158	0.498	1.217	2.523	4.673	7.973	12.77	19.46
1B	133	37.37	562	105.3	1.03	11	70.3	89.9	95.5	0.173	0.548	1.338	2.774	5.140	8.769	14.05	21.41
1BXN	143	49.19	688	130.4	1.28	14	86.0	110	117	0.132	0.416	1.016	2.108	3.905	6.662	10.67	16.26
2N	235	135.13	1150	112.3	1.10	14	144	184	196	0.048	0.152	0.370	0.767	1.421	2.425	3.884	5.920
3N	283	238.85	1688	137.1	1.34	18	211	270	287	0.027	0.086	0.209	0.434	0.804	1.372	2.198	3.349
4N	330	398.31	2414	170.8	1.68	23	302	386	410	0.016	0.051	0.126	0.260	0.482	0.823	1.318	2.008
5	311	492.62	3168	236.9	2.32	24	396	507	539	0.013	0.042	0.101	0.210	0.390	0.665	1.065	1.624
Larssen Sheet Piling																	
1A	130	24.96	384	84.1	0.83	6	48.0	61.4	65.3	0.260	0.821	2.003	4.154	7.696	13.13	21.03	32.05
1B	178	49.98	562	89.1	0.87	9	70.3	89.9	95.5	0.130	0.410	1.000	2.074	3.843	6.556	10.50	16.01
1U	130	31.84	489	106.0	1.04	9	61.1	78.2	83.1	0.204	0.643	1.570	3.256	6.033	10.29	16.48	25.13
2	200	84.94	850	122.0	1.20	14	106	136	145	0.076	0.241	0.589	1.221	2.261	3.858	6.179	9.414
2B	270	136.63	1013	116.8	1.15	17	127	162	172	0.047	0.150	0.366	0.759	1.406	2.398	3.842	5.855
3	247	168.39	1360	155.0	1.52	18	170	218	231	0.039	0.122	0.297	0.616	1.141	1.946	3.117	4.751
3B	298	239.10	1602	155.2	1.52	21	200	256	272	0.027	0.086	0.209	0.434	0.803	1.370	2.195	3.346
4A	381	451.60	2371	185.1	1.82	23	296	379	403	0.014	0.045	0.111	0.230	0.425	0.726	1.162	1.771
5	343	507.77	2962	237.7	2.33	24	370	474	504	0.013	0.040	0.099	0.204	0.378	0.645	1.034	1.576
6	440	922.98	4200	290.0	2.84	26	525	672	714	0.007	0.022	0.054	0.112	0.208	0.355	0.569	0.867
10B/20	171	60.54	706	130.7	1.28	14	88.3	113	120	0.107	0.338	0.826	1.713	3.173	5.413	8.670	13.21
1	2	3	4	5	6	7	8	9	10	11	12	13	14	15	16	17	18

*All Normal Profiles are obtainable in 'Rolled down' or 'Rolled up' versions by 0.8 mm. Information in columns 1 to 5 and 7 is extracted from *Steel Piling Products* by courtesy of the British Steel Corporation.

Timber piles are difficult to drive in compact soils, they are not very strong and efficient interlock is difficult to achieve; when impregnated with preservatives their durability can be very good.

Because of the large volume displacement, both precast prestressed and reinforced concrete piles are difficult to drive in compact soils and may require jetting; they are much heavier than steel and therefore require heavier plant for driving. Prestressed piles can be handled in longer lengths than reinforced concrete ones. Interlock is difficult to achieve but can be improved by inclusion of steel sections (Fig. 6.37b). Tie rods are usually anchored in an *in situ* reinforced concrete capping beam.

The use of *in situ* reinforced concrete sheet piling, originally limited to the dredge bulkhead type of construction, is now very wide; its main advantage over the other forms of construction is the fact that it can be installed in almost any subsoil conditions. It comprises either a row of individually bored piles (contiguous piles) or diaphragm wall panels cast in slurry-filled trenches. In the former type of construction, piles must be truly vertical and in contact with each other. To achieve good contact two tubes are first driven, concrete is placed in one and the tube is withdrawn; the tube for the third pile is driven before concrete is placed in the second. In construction of diaphragm walls, steel tubes are used as stop ends in alternate panels and are withdrawn prior to concreting of the infilling panels to provide clean contact surfaces between them. In both of the above types of construction, reinforcement is prefabricated, and concrete, with the slump of the order of 200 mm, is placed using a tremie pipe. Single or multiple rows of tie bars can be used and are usually anchored in *in situ* reinforced concrete capping beams or walings.

Steel sheet piling is usually more costly in material than reinforced or prestressed concrete but driving should be cheaper. Because of the small volume displacement, steel sheet piling can be driven into any subsoil except rock; it is ideal for temporary structures as it can be reused several times, depending on the care with which it is handled and driven. The effective life of well-maintained steel sheet piling in fresh water is likely to be of the order of 50 to 150 years, depending on the thickness of the section; in seawater these figures are reduced by about 40%. The average rate of corrosion in the British Isles has been quoted as 0.08 mm in sea water and 0.05 mm per year in fresh water and an allowance should be made for this, or for expected abnormal rates, when calculating sections. Protection against corrosion is obviously one of the main considerations in the design of steel sheet piling and much research has been done in this field, e.g. see Fancutt and Hudson (1960) and Ballard (1962). Costs of maintenance were studied by Kavanagh and Johnson (1966); continual developments and improvements of protective materials and processes and variations in relative costs of material and labour make economic studies an essential part of the design of new major projects.

Where steel sheet pile sections are of the type which alternate about the wall centre line, with the interlock in the centre (e.g. some of the Larssen sections), consideration should be given to the fact that full combined section modulus may not be developed owing to the lack of continuity at the interlock where longitudinal shear stress is a maximum. Lee (1950) states that tests show that about 60% of the full modulus is developed when the interlocks are greased but that with normal friction the value probably approaches 100%; when penetration is shallow, it is prudent to reduce the value of the section modulus to about 75% unless welding at the interlock is adopted. To correct for leaning of piles, developed during driving, taper piles can be supplied or fabricated on site by cutting two plain piles along opposite diagonals and connecting one-half from each pile.

Maximum permissible stresses and material specifications for all the above-mentioned types of piles are given in CP 2004 (BSI, 1972d). Details of steel sheet pile sections and other useful information can be found in British Steel Corporation publications (See BSP, 1971).

The most common causes of failure of sheet pile walls appear to be insufficient penetration to develop required earth resistance, anchorages located too near the wall and tie rods of insufficient strength or inadequate protection of piles and/or tie rods against corrosion.

Burland *et al.* (1981) have discussed the stability of free and propped cantilever retaining walls constructed, for example, of bored piles which are similar to sheet pile walls.

(a) Cantilever walls

Cantilever sheeting should be adopted only for relatively unimportant walls in medium to dense clean sand. The design loading should be taken as outlined in Section 6.1.3c with a factor of safety of two applied to the passive thrusts E_{p1} and E_{p2} (Fig. 6.13). The maximum BM occurs at the point of zero shearing force, generally just below ground level in front of the wall and is the governing factor in selection of the appropriate sheet pile profile (Fig. 6.37 and Table 6.11).

Rowe (1951) tested a number of model cantilever walls in loose and dense sand. In the case of very stiff sheeting it was found that, at the instant of failure, the observed stress distribution approached the triangular pattern but that this pattern was not obtained when the depth of penetration D was increased to a safe value. The observed value of K_p just below the dredge level was found to be of the same order as the theoretical value, with full wall friction taken into account, for the case of $H \geqslant 0.45H_p$, where H_p is the overall length of pile ($H_p = H + D$). As the dredge level was lowered the depth at which the passive resistance was fully mobilized increased. Safe depths of penetration in both loose and dense sands ranged from about $0.60H_p$ when $\phi' = 25°$ to about $0.33H_p$ when $\phi' = 45°$; the corresponding values of D at which failure was likely to occur were $0.44H_p$ and $0.16H_p$. With flexible sheeting it was found that the safe depth of penetration was the same as for stiff sheeting but the maximum bending moment was less because of the fact that d, defining the point of rotation O, decreased with increasing flexibility. On the basis of these findings, Rowe (1951) suggested a method of design in which advantage was taken of the reduction in bending moment arising from flexibility of the sheeting, leading to a saving of up to 20% by weight of the wall. Since cantilever walls are seldom used where the retained height H exceeds 3 to 6 m, the saving may not be appreciable in some projects.

(b) Anchored sheet pile walls

The cantilever sheet pile wall has a limited application as the introduction of a single row of ties reduces the maximum bending moment in the sheeting to less than 50% of that of the cantilever. A single row of ties is economic for retained heights up to about 10 m and where the soil is capable of developing sufficient resistance for anchorage. For heights greater than this or in the case of poor soil conditions, a relieving platform is usually added. Figure 6.38 illustrates an unusually complex design of a waterfront retaining structure in which two rows of ties and one relieving platform were used.

All the existing methods of design of the anchored sheet pile walls are based on a number of simplifying assumptions which attempt, with varying degrees of success, to reproduce the real loading and support conditions of the sheet piling in the field. In most of these methods, the simplified loadings shown in Fig. 6.14 are accepted but the assumptions regarding support conditions vary considerably: at one extreme the 'free earth support' condition is advocated (in the method of the same name) while at the other, full fixity with the point of contraflexure at the dredge level is assumed (Tschebotarioff's simplified equivalent beam method; Leonards, 1962). Terzaghi (1954) pointed out that, of all the existing methods of design, the semi-empirical method proposed by Rowe (1952) is the only one which properly takes into consideration the influence of the flexibility of the sheet piling and relative density of the soil on the amount of fixity

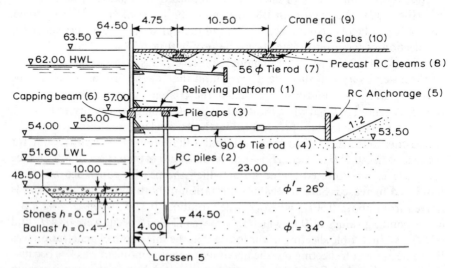

Figure 6.38 Sheet piling retaining wall on River Kame, USSR, (1) relieving platform, (2) RC piles, (3) pile cap, (4) 90ϕ tie rod, (5) RC anchorage, (6) capping beam, (7) 56ϕ tie rod, (8) precast RC beams, (9) crane rail, (10) RC slab. (After Shevchenko, 1972).

that is actually generated within soil and, therefore, on the magnitude of the tie force and bending moments in the wall. This does not invalidate completely other methods of design which, over the last four or five decades, have been successfully used in the design of retaining structures, but it highlights their limitations and hence restricts their application today to specific conditions, whereas the method proposed by Rowe is applicable over the entire range of practical values of the flexibility of piles and relative densities of soil*. Analytical investigations of the problem by Bjerrum *et al.* (1972), using finite element methods, confirm the validity of this approach. Recent observations on anchored bulkheads at Tilbury Docks on the Thames Estuary have been reported by Page *et al.* (1981).

A comprehensive comparison of quay wall design methods is given in CIRIA Technical Note 54 (1974).

In order to isolate the effects of flexibility of the sheet piling on the fixity (and hence on the maximum bending moment M developed in sheet pile walls) Rowe (1952) introduced a parameter known as the flexibility number:

$$\rho = \frac{H_p^4}{EI}, \tag{6.72}$$

where H_p is the total length of the piles and EI their flexural stiffness about an appropriate axis. The parameter ρ is not a dimensionless quantity. In imperial units it is usually quoted in $\text{ft}^4/(\text{lbf in}^2.$ per ft of wall). In SI units it is convenient to express ρ in $\text{m}^4/(\text{MN mm}^2$ per m of wall) as shown in Fig. 6.39 and Table 6.11. In the case of sheeting embedded in clay, a dimensionless parameter is also required and is known as the stability number (P.W. Rowe, 1957) which is expressed in the general form

$$S = \frac{c}{\gamma H} \frac{\sqrt{(1 + c_w/c)}}{1 + q/\gamma H}, \tag{6.73}$$

where q is a uniformly distributed surcharge loading on top of the banking and H is the free-standing height of the wall.

The influence of the flexibility number on the maximum bending moment M developed in the sheet piles is illustrated by the generalized curves (based on Rowe's extensive experimental work) shown in Fig. 6.39 and also presented in a tabular form in Table 6.12. For stiff sheeting, with relatively low flexibility number, the thrust from the backing tends to cause the sheeting to rotate about the point of attachment of the tie and the foot moves outwards as in the free earth support condition; hence the maximum moment is equal to the absolute maximum value $M_{\max}$ calculated for the free support condition. When ρ exceeds a certain critical value ρ_c, which depends on the relative density of the soil, the maximum bending moment in the sheeting becomes less than the free earth

---

*Rowe's semi-empirical method is based on the results of model tests in clean, homogeneous, sands and one has to accept that its use for real, variable, soil conditions is an extrapolation. However, in the present author's opinion there is more justification for this than for the use of some of the other methods based on purely arbitrary assumptions.

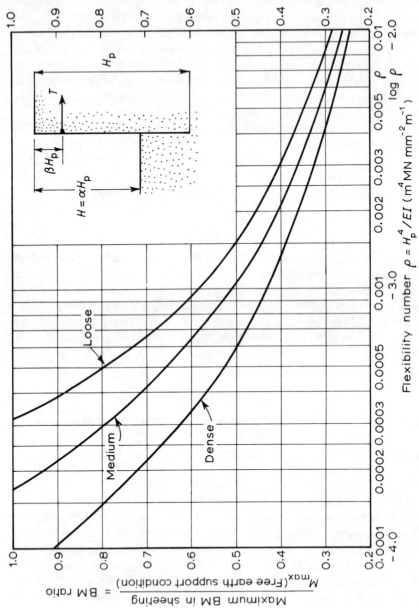

Figure 6.39 Design data for sheet pile walls in sand (after Terzaghi, 1954).

Table 6.12 Relationship between the flexibility number ρ of sheet piles and bending moment ratio (see Fig. 6.39)

$\rho\left(\dfrac{m^4}{MN\,mm^2\,m^{-1}}\right)$		Bending Moment Ratio M/M_{max}		
	$\log \rho$	Dense	Medium	Loose
0.0001	− 4.000	0.90	1.0	1.0
0.00015	− 3.824	0.78	1.0	1.0
0.000175	− 3.757	—	1.0 (ρ_c)	—
0.0002	− 3.699	0.71	0.93	1.0
0.00025	− 3.602	0.66	0.85	1.0
0.0003	− 3.523	0.62	0.79	1.0
0.00032	− 3.495	—	—	1.0 (ρ_c)
0.0004	− 3.398	0.57	0.71	0.87
0.0005	− 3.301	0.53	0.65	0.79
0.0006	− 3.222	0.50	0.61	0.72
0.0007	− 3.155	0.48	0.57	0.67
0.0008	− 3.097	0.46	0.54	0.63
0.0009	− 3.046	0.44	0.52	0.61
0.001	− 3.000	0.43	0.50	0.58
0.0015	− 2.824	0.39	0.44	0.49
0.002	− 2.699	0.36	0.40	0.45
0.003	− 2.523	0.32	0.36	0.39
0.004	− 2.398	0.30	0.33	0.36
0.005	− 2.301	0.28	0.31	0.34
0.006	− 2.222	0.27	0.30	0.33
0.007	− 2.155	0.26	0.29	0.31
0.008	− 2.097	0.25	0.27	0.29
0.009	− 2.046			
0.010	− 2.000	0.24	0.26	0.28

Note $\rho\left(\dfrac{m^4}{MNmm^2\,m^{-1}}\right) = 0.91555 \times \rho\left(\dfrac{ft^4}{lbf\,in.^2\,ft^{-1}}\right)$

support value. In the case of dense soils, as ρ increases further, the fixity approaches the value corresponding to the fixed earth support condition. It is important to note that it was only possible to generalize the results presented in Fig. 6.39 because the effects of other variables such as α, β and q were relatively small. The experimental work was done within the following limits: $\alpha = 0.6$ to 0.8, $\beta = 0$ to 0.2 and $q = 0$ to $0.2\ \gamma H_p$.

Rowe (1952) also found that the force in the tie rod decreased with an increase in ρ and relative density of the soil but the reduction was significantly dependent on variables α, β and q and hence it was not possible to generalize to the same degree as in the case of the maximum bending moment. Furthermore, the tie force was found also to be very sensitive to differential yield of the anchors and therefore it is recommended that, for practical purposes, it should be taken as evaluated from the simple free earth support condition. It is, in fact, prudent to increase this value further, depending on soil conditions, to allow for uncertainties in soil conditions and creep effects as outlined in the following sections.

Sheet piles driven into sand

The recommended procedure for the design of anchored sheet piles in clean sands of known relative density is summarized below and is illustrated in the example that follows. The depth of penetration D, the tie force T and the maximum bending moment M_{max} are determined for the free earth support condition. In the case of waterfront bulkheads, regardless of the soil type, the calculated depth of penetration is usually increased by 20% as an insurance against the effects of unintentional excess dredging, local scour and weak zones in the strata not revealed by the borings. However, the calculated depth of penetration is employed to determine bending moments in the sheeting and the force in the tie rods (see Section 6.1.3d).

A moment reduction diagram is now plotted with flexibility numbers as abscissae and bending moments as ordinates. The appropriate curve in Fig. 6.39 for the given degree of compaction of the sand is employed to plot the moment reduction curve. In practice, it is easier to use the values given in Table 6.11 which are simply multiplied by M_{max} to give the appropriate ordinates of the moment reduction curve. Terzaghi (1954) recommends that in order to compensate for the uncertainties in subsurface conditions between exploratory boreholes, if site investigations reveal that the sand below the dredge line can be classified as dense, the curve for medium sand in Fig. 6.39 should be employed for calculations, and the curve for loose sand should be employed where the sand is classified as other than dense. A pile section is then selected to cater for the M_{max} and the flexibility ρ_{min} is calculated. If ρ_{min} is less than the critical value ρ_c, no moment reduction is permissible. If ρ_{min} exceeds ρ_c, the bending moment in the sheeting will be less than M_{max}. Other lighter pile sections are then selected and the corresponding resistance moments and flexibility numbers are calculated. These values are plotted on the moment reduction diagram and the intersection of this curve (known as the structural curve) with the first gives the allowable reduction in bending moment – the nearest appropriate pile section is selected to cater for the reduced bending moment.

For design purposes, the calculated value of the tie force is increased by 40%[*] (Lee, 1968) to allow for uncertainties in evaluation and for possible differential yield of anchors and direct transfer of surcharge loads to the tie rods as the result of settlement of the retained soil.

All calculations must, of course, include for the effects of the ground water on the active thrust and passive resistance of the soil in accordance with Section 6.1.4c.

---

[*]To prevent excessive yield of unchorages, Casagrande (1973) recommends an increase of at least 100%.

Sheet piles driven into loose silty sand should be designed for the free earth support condition because the compressibility of such sands may be very high; the same practice should be followed for the design of piling driven into very variable (non-homogeneous) deposits or where site investigation data are sparse. In both cases the design tie force should be taken as 1.4 times the force obtained from the above calculation.

EXAMPLE 6.5 (DETAILS IN FIG. 6.40)

Computation of forces and moments about B for the free earth support condition (negative sign indicates forces from left to right or upwards); a is moment arm measured from B (clockwise moments are positive).

Forces (A and W) (kN)		$\tan \delta$ (*in layers*)	$A \tan \delta$	$a\,(m)$	$M_B(kNm)$
(a) Forces above the dredge level (independent of D)					
$A_1 =$	20.75 ⎫			−0.4	−8.30
$A_2 =$	39.12 ⎭ 59.87	0.397	23.77	0.3	11.74
$A_3 =$	47.86 ⎫			2.6	124.44
$A_4 =$	11.63 ⎪			2.9	33.72
$A_5 =$	59.27 ⎬			4.25	251.88
$A_6 =$	3.85 ⎭ 122.61	0.268	32.86	4.50	17.31
$A_7 =$	88.55 ⎫			6.25	553.44
$A_8 =$	6.14 ⎭ 94.69	0.397	37.59	6.67	40.94
$W_1 =$	4.91 ⎫			4.17	20.45
$W_2 =$	29.43 ⎭ 34.34	0	0	6.00	176.58
$\sum A + \sum W =$	311.51		$\sum A \tan \delta = 94.22$		$\sum M_B = 1222.20$
(b) Forces below the dredge level (dependent on D)					
1st trial: $D = 6.5\,m$					
$A_9 =$	261.15 ⎫			10.75	2818.06
$A_{10} =$	41.62 ⎭ 303.77	0.397	120.59	11.836	492.57
$W_3 =$	31.85 31.85	0	0	9.665	307.83
$\sum A + \sum W =$	647.13		$\sum A \tan \delta = 214.81$		$\sum M_B = 4840.66$
$P =$	−365.04 ⎫	0.397	−144.92	11.836	−4320.61
$T_s =$	−25.11 ⎭ −390.14	—	—	11.836	−297.20
$\sum A + \sum W$					
$+ P + T_s =$	256.99	$P \tan \delta = -144.92$			$\sum M_B = 222.85$
2nd trial: $D = 6.8\,m$					
$A_9 =$	274.24 ⎫			10.90	2989.26
$A_{10} =$	45.54 ⎭ 319.78	0.397	126.96	12.033	548.06
$W_3 =$	33.32 33.32	0	0	9.767	325.44
$\sum (A + W)$	664.61		$\sum A \tan \delta = 221.18$		$\sum M_B = 5084.96$
$P =$	−399.51 ⎫	0.397	−158.61	12.033	−4807.35
$T_s =$	23.20* ⎭ −422.71	—	—	12.033	−279.20
$\sum A + \sum W + P$					
$+ T_s =$	241.90	$P \tan \delta = -158.61$			$\sum M_B = -1.59$

*$T_s = [W_p + \sum A \tan \delta + \sum P \tan \delta] (\tan \delta / F)$ where $W_p =$ weight of pile taken as $25\,kN\,m^{-1}$.

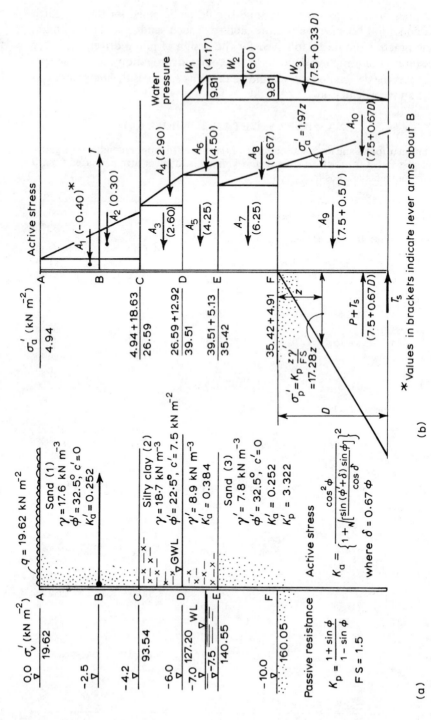

Figure 6.40 Design example details. (a) Problem details: soil (1), $\gamma = 17.6\,\text{kN m}^{-3}$, $\phi' = 32.5°$, $c' = 0$, $K_a^1 = 0.252$; soil (2), $\gamma = 18.7\,\text{kN m}^{-3}$, $\gamma' = 8.9\,\text{kN m}^{-3}$, $\phi' = 22.5°$, $K_a' = 0.384$; soil (3), $\gamma' = 7.8\,\text{kN m}^{-3}$, $\phi' = 32.5°$, $c' = 0$, $K_a' = 0.252$, $K_p' = 3.322$; (b) active and passive stresses and unbalanced water pressure.

The required calculated penetration of piling D is 6.8 m
The overall calculated length of piling H_p is 16.8 m.
The tie force T is obtained by equating horizontal forces

$$T + \sum A + \sum W + P + T_s = 0$$

$$\therefore T = 241.9 \,\text{kN}\,\text{m}^{-1}.$$

Maximum bending moment occurs at the section where the shear force is equal to zero; this is conveniently located by trial and error

Section E		Section F	
$\displaystyle\sum_1^6 A + \sum_1^2 W =$ 192.3 kN		$\displaystyle\sum_1^8 A + \sum_1^3 W =$ 311.5 kN	
$T =$ -241.9 kN		$T =$ -241.9 kN	
$S_E =$ -49.6 kN		$S_F =$ 69.6 kN	

By interpolation zero shear force (and hence M_{max}) occurs between these two sections at -8.54 m.

Force (kN)			Lever arm (m)	Moment (kNm)
$\displaystyle\sum_1^6 A + \sum_1^2 W$	$=$	192.3	3.57	686.5
35.42×1.04	$=$	36.8	0.52	19.1
$0.252 \times 7.8 \times (1.04)^2/2$	$=$	1.1	0.35	0.4
9.81×1.04	$=$	10.2	0.52	5.3
T	$=$	-241.9	6.04	-1461.1
S	$=$	-1.5	$M_{max} =$	-749.7

Maximum bending moment in piling $= 750 \,\text{kN}\,\text{m}^{-1}$.
With $\phi' = 32.5°$ it is assumed that the sand is loose and therefore ρ_c and the moment reduction curve is obtained using the last column in Table 6.12

$$\rho_c = 0.000\,32 \ (\text{in SI units})$$

The required section modulus Z for $M_{max} = 750 \,\text{kN}\,\text{mm}^{-1}$ and working stress of $160 \,\text{N}\,\text{mm}^{-2}$ is

$$Z = \frac{750 \times 10^6}{160} = 4.69 \times 10^6 \,\text{mm}^3\,\text{m}^{-1}.$$

Larssen 6 ($312.2 \,\text{kg}\,\text{m}^{-2}$) has a section modulus $Z = 4.62 \times 10^6 \,\text{mm}^3\,\text{m}^{-1}$ and $I = 1017 \times 10^6 \,\text{mm}^4\,\text{m}^{-1}$; it can be used for evaluation of ρ_{min}

$$\rho_{min} = \frac{H_p^4}{EI}\left(\frac{\text{m}^4}{\text{MN}\,\text{mm}^2\,\text{m}^{-1}}\right) = \frac{16.8^4}{0.2 \times 1017 \times 10^6} = 0.000\,39.$$

Since $\rho_{min} > \rho_c$, bending moment in the sheeting will be less than M_{max} and therefore reduction is permissible.

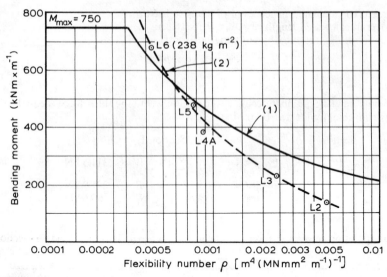

Figure 6.41 Moment reduction curve for the example: (1) BM reduction curve for loose sand; (2) structural curve for Larssen sheet piles.

The ordinates of the moment reduction curve in Fig. 6.41 are obtained by multiplying figures in the last column of Table 6.12 by M_{max}; the abscissae are obtained directly from the first column and are plotted on a logarithmic scale or in the form of logs (from the second column) on a linear scale. Other lighter Larssen sections are now selected from Table 6.11 and corresponding maximum working bending moments (for working stress of $160\,\text{N mm}^{-2}$) and flexibility numbers (for $H_p = 16.8\,\text{m}$, either obtained by linear interpolation from the table or by calculation) are plotted as individual points on the moment reduction diagrams. A curve is drawn to fit these points as closely as possible to give the structural curve shown by the dotted line in Fig. 6.41. Larssen 6 ($290\,\text{kg m}^{-2}$) section is the nearest above the reduction curve but Larssen 5 is only just below it and, therefore, can be taken as suitable. The corresponding design moment is equal to $473.9\,\text{kNm m}^{-1}$: this represents approximately $63\%\,M_{max}$, a considerable reduction in the bending moment due to fixity of the piling below the dredge level.

Summary of results (Factor of Safety $= 1.5$)
    Calculated length of piling 16.8 m
    Actual length of piling $10 + 1.2 \times 6.8 = 18.2\,\text{m}$
    Pile section: Larssen no. 5 ($237.7\,\text{kg m}^{-2}$)
    Design tie force $1.40 \times 241.9 = 339\,\text{kN m}^{-1}$

Sheet piles driven into cohesive soils

In the case of piling driven into cohesive soils it is necessary to differentiate between the 'immediate' undrained and 'final' fully drained condition; creep of the soil and progressive softening must also be considered. For the drained condition in soft clays the cohesion c' can be ignored and the analysis then becomes identical to that of sheet piling driven into sands. For the undrained condition the depth of penetration D, the tie force T and the maximum bending moment M_{max} are usually determined from the free earth support condition (see Fig. 6.14b to d); a suitable factor of safety is applied to the passive resistance (usually between 1.5 and 2.0). Although experimental evidence (P.W. Rowe, 1957) indicates that the influence of relative flexibility of piling and soil on the reduction of bending moments in the piling is similar to that for piling driven into sands a convenient generalization of these results is not possible. The common practice, therefore, is to design sheet piling driven into cohesive soils using the free earth support condition, without any reduction of bending moments, and with an increase of the tie force of up to 30% (Lee, 1968). Such a conservative approach is justified in the case of the soft cohesive soils or if there is any doubt about the reliability of the values for soil properties but for piling driven into properly investigated firm to hard clays or stiff silts, the fixity, even in the fully drained case, will remain close to that defined by the fixed earth support condition and therefore its effects on the reduction of the bending moments should be taken into account. This is done by using either the fixed earth support method or Rowe's semi-empirical method (P.W. Rowe, 1957; Lee, 1968; CIRIA Technical Note 54, 1974).

Anchorages, ties and walings

Anchorages may consist of independent ground anchors, plates or blocks, continuous beams, sheet pile or concrete anchor walls or raking piles; sheet pile anchorages may be of the balanced or cantilever type (Fig. 6.42b). A raking pile anchorage is of a doubtful efficiency unless it is stabilized by a heavy vertical load (Fig. 6.42c). A relieving platform (Fig. 6.42d) stabilizes the anchorage and also relieves the sheeting of a proportion of the lateral thrust. Particular care is required to secure a high degree of compaction of soil in front of anchor plates, blocks and beams. It may be necessary to design these units to resist bending moments and shearing forces and, in the case of steel pile anchor walls, walings are necessary for the transfer of the horizontal load to the tie rods. The force polygon shown in Fig. 6.42c for the analysis of a raking pile anchorage is based on the assumption that each pile is pinned at the head and foot.

The resistance of a continuous anchor beam or wall is based on the passive resistance less active thrust, without including skin friction, unless the weight of the anchorage is sufficient to balance the upward component of passive resistance. If $D_a/d_a < 2$ (Fig. 6.42a), it is permissible to assume that the resistance is developed by the full depth D_a (CP 2, ISE, 1951). The resistance of isolated anchorages is augmented by the shearing resistance at the sides of the wedge of soil in front of the anchorage and, in cohesionless soils, the total shearing resistance in both sides of the wedge is estimated to be $\frac{1}{3}D_a^3 K_a \tan (45 + \phi/2) \times \tan \phi$ (CP2, ISE, 1951). In cohesive soils the shearing resistance on the two sides of the wedge is $c_u \times D_a^2$ (CP2, ISE, 1951) but this should not be included where shrinkage cracks may develop in planes parallel to the tie rods, as may occur with extensive backfilling. The total shearing plus passive resistance of a

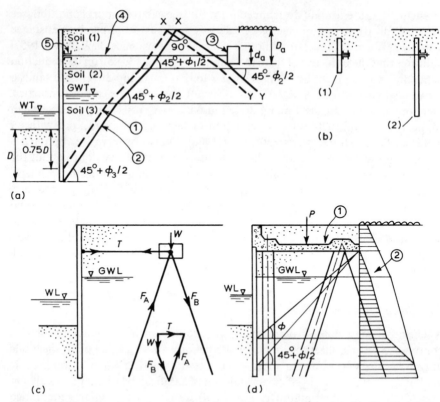

Figure 6.42 Details of anchorages of sheet pile retaining walls. (a) Positioning of anchorages. (1) for full fixity, (2) for free ES condition, (3) anchorage, (4) tie rod (with turnbuckle if length > 12 m), (5) walings; (b) types of steel sheet pile anchorages, (1) balanced anchorage, (2) cantilever anchorage; (c) raking pile anchorage, (1) W, weight of pile; (d) raking pile anchorage with (1) relieving platform, (2) reduction of active thrust.

number of isolated anchorages must not exceed the resistance of a beam of equal section. For deep continuous anchorages (D_a/d_a between 1.5 and 5.5) in sands the passive resistance can be taken as approximately equal to $1.5\gamma D_a^2$ (Buchholtz, 1930-31). The behaviour of anchorages in both cohesive and cohesionless soils have been studied by a number of investigators (Hansen, 1953; Mackenzie, 1955; Biarez *et al.*, 1965; Douglas and Davis, 1964; Hueckel, 1957; Hueckel and Kwasniewski, 1961; Hueckel *et al.*, 1965; Verdeyen and Nuyens, 1965). All anchorages should lie beyond the plane XY (Fig. 6.42a). This satisfies the condition that the underside of the passive wedge in front of the anchorage should not intersect the underside of the active wedge behind the sheeting. In all the above cases a factor of safety of between 1.5 and 2.0 should be applied to the passive resistance of the soil to limit the yield of the anchors.

Tie rods are normally spaced at 3.0 to 4.5 m intervals and are usually designed with a generous margin of safety to allow for unforeseen increases in the tie force. If the backing, or any stratum beneath backfill, is very compressible, the tie rods should be housed in relatively large section conduits in order to avoid downward drag on the rods which would increase the tension.

Walings in steel sheet pile construction commonly consist of two rolled channel sections placed back to back with a space between for the tie rods; they are fixed usually at the front of the piles in the retaining wall. The loads in continuous walings should include any arbitrary increase made in tie loads and the bending moments taken as $\pm 0.1\,pl^2$. The end spans should be reduced to about 70% of the intermediate spans l to maintain fairly constant values of maximum moments. Joints in walings should be designed to carry full shearing forces and bending moments, unless located at the points of contraflexure (about $0.2l$ from ties) where resistance to shear only is required.

6.3 LATERAL STRESSES ON THE LINING OF SHAFTS

In the case of circular shafts, both theory and experience indicate that radial stresses on linings are less than the values derived from an assumed state of active plastic equilibrium. This is due to the fact that soil surrounding the shaft functions in the manner of a ring in which radial stresses are resisted by circumferential compressive stresses. Furthermore, the presence of soil beneath the floor of the shaft leads to a further reduction in radial stresses on the lining – near the bottom only in the case of sand, although according to Terzaghi (1943), the relief may be experienced for nearly the full depth of the shaft in the case of clay. At depths greater than about four times the diameter of the shaft, radial stresses on the lining in a clean dry sand increase very slowly with depth. When a shaft is below the phreatic surface, hydrostatic pressure on the lining greatly exceeds the stresses exerted by soil. The conditions under which stresses on the timbering of a shallow, circular shaft should be calculated on the assumption that the adjacent soil yields are not known exactly. As a general guide it is prudent to assume this condition for excavations with a ratio of depth to radius less than one. For higher ratios reference should be made to the analysis given by Terzaghi (1943). It can be expected that relief of stress is likely to be developed to some extent by ring action in the case of a square shaft. For both square and rectangular shafts, relief of lateral stress is derived by shearing stresses coupled with compression on the ends of the yielding mass of soil adjacent to each side.

Peck and Berman (1948) observed lateral stresses in a shaft, 4.7 and 3.7 m diameter, for the Chicago subway sunk to a depth of about 21 m through 3.7 m of sandy and silty fill overlying soft glacial clay. The stress on the upper part of the shaft was less than the lateral stress at rest and appears to conform roughly to the active stress but on the lower half the distribution falls steadily to almost zero at the shaft bottom. For consideration of stability of the floor of shaft excavations in clays see Section 6.2.7b.

6.4 DESIGN AND CONSTRUCTION OF CULVERTS

This section is devoted principally to box section culverts which have a diverse range of application in the solution of the present day rural and urban traffic problems. *In situ* cast reinforced concrete culverts (invert slab monolithic with walls and roof slab) are the most common form of construction. Box section culverts have also been constructed of prestressed precast inverted T-walls with invert and roof slabs fixed in position with dowels or bolts and L-walls with roof slabs; the flanges of the walls serve as footings and also as part of the invert. The box section is not as economical in material as, for example, an arch section but is simpler in construction and is commonly used for spans up to 5.0 m under shallow and medium depth fills.

6.4.1 Determination of loads on culverts

Loads on culverts can be considered under four headings, depending on the method of construction: (a) in tunnel, (b) in open cut, (c) in a vertical-wall cofferdam, and (d) constructed at ground level and subsequently covered with an embankment.

The theory presented briefly below for loads on culverts constructed in tunnel is based on the assumption that arching is developed in the mass of soil above the tunnel and that shearing resistance is evoked at the surfaces MN_1P_2 (Fig. 6.43a). It is assumed that the water table is beneath the tunnel floor. Part of the yield of the soil toward the tunnel occurs as the working face passes the cross-section under consideration and the remainder after the timbering or lining is inserted. Although it is assumed here that movement takes place to an extent sufficient to reduce the vertical and lateral thrust on the lining to the value corresponding to the state of incipient shear failure in the soil, it is possible that with shield driven tunnels (usually of circular section) the movement is insufficient. Before excavation takes place the vertical and horizontal stresses are those of the earth at rest but when yield occurs, the total vertical force at the level of the roof is reduced by the vertical component of shearing resistance at the sliding surfaces.

Terzaghi (1943) makes the simplifying assumption that the sliding surfaces are MN_1P_2 and the width of the yielding strip (Fig. 6.43a) at the level of the tunnel

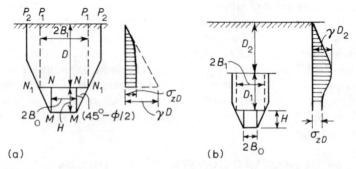

Figure 6.43 Earth on culverts in tunnel construction: (a) medium deep tunnel; (b) deep tunnel.

roof is thus

$$2B_1 = 2[B_0 + H \tan (45 - \phi/2)].$$
(6.74)

Then for a tunnel driven at a relatively shallow depth through sand the vertical stress on N_1N_1 is

$$\sigma_{ZD} = \frac{\gamma B_1}{K \tan \phi}(1 - e^{-K \tan \phi D/B_1}).$$
(6.75)

When driven at a greater depth, the vertical stress on N_1N_1 reaches an upper limiting value

$$\sigma_{ZD} = \frac{\gamma B_1}{K \tan \phi},$$
(6.76)

since arching does not extend beyond a certain elevation D_1 (Fig. 6.43b) above the tunnel roof, when the soil above this elevation acts on the zone of arching in the form of a surcharge γD_2. In these expressions K is an empirical constant which direct measurements indicate is approximately unity. The lateral thrust on the tunnel is calculated by the Rankine or Coulomb theory, treating the walls as retaining walls with a surcharge σ_{ZD}. A similar expression for a tunnel in cohesive soil is

$$\sigma_{ZD} = \frac{B_1(\gamma - c/B_1)}{K \tan \phi}(1 - e^{-K \tan \phi D/B_1})$$
(6.77)

from which it is apparent that $\sigma_{ZD} = 0$ when $B_1 \leqslant c/\gamma$.

Thrust on tunnel linings in cohesive soils may depend almost entirely on squeezing and swelling conditions. Observations on circular tunnels constructed in the London Clay show that the contact stresses on the lining depend almost entirely on factors associated with construction and the final distribution of stresses is irregular (Tattersall *et al.*, 1955); on one tunnel, 2.6 m i.d., driven at a depth of 27.5 m the average radial stress was between 200 and 300 kN m^{-2}.

Studies of subsidence above shallow tunnels in soft ground have been reported by Atkinson and Potts (1977).

Although in open cut it is conceivable that arching may be absent when the backfilling is well compacted, owing to the inability of the soil to yield in the zone surrounding the culvert, it may be argued that even the best compacted fills settle with time, and Steedman (1940) quotes experiments in which the vertical load on the culvert roof is only a portion of the total weight of the filling. Alternatively it seems possible that arching may be absent in poorly compacted backfilling owing to low resistance at the sliding surfaces. If full arching effect is likely to be established, the values of σ_{ZD} are determined by Equations 6.75, 6.76 and 6.77 given above. In order to establish arching effect, the layers of backfilling adjacent to the culvert should be lightly and uniformly compacted and the degree of compaction increased to the optimum value at some distance above the culvert.

This method of backfilling is desirable in any case to avoid damage to the culvert. R.R. Rowe (1957) describes the application of this technique in practice, which has proved satisfactory provided construction is carefully supervised. For an arch culvert the fill is lightly compacted for a height above the crown equal to one-quarter of the span and for a box culvert for a height above the roof equal to the span. It should be noted that claybackfilling is a remoulded material, the characteristics of which depend on its sensitivity.

When a culvert is constructed at ground level and subsequently buried under an embankment, the stress on the culvert roof may exceed that of the entire weight of the column of earth directly above it on account of the downward drag accompanying consolidation in the filling behind the culvert walls; Marston (1930) suggests that the upper limiting value of the stress on the roof of such a culvert is given by

$$\sigma_{ZD} = \frac{B_0(\gamma + c/B_0)}{K \tan \phi}(e^{K \tan \phi D/B} - 1). \tag{6.78}$$

Alternatively, it may be possible to determine values for settlement over the culvert and at the sides from laboratory consolidation tests, taking into account the fact that the soil directly above the culvert is subjected to a consolidation load exceeding the weight of soil, and that the soil immediately at the culvert sides is relieved of load to some extent, and hence to estimate a height above the roof, D_e, where the settlement directly above the culvert is equal to the settlement over the sides. Equation 6.78 is valid for the case $D < D_e$.

Marston (1930) considers $K = K_a$, and in support of this it is stated that measurements indicate the horizontal stresses on pipes beneath embankments may vary from 25 to 35% of the vertical stresses and extremes of 18 to 40% have been recorded. It appears that, at least with granular fill materials, the loads on culverts constructed in open cut or under embankments increase over a period of two years or more and the ultimate load is likely to exceed the initial load by up to 20 to 25%. Marston states that the load on a culvert constructed in a trench with sloping sides is the same as that constructed in a trench with vertical sides and of width equal to the width in the sloping sided trench at the level of the top of the culvert. The term c/B_0 does not appear to be included in Marston's theory, which was developed principally for granular fill materials, but it is incorporated for the purpose of comparison with other theoretical expressions. Where a culvert is constructed in a wide trench, the relief of contact stress on the roof arising from resistance at the sides of the trench may not be experienced. Instead, a condition similar to that beneath an embankment may be obtained and compression of the soil at the sides of the culvert may lead to high stresses on the roof. Whether the stress will be relieved or augmented depends on the relative proportions of trench and culvert widths and the cover depth of the culvert beneath the ground surface. The wide trench condition has been investigated by Clarke (1963).

Gould (1970) gives a summary of practical recommendations for the design of 'cut-and-cover' rigid reinforced concrete twin box culverts built inside rigidly braced vertical-wall cofferdams. It is generally assumed that ultimately the lateral earth stresses return to the 'at rest' values because of the relaxation of the shear stresses mobilized during excavation; the possibility of much larger stresses applied by overconsolidated clays is discounted. The values of lateral stress coefficients used in the design of subways in various American cities range from 0.4 for alluvium and soft Tertiary deposits, through 0.5 for glacial materials and hard surface strata of the Mexico City clays to 0.7 for the plastic Mexico City clay. For the criteria used in the design of the Rapid Transit System in Washington D.C. it was assumed that the 'at rest' lateral stress coefficient was equal to $(1 - \sin \phi')$, utilizing laboratory angles of drained shearing resistance for materials finer than medium sands and estimated values, based on the Standard Penetration Test results, for coarser soils. In all cases the vertical contact stresses acting on the roof slab were taken as equal to the full weight of the overburden plus surcharge, neglecting vertical shear on the sides of the block of heavily compacted backfill. Various plausible combinations of the above loadings were considered, together with possibilities of future excavations adjacent to the culverts and fluctuations of the ground water table.

A culvert under an embankment is liable to differential settlement in the form of a sag along its longitudinal axis. This consists of the characteristic central sag of a uniformly loaded flexible body bearing on soil, which is accentuated by the particular form of an embankment with the maximum load in the centre falling to zero at the toes of the banks. Nominal longitudinal roof and floor reinforcement is provided to cater for this effect. If the culvert gradient is slack, consideration may be given to the provision of slight initial hogging to eliminate possible backfall in the invert. If the culvert and head walls function as a monolithic unit, the earth thrust on the head walls induces longitudinal tension in the culvert. High embankments may lead to large tensile forces on account of the tendency of the fill to spread laterally, involving both elastic and plastic displacements. R.R. Rowe (1957) reports that joints in pipe culverts have opened under the influence of these forces: the largest displacements being incurred near the ends of the culverts.

Steedman (1940) shows that a uniformly distributed surcharge q at ground level produces on the roof of a culvert in open cut a stress

$$\sigma_{qD} = qe^{-K \tan \phi D/B_1}, \tag{6.79}$$

and this is applicable also to the case of a culvert in tunnel. The value of σ_{qD} is usually less than 10 to 20% of q when $D > 16B_1$.

Where a culvert is at a very shallow depth, the local effect of a knife-edge load or a point load may be considerable. In general, wheel loads can be treated as point loads except where the cover is less than about 1.0 m, or where the maximum dimension of the area of contact between wheel and ground surface exceeds one-half of the cover depth. The stress transmitted to the roof of a culvert can be obtained, with sufficient accuracy for practical purposes, by the Boussinesq equation for stresses beneath a point load on the surface of an elastic medium;

practical studies of this problem were carried out by Clarke and Young (1962). Recent experimental work indicates that, pending more extensive observations, an impact factor of not less than 2 is necessary at any depth of cover for road wheel loads directly above a culvert when the road surface is uneven.

Table 6.13 Bending moments for single cell box culvert

$I_1 \doteq I_3$ for expressions given below $K = \dfrac{H}{B} \times \dfrac{I_3}{I_2} \doteq \dfrac{H}{B}\left(\dfrac{t_3}{t_2}\right)^3$ Midspan BM are evaluated by deducting end BM from free span BM or by proportions		

Loading	BM diagram	BM per unit length of culvert	
		$M_A = M_B$	$M_C = M_D$
1 σ_{zD} + weight of roof q		$-\dfrac{q\,B^2}{12\,(K+1)}$	M_A
2 Line loading $\downarrow W$ $\dfrac{W}{B}$		$-\dfrac{WB}{12}\left[\dfrac{2K+4.5}{(K+3)(K+1)}\right]$	$-\dfrac{WB}{24}\left[\dfrac{K+6}{(K+3)(K+1)}\right]$
3 Weight of walls W_w W_w $w = \dfrac{2W_w}{B}$		$+\dfrac{wB^2}{12}\left[\dfrac{K}{(K+3)(K+1)}\right]$	$-\dfrac{wB^2}{12}\left[\dfrac{3+2K}{(K+3)(K+1)}\right]$
4 Earth thrust p_1 p_1		$-\dfrac{p_1 H^2}{60}\left[\dfrac{K(2K+7)}{(K+3)(K+1)}\right]$	$-\dfrac{p_1 H^2}{60}\left[\dfrac{K(3K+8)}{(K+3)(K+1)}\right]$
5 Earth thrust p_2 p_2		$-\dfrac{p_2 H^2 K}{12\,(K+1)}$	M_A

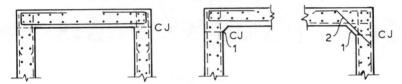

Figure 6.44 Reinforcement for box culverts: (1) 150 mm × 150 mm fillet; (2) tension reinforcement, if necessary.

6.4.2 Design of box section culverts

Appropriate combinations of the loads described above are considered in the design to obtain maximum span and support bending moments; the moment distribution method or readily available design charts (e.g. Table 6.13) or computer programs are used in the analysis of the statically indeterminate single or multicell box sections. The weight of water in the culvert and the weight of invert slab are carried directly by the soil and do not contribute to bending moments. If the water is under hydrostatic head owing to flooding at the culvert inlet, a uniformly distributed pressure is applied to the roof, invert and walls, and loadings (1) and (5) can be applied with signs of the bending moments reversed. When the flow is up to roof level without excess head, loading (4) can be applied to the walls with signs reversed. Fig. 6.44 illustrates the arrangement of reinforcement in box culverts; for speed of construction fillets are often neglected and simple U-bars are used to provide reinforcement against support bending moments. Nominal longitudinal reinforcement is provided in the top and bottom of the culverts to resist stresses induced by differential longitudinal settlement. Longitudinal distribution steel should be 0.2 to 0.3% but temperature steel is not normally provided. In long culverts, expansion joints are necessary and should be properly designed, particularly if differential settlement is likely to occur.

Chapter 7

COFFERDAMS AND CAISSONS

I. Greeves

INTRODUCTION

This chapter is an introduction to the design and construction of cofferdams and caissons used for foundations below water level in the bed of the sea or rivers, in waterlogged or unstable ground or where lateral support is necessary.

As many cofferdams are a temporary measure to enable permanent foundations to be constructed within them, the conflicting demands of economy and safety need careful balancing in this class of work. Practical experience should form a large part, though by no means the whole, of the engineering approach to these problems.

7.1 GENERAL REVIEW OF THE TYPE OF FOUNDATION TO BE USED IN BAD GROUND OR UNDERWATER

7.1.1 Preliminary investigation

It is important to carry out a comprehensive soil investigation of the site and surrounding areas in order to select the most suitable type of foundation. A number of borings should be put down to sample the soil carefully, and tests should be made with regards both to the temporary works and permanent structure.

If the work is extensive, and is to be constructed in the estuary of a river, every endeavour must be made to determine the presence of a hidden channel filled with recent alluvium. If undiscovered this soft and loose material could cause a 'blow', i.e., upward failure of the excavated surface, thus endangering the works.

The permeability and strength of the soil must also be considered to determine the depth at which the foundation should be made. In rivers and tidal waters, heights of tides, flood levels and velocity of current should be studied. In open sea the effect of wave action must also be taken into account.

7.1.2 Choice of foundation

The main difference between cofferdams, cylinders, caissons and monoliths is that cofferdams are usually temporary structures, parts or all of which are removed after construction, whereas the others are primarily permanent structures to be incorporated in the main works. The factors which affect the choice of type of construction are the depth to reach sound material, the strength of the ground being supported and size of foundation.

7.1.3 Cofferdams

Cofferdams of steel sheet piles are well suited to temporary works in which permanent foundations are constructed, the piles being either cut off or withdrawn for future use. Steel sheet piling, however, is usually limited to a maximum depth of between 20 and 30 m, although in some cases pile lengths have exceeded 40 m. The depth that water is lowered inside a cofferdam should not exceed 20 m, greater depths being excavated and concreted underwater.

Where there is a relatively impermeable stratum, such as clay, at foundation level into which the piles can penetrate and provide an effective cut-off against inflow of water, cofferdam piling should be used. However, if the foundation is of bedrock into which the piles will not penetrate without the use of special techniques, a caisson may be preferable. The presence of boulders or large flints may well cause piles to become declutched, torn or buckled. Flints in chalk and stones in clay may be carried below the toes of the piles leaving a waterway down the outside, thus causing a blow when pumping out is attempted. Caissons might then be preferred, as these major obstructions to piling can be dealt with during excavation.

7.1.4 Caissons

As the long term effects of compressed air on workmen become known, there is an increasing reluctance to use it in caisson sinking. However, there are occasions when compressed air working has advantages over open well sinking. For example, when large boulders occur under the cutting edge, when fine silts and sands overlie an uneven bedrock surface, or when other structures in the vicinity make it important that there should be no loss of ground.

A compressed air caisson is relatively easy to control during sinking as the material is excavated under the cutting edge. Lowering is controlled by 'blowing-down': this involves lowering the air pressure inside until the chamber sinks into the ground a sufficient distance. Equilibrium is then restored by raising the pressure. Provided the miners are taken out, blasting can be used to prepare the bedrock or to split boulders encountered during sinking. The depth to which a compressed air caisson can be sunk is limited by the maximum permissible air pressure under which men can work in the chamber – usually about 33 to 36 m head of water.

Open caissons are not limited by depth restrictions; they have been sunk to more than 60 m below the surface. As they are excavated by grabbing out the material inside wells, it is important that the soil is capable of being grabbed satisfactorily. Soft alluvial clays are good but peat layers can be difficult since the fibres compress under the weight of the cutting edge. Boulders can cause delay to sinking and upset the verticality of the caisson. To reduce the skin resistance during sinking the cutting edge may be about 100 mm greater in diameter than the body so that the annular space can be kept filled with bentonite pumped through special pipes built into the shaft.

7.1.5 Cylinders

Cylinders are often used for bridge pier foundations. They have a much smaller cross-sectional area than caissons and thus are more difficult to keep plumb during sinking. To overcome this they are lowered between guides until sunk sufficiently into the soil to gain lateral support. Small cylinders may be sunk aided by hammer blows but, in the case of larger ones, either kentledge is used or skin resistance is reduced by bentonite.

Sometimes when the cutting edge becomes sealed into an impervious stratum, for example London Clay, excavation may continue below the cylinder using shaft sinking techniques, with cast iron or concrete segmental linings. The pier shafts for the new London Bridge (Brown and Mead, 1973) were sunk in this way and were belled out at the bottom to reduce bearing stresses.

Often when the bedrock is not horizontal it may be preferable to sink 3 or 4 small cylinders to differing depths rather than sink a larger caisson which will get into difficulties when one side reaches bedrock and the opposite side may still be in soft material.

7.1.6 Monoliths

Monoliths are heavy open caissons, usually of mass concrete, provided with a cutting edge and sunk by grabbing from inside cells in a similar way to open well caissons. They are suitable for quay walls where a great depth of soft overburden overlies a firm stratum. They become economical compared with steel sheet piling when the depth to be retained exceeds 25 m. When they reach a suitable founding level, the bottoms are cleared out by a diver and sealed with a plug of concrete some 2.5 m thick. Having done this the cells can be pumped out and the rear row filled with concrete to resist the overturning moment when the soil is excavated in front of them. Monoliths with a plan area of about 10 m × 10 m are often sunk with a gap of about 1 m between them to allow for any inaccuracy in verticality during sinking. The space is later closed with piles driven close to each other, thus filling the gap.

7.2 TYPES OF COFFERDAM

7.2.1 Gravity types

These may be made of a bank of rock or hardcore with an infilling or a surfacing of impermeable material. In some cases a single row of sheet piles may form the cut-off, supported by a bank of hardcore tipped in front and behind.

Small gravity dams may be of *in situ* concrete or of grouted aggregate or bagwork. Gravity dams employed to close dock entrances are made of large interlocking concrete blocks, each weighing about 12 tonne. They are placed on the lock sill or apron and built up underwater by divers.

7.2.2 Sheeted cofferdams

(a) Timber sheeting

In the simplest form sheeted cofferdams can be constructed of a single row of timber sheet piles interlocked by a metal tongue. They can also be composed of a double wall of timber piles, with puddle clay infilling, strengthened by king piles and raking struts.

(b) Concrete sheeting

Cofferdams can also be constructed of reinforced concrete sheet piles, with a vee on one side and groove on the other to give a simple method of interlocking. They are used for shallow work and are usually left in position. Concrete king piles with horizontal concrete panels between may also be used.

In cases of limited headroom or in close proximity to existing buildings, walls of contiguously cast *in situ* bored piles can be employed (Section 6.2.8).

Recent developments have seen the introduction of diaphragm walling, for which a trench is excavated, the sides being supported by a bentonite slurry. When each section is excavated to full depth, reinforcement is placed and concreting effected by tremie pipe. In this way the bentonite is displaced as concreting proceeds. If carried out under careful supervision, good results have been achieved.

(c) Steel sheet piling

Cofferdams of steel sheet piling may be either single skin or double skin. Single skin cofferdams can be used to close the end of structures such as an entrance lock. They are usually driven on a circular arc supported by circular arc walings. If the depth is not great they may be cantilevered.

Single-skin sheet steel piling is also convenient for the construction of circular cofferdams for bridge foundations, gas holders or cooling water intakes for power stations. They have the advantage of being simple to construct and are supported by internal ring walings cast *in situ* of reinforced concrete. Single skin cofferdams can be supported by one, or a combination, of the following constructions: by internal struts and walings of timber, concrete or, more usually, steel, in one or more frames according to the depth to be supported; by horizontal or raking struts or raking piles used as struts; by external rods tied to anchor piles or anchor blocks or by external ground anchors or rock bolts. These latter methods give clear working space for excavation.

(d) Double-skin sheet piled cofferdams

These basically comprise two rows of sheet piles held together by tie bolts at various levels and the space between is filled with broken rock, rubble or gravel. They are self-supporting gravity structures suitable for depths of water up to 10 m, beyond which cellular cofferdams may be preferable. Both types depend on the properties of the filling and the soil at foundation level for their stability. The arrangement and type of sheet piling is also important.

Double-walled cofferdams comprise two parallel lines of sheet piles connected by walings and tie rods at one or more levels. When all soft material and mud has been removed between the rows of piles, the space is filled with cohesionless material such as sand, gravel, broken rock or clean hardcore; clay and silt must not be included in the filling. The inner line of piles acts as an anchored retaining wall, whilst the anchorage is provided by the outer line of piling. Ordinary trough type piling can be used and the width of the dam should be not less than 0.8 of the height of retained water.

It is important that the filling is adequately drained in order to reduce the pressure on the inner row of piles and to prevent loss in shear strength of the filling. This can best be done by providing weep holes at the lowest level in the inner line of sheet piling. These weep holes can be connected to a horizontal porous pipe surrounded by a bed of filter gravel in the body of the dam to give the maximum draw-down.

Before filling is commenced, vertical wells should be installed to observe the water level in the filling during the crucial dewatering operations. If the observation wells are not less than 300 mm diameter, then submersible pumps can be lowered inside them to aid dewatering. Ashes or cement can be dropped down the outside face to seal the clutches of the piles. On large dams, the face has been observed to deflect inwards by as much as 250 mm at the centre during dewatering as the full load is taken up.

The piling must be driven into the soil below the level of excavation to a depth sufficient to enable the necessary passive resistance to be developed. Should it be water-bearing granular material, then the effect of seepage has also to be considered. Should there be hard rock at formation level into which the piles cannot penetrate, then this type of cofferdam will require special measures to ensure sufficient pile penetration for stability.

7.2.3 Cofferdams of rock-filled cribs

These can be used on exposed bedrock or where river currents are swift, and have been used in water at least 15 m deep. The base of a timber cribwork is constructed on shore and floated out to the site, the upper part being assembled as sinking proceeds. Sinking is effected by placing rock and gravel in pockets in the cribwork and the remainder of the crib is filled before the cofferdam is pumped out.

To make the dam watertight a clay can be provided between close boarding at the water face. The conformation of the river bed is surveyed before starting so that irregularities can be catered for in the shaping of the cribs. This limits the amount of divers' work in preparation and packing of the foundation. Watertightness in fast flowing rivers can be achieved by hanging thick polythene sheeting down the waterside of the dam and holding it on the river bed by concrete in bags or by rock fill. The cribwork is designed as a gravity dam and must be braced by diagonal members against lateral distortion.

7.2.4 Cellular cofferdams

Cellular cofferdams are constructed using straight web sheet piling in circular, segmental or other shaped cells filled with cohesionless soil such as sand, gravel, hardcore or broken rock. High circumferential tensile forces are developed in the piling due to the outward thrust of the filling. The straight web piling has a number of significant features: a strong interlock designed to take a tensile force of not less than $2.8\,\text{kN}\,\text{mm}^{-1}$ length of pile, a straight web and an angular deviation between one pile and the next so that circular cells or arcs can be constructed. The low flexural strength of these sections about one axis leads to special problems in transport, handling, pitching and driving. Trough-shaped sections of steel sheet piles are not employed for this work as the troughs and interlocks would open under the circumferential tensile forces.

Cellular cofferdams are entirely self-supporting and enable a considerable height of water to be retained. Stability depends on the section of pile resisting bursting, the size and shape of the cells and the properties of the filling and material in the foundations. Because they are self-supporting, they are well suited for use when rock is covered with no more than a thin veneer of soils. Some penetration of the piles below formation level should be achieved in order to provide an adequate cut-off to retard seepage. As with double-skin cofferdams, it is necessary that the filling be free-draining sand, gravel, hardcore or broken rock. Any overlying soft clays and silts should be removed before filling commences. Adequate drainage on the inner side of the cofferdam must be provided by weep-holes and filters. The ratio of width to height of water upheld should be not less than 0.8 to 1.0 in order to provide an adequate factor of safety against tilting or sliding. If it is not possible to provide complete drainage of the filling, allowance should be made for any water pressure acting on the piling. Cellular cofferdams have the advantage of providing a working space clear of bracing and, when filled, the cells provide platforms for storage of materials, providing the design of the cells takes into account the surcharge loads from such materials.

In the new entrance lock cofferdam to Leith Harbour, floating craft were used to drive the piles working from both ends simultaneously (Godden *et al.*, 1970).

To maintain the shape and provide the necessary support during driving, a floating template with eight spud legs was used. This template was fitted with sections of pile clutch as a guide at each of the locations were junction piles would be pitched and could be split in half for removal from a completed cell. The piles were driven by floating derricks using diesel pile hammers. After a complete cell had been pitched, starting from the four junctions piles located in the clutch guides, all but two piles were driven completely. The last two piles were left suspended about 600 mm above the bottom to allow the passage of water under their toes until filling was ready to be placed.

On other sites more elaborate temporary spiders have been erected to support pile frames in addition to temporary gates to the piles. Compared with segmental cells, circular cells have the advantage that each cell is individually stable as soon as it has been filled. This is particularly significant if construction has to take place during difficult wind and wave conditions. It is important to note that the failure of one cell in a segmental structure can lead to progressive collapse of the whole structure. Fully circular cells should therefore be included at intervals to create strong points to contain any such effect.

7.2.5 Construction of cofferdams

Even when a cofferdam is driven to an impermeable stratum or provided with a concrete floor seal, some inflow of water must be expected. Leakage through the interlocks can sometimes be prevented by ashes, cement or sawdust tipped against the outside of the sheeting as water is pumped from the interior. Caulking the interlocks with tarred yarn has also been adopted. Leakage is aggravated in tidal situations owing to movement of the sheeting as the tide rises and falls.

The Runcorn–Widnes Bridge is founded on soft sandstone which occurs at a shallow depth below the bed of the River Mersey. A seal between cofferdam sheeting and sandstone was effected by excavating a trench 1 m deep in the rock which was filled with clay and covered with a skin of concrete. The sheeting was then driven into the clay. Two major leaks developed through a fault zone and, to overcome these, the sheeting was driven deeper, grout was injected into the rock and gunite was applied to rock surfaces.

Piping and blowing is most likely to occur where the hydraulic gradient is greatest; that is, close to the sheeting. The length of the flow lines can be increased, thus reducing the hydraulic gradient, by placing a blanket of clay, say 1.50 to 2 m thick, on the river bed outside the sheeting and also a bank of permeable material on the cofferdam floor.

Sub-drains are provided in cofferdams of large area to a distance of, say, 3 to 5 m from the toe of the bank and are connected to a sump. The depth of water in the pumping sump should be about 0.5 m in order to maintain a low velocity and reduce the possibility of sediment being carried to the pumps. Alternatively, infiltration water can be intercepted by pumping from well-points sunk within the cofferdam to approximately the sheeting toe level. The rising main from pumps can be arranged as a syphon over the top of the sheeting, with the

Table 7.1 Minimum driving depth for sheeting in cohesionless soils

Width of cofferdam W	Depth of cut-off D (cohesionless soils)
2 H or more	0.4 H
H	0.5 H
0.5 H	0.7H

discharge 1 or 2 m below water level, thereby decreasing the pumping head. The end of the pipe should be gradually increased in diameter in order to reduce the velocity and hence reduce the losses due to impact in the water outside the end of the pipe. A thin impermeable stratum or a concrete seal may be subjected to uplift, and this tendency can often be relieved by pumping from beneath the layer.

As a guide, the minimum depth D to which sheeting should be driven in non-cohesive soils to restrict the exit gradient to approximately unity varies as shown in Table 7.1, where H is the height from excavation floor to external ground water level. These values (CP2004, BSI, 1972d) are confirmed by experience. However, the stability of the bottom of a cofferdam excavation in water-bearing non-cohesive soils depends on its dimensions in addition to soil conditions. For homogeneous soils, both theory and practice show that stability decreases with the decrease in width of excavation. Although stability can be increased by increasing the depth of penetration of the sheeting, the worst conditions obtain where relatively fine-grained soil is underlain by soils having greater permeability. It may be necessary in this case to employ ground water lowering outside the excavation or to carry out excavation partly or wholly under water.

7.2.6 Work carried out in tidal waters

The first decision to be made is whether the work can be carried out using a half-tide cofferdam. This gives protection during the lower part of the tide and the cofferdam is flooded as the tide approaches high water level. Alternatively, the work can be protected by a whole-tide cofferdam in which the construction can be carried out continuously. With the reluctance of labour to work unsocial tidework-hours and the high cost of paying standby time when washed out by the tide, whole-tide dams have become more attractive economically. On some coasts, extreme low water on both morning and evening Spring tides occur in hours of darkness in wintertime, so that in these circumstances half-tide cofferdams should not be considered.

It is necessary to consult tide tables when deciding the height to which a whole-tide cofferdam is built. It must be recognized that the direction and strength of prevailing wind, wave height, flood levels and the effects of flood water can all affect the levels predicted. Direction and strength of currents during a complete tide cycle need to be considered in designing the cofferdam bracing.

A cofferdam in a river reduces the waterway and the local river authority must be consulted before any work is contemplated. In some cases, if the restriction to flow is considered to be excessive, orders may be given to construct the cofferdam in sections and the temporary piling for each completed section is removed before the next section is commenced.

The natural regime of the river is altered by construction works and scour will almost certainly take place if the current is swift. Work on hydraulic models can indicate the magnitude of the problem and on major projects a model study is desirable. Apart from scour to the river banks caused by increased velocity, scour is likely to take place in the river bed in the immediate vicinity of the cofferdam. Protective measures for both include rip-rap pitching, stone gabions, fascine mattresses and concrete tetrahedrons.

The closure of a cofferdam in swift currents is difficult and substantial falsework is required to maintain the sheeting in position during driving. Scour is likely to take place whilst closing the gap. This should be done quickly and during this time sluices or valves must remain open to equalize the water pressure on each side of the piles. When all cofferdam bracing is completed, the valves can be closed and pumping out commenced.

7.3 DESIGN OF COFFERDAMS

7.3.1 General design considerations for cofferdams

The distribution of earth loading on cofferdams is discussed in Section 6.1.3(b). The stability of the floors of excavations is treated in Sections 6.2.7 and 2.7. Sections 7.3.2, to 7.3.5 below are concerned mainly with cofferdams in water, where the external load distribution is triangular. Packshaw (1962) has reviewed the construction of cofferdams, including practical aspects of the design of sheet pile cofferdams and further pertinent information has been given by Cornfield (Packshaw, 1962).

The factor of safety for stability of cofferdams is usually 1.25 to 1.50 and Bjerrum and Eide (Packshaw, 1962) state that in Norway the factor of safety against bottom heave is taken as 1.3. Bearing in mind failures of various temporary works which have occurred within recent years, it may be considered that such low margins of safety lead to excessive risk. A major influence governing the choice of magnitude of factor of safety in any particular case is the degree of accuracy with which the applied loading can be predicted. In addition to the obvious systems of loading on cofferdams, allowance should be made for waves and currents and, in some countries, ice. Large thrusts may be imposed by frost heave in soil behind sheeting during intensely cold weather.

Permissible stresses for certain constructional timbers are given in CP 2004 (BSI, 1972d): in the case of other timbers, reference should be made to Table 10 in CP 112 (BSI, 1971). Basic permissible stresses may have to be multiplied by a factor between 1.0 and 0.7 to take account of deterioration of timber during service. Under temporary loading, at the discretion of the designer, permissible stresses specified in CP 114 (BSI, 1957; presumably this will apply also to CP 110, BSI, 1972c) for concrete can be increased by up to $33\frac{1}{3}\%$ and those specified in BS 449 (BSI, 1969a) for steel by up to 10%, except where loads result primarily from water pressure, in which case no increase is permissible.

7.3.2 Cofferdam of one row of sheeting internally braced

It is desirable that the top frame of the cofferdam should be so located that negative and positive moments are balanced. Under triangular load distribution, where frames can be installed under water prior to dewatering and excavation, the cofferdam can be designed with the walings spaced so that the bending moment in each span of the sheeting is constant, in which case the strength of the walings must be increased successively downwards. Alternatively, the walings can be spaced so that the loading is equal in each frame, with the exception of the top one, in which it is less, and the bending moment in the sheeting decreases successively downwards. The distance h between the top frame, assumed at water-level, and the second frame is determined from the given triangular loading and the selected sheet pile section and usually lies between 3 and 7 m. The distances $h_1, h_2 \ldots h_{10}$ between successive walings are then determined from Table 7.2. The methods of design discussed below are based on elastic theory.

Consider a 1 m run of cofferdam sheeting. The bending moment in the sheeting between the first two frames is taken as $(1/7.8) Wh$ (theoretical value for triangular distribution of load and simply supported condition) and between lower frames as $(1/10) Wh_n$, where partial continuity over supports is assumed. W is the total load on the span under consideration. It is generally more economical to use heavy piling and less bracing than light piling and more bracing. The former arrangement generally necessitates steel bracing because of the heavy loads on the frames. In deep cofferdams, when the water level is lowered before each successive frame is installed, calculations should be made to check whether or not the maximum permissible bending moment in the sheeting

Table 7.2 Distance between successive walings for triangular loading. (Values given by United States Steel Export Co.)

	Equal BM in sheeting	*Equal loads on frames*
h_1	$0.691\,h$	$0.565\,h$
h_2	$0.570\,h$	$0.404\,h$
h_3	$0.505\,h$	$0.334\,h$
h_4	$0.463\,h$	$0.291\,h$
h_5	$0.432\,h$	$0.261\,h$
h_6	$0.408\,h$	$0.239\,h$
h_7	$0.388\,h$	$0.222\,h$
h_8	$0.372\,h$	$0.208\,h$
h_9	$0.358\,h$	$0.196\,h$
h_{10}	$0.346\,h$	$0.186\,h$

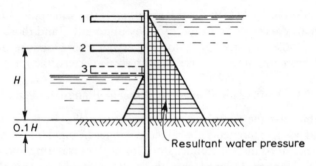

Figure 7.1 Arrangement of the frames in internally braced cofferdams.

is exceeded as the water-level is lowered and before the second or third frame is placed (Fig. 7.1). It is reasonable to assume that the sheeting acts as a simply supported beam of span $1.1H$ in firm ground but a larger value should be assumed in weak ground. The load consists of the difference in hydrostatic pressures. Stresses in walings and struts and the thrust on the soil should be investigated under this loading. Upward flow of water at the foot of the sheeting, due to unbalanced hydrostatic heads as water is pumped out, reduces the resistance of the soil to lateral thrust. If boiling or blowing occurs, the thrust normally resisted by the soil is thrown onto the bottom frame. There are numerous examples of failure at the foot of sheeting due to inadequate shear strength of soil or rock, either intrinsic or occasioned by boiling or blowing. During the construction of the George Washington Suspension Bridge, New York, a collapse occurred in one cofferdam owing to failure of the fissured sandy shale and sandstone to carry the reaction from the toe of the sheeting, which bent inward about the bottom waling. Fluctuations of stress in soil over many tidal cycles may reduce the passive resistance of the soil by fatigue.

When the design is based on equal bending moments in the sheeting, the load per metre run on the top waling is taken as $\frac{1}{3}W_1$ and on the second waling $(\frac{2}{3}W_1 + \frac{1}{2}W_2)$ where W_1 and W_2 are the total thrusts on the first and second panels, respectively. The load on any lower waling is taken as the pressure at that waling multiplied by one-half of the sum of the distances between adjacent walings above and below. When the design is based on equal loads on the frames, the load on the top waling is taken as $\frac{1}{3}W_1$ and on lower walings $1.4216W_1$. The walings are commonly designed as simply supported ($M = Wl/8$) where the number of spans is 3 or less, but for more than 3 some continuity is assumed ($M = Wl/10$). These notes on waling loads and bending moments are summarized from a publication of the United States Steel Export Co. The span l is the distance between centres of struts and for timber struts is usually 2 to 4 m. A wide spacing of struts is desirable to facilitate hoisting materials and plant. Before each frame is placed and wedged as water lowering proceeds, the last frame to be placed is subjected to thrusts somewhat greater than the design value. Timber walings should be of square section to reduce to small values the bending stresses due to self-weight. Each waling should be hung from the next above and the top one should be suspended from the top of the sheeting. Walings act also as struts carrying a thrust from the end walls. Shearing stresses in walings should be investigated and connections given careful attention.

The compression in any strut is taken as one-half of the sum of the thrusts on the spans of walings on each side of that strut. Struts and bracing should be designed to carry loads from staging or from impact of falling equipment; an arbitrary static concentrated force of 5 kN is sometimes assumed for the latter. Unless struts are rigidly braced by other struts in a horizontal plane and by struts and diagonals in a vertical plane, the effective length of strut is the distance between the walings. All connections should be secured by cleats, packings and bolts as appropriate. Excessive loads my be induced inadvertently in some members when wedges are driven.

In some situations it may be desirable to consider the possibility of water level inside the cofferdam being inadvertently higher than outside. The method of dismantling the cofferdam and the systems of loading imposed as a result of this must be taken into account in design.

Horizontal and vertical movements of the ground inevitably occur around cofferdams as water lowering proceeds: horizontal displacements can be largely nullified by jacking against the struts but vertical displacements can rarely be eliminated. It is important in some projects to observe the magnitude of these movements as the work progresses.

It is useful to examine briefly the implications of ultimate load theory in relation to the design of cofferdams. The complete objective would be to secure concomitant failure of sheeting, walings, struts and soil. Whether or not this is a rational approach to design is a matter of debate: undoubtedly other factors would enter the problem, apart from the difficulties involved in attempting to achieve balance of all the components at failure. For example, ultimate load theory might lead in some cases to unacceptable deflection of sheeting. Again, for ease of fabrication it might not be desirable to alter sections and arrangements of walings and struts in order to satisfy theory. It can be noted in passing that most sheet pile sections are thin-wall sections and local buckling of any of the section elements may prove an important criterion of design. These few comments should not deter engineers from employing ultimate load theory for cofferdam design: as yet the field is largely unexplored.

7.3.3 Circular cofferdams

A circular cofferdam is a special application of a single row of sheeting. The circular walings can be designed as rings in compression and internal bracing is not required except when the diameter exceeds 45 to 60 m. In open water, steel rings are the most convenient but reinforced concrete rings are suitable for land cofferdams and, since they can be cast against the sheeting, wedges are not necessary. If convenient, the top ring can be cast outside the sheeting, which is attached to the ring by bolts or hooks. Since there is a tendency for the upper length of sheeting to deflect outwards under certain conditions of loading as excavation proceeds, the top ring should be designed to resist some measure of tension, although the magnitude of the tensile force will probably be decided arbitrarily. An advantage of ring walings is the clear working space in the cofferdam.

Each ring must be designed to resist direct thrust and to avoid collapse from elastic instability or buckling. The critical external radial uniform load p_c per unit length of centre line of the ring required to cause collapse is $p_c = KEI/R^3$, where E is the modulus of elasticity of the material of the waling, I is the second moment of area of the cross-section of the waling about the neutral axis, and R is the radius of the ring and K is a factor dependent on the stiffness of the supported material, i.e., 6 for solid rock, 5 for stiff soils, 4 for loose soils and 3 for water (Timoshenko, 1941). The distances between walings can be taken from Table 7.2 and the load p per unit length of ring can be then estimated. The section required to resist collapse is generally larger than the section required to resist direct compression. A factor of safety F not less than 2 should be applied to the critical load and the section is then determined from $p = KEI/(FR^3)$. The spacing of circular walings may need to be closer than the normal spacing of straight strutted walings in order to avoid the tendency to collapse. Bending moments of unknown magnitude will arise from variations in the loading, unavoidable malformation of the waling and other sources. Since lives will be at stake, the risk of collapse must be carefully assessed and it may be desirable to increase the factor of safety to 3 or 4. The longitudinal reinforcement in a concrete ring subjected to direct compression should be designed in accordance with CP 110 (BSI, 1972c). The amounts of steel may be altered to cater for unusual conditions. For reinforced concrete walings a common rule is that the ratio of cofferdam diameter to waling width should not be more than 35. Ring walings can be suspended in position by rods fixed to the top of the sheeting. Consideration should be given to the tendency for walings subjected to direct compression to buckle in a verticle direction. Some resistance to buckling is afforded by friction between waling and sheeting and additional resistance can be provided by struts. Circular cofferdams are not suitable where the loading varies around the periphery due, for example, to a surcharge of fill on one side of the cofferdam.

The collapse of a cofferdam at Lake Keowee, South Carolina, USA, in 1978, leading to the loss of seven lives, was attributed to buckling of internal ring walings. Vertical small diameter steel tubular struts were inserted between the walings but the waling section was inadequate to prevent collapse in a horizontal plane.

7.3.4 Worked example of cofferdam of double row of sheeting

General description

The plan dimensions of the cofferdam are shown in Fig. 7.2. A semi-circular nosing formed by sheet piling at each end of the cofferdam, filled with gravel, provides a protection against shipping. The method of constructing the cofferdam is shown in Fig. 7.3 (a) to (d).

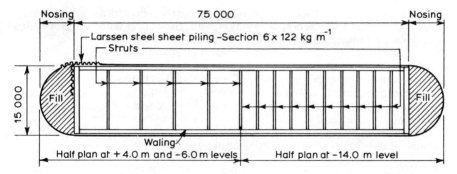

Figure 7.2 Cofferdam of double row of sheeting.

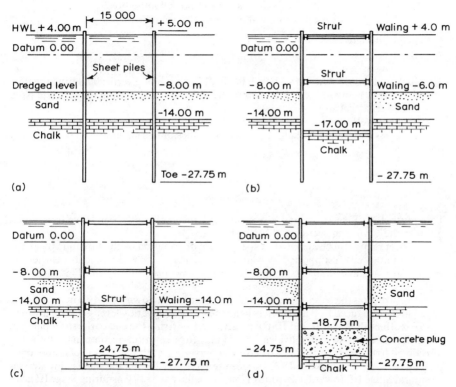

Figure 7.3 Cross-sections through cofferdam showing method of construction. (a) Stage I, drive sheet piling to level. (b) Stage II, fix waling frames at $+4.0$ m and -6.0 m levels with struts at 7.5 m centres. Excavate to -16.5 m level. (c) Stage III, fix waling frame at -14.0 m level with struts at 7.5 m centres. Excavate to -24.75 m level. (d) Stage IV, place concrete under-water up to -18.75 m level. Fix additional struts at -14.00 m level to reduce span of waling to 3.75 m. Dewater the cofferdam ready to construct new foundation up from 18.75 m level.

The sheet piling is driven to level and excavation within the cofferdam carried out through water with successive frames of strutting lowered and fixed under water. A mass concrete plug is placed through water by tremie at the base of the excavation in chalk. The plug replaces weathered chalk at the upper levels and provides a firm base for the new foundation to be constructed inside the cofferdam after dewatering. The plug also prevents water entering the dewatered cofferdam through the chalk.

In the following calculations it is assumed that the level of water inside and outside the cofferdam is the same and water pressure only will be considered in the final stage of dewatering the cofferdam.

Ground investigations in the vicinity of the cofferdam reveal a soil profile, typical of the Thames estuary, of dense sand overlying chalk.

Parameters for sand:

$$\gamma^1 = 10\,\text{kN}\,\text{m}^{-3}$$
$$\varphi = 35°$$
$$\delta = 10°$$

$K_a = 0.27$ with the effect of wall friction ignored.
$K_p = 4.8$ taking wall friction into account.

Parameter for chalk:

$$\gamma^1 = 8.8\,\text{kN}\,\text{m}^{-3}$$

The chalk at depths below $-19.0\,\text{m}$ will be capable of standing vertically during the time period involved between excavation within the cofferdam and placing the concrete plug, and no active lateral stress will occur.

At higher levels where the chalk is variably weathered, some active lateral stress is allowed for by assuming the chalk to behave as a cohesionless material with the following properties.

$$\varphi = 40°$$
$$K_a = 0.2 \text{ for } \delta = 15°$$
$$K_p = 6.5 \text{ for } \delta = 10°.$$

Lateral stress coefficients K_a and K_p for the sand and chalk have been taken from tables given in the Civil Engineering CP 2 (1951).

The sheet piling will be analysed by the moment distribution method treating the piling as a continuous beam supported by walings and struts. Fixed end moment coefficients c for partial loads and moments acting on the pile have been taken from Figs. 7.4 to 7.8 inclusive.

Active stress on the sheet piling in chalk has been taken down to $-19.0\,\text{m}$. In Stage II active stress is reduced by passive stress acting from the excavation level $-17.0\,\text{m}$ down to $-19.0\,\text{m}$. In both Stages II and III the passive stresses fixing the end of the sheet piling in chalk are assumed to be equivalent to a knife edge reaction accompanied by a full directional fixing moment acting $2.0\,\text{m}$ below excavation level. The casting of the concrete plug up to $-18.75\,\text{m}$ in Stage IV is assumed to have no effect on the sheet piling stresses existing at Stage III. Inward deflection of the sheet piling at $-19.0\,\text{m}$ during Stage III gives rise to free support for the sheet pile against the concrete plug at $-19.0\,\text{m}$ for the external water pressure caused by Stage IV dewatering of the cofferdam.

Sign convention

(a) Sheet pile bending moments
    Positive moments produce tension on the outside face of the sheet pile cofferdam and negative moments produce tension on the inside face.

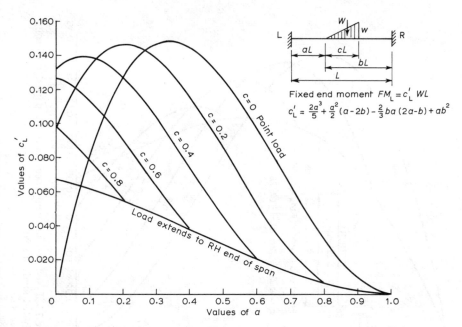

Figure 7.4 Fixed end moment coefficients for partial triangular load on span.

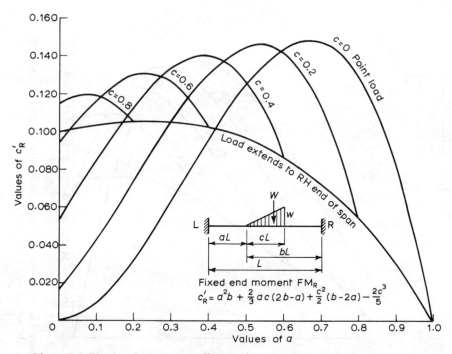

Figure 7.5 Fixed end moment coefficients for partial triangular load on span.

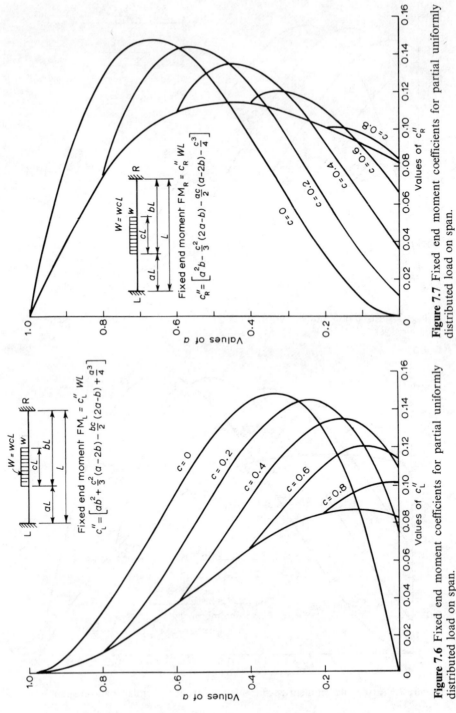

Figure 7.6 Fixed end moment coefficients for partial uniformly distributed load on span.

Figure 7.7 Fixed end moment coefficients for partial uniformly distributed load on span.

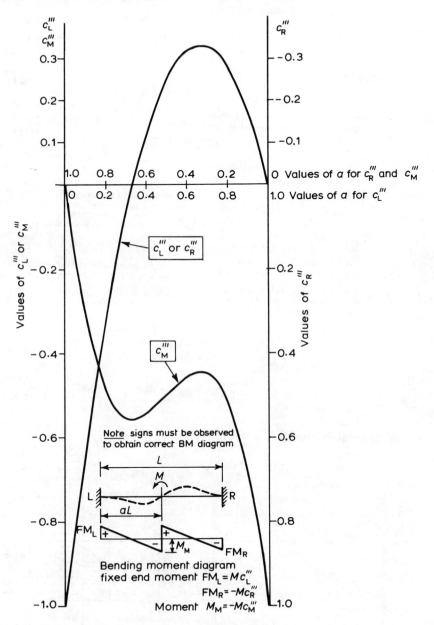

Figure 7.8 Fixed end moments for applied moment anywhere on span.

(b) Moment distribution

Moments acting on the ends of individual members are considered positive if acting clockwise and negative if acting anticlockwise.

(c) Loads on walings

Loads from sheet piles acting inward on the walings are considered positive and if acting outward are considered negative.

List of symbols		Units
σ'_v	Intensity of vertical stress	$\mathrm{kN\,m^{-2}}$
σ'_a	Intensity of active lateral stress	$\mathrm{kN\,m^{-2}}$
σ'_p	Intensity of passive lateral stress	$\mathrm{kN\,m^{-2}}$
h	Height of soil or water above any level under consideration	m
W_1, W_2	Total loads	kN
$\mathrm{FM_{AB}}$	Fixed end moment at end A of member AB	kN m
$\mathrm{FM_{BA}}$	Fixed end moment at end B of member AB	kN m
c'_{AB} or c_L	Fixed end moment coefficient for left-hand end A of member AB for partial triangular load	—
c'_{BA} or c'_R	Fixed end moment coefficient for right-hand end B of member AB for partial triangular load	—
c''_{AB} or c''_L	Fixed end moment coefficients for partial	—
c''_{BA} or c''_R	uniformly distributed loads	
c'''_{AB} or c'''_L	Fixed end moment coefficients for a moment applied to a member	
c'''_{BA} or c'''_R	acting anywhere in the span	—
K_{AB}	Stiffness of member AB at end A	—
K_{BA}	Stiffness of member AB at end B	—
D_{AB}	Moment distribution factor for member AB at end A	—
D_{BA}	Moment distribution factor for member AB at end B	—
I or I_{AB}	Second moment of area of member AB	$\mathrm{m^4}$
L or L_{AB}	Span of member AB between supports	m
S or S_{AB}	Shear at end A of member AB	kN
S or S_{BA}	Shear at end B of member BA	kN
R or R_A	Load on waling A	$\mathrm{kN\,m^{-1}}$

Sheet piling between nosing

(a) Stage II

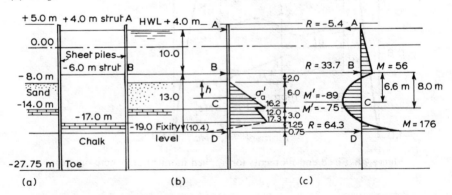

Figure 7.9 Stage II. (a) Cofferdam; (b) net soil stress diagram $(\mathrm{kN\,m^{-2}})$; (c) final moments and loads. Pile bending moments, $M(\mathrm{Kn\,mm^{-1}})$. Loads on walings, $R(\mathrm{kN\,m^{-1}})$.

(i) Soil stresses

Table 7.3 Stage II soil stresses (Fig. 7.9)

Soil	Level (m)	Vertical $\sigma'_v = \gamma'h$	Lateral $\sigma'_a = \sigma'_v K_a$	Vertical $\sigma'_v = \gamma'h$	Lateral $\sigma_p = \sigma'_v K_p$	Net stress $(kN\,m^{-2})$
		Active stress		*Passive stress*		
Sand	−14.00	60.0	16.2	—	—	16.2
Chalk	−14.00	60.0	12.0	—	—	12.0
Chalk	−17.00	86.4	17.3	—	—	17.3
Chalk	−19.00	104.0	20.8	17.6	−114.4	−10.4

(ii) Moment distribution (Fig. 7.10)

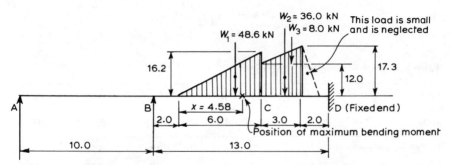

Figure 7.10 Sheet pile loading diagram, kN per m run.

$$\text{Load } W_1 = 16.2 \times \tfrac{6}{2} = 48.6\,\text{kN}$$
$$a = \tfrac{2}{13} = 0.15$$
$$c = \tfrac{6}{13} = 0.46$$
$$c'_{BC} = 0.127$$
$$c'_{CB} = 0.110$$

$$\text{Load } W_2 = 12.0 \times 3 = 36.0\,\text{kN}$$
$$a = \tfrac{8}{13} = 0.62$$
$$c = \tfrac{3}{13} = 0.23$$
$$c''_{BC} = 0.054$$
$$c''_{CB} = 0.139$$

$$\text{Load } W_3 = (17.3 - 12.0) \times \tfrac{3}{2} = 8.0\,\text{kN}$$
$$a = \tfrac{8}{13} = 0.62$$
$$c = \tfrac{3}{13} = 0.23$$
$$c'_{BC} = 0.041$$
$$c'_{CB} = 0.132$$

Fixed end moments

$$FM_{BD} = (48.6 \times 0.127 + 36.0 \times 0.054 + 8.0 \times 0.041)13 = 110\,kN\,m$$
$$FM_{DB} = (48.6 \times 0.110 + 36.0 \times 0.139 + 8.0 \times 0.132)13 = 148\,kN\,m$$

Joint B, distribution factors

$$K_{BA} = \frac{3}{4}\frac{I}{L} = \frac{3}{4} \times \frac{1}{10} = 0.075$$

$$K_{BC} = \frac{I}{L} = \frac{1}{13.0} = 0.077$$

$$\overline{0.152}$$

$$D_{AB} = \frac{0.075}{0.152} = 0.49$$

$$D_{BC} = \frac{0.077}{0.152} = 0.51$$

Table 7.4 Stage II moment distribution

	A	B		D
Joint	AB	BA	BD	DB
Distribution factor D	—	0.49	0.51	—
Fixed end moments (kN m)	0	0	− 110	+ 148
	0	+ 54	+ 56	+ 28
Final moments (kN m)		+ 54	− 54	+ 176

(iii) *Final shearing forces and bending moments*

$$S_{AB} = S_{BA} = \frac{54.0}{10} = 5.4\,kN$$

$$-S_{BD} = (48.6 \times 7 + 36.0 \times 8.0 \times 3 - 176 + 54) \times \frac{1}{13.0} = 28.3\,kN$$

$$S_{DB} = (48.6 \times 6 + 36.0 \times 9.5 + 8 \times 10 + 176 - 54) \times \frac{1}{13.0} = 64.3\,kN$$

Find the maximum bending in Span BD
Distance of zero shear from B $= x + 2$

Equating shears in Span BD. $x^2 = \dfrac{28.3 \times 6 \times 2}{16.2}$

$$x = 4.58\,m$$

Maximum $M'_{BD} = -28.3\left[(4.58+2) - \dfrac{4.58}{3}\right] + 54 = -89\,\text{kN m}.$

At C $\quad M'_{BD} = -28.3 \times 8 + 48.6 \times \dfrac{6}{3} + 54 \quad = -75\,\text{kN m}.$

Soil reaction at D, $R_D = 64.3\,\text{kN m}^{-1}$
The final bending moments are shown in Fig. 7.9(c).

(*iv*) *Loads on walings*

Waling A at $+4.0\,\text{m}$ level $R_A \qquad\qquad = -5.4\,\text{kN m}^{-1}$
Waling B at $-6.0\,\text{m}$ level $R_B = 5.4 + 28.3 = 33.7\,\text{kN m}^{-1}$

(*b*) *Stage III* (*Fig. 7.11*)

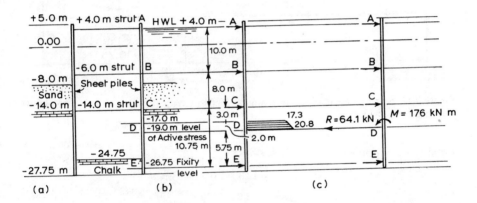

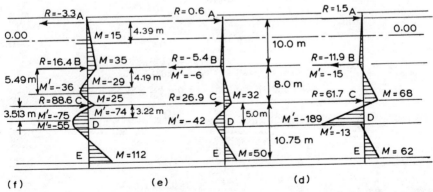

Figure 7.11 Stage III. Bending moment diagrams and waling loads. Pile bending moment, M (kN m per m run). Loads on walings, R (kN per m run). (a) Cofferdam; (b) net soil stress diagram (kN m^{-2}). Additional to Stage II stresses; see Stage II soil stress table. (c) Remove stage II soil relations at D. (d) Remove Stage II moments and loads. (e) Net soil moments and loads; (f) Stage III final moments and waling loads.

(i) Soil stresses
See Table 7.3 for stage III soil stress at D (-19.0 Level).

(ii) Moment distribution

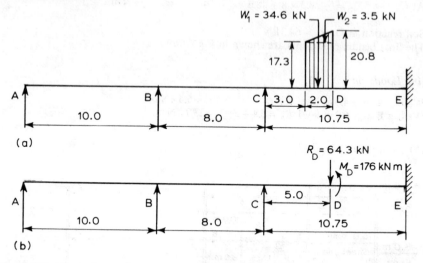

Figure 7.12 Sheet piling loading diagram. (a) Net soil loads (kN per m run): (b) Stage II soil reactions at D.

Fixed end moments for net soil loads (Fig. 7.12(a))

$$\text{Load } W_1 = 17.3 \times 2.0 = 34.6 \text{ kN}$$

$$a = \frac{3.0}{10.75} = 0.28$$

$$c = \frac{2.0}{10.75} = 0.28$$

$$c''_{CE} = 0.144$$
$$c''_{EC} = 0.088$$

$$\text{Load } W_2 = (20.8 - 17.3) \times \tfrac{1}{2} = 3.5 \text{ kN}$$

$$a = \frac{3.0}{10.75} = 0.28$$

$$c = \frac{2.0}{10.75} = 0.19$$

$$c'_{CE} = 0.142$$
$$c''_{EC} = 0.098$$

Fixed end moments

$$FM_{CE} = (34.6 \times 0.144 + 3.5 \times 0.142)10.75 = 59 \text{ kN m}$$
$$FM_{EC} = (34.6 \times 0.088 + 3.5 \times 0.098)10.75 = 37 \text{ kN m}$$

Fixed end moments for removal of stage II soil reactions at D (Fig. 7.12(b))

$$\text{Load } R_D = 64.3\,\text{kN} \quad a = \frac{5}{10.75} = 0.47$$

$$c'_{CD} = 0.132 \text{ and } c'_{ED} = 0.117$$

$$\text{Moment } M_D = 176\,\text{kN m} \quad a = \frac{5}{10.75} = 0.465 \quad b = 0.535$$

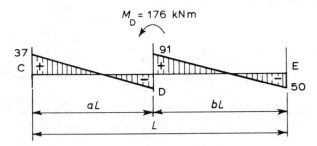

Figure 7.13 Fixed end moments (kN m).

$$c'''_{CD} = 0.211 \text{ and } c'''_{ED} = -0.281$$
$$c'''_{DC} = -0.483$$

Fixed End Moments (Fig. 7.13)

$$\text{FM}_{CD} = 64.3 \times 10.75 \times 0.132 + 176 \times 0.211 = 128\,\text{kN m}$$
$$\text{FM}_{ED} = 64.3 \times 10.75 \times 0.117 - 176 \times 0.281 = 31\,\text{kN m}$$

Distribution factors

$$\text{Joint B: } K_{BA} = \frac{3I}{4L} = \frac{3}{4} \times \frac{1}{10} = 0.075 \quad D_{BA} = \frac{0.075}{0.2} = 0.38$$

$$K_{BC} = \frac{I}{L} = \frac{1}{8.0} = 0.125 \quad D_{BC} = \frac{0.125}{0.2} = 0.62$$

$$\underline{0.200}$$

$$\text{Joint C: } K_{CB} = \frac{I}{L} = \frac{1}{8.0} = 0.125 \quad D_{CB} = \frac{0.125}{0.218} = 0.573$$

$$K_{CE} = \frac{I}{L} = \frac{1}{10.75} = 0.093 \quad D_{CE} = \frac{0.093}{0.218} = 0.427$$

$$\underline{0.218}$$

Table 7.5 Stage III moment distribution

Joint	A	B		C		E
	AB	BA	BC	CB	CE	EC
Distribution factor D.	—	0.38	0.62	0.573	0.427	—
Fixed end moments for	0	0	0	0	− 59	+ 37
net soil loads (kN m)			+ 17	+ 34	+ 25	+ 12
		− 6	− 11	− 5		
				+ 3	+ 2	+ 1
Final moments (kN m)	0	− 6	+ 6	+ 32	− 32	+ 50
Fixed end moments for	0	0	0	0	− 128	+ 31
soil reactions at D (kNm)			+ 37	+ 73	+ 55	+ 28
		− 14	− 23	− 11		
			+ 3	+ 6	+ 5	+ 3
		− 1	− 2			
Final moments (kN m)	0	− 15	+ 15	+ 68	− 68	+ 62

(iii) Shearing forces and bending moments for net soil loads

$$-S_{AB} = -S_{BA} = \frac{6}{10} = 0.6\,\text{kN}$$

$$S_{BC} = S_{CB} = \frac{6+32}{8} = 4.8\,\text{kN}$$

$$-S_{CE} = (34.6 \times 6.75 + 3.5 \times 6.42 - 50 + 32)\frac{1}{10.75} = 22.1\,\text{kN}$$

$$S_{EC} = (34.6 \times 4 + 3.5 \times 4.33 + 50 - 32)\frac{1}{10.75} = 16.0\,\text{kN}$$

$$\text{At D, } M'_{CE} = -16 \times 5.75 + 50 = 42.0\,\text{kN m}$$

Bending moments shown in Fig. 7.11(e).

(iv) Load on walings for net soil loads

Waling A at $+4.0\,\text{m}$ R_A $= 0.6\,\text{kN m}^{-1}$
Waling B at $-6.0\,\text{m}$ $R_B = -0.6 - 4.8 = -5.4\,\text{kN m}^{-1}$
Waling C at $-14.0\,\text{m}$ $R_C = 4.8 + 22.1 = 26.9\,\text{kN m}^{-1}$

(v) Shearing forces and bending moments for removal of Stage II soil reactions at D

$$-S_{AB} = -S_{BA} = \frac{15}{10} = 1.5\,\text{kN}$$

$$S_{BC} = S_{CB} = \frac{15+68}{8} = 10.4\,\text{kN}$$

$$- S_{CE} = (64.3 \times 5.75 + 68 + 176 - 62)\frac{1}{10.75} = 51.3\,\text{kN}$$

$$S_{EC} = (64.3 \times 5 - 68 - 176 + 62)\frac{1}{10.75} = 13.0\,\text{kN}$$

To left of position D, $M'_{CE} = -51.3 \times 5 + 68 = -189\,\text{kN\,m}$
To right of position D, $M'_{CE} = -13.0 \times 5.75 + 62 = -13\,\text{kN\,m}$
Bending moments shown in (Fig. 7.11(d))

(vi) Loads on walings for removal of stage II soil reactions at D

Waling A at $+4.0\,\text{m}$ R_A $= 1.5\,\text{kN\,m}^{-1}$
Waling B at $-6.0\,\text{m}$ $R_B = -1.5 - 10.4 = -11.9\,\text{kN\,m}^{-1}$
Waling C at $-14.0\,\text{m}$ $R_C = 10.4 + 51.3 = 61.7\,\text{kN\,m}^{-1}$

(vii) Final bending moments and loads on walings for Stage III

The final bending moments and loads on walings Stage III (Fig. 7.11(f)) are obtained by adding (i) final moments and loads for Stage II (Fig. 7.9c) (ii) net soil moments and loads for Stage III (Fig. 7.11e) and (iii) moments and loads due to the removal of Stage II soil reactions at D (Fig. 7.11d).

Span AB at 4.39 m from A $M'_{AB} = \dfrac{35 \times 4.39}{10} = 15\,\text{kN\,m}$

Find the maximum bending moment in span BC (Fig. 7.14).
$S_{BC} = -28.3 + 4.3 + 10.4 = -16.45\,\text{kN}$
Distance of zero shear from $B = x + 2$
Equating shears in span BC

$$x^2 = \frac{16.45 \times 6 \times 2}{16.2}$$

$$x = 3.49$$

$$\text{Maximum } M'_{BC} = -16.45\left[(3.49 + 2) - \frac{3.49}{3}\right] + 35 = -36\,\text{kN\,m}.$$

$$\text{At 4.19 m from B, } M'_{BC} = -16.45 \times 4.19 + \frac{16.2}{6} \times \frac{2.19^3}{2 \times 3} + 35 = -29\,\text{kN\,m}.$$

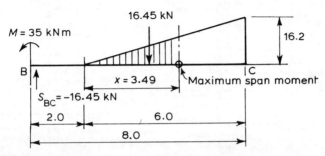

Figure 7.14 Determination of maximum bending moment in span BC.

Find the maximum bending moment in span CE (Fig. 7.15).

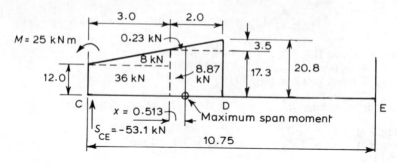

Figure 7.15 Determination of maximum bending moment in span CE.

$$S_{CE} = 20.3 - 22.1 - 51.3 = -53.1 \,\text{kN}$$
$$\text{Distance of zero shear from C} = x + 3$$

Equating shears in span CE:

$$\frac{3.5}{2} \times \frac{x^2}{2} + 17.3x - (53.1 + 36 + 8) = 0$$

$$0.875x^2 + 17.3x - 9.1 = 0$$
$$x = 0.513$$

Partial loads $0.875x^2 = 0.23 \,\text{kN}$ and $17.3x = 8.87 \,\text{kN}$

Maximum $M'_{CD} = -53.1 \times 3.513 + 36 \times 2.013 + 8 \times 1.513 + 8.37 \times 513/2$
$+ 0.23 \times 0.513/3 + 25 = -75 \,\text{kN m}$

At 3.22 m from C, $M'_{CD} = -53.1 \times 3.22 + 36 \times 1.72 + 8 \times 1.22 + 17.3 \times (0.22)^2/(2)$
$+ 0.875 \times (0.22)^3/3 + 25 = 74 \,\text{kN m}.$

(c) Stage IV

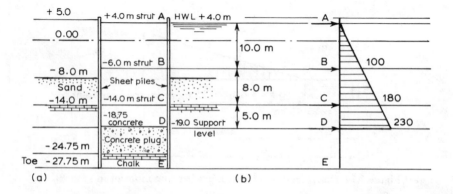

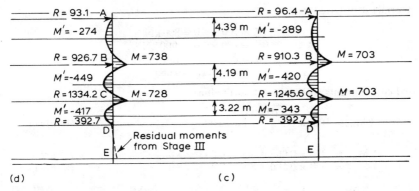

(d)   (c)

Figure 7.16 Stage IV. (a) Cofferdam; (b) external water pressure $(kN\,m^{-1})$; (c) Stage IV final moments and waling loads; (d) moments and loads due to external water pressure. Pile bending moments, M $(kN\,m$ per m run$)$. Loads on walings, R, $(kN$ per m run$)$.

(i) External water pressure

Unit weight of water $\gamma_w = 10\,kN\,m^{-3}$
High water level $+4.0\,m$
External water pressure

$$\begin{aligned}
\text{At} -6.0\,\text{m} \quad \sigma'_B &= 10 \times 10 = 100\,kN\,m^{-2}\\
\text{At} -14.0\,\text{m} \quad \sigma'_C &= 10 \times 18 = 180\,kN\,m^{-2}\\
\text{At} -19.0\,\text{m} \quad \sigma'_D &= 10 \times 23 = 230\,kN\,m^{-2}
\end{aligned}$$

(ii) Moment distribution for external water pressure

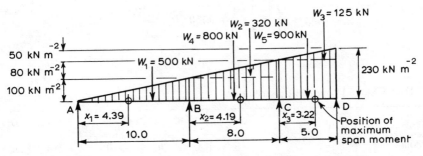

Figure 7.17 Sheet pile loading diagram. Total water pressure is $2645\,kN$ per m run.

Triangular load

$$\begin{aligned}
W_1 &= 100 \times \tfrac{10}{2} = 500\,kN\\
W_2 &= 80 \times \tfrac{8}{2} = 320\,kN\\
W_3 &= 50 \times \tfrac{5}{2} = 125\,kN\\
a &= 0 \qquad c = 1.0\\
c'_L &= 0.067 \quad c'_R = 0.100
\end{aligned}$$

Uniformly distributed load

$$W_4 = 100 \times 8.0 = 800\,\text{kN}$$
$$W_5 = 180 \times 5.0 = 900\,\text{kN}$$
$$a = 0 \qquad c = 1.0$$
$$c_L'' = 0.083 \quad c_R'' = 0.083$$

Fixed end moments with ends A and D pinned

$$FM_{BA} = 500\left(0.1 + \frac{0.067}{2}\right) \times 10.0 \qquad\qquad\qquad = 668\,\text{kN m}$$

$$FM_{BC} = (320 \times 0.067 + 800 \times 0.083) \times 8.0 \qquad = 703\,\text{kN m}$$
$$FM_{CB} = (320 \times 0.100 + 800 \times 0.083) \times 8.0 \qquad = 787\,\text{kN m}$$

$$FM_{CD} = \left[125\left(0.067 + \frac{0.1}{2}\right) + 800\left(0.083 + \frac{0.083}{2}\right)\right] \times 5.0 = 571\,\text{kN m}$$

Distribution factors
Joint B

$$K_{BA} = \frac{3I}{4L} = \frac{3}{4} \times \frac{1}{10} = 0.075 \quad D_{BA} = \frac{0.075}{0.2} = 0.38$$

$$K_{BC} = \frac{I}{L} = \frac{1}{8.0} = 0.125 \quad D_{BC} = \frac{0.125}{0.2} = 0.62$$
$$\underline{0.200}$$

Joint C

$$K_{CB} = \frac{I}{L} = \frac{1}{8.0} = 0.125 \quad D_{CB} = \frac{0.125}{0.275} = 0.46$$

$$K_{CD} = \frac{3I}{4L} = \frac{3}{4} \times \frac{1}{5} = 0.150 \quad D_{CD} = \frac{0.15}{0.275} = 0.54$$
$$\underline{0.275}$$

Table 7.6 Stage IV moment distribution

Joint	A	B		C		D
	AB	BA	BC	CB	CD	DC
Distribution factor D	—	0.38	0.62	0.46	0.54	—
Fixed end moments (kN m)	0	+ 668	− 703	+ 787	− 571	0
			− 50	− 99	− 117	
		+ 33	+ 52	+ 26		
			− 6	− 12	− 14	
		+ 2	+ 4	+ 2		
				− 1	− 1	
Final moments (kN m)	0	+ 703	− 703	+ 703	− 703	0

(iii) Shearing forces and bending moments for external water pressure

$$-S_{AB} = \frac{500}{3} - \frac{703}{10} \qquad = 96.4 \, \text{kN}$$

$$S_{BA} = 500 - 96.4 \qquad = 403.6 \, \text{kN}$$

$$-S_{BC} = \frac{800}{2} + \frac{320}{3} \qquad = 506.7 \, \text{kN}$$

$$S_{CB} = \frac{800}{2} + \frac{2}{3} \times 320 \qquad = 613.3 \, \text{kN}$$

$$-S_{CD} = \frac{900}{2} + \frac{125}{3} + \frac{703}{5} \qquad = 632.3 \, \text{kN}$$

$$S_{DC} = \frac{900}{2} + \frac{2}{3} \times 125 - \frac{703}{5} = 392.7 \, \text{kN}$$

Find the maximum bending moment in span AB.
Distance of zero shear from $A = x_1$.

Equating shears in span AB $x_1^2 = \dfrac{96.4 \times 10 \times 2}{100}$

$$x_1 = 4.39 \, \text{m}$$

Maximum $M'_{AB} = -96.4 \times \left(4.39 - \dfrac{4.39}{3}\right) = -289 \, \text{kN m}$

Find the maximum bending moment in span BC
Distance of zero shear from $B = x_2$
Equating shears in span BC:

$$\frac{80x_2}{8} \times \frac{x_2}{2} + 100x_2 - 506.7 = 0$$

$$5x_2^2 + 100x_2 - 506.7 = 0$$
$$x_2 = 4.19 \, \text{m}$$

Partial loads

$$5x_1^2 = 5 \times 4.19^2 \quad = 87.7 \, \text{kN}$$
$$100x_1 = 100 \times 4.19 = 419.0 \, \text{kN}$$

Maximum $M'_{BC} = -506.7 \times 4.19 + 87.7 \times \dfrac{4.19}{3} + 419 \times \dfrac{419}{2} + 703 = -420 \, \text{kN m}$

Find the maximum bending momentum span CD.
Distance of zero shear from $C = x_3$
Equating shears in span CD

$$\frac{50}{5}x_3 \times \frac{x_3}{2} + 180x_3 - 632.3 = 0$$

$$5x_3^2 + 180x_3 - 632.3 = 0$$
$$x_3 = 3.22 \, \text{m}$$

Partial loads

$$5x_3^2 = 5 \times 3.22^2 \quad = 52\,\text{kN}$$
$$180x_3 = 180 \times 3.22 = 580\,\text{kN}$$

Maximum $M_{CD}' = -632.3 \times 3.22 + 52 \times \dfrac{3.22}{3} + 580 \times \dfrac{3.22}{2} + 703 = -343\,\text{kN m}$

(iv) Loads on walings for external water pressure

Waling A at $+4.0\,\text{m}$ R_A $= 96.4\,\text{kN}$
Waling B at $-6.0\,\text{m}$ $R_B = 403.6 + 506.7 = 910.3\,\text{kN}$
Waling C at $-14.0\,\text{m}$ $R_C = 613.3 + 632.3 = 1245.6\,\text{kN}$

(v) Final bending moments and loads on walings for stage IV

The final bending moments and loads on walings for Stage IV (Fig. 7.16d) are obtained by adding the final moments and loads for stage III (Fig. 7.11f) to the moments and loads due to the external water pressure (Fig. 7.16c).

Sheet Pile Section

The worst bending moment occurring in the sheet piles during the various stages of construction is $738\,\text{kN m}$ at waling support B for Stage IV (Fig. 7.16d). The width of the walings will have the effect of reducing the peak support moment.

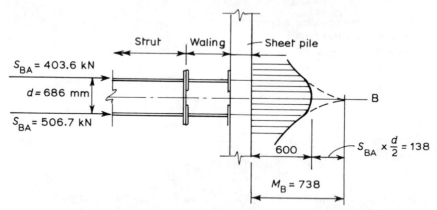

Figure 7.18 Waling support B – Stage IV bending moment and diagram (kNm m^{-1}).

At waling support B (Fig. 7.18) it is assumed that the peak support moment M_B can be reduced by an amount equal to $S_{BA} \times d/2$. Therefore,

$$\text{design moment} = 738 - 403.6 \times \dfrac{0.686}{2} = 600\,\text{kN mm}^{-1}$$

Use of Larssen steel sheet pile section No. $6 \times 122\,\mathrm{kg\,m^{-1}}$ (Fig. 7.19).

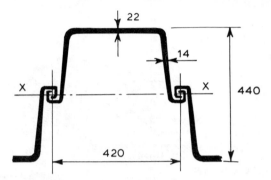

Figure 7.19 Larssen steel sheet pile section No. $6 \times 122\,\mathrm{kg\,m^{-1}}$.

Sectional area A $= 370\,\mathrm{cm^2\,m^{-1}}$
Section modulus $Z_{xx} = 4200\,\mathrm{cm^3\,m^{-1}}$

Maximum stress f_B $= \dfrac{600 \times 10^3}{4200} = 143\,\mathrm{N\,mm^{-2}}$

Permissible stress in steel sheet piling used for temporary works is $155\,\mathrm{N\,mm^{-2}}$ for steel Grade 43A (BS 4360).

7.3.5 Cofferdam of double row of sheeting with filling (Fig. 7.20)

The following notes are a summary of an exposition by Packshaw (1945). The inner line of sheeting is designed as an anchored retaining wall subjected to the stresses imposed by the filling and to the pressure of the water below the saturation line. The outer line of sheeting is designed as an anchor wall subjected to the pull of the tie rods and to the hydrostatic pressure due to the difference in head between the water in front of and behind the sheeting. The mass of earth A′DCE affords the necessary passive resistance. Cross-walls contribute very little resistance to thrusts. The saturation line is determined as indicated in the notes which follow on cellular cofferdams.

The wedge A′BE, with surface A′E assumed plane, is selected for the design of the inner sheeting. The width of the dam is generally sufficient to permit the formation of the complete wedge. If the sheeting is fixed at the toe in a rock trench, A′ is at A″. If the sheeting is freely supported in the soil, A′ is at A, and, if the shetting acts as a beam fixed in the soil, Packshaw suggested that A′A″ ≃ 0.75 AA″. It is convenient to ignore wall friction, thus leading to a slight increase in factor of safety, and the lateral earth stresses are then equal to the active Rankine values. The analysis of the forces on the sheeting is performed as

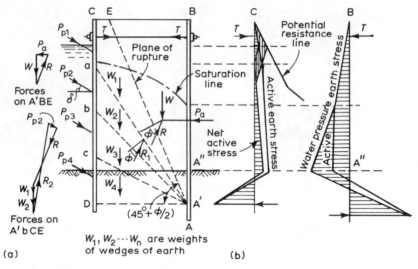

Figure 7.20 Analysis of cofferdam of double row of sheeting with filling.

indicated in Chapter 6 and the pressure and stress distributions which can be expected are indicated in Fig. 7.20

It can be shown that, for any point such as a or b on the outer sheeting, the least resistance to lateral forces large enough to evoke passive resistance is developed by failure approximately along the planes such as aA' or bA'. Area $A'DCE$ is divided into a number of wedges radiating from A'; only a few are shown in Fig. 7.20 for convenience. Wedge $A'aCE$ is in equilibrium under W_1, R acting as a surcharge, reaction R_1, and passive resistance E_{p1}. A polygon of forces is drawn to determine the magnitude of R_1 and E_{p1}. The construction is repeated to obtain the resistance due to each wedge. The potential resistance line is shown in Fig. 7.20 and is derived, for example, from E_{p2} divided by (ab) which is assumed to be the average resistance occurring at the mid-point of ab. As the depth below the top of the dam increases, the potential resistance increases rapidly, and generally only a portion is mobilized. The least outward stress on the sheeting is the active earth stress, taking into account the effect of buoyancy, and this may be partly, wholly, or more than balanced by the inward water pressure due to the difference in hydrostatic heads. If the design is based on the net diagram obtained from active stress and hydrostatic pressure, the value of T for the inner wall will not equal the value of T for the outer wall. The latter will most probably be less than the former, so that the excess tension draws the top of the outer sheeting towards the filling and some of the potential passive resistance is mobilized to effect a balance. For design purposes the extra pressure area corresponds roughly to the vertically hatched area. Several trial analyses may be required before the algebraic sums of forces and moments are reduced to zero. An alternative method of analysis for the stability of double-wall cofferdams has been described by Bishop (Packshaw, 1962).

7.3.6 Cellular cofferdams

A cellular cofferdam is constructed of a series of sheet pile cells, linked together as shown in Fig. 7.21 to enclose the excavation area, and filled with sand or gravel. Circular cells are connected by arcs or fillets and in segmental cells the diaphragms may be straight or curved. The cells are designed to be stable against sliding on the base, shear deformation created by tilting and failure in the foundation soil when not driven to rock. Furthermore, the tension in the interlock must not exceed the safe permissible value. Failure by bursting of the cells during or immediately after filling is generally due to defective driving and there are no records of failure by sliding. Experience suggests that the watertightness of cellular cofferdams is due primarily to tension in the interlocks. Since there is tension only above the level of the base of the filling, that part of the sheeting which penetrates soil is likely to be less watertight than the upper part. Particular attention should be given to this problem where the soil is silt or fine sand in order to prevent boiling or blowing. The following notes are a summary of a study by Terzaghi (1944); the original paper and discussion contain a wealth of information on cellular cofferdams. Further contributions to this problem have been presented by Descans (1954), Cummings (1960), Ovesen (1962) and Cushing and Moline (1975). Swatek (1967) has discussed practical aspects of design. Field and model studies have been reported by Schroeder *et al.* (1977), Maitland and Schroeder (1979), Schroeder and Maitland (1979), Sorota and Kinner (1981) and Sorota *et al.* (1981).

Experience indicates that the sheet pile walls of cellular cofferdams can withstand a horizontal deflection of 10% of the height without injury to the interlocks. The elastic properties of the filling have a marked influence on the

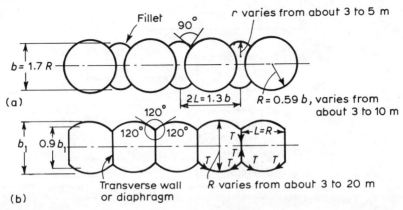

Figure 7.21 Arrangement of cellular cofferdams. (a) Circular cells; (b) segmental cells.

horizontal deflection of the sheeting. Tests show, for example, that considerably smaller deflections are obtained for the same lateral thrust when sand filling has a porosity of 40% than when the porosity is 44%. Alternatively, for equal deflections, a well-compacted filling enables the dam to resist a thrust of magnitude two or three times that resisted when the filling is less well compacted. Obviously the elastic properties of the soil depend to a large extent on the method of filling. In the analyses which follow, the filling is assumed to be sand unless otherwise stated.

Consider first the case of a cofferdam founded on rock (Fig. 7.22). The sliding resistance for well-drained cells is given to a close approximation by

$$F_b = 6(H\gamma - H_s\gamma_w)\tan\phi'. \tag{7.1}$$

The coefficient of friction between sand filling and smooth rock suface is commonly taken as 0.5 but the value is approximately equal to $\tan\phi'$ if the rock surface is rough. Calculations for stability are based on the effective width b (Fig. 7.22). If the rock surface is not very rough, efficient drainage is required for Equation 7.1 to be valid, and this can be provided by a filter layer. The efficiency of the drainage system should be checked on the site by observation pipes. The resistance is increased by penetration of the sheeting into the layer of decomposed rock which frequently occurs at the top of bedrock. When the possibility of

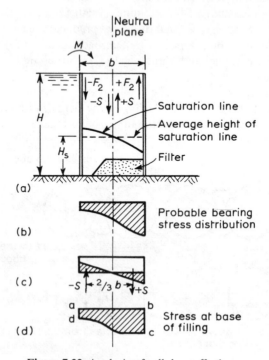

Figure 7.22 Analysis of cellular cofferdams.

overturning is being considered, the cellular cofferdam should not be treated as a gravity wall, since the bearing stress distribution does not conform to this assumption. Zero stress cannot exist at the heel of a dam filled with sand, even if arching does develop at the base, and the stress distribution is likely to be of the form indicated in Fig. 7.22(b). However, before failure by overturning occurs, it is probable that the cofferdam will fail in the manner discussed next.

Consider a unit length of dam. For simplicity it is assumed that the bearing stress distribution is linear (Fig. 7.22c) and shearing stresses between filling and sheeting acting before the overturning moment M is applied are disregarded. The shaded areas in Fig. 7.22(c) indicate the stresses on the base due to the overturning moment M. Then

$$M = \frac{2bS}{3} \quad \text{or} \quad S = \frac{3M}{2b}. \tag{7.2}$$

For stability, the shearing resistance on the neutral plane must be not less than S. Now the shearing resistance in the soil

$$S' = (\text{total thrust on neutral plane}) \times \tan \phi'$$
$$1 = \tfrac{1}{2} K \gamma H^2 \tan \phi' \tag{7.3}$$

where K is an empirical constant relating to lateral earth stress. The total tension in one interlock on the neutral plane is $\tfrac{1}{2} K \gamma H^2 R$ and the shearing resistance per unit length of dam in the interlock

$$S'' = \tfrac{1}{2} K \gamma H^2 (R/L) \mu \tag{7.4}$$

where μ is the coefficient of friction in the interlock, which averages about 0.3, and R and L are shown in Fig. 7.21. It is assumed that the arcs and transverse walls in a segmental cell intersect at $120°$, in which case the tension in the transverse walls is equal to that in the arcs, and Equation 7.4 is valid for both segmental and circular cells. The ratio R/L is close to, or equal to, unity and hence

$$(S' + S'') = \tfrac{1}{2} K \gamma H^2 (\tan \phi' + \mu) \tag{7.5}$$

and the factor of safety against failure by shear is

$$F = \frac{(S' + S'')}{S} = \frac{K \gamma b H^2}{3M} (\tan \phi' + \mu) \tag{7.6}$$

In Fig. 7.22, $M = \tfrac{1}{6} \gamma_w H^3$ and hence

$$F = \frac{2K \gamma_w b}{\gamma_w H} (\tan \phi' + \mu). \tag{7.7}$$

Substitution of common values in Equation 7.7 reveals that the factor of safety with respect to shear is considerably less than that with respect to overturning when the dam is treated as a wall. Since the average width of cofferdams with circular cells on rock is not greater than about $0.85H$ and since dams of these proportions have not failed, it is apparent that the value of K is greater than the Rankine value K_a.

Although in this discussion the value of K is assumed constant throughout the filling, observations of deflections of sheeting indicate that K varies in practice. The value in the middle of a fill of sand or gravel appears to be roughly 0.4 to 0.5. It will be observed that the interlock friction considerably increases the stability of the dam and it may be desirable to take advantage of this by increasing the interlock friction mechanically. An inner bank is a disadvantage with a cellular cofferdam since the tension in the sheeting is reduced and hence also the stability of the dam. Sill beams with dowels penetrating solid rock can be used as an alternative to banks to increase resistance to sliding.

Esrig (1970) has reviewed the problem of shear along the neutral plane and concludes from available evidence that failure in this manner is unlikely and that the coefficient K can be taken as unity.

For tension failure in the interlock, the maximum tension per metre run of interlock is

$$t = RK\gamma H \tag{7.8}$$

and for design it is commonly assumed that $K = K_a$, although the conclusion of Esrig noted above should be borne in mind. The value t must not exceed the safe permissible tension specified by the manufacturers and, for steel with ultimate tensile strength $500\,\text{N}\,\text{mm}^{-2}$, this is about $500\,\text{kN}\,\text{m}^{-1}$ for slightly arched sections and $1400\,\text{kN}\,\text{m}^{-1}$ for straight web sections. The strength of sheet pile interlocks has been discussed by Bower (1973) and Kay (1975). When it is assumed that $K = K_a$, the factor of safety is about 2.5 but, since in reality $K > K_a$, the factor of safety is probably less than two. Furthermore, when the dam is subjected to an overturning moment, the stress distribution at the base follows the pattern shown in Fig. 7.22 and the interlock tension at the toe is likely to be so high that the factor of safety becomes less than unity. Finally, if the cells are not almost completely drained, the tension in the interlock will be increased and the factor of safety reduced further.

Nevertheless, experience shows that, if the interlocks are strong enough to withstand the stress of the filling alone, they are strong enough to withstand the stress after the overturning moment is applied and there is no evidence that the factor of safety is appreciably less than two. Clearly, additional forces must be evoked to counteract the effects discussed above. Consider the inner row of sheeting. If it is assumed that no shearing stress exists between filling and sheeting, the vertical stress at a depth H is γH. Any increase in stress necessitates a slight reduction in H and this would be sufficient to mobilize the frictional resistance between filling and sheeting. Consequently, the vertical stress cannot be increased until the shearing stress becomes equal to the maximum frictional resistance. When this occurs, the vertical stress distribution at the base of the filling is abcd (Fig. 7.22d) and the vertical frictional force per metre width of sheeting at the contact surface of the inner sheeting is

$$F_1 = \tfrac{1}{2}K\gamma H^2 \tan \delta. \tag{7.9}$$

The total perimeter of sheeting in a length $2L$ of circular cells is approximately $2(2L + b)$ and the equivalent length of sheeting on one side of the centre-line is $(1 + b/2L)$ per unit length of cell. Hence the vertical frictional force per unit length of dam

$$F_2 = \tfrac{1}{2}K\gamma H^2(1 + b/2L)\tan \delta. \qquad (7.10)$$

Now $b/2L \simeq 0.77$ (Fig. 7.21) hence

$$F_2 = 1.77 \times \tfrac{1}{2}K\gamma H^2 \tan \delta. \qquad (7.11)$$

This vertical force must be opposed by $-F_2$ at the contact surface of the outer sheeting. If it is assumed that the lever arm of the couple constituted by F_2 and $-F_2$ is roughly $\tfrac{2}{3}b$, then the overturning moment required to establish the stress distribution abcd is

$$M_1 = \tfrac{2}{3}bF_2 = 1.18b \times \tfrac{1}{2}K\gamma H^2 \tan \delta. \qquad (7.12)$$

The moment due to the water pressure in $M = (1/6)\gamma_w H^3$ and hence the ratio

$$m = \frac{M_1}{M} = \frac{3.54Kb\gamma \tan \delta}{\gamma_w H}. \qquad (7.13)$$

Substitution of common values in Equation 7.13 gives $m = 0.85$. Hence 85% of the applied moment M is carried by the couple created by F_2 and $-F_2$, and the bearing stress does not begin to increase until about 85% of the overturning moment is applied. The increase in stress due to the remaining 15% is well within the capacity of the interlocks, but the danger of failure by bursting increases rapidly when $M > 1.2$ to $1.3\,M_1$. The outward deflection, and hence also the maximum interlock tension, at the base of the inner sheeting is reduced by the frictional force between the toe of the sheeting and the rock surface, induced by the downward pull F_1. Figure 7.23 shows a section of a fictitious lock. If T is the interlock tension per linear metre, the sum of the forces on the two contact lines is $T/\cos \alpha$ and the interlock friction is $f = \mu' T/\cos \alpha$ per linear metre, where μ' is the

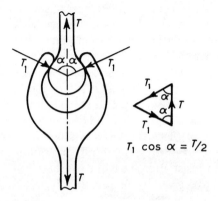

Figure 7.23 Forces in sheet pile interlocks.

coefficient of friction between steel and steel. When $\alpha = 0$, f is a minimum. Although it is almost a line contact, experiments show that μ' is the same as for surface contact. For dry steel $\mu' = 0.15$ to 0.20 and for wet steel $\mu' = 0.30$; a trace of moisture is sufficient to produce an antilubricating effect. It is apparent, therefore, that a conservative value for f is $0.3 T$; that is, assuming $\alpha = 0$, which is never the case.

The above discussion discloses some of the fallacies in the design of cellular cofferdams and the potential sources of danger but, owing to lack of precise information on some of the factors involved, no rigorous design procedure can be formulated. Nevertheless, certain reasonable safe rules can be derived. The width $b = 0.85H$ is an acceptable basis for design but the factor of safety with respect to shear along the neutral plane is certainly not greater than 1.5: $b = 0.8H$ to $1.0H$ is recommended in CP 20004 (BSI 1972d). The maximum interlock tension can be taken as $t = 0.4\gamma\ RH$ for sand, or sand and gravel, filling, although it will be realized from earlier discussion that this may underestimate the magnitude of tension. It is desirable to observe deflection of the crest of the dam during and after pumping out the cofferdam, as a safety precaution. Terzaghi (1944) points out that, in design 'Every scrap of available information should be utilized; and every known theoretical approach, including the stability computation based on the gravity concept, should be made before final decisions are reached'.

The cases of cellular cofferdams founded on a thick bed of sand or clay are less common than on rock and will be dealt with very briefly. The general approach to the analysis of problems is similar to that for a rock foundation. Three conditions must be satisfied in the case of a cofferdam on sand: (a) the sand at the outer sheeting must be protected against erosion, (b) the dam must be capable of withstanding the overturning moment, (c) the subsoil must be able to resist the stress under the toe, taking into account the tendency of the sand to liquefy under seepage forces. The sheeting must be driven into the sand to a depth sufficient to prevent the inner sheeting sinking under the force F_1. This depth is at least $\frac{2}{3}H$. If the inner sheeting sinks, the cofferdam is likely to fail by bursting of the interlocks. Resistance of the sand to F_1 is reduced by boiling and whereas, for example, a single sheet pile may sustain a load of about $800\ \text{kN}$ when driven into compact sand, lowering the water level in the cofferdam by, say, $3\ \text{m}$ may induce seepage leading to the piles becoming free. Protection against boiling can be effected with either a loaded inverted filter or a plain sand bank with a broad base.

The primary consideration for a cofferdam on clay is that the clay must be strong enough to sustain the weight of the cofferdam and danger of a foundation failure can be disregarded only if the clay is stiff or if the sheeting is driven to a lower stratum of sand or stiff clay. The stability of the dam depends to a considerable extent on the compressibility of the soil between the base of the fill in the cells and the level of the lower edge of the sheet piles. In the case of a foundation of either sand or clay, depth of penetration must be sufficient to enable the sheeting to resist a force equal to at least $1.5F_1$.

Consideration must be given to the permeability of the dam before stability

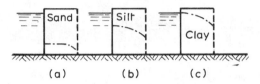

Figure 7.24 Seepage in cellular cofferdams.

calculations are made. In cells provided with weep holes in the inner row of sheeting and filled with clean sand and gravel, the loss of head across the dam is almost exclusively caused by the outer row of sheet piles. On the other hand, if the filling is clay, the loss of head caused by the outer row is negligible compared with the loss of head through the clay. The position of the seepage line for different filling material, on a rock foundation, is indicated in Fig. 7.24. The condition depicted in Fig. 7.24(a) is rarely encountered in practice, since the silt content of the filling is usually sufficient to create a condition approaching that depicted in Fig. 7.24(b). The seepage across a dam on rock affects only the tension in the interlocks. When it is founded on soil, the effect of seepage at and under the base tends to raise the positions of the seepage line-shown in Fig. 7.24 and, in silt and fine sand, boiling may occur at the toe of the inner row of sheeting.

A description of shipways constructed with cellular walls on a marl foundation has been given by FitzHugh *et al.* (1947). Jennings (1953) has described the use of cellular cofferdams, each consisting of ten circular cells with connecting arcs, for the construction of piers for a bridge across the River Ohio at Wheeling, West Virginia.

A field study of the behaviour of a bulkhead for a pier at Long Beach Harbor, California, has been reported by White *et al.* (1961). The bulkhead comprises 19 m diameter circular cells with connecting arcs and retains 17 m dredged silty sand fill. The cells are founded mainly on sand and are filled with dredged sand. The coefficient of lateral earth stress K in the cells was found to be 0.66 during filling, 0.54 shortly after the cell was filled and 0.53 two years later after application of a 4 m surcharge over cells and backing. The maximum hoop tension generally occurred near the original ground level, although the absolute maximum observed was 1.24×10^3 kN m^{-1} about 3.5 m above that level. Excess pore pressure should be included in the calculation of hoop tension. Ten days elapsed before the filling in the cells attained 90% consolidation and the maximum pore pressure recorded was of the order 3.5 kN m^{-2}.

7.4 SHEET PILES

7.4.1 Driving of sheet piles

Steel sheet piling is rolled in a variety of sections, each designed to be the most suitable for the particular purpose required. The manufacturers' handbooks give

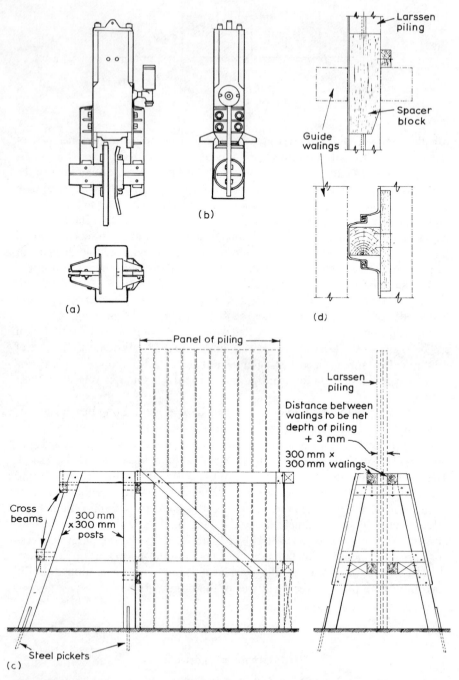

Figure 7.25 Driving sheet piling. (a) Leg guides for hammer driving single pile; (b) driving cap for single pile; (c) guide trestle; (d) spacer block.

weights, sections, moments of inertia and other relevant details. Corner sections and junctions can be supplied by the makers, fabricated from pile and structural steel sections as required. Figure 7.25 shows details of some equipment and techniques employed in driving steel sheet piling.

Automatic and single-acting steam hammers are seldom used these days due to the expense of boilers and coal. Automatic double-acting hammers are for use with high pressure compressed air. These are usually employed for driving light piles.

Self-contained single and double acting diesel hammers are now almost universally accepted as they have proved merit and efficiency. They are made in many sizes, as can be seen in Table 7.7. The heavier hammers are used for driving the largest pile sections in hard ground. Hydraulic powered vibrating pile drivers are also to be recommended in granular soils.

It is advantageous to drive steel sheet piles in pairs since they are easier to guide than when driven singly. The practice saves time and is more economical. For permanent work Frodingham type piles may be preferred but, in the case of temporary works, the extraction of piles has to be considered. If an attempt is made to pull the Frodingham piles in pairs, they tend to become declutched during pulling, whereas Larssen piles can be pulled singly.

In driving a row of Frodingham piles, the tongue should always be leading to avoid clogging the clutch with compressed soil, thereby increasing interlock

Table 7.7 Diesel hammers in common use

Hammer		Energy per blow (kgm)
Delmag	D 5	1250
	D 8/22	1305–2400
	D 12	3125
	D 16	2610–5440
	D 22	5500
	D 30	3300–7500
	D 36	4200–10200
	D 46	7300–14600
	D 62	11160–22320
BSP	DE 20	1658
	DE 30C	3730
	DE 40	3317
	DE 50C	6220
	B 15	3790
	B 25	6320
	B 35	8850
	B 45	11400
Kobe	K 13	3703
	K 25	7500
	K 35	10503
	K 45	13506

resistance. Greasing the interlock before driving helps to reduce friction and tends to make the seal more watertight.

There may be a tendency for piles to lean in the direction of driving when each pile is driven consecutively to full penetration. This occurs particularly when the toe is driven into hard ground and the pan of the pile deflects inwards, 'hunching its shoulders'. Another cause may be friction in the interlock. This develops on one side of the pile only, creating an unbalanced moment which leads to tilting. To avoid this, work is commenced by driving a pair of piles to part penetration and as near to vertical as possible. Between ten and twenty pairs of piles are then interlocked and pitched ready for driving. The last pair of piles in the panel are driven to part penetration and the remaining piles driven to finished level. If the piles are long, this should be done in stages. The process is repeated with a further panel of piles, the half-driven last pair of the first panel forming the first pair of the second panel.

Care should be taken to prevent the piles in the middle of the panel becoming ironbound. Tirfor winches and straining wires can also be used to restrain piles during driving, as trouble taken in preventing the piles getting out of line in the initial stages can prevent the use of taper piles in the last resort.

In soft ground or where high frictional resistance develops in the interlock, a driven pile may be drawn down as the next one is being driven. When this occurs it is preferable to weld an additional piece to the drawn down pile rather than attempt to jack it back in position. The problem can be overcome by bolting piles to the permanent waling when they have been driven. When it is necessary to drive the head of a pile below water level, a long dolly or follower, fabricated from a short length of pile and provided with leg guides, must be employed between the hammer and pile. Alternatively some steam or compressed-air hammers will work under water up to a maximum depth of 20 m, provided the exhaust is taken above water level.

Should a pile meet an obstruction and refuse to penetrate further, the next pile should be pitched and driven. This may break up the obstruction and, in any case, affords greater support to the obstructed pile if harder driving is necessary.

7.4.2 Water jetting

Water jets can be used to minimize resistance to driving at the toe of the pile and may also help to reduce friction along the sides of the pile. Jetting affords the greatest advantages when driving in sands, gravels and fine grained silts, provided there is little clay: little or no benefit is derived from jetting in clays. Jetting should not be used where it is likely to impair either the bearing capacity of piles which have already been driven or the stability of the ground and nearby buildings. It is best used when pile driving is over water where the disposal of water as it emerges from the ground is no problem.

Jetting equipment comprises 25 mm diameter nozzles fitted to 75 mm tubes coupled to a flexible hose pipe. As the pile is driven two pipes, one on each side of the pan, are pushed into the ground so that the jets follow the toe of the pile. The consumption of water is about 3×10^4 litre h^{-1} at a pressure of 0.5 to 1.0 N mm^{-2}. Jetting should be stopped about 1 m from final toe level and the piles driven to their final set after the jetting pipes have been removed.

7.4.3 Cofferdam closure

The closure of a cofferdam is best effected by driving a final panel which includes a corner pile, thereby affording greater latitude in adjustment. On land, and in water where the current velocity is low, all the piles in a small cofferdam can be pitched before driving is commenced. It is often convenient to erect the top frame of the cofferdam bracing before pitching piles, using it as an inner guide waling.

It is important to allow clearance between the theoretical internal dimensions of the cofferdam and the line upon which the piles are actually pitched. In a deep cofferdam necessitating four or five frames, where obstructions are likely to deflect the piles off line, as much as 150 mm clearance should be given in setting out the piles on the ground. It is most embarrassing to find that, when excavation reaches the bottom of the cofferdam, there is insufficient room left in which to build the permanent structure; on account of an inadequate allowance for clearance when setting out. The space between the waling and the pan of the pile can be filled by dry concrete packed in bags, by hardwood folding wedges or, if in water, by the deflection of the piles as pressure builds up during dewatering.

In the sea tideway or rivers, all cofferdams must be provided with valves or sluices on the upstream and downstream sides. These must be left open to balance the water pressure inside and outside the piles during closing and until bracing is complete. They should be closed at the lowest water level, before pumping out commences, and opened for flooding when construction is complete. In an emergency when there may be danger of overtopping due to floods or abnormal tides, work must be made safe and the cofferdam deliberately flooded. The valves or sluices should be provided with long spindles connected to hand wheels operated from a properly constructed and guarded staging.

7.4.4 Circular cofferdams

Certain types of piles must comprise an even number in order to secure closure. They may be driven to a circular pattern from a minimum of 4.7 m diameter solely by taking up the movement in the interlocks and without initial bending. Some increase in interlock friction is inevitable, so care must be taken in pitching the complete circle before driving. In order that the individual piles remain vertical and do not become ironbound, driving should be carried out in stages alternately clockwise and anti-clockwise around the circle.

7.4.5 Extraction of piles

If concrete is to be placed against piles which are later to be withdrawn, it is necessary to ensure that no bond takes place between the two materials. This can

Table 7.8 Output of BSP Zenith extractors

Size	Energy (kN m)
No. 20	0.84
No. 80	3.50
No. 120	5.60
HD 7	0.77
HD 10	1.10
HD 15	16.60

be achieved by placing sheets of hardboard against the face of the piles, or alternatively two sheets of waterproof paper can be used if it is required for the concrete to fill the pans of the piles.

Extraction can be carried out by direct pull from a heavy-lift crane, but more usually in combination with an extracting hammer or vibrator. If the piles are ironbound and do not pull readily, then after a few minutes on one pile which does not move, the extracting hammer should be attached to others in turn until a pile is found to move. This usually breaks the bond between other piles and allows them to be extracted more easily. Most extracting hammers are purpose-made and are powered by compressed air: details of BSP Zenith extractors are given in Table 7.8. Vibrating extractors are the same machines as those used for driving. The cable from the pulling crane should be plumb over the pile and be taut before extraction starts. The pull on the cable should exceed the weight of the pile and extractor by between 10 and 20 kN but it is most important that the minimum and maximum recommended crane pulls are complied with when operating extractors. The HD 15 is designed to meet the demand for a powerful extractor capable of pulling sheet piles in the most difficult conditions, when other types have proved to be ineffective. It requires a crane capable of exerting a steady pull of between 15 and 18 tonnes and a boiler pressure of $74\,\text{kN}\,\text{m}^{-2}$.

The extraction of piles depends upon the length of time the piles have been in the ground, the size of hammer used to drive them, the soil conditions and, most important, whether they have been buckled or split in overcoming obstructions during driving. In such cases it may not be possible to extract them and the piles have to be abandoned or the accessible parts are cut off and removed.

7.4.6 Cutting sheet steel piles

Sheet steel piles are best cut by oxy-acetylene equipment in free air. When they have to be cut under water, oxy-hydrogen cutting gear should be used. Divers using oxy-hydrogen gear can cut piles at depths of 40 m, as hydrogen can be supplied at the high pressure required at such depths. For underwater cutting, in addition to the oxygen and the hydrogen jet, an enveloping annular nozzle, passing compressed air or oxygen, forms part of the blowpipe, providing a water-free zone in which combusion takes place. An electric igniter is used to light the torch underwater. Before cutting piles underwater, the section should be attached to a crane hook and a slight pull exerted. The cable should be plumb over the piles to ensure that when the cut is almost complete, the pile does not sway or fall and trap the diver. Cutting and drilling of steelwork underwater can also be carried out by diver using saws and drills powered by compressed air. Steel can be welded underwater electrically.

7.4.7 Silent pile driving

With the upsurge of public demand for civil engineering works to have less impact on the environment, there has been an increasing demand for 'silent' pile driving in urban areas. Three main methods so far have been used with some success: these involve hydraulic pile drivers, vibration equipment and a drop hammer working in an insulated box. The hydraulic pile driver consists of a steel cross-head which houses eight double-acting hydraulic rams connected to a hydraulic power pack, each capable of exerting a downward thrust of 200 tonnes or an upwards pull of 150 tonnes. Piles are driven in groups of eight; two rams are pressurized to thrust downwards on to the piles to which they are connected, whilst all other rams are hydraulically locked. The reaction against the driving load is taken by the resistance to uplift of the other piles held by the machine, the reaction being transferred through the locked rams and the cross-head. The initial force available to drive the first two piles is the weight of the machine plus the weight of all the piles connected to it. After the first two piles have been driven, the rams adjacent to them are pressurized and the driving force now available includes the skin friction and adhesion on the piles just driven. As the piles are driven into the ground, skin resistance increases, giving additional reaction for the piles to be driven further. By reversing the procedure, piles can be extracted in a similar manner. This system has been developed by Taylor Woodrow Construction Limited and overcomes the problem of noise and vibration. Piles can therefore be driven safely close to occupied buildings without disturbance to the occupants. This type of machine does not work very well in granular material but gives excellent results in clays.

It has been found that by the use of high frequency vibrations transmitted to the pile head by a motive force, a pile can be shaken into the ground with considerable success, provided the soil is sand or gravel with standard penetration test N values below 50. The machine is much quieter than a normal pile hammer as the motor is usually electrically or hydraulically driven with power supplied from the mains or by an independent generator situated away from the piling site. The motor drives two out-of-balance weights, giving a moment of eccentricity of 40 000 kg mm at 1100 vibrations per minute, which is transmitted to the pile head through a robust hydraulically operated clamping system. The vibration method can be used also for extracting piles.

Other methods have been devised to silence pile driving operations, the most successful is the 'hush' system which consists of a drop hammer working in guides surrounded by a 'box' of sound-proofing material. The front of the box is opened, the pile pitched inside, and when ready the box is closed and piling proceeds. The head of the pile is enclosed in the box during driving. A similarly successful alternative is the Dawson quiet driver – a drop hammer with noise eliminated at source by inserting a plastic cushion between the ram and impact block.

7.5 OPEN AND PNEUMATIC CAISSONS

Caissons are constructed of steel or reinforced concrete. The latter has the advantage of considerable weight to assist sinking, although a steel shell can be filled with concrete as sinking proceeds. The former has the advantage of less weight for flotation and of being flexible and resistant to torsional stresses. Steel caissons are fabricated most conveniently in shipbuilding yards. Caisson size is usually governed by the required foundation bearing area and the plan area of the bearing under the superstructure. A margin of about 0.5 m is sometimes allowed between the plan dimensions of the caisson and the base of the pier proper in order to allow for adjustment if the caisson is located out of position. The structure should be designed to achieve continuous and steady sinking; thereby permitting good control of the process and eliminating repeated additional effort to overcome static resistance. This is most likely to be achieved in a caisson in which the sides are not tapered and when the processes of fabrication and sinking advance at the same rate. It is desirable that the fabrication cycle should be completed in a multiple of continuous 8 h shifts. Experience indicates that a convenient height for a caisson strake is 1.25 m. Vertical sides assist also in preventing blows in pneumatic caissons by ensuring that the surrounding ground is in contact with the caisson skin.

The interior of rectangular open caissons is generally divided into cells 5 to 6 m square in plan or alternatively 3 to 5 m diameter. Structural anaylsis of the external walls under water and earth loading can be made by the moment distribution method, taking into account the corner moments and also direct thrust. On completion of construction, the upper portions of the external and internal walls usually carry the whole of the load from the superstructure although, at the base, the greater part of the load is transmitted to the concrete filling and seal. With square cells the bending moment in the internal walls due to earth and water loading is theoretically zero although, in practice, non-uniform earth stresses lead to bending moments which should be catered for by consideration of arbitrary conditions. Torsional stresses can be determined for a rectangular caisson by assuming it to be supported on two diagonally opposite corners only. In designing the steel curbs of monoliths for the New Howrah Bridge, Calcutta (Ward and Bateson, 1947) bending stresses were determined assuming a maximum variation of 2:1 between the cutting-edge resistance at the ends and the centre, and vice versa. The possibility of the weight of a caisson being entirely supported by skin resistance at the top of the caisson, leading to vertical tension in the walls should be considered. The analysis of cellular structures containing web openings has been studied by Evans and Shanmugam (1979).

Progress in sinking should be observed daily and records maintained of caisson plan position, elevation of cutting edge, inclination and amount of rotation. There is often a tendency for caissons located near the banks to drift towards the centre of a river. The caisson position should be corrected when it reaches 50 to 100 mm off centre. On the Hawkesbury River Railway Bridge, New South Wales (Fewtrell, 1949) the positions of the centres of five of the cassions, along the two major axes, at heights of 52 to 55 m above the cutting edge were in error by the

Table 7.9 Skin resistance in sinking cylinders

Material	$k\,N\,m^{-2}$ of surface area
Soft clay	15
Sand and gravel	20 to 35
Sand and clay	40
Stiff clay	40 to 50
Fine silt and sand	50 to 60

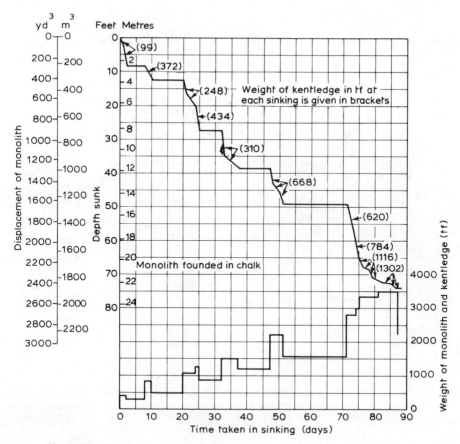

Figure 7.26 Typical record of the sinking of a monolith, at Tilbury Dock.

following amounts: 206 mm, 237 mm; 394 mm, negligible; 219 mm, 365 mm; 165 mm, negligible; 1143 mm, 360 mm. The maximum rotation was about 1°. On the New Howrah Bridge (Howorth and Smith, 1947) the monoliths for two main piers and the two anchorages were sunk to depths of 24 to 30 m below ground level and the positions in two directions were within 50 to 80 mm of the true positions. Position and inclination can be corrected by attaching cables to the shore or to dolphins and applying tension as sinking proceeds, or by dredging more on one side of the caisson than on the other, or by struts, or sliding wedge

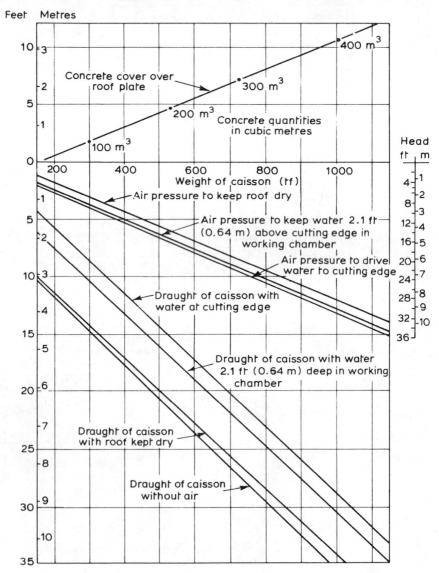

Figure 7.27 Flotation chart for caisson No. 7, Kafr el Zayat railway bridge, Egypt.

plates fixed at the cutting edge by divers, or a combination of these methods.

A sinking-effort diagram for bridge foundations sunk in sand was produced in the correspondence on papers concerning the Lower Zambezi Bridge, Africa (Howorth, 1937), and was based on observations on open wells of brick, stone and concrete for several bridges at depths below ground level from 18 to 52 m. The curve – which fits the observed values very well – is not quite linear; values taken from it are given in Table 7.9. Sinking-effort is calculated from the submerged weight and external surface area of the well and is not the same as skin resistance because of the resistance under the cutting edge – this was estimated to be equivalent to $4.8\,\text{kN}\,\text{m}^{-2}$ on the Zambezi Bridge (see Table 7.10, p. 638). Records of sinking a monolith and a caisson are shown in Figs. 7.26 and 7.27 respectively.

7.6 CONSTRUCTION OF CYLINDERS, ETC.

7.6.1 Cylinders on land

The sinking of a cylinder on land is comparatively simple compared with one in deep water. The method of sinking is governed by the nature of the soil and the depth to which the cylinder has to be sunk. The simplest case is that of sinking through soft alluvium and founding on a stratum of high bearing capacity material, as for bridge piers. For this method steel strakes are set up, with a cutting edge on the bottom of the first unit. The material is then grabbed out inside and the cylinder sinks the first few feet under its own weight. To ensure verticality, a guide frame is necessary until the cylinder has become self-

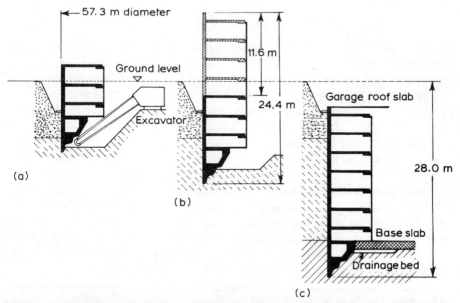

Figure 7.28 Stages in sinking a land caisson under its own weight at Geneva. (a) Concrete placed for cutting edge and three storeys of basement; (b) concreting, excavation and sinking proceed simultaneously; (c) concrete base slab placed.

supporting in the ground. Excavation continues with additional sections being bolted on to those previously sunk until the required depth is reached. However, as sinking proceeds, the skin resistance builds up and has to be overcome by putting a kentledge frame on the top section and loading it with kentledge blocks until the cylinder begins to sink again under the additional weight. A rough guide to the magnitude of skin resistance in the sinking of cylinders is given in Table 7.9.

Skin resistance can be reduced by making the diameter of the cutting edge about 100 mm greater than the rest of the shaft, thus forming a space into which bentonite slurry can be injected. This provides lubrication and enables the cylinder to be sunk to a considerable depth under its own weight without the application of kentledge.

Stages in sinking a large cylinder or caisson on land are shown in Fig. 7.28.

7.6.2 Cylinder sinking by well sinking methods

This method is used where the diameter of the cylinder is large. Segmental rings of steel, cast iron or reinforced concrete are employed for this type of construction and initially are built in a small cofferdam. Excavation is continued inside the cylinder by miners or small hydraulic excavators, additional rings being fixed underneath those already in position as excavation proceeds. This method is not difficult until ground water is encountered. It has the advantage over sinking by kentledge described above in that boulders or other obstructions can be removed by the miners when encountered. When water is reached, the provision of a compressed air deck enables further work to be carried out under compressed air.

7.6.3 Cylinder sinking over water

This is often necessary when constructing jetties or cooling water intake shafts over water. Before construction can commence it is necessary to build a temporary jetty strong enough to carry the cylinder until it becomes supported in the ground, together with a kentledge frame and all excavating machinery.

When a suitable jetty has been constructed, it is necessary to erect lowering gear, comprising a framework and either screw or hydraulic jacks, which will support the cutting edge and first few strakes of the shaft. As the weight of the sections being built up is taken by the jacks, the shaft is gradually lowered. When the full travel of the jack has been reached the load is transferred to the framework whilst the full travel of the jack is re-established. The procedure is repeated until the cutting edge is embedded in the ground and becomes capable of taking the weight of the cylinder sections above it. The lowering gear can then be disconnected and sinking continued. As with other methods, provision can be made for the installation of compressed air working should it be necessary when the cylinder is well sealed into the sea bed.

Table 7.11 gives values of skin resistance based on tests by Potyondy (1961) who states that, where the clay content of a sand is less than 15% the adhesion component c_w can be ignored and that, where the sand content of a cohesive soil does not exceed 5 to 8%, the friction component tan δ can be ignored. Potyondy considers that four factors determine skin resistance: moisture content of the soil, roughness of the surface, composition of the soil and intensity of the normal load. However, the conditions of test (undrained, consolidated-undrained or drained) influence the results to some extent and, in respect of these tests, Bishop (1962) points out that there is little doubt drained conditions obtained in the case of the sand and the rock flour, that undrained conditions probably obtained in the case of the clay and that the sand–clay mixes either were not fully saturated or were drained to varying extents.

The layout of pneumatic caisson is shown in Fig. 7.29. The height of the working chamber should not be less than 2.5 m. The depth of a caisson shoe is commonly about 4.5 m. Fig. 7.30 shows typical reinforced concrete caisson shoes (Wilson and Sully, 1949). When constructed in steel, the general outline and the cutting edge are the same but the skin is stiffened by vertical angles, say, 150 mm × 75 mm × 5 mm for the full height of the shoe, and by bracing between inner and outer skins connected to the trusses in the working chamber roof. The working chamber roof should be designed to resist upthrust due to air pressure or bearing stress when the caisson is completed, and downthrust due to the weight of concrete. The design of the walls and roof of the working chamber in a reinforced concrete caisson is analogous to that of a circular or rectangular tank, as appropriate with external soil and water loading varying with depth and of trapezoidal distribution, together with uniformly distributed internal air pressure varying from zero to working pressure. The practice in rectangular tank design is to assume that the wall functions as a vertical cantilever if its unsupported length exceeds about 1.5 times its height but with some fixity at the junctions with internal and end transverse walls. If the unsupported length is less than this, it is assumed that some, if not all, of the load is carried by the wall functioning as a slab spanning longitudinally, with some fixity at the junctions with transverse walls and with the floor. The working chamber roof is designed as a slab supported on four sides with some fixity at the edges. Direct compression or tension in walls and roof must also be taken into account.

The air lock should be located above water-level, as a safety precaution. Man locks usually hold four men. An air shaft is generally oval or figure-of-eight in section and is divided by a ladder into two shafts, each roughly 1 to 1.5 m diameter; one for men and one for buckets. The shaft is constructed in 1.5 or 3 m lengths with rubber gaskets at the joints. The shaft plate thickness should be not less than 10 mm. A sealing plate is provided at the shaft bottom to retain air pressure in the working chamber when the air lock is removed for extending the shaft. The shaft is designed to cater for circumferential and longitudinal tensile stresses due to air pressure and longitudinal compressive stresses due to the weight of the airlock. The hoist rope passes through a stuffing box in the muck-lock lid and the bucket must be unhooked in the lock before hoisting into the open air. Wilson and Sully (1949) state that one air lock should be provided for every 100 to 110 m² of excavation area and that experience indicates that the optimum labour force at one shaft is ten men. Wilson and Sully (1952) give details of the design, construction and sinking of the 50 m × 34 m × 24 m deep caisson

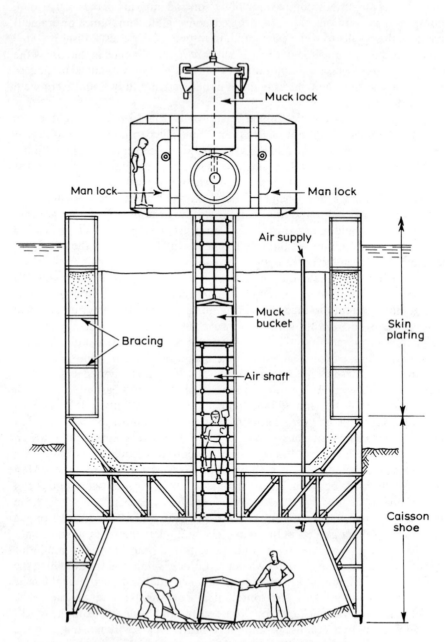

Muck lock

Man lock

Man lock

Air supply

Muck
bucket

Bracing

Skin
plating

Air shaft

Caisson
shoe

Figure 7.29 General arrangement of a pneumatic caisson.

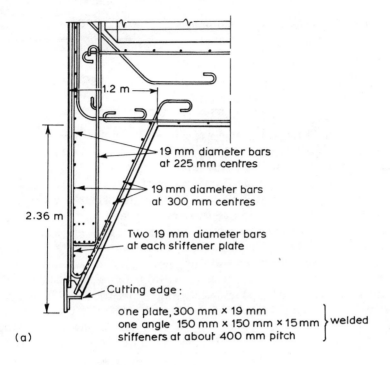

1.2 m

19 mm diameter bars
at 225 mm centres

19 mm diameter bars
at 300 mm centres

Two 19 mm diameter bars
at each stiffener plate

2.36 m

Cutting edge:

one plate, 300 mm × 19 mm
one angle 150 mm × 150 mm × 15 mm } welded
stiffeners at about 400 mm pitch

(a)

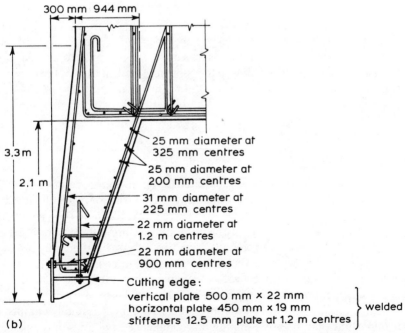

300 mm 944 mm

25 mm diameter at
325 mm centres

25 mm diameter at
200 mm centres

31 mm diameter at
225 mm centres

22 mm diameter at
1.2 m centres

22 mm diameter at
900 mm centres

3.3 m

2.1 m

Cutting edge:

vertical plate 500 mm × 22 mm
horizontal plate 450 mm × 19 mm } welded
stiffeners 12.5 mm plate at 1.2 m centres

(b)

Figure 7.30 Caisson shoes. (a) Cutting edge for pump-house caisson sunk through sand to rock; (b) cutting edge for a caisson forming part of a dock wall, designed for use in boulder clay.

Table 7.10 Sinking effort: Lower Zambezi Bridge

Depth of cutting edge below ground (m)	0	9	18	27	36	45	54
Average sinking effort $(kN\,m^{-2})$	0	4.8	10.8	17.5	24.2	31.6	39.7

for the foundation to the circulating-water pump-house for the Uskmouth, Wales, generating station, sunk principally under compressed air. Smith (1954) describes the welded construction of the shoe for this caisson.

Where the stability of a caisson is endangered during the initial stages of sinking by, for example, a considerable depth of soft material in the river bed, the sand-island method of sinking has advantages. The sand-island is normally contained in a large diameter steel cylinder and the caisson is erected on, and sunk through, the sand. The method is relatively expensive, but facilitates sinking in some situations, avoids the necessity of towing caissons and provides a working platform. The container can be dispensed with only in shallow, still water. Clay-islands protected by sheet pile cutwaters have been used in Malaya for bridge piers constructed by well-sinking through soft material to rock.

The sand-island method was employed on the New Orleans Bridge across the Mississippi River (Anon, 1937). The depth of water was from 12 to 24 m. The river bed consisted of unconsolidated silt and was followed by sand, clay and silt. The piers were founded at a maximum depth of 52 m below mean water-level. The sand-islands were confined within steel shells in deep water and within cofferdams in shallow water. The largest shell was 37 m diameter and 30 m high. The maximum ebb flow velocity was at least $2.7\,m\,s^{-1}$, and, to prevent scour, 7.6 m × up to 137 m woven willow mattresses were sunk at each of the island sites and held in place by riprap.

7.6.4 Sealing of caissons

(a) Open caissons

When the caisson is sunk to the required level, all loose material should be removed from the excavation but generally the grab will not reach all the corners of the cells and these have to be cleaned out by hand. If the caisson is in waterbearing ground then the problem becomes more difficult. High pressure water jets may be used but generally it will be necessary to employ a diver to remove any soil adhering to the side of the caisson cells. The bottom of the caisson should then be sealed with a concrete plug, preferably placed in one continuous operation to avoid laminations. The thickness of concrete should be sufficient to resist any upward water pressure when the cell is finally pumped out. It is important that, before pumping out commences, the caisson is sufficiently heavily loaded to prevent flotation.

Table 7.11 Proposed values of coefficient of skin resistance between soils and construction materials

$$[f_\phi = \delta/\phi,\ f_c = c_w/c,\ f_{c,max} = c_{w,max}/c_{max}]$$
$$\text{Ultimate skin resistance} = f_c c + \sigma \tan(f_\phi \phi)$$

Construction material	Surface finish		Sand (particles within range 0.06 to 2.0 mm)		Cohesionless silt (rock flour) (particles within range 0.002 to 0.06 mm)			Cohesive granular soil (50% clay + 50% sand)		Clay (particles not exceeding 0.06 mm)		
			Dry	Sat.	Dry	Sat.		Consistency index		Consistency index		
			Dense	Dense	Dense	Loose	Dense	1.0 to 0.5		1.0 to 0.73		
			f_ϕ	f_ϕ	f_ϕ	f_ϕ	f_ϕ	f_ϕ	f_c	f_ϕ	f_c	$f_{c,max}$
Steel	Smooth	Polished	0.54	0.64	0.79	0.40	0.68	0.40	—	0.50	0.25	0.50
	Rough	Rusted	0.76	0.80	0.95	0.48	0.75	0.65	0.35	0.50	0.50	0.80
Wood	Parallel to grain		0.76	0.85	0.92	0.55	0.87	0.80	0.20	0.60	0.40	0.85
	At right angles to grain		0.88	0.89	0.98	0.63	0.95	0.90	0.40	0.70	0.50	0.85
Concrete	Smooth	Cast in iron form	0.76	0.80	0.92	0.50	0.87	0.84	0.42	0.68	0.40	1.00
	Grained	Cast in wood form	0.88	0.88	0.98	0.62	0.96	0.90	0.58	0.80	0.50	1.00
	Rough	Cast on soil	0.98	0.90	1.00	0.79	1.00	0.95	0.80	0.95	0.60	1.00

It may be desirable to apply a factor of safety to the ultimate skin resistance in certain problems. For purely cohesive soils in the range where shearing resistance is independent of normal stress, the coefficient $f_{c,max}$ should be employed since cohesive soils have constant cohesion at maximum density.

(b) Pneumatic caissons

When the cutting edge has finally reached founding level, it should be trimmed and cleaned out by the miners. The area should be taken to the outside face of the cutting edge, thus giving the largest bearing area possible. To do this it may be necessary to undercut and underpin the cutting edge from the formation, an operation which is carried out in large sections. On occasion the whole working chamber may form the foundation of a permanent structure and the concrete filling should be placed in such a manner that the chamber is completely filled. When this has been done the air deck and compressed air plant can then be removed and construction of the permanent work continued. It is advisable to provide grout pipes in the top of the concrete so that any shrinkage or spaces which have not been completely filled can be grouted with cement before removing the air deck.

(c) Use of explosives in caissons

When large boulders are encountered in the sinking of open or pneumatic caissons they can be broken up by explosives. This method is used also for levelling the formation when the caisson has to be sunk on to an uneven rock foundation. Charges should be as light as practicable and should not exceed 0.7 kg of gelignite per hole. In exceptional circumstances several charges may be fired under expert supervision, provided millisecond delay detonators are used. These reduce the shock effect of a number of charges being fired simultaneously.

In soft rock the charges should be lighter than for hard rock, but in no case should an attempt be made to blast to a depth greater than 1 m and the total amount of explosive used at any one time should be limited to about 4 kg. There are, however, important precautions which must be taken when explosives are used in confined areas such as the working chamber of a caisson. These can be summarized as follows:

(a) One person only, well experienced in the use of explosives, must be in charge at all times, and have the key to the exploder in his possession before any charging of shot holes is made.

(b) No men other than those directly involved in laying the charges should be allowed in the working chamber.

(c) Before detonation all men must be withdrawn from the working chamber and a danger notice hung on the air lock door.

(d) The charge should be fired electrically from outside the working chamber. The cables should enter the working chamber by a duct which should be reserved for that purpose only.

(e) All electric supplies to the chamber must be disconnected and charging carried out by the use of battery-operated hand lamps.

(f) After the shots have been fired all fumes must be cleared by opening a relief valve. The time taken for this depends upon the size of the working chamber and the capacity of the air compressors but generally 20 min should be the minimum time allowed before re-entry.

(g) Tests should be made with a safety lamp to ensure that there is no carbon dioxide or carbon monoxide remaining before work continues.

In addition to the above notes, all Statutory Regulations governing the use and handling of explosives must be enforced.

7.6.5 Buoyant foundations

In some parts of the World, particularly in the estuaries of tropical rivers, the depth of mud and soft alluvium clay may be 100 m or more. This makes the sinking of normal caissons costly or in some cases impractical.

For the construction of 'La Bonne Intention' sugar factory on the Demerara river in Guyana it was found that, owing to the presence of mud and soft alluvial clay in excess of 100 m deep, buoyant foundations provided the answer to the problem of supporting heavy plant without differential settlement taking place. The whole factory was constructed on five cellular open caissons spaced 5 m apart and consisting of between 12 no. 4 m × 4 m cells for the boiler house up to 40 no. 4 m × 4 m cells for the main mill house. The diaphragms of the cells were of precast concrete, the outside walls of *in situ* concrete and the whole was made monolithic by cables in ducts passing through the walls. Each monolithic caisson was then sunk to the required depth by excavating in the cells. On reaching the required level, the bottoms of the cells were sealed with reinforced concrete floors. The weight of the caisson and the heavy plant installed upon it was only slightly heavier than the weight of material excavated inside. Settlement of 12 mm was observed during erection of plant. The ultimate settlement was approximately 30 mm. Since each building was completely independent of the others, differential movement during plant erection was immaterial.

7.6.6 The construction of monoliths

The development of docks often takes place in estuarine marshland where the adoption of monolith wall design for quays is advantageous. At Tilbury, for instance, the marshland consists of 1.5 m of reasonably firm, brown clay beneath which lies 10 to 15 m of very soft, blue-grey clay and peat. This overlies dense gravel, generally about 8 m thick, followed by chalk of indefinite thickness. The Port of London Authority therefore adopted the monolith construction for the new branch dock. For design purposes a uniform undrained shear strength in the soft clay of $19 \, \text{kN m}^{-2}$ from the surface to a depth of 8 m was adopted, increasing uniformly to $38 \, \text{kN m}^{-2}$ at 15 m. The majority of monoliths were founded in the dense gravel but a few penetrated into the underlying chalk where the gravel layers were thin. Bentonite was employed to facilitate sinking but this entailed some residual loss of skin resistance at the front and back of the wall after sinking. Smeardon *et al.* (1967) describe the work at Tilbury Dock, where 231 monoliths were constructed, each comprising 4 cells and having a plan area of 9 m × 9 m and walls approximately 1.0 m thick. They were sunk 1.4 m apart. The shoes were constructed on a thin concrete mat and were reinforced with high tensile steel to a height of 3 m with 28 no. 25 mm vertical bars extending a further 5 m. The height of the first lift was 1.5 m but subsequent lifts were 3 m. The final lift varied according to the depth sunk. After the initial stages, modifications were made to the cutting edge. The bottoms of the internal walls were unprotected and started 0.5 m higher than the outer ones. This increased the bearing stress on the outside edges and improved stability during sinking, particularly during the initial stages.

Sinking was started by breaking out the thin concrete base after the first two lifts had been concreted and was continued by grabbing from the four walls, the amount taken out from each being controlled to ensure even sinking. When the second 3 m lift had been cast and gained sufficient strength, the monolith was sunk a further 2.5 m below ground level. This left about 1.5 m above ground to avoid over-sinking. When the monolith had penetrated about 8 m, it was found possible to build up about 8 or 9.5 m above ground level before restarting sinking.

Bentonite was pumped through 12 no. 20 mm diameter pipes cast three to each wall with outlets above a 50 mm wide ledge formed at shoe level. It was injected continuously during sinking until it was seen to come out at ground level. This is important as loss of bentonite, particularly in an open stratum, can cause overlying soft clay to squeeze the side of the monolith, thus destroying the advantage of lubrication. On previous monolith sinking at Tilbury, up to 1600 tonnes of kentledge had been used to overcome skin resistance of 44 kN m^{-2}. By using bentonite, the maximum weight of kentledge employed was only 100 tonnes and this was used mainly to correct verticality during initial sinking but generally no kentledge was needed, even for monoliths penetrating a limited depth into chalk. When monoliths had to penetrate a greater depth into chalk, it was found possible to pump out the water within the cells using 150 mm diameter pumps. The removal of the flotation effect gave the equivalent of a loading of 500 tonnes, thus allowing sinking to continue at the rate of 300 mm day^{-1}. The average rates for sinking given by Smeardon *et al.* (1967) are shown in Table 7.12.

After the cells had been cleaned out by diver, the concrete seal in each cell was

Table 7.12 Monolith sinking rates: Tilbury

Site	Sinking aid	No. sunk	Total depth (m)	Av. depth (m)	Rate of sinking per week (m)
Berth I New Dock,	Kentledge	30	566	19	10
Stage I New Dock,	Bentonite	40	615	15.5	14.4
Stage II	Bentonite	159	2650	16.5	38.8

replaced by a $4\,m^3$ bottom-opening skip. The $2\,m$ thickness of concrete originally specified was found to be insufficient to provide an adequate seal and was increased to $3.75\,m$. The depth of laitence on this increased thickness of concrete was about 1 m and an analysis showed it to be composed of 74% water, 18% cement and 8% silt. This was removed before hearting the rear cells.

7.6.7 The diving bell

The ordinary or 'Smeaton' diving bell is useful for the examination of a dock bottom, entrance lock sills or for the removal of obstructions on the sea bed. The main advantages are that men working inside are not encumbered by diving suits and are not affected by the lack of visibility in turbid waters. Engineers are able to supervise the work at first hand, without having to rely on divers' reports, provided they have been passed medically fit for work in compressed air. Four men can work in comfort in a rectangular bell 3.5 m long 2 m wide and 2 m high. Such a bell weighs about 15 tonnes in free air and has a negative buoyancy of 5 tonnes when submerged. A typical bell is made of mild steel, weighted by cast iron slabs of kentledge bolted to the walls to ensure that it is stable on the dock bottom. It can be raised and lowered by a suitable floating crane or, if the work is near a quayside, by a derrick crane of adequate capacity.

When on the deck or shore, the ends of the bell are supported on baulks of timber allowing access under the sides. In this position men climb inside and sit on seats provided. The bell is then swung out and slowly lowered into the water. Compressed air is forced into the bell by a suitable air line at sufficient pressure and volume to keep the water out and the air pure. Surplus air is allowed to escape below the bottom rim of the bell.

Regulations affecting the use of the diving bell

The regulations mentioned in Section 7.8 (Use of compressed air in excavation work) apply to work in the diving bell. In addition, it is necessary to comply with Factories Act (1961) and with special regulations for diving operations and work in compressed air.

The diving bell should be provided with an ample supply of clean cool air at a pressure sufficient to more than balance the water pressure outside at the maximum depth likely to be achieved. The rate of supply of air must be not less than $0.3\,m^3$ per person per minute. An 80 mm diameter armoured flexible hose should be connected to the roof of the bell to supply the air. Separate high-pressure air connections should be provided for compressed air hand tools to be used in the bell; it should be noted that air tools lose their efficiency when working in compressed air. Standby compressor plant must be provided in case of power failure.

Electricity supply to illuminate the chamber should be at a maximum of 50 V, and fittings should be flameproof. A telephone in the bell should be connected directly to the crane driver and a loudspeaker should be provided and connected for others to hear the instructions at the same time. A life line controlled by a handle and rod inside the bell and passing through a watertight gland in the roof to a linesman close to the crane should be provided in case of breakdown in telephonic communication. All men working in the bell must be medically examined and passed fit before starting. The time taken for decompression laid down in the regulations applies to diving bells and the rate of raising the bell from the sea bed must be such as to conform to the rules.

7.7 UNDERWATER CONCRETING

In the past, concrete placed underwater was considered only in the last resort as it was considered that its strength and consistency were unreliable. However, in recent years considerable advances have been achieved in the quality of concrete placed underwater using modern techniques. The main uses are in locations where it is impracticable or impossible to place concrete in the dry, such as in the bottoms of cofferdams, monoliths or caissons where a watertight seal is required but, owing to the nature of the ground, pumping out cannot be employed. Other major uses are for levelling beds ready to receive blockwork dams, for making underwater foundations at depth, and for repairs to underwater structures such as dock entrances.

7.7.1 Concrete

Most rules, recommendations and principles for normal work in the dry apply, with commonsense, to underwater concrete. However, the following special points should be observed, The slump of the concrete should be not less than 125 mm and the minimum cement content should be about $350\,kg\,m^{-3}$.

In designing the mix it is important that the concrete should flow without segregation by producing a good plastic mix with high workability. Rounded aggregates, the addition of plasticizers and a relatively high percentage of fines should help to obtain the required consistency. Ordinary Portland cement should be used since it sets as well underwater as in air. However about 15 to 20% more cement should be used than for similar mix designs for use in the dry. Excess of cement leads to excessive laitence, as also does too finely ground cement. No attempt should be made to use vibrators underwater.

7.7.2 Preparation to receive concrete

The general aim should be to place the concrete in its final position with the minimum of disturbance. When placed, precautions should be taken to prevent cement being washed out by covering with plastic sheeting well weighted down, especially when there is a chance of water flowing over the newly completed work.

The formation should be cleaned by diver and all loose material removed before placing concrete. An air lift is convenient for doing this. The shuttering should be of steel or weighted timber and be of simple design with strong, but easily removed, fixings. Intricate shapes, complicated bracings and construction joints should be avoided. The concrete should be placed in one continuous pour whenever possible. If placed in more than one lift, it is necessary to remove any laitence which has formed or sediment which has been deposited on the concrete by high pressure air jet or by air lift. Care should be taken to ensure that no sediment or laitence is trapped in the corners.

7.7.3 Reinforcement

Any reinforcement incorporated in the work should be as simple as possible, using large diameter bars made up into cages which are lowered to the bottom by crane and carefully handled into position by diver just before concreting commences. It should be borne in mind that any reinforcement will impede the flow of concrete to its desired position when being placed. It will also cause difficulty in moving tremie pipes.

A cover of between 70 to 100 mm of dense concrete is essential but, if possible, this should be increased. When it is necessary to finish the surface of the concrete to a level, for instance in preparing the foundations for a block dam, carefully aligned flat bottom rails should be used as screeds. A further rail worked across the surface of the concrete by two divers can be employed to obtain a finished surface. It is better to over-fill the shutter and to allow the divers to screed off any excess than to under-fill and try to top up. During screeding, levels should be carefully checked by engineers using long light tubes with flat plates on the bottom as measuring staffs.

7.7.4 Methods of placing concrete

(a) Concrete pump

The development of the mobile concrete pump, with variable radius boom, fed with ready-mixed concrete, has facilitated the work of placing concrete under-water. Several concrete pumps fed by a number of ready-mixed concrete trucks enable large quantities of concrete to be placed at one time. By using a flexible delivery pipe and adjusting the boom a diver can direct the flow of concrete into the work. All divers should have telephonic communication with the surface and with each other. The concrete pump line remains full of concrete and is less susceptible to inadvertent emptying than the standard tremie pipe. It is important, of course, that there should be a steady supply of concrete to feed the pumps.

(b) Tremie pipe

Watertight smooth-bore vertical steel pipes, with quick release couplings, are erected over the work and attached to equipment for rapid raising and lowering.

A hopper is fitted on top of the pipe to receive concrete. The diameter of pipe is commonly between 150 and 225 mm, according to the volume of concrete to be placed. With the smaller diameter pipe the maximum size of aggregate should not exceed 20 mm but with the larger diameter pipe it can be increased to 40 mm.

The tremie pipe is charged by inserting a plug into the top to force out the water as the concrete follows it. It should fit sufficiently closely to enable the weight of concrete filling the pipe from the hopper to push it forward until it comes out at the bottom and floats to the surface. The hopper should be kept full and the concrete allowed to flow out through the pipe until it builds up around it. Raising the tremie pipe allows the concrete to flow, lowering slows or stops the flow. The bottom of the tremie pipe must always remain buried in the concrete until the pour is completed. Should contact be lost and the pipe become filled with water, the initial process of charging must be repeated. When the depth of concrete around one tremie pipe has filled the shutter to its correct level, the hopper is moved to a new pipe already positioned and concreting continued.

(c) Underwater skips

Volumes of concrete varying from small amounts to quite large quantities can be placed underwater using specially constructed skips. The plant required is less cumbersome than the more elaborate tremie gear. A crane of suitable capacity, equipped with bottom dump skip, and a supply of concrete is all that is required.

Underwater skips vary in shape and capacity from about $0.20\,m^3$ for small quantities to fill corners, to 3 or $4\,m^3$ for large pours. Small ones have bottom-opening doors, operated by diver or by trip line from the surface. The larger ones usually open by a self-tripping device which operates when the skip lands on the bottom or on the surface of the previously placed concrete. The top of the skip should be closed either by a canvas cover or by top doors to prevent interference with the surface of the contained concrete by the swirl of water whilst it is being lowered. Two divers should be used to guide the skip into position. When tripped it should be raised very slowly so that the concrete flows onto the surface of the previously placed batch with the minimum of disturbance.

(d) Toggle bags

These consist of re-usable canvas bags about 0.5 m in diameter and 1 m long and open at both ends. The bottom end is secured by a cord and slip knot, or chain and toggle. The bag is filled with concrete and the top secured in a similar manner. It is lowered to a diver who unfixes the lower end, thus allowing the concrete to flow gently into the works. This method is generally used for small quantities in corners where skips cannot be used. The bottom end of the bag should be opened close to the work, since concrete must not be allowed to fall through water.

(e) Bagwork

Bagwork is not used very much these days, since it is slow and therefore expensive. It should be considered only where the quantity is small, or where bags can be used for filling gaps between, for instance, the end of a block work dam and the side of an existing dock wall or other temporary works. Hessian sandbags should be half-filled with a plastic mix of concrete, as described in Section 7.7.1, and the mouth tied. They are lowered to the diver in a sling in batches of about 40. The bags are bonded together to form a wall, with the mouths of the bags facing inwards. The diver should well tread the bags into position. since the bond is strengthened by adhesion of cement between them. Dry concrete should not be used and large or sharp aggregates should be avoided.

(f) Grouted concrete

Grouted concrete has advantages over other methods of placing concrete underwater. It is more successfully placed in flowing water, in difficult situations and by floating craft in tidal waters or on the sea bed. Reinforcement can be used more freely than with tremie concrete. The development of colloidal grout mixing techniques and of wetting and dispersing agents has greatly improved the quality of grouted concrete.

The minimum size of aggregate to be grouted should be 40 mm. A suitable grout consists of ordinary Portland cement, with PFA and special workability agents, mixed with well-graded fine sand in the proportions of $1:1\frac{1}{2}$ by weight and prepared with a water/cement ratio of 0.5.

Two colloidal mixers, coupled in tandem to give a continuous flow, should be used to prepare the grout, each equipped with a 7 mm screen over the discharge hopper to ensure that no lumps of unmixed material can enter the pipeline and cause an obstruction. It is essential that the consistency is checked regularly to ensure a uniform creamy flow.

The shutters should be stout, firmly fixed and grout-tight. Reinforcement can be used to support grout pipes and the large aggregate is carefully packed around them. Grouting should commence as soon as possible after placing the aggregate, before any sediment settles on the work.

The first batch of grout should be without sand in order to lubricate the pipes. Work should start at the lowest point and the grout be allowed to work its way upwards (grouting downhill leads to the formation of voids). The pump pressure should be watched to ensure that it builds up steadily.

7.7.5 Use of the diving bell for underwater concreting

Diving bells, when available, can be used for underwater concrete repairs, usually in the entrance locks to impounded docks damaged when ships drag their anchors. The concrete is carried in a bottom dump skip suspended from the roof of the bell. A good plastic mix should be used but the slump can be less than for the methods discussed above, say 75 mm. When the formation is level, the bell sits evenly on the bottom and it is possible to exclude all water so that concreting can be done in the dry. If the bottom is uneven or on a slope, then it may not be possible to exclude all the water. Rapid-hardening cements can be used to advantage in these conditions.

There are two main disadvantages in using a diving bell for concreting. Firstly, the surface area to be concreted is limited by the plan area of the bell (usually not more than $3 \text{ m} \times 2 \text{ m}$). Secondly, the maximum amount carried in a skip is only about $\frac{1}{2} \text{ m}^3$ and since the men travel in the bell with the concrete, they will be compressed and decompressed with each batch; this introduces a medical hazard if repeated frequently.

7.8 THE USE OF COMPRESSED AIR IN EXCAVATION WORK

Compressed air has been used for many years for the construction of shafts, tunnels, caissons and cofferdams. By increasing the pressure within the chamber, water is forced back into the ground and excluded from the working face. Its use has been the traditional method of supporting difficult ground in the construction of London's underground system since the beginning of the present century. The Bow–Leyton section of the Central Line extension through the Lea Valley was at the time (1937–39) the longest compressed air tunnel in the World. Engineers, skilled in its use, turned to compressed air immediately they ran into difficulties with their excavations. Recently there has been a reluctance to use compressed air in civil engineering construction since reports have been made on the delayed effect of aseptic necrosis of the bone associated with prolonged working under pressure. The development of geotechnical processes, particularly those involving bentonite, grouts, cement slurries and chemicals, have gone a long way to replacing compressed air working in recent years (Haswell, 1969).

The use of compressed air is often necessary in fine silts, peats and soft clays which do not yield groundwater readily. They are difficult to drain, although electro-osmosis has been used on some silts. They are equally difficult and expensive to treat by injection processes. When excavating through waterbearing fissured rocks the inflow of water to an excavation may be too great to be removed economically by pumping and compressed air working may be the only practical solution, provided the fissures are not too open, a condition which could lead to excessive loss of air.

The ground conditions must be recognized from site investigations and the use of compressed air anticipated before the cofferdam, cylinder or caisson is designed. Provision must be made for air decks, air locks and compressed air plant to avoid delays should conditions worse than those originally anticipated be encountered during the course of sinking. Considerable trouble was experienced during the construction of Ford's Power House, Dagenham, due to unexpected difficulties from groundwater.

However, compressed air has its limitations: the health risk on the men working inside the chamber increases significantly when the pressure is greater than $100 \, \text{kN m}^{-2}$ above atmospheric pressure. For physiological reasons the maximum pressure to be considered is about $350 \, \text{kN m}^{-2}$ above atmospheric pressure. This corresponds to a head of water of 35 m.

7.8.1 The effect of compression upon workmen

The act of compression has no ill effects upon man provided he is healthy, has a strong heart and is free from catarrh. Men aged over 40 or who are fat do not take kindly to compressed air working if not accustomed to it. On decompression, however, compressed air sickness or 'the bends' may occur if pressure reduction is too rapid. The reason for this is that on compression the blood and other body fluids absorb an additional quantity of air through the lungs. The extra oxygen is dealt with chemically by the blood in the normal way but the nitrogen remains unaltered, thereby saturating the blood. It may take two or three hours before every part of the body reaches 70 or 80% saturation. Thus an engineer visiting the working face for a short time is at much less risk than the miner working physically during the whole shift.

On decompression the blood becomes supersaturated and, if the rate of decompression is low, the nitrogen will be expelled by the lungs. If the rate of decompression is too rapid, bubbles of appreciable magnitude are formed in the blood. They collect in certain parts of the body, usually the muscles of the arms and legs, causing severe pain known as the 'niggles' or 'bends'. In extreme cases, interference with the circulation, paralysis or death may occur. Because of these risks, special regulations are in force applicable to work in compressed air. The *Medical Code of Practice for Work in Compressed Air* (CIRIA Report No. 44, 1975) gives full details of rules, regulations and decompression (Blackpool) tables and these show that the 1958 regulations were inadequate, particularly with reference to decompression times.

It is essential that those responsible for work in compressed air are competent and experienced in the procedures relevant to such work and have full knowledge of the regulations. All men to be employed under compressed air must have a medical examination and should undergo a test compression and decompression, accompanied by a competent instructor so that they can obtain confidence. A new starter is someone who has never worked in compressed air before and requires acclimatization of approximately 10 days. When the working pressure exceeds $100 \, \text{kN m}^{-2}$, new starters should work a shift not exceeding 4 h. This should be the latter part of a normal shift, so that new starters are decompressed with other experienced men. Compressed air sickness in men who have been passed medically fit is unlikely with pressures less than $100 \, \text{kN m}^{-2}$.

If it is anticipated that pressures considerably in excess of $100 \, \text{kN m}^{-2}$ above atmospheric are to be used, the Medical Research Council's decompression sickness panel should be consulted. The regulations and decompression tables applicable to the work should be included in the contract documents. The decompression (Blackpool) tables, referred to above, relate the time of decompression to the maximum working pressure and time during which the man has been exposed to compressed air. Bennett and Elliott (1975) have produced a comprehensive treatise on the physiology and medicine of diving and compressed air work.

Should a man, after decompression, suffer physical pain the immediate cure is recompression in a medical lock, followed by a more gradual reduction in pressure. A medical lock must be provided when working pressures exceed $100 \, \text{kN m}^{-2}$. It should be provided with electric light, heating and a bed with plenty of blankets. The patient should be kept warm during decompression since this is always accompanied by a feeling of coldness. Chilling is experienced also in the main air lock, which should be heated. The men should be supplied with hot drinks and warm clothes on emerging. As a precaution, all men should be provided with a disc inscribed with a note to the effect that he is a compressed air worker and, in the event of illness away from the site, steps can be taken immediately to admit him to a medical lock.

7.8.2 Use of compressed air plant in relation to ground conditions and safety

The installation of compressed air plant should be carried out under the supervision of qualified engineers experienced in this work. Adequate arrangements must be made for the provision of standby plant to ensure that air pressure is maintained at the working face in the event of a power failure. An adequate supply of air will be needed at all times during the sinking of caissons. This must allow time for men to leave the working chamber in the event of flooding.

As the nature of the ground to be encountered is likely to vary, it is important that the total compressed air requirement is assessed by a supervising engineer experienced in working under similar conditions. The greatest risk of sudden loss of air pressure occurs, for example, when the caisson has been sunk through soft clays, with little loss of air, and suddenly reaches open gravel through which the air escapes at an alarming rate. When reaching the anticipated level of such ground conditions, it is essential that the compressor drivers are alert to the need of increasing the air supply. Telephone communication between the working chamber and the compressor house should be provided, together with an alarm bell to attract the attention of the compressor drivers.

A good standard of lighting is necessary so that the men in the working chamber have confidence and can see what they are doing. As with compressed air plant, an emergency reserve supply of lighting should be available, possibly by independent sources such as batteries or emergency generators.

There is always a risk of fire when working under compressed air, due to the concentration of oxygen, which can cause ignited timber to burn fiercely. It is important, therefore, that inflammable rubbish and other materials should be removed from the working chamber. If timber is required, it can be treated to render it non-inflammable. Beds of peat have been known to catch fire below ground and to burn furiously in the stream of compressed air escaping from the working face. Smoking should be prohibited and men should not be allowed to take matches with them into the chamber. Should a fire occur anywhere in the working chamber, smoke fumes will endanger the men. Provision should be made in the air deck for fire mains to be installed in agreement with the local Fire Brigade.

Care should be taken by the engineer to check the soil, through which the excavation is to take place, for the presence of vegetable matter which gives off noxious gases when in contact with compressed air. Such gases can be dangerous to men working in the vicinity. Escaped compressed air may travel a considerable distance if the ground is permeable and enter excavations in which other men are working. During its passage through the ground it can lose much of its oxygen in oxidizing organic matter and so become unable to support life. At Dagenham in 1931 six men were killed when climbing down a shaft into which deoxidized air had seeped from adjacent workings.

The pressure of air in the working chamber should be controlled so that it exceeds only slightly the value required to keep the works free from inflow of water. A check on the correct air pressure can be made by keeping a hole ahead of the general excavation in one corner of the floor. The air pressure can then be adjusted so that water can be seen standing in the hole. If the air pressure is too great, power is wasted, the men are caused inconvenience and the ground may become unstable. On the other hand, in cohesive soils, the amount of air escaping from the working chamber may be so small that a fresh-air supply must be piped to the face and a blow-off valve provided in the roof of the chamber to allow sufficient air circulation.

The principles of safe air diving practice are given in an Underwater Engineering Group publication (UEG Report UR 2, Part 1, 1975). A general account of diving has been presented by Twine (1980). Further guides to save diving practice are presented in publications by CIRIA, UEG (1974) and HMSO (1981).

Chapter 8

BEARING PILES AND PILING

I.A. Rennie

INTRODUCTION

Piles and the art of piling have been used for centuries by man but it is only in the last two or three decades that the methods of assessing the behaviour of piles and of pile groups have been subjected to detailed analyses. This can be attributed to the growth in the subject of soil mechanics and to economic and commercial pressures. In 1954, the Institution of Civil Engineers published CP 4 which provided a guide for good piling practice. This document was superseded in 1972 by CP 2004 (BSI, 1972d). Over the last 20 years a considerable volume of literature has been published on all aspects of piling.

The principal function of a bearing pile is to transfer load to lower levels of the ground which are capable of sustaining it with an adequate factor of safety and without settling under normal working conditions by an amount detrimental to the structure. The carrying capacity of a pile is a combination of the resistance mobilized over the surface area of the shaft, usually referred to as friction or adhesion, and the end resistance mobilized at the pile point or base; the nature of the ground determines the proportions of shaft and base load.

Piles can have many uses; when driven into loose cohesionless soil they decrease the void ratio and increase the bearing capacity; when used in river piers they carry loads to below scour level; they form protective frameworks or fenders round structures in navigable waters; they can act against lateral thrust loads; they can be installed to form retaining walls; they can be used to resist overturning moments and they can be used as anchors. They can also be installed either vertically or at an angle of rake to cater for horizontal forces.

There are many different types of pile in use today, with varying geometry which depends upon imposed loading and soil conditions. The main types are dealt with in Section 8.1. The choice of pile type for a particular job depends upon the combination of all the various soil conditions and the magnitude of the applied load. As examples, a driven pile may be placed through weak and compressible soil to rock-head to carry a concentrated load, and a large diameter bored pile may be used in London Clay to carry a point load of many hundreds of tonnes. As with any type of foundation, a piled foundation should be judged on its technical as well as its economic merits; for the same superficial area of pile surface, a small number of long piles are more effective and will, in general, support the load with less settlement than a large number of short piles.

Where possible the applied load should be concentric with the pile axis or centre of gravity of the pile group, with due allowance for tolerance in position and verticality, especially with single piles and pairs of piles. Eccentric and horizontal loads must be catered for specifically in the design of the pile or pile group. Care must be taken when assessing the passive resistance of the soil near the surface in resisting lateral loads. Interconnecting beams or caps can be used to spread applied load over several piles.

Extraneous factors should also receive consideration when designing piled foundations; for example, sulphate attack on the concrete shaft, damage to the shaft due to obstructions during driving, damage to adjacent property, unacceptable noise and/or vibration and infiltration of artesian water.

A prerequisite to the design of any piled foundation is an investigation of ground conditions by a competent firm preferably experienced in piling. Undisturbed or disturbed soil samples must be taken and *in situ* or laboratory tests, where applicable, carried out to yield design parameters. Boreholes must be sufficiently deep and located so as to examine the soil both around and below the proposed foundation. Meticulous observations regarding ground water strike level, flow rate, standing level and chemical composition are necessary. The preliminary investigation should also cover an appraisal of adjacent property, underground workings, pits, potential subsidence, possible or past slips and similar contingencies which may affect the choice of pile (Marsland, 1973a).

8.1 TYPES OF PILE

There are two basic kinds of pile; those that displace the ground and those that replace the ground in which they are installed. The former method is the older of the two techniques and has been in existence since man first hammered a stake into the ground. Timber piles were the first displacement pile to be employed and are still used extensively under certain circumstances. The city of Venice sits almost exclusively on driven timber piles. Today, however, the vast majority of piles in industrialized countries are composed of steel and concrete in various combinations. There is a subdivision in displacement pile types: some displacing large and some only small volumes of soil.

The large-displacement steel pile in general consists of a closed-ended cylinder which can be left in the ground (permanent) and the internal void filled with concrete, or alternatively the tube can be withdrawn (temporary) for re-use as concrete is poured into the remaining void. The large-displacement concrete pile is often precast and can be either segmental or full-length. It may be hollow, with the core later being filled with concrete, or it may be solid. Its plan shape may be circular, square or polygonal.

The small-displacement pile usually consists of steel in the form of an H section or an open-ended box or tube. Because it has a small cross-sectional area, it displaces only a small volume of soil. Since there is no concreted void, the load is carried exclusively on the steel cross-section. A screw pile falls between the large-

and small-displacement piles and consists of a helix attached to the base of a smaller diameter shaft: it can be manufactured in either steel or concrete. However this type of pile is now rare.

The disadvantages of a driven pile lie principally in the noise and vibration which results from impact of the hammer during driving and the displacement of the soil. These nuisance features are largely overcome by the replacement type of pile.

For many years, replacement piles were installed using percussive drilling rigs. In cohesive soils using this method of installation, an open-ended heavy tube, no more than 2 m long, with a cutting edge, is dropped repeatedly to cut a hole, which is then filled with concrete. Cohesionless soils are removed by using a tubular cutting tool with a bottom opening flap. Sections of casing are inserted as necessary to support the sides of the opened hole as boring progresses downwards; the void is then filled with concrete and the casing withdrawn. This pile is commonly called a bored pile but should more correctly be called a percussive bored pile.

Like the driven pile, the percussive bored pile has a limitation on size because of its method of installation. Both types are normally restricted to a maximum load of about 1200 kN, a diameter of 600 mm and a length of 20 to 25 m. Occasionally special conditions prevail where these limits are surpassed, as at the M 62 River Ouse Bridge where 'Big R' Raymond Steel tube piles, 20 m long by 1530 mm diameter, were jacked into position to carry the river piers. However, in the past few decades, some buildings and structures have become heavier and require a more efficient method of carrying large concentrated loads than on clusters of small capacity piles. Development of the caisson type of foundation resulted and brought about a revolution in piling practice. By mounting an auger on a long driving shaft which is suspended from a crane jib and rotated by a powerful diesel engine, large diameter holes may be formed and these can be filled with concrete to provide substantial piles. In cohesionless soils, the sides of the hole are supported by temporary steel casing. Rotary bored piles can be bored to 60 m in depth and up to about 2.5 m in shaft diameter.

Two additional and important developments have taken place within the large diameter rotary field. Firstly, by using a specially designed tool, the base of a pile bored in a cohesive soil can be enlarged to make it up to three times the shaft diameter, so increasing the base carrying capacity (under-reamed or belled pile). Secondly, by using bentonite drilling mud, large diameter holes can be supported without the use of temporary steel casing, except for an initial length of about 1.5 m casing at ground level, to very substantial depths.

A preformed displacement pile (timber, steel or concrete) can be manufactured under factory conditions and prior to driving can be inspected for its integrity as a load-carrying element; nevertheless, it can be severely damaged during driving. The bore of a replacement pile can be examined, as can the concrete used to form the shaft, but the final cured shaft cannot under normal contract conditions be inspected for correctness of diameter and integrity. These advantages and

Table 8.1 Types of displacement piles

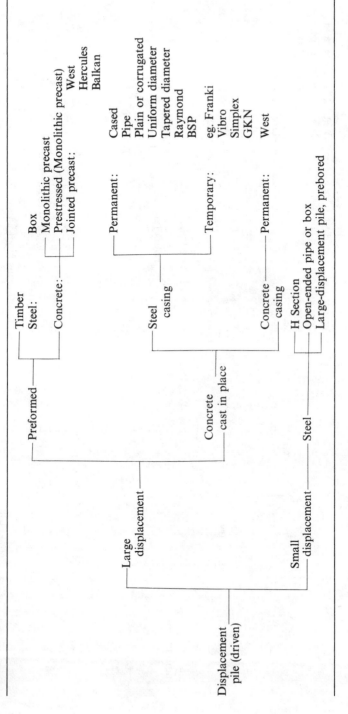

Table 8.2 Types of replacement piles

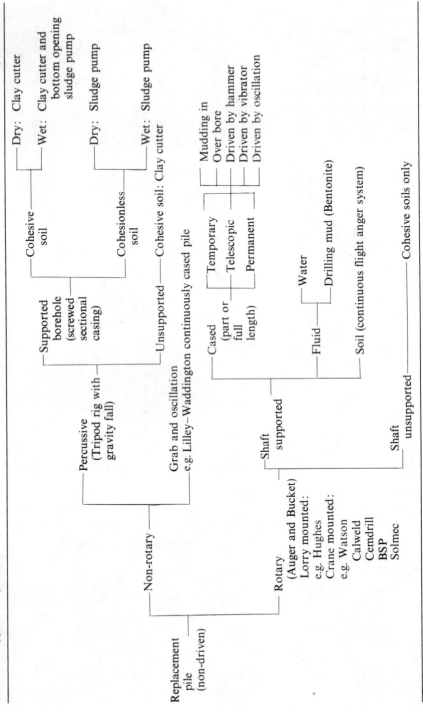

disadvantages must be realized during the pile selection procedure. However modern sonic and other integrity testing methods can supply important evidence about the continuity and shape of finished piles in the ground.

Although hewn trees may still be used for piles in some regions of the world, the modern pile is a sophisticated piece of engineering merchandise which is marketed by a highly competitive specialist industry. Concurrent with development in piling practice has been improvement in the understanding of the theory behind these developments, and there has been much progress through research into the mechanics of shaft and base support. The theory of support, using both dynamic and static methods, is covered in Section 8.4.

The expansion of the piling market has resulted in certain types of pile and piling practice becoming synonymous with a brand name, especially in the driven range. For example, West produce a driven segmental concrete shell displacement pile, Raymond a tapered permanent steel driven pile and Cementation Franki a driven bulb-ended cast in place pile. In the percussive bored and auger bored market the same basic equipment is used by most of the specialist contractors, each one offering particular variations in diameter, size, depth of bore, method of casing driving and extraction, speed of drilling, etc. Tables 8.1 and 8.2 show available types of piles, the principles of construction and typical brand names.

Developments in the design and construction of piles are discussed in the proceedings of conferences (ICE, 1980 and ICE, 1984), the behaviour of piles in an earlier publication (ICE, 1971) and piling specifications in another (ICE, 1978). Problems associated with the installation of piles have been discussed by Thorburn and Thorburn (1972) and by Healy and Weltman (1980). Ground vibrations from piling operations are the subject of two reports (Martin, 1980; Weltman, 1980). It should be understood clearly that piling is an art requiring a high level of skill on the part of all concerned, and inexperience can prove costly in the long run.

8.2 DISPLACEMENT PILES

8.2.1 Timber piles

From the earliest time man has used timber to his advantage. The wooden stilts which supported his riverside hut and the stakes with which he constructed his jetty are two examples. Pitch pine was used almost exclusively at the turn of this century but it has now largely been replaced by Douglas fir and greenheart. The main problem with timber lies in its organic structure which can be attacked by air-borne pests and fungi or by marine organisms which bore into it below water level, causing decay and eventually structural failure. The length of pile over which the water level fluctuates is subject to serious and often rapid decay. Careful examination of ground water and its seasonal levels is of paramount importance prior to pile selection.

The following timbers are suitable for piles under the stated conditions:

Totally dry   American yellow and red pine, pitch pine, Norway fir, spruce, beech.

Totally wet   Northern pine, pitch pine, spruce, larch, elm, oak, teak.

Dry/wet   Northern pine, larch, Lebanon cedar, greenheart, jarrah, teak, oak.

All timber piles should be free from natural defects and the centre line of a sawn pile should not deviate from the straight by more than 25 mm throughout its length, but for round piles a deviation of up to 25 mm on a 6 m chord may be permitted, (CP 2004; BSI, 1972d). All timber used for piles should, where possible, be treated with preservative in order to prolong its working life. Most piles not so treated quickly deteriorate. When possible, pressure impregnation should be used, but if this is not available then alternative, though inferior methods are possible (BS 913, BSI, 1973a and BS 4072, BSI, 1974a). Further information can be obtained from CP 2004 (BSI, 1972d), the *Handbook of Softwoods* (HMSO, 1957b) and the *Handbook of Hardwoods* (HMSO, 1957a). Care must be exercised during handling and driving to cause minimum surface damage to the timber. When splicing or finishing the head of a pile, the timber should be coated liberally with a preservative.

The working stress of a pile must not exceed that specified in CP 112 (BSI, 1971) for the grade of timber used. The driving stresses must be checked to ensure that failure of the pile does not occur. The head of the pile should be fitted with a tight fitting steel or iron ring to prevent 'brooming' during driving. The overall length of the pile should be sufficient to allow for the damaged end to be removed after driving. Reduction in both the drop and frequency of blows of the hammer can limit damage to the fabric of the pile and as a general rule the weight of the hammer should be equal to the weight of the pile for hard driving and half the weight of the pile for soft driving. Working loads are normally restricted to between 150 and 500 kN per pile, depending on cross-sectional area.

When splicing is necessary to increase the length of a pile, the two abutting surfaces must be square to their length to ensure total surface contact. To avoid splitting or brooming the two adjacent ends must be securely restrained either by hoops for round piles or by angles for square piles.

Timber piles are used where they are cheap and plentiful or where their lightness and flexibility make them ideal for absorbing shock loads, as in jetties and temporary land and marine works.

8.2.2 Precast concrete piles

Precast concrete piles come in two basic designs: those cast as a single unit of approximately the correct driven length, up to about 20 m long, and those cast as small units, 1.0 to 10.0 m long, connected together during driving to attain the final pile length. The full-length precast pile has its best use where it functions essentially as a column, obtaining poor lateral stabilization from the ground and subjected to moments and lateral forces. However if piles are to be longer than 10 m, a substantial contract is required to justify the use of piles cast in one long length.

Except when factory produced, these piles are usually fabricated in specially prepared casting yards set up by contractors on or near site, where supervision and control are good and where a high quality product can be economically produced. They must be able to withstand the stresses imparted to them during handling and driving and generally these are substantially larger than those derived from carrying the finally imposed loads.

For small loads and short lengths, the pile section is usually square with chamfered corners but when the side dimensions are greater than 400 mm, an octagonal section is to be preferred. For exceptionally long piles, a hollow concrete section can be used which provides high stiffness for relatively low dead weight. The high stresses imposed during driving and lifting necessitate accurate design of the pile as a reinforced concrete member subjected to bending and shear. Figure 8.1 shows four different methods of lifting a pile; two by using a set of brothers and two employing a single lifting point. Design of the pile to a maximum static bending moment based on self weight will adequately cope with most lifting conditions. It is advisable to fit toggles at the designed lifting points or to mark the freshly cast pile at these points so that slinging is carried out correctly and safely.

The concrete should be designed in accordance with CP 110 (BSI, 1972c). Where easy driving is expected, a nominal 1:2:4 mix can be used, but if hard

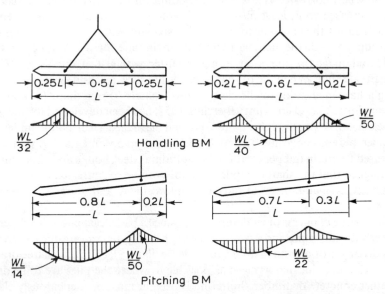

Figure 8.1 Handling and pitching piles. *W* is the total weight of the pile.

driving is even only a possibility, the mix should be richer at $1:1\frac{1}{2}:3$ or $1:1:2$. CP 2004 (BSI, 1972d) Table 8, states that for hard and very hard driving conditions and in marine works, minimum cement quantity should be $400 \, kg \, m^{-3}$: giving a cube works strength within 28 days after mixing of $25 \, N \, mm^{-2}$. For normal and easy driving the corresponding figures for minimum cement and strength are $300 \, kg \, m^{-3}$ and $20 \, N \, mm^{-2}$. In general ordinary Portland cement should be used, but where sulphates may be present in the ground water or soil the recommendations of BRS Digest No. 174 (Article 1.3) should be implemented. High alumina cement concrete should be avoided due to possible substantial strength losses after curing.

Longitudinal and lateral reinforcement must comply with CP 110 (BSI, 1972c) and be sufficient to resist the stresses generated from lifting, storing, transporting and driving as well as those transmitted from the superstructure. Joints in longitudinal bars should be avoided if possible but where they are necessary, it is usually recommended that they should be continuously butt welded. Lateral reinforcement, comprising hoops or links of not less than 6 mm diameter, is vitally important for resisting the driving stresses. The head and toe of the pile usually have to sustain the highest driving stresses, hence for a distance of 3 to 4 equivalent diameters the 'volume of lateral reinforcement should be not less than 0.6% of the gross pile volume. In the body of the pile, the links should represent not less than 0.2% of the pile volume, and be spaced at not more than half the width of the pile', (CP 2004, BSI, 1972d). Changes in lateral reinforcement should be gradual over a length of at least three times the width. For most normal land-based piles, cover to all reinforcement should at least comply with CP 110 (Table 19), but where corrosion is possible, for example in sea water, the cover may have to be increased to at least 50 mm. The reinforcement cage must be assembled tightly against the spreader forks at 1.5 m spacing, with no tying wire ends left protruding outside the main bars which might initiate corrosion. When driving is liable to damage the pile point, it must be fitted with a steel, ductile iron or cast iron driving shoe. When the toe has to penetrate rock, a sound key is obtained by using a hardened point to the shoe, approximately 75 mm diameter and up to 200 mm long. The 'Oslo Point' (Bjerrum, 1957) has been developed for use where a rock head is sloping; the hollow-ground hardened steel point, under light hammer blows, penetrates the rock without slipping and the blows are then increased to attain full penetration. Longitudinal steel, hoops and shoes must be so designed that the shaft of a pile does not become overstressed and so that its continuity is maintained. Details of reinforcement and shoes are shown in Fig. 8.2.

The casting of piles is of vital importance since a badly cast pile will not perform properly during lifting and driving and will probably deteriorate due to corrosion after driving. For on-site casting yards a large, flat, well compacted, free draining area is essential, located as close as possible to where the piles are to be driven. Whether concrete or timber shuttering is used, it must be meticulously cleaned and oiled before the pre-assembled reinforcement cage is positioned in the

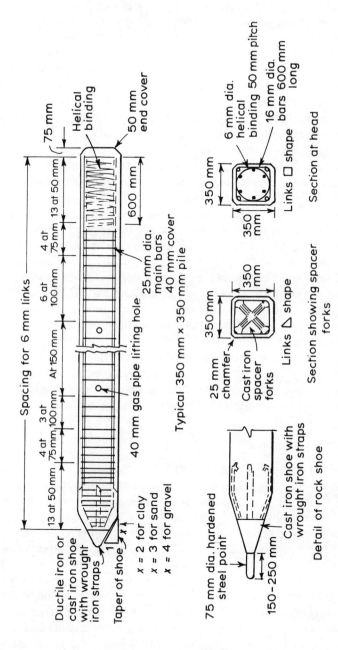

Figure 8.2 Arrangement of reinforcement in precast concrete piles.

mould and supporting arrangements should be sufficiently stiff to prevent the steel reinforcement from sagging, since this could reduce the cover. The head end-plate of the pile formwork should be truly square to the shaft axis. Concrete should be placed in one continuous operation and thoroughly compacted by a hand held poker-vibrator. The exposed face of the wet concrete must be finished with a wooden float to produce a smooth dense surface. All faces of the pile should be of well compacted concrete, free from roughness and pitting. The casting date and a reference number should be marked upon the upper face of each finished unit, as well as the locations of lifting points.

The side shutters can be removed as soon as practicable but not sooner than 12 hours after casting. During curing the piles should be kept wet for at least 4 to 5 days by a combination of spraying and wet hessian, preferably up to the time when they can be lifted for transfer to the stacking yard; this is usually after 7 to 10 days. Alternatively an appropriate steam curing system may be used. Three test cubes cast at the same time as the pile can be used to ensure that the necessary concrete strength has been achieved prior to lifting and driving. The test cubes should of necessity be stored in similar conditions to the pile and not in artificial conditions. Piles should be stacked on supports placed at the predetermined, marked stacking points. The stack should be uniform and not excessively high, because the piles at the bottom are the ones that have cured for the longest time and should be the first to be driven. After curing but prior to driving, the pile should be closely inspected for cracks and defects which could cause problems after installation. A 'state of the art' report on the requirements of design, loading and inspection of high quality precast concrete piles has been given by Fellenius (1974). A report on the design, manufacture and installation of precast and cast-in-place piles in the United States has also been published (Talbot, 1973).

Precast segmental driven pile

The Hercules pile consists of up to 6, 9 and 12 m lengths of precast hexagonal concrete shaft connected by a patent bayonet-type coupling. The shaft units are manufactured under factory conditions from dense concrete having a minimum compressive cube strength of 48 N mm^{-2} at 28 days, a water:cement ratio of 0.39 and a minimum cement content of 480 kg m^{-3}. Each unit is reinforced with six No. 16 mm diameter high tensile deformed bars having a yield stress of not less than 590 N mm^{-2}. The lateral reinforcement consists of 5 mm diameter spiral binders of cold drawn steel at 100 mm centres, reducing to 50 mm at each end of the unit. The joint consists of a central locating spigot and a male and female steel fitting, connecting together in the form of a bayonet coupling. The reinforcement bars are screwed into the end plates to ensure structural continuity and this joint is able to form a splice as strong as the shaft itself. To ensure a well finished surface at each joint a short, thin sheet steel 'cuff' is provided around each concrete element. The specially designed 30 mm cylindrical pile point is made from special oil-tempered steel and has a concave leading face like the 'Oslo point'. The end

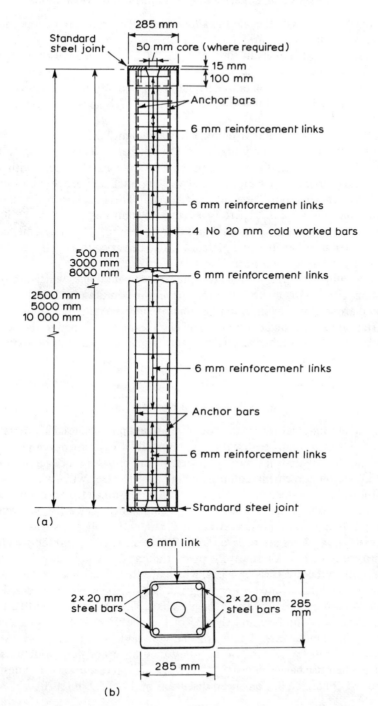

Figure 8.3 West Hardrive pile (a) longitudinal section; (b) cross section.

plate is cast integral with the pile and allows the driving energy to be concentrated into the very small point area.

The Hercules pile comes in three sizes; a 700 kN pile, a 1500 kN pile, and a 2000 kN pile. The Herklid 625 is a smaller version of the Hercules pile of about 500 kN capacity. It is manufactured from the same quality concrete, has a square cross-section (250 mm sides) and uses 4 No. high tensile deformed steel reinforcing bars.

The West Hardrive pile is employed where hard driving is encountered. Standard 285 mm × 285 mm units are cast in 2.5, 5 and 10 m lengths in concrete having a crushing strength of $50 \, \text{N} \, \text{mm}^{-2}$. The units are reinforced with 4 No. 20 mm cold-worked bars, 6 mm links at varying pitch and are jointed together with 4 No. locking pins, one at each corner. The pile is top driven by a conventional piling frame. There is a central 50 mm inspection hole which, if required, can be used for post tensioning or anchoring. The shoe units can be manufactured in either steel or concrete and can be varied to suit soil conditions. Details of the pile are shown in Fig. 8.3.

Balken Piling Ltd also produce two precast segmental piles which are square in section, either $275 \, \text{mm}^2$ or $300 \, \text{mm}^2$, have a maximum unit length of about 13 m and are reinforced with either 12 mm or 16 mm diameter bars. The concrete has a strength exceeding $56 \, \text{N} \, \text{mm}^{-2}$ and each segment is joined to the next one by a special locking mechanism. This segmental piling system has been operated in Europe for more than 20 years, especially in the Scandanavian countries.

8.2.3 Prestressed concrete piles

Both concrete and steel are expensive commodities and when used in precast piles are relatively inefficient, because the maximum pile stresses are developed during lifting and driving and not during service. In the prestressed concrete pile the longitudinal reinforcement comprises high tensile steel wires which can be pretensioned or post-tensioned. The wires can be tensioned to the required stress while in the mould prior to the concrete being added. Once the concrete has reached its required compressive strength (2.5 times the concrete stress due to the prestressing), the wires are released from the stressing jacks and the concrete is put into compression. Alternatively, precast units having holes cast in them are threaded onto the prestressing wires, which are then tensioned to the required stress and the holes grouted. After the grout has set, the wires are released, thus producing compressive stresses in the pile shaft. The prestressing wires are then cut off flush with the face and the toe of the pile. High control of quality is necessary in manufacture. There are many advantages in using precast pre-stressed piles but for reasons of economy they have not been widely used in practice. Their use has been restricted to a few special conditions. The design of prestressed concrete piles has been discussed by Li and Liu (1970).

8.2.4 Driven cast in place piles

Monolithic precast piles have two major disadvantages: they are difficult to extend or shorten in length during a contract, especially if the contract is not very large and their handling and transportation is expensive. To circumvent these difficulties a range of driven piles exists in which the load bearing column or core is cast 'in place' and can be varied in length. These piles can be divided into two general categories: either a permanent steel tube or a hollow segmental concrete cylinder is driven into the ground and filled with concrete, or a temporary steel tube with a detachable shoe is driven into the ground and filled with concrete, following which the tube is withdrawn, leaving the concrete in contact with the displaced soil. With the former method, the required length of pile is estimated and, after driving the tube, any excess length is cut off, or alternatively, the casing can be lengthened by welding. With segmental hollow cored concrete cylinders or shells, short lengths of precast concrete are threaded on to a mandrel to make up the required length of pile. When installation is effected by temporary steel tubes, the length of the casing is chosen to exceed that required for the average pile and is used for both longer and shorter piles. After the tube has been driven, concrete is discharged into it through a hopper. The tube is then slowly withdrawn and reused on the next pile; only the expendable detachable shoe remains at the toe of the pile. A brief description of some of the better known proprietary brands of driven cast in place pile is given below.

(a) Permanent steel cased pile

In the Raymond pile the permanent casing consists of either a uniformly tapered or a step tapered light corrugated steel tube, which is usually assembled on site from 1.25, 2.5 and 5 m lengths of casing; the foot section has a closed end. The individual lengths of casing are screwed or 'spun' together to produce, when assembled, a tube which is almost completely watertight. An internal mandrel is used to drive the casings: with the uniformly tapered casing, the mandrel is composed of two halves for ease of extraction, while in the step tapered casing the shoulders of the mandrel bear against the shoulders of the steps in the casing.

To facilitate easier working on site, the tube to be driven is assembled and then lifted and lowered onto the previously driven tube, prior to it being concreted. This allows the long mandrel to be fed into the tube. The tube and mandrel are then lifted out together and positioned ready for the next drive. If ground heave is suspected, a continuous or single flight auger is often used to prebore the initial part of the pile; the tube can then be driven to the required set in the bearing stratum. At the Isle of Grain Power Station in Kent the Raymond Concrete Pile Company installed nearly 18 000 piles up to 40 m in length, yielding a total length of more than 600 000 m in just 60 weeks. In order to minimize surface heave, the top ground of softer brown London Clay was prebored.

The uniformly tapered pile is manufactured in 2.5 m lengths of light gauge sheet steel, reinforced internally with a spirally wound wire. The toe diameter is 200 mm and the tube tapers uniformly at a rate of 1 mm in 30 mm, with a maximum length of 11 m.

The step taper pile is fabricated from 2.5, 3.75, 5.0 and 7.5 m tube lengths and the diameter can be increased by 25 mm per section. Thus a toe diameter of 220 mm can be increased to a cap diameter of 445 mm over 10 sections. By varying the length of each individual section, a pile of variable geometry can be obtained to optimize use of available ground strengths. After driving, the empty casing is checked for verticality but concreting is usually postponed until the driving rig is at a distance sufficient to avoid disturbance of freshly poured concrete. Davis (1970) carried out experimental drives on various pile geometries and found that for two piles of equal length, the pile containing the greater number of short lengths of step taper carried 33% more load, although the pile point was 220 mm as opposed to 240 mm in diameter.

(b) Permanent concrete shell pile

In the West shell pile, precast concrete shell units 1 m long and reinforced with polypropylene are threaded onto a straight steel mandrel which has a detachable expendable conical concrete shoe. Tightly fitting steel bands are located in circumferential recesses to join one shell to another. The concrete shells are threaded onto the mandrel to give a total pile length slightly in excess of that required. The hammer blow is shared between the mandrel and the concrete shells. Most of the blow travels down the mandrel to the shoe, resulting in the pile being driven from the bottom; the rest of the blow is transmitted to the outer concrete shells preventing them from separating during driving. After driving to the required set or length, the excess shells at the surface are removed and, if required, a longitudinal reinforcement cage is inserted. Concreting can take place as and when convenient but, if delayed, the head of the pile must be covered to prevent ingress of water or foreign matter.

These piles are generally supplied in four nominal diameters, 610, 510/515, 445 and 370/380 mm, carrying up to 1700, 1200, 800 and 600 kN, respectively, depending on length and ground conditions. By using a mandrel formed from connected short lengths, piles can be driven in conditions of restricted head room. They also can be lengthened by fitting an extension to the mandrel and adding several more sections of shell. Difficulties may be encountered in variable ground with the West pile which, like the Raymond pile, is driven with a mandrel. In such conditions the shaft may deviate, thus causing the mandrel to jam in the pile and, during subsequent extraction, the outer skin may be damaged. Since part of the load in the West pile is carried by the outer shells, they must fit tightly and about one another properly. As with all driven piles, heave must be closely checked and prevented by preboring. Details of the pile are shown in Fig. 8.4.

The West cylindrical segmental pile is similar in some respects to the shell pile.

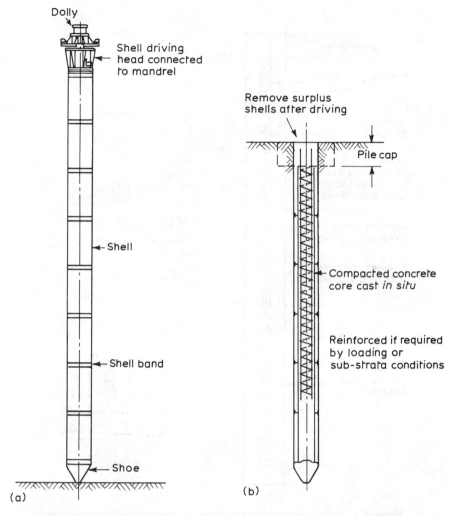

Figure 8.4 West shell pile (a) pile ready for driving, shells threaded on steel mandrel; (b) completed pile.

It consists of 280 mm diameter, polypropylene reinforced, 1 m long shells of concrete, having a crushing strength of 55 N mm^{-2}. The segments are jointed internally by a spigot and socket joint and externally by flush fitting steel bands. The pile is driven by a small drop hammer and a new segment is placed on top of the last segment until the required length has been obtained. A 70 mm diameter central hole allows inspection of the driven pile or a single bar of reinforcement for post-tensioning the pile if required. It is then filled with grout to complete the pile. It is used for light loads: the maximum working load is about 300 kN with a maximum length of around 15 m. Details of the pile are shown in Fig. 8.5.

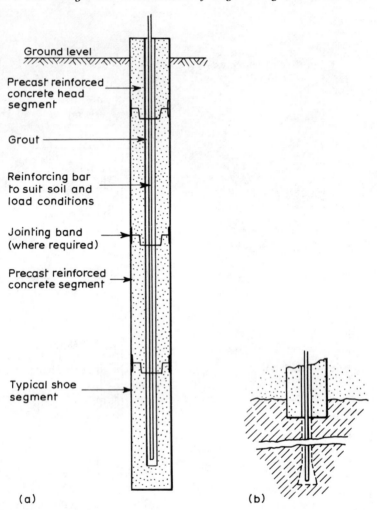

Figure 8.5 West segmental pile (b) shows alternative anchored toe unit.

(c) Driven cast in place pile (temporarily cased)

The temporary steel driving casing of the Franki pile is positioned in the driving frame and a charge of dry ballast or gravel inserted to form a plug at the foot of the open-ended tube. The hammer is then repeatedly dropped on the plug which is compacted; further hammer blows striking the plug thrust the tube downwards. When the bearing stratum has been met or when the set has been reached, lifting ropes arrest the downward motion of the casing and the gravel plug and additional concrete is then hammered out of the tube to form a bulb-end. A reinforcement cage is then lowered into the tube and charges of concrete added and rammed with the hammer as the tube is slowly withdrawn. This ensures that the shaft is concreted under dry conditions.

A semi-dry mix is used since this resists the squeezing force imposed by the displaced soil on withdrawal of the casing and furthermore it can be compacted to a high density using the gravity hammer. The enlarged end to the pile gives it a greater end bearing area than a conventional straight shafted pile and the end ramming compacts the soil adjacent to the toe. This results in an increase of the permissible end bearing stress, while the shaft compaction increases the frictional resistance between the concrete and the surrounding ground. Because the hammer strikes the plug at the bottom of the casing, surface vibrations are reduced during driving. Details of the driving sequence are shown in Fig. 8.6.

Franki piles are formed in five sizes (Table 8.3). The design bearing capacity of the Franki pile is best estimated by a combination of a static analysis and a check an driving data. As for other piles representative pile tests give the best information on capacity. In many instances the Franki pile can be modified to carry out the base enlargement operation as described, but the pile shaft may then be completed using normal concrete as for other cast in place piles. Franki piles are constructed in United Kingdom by Cementation Piling and Foundations Limited.

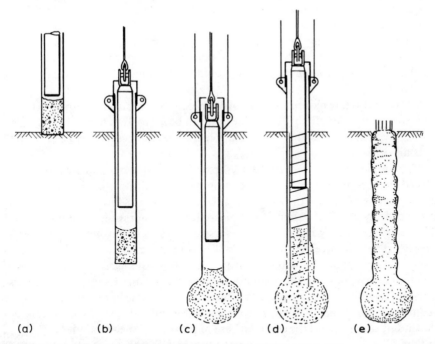

(a)   (b)   (c)   (d)   (e)

Figure 8.6 Bottom driven Franki pile (a) consolidating the 600 to 900 mm of aggregate to form driving plug, (b) driving the tube, (c) forming the base, (d) forming the shaft, (e) completed pile.

Table 8.3 Franki pile dimensions and capacities (Cementation Frankipile Ltd)

Nominal shaft diameter (mm)	Nominal maximum working load (kN)*
275–300	320
355–380	500
405–455	700
510–560	1100
560–635	1400

*Capacities may be increased by use of high strength concrete mixes.

The Vibro pile falls within the same category as the Franki pile, the difference being in the details of installation. An open-ended 20 mm thick steel tube is fitted with an expendable steel or cast iron shoe (either the diameter of the pile or oversize). Using a hammer operating at 40 to 50 blows per minute the tube and shoe are driven to the correct length or required set. The reinforcement cage, as required, is then placed and by means of a mobile hopper attached to the pile-frame leaders, the shaft is charged with concrete. Extraction links are then fitted to the head of the tube and the casing withdrawn by reverse motion imposed by the hammer operating at about 80 blows per minute. During the upward movement, concrete is drawn under the bottom flange of the casing and is then partially compacted when the direction of motion changes. Details of the pile are shown in Fig. 8.7.

Vibro piles are formed in a variety of diameters ranging from 350 mm–600 mm. Tubes of various lengths are available and can be up to 42 m long. Several Vibro piles have been unearthed after installation and the surfaces found to be ribbed as a result of the compaction process.

The Simplex pile is similar to the Vibro pile. A cast iron or steel plate shoe is placed on the ground in the pile position and a steel tube lowered on to it. A driving cap is placed on the top of the tube and the pile is formed by top driving using a 'drop' or diesel hammer until the required penetration or set is obtained. The seal between tube and shoe is effective in keeping out any ground water present. The reinforcement cage is then lowered into the tube and the pile is concreted and the tube withdrawn in one continuous action. When ground conditions are poor, the pile may not achieve the required set and under these circumstances it can be redriven. The initial drive is carried out and ceased at the required depth. A mix of semi-dry concrete is then placed in the tube, the amount depending on the size of the enlarged base required, and the remainder of the tube is then filled to ground level with mixed fine and coarse aggregate. The tube is then withdrawn and redriven in the same position. During the redrive process the aggregate mix and the semi-dry concrete are displaced laterally into the

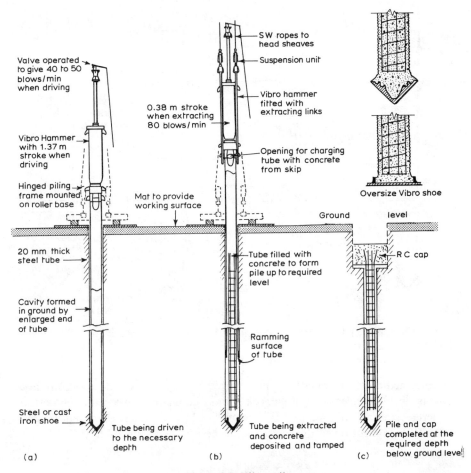

Figure 8.7 Vibro pile

surrounding ground, forming an enlarged pile which is then completed as previously described. Details are shown in Fig. 8.8. The pile can be formed in diameters of 400, 450 and 500 mm with a maximum depth of up to 20.0 m; loads of up to 800 kN can be carried, depending upon ground conditions, and can be increased to 1200 kN by redriving.

Many inventive ideas have come forward in the field of cast in place driven piling throughout the last fifty years and there is no doubt that other new developments will take place in future years. Apart from the specific pile types mentioned above, other systems of a generally similar nature are available from various companies.

Both driven precast and driven cast in place piles can suffer from heave. Unlike piles with a permanent external casing of either steel or concrete, cast in place piles may in poor ground suffer locally from a reduction in the cross-sectional

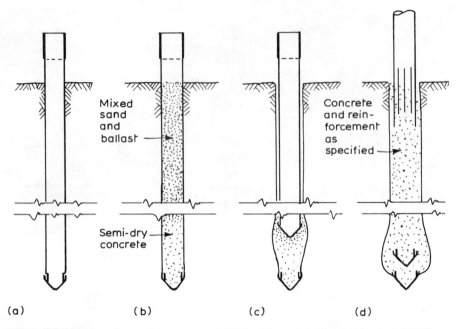

Mixed
sand
and
ballast

Concrete
and rein-
forcement
as
specified

Semi-dry
concrete

(a)   (b)   (c)   (d)

Figure 8.8 Simplex pile – redrive process. (a) Tube driven on initial drive to 'set'; (b) filling and withdrawal of tube; (c) redrive with enlargement of pile section; (d) reinforcement and concrete placed and final tube withdrawal.

area, termed waisting or necking, due to rapid changes in ground conditions. For all cast in place piles, a careful check should be made between the theoretical concrete volume required and that actually placed. Should these differ significantly, investigation of the cause should be initiated immediately.

8.2.5 Steel bearing piles

The strength of steel makes it an excellent material from which to fabricate piles which are subjected to heavy loads during handling, driving and service. For many foundations the use of steel bearing piles is the best technical answer but not always the most economical. Today steel is expensive and the engineer must be sure that the technical requirements justify the more expensive material. Steel is used to form four basic types of bearing pile; cased, pipe, box, and H or universal. Steel piles can easily be lengthened by welding and shortened by cutting. Special feet are available for hard driving, and when uplift in service is anticipated, due for example to excess hydrostatic pressure beneath thick ice, barbs may be fitted along the buried length of piles.

(a) Cased piles

These are manufactured by the continuous Driam process in which a strip of mild steel, of minimum yield strength $240 \, \text{N} \, \text{mm}^{-2}$, is helically twisted to form a cylinder and then internally and externally butt welded. This continuous process allows a variety of pile lengths to be available, the longer lengths being limited by transportation problems. The casings are manufactured in increments of diameter of 50 mm, from 250 mm to 700 mm. The casing thicknesses range from 3.25 mm for the smallest diameter to 9.53 mm for the largest.

They are installed using the minimum of equipment. Once the tube is set up on a frame or leaders, it is fitted with a welded steel plate shoe, 9 to 18 mm thick. A plug of 1 : 2 : 4 concrete, or richer, having a water : cement ratio not exceeding 0.25 by weight, is placed in the casing and, when compacted, its height should be not less than $2\frac{1}{2}$ times the pile diameter. An internal drop hammer is then used to drive the casing to its required depth or set, after which it is filled with concrete. By driving from the base, the thickness of the casing can be kept to a minimum because buckling stresses are eliminated and the noise and vibration are significantly less than with a top driven casing. The hammer weight, length and diameter are suited to the various sizes of piles, ranging from 0.75 tonne, 2.7 m and 230 mm, to 6 tonne, 3.3 m and 590 mm.

The capacity of the pile can be up to 1500 kN if it bears upon rock, while in poor ground or in estuarine conditions the casing acts as a permanent shutter and can permit the concrete to be taken above normal ground level. This procedure results in a cast in place shaft, unaffected by driving stresses. Short lengths of reinforcement may be inserted to bond the pile shaft to the pile cap or a steel plate can be welded to the top of the casing. Typical casing and hammer specifications are given in the BSP *Pocket Book* (1969).

(b) Pipe piles

These are similar in many respects to cased piles; they are manufactured by the same process, may be driven in the same manner and carry approximately the same loads. They are, however, manufactured from thicker plate, 6.35 to 12.70 mm, and hence can withstand slightly harder driving. The diameter ranges from 400 to 600 mm in 50 mm increments. The additional wall thickness allows them to be driven also from the top with, for example, a diesel or hydraulic hammer. These pipes can be driven either open-ended, later to be internally cleaned out, or they can have a 25 to 30 mm flat shoe welded on the base. Both kinds of pile can be installed at a rake although in heterogeneous ground conditions, the pile may deflect sideways. After driving, the deflection can be checked by lowering a lamp down the pipe.

(c) Box piles

These are heavy duty steel piles fabricated either from two sections of plate welded longitudinally to produce an eight- or twelve-sided cross-section or from special sections of steel sheet piling. For example, Frodingham box piles are available in three sizes of constant wall thickness. At their greatest diameter they are 419, 533 and 673 mm respectively, with a wall thickness of 15.9 and 17.8 mm.

Their respective weights are 167, 233 and 313 kg m^{-1}. They are normally fitted with a flat shoe prior to driving and when driven to rock, can carry loads ranging from 2000 to 4000 kN.

When driven open-ended into a cohesive soil, a plug of clay is formed shortly after driving at the toe of the pile and this is carried down to founding level: the remainder of the cylinder is then filled with concrete. In clean well graded cohesionless soil, grout can be introduced through a special tube and aggregate, enclosed within the tube, can thus be grouted. Under such circumstances preliminary aggregate-grouting tests are essential.

The main application of box piles is in situations where the upper ground conditions are poor, the lower ground conditions good and where large horizontal loads occur. The flat faces of the pile are eminently suitable for the attachment by welding of bracing and other structural elements. Because of their large cross-sectional area, soil displacement may be considerable when they are driven in groups. Under normal manufacturing conditions they are supplied in lengths up to about 35 m. Larssen and Rendhex also manufacture a wide variety of box steel piles.

(d) Universal bearing piles

These piles are manufactured by rolling into H-sections, with the flange width equal to the web depth. They are relatively stiff and can withstand considerable driving stresses. They have a low displacement volume and can be driven at close centres or near existing buildings provided the latter can withstand the accompanying vibration. Large horizontal forces can be carried perpendicular to the XX-axis. The piles are generally available in three serial sizes, 300×300, 250×250 and 200×200 mm. Standard sections vary from 280 to about 44 kg m^{-1}, although special sections can be rolled for particular applications. The piles are generally driven by single or double acting hammers, but in certain ground conditions they can be driven also by a drop hammer or a vibrator. As an example of their use, Lanarkshire Universal steel bearing piles of varying length were used for the pier foundations of the 2300 m long Tay Road Bridge.

(e) Corrosion

Having selected a system of piling using steel piles and having justified the high capital cost of the piles, the problems of corrosion must be understood and preventative measures taken. Corrosion is an electro-chemical reaction which occurs mainly when metal is buried in soil or water or a mixture of the two. Owing to the manufacturing processes, areas of different electropotential exist on the surface of the pile and produce the necessary components of rusting. Current flows from the cathode to the anode through the soil and/or water which acts as the electrolyte, corrosion taking place at the anode because it has the higher electropotential. The presence of oxygen is essential before corrosion can take place, although it may occur also as a result of bacterial activity or where stray electric currents are found in the ground. Corrosion can also arise from atmospheric pollution involving dust particles containing salts and acids. Electrolytes which contain dissolved salts and acids have high conductivity and can cause, under certain conditions, rapid corrosion of metal.

The first priority with a corrosion problem is to analyse the ground and atmospheric conditions to ascertain the corrosion sources. Measures can then be taken to retard rather than totally prevent corrosion. For piles driven into virgin undisturbed soil, corrosion below the permanent ground water level will be extremely small and generally can be ignored. Above the water line, where the soil may be alternately saturated or moist, corrosion can occur and preventative measures should be adopted. Research carried out by the Sea Action Committee of the Institution of Civil Engineers showed that the average reduction of wall thickness for piles immersed in sea and fresh water was 0.076 and 0.051 mm, respectively, per annum. These figures will be increased if immersion is followed by a period of drying, as within the tidal range or in wave splash zones.

CP 2004 states six further cases where corrosion may occur: (1) in tropical conditions: (2) where aggressive chemicals are present; (3) where sulphate reducing bacteria occur; (4) where steel is in contact with other metals; (5) where steel is in contact with wood, especially oak; and (6) where used with unsuitable fill, such as unweathered ashes.

Four methods are commonly used to retard corrosion:

(1) Where atmospheric corrosion is prevalent, steel containing 0.25 to 0.35% Cu is slightly more resistant than ordinary steel, although larger percentages than these produce no increase in resistance. Corrosion below ground surface is not arrested by this process.

(2) The most positive way to prevent rapid corrosion below ground level or where maintenance is impossible, is to coat the steel with, for example, paint, coal-tar pitch, bitumen, epoxy resin or polyurethane. These coverings must be applied to clean, oil-free surfaces if they are to have any chance of preventing corrosion. It is thus advisable to ensure that the manufacturer's instructions for the coatings are strictly followed, if possible under factory conditions. Such treatment is wasted if the piles are carelessly handled during transit and the coatings damaged, although it is sometimes impossible to prevent cracking and scraping during driving, for example, in coarse granular soils. Here the cost of thicker protective coatings must be balanced against the cost of choosing a thicker steel section which will have sufficient cross-sectional area to carry the applied load at the end of the life span of the structure. It may be more economical to replace steel sheet piles forming certain structures, such as jetties and sea walls when corroded, than to undertake coating the piles. Where steel sheet piling is used, the section is generally chosen to carry the maximum bending moment, but the point at which this occurs may not be the portion of the pile subject to the most aggressive corrosion. By calculating the critical stress at a point of lower bending moment, the life span of the pile may be increased by many years.

(3) There are certain conditions when protective coatings are not suitable; for example, in positions of river scour. Here the best solution is to case the steel in good quality concrete which can withstand the abrasion of stones and sand. Alternatively, timber or non-corrosive metals, such as aluminium or zinc, can be used as a shroud.

(4) Cathodic protection is an expensive but positive way of preventing steel piles from decay and is described in Section 1.3.

8.2.6 Driving plant and hammers

Just as there are a large number of types of driven piles, similar in principle but different in detail, so there are many types of frames and hammers, again reasonably similar in principle but adapted to suit the requirements of the individual pile.

Piling frames are of two basic kinds. The first and simpler is a frame or trestle, usually manufactured in wood, into which the pile shaft is placed. The frame is located over the pile position and the hammer mounted on to the pile head or suspended from a crane. Once the pile has been driven, the frame is moved to the next position. This type of frame is economical for piles driven in a line at close centres or where only a few piles have to be driven; it is not economical where many piles have to be installed over a wide area. A handling crane is necessary to lift the piles, hammer, frame, etc.

The second type of plant is more mobile and adaptable. The basic unit is either a conventional track laying crane or a walking or sliding frame. The conventional crane is fitted with a set of leaders which are pivoted at the end of the jib and secured by hydraulic rams to the base of the crane turntable. The rams are used to adjust the rake of the leaders. It is normal practice when using a drop hammer to have two drums and sheaves: one is used exclusively for the hammer, the other is a general purpose lifting line. This piece of plant has great mobility and can travel safely with the leaders upright and in a driving position.

The walking or sliding plant is designed for efficient driving at the expense of site mobility. A set of leaders, mounted upon a heavy duty frame at the front, can be adjusted for driving vertical and raking piles. At the rear of the frame is a diesel engine which is used to drive the various winches and to operate the hydraulic devices. These machines have greater pulling power than the equivalent crane mounted rig because the pulling energy is usually better distributed within the frame and over the ground. The walking rig consists of a frame, upon which all the driving equipment is mounted, a turntable and four hydraulic feet. The frame moves along slides until it reaches the end of its travel. The legs are raised and progressed along the frame while it sits upon its base. The legs are then extended and the operation repeated as necessary. With the sliding rig, the frame is mounted fore and aft on two large, long cylindrical pipes. By winching on a block and tackle, the frame slides on shoes which fit over the pipes. Slewing is accomplished by winching on the block while the head of the frame is restrained, usually by resting it against a tube lightly driven into the ground. The pipes normally rest upon a grid of sleepers which disperses the weight and the extraction loads. Sliding rigs are cumbersome to move and are no longer in common use.

The simplest form of hammer is the cylindrical drop hammer which is used to drive tube piles. These come in various diameters, lengths and weights to suit the various pile tube sizes. The hammer is not located in guide rails but is simply lifted inside the tube to the required height and dropped. These drop hammers weigh up to 6 tonnes. They are normally lifted by winch and rope and hence their energy output is not accurately controlled.

When the pile is top driven, guides are located on the front faces of the outer structural members of the leader in order to ensure that the hammer maintains its vertical or raking fall and strikes the packing at right angles, otherwise severe damage can occur, especially with concrete piles. The hammer is difficult to control accurately with this arrangement also, but trip monkeys can be fitted to ensure a constant drop, although these can be troublesome and are best avoided if possible.

Steam, air, diesel and hydraulic hammers are more flexible and efficient than drop hammers. The steam hammer is powerful but requires a large amount of auxiliary equipment to produce a sufficient quantity of steam. The air hammer is more compact and requires only a mobile compressor but it cannot produce as much driving energy as the steam hammer. The diesel hammer combines compactness with high energy output while the hydraulic hammer is cleaner and of more constant efficiency. Each of these hammers has its own virtues and is best suited to certain types of driving.

Steam of air single-acting hammers are of two types. In one type the hammer is rigidly fixed to the internal piston rod which is raised by steam or air under pressure entering the hammer cylinder. The steam or air escapes through a valve at the top of the cylinder, either at the end of its stroke or when released by a rope operated from ground level. The falling cylinder imparts the necessary energy and is emptied of steam or air during its fall. In the other type of hammer, the internal piston is fixed and the external cylinder forms the hammer. Depending upon the length of stroke, the hammer can operate at up to 40 blows per minute. In the double-acting steam hammer, additional driving energy is obtained by assisting the gravity blow of the hammer with steam.

The diesel hammer requires none of the attendant plant of the steam or air hammer. The hammer must be started manually. An internal drop piston, having a hemispherical lower face, is raised inside the cylinder until it is released by an automatic trip. As the piston falls, a small quantity of diesel oil is pumped into a cup at the base of the cylinder. The falling piston strikes the base of the cylinder, thus imparting a blow through the anvil to the pile. Simultaneously, the hemispherical end splashes the diesel oil into the compressed air where it ignites. The explosion firstly imparts an additional small downwards blow immediately after the gravity blow, and secondly returns the piston to the top of the cylinder, during which time the cylinder is exhausted of burnt gases and fresh air is admitted. If little pile resistance is met during the impact of the piston on the diesel oil, combustion may not take place, rendering the diesel hammer practically impotent. In the hydraulic hammer, powerful rapid acting rams raise a heavy steel cylindrical weight, which is then released to fall back on to a striking head fitted onto the top of the pile.

8.2.7 Pile driving by vibration

Vibration is used in one of two ways to drive piles and casings. The first method employs either low or high frequency vibrations to reduce the surface resistance on the pile shaft. The second method uses an impact technique where the blows originate from rotating eccentric weights. These two methods originated in Russia, where they have been used since 1935 (Barkan, 1957). More recently, vibrators have been developed in Germany, Poland, France, Japan and Great Britain. The state of the art in Russia was described by Smorodinov *et al.* (1967).

High and low frequency vibrators consist usually of two or four counter-rotating eccentric weights driven by an electric motor. The whole assembly is rigidly clamped to the pile shaft or casing and imparts only a longitudinal vibration as the eccentric weights revolve in counter-clockwise directions; forces at right angles to the longitudinal axis cancel each other. The vibrator, it is claimed, destroys the friction between adjacent grains of soil along the shaft of the pile, thus concentrating the whole assembly load at the pile point and this implies that they are most efficient in cohesionless soils but inefficient in cohesive ground. In stiff clays they are relatively ineffective unless very powerful. In saturated cohesionless soils, the oscillations generate positive and negative alternation of pressure under the pile point, which causes liquefaction in the saturated soil and allows the pile to sink under its own weight as if in a heavy fluid. Low frequency vibrators operate at between 15 and 40 cycles s^{-1} and are effective only in cohesionless soils. The precise operating frequencies and weights of assembly (up to 20 tonnes) for various ground conditions and densities have been found by experience. In any one system the exciting force must be great enough to overcome the ultimate value of skin resistance over the total length of pile embedment. If the total weight of the vibrator assembly is not sufficient to give the necessary rate of penetration, an additional load must be applied to the overall assembly or a downward pull exerted. The vibrator can be used also for extraction purposes, where the vibration reduces the skin resistance and the extractor force is supplied by the pull of a crane rope.

Pile driving by vibrating impact is little used in the UK, although it is based on the principles of the high and low frequency method. Instead of the vibrator being rigidly clamped to the casing and producing a longitudinal vibration, it is fitted loosely to the head of the pile. During each revolution of the weights, an upward and downward force is produced which acts as a hammer on the head of the pile. Whitaker (1970) quotes frequencies of between 480 and 1450 blows min^{-1} from motors of between 2 and 28 kW rating, For further details reference should be made to Barkan (1957), Smorodinov *et al.* (1967) and Whitaker (1970).

8.3 REPLACEMENT PILES

In this form of piling the concrete pile shaft replaces the ground in which it is installed.

8.3.1 Hand augered piles

Hand augering is the simplest method of constructing replacement piles. Because the hole is unlined, the technique can be employed only in soils which stand unsupported above the water table. The augered hole is then filled with concrete. Under normal conditions and unless clay heave is expected following tree cutting, it is not necessary to install a reinforcement cage; a few starter or dowel bars can be pushed into the wet concrete to bond it to the pile cap or ground beam. Because of the method of installation, hand augered piles are usually restricted to a diameter of about 300 mm and a length of 5 m. Their main application is where the loading is relatively light, as in the case of houses or single storey framed structures, and where soil of good bearing capacity lies at a shallow depth beneath softer soils. They are economic only if a few are required and where the transportation costs for alternative methods are high.

8.3.2 Percussive bored piles

The traditional method of installing a small-diameter replacement pile is by using the percussive boring technique, producing what is colloquially known as a 'bored' or 'tripod' pile. The apparatus consists of a set of shear legs, an air, petrol or diesel driven winch, a digging tool and say 1.2 m lengths of threaded steel casing. The shear legs are set up with the pulley block directly over the position of the pile. The digging tool, which is shackled to a wire rope attached to the winch, is raised and dropped on to the ground. It is normal to commence operations with a clay cutter, which consists of a 1.5 m long cylinder with a lower cutting edge and an upper weighted portion. By repeated raising and dropping, granular and/or cohesive soil is forced up into the tube. The soil is subsequently removed from the tool through two slots cut longitudinally along the middle portion of the cylinder. When about 1 m of ground has been removed, a length of steel casing is inserted into the hole, with the male thread uppermost. A driving head is then screwed on to the casing, a driving bar inserted through the slots in the clay cutter and through two holes in the driving head and the weight of the clay cutter used to drive the casing into the ground. The driving head is then removed, another length of casing screwed on, the driving head replaced and the casing driven until the upper edge is approximately level with ground level. The digging tool is then used to remove soil from within the casing.

It is normal practice to use casing only where the ground will not stand unsupported or where water is encountered. When boring through granular soil into cohesive soil, the casing must be advanced ahead of the digging tool and should be driven into clay when encountered, by about 0.5 to 1 m; this is sufficient in most cases to form a seal against the ingress of water.

In granular materials, especially under water, a 'sludge pump' is used to extract the soil. It is usually a 1.5 m long cylinder, weighted at the upper and, fitted with a cutting edge and a hinged flap-valve inside the lower end. On being dropped, the flap opens to permit the passage of soil into the cylinder. When the 'pump' is lifted the flap shuts, retaining the soil and water inside. It is emptied by being upended and allowing the soil to flow out the upper end.

In another type of excavating tool known as the 'cruciform clay cutter', the main body comprises two steel plates 1 to 1.5 m long, welded to form a cross in section, a weighted top cap and a cutting ring welded around the lower end of the plates. When dropped into a cohesive soil, the clay is compressed through the ring and expands between the vanes. On the tool being removed from the borehole, the clay virtually falls away from the cutter.

Once the borehole has been excavated to the required depth, a reinforcement cage, if required, is inserted and the pile concreted. Because the concrete is cast in place and the shaft cannot be inspected once cured, it is essential that every care be taken during concreting operations. The concrete must be rich enough to give the required design strength and have sufficient workability to become densely compacted around the reinforcing bars and helix under its own weight. It should be discharged into the pile borehole through a funnel with a short length of tremie pipe. This ensures that the concrete drops vertically through the longitudinal reinforcement and does not ricochet from one side of the borehole to another, depositing cement on the sides and reinforcement cage. The above precaution is imperative where contiguous retaining wall piles are constructed with a high percentage of reinforcement. A slump test should be taken at frequent intervals and a set of test cubes once or twice daily.

When a pile is bored under water, the concrete must be placed by tremie pipe extending initially to the foot of the bore. A Vermiculite plug is first placed in the tremie to form a barrier between concrete and water, and concrete pouring is commenced. Construction should then be continuous until completion of the pile, apart from necessary interruptions to unscrew and shorten the pipe as concrete rises within the borehole. The base of the tremie must remain embedded in concrete by one or two metres at least throughout the process, because if water enters the tube, the continuity of the operation will be broken. The first concrete placed eventually returns to the head of the pile and as it may contain debris, it is usually over-spilled. It is unwise however to rely on the scouring action of concrete to fully clean the base of a borehole and where end bearing is important, a separate cleaning and checking procedure should precede the concreting phase of construction. Temporary casing is usually withdrawn from all piles of this type in stages during concreting, ensuring that the level of the concrete is above the bottom of the casing, taking account of the need to compensate for the volume occupied by the casing and any overbore. Starter bars can be inserted at the top if a reinforcement cage has not been installed. A pile containing a reinforcing cage approximately the same length as the temporary casing should be avoided because extraction of the casing may cause the cage to rise inside the pile. The problem can be overcome by ensuring that the cage projects below the casing by several metres and has a concentration of helical binding at the lower end; this anchors it to the concrete below the casing. Alternatively, if the cage is short and the casing long, two or three bars are continued below casing bottom level and bound with a concentration of helical reinforcement. When a deep cut-off level pile has to be concreted it is better to ensure that too much concrete is placed rather than too little. It is easier to cut back a pile than to build it up. The Federation of Piling Specialists recommends the concrete standards given in Table 8.4.

Tripod piles are commonly of diameter 400, 450, 500 or 600 mm, carrying up to 350, 500, 600 and 1000 kN, respectively, depending upon ground conditions. They are usually reinforced nominally with 4 or 5 No. 12 mm, or 4, 5 or 6 No. 16 mm bars respectively with 6 mm helical binders at 200 to 250 mm spacing.

When piles are designed to resist bending moments, the diameter of the cage should be formed accurately, since even small errors in diameter have a relatively large percentage effect on the moment of resistance. The cage must also contain sufficient spacers to ensure that it is located centrally within the bore.

Table 8.4 Federation of Piling Specialists recommended concrete workability for cast in place piles

| Piling mix | Slump | | Typical conditions of use |
	Minimum (mm)	Range (mm)	
A	75	75–125	Poured into water-free unlined bore. Widely spaced reinforcement leaving ample room for free movement between bars.
B	100	100–175	Where reinforcement is not spaced widely enough to give free movement between bars. Where cut-off level of concrete is within casing. Where pile diameter is less than 600 mm.
C	150	150 to collapse	Where concrete is to be placed by tremie under water or drilling mud.

The piles can be installed normally to a depth of 25 m but, by using specially adapted machines, depths of 35 m can be achieved. If obstructions are met during boring, a heavy chisel is employed to break them. Boring in heterogeneous soils like boulder clay can cause problems with verticality when the digging tool is deflected by a rock or boulder; a borehole is maintained vertical most effectively by using continuous casing. A long straightening tool can sometimes be used to correct alignment but once the bore is deflected, it is very difficult to get it back on line. It is preferable to proceed slowly and to take remedial action at the first sign of deviation.

The Prestcore pile is a bored segmental pile, constructed in the same way as the standard tripod or bored pile up to the point when the borehole is complete. Precast concrete cylindrical units, each with a square central hole, are threaded on to a specially designed square section mandrel. As the units are threaded on and interlocked, the assembly is lowered into the hole. A reinforcement rod is then threaded down each of the pre-aligned holes in the units. The temporary casing is then withdrawn a short distance in order to free it prior to grouting.

A rich cement and water grout is then pumped from a mixing tank down the hollow mandrel which has a specially designed locking and release machanism at its lower end. The grout penetrates the void left between the units and the soil as the casing is withdrawn, the small joint between one unit and another and the voids around the reinforcement. Although the grout effectively seals all the spaces and joints, it is important to ensure that one unit abuts squarely its neighbours, because the load is transferred from one unit to another by the adjacent faces.

The Prestcore pile is used under conditions where a standard tripod bored pile cannot be employed, for example, when water and/or poor ground makes concreting an unlined hole a precarious operation. Where a Prestcore pile is used the cross-sectional area of the pile shaft can be guaranteed to be at least that of the diameter of the concrete segments. Its bearing capacity will again depend upon the various strata through which it passes.

The piles are manufactured with nominal diameters of 350, 450 or 650 mm, having 4, 5 and 6 reinforcement holes respectively which can accommodate up to 25 mm diameter bars. The Prestcore pile is constructed by Cementation Piling and Foundations Ltd.

8.3.3 Auger bored piles

The size, and hence load bearing capacity, of tripod bored piles is limited because the piling plant is hand operated. By replacing the gravity cutter of the tripod rig with a mechanically driven auger, outputs are increased and piles of considerably greater depth, diameter and capacity can be produced. The rotary bored pile has seen many improvements in recent years and at present a wide range of pile types is commercially available. It has for many uses replaced the tripod bored system. Because the installation of rotary bored piles requires a large amount of costly production plant, large diameter rotary bored piles are not economical when only a few are required. Like any operation involving complex mechanical plant, breakdown time is vitally important and the profitability of a contract can vanish if outputs drop.

The rotary or auger bored large-diameter pile is installed by a number of specialist contractors using the same basic principles of operation. A digging tool, either an auger, cleaning bucket or under-reamer, is attached to the foot of a long kelly bar which is suspended from a swivel connection. In section the kelly bar is either square or circular with protruding lugs at diametrically opposite points and comprises one single length or up to three lengths which fit one inside another. At about 2 to 3 m above ground level the kelly bar passes through a ring gear which converts the drive from an auxiliary diesel engine into a rotation of the kelly bar. By a system of gears, the speed of rotation and torque can be varied to cope with different digging conditions.

On crane mounted machines, the auxiliary boring engine usually sits on a raised platform in front of the crane unit. The platform is hinged at its lower inward end and connected by adjustable hydraulic pistons at its upper inward end, thus allowing the platform to be adjusted independently of the crane unit. The kelly bar is suspended from the top of the mast by a swivel shackle and rope, which is independent of the mast ropes, and allows the kelly bar to be raised or lowered.

On lorry mounted rigs, the boring unit and mast are usually mounted on a fixed eight-wheel chassis. The mast is hinged about its rear end, lies flat along the length of the vehicle while in transit and is hoisted into the working position by means of hydraulic rams. The boring assembly is mounted on a set of parallel slides which in turn sit upon a 270° revolving turn-table.

Raking piles are bored by inclining the mast. On the crane mounted rig this is done by slightly lowering the crane jib and tilting the boring unit about the hinged lower inward end, digging taking place towards the tracks of the crane. On the lorry rig, the mast is not fully extended to the vertical and the hole is bored away from the rig. With either system it is not possible to bore in the opposite raking direction; the machine has to be turned about.

The auger on a crane rig is aligned over the pile position by movement of the crawler tracks and use of the turn-table; the machine is then stable enough for drilling. On a lorry-mounted rig, the lorry manoeuvres as close as possible to the pile position with the mast in the down position. The outriggers, situated on the chassis below the corners of the boring unit, are extended at the front and at the rear. The mast is then hoisted and the slides and turn-table are used to make the final adjustments to bring the auger directly over the pile position. Once a machine has been positioned correctly over a pile centre the kelly bar is plumbed to the vertical, or the correct amount of rake is applied.

Where a shallow depth of fill or granular material overlies cohesive soil, it is often possible to auger out sufficient material to allow placement of a short length of steel casing, normally called the 'collar' casing. This is driven, say, 1 m into underlying cohesive soil and prevents collapse of the upper portion of the borehole during further boring in the clay. Where the above technique is not possible, three other methods are available to deal with unstable ground.

First, casing can be inserted into the ground prior to boring. In very loose ground this may be possible under its own weight, the weight of the kelly bar and the screwing action of the torque-bar connecting the kelly bar to the casing. If the ground is more compact and predominantly granular, the casing can be sunk into it by the use of a vibrator (Section 8.2.7). The whole vibrator assembly, weighing several tonnes, is mounted on a frame and lowered by an auxiliary handling crane on to the top of the casing. The vibrations are parallel to the longitudinal axis of the casing and greatly reduce the skin resistance between the ground and the casing. The casing may also be extracted by vibration, which assists in compacting the freshly placed concrete.

The second method can only be employed in cohesionless soil. The auger is used to agitate the material, without removing it from the borehole, and dry bentonite powder and water, where necessary, are introduced as the auger loosens the ground. When thoroughly loosened, it is relatively easy to insert the casing under its own weight and excavate from within it. This technique is only successful over fairly small depths, say up to 8 metres, and with machines which can develop a quick and lively torque in the auger.

In the third method of supporting unstable ground, drilling mud, normally a fully hydrated bentonite suspension containing 5 to 7% of bentonite powder, is employed. After a short length of collar casing has been placed, bentonite slurry is pumped into the borehole as soil is removed. The slurry works in two ways. It forms a membrane around the surface of the borehole, this being more pronounced with coarser soil grain sizes, and it acts hydrostatically against this membrane to maintain borehole stability. Once the required depth has been reached, a high slump concrete is placed by tremie (Section 8.3.2).

Two main problems are associated with the bentonite method. First, it requires a large amount of plant to operate an efficient system and it involves frequent monitoring of the suspension to ensure that it conforms to the required standards. It is often rapidly contaminated and must be screened mechanically or treated chemically to remove impurities before it begins to lose its thixotropic properties, thereby leading to pumps and pipes becoming blocked or to local collapse in the borehole. Modern hydrocyclones do however provide a rapid and effective means of removing solid impurities. The fluid is also a bulky material to dispose of and generally it is removed in tankers.

The second main problem with bentonite is associated with one of its virtues. If the membrane or filter cake builds up on the borehole surface of a granular layer over a long period of time and becomes static and contaminated, it can remain there after concreting and reduce the friction between the concrete shaft and the sand or gravel. This problem has been studied by Fleming and Sliwinski (CIRIA Report PG 3) who concluded that it was necessary to keep solid contaminants in a suspension below certain levels and to concrete piles without undue delay. Provided these requirements are observed and construction procedures are carried out to good standards, the effect on friction development is minor and can be regarded as insignificant when using normal design procedures. In cohesive soils filter cake is either non-existent or very thin because of the low soil permeability. The use of bentonite suspension is described in the CIRIA report PG3, where many practical details will be found. The properties and use of bentonite drilling muds are also covered in the FPS* specification for cast in place piles formed under bentonite suspension.

The use of a bentonite suspension also prevents an exposed bore wall from being subjected to zero radial total stress, but little is known about the actual influence of the fluid in restricting stress relief and research is necessary before cognizance can be taken of any beneficial effect in the design of piles.

It is not normal practice to construct piles of diameter less than 600 mm with bentonite because smaller diameter bores become too congested with tremie pipe, steel and incoming concrete to ensure a free outflow of the displaced bentonite. If the reinforcement cage is heavy and contains a large number of bars and links or the concrete is insufficiently workable, the bond between the steel and concrete may not be fully developed because the concrete rising up the borehole is unable to scour the bentonite completely from the reinforcement, especially at the intersection of the bars. The same problem may be found in diaphragm wall construction as reported at the conference on 'Diaphragm walls and anchorages', ICE (1974). In favour of the bentonite construction method it should be pointed out that difficulties with the extraction of long casings and associated integrity risks are avoided.

The most efficient pile is in general one which mobilizes the full permissible concrete stress in its shaft, but in most soils this is not possible unless the length of the shaft becomes excessive. In non-waterbearing cohesive soils however, it is possible to increase the base diameter over the shaft diameter, thus shortening the pile length, increasing the bearing capacity and improving the working stress in the shaft.

Widening of the base, usually called belling or underreaming, is accomplished by replacing the auger with a belling bucket. The latter is a cylinder of the same diameter as the shaft auger, between 2 and 4 m in height, which has a pair of arms opening out from it containing digging teeth. As the bucket is rotated, the downward movement of the kelly bar pushes out the arms, which are hinged about their upper end. Cut lumps of clay fall to the foot of the bore where the folding arms draw them into the base of the bucket. When the kelly bar is lifted, the arms of the belling bucket fold back and allow the bucket to be withdrawn and emptied. It is normal practice not to exceed a ratio of 3 to 1 for the relationship of base to shaft diameter, with the angle of the bell side being about 55° to 60° from the plane of the pile base. The bell geometry is usually completed by a 150 mm vertical upstand on the outside edge of the base and a 150 mm deep socket, of the same diameter as the shaft, below the base. The socket has three

*Federation of Piling Specialists, 15 Tooks Court, London EC4.

purposes: it locates the bottom of the underreamer during digging, it houses part of the mechanism for opening and shutting the arms and it collects the debris cut from the roof of the bell. Once a bell has been completed it is common practice to visually inspect the roof and base prior to concreting and, if necessary, to hand-clean the bottom. Elaborate safety precautions must be taken during this exercise; these are dealt with in Section 8.7. In a certain contract in London, Cementation Piling and Foundations Ltd installed 5.5 m diameter bells on 2.2 m diameter shafts at a total depth of 47 m below ground level.

Large diameter piles can be constructed in the following nominal diameters: 600, 750, 900, 1050, 1200, 1350 and 1500 mm. The maximum load depends on ground conditions. It is economical to found all piles at the same depth and to employ as few shaft sizes as possible, since the cost of transporting casings, augers, belling buckets and cleaning tools is high. A large diameter pile must never be belled in conditions where the bell roof might collapse; for example, when a water-bearing stratum occurs within 2.5 to 3 m of the base of the bell, or in a highly fissured clay or soil which does not possess good cohesion. Using standard equipment, piles up to 35 m long can be constructed but if extensions and modifications are used, piles up to 2.5 m in shaft diameter, and 5.5 m in bell diameter can be extended to about 55 m. The design of large diameter piles is dealt with in Section 8.4.3. It is desirable to discuss with piling contractors all projects in which piles are likely to be employed, at an early stage in planning and design: lists of contractors, including those concerned with large diameter piles, can be obtained from the Federation of Piling Specialists.

Piles constructed by means of hollow stemmed continuous flight augers come under the category of auger bored cast in place piles. There are two basic pile types, each constructed using a hollow shaft single flight auger mounted on a crane or similar unit; the differences occur during the concreting stage. The first involves the use of grout to form the pile and can be used for small diameter piles in the range 300 to 450 mm. Once the continuous flight auger has penetrated to the required depth a grout is injected through the hollow stem, consisting of sand, cement and an intrusion additive. It is pumped into the hole starting at the bottom and working up as the auger is extracted. Reinforcement, if required, is added immediately after grouting, but must be of limited length to ensure complete embedment. In the second type the material used to form the pile is a concrete of high workability and this type of pile is available in diameters up to 750 mm. The auger stem has usually an internal diameter of about 100 mm. After the required depth has been reached with the continuous flight auger, concrete is pumped through the hollow stem as it is withdrawn. This process continues until the auger with the spoil has been removed from the bore, when reinforcement is inserted. The maximum length of reinforcing steel which can be provided is usually about 6 m, but such piles have been reinforced to greater depths, for example using steel joists in place of conventional bars. It is self evident that the reliability of this type of pile depends on matching the rate of grout or concrete supply to the rate of auger withdrawal. Monitoring equipment to ensure that supply and demand can constantly be checked has been described by Derbyshire (1984). A specification for this type of work is available from the Federation of Piling Specialists, London. During construction these piles are relatively free from vibration and noise and can be constructed in ground conditions where, under normal circumstances, casing would have to be inserted and withdrawn.

8.4 BEARING CAPACITY AND SETTLEMENT OF PILES AND PILE GROUPS

8.4.1 Bearing capacity of single piles

It is an essential prerequisite to the design of a piled foundation that the ultimate bearing capacity of the piles be known either before or after installation. For many years, this problem of assessing bearing capacity has taxed and frustrated the minds of engineers. In the early days of piling, the normal method of installation was by striking the head of a pile with a hammer. It was assumed that the harder the driving, the greater the capacity, and generally this has led to piles carrying their loads safely. This dynamic method of estimation has been much considered over the years and many formulae exist in which the pile capacity is calculated from a knowledge of the input hammer energy. This method was seen to be often inaccurate and as the science of soil mechanics grew, more reliable static methods of capacity calculation were evolved, based on the shear strength of the soil in which the piles were installed. The bearing capacity of piles will be considered under the separate headings of dynamic methods and static methods.

8.4.2 Dynamic methods of calculating bearing capacity using driving equations

The simplest and one of the earliest formulae to be proposed (*ca* 1850) for the installation of driven piles, equated the input energy of the blow to the resistance of the ground, that is

$$WH = RS \qquad (8.1)$$

where W is the weight of the hammer; H is the stroke or fall of the hammer; R is the resistance of the ground to the pile (capacity); S is the set of the pile under the blow. This formula was based upon three assumptions: (a) the hammer and pile can be considered as impinging bodies; (b) on impact the hammer releases its entire energy; (c) on hammer impact, the resistance R instantaneously rises to a constant value and remains so while the pile is in motion.

These assumptions are somewhat unrealistic and have been modified over the years; for example it is often now the case that, the pile is treated as an elastic member, Newton's laws of motion relating to impinging bodies are applied, account is taken of losses in energy due to hammer friction, of energy losses at impact, of elastic compression at impact and the soil is treated as a partially elastic medium. Two formulae incorporating these modifications are commonly used; they are the Engineering News formula and the Hiley formula.

In the Engineering News formula for gravity drop hammers

$$WH = R(s + 1.0) \qquad (8.2)$$

and for single acting steam hammers

$$WH = R(s + 0.1) \qquad (8.3)$$

In the simplified Hiley formula

$$R_u = \frac{WH\eta}{(s + c/2)} \qquad (8.4)$$

in which

$$\eta = \frac{K(W + e^2 P)}{(W + P)}, \qquad (8.5)$$

where R_u is the ultimate driving resistance (tonf); W is the weight of the ram or moving part of the hammer (ton); H is the effective fall of the ram (in.), equal to K times the actual fall of the ram, where K is a coefficient which depends on the type of hammer; η is the efficiency of the blow, dependent upon the coefficient of restitution e and the ratio P/W, where P is the total weight of pile, dolly, etc. (tonf); s is the set or penetration of the pile under the blow (in.); c is the sum of the temporary compression of the dolly and packing c_c, compression of the body of the pile c_p and compression of the ground (or quake) c_q (in.).

The BSP *Pocket Book* (1969), for example, contains graphs and tables from which the above constants can be found for various types of piles, hammers and ground conditions. Note that the formulae are here quoted in imperial measure since this conforms with most of the data available.

The Hiley formula is not applicable to base driven piles; the approximate bearing capacity of these can be calculated using the empirical dynamic formula proposed by Cornfield (1968)

$$R_u = \frac{3.6W(3.0 + h)}{(s + 0.5)} \qquad (8.6)$$

where R_u is the estimated ultimate driving resistance (tonf); W is the weight of internal drop hammer (tonf); h is the actual hammer drop at final set (ft); and s is the final set (in./blow).

The formula applies only when the hammer drop at final set is between 1.3 and 1.8 m and when the final set is less than 5 mm/blow. Cornfield states that the formula should be used only for piles driven into soils such as sand, gravel, rock, hard marl or very stiff clay, that the internal drop hammer must strike a plug of concrete inside the base of the pile $2\frac{1}{2}$ times the pile diameter in height after compaction and that the concrete should be 1:2:4 with a water:cement ratio not exceeding 0.25 by weight.

A more recent technique of assessing the static bearing capacity of a driven pile is to use the 'Wave Equation Method', in which a value of Soil Resistance at the time of Driving (or SRD) is obtained from the interaction of several input parameters. An extensive description of the wave equation as a pile driving aid can be found in papers by E.A.L. Smith (1962) and Lowery *et al.* (1969). Some aspects of the method are discussed below. There are three sets of input parameters, one for each of the hammer, pile and soil; a schematic diagram is shown in Fig. 8.9. The concept differs from earlier driving formulae in that Newton's law of motion relating to impact, which was based on the concept of colliding spheres, is no longer applied, but rather the wave motion in the pile is examined. Because digital computer techniques are employed, more complex and variable input parameters can be used.

The energy delivered by a single hammer blow initiates a compressive stress wave which travels down the body of the pile, with a velocity of propagation $v = (E/\rho)^{1/2}$. Along its path the wave energy is dissipated through pile shaft

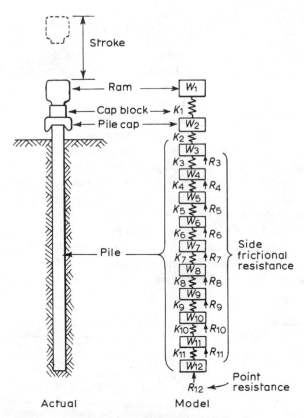

Figure 8.9 Diagrammatic representation of wave equation parameters: W is mass, K is spring force, R is soil resistance (after Smith, 1962).

resistance against the soil, heat (due to the hysteresis of the pile material) and variations in cross-section. If at the pile tip there is any remaining energy, it mobilizes end-bearing soil resistance (causing penetration of the pile if the energy exceeds the tip resistance), and initiates a reflected stress wave.

A wave equation computer program calculates for all elements in the system per time increment, velocities, displacements and forces as generated by the impact blow. Computations are continued until the permanent set of the pile tip is achieved. The time interval selected must be such that the travelling distance of the force wave per time step is significantly less than the length of a segment. Hence, for a segmental length of, say, 2.0 m, a suitable time step would be

$$0.5(2.0 \div \text{velocity of propagation})$$
$$= 0.5(2.0 \div 5000)$$
$$= 0.2 \,\text{ms.}$$

The soil model used in the wave equation program is usually elasto-plastic, and damping values are introduced to compute the instantaneous values of soil resistance during driving,

$$(\text{SRD})_i = \text{SRD}(1 + Jv)$$

where $(\text{SRD})_i$ is the instantaneous value of SRD, SRD is the input value of SRD during driving, J is the damping value of the soil, and v is the velocity of the pile segment relative to the soil. Typical soil values are given in Table 8.5.

The program thus allows a relationship to be established between (1) the pile driver, (2) the pile, (3) the soil resistance along the shaft and below the pile tip, and (4) the permanent set (if any). If constant input values are assumed for items (1) and (2) above, and variable values of (3), then successive values of set (4) can be achieved as output. The corresponding values of (3) and (4) are then plotted to obtain a relationship between soil resistance at the time of driving (SRD) and set: the inverse of set is normally used, which is expressed as 'blows per 25 mm'. Such a graph is shown in Fig. 8.10 for several values of hammer efficiency (0.4 to 0.8).

During pile installation, the SRD can be assessed from a knowledge of the pile driving records (blows per 300 mm) and the corresponding SRD against blow count curve for the particular hammer and soil profile. This is not, however, a direct measure of ultimate bearing capacity, because 'set-up' and/or refusal

Table 8.5 Typical quake and damping values for driven piles

	Quake (mm)		Damping ($s\,m^{-1}$)	
	Side	Point	Side	Point
Sand	2.54	2.54	0.164	0.492
Clay	2.54	2.54	0.656	0.033

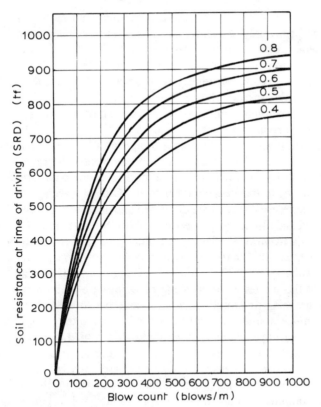

Figure 8.10 Blow count against soil resistance at time of driving (SRD). Note: Conductor: 662 mm × 25 mm; Hammer: Delmag 55; Energy: 18.4 tfm; Penetration: 39 m; Pile Tip in sand; Friction pile – Hammer Efficiency: 0.8, 0.7, 0.6, 0.5, 0.4.

conditions may prevail. When a pile is driven into a cohesive soil, excess pore water pressure will probably be generated. This phenomenon reduces the SRD relative to the long-term ultimate bearing capacity. If a redrive test is carried out some time after installation, then the set-up can be quantified; it is the difference between the restart SRD and the SRD at the end of the previous drive. It is important to note that this value of set-up will correspond to the delay time between the end and restart drives, which will be considerably shorter than the time between installation and when full load is applied. If a pile is driven into a sand stratum, then refusal will most probably occur: refusal is normally defined as a blow count per 25 mm of more than about 8. The ultimate bearing capacity will be in excess of the refusal SRD obtained from the appropriate driving curve.

The wave equation method of analysis is used exclusively in the offshore oil industry, where small numbers of very large piles (up to 1.8 m diameter) are driven with extremely large hammers (approximately 200 m tonnes) to provide loads up to 3000 tonnes. In this context, considerable time and money are spent in the design stage, principally to ensure that those very large piles can be driven to the required penetrations with the available hammers. Failure to achieve the pile design penetration in the offshore environment can be financially punitive, because extremely costly remedial measures will have to be adopted (Rennie and Fried, 1979), with consequent cash-flow penalties for the development of the oil field. A separate static bearing capacity assessment must always be carried out in tandem with a driving analysis.

The use of the wave equation as a design tool for on-land piles is limited, although as post-installation analyses are completed, the required amount of confidence will be obtained to make its use more widespread. Numerous references can be found in the papers of the Offshore Technology Conference, Houston, where case histories are described and analytical parameters evaluated and discussed and also in the proceedings of two Stress-Wave conferences held in Stockholm. Conventional driving formulae are not generally applicable to deposits such as silts, marls, clays and other soils which can be classed as possessing cohesion. CP 2004 (BSI, 1972d) states that if driving formulae are restricted to those piles whose bearing capacities are predominantly derived at the toe in gravels, sands and other non-cohesive soils, then one of the more reliable formulae should give a calculated result within the range of 40 to 130% of the ultimate bearing capacity that would be determined by test load. Whenever possible, a load test on a driven pile should be carried out. Every care must be taken with the test pile to ensure it will be representative of the working piles. If the estimated load and the test load differ significantly, then other formulae should be examined to find the most appropriate one. It can subsequently be adapted by an overall coefficient dependent upon all those points peculiar to the site in question. Should any of these characteristics alter, then a re-appraisal must be made between the test pile and the driving formula: factors include for example, change of the winch operator, the depth of overburden to the bearing stratum, the position of the water table and the type of hammer.

8.4.3 Static methods of calculating bearing capacity

The bearing capacity of a pile is a summation of shaft resistance and base resistance. When a load is applied to the head of the pile, it settles by an amount governed by the load, the pile geometry and the soil characteristics. Downward movement of the pile mobilizes the shear strength of the soil at the base and a force around the shaft which is frictional and/or adhesive. Prior to the installation of the pile, the ground was in equilibrium and supported a column of soil of weight equal to the unit density of the soil times the pile volume; it now supports a force equal to the weight of the pile body. Equating these forces allows the ultimate bearing capacity of the pile to be assessed as,

$$Q_f + W = f_f A_s + q_f A_b + \gamma D A_b \tag{8.7}$$

where Q_f is the ultimate bearing capacity; W is the weight of pile body below ground level; f_f is the ultimate tangential force per unit area of the shaft due to friction and/or adhesion; A_s is the area of the shaft over which f_f is considered operative; q_f is the ultimate resistance of the base per unit area due to the shear strength of the soil at the base; A_b is the area of the base; γ is the average unit weight of soil replaced by the pile body; D is the length of the pile shaft below ground level.

The difference in unit weights between the displaced soil and the concrete shaft of the pile is approximately $450\,\text{kg}\,\text{m}^{-3}$. Considering the magnitude of the capacity of most piles, this represents an error which, although on the unsafe side, is usually neglected with little practical loss in accuracy. The ultimate bearing capacity equation thus becomes

$$Q_f = f_f A_s + q_f A_b \qquad (8.8)$$

It is unusual for a pile to be founded in a soil which possesses cohesion and friction to the same degree; usually one or other predominates in terms of the important short term consideration of capacity. If a pile shaft passes through alternate layers of cohesionless and cohesive soils, the contribution to the bearing capacity in each layer is found separately. Caution is necessary to ensure that in the summation, the strains at working loads are compatible for the different soil types and that the proportions of cohesive and cohesionless strata are correctly represented.

(a) Bearing capacity in granular soils

In granular soils, the cohesion $c = 0$ and the terms relating to cohesion and adhesion are omitted from the expressions for bearing capacity discussed in Section 3.3.

The Standard Penetration Test

The bearing capacity coefficients referred to in Section 3.3 depend partly upon ϕ, but it is impossible to measure the *in situ* value of ϕ in laboratory tests because of inevitable disturbance in sampling. During driving, displacement piles usually compact cohesionless soil, thus altering the *in situ* shear strength, the unit weight and the angle of shaft friction. Generally, the value of ϕ is assessed by correlation with *in situ* density derived from the results of standard penetration tests (Section 1.2). Relevant correlations given by Peck *et al.* (1953) are shown in Fig. 8.11. A widely used relationship between ϕ and N_q proposed by Berezantsev *et al.* for deep driven pile foundations, is shown in Fig. 8.12. Meyerhof (1956) proposed the following approximate empirical equation for the ultimate load on a single displacement pile in saturated sand:

$$Q_f = 400\,N A_b + 2\bar{N} A_s, \quad (\text{unit skin friction } f_s \not> 100\,\text{kN}\,\text{m}^{-2}) \qquad (8.9)$$

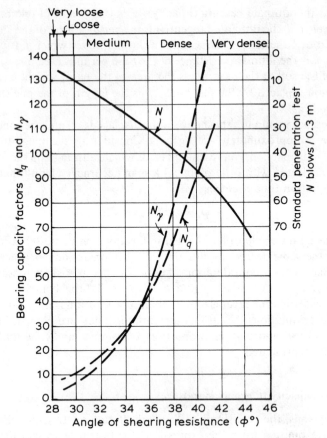

Figure 8.11 Relationship between ϕ, bearing capacity factors and N values from the Standard Penetration Test (after Peck, Hanson and Thornburn, 1953).

For small-displacement piles, this becomes:

$$Q_f = 400\,NA_{nb} + \bar{N}A_{gs}, \quad (f_s \not> 50\,\text{kN}\,\text{m}^{-2}), \tag{8.10}$$

where N is the Standard Penetration Test result at the base; $\bar{N}$ is the mean N value along the shaft; A_b is the area of pile base; A_s is the area of pile shaft; A_{nb} is the net area of pile base; A_{gs} is the gross surface area of pile including web and flanges. Note that all units in the above equations are in kN and metres.

Two corrections are often applied to the N values. In silt and fine sand at and below the water table, Terzaghi and Peck (1967) and Peck and Bazarra (1969) suggest the value of N should be taken as $N = 0.5\,(N' - 15) + 15$ where N' is the actual value obtained during the test; the correction is applied only if $N' > 15$. Meyerhof (1965), D'Appolonia *et al.* (1970) and Parry (1971) contend, however, that the effect of the water is automatically reflected in lower test values of N and

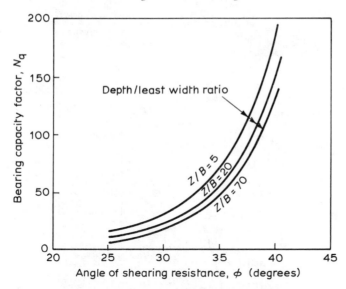

Figure 8.12 Values of bearing capacity factor plotted against angle of shearing resistance. (after Berezantsev *et al.*).

that no further reduction is required. The correction for overburden stress was introduced by Gibbs and Holtz (1957) and values of this are shown in Fig. 8.13. Unless the piles are short, this correction can be neglected.

The skin resistance of a pile driven or bored in a granular soil depends upon several factors, including the material of the pile, the method of driving, the initial soil density, the grading of the soil, the position of the water table and the length of embedment. If the pile does not found in a resistant bed, the total pile capacity should be calculated only upon the shaft load. If, however, the pile point bears in dense sand or gravel, or upon bedrock, the shaft friction may not be fully mobilized under working conditions. However since factors of safety are applied to ultimate loads which can only involve significant pile displacement and which would mobilize full friction, it is usual to include in calculations for all piles, both shaft friction and end bearing. Potyondy (1961) found values of the ratio of the angle of foundation shaft friction (δ) to the soil internal angle of friction (ϕ) from shear box tests using wood, steel and concrete as the foundation surface: these are given in Table 7.11.

Using the average value of N along that part of the shaft considered as load bearing, the corresponding ϕ angle can be found from Fig. 8.11 and hence $\tan \delta$. From the various charts and figures mentioned above, the ultimate pile capacity can be found to which an appropriate factor of safety must be applied. Typical calculations for pile capacity are illustrated in Example 3.7.

It is often desirable not only to calculate the ultimate bearing capacity of a pile,

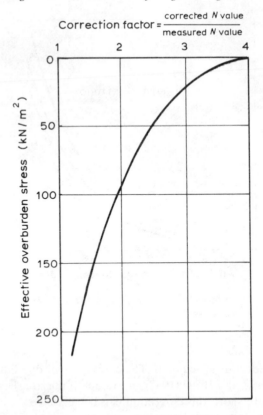

Figure 8.13 Correction factors for Standard Penetration Test results (after Gibbs and Holtz, 1957).

but also to check anticipated settlement under the design loading. As a first approximation it can be assumed that full ultimate shaft friction will be mobilized at 0.75 to 1% of the pile diameter, with friction increasing linearly to this point and then remaining constant. The settlement of a circular pile base can be calculated from the equation $E = \pi/4.q/\rho.D(1 - v^2) f$ where E = Modulus of soil elasticity, q = applied base pressure, ρ = allowable settlement, D = pile diameter, v = Poisson's Ratio (usually 0.3) and f = a depth factor (in the range 0.5 to 0.75). In undisturbed sands or gravels E is approximately 750 N to 1000 N where N is a representative Standard Penetration value in the region of the pile toe. Taking elastic shortening of the pile into account, an approximate load/settlement graph can then be drawn and the approximate pile behaviour may be checked against structural requirements and the Factor of Safety. For driven piles the E/N ratio will be enhanced if the soil compacts whereas if a bored pile base is loosened the bearing capacity will be reduced.

The static (Dutch) cone penetrometer

The static cone penetrometer is widely used for site investigation in Europe: the apparatus and procedure are described in Section 1.2. As the cone is advanced into the ground at a constant rate of up to 20 mm s^{-1}, readings of base resistance q_c and shaft resistance f_s are taken at regular intervals or, in the case of the electric cone, continuously (Fig. 8.14). Sanglerat (1972) gives a detailed account of the penetrometer. Experience has shown that the penetrometer can be considered as a model pile and the recorded values of q_c and f_s, subject to minor modifications, used directly to calculate ultimate pile bearing capacity.

When the approximate founding level of the pile has been selected, the method suggested by Van der Veen (1957), based on field test results, is commonly used to allow for the scatter of results in and around the cone point and to allow for the shear zone which is above and just below the pile point. The average cone

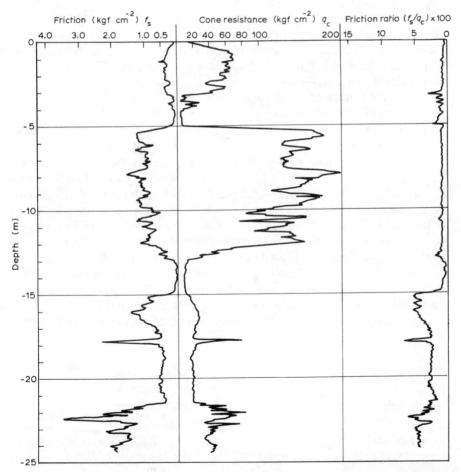

Figure 8.14 Cone penetration test for temporary dry dock on Seal Sands, Teesside.

resistance is taken over a distance of $3.75B$ above and $1B$ below the point level. This value of q_c is used directly to calculate ultimate base bearing stress q_u. The procedure does not take into account the difference in scale between the cone point and the pile point.

An alternative procedure has been described by Fleming and Thorburn (1984) who consider that a pile may be treated as deeply embedded when its depth into a fine cohesionless stratum is more than 8 diameters. Procedures for calculating ultimate base resistance are given, also based on the cone resistance both above and below the pile base level.

Whatever the founding level, the penetration test should be taken to a suitable depth below its estimated position. The allowable bearing stress is obtained by dividing q_c by a factor of safety, which should be not less than 2 and may be between 2 and 5, depending on the depth of embedment, the higher factors being applicable to shallower depths.

Where side resistance has been measured using a short movable sleeve just above the cone point, the f_s value can be used directly to calculate the ultimate shaft capacity. Side resistance values obtained from penetrometers not equipped with a short sleeve should be treated with caution (Sanglerat, 1972). Although the measured resistance is of steel to soil, while that of the pile to soil depends upon the method of construction, the elapsed drive time, the pile spacing, the depth of embedment and the soil characteristics, Belgian and Dutch engineers have found that there is a reasonably linear relationship between the measured value and the pile value. A safety factor of between $2\frac{1}{2}$ and 3 is usually recommended.

It is not always possible to obtain values of f_s directly and various attempts have been made to correlate f_s with q_c. The ratio depends in sands on the *in situ* density, with f_s/q_c varying between 2 and 0.5%. Schmertmann (1969) found for sands in Florida, USA, the values shown in Table 8.6.

Begemann (1965) and Schmertmann (1969) have classified various soil types according to the relationship between the cone base resistance and the ratio of the shaft to base resistance, as shown in Fig. 8.15. Sanglerat found similar relationships for soils tested in France. These diagrams should be used as a guide only to soil type identification and, where possible, deduced soil characteristics should be verified by correlation between cone tests and borehole samples.

Many authors have attempted to compare SPT N values and cone q_c and f_s

Table 8.6 Relationship of cone friction to cone end bearing for sands (after Schmertmann 1969)

	Loose	Medium	Dense
Relative density (%)	30	60	90
f_s/q_c	1/60	1/120	2/330
%	(1.7)	(0.8)	(0.6)

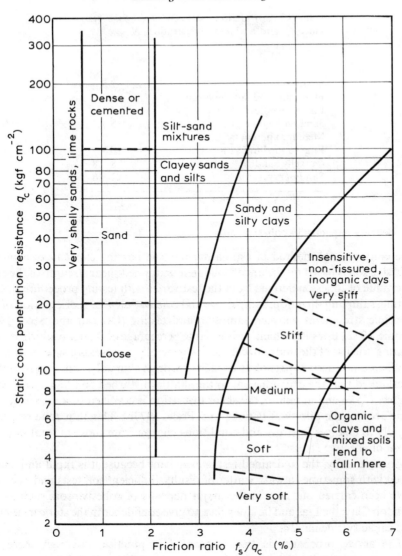

Figure 8.15 Guide for identifying soil type from Dutch friction-cone ratio (Begemann mechanical tip) after Schmertmann, 1969.

values but correlations should be treated with caution (Meyerhof, 1956; Meigh and Nixon, 1961; Rodin, 1961; Sutherland, 1961; Sanglerat, 1972; Belshaw, 1973). Sutherland (1974) has given the following correlation for various soils (Table 8.7)

In situ testing was the subject of a conference in Stockholm (1974) and reference may be made to its proceedings entitled 'Symposium of Penetration Testing'.

Table 8.7 Correlation between cone base resistance, q_c and Standard Penetration, N

Soil	q_c/N
Sandy silt	2.5
Fine sand and silty fine sand	4.0
Fine to medium sand	4.8
Sand with some gravel	8.0
Medium and coarse sand	8.0
Fine to medium sand	10.0
Gravelly sand	8–18
Sandy gravel	12–16

(b) Bearing capacity in cohesive soils

The conventional method of estimating the safe bearing capacity of a pile in cohesive soil is based on the unconsolidated undrained shear strength of the clay. The value of cohesion derived from the test varies with testing procedure, rate of testing, sampling direction, sample size and orientation, and reorientation of the principle stress axes between sampling and testing (Duncan and Seed, 1966; Rennie, 1972). For piles which derive a large percentage of their capacity from end bearing, the use of the undrained cohesion generally represents a safe lower limit.

When a pile derives part of its capacity through skin adhesion, it must not be overlooked that the clay adjacent to the shaft is disturbed to a lesser or greater degree. Up to and at failure, the shear distortion is confined to a relatively thin zone around the pile shaft (Cooke and Price, 1973b). This thin zone responds rapidly to changes in stress and conditions change from those of total stress to those of effective stress.

Commercially, the undrained test is expedient because it is rapid and cheap, albeit with sometimes a wide scatter of results. Sufficient soil tests and pile tests have been carried out now in the major deposits of cohesive soil, such as the London Clay, the Lias and boulder clay, to give confidence in the static method of pile capacity calculation.

The above procedure is the standard UK practice, although there are alternative methods, such as the standard penetration test and the static cone penetrometer.

Bearing capacity based on undrained cohesion

From the general equation

$$Q_f = f_f A_s + q_f A_b,$$ (8.11)

it has been found that

$$f_f = \alpha \bar{c}_s$$

and

$$q_f = N_c W c_e$$

where c_s is the average undrained cohesion of the soil surrounding the shaft, c_e is the average undrained cohesion at the base, α is the adhesion factor, N_c is the pile bearing capacity factor, assuming that the rupture zones are entirely below ground surface, and W is a bulk soil strength factor.

The bearing capacity factor N_c for deep piles has been examined by several authors. Meyerhof (1951) analytically found $N_c = 9.3$ and 9.8 for smooth and rough bases respectively, while Whitaker and Cooke (1966) quote values of 8.0 (Wilson) and 8.5 (Gibson). After examining various analytical and experimental values, Skempton (1951) concluded that $N_c = 9$ was accurate enough for most practical problems. Whitaker and Cook (1966) accepted a value of $N_c = 9$, but introduced a bulk strength factor W to account for a value of $N_c = 6.75$ derived from field observations on large-diameter augered piles.

In cohesive soils a distinction must be made between driven and bored piles.

Driven piles

When a pile is driven into cohesive soil, excess pore-water pressure can be set up adjacent to the shaft and, because of the very low permeability of such soils, the excess pressure takes time to dissipate. As consolidation proceeds, the shear strength gradually increases until it approaches the maximum *in situ* value. This time-dependent phenomenon influences the carrying capacity of the pile and as much time as possible should be left between installation and testing. High pore-water pressure is not normally induced during driving in stiff over-consolidated clay which, because of its low *in situ* water content and over-consolidated nature, does not significantly consolidate under the increased radial stress. The clay usually adapts to the driven pile by heaving at the surface by an amount dependent on the displacement volume and the soil characteristics such as over-consolidation ratio, water content and frequency of fissures. High strains occurring during pile driving can result in the shear strength of over-consolidated clays falling to a residual value as low as one-half of the peak strength.

Table 8.8 Adhesion factor for piles driven into clay soils of various undrained shear strengths (after Tomlinson, 1975)

Adhesion factor α	Unconsolidated undrained cohesion $c_u (kN\,m^{-2})$
0.75	0–25
0.60	25–50
0.50	50–75
0.40	75–125
0.33	125–175

The relationship between adhesion and undrained clay shear strength is referred to as the adhesion factor and is a complex parameter to define. Table 8.8 contains for various values of undrained cohesion, the corresponding value of the adhesion factor. These must be considered average design values to be used when information about a particular case does not exist. When tapered piles are used, the values in Table 8.8 can be increased by 20% (Tomlinson, 1975). Tomlinson (1970) further concluded that the mobilized adhesion is influenced by the succession of strata into which the pile is driven, as indicated in Table 8.9 and Fig. 8.16. When piles are driven into fissured clay the fissures tend to close, hence the parameter W assumes the value of unity.

Hence for a driven pile the ultimate load may be expressed

$$Q_f = \alpha c_s A_s + 9 c_e A_b \tag{8.12}$$

The American Petroleum Institute have proposed the following relationship α between f_f and c_u:

'f_f shall not exceed c_u or the following limits:

(1) For highly plastic clays, or for under-consolidated and normally con-solidated clays, f_f may be equal to c_u. For over-consolidated clays, f_f shall not exceed $50 \, \text{kN m}^{-2}$ for shallow penetrations or c_u equivalent to a normally consolidated clay for deeper penetrations. The normally con-solidated value of c_u can be determined for most marine clays from the relationship $c_u/\sigma_0' = 0.11 + 0.0037 \, I_p$ where σ_0' is the consolidation over-burden pressure at a particular level and I_p is the plasticity index.

(2) For other types of clays, f_f shall be taken equal to c_u for c_u less than or equal

Table 8.9 Design values of adhesion for piles driven to various penetrations in stiff cohesive soils (after Tomlinson, 1970)

Case	Soil conditions	Penetration ratio*	Adhesion factor α
I	Sands or sandy gravels overlying stiff to very stiff cohesive soils	< 20 > 20	1.25 $c_u \not> 70 \, \text{kN m}^{-2}, \alpha = 1.25$ $70 < c_u < 175, \alpha = 1.75 - 0.007 c_u$
II	Soft clays or silts overlying stiff to very stiff cohesive soils	< 20 (but > 8) > 20	0.40 0.70
III	Stiff to very stiff cohesive soils without overlying strata	< 20 (but > 8) > 20	0.40 $c_u \not> 90 \, \text{kN m}^{-2}, \alpha = 1.0$ $90 < c_u < 150, \alpha = 1.9 - 0.01 c_u$

$*\text{Penetration ratio} = \dfrac{\text{depth of penetration into stiff to very stiff cohesive soil}}{\text{diameter of pile}}$

Note: adhesion factors shown are not applicable to H-section piles.

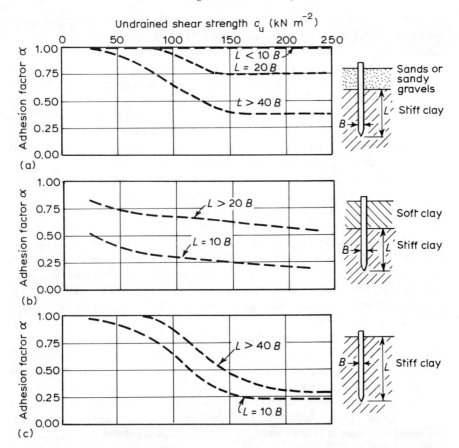

Figure 8.16 Adhesion factor against shear strength for various penetrations into stiff clay. (a) piles driven through overlying sand or sandy gravels; (b) piles driven through overlying soft clay; (c) piles without different overlying strata. (After Tomlinson, 1970)

Note: These curves are not applicable to H or cruciform sections or to bored or driven cast in place piles; safety factor should be not less than 2.5 except for design based on adequate loading test data.

to $25 \, \mathrm{kN \, m^{-2}}$. For c_u in excess of $25 \, \mathrm{kN \, m^{-2}}$ but less than or equal to $75 \, \mathrm{kN \, m^{-2}}$, the ratio of f_f to c_u shall decrease linearly from unity at c_u equal to $25 \, \mathrm{kN \, m^{-2}}$ to one-half at c_u equal to $75 \, \mathrm{kN \, m^{-2}}$. For c_u in excess of $75 \, \mathrm{kN \, m^{-2}}$, f_f shall be taken as one-half c_u.'

Bored cast in place piles

Cast in place bored piles can be divided into percussive bored piles (normally not greater than 600 mm diameter) and auger bored piles (up to 2.5 m shaft diameter and perhaps with up to 5.5 m bell diameter). When a pile borehole is opened in cohesive ground, stress relief occurs which is accompanied by migration of pore water to the open bore and opening of any adjacent fissures. The longer is the time taken in boring, the greater the deterioration in the sides of the bore, although the action of the boring tool does have a smearing effect on the clay and can partially cover small fissures. Boring under bentonite, seldom executed in cohesive soils, can be beneficial because it prevents a fall of the total radial stress to zero and the development of negative pore-water pressure in the clay borewall (Kirkpatrick and Rennie; 1975). The hydrostatic bentonite pressure may be sufficient to equal $K_0 \sigma_1$ for a normally consolidated clay or a fraction (say 1/6) of σ_3 for an over-consolidated clay. The clay may take water from the concrete when the pile is poured but there is unlikely to be any water content change in the clay beyond about 60 mm from the wall, depending largely on the extent of fissures. In certain heavily fissured clays the sides of the bore can deteriorate rapidly and the strength falls to low values which can be found from appropriate laboratory tests. Shaft adhesion may be estimated from these values and because this will be a lower limiting value, no significant further safety factor need be applied. Whitaker and Cooke (1966) found that the London Clay at Wembley had a constant residual softened shear strength of about $16 \, \mathrm{kN \, m^{-2}}$; the original shear strength varied from 100 to $250 \, \mathrm{kN \, m^{-2}}$.

Care must be exercised in ensuring that adhesion is confined to that part of the shaft which will always contribute adhesion. The seasonal desiccation zone, fill and ground which may subsequently be removed must be omitted in the calculation of shaft resistance.

Skempton (1959) examined for London Clay the value of the adhesion factor α as the ratio of unit skin adhesion to the average undrained shear strength, $f_\mathrm{f}/\bar{c}_\mathrm{u}$. He found that α lay between values of 0.3 and 0.6 and suggested a design value of 0.45 for values of undrained cohesion of up to about $220 \, \mathrm{kN \, m^{-2}}$ and furthermore that f_f should be limited to about $100 \, \mathrm{kN \, m^{-2}}$. Pile test results in other types of clay seem to indicate that the value of $\alpha = 0.45$ is uniformly acceptable. Whitaker and Cooke (1966) found from scale tests in London Clay that $\alpha = 0.45$ lies close to their mean shear strength profile.

A detailed analysis of many tests on piles constructed in cohesive Glacial till has been carried out by Weltman and Healy (1978) who show that the adhesion factor α varies from near 1.0 at shear strength $c_\mathrm{u} = 60 \, \mathrm{kN \, m^{-2}}$ to about 0.4 when $c_\mathrm{u} = 200 \, \mathrm{kN \, m^{-2}}$. They suggest that in this type of ground a good first approximation is to assume an adhesion value of $70 \, \mathrm{kN \, m^{-2}}$ throughout the strength range.

In regard to end bearing the value of N_c can be assumed equal to 9 although a bulk soil strength factor is now generally applied, especially where large-diameter piles are concerned. As mentioned earlier, Whitaker and Cooke (1966) found from their large scale field tests that $N_\mathrm{c} = 6.75$ but it is preferable to accept the value of $N_\mathrm{c} = 9$ for all clays and to vary the bulk soil strength factor W for various combinations of pile size and clay type. For large diameter piles without and with bells in London Clay, they suggest that $W = 0.75$. This factor accounts for change in stress within the clay when the bore is drilled, left open and refilled with concrete. For percussive bored piles less than 600 mm diameter, W is normally assumed to be equal to 1.0.

Hence for bored piles not exceeding 600 mm diameter in stiff fissured clay

$$Q_f = 0.45\bar{c}_s A_s + 9c_e A_b \qquad (8.13)$$

and for bored piles greater than 600 mm diameter

$$Q_f = 0.45\bar{c}_s A_s + 9(0.75)c_e A_b \qquad (8.14)$$

where $0.45\bar{c}_s \not> $ approx. $100 \, \text{kN m}^{-2}$. Factors of safety will be considered later.

Modifications are sometimes applied to the above equations but they are in detail and not principle. In the absence of test pile data, these equations are adequate. When calculating the shaft capacity of a belled pile, a length of shaft above the top of the bell is generally neglected to compensate for any relaxation in that part of the clay due to the presence of the bell. Estimates are usually about 1.5 times the height of the bell, measured from the base of the bell. A detailed account of load transfer has been reported by Whitaker and Cooke (1966).

Bearing capacity based on in place measurement

This method is used where piles have previously been installed and tested, and have yielded some design parameters or where SPT results or Dutch cone penetrometer results are available. For driven concrete, steel or timber bearing piles, Tomlinson (1957) gave the values for adhesion shown in Table 8.10. For tension piles it is usually assumed that adhesion is the same as for a compression pile, though there is some disagreement on this subject and values of friction reduced by up to 50% have sometimes been advocated particularly for load cases involving cyclic loads and load direction reversals.

Table 8.10 Relationship between cohesion and adhesion for concrete, steel and timber piles in cohesive soil (after Tomlinson, 1957)

Pile material	Soil consistency	Cohesion $(kN\,m^{-2})$	Adhesion $(kN\,m^{-2})$
Concrete or timber	Very soft	12	12
	Soft	12–24	12–23
	Medium stiff	24–48	23–36
	Stiff	48–96	36–46
	Very stiff	96–192	46–62
Steel	Very soft	12	12
	Soft	12–24	12–22
	Medium stiff	24–48	22–34
	Stiff	48–96	34–35
	Very stiff	96–192	35–37

For a simple type of static cone penetrometer without a sleeve, Sanglerat (1972) states

$$c_u = q_c/10,$$

where q_c is the point resistance ($kgf\,cm^{-2}$). Where there is a 100 mm long sleeve above the point,

$$c_u = q_c/20.$$

The average local side resistance, obtained from total side resistance measurements, can be expressed as

$$f_s = q_c/50,$$

but where local side resistance is measured on the sleeve,

$$f_s = 3q_c/50.$$

The standard penetration test (SPT) is not very widely used in cohesive soils and in the United Kingdom at least, the undisturbed sampling technique has generally been preferred. However a detailed study by Stroud (1974) shows that in insensitive stiff clays such as those of the London Clay, Bracklesham Beds, Oxford Clay, Kimmeridge Clay, Woolwich and Reading Beds and Upper Lias and in stiff boulder clay and laminated clay reasonable correlations can be used, provided the standard penetration test is carried out carefully using a standard Raymond split spoon sampler in a 150 mm or 200 mm diameter borehole. His work was based on the use of an automatic Pilcon trip monkey (63.5 kg) falling through a height of 760 mm. In general terms it was found that the shear strength $c_u = 4.5\,N$ where N is the standard penetration result, though for soils with plasticity indices less than about 30% the relationship may change to $c_u = 6\,N$ or more.

Design factors of safety
For replacement piles in granular and cohesive soils the factor of safety is a very complex parameter. Whitaker and Cooke (1966) state that 'the derivation of a factor of safety for the calculation of working load from the ultimate bearing capacity to ensure that the latter is not exceeded demands two decisions: one concerning the level of known risk to be accepted and the other to cover contingencies and the unknown effect of using parameters derived from one test site as representative of others'. They also state that a factor of safety must limit the pile settlement at working load; this will be considered further in Section 8.4.4.

A factor of safety of not less than 2 based on a reliable calculation method should be used for piles which derive the majority of their load from the shaft, that is, piles in granular soils and long straight-shafted piles in cohesive soils. Where there is doubt about the accuracy of the design parameters or the certainty of the calculation method, the safety factor should be increased to at least 2.5 if not 3.0. When a pile is designed so that the load is shared equally by the shaft and base, as in the case of long underreamed piles in cohesive soils and short straight piles in granular soils, cognizance must be taken of the different load/settlement characteristics of each (see Section 8.4.4). For large diameter bored piles in cohesive soil Tomlinson (1975) suggests that a factor of safety of $2\frac{1}{2}$ on the ultimate load as given by the sum of the base resistance and skin resistance should ensure that settlement at working load will not exceed a tolerable value. Whitaker and Cooke (1966), Burland et al. (1966) and Skempton (1966) differentiate between the factor of safety for the shaft and the base.

Burland *et al.* (1966) suggest that the maximum safe working load (SWL) can be taken as the lesser of the two values given by

$$\text{SWL} = \frac{Q_f}{2} \tag{8.15}$$

$$\text{SWL} = Q_f \text{ (shaft)} + \frac{Q_f(\text{base})}{3} \tag{8.16}$$

Skempton (1966) states that for a base diameter of less than 1.8 m

$$\text{SWL} = \frac{Q_f}{2} \tag{8.17}$$

$$\text{SWL} = \frac{Q_f(\text{shaft})}{1.5} + \frac{Q_f(\text{base})}{3.0} \tag{8.18}$$

whichever is the lesser, and for bases greater than 1.8 m the working load should be evaluated from settlement calculations.

The designer of piled foundations should treat the factor of safety with respect and not necessarily apply the same value to all the piles irrespective of geometry. Analysis of load distribution on large rigidly capped and centrally loaded pile groups in an elastically deforming soil mass shows that for piles of uniform length the loads on peripheral piles increase while those in the centre of the group diminish. The computer programs PGROUP, DEFPIG and PIGLET (Randolph and Wroth, 1979, Banerjee and Driscoll, 1978 and Poulos and Randolph, 1983), which may be used to compute pile loads in these circumstances, demonstrate this effect clearly. It is not generally necessary to take this load variation into account specifically in the design of such pile groups and it is sufficient to ensure that the average factor of safety is sufficient, except in so far as the material stresses in peripheral piles may be exceeded and the bending stresses in the cap may require additional consideration.

One approach to this problem is for example to use longer, and hence apparently stiffer piles beneath the centre of a raft foundation, while using shorter piles, perhaps of a larger diameter around the periphery. Thus the support may be more evenly distributed.

(c) Bearing capacity in shale, chalk and other rocks

Table 3.6 gives presumed bearing values for rocks. When a pile founds upon rock, even a soft rock, it is likely to gain most of its capacity from end bearing. The settlement at working load will be extremely small, even for large diameters and loads, provided that for bored piles either a socket to penetrate through weathered rock is formed or that the rock is so hard as to make further penetration impractical. The same is true of driven piles where virtual refusal is attained, although care must be taken not to damage any precast pile shaft. The safe pile working load for many rocks is often limited by the safe working stress of the shaft material. With rock classes 1, 2 and 3, Table 3.6, refusal is usually met after only small penetration. With the softer rocks, like shale, marl and chalk, bearing capacity is frequently estimated from the results of standard penetration tests.

Marls and shales can be loosely classed together and vary from hard, intact deposits with high bearing capacity to soft, weathered and considerably jointed deposits with poor bearing capacity. When the SPT results are relatively low, say less than 50, the bearing values given in Table 3.6 can be used and verified by test loading, but when the SPT results are greater than 50, a better quality rock is indicated. An approximate design guide is to take the safe shaft resistance equal to $0.9\,\bar{N}_s$, with an upper limit of $150\,\mathrm{kN\,m^{-2}}$ and the safe end bearing equal to $15\,N_b\,\mathrm{kN\,m^{-2}}$, where $\bar{N}_s$, is the average of the SPT results over the entire length of the shaft and N_b the average value at and below the base.

Chalk as a material has received a great deal of study in recent years. Ward et al. (1968) published the results of research into the chalk at Mundford, Norfolk, and graded chalk in five distinct classes as shown in Table 1.21. Wakeling (1970) developed the classification further by introducing SPT results for the 5 grades (Table 1.21). Chalk sampled in a standard site investigation hole seldom resembles the descriptions given in Table 1.21, due to disturbance during extraction. SPT results obtained using modern equipment and good techniques should, however, give a reasonably good correlation but should be examined carefully in the light of geological information. It is not generally considered good practice to found piles in chalk where $N < 10$.

A report on piling in chalk by Hobbs and Healy (1979) provides a good basis for the design of various types of driven and bored piles. For bored piles approximate ultimate shaft resistance in $\mathrm{kN\,m^{-2}}$ is as follows:

SPT result N: 10, 15,   20,   25,   30, > 30
Unit ultimate shaft resistance: 35, 70, 105, 170, 250,   250

The ultimate shaft resistance is limited at a value of $250\,\mathrm{kN\,m^{-2}}$.
Ultimate base resistance is given by the following relationship:

For $N_b < 30$, Ultimate bearing stress $= 240\,N\,\mathrm{kN\,m^{-2}}$
For $N_b > 40$, Ultimate bearing stress $= 200\,N\,\mathrm{kN\,m^{-2}}$

For values of N_b between 30 and 40 a linear increase in the factor applied to N_b may be assumed.

In the case of driven piles, either preformed or cast in place, the ultimate friction values vary, depending on the method of installation, and reference should be made to the report for further information.

Factors of safety are generally taken as 2 for safe shaft resistance and 3 for end bearing. The report suggests other ways in which factors of safety may be used but in general terms it is sufficient to use the values stated, with the proviso that in order to limit settlement, it is often prudent to ensure that the design load on a pile does not exceed the ultimate shaft resistance load.

In chalks, the danger of 'swallow holes' and large fissures is always present and it should be borne in mind at the site investigation stage that it is important to recognize such features near which the level and quality of chalk may vary rapidly. Rennie and McDonald (1976) report a case of concrete which had been placed by the tremie method, flowing from pile bores into adjacent fissures and cavities. When this occurred, large quantities of concrete were involved, which were traced by ground cores to be as much as 2.5 m from the piles.

The testing of a pile in chalk is a critical operation and must be carried out correctly if a representative result is to be obtained (see Section 8.6).

Meigh (1970) describes a series of tests on driven cased piles in chalk near Chatham. He found that if an allowance for the skin resistance is deducted from the estimated ultimate load of the test piles, the resulting values are broadly in agreement with the ultimate values calculated from the final driving resistance using the formula devised by Cornfield (1968) (see Section 8.4.2). Meigh observed the skin resistance and end bearing to be time-dependent parameters and that the values of $29 \, \text{kN} \, \text{m}^{-2}$ skin resistance and $2000 \, \text{kN} \, \text{m}^{-2}$ end bearing indicated a safety factor of 3. In chalk it is usually advisable to drive piles to a length and not to a set, since under impact the chalk may lose a large part of its strength and give a poor set. After allowing a suitable recovery period the pile may be load-tested in order to verify the design drive length. Should the ground conditions, the hammer weight or the pile geometry change, a further pile test is advisable.

(d) Bearing capacity based on effective stress principles

Over the last 15 years several attempts have been made to assess shaft resistance in clays, based on effective stress principles. Burland (1973) proposed that

$$\tau_s = \sigma'_h \tan \delta,$$

where τ_s is the shaft friction at any point, σ'_h is the horizontal effective stress on the pile at that point and δ is the effective angle of friction between clay and pile shaft. Assuming that

$$\sigma'_h = K\sigma'_0$$

where σ'_0 is the effective overburden stress,

$$\tau_s = K\sigma'_0 \tan \delta,$$

or

$$\tau_s/\sigma'_0 = K \tan \delta = \beta,$$

where β is analogous to the adhesion factor α ($= f_f/\bar{c}$). Both K and δ are fundamental parameters which can be evaluated and the ratio τ_s/σ_0' or β can be compared with values found experimentally. At failure in soft clays the soil close to the pile shaft is in a remoulded state, hence, according to Tomlinson (1970) $\delta = \phi_d$, the drained angle of friction of the remoulded soil. For a normally consolidated clay $K_0 = (1 - \sin \phi_d)$, but for a driven pile $K > K_0$, hence $K = K_0$ will be a lower limit. Thus the lowest possible value for β can be given as

$$\beta = \frac{\tau_s}{\sigma_0'} = (1 - \sin \phi_d) \tan \phi_d.$$

Burland demonstrated that for values of ϕ_d in the range 20° to 30°, β has a small range of 0.24 to 0.29. Field test results on several different clays resulted in β values of between 0.25 and 0.4 (average 0.32) while the corresponding α values are between 0.43 and 0.79 respectively. From a large number of other pile tests β was found to be between 0.25 and 0.4 ($\alpha = 0.5$ to 1.6) and Burland concluded that a reasonable design value is 0.3.

For stiff over-consolidated clays these considerations are more complex but

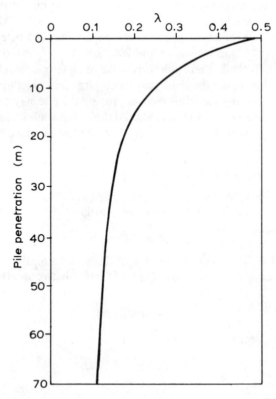

Figure 8.17 Design criteria for axial pile capacity in clays (λ method) (after Vijayvergiya and Focht, 1972).

Burland found that for bored piles an average value of β is about 0.8; for driven piles the scatter of results precluded any conclusion.

Vijayvergiya and Focht (1972) derived the following equation for the total frictional capacity Q_s of a driven pipe pile embedded entirely in clay.

$$Q_s = \lambda(\bar{\sigma}'_m + 2c_m)A_s$$

where $\bar{\sigma}'_m$ is the mean effective vertical stress between the ground and pile toe, c_m is the mean undrained shear strength along the pile, A_s is the pile surface area and λ is a dimensionless coefficient derived from pile tests. They found from 47 tests that the coefficient λ had a close correlation with the embedded length (Fig. 8.17).

8.4.4 Settlement of single piles

When a load is applied to a pile it settles by an amount which is a function of the soil characteristics and pile geometry. It is not practical to determine accurately the soil characteristics for each pile and thus part of the usually applied factor of safety is used to overcome the scatter in the soil parameters which are used to evaluate typical pile loads. It is believed that for driven piles installed using a reliable driving formula, a factor of safety of 2 will result in about 5% of the piles having working loads in excess of their rated capacities. Hence where loading tests are not carried out, a factor of safety exceeding 2 should be used.

A test on a typical pile is almost always worthwhile because it allows the design to be assessed, albeit for only one small part of the site. Pile settlement varies with soil type, with piles driven or bored to rock head perhaps producing settlements of least doubt at working load, that is, little more than the elastic shortening of the shaft. Settlement of piles installed in free draining or dry granular soils is usually completed as the load from the structure accumulates. In general it is only when cohesive soils occur below founding level that long term settlement will occur (Section 8.4.5).

The settlement of a pile under load in a cohesive soil is a complex phenomenon and may best be analysed in terms of shaft settlement and base settlement. Sufficient field tests have been carried out to verify that the shaft and base have distinct load/settlement characteristics. Consider a pile having an idealized shaft and base settlement as shown in Fig. 8.18. It can be seen that considerably less settlement is required to mobilize the ultimate shaft resistance than the ultimate base resistance. This diagram varies with the geometry of the pile: thus when the pile is long and straight the shaft contributes most of the load with a relatively small settlement, but when it is short and under-reamed, most of the capacity is derived from the base, resulting in a larger relative settlement.

If a factor of safety of 3 is applied to the ultimate load shown in Fig. 8.18 then EH represents the working load and OE the corresponding settlement: EF and EG represent the base and shaft contributions, respectively. These give corresponding factors of safety of approximately 6 and 1.6 on Q_{fb} and Q_{fs}. If a constant factor of safety of 3 is applied to shaft and base alike, then the contributions are JK and

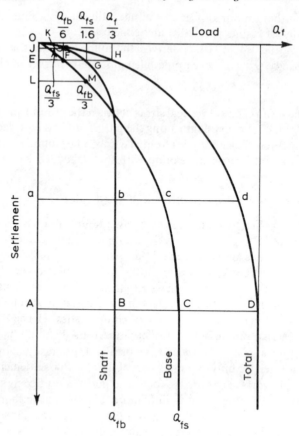

Figure 8.18 Idealized relationship between load and settlement.

LM respectively. These two are obviously incompatible as they involve different settlements, hence although empirical and arbitrary factors of safety are commonly used, there is merit in choosing differing values which correspond approximately to the same settlement value. Thus one may write

$$\frac{Q_f}{F} = \frac{Q_{fs}}{F_s} + \frac{Q_{fb}}{F_b}. \tag{8.19}$$

where F_s and F_b are factors of safety applied to shaft and base ultimate loads respectively. Whitaker and Cooke (1966) studied the settlement characteristics of large diameter straight and under-reamed piles bored in London Clay. They found that for a given degree of mobilization of shaft resistance, settlement increased with shaft diameter and is fully mobilized at between 0.5 and 1% of the diameter. Similarly the degree of base resistance mobilization increased with settlement up to a maximum of between 10 and 20% of the base diameter.

The settlement of pile bases can be predicted from plate bearing tests. Burland *et al.* (1966) found an almost unique dimensionless plot for tests carried out on London Clay at Moorfields. By plotting the ratio (q/q_f) of plate bearing stress q at any settlement ρ_i up to failure (q_f), they found that

$$\frac{\rho_i}{B} = \kappa \frac{q}{q_f},$$

where B is the plate diameter. From the test curves, values of κ were found for $(q/q_f) = \frac{1}{3}$, enabling ρ_i to be calculated for other values of B. The value of κ generally lay between 0.009 and 0.02; the latter figure, if used in the absence of plate bearing tests, gives conservative predictions of settlement.

Various attempts have been made to predict the settlement of individual isolated piles on a more generalized basis, mainly in order to set reasonable standards for the assessment of pile test results. For many years such assessments have been based on specialist experience involving land based piles in known soil conditions. However the use of long and often slender piles for offshore work has stimulated new studies of pile behaviour. Settlement at working loads, where a factor of safety in the range 2 to 3 is used may, for the case of reasonably rigid piles, reasonably be assumed to be elastic and hence, in order to assess probable settlements, analysis based on elastic methods may usually be taken as appropriate.

Using numerical methods, based on such elastic principles, the load deformation behaviour of piles has been studied extensively. Poulos and Davis (1980) have compiled charts which allow the load/settlement behaviour to be estimated with reasonable confidence. Similarly Randolph and Wroth (1978) and Randolph (1983) have developed computational methods which yield good results and are convenient to use, especially with the aid of a computer.

If the predicted settlement of a pile exceeds a tolerable magnitude then either the column load must be reduced or the proportion of shaft load increased by increasing the shaft diameter and/or increasing the length of the pile. Reducing the pile load by increasing the number of piles reduces the settlement of a single pile but probably does little to reduce the total settlement of the structure. Increasing the length or diameter of the piles usually provides the better technical answer, although it is not necessarily the most economical. Pile lengths may generally be varied within a structural foundation, but not if it means founding piles in strata with differing long term settlement characteristics. Where this is inevitable, perhaps because of a fault in rock beneath a building, particular care should be taken in structural detailing to minimize the effects of differential pile behaviour.

8.4.5 Bearing capacity of pile groups

Before the advent of large diameter augered piles, most heavy column loads were carried on groups of small section piles. Group behaviour differs from individual

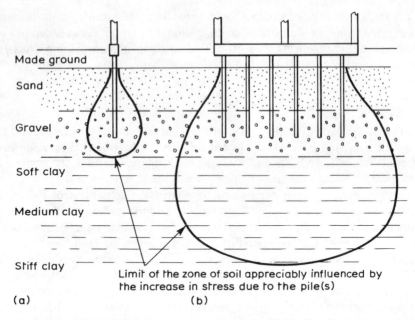

Made ground

Sand

Gravel

Soft clay

Medium clay

Stiff clay

Limit of the zone of soil appreciably influenced by
the increase in stress due to the pile(s)

(a)   (b)

Figure 8.19 Bulbs of stress around a single pile and a pile group.

behaviour. When piles are installed in a group and derive their capacity from end-bearing on rock, it can generally be assumed that the group capacity will be equal to the summation of the individual capacities and that group settlement will be little more than the elastic shortening of each pile shaft. In all other circumstances, an analysis should be undertaken to examine group action.

An individual pile generates its capacity from a combination of shaft resistance and end bearing. This may be considered as producing a bulb of stress around and below the pile (Fig. 8.19a). When several piles are combined to form a group, the bulbs of stress interact and merge to form a group bulb of stress (Fig. 8.19b). The ground below a pile group can be appreciably stressed to a depth of up to 2 or $2\frac{1}{2}$ times the width of the group and for a large group of, say, 100 piles the effects on settlement can be considerable. Frequently the succession of strata is such that the soil increases in shear strength and decreases in compressibility with depth and this fortunately has a slight compensating effect. If, however, the reverse should be the case, as in Fig. 8.19, where sand and gravel are underlain by soft clay, the use of piles may lead to excessive settlement or even failure in a block fashion.

(a) Pile groups in cohesive soils

The behaviour of a pile group in clay is a function both of the stresses around an individual pile and of the increased stresses set up in the whole soil mass when a series of piles in close proximity are loaded simultaneously.

The modes of potential failure of a group may involve the failure of individual piles as they reach a load where they slip relative to the soil mass in which they are embedded, or failure of the whole block of soil involved within the group, or failure of some part of it.

It is evident that as more load is concentrated into any block of soil the settlement must increase, and so for a given load per pile the settlement of groups must be expected to be greater than that of individual piles. Experimental studies by Cooke, Price and Tarr (1980) at the Building Research Establishment in the United Kingdom have demonstrated that at normal design loads, the strain pattern in the soil around an individual pile may be combined with the strain pattern of adjacent piles, using the principle of superposition, to represent the behaviour of a series of piles simultaneously loaded in a group configuration. The effect of this is to produce a dishing of soil layers around and within a pile group so that for example, when a group is rigidly capped, the peripheral piles are forced to move more relative to the soil than are piles in the group interior. This means that loads on piles at the periphery of a group, and particularly at the corners of a group, will be higher than those in the group centre.

In the case where soil deformations are substantial and the capping or the structure erected on a series of caps is rigid, it is possible for peripheral piles to have reached an ultimate load condition while the loads on internal piles are still quite low. For piles in stiff clay the load/settlement behaviour is such that in general large strains of any individual pile relative to the surrounding soil do not mean a diminution of frictional reaction, but in the case of soft sensitive clays the resistance to individual pile penetration may reach a peak value and then diminish as further strain takes place. Thus in certain soft clay soils the total bearing capacity of a group may not be equal to the sum of the maximum bearing capacities of individual piles. However, to produce a true failure of any large pile group will involve considerable settlement, so that there is very little available information on the true comparison between the failure of individual piles and groups.

The real problem associated with pile groups, provided the piles are not so closely spaced that a block failure is possible, is one of settlement and differential settlement. It has been practice in the past to think of groups of piles in terms of 'efficiency'; this being defined as the ratio of the ultimate load of a pile group to the ultimate capacity of the same number of individual widely separated piles. Various formulae purporting to relate this 'efficiency' to group geometry have been suggested but such treatment of groups has little or no justification and is more of a hindrance than a help in the process of pile design. Such methods lead engineers to imagine that they have taken group effects fully into account, when this is clearly not so. They are usually applied to piles in cohesive soil only, they offer no way to take soil characteristics nor pile length into account and when they are applied to piles in stiff clays, their sole effect is often to lengthen piles and increase costs while conferring minimal benefits in terms of settlement control.

The code of practice CP 2004 Foundations states that pile spacings measured centre to centre should be not less than the pile peripheral length and this is usually taken as meaning three times the pile diameter for piles of circular section. At this spacing in stiff clays it will usually be found that the factor of safety against block failure is higher than for corresponding individual piles, and an analysis of block behaviour will demonstrate that this spacing can often be reduced to two and a half times the pile diameter or less, when conditions of layout make this necessary.

In cases where the factor of safety against block failure exceeds say two and a half, and exceeds also the factor of safety on the same number of individual piles, the principle of superposition may be applied to calculate the total group settlements. A block failure is more likely to take place with a large number of long and slender piles than with similar groups consisting of short piles.

For a group of piles with a cap which bears on the same founding stratum, a proportion of applied load is transmitted to the ground through the cap, thus ensuring that the group will in general fail as a whole block at larger pile spacings than if there were no such cap. Where a compressible fill or weak soil underlies the cap then such a benefit would not be expected.

Using numerical procedures and the principle of superposition, various workers have produced methods for estimating the stiffness of pile groups and thus calculating group deformations. Charts have been produced by Butterfield and Douglas (1981) which enable the designer to estimate settlements, and Randolph and Wroth (1979) have produced for the same purpose a method using interaction factors based on the deformation field around an individual pile, which treats the strains due to pile shafts and bases independently. While this latter method may be used by hand, based on graphs, it is generally more convenient to perform the calculations using a computer and this may be done using the program PIGLET (Randolph, 1980).

In multi-layered soils some difficulty will be experienced with all methods since the soil shear modulus must be taken as uniform or increasing linearly with depth to simplify calculation; thus some approximations are inevitable. However the methods are very much better than the somewhat rudimentary earlier methods for calculating deformation in this sort of problem.

Driven piles in cohesive soils may cause problems when they are installed in groups, because of the fact that the displaced volume of soil leads to an upward movement of the ground surface. This effect can cause damage to adjacent structures and services, and in the case where the piles pass through clay to rock or some other hard bearing stratum, the piles may be lifted up so that the important contribution to capacity from end bearing may be lost. If this should be the case, precast piles may be redriven to restore the seating, but this is not normally practicable with cast in place piles. Fortunately this occurrence is fairly unusual and where the piles are carrying most of their load by shaft resistance, the effect is not normally very significant although it is still worth considering the possibility of damage to adjacent piles and the best sequence of driving to minimize any such problems.

Groups of piles in clay may be designed to carry lateral loads and moments, as well as purely vertical loading, and often the designer elects to use raking piles either to resist the lateral loads or for some other reason. When raking piles are used in a group, whether in conjunction with vertical piles or not, certain difficulties may be encountered and these should be borne in mind. The following problems are worth noting:

(a) If by reason of additional applied loading to the soil mass, or excavation in the vicinity of a group, clay is caused to settle or to heave, then a secondary system of bending is induced within the raking piles and they may be damaged as a result. In these circumstances vertical piles only are to be preferred, designing them to resist any lateral forces which may be applied.

(b) The lateral loading applied to groups of piles in clay may be transitory or may vary in magnitude. If the continuously applied force is vertical and the group as a whole is forced to settle in a vertical direction, then raking piles will be subject to bending and where the bases of these piles extend well clear of the vertical piles in the group, their axial loading may also increase due to the soil dishing effects. The combined bending and direct stresses induced may lead to damage and sometimes premature pile failure.

(c) To maintain accuracy of direction in placing raking piles is more difficult than in the case of comparable vertical piles and this is recognized in many specifications such as in 'Piling: Model procedures and specifications' published by the Institution of Civil Engineers (1978). This gives an alignment tolerance of 1:25 for raking piles as compared with 1:75 for vertical piles. In particular the larger permitted tolerances may lead to variations of load per pile from those assumed in an idealized design condition.

For these reasons considerable research has taken place into the lateral load carrying capacity of vertical piles and pile groups in recent years and many engineers now seek to avoid raking pile arrangements where possible. Analysis of laterally loaded piles and pile groups may now be carried out with confidence using the work of Broms (1964), Poulos (1971) and Randolph (1980, 1981).

(b) Pile groups in cohesionless soils

Driven piles are commonly used where the majority of load is to be carried onto or into sands or gravels. Often in these circumstances the reason for piling is because of overlying soft clay or fill. Research and experience shows that in such cases, much of the load is carried by end bearing.

If the sands or gravels are initially loose or only moderately dense, then the process of pile driving leads to an increase of soil density and in turn produces increased load capacity. The sequence of pile driving usually has to be chosen so that one is not attempting to install piles between others which have already been driven, particularly if one is to obtain piles of similar length throughout a group. The closer the pile spacing, so the more noticeable are the effects of soil densification. A common procedure is to commence driving at the centre of a group and to progress outwards from this point, bearing in mind that damage may be inflicted on freshly placed piles by close driving before concrete has set and that tracking of machines across the tops of piles already constructed could lead to breakage of the pile heads. Should the granular soil into which piles are being driven be already in a dense state, then the compaction effect will be relatively less and some displacement or heave may be expected. However in these conditions it is not usual to penetrate the dense layer to any significant depth and therefore the heave problem is minimized. The ultimate bearing capacity of pile groups driven into sand is greater than the capacity of the same number of isolated piles both because of compaction effects and because, if the piles are of the same length, interference takes place between the failure mechanisms at the bases which results in an increase of pile capacity.

Where piles are largely end bearing in a dense sand, whatever type of pile is considered, spacing is not an important factor in so far as ultimate bearing capacity is concerned and piles can be placed as close together as practical considerations of tolerance and potential damage to adjacent piles will allow. Bored piles in such circumstances are often placed at centres equal to two pile diameters.

Where piles are founded in sands and gravels, and where there are no underlying clay layers, the possibility of a block failure mechanism is usually remote. If there are underlying clay layers the possibility of a block failure must be checked and care should be taken that the intensity of loading is not such as to cause large deformations in the clay.

As in the case of piles founded in clay soils, settlement and differential settlement may become overriding considerations, and because factors of safety are usually three or more, an elastic analysis using one of the methods previously referred to for groups in clay soils may be used. For driven piles some difficulty may be experienced in such a calculation due to the improvement of the soil shear modulus with compaction, but since the settlements are often more associated with soils below rather than within a group, it is probably best to carry out the calculations using the pre-piling soils data, bearing in mind that some improvement is probable.

For groups of bored piles bearing on sand, it is important as in the case of a single pile, to ensure that the process of installation does not permit fundamental disturbance of the founding stratum, because this can lead to substantial increase in settlement. If piles are installed in wet ground, the danger of boreholes 'blowing' due to inflows of water must be a primary consideration.

8.4.6 Laterally loaded piles and pile groups

Piles, whether installed singly or in groups, are frequently called upon to support lateral loads in addition to vertically applied loads. The reasons may be because of structural reactions, retained earth forces, anchorage pulls, wave loadings, reactions from machine bases, earthquakes, wind, braking, impacts, thermal movements in the supported structure, or creep and shrinkage. Some of these effects are difficult to quantify and particular care needs to be exercised in assessing the effects of major cyclic loads, since the deflections suffered by the piles may increase gradually to several times the value which would apply to a similar loading condition of the static kind.

Simple design procedures often involve the assumption that upper soft soils contribute nothing to pile capacity to carry lateral loads and a point of fixity is assumed arbitrarily say a short distance into the first firm or stiff stratum that is encountered. The pile is then designed either as a free or fixed headed cantilever depending on circumstances. The method is in general uneconomic because it fails to take into account the relative stiffnesses of the pile and upper soil system.

The work of Broms (1964a and b) on the ultimate lateral resistance of piles represents an advance on such simple considerations. His analysis shows that the ultimate lateral resistance to displacement in cohesionless soils at any level is approximately

$$3 \times K_p \times \text{overburden pressure}$$

Where K_p is the normal two-dimensional coefficient of passive earth pressure derived by the Rankine or Coulomb methods.

In cohesive soils the available lateral resistance rises from two times the undrained shear strength at the ground surface to about nine times the undrained shear strength at a depth of three pile diameters below the ground surface. For simple cases where the lateral loading is small it is often sufficiently accurate to reduce the ultimate lateral stresses by a factor of about three and use the resulting stress diagram to derive bending moments in the piles. This kind of analysis is sufficient for assessing requirements for steel reinforcement to cope with tolerance effects for example. However for larger loads the effects of pile and soil stiffness have to be taken into account in order to estimate the bending moments and deflections which will result in piles.

Various methods have been proposed for this purpose including a subgrade modulus approach. An alternative is to use the so-called 'p-y' methods which are variants on the Winkler spring analysis system. These like the subgrade modulus methods, cannot predict interaction effects between piles in a group and are therefore at some disadvantage. A related empirical procedure for simple pile groupings is however proposed in the United States Navy (1971) Manual NAVFAC DM7 to overcome this difficulty in part. The most satisfactory theoretical solution to pile bending and deflection problems lies in the elastic continuum approach and three computer programs based on this concept are available for dealing with individual piles and groups.

PGROUP 3.   This program is available from the Highway Engineering Computing Branch of the Department of Transport in London. It is based on a boundary element analysis, the soil being modelled as an elastic medium in which linear variations of soil modulus with depth can be accommodated. The effect of a pile cap bearing on the ground can be included and piles may be raked in the plane of loading. The program calls for a considerable computing facility and the selection of element lengths is limited, so that results for laterally loaded piles may be relatively coarse.

DEFPIG   Poulos (1980). This program is based on superposition and interaction factors between individual piles. Variations of soil stiffness with depth are possible and piles may be vertical or raked in the plane of loading. Cap effects in relation to vertical loading may be taken into account. Considerable computing power is required to cope with the pile stiffnesses and the interaction factors.

PIGLET   Randolph (1980). This program is based on power curves fitted to the normalized results of a large number of finite element analyses. Piles of equal length may be considered raked in any direction. The soil is assumed to be elastic and the soil modulus may be taken as uniform or varying linearly with depth. No provision is made for the effect of a cap bearing on the ground but, because of the method of operation, the computer requirements are relatively modest.

Comparison of the program results for some typical cases has been carried out and presented by Poulos and Randolph (1983), showing that there is reasonable agreement. All the programs mentioned will take into account combined vertical and lateral loading effects.

A summary of the various methods available for the design of laterally loaded piles and pile groups by Elson (1984) is published in CIRIA Report 103. The analysis of forces and deflexions in pile groups subject to a combination of lateral and vertical loads is a complex problem and is normally too demanding of time and effort to be done by hand, especially in those cases where the best arrangement of piling is sought through a series of trials. However in some simpler groupings and in circumstances, where for example piles are founded on a rock stratum and have perhaps a considerable part of their length free standing above a river or harbour bottom, a graphical or numerical solution may be sufficient for design. The important advantage of a hand calculation method is that it gives the engineer an intimate knowledge of the way in which the geometry influences forces and deformations and can provide the feeling with which later to solve more complex problems by computer methods. If one takes the step of ignoring soil structure interaction, then the problem resolves itself into one of structural analysis, and slope deflection or strain energy solutions are often convenient. Two analytical techniques to such problems are given in the Appendix.

8.4.7 Piles in fill and soft soils

(a) Buckling of axially loaded piles

Investigations carried out into piles buckling in soft soil under axial loads lead to the general conclusion that the soil provides sufficient restraint to prevent buckling except under those rare circumstances when it behaves as a fluid. Francis *et al.* (1963), Lee (1968) and Whitaker (1970) give detailed analyses of the critical load which a pile can safely carry. When an axially loaded pile has initial imperfections it will have the same buckling load as an initially straight pile, although the deflection of the pile prior to this being reached will probably be larger. Under eccentric loading, however, the buckling load may be only 25% of that of a perfect axially loaded pile.

If part of a pile projects above ground level, that part must be treated as a column and consideration given to column reduction factors (BS 449, BSI, 1969a and CP 110, BSI, 1972c) and the degree of fixity. In good ground it is suggested that the lower point of contraflexure can be taken as 1 or 2 m below ground level, while in poor ground it can be as much as 4 m.

(b) Piles subject to soil downdrag or negative skin resistance

If a pile is carrying load into a firm stratum overlain by compressible soils or fill, then the possibility of settlement of the upper soil layers contributing to the applied pile load must be considered. This occurs because under the new loading conditions, the upper soils may be subject to settlement exceeding the anticipated settlement of the associated pile or pile group.

The accurate assessment of downdrag forces is not an easy matter since soil displacements depend very much on the history of a site and on features connected with new construction, such as regrading of the site surface and the use of non-suspended floor slabs.

One of the main problems with fill is its organic content and whether the organic matter will degrade in time, leading to volume changes. Domestic refuse, for example, now contains a considerable quantity of variable plastic material; before this existed degradation was more predictable. Old fills, not subject to load increases, will rarely produce large downdrag forces.

Soft clays which exist in a natural condition, as opposed to those which have been remoulded and redeposited, may also be subject to little settlement unless additional load is applied. However many cases exist where downdrag is a real possibility and steps must be taken to quantify the depth of material subject to settlement greater than that of the pile and the magnitude of the forces which will result. Relative movement will be greatest near the pile head, diminishing to zero, usually at the interface with a firm bearing stratum, so that in general ultimate downdrag values will be possible to within a short distance of perhaps a metre or so from the interface.

Unit downdrag values are normally calculated in exactly the same way as if the calculation was being done for positive resistance, with the normally applied proviso that within a large group the total downdrag force cannot exceed the self weight of the soils contained within the group in the case of soil degradation, or the newly applied load at the ground surface in the case of old fills or soft natural soils subject to surcharge.

In practice many engineers have in the past provided a pile in which ultimate downdrag force is equated to part of the ultimate load available below the zone of downdrag, applying the factor of safety to the remainder of the ultimate capacity as against the structural load applied at the pile head. This approach may be acceptable where total downdrag forces are small in relation to applied load, but it can be very unsatisfactory where downdrag loads are high, bearing in mind that calculation may be inexact and downdrag may be underestimated, whereas the capacity of the pile in the bearing strata may be overestimated. Normal pile tests, unless elaborately instrumented, will do little to provide design data in such cases since, in the short term, both the upper and lower soil layers may contribute to apparent pile capacity. The best procedure is to add ultimate downdrag force to applied structural load and then to provide a realistic factor of safety to the pile bearing calculation which should reflect the available knowledge of pile behaviour in the particular soil conditions concerned and the uncertainties which may exist. It should be borne in mind that the material stress in a pile is also increased by downdrag forces and this feature may be all the more serious because of the way in which pile loads are redistributed within some pile groups. Large forces may have to be catered for at the bottom of the downdrag zone.

8.5 PILE CAPS

A pile cap must be capable of safely carrying the imposed bending moments and shearing forces, be deep enough to provide adequate bond length to the pile steel and the column starter bars and be sufficiently large in plan to accommodate the tolerance in pile position. The pile concrete shaft should be broken back until sound concrete is met and the upper face of the pile buried at less than 150 mm in the pile cap. When H-section piles are used, it is normal practice to weld a bearing plate onto the upper section of each pile to distribute the pile reaction.

Single piles must be treated with caution especially when they are of small diameter and where the pile heads are surrounded by soft ground, because eccentric loadings resulting from tolerance effects in placement may introduce bending stresses and deflexions which are greater than can be accommodated by the pile section or structure. In such circumstances it is usually advisable that they be linked together by a series of ground beams. This is also true of pairs of piles where the column may be displaced laterally from the axis connecting the piles. A large diameter pile suitably reinforced needs only a square cap into which the pile steel and column starter bars are bonded; it is generally prudent to allow at least 300 mm over the pile diameter for the cap dimensions. When single piles or pairs of piles are arranged to cater for moving loads, as imposed by a crane track, each cap must be carefully designed in conjunction with the piles to resist the worst loading case.

There are two approaches in present use for the design of pile caps involving two or more piles, one being to assume that the pile cap acts as a beam spanning between the piles while the other assumes that the pile cap acts as a truss or space frame wherein the steel reinforcement performs the role of tension members and the concrete acts as strutting.

It seems to be generally the case that for conventional pile caps, the truss approach is somewhere between 5% and 20% more economical in the use of steel reinforcement than the comparable beam analysis, some small advantage being lost if the column applied to the cap is treated as a point load rather than a loaded area.

The basis of the truss method is outlined in Fig. 8.20 which displays the force line assumptions involved in a four-pile cap.

A summarized method of design as given by Clarke (1973) is as follows:

(1) Fix the dimensions of the pile cap. The most compact arrangement of piles in general gives the most economic cap bearing in mind that the spacing of piles will generally lie between 2 and 3 pile diameters measured centre to centre, the former spacing relating to end bearing piles in general and the latter referring to friction piles. The total depth of the pile cap will usually be in the range 2.2 to 2.4 times the pile diameter.

(2) The area of steel required in a given direction along a tensile force line is

$$A_s = K \frac{1}{0.87 f_y} \frac{k h_p}{d} \left(1 - \frac{h_c}{2 k h_p} \right) N$$

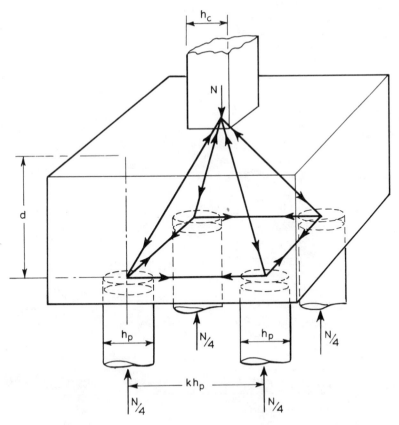

Figure 8.20 Truss method of analysis of four-pile group.

The Design and Construction of Engineering Foundations

where A_s = area of tensile steel
  K = steel force coefficient (as given in the following table)
  f_y = characteristic strength of reinforcing steel
  k = pile spacing factor (pitch of piles = kh_p)
  h_p = pile diameter
  d = effective depth of cap of reinforcement
  h_c = column breadth
  N = total load on cap

Table 8.11 Steel force coefficient for various pile group arrangements

Number of piles	Plan arrangement of piles	Steel force coefficient, K		
		AB	BC	AD
2		1/4		
3		1/9	1/9	
4		1/8	1/8	
5		1/10	1/10	
6		1/6	1/12	1/12

Note: piles are at uniform spacings of kh_p

(3) Shear strength on plane sections adjacent to the column. Consider the steel banded over the piles in the form of straight bars and calculate the shear on a critical plane across the cap at the face of the column (Fig. 8.21).

$$\text{The percentage of steel over piles} = 100 \ \frac{\text{area steel over piles}}{\text{Zone 1 area}}$$

Hence determine allowable shear stress v_c from Table 5 of CP110. Enhanced shear stress $= (2d/a_v)v_c$ (must be less than values in Table 6 of CP110) can be taken on Zone 1 areas, see Fig. 8.21 where $a_v =$ distance from edge of column to support. Thus

$$a_v = \tfrac{1}{2}[(k-1)h_p - h_c] + 0.2h_p$$

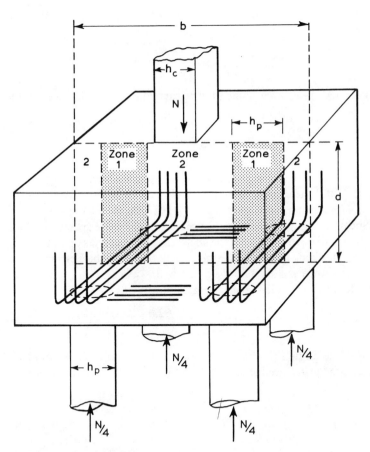

Figure 8.21 Method for calculating shear on a vertical plane surface at the edge of the column position.

and, where v_c = allowable shear stress, the total shear force on the critical plane is

$$(\text{Zone 1 area})\, v_{c1} \frac{2d}{a_v} + (\text{Zone 2 area})\, v_{c2}$$

Thence derive total shear capacity of the cap V_B.

If $V_B \geqslant N$ the design is satisfactory but if $V_B < N$ distribute the steel in a uniform grid across the cap width; provide full anchorage lengths and recalculate the shear capacity.

Where shear capacity must be recalculated, the percentage of steel is now

$$100 \frac{A_s}{bd}$$

The enhancement factor, $2d/a_v$ applies only to steel passing over the piles. Thus total shear force on the critical plane is

$$v_c[(\text{Zone 1 area}) \times \frac{2d}{a_v} + (\text{Zone 2 area})]$$

The additional shear force which may be carried by the anchorage lengths acting as links is

$$2 \times 0.87 f_y \frac{d}{s_v} A_{sb}$$

where s_v is the pitch of the bars and A_{sb} is the cross-sectional area of one bar. Hence, the total shear capacity of the whole cap V_G is derived and if $V_G \geqslant N$, the design is satisfactory. If however $V_G < N$, the A_s must be increased until $V_G = N$.

(4) *Shear strength – punching.* For pile caps where the spacing of the piles is 4 times the pile diameter or more, the shear capacity in terms of a punching failure should be checked. In this case, the calculation may be performed using the diagram as shown in Fig. 8.22 for a cap with 4 piles.

For a case with banded steel, the punching shear capacity

$$V_{BP} = 4\beta r v_c \left(\frac{2d}{a_{vp}} \right) d$$

where β is the angle subtended by a pile at the centre of the column, r is the distance from the inside edge of the pile to the centre of the column plus 20% of the pile diameter and

$$a_{vp} = r - \frac{1}{\sqrt{\pi}} h_c$$

In the case where the steel is placed in the form of a grid, the total shear capacity

$$V_{GP} = [4\beta \frac{2d}{a_{vp}} + (2\pi - 4\beta)] r v_c d$$

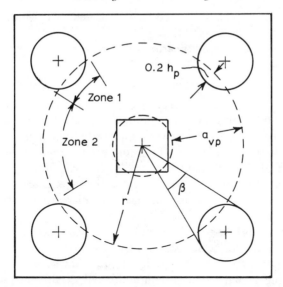

Figure 8.22 Method of calculating punching shear on a cylindrical surface.

The shear capacity of the cap should be taken as the lesser of V_B and V_{BP} if banded steel is used and of V_G and V_{GP} if grid steel is used.

Further information on this method and its basis can be obtained from the paper by Clarke (1973).

The method given by Whittle and Beattie (1972) is similar in many respects and suggests standard pile caps for groups containing up to 9 piles. It may be slightly less economic in use since the column load is taken as a point load in calculating the area of tensile reinforcement. However, there are significant practical merits in standardization and tables are given which simplify routine design procedures in this paper.

It should be borne in mind that where groups are loaded eccentrically or where the pile loads within a group vary, the conditions should be treated on an individual basis and steel requirements derived from first principles after the manner indicated above.

8.6 PILE TESTING

8.6.1 Functions of a pile test

The main functions of a pile test may be as follows:

(a) To determine at one specific pile location only the relationship between load and settlement, particularly up to working load.
(b) To prove, again at one specific location only, that failure does not happen

before the ultimate design load has been reached, hence verifying the minimum design factor of safety.

(c) To determine the ultimate bearing capacity of the pile and so define the maximum design factor of safety.

(d) To check that the workmanship of any randomly selected pile is satisfactory.

Because a precast pile can be inspected prior to installation, there should be little doubt that the shaft is of stated minimum dimensions. However, no check can be made on alignment after driving unless a tell-tale pipe is driven with or within the pile. In the case of a steel H pile for example, this is welded or attached to the pile and enables an inclinometer to be lowered after driving, thus indicating the verticality and continuity of the pile. One of the advantages of a driven cast in place pile over a precast pile is that the verticality can be verified prior to concreting.

The question of workmanship is of vital importance, especially with deep large diameter bored piles for which the design load may exceed 10 000 kN. One important danger with a bored cast in place pile in poor ground without a permanent casing, is that a shaft discontinuity may occur which will reduce the design factor of safety. Prevention in this case is better than cure and a careful check should be kept metre by metre on the volume of placed concrete with respect to the theoretical volume and on any subsidence of the ground surface. In addition the use of relatively inexpensive sonic or vibrational integrity testing is now possible to check for the presence of such faults in most cast in place piles, and this, judiciously combined with load testing, can give a good level of confidence that piles are sound and sufficient (See 8.6.7).

8.6.2 Methods of test loading

There are four methods by which a pile can be test loaded.

(a) Kentledge, in the form of blocks of concrete or cast iron, is loaded on to a framework which rests directly upon the pile head. The load is measured by the mass of kentledge placed at any time but this technique affords only clumsy control of loading and unloading and the method is not normally adopted for safety and other reasons.

(b) Kentledge, of mass not less than about 1.15 times the estimated test load, is placed on a test frame and cribs which rest upon the ground. A hydraulic jack is then used to jack the pile against the kentledge. It is desirable to have a dual load measuring system comprising a load cell between the jack and the kentledge framework and a pressure gauge linked to the hydraulic pump. This can help resolve uncertainties in the event of a malfunction of one part of the measuring system.

(c) Kentledge is not used and instead the reaction is provided by tension piles which are linked by a beam and yoke system. The load, which is applied by a jack between the yoke and the test pile, is measured by a load cell and a hydraulic gauge. For safety reasons it is advisable to have at least two tension piles per side of the test pile, especially when high test loads are required.

(d) High tensile steel cables or rods will in this case replace the tension piles of the above system and are usually anchored into sound bedrock.

Whichever loading system is used, great care must be exercised to ensure that the arrangement is symmetrical and that the thrust is axial to the pile, the pile cap, the load cell and the test frame. Failure to ensure this can lead to poor results or to accidents. In pile testing there is no substitute for careful and methodical work; once a test has started, danger may be only seconds away and the test cannot easily be stopped for adjustment of equipment without consideration of the procedure to be followed and of possible effects on the accuracy of the results.

The reaction system should not interfere significantly with the test. For a kentledge test, the supporting crib or framework should be sufficiently far from the test pile and have a relatively low bearing stress, so as not to influence the ground around the test pile. Similarly, tension piles and anchors must be kept well clear of the test pile; usually not nearer than 3 pile diameters centre to centre. When under-reamed piles are test loaded, 1.5 times the bell diameter is frequently taken as the minimum spacing.

When a hydraulic jack and linked pressure gauge are used, a current test calibration certificate should be available, since large errors can occur in a jack due to piston friction resulting from wear and tear. The jack should be centred axially over the pile and under the cross-head because eccentricity will produce errors in the measured load. The beams used to transmit the load to the tension piles or anchors should be in good condition and given ample lateral support. They should be marked at the load application and support points and exhibit the maximum permissible point load. They should be employed always with the same flange uppermost, since repeated turning around of a test beam tends to cause fatigue in the material, thus eventually reducing the safe working load.

Safety is an important consideration in test loading. The welds attaching tension bars to the loading frame should be inspected prior to the commencement of loading and during the test where a tension pile system is used. Site personnel should be kept well clear of the loading structure during periods when load is being increased. The reading of instruments can often be carried out by remote or electrical means in order to minimize risks.

8.6.3 Methods of measuring vertical displacement

There are three reliable ways of measuring the vertical displacement of a pile. In the first method, a surveyor's level is used in conjunction with a fixed datum and a scale. The instrument should be set up over a datum point which will be unaffected by ground movements, but near enough to give the shortest possible sight. A vernier scale should be securely attached to the reference part of the test pile and should be read to an accuracy of at least 0.1 mm.

The second method involves the use of dial gauges or strain transducers and a reference beam and is to be preferred to the first system, although there are possible sources or error unless care is taken during installation. The reference beam should be securely attached to at least two long, rigid posts which are driven to a firm stratum. Usually four dial gauges or strain transducers should be securely attached to the beam and their plungers should rest upon pieces of glass set flat in a bed of adhesive on top of the test pile cap. By placing the gauges in diametrically opposite pairs, differential settlement between gauge positions can be detected and an average deflection calculated. The underside of the test pile cap should be clear of the ground by at least 150 mm in order to prevent the cap transmitting load directly to the ground. The pressure gauge, the load cell and the dial gauges should be readily accessible and easily read. The reference beam and dial gauges should be protected from the effects of wind, rain and sun as these can affect the readings. Where large changes in temperature are suspected over the duration of the test, a max./min. thermometer should be used and the ambient temperature recorded with each test reading. The dial gauges should have more than enough travel to accommodate the maximum settlement, say 10% of the pile base diameter, and be capable of being read to an accuracy of say 0.05 mm. Machined blocks of appropriate and exactly known thickness may be used as inserts if the movement of the pile is expected to exceed available gauge travel.

The third system replaces the reference beam by a strained high tensile wire which is read against a fixed scale to an accuracy of 0.5 mm. If possible, the dial gauge reference beam or wire system should be backed up by the instrument levelling method.

8.6.4 Testing procedure

Where the primary interest lies in the settlement of a pile at its working load or where the test is to be carried out as a functional proof up to say 1.5 times working load, a 'maintained load' test is usually carried out. This type of test is not well suited to the determination of ultimate loads, and hence in such cases the 'constant rate of penetration' test is to be preferred.

(a) Maintained load (ML) test

In this test an increment of load, usually not more than 25% of the working load, is applied to the pile and held constant until either the pile ceases to settle or settles at a limiting rate, which may be arbitrarily defined as 0.1 mm in 20 min. At this point the next increment of load is applied and the procedure is repeated until the maximum load of the cycle has been achieved: similar measurements are made during off-loading cycles. By holding the load constant at the working load and at 1.5 times working load for longer periods, indicative settlements are recorded which are often used as the basis for performance specifications.

Graphs of load against time, settlement against time, and settlement against

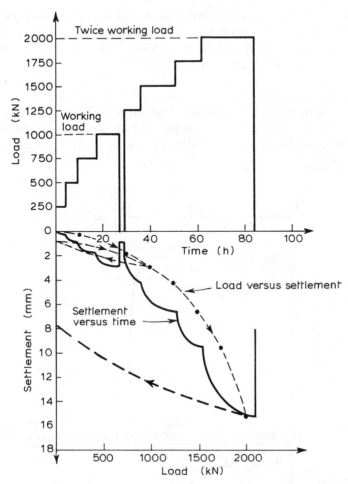

Figure 8.23 Maintained load test: Load/time, time/settlement and load/settlement curves.

load can be prepared as shown in Fig. 8.23. From the maintained load test the ultimate bearing capacity cannot easily be established, partly because of the need to use reasonably large load increments, but also because as ultimate load is approached, particularly for piles in clay soils, the time taken to reach low settlement rates becomes excessively long.

(b) Constant rate of penetration (CRP) test

This test is described by Whitaker (1957) and has significant advantages if the requirement of a test is to determine ultimate loads. The load is adjusted so that the pile penetrates the soil at a fixed, pre-determined rate. Readings of load are recorded with increasing settlement. The pile can in this case be regarded as a device for testing the soil, analogous with the constant rate of deformation of the triaxial shear test. Whitaker defines the ultimate bearing capacity for this test as the load at which the full resistance of the soil becomes mobilized. The rate of testing is usually chosen to be comparable with that used in the undrained shear test in the laboratory. For friction piles in clay CP 2004 (BSI, 1972d) recommends a rate of penetration of 0.75 mm min^{-1} while for end-bearing piles in sand or gravel the rate should be about 1.5 mm min^{-1}. During the test, several operatives should be available, to read the dial gauges, to apply the load at a rate

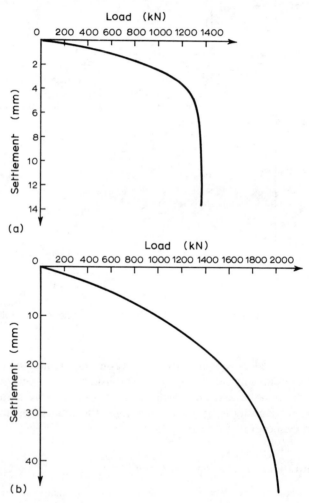

Figure 8.24 Constant rate of penetration test: Load/settlement curves (a) for a friction pile and (b) for an end bearing pile.

appropriate to maintain a constant rate of penetration and to plot the load/settlement curve.

Figure 8.24 shows the results of two typical CRP tests for a predominantly friction pile and an end-bearing pile in clay. Failure of the end-bearing pile can be taken when the settlement of the head is 10% of the base diameter; for very long piles the elastic shortening must be taken into account. With large capacity piles it may be necessary to extrapolate the CRP test curve if failure is not approached at the ultimate capacity of the testing frame.

In sands and other free-draining soils, load testing takes place under drained conditions because the pore-water pressure readily dissipates. In cohesive soils, load testing usually takes place under undrained conditions due to the much lower permeability of the soil. Under actual loading conditions during construction in cohesive soil, drainage takes place around the pile due to the relatively slow build-up of the pile load, hence the test conditions do not truly represent actual service conditions and in most cohesive soils, the test results will be on the conservative side.

Over recent years design figures for the safe bearing capacity of chalk have increased because of research into this particular material and because the testing procedure has been improved. The best method is considered by some to be a cross between a maintained load test and a slow CRP test, using increments of about 15% of working load. Each increment is applied rapidly and maintained constant while the pile settles. During the test the settlement is plotted to a base of the logarithm of time and the next increment is not applied until primary settlement has ceased: this is indicated by the change in curvature of the log. time/settlement curve.

The testing procedures outlined above are described in CP 2004 (BSI, 1972d) and should be followed save where experience dictates otherwise. A comprehensive review of pile load testing has recently been published (CIRIA, 1980), and the Institution of Civil Engineers publication 'Piling: Model Procedures and Specifications' (1978) will provide a useful framework for the control of testing on sites.

8.6.5 Interpretation of results

There is often a tendency to extract more information from a pile test than is justified. Unless a pile is well instrumented, a successful test usually only vindicates the design to some factor of safety at some particular part of the site. It also indicates that the workmanship has been to an approved standard and if repeated on every pile should eliminate a possible source of concern to the engineer. Nevertheless, sudden changes in ground conditions usually merit another test to check the influence of the changes.

Extrapolation of pile results should in general by avoided between piles of different geometry and definitely not attempted between piles of different construction: each type of pile produces different construction soil stresses and each has different shaft-soil and base-soil contact characteristics. One of the main criticisms of pile testing in the past has been that it is non-standard, making comparison of results virtually impossible, but the procedures laid down in the Piling: Model Procedures and Specifications may lead to some improvement in this situation. The analysis of results from pile load tests has been discussed by Fellenius (1980).

8.6.6 Lateral load and vertical pulling pile tests

Piles subjected to horizontal or inclined loads can be tested, either to working load or failure, to determine the deflection/load characteristics. The apparatus is basically the same as that used in the vertical loading pile test, except that the configuration of the piles is different.

For lateral pile tests the usual practice is to jack two identical piles apart; the test layout must resemble as closely as possible the conditions in which the piles will finally carry the working loads. Dial gauges may be used to monitor the horizontal movement of the pile shaft at various points near to or a little below ground level, while on more sophisticated tests, strain gauges can be fitted to the pile surface and the stresses thus calculated up to failure (Matlock and Ripperger, 1956; Reese *et al.* 1974). It is normal to test such a pile in a free-headed mode, but under exceptional circumstances a fixed-headed pile which is vertically loaded can be tested to examine horizontal deflection characteristics. Further references to lateral pile test arrangements and field results have been quoted by Broms (1964 a,b).

Estimates are often required of the pulling capacity of vertical piles. Sowa (1970) investigated the pulling capacity of cast in place bored piles in both sandy and cohesive soils and concluded that the pulling capacity of piles in cohesive soils can be estimated approximately, but that similar calculations in sandy soils are much more difficult. The most practical way of testing a pile for pulling capacity is to reverse the anchor pile arrangement described in Section 8.6.2 and jack up the test pile against two vertical reaction piles or ground bearing pads. During a conventional pile test, vernier scales can be put on tension piles and an approximate relationship found for pile lift under tension.

8.6.7 Non-destructive pile testing

Many techniques are available whereby a pile can be tested non-destructively to prove its continuity. These are described below.

(a) Borehole camera method

In this method a borehole video camera is lowered down a drill hole and a continuous record of the walls of the borehole can be obtained. Interpretation requires considerable skill.

(b) Caliper method

A caliper, fitted with three arms, each at 120° is lowered down the drill hole. When a defect such as a crack or void is encountered, one or all of the arms registers a change in diameter which can be investigated further. By incorporating an inclinometer in the caliper the verticality of the drill hole can be simultaneously checked.

(c) Coring method

A diamond bit is used to extract one or more continuous concrete cores from the pile. Cracks and discontinuities are detected either by the driller noting a change in resistance to the bit or a change in colour of return water or by checking the drilled length with the core length. Cracks in the cross-section of the pile can be detected by filling a drill hole with water and noting any change in level of the water or the movement of water between drill holes. Care must be taken to check that the core does not pass to the outside face of the concrete shaft.

(d) Electrical resistance method

In this method the pile reinforcement is used as an electrode and the resistance and change in resistance measured between the pile and other electrodes inserted in the soil surrounding the pile. It has been shown by Forde and Whittington (1983) that various defects can be identified, although the method is more cumbersome in operation than for example the sonic systems. It can however be used where the pile concrete has only recently been cast.

(e) Percussive drilling method

A percussive drill is used to bore a hole down the centre of the pile. The non-verticality of such a hole can be a problem and defects are only noted by the skill of the driller or by periodic water level checks.

(f) Prestressing method

A continuously sheathed U-shaped high tensile strand is cast throughout the length of the pile and if defects are thought to exist the strand is tensioned. If the head of the pile moves downwards, a defect near the surface is implied while a large elongation of the wires indicates a deep seated shaft discontinuity. An alternative to casting a prestressing strand into the pile is to use sleeved Macalloy bars anchored at the base of the pile. However the base plate must be sufficiently large not to initiate a failure of the concrete of the pile.

(g) Radiation method

In this method either gamma ray transmission or neutron back-scatter is used (Preiss, 1971). In the former the central area of the pile is monitored from between two cast *in situ* tubes down which are lowered a gamma ray source and a detector. With the neutron back-scatter, the variation in total water content is measured with a single probe up to a thickness of not more than 200 mm of concrete. Several probe holes are necessary to ensure that a reasonable assessment of the whole pile section can be made.

(h) Sonic coring method

This method has been developed by CEBTP (Centre Expérimental de Recherches et d'Etudes du Bâtiment et des Travaux Publics, Paris). Three holes are usually necessary and these can be either drilled after construction or installed as tubes tied to the reinforcing cage: they are about 50 mm in diameter. The tubes are filled with water to act as a contact for the transmitter which is lowered down one hole, while the receiver is lowered down another at the same rate. Defects are detected by apparent lengthening of the wave path or a fall-off in velocity of the transmitted signal. A disadvantage of the method is the cost of tubes and their installation.

An alternative is to use one tube in the pile and two tubes in the soil. The resistance of the concrete/soil pair is compared with the soil/soil pair, similarities indicating changes in concrete thickness. Further reference can be made to, Gardner and Moses (1973) and Davis and Dunn (1974).

(i) Vibration method

By mechanical or electrical means a constant force vibration is applied vertically to a pile head at a frequency which may be varied between about 10 and 100 cycles per second. A velocity transducer placed at the top of the pile measures the impedance of the pile as a relationship between the applied frequency and velocity. An undulation in the velocity/frequency curve is an indication of the effective length of the pile while variations in the output will indicate various forms of defect.

(j) Sonic echo method

The sonic echo method was first developed by TNO in the Netherlands and has proved a useful method for investigating significant problems in the construction of driven precast piles, apart from the segmental type, driven cast in place piles and bored piles. A blow, from a hand-held hammer, is delivered to the head of the pile and is reflected from the base of the pile back to an accelerometer at the pile head (Fig. 8.25). A trace is displayed on an oscilloscope screen and several traces may be compared for consistency in the record produced for any one pile. The wave is also reflected in characteristic ways from reductions or enlargements of the pile section.

This method of integrity testing is probably the least costly and most widely used at the present time but it requires considerable skill on the part of the

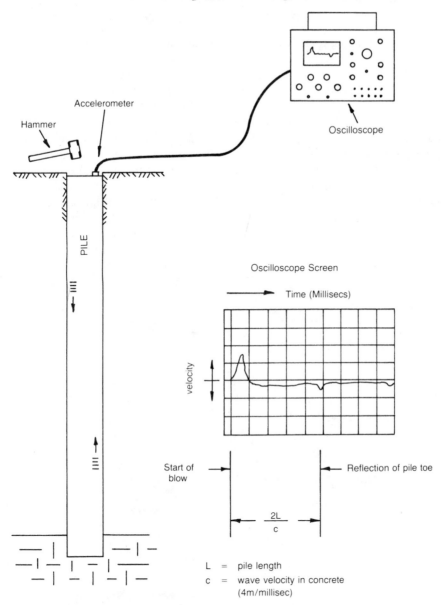

Figure 8.25 TNO sonic integrity testing method.

interpreter, who should have considerable experience based on the comparison of signals and exhumed piles. He should be aware of the effects of soil conditions and the specific features of the construction methods used. For further and more detailed information on this method the reader is referred to papers by Sliwinski and Fleming (1984), Fleming (1984), Reiding *et al.* (1984).

8.7 SAFETY PRECAUTIONS

Safety on site can be considered under two headings: plant and men. First, there are statutory safety requirements for all the major items of equipment used on site and, second, there are safety procedures which must be adopted when men are involved in certain tasks. The most obvious sign that personnel on sites are now more safety conscious is the almost universal use of the safety helmet.

CP 2004 (BSI, 1972d) contains a wealth of information about general site safety during piling operations and about specific aspects relating to driven precast concrete, steel, sheet and bored piling. Piling contractors are now paying more attention to all aspects of safety and most of the larger companies have safety officers. Safety should always be kept in mind when on site and if anything looks suspicious or dangerous, it should be brought to the attention of senior site personnel without delay.

Chapter 9

STRUCTURES
LIABLE TO THE EFFECTS
OF MINING SUBSIDENCE

G.S. Littlejohn

INTRODUCTION

Mining subsidence is the result of the action of gravity on strata rendered unstable by the withdrawal of their natural support over a sufficiently large area. Subsidence, starting at seam level, extends vertically and horizontally with varying time lags to the surface where disturbed ground sinks slowly at different rates until, in the course of time, stability of the whole mass is restored at a lower level. This chapter deals exclusively with subsidence due to modern longwall mining since it is amenable to prediction and analysis unlike surface stability in areas of old coal workings formed by room and pillar methods (Gray & Meyers, 1970; Littlejohn, 1979a, b).

At the beginning of the present century, it was known that mining subsidence varied with the size and shape of the underground workings but no generally acceptable theory had been proposed. Since then, considerable progress towards a better understanding has been made by the recognition of basic geometrical relationships between some of the principal factors in mining subsidence, although unsolved problems remain to confuse the prediction of events in specific physical conditions.

With regard to the mining precautions which can be adopted to minimize surface damage, the development of surface curvatures and slopes, with their complementary lateral movements, is not beyond control by subsidence techniques and may be predicted with reasonable accuracy. Structural precautions, on the other hand, have not been wholly successful and many problems remain to be tackled, especially in the fields of structure–soil interaction and soil mechanics, before structures can be adequately designed to accommodate or resist ground movements.

739

9.1 MECHANICS OF MINING SUBSIDENCE

9.1.1 History

The original idea regarding the occurrence of subsidence was that the strata subsided vertically above the excavation as shown in Fig. 9.1(a). Potts (1949) states that this view was followed in 1825 by the normal theory proposed by Ganot, a Belgian engineer. He presumed that the breaking lines on each side of an excavation were perpendicular to the plane of the seam (Fig. 9.1b). However, this theory had limitations in inclined measures, as was pointed out by Schulz, a German engineer, in 1867 (Potts, 1949). He suggested that the normal theory could be applied to workings to the dip but that a vertical line of break was more probable when working to the rise. Following this work Hausse, in 1907, concluded that the angle of break in horizontal strata lay between the angle of repose and the vertical (Fig. 9.1c). It was not until 1927 that the concept of 'draw' was established: O'Donahue stated that in horizontal strata, if a large enough pillar was removed, the subsidence profile would be as shown in Fig. 9.1d. He suggested there was a 'pull' or 'draw' beyond the edge of the workings and, since lateral support was not entirely removed, the angle of repose was not fully developed.

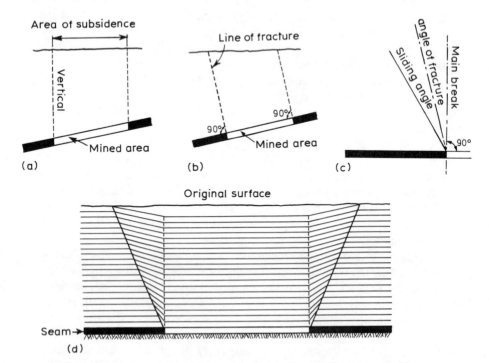

Figure 9.1 Early concepts concerning the mechanics of mining subsidence.

With the accumulation of large numbers of observations during the last two decades in mining areas, empirical rules have been established which relate surface movements to such factors as (a) seam thickness, (b) type of extraction, i.e. mining method, and degree of packing, if any, (c) depth of overburden, (d) width of working and (e) rate of advance. However, surface movements are governed also by less determinate factors which include (a) character of strata between seam and surface level, (b) presence of goaf (voids) in old workings above the new seam and (c) presence of faults.

Nevertheless, as a result of many observations, the basic principles governing mining subsidence have become widely accepted and the current concepts are now discussed.

9.1.2 Mining methods

There are two distinct methods of working: longwall, and pillar and stall. The longwall method shown in Fig. 9.2 is employed in the majority of coal mines in Britain and consequently has been more widely investigated than the latter. In longwall mining, large areas of coal are completely extracted by a number of more-or-less broad working fronts as the mine face advances. In pillar and stall mining (Fig. 9.3), coal is extracted in rooms and only 30 to 50% may be removed in the first operation. The walls may be extracted later or left as pillars to support the roof.

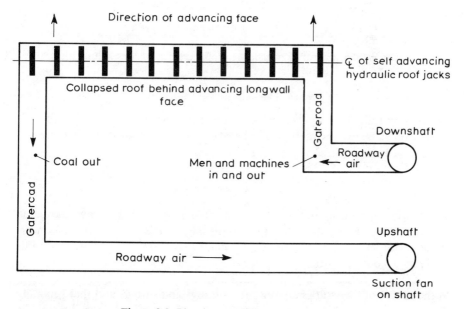

Figure 9.2 Plan layout of longwall working.

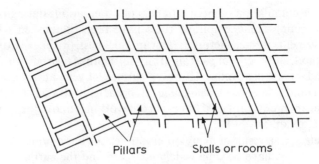

Figure 9.3 Plan layout of pillar and stall workings.

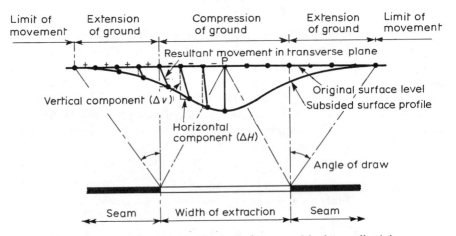

Figure 9.4 Basic form of subsidence profile created by longwall mining.

9.1.3 The subsidence diagram

In planning and recording their work, mining engineers construct graphs relating surface subsidence to extraction. The basic form of the subsidence profile created by longwall mining is shown in Fig. 9.4.

9.1.4 Angle of draw

The extent of surface subsidence is defined by the angle of draw, measured between the vertical and the line joining the edge of the movement zone on the surface with the edge of the working (Fig. 9.4). Wardell (1954) states that this angle is not constant, although a value of 35° repeats itself with a high measure of consistency in British coalfields. Wardell suggests that the angle varies according to the nature of the strata between the workings and surface and that generally the angle tends to decrease with increase in resistance to deformation of the

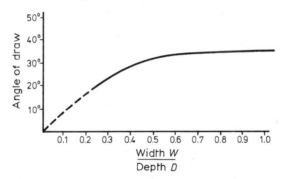

Figure 9.5 Relationship between angle of draw and width/depth ratio (after Marr, 1958).

strata. In addition, the maximum angle is not developed under all mining conditions in level strata. As confirmed in practice, the smaller the width of face, the smaller the amount of subsidence produced at the surface. By considering faces of progressively reduced width, a stage is eventually reached when a longwall face is too narrow, for example a single roadway, to produce subsidence at the surface. At this point there is no subsidence and hence no angle of draw. From field observations, Marr (1958) prepared a graph of angle of draw against the ratio of width of working W to depth of working D. Figure 9.5 shows that the maximum angle of 35° is produced only in undisturbed strata when W/D is 0.7 or more.

9.1.5 Critical area of extraction (area of influence)

If from a surface point P (Fig. 9.4) angle of draw lines are plotted downwards on all sides, the area subtended in the seam is called the critical area of extraction. In horizontal measures, P subtends a circular area of radius (R) equal to the depth of the seam (D) × the tangent of the angle of draw $(R = D \tan \alpha)$ and no workings outside this circle will cause subsidence at the point P. It follows that if workings outside this critical area cannot affect the surface point P, then the point must undergo its maximum subsidence when its full critical area has been worked. As the dimension of the critical area in any particular case depends upon the depth of the working and the value of the angle of draw, it will be appreciated that, all other things being equal, the area of influence increases with depth.

The area of extraction affects the subsidence profile considerably as shown in Fig. 9.6. Since point P suffers its maximum possible subsidence when the whole of the critical area is extracted, it is convenient to call this 'full subsidence'. With supercritical areas, where the goaf (extraction) is wider than the critical value, not only does point P suffer full subsidence but other points nearby subside equally thus forming a flat-bottomed basin. On the other hand, when a panel is worked with a width less than the critical value, this subcritical area causes a degree of

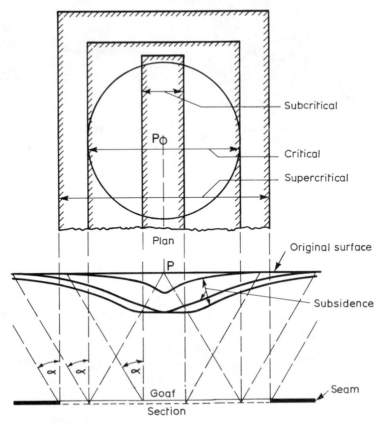

Figure 9.6 Relationship of subsidence to different widths of extraction in a seam of given depth.

subsidence less than the maximum possible and is referred to as 'partial subsidence'.

9.1.6 Seam thickness (*t*)

The greater the thickness of seam extracted, the greater the subsidence, and it is common practice to express subsidence as a percentage of the seam thickness. Care is required, however, to ascertain the true average seam section extracted.

9.1.7 Width of working (*W*)

The effect of the area of working has already been discussed but the width of the advancing face itself can also be a relevant factor in estimating subsidence. When the width of a panel is small, the area occupied by supporting gates (roadways) and gateside packs is a greater proportion of the goaf than when the panel is wider. The convergence of roof and floor and, correspondingly, the amplitude of the surface subsidence may be much reduced, compared with a panel of similar width/depth ratio but greater width.

9.1.8 Stowage

Subsidence may be reduced by replacing the extracted coal with waste material. This operation is called stowage. The material is packed tightly in place before appreciable settlement has taken place and may be done by hand or, more effectively, by pneumatic, hydraulic or mechanical means. It is obvious that any material imported into the goaf must, to some extent, reduce the amount of convergence of the roof and floor and thus reduce subsidence at the surface. However, stowing cannot eliminate surface subsidence completely, since in almost all cases the roof subsides 5 to 10% of the thickness worked before any packing material can be introduced. The stowing is invariably more compressible than the coal removed, so it is impossible to prevent some degree of surface subsidence over longwall workings using present day techniques. Whetton (1957) has shown that the initial convergence that occurs at a face before packs are built or stowing material placed in position is sometimes as high as 30% of the seam thickness. Thus, other things being equal, the speedier the insertion of the packing that follows coal extraction, the less will be the subsidence.

The most efficient, i.e. tightly packed, stowing observed in Britain has been pneumatic stowing and this has the effect in some cases of approximately halving the amount of subsidence which would have occurred had the more conventional method of strip packing been employed. Hydraulic stowing is often considered the most effective method of filling the goaf and the technique has been used with considerable success in the Polish coalfields. In these collieries, where thick seams are worked under towns, sand mixed with water is delivered underground

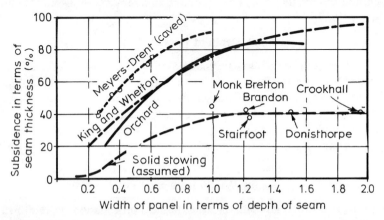

Figure 9.7 Comparison of subsidence under various stowing conditions.

Table 9.1 Maximum subsidence related to type of stowing

Type of stowing	Percentage subsidence	Percentage subsidence	Percentage subsidence
Solid pneumatic stowing	45–50 (German, see Potts, 1949)	55 (Loffler, see Potts, 1949)	47 (Aynsley and Hewitt, 1960)
Solid mechanical stowing	45–50 (German, see Potts, 1949)		45–50 (Potts, 1949)
Solid manual stowing	50–60 (German, see Potts, 1949)	66 (Loffler, see Potts, 1949)	58 (Aynsley and Hewitt, 1960)
Strip packing	75–85 (Flaeschentraeger, 1958)	58–74 (Schultz, see Potts, 1949)	61–85 (Aynsley and Hewitt, 1960)
Hydraulic stowing		23 (Orchard, 1961)	10–30 (Potts, 1949)
Caving	85–95 (German, see Potts, 1949)	95 (Loffler, see Potts, 1949)	95 (Potts, 1949)

through pipes and deposited at the faces, where the water drains off through hessian screens and the sand becomes very compact. Sand would be expensive in Britain and the effect of water in the seam floor would, in many cases, render the seam unworkable.

One of the major difficulties in comparing different systems of stowing is that rarely is the full critical area of coal extracted. Ideally, comparisons should be made with cases having similar dimensions, such as width and depth. However, Fig. 9.7 indicates the partial subsidence curve deduced by King and Whetton (1958) from laboratory experiments under ideal conditions and, in addition, three curves for strip packing, caving and solid stowing (Orchard, 1961) have been included for comparison. Typical values of the percentage of maximum subsidence for different types of stowing are given in Table 9.1 and it is apparent that, in general, there is fair agreement between observers. In the case of strip packing, the discrepancy is probably due to different materials and methods.

9.1.9 Depth of working (*D*)

Earlier in this century, it was considered that the amplitude of subsidence decreased with depth and in 1929 Briggs published the chart (Fig. 9.8) comparing his observations with those of Menzel and O'Donahue (see Potts, 1949). This information suggests that the ratio maximum subsidence (S)/thickness extracted (t) decreases with increasing depth but no data are presented on the width of working pertaining to each field case.

9.1.10 Width/depth ratio (*W/D*)

For a given W/D ratio and seam thickness (t), field observations have shown that the maximum subsidence is practically the same irrespective of depth (Fig. 9.9)

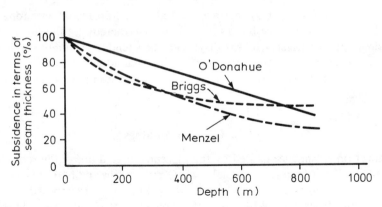

Figure 9.8 Relationship between amplitude of subsidence and depth of workings (after Briggs, 1929).

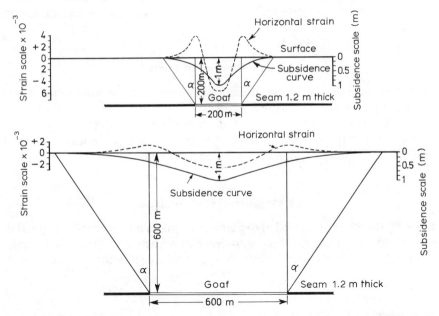

Figure 9.9 Effect of width and depth of extraction on subsidence curve and strain intensity.

but ground strains (to be discussed later) are affected by amplitude of subsidence and depth.

9.1.11 Geology of the overburden

In order to predict the effect of geological characteristics, an intimate knowledge of the strata comprising the overburden is essential. McTrusty (1959) has observed in the field that strong massive beds of sandstone or limestone may subside erratically in large blocks between joints, thereby leading to irregular subsidence at the surface, whereas shales and clays tend to yield more uniformly. A thick bed of sand or sandy clay may, by internal adjustment of its particles, absorb erratic displacements in underlying beds and thus reduce irregularities at the surface.

9.1.12 Old coal workings

The effect on subsidence caused by the presence of goaf above the seam being worked is difficult to assess, especially where pillars, ribs and barriers have been left in old workings. Movement of the old workings is superimposed upon the normal subsidence and Whetton (1957) considers that the major problems concern prediction of the magnitude of the strain on the surface, both in position and direction. The subjects of surface stability and remedial treatments in areas of old coal workings are described by Gray and Meyers (1970) and Littlejohn (1979a, b).

9.1.13 Presence of faults

The most erratic surface changes occur in the presence of a geological fault. Peate (1956) has shown that this natural plane of shear modifies the normal effects of draw and may cause abrupt changes of surface level with spectacular effect, as shown in Fig. 9.10. All such problems, however, are unique in time, place and circumstance, and the appropriate answers vary with the facts.

9.1.14 Estimation of subsidence

When all the effects discussed above have been taken into account, it is usual to express the estimated maximum subsidence as a fraction of the seam thickness extracted. This fraction is known as the subsidence factor. An estimate of

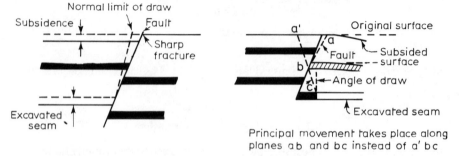

Figure 9.10 Influence of a fault on the subsidence profile (after Peate, 1956).

subsidence factor can be made from Fig. 9.11 for longwall workings with strip packing, provided the seam is level, the width of panel is constant and the face has advanced through the critical area for the point considered (Orchard, 1953). Whilst the greatest subsidence factor is 0.84, the value may be higher for full caving but is certainly lower for solid stowing. Thus the method is strictly limited

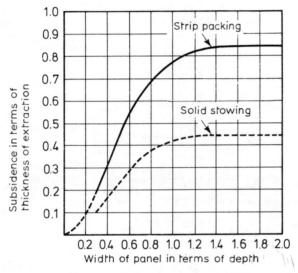

Example : 150 m face, 1.5 m seam, depth 250 m, strip packing.
W/D = 0.6, hence subsidence factor = 0.53. Therefore estimated maximum subsidence = 0.53 × 1.5 = 0.8 m

Figure 9.11 Estimation of subsidence factor.

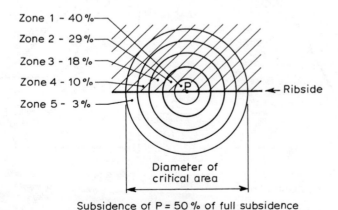

Figure 9.12 Prediction of subsidence by zones in Holland.

in its scope and, even when applied to strip packing, there is no doubt that the factor is subject to variations, according to local conditions.

Subsidence can be predicted also by a method of zones and, although the system is not common in Britain, it has been widely adopted in Europe, especially Holland. In order to predict the subsidence over an irregular goaf, or at some distance from the workings, the critical area is divided into a number of annular zones and the percentage which the extraction of each zone contributes towards the full subsidence is based on observational data collected over many years (Fig. 9.12). The diagram is analogous to Newmark's chart for foundation stresses

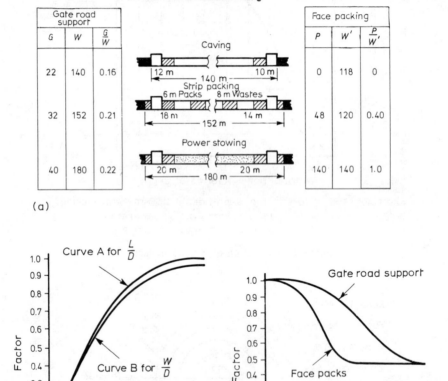

Figure 9.13 Estimation of subsidence with particular regard to degree of packing. G = total width of Gateside packs; W = full width of panel, P = total width of face packs; $W'' = W - G$; L = face advance into critical area. (After Marr, 1958.)

and in the Dutch coalfields it is fortunate that half subsidence is normally observed over the ribside. Thus when a series of zones are described round a point over a ribside, half subsidence is inevitably predicted.

Marr (1958) has described a method which attempts to correlate the known mining parameters with particular regard to the degree of packing applied to the face (Fig. 9.13a) and it appears that his empirical analysis is fairly accurate. To obtain the maximum subsidence the various factors for W/D, L/D, G/W and P/W' and taken from graphs in Fig. 9.13(b) and (c) and their product is multiplied by the seam thickness.

EXAMPLE 9.1

For the strip packing situation shown in Fig. 9.13(a) where the depth is 400 m, length of advance 450 m, width of panel 152 m and seam thickness 1.5 m, the maximum subsidence can be determined as follows:

Subsidence factor = product of factors for W/D, L/D, G/W and P/W'

$$W/D = \frac{152}{400} = 0.385 \quad \text{therefore} \quad F = 0.51$$

$$L/D = \frac{450}{400} = 1.125 \quad \text{therefore} \quad F = 0.97$$

$$G/W = \frac{18 + 14}{152} = 0.21 \quad \text{therefore} \quad F = 0.97$$

$$P/W' = \frac{8 \text{ no. packs of 6 m}}{152 \text{ m} - 18 \text{ m} - 14 \text{ m (gate packs)}} = \frac{48}{120} = 0.40 \quad \text{therefore} \quad F = 0.66$$

Therefore subsidence factor $= 0.51 \times 0.97 \times 0.97 \times 0.66 = 0.32$
Therefore maximum subsidence $= 0.32 \times 1.5 \text{ m} = 0.48 \text{ m}$.

In a similar way, the ribside subsidence can be calculated using additional graphs.

In the past decade greater interest has been expressed in design and analysis in the United States, and the state of the predictive art is described by Voight and Pariseau (1970).

9.1.15 Strains and displacements

When a trough of subsidence is formed as shown in Fig. 9.4, the centre part subsides vertically only; the remainder moves inwards in addition to moving downwards. This fact was not always recognized and the use of the term 'subsidence' – implying vertical movement – probably tended to obscure horizontal components in the early days. Whetton (1957) observed that, although Stewart published records of lateral displacement as early as 1885, the neglect or inability to recognize surface strains persisted in this country until 1930 when, in Scotland, Briggs and Ferguson devised a scheme for the measurement of the component. Observations show that the horizontal components of the movement at individual surface points in the subsiding area are not equal in amount and the differential displacements cause tensile and compressive strains according to whether the inner or outer point is displaced the more. The manner in which the displacements and accompanying strains are developed is indicated in Fig. 9.14. The tensile strains occur mainly around the edges of a subsiding basin and the compressive strains more or less directly over the worked-out area. The horizontal displacements reach a maximum at points approximately vertically over the edge of the worked-out area and are zero at the extremities and at the centre of the subsiding basin. One phenomenon, which is a direct outcome of tensile strain, is the occurrence of breaks or fissures in the ground surface, usually at points of maximum tensile strain. These fissures tend to appear where a stratum of hard rock is present, at or near the surface, but may occur also in softer strata if the tensile strain is great.

The horizontal displacement and, therefore, the strain, is proportional to the subsidence and varies roughly inversely with depth of seam. Figure 9.9 shows the manner in which a similar amplitude of subsidence can be caused by workings at different depths, provided the seam thickness is similar and the width/depth ratio is the same. This is an important relationship in itself but, in addition, it can be seen that the strains reduce in value with increase in depth. With shallow workings, the curvature of the profile of the subsidence basin is greater, owing to the smaller extent of the draw. This increased curvature accompanies greater differential displacement and therefore greater strain, and is the reason why shallow workings cause more damage to surface structures than deep workings. The linear relationship $\pm E \propto S/D$ was pointed out by King and Smith (1954), where E is the maximum strain, S the maximum subsidence and D the depth of the seam. Since this date various relationships have been evolved by different investigators but the most widely known are those of Orchard (1956) where

$$+ E(\text{tensile strain}) = 0.75 \, S/D$$
$$- E(\text{compressive strain}) = 1.5 \, S/D.$$

The compressive strain equation applies only to cases where the width of extraction does not exceed the radius of the critical area. With greater widths, the maximum compressive strain reduces in value (Fig. 9.14). To obtain full subsidence the width of working must be equal to the diameter of the critical area but it has been observed that the maximum strains are reached when the width is approximately equal to the radius of the critical area. As the width of working increases, the tensile strain remains constant at its maximum intensity whilst the compressive strain reduces in value and gradually spreads out into two zones when the full critical area has been worked. Under these conditions it is commonly observed that the compressive strain tends to reduce to the value of the tensile strain, in accordance with the long established principle that the amount of tensile strain in two zones must more or less balance the compressive strain over the goaf, whether in one or two zones. For a critical area of extraction there exists a point which is strain free (Fig. 9.14b) and when a supercritical width is worked this strain free point becomes a zone (Fig. 9.14c).

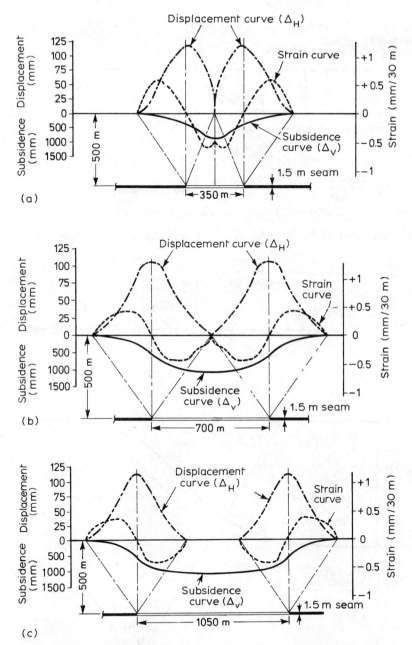

Figure 9.14 Development of horizontal displacements and strains with respect to subsidence profile. (a) Subcritical condition; (b) critical condition; (c) supercritical condition.

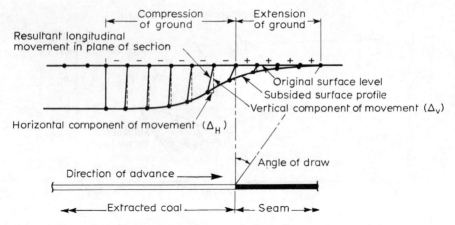

Figure 9.15 Longitudinal section on line of advance of working face.

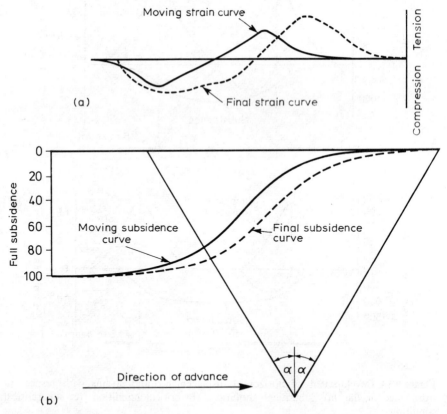

Figure 9.16 Comparison of transient and final subsidence movements (after Beevers and Wardell, 1954).

9.1.16 Travelling or dynamic movements

The foregoing diagrams illustrate the final position which is reached only when the movement is complete. Development of subsidence is a three-dimensional process and as the face advances it initiates a wave of subsidence at the surface where tensile and compressive strains are experienced alternately as shown in Fig. 9.15. Normally these transient strains are somewhat less than the final strains because of the differences in subsidence profile, shown in Fig. 9.16(a) and (b). Beevers and Wardell (1954) found that the intensity of the travelling strain bears a direct relationship to the amount of subsidence and thus, for a face equal in width to the diameter of the critical area, the maximum travelling strain develops over the centre of the face, where the greatest subsidence also is developed. To right and left of the centre, both the subsidence and intensity of the travelling strains decrease, as shown in Fig. 9.17. Similarly, both subsidence and travelling strains are reduced when subcritical widths are worked, as shown in Fig. 9.18. Consequently, travelling strains are greatest when the width of the advancing face equals or exceeds the diameter of the critical area. Thus a surface structure placed in the central zone of a supercritical area of extraction would be subjected to maximum travelling strains, although eventually it would return to a strain-free condition at a lower level than originally.

Much attention is now being paid to the development and effect of these strains and to ways and means of minimizing damage to buildings; reference to these topics will be made later.

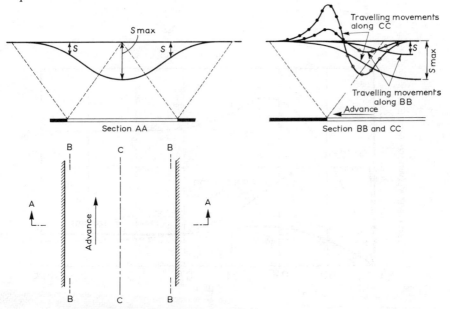

Figure 9.17 Development of travelling movements across the advancing face, where ——— is subsidence and ••◦◦ is strain, (after Beevers and Wardell, 1954).

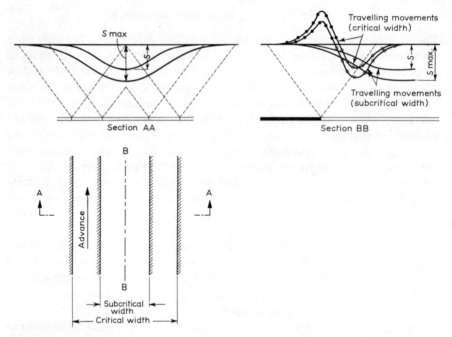

Figure 9.18 Influence of width of working on travelling movements, where ———— is subsidence and -•-•-○-○- is strain, (after Beevers and Wardell, 1954).

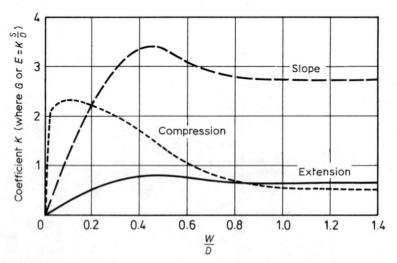

Figure 9.19 Relationship of slope (*G*) and strain (*E*) to maximum subsidence and depth, at various width/depth ratios (after NCB, 1975).

9.1.17 Slope

An estimate of surface slope or tilt is required to determine the size of structural units and the need for expansion gaps to minimize the risk of damage. Mining engineers calculate slope between adjacent subsidence stations (15 to 30 m apart) by dividing the differential subsidence by the horizontal distance between stations. Based on field observations of this type, the maximum slope (G) which extraction is likely to cause is 2.75 S/D when full subsidence develops, although the maximum coefficient in terms of S/D occurs when $W = 0.45\ D$ (Fig. 9.19). Figure 9.20 is based on field observations and illustrates the effect of depth and thickness of seam on maximum slope produced at the surface. Both Figs 9.19 and 9.20 have been produced by the National Coal Board and, although the results are obviously limited in their application, they give the structural engineer a guide when determining the limiting dimensions of construction.

9.1.18 Inclined seams

In the various methods that have been used for the estimation of ground movements, the seams are usually assumed to be horizontal or nearly so. The effect of inclined measures has been investigated in several coalfields with some success, although the cases examined cover seams ranging only from horizontal to a dip of approximately 20°. Orchard (1953) found that inclination of strata displaces the subsidence trough and the strain zones in the direction of dip. Generally, points on the subsidence curve are related to relevant points in the seam by imaginary lines drawn normal to the seam and not by vertical lines. Thus a point such as that of maximum subsidence, which lies vertically above the

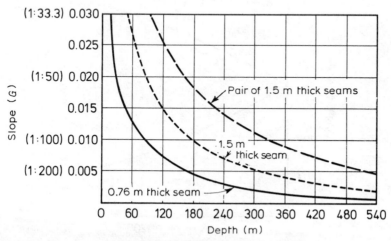

Figure 9.20 Effect of depth and thickness of seam on maximum slope at the surface (after NCB, 1975).

centre of the workings in horizontal measures, occurs perpendicular to the centre of the goaf when the seam is inclined, as shown in Fig. 9.21.

The normal angle of draw reduces to the rise and increases to the dip with inclined seams, as shown in Fig. 9.22 (Marr, 1958) in which the normal angle of draw is 35° and, therefore, the data apply only to cases where $W/D > 0.7$. When used for ratios less than 0.7, a lower value of angle of draw is selected from Fig. 9.5 and the curves shown in Fig. 9.22 are displaced accordingly.

The best layout of working for the protection of surface property is difficult to determine where the seams are inclined. The differences in stress arising from the depth differential can result in greater consolidation of packs on the dip side of faces. Consequently, there is a greater degree of vertical convergence at the greater depth, which modifies the pattern of movements to be found over inclined

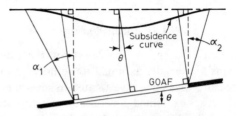

Figure 9.21 Effect of inclination of seam on position of subsidence profile (after Orchard, 1953).

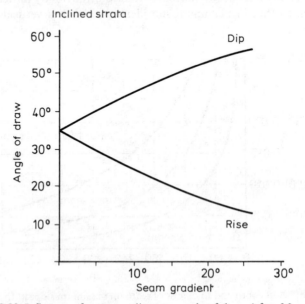

Figure 9.22 Influence of seam gradient on angle of draw (after Marr, 1958).

workings. There is still a considerable amount of field research to be done where the opportunities exist to take observations over semi-steep and steep seams.

9.1.19 The time factor

From the point of view of surface damage the duration of ground movement is a factor of concern to mining engineers and surveyors, who have to deal with damage claims and repairs. It is important to know when a particular structure is likely to be initially affected, when the danger will be most acute, and when the movement will be virtually complete. Workings must be properly phased to avoid cumulative strains. Hence a knowledge of the relationship between time and subsidence is essential.

For many years subsidence of a point expressed as a ratio of the total subsidence occurring in unit time was referred to as the time factor. Bals (1931) was probably the first to point out that this method was incorrect and that the time factor was a function of the area of seam worked in unit time. In spite of this criticism the original method continued unchanged until Wardell (1953) clarified the situation by introducing the 'subsidence development curve'. This curve (Fig. 9.23) indicates the subsidence of a given point on the surface for successive positions of a face passing through the critical area of the point. Wardell represented the time factor by the rate of advance expressed in terms of the radius of the critical area and, from eleven case histories, concluded that the development of subsidence depended primarily on the position and rate of advance of the face within the critical area.

As Fig. 9.23 shows, the point P begins to subside as soon as the working enters the critical area and therefore it must be assumed that the extraction which is going to influence point P has already been in operation for some time when it reaches the zone of influence, that is, the origin of the diagram. Subsidence commences slowly during a period corresponding to an advance of about $0.6\,R$. In this zone the basin is very slightly inclined and the fall of point P amounts to about 5% only of the final subsidence. A period of acceleration follows situated between 0.6 and $1.1\,R$ and this corresponds to a zone characterized by heavy surface damage. At $0.9\,R$ the slope at point P is greatly increased and the tensile ground strain reaches its maximum. When the face passes vertically below point P the subsidence is then about 15% of the total and the tensile ground strains diminish gradually to zero at $1.1\,R$. The rate of subsidence development reaches its maximum and maintains it between 1.1 and $1.6\,R$ and, during this stage of the advance, point P effects about two-thirds of its descent. At $1.6\,R$, compressive ground strains reach maximum values, after which they steadily decrease to negligible values at $2.0\,R$. It is commonly observed that about 95% of the ultimate subsidence has taken place when the extraction moves out of the critical area. The time taken for the residual 5% to develop appears to depend on the dimensions of the critical area and the rate of advance. Wardell indicates that it is necessary for the extraction to advance a further $0.5\,R$ before subsidence is complete, and this

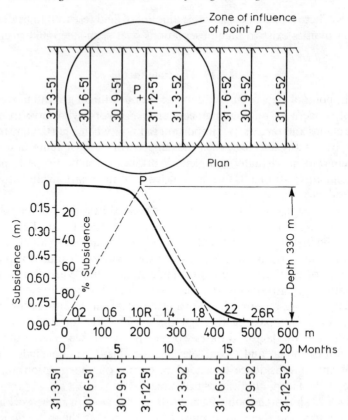

Figure 9.23 Subsidence–time relationship (after Wardell, 1953).

may take only a few months. However, Orchard (1956) has shown that cases are not unknown where a residual subsidence of 5% has taken 12 months or more to occur. In spite of its small percentage value, residual subsidence is a factor of some concern and certain types of structure can be seriously affected by it.

The subsidence development curve is invaluable in producing time/subsidence curves for any situation in which the final subsidence is known. The following example (Wardell, 1953) illustrates the method of application.

EXAMPLE 9.2

Given a depth of seam of 500 m, angle of draw 35°, rate of advance 350 m year^{-1} and a total anticipated subsidence at a point P of 1.0 m, the subsidence at intervals of 6 months is determined in the following manner.

The radius of the critical area = depth × tan angle of draw = 500 × 0.7
$$= 350 \text{ m}$$

Hence, the rate of advance is 1.0 R year^{-1}.

In 6 months the working will have advanced 0.5 R, in 12 months 1.0 R, in 18 months 1.5 R, in 24 months 2.0 R and in 30 months 2.5 R. The percentage of the total subsidence, measured from the original plot for Fig. 9.23 for each of these positions, is 1.25; 14.5; 69.0; 94.0 and 100, respectively. Thus at the end of 6, 12, 18, 24 and 30 months the magnitudes of subsidence will be 0.0125, 0.145, 0.69, 0.94 and 1.00 m, respectively.

The relationship between the rate of advance and the time factor has also been investigated by Wardell. It is clear that the slower the working traverses the critical area, the longer it takes for the subsidence to develop, and Wardell has stated that, for any case, the time taken for the subsidence at the surface to be completed is more or less inversely proportional to the rate of advance of the workings. In Fig. 9.24 for example, the rates of advance are expressed in terms of the radius of the critical area and the curves show that the greater the rate of advance, the less time is required for the full development of subsidence.

Depth is also an important factor, partly because the diameter of the critical

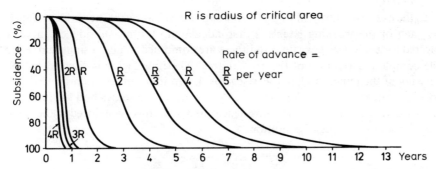

Figure 9.24 Percentage subsidence in relation to time and rate of advance (after Wardell, 1953).

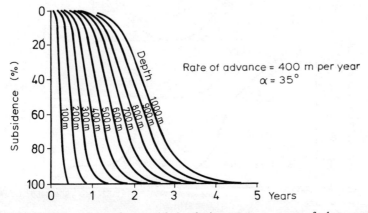

Figure 9.25 Variation of time factor with depth given a constant rate of advance and angle of draw, where Δ is total settlement and δ is differential settlement, (after Wardell, 1953).

area, and therefore the time taken for a working with a given rate of advance to traverse it, increases with depth, and partly because at greater depths several separate but contiguous workings may be necessary before the critical area is completely worked out. Wardell plotted a family of curves, shown in Fig. 9.25, assuming an angle of draw of 35° and an annual advance of 400 m, and the influence of depth can be clearly seen.

With regard to the effect of width of working, the time factor discussed thus far has been considered in relation to the subsidence caused at a surface point by a single working of uniform width. At shallow depths it is quite possible for a working greater in width than the diameter of the critical area to be extracted in one operation and consequently for the time–subsidence relationship already discussed to represent also the total time which a point will take to subside from the working of its critical area. At greater depths, however, the full critical area of a surface point P will rarely, if ever, be completely extracted by a single working panel and the effect of working in strips or panels at different times must be considered.

In the example shown in Fig. 9.26, the subsidence caused at the surface point P by each of the working panels is first calculated. A time–subsidence curve is plotted for each one separately and these are combined to give a single curve for the complete extraction of the critical area. The time factor is affected by the phasing of the panels and, in the case shown, two time curves are produced, the former representing the phasing indicated by the earlier dates and the latter by the more recent dates. A time–subsidence curve can thus be deduced in the same way for any given set of conditions.

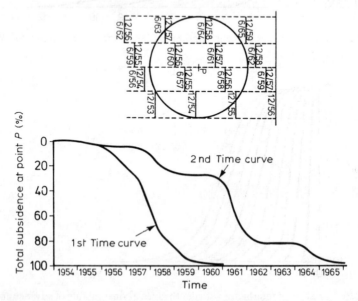

Figure 9.26 Time curves illustrating variations with different phasing of extraction.

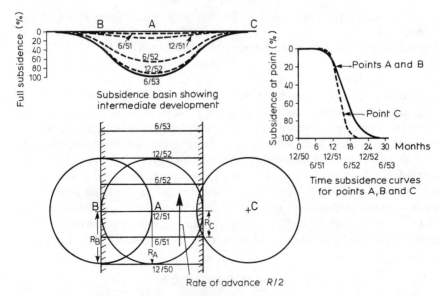

Figure 9.27 Variations in time–subsidence development at different points with respect to a working (after Wardell, 1953).

So far, all the points used in producing subsidence curves have been situated over the working area. However, it is necessary to consider also the development of subsidence at points remote from, but still affected by, the workings. In Fig. 9.27 point A represents the cases already dealt with, and the foregoing method can be similarly applied to all points overlying the goaf, since working will continue within their critical areas for exactly the same time. Thus the time factor for all points between A and B will be identical, although the amount of subsidence will decrease from A to B. At point C, however, the working enters its critical area for only half the time relating to points A or B. Clearly point C cannot begin to subside until the working enters its critical area and will probably cease to do so after the working has advanced beyond it. In this situation Wardell assumes that the chord R_C represents the full critical radius R for point C. Although this assumption has not been fully substantiated, observations indicate that the further a point is away from the working, the less time it takes to undergo the maximum subsidence caused by that working. This can be seen from the time/subsidence curves for points A, B and C shown in Fig. 9.27, and these curves are particularly valuable in determining the intermediate stages in the development of the final subsidence basin.

9.1.20 Conclusions

The purpose of this section has been to describe the principles of subsidence due to longwall mining, since it is essential that the engineer who sets out to design a structure in a mining area should be familiar with the basic concepts of the problem.

The subsidence prediction procedures of the National Coal Board (NCB) are based primarily upon surface and seam level survey data and, in spite of the fact that these prediction procedures are purely geometric, an accuracy of $\pm 10\%$ is claimed in the majority of cases. In the NCB *Subsidence Engineer's Handbook* (1975), the empirical rules and curves quoted earlier have been employed to produce contour maps of subsidence and strain plotted in relation to position of mine panel for different width/depth ratios. These charts have proved valuable to practising engineers but, since the rules do not take account of geology, topography and material properties of the subsiding rock and soil masses, all of which influence subsidence development, there is risk in applying empirically based procedures or charts too generally. Nevertheless, the accumulation of data by the NCB represents the largest systematic study of subsidence to date and all engineers concerned with mining subsidence should be familiar with that work.

9.2 SURFACE DAMAGE

9.2.1 General aspects

The effect of subsidence on structures and services hardly needs describing. Minor effects include cracking of plasterwork and slight jamming of doors and windows. More serious effects are fractures in brickwork and distortion of steel or concrete frameworks which may lead to structural failure.

Roads, railways, rivers and canals have subsided as much as 6 m where a number of seams overlying each other have been worked. This has necessitated the building up of banks of rivers and canals, followed by the raising of bridges to maintain headroom. Areas of low-lying land become inundated with water to form flashes and meres, thereby detracting from land values and rendering existing buildings untenable. Backfalls may be created in pipelines due to variations in vertical subsidence.

In zones of compression, the ground has been observed to form ridges some 1 m wide and 200 mm high, whilst in tension zones cracks of width exceeding 150 mm have been measured. As a result of these horizontal ground displacements, sewers and other pipelines may be buckled and fractured and basements and manholes distorted and crushed.

With regard to surface structures, it is well established that the tensile and compressive ground strains are the main cause of damage. This is confirmed by the fact that the greatest damage does not necessarily take place where maximum subsidence occurs but rather where the curvature of the subsidence basin is greatest. The degree of damage suffered by the structure, however, is not simply dependent on the intensity of compressive or tensile ground strain, but also on the dimensions and orientation of the affected structure with respect to the advancing face, the subgrade forces at the foundation interface and the flexibility of the superstructure. If the ground is being compressed at uniform strain of, say, 1 mm m^{-1}, the damage resulting from this compressive strain may be slight in the case of a small building but, if the whole of the shortening takes place at one weak place in the structure, a long building, such as a terrace of houses, may suffer severe damage.

In addition to horizontal strains, tilt can also adversely affect buildings, especially tall structures such as chimneys. Here the structure may lean to such an extent that it becomes unstable. This occurrence, however, is rare and nearly always due to several seams being worked on one side only of the structure. The remedy is to work coal on the other side, although this is not always convenient. Slight or moderate tilting, unaccompanied by appreciable strains, is often tolerable or even unnoticeable but certain industrial processes can be upset by slight departures from the horizontal and provision for jacking is sometimes made where the possibility of levelling by underground extraction is impracticable.

In general, although damage due to mining can be very severe, it appears that aesthetic damage is a prime concern. Although structurally sound, a house which leans or contains plaster cracks can cause serious distress to its occupants. For this reason a knowledge of ground strains and settlements present at the first appearance of cracks is of particular importance.

Although surface subsidence due to mining is three-dimensional in character, in order to study the threshold or limiting movements in relation to degree of damage, it is convenient to consider the vertical and horizontal components separately, in spite of the fact that they occur simultaneously.

9.2.2 Vertical settlements

Curvature of the ground surface usually gives rise to differential settlement between parts of a structure and, unless the structure is statically determinate or perfectly flexible, the distribution of the supporting forces will be changed. This will, in turn, produce stress changes within the structure and these may ultimately produce structural damage. Bjerrum (1963) pointed out that, in ground settlement problems, the engineer has to answer two questions: firstly he has to evaluate the allowable differential settlement the structure can withstand without experiencing damage and secondly predict what total and differential settlements can be expected at a given site.

Field observations show that the magnitude of differential settlement which will cause damage to a building cannot be predicted from a theoretical computation, since the behaviour of the building is influenced by a number of factors which are difficult or impossible to take into account in theoretical calculations such as interaction between frame and cladding, the varying rates of subsidence and the redistribution of the loads. Allowable settlements have, therefore, to be decided on the basis of experience derived from observations on. similar types of structures. From a study of deformation caused by settlement,

Skempton and Macdonald (1956) concluded that allowable settlement is governed more by the avoidance of cracking in panels and finishes than by the overstressing of stanchions and beams. Consequently, the allowable distortion will in almost every case be governed by a consideration of factors which are related to the practical use of the building. It follows that there is no single damage criterion which can be specified and in any particular case it must be decided what type of damage can be considered permissible, and this obviously varies from building to building. The allowable settlement will be different for a warehouse, where the occurrence of minor cracks in partition walls would not justify an expensive foundation, whereas for a hospital or school, unsightly cracks should be prevented.

The results of Skempton and MacDonald's investigation briefly summarized are:

(a) Cracking of the panels of framed buildings of the traditional type or of the walls in load bearing wall buildings is likely to occur if the angular distortion δ/l exceeds 1/300 (Fig. 9.28a).
(b) Structural damage to stanchions and beams is likely to occur if the angular distortion exceeds 1/150.

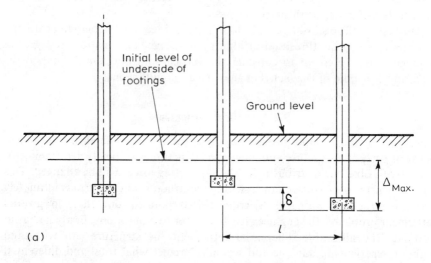

(a)

Criterion		Isolated foundations	Rafts
Angular distortion δ/l		1 / 300	
Maximum settlement $\Delta_{Max.}$	Clays	75 mm	75 - 125 mm
	Sands	50 mm	50 - 75 mm

(b)

Figure 9.28 Settlement limits (after Skempton and MacDonald, 1956).

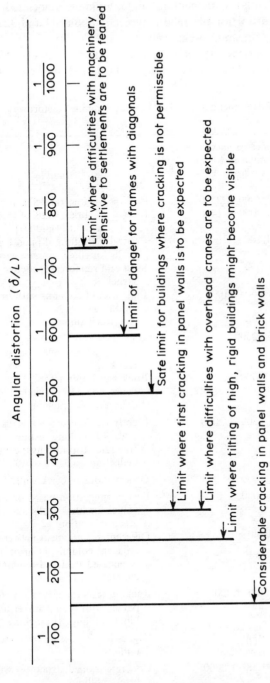

Figure 9.29 Damage criteria. Height (*h*); length (*L*) (after Bjerrum, 1963).

(c) The greatest differential settlement (δ) in a building associated with a maximum angular distortion of 1/300 is typically about 44 and 32 mm for foundations on clay or sand, respectively.

(d) Figure 9.28(b) summarizes the recommended settlement limits. Although tending to err on the side of safety, these limits should not be exceeded in

Table 9.2 Summary of Recommendations for limiting angular distortions of brick structures

Reference	Suggested maximum angular distortion (δ/L)	Details
Polshin and Tokar (1957)	1/3000–1/2500	($L/H \leqslant 3$) Central deflection of plain
	1/2000–1/1400	($L/H \geqslant 5$) brick walls of length (L) and height (H)
	1/1000	Central deflection of plain brick walls for single storey factories
	1/500	Steel and reinforced concrete frame structures
	1/150–1/1000	End rows of columns with brick cladding
	1/300	Crane track slopes
Wood (1952)	1/20000	Solid brick wall 2.44 m high, span 3.2 m; central deflection at cracking
	1/4000	Brick walls on reinforced concrete beam, span 3.2 m; central deflection at cracking
Rigby and Dakema (1952)	1/4500–1/7000	Lightly reinforced brick walls 3.05 m high, 9.15 m long resting on concrete strip footings; central deflection
Terzaghi (1935)	1/1000	Observations of brick walls
Meyerhof (1953)	1/2000	Design suggestion based on above values for brick walls allowing for creep relaxation
	1/300	Differential settlement between adjacent columns of open frames of encased steel and reinforced concrete; at cracking
Thomas (1953)	1/300–1/400	Racking tests on 3.05 m × 3.66 m panels with brick, clinker and clay block infilling; at cracking
Meyerhof (1953)	1/1000	Suggestion for design of frames with infilling
Fjeld (1963)	1/1200–1/2000	Concrete framed structures with brick infilling

design without special investigation or knowledge from local experience.

(e) For box foundations on clay the allowable maximum settlements may exceed 100 mm provided that such settlements will not give rise to visible tilt in the building.

Bjerrum (1963) also studied the problem of differential settlement and determined the values of limiting angular distortions related to type of damage or difficulty to be expected, shown in Fig. 9.29. Although Skempton and MacDonald's limit for δ/l of 1/300 has been generally accepted for many years, the design limits given in Figs. 9.28 and 9.29 must not be regarded as a set of rigid rules, because the ground conditions are not described in detail and the load distribution within the structure is not specified. For conventional settlements it is unlikely that any damage to a building will occur provided these limits are not exceeded but in some cases the engineer must use judgement and experience which may lead him to adopt very different criteria. For example Table 9.2 shows that in brick structures crack initiation has been observed over a wide range of angular distortions. It should also be remembered that methods of construction have changed since 1956 and some modern structures are outside the scope of Skempton and MacDonald's investigation (ISE, 1978).

The second problem which the engineer faces concerns the prediction of differential settlement at a given site. Settlements occurring as a result of mining operations are, of course, additional to settlements existing before mining operations commence. The factors affecting the estimation of magnitude and rate of subsidence have been discussed in Section 9.1 and normal settlement is treated in Sections 2.4 to 2.6. It should be noted that, in addition to vertical settlements, horizontal displacements must be taken into account in connection with mining subsidence: the latter are treated in the following section.

9.2.3 Horizontal displacements

Ground strains at the surface arising from differential horizontal displacements are not transferred directly to the structure: the ground movements impose a force on the structure which, in turn, causes structural strain. The fact that ground strains do not reproduce identical strains on the structure sited on or in the ground was not generally appreciated by planning and mining engineers until the late 1950s.

At present the only guide to estimating damage is based on many years of observing countless buildings and, in spite of the fact that several factors such as soil type, subgrade force and structural flexibility have not been included in the observations or in the empirical analyses, simple relationships between ground strain and damage have proved to be of value. When new extractions are planned, a fair idea can be obtained of the damage likely to be caused to overlying surface properties. Any particular precautions indicated can then be taken, whether this be in the form of modifications to the structure or to the proposed mine workings.

Table 9.3 Classification of structural damage

Classification of damage	Description of typical damage	Strain (%)
Negligible	Hair-line cracks in plaster, crack width $\not> 0.1$ mm	0–0.05
Very slight	Hair-line cracks in plaster, perhaps isolated fracture in building not visible outside, crack width $\not> 0.5$ mm	0.05–0.1
Slight	Several slight fractures, width $\not> 2$ mm, showing inside building, doors and windows may jam slightly, repairs to decoration probably necessary	0.1 –0.2
Appreciable	Slight fractures showing on outside of building (or one main fracture, width $\not> 6$ mm), doors and windows jamming	0.2 –0.35
Severe	Open fractures, width $\not> 15$ mm, requiring rebonding and allowing weather into structure, window and door frames distorted, walls leaning or bulging noticeably, some loss of bearing in beams	0.35–0.6
Very severe	Open fractures, width > 15 mm, and repairs involving partial or complete rebuilding, roof and floor beams lose bearing, walls lean badly and require to be shored up	> 0.6

Most buildings can withstand a certain degree of tensile or compressive ground strain without apparent signs of damage, and observations indicate that cracking of internal plaster and jamming of doors and windows tend to occur when tensile and compressive ground strains exceed 0.4 and 0.8 mm m^{-1}, respectively. These figures lead to the commonly held view that most buildings appear to be about twice as strong in compression as in tension. Compressive ground strains up to 6 mm m^{-1} have been recorded and these would, without doubt, cause severe surface damage. Ogden and Orchard (1959) compared several cases where the ground strains have been measured precisely and the structural damage has been photographed and classified according to an accepted scale of damage. Table 9.3 indicates six classes of damage related to ground strain and the present author has included crack measurements to further quantify degree of damage. It is noteworthy that Burland and Worth (1975) judge that 0.1 mm is the width of crack likely to be first discerned by a casual observer and Pryke (1975) states that cracks seem to worry people when the width exceeds 3 mm.

Orchard and King (1959) published data represented in Fig. 9.30 in an attempt to show the relationship between the size of structure, ground strain and the severity of the resulting damage.

Bearing in mind the importance of differential settlement, as described in Section 9.2.2, it is unfortunate that the combined movements have not been

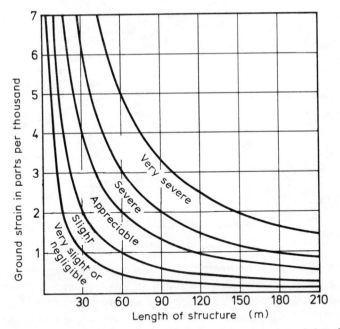

Figure 9.30 Relationship between severity of damage, ground strain and size of structure (after Orchard and King, 1959).

monitored in relation to damage. It is considered that more detailed observations of structures in areas of mining subsidence are required since the limiting angular distortions and threshold ground strains already discussed may still underestimate the likely degree of damage. A field investigation (Littlejohn, 1975a) has indicated that aesthetic damage to unreinforced brickwork occurred at an angular distortion (δ/l) of 1/6000 and the tensile ground strain monitored was only 0.22 mm m^{-1}.

9.3 MINING PRECAUTIONS AGAINST THE EFFECTS OF SUBSIDENCE

9.3.1 Sterilization

If a valuable structure, such as a public building or power station, requires protection, part of the coal seam underneath may be left unworked in order to form a pillar of support. The size of this pillar (Fig. 9.31) depends on the seam depth, the angle of the draw and the margin of safety required. With increasing depth, the amount of sterilized coal increases appreciably and obvious limitations ensue where the area to be protected is extensive. Modern ideas of domestic and industrial development, since the advent of motor transport, have tended to destroy the previous state of affairs where there were compact and densely built-

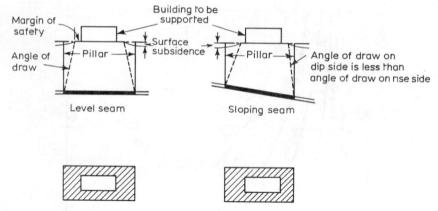

Figure 9.31 Relationship between area of sterilized coal, seam depth, angle of draw and safety margin.

up industrial areas in the towns with a relatively much larger space of open country outside. This trend has resulted in much interference between coal mining and built-up areas. The trouble is accentuated by the fact that the steady exhaustion of the better seams has increased the necessity of finding some method of extracting the remaining portions of the seams even though they may lie under industrialized areas. Hence, the adoption of protective pillars is more or less regarded as an extreme measure.

9.3.2 Harmonious mining

Considerable success has been achieved in Holland and Germany in protecting particular buildings from major damage by careful planning of workings. Broadly, this involves arranging a particular working so that the building to be protected lies in an area of minimum strain or, by working two seams concurrently, surface strains are cancelled. In the past it was often suggested that the best way to minimize surface damage was to work beneath the structure with a wide and quickly advancing face. At a depth of 200 m, an advancing face 350 m wide would have a zone some 70 m wide in the centre which would undergo uniform maximum subsidence and be practically free of strain after completion of subsidence. A similar idea to this was advocated by Lehmann as early as 1938, when he showed that supercritical areas of extraction could be achieved by simultaneous working of three or more seams: the reduction in strain for two seams only is shown in Fig. 9.32. Unfortunately Lehmann's description of 'harmonious mining' referred only to the elimination of transverse ground strains which would be ideal but for the three-dimensional nature of subsidence development. In the case quoted above, the travelling movements would be maxima because maximum subsidence would be developed over the centre of the working.

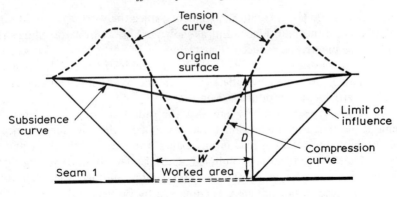

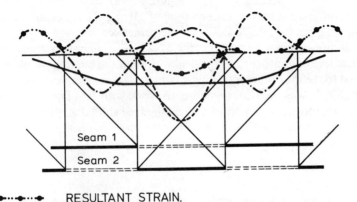

-•--•--• RESULTANT STRAIN.

Figure 9.32 Simultaneous working in two seams with width and position of faces harmonized, where --- represents the first seam working and strains related thereto; –·–·– represents the second seam working and strains related thereto and -•·· -•- ··-•- represents the resultant strain (after Grond, 1950).

Grond (1950) described how the anticipated tensile ground strain from the workings in one seam can be practically offset by the compressive ground strain caused by the extraction in another seam. This method, however, is difficult to adopt in British mining because suitable access to two or more seams for simultaneous working in the same locality does not often occur. Nevertheless, the influence of travelling strains is a prime concern in Britain and since continental practices cannot always be imitated, alternative methods have been adopted to minimize damage. A face of the same width (350 m) as in the foregoing example at a depth of 800 m would be too narrow to develop either maximum tensile or compressive ground strain in a direction parallel to the face and the danger of damage to a centrally disposed building would not be great. Nor would there be much danger from the travelling movements, since the workings would not be developing maximum subsidence and, as shown in Fig. 9.18, the travelling strains would, consequently, be reduced. This technique of partial extraction has been adopted with considerable success in Yorkshire and it simply consists of narrow longwall panels separated by pillars. Beevers and Wardell (1954) suggest that negligible settlement is produced when the widths of the panels and pillars are $R/6.5$ and $R/4.4$ respectively, where R is the radius of the critical area. This technique is primarily applicable to seams at considerable depth, since for shallow seams the working face in any panel becomes unduly short and this retards efficient production. Combined with the fact that travelling strains increase with decrease in depth, this necessitates another approach to the problem. A successful technique was employed by Beevers and Wardell (1954), who adopted stepped faces in an attempt to attenuate the travelling subsidence curve and hence reduce the strain. Figure 9.33(a) shows the situation which confronted these authors and the method of solving the problem is discussed in the following paragraphs to illustrate the principles involved.

The circumstances were such that the workings could have been carried out by one single face of approximately critical width but the building would have been subjected to the maximum travelling strains and this was considered to be highly undesirable. It was decided, therefore, to work with two faces of more or less equal width and to arrange that the leading face should advance at a constant distance of 150 m in front of the other. The authors anticipated that in the locality of the building:

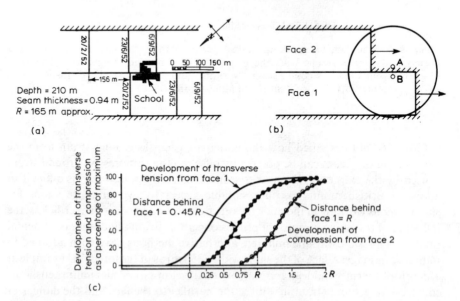

Figure 9.33 Harmonious mining using stepped faces (after Beevers and Wardell, 1954).

(a) The travelling tensile ground strain from the leading face would be about 30% only of the maximum.
(b) The travelling compressive zone from the leading face and the travelling tensile zone from the second face would more or less coincide and that this would result in a net reduction of strain.
(c) The travelling compressive ground strain from the second face would be about 70% of the maximum.

Observations showed that this was effectively the case, although the distance between the faces had been slightly overestimated. Wardell (1957) states that the most important factors in this method are (a) the position of the common working edge between the faces, and (b) the distance between the faces. Ideally, the protected structure should lie just inside the edge of the first working face. The distance between working faces is governed largely by whether the most urgent problem is to minimize the travelling strains or prevent the full development of transverse strain. In the case of travelling strain, Wardell suggests that the distance between faces should be equal to the distance between the peaks of maximum travelling tensile and compressive strain and this would be determined from site observations. The development of transverse strain with time is not fully understood but Wardell suggests that it may be proportional to the development of subsidence and so the subsidence development curve (Fig. 9.23) would represent also the development of strain. In the case of the two surface points A and B (Fig. 9.33b) it has been shown that the subsidence at a point between them develops according to the advance of the working through its critical area. In accordance with Wardell's assumption, the strain between points A and B would develop in the same way. The solid curve in Fig. 9.33(c) therefore represents the development of tensile strain between A and B which would be caused by the working face 1. In the same way face 2 advancing some distance behind causes compressive strains between A and B and the development of this strain can be represented by the same curve. The degree of strain cancellation is governed by the distance between the two faces. If the distance is equivalent to $0.45\ R$, then the tensile strain from the leading face should not develop beyond 50% of its maximum. If the distance is increased to R, then 85% of the tensile strain from the first face would be developed. Wardell admits that this analysis is open to criticism but at least it gives an idea of the principle involved.

The importance of the travelling movements has thus been demonstrated and, although every case has to be considered entirely on its merits, it appears, in the case of shallow workings, that the method of staggered faces is the best solution at present.

9.3.3 Stowing

As shown earlier, stowing may be applied to restrict subsidence and thus prevent surface damage. Whilst there appears to be no justification for wholesale solid stowing in this country, unless for reasons other than subsidence reduction, there are definite advantages to be gained from selective and local application. Orchard (1961) has shown that, by increasing the amount of stowing near a ribside, it is possible to move a zone of maximum tensile strain further away from a particular structure and that, increasing the packing in the centre of a panel, the subsidence profile may be flattened, thus reducing the curvature and hence the possibility of damage. This method is particularly valuable where the width/depth ratio of the panel is such that a zone of 'double compression' (Fig. 9.14) may be formed. Intensive hand packing has been successful in this connection but it is for this purpose that pneumatic stowing or grouted mine packs could probably be used to the best advantage. Solid stowing could be used in a complete panel at the edge of a colliery in order to ease the gradient of the subsidence basin at the boundary, although the same effect can be gained by staggering the stop lines of successive seams.

In certain circumstances, therefore, mining precautions can be employed to avoid damage; however, in general, the mining engineer must devise methods of working and roof control which will be efficient and economical from the point of view of coal production, and these considerations may dictate that limited damage and disruption at the surface should be accepted. In addition, it should be appreciated that situations exist where, in spite of correct roof support, damage is inevitable.

9.4 STRUCTURAL PRECAUTIONS AGAINST THE EFFECTS OF SUBSIDENCE

9.4.1 Design for vertical movements

So far as buildings are concerned, structural precautions against mining damage in Britain have been aimed largely at strengthening foundations by reinforcement. The effects caused by vertical and horizontal movements are usually calculated separately and a method of analysis was published by Mautner (1948). In areas of mining subsidence, Mautner concludes that complete protection from damage can be guaranteed only if the structure is limited in plan size and of great rigidity. With regard to vertical movements, Mautner considers that in the centre of the subsidence basin the structure acts as a simply supported beam (Fig. 9.34a) whilst at the edge of the basin there is a possibility of part of the structure functioning as a cantilever, as shown in Fig. 9.34(b). Considering unit width of the structure and assuming a linear distribution of bearing stress beneath the contact areas, the equilibrium in both cases gives the maximum stress.

$$q_z(\text{max}) = \frac{4q_z}{3(1 - 2k)} \text{ for cantilevering} \tag{9.1}$$

$$q_z(\text{max}) = \frac{q_z}{(1 - k)} \text{ for simple support} \tag{9.2}$$

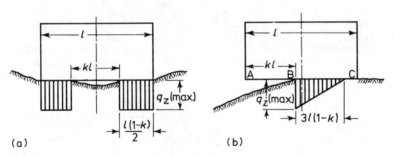

Figure 9.34 Foundation stress distributions at centre and edge of subsidence basin (after Mautner, 1948).

where q_z is the assumed uniform design bearing stress and, if l is the length of the structure, kl is the effective unsupported length. The latter depends on the ratio $n = q_z(\text{yield})/q_z$ where $q_z(\text{yield})$ is the yield stress of the subgrade, that is, the stress under which settlement increases considerably compared with the increasing load. From Figs. 9.34(a) and (b) it can be seen that, if yield stress is reached, any further increment in loading will cause kl to decrease. The value of kl increases with increase in value of yield stress and this leads to increases in the shearing forces and bending moments imposed on the foundation. It follows that, in order to limit bending moments, the design bearing stress should be as high as possible, even if the initial settlement is likely to be great. Clearly, however, continuous settlement due to yield must be prevented, and the Report of the Institution of Structural Engineers (ISE, 1947) specifies that q_z should not exceed half the ultimate bearing capacity of the soil. Little or no relief of bending moment will be gained by using relatively high design stresses in ground with a very high yield strength.

If the direction of advance is parallel to one of the principal axes of the foundation, the bearing stress has the values given in Fig. 9.35(a) and (b) for the cantilever condition. Applying conditions for static equilibrium to the foundation it can be shown that when $k = 0.25$, $q_{z2} = 0$ and when $k = 0.5$, $q_{z1} = \infty$. Bending moments and shearing forces are readily determined from the distribution of applied load and soil bearing stress. The maximum bending moment (M_{max}), not necessarily in the centre of the foundation, when $k = 0.25$ is $0.053 \, q_z l^2$ and when $k = 0.33$ is $0.0785 \, q_z l^2$. This approach to the problem can be adapted to other systems of loading on a foundation and a general expression for bending moment derived, from which maximum moments can be determined by differentiation.

For the simply supported beam condition, it is reasonable to assume that the distribution of bearing stress may be either triangular or uniform (Fig. 9.35c) and in practice it probably lies between the two limits. When $k = 0.33$, the maximum bending moment, which occurs at mid-span, is $0.069 \, q_z l^2$ for the triangular distribution and $0.042 \, q_z l^2$ for the uniform distribution. A mean value

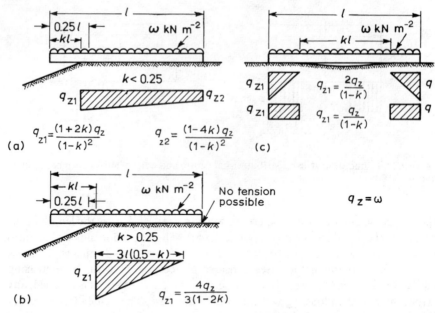

Figure 9.35 Foundation stress distribution during development of mining subsidence for different effective spans (kl).

for design is $M_{max} = 0.055\, q_z l^2$. As already indicated, the value of k depends on the ratio n and the moment will be a minimum when k is as small as possible, that is when n is unity. A structure should be rigid enough to resist the bending moments and shearing forces in all directions and, depending on the orientation of the structure concerned, torque may also have to be considered. If a structure is very long in one or both directions in plan, it may not be economical to design it to withstand the imposed forces and the structure must be sub-divided into sections which will in themselves be rigid.

The effect of twist may be eliminated by supporting the structure on three suitable bearings. Except in very hard ground, Mautner considers that the adoption of this method is an extreme measure since, although it is theoretically immune to all differential subsidence, the three-point support system necessitates the provision of large spherical contact joints of hardened steel between foundations and superstructure to permit large angles of rotation and, in general, it appears that there are practicable difficulties in construction.

In practice, Mautner's method of foundation design appears to be justified only in rare instances where there is a possibility of a discontinuity or sudden collapse over a small area, as in the case of a crown hole. In cases of normal mining subsidence where ground deformations are most frequently caused by general bending of strata, the foregoing method of design gives excessive values of moments and shearing forces. In addition, this method involves considerable

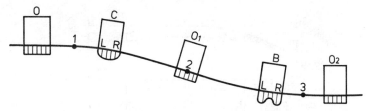

Figure 9.36 Contact stress distribution cycle for a travelling wave (after Wasilkowski, 1956).

expense, and Dobson *et al.* (1959) have shown that the use of a 150 mm thick concrete raft on sand is more economical for low rise buildings. The initial cost of construction is much less although the houses are more susceptible to damage. It appears, however, that the initial saving in construction more than compensates for the repair costs.

From the foregoing remarks it will be appreciated that the use of Mautner's method of design can be uneconomical, especially for low rise building, where foundation costs may represent a significant proportion of the overall cost. An alternative design procedure has been published by Wasilkowski (1956), who considers that, in practice, conditions are more favourable than those assumed in the analysis outlined by Mautner due to (a) plastic deformation of the soil at stresses less than the nominal yield value, (b) the large radius of curvature of the ground surface and (c) the flexibility of most structures.

For a travelling subsidence wave, Wasilkowski considers that the contact stress distribution cycle is as shown in Fig. 9.36. The effect of tilting is assumed negligible and the stress distribution is symmetrical about the neutral axis. The most critical conditions are obviously at points B and C, since the bending moments and shearing forces then acting on the structure are maxima and, at the same time, horizontal forces are developed. Wasilkowski analyses the forces and moments in stage C (Fig. 9.37) as follows.

Flexing of the ground is assumed to increase the ground reaction under the middle portion of the foundation which, in turn, cause deflections of the foundation. If the load from the structure is uniformly distributed on the foundation, the increase in contact stress in the middle and the decrease at the edges causes bending (Fig. 9.37). When calculating the magnitude of the bending moments and shearing forces acting on the structure, Wasilkowski considers the following important factors: stiffness of the foundation EI; length of foundation l; breadth of foundation B; radius of curvature of ground R_g; stiffness of ground k; initial contact stress q_z; ultimate ground bearing stress q_z (yield).

Referring to Fig. 9.37, the deflections of the ground and the foundation are both very small and it is assumed that the contact stress distribution is parabolic. Considering vertical displacements

$$\Delta_2 + \delta_f = \Delta_1 + \delta_g \qquad (9.3)$$

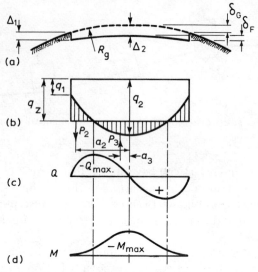

Figure 9.37 Distribution of forces and moments at stage C (after Wasilkowski, 1956).

and vertical stresses

$$q_z = q_1 + \tfrac{2}{3}(q_2 - q_1) \tag{9.4}$$

where q_1 is minimum contact stress, q_2 is maximum contact stress, δ_g is deflection of ground over the length l, δ_f is deflection of foundation over the length l, Δ_1 is settlement of foundation into ground at point 1, Δ_2 is settlement of foundation into ground at point 2, M_{max} is maximum bending moment in the foundation.

If in Equations 9.3 and 9.4 the following substitutions are made

$$\Delta_1 = \frac{q_1}{k}; \quad \Delta_2 = \frac{q_2}{k}; \quad \delta_g = \frac{l^2}{8R_g}; \quad \delta_f = \frac{M_{max}l^2}{12EI}$$

then

$$\frac{q_2}{k} + \frac{M_{max}l^2}{12EI} = \frac{q_1}{k} + \frac{l^2}{8R_g}$$

and three unknowns are left, namely q_1, q_2 and M_{max}.

Now

$$M_{max} = -P_2 a_2 + P_3 a_3 \quad \text{(Fig. 9.37b)}$$

where

$$P_2 = P_3 = Q_{max} = \tfrac{2}{3}(q_2 - q_z)\frac{bl}{2\sqrt{3}}$$

and $a_2 = 0.108l$, $a_3 = 0.429l$.

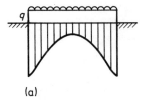
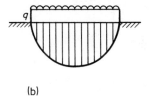

(a)   (b)

Figure 9.38 Contact stress distribution under a rigid foundation loaded with a uniform stress, q. (a) Cohesive soil; (b) cohesionless soil.

Hence,

$$M_{max} = \tfrac{2}{3}(q_2 - q_z)\frac{bl}{2\sqrt{3}}(0.429l - 0.108l)$$
$$= 0.062\,(q_2 - q_z)bl^2. \tag{9.5}$$

Substituting for M_{max} in Equation 9.5 gives

$$\frac{q_2}{k} + \frac{(q_2 - q_z)bl^4}{200\,EI} = \frac{q_1}{k} + \frac{l^2}{8R_g}. \tag{9.6}$$

Thus Equations 9.4 and 9.6 with two unknowns q_1 and q_2 are obtained, and so the solution for stage C can be found. The solution for stage B (Fig. 9.36) is found using a similar procedure.

In the foregoing discussions it has been assumed that the contact stress value q_z is uniform over the whole base of the foundation prior to the start of subsidence, but a uniformly loaded foundation will not necessarily transmit a uniform contact stress to the soil. This is possible only if the foundation is perfectly flexible. As discussed in Section 2.3, the bearing stress distribution for uniformly loaded rigid foundations on sand is roughly parabolic and on clay roughly elliptical (Fig. 9.38). In the case of clay, for example, the foundation may have settled with a convex upwards curvature prior to subsidence due to the contact stress distribution shown (Fig. 9.38a). As subsidence develops and tensile ground strains are mobilized, the ground becomes weaker and further settlement takes place, and with the stress distribution of Fig. 9.38(a), the convex curvature can also increase to an extent such that the foundation curvature is greater than the ground curvature. Field observations on a simple brick wall are shown in Fig. 9.39 in which deflections are plotted for different positions of the advancing face in the critical area (Littlejohn, 1975a). The greater magnitude of the structural deflection can be explained by reference to Fig. 9.40. During initial ground movement (0.4–0.5R) the foundation penetrated the ground at the ends and separated under the central section of the structure. As the tensile ground strain increased, the entire foundation penetrated the ground, although the former pattern remained. Maximum penetration and tensile ground strains were recorded at the same time (1.1–1.2R). It can be concluded that the development of tensile ground strain altered significantly the load–settlement properties of the

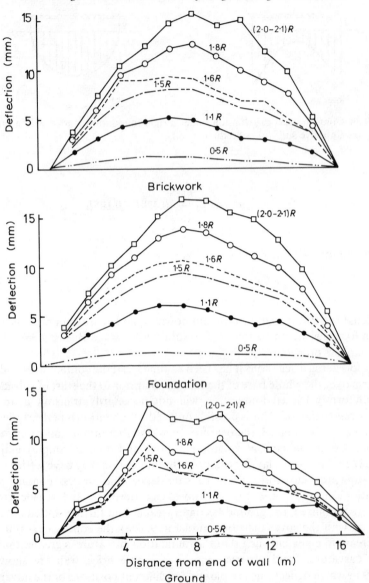

Figure 9.39 Ground and structural deflections related to face advance (16 m long brick wall).

ground and the distribution of the supporting forces. As anticipated, the development of compressive ground strain caused a reduction in penetration. No current foundation design procedure takes account of the changing soil properties and, bearing in mind the difficulty in specifying an overall stiffness factor for a full scale structure consisting of a range of materials and claddings, it

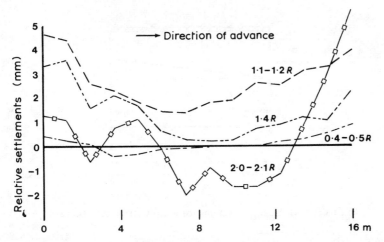

Figure 9.40 Relative settlements at ground/foundation interface related to face advance (16 m long brick wall).

will be appreciated that much research in the field of structure–soil interaction is required and, in the future, field investigators should monitor relative movements at the soil–foundation interface.

Bearing in mind that stress distributions with discontinuities are unlikely in areas of subsidence due to longwall mining, it is noteworthy that observations of structures, having small dimensions compared with the subsidence wave, are often confined to differential settlements, and little or no data are available regarding the ground strain under the foundation. As a result, few attempts are made to analyse the structural deformations. To make better use of observed foundation deflections the following method of analysis is presented for the approximate estimation of the maximum bending moment in the foundation.

(a) Strip footings

For simply supported beams subjected to different loading systems (Fig. 9.41) the centre line bending moments and deflections are as follows:

$$M_a = \frac{wl^2}{8}; \quad \delta_a = \frac{5wl^4}{384EI} \tag{9.7}$$

$$M_b = \frac{w_0 l^2}{16}; \quad \delta_b = \frac{5w_0 l^4}{16 \times 48EI} \tag{9.8}$$

$$M_c = \frac{Pl}{4}; \quad \delta_c = \frac{Pl^3}{48EI} \tag{9.9}$$

$$M_d = \frac{P_1 l}{8}; \quad \delta_d = \frac{11P_1 l^3}{6 \times 128 EI}. \tag{9.10}$$

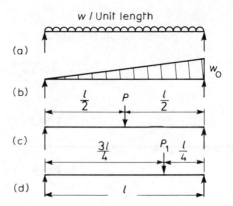

Figure 9.41 Loading systems for simply supported beams.

If it is now assumed that the observed centre line deflection is δ, which might be caused by any one of the four loading systems, the bending moments induced by the deflection can be compared by assuming $\delta_a = \delta_b = \delta_c = \delta_d = \delta$.

$$\text{From } \delta_a = \delta_c, \quad w = \frac{8P}{5l} \tag{9.11}$$

$$\text{From } \delta_b = \delta_c, \quad w_0 = \frac{16P}{5l} \tag{9.12}$$

$$\text{From } \delta_d = \delta_c, \quad p_1 = \frac{16P}{11}. \tag{9.13}$$

Incorporating Equations 9.11, 9.12 and 9.13 in moment Equations 9.7, 9.8 and 9.10, respectively,

$$M_a = \frac{Pl}{5}; \quad M_c = \frac{Pl}{4}$$

$$M_b = \frac{Pl}{5}; \quad M_d = \frac{2Pl}{11}.$$

The bending moments expressed in terms of P do not vary greatly for the same centre-line deflection and hence, for an approximate determination of the moment, a simple loading distribution can be assumed to represent the contact stress under the footing. The intensity of this loading required to produce the observed deflection is calculated and hence the bending moment can be found.

(b) Concrete rafts

The theory of flat plates offers considerable possibilities for the estimation of moments in concrete slabs. Measured deflections of floor slabs indicate that

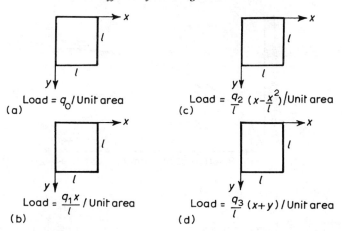

Figure 9.42 Loading systems for square plates.

agreement between design and practice is closer with this treatment than by normal semi-empirical methods.

For the loadings on square plates shown in Fig. 9.42, the centre-line moments and deflections, calculated using Navier's solution, are as follows:

$$M_a = \frac{52q_0 l^2}{15\pi^4}; \quad \delta_a = \frac{292q_0 l^4}{75\pi^6 EI} \tag{9.14}$$

$$M_b = \frac{148q_1 l^2}{75\pi^4}; \quad \delta_b = \frac{148q_1 l^4}{75\pi^6 EI} \tag{9.15}$$

$$M_c = \frac{292q_2 l^2}{75\pi^6}; \quad \delta_c = \frac{532q_2 l^4}{135\pi^8 EI} \tag{9.16}$$

$$M_d = \frac{17q_3 l^2}{15\pi^4}; \quad \delta_d = \frac{78q_3 l^4}{75\pi^6 EI}. \tag{9.17}$$

If it is assumed that $\delta_a = \delta_b = \delta_c = \delta_d = \delta$ and that $M = \alpha K$, it can be shown by using the same procedure as before that for loading condition a, b, c and d, $\alpha = 1.00, 1.00, 1.10$ and 1.09 respectively, where

$$K = \frac{292q_0 l^2}{75\pi^4}.$$

Thus the technique suggested for strip footings can be applied also to concrete rafts, although it should be noted that these approximate methods neglect the horizontal subgrade forces which may be present. Bearing in mind the current dearth of information on this subject, however, at least some assessment of the magnitude of bending moments mobilized as a result of subsidence can be made available.

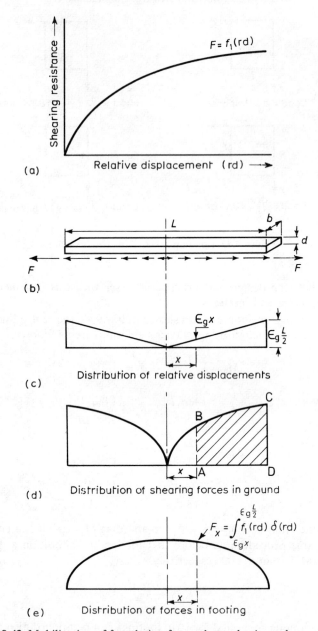

Figure 9.43 Mobilization of foundation forces due to horizontal ground strain.

9.4.2 Design for horizontal displacements

Differential horizontal displacements are the components of ground movement considered most liable to cause damage to surface structures. These movements create shear forces on the underside of foundations and can produce high lateral stresses on structures sited below ground level in compression zones. When relative displacement occurs between a surface foundation and the soil, the shearing stresses increase with the magnitude of movement up to a maximum value and thereafter subsequent relative displacement takes place against this ultimate shearing resistance (Fig. 9.43a).

Thus, for a rigid foundation subjected to tensile ground strain (Fig. 9.43b) the relative displacement increases linearly from zero at the neutral axis (in this case the centre line) to a maximum at the ends (Fig. 9.43c). As a result the shearing force increases from zero at the centre line to a maximum at the ends (Fig. 9.43d). Consequently, the stresses produced in the foundation by the shearing forces in the ground increase from zero at the free ends of the foundation, where there is no restraint to movement, to a maximum at the centre line, where the internal stresses must balance the shearing forces created by the relative displacement under each half of the foundation (Fig. 9.43e).

EXAMPLE 9.3   RAFT FOUNDATION

For a strip of unit width subjected to ground strain ε_g (Fig. 9.43b) the stress distribution at any point at horizontal distance x from the centre line is of the form

$$\sigma_x = F/A \pm My/I \tag{9.18}$$

where $M = Fd/2$, bearing in mind that the force F acts at the underside of the slab and y is any distance measured vertically from the neutral axis of the cross section of the footing.

Assuming that the relationship between shearing resistance and relative displacement for the footing can be represented by

$$F = f_1(\mathrm{rd}), \tag{9.19}$$

the manner in which the subgrade force acts on the footing is shown in Fig. 9.43(c).

The force F at a point x measured horizontally from the centre line is given by

$$F_x = \int_{\varepsilon_g x}^{\varepsilon_g(L/2)} f_1(\mathrm{rd})\partial(\mathrm{rd}) \tag{9.20}$$

where the footing is assumed rigid in a horizontal direction. In other words, F_x is equivalent to the area ABCD in Fig. 9.43(d).

The maximum force occurs at the centre line

$$F = \int_0^{\varepsilon_g(L/2)} f_1(\mathrm{rd})\partial(\mathrm{rd}). \tag{9.21}$$

From Equations 9.18 and 9.20, the stress distribution within the footing can be determined for any position of x.

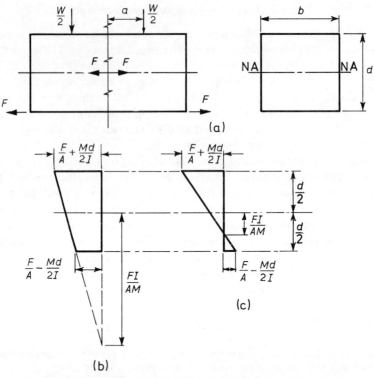

Figure 9.44 Stress distribution over cross-section of beam.

The possible stress distribution due to the combined effects of M and F are shown in Figs. 9.44(b) and (c), the actual form for a given problem depending on the relative magnitudes of $Md/2I$ and F/A. The internal foundation force acts along the base of the footing at the ends, but axially at the middle section, since the footing has no freedom and the eccentric moment Fe ($e =$ distance from line of action of F from neutral axis) is cancelled by part of a loading moment (e.g. $\frac{1}{2}W/a$ in Fig. 9.44(a)). In lightly loaded low rise buildings founded on thin rafts or strip footings the eccentric moment is ignored and the internal forces are assumed to act axially throughout.

In the above example the analysis is concerned with an individual unit. However, where structures are supported on several independent footings it is important to understand the factors which influence the transmission of horizontal forces from one footing to another when they undergo different relative displacements. To illustrate this problem, consider the case of a beam resting on two pad footings as shown in Fig. 9.45. Assuming that the graphs (Fig. 9.45b) of horizontal force against relative displacement for the two loaded footings can be represented by

$$F_1 = f_1 \, (\mathrm{rd}_1) \qquad (9.22)$$

and

$$F_2 = f_2(\mathrm{rd}_2). \qquad (9.23)$$

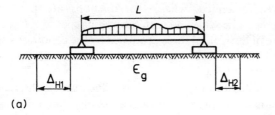

(a)

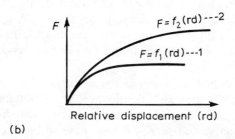

(b)

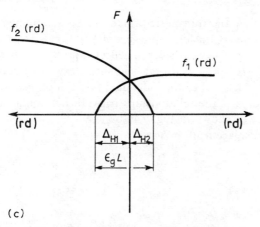

(c)

Figure 9.45 Transmission of horizontal forces between rigidly connected footings.

Then for static equilibrium

$$f_1(\mathrm{rd}_1) = f_2(\mathrm{rd}_2). \tag{9.24}$$

Assuming the system is rigid, the total relative displacement occurring beneath the foundations equals $\varepsilon_g L$, where ε_g is the ground strain and thus

$$\Delta_{\mathrm{H1}} + \Delta_{\mathrm{H2}} = \varepsilon_g L. \tag{9.25}$$

To solve Equations 9.24 and 9.25 the graphical construction shown in

Fig. 9.45(c) is recommended, where the functions are plotted in opposite directions using the same scale and with the origins spaced $\varepsilon_g L$ apart. The curves cross at the value of horizontal force produced by the conditions specified above and by drawing an ordinate through this point the relative displacements for the two footings can be obtained.

With equally loaded footings of the same dimensions, the relative displacements for the pads are approximately equal for horizontal forces less than the maximum shearing resistance. Where the pads are not identical, all sliding will ultimately occur beneath the pad with the lower shearing resistance when the transmitted force reaches that value. For footings carrying unequal loads, sliding ultimately occurs under the footing carrying the lesser load, when its limiting resistance is attained, provided the source of that resistance is friction. Where the source of the resistance is adhesion, the foundation having the smaller plan area may slide first.

For any system of connected footings the horizontal force due to relative displacement between a footing and its subgrade can be obtained from simple field shear tests on single footings. The problem is complicated in the case of a buried footing, since active and passive thrusts have to be taken into account together with friction and adhesion on two sides developed under the earth at rest condition.

To illustrate how the transmitted force can be determined when structural flexibility is considered, the following general solution is proposed for the portal frame shown in Fig. 9.46(a), where the deflected profile has been produced by a ground strain ε_g. It is assumed that field shear tests have been carried out on single slab units, identical to the individual portal footings, with regard to size and loading, and that the relationships between horizontal force and relative displacement are as shown in Fig. 9.46(b).

The relative displacement at A is

$$\Delta_{H1} = \frac{\varepsilon_g l}{2} - p \tag{9.26}$$

and at B is

$$\Delta_{H2} = \frac{\varepsilon_g l}{2} - q. \tag{9.27}$$

Therefore,

$$F_A = f_A\left(\frac{\varepsilon_g l}{2} - p\right) \tag{9.28}$$

and

$$F_B = f_B\left(\frac{\varepsilon_g l}{2} - q\right). \tag{9.29}$$

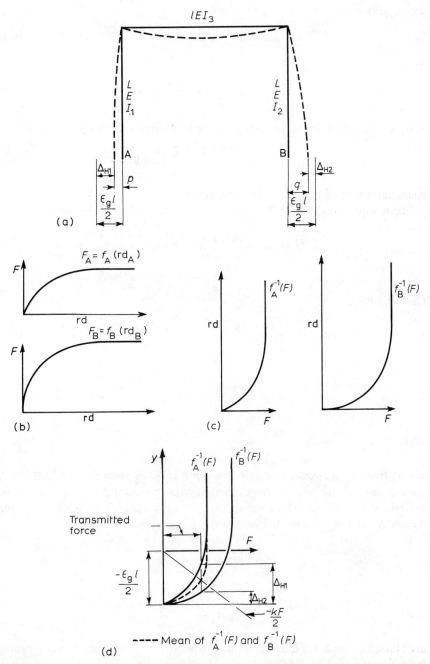

Figure 9.46 Transmission of horizontal forces between footings connected by a flexible structure.

For equilibrium

$$F_A = F_B = F, \tag{9.30}$$

hence,

$$f_A\left(\frac{\varepsilon_g l}{2} - p\right) = f_B\left(\frac{\varepsilon_g l}{2} - q\right). \tag{9.31}$$

It can be shown by slope–deflection analysis that for a fixed-base portal

$$p + q = \frac{L^3 F}{6E}\left(\frac{1}{I_1} + \frac{1}{I_2}\right) = kF \tag{9.32}$$

when extension of the portal beam is negligible.

From Equations 9.28, 9.29 and 9.32 and referring to Fig. 9.46(c)

$$f_A^{-1}(F) + f_B^{-1}(F) = \left(\frac{\varepsilon_g l}{2} - p\right) + \left(\frac{\varepsilon_g l}{2} - q\right)$$

$$= \varepsilon_g l - (p + q) = \varepsilon_g l - kF.$$

Dividing by 2,

$$\frac{f_A^{-1}(F) + f_B^{-1}(F)}{2} - \frac{\varepsilon_g l}{2} + \frac{kF}{2} = 0. \tag{9.33}$$

If

$$y = \frac{f_A^{-1}(F) + f_B^{-1}(F)}{2} - \frac{\varepsilon_g l}{2}$$

and

$$y = \frac{-kF}{2}$$

are plotted on the same F axis and to same scale, as shown in Fig. 9.46(d), the intersection of the two curves determines the horizontal force for the given boundary conditions. The relative displacement occurring under each foundation Δ_1 and Δ_2 can be measured from the graphs.

If

$$y = \frac{f_A^{-1}(F) + f_B^{-1}(F)}{2}$$

and

$$y = -\frac{kF}{2}$$

are plotted on separate sheets of tracing paper to the same scale, then the solution for different ε_g and l values can be obtained simply by sliding one sheet down the y-axis relative to the other until the correct value for $\varepsilon_g l/2$ is obtained. It can be observed from Fig. 9.46(d) that, for the infinitely stiff connection, $-k \to 0$ and

$y \rightarrow 0$. Consequently, the transmitted force is determined from the junction of the mean curve with the F axis.

For the engineer concerned with the design of frame structures in mining areas, it is suggested that a family of curves of the type $y = -kF/2$ be plotted on one tracing sheet to assist in determining the best solution for a given problem. For example, if the portal frame treated above is pinned at the bases

$$k = \frac{L^3}{3E}\left(\frac{1}{I_1} + \frac{1}{I_2}\right)$$

which is double the previous value. This reduces the horizontal force in the frame considerably. Needless to say, whilst k values for open frame structures can be estimated in practice, it is virtually impossible at present to establish k values for composite structures with a variety of claddings (ISE, 1978).

Where foundations or other structures are in direct contact with the soil, the considerations discussed above indicate that it is important to understand the load–displacement behaviour at the foundation–soil interface.

The relationships between shearing resistance, normal stress and shearing movements in soils are well known phenomena, but field shear tests on foundations of direct use to the designer are surprisingly rare. A summary of interesting tests now follows and, although the results refer to the sliding of concrete slabs over soil, it is logical to assume that the reverse, which occurs in mining subsidence problems, produces similar effects.

As early as 1924, Goldbeck carried out shear tests using concrete slabs

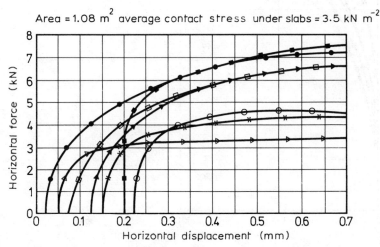

Figure 9.47 Displacement of concrete slabs on various subgrades due to a horizontal force; where —•—•— represents clay, smooth surface; —▷—▷— represents broken stone, 18 mm to dust; —□—□— represents loam; —▶—▶— represents clay, rough suface; —✕—✕— represents gravel, 18 mm to 6 mm; —■—■— represents broken stone, 75 mm; —○—○— represents sand, smooth surface (after Goldbeck, 1924).

1.83 m × 0.61 m × 150 mm thick cast on various subgrades. The forces required to slide the specimens, together with their corresponding movements, are shown in Fig. 9.47.

Mautner (1948) suggests that if a foundation is constructed on a layer of oiled paper, underlain by a concrete blinding layer, the coefficient of friction can be taken as 0.4 to 0.6. In the present author's experience this construction procedure may not provide a long term solution since oiled paper can abrade quickly in sliding over the subgrade. Van der Veen (1953), quotes a coefficient of friction as low as 0.25 for a 2 mm thick bituminous layer inserted between a 140 mm thick prestressed runway slab and a fine sand subgrade. In the field of concrete footings, Littlejohn (1966) produced the curves shown in Fig. 9.48, where the coefficient of subgrade resistance is plotted against relative displacement for a dry sand/rough concrete interface. The limiting coefficient for any given normal stress is not reached until a significant relative displacement has taken place, in this case in the range 8 to 16 mm. Hence unless the critical displacement for any soil is exceeded

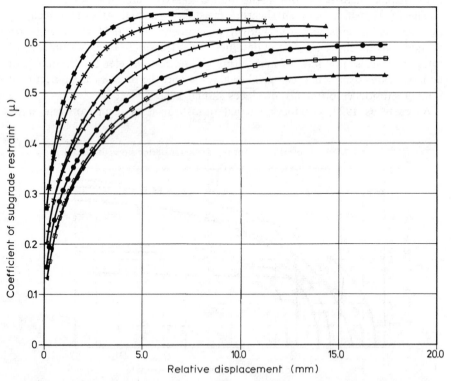

Figure 9.48 Effect of normal stress on the development of subgrade restraint with respect to relative displacement. The dimensions of the slab are 915 mm long × 305 mm wide, the surface is rough dune sand and μ = (horizontal stress)/(normal stress). The normal stress (kN m^{-2}) is given as follows: ▬■▬, 1.8; ✕✕, 12.6; ▲▲, 25.2; ─┼─ 38.7; ─⊖⊖, 50.4; ─⊟⊟, 63.0; ▲▲, 75.6.

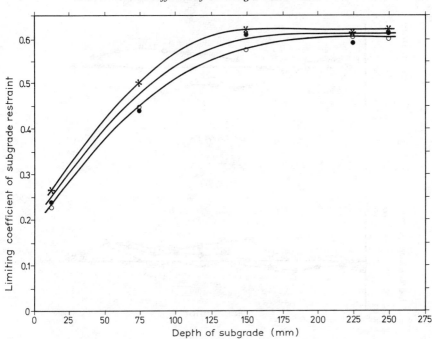

Figure 9.49 Effect of subgrade depth on subgrade restraint. The dimensions of the slab are 610 mm long × 610 mm wide, the surface is rough dune sand. The normal stress (kN m^{-2}) is given as follows: —×—, 6.3; —●—, 12.6; —○—, 25.2.

the ultimate shear values, often employed in practice, may not be fully mobilized. It is, therefore, important to know how the subgrade shearing coefficient increases with relative displacement to avoid over-estimating the horizontal ground forces. Experiments carried out to investigate the effect of subgrade depth on shearing resistance are also relevant (Fig. 9.49).

From these observations it seems reasonable to conclude that for light structural loads the 150 mm thick layer of sand adopted by Dobson *et al.* (1959) is adequate as a sliding layer. For heavier loads, bearing in mind the desire to reduce the magnitude of the accompanying ground curvature in the case of mining subsidence, the author's field observations suggest that a 600 mm layer of loose to medium dense, fine to medium sized sand should be employed (Fig. 9.50). Dense sand is not recommended since it gives a higher ultimate shearing resistance and a smaller critical displacement.

Whilst sand is widely used as a sliding layer, it may not always be the best available subgrade to minimize the horizontal forces invoked by the anticipated ground movement. Based on results obtained by Potyondy (1961), Figure 9.51 shows that, if ultimate shear forces are mobilized, then only in the case of light loads will sand give a smaller shear force compared with cohesive soils. For the

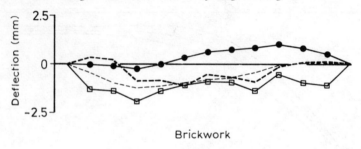

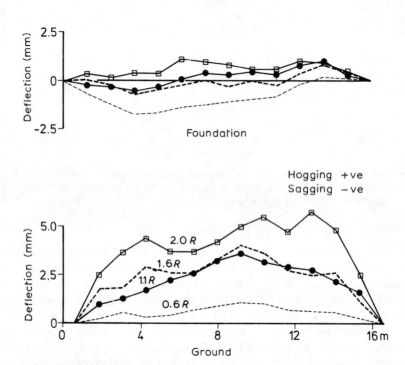

Figure 9.50 Reduction of structural deflections using a 600 mm thick layer of sand subgrade.

clay (3), the sand need not replace it as the immediate subgrade when the normal stress exceeds $36\,\text{kN}\,\text{m}^{-2}$ approximately; the equivalent stress for the cohesive-granular soil (2) being $60\,\text{kN}\,\text{m}^{-2}$.

It is suggested that, prior to the building of new structures, field shear tests should be carried out where prototype footings or rafts are pushed horizontally over (a) the natural subgrade, and (b) an alternative artificially placed soil. Ideally full scale foundations should be tested but, if this is impracticable, smaller

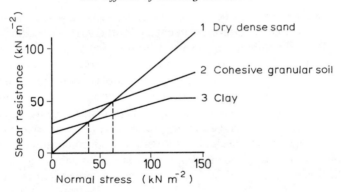

Figure 9.51 Relationship between shear resistance and normal stress for smooth concrete on various types of subgrade (after Potyondy, 1961).

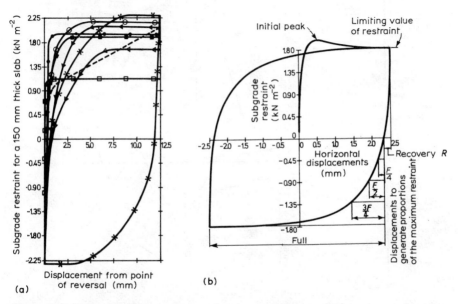

Figure 9.52 Force–displacement curves for concrete slabs cast on various subgrades; where ✳ represents gravel; ⊖ represents dune sand; ◄ represents sharp sand B; ● represents concreting paper on smooth mortar base; ✛ represents sharp sand A; ◁ represents limestone chippings; --- represents asphalt; —▫— represents smooth mortar base (after Stott, 1961).

versions may be acceptable provided that for rafts the minimum side dimension in plan is not less than 1 m and for strip footings the slenderness ratio (L/B) is not less than 4, otherwise scale effects and tilting may distort the results.

The shearing force/relative displacement curves discussed so far have been obtained by sliding a block of material over the subgrade in one direction. However, when a structure is subjected to a dynamic subsidence wave, the foundation has to resist tensile and compressive ground strains as shown in Fig. 9.17. In this connection the experiments of Stott (1961) in the laboratory are interesting. Besides obtaining the normal type of force–displacement curves for different materials (Fig. 9.52a), Stott produced curves of hysteretic form (Fig. 9.52b). These were formed by pushing a block of concrete horizontally over the subgrade until the limiting or steady force was attained. The direction of the applied force was then reversed. This reversal of displacement was repeated several times and it was observed that the initial limiting shear force diminished slightly during the first few cycles but ultimately attained constancy. From the point of view of mining subsidence both types of graphs shown in Fig. 9.52 are required for different materials although it should be noted that the particular

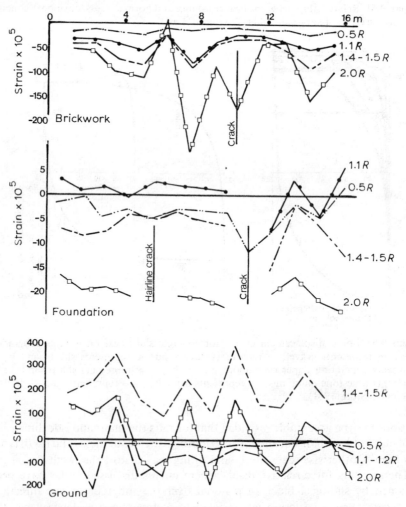

Figure 9.53 Distribution of horizontal strains with respect to face advance.

figures published by Stott were obtained using a normal stress of $3 \, \text{kN m}^{-2}$ and very shallow subgrade layers.

From this brief review of research concerning slabs subjected to horizontal forces, it will be appreciated that there is a lack of fundamental knowledge concerning the manner in which structures and soils interact and, since the horizontal movements are often the most damaging to a structure, ground strains in particular require detailed study. It is interesting to note that, even when modern flexible structures, such as CLASP (Section 9.4.3(c)) have been con-structed, damage in the foundation slab due to horizontal ground strains has occurred, as reported by Jones (1963) and Heathcote (1965). The results of field observations shown in Figs. 9.53 and 9.54 (Littlejohn, 1975a) illustrate that useful results can be obtained by careful monitoring of interface movements and analysis of these results indicates that the longitudinal slip H_1 at any point l m from the neutral axis of the footing can be estimated from the following equation, provided that there is no damage.

$$H_1 = 1.0 l E_g \qquad\qquad (9.34)$$

Since subgrade shear force is a function of slip, a knowledge of this function from a foundation/soil shear test combined with Equation 9.34 would enable foundation stresses to be calculated along the length of the structure. It should be noted that the maximum tensile force in the footing occurs at the position of zero slip (Fig. 9.43c). Field observations of a 15.8 m long × 1.2 m high brick wall show that the main crack (Fig. 9.53) formed at a point 9 m along the wall, whilst at a later stage a hairline crack occurred at 5 m. Both cracks formed at positions of zero slip and the results confirm that the main cause of damage was tensile ground strain.

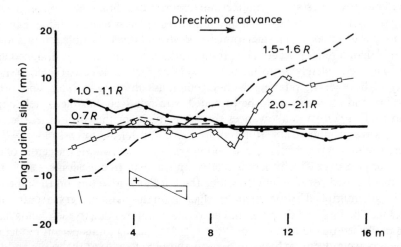

Figure 9.54 Distribution of longitudinal ground slip beneath a strip footing with respect to face advance.

9.4.3 Constructional considerations

(a) General aspects

With regard to construction in general, the size of the building is the first important consideration, not only its length and breadth but also its height. Long continuous buildings should be divided into units, so that each can settle independently, and should be provided, where appropriate, with flexible connections and services and a gap of 75 mm or more, depending on the height of the building, between units to cater for compression and differential tilting. In tall buidings a knowledge of the expected tilt, as well as compressive ground strain, is essential (Section 9.1.19). Careful town planning can minimize damage considerably and it is preferable to orientate the buildings such that the shorter axis coincides with the direction of the anticipated maximum curvature of the ground surface. Orchard (1956) found that weak zones in the subsoil absorb considerable amounts or horizontal strain, whereas in fairly homogeneous ground, the strain is spread evenly. Advantage of this has been taken on some sites by cutting deep trenches close to the walls of structures where the trenches are subsequently backfilled by a compressible medium such as coke. Engineering opinion is that this has saved a considerable amount of damage but clearly the trenches must be deeper than the adjacent foundations to reduce lateral thrusts effectively.

More recently, expanded polystyrene blocks have been employed around basement walls (Fig. 9.55a) to prevent the mobilization of full passive thrusts during the period of compressive ground strain. A compression curve for this low density material is shown in Fig. 9.55(b) and the compressibility can be further increased, according to the design requirements, by drilling holes into the blocks (Fig. 9.55c).

As far as possible, statically indeterminate structures should be avoided. Thus a bridge consisting of a series of simply supported spans is more suitable than one of continuous spans, and a steel framework with relatively flexible joints is more suitable than a reinforced concrete framework, unless the latter is designed as a small single completely protected unit. Roads are preferably constructed in short concrete bays or in pitching with bituminous surfacing. Sewers and other pipelines and services should be provided with relatively flexible or telescopic joints (ICE, 1977b) and allowances should be made for the effect of tensile and compressive ground strain. Both compressive and tensile forces, mobilized at different stages in the subsidence development, increase towards the centre of the length of pipe over which there is relative movement. In cohesionless soils, the incremental load per unit length will be the normal load acting on the pipe per unit length multiplied by the friction coefficient. In this manner, very large friction forces may be imposed on pipes and accounts of failures due to such loads have been published by the Ministry of Works (1951). Sliding joints permit reduction of the relative movement between the soil and pipe to zero at the joints and the maximum load on a pipeline is, therefore, limited to the maximum load on an individual pipe length. Alternatively, welding of a pipeline is often sufficient to

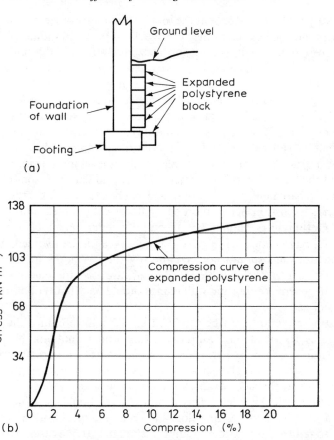

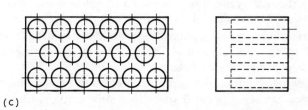

Figure 9.55 Use of polystyrene blocks to prevent the mobilization of full passive thrusts during the period of compressive ground strain. (a) Position of proposed blocks relative to foundation; (b) compression curve of expanded polystyrene; (c) design of expanded polystyrene blocks.

prevent disruption: while the line may be heaved out of the ground or stressed beyond its elastic limit, failures are rare.

Wall footings for ordinary domestic dwellings should be reinforced and tied together to cater for horizontal tension. A disadvantage of a raft for domestic houses is that it reduces the contact-bearing stress to a value much less than that under strip footings and lessens the remote possibility of yield taking place, but it was recommended by the Ministry of Works in 1951 and should be designed with sufficient reinforcement to cater for horizontal strains. Additional resistance to flexure and direct tension can be provided in domestic dwellings by 1 m deep horizontal bands of reinforced brickwork at ground-floor, first-floor and roof levels, together with similar vertical bands at corners. Reinforced brickwork should be constructed in cement or cement–lime mortar, although the enclosed panels of plain brickwork are more suitably constructed in lime mortar, provided the pointing can be matched. If lime mortar is used in brickwork instead of cement mortar, the brickwork is able to adjust itself more readily to subsidence without the formation of wide cracks. Floors and flat roofs should be tied to all walls and a basement should be designed as a box to cater for vertical movements, resistance to torsion being afforded by floors constructed monolithic with the walls. Outbuildings should be structurally independent of a house. Corner windows, bay windows, arches and porches should be avoided, and door and window openings should not be close together. Window frames should be of wood with oversize rebates or, if steel frames are used, they should be mounted in wood. In general, ceiling and wall linings should be of plaster board in preference to plaster.

Existing structures can be modified in order to minimize the possibility of damage, as described in the Subsidence Engineers Handbook (NCB, 1975). Such modifications comprise permanent strengthening, temporary supports, or removal of structural elements. Since extension is commonly followed by compression as the face advances, fractures which have been opened should not be filled immediately and debris within cracks should be removed before the onset of compression. Cutting slots and removing brickwork and window glass may also reduce damage due to compression. In extreme cases, part of a house in a long terrace has been cut out, or even a complete house removed, to allow for compression. Use of tie rods is justified only if the predicted extension of the ground is sufficiently severe to create serious loss of bearing for roof trusses when the supporting walls move apart. In some cases extra bearing has been provided by means of temporary corbels.

Bridges generally require complete protection against the effects of mining subsidence. Simply supported spans should be employed and skew spans and continuous structures should be avoided. Cantilever and suspended span designs are suitable, provided adequate bearing is allowed for the suspended spans to guard against collapse due to horizontal tension and provided an adequate gap is left to permit horizontal compression without fouling the cantilever spans. Wide bridges are preferably split along the centre of the superstructure and abutments. Considerable passive earth resistance may be developed behind abutment walls during the period of compression and this should be catered for at the design stage by incorporating compressible materials to accommodate the movements or providing sufficient strength to resist full passive thrusts.

Provision should be made on all bridges for jacking the superstructure to grade levels. This is usually afforded by jacking pockets or brackets. A sill beam should be divided into convenient lengths, say 6 to 10 m, depending on dead load per jack and the magnitude of differential subsidence anticipated. Two pockets should be provided per sill beam, positioned approximately at the quarter points.

(b) Examples of rigid construction

With regard to the design of rigid structures, Mautner's method was considered to be the answer to mining subsidence for some years after publication in 1947. At this time a large school-building programme was under way in Britain and the building industry developed a method of rapid construction called the modular system. When this technique was employed the framework of the superstructure was built using a basic unit or module, hence any room or section of the building had a height, length and breadth which were simple multiples of the module. It is, therefore, not surprising that this method of construction was adopted also in areas of mining subsidence where the ground movement problem was catered for by special foundations. For schools in the Midlands, Geddes and Cooper (1962) describe a scheme, devised by Ove Arup & Partners, using a raft of beams, designed according to Mautner's conditions, which ran longitudinally and transversely, and a single storey building of timber construction was erected on it. Similar rigid foundations can be constructed using a grid of post-tensioned concrete beams, and the foundation for a six-storey teaching block, designed by Donovan Lee, is illustrated in Fig. 9.56. The concrete beams, post-tensioned by Lee McCall bars, are 1.83 m deep, covering an area approximately 28 m × 18.3 m. The grid is supported on three foundation pads to eliminate torsion and the six-storey superstructure is built in the normal way.

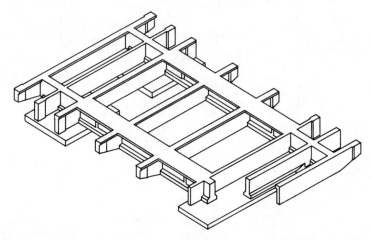

Figure 9.56 Foundation of building at College of Ceramics and Pottery, Stoke-upon-Trent.

A more recent building to be constructed in this country to overcome subsidence is the Atherton–Tyldesley County Secondary School, Lancashire, completed in 1964 (Fig. 9.57). Modular construction, based on a unit of 1 m, was used for the construction of the superstructure of the two-storey blocks with a special raft foundation, 19.5 m × 19.5 m, supported on a reduced bearing area. The load was concentrated on the reduced bearing area which was in the form of an annular rectangle, in effect a rectangle of strip foundations 1.67 m wide which surrounds an area 11 m × 11 m × 1.67 m deep.

How far such precautions are generally effective is not known with any certainty but observations have revealed a number of buildings with strongly reinforced foundations which have still been damaged. As a principle, it appears preferable to construct buildings in a way such that they can accommodate movement.

In this connection, the construction of the Reinforced Concrete Research Centre at Ougrée in Belgium is of interest (Venter *et al.* 1954). The normal

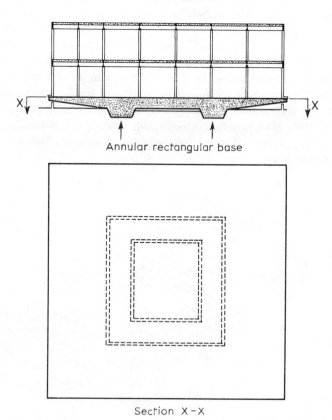

Annular rectangular base

Section X - X

Figure 9.57 Special base construction for two-storey blocks at Atherton–Tyldesley School.

procedure of placing reinforced concrete beams underneath the walls of buildings was rejected because of weakness of standard beams when subjected to torsion. The direction of the resultants of the tensile and compressive forces are generally orientated at random to the direction of the walls or columns and the beams cannot resist these torsional forces unless strengthened at prohibitive cost. This results in a general distortion of the foundation, as shown in Fig. 9.58(a). The magnitude of the forces acting upon the lower portion of the building, as a result of horizontal ground movement, was estimated to be approximately one-third of the total weight of the building, and since the main experimental hall was 94.5 m long, 14.2 m wide and 14.2 m high it was decided to separate the columns from their foundations by means of a sliding joint (Fig. 9.58b). The bracing slab was placed above the sliding joint so that the foundations could move with the ground whilst rigidly braced columns maintained their positions relative to one another. This plan was abandoned, however, because of the difficulty of reducing the coefficient of friction below 0.4. Consideration was therefore given to the possibility of inserting a rocking column between the foundation and the base of the column (Fig. 9.58c). This system is ideal for small horizontal displacements but

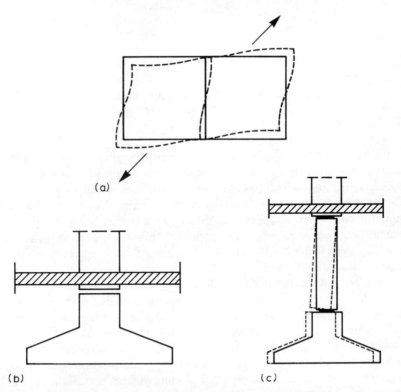

(a)

(b)   (c)

Figure 9.58 Foundation proposals for reinforced concrete research centre at Ougrée, Belgium (after Venter *et al.*, 1954).

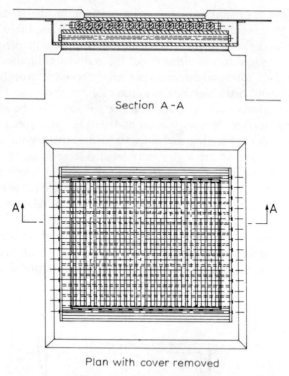

Section A-A

Plan with cover removed

Figure 9.59 Sliding roller joint (after Venter *et al.*, 1954).

in the building being considered the maximum design value was 100 mm and so these proposals were also abandoned. Since column loads were of the order of 3000 kN, it is not difficult to imagine the effect of an eccentricity of 100 mm and the solution finally adopted consisted of carrying the column base on two superimposed layers of rollers, the directions of rolling being at right angles to each other. The rollers were supported by 12 mm thick steel plates as shown in Fig. 9.59, there being 21 rollers in each row under the columns carrying the higher loads. Using this method the coefficient of friction was reduced to 0.02 and, consequently, it was feasible to employ cross-bracing at the feet of the columns.

This leads to the problem of designing the cross-braces, for which it is necessary to determine the maximum rotation of the building. For the design of the testing hall, Venter *et al.* (1954) assumed that ground movement takes place along a line running at an angle of 45° to the principal axis of the hall (Fig. 9.60). The forces acting at the feet of the columns and possessing a line of action also inclined at 45° will be felt in one direction or the other of a line which is termed the neutral line of movement of the building in relation to the ground. Venter *et al.* state that the resultants of the horizontal forces at the feet of the columns on both sides of this

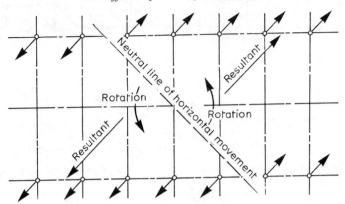

Figure 9.60 Overall rotation of building due to column forces acting at 45° (after Venter *et al.*, 1954).

neutral line will not, in general, have the same line of action and this creates a torque which will produce an overall rotation of the building.

(c) Examples of flexible construction

In general it is not easy to incorporate full articulation in reinforced concrete structures and this has led to the choice of flexible steel frames for construction of buildings in areas subjected to mining subsidence. At Peterlee New Town, County Durham, where a minimum vertical movement of ± 450 mm accompanied by a horizontal movement of ± 300 mm was predicted, a factory for clothing production was erected in 1954. The structure was designed to accommodate this movement without any change in stress and without damage to glazing and finishes. The main workshop area is 3500 m² and it is spanned by welded steel frames constructed and erected on independent footings, the whole structure being covered with light-weight roofing and wall panels. Geddes and Cooper (1962) state that the frame is fully articulated and covers a clear span of 36.5 m, as shown in Fig. 9.61.

Another example of this type of flexible factory is the Krupps Works in Westfalenhutte, which was designed in 1957. It was originally proposed to employ a fully articulated frame which gained its stability from one main foundation (Fig. 9.62a). However, the presence of heavy overhead cranes with accompanying surge loads brought about an overturning moment of about 50 000 kN m to bear upon this single foundation and so the scheme was abandoned. A structural system, similar to that of the Peterlee Factory, was finally adopted, the structural rigidity being achieved by one rigid joint at eaves level (Fig. 9.62b).

Probably the most widely adopted method of construction in this country in

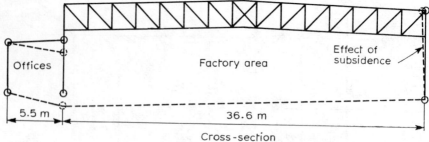

Cross-section
Bays, 7.6 m x 36.6 m, act independently

Figure 9.61 Flexible superstructure of Peterlee factory (after Geddes and Cooper, 1962).

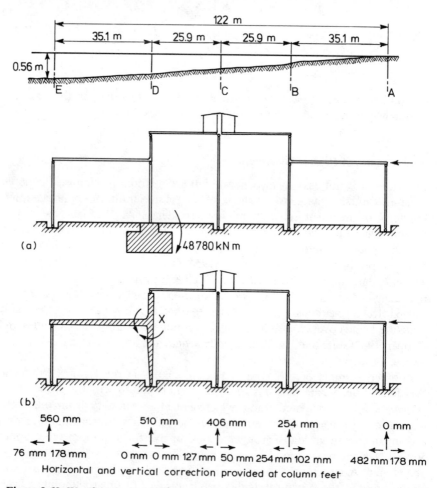

Horizontal and vertical correction provided at column feet

Figure 9.62 Westfalenhutte melting shop (after Geddes and Cooper, 1962). (a) Conventional hinged frame system. (b) Hinge system finally adopted; lateral stability depends on the moment of resistance at X.

the CLASP or Nottinghamshire system (Fig. 9.63a). This method was developed by Gibson (1957) to meet the requirements of a large school-building programme on sites overlying coal workings. Buildings designed according to the system are erected on a flat base which is founded on a subgrade, such as sand, which facilitates relative horizontal displacements between base and subgrade. Vertical and horizontal displacements are catered for separately; the horizontal

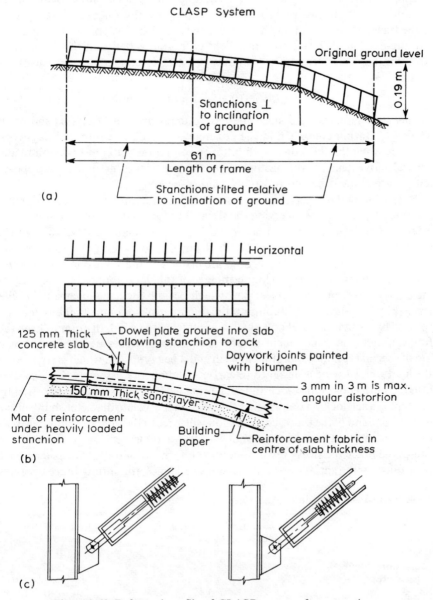

Figure 9.63 Deformed profile of CLASP system of construction.

movement by the base and subgrade and the vertical component by the design of the superstructure. The entire foundation is a simple concrete slab 125 to 150 mm thick containing a continuous steel mesh which is usually placed at mid-depth. This reinforcement is sometimes called the subsidence mesh and its function is to prevent the slab being split in two by the ground strain developed during mining. The force it has to resist is directly proportional to the least total weight of the building to the left or right of any section, assuming a constant coefficient of friction between the slab and soil. Gibson shows that this force can be expressed as the least $(W + \frac{1}{3}LW)$ at any line, where W is the dead load of the building and LW the live load. Thus, if the coefficient of friction is 0.67 and the permissible stress for the steel is 310 MN m^{-2}, the area of steel A_s required per metre width of slab is

$$A_s = \tfrac{2}{3}(W + \tfrac{1}{3}LW)/310 \times (\text{width of slab in metres}).$$

This steel mesh is light since the design loads are not heavy, as shown in Table 9.4. During construction the concrete is poured in controlled day-work areas, painting the slab edges with bitumen so that the slab units will follow the vertical movements of the ground. Thus the subsidence mesh simply opposes tensile stresses and has no resistance against bending.

The design of the superstructure calls for a light building, to facilitate slipping, and not only capable of following the slope of the ground, but of re-assuming its original shape if necessary and hence the framework is fabricated of the lightest possible steel sections. There is no rigid junction between any beam and its supporting column but all connections are made with the aid of one or two large-diameter bolts, which permit a limited degree of angular rotation. The majority of the columns rest on simple steel pins which permit angular movement to take place (Fig. 9.63b) but the columns on the outer perimeter of the building may be secured by a single nut and bolt against uplift from the wind. In order to keep the frame upright and in alignment, a number of diagonal braces are employed which are of special type, incorporating at each end a heavy spring, which is stressed to the load which the tie is designed to carry. A sliding fork assembly (Fig. 9.63c) is fitted to both ends of the brace and each is loaded by a coil spring compressed by tightening a self locking nut which is threaded to the 32 mm diameter shank of the fork. At one end of the brace, compression will force in the fork assembly, whereas the self-locking nut and sleeve prevent withdrawal under tension. At the opposite end of the brace the sleeve resists compression, and withdrawal under tension is controlled by the spring. Prior to ground subsidence, the sprung brace operates

Table 9.4 Loads applied in CLASP system

Structure	Dead load $(kN\,m^{-2})$	Live load $(kN\,m^{-2})$
Roof	1.05	0.72
Multi-storey floor per storey	1.58	2.87
Site slab	3.45	2.87

as a fixed brace but as soon as the load exceeds that for which the tie was designed, the springs compress or extend and permit the building to deform (Fig. 9.63a).

9.5 MONITORING FOUNDATION MOVEMENTS

In recent years there has been a growing awareness of the need to increase the number of field monitoring studies concerned with the performance of structures subjected to ground movement. Significant advances have been made in the development and use of field instrumentation for the accurate measurement of ground or structural deformations but relatively little attention has been directed towards monitoring relative movements at the soil–foundation interface (Littlejohn, 1973).

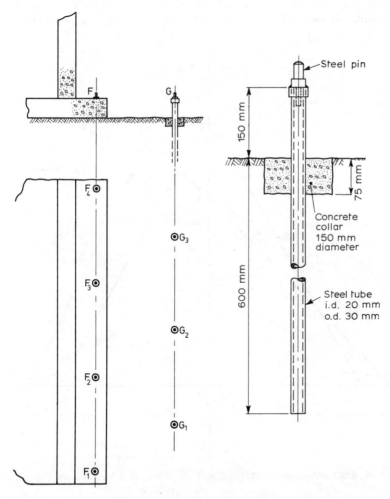

Figure 9.64 Layout of monitoring stations.

9.5.1 Layout of monitoring stations

To study in detail the behaviour of foundations and the adjacent ground during mining subsidence, stations inserted at 1 to 1.5 m intervals are recommended for the foundation and ground. A useful layout of stations is shown in Fig. 9.64, where it can be observed that the foundation pins are staggered with respect to the ground pins. In this way it is possible for two consecutive foundation pins to form an equilateral triangle with the adjacent ground pin. Thus, with a knowledge of the horizontal strains along each side of this triangle, the displacement of the ground pin relative to the point on the footing opposite can be calculated parallel and normal to the major axis of the footing.

The relationship between these strains and slip can be found by co-ordinate geometry by reference to Fig. 9.65. The original positions of concrete stations F and ground station G form an equilateral triangle with lengths as shown. These points are then displaced due to straining to form triangle F'G'F'. Assuming that

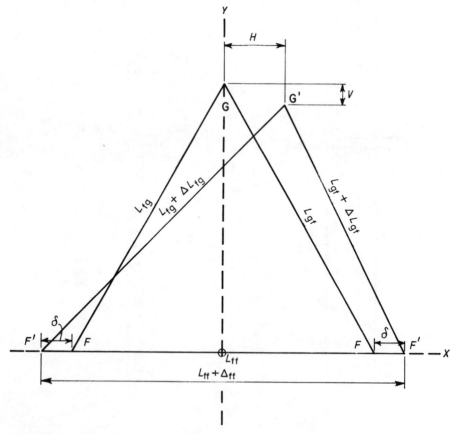

Figure 9.65 Relationship between strain and slip where H is longitudinal slip, and V is lateral slip.

all three stations remain in the same plane, longitudinal slip is given by

$$H = \frac{\Delta L_{fg} - \Delta L_{gf}}{(1 + \Delta L_{ff}/L)} + \frac{\Delta L_{fg}^2 - \Delta L_{gf}^2}{2(L + \Delta L_{ff})} \tag{9.35}$$

and lateral slip by

$$V = \frac{-1}{\sqrt{3}}(\Delta L_{fg} + \Delta L_{gf} - \Delta L_{ff}/2). \tag{9.36}$$

Neglecting the second term in Equation 9.35, since it is extremely small, then for most practical purposes

$$H = \frac{\Delta L_{fg} - \Delta L_{gf}}{(1 + \Delta L_{ff}/L).} \tag{9.37}$$

9.5.2 Field measurements

Differential settlements can be plotted, as shown in Fig. 9.66, by assuming that the end pin on each set of stations is held fixed throughout the duration of the study.

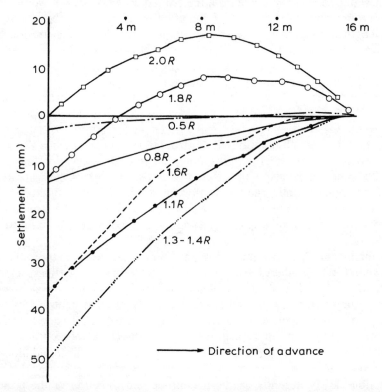

Figure 9.66 Differential settlements of a strip footing related to face advance.

The settlements of adjacent pins for each line are evaluated for any mine face position with respect to the end pin and the results plotted along the length of the structure. The manner in which the end pins subside with time can be shown on a graph of absolute settlements, observed using precision levelling equipment.

When comparing differential settlements for the foundation and ground, study of the results is facilitated by the production of deflection diagrams (Fig. 9.39) which indicate the bending of the foundation and ground with reference to a straight line joining the end stations.

In Fig. 9.39 it is interesting to note that, for the case of a clay subgrade, the hogging deflections of the foundation are greater than the corresponding ground deflections. This is the reverse of the condition often assumed by structural designers and the situation can best be explained using a diagram of relative settlement (Fig. 9.40). These graphs represent the relative settlement of the foundation with respect to the adjacent clay subgrade and the points are evaluated from a comparison of the differential settlement readings of the ground and foundation pins, assuming both lines level at the start of the investigation.

During the initial period of ground movement, when the ground strain is very small, the foundation penetrates the ground at the ends where the contact stress is highest and tends to separate from the ground at the centre. However, as the tensile ground strain increases, the foundation penetrates the subgrade over its entire length although the former trend is maintained, a higher penetration being recorded at the ends compared with the centre. Thus the distribution of support from the ground is altered and a greater proportion of the total vertical load is taken at the ends. Since a change in the distribution of the supporting forces alters the stresses in the foundation, it is clear that readings from pressure cells inserted into the foundation combined with the simple measurements above will facilitate study of the variation of vertical stress with change in relative settlement.

The results of field observations for horizontal strains are shown in Fig. 9.53. It can be observed that strain patterns along the ground observation line do not show the same uniformity as the differential settlement curves. This situation is common in practice and structures in close proximity may be subjected to quite different strain distributions. Detailed observations around each structure are, therefore, required if subsequent behaviour is to be properly analysed and understood.

The manner in which ground strains induce horizontal forces in the foundation is shown in Fig. 9.54, where the longitudinal slip of the ground with respect to the foundation is plotted in relation to foundation length. During the development of tensile ground strain, the ground moves outwards from the neutral axis of the foundation, thus inducing tensile forces in the structure. With the development of compressive ground strain, slips in the opposite direction are recorded.

Figure 9.67 shows the lateral slip of the ground relative to the foundation. The magnitude and distribution of this slip with respect to position along the foundation clearly depends on the magnitude and direction of the principal strains. Except in special cases the magnitude and direction of the principal

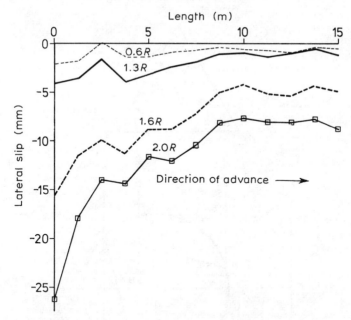

Figure 9.67 Distribution of lateral ground slip beneath a strip footing with respect to face advance (16 m long brick wall).

strains cannot be obtained by the use of a single line of stations, since measurements taken along such a line indicate component values in the direction of the line.

The following procedure is suggested as a means of deriving the maximum instantaneous values in direction and magnitude with three stations A, B and C (Fig. 9.68a). Strains along AB, BC and CA are measured and the magnitude and direction of the principal strains can be evaluated if the following assumptions are made.

(1) The strains along the axis parallel to the measured axes are sensibly constant in magnitude within the triangle.
(2) The measured strains are considered to prevail at the centroid of the triangle.
(3) Straight lines existing in the unstrained state remain straight after deformation.

In Fig. 9.68(b), strains Oa, Ob and Oc are scaled and ordinates aa', bb' and cc' drawn. Since triangle ABC is equilateral, the centroid D coincides with the centre of the circumscribed circle and if the strain value at D is denoted by m, then the following relationships are obtained.

$$m = \tfrac{1}{3}(e_a + e_b + e_c) \tag{9.38}$$

$$r^2 = \tfrac{2}{9}(e_a - e_b)^2 + (e_b - e_c)^2 + (e_c - e_a)^2 \tag{9.39}$$

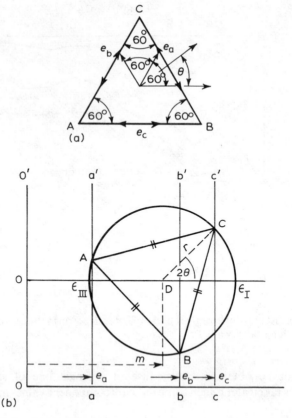

Figure 9.68 Relationship between principal strains and side strains of a triangular unit.

Maximum principal strain $\varepsilon_I = m + r$
Maximum principal strain $\varepsilon_{III} = m - r$

and
$$\theta = \frac{1}{2}\cos^{-1}\frac{e_c - m}{r}.$$
(9.40)

With this information, the magnitude and direction of principal strains can be evaluated and, since the triangular units need have only small dimensions, the instruments described above can be adopted for the measurement of differential settlement and horizontal strain. It is recommended that each unit should consist of four pins, forming two equilateral triangles (Fig. 9.69) and, since the measurements in each unit are independent, there is no siting problem in built-up areas. Consequently units of 'double triangles' can be placed anywhere provided space is available and, by interpolation of the results, contours of change of slope and horizontal strain can be produced over the area concerned.

In conclusion, it is hoped that engineers, when planning monitoring studies

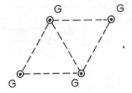

Figure 9.69 Double triangle monitoring unit.

concerned with mining subsidence, will consider both the detailed measurement of horizontal strains, in addition to differential settlements, and the estimation of relative movements at the soil-foundation interface. More information on relative movements, when analysed in conjunction with the resulting structural deformations, would lead to a much better understanding of how buildings perform when subjected to ground movements.

9.6 GENERAL CONCLUSIONS

In areas of mining subsidence all buildings which suffer damage (and to a lesser extent those which do not) are potential sources of information on which improved design techniques could be based. Further observation of building performance and better access to existing records, including related damage, should lead to a closer correlation between predicted and real behaviour in the future. International co-operation in such work would be valuable. These case studies should, however, be backed up by a number of studies where the real behaviour of buildings is observed in detail (Section 9.5) and in which attempts are made to correlate the behaviour with analysis.

Since current damage criteria may seriously underestimate the degree of damage, improved guides to acceptable damage levels related to building type and use are required. In this connection a better classification of building types would facilitate correlation of case records. Figs 9.70(a) and (b) include simple categories of superstructure and substructure which have been produced by the Special Study Group (Institution of Structural Engineers) for Structure – soil Interaction (ISE, 1974). Figure 9.70 suggests a further refinement by including aspect ratio, since this could be important in reflecting differences in overall behaviour of buildings.

Use of more relevant parameters to define movements is also required for the beam shown in Fig. 9.71, $[(\Delta_2 - \Delta_1)/l]$ represents the widely accepted angular distortion but (δ_3/l) could be more important since it represents curvature, and therefore strain, more closely. Differential settlement can be related also to upward convex (Fig. 9.71a) or concave (Fig. 9.71b) curvature where the ground and/or structural strains are tensile and compressive respectively. It is likely that different damage criteria apply to these situations for the same building since

Aspect ratio (height/ width)	Main type	Frame							Wall		
	Material	Steel			Concrete			Timber	Brick	Concrete	
	Cladding	1	2	3	1	2	3		Block	*In situ*	Precast
>10											
10 - 5											
5 - 2											
2 - 1											
1 - 0 5											
0.5 - 0.2											
0.2 - 0.1											
< 0.1											

Figure 9.70 Classification of building types. (a) Superstructure; (b) substructure; (c) superstructure including aspect ratio.

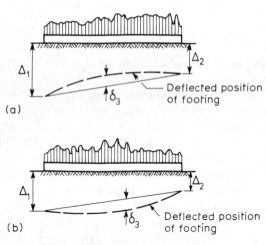

Figure 9.71 Distortional settlement.

structures are generally weaker in tension than in compression. To clarify this situation, consideration should be given not only to total settlement across a structure but also to horizontal and vertical movements of foundations relative to adjacent strata.

Finally, it should be remembered that there is a demand for houses, blocks of flats, offices and factories, and a shortage of land fit for buildings. In many districts where building sites are most wanted (Fig. 9.72) the ground has been disturbed by mining and is often regarded as unsuitable, and indeed dangerous,

Figure 9.72 The coalfields of Britain in relation to the main centre of population (after Price, 1971).

for building. Unlike civil engineers, mining engineers are continually faced with the problems of subsidence and surface damage costs are rising annually (Fig. 9.73) with the increase in numbers of buildings and their spread over coalfields. The greater part of these costs is spent on houses whose design is persistently traditional and delays occur which often are not included in costs. With reference to the construction of offices and domestic multi-storey blocks alone, where the foundations are underlain by coal measures, it was estimated by Price *et al.* (1969) that this work represented a national investment of about

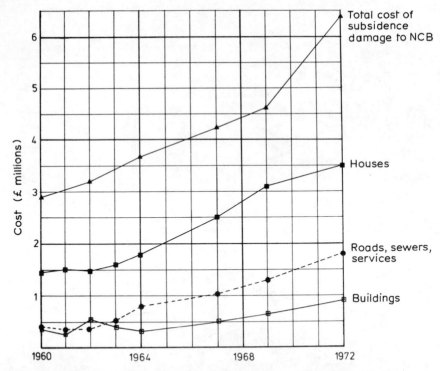

Figure 9.73 Estimated expenditure on main types of claim (Great Britain).

£70 million annually. It is therefore clear that co-operative research between structural, geotechnical and mining engineers should be encouraged on a long term basis with the objective of reducing damage costs pertaining to new buildings by improving foundation design and planning methods.

Chapter 10

BRIDGE ABUTMENTS AND PIERS

W.J. Walley and J.A. Purkiss

INTRODUCTION

Many of the factors taken into account in the design of retaining walls are applicable also to bridge abutments, and Chapter 6 should be studied in conjunction with this chapter. Chapter 7 (Cofferdams and caissons) and Chapter 8 (Piles) also contain material which is relevant to the design of bridge abutments and piers.

Hambly (1979) deals with bridge abutments and their foundations from site investigation through to design and construction and provides a very useful guide to the problems that may be encountered.

In 1953 the European Committee for Concrete (CEB) was set up with the object of providing a more soundly based philosophy of design. Since that time considerable development has taken place and a British Standard Code of Practice (CP 110, BSI, 1972) has been produced, based upon 'limit state' philosophy. This concept was discussed by Rowe *et al.* (1965) and the implications of its use in bridge design were investigated by Flint and Edwards (1970).

British Standard BS 5400*: *Steel, Concrete and Composite Bridges*, which has now been published in full (1983), deals with the design and specification of all types of bridges and is based on limit state principles.

10.1 PRELIMINARY CONSIDERATIONS

10.1.1 General considerations

In general the most economical design of a multi-span bridge is achieved when the cost of the piers is equal to the cost of the superstructure (Henry, 1956). That is

---

*Extracts from BS 5400 are reproduced by permission of the British Standards Institution, 2 Park Street, London WIA 2BS, from whom complete copies of the standard can be obtained.

to say, a pier spacing should be chosen such the cost of spanning the gap is equal to the average cost of the piers. However, two and four-span river bridges should be avoided on aesthetic grounds (see Section 10.1.4). For a given type of construction, the cost of a bridge superstructure can be assumed to be approximately proportional to the square of the span. Substructure costs for short-span bridges are relatively high but substantial economy can often be achieved in the design of the foundations provided adequate site investigations are made. Both costs and illustrations of the numerous types of bridge used on the Hamilton By-Pass are given by Paton *et al.* (1968).

Tomlinson and Francis (1972) recommend a flexible approach to site investigation and stress the importance of adequate communication between geotechnical engineers and design engineers at an early stage in the design operations. Inadequate or badly interpreted site investigation can have serious consequences, as mentioned in Section 1.1.3 in connection with a bridge over the St Lawrence and the construction of cofferdams in glacial drift in Massachusetts.

The most acceptable site for the construction of a bridge is one where firm compact gravel or intact solid rock, free from faults, is either exposed at the surface or occurs at a shallow or moderate depth below ground, and preferably where no difficulty due to groundwater during construction is anticipated. The choice of bridge type for a particular site may be greatly influenced by the nature of the ground. For instance, arch bridges and portal frames require firm foundations which can withstand both vertical and horizontal forces while suffering negligible displacement. Special consideration must to given to the design of bridges for areas which suffer from subsidence due to coal mining and brine pumping (Section 10.3.1(h)).

In the early stages of design the overall appearance of the bridge is an important consideration especially when the bridge is to occupy a prominent position in the urban or rural scene. This topic is considered in more detail in Section 10.1.4.

10.1.2 Special considerations relating to river bridges

By their very nature river bridges are often sited on flood plains where ground conditions are far from ideal, the soils often being highly variable: geologically recent deposits having high compressibility and low strength. Where a site for a relatively short-span bridge over a stream is underlain by a considerable thickness of soft or loose deposits, consideration should be given to a box culvert of two or three spans as an alternative form of construction. The bearing stress beneath the culvert is relatively low and the monolithic nature of the structure reduces differential settlement. A natural blanket of clay is desirable for river piers in order to provide a seal for an economical sheet-pile cofferdam.

A river bridge should, wherever possible, be sited where the flow is free from scour-provoking currents due to bends, tributaries or other causes. Scour may be considered to be of two kinds; general scour and local scour. General scour is the most significant and results from the high current velocities which occur during floods. A sharp bend immediately upstream is particularly undesirable. Local scour is of minor importance compared with general scour, except that it adds to the effects of general scour. It is caused by the increased velocities resulting from the disturbance of flow around the piers. Wherever possible piers should have a streamline shape and be placed with their major axes parallel to the direction of flow so as to minimize the effects of local scour. A fuller treatment of scour is given in Section 10.2. Straight reaches are, of course, essential for multi-short-span and movable bridges over navigable rivers. A right-angled crossing between high banks, which reduces approach works, and a narrow waterway are desirable features of a bridge site.

Adequate waterway opening must be provided and should be determined on historical evidence, gaugings and calculations based on run-off from the catchment area. Mosonyi (1973) and Dalrymple (1964) have described methods of evaluating design flood. The hydraulics of river flow through bridge constructions has been discussed by Biery and Delleur (1962), Pagan (1963) and HRA (1983). Normal headroom allowances between maximum flood level and soffit of superstructure varies between 0.3 m and 1.50 m, depending on the reliability of calculations for waterway opening, the amount of adjacent property affected by flooding and the likelihood of driftwood being brought downstream. Bridges over navigable waters must provide the headroom required by the navigation authority.

10.1.3 Special considerations relating to land bridges

During the last two decades there has been a rapid growth in the number of land bridges due largely to grade separation of major highways. The main purpose of these bridges is to remove the directional conflicts between vehicles at major junctions and the pedestrian–vehicle conflict in urban areas. This has led to the development of four basic bridge types:

(a) Urban flyovers or viaducts, often curved in plan.
(b) Motorway bridges with up to four spans which are generally straight but sometimes markedly skewed.
(c) Footbridges which are often long and slender.
(d) Subways and underpasses.

One might expect the foundation conditions at sites for land bridges to be less variable than those encountered on flood plains, but this is often not so, since much of the British Isles is covered by highly variable drift deposits. There are so many constraints governing the location of modern highways and so many bridges associated with them, that there is rarely any freedom of choice as regards site or even angle of skew. Two-span and single-span bridges with cantilever abutments are common forms of construction where approach roads are on embankment (Fig. 10.1a). For bridges in cuttings, however, it may be preferable

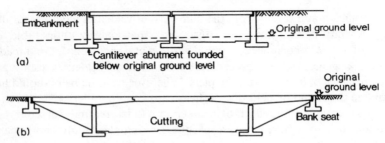

Figure 10.1 (a) Two-span bridge with cantilever abutments retaining embankments. (b) Three-span with bank seats constructed in cutting.

on cost and other grounds to use three- or four-span bridges with bank seats (Fig. 10.1b). Pedestrian subways are preferable to footbridges in that they involve the pedestrian in a smaller change of level. However, their construction interrupts existing traffic to a far greater extent and often necessitates extensive diversions to sewers, water mains and other underground services. It should be remembered that pedestrian footbridges and subways are intended for the pedestrians' convenience and should be located and designed accordingly. The high price of urban land often necessitates the use of viaducts where one would otherwise use embankments. Shortage of space sometimes necessitates the use of piers which cause minimum obstruction at ground level. This is best achieved by the use of suitably proportioned reinforced concrete T-shaped piers or portal frames.

The fact that urban life must be permitted to continue without unnecessary interruption due to noise or traffic delays is an important consideration when deciding on the form of construction. Noisy piling operations and the transport of very large components should be avoided if possible. Noise due to traffic in urban areas is becoming a major problem and consideration should be given to the use of underpasses as opposed to flyovers. This is particularly relevant now that householders are able to claim compensation for increased traffic noise resulting from the construction of new roads.

The overall cost of a grade-separated junction is greatly influenced by the headroom requirements. Minimum headroom standards are specified by the Department of Environment in BE 14, Ministry of Transport, 1968) for subways and bridges over highways. Basically, they require 5.10 m headroom over highways plus additional allowances to compensate for vertical curves and deflection of the deck, and 2.25 m in pedestrian subways. Minimum side clearances from the kerbside to abutments and piers are specified by the Ministry of Transport *et al.* (1966). On curved roads, however, additional side clearance may be required in order to provide drivers with the necessary visibility distance. Consideration should be given to spanning dual carriageways with a single span where the use of a central pier would obstruct drivers' visibility.

Bridges over railways should be designed to minimize track possession during construction. The use of side clearances greater than the required minimum avoids delays to rail traffic and permits better continuity of work construction.

10.1.4 Aesthetic considerations

The overall appearance of a bridge can either enhance or degrade the environment, urban or rural, for a century or more. The bridge designer, therefore, has a great responsibility to society to ensure that his bridges not only perform their function safety and at reasonable cost, but are also visually acceptable in their particular environment. Unfortunately, there are many railway bridges and some early motorway bridges which are ugly products of purely functional and economic design. This is indeed a great pity, since economic and aesthetic requirements need not conflict to a significant extent, in fact, they are often complementary. For example, the final design of the Erskine bridge was chosen from many alternatives on the grounds that it was the most economic and aesthetically attractive (Kerensky *et al.*, 1972).

To be truly successful the bridge designer must be prepared to devote some time to developing his creative and artistic capability (Siegel, 1962; Robinson, 1964). Probably the most valuable exercise, even for those of very limited artistic ability, is to study closely the good and bad features of existing bridges. The Ministry of Transport (1969) consider the general principles of good design to be:

(a) The expression of function without any attempt to deceive.
(b) The development of character and individuality.
(c) The use of proportions and apparent weights which please the eye.
(d) Simplicity of form.

It is beyond the scope of this chapter to elaborate on these points and the interested reader should consult the original publication.

When choosing the number of spans for a multi-span bridge, consideration should be given to the nature of the obstacle being bridged. A river flowing in a single channel should not be bridged by a two- or four-span bridge because the dual arrangement of the spans, in this case, is aesthetically undesirable. Such bridges are, however, quite acceptable on motorways because of the dual nature of the carriageways. Even so, the inherent charm of the three-span cantilever and suspended span bridge is difficult to match. The construction of two bridges close together is generally aesthetically undesirable but to attempt to match the style of a new bridge to that of an older one is often more undesirable. There is a general feeling that bridges which are associated with the high speeds of modern transport should be as light and slender as possible. Continuous slabs, portal frames, prestressed concrete and steel beams are all useful forms of construction in this respect. The position and basic proportions of the abutments and piers will be governed by the considerations relating to overall appearance but their final shape and surface texture warrants detailed consideration of function, appearance and durability. In fact, the piers of many modern urban flyovers and motorway bridges are a main feature of the design and add greatly to the overall appearance. A few examples of such piers are shown in Fig. 10.2.

Although there is no excuse for designing ugly bridges anywhere, there is clearly no justification in spending extra money on beauty that cannot be appreciated. Many motorway underbridges are rarely seen by anyone, whereas the side elevation of all the overbridges are seen by thousands of motorists daily. These elevations should be pleasant and varied so as to leave the motorists with a good impression of the route. They should not be so spectacular as to distract the motorist or so drab as to bore him. A highway bridge which passes under a motorway is of necessity long and can produce a tunnel effect. This can be overcome by the use of three-span bridges with open piers. The streamline form required of river piers should be expressed in their appearance. A bridge can be

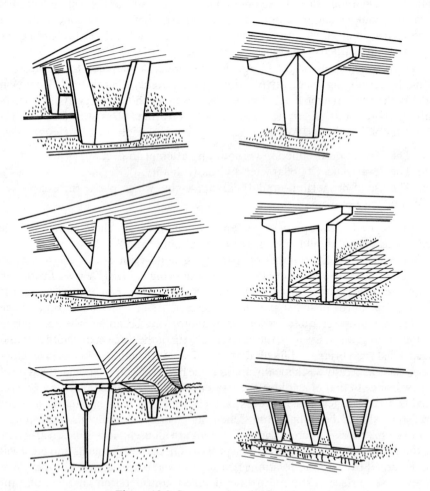

Figure 10.2 Some modern bridge piers.

given a character which is typical of its locality by facing the abutments and piers in the same material as local buildings. This is particularly effective in areas which have highly characteristic materials.

More extensive and general treatments of bridge aesthetics are given by Allen (1971, 1976) and a booklet published by the Cement and Concrete Association (1977) on *The Appearance of Highway Structures.*

10.2 SCOUR AT RIVER BRIDGES

Most failures of bridges arising from floods are due to scour. An accumulation of drift-wood near the base of a pier may aggravate scour and an accumulation at higher levels may lead to the development of hydrodynamic forces sufficient to overthrow the superstructure. In order to obtain the best estimate of the extent of scour, conditions at existing bridges should be examined and the results of empirical calculations, mathematical analyses (usually based on simplified premises) and model investigations should be studied. Various approaches to this problem have been described by Moulton *et al.* (1958), Laursen (1960, 1963), Neill (1965), Carstens (1966), Shen *et al.* (1969), Jain and Fischer (1980), Qadar (1981) and Zwamborn (1981). Experiments can be made on model piers formed of transparent plastic sheets on a framework and changes in the extent of scour with varying discharges can be observed from inside each pier by means of a mirror (Posey, 1949).

Scour is generally considered to take place during a rapid rise in flood level and not necessarily when the discharge is a maximum. Figure 10.3 shows the changes which occurred in the bed of the Rio Grande at the Santa Fé Railroad Bridge at San Marcial, New Mexico, in (1929) (Lane and Borland, 1954). This supports the recommendation of Terzaghi and Peck (1967) that, if the depth to bedrock is great, it is advisable to found piers at a depth below the low-water bed not less than four times the greatest known rise in water level. Niell (1965) carried out field measurements of scour at bridges in Alberta and showed that deep scouring occurred beneath two bridges and that it refilled as the flood subsided. Scoured depths at certain sharp bends in the river were found to be considerably greater than at the bridges which were sited within fairly straight reaches. Niell suggests that scour at bridge foundations is due to at least three separate causes: (a) constriction of flow through a waterway narrower than the flood channel, (b) local obstruction of flow at piers, and (c) local curvature and separation of flow at abutments. Field measurements at bridges in South Africa have been reported by Kuhn and Williams (1961).

Scour can be diminished by flood control works upstream which are designed to reduce maximum velocities beneath a bridge. Posey (1963) suggests that scour can be prevented by placing on the river bed a layer of material graded from sand at the base to cobbles and boulders at the top according to the rule for filter layers given in Section 2.7.3 and this should prevent the uplift of fine bed material by rising currents. In emergencies, long rolls of sand or gravel wrapped in nylon

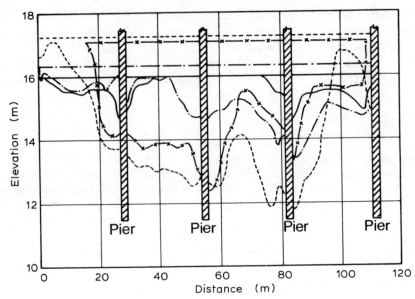

Figure 10.3 Changes in Rio Grande bed due to scour, where --- is 24 September,— × —is 14 August, —·—· is 13 May and—is 17 December.

skins or wire mesh have been placed around the nose of a pier so that the two ends are carried alongside the pier by the current. More permanent measures are effected by underpinning with piles or lithifying the ground by geotechnical processes. Regular inspection of piers and abutments is essential for effective maintenance.

Extensive laboratory investigations were carried out by Tison (1940) on the problem of scour in cohesionless soils but the generalizations given below should be regarded only as a guide. It is apparent that for a single pier of length L and breadth B, with its longitudinal axis parallel to the current, a rectangular plan profile leads to a maximum depth D of scour at the upstream end and a minimum at the downstream end where a region of backflow tends to reduce the depth of scour. The maximum depth is reduced to about $\frac{3}{4}D$ by rounding the corners of a rectangular profile to a radius of about $\frac{1}{3}D$. A rectangular pier with triangular (nearly equilateral) cut-waters reduces the maximum depth at the upstream end to slightly more than $\frac{1}{2}D$. Since the extent of scour depends partly on the total angle through which the stream lines are deviated at the upstream end, an aerodynamic profile does not lead to a marked reduction in depth of erosion and the maximum is about the same as that for triangular cutwaters. An example of experimental results obtained by Tison is given in Fig. 10.4.

Whereas the maximum scour for a rectangular profile is localized at the upstream face, it extends for increasing distances along the sides for the other three profiles mentioned below. A symmetrical lenticular profile leads to a

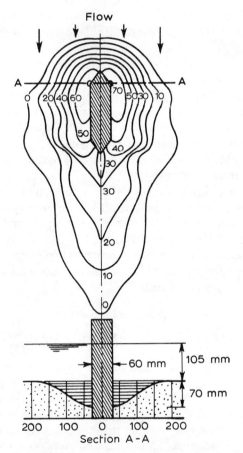

Figure 10.4 Experimental results of scour obtained by Tison (1940).

maximum depth of less than $\frac{1}{2}D$ and the region of maximum scour occurs alongside the upstream half of the pier but does not extend to the upstream tip. The length L of a pier does not affect the magnitude or extent of scour at the upstream face but does affect scour along the sides. If the base of a lenticular pier is widened to merge into a rectangular foundation block with its top surface at river-bed level and front face upstream of the pier by a distance equal to about B, the vertical distribution of current velocity is made more uniform. The scour round the pier is thus reduced to a negligible amount but some scour occurs well beyond the downstream tip.

A single vertical pile, placed upstream on the longitudinal axis of a pier of lenticular profile at a distance from the pier equal to about B, reduces the maximum erosion to about $\frac{1}{3}D$ partly by creating a region of backflow downstream of the pile and, in effect, increasing the length of the pier. A pair of piles placed at a distance equal to about B upstream of the pier with a single pile placed a distance equal to about B upstream of these reduces the maximum depth of scour to about $\frac{1}{4}D$.

If the longitudinal axis of a pier is inclined at an angle ($6°$ to $14\frac{1}{2}°$ in Tison's experiments) to the direction of the current, the projecting longitudinal face causes an increase in the total deviation of the current, and hence the scour on that face is greater than when the axis is parallel to the current. For the larger angles of inclination, any system of piles which is to be effective in reducing scour must be more elaborate than when the axis is parallel to the current and must be arranged asymmetrically. Piers which are set skew to the direction of the current may deflect the current towards the bank which consequently may suffer erosion.

Where a multi-span bridge across a river is supported on a number of piers arranged with their longitudinal axes parallel to the current, the depth and extent of scour at the upstream ends of the piers is the same as that for a single pier but the stream lines are brought parallel to the longitudinal axes of the piers in a relatively short distance, the velocity is increased and there is a tendency for the region of maximum scour to extend progressively along the sides of the piers as the spacing of the piers is reduced.

10.3 STANDARD BRIDGE LOADING

Loads applied to bridge superstructures are transmitted to the substructures either directly, as in the case of fixed arches, or indirectly via bearings. In the latter case bending moments are not transmitted and the load is transferred as normal and transverse forces at the bearings. Considerable distribution of superstructure loads occurs before their effect reaches the foundations, and this fact should be borne in mind when designing abutments and piers on a 1 m strip basis.

A new British Standard (BS 5400) for the design of steel, concrete and composite bridges has now been published. The constituent documents are based upon limit state design philosophy and include sections dealing with loading (Part 2) and the design of concrete bridges (Part 4). This section summarizes Part 2: *Specification for Loads*, and comments on its application to the design of bridge abutments and piers. Part 4: *Code of Practice for Design of Concrete Bridges* is similar to Part 1 of the *Code of Practice for the structural use of concrete*, CP 110 (BSI, 1972c). For full details the reader should refer to the original documents.

Part 2: *Sepcification for Loads* defines two load factors (the term 'partial safety factor' is more in keeping with limit state terminology and will be used hereafter); one (γ_{fL}) provides for variability of each type of load (or more correctly each characteristic force) and may be applied to increase or decrease the effect of the load as appropriate, while the other (γ_{f3}) provides a global margin of safety. For example, when considering overturning of a pier due to wind forces, γ_{fL} may be used to reduce the dead load and increase the wind load, while γ_{f3} serves to cause an overall increase in the net effect. Two limit states are considered: ultimate and serviceability. The global partial safety factor (γ_{f3}) varies with the constructional material and the method of analysis, but is typically 1.10 to 1.15 for the ultimate limit state and 1.0 for the serviceability limit state.

The value of γ_{fL} varies with the type of loading both for the ultimate and serviceability limit states. Table 10.1 summarizes the values of γ_{fL} appropriate to both limit states.

Design forces are determined by multiplying the characteristic forces by the two appropriate partial safety factors. Characteristic forces may be considered to be either permanent (i.e. dead load, superimposed dead loads and differential settlement where no provision is made to alleviate this effect) or transient (all other forces).

A total of five combinations of loading must be considered, not all of which will be critical.

Table 10.1 Loads to be taken in each Combination with appropriate γ_{fL}

Load	Limit state*	γ_{fL} to be considered in combination				
		1	2	3	4	5
Dead: steel	ULS†	1.05	1.05	1.05	1.05	1.05
	SLS	1.00	1.00	1.00	1.00	1.00
concrete	ULS†	1.15	1.15	1.15	1.15	1.15
	SLS	1.00	1.00	1.00	1.00	1.00
Superimposed dead	ULS	1.75	1.75	1.75	1.75	1.75
	SLS	1.20	1.20	1.20	1.20	1.20
Reduced load factor for dead and superimposed dead load where this has a more severe total effect	ULS	1.00	1.00	1.00	1.00	1.00
Wind: during erection	ULS		1.10			
	SLS		1.00			
with dead plus superimposed dead load only, and for members primarily resisting wind loads	ULS		1.40			
	SLS		1.00			
with dead plus superimposed dead plus other appropriate combination 2 loads	ULS		1.10			
	SLS		1.00			
relieving effect of wind	ULS		1.00			
	SLS		1.00			

(*Contd.*)

Table 10.1 (*Contd.*)

Load	Limit state	γ_{fL} to be considered in combination				
		1	2	3	4	5
Temperature: restraint due to range	ULS			1.30		
	SLS			1.00		
frictional bearing restraint	ULS					1.30
	SLS					1.00
effect of temperature difference	ULS			1.00		
	SLS			0.80		
Differential settlement	ULS⎫ SLS⎭	to be assessed and agreed between the engineer and the appropriate authority				
Earth pressure: retained fill and/or live load surcharge	ULS	1.50	1.50	1.50	1.50	1.50
	SLS	1.00	1.00	1.00	1.00	1.00
relieving effect	ULS	1.00	1.00	1.00	1.00	1.00
Erection: temporary loads	ULS		1.15	1.15		
Highway bridges live loading: HA alone	ULS	1.50	1.25	1.25		
HA with HB	ULS	1.30	1.10	1.10		
or HB alone	SLS	1.10	1.00	1.00		
Centrifugal load and associated primary live load[†]	ULS				1.50	
	SLS				1.00	
Longitudinal load: HA and associated primary live load	ULS				1.25	
	SLS				1.00	
HB and associated primary live load	ULS				1.10	
	SLS				1.00	
Accidental skidding load and associated primary live load	ULS				1.25	
	SLS				1.00	
Vehicle collision load with bridge parapets and associated primary live load	ULS				1.25	
	SLS				1.00	
Vehicle collision load with bridge supports	ULS				1.25	
	SLS				1.00	
Foot/cycle track bridges: live load and parapet load	ULS	1.50	1.25	1.25	1.25	
	SLS	1.00	1.00	1.00	1.00	
Railway bridges: type RU and RL primary and secondary live loading	ULS	1.40	1.20	1.20		
	SLS	1.10	1.00	1.00		

*ULS: ultimate limit state; SLS: serviceability limit state.
[†]Each secondary live load shall be considered separately together with the other combination four loads as appropriate

Combination 1   All permanent loads together with the live load due to traffic (for railway bridges associated *or* secondary loads must be included).

Combination 2   As Combination 1 together with wind forces, and where appropriate erection forces.

Combination 3   As Combination 1 together with loading from temperature restraint and erection loads.

Combination 4   Accidental loading, e.g. vehicle collision on bridge supports.

Combination 5   Permanent loading together with friction forces at bearings.

The following items are the characteristic loads as specified in BS 5400 and comments are made on their importance in the design of abutments, piers and foundations.

10.3.1 Forces common to all bridges

(a) Dead load

The characteristic dead load includes the weight of all structural elements. Fairly precise loads will normally be available for the bridge deck since this is usually designed first. Loads for the remainder of the structure must be estimated and their accuracy checked on completion of the design.

(b) Superimposed dead load

This includes the weight of all non-structural elements which are of a permanent nature. They are allowed for separately from dead load because their weights are far less predictable than those of the structural elements. Road surfacing and railway ballast fall into this category. Where appropriate, consideration should be given to the possible removal of all or part of the superimposed dead load. Forces due to retained earth are also regarded as superimposed dead load. These forces should be evaluated from the properties of the retained material using the principles described in Chapter 6.

(c) Wind load

The overturning effect due to wind forces is generally of insignificant magnitude for short and medium span bridges with low height/breadth ratios, but may be quite considerable for tall slender piers supporting relatively light superstructures having a large side area. Footbridges and long-span bridges over navigable water often fall into this category. The maximum probable wind force to which a bridge will be subjected depends upon its geographical location, local topography, the overall size and shape of the bridge and its height above ground level. The highest wind pressures are due to local gusts of very limited impact area and duration. Hence, the bigger the bridge the less likely is the chance of the whole structure being subjected to the full force of the gust. These factors are taken into account when determining the characteristic wind speed, v_c:

$$v_c = v K_1 S_1 S_2, \tag{10.1}$$

where v is the hourly mean wind speed for the location $(m\,s^{-1})$ likely to be exceeded only once in 120 years at a height of 10 m; S_1 is a funnelling factor which is normally taken as 1.0 but may be increased to 1.1 where it is thought that the local topography tends to funnel the wind; S_2 is a gust factor which depends upon height above ground level and the length of a bridge or component under consideration; K_1 is a factor which is normally 1.0 but which may be less than 1.0 when wind forces are considered to act in conjunction with temperature or erection forces.

The characteristic transverse wind force, P_T (Newton), is given by:

$$P_T = 0.613 v_c^2 C_D A_1 \tag{10.2}$$

where C_D is a drag coefficient, which depends on the shape of the element and, in the case of bridge decks, the form of construction, the amount of superelevation, and the presence or absence of live load; and A_1 is the net exposed area (m^2) of the element, which in the case of bridge decks depends upon the type of parapet and presence or absence of live load. A characteristic longitudinal load, P_L, and vertical uplift, P_V, are defined in a similar manner to P_T. Three possible combinations of the three characteristic wind forces are specified. They are: P_T alone or in combination with $\pm P_V$; P_L alone; and 0.5 P_T together with P_L and $\pm 0.5\ P_V$. Where overturning effects are being investigated, wind loads may be considered to act in conjunction with normal traffic loading acting eccentrically or a specified light traffic loading if this produces a worse effect.

(d) Temperature effects

For bridges in the British Isles the extremes of shade air temperatures can be obtained from a map of isotherms which correspond to a return period of 120 years. Adjustments can be made to give temperatures corresponding to a return period of 50 years. The effects of change in air temperature are two-fold. Firstly, they cause continual changes in the overall length of structural members but the variation of the mean temperature of the structure is less than that of the surrounding air. The amount of this difference depends upon the form of construction, the type of material and the mass of the components. Secondly, continual change in air temperature causes temperature differences within the structure which result in non-uniform strain distributions. Severe temperature gradients may occur within structural components following large and rapid changes in surface temperature. The maximum variation in mean bridge temperature and the maximum temperature gradients corresponding to return periods of 120 and 50 years are specified for various types of bridge deck. Temperature effects should, in general, be determined for a return period of 120 years (i.e. the design life of the bridge) but where they are being considered in conjunction with erection forces or the design of components such as carriageway joints, a return period of 50 years is considered adequate. For the purpose of evaluating temperature effects the coefficients of thermal expansion for both concrete and steel can, in general, be taken as $12 \times 10^{-6} \,°C^{-1}$.

In the design of foundations for fixed and two-pinned arches and portal frames, temperature forces form a significant part of the loading. For most free standing abutments and piers, however, temperature forces due to expansion and contraction of the superstructure are limited to that value which is capable of being transmitted by the bearings. Temperature differences between the exposed faces and earth retaining faces of abutments may be large but their effect is relatively insignificant.

(e) Restraint at bearings

Abutments and piers supporting expansion bearings only must be designed to withstand the frictional forces developed at bearings based upon a vertical load equal to the dead load plus superimposed dead load. Where a pier supports two spans the frictional resistance from the two sets of bearings are not necessarily equal and opposite at any given time and account should be taken of this fact by assuming the resistance from one set to be zero. A pier or abutment which supports a set of fixed bearings will be subject to a force which is equal and opposite to that which exists at the expansion bearing. However, this will not appear in the design calculations unless it exceeds the force due to traction and braking (Section 10.2.2). The frictional coefficients for various types of bearing are given in Section 10.8.1. In the case of rubber bearings expansion and contraction of the superstructure is accommodated by shear deformation of the rubber. The maximum horizontal force developed at the bearing is therefore equal to the shear stiffness of the bearing multiplied by the maximum shear displacement. Where rubber bearings are not set to have zero shear displacement at the mean of the appropriate specified temperature range, the maximum displacement due to the temperature should be based upon the normal temperature range extended by 15°C at either end of the range.

(f) Effects of shrinkage and creep in concrete

Considerable forces may be induced in hyperstatic structures due to the shrinkage of concrete members. Fixed arches, portal frames and mutli-column piers with a capping beam should all be checked for the effects of shrinkage. When evaluating shrinkage forces, consideration must be given to the order in which the concrete is placed and allowance should be made for the relieving effect of creep.

(g) Erection forces and effects

There have been several bridge disasters during the last decade which occurred during the erection stage. These may not have been entirely due to inadequate assessment of erection forces, but they have drawn attention to the danger of collapse during erection. If the method of construction is defined by the designer, he must assess the erection forces at each stage of construction and allow for them in the design. Alternatively, if the method of construction is agreed after completion of the design, the chosen method should be checked for suitability of both the permanent and temporary work at each stage of erection. This may prove to be a greater task than that of producing the basic design but it is equally important and must be executed with the same thoroughness. The weight of all permanent and temporary materials, together with all other forces and effects which can operate any part of the structure, must be taken into account.

(h) Differential subsidence and settlement

Subsidence from coal mining and brine pumping present serious problems which require careful assessment in the light of local records. Differential settlement of piers and abutments may cause serious redistribution of support reactions if the superstructure is continuous over two or more spans. Rotation of one support relative to another will cause overloading of bearings at diagonally opposite corners if the superstructure is torsionally rigid. In both of these cases, serious damage to the superstructure may occur if prompt remedial action is not taken. Consequently, simply supported decks having relatively low torsional rigidity are now a common form of construction in areas which suffer from subsidence. It is also prudent to provide wide bearings with provision for jacking. Where it is thought that a small amount of subsidence may occur before remedial measures are taken, the forces so induced should be allowed for in the design.

Sims and Bridle (1966) discussed the problems of bridge design in areas of mining subsidence and outlined methods employed in the West Riding of Yorkshire. The Calder Bridge (Gifford *et al.*, 1969) was made structurally immune to the effects of subsidence by supporting each bridge deck on a three-point bearing system.

(i) Exceptional forces

The effects of stream flow, waves, ice packs, floating objects and earthquakes are considered under this heading.

A pier standing in a river is subject to a hydrodynamic force resulting from the change in momentum imposed on the water which is deflected by it. Under normal flow conditions this force is small and can generally be ignored, but during flood conditions it may be quite significant. The hydrodynamic force (P) on a 1 m deep strip of the submerged upstream face of a river pier which stands parallel to the direction of flow is given by:

$$P = Kb\gamma_w v \, \text{kN m}^{-2} \qquad (10.3)$$

where K is a dimensionless constant which depends on the shape of the pier and the Reynolds number; b is the breadth of pier (m); γ_w is the unit weight of water (kN m^{-3}); and v is the mean approach velocity of current (m s^{-1}).

The total equivalent hydrostatic force acting on a pier is commonly assumed to be $0.08 \, Av_m^2$ (kN) for flat fronted piers and $0.04 \, Av_m^2$ (kN) for rounded piers, where v_m is the mean approach velocity over the depth of the pier and A (m²) is the projected submerged area of the upstream face. The force is applied at a depth below water level of one-third the overall depth. Minimum values for the above two conditions are often taken as $7.4 \, A$ (kN) and $3.7 \, A$ (kN), respectively, for flood flows and $2.4 \, A$ (kN) and $1.2 \, A$ (kN) for tidal currents (Robinson, 1964). Tentative modifications, based on discussion by Gibson (1952), for the case where the current strikes the pier obliquely are illustrated in Fig. 10.5. Hydraulic aspects of bridge design have been studied by Farrady and Chartton (1983) and hydrodynamic force coefficients by Apelt and Isaacs (1968).

Forces due to waves should be taken into account where substructures are exposed to the open sea. In deep water the advancing wave does not break but contact with the vertical wall of an abutment or pier deflects the water upwards as a clapotis or standing wave having an amplitude of about twice that of the component waves. In this case the pressure on the wall at any level is merely the hydrostatic pressure corresponding to the height of the clapotis. The pressures produced by this type of wave are, therefore, relatively small but they may exist for several seconds. In shallow water, however, the oncoming wave may strike the wall at the moment of breaking when the water face is near vertical. In this case the water is not only deflected upwards, thus producing a long period of hydrostatic pressure of low intensity, but it also produces a shock pressure of high

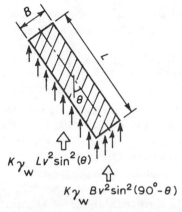

Figure 10.5 Hydrodynamic forces acting on a bridge pier due to an oblique current.

intensity but short duration. Shock pressures up to $700 \, \mathrm{kN \, m^{-2}}$ have been recorded in full-scale tests at Dieppe. These pressures tend to occur in the region between mid-height and the top of the wave. For further information on this subject reference should be made to Bagnold (1939), Minikin (1950), Denny (1951), Henry (1956), Draper (1963) and Nagai (1968).

In Great Britain it is not necessary to take into account the pressure exerted by sheets of ice but in North America the crushing of ice against the end of a pier is allowed for by empirical loads of up to $150 \, \mathrm{kN \, m^{-1}}$ of pier width at the water line. Other loading allows for ice jams between piers and the impact of moving flows. In order to break up ice accumulating at a pier, triangular starlings may be used with steel angle noses raked at 45° or less to the horizontal. The ice rises up the starling and breaks under its own weight. The formation of ice covers and ice jams on rivers has been studied by Pariset *et al.* (1966).

Possible blows from vessels in navigable rivers should be taken into account, although the whole or major part of the impact is best carried by fenders constructed as an independent framework clear of the main structure. The probable magnitude of such a blow can be estimated from maximum vessel weights and speeds. However, head-on collisions are not normally possible because the fenders are tapered at the ends.

10.3.2 Highway bridge live loading

Two types of live loading are specified for highway bridges. Type HA loading represents normal traffic loading in Great Britain and consists of a uniformly distributed lane loading together with a transverse knife edge load, while HB loading comprises an abnormal load applied to a standard trailer. Both types of load include an allowance for impact.

(a) Type HA loading

The magnitude of the uniformly distributed load is constant at $30 \, \mathrm{kN \, m^{-1}}$ per lane for loaded lengths up to 30 m, but for greater lengths it decreases in accordance with:

$$W = 151 \times \left(\frac{1}{L}\right)^{0.475} \, \mathrm{kN \, m^{-1}},$$

but not less than $9 \, \mathrm{kN \, m^{-1}}$, where L is the loaded length in metres. The knife edge load consists of 120 kN per lane. Both of these loads occupy one traffic lane and are uniformly distributed over the width of the lane. The position of the knife edge load and the position and length of the distributed load are chosen, after studying the appropriate influence line, to give the worst possible effect in the member being considered.

Only two lanes may be occupied with the full HA loading, the remainder being occupied by one third HA loading, except that when the superstructure supports eight or more traffic lanes one more lane of full HA loading may be applied, provided that no carriageway has more than two lanes of full HA loading. Where a substructure supports more than one superstructure, as may occur at multilevel junctions, the load on the substructure shall be based on a loaded length equal to the sum of the loaded lengths for each superstructure.

(b) Type HB loading

This consists of a standard trailer having four axles, each having four wheels and set at a series of standard spacings (Fig. 10.6). Unit HB loading comprises 10 kN per axle giving a total load of 40 kN. Most trunk and principal road bridges in Great Britain are required to be capable of carrying 45 units of HB loading. Other bridges are required to carry $37\frac{1}{2}$ or 20 units of HB loading, depending on their importance. Only one HB vehicle may be considered, and its application to any lane displaces the HA loading in that lane over a length of 25 m in front to 25 m behind the HB vehicle.

(c) Braking force

This force is assumed to act horizontally and may be generated by one full lane of HA vehicles or by one HB vehicle. The force corresponding to HA loading is 200 kN plus 8 kN per metre of loaded length, but subject to a maximum of 700 kN, and is assumed to be uniformly distributed over one notional lane width extending over the full loaded length. The force due to braking of the HB vehicle is 25% of the total load equally distributed between the eight wheels of the front or rear pair of axles. Most of the resistance to longitudinal force is provided by the fixed bearings although some resistance, not exceeding the frictional resistance, may be provided by sliding bearings. Abutments and piers supporting fixed

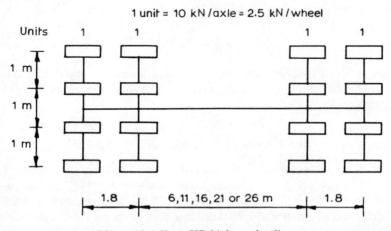

Figure 10.6 Type HB highway loading.

bearings should be designed to provide the full resistance to longitudinal forces.

In the United States of America, loading from *The Standard Specification for Highway Bridges* published by the American Association of State Highway Officials (AASHO) is generally used. The code allows two types of loading:

(a) HS20-44 which is a truck loading based on two 160 kN loads spaced from 9.144 to 4.267 m apart with a leading load of 40 kN at a distance of 4.267 m from the leading 160 kN load.
(b) H20-44 which comprises the 40 and 160 kN loads from HS20-44.

Both of these loads must be considered in conjunction with a lane loading of 9.34 kN m^{-1} per lane plus a concentrated load of 116 kN (for shear) or 80.4 kN (for moments). These loads must be positioned to give the worst case for the effect being considered. Only one truck may be used. In general, the effects of the American standard loading will be less than those of the British loading system.

(d) Centrifugal force

This is generally of minor importance but its influence may be significant in the design of tall slender piers supporting a sharply curved superstructure. Such structural arrangements often occur at grade separated interchanges in urban areas. The characteristic force is based upon the situation where 300 kN vehicles follow at 50 m intervals in each of two lanes. The centrifugal force (F) exerted by each vehicle is given by:

$$F = \frac{30\,000}{r + 150}\,\text{kN}$$

where r is the radius of curvature (m). This corresponds to vehicle speeds of 88 km h^{-1} (70 mph) for a large radius curve and 62 km h^{-1} (50 mph) for a radius of 150 m. Centrifugal force should not be considered to act in conjunction with braking force.

(e) Other forces

Fatigue investigations and the application of accidental forces due to skidding and vehicular collision with parapets are not relevant to the design of abutments and piers.

10.3.3 Footway and cycle track live loading

Structural elements supporting footways and cycle tracks only are designed for a live load intensity (kN m^{-2}) equal to one-sixth of the HA distributed load (kN m^{-1}) for the appropriate loaded length. If, however, the element also supports road or rail traffic, a reduction of 20% in the footway and cycle track loading is permitted provided crowd loading is not expected. Further reductions are permitted for footways which exceed 2 m in width. Special highway wheel loadings are specified for footways which are not protected from traffic but these are not relevant to the design of substructures.

10.3.4 Railway bridge live loading

Two types of live loading are specified for railway bridges. Type RU allows for all combinations of vehicles running or projected to run on European railways and is intended for use in the design of main line railway bridges. Type RL loading is a lighter loading which represents the loading on passenger rapid transit railway systems where main line locomotives and rolling stock do not operate. Both types of loading are equivalent static loadings and are multiplied by the appropriate factor to allow for the effects of impact, oscillation and other dynamic effects. The derivation of both load types is given in Appendix D of the British Standard.

(a) Application of type RU and type RL loading

Type RU loading consists of four 250 kN concentrated axle loads preceded, and followed, by a uniformly distributed loading of 80 kN m^{-1} (Fig. 10.7).

Type RL loading consists of a uniformly distributed load of 50 kN m^{-1} of maximum length 100 m; which then reduces to 25 kN m^{-1}, together with a single concentrated load of 200 kN which may act at any point along the distributed load. An alternative two-axle bogey loading, consisting of 300 kN and 150 kN axle loads spaced at 2.4 m centres with no allowance for dynamic effects, shall be used for the design of deck type elements where this produces a worse condition than type RL loading.

To allow for dynamic effects, type RL loading is multiplied by a dynamic factor of 1.2. In the case of type RU loading separate dynamic factors are specified for bending moments and shearing forces. They are defined in terms of a length L, which is a function of the type of element being considered, and are subject to a minimum value of 1.0 and maximum values of 2.0 and 1.67 for bending moment and shearing force, respectively.

Standard railway loading is a track loading and is applied to each and every track in such a way as to cause the most severe effect in the structural element under consideration. The concentrated axle loads can be applied only once per track, but any number of lengths of the distributed loads can be used.

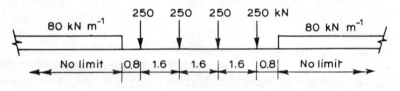

Figure 10.7 Type RU railway loading.

American railway loading is based on Cooper's conventional loading which comprises a series of axle loads and a trailing uniformly distributed load. The heaviest load is E60 which comprises:

(i) A load train of 133 kN followed by 267 kN, 2.44 m behind then 3 No. 267 kN loads at 1.52 m centres.
(ii) A 173 kN load, 2.74 m behind load train (i), and then three 173 kN loads spaced 1.52 m, 1.83 m and 1.52 m, respectively.
(iii) A load train similar to (i), spaced 2.44 m behind (ii).
(iv) A load train similar to (ii) following (iii).
(v) A line load of 87.6 kN m starting 1.52 m from (iv).

British loading, in comparison, is equivalent to approximately E50, since British locomotives and trains tend to be less massive.

(b) Lurching and nosing

Lurching causes a temporary transfer of live load from one rail to the other and is allowed for in the case of type RL loading by assuming that 0.56 of the track loading is carried by one rail and 0.44 by the other. No allowance for lurching is necessary in the case of type RU loading since this effect is taken into account in the dynamic factor mentioned above.

The effect of nosing is defined as being equivalent to a 100 kN force, acting horizontally at rail level in either direction at right angles to the track. This force should be applied to one track only at that point which causes the most severe effect in the member being considered. Consideration should also be given to the vertical effects of this force on secondary elements such as rail bearers.

(c) Braking and traction forces

Longitudinal forces due to braking or traction are considered to act at rail level in a direction parallel to the track. For type RU loading the specified forces due to traction and braking are equivalent to 30% and 25%, respectively, of the maximum live load corresponding to the given loaded length. In the case of type RL loading, the specified braking force is $8 \, \mathrm{kN \, m^{-1}}$, subject to a minimum value of 64 kN and a maximum value of 800 kN. The corresponding traction force is $10 \, \mathrm{kN \, m^{-1}}$ subject to a minimum value of 80 kN and a maximum value of 300 kN where the loaded length is less than 60 m. For loaded lengths greater than 60 m the force due to traction should be taken as $5 \, \mathrm{kN \, m^{-1}}$. Where a bridge supports two or more tracks, longitudinal forces due to braking or traction may be applied to no more than two tracks. If the two tracks carry traffic in opposite directions the force due to braking shall be applied to one track and the force due to traction to the other. For bridges supporting welded ballasted track, up to one-third of the longitudinal forces may be assumed to be transmitted by the track to resistances outside the bridge structure provided that there are no rail discontinuities on or near the bridge.

(d) Centrifugal force

The centrifugal force imposed on bridges carrying curved tracks is considered to act radially at a height of 1.8 m above the track. The magnitude of this force is defined in the Standard by an equation which would appear to be far more complicated than the problem warrants. Admittedly there is no real difficulty in mechanically substituting the appropriate values in the equation but most engineers are averse to such practices when there is no explanation or even indication of its derivation.

(e) Derailment forces

Three requirements are specified with respect to accidental forces due to derailment but only one of these is likely to influence the design of abutments or piers. This states that the structure as a whole, and its constituent elements, shall be safe against overturning effects due to a specified static force resting on the parapet or outermost edge of the bridge. This force is intended to represent the effect of a locomotive and one wagon being momentarily balanced on the parapet and is specified as $80 \, \text{kN m}^{-1}$ and $30 \, \text{kN m}^{-1}$ for types RU and RL loadings, respectively. For spans over 20 m, however, the total load spread over a 20 m length is limited to 1600 kN in the case of type RU loading and 600 kN for type RL loading. Derailment forces do not qualify for inclusion in Combination 1 as defined earlier (Section 10.3). These forces will, however, greatly influence the overturning stability of piers, particularly slender ones.

(f) Other forces

A 120 year load spectrum for standard railway loading is given for the purpose of checking the effects of metal fatigue. However, such calculations are unlikely to be required in the design of the substructures since few are constructed of steel. Similarly, the dispersion of railway wheel loads is a topic which is more relevant to superstructure design than substructure design.

10.4 ABUTMENTS FOR SIMPLY SUPPORTED BRIDGE DECKS

10.4.1 General considerations

The first steps in the design of an abutment have traditionally been:

(a) The choice of foundation level and determination of allowable bearing stress.
(b) The determination of the overall dimensions necessary to provide adequate factors of safety against overturning and sliding.

When determining (a), account is taken of the fact that shallow foundations are more prone to failure due to rupture of the soil than are deep foundations and that deep foundations may settle excessively if their load-bearing capacity is fully exploited. In 'limit state' terms this is equivalent to a 'collapse' criterion for shallow foundations and a 'serviceability' criterion for deep foundations. Obviously both of these limit states would have to be satisfied. It is unfortunate that the recently revised code of practice CP 2004 *Foundations* (BSI, 1972d) was not based upon limit state principles, because this now involves the designer in two different philosophies in one design; a potential cause of confusion and hence error. When traditional methods are used for checking the adequacy of foundations the loads used must always be the characteristic loads applied without any modification.

The determination of overall proportions which satisfy the requirements is, in effect, the application of a limit state of collapse criterion. Overturning should be treated at ultimate limit state calculated for rotation about the toe. However, since bearing failure will occur beneath the toe before overturning occurs, it is obvious that rotation will not occur about the toe but about some point within the base. In view of this apparent anomaly and the separate treatment of sliding, the authors suggest that both sliding and overturning be checked by the traditional method (CP 2, ISE, 1951) using characteristic loads without modification.

In many cases certain other factors, which have often been forgotten, are of importance when choosing overall proportions. These are, in effect, limit state of serviceability criteria. One such factor is the shape of the bearing stress distribution under the permanent and construction loadings. These distributions determine whether or not the abutment will suffer rotational displacement about its base. Such rotation may cause a 30 mm or more horizontal movement of the stem at the level of the bearings (see Fig. 10.8) in addition to that caused by flexure

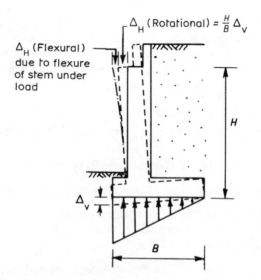

Figure 10.8 Rotation of abutment due to triangular bearing stress distribution.

of the stem. This could cause severe overstressing of rubber bearings and complete closure of the expansion joints between the deck and abutment or, in the case of wing walls, it may cause a discontinuity in the line of the coping where it joins the abutment. To overcome these difficulties the base should be proportioned so that the bearing stress distribution due to permanent loading is as near rectangular as possible. This is not difficult or uneconomic since, although a larger base is required than for a triangular distribution, the amount of steel reinforcement in the toe and heel is reduced because the net load on these components is less. When designing an abutment, sufficient load combinations must be considered to ensure that each component is designed for the worst effects of all possible loadings. Design of the base slab will generally involve determining two or three bearing stress distributions. Traffic loading is most effectively distributed over the foundations if the abutments are stiff and for this reason abutments are not generally designed to minimum proportions.

Wing walls to abutments may be in line with the abutment face, returned parallel to the longitudinal axis, splayed at an angle or curved from a tangent with the longitudinal axis. Due consideration should, of course, be given to appearance before choosing a type. Wing walls which are short in length may be constructed to be monolithic with the abutment and cantilevered from it, provided that the abutment is designed to withstand the cantilever bending moment and shearing force induced by the retained earth. Where wing walls are independent of the abutment they should be separated by a joint which permits differential settlement and a small amount of differential lateral movement. Differential settlement can be minimized by suitably proportioning the wing wall and abutment bases to give the same bearing stress.

Expansion joints are sometimes provided in abutment walls exceeding about 12 m in length and in such cases a longitudinal joint in the superstructure, or other provision for expansion, is required. If the abutment wall is shielded to a great extent from the rays of the sun, it is doubtful if expansion joints are necessary, although a contraction joint may be desirable to cater for concrete shrinkage, but on wide bridges provision must be made either in the superstructure or at the bearings for lateral expansion of the superstructure. Expansion and contraction joints should include a waterproof seal.

Differential settlement between abutment and backfilling is difficult to avoid and can result in a fairly abrupt step in the road surfacing if measures are not taken to distribute this step over a longer length. Figure 10.9 shows two methods which have been used to reduce the effect of differential settlement. The first method replaces the normal fill material at the rear of the abutment with a material, such as granular fill or pulverized fuel ash, which is not prone to settlement. By constructing this in the shape of a wedge with its apex extending back into the embankment, any differential settlement that does occur will be distributed over the length of the wedge. The second method has a similar effect and consists of a reinforced concrete relieving slab resting on the embankment and supported on the abutment at one end.

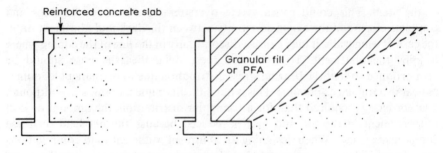

Figure 10.9 Two methods of overcoming differential settlement behind abutments.

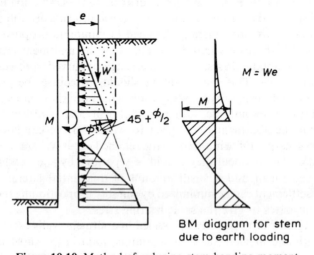

BM diagram for stem
due to earth loading

Figure 10.10 Method of reducing stem bending moment.

The maximum bending moment in the stem of a cantilever abutment can be reduced by constructing a cantilever relieving slab about half way down the back of the wall (Fig. 10.10). This slab not only relieves the stem of some of the horizontal force from the retained earth as illustrated, but also applies a bending moment to the stem which is opposite in effect to that applied by the retained earth. Figure 10.10 also illustrates a typical bending moment diagram for the stem of this type of wall.

To prevent hydrostatic pressure developing behind abutment and retaining walls it is usual to provide a 300 mm thick drainage layer down the back of the wall. Such layers are normally constructed of broken stone or rubble – strictly they should be designed as filter layers (Section 2.7.3) and are connected to a piped drainage system, weepholes may be constructed through the bottom of the wall, but where these are provided they should be inclined towards the back of the wall to prevent discolouration of the concrete due to continual weeping. Severe discolouration of the face of an abutment may occur also as a result of water weeping down the face from the bearing platform. This may be avoided by placing a small upstand at the front edge of the bearing platform, sloping the surface of the platform backwards and providing a drainage pipe through the curtain wall to the porous layer at the rear of the wall. Much of the discolouration of wing walls can be prevented by sloping the top surface of the coping away from the front face of the wall. The rear surface of all walls should be waterproofed with bituminous paint or other suitable material; special attention being given to all construction joints. This helps to protect steel reinforcement from corrosion and to eliminate staining due to leakage of water through badly formed construction joints.

Bearings should be set sufficiently far back from the face of the abutment to avoid possible spalling of the concrete, due to the development of high stresses near the face. Where the superstructure of a short span bridge rests directly on the abutment walls, a 50 mm chamfer should be formed around the edge of the bearing surface to prevent the concrete from spalling. In order to increase the stability of the abutment against overturning, the bearings should be placed as near to the back of the wall as possible but the benefit thus gained should be weighed against the increased cost due to lengthening the span. It is also wise to place a mat of reinforcement (say 12 mm bars at 150 mm centres) underneath the bearings to help the concrete resist the stress concentration in this region.

10.4.2 Mass concrete, brick and masonry abutments

Mass concrete abutments and retaining walls are commonly considered to be cheaper than reinforced concrete walls for heights up to about 5.5 m, or perhaps rather less, but the economic height varies with local conditions and changes in relative costs between steel, concrete, formwork and labour. There are, however, certain other relevant factors which, if considered, make mass construction more attractive. At the present time, with the shortage of skilled labour there is a lot to be said for mass construction quite apart from what economic consideration might reveal. Mass abutments can be designed and detailed quickly with the minimum of qualified staff. Amendments to the initial design can be quickly carried out without fear of introducing errors in the bending schedule. Construction can be achieved with the minimum of skilled labour. If the wall is to be brick faced the brickwork can be used as the front shutter provided lifts are restricted to about 0.6 m. Stability against overturning can be improved by a toe extending forwards and constructed in reinforced concrete.

10.4.3 Reinforced concrete cantilever abutments

Cantilever abutments are similar to cantilever retaining walls and the design methods are practically identical. The base of an abutment is normally made slightly wider than that of a retaining wall since the abutment carries a greater vertical load. Provision must be made to permit relative movement between the bridge deck and the abutment. This movement may be due to thermal expansion of the deck or lateral deflection of the cantilever wall. The latter effect is minimized if the stem is relatively thick and if the foundation is designed to have a rectangular bearing stress distribution when subject to the permanent loads.

The stem of an abutment is sometimes made the same thickness throughout its height, but some economy may be effected, at the expense of increased deflection, if the wall is tapered. The thickness at the bridge seat should not, however, be less than about one-fifteenth of the height of the wall. Cantilever abutments are generally considered to be unsuitable for height in excess of about 6.5 m because both cost and deflection become excessive. The counterfort abutment is recommended as a possible alternative. Reinforced concrete walls supporting vertical loads are classified as short or slender and braced or unbraced. A short wall is one which has an effective height-to-thickness ratio less than or equal to 12. A cantilever wall must, therefore, have an actual height-to-thickness ratio of 6 or less in order to qualify as a short wall. In general, cantilever abutments are both slender and unbraced and are subject to bending moment and axial force. For design purposes the portion of the wall subjected to the highest intensity of vertical load should be considered as a slender column of unit width. The stem reinforcement must also be checked for the load conditions where bending moment is high and vertical load low. These conditions include:

(a) Lateral earth loading prior to placing the deck, if this condition is likely to exist during construction.
(b) Maximum lateral force at the bearing in conjunction with the minimum compatible vertical force the maximum lateral earth stresses. For abutments supporting fixed bearings this condition will generally obtain when the HB braking force is applied with only two axles of HB vehicle on the superstructure. The corresponding condition for abutments supporting expansion bearings is when the frictional resistance or shear resistance of the bearing is a maximum.

The Standard BS 5400, Part 4, specifies the following minimum areas of reinforcement in members. The area of vertical reinforcement in any cross-section of a reinforced concrete wall should not be greater than 4% or be less than 0.4% of the gross cross-sectional area. If the main vertical reinforcement is used to resist compression, horizontal reinforcement, equivalent to 0.25% in the case of mild steel and 0.3% in the case of high yield steel, must be provided. If the amount of vertical compression reinforcement exceeds 2%, horizontal links must be provided. The amount of secondary reinforcement required when the main reinforcement is not used in compression is not specified but the minimum area of secondary reinforcement specified for suspended slabs is 0.15% mild steel and 0.12% high yield steel.

10.4.4 Reinforced concrete counterfort abutments

Counterfort construction is used for abutments which are particularly high, say exceeding 6.5 m. The counterforts are designed as cantilever beams which resist the transverse forces transmitted to them from the wall slabs. The vertical loads from the superstructure are usually supported by a capping beam spanning the tops of the counterforts. The counterforts should be spaced closer than for a retaining wall and, ideally, one should be placed under each concentrated load. The capping beam should be substantial so as to relieve the wall slab of vertical load and also minimize differential deflection of the bearings. The wall slabs are designed as two-way slabs supported on four sides and subject to a distributed load which increases linearly downwards. The slab is assumed fixed along the bottom edge, continuous over the two vertical edges and either fixed or simply supported along the top edge depending upon the rigidity of the capping beam. The applied load is treated as a combination of rectangularly distributed and triangularly distributed loads. Bending moment coefficients for two-way slabs of uniform thickness subjected to rectangularly distributed and triangularly distributed loads are given by Reynolds and Steedman (1974).

10.4.5 Reinforced concrete strutted abutments

Strutted abutments have been used extensively for square and slightly skewed bridges up to 12 m span. In this type of bridge no allowance is made for the thermal expansion of the superstructure and thus the deck will act as a strut between the abutments. The abutment wall may be monolithic with the base or hinged by means of a simple joint with a shear key as illustrated in Fig. 10.11.

In the case of the monolithic wall, the wall slab can be considered to act as a propped cantilever, fixed at the base and simply supported at the top, or as a beam simply supported at both ends. Clearly neither assumption is strictly true and the wall should be designed to satisfy both cases. Hinged walls are designed

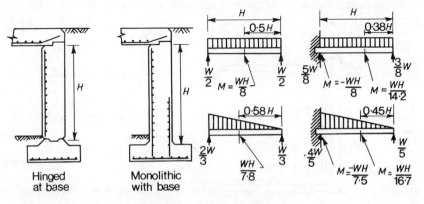

Figure 10.11 Strutted abutments.

as simply supported beams and overall stability of the structure is provided by the retained earth. The joint between the wall and the foundation should have chamfered edges to prevent spalling of the concrete. With both types of wall the foundations are assumed to be axially loaded and the backfilling is placed simultaneously behind both abutments after erection of the superstructure. This latter point must be specified in the Contract Documents. The stems of hinged walls are normally made quite stocky in order to provide adequate stability against overturning during construction. It is also very important to check the adequacy of the shear keys at the top and bottom of the stem.

The lateral load distribution due to retained earth is normally assumed to be that due to earth at rest. However, allowance should also be made for the passive resistance induced by thermal expansion of the superstructure. A rule of thumb method of allowing for this, suggested by the authors, is to use a characteristic uniformly distributed load having an intensity corresponding to the earth stress at rest existing at the bottom of the wall.

10.4.6 Reinforced concrete U-shaped and T-shaped abutments

The U abutment has been used increasingly in recent years because it has proved to be an economic form of construction. Basically, it consists of a cantilever abutment wall constructed monolithic with the wing walls which are normally cantilevered, both vertically and horizontally, off the abutment wall (Fig. 10.12). The wing wall thickness at any vertical cross-section is governed by the magnitude of the cantilever moment caused by the lateral force from the retained earth. In practice, the wall thickness is normally made to vary linearly from a maximum at its junction with the abutment to a minimum of about 200 mm at the free end. The cantilever effect in the vertical direction is not so critical since the section depth is very large. When determining bearing stresses beneath the foundation and checking stability against overturning, the abutment should be considered as a whole, including the wing walls, since the weight of the latter has a significant effect upon the restraining moment, and greatly influences the design of the foundation.

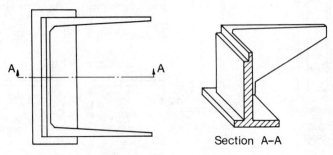

Section A–A

Figure 10.12 U abutment.

Alternatively, the whole structure, abutment and wing walls, may be founded on a U-shaped strip footing provided that the abutment is not so wide as to be inadequately restrained against overturning. In this case the footings are assumed to be axially loaded, and may be stepped for economy provided they are founded in undisturbed soil.

If the abutment is narrow, it may prove economic to tie the wings walls to each other by means of tie beams, or, more preferably, a single tie wall. A wall, as opposed to beams, has fewer constructional difficulties and is less vulnerable to damage due to settlement of the fill. In addition to direct tensile force, ties must be designed for earth loading from either side and, in the case of tie beams, vertical loading from the soil and self weight. In order to minimize the risk of corrosion of the tensile reinforcement in the ties, the concrete cover should be at least that specified for very severe conditions. Alternatively, the tendency to crack may be minimized by prestressing the ties. To help prevent failure due to pulling out of ties, the tie reinforcement should always be hooked around the main reinforcement in the walls.

Wing walls which are tied by a single tie wall can be designed as propped cantilever slabs, fixed at one end and simply supported part way along their length. The bending moment within the span should, however, be increased to allow for rotational yielding of the 'fixed' end: 10% to 20% redistribution of the support moment, depending on relative stiffness, appears appropriate. Allowance should be made for the forces due to differential thermal expansion and contraction of the abutment wall and tie wall. The tie wall should be so positioned that the wall thickness can economically be made uniform throughout and the reinforcement made continuous in both faces.

In all types of construction the abutment wall acts not only as a cantilever off the foundation but also as a support and tie wall for the wing walls. Consequently, horizontal reinforcement must be provided:

(a) Throughout the whole length of the wall to withstand the horizontal tensile force transferred from the wing walls.
(b) In the back face at the junction with the wing walls to withstand the bending moment applied by the wing walls.

If the abutment is wide the vertical steel should be designed on the same basis as for a cantilever abutment. Alternatively, if the abutment is narrow (i.e. width less than about twice the height) the wall should be designed as a two-way slab supported on three sides and subjected to a loading which increases linearly downwards due to earth and surcharge stresses. Bending moment coefficients for these conditions are given by Reynolds and Steedman (1974). The heel of the foundation can be designed in a similar manner. Tests on a model U-shaped abutment have been reported by Lindsell and El-Dharat (1979).

The general form of a T abutment is illustrated in Fig. 10.13. These abutments are very economical for narrow superstructures such as those of footbridges and farm accommodation bridges. The abutment wall does not retain earth but acts

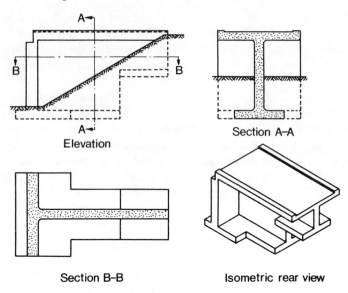

A→

B   B

A→|

Elevation   Section A–A

Section B–B   Isometric rear view

Figure 10.13 T abutment.

purely as a load-bearing wall braced by the spine wall. The spine wall supports an approach slab and retains an equal amount of earth on each side. The foundation can be all at one level or, if founded in undisturbed ground, it can be stepped for economy. The bearing stresses should be evaluated from the overall effect of the loads on the entire structure.

10.4.7 Skeleton abutments and bank seats

The general form of skeleton abutments and bank seats is illustrated in Fig. 10.14. An abutment wall to retain earth is not required but the length of the superstructure required is greater than when earth-retaining abutments are used. Three- and four-span bridges with bank seats or skeleton abutments have been used quite extensively as motorway overbridges. In certain circumstances their use may be justified on purely economic grounds but they are often used on aesthetic grounds or as a means of adding variety to the highway landscape. In order to be more economic than the equivalent two-span or single-span bridge with earth-retaining abutments, the saving in the cost of substructures must be greater than the additional cost of the superstructure. Bank seats are cheaper than skeleton abutments but, if they are to be founded in undisturbed soil, they are suitable only for bridges in relatively deep cuttings.

A bank seat is designed as a cantilever abutment with a very short stem. In addition to the vertical and horizontal forces from the superstructure, each column of a skeleton abutment is designed for a lateral earth stress which, arbitrarily, is 50% to 100% greater than the active stresses. If the columns are

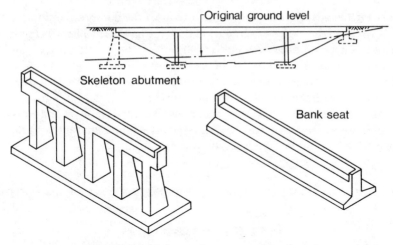

Figure 10.14 Skeleton abutment and bank seat.

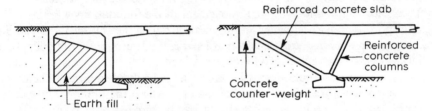

Figure 10.15 Two examples of counterbalanced cantilever and suspended span construction.

closely spaced, it is likely that, owing to arching, the total thrust on the columns will be almost the same as for a continuous wall. The lateral forces due to the earth in front of the columns are ignored in design. A variation of the skeleton abutment is a bank seat founded on piles. This type of construction has distinct advantages when used for river crossings. It avoids excavating foundations in flood plain conditions and the foundation cannot be undermined by severe floods.

10.4.8 Counterbalanced and cellular abutments

Figure 10.15 illustrates two types of abutment which can be used for cantilever and suspended span construction. The cellular abutment supports a relatively short cantilever which is counterbalanced by earth filling. The counterbalanced abutment with raking columns shown in Fig. 10.15 has been used extensively on British motorways. In this case the concrete counterweight not only provides support for the suspended span but also reduces the compressive force acting upon it. Many of the motorway bridges of this type have used a heavily reinforced Mesnager concrete hinge (see Section 10.8) between the cantilever span and the suspended span as opposed to the simple support shown in Fig. 10.15. However, the authors do not recommend this system, since they feel that the steel reinforcement in the throat of the hinge is close to the surface of a road, and is vulnerable to corrosive attack. It is well known that all water bars and seals have their failures, so even the best sealing systems cannot guarantee protection against the salt solutions produced by winter salting.

10.5 ABUTMENTS FOR PORTAL FRAME AND ARCH BRIDGES

10.5.1 General considerations

Statically indeterminate structures such as portal frames and arches may be seriously damaged by differential settlement of the foundations. Fixed portals and arches are statically indeterminate to the third degree and are sensitive to vertical, horizontal and rotational movement of the foundations whereas two-pinned portals and arches suffer serious damage only if the abutments spread. In fact, even three-pinned arches, which are statically determinate, will collapse if abutment yield is progressive. It follows that fixed arches and portals are suitable only for sites where foundation movements will be absolutely minimal. Therefore, fixed ended structures should be adopted only where the foundations can be constructed on solid rock. Where the bridge can be founded direct on firm sand or gravel or where rock or firm gravel can be reached by piling, a two-pinned structure is feasible provided adequate resistance against spread of the abutments is provided.

The bearing stress distribution beneath a foundation to a fixed arch or portal may vary considerably due to variations in live load. In order to minimize the possibility of rotational settlement, the foundation should be designed, wherever possible, to have a uniform bearing stress distribution when subject to the permanent loads. For this purpose a small proportion of the traffic load may be considered to be permanent. In the case of two-pinned structures the resultant reaction at the underside of the base is maintained within fairly close limits. When designing a foundation it is necessary to examine, in addition to the permanent loading, those combinations of loading which, by careful inspection, are seen to produce extreme bearing stress distributions. These may include temperature and shrinkage effects. There is some difference of opinion about the resisting forces which should be included when evaluating the factor of safety against sliding of the foundation but normally these should never exceed the base frictional force plus 50% of the passive resistance of the earth. It must be remembered also that hydrostatic uplift reduces the frictional resistance to sliding.

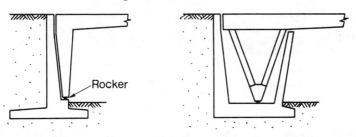

Figure 10.16 Portal frame abutments with independent retaining wall.

10.5.2 Portal frame abutments

Abutments to portal frame bridges take many forms and their foundations can be designed as strip footings, retaining wall bases, column bases or pile caps depending on the arrangement of the superstructure. Where the legs of a portal frame bridge retain earth, they should be designed to cater for earth stresses corresponding to the 'at rest' condition, or in some cases, a proportion of the stresses corresponding to the passive state. The proportions of the bridge, method of backfilling, method of resisting braking forces and sensitivity to temperature change are all factors which influence the magnitude of the lateral earth stresses. The designer must take all such factors into account when deciding on the applicable earth loading. If, however, the earth is retained by a wall which is independent of the portal frame (see Fig. 10.16), the leg is relieved of the earth forces and the wall can be designed as a mass, cantilever or counterfort retaining wall. If the foundation is common to the retaining wall and portal frame, then the wall cannot move forward to invoke the active condition and design should be based on 'at rest' stresses. Hinges suitable for two-pinned portal frames are described in Section 10.8.

10.5.3 Arch abutments

The stability of arch abutments can be examined more closely in the light of soil mechanics investigations and can be improved with the assistance of geotechnical processes. The Runnymede Bridge (Cracknell, 1963) over the River Thames comprises a series of 54 m span steel spandrel-braced two-pinned arch ribs with a rise of 5.5 m. After soil mechanics investigations, the horizontal displacement of each abutment, bearing on the London Clay, allowed for in design was 25 mm, taking into account elastic deformation and consolidation of the soil. The resulting 50 mm increase in span greatly influenced the stress distribution in the ribs. It is worth remembering that, in the extreme, a two-pinned arch can be designed as a simply supported curved beam. Undoubtedly, many spandrel filled arch bridges founded on yielding material derive considerable strength from the fill.

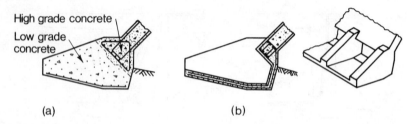

Figure 10.17 Mass and cellular arch abutments. (a) Mass abutment; (b) cellular abutment.

Preloading was employed to eliminate horizontal movement of abutments for a bridge across the Fraser River, British Columbia (Golder and Sanderson, 1961). In the site investigations, boulders were mistaken for bedrock and, during excavation, it was found necessary to change the original design for a fixed-ended arch to a two-pinned arch because of the depth of sand and gravel overlying bedrock. The bearing stress beneath the abutments was about $500 \, \text{kN m}^{-2}$ but the safe bearing capacity of the sand and gravel was adequate. Settlement was therefore the criterion and, after considering several alternatives, it was decided to preload the soil thereby eliminating further settlement when the arch was constructed. Since the soil was largely cohesionless, with one thin layer of clay, settlement would be practically instantaneous when the load was applied. Vertical preloading of $110\,000 \, \text{kN}$ was effected by jacks operating in opposition to cables anchored in bedrock. Horizontal preloading of $18\,000 \, \text{kN}$ was effected by jacking between the back of the abutment and soil, after which concrete was placed in the jacking chamber. The frictional resistance at the base of the abutment developed under the vertical load was more than sufficient to provide the reaction. The maximum vertical settlement observed in the preloading was about 13 mm and the maximum horizontal movement about 2.5 mm. In fact the arch would not have been stressed excessively by a differential settlement of 50 mm and a spread of 25 mm but, although the observed displacements were less than these, it would have been difficult to predict displacements because of possible variations in the degree of compaction of the soils. Arch abutments may be mass concrete or cellular reinforced concrete as shown in Fig. 10.17.

(a) Mass concrete abutments

Abutments for arch bridges are usually constructed in mass concrete because they are very rigid, simple to design and construct and their great weight helps to deflect the line of thrust from the arch into a more vertical direction. They are also easily modified to suit conditions on site or to accommodate changes in design or construction. They are, of course, expensive in material, but this has to be weighed against the extra cost of skilled labour required for the design and construction of cellular abutments. Mass concrete abutments are invariably used for short-span arches, for highly skewed arches and for some longer span structures such as the 245 m span two-pinned arch of the Volta River Bridge, Ghana (Scott and Roberts, 1958).

Opinion differs as to the relative merits of a horizontal bed to an abutment and a sloping (usually 15° to 30°) or a stepped bed. In practice, careful study of the specific problems involved should reveal which type is the most suitable, consideration being given to the nature of the rock, the provision of horizontal resistance and the effective distribution of bearing stress. Meyerhof (1953b) found that a base which is inclined so that it is normal to the resultant thrust leads to a greater bearing capacity than a horizontal base. The bearing capacity beneath foundations subjected to inclined loads is discussed briefly in Section 6.2.1. A sloping bed is normally terminated by a short horizontal toe and, provided the slope is not excessive, it is sufficient in most cases to project the horizontal portion on to the inclined plane in order to determine the bearing stress distribution.

Where piles are used, both raking and vertical piles are normally required but in some cases it is possible to rake the majority, if not all, in the average direction of thrust. The spacing across the width between rows of piles can be varied to suit the bearing pressure distribution. Reinforcement may be placed in the base to prevent tension cracking and around the skewback to prevent bursting (Fig. 10.17a); this is conservative practice and is not always adopted. Since the loads imposed by the superstructure on the abutment are rapidly distributed, stresses within the abutment are generally relatively low and it may be economic to consider using two grades of concrete as illustrated in Fig. 10.17(a).

(b) Reinforced concrete cellular abutments

Reinforced concrete cellular abutments (Fig. 10.17b) are more economical than mass concrete abutments for long-span arches. They are, however, suitable only for two- or three-pinned arches since they are insufficiently rigid to withstand the end moments without suffering significant rotational deformation. The arch bears on a substantial horizontal beam, known as the skewback, which is supported at about 2.5 m intervals by buttress walls. These walls bear on a thick foundation slab and also act as counterforts for the face wall which retains the earth filling. Buttress walls are generally about 350 to 500 mm thick but they should be sufficiently thick to avoid buckling. Sometimes a thinner base slab is used which is then stiffened by beams spanning between the buttresses. In order to reduce concentrations of stress, adequate fillets, say 300 mm, should be provided at the junctions of all members. The foundation slab and face wall are designed as two-way slabs. If the bridge is slightly skewed the buttress walls must be skewed to the same angle so that they are in line with the longitudinal axis of the bridge. Cellular abutments should not, however, be used for highly skewed bridges, because such bridges tend to span between the obtuse corners thus making the determination of the size and direction of the buttresses extremely difficult. For this type of bridge the mass concrete abutment with its superior rigidity is much to be preferred.

10.6 BRIDGE PIERS

10.6.1 Types of piers

Bridge piers take many forms and may be land-based piers or river piers. Prior to the mid-1950s most bridge piers were associated with rivers or railways and were constructed of brick, masonry and reinforced concrete. However, since then there has been a rapid expansion of grade separated highways having many multispan bridges. This has involved the design and construction of land piers of many kinds constructed mainly in reinforced concrete but occasionally in steel.

In the transverse direction piers act either as cantilevers, constructed monolithic with the foundation or superstructure, subject to axial and transverse end loads and moments, or as axially loaded pin-ended struts (Fig. 10.18). They are normally vertical but may occasionally be inclined. Piers that are pinned at the foot require relatively narrow bases because they produce axial or near axial loads on the foundations, whereas piers which are cantilevered from the base

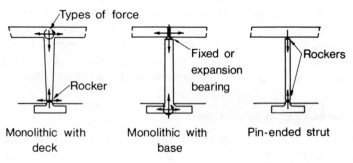

Figure 10.18 Basic types of pier (transverse direction).

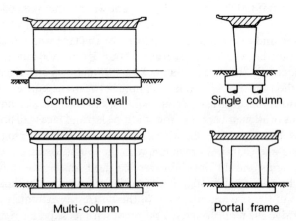

Figure 10.19 Basic types of pier (longitudinal direction).

require broader bases to resist the overturning moment, especially if designed to resist braking forces.

In the longitudinal direction piers may be continuous walls, independent columns, portal frames or multi-column structures with a continuous capping beam (Fig. 10.19).

Continuous wall piers are well suited to river crossings because they can be constructed to slender and streamline proportions, thus causing minimum resistance to flood flows. Portal frame and single-legged T-shaped piers are well suited to urban environments because they can be made attractive and occupy the minimum amount of ground space. An economic form of construction is the multiple circular column pier with rectangular capping beam, although this type of pier may produce undesirable aesthetic effects if used indiscriminately. For example, when used for viaducts it produces a 'concrete jungle' effect.

10.6.2 Design considerations

Once the general form of a pier has been decided, design involves consideration of the following criteria:

(a) Safety against bearing failure of the soil.
(b) Limitation of settlement.
(c) Safety against overturning.
(d) Safety against sliding.
(e) Strength and serviceability of the structural elements.

Items (a) (c) and (d) are, in fact, collapse criteria while (b) is a serviceability criterion. However, the Code (BS 5400) recommends that items (a) (b) and (d) should be evaluated in accordance with CP 2004 using unmodified characteristic forces since this document is not based upon limit state principles. Although the code specifies that item (c) should be treated as a limit state collapse, the authors believe that the point about which rotation occurs has not been satisfactorily defined and recommend that this also be treated in the traditional manner using unmodified characteristic forces.

In order to determine the most critical load condition for each of the above criteria, it is necessary to inspect all possible compatible combinations of the vertical, longitudinal and transverse forces outlined in Section 10.3. Under normal circumstances the only longitudinal forces acting on a pier are those due to frictional or shear resistance of the expansion bearings and longitudinal wind force. Occasionally, however, a pier may be required to resist braking forces but this is a practice to be avoided wherever possible since it considerably increases the internal bending and overturning moments. The transverse forces include wind and, where the superstructure is curved in plan, centrifugal force. In some circumstances it may be necessary to apply a vertical uplift force due to wind effects on the superstructure. This has the effect of reducing the stabilizing influence of dead load and may be significant when checking resistance to overturning and sliding. Permissible combinations of transverse, longitudinal and vertical uplift forces due to wind are given in Section 10.3.1(c).

Calculations for settlement and bearing capacity of piers should be made in accordance with the methods discussed in Chapters 2 and 3. When the depth of founding is considerable, increase in bearing capacity arising from overburden stress and skin resistance may be appreciable and, furthermore, settlement can be expected to be less than for a surface footing. Account should be taken of the anticipated depth of scour which will lead to a reduction in these benefits. Skin resistance should not be included in bearing capacity when the base is founded on a resistant stratum and the shaft is surrounded by relatively compressible soil.

The response of piers to lateral and longitudinal forces may involve translation only or, more likely, will include rotation about some point between bed level and founding level. In some cases the stability of the pier will have to be investigated, taking into account active and passive stresses. On the other hand, in many cases it is likely that deformation of the soil, theoretically within the elastic range, will be more important since it is necessary to limit horizontal movements in order to maintain the stability of the superstructure. The horizontal modulus of subgrade reaction would have to be determined or estimated but some piers are comparatively rigid and their flexibility can probably be ignored in calculations for deformation. A method of taking into account the flexibility of piers embedded in cohesionless soil has been given by Karmalsky and Korner (1956).

Mass concrete may be used for piers provided lateral loads are relatively light but nominal reinforcement is necessary to cater for shrinkage and temperature stresses.

Hollow piers are usually constructed in reinforced concrete, with or without internal transverse walls, and are useful where foundation loads must be kept to a minimum. In the case of river piers, additional bouyancy can be provided by watertight compartments but this is at the expense of lateral stability. Piers may be hollow for full or part depth but, where it is possible for water to freeze inside the pier and perhaps burst the shell, the solid portion should be carried above high-water level and the hollow portion adequately drained. The minimum thickness of external walls is commonly 300 mm at the top and 450 mm at the bottom, while for internal transverse walls 225 mm is the usual minimum.

When analysing single or multiple bay portal frame piers constructed of reinforced concrete, account should be taken of the fact that the dimensions of the joints are relatively large. A method of moment distribution for stiff jointed frames described by Pippard and Baker (1968) is easily modified for this purpose. In this method the joints are assumed to be rigid and the moments, including those due to the shear forces at the ends of the members, are balanced about the centre points of the members. Alternatively, computer methods can be used for the analysis provided they allow for the relatively large size of the joints.

The benefits to be gained from the use of climbing formwork, or standardized formwork, should be considered at the design stage and, if adopted, the pier must be suitably proportioned but care must be taken to ensure that undesirable aesthetic effects are not introduced.

Special formwork was employed for 1.8 and 2.1 m diameter columns for the Willis Bridge (Williams, 1960) on the Texas–Oklahoma border. The cylindrical forms comprised three arcs connected by double-acting hydraulic rams and could be clamped in position under water, pumped dry, filled with concrete and stripped by reversing the movement of the rams.

Where a series of arches is carried on piers, every third or forth pier should be designed as an abutment pier to carry any unbalanced thrust from either side in the event of one of the spans collapsing. If a pier carries two arch spans of unequal length, it is usually necessary to offset the pier foundation in order to achieve a reasonably uniform distribution of bearing stress.

The Erskine Bridge over the River Clyde (Kerensky *et al.*, 1972) has several interesting features. The south abutment provides the fixed anchorage for 840 m of superstructure; the seven supporting piers are pinned to the deck, 'fixed' at the base and flex longitudinally to accommodate changes in length of the superstructure. Because of poor ground and proximity to the railway, it was uneconomic to design the north abutment to resist longitudinal forces, so the 483 m of the superstructure on the northern shore is pinned to six piers designed to resist all longitudinal forces in proportion to their stiffnesses and to flex so as to accommodate changes in length of the superstructure. The design of the piers was rather critical, since it was necessary to have sufficient cross-sectional area to carry the vertical load and yet sufficient flexibility to accommodate horizontal movements at the top of ± 260 mm without overstressing the concrete. The lozenge-shaped cross-section (see Fig. 10.20) gives a slender appearance and offers little resistance to wind. The foundations to the northern side are also of interest. The ground consisted of very soft silt overlying 17 to 20 m of stiff boulder clay, followed by a soft silt and finally very compact gravel overlying bedrock. During the site investigation it was discovered that the compact gravel was

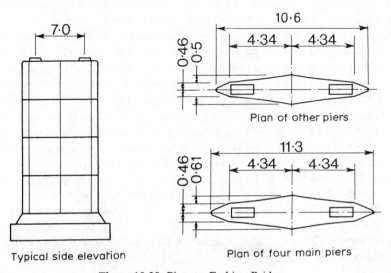

Typical side elevation   Plan of four main piers

Figure 10.20 Piers to Erskine Bridge.

subject to high artesian pressure. In order to pile to bedrock without releasing water from the artesian layer it was necessary to use driven piles since bored cast *in situ* piles were clearly unsuitable. However, the presence of the boulder clay made driving difficult, so it was decided to pre-bore the boulder clay, fill the hole with sand and drive the piles through the sand and down to bedrock.

Flexible columns in cellular piers were used on the Clifton Bridge, Nottingham, and independent hinged reinforced concrete columns were used for the Pelham Bridge, Lincoln (Reisser and Wright, 1958).

Single column, T-shaped and V-shaped piers are used extensively for viaducts, because they are economical, highly functional and, when carefully designed, have a pleasing appearance. The single-column piers used on the Western Avenue Extension are illustrated by Baxter *et al.* (1972). The foundations to these piers consisted of 1.1 m diameter reinforced concrete shafts with enlarged bases of 3.2 m diameter. The depth of the shafts in the firm to very stiff London Clay was varied to give working loads of 380 to 450 tonnes per shaft. Similar piers and foundations were used for the Mancunian Way (Bingham and Lee, 1969). Large diameter reinforced concrete cast in situ piles were chosen for the Coventry Inner Ring Road (Aizlewood and Heathcote, 1970) because of the need to suppress noise and to make possible inspection of the foundations. The piers consisted of steel portal frames in some locations and reinforced concrete single-leg and double-leg trestles in others. The base of each column pier on the Hammersmith Flyover (Rawlinson and Stott, 1964) is supported on a pair of roller bearings resting on a 8.5 m × 7.3 m reinforced concrete footing founded on gravel overlying London Clay.

Large precast concrete units were used for the construction of the piers on the Richmond–San Rafael Bridge over San Francisco Bay (Gerwick, 1954). The bridge crossed two main shipping channels where the water was some 15 to 18 m deep. The sea-bed consisted of a thick layer of very soft mud and sediments overlying rock. The depth to the rock surface varied from 10 to 90 m but fortunately the deepest section was covered by a thick stratum of dense sand and gravel. At the site of each pier a depth of about 4 m was dredged and timber falsework piles were driven in the excavation and cut off at formation level. The piers generally consisted of a pair of shafts on 250 mm thick reinforced concrete bases perforated with holes to take steel H-piles. Having accurately positioned the bases on the timber piles, the H-piles were driven through the perforations to a depth of about 60 m and then grouted. The 200 mm thick precast reinforced concrete shell units were then built up on each base and filled with concrete as construction proceeded. Developments of this type of construction have been discussed by Gerwick (1965). Precast concrete units were used for the piers of the 5 km long Oosterschelde Bridge, Netherlands (Cement and Concrete Association, 1966). Each pier was founded on three 4.3 m diameter hollow precast concrete piles placed in line with the tidal stream (Fig. 10.21). The piles were capped with a precast concrete caisson which acted as a plinth for the two inclined columns forming the inverted V-shaped piers.

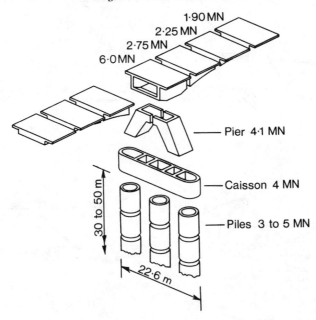

Figure 10.21 Precast concrete piers used for Oosterschelde Bridge, Netherlands.

10.7 ABUTMENTS AND PIERS FOR MOVEABLE BRIDGES

Moveable bridges are commonly used to bridge relatively narrow navigable waters such as shipping canals and entrances to docks. The design and construction of the new Manchester Road bridge in the Port of London have been described by West (1971). The various types of moveable bridge which were considered by the Port Authority for use at Manchester Road are shown in Fig. 10.22. The clearance between top water level and the bridge soffit was an important preliminary consideration in the design of the bridge. It was thought that sufficient headroom could be provided to enable the smaller vessels, such as tugs, to pass under the bridge while in the closed position, thus reducing the number of opening operations and hence delays to motor traffic. However, a shipping survey showed that the headroom required to produce a significant reduction in the number of operations was so great as to be economically unfeasible.

Basically, there are four types of moveable bridge; bascule, swing, vertical lift and retractable. Bascule bridges may be single or double leaf and, depending upon the type of lifting mechanism, may be of the Scherzer rolling, trunnion, drawbridge or Strauss type. A disadvantage to the double leaf type is that two sets of expensive lifting gear are required. The single leaf type has the disadvantage that the main abutment has to be designed to withstand the full weight of the bridge. Under certain circumstances considerable savings in cost of foundations

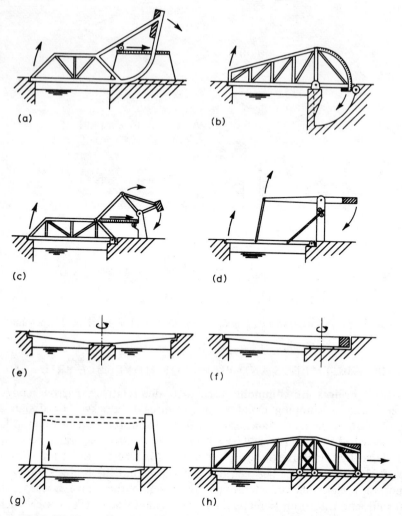

Figure 10.22 Various types of moveable bridge. (a) Scherzer rolling bascule; (b) Trunion bascule; (c) Strauss bascule; (d) drawbridge; (e) two-span balanced swing; (f) single-span counterbalanced swing; (g) vertical lift; (h) retractable.

and operating machinery can be achieved by the use of lightweight materials for the superstructure. The use of aluminium alloy for the superstructure of the double leaf trunnion bascule bridge at Sunderland Docks enabled the weight of the moving spans to be reduced to 40% of the equivalent steel spans.

The Scherzer rolling type is quite common since all parts are above ground level and the operating machinery is simple and easily installed. However, due to the reaction at the main support moving considerably during operation, the foundations have to be both heavy and extensive. Maintaining the rollers on the correct travel line and local crushing due to the high contact stresses beneath the rollers can also cause problems with this type of bridge.

The trunnion bascule bridge is often the best type for relatively high level bridges but, when installed at ground level, it suffers the disadvantage of requiring a large pit to house the tail span counterweight when open. Since the pit must be kept free of water it is generally necessary to install pumps, and the effects of bouyancy must be allowed for in design. Shock absorbers should also be provided at the limits of travel to avoid damage to the main structure.

The Strauss and drawbridge types operate on very similar principles but the drawbridge has a simpler appearance and is more acceptable aesthetically than the Strauss. Both types possess the advantage of having all moving parts above ground level. The drawbridge has been used extensively in Holland and was chosen for use at Manchester Road, Port of London, because among other things, it looks elegant and could be constructed quickly using large pre-fabricated members.

A swing bridge is generally the most economical form of moveable bridge for many sites. It may be of the two-span balanced cantilever type with centre bearing or single-span counterbalanced bobtail type. After considering various types of bascule, vertical lift and swing bridges, a balanced cantilever centre bearing swing bridge was chosen for the Cumberland Basin Scheme, Bristol (Vavasour and Wilson, 1966), because it provided the most economical and pleasing solution. When open to shipping the ends of a two span balanced cantilever swing bridge generally project over navigable water and are, therefore, highly vulnerable to damage. In order to protect the structure when in this position it is necessary to provide a system of fenders. For the Cumberland Basin Bridge these consisted of a 0.9 m thick concrete walkway supported on steel box piles with advanced protection of timber fendering supported on 0.4 m square greenheart piles. Although not requiring extensive fendering, single-span swing bridges which lie along a canal bank when open may restrict mooring operation at sites where this is necessary.

A number of swing bridges over the River Weaver in Cheshire are designed to facilitate adjustments to counteract the effects of brine subsidence. For example, the superstructure of the bridge at Acton Bridge rests on a floating steel shell pontoon contained within a hollow pivot pier filled with water. The pontoon carries 70% of the weight of the superstructure, the remainder being carried by the rollers. The provision of the pontoon not only facilitates occasional re-levelling of the roller track but also reduces the load carried by the rollers and hence the power required to operate the bridge. McNaughton (1953) gives details of the design and construction of the foundations for the Dalgrain Swing Bridge over the Forth and Clyde Canal at a site near Grangemouth, Stirling.

A vertical lift bridge is very expensive for short spans but for long spans where it is necessary to provide a wide opening for navigation, it is sometimes the only feasible solution. The foundations for the vertical lift bridge at Kingsferry, Isle of Sheppey, have been described by Anderson and Brown (1964).

Retractable bridges require expensive rolling or sliding gear, extensive and heavy foundations and a suitable approach to accommodate the superstructure when withdrawn. Consequently, this type of bridge is little used.

The number of load conditions to be considered in the design of foundations for moveable bridges is quite considerable because, in addition to the normal dead, live, wind and constructional loadings, there are wind loads in the open position, transfers of dead load during operation and dynamic loadings to be considered. Wind loading is also an important consideration in the design of the brakes which restrain the moving parts.

10.8 BRIDGE BEARINGS AND ARTICULATION SYSTEMS

10.8.1 Introduction

Basically there are four degrees of freedom which a bridge bearing may be required to provide; translation along and rotation about each of the two principle axes. Swing bridge bearings, which provide rotation about the vertical axis, are not considered here in view of their specialized nature. Most bearings, in fact, provide only one or two of these degrees of freedom (Fig. 10.23). Sliding bearings cater primarily for translation in one or two directions but, when suitably adapted, may also provide for rotation. A single roller bearing permits translation along one axis and rotation about the other, whereas a rocker bearing provides only for rotation about one axis. A concrete hinge can be designed to permit rotation about one or two axes. Rubber bearings can be arranged to provide any combination of the four basic degrees of freedom. Rotation about both axes can be provided also by a spherical steel bearing.

The choice of bearings for a bridge should be made at an early stage since it may affect many aspects of the detailed design. It may be that aesthetic considerations dictate the type of bearing to be used or that the span is so great that the magnitude of the loads and displacements necessitate the use of specially hardened high-strength rollers. In the majority of cases, however, the choice is based upon considerations involving some or all of the following factors: type and magnitude of resistance to movement, size of bearing, durability and cost. Bridge temperature for setting bearings and expansion joints is the subject of a report by Emerson (1979).

The first step is to decide upon the articulation system for the bridge and then to determine for each bearing location the type and magnitude of movements to be accommodated (Fig. 10.24). For wide bridges and skew spans, consideration should be given to the effects of lateral expansion and contraction. In the case of short-span bridges there are many types of bearing which can satisfy the specification, since both loads and movements are relatively small, and consequently the choice tends to be based almost entirely on cost. The economics of the problem may, however, be complicated since different bearings have different shapes, sizes and resistance to movement, all of which may affect the cost of the

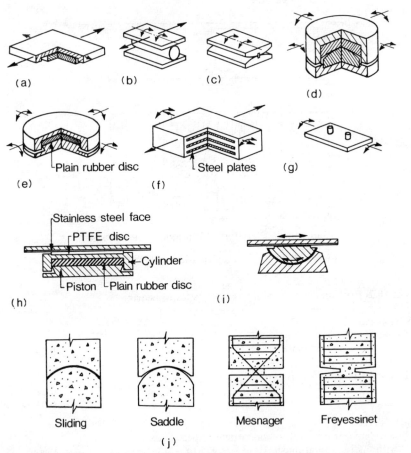

Figure 10.23 Various types of bridge bearings. (a) Sliding; (b) roller; (c) rocker; (d) spherical; (e) rubber pot; (f) laminated rubber; (g) plain rubber pad, fixed type; (h) combined rubber pot and sliding bearing; (i) combined flat and cylindrical sliding surfaces; (j) concrete hinges.

supporting structure. When a bridge is to be supported on tall piers it is possible that some or all of the movement may be more economically provided for by flexure of the piers, as on the Erskine Bridge (Kerensky *et al.*, 1972).

Durability is something which is more a function of detailed design and construction than of bearing type. For example, grease boxes undoubtedly give added protection to roller bearings and accurately placed rubber bearings will last longer than badly placed ones because of the adverse stress conditions experienced by the latter. Consideration should be given during the design stage to the maintenance and possible replacement of the bearings during the working life of the bridge. If necessary the design should incorporate special features to facilitate this work.

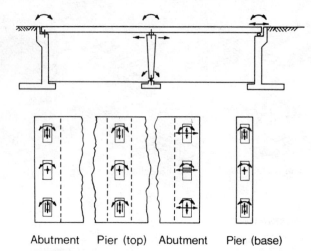

Abutment   Pier (top)   Abutment   Pier (base)

Figure 10.24 Typical articulation system for a bridge.

Table 10.2 Friction coefficients for various types of bearing

BS 5400 Part 2	
Roller bearing 1 or 2 rollers	0.03
3 or more rollers	0.05
Sliding (steel on PTFE)	
Contact stress $10 \, N \, mm^{-2}$	0.06
$20 \, N \, mm^{-2}$	0.04
$30 \, N \, mm^{-2}$	0.03
Black (1971)	
For roller bearings in through-hardened special steel with finely ground finish. Hardness 425 to 450 HB	0.006
For roller bearings with one or two rollers in mild steel to BS 15, High Tensile steel to BS 548 or BS 968 and grey cast iron to BS 1452 all with a normal as turned finish	0.03
For bearings using polished stainless steel sliding on PTFE (see Section 10.8.2(a)) and the surfaces initially lubricated	0.03–0.07
For sliding of steel on hard copper alloys or grey cast iron on grey cast iron	0.35
For sliding of steel on steel	0.80

Black (1971) reported on a field study of bridge bearings carried out by the Transport and Road Research Laboratory and concluded that the allowable values of friction in bearings given in BS 153 (BSI, 1972a) may be too low. In fact, BS 5400 Part 2 does not recommend minimum allowable values but actually states the values to be used. The coefficients of friction specified in BS 5400 together with those recommended by Black, are given in Table 10.2.

10.8.2 Bearing design and performance

This subsection deals with matters relating to the design and performance of various types of bridge bearing. Much of this material has been taken from the paper by Black (1971) and the interested reader should refer to the original document for greater detail.

The design of bearings to BS 5400 Part 9, are to 'limit state' principles. Present design methods are based on characteristic loads and various Department of Transport Technical Memoranda (BE 1/76).

(a) Sliding bearings

The early types of sliding bearing include cast iron sliding on wrought iron, mild steel sliding on mild steel and prefabricated bituminous layers. More recently, grey cast iron sliding surfaces were used and sometimes copper strips were placed between steel plates to reduce the sliding friction. In many of these bearings no provision was made for rotation. Although the coefficient of friction for sliding bearings specified in BS 5400 Part 9 (BSI, 1972a) is 0.25, there is evidence that values in practice may be between 0.6 and 0.8.

Recently, various types of bearing have been used in which translational movements are taken by a stainless steel plate sliding on a layer of low friction plastic known as PTFE (polytetrafluoroethylene). In many cases pure PTFE sheet is used, bonded or recessed into a baseplate. Other forms include PTFE lead lubricant impregnated into a sintered backing material, PTFE with various fillers and woven PTFE fibres. The merit of these bearings lies in their low constructional height and low frictional resistance to movement. In order to achieve low friction values the bearings are used at high stresses of 7.5 to 30 MN m^{-2}. Provision for rotation can be achieved in several ways but the use of confined rubber in the form of a pot bearing (Fig. 10.23e) enables the bearing height to be maintained at 100 mm or less. Laboratory investigations show that, after running in for a few cycles, the coefficient of friction of some commercial bearings is likely to lie between 0.005 and 0.05, depending on the loading, temperature, and speed of sliding and the surface finish of the steel rubbing plate. In fact, the coefficient of friction tends to increase as the stress in the PTFE decreases, as the speed of movement increases, as the temperature decreases and as the roughness of the steel plate increases. It should be noted that the initial peak value of the coefficient may lie between 0.03 and 0.18 and values of 0.25 have been quoted. With the present lack of knowledge and fixed experience of PTFE bearings it may be desirable to design for ultimate values between 0.03 and 0.06, depending on conditions of service. Initial peak friction should be eliminated by the use of an appropriate lubricant approved by the manufacturers. The results of tests involving a large number of backward and forward movements of steel on PTFE suggest that wear will be small but tolerable, provided that the sliding surfaces are adequately protected from grit. The use of PTFE in highway bridge bearings has been discussed in detail by Taylor (1972).

Figure 10.23 (h and i) illustrates two sliding bearings which provide translational movement by means of a rubber pot bearing in one case and a curved sliding surface in the other.

(b) Roller, rocker and spherical bearings

Table 10.3 summarizes the allowable working loads specified in BS 5400: Part 9 (BSI, 1972a) for cylindrical roller and spherical bearings manufactured of mild steel, high tensile steel or cast iron.

The formulae given in this table are based upon relationships derived by Timoshenko and Goodier (1970) for stress concentrations at contacts between balls and rollers and bearing surfaces. It will be observed that the unit load for three or more rollers is less than for one or two rollers. This takes account of the fact that where several rollers are used the load is not shared equally because of the inevitable small discrepancies in dimensions. Through-hardened special steel rollers having hardnesses of 425–450 HB, compared with 110–150 HB for mild steel, are often used where loads are very high. These bearings are not covered by BS 153 but the manufacturers specify an allowable load of $96D$ (N mm^{-1} where D is the diameter of the roller).

Tests have shown that the value of the coefficient of rolling friction varies with the number of repetitions of rolling over the same area. The magnitude of the effect depends on the hardness of the roller and base plate, the surface finish of the parts in contact and the relation of the contact stress to the yield stress of the material. The allowable loads on bridge rollers are sufficient to produce the first stages of plastic deformation in the contact area and for this condition the rolling friction during the first cycles of rolling may be two to four times higher than the constant value recorded after several cycles. However, if the roller traverses the

Table 10.3 Allowable working loads on cylindrical roller and spherical bearings

Type of bearing	Grade 43 steel to BS 4360	Grade 50 steel to BS 4360 or cast iron to BS 1452 Grade 23	Grade 55 steel to BS 4360
Cylindrical rollers on flat surfaces: Single or double rollers	$7.7D*$(N mm^{-1})	$10.8D.$(N mm^{-1})	$13.4D$ (N mm^{-1})
Three or more rollers	$4.9D$ (N mm^{-1})	$6.9D$ (N mm^{-1})	$8.5D$ (N mm^{-1})
Spherical bearing on flat surface	$D^2/13$ (N)	$D^2/7.8$ (N)	$D^2/5.2$ (N)
Roller or spherical bearing on curved surface	Replace D in above equations by $1/(1/D_1 - 1/D_2)$ where D_1 and D_2 are the diameters (mm) of the convex and concave contact surfaces, respectively		

$*D$ = diameter of roller or sphere in mm.

same path several times and is then rolled over the ridge formed at either end, a large resistance is again recorded. The ridge in high tensile steel amounts to about 0.01 mm in height at normal loads and 0.1 mm at 60% overload. Even at normal loads the effect of rolling over a ridge is to produce a coefficient of rolling friction many times the value measured on the well rolled surfaces. In a bridge the ridges are not likely to be so well defined as in laboratory tests since the annual range of movement varies from year to year. Therefore it is considered that the initial friction, on first rolling, will be the highest value which occurs in practice. However, because of weekly, monthly and seasonal variations in mean bridge temperature, the daily to and fro movements do not occur over the same area of base plate and the running in period is, therefore, extended to at least one year. Hence when calculating the maximum horizontal force, it appears appropriate to use the initial coefficient of rolling resistance. The coefficients of rolling friction specified in BS 5400 and those recommended by Black (1971) are given in Table 10.2. The values recommended by Black include a factor of safety of 3 to allow for bearings which are badly machined, poorly aligned, overloaded, dirty or corroded.

For long life, roller bearings should be encased in a grease box. If not protected in this way, pitting generally occurs after a few years service and corrosion debris and grit accumulate under the rollers. In recent years grey cast iron rollers have been used extensively because of the popular belief that they require no protection against corrosion and, in fact, very few of these bearings have been given any protective treatment. Unless used in the as-cast condition, when there is a corrosion-resistant skin, this belief is unfounded.

The steel spherical bearings used for the pin-ended columns on the four-level interchange between the M4 and M5 motorway at Almondsbury have been described and illustrated by Kerensky and Dallard (1968).

(c) Rubber bearings

There are basically two types of rubber bearing: the laminated bearing and the pot bearing. Both are relatively new developments, the first of the accepted laminated bearings being developed in 1955 and the pot bearing in 1959. The majority of British bearings are manufactured from natural rubber but synthetic rubber, such as neoprene, has been used extensively in other countries. The detailed design of rubber bearings is beyond the scope of this text and the interested reader should refer to the papers by Black (1962b), Lindley (1962) and Taylor (1965). Department of Transport (1976) Technical Memorandum (Bridges) BE 1/76 provides rules for checking the suitability of laminated rubber bearings for use on highway bridges.

A laminated bearing consists of a number of layers of rubber sandwiched between steel plates and bonded to them (Fig. 10.23f). This arrangement prevents the rubber from bulging excessively when subject to vertical load and thus greatly increases its stiffness in this direction. The deformation due to shear force is, however, virtually unchanged and in this respect the bearing acts as if it were a rubber block. If required, greater resistance to shear in one direction can be introduced by making the steel plates slightly V-shaped instead of flat. In this way resistance to lateral loads can be provided while at the same time permitting longitudinal and lateral thermal movements. Laminated bearings may accommodate small rotational movements by means of differential vertical deflection, provided that such movements do not cause net tensile strains at the edge of the bearing. When the shear deformation is limited to about 10°, the shear resistance is equivalent to a coefficient of friction for a sliding bearing of about 0.07. If dowel bars are not provided to resist sliding at the upper and lower contact surfaces, the coefficient of friction can be taken as 0.33 between rubber and concrete and 0.2 between rubber and steel.

A rubber pot bearing consists of a circular slab of rubber confined within a shallow steel cylindrical pot by means of a thin steel piston (Fig. 10.23e). The edge of the piston is bevelled so that it may tilt without jamming against the sides of the cylinder. To prevent extrusion of the rubber through the small gap between the piston and cylinder wall, brass piston rings are fitted. The totally encased rubber tends to behave like a confined liquid. That is, it allows the pot to tilt quite freely while permitting only fractional vertical displacement even under high loads. Since the rubber is completely confined it can tolerate bearing stresses of about 30 MN m^{-2}. Even when subjected to relatively large rotations the bearing stress distribution in the rubber is almost uniform because of the low resistance to tilting of the piston. The thickness of rubber required for a bearing of diameter d is about $d/8$ but this can be reduced to $d/15$ if teflon sliding surfaces are provided above and below the pad to permit sideways movement of the rubber when subject to differential compression. The advantages of these bearings are, therefore, high load capacity, low resistance to rocking, good load spreading characteristics and small constructional height. They are used for loads of 1 to 20 MN and the manufacturers claim that they are more economical than rocker bearings for loads greater than 4 MN. Where translation is to be accommodated in addition to rotation, a PTFE and stainless steel sliding bearing may be used in conjunction with a rubber pot bearing (Fig. 10.23h). A description of the rubber pot bearing and a guide to design are given by Black (1962).

Although rubber bearings have been used extensively in Great Britain in recent years, none have been in service long enough to test their long-term performance. In view of this uncertainty about durability it is usual to make provision for their replacement within the life of the bridge. Since excessive strain is detrimental to long life it is important that the bearings should be designed for the appropriate temperature range (Section 10.3.1(d)) and accurately bedded and positioned.

Rubber is subject to creep but it is considered that the major part of the creep deformation will be attained by the time the superstructure is completed. Oxygen and ozone have deleterious effects on rubber but inhibitors are introduced in manufacture as a preventive measure. Direct sunlight may also be injurious to rubber but most bearings are protected by the superstructure. Oil is also injurious to rubber. At low temperature, rubber, particularly some synthetic varieties, becomes stiff due to crystallization. It is, however, generally accepted that a laminated rubber bearing will have a working life of at least 25 years and that the working life of a rubber pot bearing may be considerably longer than this because of its enclosed nature.

(d) Concrete hinges

Since the start of the twentieth century four types of concrete hinge have been used in bridge design (Fig. 10.23j). The first type was a sliding hinge which was used in the construction of three hinged arches. It consisted of a member with a convex cylindrical surface registered in a concave cylindrical cavity in the adjacent member. The two cylindrical surfaces were of the same radius and were often separated by a thin layer of lead which was introduced to reduce sliding friction and to provide better fit between the surfaces. The second type, commonly known as a saddle bearing, was similar in form to the first but used unequal radii so that rotation produced a rolling as opposed to a sliding motion.

Later Mesnager introduced a hinge which made use of reinforcing bars which crossed over each other in a small gap left between the members. The bars were often protected from corrosion by surrounding them with mortar but the mortar was not considered to contribute to the strength of the hinge. Finally, it was realized that the mortar in the throat of the Mesnager hinge contributed considerably to the strength of the hinge and this led to the development of the so-called Freyssinet hinge, in which little or no reinforcement passes through the throat. Freyssinet believed that by thinning a member down to a narrow throat the concrete at the throat would crack, thus producing two perfectly fitting surfaces which would provide ideal conditions for bearing and rocking. Cracking, however, did not occur, but the hinge was nevertheless highly successful and has since been extensively investigated. Freyssinet hinges have been used for loads up to $6.5 \, \text{MN m}^{-1}$ and may feasibly be used for loads of twice this value.

Base (1965) has described the results of a test programme carried out by the Cement and Concrete Association into the performance of four types of concrete hinge; two Freyssinet, one Mesnager and one saddle bearing. It was from the results of these tests that the Ministry of Transport compiled their rules for the design and use of Freyssinet concrete hinges in highway structures (Ministry of Transport, 1966, Memorandum 577/1). Prior to the publication of this document, the West Riding County Council had developed their own set of empirical rules for the design of concrete hinges. Sims and Bridle (1964) described these rules and the considerations which led to their adoption.

In each of the three basic types of hinge tested by the Cement and Concrete Association the load is channelled through a narrow section. It is this convergence of compressive forces that gives the throat of the Freyssinet hinge its strength, because in this way the concrete in the throat tends to be placed in three-dimensional compression; an ideal condition for a material which is weak in tension. In fact the concrete in the throat can withstand stress many times greater than the cube strength.

The convergence of the compressive forces towards the throat does, however, place the members on either side of the throat in transverse tension. In order to contain this tension and thus prevent the members from splitting, it is necessary to provide transverse reinforcing mats. In the case of Freyssinet hinges Memorandum 577/1 defines 'tension zones' which extend above and below the throat for a distance equal to the width of the member. A primary tensile force equal to three-eighths of the resultant of the direct and shear forces acting at the throat is assumed to act across the width of each tension zone. A secondary tensile force is assumed to act horizontally and at right angles to the primary force with a magnitude one-third of its value.

The horizontal reinforcing mats must be contained within the tension zones and should be concentrated as far as possible in the half nearest the throat. Rectangular stirrups around the perimeter of the member should not be used since these tend to deform to a circular shape, thus permitting bursting of the member. Primary and secondary reinforcing mats should be either bent from a single length of steel, bent alternatively left and right, or constructed as an orthogonal mat with welded joints. The first mat on either side of the throat should be as close as possible to the throat. The ends of the throat should be recessed to prevent spalling at the ends and to assist the development of compression along its length.

Reinforcement through the throat of a Freyssinet hinge is generally unnecessary and sometimes undesirable. However, the shear resistance of the throat, although quite adequate for normal loading, may benefit from the inclusion of crossed bars where there is any possibility of impact shear. Such reinforcement should not, however, exceed 3% of the cross-sectional area of the throat. MOT Memorandum 577/1 recognizes the weakness that Freyssinet hinges have with respect to impact loading and requires that they shall not be used where there is risk of collision with the structure which might cause damage or displacement of the hinge.

When designing the proportions of the throat there are two basic considerations. Firstly, the cross-sectional area should be sufficient to enable the concrete to safely withstand the load. The permissible compressive stress in the concrete, as defined in Memorandum 577/1, is twice the specified works cube strength at 28 days or 103 N mm^{-2}, whichever is the lesser. Secondly, the hinge should be able to provide the required rotations with the minimum of resistance and without causing tensile stress in the throat. To achieve this, the ability of the concrete in the throat to withstand hinge compressive stresses should be fully exploited.

Fatigue tests equivalent to hundreds of years of life carried out on the four types of hinge tested by the Cement and Concrete Association (Base, 1965) did not cause significant damage to any type but spalling cracks did appear in the Masnager and saddle hinges, which in practice could have resulted in corrosion of the transverse reinforcement.

During construction, hinges should be supported and care taken to ensure that they are not subjected to rotation before completion of the structure. The Freyssinet hinges used to form the pin ends to the inclined legs of the Wentbridge Viaduct, near Pontefract, were designed as precast units and were fully tested as part of the Cement and Concrete Association research programme prior to construction (Base, 1960).

Finally, concrete hinges require no maintenance and perform well under load, particularly when there is little or no reinforcement through the throat. They are relatively simple and cheap to construct.

10.9 DESIGN EXAMPLE. CALCULATION OF DESIGN MOMENTS AND FORCES FOR A CANTILEVER ABUTMENT

The following simple example illustrates the limit state method of evaluating design moments and forces for a cantilever abutment. These calculations are in accordance with the Code of Practice for bridges (BS 5400) as outlined in the text of this chapter.

10.9.1 Basic specification

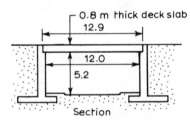

Section

Plan

Retained earth (non-cohesive):
  unit weight $1900 \, \text{kg m}^{-3}$,
  angle of shearing resistance $\phi = 35°$.

Foundation:
  allowable bearing capacity $= 250 \, \text{kN m}^{-2}$,
  coef. of friction (conc/soil) $= 0.3$.

Concrete:
  unit weight $2400 \, \text{kg m}^{-3}$.

Dead weight of deck:
  $1680 \, \text{kg m}^{-2}$ (structural),
  $240 \, \text{kg m}^{-2}$ (non-structural).

Bearings:
  stainless steel/PTFE ($\mu = 0.03$).

10.9.2 Loading (characteristic forces per metre width of abutment)

Lateral earth stresses in accordance with CP2 (ISE, 1951) (see also Chapter 6).
  $\phi = 35°$. $\delta = 0$ at actual back and virtual back of wall because of traffic vibration.

$\therefore$ $K_a = 0.27$ and $p = 0.27 \times 1900 \times 10^{-3} \times 9.81 \times h = \underline{5.03 \, h \, \text{kN m}^{-2}}$.

Dead load from deck $= 1680 \times 10^{-3} \times 9.81 \times 1 \times 12.9/2$
$= \underline{106.5 \, \text{kN m}^{-1}}$.

Superimposed dead load from deck $= 240 \times 10^{-3} \times 9.81 \times 1 \times 12.9/2$
$$= \underline{15.2\,\text{kN}\,\text{m}^{-1}}.$$

Live load.
    Max. reaction due to one lane of HA loading

$$= 30 \times \frac{12.9}{2} + 120 \times \frac{12.75}{12.6} = 315\,\text{kN}.$$

    or $315/11.3 = \underline{27.8\,\text{kN}\,\text{m}^{-1}}$.

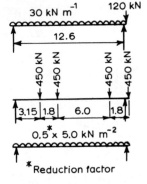

    Max. reaction due to 45 units of HB loading

$$= \frac{450}{12.9}(3.15 + 4.95 + 10.95 + 12.75) = 1135\,\text{kN}$$

    or $1135/11.3 = \underline{100.3\,\text{kN}\,\text{m}^{-1}}$.

    Max. reaction due to footway loading (both footways)

$$= \frac{12.9}{2}(0.5 \times 5.0) \times 2 \times 2 = 64.5\,\text{kN}$$

    or $\underline{5.7\,\text{kN}\,\text{m}^{-1}}$.

    Max. live load reaction (1 HA, 1 HB and 2 footways)
        $= 27.8 + 100.3 + 5.7 = 133.8\,\text{kN}\,\text{m}^{-1}$.

    Max. braking force: HA $= (200 + 8 \times 12.9)/11.3 = \underline{26.8\,\text{kN}\,\text{m}^{-1}}$.
                        HB $= 0.25(4 \times 450)/11.3 \quad = \underline{39.8\,\text{kN}\,\text{m}^{-1}}$.

    Earth surcharge to allow for live load on embankment (i.e. 0.6 m surcharge)
        $= 0.6 \times 1900 \times 10^{-3} \times 9.81 \ = \underline{11.2\,\text{kN}\,\text{m}^{-2}}$.

10.9.3 Choice of overall proportions

The next step is to choose an abutment cross-section which satisfies the stability requirements and produces an approximately uniform bearing stress distribution when subjected to the permanent loads (i.e. dead plus superimposed dead loads). These calculations are based upon CP2 as recommended in Section 10.4.1 (see also Chapter 6) and on characteristic loads without modification.

Try the followinng section:

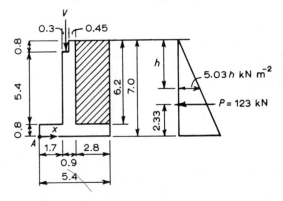

	W	x	M
Base	101.8	2.70	274
Stem	114.5	2.15	246
Curtain wall	8.5	2.38	20
Total	224.8		540
Earth fill	324.0	4.00	1296
Lat. force E			− 286
Total	549.8		1550
V (dead)	106.5	2.00	213
V (sup. dead)	15.2	2.00	30
Total	670.5		1793

$\bar{x} = 1793/670.5 = 2.68\,\text{m}$, $e = 2.70 - 2.68 = 0.02\,\text{m}$.

Factors of safety:

$$\text{against overturning} = \frac{\text{restraining moment}}{\text{overturning moment}} = \frac{2079}{286} = 7.26$$

$$\text{against sliding (friction only)} = \frac{\mu W}{P} = \frac{0.3 \times 670.5}{123} = 1.64.$$

(NB CP2 requires a factor of safety against overturning $\geqslant 2.0$ and a factor of safety against sliding, based upon base friction and passive resistance, approximating to 2. However, where resistance to sliding is assumed to be provided by base friction alone, a value $\geqslant 1.5$ is often used.)

Therefore stability is satisfactory.

Bearing stresses:

$$\text{Stress at edges} = \frac{W}{B}\left(1 \pm \frac{6e}{B}\right) = \frac{670.5}{5.4}\left(1 \pm \frac{0.12}{5.4}\right)$$

$$= 126.8\,\text{kN}\,\text{m}^{-2} \text{ and } 121.4\,\text{kN}\,\text{m}^{-2}$$

Therefore bearing stress distribution is satisfactory.

10.9.4 Adequacy of foundation

Since CP2004 *Foundations* (BSI, 1972d) is not based upon limit state principles the maximum bearing stress must be calculated using characteristic loads without modification. The allowable bearing stress as defined by CP2004 is based upon considerations of both settlement and bearing failure.

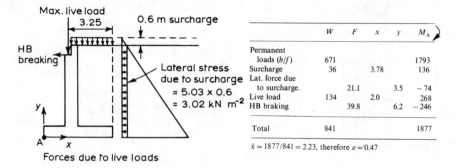

Forces due to live loads

$$\text{Bearing stresses} = \frac{841}{5.4}\left(1 \pm \frac{6 \times 0.47}{5.4}\right) = 237 \, \text{kN m}^{-2} \text{ and } 74 \, \text{kN m}^{-2}$$

These are less than $250 \, \text{kN m}^{-2}$, therefore satisfactory.

10.9.5 Determination of design moments and forces

(a) Stem (overall height 6.2 m)

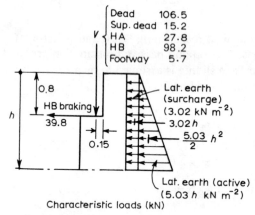

Characteristic loads (kN)

		Serviceability		Ultimate	
		γ_s	Des. load	γ_s	Des. load
Dead		1.0	106.5	1.15	122.5
Sup. dead		1.2	18.1	1.75	26.4
HA		1.2	33.4	1.5	41.7
HB		1.1	110.3	1.3	130.4
Footway		1.0	5.7	1.5	8.5
Therefore max. V			274.0		329.5
HB braking		1.1	43.8	1.3	51.7
Lat earth forces:					
(surcharge)		1.2	$3.63h$	1.5	$4.53h$
(active)		1.0	$2.52h^2$	1.5	$3.77h^2$

Ultimate limit state.

Design BM at depth $h = \gamma_{f3}(329.5 \times 0.15 + 51.7[h - 0.8] + 4.53h^2/2 + 3.77h^3/3) \, \text{kN m}$

Max. design BM ($h = 6.2$, $\gamma_{f3} = 1.15$) $= 822 \, \text{kN m}$

Max. design shear force $= 1.15(51.7 + 4.53 \times 6.2 + 3.77 \times 6.2^2) = 258 \, \text{kN}$

Max. design axial force $= 1.15(326.8 + \gamma_s \times \text{self wt of stem})$
$= 1.15(326.8 + 1.15(114.5 + 8.5))$ $= 538 \, \text{kN}$

Serviceability limit state.

Design BM at depth $h = \gamma_{f3}(274.0 \times 0.15 + 43.8[h - 0.8]$
$$+ 3.63h^2/2 + 2.52h^3/3)\,\text{kN m}$$

Max. design BM ($h = 6.2$, $\gamma_{f3} = 1.0$) = $\underline{548\,\text{kN m}}$

Max. design shear force = $1.0(43.8 + 3.63 \times 6.2 + 2.52 \times 6.2^2) = \underline{164\,\text{kN}}$

Max. design axial force = $1.0(274.0 + 1.0 \times \text{self wt. of stem})$ = $\underline{395\,\text{kN}}$

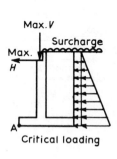

Max. V

Max. H

Critical loading

		Characteristic		Serviceability			Ultimate		
		W	M_A	γ_s	W	M_A	γ_s	W	M_A
Concrete section		224.8	540	1.0	225	540	1.15	259	621
Earth (self wt)		324.0	1296	1.0	324	1296	0.85	275	1100
Surcharge		36.0	136	1.2*	43	163	1.5*	54	204
Lat. earth forces:									
(active)			−286	1.0		−286	1.5		−429
(surcharge)			−74	1.2*		−89	1.5*		−111
Dead		106.5	213	1.0	107	213	1.15	123	245
Sup. dead		15.2	30	1.2	18	36	1.75	27	53
V HA		27.8		1.2	32.2		1.5	41.7	
HB		100.3	268	1.1	110.5	296	1.3	130.4	361
Footway		5.7		1.0	5.7		1.5	8.5	
HB braking			−246	1.1		−271	1.3		−320
Total					865	1898		919	1724

(*Assumed to be equivalent to HA live load.)

$\bar{x} = 2.19$ $\bar{x} = 1.88$
$e = 0.51$ $e = 0.82$

Ultimate limit state ($\gamma_{f3} = 1.15$)

$$\text{Bearing stresses} = \frac{919}{5.4}\left(1 \pm \frac{6 \times 0.82}{5.4}\right) = 324\,\text{kN m}^{-2} \text{ and } 15\,\text{kN m}^{-2}$$

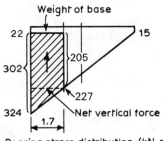

Weight of base

22

302

324

15

205

227

Net vertical force

1.7

Bearing stress distribution (kN m^{-2})

Self wt of base = $1.15(0.8 \times 2400 \times 10^{-3}$
$$\times 9.81)$$
$$= 1.15(19) = 22\,\text{kN m}^{-2}$$

Max. design BM = $1.15(205 \times 1.7^2/2$
$$+ 97 \times 1.7^2/3)$$
$$= \underline{448\,\text{kN m}}$$

Max. design shear force = $1.15(302$
$$+ 205)1.7/2$$
$$= \underline{496\text{kN}}$$

Serviceability limit state ($\gamma_{f3} = 1.0$)

$$\text{Bearing stresses} = \frac{865}{5.4}\left(1 \pm \frac{6 \times 0.51}{5.4}\right) = 250\,\text{kN m}^{-2} \text{ and } 69\,\text{kN m}^{-2}$$

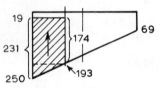

Bearing stress distribution (kN m^{-2})

Self wt of base $= 1.0(19) = 19\,\text{kN}\,\text{m}^{-2}$

Max. design BM $= 1.0(174 \times 17^2/2$
$+ 57 \times 1.7^2/3)$
$= 306\,\text{kN}\,\text{m}.$

Max. design shear force $= 1.0(231$
$+ 174 \times 1.7/2$
$= 344\,\text{kN}.$

(c) Heel (sagging moments and shears)

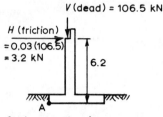

Critical loading (assumes that
deck is placed before backfill)

	Charac-teristic		Serviceability			Ultimate		
	W	M_A	γ_s	W	M_A	γ_s	W	M_A
Concrete section	224.8	540	1.0	225	540	1.15	259	621
V(dead)	106.5	213	1.0	107	213	1.15	123	245
H(friction)		19	1.0		19	1.5		29
Total				332	772		382	895

$\bar{x} = 2.32$ $\bar{x} = 2.34$
$e = 0.38$ $e = 0.36$

Ultimate limit state ($\gamma_{f3} = 1.15$).

$$\text{Bearing stresses} = \frac{382}{5.4}\left(1 \pm \frac{6 \times 0.36}{5.4}\right) = 99.0\,\text{kN}\,\text{m}^{-2} \text{ and } 42.5\,\text{kN}\,\text{m}^{-2}$$

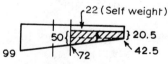

Bearing stress distribution (kN m^{-2})

Max. design BM $= 1.15(20.5 \times 2.8^2/2$
$+ 29.5 \times 2.8^2/6)$
$= 137\,\text{kN}\,\text{m}\,\text{(sagging)}$

Max. design shear force
$= 1.15(20.5 + 50)2.8/2$
$= 113.5\,\text{kN}$

Serviceability limit state ($\gamma_{f3} = 1.0$)

$$\text{Bearing stresses} = \frac{332}{5.4}\left(1 \pm \frac{6 \times 0.38}{5.4}\right) = 87.5\,\text{kN}\,\text{m}^{-2} \text{ and } 35.5\,\text{kN}\,\text{m}^{-2}$$

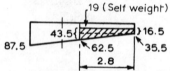

Bearing stress distribution (kN m^{-2})

Max. design BM $= 1.0(16.5 \times 2.8^2/2$
$+ 27 \times 2.8^2/6)$
$= 100\,\text{kN}\,\text{m}\,\text{(sagging)}$

Max. design shear force
$= 1.0(43.5 + 16.5)2.8/2$
$= 84\,\text{kN}$

(d) Heel (hogging moments and shears)

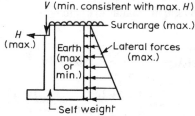

Critical loading system
(This diagram indicates both the nature
of the forces and the appropriate values of γ_{fL})

In this case it is not immediately apparent which load system is the most critical. Examination of the possible load systems shows that:

(i) The value of γ_{fL} chosen for the self weight of the earth fill for the limit state of collapse (i.e. 0.85 or 1.3) has little or no effect upon the net load,

(ii) The live load should be chosen to produce the minimum vertical reaction (V) which is consistent with H being equal to the maximum braking force,

(iii) The application of surcharge increases the net load on the heel.

The minimum value of V consistent with H being equal to the maximum braking force is obtained by applying two axles of HB loading to the expansion end of the deck. This produces a relatively small vertical reaction at the fixed end while still permitting the application of the full HB braking force.

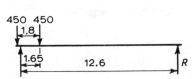

$$R = \frac{450}{12.6}(1.65 - 0.15) = 53.5\,\text{kN}$$

R or $53.5/11.3 = 4.7\,\text{kN}\,\text{m}^{-1}$

	Characteristic		Serviceability			Ultimate		
	W	M_A	γ_S	W	M_A	γ_S	W	M_A
Concrete section	224.8	540	1.0	225	540	1.15	259	621
Earth (self wt)	324.0	1296	1.0	324	1296	1.3	421	1685
Surcharge	36.0	136	1.2	43	163	1.5	54	204
Lateral earth forces:								
(active)		− 286	1.0		− 286	1.5		− 429
(surcharge)		− 74	1.2		− 89	1.5		− 111
V { Dead	106.5	213	1.0	107	213	1.15	123	245
Sup. dead	15.2	30	1.2	18	36	1.75	27	53
HB (min.)	4.7	9	1.1	5	10	1.3	6	12
HB braking		− 246	1.1		− 271	1.3		− 320
				722	1612		890	1960
				$\bar{x} = 2.23$			$\bar{x} = 2.20$	
				$e = 0.47$			$e = 0.50$	

Ultimate limit state ($\gamma_{f3} = 1.15$)

$$\text{Bearing stresses} = \frac{890}{5.4}\left(1 \pm \frac{6 \times 0.50}{5.4}\right) = 257\,\text{kN m}^{-2} \text{ and } 73\,\text{kN m}^{-2}$$

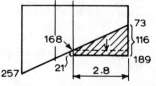

Bearing stress distribution (kN m^{-2})

Downward load on heel (kN m^{-2})

$$
\begin{aligned}
\text{Surcharge} &= 1.5(0.6 \times 1900 \times 10^{-3} \\
&\quad \times 9.81) \\
&= 1.5(11.2) &&= 17 \\
\text{Earth} &= 1.3(6.2 \times 1900 \times 10^{-3} \\
&\quad \times 9.81) \\
&= 1.3(115.5) &&= 150 \\
\text{Base slab} &= 1.15(0.8 \times 2400 \times 10^{-3} \\
&\quad \times 9.81) \\
&= 1.15(18.8) &&= \underline{22} \\
& &&\underline{189}
\end{aligned}
$$

Max. design BM $= 1.5(21 \times 2.8^2/2 + 95 \times 2.8/3) = \underline{380\,\text{kN m}}$ (hogging)

Max. design shear force $= 1.15(21 + 116)2.8/2 \quad = \underline{220\,\text{kN}}$

Serviceability limit state ($\gamma_{f3} = 1.0$)

$$\text{Bearing stresses} = \frac{722}{5.4}\left(1 \pm \frac{6 \times 0.47}{5.4}\right) = 204\,\text{kN m}^{-2} \text{ and } 64\,\text{kN m}^{-2}$$

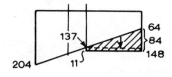

Downward load on heel
$$
\begin{aligned}
&= 1.0(18.8) + 1.0(115.5) + 1.2(11.2) \\
&= 148\,\text{kN m}^{-2}
\end{aligned}
$$

Max. design BM $= 1.0(11 \times 2.8^2/2$
$$
\begin{aligned}
&\quad + 73 \times 2.8^2/3) \\
&= 234\,\text{kN m} \text{ (hogging)}
\end{aligned}
$$

Max. design shear force $= 1.0(11 + 84)2.8/2$
$$= 133\,\text{kN}$$

(e) Curtain wall

Assume that one axle of HB braking (i.e. 225 kN) may be applied at the top of the wall in either direction.

Therefore max. characteristic BM $= 225 \times 0.8 = \underline{180\,\text{kN m}}$

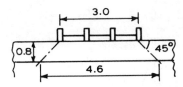

If the effect of the braking force is distributed transversely down from the outer wheels, the effective width of the section resisting the bending moment at the bottom of the curtain wall is 4.6 m. Therefore maximum characteristic BM per metre width

$$= 180/4.6 = 39.2 \text{ kN m}$$

Ultimate limit state ($\gamma_{f3} = 1.15$)
Max. design BM $= \gamma_{f3} \times \gamma_{fL} \times 39.2 = 1.15 \times 1.3 \times 39.2 = 59 \text{ kN m}$

Max. design shear force $= \gamma_{f3} \times \gamma_{fL} \times 225/4.6 = 1.15 \times 1.3 \times 49 = 73 \text{ kN}$

Serviceability limit state ($\gamma_{f3} = 1.0$)
Max. design BM $= 1.0 \times 1.1 \times 39.2 = 43 \text{ kN m}$

Max. design shear force $= 1.0 \times 1.1 \times 225/4.6 = 54 \text{ kN}$

10.9.6 Summary of design moments and shears

(a) Ultimate limit state

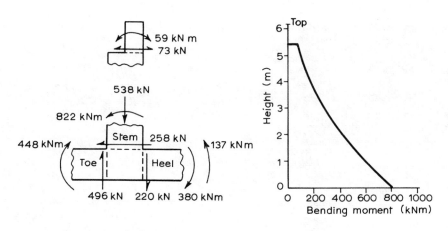

(b) Serviceability limit state

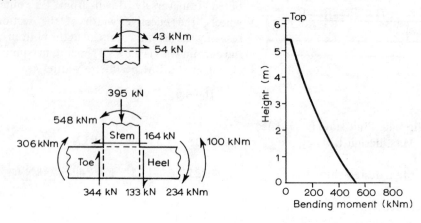

Chapter 11

MISCELLANEOUS
FOUNDATION PROBLEMS

K. Elson and D.A. Greenwood

INTRODUCTION

This chapter is concerned with problems which generally do not fall within the scope of other chapters though such problems are often linked with other parts of the text. A number of geotechnical processes are now available to facilitate construction in situations where a century ago such work would have been extremely dangerous, tedious or impossible to execute. Underpinning embraces a wide variety of foundation types, the principles of which are discussed in other chapters. Mobile foundations are required when structures are moved bodily. The depth of concrete placed in a single lift in foundations is often considerable and it is pertinent to discuss the pressure of fresh concrete on formwork. Some years ago the editor witnessed the failure of formwork for the 3 m high wall of a sunken reinforced concrete tank in Egypt. The concrete was placed in one lift but, although the formwork and bracing looked adequately strong, no analysis had been made of the forces to which it would be subjected. Accident prevention in foundation work is covered by legislation and publications dealing with construction in general. In the UK, legislation is set out in the 'Health and safety at work, etc., Act', Chapter 37, of 1974 and a guide to requirements in the construction industry has been provided by the Royal Society for Prevention of Accidents (RoSPA, 1976). General reviews of problems have been published in several papers and conference reports; for example, the Institution of Civil Engineers' 1980 conference (ICE, 1980b). Special problems, such as those associated with diving and work in compressed air and with hazardous conditions in sewers, are covered by more specific regulations and publications.

11.1 DE-WATERING

Almost all excavations require to be de-watered continuously by one means or another. There is not a great range of techniques available and some operate only under limited conditions. The methods of assessing the quantity of water flowing

885

into an excavation are discussed in Section 2.7. In practice the simplest methods are adopted where possible and, if the inflow is great, measures are taken to reduce the volume to manageable proportions.

Simple sumps are used extensively to collect water in an excavation so that it can be pumped out. Such a technique has limited applications and a careful check should be kept on the inflow to avoid damage arising from subsurface erosion. Sumps should be used only in situations where the hydraulic gradients induced by pumping are low and the factor of safety against boiling should be investigated. Should boiling occur, the resultant loss of ground can rapidly lead to collapse of the sheeting to the excavation and possible damage to adjacent structures (Marsland, 1953). Another cause of failure is internal erosion of fine soil particles and, over a period of months, the loss of material may result in substantial subsidence at considerable distances from the excavation. Subsurface erosion is difficult to detect and the likelihood of its occurrence can best be assessed from the grading of the soil. In rock excavation, joint filling may be washed out, thereby reducing the strength of the rock mass.

Placing concrete or waterproof membranes of, for example, asphalt in excavations drained by a sump is undertaken by normal techniques, although it is usual to place a drainage layer on the formation and to pipe discrete inflows to the sump. Placing structural concrete over a sump presents a problem, since pumping must continue. A commonly adopted technique is to insert pipes into the sump and rapidly fill the area with concrete; the water overflowing from the pipes. This procedure is satisfactory provided the head of water beneath the fresh concrete does not build up and wash out the cement. In situations where the inflow is large and an appreciable head may be developed, high-capacity borehole pumps can be installed in a steel casing passing through the concrete. After the concrete has set it is usual for the area around the sump to be grouted to fill any remaining cavities.

For excavations at any depth below the permanent water table, simple sumps are unsuitable unless the inflow is reduced to acceptable levels by grouting, sheetpiling or diaphragm walls. In gravel soils little can usually be done, other than to seal the area to be excavated, since the volume of water to pump is frequently unmanageable. In sands, various types of well can be used with success (Ward, 1957) but difficulties arise in attempting to de-water excavations in silt as the effective radius of any well is small and the spacing of wells becomes uneconomically close. Indeed silty soils are the most intractable.

A satisfactory technique for de-watering sands and gravels is to sink deep wells around the site and install high-capacity borehole pumps to lower the water table below the level of the proposed excavation. The capacity of such a system is limited only by the depth of the boreholes and the capacity of the pumps employed. The boreholes for such an installation are lined with steel casing and the lower slotted portion is wrapped with glass-fibre cloth to prevent internal erosion of the soil. For semi-permanent installations graded gravel packs can be used to prevent piping; such packs require large diameter boreholes. The design of gravel packs should be in accordance with the filter rules specified in Section 2.7.3.

Well-point systems are a flexible means of de-watering fine sands and silts where the quantities of water to be pumped are not large. A well-point system comprises a ring main connected to combined centrifugal and vacuum pumps. The purpose of the vacuum pump is to increase the vaccum developed by the centrifugal pump and ensure that it is self priming. A series of small diameter tubes, the well-points, are installed at 1 to 3 m intervals along the header pipe. The suction head developed in the header pipe draws water and some air from the well points. The system is particularly suited for de-watering shallow open excavations, the maximum reduction of water level that can be reliably achieved is about 5 m. Deep excavations can be de-watered by employing multiple well-point systems installed at about 5 m vertical intervals (Fig. 11.1). The success of the system depends on maintaining a high vacuum as wells drawing air or leaking joints rapidly decrease the efficiency of the whole system.

A well-point comprises a steel tube, 35 to 50 mm diameter, with a perforated sleeve about a metre long at the lower end. In silts the perforated section may be wrapped with a filter cloth to exclude soil particles. The tubes are normally driven to the required depth but jetting may be used as an alternative.

Where the flow of water per well-point is small, say less than 1 litre s^{-1}, jet ejector pumps can be used. Such pumps are relatively inefficient but can produce a vacuum of 3 m head and a lift of 30 m: hence a single deep well may be used as an alternative to a multiple well-point system. Each jet ejector can be powered by a separate high-pressure centrifugal pump or a high-pressure ring-main can be installed to serve several pumps.

An alternative system to individual vertical well points is to lay a perforated pipe in a trench around the area to be de-watered. Modern trenching machines can place pipes at depths to 4 m.

Well-point systems can be designed using the formulae given in Section 2.7.4, assuming that the system is the equivalent of a continuous slot (Example 2.17) (Chapman, 1957). The sizes of pipe work should be generous to reduce friction losses and the length of a vacuum header should not in general exceed 200 m. The

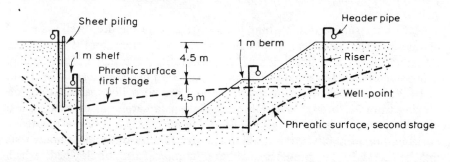

Figure 11.1 Multiple well-point arrangements with sheeted and unsheeted excavations.

well-point pump should be sited as near to the existing groundwater level as possible to increase the effectiveness of the system. It is usual to design each ring of a multiple system to cope with the entire flow from the excavation since it may be possible to stop pumping the upper well-points as the excavation is deepened but a check should be made to ensure that the water surface is well below the floor of the excavation. An advantage of the well-point system is that it can be rapidly extended to cope with variations in the soil conditions.

Vacuum wells can be used to stabilize silts, in which deep pumping wells and normal well-points are not very effective. The object of a vacuum well is not so much to pump water from the strata but to establish an area of low pore pressure and hence increase the effective stress in this region, with a consequent increase in the strength of the soil. Vacuum wells normally employ separate pumps to create the vacuum and to pump the small inflow out of the well. The lower part of the well comprises a slotted screen and sand filter, whilst the upper portion is grouted into the borehole to provide an air tight seal. Such wells may be very effective but are limited in application to a restricted range of soil types.

The effect of electro-osmosis can be used to increase the rate of drainage from fine or organic silts, which cannot be de-watered by other techniques (Casagrande, 1949). Again this process has limited application, and requires a large power consumption, consequently the technique is rarely adopted. Where possible the problem is circumvented by the use of diaphragm walls or sheet pile cofferdams. In the electro-osmosis technique, anodes are installed in the ground at 8 to 12 m centres and well-points are used as cathodes spaced equally between the anodes. The discharge from the well is small, ranging from 1 to 50 litre day^{-1}, and intermittent pumping may be all that is required to drain the wells. The applied voltage depends on the conductivity of the pore fluid and is usually in the range 30 to 100 V d.c. The current consumed varies between 10 and 30 A per well and the power consumption is therefore high. The discharge from a well can be estimated from the equation

$$Q = k_e i_e a Z \qquad (11.1)$$

where, k_e is the osmotic permeability (about 5×10^{-7} m s^{-1}) (Casagrande, 1949), i_e is voltage gradient in V m^{-1}, a is spacing of the wells, and Z is depth of soil being stabilized.

Multi-stage submersible turbine pumps are used in the deep well system of dewatering. The system can be designed to reduce the uplift on the floor of the excavation or to completely dewater the excavation. The shape of the cone of depression of the wells should be accurately assessed to enable a suitable layout to be planned. Such wells are expensive and are normally spaced at centres ranging from 20 to 50 m. For large installations a careful assessment of soil conditions must be made. In this respect it is important to correctly estimate the horizontal permeability of the strata. Field permeability and trial pumping tests may be necessary. The presence of either thin gravel lenses or clay beds, which may not be detected by normal boring techniques, can have a major effect on the discharge and drawdown characteristics of the soil. Adequate precautions should be taken to provide sufficient standby plant to cope with emergencies.

Large schemes necessarily lower the water table over extensive areas and, although normally temporary, the effect on local water abstraction may be serious. Furthermore, the reduction in water-level results in an increase in effective stress in the soil mass, with resultant settlement. The effect of likely settlement on adjacent properties should be carefully considered when planning a de-watering scheme.

11.2 DRAINAGE AND PRELOADING

With the greater intensity of development, less suitable sites are having to be used for industrial and housing projects. Many such sites are on low-lying alluvial flats or on infilled quarries. Port development, in particular, frequently involves designing heavy structures founded on weak compressible soils. The construction of high embankments and heavy buildings on deep alluvial deposits requires that a careful assessment of the effect of consolidation and secondary settlements be made. The effects of large deformations on structures are frequently unacceptable and various expedients may be adopted to minimize these. For large structures and embankments total settlements up to 1 m are not uncommon. Expedients range from piling and replacement of compressible soil, which avoid the problem, to preloading and artificial drainage. The various geotechnical methods of minimizing the effects rely on accelerating settlement such that the greater part of the movement is complete by the time the structure is put into operation.

Preloading the site to the full working ground stress by placing a temporary fill on the site is a satisfactory means of reducing post-constructional settlement and may conveniently be incorporated in land reclamation projects. The method is simple and reliable but the importation and removal of the required quantities of fill is expensive and the length of time for a reasonable proportion of the settlement to take place is crucial to the success of the operation. A typical project of this nature was adopted at Port Victoria, Seychelles (Pannett, 1974). The main quay structure was piled to bedrock at 35 m below sea-level but it was not economic to pile the associated warehouses and factories to such great depths. Consolidation of the overlying coralline silts and clays under the weight of the superimposed fill and structures would be considerable. To reduce the settlement after commissioning of the project, a 3 m high mound was built of dredged sand at one end of the warehouse area. This mound was moved forward in stages so that the entire area to be developed was preloaded for 4 months. The sand was eventually used to reclaim a further area. Storage areas where settlement could be tolerated were not preloaded. Settlement gauges were installed on the sea-bed beneath the sand fill so that measurements of the rate of consolidation could be made.

For this type of project an accurate assessment of the rate of consolidation has to be made, since preloading must be planned ahead of the main construction phase. A thorough site investigation is necessary and an examination of the soil fabric in conjunction with field permeability tests enables a reliable value of the coefficient of consolidation to be estimated. Rowe (1964) has discussed the problem of predicting the rate of settlement in alluvial soils at some length. On major projects an instrumented trial embankment may be constructed prior to the award of the main contract to prove the design predictions. Fortunately, the coefficient of consolidation is usually underestimated at the design stage, rather than the reverse. The economics of preloading depend on the time required to complete an adequate proportion of the total settlement; usually 80 to 90% settlement is the design object of preloading. To achieve a shorter preload period the foundation may be surcharged above the design bearing pressure, although on weak alluvium this may result in instability of the soil beneath the preload embankment.

Shallow layers of very soft soil or peat may be displaced by overloading the embankment, since it is commonly formed by end tipping. The process may be assisted by firing small charges beneath the fill to eject the soft material. The application of this technique is limited and is seldom used today. The alternative to displacement is total replacement of the unsuitable soil. This technique was successful at Fawley, Hampshire, for the foundation of large oil tanks (Bratchell, *et al.* 1974).

Artificial methods of accelerating the rate of consolidation of soils are of limited use in foundation construction as they are usually applicable to relatively large areas. They involve the use of drainage blankets, sand drains or sand wicks. The main problem with the use of these methods is the uncertainty in predicting the effectiveness of the system. Sand drains comprise vertical boreholes backfilled with suitably graded sand. The boreholes may require casing when drilled in soft sediments. The construction technique should be such that disturbance of the surrounding soil and smearing of the borehole surface is reduced to a minimum since the remoulded layer is less permeable than the mass and greatly reduces the efficiency of the system. Wicks are formed by driving a mandrel into the soil and placing a permeable polypropylene woven or melded sheath, stiffened with a semi-rigid plastic core or alternatively sand-filled, within the hole. Drains or wicks are usually spaced between 2 and 6 m apart on a regular grid. The drains reduce the length of the drainage path and hence accelerate the rate of consolidation. They do not reduce the amount of settlement which will occur, nor do they physically reinforce the ground as would stone columns.

11.3 DEEP GROUND IMPROVEMENT BY VIBRATION AND TAMPING

foundations to structures on weak soil, the ground on which the foundation has to be either improved or by-passed by piling. Improving the ground is most economic solution, especially for comparatively light structures and housing.

Compaction of weak soils usually gives them higher strength and greater resistance to settlement and is thus the basis of an effective system of improvement. For loose, cohesionless sandy or gravelly soils, application of continuous vibration internally destroys intergranular friction and allows gravity to cause particles to pack to a denser state. Cohesive soils do not settle when vibrated but columns of sand or gravel can be formed in them by vibrating plant to act as reinforcement, analogous to compression steel in concrete. Tamping the ground surface compresses weak soil and also imposes vibrations.

Neither process is effective unless it influences a mass of soil to a depth equal to that which would be stressed significantly under foundation loads. Both intensity and plan dimensions of foundation loading establish this depth, which can be determined by the application of Boussinesq, Westergaard or other stress-distribution theories. In practical terms, for most buildings improvement to a depth of about 10 m is sufficient. There are always exceptions where greater depths are required, and individual circumstances must be reviewed according to the capability of improvement processes relative to soil characteristics.

11.3.1 Vibratory systems

Vibrating machines in the form of vertical pokers are generally excellent boring tools for weak soils in which rapid penetration is achieved. All such machines are usually freely suspended from a mobile crane and thus penetrate vertically.

There are two major types of vibratory probe but each may take a variety of forms. The difference lies in the location of the source of vibrations. Vibroflotation processes use a poker which contains an eccentric weight mounted at the bottom on a vertical shaft directly linked to a motor contained in the body of the machine. The vibratory motion is horizontal and the body cycles around the vertical axis. Vibratory displacements are applied directly to the ground through the tubular casing of the machine and the output from the machine remains constant whatever the depth of penetration.

The second type uses a casing or pipe driven by an oscillator mounted at the top and applying essentially vertical vibrations to the casing. Thus the vibratory motions are much modified in energy, form and amplitude as the machine penetrates more deeply.

These inherent differences lead to differing modes of practical application and results.

(a) Vibroflotation processes

The vibroflot is a poker of diameter 300 to 450 mm and length 2 to 3.25 m: it weighs 2 to 4 tonne according to type and purpose.

A flexible vibration damping connection to a follower tube of about 300 mm diameter provides for extension pieces to allow deep penetration into the ground. The follower carries power and water from the surface for jets in the nose cone and sides of the vibrator. Electric or hydraulic motors are used. Power development ranges from 35 to 160 kW, the most common range being 50 to 100 kW. Lateral impact force is usually 7 to 15 tonne, ranging from 5 to 30 tonne.

Vibration frequency is normally fixed and the operating range is 30 to 50 Hz. Amplitude, when hanging free, is about 5 to 10 mm, representing half the total displacement range. When the machine is constrained by the ground, it is much less.

To achieve maximum compaction of cohesionless soils, large amplitude is desirable. The design of a machine with this result is very complex, involving the relative weights and positions of moving and static parts and the power and speed of the machine and ground response. Some machines have variable frequencies but, if variation is allowed in the field, interpretation of measured ground response is difficult. Since this is an important part of quality control, constant speed is better than fine tuning to variable ground conditions.

Vibroflot techniques

The type of soil dictates how the machine is used. Compaction of loose, non-cohesionless granular soil can be achieved simply by inserting the machine at regular intervals all over the area of proposed footings and slowly withdrawing it. Usually, however, ground variations or content of fines, even in essentially cohesionless material, dictate that imported backfill is used to fill craters formed as the mass is compacted. Broadly, imported granular backfill is used wherever sand is fine and comparatively impermeable. Soil permeability must be greater than about 10^{-2} mm s^{-1} if compaction is to be achieved during the time vibrations are applied, since water must be expelled simultaneously. The cyclic shear strains induced by the machine break the frictional strength of the soil, allowing it to settle under gravity – but only if pore-water can be forced out.

As soils become finer and the clay content increases, this gravitational mechanism is progressively inhibited and the machine simply forms a hole in the ground without achieving compaction. If the hole is then filled with coarse gravel, compacted by the vibrator, the resulting gravel column lends vertical stiffness to the composite of soil and column. Groups of such columns make the ground stiffer and more resistant to settlement than in its natural state. This technique has been widely used in silty sands and glacial soils in the UK and Germany.

Penetration of the machine may be assisted by water jets which transport soil upwards and allow the machine to sink. Backfilling of gravel may be accomplished whilst this water flows up the annulus between the soil and machine in the so-called 'wet process' of column construction. If no water is used, then the machine simply displaces the soil and must be removed from the hole to allow the stone to be tipped into it. The machine is then lowered to compact the stone *in situ*. Repetitions of this process result in a 'dry' column.

Sometimes a small volume flow of compressed air is used in dry treatment to relieve suction pressures which would otherwise tend to create instability in a tight-fitting bore as the machine is withdrawn. Its value in assisting in penetration is minimal and an excessive amount of air at high pressure can damage the fabric of soft normally consolidated soils. Air should be used only where the soil remains essentially dry and stable in the borehole during the process of column construction. The wet process is safer if there is any doubt about this although it has the disadvantage of having to remove water overflowing the bore from the surface of the site which would otherwise inhibit traffic of tracked vehicles.

Water is always used in cohesionless sandy soils since compaction induced by vibration otherwise inhibits penetration. Water saturation also reduces effective stress in the soil so that vibratory stresses are able more easily to overcome frictional strength: the radius of compacted volume is thus extended. There is little surface overflow because water drains into the sand. In practice, penetration is assisted by low-pressure high-volume jets through the nose cone whilst during compaction only side jets are used. It would be impracticable to allow a high upward velocity of flow in the annulus during compaction as for stone column construction, as this would simply remove fine sands in the water stream. Thus to compact sands the side jets must be of low volume and high pressure directed to undercut the sides of the bore but not to provide sufficient upward velocity to remove the sand from the hole. Controlled manipulation of the water jets is an important facet of the sand compaction technique. In very deep holes (20 to 30 m), it may be necessary to incorporate additional water jets at intervals along the follower tube to maintain a small upward velocity against losses from the bore.

Compaction of sands to a radius of several metres requires time. It is important not to allow the machine to sink in the bore, giving rise to increased power demand and thus a false indication of satisfactory compaction. During withdrawal the machine is raised in increments of 0.3 m and allowed to rest for 0.5 to 1 min per interval, precise time being determined by initial empirical trial on site. The fastest compaction is obviously achieved when sand is supplied quickly to the tip of the machine to replace compacted volume either by side jetting or shovelling from the surface. Otherwise the machine hangs in a suspension of fluidized sand and cannot impact on the sides of the bore to continue compaction.

The wet or vibro-replacement process of stone column construction is used in soft, relatively impermeable cohesive soil with undrained strengths in the range of 15 to 50 kN m^{-2}. Such soils are easily penetrated with low-pressure large-volume bottom jets and any softened material is removed to surface. Resistant layers are penetrated by direct impact of the machine. This flow continues during backfilling with gravel. Thus the more sensitive soils in the immediate vicinity of the machine are weakened and continually removed and replaced with gravel. The column expands until the amplitude of vibration transmitted through the gravel is attenuated such that the shearing motion is insufficient to break down the soil further. As a result the remaining soil between columns is not significantly disturbed by the process and retains its original fabric. The resulting columns are typically 0.9 to 1.1 m diameter.

The dry or vibro-displacement process is more akin to driven piling and shears the soil. It is used in stable insensitive cohesive soils with undrained strength of the order 30 to $60\,kN\,m^{-2}$. Since the machine is a tight fit in the bore, column diameters are little more than that of the machine – about 0.6 to 0.7 m.

A recently developed German machine allows backfill feed direct to the tip of the vibroflot *in situ* by means of a tube filled with backfill strapped to the vibrator to discharge through the nose cone. Penetration is assisted by a pull-down arrangement on dedicated cranes to compensate for increased machine cross-section. The tube is provided with an air lock to allow compressed air to assist discharge of backfill.

For wet treatment, rounded or subangular gravel of 25 to 50 mm sizes is preferred since these pass most easily around the machine. Excavated columns show a much greater range of grading as the water used contains a suspension of the coarser materials from the natural soils which become blocked between the gravel particles as they settle to mutual contact under vibration.

With the vibro-displacement system, a well-graded backfill from 10 to 100 mm sizes of angular hard material is preferred to allow the best mechanical interlock and reduction of void spaces which are not otherwise filled in the process.

(b) Casing drivers

Vibratory equipment for driving pile casings has been used as a simple expedient for sand compaction in lieu of purpose-built plant. However, this has in some instances been developed specifically for compaction and ground improvement. The best known of these is the Japanese Compozer system which is a relatively sophisticated version of casing driver.

A pipe, usually 600 mm diameter, extended as necessary, is driven by means of a vibrorammer mounted on the top of the casing and isolated by a spring connection from the suspending crane. Penetration to 20 m or more in hydraulic fills, soft clays and silts with SPT less than about 15 is possible. The power of the oscillator is generally 90 to 120 kW, giving a vertical force of 40 to 60 tonne and 15 to 18 mm amplitude at about 10 Hz. Below the oscillator, the tube has a side-mounted airlock and feed tube through which sand backfill may be introduced during penetration under compressed air without removing the vibrator. A constricted casing shoe or type of cactus or clack valve at the toe, together with the compressed air above the fill, keeps the natural soil out as penetration proceeds. At the designed depth the sand is forced out by compressed air as the pipes are simultaneously withdrawn whilst vibrating, always keeping a plug of sand within the tube. The casing is then redriven by vibration into the sand which is compacted to some extent as a result and forced into the surrounding soil until equilibrium is reached, allowing the column to be built upwards in a succession of similar steps. The final columns are about 0.8 m diameter is soft soils or 0.6 m in firmer frictional materials.

A mammoth variant of Compozer uses 1.2 m diameter casings to form columns in the softest hydraulic fills of about 2 m diameter. This machine is usually operated over water in barge-mounted form. To improve the strength of the columns a 30 kW mini-vibroflot is sometimes introduced immediately below the toe of the tube to assist compaction during steady withdrawal. These large Compozer columns are not so compact since they are formed without direct lateral vibration.

Water is sometimes added above the backfill to assist its release and to increase density. Whilst not so stiff as a gravel column, this disadvantage is offset to some extent by the heavy ramming and preconsolidation of the surrounding natural soil. Usually locally available pit sands are employed, with grading in the fine to coarse sand ranges, uniform or not. In the larger mammoth process a certain amount of gravel is permissible.

Another form of top drive vibratory system is the vibrorod described by Saito (1977) and by Massarsch and Broms (1983). These have a pipe approximately 300 mm diameter, expanded by various external fins or protuberances which increase the diameter to about 1 to 2 m. A particular form of this was used in Holland on the Oosterschelde barrier where it had a dimension of 2.1 m across the fins (Davis *et al*, 1981). Simple tubes and vibrorods need to be assisted by internal water pipe jets to penetrate to the requisite depth.

Various claims have been made regarding the relative merits of casing drivers and vibroflotation processes but, in fact, results depend on specific equipment and the precise mode of operation in relation to each site. Valid comparisons can be made only if they are based on economic and technical evaluations relevant to a specific project. Operator technique can affect results and experienced crews are desirable. Trials for major projects are preferable to optimize techniques. The bulk of evidence suggests that the vibroflot is a more efficient compactor than casing drivers (Saito, 1977; Brown and Glenn, 1976; Schroeder and Bynington, 1972). Generally the latter penetrate three or four times faster than vibroflots but require about four times as many penetrations to achieve the same densification. Tentatively it may be concluded also that casing drivers do not achieve as high a density, regardless of spacing.

11.3.2 Soils responsive to vibratory compaction

It is axiomatic that preliminary site investigation has established a need for compaction in loose soils or those with varying density. The objective of compaction is to achieve a high and uniform density to minimize settlement – especially differential settlements – of structural foundations. It should be noted that varying densities often exist in nominally loose sands and that the resulting differential settlements may sometimes be of magnitude similar to the total settlements of structures. Sands that have been preloaded by natural or artificial overburden may not yield as much settlement as normally consolidated sands of the same penetration resistance: in such cases vibratory treatment destroys any existing particle orientation and compacts all the sand to a uniform density and fabric, giving a uniform settlement response.

As noted above, increasing fines content in sands tends to inhibit compaction

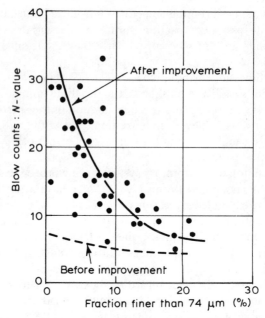

Figure 11.2 Compaction response with increasing silt content (after Saito, 1977).

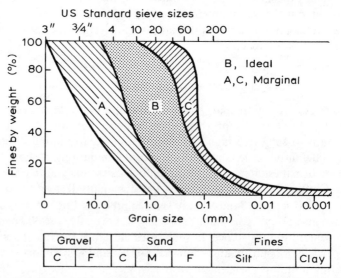

Figure 11.3 Soil grading suitable for vibro-compaction (after Mitchell, 1968).

(Fig. 11.2). Figure 11.3 shows a broadly acceptable range of gradings for vibroflotation compaction whilst Brown (1977) has defined a suitability number for vibroflotation given by

$$\text{Suitability Number} = 1.7\left[\frac{3}{(D_{50})^2} + \frac{1}{(D_{20})^2} + \frac{1}{(D_{10})^2}\right]^{1/2}, \qquad (11.2)$$

in which D_{50}, D_{20} and D_{10} are 50, 20 and 10% grain size diameters in mm. A low suitability number is better than a high one of 40 to 50, above which the ground is unsuitable. Compaction is quicker with lower numbers.

Whilst permeability below 10^{-2} mm s^{-1} inhibits compaction, a value above 10 mm s^{-1} tends to slow penetration because of loss of water. The coarser limit of grading for treatment depends on the characteristics and power of the machines. Gravels or cobbles without sand are more difficult to penetrate than a well-graded mixture. Usually, therefore, treatable soils require a sand content although, if cobbles and gravels are naturally loose, the greater power of the present generation of machines allows penetration of them for several metres. Brick demolition debris is regularly treated in this way without water. Gap grading with cobbles 'floating' in sand may present a penetration difficulty as the cobbles accumulate in the bottom of the bore to obstruct further progress.

Particle size distribution and penetration resistance are not alone sufficient for defining suitability; specific gravity and mineral composition are sometimes necessary, especially for non-siliceous materials. Low relative density is also not necessarily a sound guide if the sands are naturally cemented; weakly cemented soil with low relative density may respond by collapsing, thereby yielding a post-treatment penetration resistance lower than pre-treatment, despite an inherently more stable state. Alternatively, a strongly cemented sand may not settle at all but yet may be stiffened by stone columns, a situation which can be verified only by a loading test covering a number of columns and intervening soil.

The Compozer system was intended for stabilizing soft, natural and filled soils which are subsequently pre-loaded by earth fill prior to use. It thus precon-solidates non-cohesive silts and normally consolidated clays, which are difficult to treat by vibroflotation, even with stone columns, because of rapid loss of fabric coherence when vibrated. Such soils of low permeability would take too long to consolidate by small size drains because of insufficient initial edge stability to support the maximum pre-load necessary to achieve the requisite consolidation. Phasing of pre-load heights delays construction.

11.3.3 Arrangement of vibratory compactions

Compaction naturally attenuates with increasing distance from the treatment centre. Thus for large loaded areas, triangular grids of compaction centres are employed such that the effects overlap. For isolated footings, small groups are used, such as those illustrated in Fig. 11.4.

Depth of treatment through loose cohesionless soils need not be great.

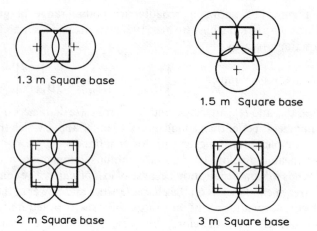

1.3 m Square base

1.5 m Square base

2 m Square base

3 m Square base

Figure 11.4 Typical arrangement of compaction points for isolated footings.

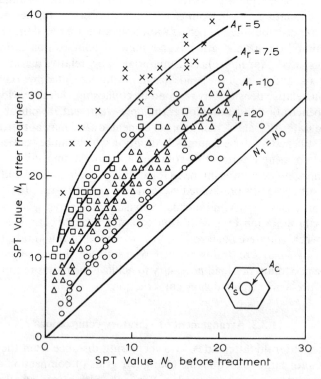

Figure 11.5 Empirical spacing/SPT relationship for Compozer.
$A = A_s + A_c$, $A_r = A/A_c$. ○ × , $A_r = 40$–15; ▲, $A_r = 15$–8; □, $A_r = 8$–6; ×, $A_r = 6$–4.
(After Fudo Construction Co., 1963)

Cohesionless materials have increasing strength and decreasing compressibility with depth because of increasing overburden weight. The depth of required penetration is limited either by the soil strength/depth profile or, for independent footings, by the geometric relation of footing width and applied stresses. It is rarely necessary to penetrate more than 8 m in sandy soils unless pre-treatment probes suggest irregular and low compaction at depth, taking account of overlapping stress distribution from adjacent footings.

Compaction spacing determines the relative density or bearing stress achievable. Since each type of machine is unique, empirical tests are necessary to establish potential achievement. The current machines can achieve better than 80% relative density with spacings of about 2.5 m in clean compactable sands. Safe bearing stress exceeding $500 \, \mathrm{kN \, m^{-2}}$ is possible on clean sands for settlement less than 25 mm. Similar empirical relationships for Compozer systems have been established in Japan, relating replacement ratio A_r to SPT and giving spacings from 1.7 to 2 m for similar relative density (Fig. 11.5).

Compaction is very sensitive to variations in permeability, related principally to silt content, in the soil and to even small changes in compaction spacing. This is very significant when planning or assessing site investigation data for compaction.

11.3.4 Earthquake resistance

Vibrating probes exert accelerations on the soil greater than do natural earthquakes. Typically the machines generate accelerations exceeding 10 g. By inducing maximum densification, liquefaction due to seismic stress can be inhibited. Depth and spacing of treatment are dictated by liquefaction potential.

If gravel columns are constructed through seismically sensitive fine sands and coarse silts, they can act as drains to inhibit excessive increase in pore pressure. Vibroflotation is a speedy and economical means of installing such columns and simultaneously helps to prevent liquefaction by compaction of the *in situ* sand. In this case the gravel columns must be well graded to preclude entry of sand and must be installed with a high volume of water flush to ensure high permeability relative to the surrounding sands.

11.3.5 Design of gravel columns and sand piles in cohesive soils

Columns of granular backfill improve foundation bearing capacity because they are stiffer than the soils which they replace. The degree of improvement naturally depends on the characteristics of the gravel and soil but, since the column is cohesionless, its stiffness depends also upon the lateral support it receives from the soil. In turn this depends on the way in which the composite of soil and columns is loaded.

The column is not designed as a pile to carry all the load but shares it with the intervening soil. Because of their greater stiffness, load tends to become concentrated on the columns and the design problem is to establish what the ratio of stresses on column and soil will be and by how much settlement of the untreated soil will be reduced.

Stress distribution also varies according to foundation width. Under widespread loading the applied stresses are essentially vertical and the composite of columns and soil is compressed in one dimensional consolidation. Stone columns under a narrow footing or near the edges of a wide one experience skin friction shear and are supported laterally only by overburden pressure without additional benefit from applied load. Thus they cannot carry as much as a central column.

Nevertheless, gravel columns are rather like piles and, if of inadequate length, may fail by overstressing skin friction and/or end bearing, which must be checked as for bored pile design. Additionally, however, because the gravel is not as stiff as concrete or steel, it experiences significant compression due to loading and tends to bulge into upper soft strata. Conventional piles are so rigid that it is not necessary to check this type of failure but it must be done for gravel columns.

Hughes and Withers (1974), considering the column as a pile, defined a critical length for an isolated column at which end bearing and skin friction are equated. Beyond this length the column contributes no extra benefit in terms of enhanced bearing capacity but it helps to reduce settlement by penetrating to a firmer stratum. Typically the critical depth is usually only about four column diameters.

The important criterion for most structures on soft ground is settlement rather than bearing capacity, so that partially penetrating columns are rarely used. For widespread loading, an array of columns penetrating to a firm stratum has virtually constant vertical stresses at every horizon so that bulging of a column rather than failure as a friction pile then becomes the main feature of design for control of settlement and ultimate strength.

To simplify consideration of the design elements the problem is divided into estimation of ultimate load capacity and settlement for both isolated columns and for widespread loading. Small groups of columns are treated similarly by rough interpolation.

(a) Isolated columns – ultimate load capacity

The assumption that a column is triaxially loaded in confined compression is common to all hypotheses. At some depth where lateral support from the ground is a minimum, a horizon of the column will reach a critical state first wherein the maximum ratio of vertical to horizontal stresses in the gravel occurs:

$$\sigma'_{vc} = \tan^2(45° + \phi'_c/2)\sigma'_{rc} \tag{11.3}$$

where in σ'_{vc}, σ'_{rc} and ϕ'_c represent, respectively, vertical and radial stress and angle of shearing resistance in the column.

σ'_{rc} must be balanced by the minimum passive support of the soil so that

$$\sigma'_{vc} = \tan^2(45° + \phi'_c/2)(Fc_u + \sigma_{ros} - u_o) \tag{11.4}$$

where F is a multiplier, c_u the undrained soil cohesion, and σ_{ros} the initial radial stress in the soil; u_o is the initial pore pressure.

In the second parenthetic term, which represents lateral stress, σ_{ros} includes any stress component of surcharge and soil weight, taking account of the initial ratio of horizontal and vertical stresses in the soil K_{os}. Because the column is assumed to be effective as a drain, excess pore pressure in the equation is taken as zero so that the calculation is based on effective stress (represented by σ') in the column and undrained strength in the soil.

Various investigators have approached the problem and have evolved different values of F. Brauns (1978) compared some of these, as shown in Fig. 11.6. The most commonly used value of $F = 4$ is that chosen by Hughes and Withers (1974) and thus the second parenthetic term of Equation (11.4) reduces to $4c_u + \sigma_{ros}$. This method is widely used because of its simplicity. Also shown on Fig. 11.6 are results of field loading tests at four different sites. The scatter indicates the difficulty of assessing the precise parameters of ground and column or possibly of forecasting the value of F. Nevertheless the benefit of the column is clearly demonstrated.

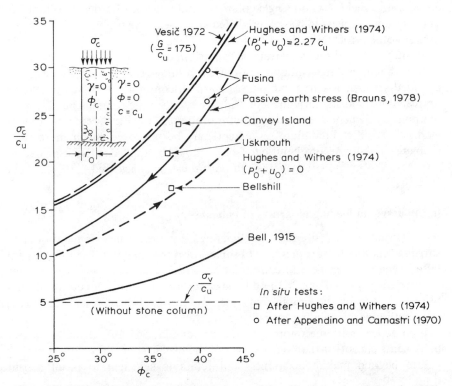

Figure 11.6 Ultimate capacity of stone columns (after Greenwood and Kirsch, 1983).

(b) Settlement of isolated columns

Immediate pseudo-elastic settlement contributes the major part of deformation. Mattes and Poulos (1969), in their elasto-plastic analysis, show that substitution of effective stress modulus for undrained soil modulus results in a difference of order only 10%. The same analysis demonstrates that Poisson's ratio is of little significance in determining stress distribution in the column and consequently also settlement. Thus their method, based on the simple elastic relationship settlement $s = (P/E_s \times L)I_P$, is easy to apply using their graphs of the value of I_P, an influence factor based on the geometry of the column and the ratio of column and soil moduli. These charts are derived from finite element analysis. P is the total load on the column and L is the length of the column. It is recommended that values of the appropriate soil modulus should be determined directly from tests, either drained or undrained according to loading circumstances.

A simpler approach is that of Priebe (1976), who also uses elastic analysis. The ratio of settlement without a column to settlement of the composite soil, defined as a soil improvement factor n, is calculated in the theoretical treatment to yield $n = (1 - v)/2 \, K_{ac}$ where K_{ac} is the active stress ratio in the column. With Poisson's ratio $v = 0.33$ this reduces to $\frac{1}{3}K_{ac}$, which depends only on the friction angle ϕ_c' of the column. This results in an improvement ranging from 1.2 to 1.9 for ϕ_c' between 35° and 45°. Field tests on single columns yield a factor of order 2, so that the improvement appears to be well predicted and insensitive to soil parameters. To obtain total settlement, the deformation of the untreated soil is estimated conventionally and the improvement factor applied.

Note that for isolated columns or a single row under a narrow strip the shape of the load/settlement curve exhibits a sharp discontinuity relative to that of untreated ground similarly loaded. This occurs when the columns bulge. Up to this point settlement is reduced by a factor of about 2 but after the columns yield it increases rapidly. Pre-failure settlement may be improved much more than ultimate strength. Accordingly, load factors should be applied to the deformation rather than ultimate strength, though both must be considered.

(c) Widespread loading on arrays of columns

As columns in this situation are confined by the surface foundation and surrounding soil they are in a state of constrained compression. The soil does not fail in shear but elastic and consolidation settlements occur which ultimately result in excessive bulging. In this dilated condition a column does not wholly fail because it is in plastic equilibrium. Internal stresses then represent the plastic state with a corresponding stress ratio.

In practice it is uneconomic to design foundations for oil tanks or embankments etc., other than on the basis of the plastic state, the columns having passed beyond maximum stiffness but retaining sufficient to be of benefit (Fig. 11.7).

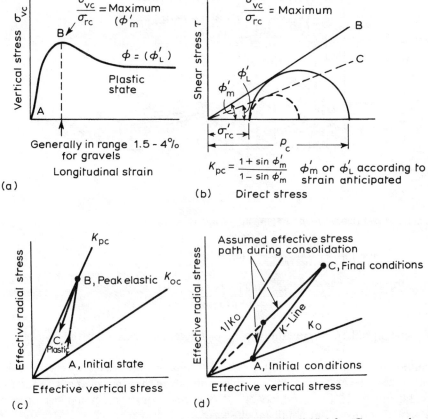

Figure 11.7 Idealized stress changes in column (a,b,c) and soil (d) (after Greenwood and Kirsch, 1983).

Analysis deals with resulting reduction in settlement of the foundation. The proportion of load carried by the columns and by the soil must be found. The simplest analyses estimate column bearing capacity according to estimates of anticipated vertical strains and calculated consolidation settlements of the soil under a stress reduced according to the total load carried on the columns. Iteration then allows compatibility between deformations assumed for columns and those calculated for soil consolidation. The more rigorous approaches derive a settlement improvement ratio, calculated without prior definition of elastic or plastic deformation on the basis of varying radial stress in the soil.

As for isolated columns, Priebe's (1976) analysis is one of the simplest to apply. The basis of this is an incompressible column material within a cylindrical elastic half space with no change in lateral stress with depth and no peripheral shear. The derived settlement improvement ratio n is shown in Fig. 11.8. As might be anticipated this depends strongly on the area ratio of the column and its

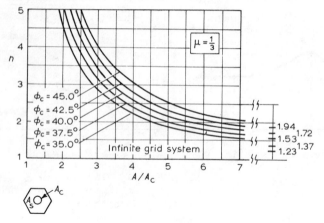

Figure 11.8 Priebe's (1976) design curves. $A = A_s + A_c$.

$$n = 1 + \frac{A_c}{A}\left(\frac{0.5 + f}{K_{ac}f} - 1\right); \quad K_{ac} = \tan^2\left(45° - \frac{\phi_c}{2}\right);$$

$$f = \frac{1 - \mu^2}{1 - \mu - 2\mu^2} \times \frac{(1 - 2\mu)[1 - (A_c/A)]}{1 - 2\mu + (A_c/A)}.$$

dependent soil, i.e. on column spacing. By reference to charts derived from elasto-plastic finite element analysis, Balaam (1978) has also provided a comparatively simple practical approach. However, this demands knowledge of the modular ratio between column and soil which is generally in the range of about 5 to 10 but is difficult to define accurately even from laboratory tests. Again the value of Poisson's ratio has insignificant effect.

A variety of other studies based on elastic theory have been made and are compared in Fig. 11.9.

The disparity between different theoretical approaches and between theory and field observations is not very important where total settlements are small. However, for structures such as oil tanks or embankments on very soft soils yielding potential settlements of 0.25 m or more, the differences are important.

It appears that at area ratios less than 3, settlements diminish significantly. With wider column spacings plastic shear displacements dominate settlement. The magnitude of settlement is also controlled by the stiffness ratio. Experimental studies (Charles, 1976) demonstrated that the elastic range of the columns extends to about 80% of the peak principal stress ratio prior to failure. Almost the whole column has to become plastic before effects on settlement are significant.

A recent theoretical study by Goughnour and Bayuk (1979) for fully penetrating columns solves simultaneous equations for plastic and elastic states to allow choice of the maximum value of vertical strain for a given stress increment on the clay. Thus a composite stress/strain plot can be constructed. The analysis allows definition of the depth of the plastic zone in the column and the ratio of column/soil stresses at any stage of loading and consolidation. The

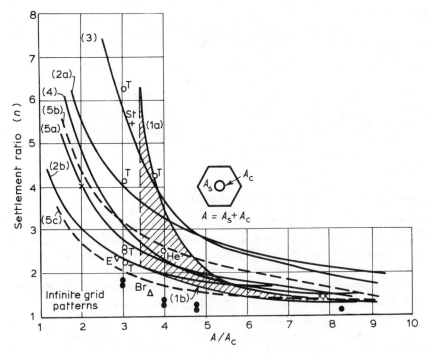

Figure 11.9 Comparison of elastic theories and field observations. (1) Greenwood (for $A_c = 1\,\text{m}^2$); (a) $c_u = 4\,\text{N cm}^{-2}$, (b) $c_u = 2\,\text{N cm}^{-2}$. (2) 'Weighted modulus' method: (a) $E_c/E_s = 10$, (b) $E_c/E_s = 5$. (3) Baumann and Bauer (1974); sandy silt, $K_s = 1.5$, $K_c = 0.6$, $E_c/E_s = 8$. (4) Priebe (1976); $\phi = 37.5°$. (5) Balaam (1978); (a) $E_c/E_s = 10$, flexible; (b) $E_c/E_s = 10$, rigid; (c) $E_c/E_s = 5$, rigid. Observed results: width foundation/depth soil > 3; $\bigcirc$. Watt *et al.* (1967, field); $\triangle$, Greenwood (1970, field); $+$, Greenwood (1974, field); $\blacktriangledown$, Kirsch (1979, field); $\bullet$, Charles and Watts (1983, laboratory); $\times$, Nahrgang (1974, laboratory), $T = $ Teesport, $Br = $ Bremerhaven, $St = $ Stanlow, $He = $ Hedon, $E = $ Essen. (After Greenwood and Kirsch, 1983.)

method is complex but is sufficiently versatile to allow introduction of the lateral stress ratio K, reflecting preconsolidation of the soil naturally or by action of the construction process. Many soil parameters are required which need to be accurately measured and therefore a well conducted site investigation is essential. Time for consolidation assisted by drainage through the columns is calculated as for sand drains using the Biot equations. To simplify application of the method, a supplementary paper has been produced (Goughnour, 1983) with various parameters in chart form.

(d) Analysis for small groups of columns

The simplest approach for small groups of columns is the use of Balaam (1978) design charts, based on finite element studies with elastic interaction factors.

These allow direct calculation based on elastic theory. Priebe (1976) also proposes a method averaging contributions of improvement ratios for internal and peripheral columns in a small group.

Both methods account for the fact that moderately large groups of columns with surface loading of intermediate width experience significant attenuation of vertical load stresses. The conventional way of dealing with this problem is to divide the various strata into layers, allocating appropriate stresses and properties after the manner of consolidation calculations using Boussinesq or Westergaard stress distributions and summing the contributions from each layer to obtain total settlements.

(e) Soil parameters employed for calculations

The various calculations can be performed with either drained or undrained soil properties, according to the rate of loading and the loading circumstances at various states of working life. However, calculation of ultimate strengths of isolated columns or small groups in terms of undrained strengths is always safe since transition to the drained condition always enhances resistance, settlement calculations ought to be made with effective stress parameters in any event.

The most important influence on the performance of the columns is the area ratio, i.e. column area in relation to its dependent soil. Small changes are significant and in practice the major point of concentration for quality control should be on the uniformity of column diameter and on its magnitude to ensure that the design assumptions are achieved.

Within the practical range of variation, the angle of shearing resistance of a chosen column material does not have a major influence. However, differences in values between differing materials, e.g. sand or gravel, can be significant.

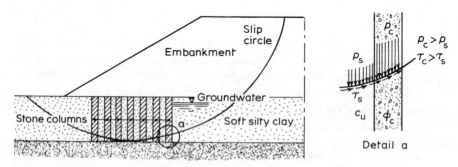

Figure 11.10 Slip circle analysis of slope on stone columns. Stress concentration on columns allowed for in composite values of strength parameters: $\tan \phi = m \tan \phi_c + (1-m) \tan \phi_s$, $\bar{c} = (1-m)c$, where $m = \dfrac{1}{A_r} \times \dfrac{\sigma_c}{\sigma}$. (After Priebe, 1978.)

(f) Gravel columns for improvement of shear resistance

For large widespread structures, such as oil tanks or embankments, resistance to rotational shearing is required under the edges. Stone columns are sometimes used to improve shear resistance. Designs are based on conventional slip circle analysis. A substantial replacement of soil by gravel is required for significant improvement, so that columns must be relatively closely spaced. Furthermore, to mobilize the frictional strength of the columns, substantial overburden stress is necessary. Liquid loads in tanks enhance frictional strength but at the same time increase the load to be carried. A method proposed by Priebe (1978) accounts for this load concentration, as shown in Fig. 11.10. A fuller treatment of design approaches for stone columns is given in a paper by Greenwood and Kirsch (1983).

11.3.6 Application in heterogeneous strata

In practice most sites have strata varying in thicknesses which are significant relative to the diameter of a compaction probe. Successive layers having radically different characteristics or lenses of inconsistent thickness may exist. Often the need for treatment arises because of such mixtures of soil types being present. The formation and history of the superficial deposits can sometimes suggest probable form of variation but not its detailed scale. Site investigation defines limits of the range of variation likely to be encountered but can never be specific except at individual boreholes.

Treatments must be designed to cope with the anticipated variations without alteration of projected structural design. The possibility of simultaneous compaction and stone column construction makes vibroflotation processes comparatively insensitive to the inclusion of thin lenses of soft material in sandy sites and provides security against unexpected variations. The columns are especially stiff if the thickness of soft lenses is not greater than the column diameter so that shearing planes through the column are constrained to directions in which resistance is greater than the minimum. Potential for lenses thicker than two column diameters should be revealed by site investigation and accounted for in design.

Where granular material of significant thickness overlies cohesive strata, compacted stiffness of the former can enhance raft effect over the latter. Foundation designs should account for such possibilities to obtain the best economic results.

11.3.7 Practical constraints

When working in cohesive soils it is important for both economy and quality of work that backfill supplies should keep the vibroflot working quickly and continuously. Delays during the cycle of column construction can adversely affect soil fabric surrounding unsupported bores in saturated silts. A single vibroflot may require 100 to 150 tonne per day of stone. The Compozer method cases the bore and so can cope better with very unstable soils.

Efficient site operation requires a sound working surface. Sufficient bearing for a crawler crane must be provided. The crane may have to pull up to five times the weight of the probe in cohesionless soils as they compact around it. While surface stability is never a problem on sandy sites, except where the water table coincides with ground surface, it is necessary to place a granular layer to support tracked equipment over clays, especially in wet weather. The vibro-replacement and other methods employing jetting water all require provision and maintenance of temporary surface drainage. Good site management should not allow water to cause deterioration of site surface. After winter working or wet treatment over clay, surface trimming is essential to remove soil affected by tracked equipment. In cohesionless soils the upper 0.5 m are not well compacted because of over-excitation by the power of vibration required for greater depths. Thus surface rolling is necessary if excavation is not planned.

Since probes are spaced between 1.7 and 3 m apart for most jobs, each provides a means of locating unsuspected zones of poor soil. The time to construct a column is generally only 10 to 30 min so any such zones located may be strengthened by additional columns without serious delay to the project. Compaction probes are also able to define outlines of buried obstructions which can then be either removed or surrounded by additional probes if sufficiently small.

11.3.8 Quality control

In cohesionless sands and silts, results of treatment are normally assessed by means of post-treatment penetration tests. From conventional analysis of these, appropriate bearing pressure/settlement properties are estimated. Commonly either SPT or Dutch cone is used. Comparative tests prior to treatment can also be beneficial.

It is important to average penetrometer results within zones potentially affected by foundation stresses. An isolated low result may arise equally from a lense of cohesive material as from bad workmanship. It is not significant unless extensive. Johnson *et al.* (1983) have shown how a site containing clay inclusions has been analysed by penetration testing.

Where widespread arrays of compactions are undertaken on sandy sites, an approximate check on relative density improvement can be made by reconciliation of imported backfill quantities and site surface depression, using the compacted volume of backfill. The accuracy of this approach is limited, except for very extensive areas, as volume reduction is usually only about 5% to 15% according to initial density.

If post-treatment settlement characteristics are very important, the expense of simulated footing loading tests may be justified. However, this type of test is more severe than structural working load because the footing is unrestrained by interconnection with the structure and the loading is usually applied more rapidly than building loads.

For granular columns in cohesive or naturally cemented soils, large-scale footing tests encompassing several columns are the only reliable guide to structural performance. Penetration tests are ineffective in gravel columns, whilst the clay soil between is substantially unaffected by vibration.

For building projects incorporating only narrow strip footings simple plate load tests can be made on individual columns. These may be a useful guide to column quality at shallow depth but otherwise do not give reliable indication of foundation performance.

The loading tests described above follow treatment and may take several days to accomplish. A useful immediate control during the course of construction is achieved by monitoring power consumption in relation to depth of treatment for each compaction. Power demand increases as the soil compacts around the machine. Amperage of electric current or pressure of hydraulic oil indicates when peak demand is achieved and no further benefit can be obtained by longer vibration. In some cohesionless silty soils and very soft clays, continuing vibration after the peak may sometimes yield a fall in demand, indicating undesirable weakening of soil fabric.

Because column diameter and its consistency with depth are so important for achieving desired results, the backfill quantity against depth must be recorded for each compaction. On completion these records should be cumulatively correlated with the total volume of backfill imported less any surface wastage, which should be measured on a regular survey grid to obtain the most accurate assessment.

11.3.9 Deep compaction by heavy tamping

This method uses weights of 15 to 20 tonne dropped from heights of about 20 m to pound the surface of the ground. It is apparently very simple but requires some experience to obtain the best results. The method may vary from site to site according to the soil strength profile within the potential depth of treatment and to some extent according to the nature of the soil to be strengthened.

The weights usually consist of toughened steel plates bolted together to give the requisite mass. This form has been found to be fairly durable although other types of weight and materials have been used. A tracked crane is most frequently employed to raise and drop the weight. The crane must be substantial to accept stresses resulting from reaction to sudden release of the weight.

If it is not present naturally, a surface blanket of unsaturated granular material 1 m or more thick is spread over the surface of the area to be tamped. The function of this granular layer is to act as a 'dolly' to lessen local impulsive shear stresses which might cause failure of the immediate surface and so inhibit effective compaction at greater depth. The presence of the granular layer also provides stability for tracked plant and lessens risk of flying stones or mud displaced by the impacts.

Tamping is generally done at points distributed over the site area on a rectangular grid 5 to 10 m apart. At each point 5 to 10 blows of the tamper may be applied. The number chosen is controlled primarily by observation of the depth of depression created. The weaker the soil the less blows will be required initially to make a crater 0.5 to 2 m deep with a weight typically 2 m square in plan. At selected grid points careful measurements are made to equate the volume of depression with the adjacent heave to ascertain the compression volume. Height of drop or plan dimensions and magnitude of weight are adjusted to optimize benefit. Subsequently, the craters are filled with surface or imported material and the site is regraded and levels checked to assess the average forced settlement before starting a further pass of tamping. The second pass of tamping points may be on the same or an intermediate grid. If the depth of treatment required is large and the loose soils are relatively deep, the same grid is worked but, if the depth of firm strata is limited, an intermediate grid would be chosen. Timing of successive passes is controlled by the need to dissipate excess pore pressure generated by the completed passes. For clay soils this may require three to four weeks. It follows that the process is best suited to treatment of large areas exceeding 10 000 m^2 so that tamping may continue on the site without interruption between passes.

Tamping is repeated until forced settlement is compatible with the expected volume changes anticipated from initial soil properties or until the requisite post-treatment *in situ* strengths have been achieved. To complete the job a smoothing pass may be given, dropping the weight from 2 to 3 m height contiguously over the whole surface area or alternatively the surface may be rolled.

Each time the weight impacts there is an impulsive shock of low frequency (up to 10 Hz) which is rapidly damped in saturated soils. The influence of such shock waves is probably significant only in coarse free-draining soil and of very little value in fine-grained saturated sands and silts or clays.

The initial passes generally require the greatest height of drop. The weight can then most easily punch through the upper soil layers and with each successive blow it carries down below it a zone of compacted material which grows with each blow. In effect this provides an increasingly larger ram, forcing the soil downwards and outwards until in plastic equilibrium with the surrounding soil. No further benefit results from continued tamping on the same spot after closure of soil voids (Fig. 11.11). Elevated pore pressure may cause shear before complete plastic closure so that dissipation is necessary before succeeding passes can complete the compaction process.

A threshold of energy per blow is needed to overcome elastic limit stress for each soil or site, up to which little compaction results. With greater impact the effect is substantial. Mascardi (1981) has inferred from inspection of published data that this may range from 1000 kJ per blow for sandy silts to 3000 kJ per blow for clayey soil. However, successful treatments have been reported with much lower thresholds but the majority of jobs have been done with energy in the range 1000 to 2250 kJ per blow. Laboratory work by Sowers and Kennedy (*ca.*

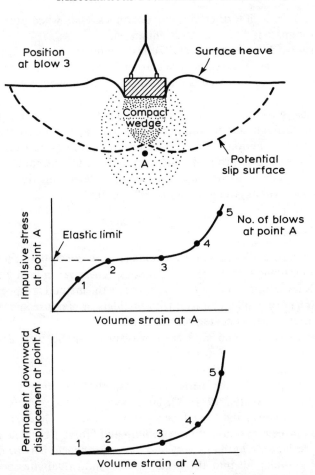

Figure 11.11 Idealized stress and displacement at a point within compaction zone (after Greenwood and Kirsch, 1983).

1953) suggests that maximum density is achieved if the requisite energy is applied with the least number of blows.

Benefit is obtained in tamping by the formation of granular columns of material carried down with the weight from surface, effectively forming short piers of gravel at each grid point.

It has been demonstrated by Jessberger and Beine (1981), using dynamic triaxial tests, that tamping initially induces a rise in pore pressure which reduces effective vertical stress. With continued ramming, lateral effective stress increases together with total horizontal stress. These reach a peak before diminishing in concert with the vertical stress as pore pressure continues to rise. When the pore pressure matches overburden pressure there is no benefit, and possibly some damage to soil fabric, if further tamping continues. Only when pore pressure is allowed to dissipate is effective stress enhanced.

Laboratory work by Fournier (1977) has demonstrated that dynamic compaction can be similar to static preloading in its effect on clay settlement characteristics. There is some support from field evidence for this result (Ramaswamy *et al.*, 1979).

There is also growing field and laboratory evidence suggesting that certain soils, ranging through silts to kaolinite clays, are capable of recovering some strength over a long period of time, up to several months as measured by *in situ* tests. Such increases are probably due, respectively, to skeletal stress relief or electrochemical action in the clays. The magnitude is sometimes significant, amounting to apparent strength gain of 10% to 50%.

(a) Suitable soils

Since pore pressure dissipation by drainage is a controlling factor in the mechanism, it is obvious that free-draining soils, generally sandy or granular, are most suitable for tamping. Owing to its simplicity the method is especially suited to compaction of rock fill and gravel with boulders or of refuse and other waste or natural heterogeneous coarse soils.

Very fine sands, silts and silty clays may not respond well, according to permeability and cohesion, but such materials in partly saturated fills have been successfully treated. Alluvial clays are sometimes successfully treated but inconclusive or unsatisfactory performance has also been reported (Bhandari, 1977; Charles and Watts, 1982). Organic soils can conceivably be treated sufficient for roadworks but the efficiency of the process is likely to depend to some extent on their strength, compressibility and fabric. Complete collapse of fabric may preclude strength recovery and the major benefit may be due only to ramming of granular fill into the organic soils. Individual sites need specific consideration with respect to mechanism and soil circumstances.

Site investigations need to reveal details of soil fabric to assess whether apparently impermeable ground may have interconnected drainage passages within it which might permit successful treatment.

(b) Design considerations

Construction design needs estimates of total energy required to compact a site, subdivided into number of passes for given equipment and time for completion. The depth of effective treatment and energy per blow are required to allow selection of plant capable of imparting the required minimum impulsive stresses to initiate compaction. Spacing of tamping points has also to be decided.

Empirical results compiled by Leonards *et al.* (1980) imply that a wide variety of soil types respond similarly in respect of total energy requirement (Fig. 11.12). In the laboratory, Charles (1978) has established a linear relationship between

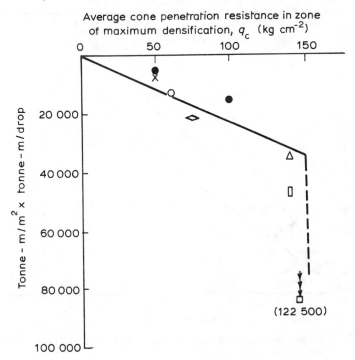

Figure 11.12 Energy against cone penetration resistance after dynamic compaction, where ○ is Belgium; △ is Sweden; ▫ is France; □ is Scotland; ◇ is Israel; × is Chicago; ● is Indianapolis (after Leonards *et al.*, 1980).

energy per unit volume treated and undrained cohesion for partly saturated clays but has not established field values. Empirical results concerning depth of effective treatment have been presented by Mitchell and Katti (1981) (Fig. 11.13). The definition of effective treatment based on penetration testing and pressure-meter testing is necessarily imprecise but nevertheless the results, although exhibiting some scatter, indicate that effective depth $\simeq 0.5(WH)^{1/2}$, where W is the weight of the tamper in tonnes and H is the height of drop in metres.

More sophisticated methods proposed by Scott and Pearce (1976) and by Jessberger and Beine (1981) require input of measured soil properties. The former requires both laboratory tests for soil strength and density, together with field measurement of peak particle velocities. The latter is semi-empirical and relies on an assumed stress distribution with depth.

Grid spacing is chosen according to the depth of soft soil requiring consolidation. For deep tamping a comparatively wide array of points is required in the early stages but if the depth of soft ground is only 3 to 4 m it cannot absorb severe shearing and may not be improved by very heavy tamping: only a small height of drop is appropriate. A grid spacing similar to drop height would then be needed, with final contiguous treatment. Grid spacing therefore depends on

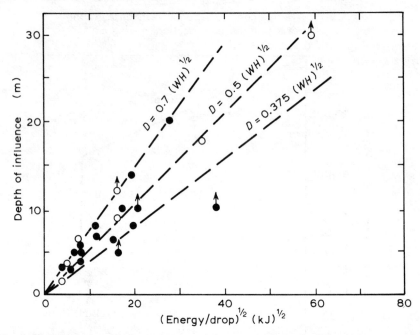

Figure 11.13 Relationship between potential energy and depth of dynamic compaction, where ● is predominantly cohesionless soil, ○ is silty soil, fills, rubbles; ↑ is full depth of improved zone undetermined (after Mitchell and Katti, 1981).

judgement of the size of compaction zones induced below each tamping point.

The volume of the imprint at individual grid points less the volume of surface heave is important. Substantial depressions must be formed if compaction is to be achieved, especially in fine grained or cohesive soils which do not compact by shock loading. On completion, the average depression of the site after levelling gives the cumulative volume change but, in the course of treatment, imprint volume is the initial guide. Graphical representation of volume against number of blows should be monitored: in successive passes imprint volume should diminish for the same energy input. Total volume change depends on the initial state of the soil but 5% to 10% is usual.

The size of the weight and its mass per unit area affect the volume of imprint. Weights normally have a mass per unit area of 4 to 5 tonne m^{-2}, giving imprints of roughly 0.3 to 2 m deep, according to the strength of the soil surface. Weights of 15 to 20 tonne are thus about 2 m^2 in area. Improved depression with the largest practicable area has been sought by using a weight in the form of an annular ring. For tamping through water, streamline shape encourages maximum impact velocity and minimum splash. Preferably, however, tamping below water should not require raising the weight above the water surface between blows.

(c) Assessment of results

Tamping is a rapid process and control by conventional soil sampling and laboratory testing is tardy and thus of little value during the course of the work. As a result, *in situ* testing by penetration methods or pressure-meter have been adopted. Because of lateral stress increases induced by tamping, penetration tests may underestimate soil bearing properties whilst pressure-meters may over-estimate them. *In situ* strength tests are performed either on a regular pattern or at known footing locations. They must be made after dissipation of pore pressure.

Changes in pore pressure should be monitored by piezometers installed prior to tamping at locations which will not be damaged by excessive disturbance. Rapid response of the system is desirable, as described by Charles and Watts (1982). Such *in situ* tests may be inappropriate in heavy rock fill or boulders. In these cases an accurate check on forced depression is essential. This is confined to the nominal treated area although in practice the plan area of treatment will extend a distance beyond the edges equal to at least the anticipated depth of compaction.

Seismic methods have been used (Dash, 1976) where other forms of test have been impracticable. Sensitivity depends on the form and change of density of the compacted ground. However, correlation with other forms of test is desirable if possible, especially the correlation of dynamically and statically determined moduli (Hansbö, 1977).

The bearing properties of foundations are determined in a conventional manner on the basis of the post-treatment tests.

11.3.10 Vibration nuisance and damage to structures

Deep compaction by vibration and by tamping both cause potential hazards of nuisance to people or damage to surrounding structures. Fig. 11.14 shows field measurements by accelerometers of vibrations induced in the ground (Fig. 11.14(a)) and on foundations of structures (Fig. 11.14(b)) by vibroflotation processes. Peak particle velocities are the resultants of independent records of vertical, horizontal and shear waves and are plotted against normalized energy and distance to source of energy. Significantly lower particle velocities are generated in the structures than on the soil surface and this is a fortuitous benefit arising from differences in constraint. Sensitivity to vibration depends on frequency so that the design range proposed by Wiss (1967) shown is not at a constant peak particle velocity. An indication of perception of vibrations by people is given on the right-hand side of the diagrams. Occasionally reasonance, which depends on the orientation, damping and fixity of structural elements and strata arrangement, can exaggerate the basic velocities. Such magnified vib-rations are preculiar to local circumstances but can be annoying. Usually a simple change in the detail of the affected element can control it.

Damage to structures may arise directly from vibrations or by settlement of

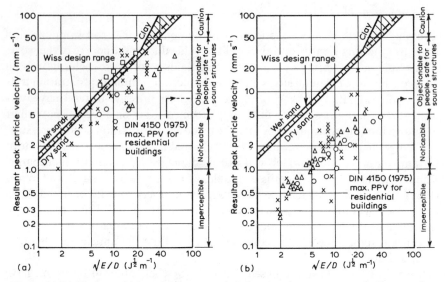

Figure 11.14 Peak particle velocity recorded for varying energy and distance from vibroflots. $D =$ distance; $E =$ energy. (a) Measurements on soil surface; (b) measurements on footings of structure. Clean sand, Δ, $50\,kW \times 30\,Hz$ 1640 J/cycle; sand $\bigcirc$, $50\,kW \times 50\,Hz$ 1000 J/cycle; silty sand, $\square$, $50\,kW \times 33\,Hz$ 1500 J/cycle; ash/rubble/clayey fill, $\times$, $90\,kW \times 30\,Hz$ 3000 J/cycle. (After Greenwood and Kirsch, 1983.)

foundations in responsive soil. The former is more sensitive. Vibrating probes of current design can safely be used at 15 m distance from sound structures whilst foundation settlements are rarely induced even in the loosest sandy soils, except within a range of 5 m. In cohesive soil vibroflots have been safely used within 1 m of existing foundations, despite strong vibrations in the structure.

Figure 11.15 shows similar results for dynamic compaction. The much greater energy of this process, compared with vibrating probes, provokes higher peak particle velocities for the same source distance. Usually this process is not operated within 20 to 30 m of surrounding structures because of potential damage and nuisance and also for safety from ejected debris.

To limit nuisance, an increase of distance from the sensitive point is much more effective than reducing energy input.

11.4 GROUTING IN GEOTECHNICS

The injection of a fluid which subsequently solidifies in place within a porous medium to change its properties is a beguilingly simple concept. But that definition contains the only common feature in a huge range of applications with a wide divergence of technique. A moment's reflection on the nature of porosity of the injected medium suggests the variety of applications: pores may be uniformly

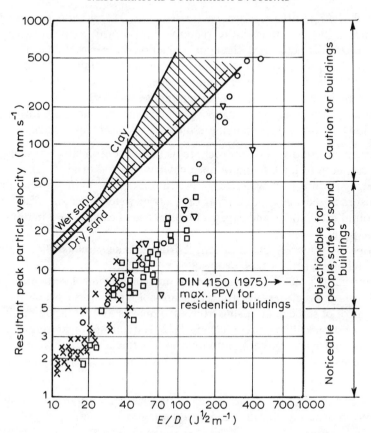

Figure 11.15 Peak particle velocity recorded for varying energy and distance from tamping point. D = distance; E = energy. Silty sand, ○, 16 T 20 m drop, 3200 kJ/drop; PFA/silty sand and clay, ×, 10 T 10 m drop, 100 kJ/drop; □, 10 T 20 m drop, 2000 kJ/drop; silty sand, ▽, 15 T 16 m drop, 2400 kJ/drop. (After Greenwood and Kirsch, 1983.)

distributed or be concentrated in simple cracks or both. Physical constraints on injection will differ significantly.

The principal objectives of grouting are to diminish deformation, increase strength or reduce the permeability of treated ground. Clearly, pore spaces must be penetrated to a substantial degree to achieve these aims. Yet differing standards of pore filling may satisfy the objective. Even nominally uniform soils have a range of pore sizes, the coarsest tortuously connected providing the main bulk of any throughput of water. A dramatic reduction in permeability can be achieved by filling only these coarsest conduits. However, such treatment may be inadequate for strengthening the soil since to resist deformation either a large majority of particles must be bonded or pore spaces filled to prevent particles from packing closer. To resist ultimate shear failure a substantial part of all potential sliding surfaces must be treated. On the criterion of pore filling, therefore, a higher standard is required for improvement in strength or deformation characteristics than for reduction in permeability. Practically, however, each project presents unique problems of technique and application to achieve acceptable values of relevant parameters.

Grouting is an extraordinarily flexible engineering tool. It is used as a part of permanent structures, as a temporary construction aid and for remedial operations. Reasons for employing grouting in geotechnics include:

(a) Settlement control of soil or rock: reduction of consolidation or shear strain in response to load, filling of open cavities, natural or artificial.
(b) Seepage control for water or gases: cut-offs below dams in rock or soils, closure of seepage around pipes or ducts in water-retaining structures and around excavations and caverns by blockage of cavities or a multiplicity of fine pores.
(c) Slope or excavation stabilization: resistance to deep-seated slides or surface sloughing in soils, screes, or unstable blocky rock.
(d) Tunnelling: groundwater and ground movement control, settlement control of surface structures over tunnels, shaft sinking.
(e) Control of settlement arising from earthquakes and vibrating machinery.
(f) Restoration of structures: jacking back settlements of footings and road slabs, underpinning both temporary and permanent.
(g) Introduction by injection of materials which produce inert gases to damp tip fires.

This flexibility emphasizes the variety of physical circumstances and processes involved in grouting and the need for differing techniques.

11.4.1 Injection constraints

Grout injection is concerned with the interaction between the properties of the ground (permeability, porosity, pore size and shape, effective stresses and pore fluid properties) and those of the fluid grout (viscosity, shear strength, particle content and sizes). This interaction controls injectability or groutability. Such properties must not be incompatible with the required set properties of the grout to give the final characteristics required of the treated ground.

Over the period of an injection the groutability must not become significantly more difficult. Immediately after injection, grout must resist displacement and eventually must develop the required working characteristics and be permanent, at least for the duration of project life. All this must be achieved with economy.

11.4.2 Site investigation

If the physical processes involved in grouting are to be not only understood but quantified to allow grouts to be designed for the job it is implicit that there is at-hand knowledge of the vital properties of both the grouts and injected medium. An adequate site investigation is essential, preferably characterizing the strata by means of permeability distribution, since this is the most important property governing resistance to injection and eventual distribution of grout. In soils this can be done only by short cell permeability tests, injecting or pumping out water to simulate the grouting operation rather than by deep wells which integrate the yield of water from all strata traversed, with a bias to the most permeable stratum or strata present (Golder and Gass, 1962). For the same permeability, uniformly graded soils have a lesser range of pore sizes than well graded ones but the average pore diameter is the same. More uniform distribution of treatment can be expected in uniform soils.

In the extreme case of cracked rock all the flow may be concentrated in very few fissures and, if the fissures are widely spaced, it is important to know their location in the bore. In such cases the concept of permeability loses validity and a measure of comparative water loss or make for fractured rocks was devised by Lugeon (1937) based on water leakage at a standard 10 atm pressure at the top of the bore (Tornaghi, 1978a). For grouting fissures, however, it is necessary to know also whether the leakage is attributable to one or two major fissures or to a band of finely fractured ground since this will affect penetrability of particulate grouts and hence the appropriate choice of grout. This is especially important in deep shafts where single borehole techniques using pressure depletion tests give the most information economically (Adamson and Scott, 1973; Chalmers *et al.*, 1979; Black *et al.*, 1982).

The better the frequency and quality of basic data, the better the chance of successful treatment. It is obvious, however, that the intent of site investigation is not to describe the ground in such detail as to define precisely the properties for each injection.

11.4.3 Technique of repetition

It follows that, if the inherent variability of the soil to be treated is large, some of the injections will be located in zones not treatable by the grout whilst others will accept it readily: the distribution of these zones is unknown. If the range of variability is very great, several grouts may have to be used in succession in the same zone, as the range of applicability of each grout is limited. Grout penetrates most easily where resistance is low. If pumping is continued, grout may travel extensively along the lines of least resistance outside the zone of intended treatment whilst simply choking the mouths of more resistant layers intersected by the grout hole. Distribution of grout in variable ground is thus irregular.

This leads to one of the simple basic practical rules – that of repetition of injection. The best result is obtained by many repetitions of injection each with limited quantities of grout in the same zone. By this means grout sets first in the most open layers and is progressively forced to enter the resistant ones to the

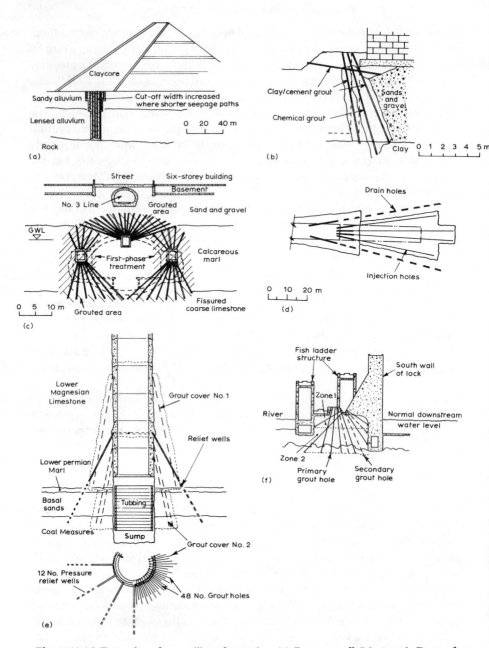

Figure 11.16 Examples of versatility of grouting. (a) Dam cut-off (Mattmark Dam, after Lossinger S.A., publicity literature); (b) underpinning for excavation; (c) multiple phase treatments for large excavations (Auber station, after C.I.F Bachy, publicity literature); (d) tunnel covers (Pollino Nord, Cementation Mining Ltd, publicity literature); (e) mine shaft covers (Riccall, after Black *et al.*, 1982); (f) foundation grouting (Locksand Dam 26, after Davidson and Perez, 1982).

limits of its physical properties. Grout thus remains closest to the injection point and distributed throughout the zone of treatment. Any scheme of injection which does not allow significant repetition will not produce a good result.

To keep the proper perspective of both technical and economic judgements involved in grouting it is useful to have in mind a concept of relative scale or width of the volume to be injected and of the pore sizes of that volume in relation to the range and shape of a single injection. A barrier of width only 1 or 2 m comprising two rows of injection holes may be more difficult to achieve even in uniform soil than a wider three or four row curtain with more widely spaced rows. The possibility for repetition in the latter is better. It can be false economy to omit a row of injection holes even whilst closing their spacing. The total grout quantity injected for a given volume of ground remains unaltered but the number of separate injections to achieve it is important: the more variable the ground the more important this repetition becomes.

Except for filling mine workings and other large caverns, the volume flow rate for grouting is restricted by the resistance of the ground to injection. Rates are not usually very large and consequently injection holes are generally of small bore, say 25 to 100 mm diameter. Flexibility of grouting stems from possibilities of arranging grout holes as access permits. Some typical arrays of holes are given in Fig. 11.16. Hole spacings generally range from just under 1 m to about 6 m, the closer spacing usually being associated with treatment of tighter ground with chemicals and the larger with very open cobbles or rock fissures. The hole arrays are arranged to cover the nominal volume of treatment required at the predetermined hole spacing to give the necessary repetition of treatment. The nominal volume and shape of grout masses in diagrammatic form must not be confused with the actual shape achieved in the ground: this is dictated by the hydraulic properties of the ground. The nominal volumes and shapes are most closely achieved by frequent repetition of small injections to give contiguous bodies of set grout. Economic balance has to be found between this technical need and cost of repetition.

11.4.4 Grout holes and stages

To place grout at correct depths the lengths of injection holes are divided into stages into which predetermined quantities are injected according to anticipated ground conditions. Here techniques diverge according to the scale of injection and the nature of the ground. Injection holes 20 to 30 m or more long clearly need more division than those 5 to 10 m long. The technique of separating stages one from another by the most economical means must vary with the stability of the hole. In sound hard rock it is possible to use packers to inject in ascending stages after drilling in one operation to full depth but, as the rock quality deteriorates, packers cannot be relied upon to prevent leakage. Descending stages clearly need redrilling with consequent additional costs in moving rigs and repeated drilling of lengths of holes. In unstable soils pipes may be driven, fitted either with an

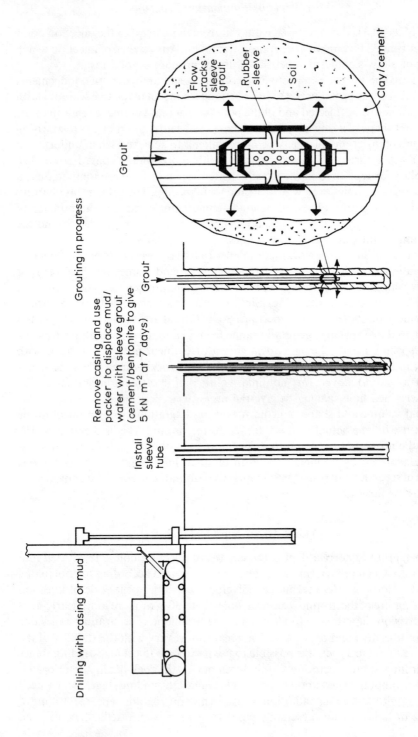

Drilling with casing or mud

Install sleeve tube

Remove casing and use packer to displace mud/water with sleeve grout (cement/bentonite to give 5 kN m^{-2} at 7 days)

Grouting in progress

Grout

Grout

Flow cracks sleeve grout

Rubber sleeve

Soil

Clay/cement

SLEEVED TUBE GROUTING METHOD FOR SOILS

Figure 11.17 Sleeved tube grouting method for soils.

expendable point allowing injection during withdrawal or with a perforated tip which it is sometimes difficult to prevent from clogging. Such driven pipes are obviously more suited to short lengths of holes. To overcome this difficulty in unstable ground requiring treatment to considerable depth, Ischy is credited with early use of the sleeve pipe system now almost universally employed (Fig. 11.17). Whilst this has the cost benefit of separating drilling from injection and allowing repetitive injection in short stages in the same borehole its establishment requires several operations to effect a seal in the ground to prevent grout escaping between stages or to surface. It is not always the most economical system for short holes or in stable ground: in these it is sometimes better to grout in a temporary rigid plastic casing as a surface seal and drive or drill an uncased hole below for stage treatment. Judgement must be made of the correct solution according to the economics of the job.

The sealing grout surrounding the sleeve tube requires bursting within a day or two of placing: otherwise pressures exceeding 10 atm may be required to break out. Providing large quantities of fluid are not pumped in the bursting process there is little risk of significant damage to the ground. Grout injection pressures require control, however, both to avoid damage to the ground and waste of grout as well for their value in recording the effects of grouting when used in a consistent and systematic manner. For delicate grouting work the pressure necessary to lift the sleeves can mask effective grouting pressure.

11.4.5 Hydrofracture and permeation techniques

The amount of pressure which can be applied to an injection to speed its progress is circumscribed by the resistance of interaction of the ground with the grout and by effective stress in the ground resulting from stress history or new construction. Ground fabric also has an influence. Too much pressure may disturb the ground structure adjacent to the point of injection (Hubbert and Willis, 1957). At lower pressures grout may penetrate by permeating in a continuously expanding rudely spherical or cylindrical volume centred on the injection pipe. If rupture occurs the physical controls on grout penetration change and preferential passages along lines or planes of rupture allow rapid grout spread with very restricted penetration between fissures. This is known as hydrofracture. If hydrofracture injections are practised repetitiously, successive grout fissures may cumulatively create surface uplift. In stratified or laminated rocks the query is raised whether hydraulic ram effects between laminae might result in greater damage than the benefit yielded by grouting (Morgenstern and Vaughan, 1963). Generally, French practice for soils favours monitoring of injections by deformations using high pressure and volume of grout. Practice elsewhere tends to favour permeation, especially for rock fissures in cut-offs below dams (Houlsby, 1982a, b).

In permeation, grout steadily advances from the source (grout pipe) in continuous flow through the pore spaces of the ground displacing pore fluids

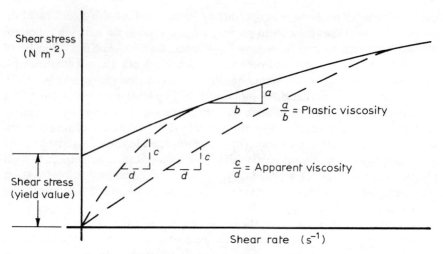

Figure 11.18 Rheological properties.

ahead. Factors controlling permeation, based on ground permeability (Scott, 1963), may be reduced to

(a) Filtration of particles at the interface with the ground and grouting suspensions.
(b) Shear resistance during flow of particulate suspensions due to continuous rearrangement of temporary particulate structures.
(c) Resistance due to viscosity of the fluid.

Sometimes the last two properties are combined and referred to as apparent viscosity or 'thickness' of a grout (Fig. 11.18). Whilst increasing viscosity and attenuation of pressures impede permeation, flow does not stop as long as pressure is applied. Shear strength and filtration may stop flow unless pressure is increased and hydrofracture supervenes.

By assuming spherical or cylindrical flow patterns from the point of injection, according to lithology of the ground, penetration distance can be calculated from grout and ground properties (Raffle and Greenwood, 1961). For spherical flow (in uniform porous media) the time to penetrate to radius R is given by:

$$t = \frac{na^2}{kh}\left[\frac{\eta}{3}\left(\frac{R^3}{a^3} - 1\right) - \frac{\eta - 1}{2}\left(\frac{R^2}{a^2} - 1\right)\right] \tag{11.5}$$

where n is porosity, a is radius of equivalent spherical source or sink, k is permeability of soil to water, h is driving head of water and η is ratio of grout viscosity to that of water.

This expression takes account of displacement of water ahead of the grout. Similar relationships can be derived for cylindrical flow: differences in result are not significantly large in practical terms having regard to unknown flow patterns in real ground.

Using such expressions, charts are drawn for various values of parameters to yield time for injection and hence economic spacings of injection holes. Values do not vary much and yield typical hole spacings of 2 to 3 m for cement grouts in open fine gravels and 0.5 to 1.5 m for chemicals in sands. Precise calculation is not justified except for large projects for which there is accurate information on soil fabric.

Using such a concept of tubular pore with mean pore diameter related to permeability, the maximum size, D_{max}, of particle which does not filter out can be determined (Raffle and Greenwood, 1961). On this basis, D_{max} in the grout is

$$D_{max} = \left(\frac{8\eta k}{n\gamma_w g} \right)^{1/2} \tag{11.6}$$

where γ_w is density of water, and g is acceleration due to gravity.

Cements typically become restricted in ground finer than $10\,\text{mm\,s}^{-1}$ permeability. Alternatively, Mitchell and Katti (1981), using empirical filtration relationships, suggest that US types 1 and 2 Portland cements are suitable only for soils coarser than 0.6 mm, type 3 for particles coarser than 0.42 mm and bentonites coarser than 0.25 mm. These results are in fair accord. Fractures of width less than about $200\,\mu\text{m}$ will not accept cement grout in practical circumstances.

The concept of mean pore diameter of granular soils related to their permeability may also be used with a spherical flow model, for example, to give penetration limit for known shear strength of grout and pressure, the radius of penetration R is given by

$$R = \frac{\gamma_w g h \alpha}{2S} \tag{11.7}$$

where S is shear strength (yield value) and α is radius of mean tubular pore with same permeability as soil.

Substitution of typical values for cement grouts (in the ranges shear strength 5 to $500\,\text{N\,m}^{-2}$; viscosity 10 to 100 cP) shows that filtration rather than shear strength is first likely to inhibit penetration except for very thick mixes. This may not always be true for cement/bentonite mixtures nor for clay/chemicals.

The accurate measurement of rheological properties of grout suspensions, even in laboratory conditions, is not easy and depends very critically on mixing efficiency, temperature, rest time after mixing and instrumental and experimental details. There is no standard method. Quoted values in the literature vary by factors of 10. In the field, however, a thorough and consistent method of mixing at high rates of shear must be beneficial.

11.4.6 Effects of consolidation ('bleed') of particulate grouts

Water/solids ratios of particulate grouts are usually higher than that above which self-weight consolidation occurs (0.4 w/c by weight for cements). The particles are effectively sluiced into voids using water as a transport medium. When filtration occurs water is expelled by pressure or, as the pressure gradient attenuates in large voids and velocity drops, particles may simply settle (Tornaghi, 1978b). Shear resistance increases and considerably greater pressure is necessary for further movement. This may affect injections in two ways: if surplus water cannot escape through micropores potentially injectable passages may not be filled (Houlsby, 1982a) or, if too rapid escape occurs, premature blockage results before complete filling. It was for this reason, when injecting fissures in finely porous rock, that Francois introduced prior injection of weak chemical gels, e.g. sodium silicate with aluminium sulphate or silicate alone to react with subsequent cement injection. These gels choked the porous surface and precluded premature bleed. Modern practice is to include a viscosity increasing agent such as cellulose or a mildly thixotropic suspension, e.g. bentonite, both of which inhibit consolidation of the relatively coarse cement particles. The use of bentonite may increase the shear resistance of the grout at both low and high proportions with cement (Greenwood, 1982): whether this is a more important factor in inhibiting injection than bleed depends on fissure size and injection radius, which control the pressure gradient at the grout front.

When grouting large cavities, such as disused mines or natural caverns, grout particles may consolidate at rest after injection, with surface bleed leaving a space between grout and roof of the void. This can occur if constraints of pumpability (shear strength and viscosity with small pipes) to reach the void dictate grouts with higher water content. Economy may preclude anti-bleed agents or strength reduction from their use may be unacceptable. In such uses re-injection is necessary, for test and precautionary purposes, using rock fissure techniques.

11.4.7 Filtration

The practice of rock fracture injection involves grouting initially with a very weak suspension, gradually thickening this if no resistance is encountered. Water/cement ratios exceeding 6:1 are sometimes used but such high values are rarely necessary and starting ratios of 4:1 or 3:1 should quickly be reduced as injection proceeds. Nevertheless, filtration rather than flow resistance is the predominant factor initially and too thick a grout used too early can preclude otherwise practicable treatment of the finer fissures.

For sealing widened joints in rock (up to about 150 mm) in which water flow makes retention of grout difficult, filtration is used to build a barrier against which to grout. Coarse materials like cotton floc, nutshells, artificial 'porcupines', etc., are introduced to the flow to try to choke the irregular cavity. Mixtures of bituminous emulsion and cement grout form instant balls and can be helpful but generally, injected fluids are useless in the face of heavy flow and high pressure in wide fissures. Once flow is staunched, grouting can proceed normally.

11.4.8 Permeation with chemical grouts

For permeation into rocks and soils with relatively fine pores, including virtually all sandy ground, resort to chemical gels is necessary. These are aqueous solutions with viscosity only two to five times that of water and practically without particles or structure until gelation occurs. Usually the reactive components are combined into two solutions which, when mixed together, produce a weak jelly-like material at a time after mixing controlled by temperature and concentration. Gel times are usually comparatively short – 10 to 90 min in practical applications. As molecules increase in size in condensation polymers there is a gradual but increasingly rapid thickening until gelation (defined by a continuous structure with shear strength) occurs. Addition polymers of the acrylamide type tend to have almost constant viscosity until virtually instantaneous gelation occurs throughout the fluid. Much has been made of the benefits of the latter characteristic in improving penetrability but in practical terms it is of little significance because of variability of soil voids and radially diverging flow from injection points, which results in the bulk of volume being filled quickly with little radial extension at the extremities for constant flow rate (James, 1963): it is more economical to reduce hole spacing a little. However, with certain chemicals the gelation is difficult to control and occurs irregularly throughout the fluid. Some chemicals have also a small proportion of solid impurities. For jobs where penetration by permeation is critically important the former grouts are unacceptable whilst the latter must be clarified by prior filtration or centrifuging.

These visco-elastic gels may have strengths as little as 50 to 100 $N m^{-2}$ and yet in fine pores are well able to withstand hydraulic gradients in excess of 100 without displacement (Scott, 1963).

The two-shot systems developed in the 1920s, based on sodium silicate and calcium chloride, are not much used now. These chemicals give an instant precipitate and the process relies on intrusive penetration of the chloride into the silicate to achieve depth of treatment (Scott, 1963). They thus require hole spacings of not more than 0.5 m and are restricted to coarse sands or sandy gravels. They have been very effective when correctly applied (Harding, 1981).

Many grouting chemicals are toxic or are irritants to personnel exposed during their use. Practically, therefore the sodium silicate group has been found to be most acceptable whilst capable of formulation to give sufficiently useful properties. There are, however, only one to two formulations of silicate grouts which combine desirable properties of low viscosity, controllable gel time, adequate set strength and long-term stability. Many formulations which are marketed commercially do not have these features. Examination and correct

Table 11.1 Chemical grouts

	Initial viscosity (cP)	Gel time range (min)	Strength in coarse sand (MN m^{-2})	Risk and toxicity	Remarks
Silicate	1.5–40	1–200	0.7–3.0	Household chemicals	Only 1 or 2 stable gels of high penetrability
Lignochromes	2.5–20	5–120	1.0–1.75	Dermatitis risk	Hexavalent chromium is accumulative pollutant. Needs clarification to remove particles
Phenolic resins	1.5–10	5–60	1.0–3.0	Respiratory irritant: caustic	Poor gel time control with high strengths. Some need clarification
Acrylic resins	1.3–10	1–200	1.0–3.0	AM-9 neurotoxic (banned in Japan)	Latest forms less toxic than AM-9 and not neurotoxic
Aminoplasts	6.0–30	40–300	1.0–3.5	Respiratory irritant to users when pure	Very viscous unless pure
Polyurethane	19.0–150	Reacts instantly with water	0.8–1.0	Irritant	Gaseous foam expands fluid

(1) All gelling chemicals are toxic if mishandled.
(2) Tabulated figures are indicative for common formulations and cannot be used without reservation.

choice of the chemical can critically affect the success of treatment.

Table 11.1 gives outline characteristics of various grout categories.

11.4.9 Permeation limits

Figure 11.19 indicates limiting permeability for penetration by various grouts. It also suggests improvement possible in terms of resultant permeability, as measured in the field. Such representations are necessarily based on generalizations but are correct for most practical circumstances. Better results can be obtained when the permeability of the natural ground is higher than the lower limit of penetrability by grout permeation.

The coarser voids are always treated first in permeation, using the thicker and least penetrating grouts. This precludes turbulent intermixing of fluid grout with pore water in coarse voids which might otherwise conduct chemicals away with no effect and is also an economy measure.

11.4.10 Hydrofracture

As injection pressures are increased to speed penetration, hydrofracture becomes a possibility. In rocks with weaknesses along bedding planes, joints or fractures, or with residual stresses from historical loading, the direction of hydrofractures is predetermined and hydraulic ram effects are possible. In recent granular soils, however, residual stresses are not so well developed and the ground structure is more uniform. Local rupture occurs without relation to soil structure, with

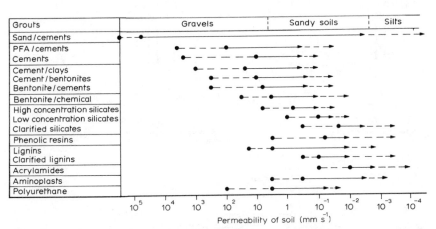

Figure 11.19 Indicative range of grouting treatments. Note: (1) If initial permeability lies at the lower end of its acceptable range, the resulting residual permeability tends to the high end of its range because of increased difficulty of treatment. (2) Permeability ranges indicated are for superficial soils only. ●--● is the usual range of minimum permeability for injection; →--→ is the usual range of residual permeability of treated soils.

direction controlled by principal effective stresses (Bjerrum *et al.*, 1972b). Heave may eventually be recorded as a result of deposition of multiple lenses of particulate grouts after many repetitions. Hydrofracture can occur with true fluids but serves then to open larger areas to permeation and more rapid absorption of grout. This technique is often deliberately used when grouting large volumes of soil and cut-offs below dams to permit effective treatment through wider hole spacing and to circumvent or cut through comparatively impervious materials which may surround the locality of an injection stage adjacent to hole. The process is effective providing the grouts selected are capable of permeating the ground comprising the bulk of the stratum to be treated.

If the grouts cannot permeate the soil and hydrofracture occurs, the shape and extent of grout fissures are determined by the rheological properties of the grout. Thus fissures 20 to 30 m long and 1 or 2 mm thick may occur if the grout has low viscosity and low shear strength. Their occurrence is easily recognized by sudden drop in pumping pressure associated with increase in pumping rate. This is undesirable as energy is used in extending the fissure with little practical benefit. With moderate viscosity and shear strength the grout sheet becomes shorter and thicker and pressure and flow rate change is less marked on fracturing. Taken to extreme, grouts with high shear strength, such as cement–silt–sand mortars, can be forced into the ground to form irregular bulbs displacing the soil structure three dimensionally (Fig. 11.20). This is so-called compaction grouting (Graf, 1969; Warner, 1982). There is thus a continuous range of grout rheological properties which controls the form of fracture.

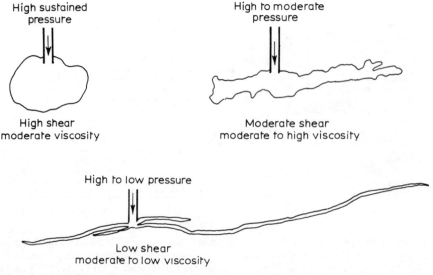

Figure 11.20 Effects of rheological properties on grout fissures in soil and on injection pressures.

11.4.11 Squeeze grouting

Fracturing with thickened grouts is adopted deliberately for stabilizing broken rock associated with fault zones in deep tunnels using squeezing and wedging to prestress and fix rock fragments ahead of excavation (Greenwood and Hutchinson, 1982). Fractures are opened and porous surfaces sealed by thickened chemical gels followed by cement grouts with anti-bleed agents. Squeeze grouting of this kind is often the only resort in finely crushed rocks alternating in fault zones with open cracks in harder strata. These cracks conduct away penetrating grouts capable of permeating the finer materials and thus, because of the large and unpredictable range of permeability, usual permeation treatment is ineffective. At the other end of the scale rock mosaics are not easily displaced by stiff mortar-like grouts which are also ineffective. The squeezing and wedging technique may also be employed with particulate grouts to reinstate original levels of footings which have settled (King and Bindhoff, 1982; Vaughan *et al.*, 1983).

In compaction grouting the formation of a bulb is engendered by using modified concrete pumps to force mortar into the ground at pressures of 30 to 40 atm deliberately inducing bleed in grout to give it shear strength to keep particles in place immediately after injection. Bleed from the pipelines must be avoided if blockages are to be averted. This treatment is used for placing injection bulbs, typically 1 m diameter, to take up potential consolidation or compaction settlement in soils or to jack back to level foundations which have settled. Multiples of bulbs are formed by successive injections as required above or alongside each other (Warner, 1982).

11.4.12 Examples of grouting techniques

The potential variety of applications of the various concepts of treatment is enormous. Numerous examples of treatment are given in the proceedings of two international conferences on grouting at ICE London, May 1963, and ASCE New Orleans, February 1982. The specific approach and methods depend on the circumstances of the job. Extensive reading beyond the scope of this section would be required to gain a perspective of accumulated experience.

Different job circumstances may dictate different grouting concepts to achieve apparently similar broad objectives. The Blackwall and Dartford road tunnel duplications beneath the River Thames were of comparable diameter and length and both were accomplished with the aid of compressed air. There was a need to restrict potential air loss through the Thames gravels, which were traversed by both tunnels, and grouting was used with the objective of reducing working air pressures (Fig. 11.21). At Blackwall the tunnel passed almost entirely through gravels or contemporary deposits, filling the river channel, with the crown only 5 to 10 m below the bed. Here permeation at low pressure was essential to contain grout within a relatively thin (3 to 5 m) annular ring of treatment without loss to

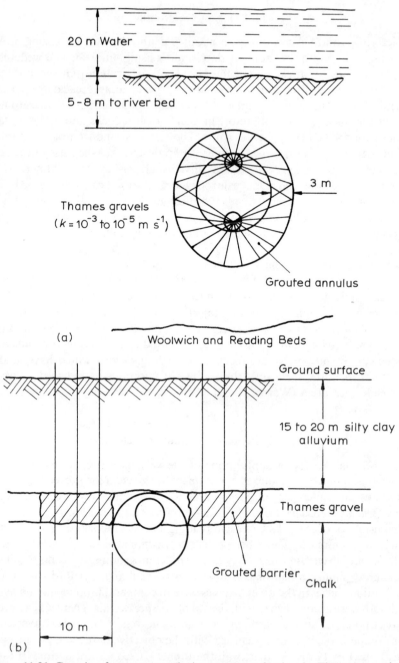

20 m Water

5–8 m to river bed

Thames gravels
($k = 10^{-3}$ to 10^{-5} m s^{-1})

3 m

Grouted annulus

(a)

Woolwich and Reading Beds

Ground surface

15 to 20 m silty clay
alluvium

Thames gravel

Grouted barrier

Chalk

10 m

(b)

Figure 11.21 Grouting for compressed air pressure reduction in Thames tunnels. (a) Woolwich and Reading beds, Blackwall tunnel. Permeation grouting from pilots (narrow annulus, close to river bed, adverse gradient). (b) Dartford tunnel. Hydrofracture grouting from surface (depth of cover for pressure, wide treatment zone, no gradient).

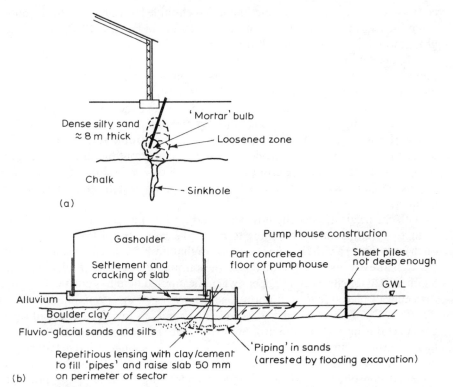

Figure 11.22 Squeeze and compaction grouting to restore disturbed ground and relevel structures. (a) Hermitage (concept 1982). (b) Rhyl gasholder (*ca* 1963).

the river (Perrott, 1965). At Dartford the under-river section of the tunnel was in chalk and only the ramps on either bank passed through the gravels which were buried by about 15 m impervious alluvium. Here sleeved injection pipes, driven vertically from surface, were used to create a box in the gravel through which the tunnel could pass. The sides of the box were about 10 m thick and grout pipe spacing was designed for hydrofracture penetration.

Similarly, raising settled footings can be achieved by squeezing or compaction grouting (Fig. 11.22); or impervious cut-off through alluvium by permeation or hydrofracture techniques (Fig. 11.16).

11.4.13 Hydraulic displacement after injection

Grouts for water-stopping must remain in place immediately after injection and after hardening must resist working hydraulic gradients. When circumstances dictate treatment from pilot headings below water, local gradients to the heading in rock fissures or permeable soil can be higher than working gradients as water converges on the tunnel. A sound lining and effective back grouting to the pilot can be beneficial. Gradients in such circumstances can be up to 10 whilst normal groundwater gradients are usually of the order of 0.1. Working gradients for dam cutoffs are restricted to about 3 or 5 but in deep mines can be over 1000, representing a completely different magnitude of problems. Grout placed above water table naturally tends to sink with a gradient of 1.

Cementitious particulate grouts are packed in place either by pressure bleed to lock them in by friction or sometimes by thixotropic thickening (reversible stiffening with decreasing flow rate). However, thixotropic thickening is often deliberately destroyed by chemical admixtures in grouting since unplanned stoppages in pumping cause grout to stiffen too early and in some instances may require pressure greater than the capacity of the pump to restart flow and trigger thinning.

With chemicals it is usually possible to reduce gel time so that sufficient shear strength develops during pumping to retain the grout. In this case pumping is often continued past the nominal gel time so that fresh grout cuts through gelled material already injected and travels furthest from the pipe. This results from the Saffman–Taylor effects and happens when the injected fluid has greater viscosity than that displaced from soil pores together with a positive pumping gradient (Scott, 1963). This is a particularly useful technique for combating adverse gradients during pumping (Karol and Swift, 1961). The resistance to displacement of grout from a soil is given by $2S/\alpha$ and can be similarly derived for plane fissures. If the fissures are wide (1 or 2 mm), account must also be taken of potential consolidation of chemical gels under direct and seepage pressure through the gel; this is because such gels have a tenuous structure and comprise mainly water. In practice, tortuosity and roughness of fissures helps to retain grout and can result in consolidation, which is generally beneficial.

11.4.14 Permanence of grout treatment

Associated with displacement of grouts is the question of permanence in the face of constant hydraulic gradient and possibly aggressive ground waters. The soundest defence is good grouting practice to ensure maximum void filling. This precludes access of aggressive water to the grout. For cut-offs in rock the technique of mix control to avoid trapping bleed water in impervious fractures and completion of each repetition of injection with rising pressure is helpful.

Chemical gels suffer more damaging dissolution by flow over the surface than through the body of the gel; this is because the gel is usually very impervious, about $10^{-7}\,\mathrm{mm\,s^{-1}}$ or less. Therefore, good grouting practice to achieve maximum penetration and choice of gels which do not suffer shrinkage nor collapse by syneresis all help to reduce decay. Only in mine shafts, where gradients retained are extremely high, is through flow the predominant factor in decay but in such cases consolidation of the gel by expulsion of water counters this effect. Continuous laboratory tests on the more stable formulations of silicate gels predicate lives exceeding 25 years and with further extension of tests may be considerably more. Such predictions are on unconsolidated gel.

11.4.15 Toxic hazards

It should be noted that considerations of rate of flow through the gel also influence assessments of toxicity of leachates. Generally speaking toxic hazard from the use of grouts is negligible after gelation since practical hydraulic gradients are incapable of removing significant quantities of material from such impervious materials. The greatest toxic hazard is to the work force and due to spillage of unmixed chemicals and cleaning plant during injection.

11.4.16 Structural design with grouted soils

The above paragraphs have been primarily concerned with placing grout in the ground. Design of grouted structures must take cognisance of the practical difficulties of distributing grout uniformly within the zone intended for treatment. Knowledge of the ground gleaned from initial site investigation and subsequent site experience and its interaction with grouts and grouting methods must be considered.

A concept which relies on uniformity of treatment, like a grouted retaining wall excavated on one side so as to support both external ground and water pressure, is inherently risky since one gap can cause undercutting of surrounding ground which quickly becomes extremely difficult to control. The same gap in a simple water cut-off not required to retain soil may be quite acceptable and can in any case be made good by further grouting on a practical time scale. Similarly, grouting undertaken from tunnels and mines below groundwater level usually has to be done in conditions of high hydraulic gradients which vastly complicate and restrict injection techniques to ensure grout stays where it is intended long enough to harden. Consideration of such factors is essential to good design philosophy.

The ultimate set strength of grout in rocks is not usually an issue since rock is usually competent to support itself in excavations providing water inflow is controlled. In highly fractured or pulverized rocks design may be based on examination of locked-in grouting pressure and the resulting frictional strength based on rock properties. Fortunately, with the tubular geometry of tunnels, a comparatively thin annular ring usually suffices to support the surrounding ground with elastoplastic deformation resulting in considerable stress relief.

Soils in the groutable range are, however, sandy and cohesionless. Strength as well as water-stopping is often required. The properties of grouted soil are similar to those of any granular material with a cohesive cement: there are both frictional and cohesion components of shear strength on the Mohr circle plot. The angle of friction of the natural material is barely altered by grouting and after injection the granular framework becomes fairly rigid with the bulk of pore-space grout-filled.

The material has high compressive and low tensile strength. The relative value and magnitude of cohesive and frictional components depends upon the material injected: being concrete-like for cement grouts injected into coarse sand-free gravel, down to weakly cemented sands with unconfined crushing strengths less than $1 \, \mathrm{MN \, m^{-2}}$ for a soft gel. Strength *in situ* depends on the value of principal confining stresses. Since the gels are visco-elastic the behaviour of chemically injected ground depends somewhat on the rate of loading and may exhibit separate peaks in the stress–strain diagram as viscous and frictional effects are mobilized (Stetzler, 1982). In the limit, if the stress level is too high, continuing creep will eventually lead to rupture within a practical time scale (Borden *et al.*, 1982). The magnitude of effective stress is the main strength factor. Sudden overstress results in brittle fracture.

11.4.17 Quality control tests

The importance of maximum penetration of grouts is reflected in field control testing during injection. This tends to concentrate on consistent rheological properties and mixture concentration. For particulate grouts, testing at the pump (or preferably at the point of injection) is primarily in the form of checks on apparent viscosity (shear strength and plastic viscosity). The most common tests are empirical and as such must be performed precisely according to the test instructions. In UK the Colcrete flow trough measures the distance travelled by a standard quantity of grout released at a standard height from a conical container along a horizontal trough. Grouts with identical apparent viscosities will travel consistently to the same distance. United States practice is to use a Marsh cone, timing the efflux of a standard grout quantity from a standard cone. Variations in mixing time or agglomerations of particles affect the results and consistent mixing control can thus be assured. In hot climates temperatures can be high enough to affect the result and adjustments must be made to the mix to achieve consistent properties at injection. Density checks in a mud balance confirm concentration.

For chemical grouts, gel-time tests are made. Usually these consist of drawing a thread of material from the surface of the liquid by means of a glass rod. Only when a continuous thread can so be drawn is there sufficient structure and the time from mixing is designated the gel time. This is extremely sensitive to temperature variation and account must be taken of groundwater temperatures in the borehole when establishing the chemical concentrations necessary for the designed gel time. Again density by hydrometer helps to control consistency of independent components.

Bleed tests may be important for particulate grouts, the most favoured being the ASTM pressure bleed test. Such tests not under pressure are inappropriate except for grouting situations where high pressure will not be applied. For the same reasons grout cube tests are usually worthless when it is known that bleed will affect grout densities *in situ*. This may occur either by self-weight consolidation or by grouting pressure.

Strength tests on pure chemical gels are likewise unrepresentative of grouted soil. The gels are usually highly elastic and very weak, with a strength of a few $kN\,m^{-2}$ only.

In the field, heterogeneous properties are anticipated following treatment. *In situ* tests to sample the grouted body with minimum disturbance are best. The sampling principle is important to obtain distribution of the measured property.

For water-stopping these can take the form of permeability tests similar to pretreatment short cell tests or it may be adequate at this stage to check by a deep well method if grouting was introduced only to seal known aquifers. For strength of grouted soil, *in situ* pressuremeter tests can be valuable since boreholes put down after treatment should stand open unsupported to allow the pressuremeter to be introduced for modulus and rupture tests. Standard penetration testing may also be appropriate but Dutch cones will not penetrate treated ground to any useful distance. Whilst absolute parameters for treated ground are needed for design correlations, such tests are usefully compared with similar ones made before treatment.

For grout jacking, testing is necessarily empirical by means of levelling on the footings or ground to be raised.

When mine-filling the objective is to make good open void but not necessarily to treat the finest fractured ground above the headings: in this case water testing may be inappropriate and injection tests under limited pressure more truly represent the practical limit of treatment and achievement of the objective.

11.4.18 Grouting equipment

Modern grouting equipment has been described in detail by Gourlay and Carson, 1982. The following commentary applies to particular aspects and is not a comprehensive review.

In most practical situations grouting in a single hole may require flow rates from 2 to 3 litre min^{-1} up to 25 litre min^{-1} or more with injection pressure ranging from 0.1 to 10 $MN\,m^{-2}$, the higher pressures being associated with slower injection rates. Except for special works, pumping rates smaller than this range are generally uneconomic and higher rates apply to open hole filling. Generally, resistance in the bore (subject to pressure limit) restricts the rate of grout acceptance. For most soils, which by their nature are generally superficial deposits, pressure is limited by restrictions on hydrofracture or on excessive waste, typically to about 0.75 $MN\,m^{-2}$ at the top of the injection hole. In mine shafts and tunnels, however, very high injection pressure may have to be used to overcome hydrostatic pressure which may range up to 6 to 10 $MN\,m^{-2}$.

There are several types of pumps in common use to meet these demands. European practice tends to ram type positive displacement pumps which are simple and easily maintained. Originally compressed-air operated, a hydraulic type is now frequently used with a relatively constant pressure characteristic over a wide range of flow rates and with the capability of controlled cut-off at any predetermined pressure. Hydraulic equipment needs a high quality of maintenance and cleanliness which is not always practicable in underground conditions.

Furthermore, compressed-air available on site is always useful for cleaning out grout lines and other purposes.

High-volume positive displacement action can be obtained with Duplex and Triplex piston pumps with synchronized strokes and high pressure capability. Diaphragm pumps can cope with large-volume low-pressure work rather more economically. Screw pumps of the moving void type (Mono or Moyno, single or multi-stage) are also popular for comparatively low-pressure high-volume work.

A typical rock grouting arrangement, as shown in Fig. 11.23 permits re-circulation of grout to the mixer, thus enabling grout-velocity to be kept high in the circulating main with little chance of blockages. The valve arrangement at the injection pipe also allows direct injection.

Inconclusive argument continues on the merits of pulsating injection pressure. This is possible only with ram pumps and direct injection. At high grout flows, pulsation is probably damped by hydraulic turbulence but, as refusal is approached, the ram effect is likely to be transmitted to the ground. The resulting momentary excess pressure may allow temporary blockages to be overcome and further beneficial grouting without risk of general hydrofracture.

Proportioning pumps are essential for chemical grouting when short gel times must be used. Large premixed batches stiffen in the tank and are either wasted or cannot be pumped with the benefits of low viscosity. In the proportioning system the two reactive components of a grout are separately mixed and pumped simultaneously to meet in a mixing head at the injection point, thus taking advantage of the maximum available gel time. Most proportioning pumps are of

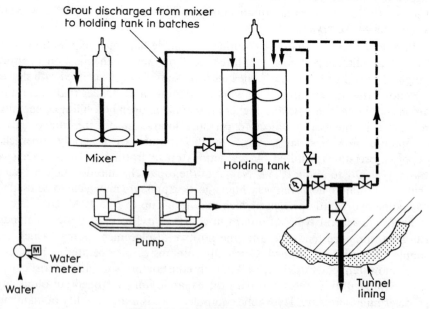

Figure 11.23 Typical grout plant.

the ram type since, by varying piston size, they can be proportioned conveniently. When using proportioning pumps it is prudent to keep the pump close to the injection point, with short lines to ensure that the resistance is the same in each line. Some gelling chemicals are quite sensitive to small variations in proportion and differential expansion of rubber hoses or unequal resistance to flow in each pipe can create sufficient disturbance of proportioning to vary the gel time irregularly. This is especially true when pumping at proportions other than 1:1. The detail of the mixing head is also important to ensure thorough intermixing of components, especially with sensitive grouts whose gelation can otherwise be 'lumpy'.

For high quality cement grouting of fine fissures, high shear-rate mixers are essential. Simple paddle mixers do not shear the grout stream and break down agglomerations of particles with thorough wetting of each particle: consequently, they produce grouts with greater bleed, sedimentation and filtration characteristics. High-shear mixers of the rotating disc or cylinder types are common but for fine particles and large discharges the Alpha Laval centrifuge type is useful, especially for clarifying 'dirty' chemicals.

In deep mining and tunnelling work, where high groundwater pressure may be encountered, it is important before drilling any injection hole to establish in the rock a standpipe grouted in over a sufficient length to resist displacement by water pressure. Such pipes should also be anchored separately to the rock (Atherton and Garrett, 1959). Standpipes are capped with a gland and valve through which drill rods can then be advanced into the hole and replaced by injection pipes, keeping full control of water pressure and inflow at all times.

For compaction grouting with mortar of less than 50 mm slump, modified concrete pumps are employed to give pressures up to $5\,\mathrm{MN\,m^{-2}}$ whilst continuously extruding the mortar. A continuous mortar mixer of the pugmill type, with helical feed to the pump to ensure continuity of mixed materials in the grout pipe, is essential.

The recent advent of microelectronics and magnetic flow meters has resulted in the introduction of multiple injection units. With these a single pump feeds a manifold linked to several individual injection pipes on each one of which is a magnetic flow meter. Such meters do not obstruct the flow passage and can handle particulate grouts or chemicals. The system can be used also whilst pumping water for test purposes on some holes and grouting on others simultaneously (Mueller, 1982). Electronic recording produces an individual record of injection pressure fluctuations and flow rates for each hole. This system is likely to lead to more economical grouting units in the future when its potential is fully realized. However, it does not allow variation of grout mix to each individual hole as required in current rock fissure grouting.

11.4.19 Mix-in-place methods

Whilst not true injection processes, mix-in-place methods involve pumping grout as a stabilizing agent into soil whose structure has been completely disturbed by the action of a mixer or jet to form a column centred on the drill pipe of thoroughly mixed grout and soil which then sets *in situ* after removal of the drill.

A Swedish adaptation of a Japanese treatment, using an 'egg-beater' mixer formed of a double helix to drill and stir soft clays, mixes them with a lime suspension which, when uniformly distributed, stabilizes and hardens the clay. Contiguous columns can build treated blocks for cut-offs or retaining functions, foundation support, etc. (Broms and Borman, 1979). This mechanical system has been used also with cement grout injection to form soil/cement columns of predetermined diameter by Intrusion Prepakt.

The current Japanese development of a UK jet grouting invention (Nicholson, 1963) uses a rotating horizontal jet to erode and disturb soil (Shibasaki and Ohta, 1982). Originally the jet comprised cement grout which was recirculated with disturbed soil from the bore through a mixer and back to the hole continuously, surplus bulked material being discharged on completion. In the Japanese development better erodability is claimed by using a water jet shrouded concentrically by compressed air with a separate grout jet about 30 cm below. Thereby disturbed soil is floated up on the grout and discharged out of the hole with very little original soil remaining in the bore.

Both these systems suffer from the disadvantage of irregular column size and irregular distribution of cement, although they can offer very useful stabilization with considerable strength. The virtue of jet grouting is that it can be used in uniform silts and silty sands which are otherwise virtually untreatable by injection processes close to surface. Sufficient strength for underpinning purposes is possible but continuity of water cut-off is difficult to achieve without uneconomically close spacing of holes.

11.4.20 Grouting contract philosophy

To achieve the best results from grouting it is important to allow a flexible approach. Significant ground variation cannot be predetermined and it may be necessary to change the detail of technique to suit. This demands an excellent rapport between the engineer and grouting contractor with day-to-day consultation on site. It implies also that details of individual grouting operations are fully recorded in respect of flow rate, quantity and pressure change and kept up-to-date. Records should be displayed in chart form for easy assimilation. Neither the specification nor pricing structure of a grouting contract should inhibit variation to improve the technical result. Grouting demands full application of the observational method of geotechnics first given expression by Terzaghi.

11.5 DIAPHRAGM WALLS

The technique of excavating a trench through a clay slurry was developed in the decade 1950 to 1960, mainly in Italy and Germany. The early excavations formed in this way were used for dam cut-off walls and other foundations. The system developed rapidly in Europe and was first introduced into Britain in 1961, when diaphragm walls were used as temporary works to support the excavation for the Hyde Park Corner Underpass. The technique is now firmly established and finds a wide application in its various forms. (ICE, 1974). Slurry wall specifications in the USA have been described by Millet and Perez (1981).

Diaphragm walls offer several advantages as a constructional technique. Basically the system offers immediate support to the ground as excavation proceeds and subsequent deformation of the adjacent soil is normally reduced to a minimum. This aspect is discussed later. Such walls can be constructed adjacent to existing foundations and plant has been developed which can work within a few decimetres of existing buildings. Furthermore, the operation is reasonably quiet and is virtually vibrationless. Excavations, with certain limitations, can be made in most soils, ranging from water-bearing sands to still clays. The technique is not cheap, particularly when reinforced walls are to be constructed. To maximize the advantages of the system it is usual to incorporate the diaphragm wall in the permanent structure whenever practicable. Watertight structures are also difficult to construct because of problems associated with forming joints between adjacent panels. Several patent joints have been developed which behave satisfactorily but it is virtually impossible to guarantee every joint watertight. Defective joints are usually grouted at a later stage in the project to reduce the inflow of water to a minimum. Frequently, a small amount of seepage is acceptable and can be channelled for pumping from the excavation. Such small seepages can be concealed behind a false wall, often constructed for a fair finish, since the surface of a diaphragm wall is generally rough; the roughness depending on the soil in which it is excavated, stiff clays giving the best appearance.

The clay slurry used to support the excavation is manufactured from bentonite, the trade name for the clay mineral sodium montmorillonite. The mineral montmorillonite is a three-layer structure comprising two silicate lattices and one aluminate lattice. The surface of the clay platelet is negatively unbalanced and absorbs metal cations to form the hydrogen, sodium, calcium or other form of the clay mineral. The activity of the clay depends on the cation absorbed, hydrogen being the most active. The bonds between individual plates of sodium montmorillonite are weak and the clay can be dispersed in water by violent shearing action. The bentonite rapidly hydrates, absorbs water around each clay platelet, and forms viscous slurry approximating to a Bingham body. The suspension also possesses the property of thixotropy and forms a weak gel when left to stand. The gel is readily broken down when disturbed. Typical properties of a bentonite slurry suitable for diaphragm wall construction are as follows (Hanna and West, 1976):

Bentonite concentration in water (%)	5
Plastic viscosity (cP)	9
Apparent viscosity (cP)	18
Gel strength after 10 min ($N m^{-2}$)	40
Density ($t m^{-3}$)	1.02

The bentonite must be thoroughly mixed before use and it is usual to allow it to hydrate for several hours. Equipment for mixing the clay with water should apply a high rate of shear to the slurry. High-speed purpose-made mixers are available, or alternatively a centrifugal pump can be used. During use the slurry rapidly becomes contaminated with soil and its density increases. For satisfactory construction a typical specification of the slurry would be as follows:

Density	less than 1.1 t m^{-3}
Viscosity (Marsh cone test)*	30 to 90 seconds
Shear strength	1.4 to 10 N m^{-2}

The problem of contamination is particularly severe in silts and fine sands where the slurry may have to be discarded after use in one excavation. In clay soils the slurry is normally used two or three times.

For trenches excavated in saline conditions it is desirable for the slurry to be mixed with fresh water before it is supplied to the trench since Bentonite will not hydrate satisfactorily in salt water. In special conditions the slurry may be treated with de-peptizing agents, such as sodium hexametaphosphate, to reduce the liability of flocculation, or a slurry manufactured using the clay mineral attapulgite which will hydrate in salt water may be used. It is possible to greatly increase the density of the slurry by the addition of powdered barytes or haematite, although such action is justified only if an exceptional constructional problem is encountered.

A clay slurry will stabilize excavations in a wide range of soil types. In permeable material bentonite slurry penetrates the pores and rapidly gels to form a barrier to prevent further loss. A filter cake of slurry will, in time, consolidate against the sides of the excavation and may form a skin of substantial thickness. The stability of the excavation is maintained by the hydrostatic pressure of the slurry acting on the soil structure through the filter cake. The hydrostatic pressure of the slurry is equivalent to an earth stress coefficient in the range 0.5 to 0.6 and is sufficient to stabilize most excavations. Typically, $K_a = 0.5$ in granular soils in the dry state. If the soil is saturated with water the equivalent earth stress coefficient increases to about 0.8 and failure of the excavation is liable to occur. In granular soils it is usual to ensure that the head of slurry is at least 2 m higher than the ground water level. In cohesive soils slurry-filled excavations are almost without exception stable. It is thus seen that normal wedge or slope stability theory may be applied to estimate the stability of an excavation filled with bentonite slurry. However, the above simple assumption of hydrostatic pressure slightly underestimates the stabilizing force exerted by the slurry. Other minor factors are the passive resistance of the slurry and the increase in shearing resistance of the zone of soil saturated with slurry due to the development of negative pore pressure in granular soils. These factors may increase the stability of the excavation by 10% to 25% (Elson, 1968).

The pressure exerted by the bentonite slurry is normally less than that existing

*Time for given volume of slurry to drain from standard cone.

in the soil before excavation (K_0 is frequently in the range 0.5 to 2.0), hence the soil will deform as excavation proceeds. The resultant movements are usually within the elastic range of the soil and can be estimated by a suitable elastic continuum calculation. Movement sufficient to cause structural damage does not normally occur unless the factor of safety for the excavation falls to a low value and a wedge failure starts to develop. The stability and deformation of heavily loaded footings adjacent to an excavation require careful consideration. The deformation of a diaphragm wall in London Clay was investigated by Cole and Burland (1972).

The above comments apply to linear excavations but in practice only relatively short panels are open at any one time. Arching of the soil in such cases is a significant factor in increasing the stability of the excavation. Schneebeli (1964) has treated the case for sands and Meyerhof (1972) has presented a simplified analysis for trenches in clay soils.

For rectangular cuts

$$F = \frac{Nc_u}{K_0\gamma h - \gamma_s h_1} \tag{11.8}$$

where h and h_1 are the depth of the trench and of the slurry, respectively. The stability factor, N, varies from a value of 4.0 for shallow excavations to a value,

$$N = 4(1 + B/L) \text{ for deep trenches}$$

where B is the width of the panel and L is depth of the panel.

The equipment developed to construct diaphragm walls has changed markedly since the first walls were excavated with a drop chisel (Veder, 1963). This method is still occasionally used if boulders are encountered or if the toe of the wall is to be located in rock. Trenches for most modern structural walls are excavated with a grab. Such rigs can excavate a wide range of soil types and give a relatively smooth finish to the wall. The rigs may be either cable operated or hydraulically operated and guided by a kelly bar. The various contractors undertaking this work have usually developed a machine particular to their needs. An alternative technique is to use a rotary drill with reversed circulation. Such machines have a high output but are limited to relatively uniform soil conditions. A reinforced concrete guide trench about 3 m deep is constructed. The trench acts as a guide for the grab and as a reservoir for the bentonite slurry. It is reasonable to expect the verticality of the wall to be within about ± 1 in 80. The operator does not have a great deal of control over the alignment of the wall but the heavy grabs used are normally sufficient to ensure the wall is sensibly vertical. For shallow excavations a standard hydraulic excavator may be suitable. Alternatively the EFTE process for constructing impermeable diaphragms can be used. In this method a series of piles are driven adjacent to each other and the space resulting on the withdrawal of the piles is filled with an impermeable grout.

To form impermeable diaphragms it is not necessary to backfill the trench with concrete. The cut-off wall of the Wanapum Dam was formed in river alluvium (La Russo, 1963). The 3 m wide trench was excavated at depths down to 25 m with a dragline. The trench was backfilled with excavated material comprising sandy gravel mixed with 20% silt. Slurry was added to the backfill to give it the consistency of high slump concrete. The material was blended in windrows with bulldozers before being dozed back into the trench.

The construction of reinforced concrete retaining walls in slurry trenches is now standard practice; panels up to 4 m long are excavated with a suitable machine and the trench floor carefully trimmed. The length of the trench is limited to that which can be backfilled with one tremie pipe. More than one pipe can be used in special circumstances but great care is required to avoid inclusion of zones of slurry or uncompacted conctete. The prefabricated reinforcing cage is lifted into position and the excavation backfilled with high slump concrete placed by tremie (FPS/BRMCA, 1977). The wall is usually concreted 200 to 300 mm higher than the design level and this concrete is subsequently removed because of contamination of the uppermost concrete with slurry and soil.

Innumerable shapes of wall are possible with the diaphragm technique, including plain curtain wall, Y or W section walls and cellular structures. At Seaforth Dock, Liverpool, a cantilever wall comprising wine-glass shaped segments was constructed to provide the required water depth of 15 m (Agar and Irwin-Childs, 1973). The structural continuity of panels is a problem in these composite structures and a method of positively connecting adjacent panels based on the use of a pile clutch has been patented. Reinforcement in walls can be made continuous with floor slabs placed after the main excavation is completed by fixing polystyrene blocks around bars that can subsequently be recovered and bent out. Polystyrene blocks fastened to the reinforcing cage can also be used to form box-outs or other openings in the diaphragm walls but due allowance should be made for the buoyancy of such blocks.

A recent development is the placing of precast panels in slurry trenches. Problems of placing tremie concrete and poor finishes to the wall are thus overcome but the problem of forming watertight joints between panels remains.

EXAMPLE 11.1 STABILITY OF SLURRY-FILLED TRENCH

Although this problem concerns a model trench in a laboratory – one of a number of experiments carried out by W.K. Elson – the method of analysis is applicable to full-scale excavations. The model trench 0.102 m wide and 0.912 m deep was formed in cohesionless sand and filled with a liquid suspension of 7.25% bentonite. When the side of the trench collapsed, the slurry level and the phreatic surface were both at ground level. The slightly curved rupture surface shown in Fig. 11.24 can be represented by a plane surface inclined at 67° to the horizontal. The buoyant unit weight of the sand was $10.54 \, kN \, m^{-3}$ and the unit weight of the suspension was $10.20 \, kN \, m^{-3}$. The unit weight of water is $9.80 \, kN \, m^{-3}$. The angle of shearing resistance of the sand was 33.5° and the apparent cohesion of the suspension at maximum gelation was $0.100 \, kN \, m^{-2}$. It is necessary to employ in these calculations Equation 3.79 for the yield pressure of a thin block of material between platens: in this case the width of the block is twice the depth of the trench since mass displacement of the block is restricted to upward movement from the bottom of the trench.

Weight of wedge of soil $W = 0.5 \times 10.54 \times 0.912^2 \tan 23° = 1.863 \, kN$.

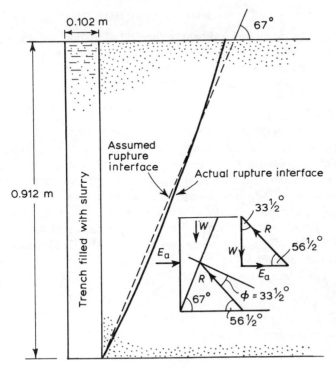

Figure 11.24 Model trench.

Horizontal thrust from wedge at failure $E_a = 1.863 \tan 33.5°$ $= \underline{1.23\,\text{kN}}.$
Horizontal hydrostatic thrust from groundwater $P_w = 0.5 \times 9.80 \times 0.912^2 = \underline{4.07\,\text{kN}}$
          Total disturbing thrust $= \underline{5.30\,\text{kN}}.$

Horizontal resistance of slurry functioning as block between platens from Equation 3.79,

$$q_f = c_u \left[\frac{B}{2H} + \frac{\pi}{2} \right] = 0.912 \times 0.100 \left[\frac{2 \times 0.912}{2 \times 0.102} + \frac{\pi}{2} \right] \quad = \underline{0.96\,\text{kN}}.$$

Horizontal hydrostatic thrust from slurry $P_s = 0.5 \times 10.20 \times 0.912^2 = \underline{4.24\,\text{kN}}.$
          Total resisting thrust $= \underline{5.20\,\text{kN}}.$

The small discrepancy of 0.10 kN is within the range of experimental error. Possible sources of small errors may arise from the assumption of a plane rupture surface and from inconsistency in the value of ϕ: a slight increase in the mean value of ϕ would lead to a reduction in disturbing thrust. Cohesion arising from penetration of the gel into the sand would also lead to a reduction in disturbing thrust.

For design purposes the two components of resisting thrust and the hydrostatic component of disturbing thrust can be calculated as shown above but the active thrust of the soil would be calculated from the relationship $P_a = 0.5 K_a \gamma' z^2 = 0.5 \times 0.29 \times 10.54 \times 0.912^2 = 1.27\,\text{kN}$, since $K_a = 0.29$ when $\phi = 33.5°$ and $\delta = 0$. This value is sufficiently close to the experimental value for design. A factor of safety should be applied, of course, to the resisting thrust which would accordingly have to be increased.

11.6 GROUND ANCHORS

A comparatively recent development in the field of foundation engineering has been the use of ground anchors or tie-backs to provide temporary support to excavations and permanent support for a variety of structures. Construction and foundation engineering specialists have introduced a number of patent anchorage systems.

Ground anchors basically comprise a steel tendon, the main structural element, which is anchored in the soil, usually by grouting, and an anchor block fixed to the structure. The normal upper limit for the working life of temporary ground anchors is considered to be 2 years. Permanent anchors require careful design and particular attention must be paid to the creep characteristics of the system. Such effects are not well understood at present and basic research is in progress. For this reason the use of ground anchors in permanent situations is limited to the best soil conditions. The working capacity of anchors depends mainly on the type of ground in which they are installed and varies from 20 tonnes in clays to over 300 tonnes in soft rocks.

The principal categories of anchors are:

(a) Straight shafted anchors.
(b) Under-reamed anchors.
(c) Displacement anchors.
(d) Mechanical anchors.

Straight shafted anchors were the original type that was developed and may be installed in either cohesive or cohesionless soils. The anchors rely on the development of adhesion along the shaft to provide the stabilizing force. In clays the anchors are simply constructed by augering a hole to the required depth, inserting the tendon and filling the annular space with high-strength grout. In cohesionless soils the pocket in which the anchor is accommodated is formed by drilling or driving a casing into the soil. For high-capacity anchors in suitable soil, tendons may be made up with several 5 to 7 mm diameter steel wires. Anchors with capacities up to 200 tonnes may be constructed by this technique.

Under-reamed anchors have largely superseded the straight shafted type in cohesive soils. One or more conical cavities are cut in the soil along the length of the borehole using an expanding cutter. The cones increase the effective diameter of the anchor and offer an increase in end-bearing capacity. In multi-under-reamed systems the capacity does not depend solely on adhesion, which may be only 30% to 35% of the undrained shear strength, but also on cohesion since the failure surface is mainly in undisturbed clay. However, the shear strength of the clay should be not less than about $100 \, \text{kN m}^{-2}$ as the hole may collapse during the under-reaming operation.

Displacement anchors are formed by filling the auger hole over the required anchor length with gravel. A hollow mandrel fitted with a loose shoe is driven into the hole to displace the gravel into the soil and the tendon placed inside the mandrel. The tendon and gravel are grouted while the mandrel is withdrawn. This technique is employed in fine-grained soils in which cement grout will not penetrate to form a larger anchor zone.

Mechanical anchorage systems are of various patent designs and usually involve the insertion, and subsequent expansion, of a metallic anchor plate in the soil.

11.6.1 Anchor design

Most anchors depend on a cement-based grout to transfer the load from the tendon to the soil, hence the properties of the grout and the adhesion between the soil and the grout are important factors to consider in design. The limiting factor is normally failure or excessive deformation of the soil. It is therefore necessary to obtain adequate geotechnical data, including pH and sulphate content of the soil.

In granular soils, a displacement technique is normally used to install the tendon and grouting is employed to consolidate the anchor zone. Compaction of the soil around the anchor by the installation technique is advantageous in that it increases the shearing resistance of the soil, and lithification of the anchor zone by suitable grouting techniques greatly increases the diameter of the anchor.

The pull-out capacity T_f of the anchor may be estimated from a consideration of the shear strength of the soil and may be assumed to be the sum of the side shear on the anchor and the end bearing resistance (Littlejohn *et al.*, 1972).

$$T_f = A\gamma\left(h + \frac{L}{2}\right)\pi DL \tan \phi + B\gamma h \frac{\pi}{4}(D^2 - d^2). \tag{11.9}$$

where A is the ratio of the contact stress at the fixed anchor/soil interface to the effective stress of the overburden; B is a bearing capacity factor; γ is the unit weight of overburden (submerged unit weight beneath water table); h is the depth of overburden to top of fixed anchor; L is the length of the fixed anchor; D is the effective diameter of the fixed anchor; d is the effective diameter of the grout shaft or column.

The value of the factor A depends largely on the installation technique adopted and normally ranges from the earth stress coefficient at rest, K_0, (about unity), where no compaction of the soil occurs during installation, to 2 where heavy compaction takes place or grout pressure is maintained. The factor B is related to the bearing capacity factor N_q, although experience has shown that the value of B is less than the value of N_q given by Terzaghi, on account of the slender shape of the anchorage and the disturbance of the soil during construction. Littlejohn (1972) estimates the ratio of N_q/B to be about 1.4 when $\phi = 35°$, rotary percussion drilling is employed and the grout injection pressure is $350 \, \text{kN m}^{-2}$. As an alternative to the theoretical Equation 11.9 the bearing capacity of the anchor may be estimated from the empirical relationship

$$T_f = Ln' \tan \phi \tag{11.10}$$

where $n' = 13$ to 16.5 tonne m^{-1}, based on observations for $h = 6.1$ to $9.2\,m$, $D = 180$ to $200\,mm$ and $L = 0.9$ to $3.7\,m$. This rule is particularly useful where previous experience in the area is available on which to base the choice of the constant n'.

Cement grouts will not permeate soils with permeability of less than $1 \times 10^{-4}\,m\,s^{-1}$ and a cylindrical body of grout is formed in the soil. The application high grout pressure is used to displace the soil and hence enlarge the anchorage. In particular cases the grout pressure may be sufficiently high to cause hydraulic fracture of the soil. Equation (11.10) may be used to estimate the bearing capacity of such anchors, the factor n' being in the range 13 to 20 tonne m^{-1}. The resistance of anchors in this type of soil depends mainly on friction on the surface of the grout body. It is therefore reasonable to assume that this resistance is related to the grouting pressure used, provided the pressure is less than that required to cause hydraulic fracture, hence the ultimate pull-out resistance may be estimated from the equation

$$T_f \simeq p_i \pi D L \tan \phi \qquad (11.11)$$

where p_i is the pressure of the grout.

In cohesive soils the pull-out resistance of an anchor may be estimated in a manner similar to that adopted for the design of piles. The pull-out capacity is taken as the sum of the end-bearing resistance plus adhesion on the perimeter of the anchor body. A small contribution will also be made by adhesion on the shaft of the anchor. Hence the pull-out resistance can be estimated from the formula

$$T_f = \alpha_a c_u \pi D L + \frac{\pi}{4}(D^2 - d^2)N_c c_u + \alpha_s c_u \pi d L \qquad (11.12)$$

where α_a is the adhesion factor applicable to the anchor and α_s is the adhesion factor applicable to the shaft.

The value of the adhesion factor varies with the type of anchor employed. For straight shafted anchors the value of the adhesion factor corresponds to the fully softened strength of the soil (0.3 to 0.4). For gravel displacement anchors the adhesion factor is in the range 0.6 to 0.75 (Littlejohn, 1970), presumably on account of the increased roughness of the anchorage and reduced contact of the soil with grout. For multi-under-reamed anchors the failure surface passes, to a large extent, through undisturbed soil. The rupture surface between under-reams is not, however, perfectly cylindrical and the adhesion factor is thus slightly less than unity. A probable range 0.7 to 0.9 is suggested by Basset (1970). The value of N_c is normally considered to be 9, the value corresponding to deep foundation. A comprehensive review of the design of ground anchors has been published by Littlejohn (1980).

In clay soils a conservative approach should be adopted for design of permanent anchors. Excessive creep of such anchors has occurred, especially in soft clay, and it is suggested that soil with an undrained shear strength of less than $150\,kN\,m^{-2}$ is an unsuitable foundation for such anchors.

For anchors in soft rocks such as marl or chalk, very high anchor loads may be developed according to the state of weathering of the material. Indeed, in many cases the strength of the anchor is governed by the failure of the steel tendon or by bond stress limitations. Design of anchors in such materials is usually by empirical rules based on previous experience, since most laboratory or common *in situ* tests, such as the Standard Penetration Test, are not suitable for assessing the relevant strength parameters of the rock; this problem is particularly intractable in weathered material.

To confirm the design assumptions and guarantee the required performance of the anchors, it is normal practice to test load several trial anchors to destruction. A proportion of the working anchors are normally tested to about 1.5 times the working load as a quality control measure and as a means of confirming the satisfactory behaviour of the anchorage system. Care should be taken that the test load does not exceed the limit of proportionality of the steel tendons and it may be desirable to construct the test anchors with an additional tendon to overcome this problem. The load/deformation curve for the production anchors may be compared with that of the initial trial anchors and an estimate made of their actual factor of safety. The load test is easily carried out with the hydraulic jack used to tension the cables and a datum beam from which the deformation can be measured. The jack requires calibration before use or, preferably, a dynamometer should be used to record the applied load.

11.6.2 Other design considerations

Reasonably generous factors of safety should be adopted for the design of the soil anchor to limit prestress loss and to allow for uncertainties in the values of soil properties adopted for the design. For temporary anchors and those in granular soil, a factor of safety of 2 to $2\frac{1}{2}$ is commonly employed. In clay soils, particularly when soft, a factor of safety in the range 3 to 3.5 may be desirable in order to limit creep to an acceptable figure. The steel components are normally designed to comply with the relevant code of practice (CP 110, BSI, 1972c). It is normal practice to load each anchor to 1.25 times the working load during installation as a check and to reduce creep of the anchor in service.

After the safe bearing capacity of the individual anchors has been determined, the overall design of the retaining wall should be considered. The load distribution on the wall may be estimated by the methods described in Chapter 6 and reference can be made, for example, to Jack (1971) and James and Jack (1974). However, intermediate stages of construction should also be taken into account since, depending on the construction method, the wall may be subjected to stress reversal and the maximum permissible load in any line of anchors should not be exceeded. The spacing of the anchors is dictated by the wall height, wall flexibility, anchor working load and construction limitations. Littlejohn (1970) recommends that the minimum cover to the top anchorage from the ground surface should be 5 m. The anchors may be inclined but ideally they should be near horizontal. Inclinations as high as 60° may be adopted with advantage if the underlying strata have a high anchorage capacity. The minimum horizontal spacing of the anchors in UK practice is taken as four times the diameter of the anchor. This spacing can be reduced if adjacent anchors are splayed and are of different lengths. The length of the anchor is usually determined by estimating the location of the potential failure plane behind the wall and ensuring the anchors are placed well beyond it. A minimum projection of 5 m beyond the potential failure plane is suggested by Littlejohn (1970). If inclined anchors are used, a check should be made that the vertical component of the anchor force does not lead to a total downward force exceeding the bearing resistance of the soil beneath the wall.

Ground anchors are normally constructed using a cement grout for the anchor; rapid hardening cement is commonly used. Typical times between grouting and stressing the anchor for various cements are given below:

Ordinary Portland cement	21 days
Rapid hardening cement	7 days
Epoxide or polyester resin	a few hours.

In general, the strength of the grout should exceed 20 MPa at the time of stressing and to this end low water/cement ratios, in the range 0.4 to 0.6, are used, frequently with the addition of plasticizers. Flow meters and grout cubes are used to control the quality of the grout. In aggressive environments it may be necessary to use sulphate resisting cement. For anchors used in temporary works, sophisticated corrosion protection is not necessary unless the ground is particularly aggressive. Temporary anchor cables are normally protected with a greased tape decoupling sheath over the elastic length and the entire hole filled with cement grout. For permanent anchors a system developed by the Cementation Company is typical. The anchor cable is made up of strands of wire which have been greased and sheathed in polypropylene 1 mm thick and bound with PVC tape. The anchor length of the cable is stripped, degreased and cased in a corrugated block of epoxide resin. The cables are inserted in the drill hole and grouted over the anchor length, stressed and the hole filled with grout. There is sufficient freedom between the sheathing and the tendons to allow the cable to be restressed at any time.

11.7 PRESSURE OF CONCRETE ON FORMWORK

An example of the collapse of formwork under pressure of concrete was quoted in the preamble to this chapter but failures occur even today in spite of the greater care exercised generally in the design of temporary work. The estimation of lateral pressure of freshly poured concrete on vertical formwork is important in foundation construction where the depth of concrete placed in a single lift is considerable and the tendency is to increase rates of pouring. Since formwork often accounts for 30% or more of the cost of a structure, designs must be economical as well as safe. Rodin (1952) reviewed the work of a number of investigators and examined factors affecting the pressure of fresh concrete. Experimental and other studies have been reported by Ritchie (1962), Fleming and Wolf (1963), Peurifoy (1965) and Ore and Straughan (1968). A field investigation of pressure on formwork for prepacked concrete has been described by Akatsuka (1968).

Thirteen variables entering the problem of pressure on formwork are discussed in a report by ACI Committee 622 (ACI, 1958) and recommendations for design of formwork are given in ACI Standard 347 (ACI, 1978). The results of one of the most comprehensive investigations on concrete pressure, sponsored by the Construction Industry Research and Information Association and carried out by the Cement and Concrete Association, are described in CIRIA Research Report 1 (CIRIA, 1965) and are outlined below. An instrument termed a formwork pressure balance (Kinnear, 1963) was designed especially for the investigations described in Research Report 1 since it was considered that discrepancies between the results of previous studies may have been due to difference in measuring techniques. Observations were made both in the laboratory and at a number of sites. It was found that fresh concrete behaves almost as a liquid under the influence of vibration but that stiffening of the concrete and arching between forms lead to deviation from the equivalent linear hydrostatic distribution of pressure and, furthermore, that the impact of concrete discharged into formwork may lead to an increase in pressure. The maximum values of pressure recorded in these investigations ranged between 2.5 and $134.0\,\text{kN}\,\text{m}^{-2}$. A typical record of pressure variations at a given level on formwork as placing proceeds is shown in Fig. 11.25(a).

Stiffening of concrete involves progressive increase in resistance to mobilization under vibration. This is due partly to chemical changes in the cement matrix and these depend upon time, temperature and the type and fineness of the cement. It is due also partly to mechanical interlocking between aggregate particles and this is a function of pressure, workability (including water/cement ratio, cement content, and aggregate size, shape and grading) and, in addition, the history of vibration. Workability involves both compactability (the ease with which dense concrete can be formed) and mobility (the ease with which concrete can be made to flow under the influence of vibration). As stiffening develops, the concrete becomes capable of supporting additional surcharge without increase in lateral pressure.

The phenomenon of arching, generated by frictional forces developed between the concrete and the form faces, leads to a reduction in vertical load at any horizontal section. These frictional forces assume particular importance in narrow sections where the surcharge volume is comparatively small. Arching depends on the minimum dimension of the section (Fig. 11.25(b) and (c)), the profile and slope of the form surface, the distribution of lateral pressure over the form, and the variation of the coefficient of friction between the concrete and the form faces. The presence of reinforcement does not appear to influence arching,

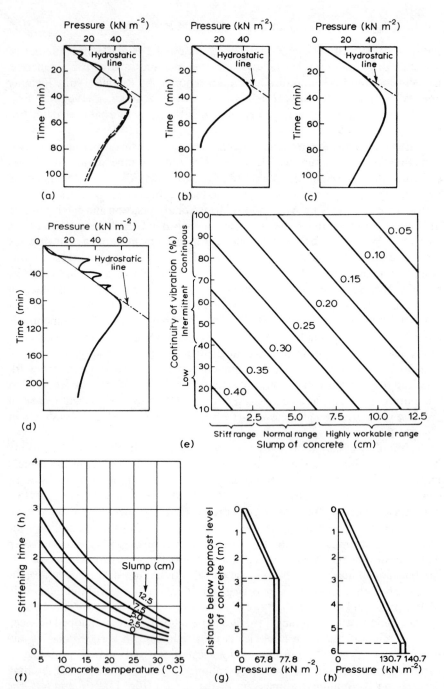

Figure 11.25 (a) Superposition of idealized pressure curve upon test result. (b) Typical idealized pressure curve for narrow section. (c) Typical idealized pressure curve for large section. (d) Pressure surcharges due to impact. (e) Values of factor C. (f) Stiffening time related to concrete temperature and workability. (After CIRIA, 1965.) (g) Pressure on formwork for wall. (h) Pressure on formwork for column.

probably because of the transmission of vibrations through the steel. In columns of small section, arching may develop in two directions but, since the compactive effort in columns less than 0.3 m × 0.3 m. is generally very concentrated, columns are treated as short walls for design purposes.

The effect of impact of discharge is to increase the pressure above that due to the static surcharge. It is reasonable to assume that the pressure developed depends upon the impact energy applied and its distribution in the mass of concrete. In fresh concrete, which behaves as a liquid, impact energy is absorbed as a surge of pressure which is relieved by continuing vibration (Fig. 11.25d). On the other hand, concrete which has assumed initial stiffening may become wedged in the formwork under impact and is more resistant to mobilization; in this case the pressure surcharges due to impact are generally permanent and cumulative. However, concrete which has developed initial stiffening is generally more remote from the point of impact and its inherent stiffness affords partial protection against pressure surge; permanent surcharges are consequently small in scale.

The five factors which have the greatest influence on concrete pressure are temperature of concrete, rate of placing, minimum sectional dimension between forms, workability and degree of vibration. Pressure on formwork can be estimated from a chart given in Research Report 1 or from the equations quoted below. Calculated pressure can be expected normally to be accurate to within about $\pm 10\%$. The findings published in Research Report 1 agree closely with those reported by ACI Committee 622 for concrete placed in columns at low temperatures but it appears that the ACI figures are conservative both at temperatures exceeding 20°C and in wall sections where arching is appreciable.

It was found that the variation of pressure at a given level on formwork with time after placing concrete at that level and as placing continued above it can be represented by the equation

$$p = \frac{\gamma_c R t}{1 + C(t/t_{max})^4} \tag{11.13}$$

where p is the pressure (kN m^{-2}), γ_c the unit weight of the concrete (kN m^{-3}), R the rate of placing (m h^{-1}), t the time from commencement of placing (h), t_{max} the stiffening time of the concrete (h) (Fig. 11.25f), and C a factor depending on the workability of the concrete and the continuity of vibration (Fig. 11.25e).

When $t = t_{max}$,

$$p_{max} = \frac{\gamma_c R t_{max}}{1 + C}. \tag{11.14}$$

The stiffening time is the time at which maximum pressure occurs and this depends on the concrete temperature, since the rate of chemical reaction depends on it, and on the workability of the concrete. The relationship between stiffening time, temperature and workability, as expressed by slump, is shown in Fig. 11.25(f). When agents are added to a mix to accelerate or retard setting time, the value of C must be adjusted accordingly. The degree of continuity of applied vibration is expressed as a percentage of the total placing time, as shown in Fig. 11.25(e). It should be noted that p_{max} cannot exceed the corresponding equivalent hydrostatic pressure $\gamma_c H$, where H is the head of concrete.

In order to take account of the withdrawal of the source of vibration which normally occurs with very high rates of placing, a term is added to Equation (11.13). In such instances the vibrator may be some distance above the fresh concrete before any appreciable stiffening due to chemical reaction occurs and the mechanical shear strength quickly reaches its maximum value. The term $1.9\ (2.4 - R)$ has the effect of adding small values of pressure at $R < 2.4$ and subtracting comparatively large values at $R > 2.4$. The pressure equation now becomes

$$p = \frac{\gamma_c Rt}{1 + C(t/t_{max})^4} + 1.9(2.4 - R). \tag{11.15}$$

Where the rate of placing is very high, the total time of placing may be less than necessary for t_{max} to apply and in such cases the maximum pressure is determined by use of the full formula, substituting the total time of placing for t. Impact pressure surcharges may be up to $20\,\mathrm{kN\,m}^{-2}$ in the early stages of a lift where large heights of discharge occur. Under conditions of vibration, however, these pressure surcharges are of short duration and the pressure returns to the original hydrostatic level within a few minutes (Fig. 11.25b). As the depth of concrete increases, the height of free discharge is reduced and the severity of impact is also diminished. In addition, the impact energy is dissipated throughout the fresh concrete and rarely extends over depths greater than 2 m. As the concrete develops, higher internal shear strength and consequent resistance to mobilization under vibration, it becomes equally resistant to impact surcharges. Nevertheless, those surcharges which do penetrate the concrete lead to permanent cumulative increases in pressure but these have never been observed to exceed $10\,\mathrm{kN\,m}^{-2}$ and it is recommended that an allowance of this magnitude be made in design to account for impact.

It was found that when the maximum influence of arching between forms is taken into account

$$p = 15 + d + 3.2R, \tag{11.16}$$

where d is the minimum section dimension in centimetres. This holds for $d = 15$ to 45 cm and may be applicable for $d > 45$ cm although this has not yet been confirmed by tests. For design purposes, pressure determined from Equation (11.16) should be compared with pressure determined from Equation (11.13) or (11.14) and the former is employed for design if it is less than the latter. The influence of arching is destroyed if extremely heavy or external vibration is used with highly workable mixes and Equation (11.16) is not then valid.

The following relationships are given in ACI Standard 347 (1978b) for the design lateral pressure $p\ \mathrm{kN\,m}^{-2}$ for formwork.

Ordinary work with normal internal vibration, $\gamma_c = 23.56 \, \text{kN} \, \text{m}^{-3}$ (150 1bf ft^{-3}), slump $\ngtr 100 \, \text{mm}$.

Columns $\quad p = 7.2 + \dfrac{785 R}{t + 17.8}$, $(\ngtr 144 \, \text{kN} \, \text{m}^{-2} \text{ or } 23.5 H)*$

Walls $\quad R \ngtr 2 \, \text{m} \, \text{h}^{-1} \quad p = 7.2 + \dfrac{785 R}{t + 17.8}$, $(\ngtr 95.8 \, \text{kN} \, \text{m}^{-2} \text{ or } 23.5 H)*$

Walls $\quad R = 2 \text{ to } 3 \, \text{m} \, \text{h}^{-1} \quad\quad\quad p = 7.2 + \dfrac{1156}{t + 17.8} + \dfrac{244 R}{t + 17.8}$,

$\quad\quad (\ngtr 95.8 \, \text{kN} \, \text{m}^{-2} \text{ or } 23.5 H)*$

Walls $\quad R > 3 \, \text{m} \, \text{h}^{-1} \quad p = 23.5 H$

where

γ_c is the unit weight of fresh concrete in $\text{kN} \, \text{m}^{-3}$,
R is the rate of placement of concrete in $\text{m} \, \text{h}^{-1}$,
t is the temperature in °C of the concrete in the formwork,
H is the height in m of fresh concrete above point considered.

Modifications to pressures are suggested in ACI Standard 347 (1978b) when special cements or admixtures or external vibrators are employed. Higher values of pressure are developed when the aggregate is preplaced and design recommendations are given in Standard 347. The relationships given above for walls can be used also for mass concrete.

Concreting in slip forms

$$p = 4.79 + \frac{524 R}{t + 17.8}$$

The relatively low value of 4.79 for the first term is justified by slight vibration only in slip formwork, by absence of re-vibration and by placing concrete in shallow layers of 150 to 250 mm. Where these limitations are exceeded, the first term should be increased to 7.2.

The ACI equations for walls lead to higher values of absolute maximum pressure than the CIRIA equations. Studies of formwork for walls by Gardner and Ho (1979) confirmed the hydrostatic nature of pressure up to a maximum value but this was less than the CIRIA value in the temperature range (12 to 22°C) of the investigation. Further conclusions derived from this study were that higher pressures are created by higher rates of pour, by higher values of slump and by wider forms. Variations in maximum aggregate size and in concrete strength had no significant influence on lateral pressure. Yielding of formwork leads to a decrease in pressure. Douglas et al. (1981) and Harrison (1983) have reported further investigations.

*Whichever is least.

EXAMPLE 11.2 ESTIMATION OF PRESSURE OF CONCRETE ON FORMWORK

Consider the formwork for a wing wall and for a column for a bridge. The concrete will be mixed on site and poured from a skip. The rate of placing will be $4\,\mathrm{m\,h^{-1}}$ and the estimated temperature of the concrete is about 15°C. Unit weight of fresh concrete $23.5\,\mathrm{kN\,m^{-3}}$. A poker vibrator will be employed and the continuity of vibration will be about 60%.

Wing wall. 6 m high × 40 cm wide. Slump 5 cm.
Time required to complete concreting is $6/4\,\mathrm{h} = 90\,\mathrm{min}$. From Fig. 11.25(f), t_{max} is 80 min. From Fig. 11.25(e), C is 0.22.
Stiffening criterion (Equation 11.15), noting that the stiffening time is less than 90 min:

$$p_{max} = \frac{23.5 \times 4 \times 1.33}{1 + 0.22} + 1.9(2.4 - 4) = 99.5\,\mathrm{kN\,m^{-2}}$$

Arching criterion (Equation 11.16)

$$p_{max} = 15 + 40 + 3.2 \times 4 = 67.8\,\mathrm{kN\,m^{-2}}.$$

Adding the impact allowance, maximum design pressure is $67.8 + 10 = 77.8\,\mathrm{kN\,m^{-2}}$.
The impact allowance would not be required if concrete were placed relatively gently through trunking.

$$p_{max}/\gamma_c = 67.8/23.5 = 2.9\,\mathrm{m}\ (\text{see below}).$$

Column. 75 cm × 120 cm × 6 m high. Slump 10 cm. Time required to complete concreting is 90 min. From Fig. 11.25(f), t_{max} is 110 min and since this exceeds 90 min, the maximum pressure will be developed at completion of placing. From Fig. 11.25(e), $C = 0.12$.
Stiffening criterion (Equation 11.15)

$$p_{max} = \frac{23.5 \times 4 \times 1.5}{1 + 0.12(1.50/1.83)^4} + 1.9(2.4 - 4) = 130.7\,\mathrm{kN\,m^{-2}}.$$

Whether or not the influence of arching is experienced in a section of width 75 cm is not yet known. Hence, adding the impact allowance, the maximum design pressure is $140.7\,\mathrm{kN\,m^{-2}}$

$$p_{max}/\gamma_c = 130.7/23.5 = 5.6\,\mathrm{m}\ (\text{see below}).$$

The maximum possible equivalent hydrostatic pressure in both wall and column is $(23.5 \times 6) = 141\,\mathrm{kN\,m^{-2}}$: if this were less than the values calculated above, then it would be employed in design unless conditions dictated otherwise.
Pressure diagrams for the wall and column formwork are shown in Fig. 11.25(g) and (h): the maximum pressure is developed below a level equal to p_{max}/γ_c from the topmost level of the concrete.

11.8 GENERAL REVIEW OF UNDERPINNING

Underpinning is the process of taking out existing defective foundations of a structure and replacing them with new and often deeper foundations or of carrying existing foundations to a greater depth by reason of new and deeper construction in the vicinity. Adequate site investigations are as necessary for underpinning as they are for all other foundation work. Live and dead load should be reduced as much as possible before underpinning operations are commenced. The structure to be underpinned, together with adjacent structures, should be examined for cracks both before work is commenced and during construction. Sketches should be made and photographs taken of existing cracks, room by room, and notes kept of further developments. Paper or glass strips (tell-tales) fixed with adhesive across cracks indicate whether or not movement takes place. These can be supplemented by horizontal pencil lines a few centimetres long, with short vertical lines at the ends, carefully drawn on the wall on both sides of the crack. Vertical and horizontal movements can then be measured. More refined measurements can be taken by inserting in walls reference plugs for portable strain gauges, arranging one plug on one side of each crack and, on the other side, one horizontally opposite it and one vertically below; so that both horizontal and vertical displacements can be observed.

New foundations for underpinning are designed by normal methods. Underpinning is nearly always followed by settlement, although it is generally small in magnitude. The soil within the range of influence of new foundations should already be consolidated under the load from the structure and, unless conditions are altered considerably, settlement should not be very much in excess of the amount required for the structure to bed down on the new foundations and on the soil at the contact face. Pre-loading reduces settlement.

There are numerous reasons for failure of existing foundations. Failure of timber piles may be due to marine borers, insufficient initial penetration or deterioration of the heads owing to a fall in water table. The corrosive action of water-quenched boiler ashes has, in at least one instance, caused failure of grillage beams encased in relatively porous concrete. At Brighton, Sussex, liquid employed to clean railway carriages attacked both the concrete foundations of the washing plant and the chalk bedrock. Evidence points to poor workmanship or unpredicted occurrences as the most probable cause of defects in foundations. It follows that consideration should be given in the design stage to possible extraneous causes of failure.

Underpinning operations should be planned to cause a minimum of interference with traffic and the inhabitants of buildings. Foundations for some new buildings in the USA have been constructed before the old building has been demolished. Deeper basements can be constructed and the new foundations utilized to carry the old superstructure temporarily.

It is convenient to consider moving structures bodily in this section since the operations can be regarded as a form of mobile underpinning.

Shoring is commonly required to give temporary support to a structure which is to be underpinned or where new and deeper foundations are being constructed at an adjacent site. Loads on shoring can seldom be subjected to rigorous analysis and, although approximate estimates of forces in members can be made, the dimensions of shores are determined largely by experience. In practice, the force in a shore, or the force imposed by a shore on a structure, depends on the extent to which wedges are driven or the extent to which members are levered or jacked into position. Overdriving may damage a structure or cause an accident. Shoring should be inspected at least once daily to ensure that fastenings have not worked loose or that swelling of timber in wet weather does not lead to the development of excessive forces. Safe basic stresses for stress-graded timber, together with allowable loads on joints fabricated with nails, screws, bolts and connectors, are specified in CP 112 (BSI, 1971). Reference should also be made to CP 2004 (BSI, 1972d). As a guide for preliminary studies only the following basic stresses can be employed: bending and tension parallel to grain (7 to 9) 10^3 kN m^{-2}, compression parallel to grain (5 to 6) 10^3 kN m^{-2}, compression perpendicular to grain (13

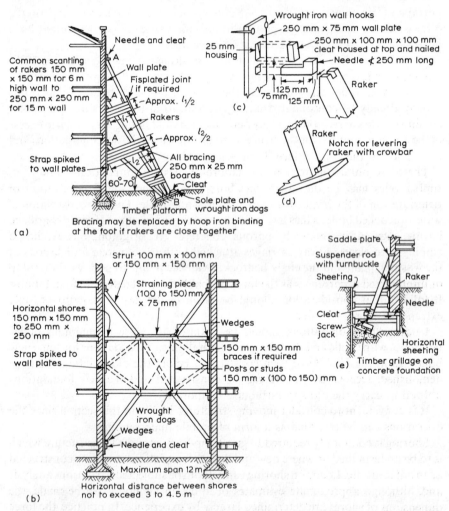

Figure 11.26 Details of shoring. (a) Raking shore; (b) multiple flying shore; (c) detail at A; (d) detail at B; (e) shoring for underpinning; (f) (not shown) similar to multiple flying shore but one horizontal shore and the two posts are omitted. It is not used when the distance between walls exceeds about 8 m. The horizontal distance between shores should not exceed 3 to 4.5 m.

to 15) 10^2 kN m^{-2}, shear parallel to grain (7 to 9) 10^2 kN m^{-2}. Mean modulus of elasticity 9×10^6 kN m^{-2}. The safe permissible or working stress is equal to the basic stress reduced by factors which take into account service conditions, slenderness ratio, deterioration and other factors. Owing to imperfections in constructional timber, which may not be stress graded, and to varying service conditions, permissible stresses are commonly restricted to relatively low values.

Figures 11.26 and 11.27 show details mainly of the traditional kinds of shoring developed by long experience, although little attention was given to the loads to

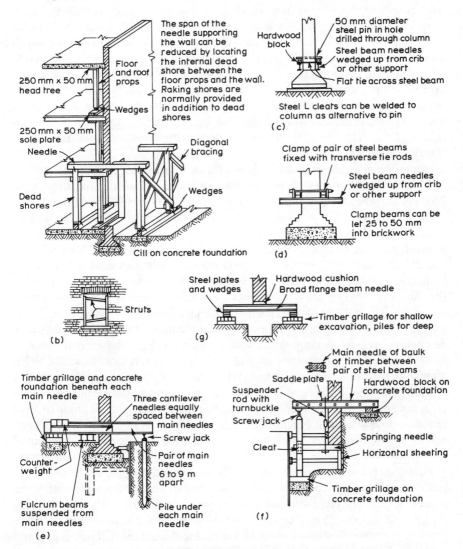

Figure 11.27 Details of shoring. (a) Dead shore; (b) strutting openings; (c) temporary support for steel column; (d) temporary support for brick pier; (e), (f), (g) shoring for underpinning.

be supported. If the nature of the forces are considered carefully, it may be found that alternative methods of achieving stability may be more satisfactory or less costly than traditional methods.

When it is necessary to remove old foundations of brickwork, masonry or mass concrete, they can be shattered by hydraulic cartridges inserted in holes drilled in the mass. Application of hydraulic pressure bursts the cartridges and the mass is disrupted. Thermic drilling can be employed to split similar masses and also reinforced concrete foundations. Disintegration is effected by differential thermal stresses induced by intense heat which is applied by a lance comprising a steel tube, 1.5 m long and about 20 mm diameter, housing a number of 3 mm diameter steel rods and with a handle at one end. Oxygen is passed through the tube which is ignited at the open end. The lance is consumed as holes are burnt in the mass to effect disruption and requires renewal at intervals. Water-cooled diamond saws operating on frames fixed to structures are now available for dismantling buildings. The demolition of buildings has been discussed by Musannif (1980), in BS 6187 (BSI, 1982) and in a publication by the Cement and Concrete Association (1982). The use of explosives for demolition is described by Brook and Westwater (1956).

Hydraulic or screw jacks are required for shoring, underpinning and moving structures. Hydraulic jacks are commonly available with capacities up to 2000 kN and the travel of the ram varies between 150 and 300 mm. The overall efficiency of hydraulic jacks approaches 70% but that of screw jacks seldom exceeds 30%. Hydraulic jacks are more easily controlled than screw jacks and in an emergency the load can be quickly released but, whereas a screw jack will hold a load for an unlimited period, an hydraulic jack may slacken. The load on an hydraulic jack can be measured approximately by a pressure gauge connected to the fluid circuit but accurate measurements should be made with a dynamometer or a proving ring. A flat diaphragm hydraulic jack is useful for lifting heavy loads but the displacement generally does not exceed 25 mm. The diaphragm jack may be circular, oval or rectangular. A 900 mm diameter jack can exert a thrust of 8500 kN but a 75 mm diameter jack will probably not exert more than 30 kN. If a number of hydraulic jacks are in use at the same time, they can be connected to a central pump and accumulator or they can be operated independently. When jacks are operated independently at the same time, the men should pump or lever in unison to a whistle in batches of twenty or thirty strokes. At the end of each batch the extensions of the rams are measured to determine differential movements and those are commonly corrected when they reach about 3 mm.

New underpinning foundations can be pre-loaded by placing one or more jacks in a 0.5 to 1 m gap between the new foundations and the underside of the superstructure. The jacks are loaded, where possible, to about 50% more than the design load but care must be taken not to damage the superstructure in the process. Short steel columns are then inserted and wedged tight in the gap and encased in concrete after the jacks are removed. The process eliminates nearly all settlement of the superstructure on sand but only a smaller proportion of the

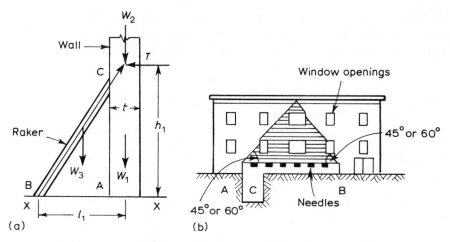

Figure 11.28 Loads on shoring.

settlement on clay. Where the bulbs of stress of a group of underpinning piles overlap, all the piles should be loaded simultaneously if possible.

If sections of the base of a wall are to be removed, consideration must be given to the influence of arching. The effect of arching can be observed when a beam or floor is test-loaded with a stack of bricks. In order to secure a uniform distribution of load it is necessary to subdivide the stack into a number of independent sections. Increased arching and beam effects are brought into play if the bricks are cemented together. The hatched area in Fig. 11.28 indicates the approximate load which can be expected on vertical shores; the work of Wood (1952) shows that this is conservative. Owing to arching, the load on the brickwork and soil at A and B is high and local failures are possible. The timbering for an excavation such as C must be adequate to cater for these high local loads. If a wall is thin, it can be considered as a point load on a needle, but, if it is thick, it should be treated as a uniformly distributed load.

Wood (1952) tested a number of brick panels supported on a reinforced concrete beam of effective span 3.2 m and clear span 3 m. The 2.4 m × 3.6 m panels consisted of plain 230 mm solid and 280 mm cavity walls and 280 mm cavity walls with door and window openings. The mortar for the brickwork was a cement–lime–sand mix in the proportions 1:2:9 by volume.

A plain solid panel of 230 mm brickwork of depth 2.4 m, clear span 3 m and weight 40 kN was also tested without a supporting beam and showed a central deflection of 0.1 mm under a superposed distributed load of 60 kN (total 100 kN). Although cracks appeared at the bottom of the panel at a total load of about 120 kN, the panel withstood eventually a total load of 320 kN, thus demonstrating that the strength of the panel was derived principally from arching within the brickwork and much less from the resisting moment of the panel acting as a beam.

When a wall is being constructed in brickwork on a beam, arching is negligible until the mortar begins to set and harden, generally when about five courses have been laid, after which arching becomes more apparent. In addition, it is reasonable to except that arching increases as the depth of the panel increases. In Wood's tests on completed panels the supporting beam was wholly in tension and thus acted as tensile reinforcement for the composite panel which functioned as a deep beam. Furthermore, the stresses so developed were low and it is unlikely that cracking developed in the concrete of the beam. The measured steel stresses in the beam for the highest loads in wall types 1, 2, 3 and 4 (Table 11.2) were less

Table 11.2 Load distribution on beams supporting brick walls

Type of wall	Applied* uniformly distributed load (kN)	Equivalent[†] bending moment (WL÷)	Typical distribution of load at top of beam
1. 230 mm solid wall	60	274	
	150	248	Almost certainly
	185	218	similar to 280 mm
	258	130	cavity wall
2. 280 mm cavity wall	60	800	
	96	730	
	192	430	
	288	330	
3. 280 mm cavity wall, central window	60	530	
	120	510	
	200	440	
	280	415	
4. 280 mm cavity wall, central door	60	320	
	192	310	
	288	270	
5. 280 mm cavity wall, end door	60	50	
	86	48	
	144	51–44	

*Excluding weight of wall. Wall weights: types 1 and 2, 40 kN; types 3, 4 and 5, 30 kN.
[†]The equivalent bending moment is that bending moment which would produce on the same beam, when free and simply supported, steel stresses equal to those produced in the tests when carrying a total load W. Compare with the standard B.M. $= WL/8$. The value of the equivalent B.M. is WL divided by the amount given in the column.

than $35 \times 10^4 \, \text{kN m}^{-2}$, at which cracking may be expected to develop, but reached $7.6 \times 10^4 \, \text{kN m}^{-2}$ with type 5. Since arching throws the major part of the load towards the ends of the beam, there is a tendency for a degree of fixity to be established at the ends.

Table 11.2 shows the equivalent bending moment on the beam for various wall types and also the general pattern of load distribution. The bending moment for type 2 is lower than for type 1, although the total thickness of brickwork is the same. This is attributed to the fact that the cavity wall was built wholly in stretchers whereas the 230 mm wall was in English bond; the former probably developing a greater degree of arching. Slight asymmetry of the load distributions is due to the variable nature of this type of construction.

Some creep was detected in the 230 mm solid wall without supporting beam when under load but its magnitude was very small and the brickwork exhibited a remarkable degree of elasticity. Up to the development of tensile cracks, the assumption of an elastic medium should be fairly satisfactory. The mathematical analysis of stress distributions in deep walls confirms the existence of high stresses at the vertical ends of panels as observed in the tests. The simple theory of bending assumes zero stresses along these edges.

The tests and analyses outlined above are fundamental to underpinning and settlement problems. A preliminary method of design for a freely supported deep wall, the depth of which is not less than 0.6 times the span, is to employ a moment arm of 0.67 times the depth with a maximum of 0.7 times the span. For deep panels the reinforcement is considered effective up to 0.1 times the span from the lower edge and for relatively shallow panels up to 0.25 times the span. It is recommended that steel stresses should be maintained fairly low; the values are $11 \times 10^4 \, \text{kN m}^{-2}$ when the beam is propped during bricklaying and $8 \times 10^4 \, \text{kN m}^2$ when unsupported during bricklaying.

A number of investigations have been reported concerning deep beams, composite action between brick panels and supporting beams and other structural forms: one or more of these may be relevant to particular underpinning problems. The following deal with deep beams: Ramakrishnan and Ananthanarayana (1968), Kong *et al.* (1970), Manuel *et al.* (1971), Kong *et al.* (1972a, b), Kong and Sharp (1973), Kong *et al.* (1975), Kong and Sharp (1977), Kumar (1977), Kong *et al.* (1978) and Kubik (1980). Although these studies are concerned with reinforced concrete deep beams, steel deep beams have also been employed, inserting the beams in sections to pre-determined deflections by leaving slight gaps at the tension edge of the joints which are closed by tightening high-strength bolts, thereby eliminating the deflections. Many studies have been made on composite action between brickwork, beams and frames but the following should enable investigations to commence for specific problems: Wood and Simms (1969), Smith and Rahman (1972), Mainstone (1974), Suter and Hendry (1975), Liauw and Lee (1977), West *et al.* (1977), Haseltine *et al.* (1977), King and Pandey (1978), Smith and Riddington (1978), Page (1979), Hulse *et al.* (1980), MacLeod and Abu-el-Magd (1980), Hulse *et al.* (1981) and Page (1981).

11.8.1 Underpinning methods

Underpinning methods can be classified as follows: (a) protection wall, (b) pit, (c) cylinder, (d) pile, (e) geotechnical. In order to avoid vibration, most pile underpinning is done with bored piles. The difference between bored piles and small-diameter cylinders is assumed to be solely the fact that a cylinder must be large enough for a man to work in; that is, not less than about 1 m diameter. Loss of ground must be avoided if possible in any underpinning operations since, if it is incurred, it may lead not only to settlement of a structure but also to high stresses on timbering due to destruction of arching action within the soil, although slight yield is necessary to develop arching.

Pre-loading is a technique employed to reduce settlement of the superstructure following underpinning. A gap of 0.5 to 0.75 m is left between old and new foundations. Jacks are placed in the gap and the new foundation is loaded, where possible, to about 50% in excess of the design load. Care must be taken not to damage the superstructure by overloading. While the jacks are holding the load, a short steel wedging column is placed in the gap and wedged tight with steel plates. On removal of the jacks, wedging columns can be encased in concrete. An advantage of this method is that all or part of the settlement which takes place when a foundation is loaded is not transferred to the superstructure. The process has been applied to new buildings during construction to reduce settlement. The Pynford stool can be used as an alternative to a wedging column. This is about 0.75 m high and comprises seat and base plates connected by four legs of steel rods braced laterally.

(a) Protection wall

Where a permanent excavation, for example a road cutting, is made in proximity to a light structure, sufficient support for the structure is sometimes secured by means of a retaining wall designed to cater for the loads from the foundations in addition to earth load. Some settlement of the structure usually occurs with this method, owing to either loss of ground during construction or slight yield of the wall under load. Permanent struts across the top of the excavation between retaining walls on each side reduce yields. Provision of temporary support during construction is usually difficult and calls for considerable ingenuity. For the construction of a subway in Toronto, Canada, steel I-beam soldier piles were driven at 2 m centres along the sides of the excavation to a depth of 2 to 2.5 m below formation level. Steel I-beam braces were provided across the top and wood sheeting was placed horizontally between the flanges of the soldier piles as excavation proceeded. To prevent movement of the ground, sand was packed behind the sheeting and, in clay soils, the sheeting was wedged back from the flanges. The permanent structure is of reinforced concrete. It is sometimes possible to incorporate part of temporary work in the permanent structure. In built-up areas, exploratory borings, excavations and geophysical surveys are required to locate sub-surface structures and services, in order to reduce damage when driving the soldier piles. An example of protection wall underpinning in Pittsburgh has been described by Swiger (1957). Many temporary and permanent protection walls have been constructed within recent years employing multiple inclined ties anchored well beneath the foundations of existing structures: an example is afforded by construction of a bank in Seattle, USA (Shannon and Strazer, 1970). A tilting masonry retaining wall, built in 1821, in the precincts of Durham Cathedral, England, was stabilized with 10 m long stainless steel rods anchored in sandstone bedrock beneath an existing building (Fondedile, 1975).

(b) Pit

A well-known example of pit underpinning is the method of constructing foundations beneath an existing wall in sections about 3 m long. Not more than about 15% of the length of the wall should be excavated at one time unless dead shoring is provided and sections worked concurrently should be at least 4 m apart. On a wall divided into twelve sections (numbered consecutively) a suitable order of excavation and construction in pairs of sections is (5, 11), (1, 7), (3, 9), (6, 12), (2, 8), (4, 10). The new brickwork is built to within about 50 mm of the old foundation or brickwork and the gap is packed with stiff fine aggregate concrete. Sections exceeding 1 m can be taken out if a wall is wholly or partly supported on dead shores at 1 to 3 m centres. Particular attention must be given to the foundations of the shores, as settlement of the superstructure must be avoided. Shores at the corners of structures are most important.

The Chicago well method of excavation can be used for the construction of piers to support foundation beams or individual columns. Support for the soil is provided by vertical sheeting of 150 mm × 50 to 75 mm boards braced by inner steel rings of 100 mm × 25 mm section and 1 to 4 m diameter, spaced at 0.5 to 2 m centres, depending on the character of the soil. The method is used exclusively in clay, which should be firm enough to stand between sets.

(c) Cylinders

Cylinders 1 to 1.5 m diameter of 10 to 20 mm plate are driven by a pair of compressed-air hammers or by jacking down from the superstructure. Sections are generally no longer than 1 m and shorter lengths are used in confined headroom. Connections between sections are internal sleeves. Joints must be caulked where aquifers are encountered in order that the cylinder can be pumped dry before concreting. The excavation processes are similar to those for larger caissons and working under compressed air can be adopted. A supply of air is required for a man working in a small-diameter cylinder. The following tolerances for position and verticality of small-diameter cylinders are considered reasonable: position at the top 75 mm, out of plumb $1\frac{1}{2}\%$ of the length, bowing $1\frac{1}{2}\%$ of the length. On completion of excavation, cylinders are filled with concrete.

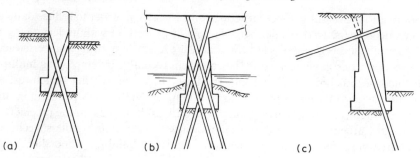

Figure 11.29 Examples of underpinning with piles (Pali Rodice). (a) Building wall; (b) bridge pier; (c) retaining wall.

The bearing area can be increased by forming a bell-shaped excavation under the cylinder with a side slope not less than 60° to the horizontal.

(d) Piles

Jacking down from the superstructure is frequently adopted for the installation of underpinning piles and the piles are often constructed in short lengths where headroom is limited. If pre-loading is adopted, the wedging columns of a pile group are encased in a reinforced concrete block. Figure 11.29 illustrates examples of the Fondedile (Pali Radice) method of underpinning in which piles are formed in holes drilled through the substructure into the underlying strata.

(e) Geotechnical processes

Geotechnical processes can be employed, in appropriate circumstances, to underpin a structure by stabilization of the soil. An example of the application of chemical injections has been described by Golder *et al.* (1961).

An important feature of underpinning operations is the vertical and horizontal control which must be established. For example, the roof of a subway in clay in Chicago, USA, is about 6 m below the underside of piers carrying the elevated railroad tracks of La Salle Street Station. Six to ten level readings were taken on each pier during each 8 h shift and any settlement was compensated by jacks. The labour expended on these observations was justified by one pier in particular which settled 0.5 m in 24 h but the tracks were maintained at grade level by jacking. When the tunnel was completed, level observations and jacking were continued until the records showed no settlement for a 60-day period. When observations for verticality of tall structures are made, account should be taken of deflection arising from wind or from differential heating in sunshine and it may be preferable to restrict observations to calm overcast or night conditions. Several specialized types of apparatus are available to control movements to within close limits. The accuracy and frequency of control observations depend on local circumstances.

Expanding cement has been used in underpinning work to eliminate the necessity for using jacks and struts. Various compositions of cements with controlled expansions have been developed: one form comprises essentially (i) Portland cement (basis), (ii) sulpho-aluminate cement (expanding agent), (iii) commonly slag cement (stabilizing agent). These cements are liable to deteriorate in storage and may not be resistant to sulphates or sea water. A comprehensive review of expanding cement has been presented by a technical committee (Kesler, 1970). When employed for underpinning, a gap of about 1 m is left between the new and the old foundations, suitable formwork is provided and the gap is filled with expanding cement concrete which is vibrated. Holes are formed in this concrete by passing 30 mm diameter bars at 200 mm centres through the mass and withdrawing them as the concrete sets. Water is then supplied to the holes, under a few centimetres head, in order to maintain the expansion process. When the supply of water is stopped, the expansion process is retarded and ceases 24 to 48 h later, thus permitting control of the process. The full expansion must not be developed until the concrete has attained sufficient strength to resist the consequent compressive stresses.

11.8.2 Examples of underpinning

One of the best-known examples of underpinning is that of grain silos at Winnipeg, Canada, in 1914 (Peck and Bryant, 1953; White, 1953). The bin house consisted of sixty-five circular bins on a concrete raft, forming a structure 59.5 m × 23.5 m in plan and 31 m high. It was founded on 16.5 m clay, with a high moisture content, overlying limestone bedrock. During the initial charging with grain, the structure commenced to rotate about its foundation when the bins were three-quarters full and finally halted at an inclination of 29° 53′, with one side 9 m below its original level and the other side 1.5 m above. There was also a longitudinal difference in level of 1.2 m. The bins were emptied of grain and blocks and shores were installed. Five rows of fourteen piers 2 m in diameter were sunk to bedrock beneath the structure by the Chicago well method. The structure was then rotated by jacks about the centre row of piers, using oak rockers as fulcrums. It was then rotated about the next highest row of piers in order to lift it in addition to plumbing. Jacks and blocks were then removed as the piers were concreted to the underside of the raft.

During the 1939–45 war it was necessary to increase the number of spans of some bridges over railways in Great Britain in order to accommodate additional tracks. When it was essential to interfere as little as possible with road traffic the following procedure was adopted. Bored piles were sunk on the line of the new abutment for half the road width and a reinforced concrete slab constructed to span from the old abutment to the piles. This was repeated on the other half of the road width. New wing walls were constructed and the dumpling between piles and old abutment excavated. Poling boards were inserted behind the piles to retain the earth, the piles having additional reinforcement to cater for earth thrusts. The new abutment was then constructed round the piles and the back of the old abutment was cut or thickened as necessary. Temporary struts were provided between the piles and the old abutment to assist the former in resisting earth thrust and, in the case of arches, additional temporary support was provided to assist the old abutment in carrying the horizontal thrust. An interesting example of underpinning a viaduct at London Bridge Station has been described by Cantrell (1953).

When the timber piles supporting a 53 m high factory chimney at Kearny, NJ, USA, were found to be partly rotten, underpinning was carried out without interruption to service. Rotting had taken place in the 0.4 m between the underside of the foundation and the water table. Vertical sheeted pits 1 m × 1 m × 3 m deep were sunk at four points round the base, filled with gravel and the sheeting removed. These served as drainage sumps and, when they were connected to 75 mm pumps, the water level was lowered by 0.8 m. Without a drainage system the fine sand beneath the base would have flowed into the excavations. The piles were sound from a few centimetres below permanent water table and test loads of 250 kN per pile indicated no permanent set. The pile heads were cut off 0.5 m below permanent water table level one at a time and steel struts and plates inserted with wedges driven by a 4 kg hammer. The entire excavation from 150 mm below the point of cut-off of the piles to 50 mm beneath the underside of the foundation was filled with concrete and the gap finally packed with cement mortar. The foundation was divided into sixteen sections and opposite pairs worked concurrently for underpinning. Level marks were established at eight points on the base but no movement was observed. Daily observations for tilt were made to the chimney top from one theodolite station north of the stack and from one west of the stack. Sights on the top of the stack were brought down and marked on the ground but no change was noted. Differential settlement of a chimney can be arrested and corrected by insertion of a temporary needle at the base supported on piles and by raking shores on jacks. New permanent foundations can then be constructed.

In Sao Paulo, Brazil, a 90 m high reinforced concrete frame building of twenty-six storeys supported on piles settled when construction was almost completed (Dumont-Villares, 1956). In one direction it was 650 mm out of plumb and in a direction at right-angles 600 mm out of plumb. Settlement was activated at one corner of the building by the movement of fine wet sand when an adjacent excavation was made. Both cement grouting and chemical injection were unsuccessful in halting settlement. Finally the ground temperature was reduced to $-20°C$ by geotechnical freezing during a period of eight months. Pits about 1 m diameter were sunk to firm ground and concrete piers were constructed. The concrete contained calcium chloride to accelerate setting at the low temperature of the ground. The structure was then righted to within 5 mm of plumb by jacks bearing on the piers and the underpinning completed.

In Paris, six-storey underground car park, beneath which is a new railway tunnel, was constructed by underpinning storey by storey; the whole being partly suspended by a slab at road level supported at the edges on bored piles put down beyond the limits of the main structure (Anon., 1968).

Pipe-jacking is a convenient method of constructing small diameter tunnels for some underpinning projects. For construction of a station on the Antwerp, Belgium, underground railway, contiguous horizontal pipes were jacked into position and eventually formed the roof of part of the station complex (Musso, 1979). Pipe-jacking is discussed in several publications (PJA, 1975, 1981; Craig, 1983). Artificially induced intense cold or heat can produce foundation problems as the effects accumulate over a period of years. In Bermondsey, London, a six-storey cold store suffered the effects of frost heave developed over a period of 50 years: underpinning with deeper foundations taken to unfrozen strata prevented ultimate collapse of the structure (Pynford, 1977). The underpinning of a steel tank has been described by Mohan *et al.* (1978).

Ancient buildings are often preserved as national monuments. Occasional remedial measures are called for as the superstructure decays or because the foundations are defective. Special attention must be given to the preservation of aesthetic qualities. The superstructure must be rendered safe before underpinning operations are commenced and shoring and centering are erected as necessary. Cracked walls can be tied by steel rods inserted in drilled holes. The rods are tensioned by nuts bearing on end plates. The holes, together with masonry or brick joints, are then grouted under a pressure of, say, $700 \, kN \, m^{-2}$, commencing at the foundations and working upwards in stages. Before grouting, defective mortar is raked out of joints and both joints and holes are blasted with compressed air to remove dust. Washing joints out with water under pressure often reveals the extent of a fracture. Prestressed cables can be employed instead of steel rods for tying walls both horizontally and vertically. Many old buildings were not constructed on proper foundations and consequently reinforced concrete strip footings or ring beams must be provided. Precast or cast *in situ* concrete sections can be inserted under a wall to form a prestressed beam. The construction of a ring beam may be necessary before the superstructure is capable of withstanding jacking thrusts during piling. The preservation of ancient monuments has been discussed by Heasman (1936) and by Gifford and Taylor (1964).

Magnel (1949) describes the raising of two mediaeval towers at Tournai, Belgium. Each tower weighed about 28 000 kN and the total lift was 2.5 m. Two tiers of beams, one 200 mm above the other, were placed in the base of each tower. The beams of one tier were at right angles to the beams of the other. Each beam was placed in sections and then prestressed. Twenty-six jacks, each of 1500 kN capacity, were placed between the two tiers and each tower raised 200 mm at a time by two motor-driven pumps. After each lift, blocks were inserted for the next lift. The application of prestressed concrete to this project considerably facilitated the work.

Prestressing was applied to the base of church tower in Staffordshire which tilted and cracked under the influence of mining subsidence (CCE, 1948). Part of mediaeval St Swithin's Church, Winchester, England, exhibited distress and remedial measures comprised underpinning foundations and a reinforced concrete frame incorporated within the flint walls of the superstructure (Pynford, 1977). Problems associated with ancient buildings have been discussed at length by Kerisel (1975).

Underpinning was carried out between 1906 and 1912 on Eleventh Century Winchester Cathedral, England, which showed signs of excessive settlement. The structure was founded on clay extending to a depth of 2 m, beneath which lay 2.5 m peat, 0.1 m silt and sand, a few centimetres of gravel and flints and then solid chalk. The water table was at the underside of the foundations. Pumping from excavations was not possible since this would have drawn out the sand and silt. Piling was not adopted because of vibrations. Therefore the new foundations were constructed of walling in 1.5 m lengths and were taken to dense gravel. Excavation was carried out by a diver who worked solely by touch since a torch was useless in the peat-discoloured water. Four courses of concrete in bags, slit open in position, were sufficient to seal the bottom of the excavation from water percolating through the gravel. The excavation was then pumped dry and the underpinning completed in concrete and bricks. The work reflects great credit on the patience and perseverance of the diver. Today this work could undoubtedly be facilitated by the application of geotechnical processes.

The use of prestressing in foundation strengthening at York Minster has been described by Dowrick (1970). Further examples of underpinning are described by Green (1954), Goldfinger (1960), Pryke (1960), McNulty and O'Brien (1961), Wilson (1962), Gifford and Butler (1966) and Bethel (1968).

11.8.3 Moving complete structures

Within the last four decades, to permit new construction, whole structures have been moved bodily as an alternative to demolition. In some cases the cost of moving a structure has been less than 40% of the cost of demolition and erection of a new building and often there is also a saving in time. Services to buildings can be maintained by flexible connections and interference to inhabitants can be limited to a few hours. The initial lifting, horizontal movement and final lowering on to new foundations generally occupies a few days only although the preparatory work may take several weeks or months.

The procedure is generally as follows. A track is laid from beneath the existing structure to the new site and may consist of steel beams, flat-bottom rails, steel channels or bull-head section rails on their sides. These tracks may be supported on concrete strip foundations but, where the bearing capacity of the ground is low, deeper foundations such as piles have been employed. Runner beams must be fitted to the base of the structure and, for horizontal movement, 50 to 75 mm diameter roller or ball bearings are inserted between the runners and the tracks. The structure may be severed from its existing foundations and jacked up either before or after the runners are fitted. Adequate cross-bracing between runners and between existing columns must be provided to maintain stability during the operations. With some structures, particularly ancient buildings, it may be necessary to provide considerable temporary or permanent support, such as enveloping frames or prestressed cables and rods.

Vertical movement is generally effected by 20 to 60 hydraulic or screw jacks. Where the lift exceeds the travel of the jacks, packing blocks must be inserted between the stages of lifting. Each lift is made in a series of increments, generally less than 25 mm, and it is usual to check the rise of each jack after each increment is made. The load distribution in a framed structure can be checked throughout the operations by fitting extensometers to the columns. The initial load on a column can be determined by fitting extensometers to that part of the column below the runner where the force falls to zero as jacking proceeds.

Horizontal movement can be effected by tractor, winch or turnbuckles, although screw or hydraulic jacks commonly afford the best control. The jacks are generally fixed to the runners and bear against clamps attached to the track. The clamps are moved forward in stages as the travel of the jacks is reached. Setting the structure on new foundations can be effected by lowering on jacks and by wedging and packing with stiff mortar. Throughout all operations it is necessary to maintain strict control of line and level and instrument and reference stations should be carefully established at the commencement of the project. The movement of two 700 kN tanks at Manchester, using the hovercraft principle, has been described by Pryke (1967).

Among the structures which have been moved are a steel-framed hospital block (23 500 kN) (Spencer, 1941), a high pressure boiler (7500 kN) (Flay, 1944), an ancient wine cellar in Whitehall, London (White and Gardner, 1950), a masonry building (The Oblate Fathers' Monastery, Quebec, 31 000 kN), a 24 m high water tower (19 000 kN) (CRME, 1953), a 300-year-old house at Hereford (Pryke, 1967), the 350-year-old Ballingdon Hall, Suffolk (Anon., 1972) and 15th Century timber-framed buildings in Old Shambles, Manchester (Charge, 1972). Other examples are described by Olsen (1958). A classic example is the raising of the Temples of Abu Simbel alongside of the River Nile.

EXAMPLE 11.3

Two timber beams are arranged to support a load of 2.5 kN over a pit excavated for underpinning work, as shown in Fig. 11.30. If the cross-sections of the beams are identical, calculate the proportion of the load carried by each beam assuming that the supports to both beams are unyielding. Also calculate the load carried by each beam if the supports of beam 1 are unyielding but the supports of beam 2 settle elastically.

The modulus of elasticity of the timber is 8×10^6 kN m^{-2}. The cross-section of each beam is 0.15 m $\times$ 0.10 m and the effective spans l_1 and l_2 are respectively, 2.00 and 1.50 m.

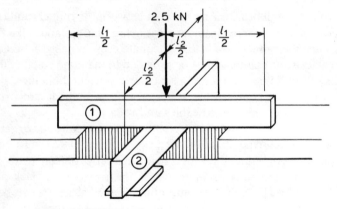

Figure 11.30 Underpinning supports over excavation.

The bearing area at each support of beam 2 is $0.15\,\text{m} \times 0.10\,\text{m}$ and the modulus of subgrade reaction appropriate to the shape and size of the bearing plates is $5 \times 10^4\,\text{kN}\,\text{m}^{-3}$.

$$I = \frac{bd^3}{12} = 0.10 \times 0.15^3/12 = 2.81 \times 10^{-5}\text{m}^3$$

$$\delta_1 = \frac{WL^3}{48EI} = \frac{W_1 2^3}{48 \times 8 \times 10^6 \times 2.81 \times 10^{-5}} = \delta_2 = \frac{W_2 1.5^3}{48 \times 8 \times 10^6 \times 2.81 \times 10^5}$$

or

$$W_1 2^3 = (W - W_1)1.5^3 \quad \text{or} \quad W_1 = \frac{1.5^3 W}{(2^3 + 1.5^3)} \quad \text{and} \quad W_2 = \frac{2^3 W}{(2^3 + 1.5^3)}$$

whence

$$W_1 = 0.30W = 0.75\,\text{kN} \quad \text{and} \quad W_2 = 0.70W = 1.75\,\text{kN}$$

$$\delta_1 = \frac{W_1 2^3}{48 \times 8 \times 10^6 \times 2.81 \times 10^{-5}} = \delta_2 = \frac{W_2 1.5^3}{48 \times 8 \times 10^6 \times 2.81 \times 10^{-5}}$$

$$+ \frac{W_2}{2 \times 0.15 \times 0.10 \times 5 \times 10^4}$$

or

$$W_1 7.414 \times 10^{-4} = W_2 3.128 \times 10^{-4} + W_2 6.667 \times 10^{-4}$$

or

$$W_1 7.414 = (W - W_1) 9.795 \quad \text{or} \quad W_1 = \frac{9.795W}{17.209} \quad \text{and} \quad W_2 = \frac{7.414W}{17.209}$$

whence

$$W_1 = 1.42\,\text{kN} \quad \text{and} \quad W_2 = 1.08\,\text{kN}.$$

Appendix

THE STRUCTURAL
ANALYSIS OF
PILE GROUPS

F.D.C. Henry

The distribution of forces in groups of piles can be investigated by the simple graphical and numerical methods – termed static methods in this text for convenience. The term static is employed to distinguish the techniques of analyses from those discussed in this appendix and termed elastic methods. For very important structures, such as oil-production platforms, it is necessary to employ comprehensive and sophisticated three-dimensional analyses which are most conveniently solved through computer programs. Nevertheless, there are many piled structures, such as bridge abutments, where two-dimensional force systems are dominant. Apart from static methods, such problems can be investigated by methods which take account of elasticity in the piles. The graphical and numerical methods discussed below are two of the numerous techniques published during the last three or four decades. Although the numerical method is convenient for manual computation, it can be expressed as a program for automatic computation. In fact, most methods of structural analysis can be adapted to evaluate forces in pile groups. Manning (1933) employed slope–deflection and also discussed in an interesting manner various aspects of the design of pile groups. In the first edition of this book (Henry, 1956), strain energy was employed in a simple worked example.

Some engineers regard graphical methods as archaic, indeed they are often tedious, but they have the merit of readily demonstrating the influence of changes in forces and deformations. Though not employed in the following methods, graphical integration, and with less facility differentiation, can be employed to advantage in some structural problems, particularly where the loading is very irregular. This is illustrated, for example, by the analysis of anchored bulkheads by successive graphical integration, commencing with loading, to determine the

973

974   *The Design and Construction of Engineering Foundations*

deformation of the sheeting, taking account by estimation of translation of foot and anchorage. The technique is discussed briefly by Terzaghi (1943).

The use of any particular method is largely a matter of preference by the engineer, based often on familiarity and dexterity with the technique. Because of the imponderables in pile group analyses, it is recommended that the static method and at least one more sophisticated method, such as either of the techniques discussed below, be employed. Forces in piles determined by these methods should be compared and judgment invoked to assess their validity, bearing in mind indeterminate factors such as compression in soil or rock beneath pile feet and the influence of lateral restraint of soil or rock.

It should be emphasized that the following notes are concerned with structural analysis and soil-structure interaction is ignored. The response of soil to loading of pile groups is discussed in Section 8.4.5 and several references are given in Examples 2.9 and 3.10. Since any unpredictable change in length of a pile can markedly influence load distribution in a group, it is doubtful whether any analysis can predict pile loads accurately. Nevertheless, design requires analytical estimates and, as mentioned above, judgment must play a part in final assessments. A comprehensive analysis of pile groups has been published by Banerjee (1978) and by Randolph (1980) as discussed in Chapter 8.

A.1 GRAPHICAL ANALYSIS OF PILE GROUPS

The static method provides an analysis for up to three rows, or groups of rows, of piles, but for more complicated systems, or where the piles must be considered as having fixed-ends, the elastic method is used. In this method it is assumed that the piles are wholly elastic, the soil under the caps carries no load, the cap is rigid and the resistance of a point-bearing pile is at the foot and that of a friction pile is concentrated at some point along the length of the pile. The effective length L_e of a point-bearing pile is taken as the full length of the pile, and it can be shown that the effective length of a friction pile with uniform frictional resistance is two-thirds the length of the pile. The effects of movement of the pile feet and of lateral restraint by the soil, except in strata where fixed-end conditions are assumed, are ignored. When a load is applied to the cap, the displacement of cap and piles is a combination of rotation and translation which can be considered as rotation about the centre of rotation. Where the displacement is pure translation, the centre of rotation is assumed to be at infinity, and the resultant of the applied forces passes through a fixed point termed the elastic centre. The elastic centre is located by the intersection of the resultants for two translations in different directions. If the applied forces constitute a pure couple, the centre of rotation coincides with the elastic centre. Westergaard (1918) and Vetter (1938) have developed elastic methods of analysis, from which the method advocated in CP 2 (ISE, 1951) and described below has been evolved. Part of the calculations in Vetter's method have been replaced by graphical techniques. Vetter states that in many practical cases, such as bridge piers, the error introduced by the assumption of hinged ends is insignificant but in some structures, such as pile bents for trestles, the error becomes of considerable magnitude.

A.1.1 Hinged-ended piles

The elastic centre is located by applying to the pile cap a unit vertical translation and analysing the deformation of the piles by a Williot deflection diagram, followed by a unit horizontal translation with a similar graphical analysis. The process of constructing the deflection diagrams is the reverse of that normally used for frameworks. In the Williot and Williot–Mohr graphical methods for the analysis of deflections of frameworks, it is assumed that, when a load W is applied to a framework, each member expands or contracts along its length a distance δl_1, δl_2 etc., according to Hooke's law, and that one end of each member rotates about the other end but, as the rotation is small, the movement can be represented graphically by a tangent drawn perpendicular to the initial position of the member. The values δl_1 etc. are plotted to an exaggerated scale, as in Fig. A.1. The vertical, horizontal and resultant deflections can be measured as required.

In Fig. A.2(b) the unit vertical translation is plotted first, lines are drawn from v perpendicular to the direction of the piles and the extension of each of the piles is given by lines from O drawn parallel to the direction of the piles. A deflection diagram (Fig. A.2 (d)) is drawn in a similar manner for unit horizontal translation. Let A be the cross-sectional area, L_e the effective length and E the modulus of elasticity of each pile. Since the deformation of a pile multiplied by EA/L_e represents the axial force (F_v or F_h) exerted on the pile when unit vertical or horizontal translation occurs at the pile cap, a force polygon for axial loads in the piles can be drawn in the usual way and resultants R_v and R_h determined (Figs. A.2 (c) and (e)). If the values A, L_e and E are the same for all piles, the deformation alone can be plotted to represent the forces. A link or funicular polygon (broken lines in Fig. A.2 (a)) is constructed in conjunction with the force polygon, on which a pole is established at any convenient location, thus enabling

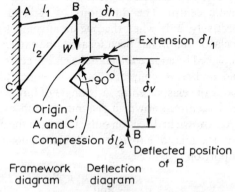

Figure A.1 Principles of Williot–Mohr diagrams.

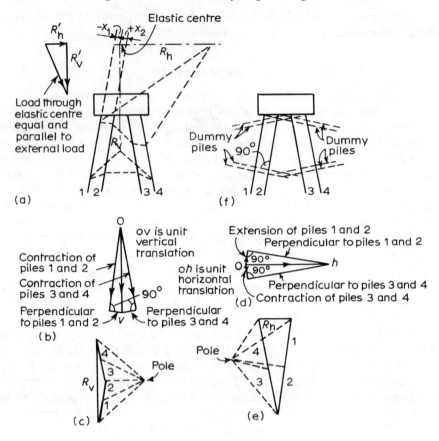

Figure A.2 Elastic analysis of pile groups; where --- dummy piles.

R_v and R_h to be plotted on the pile diagram; the intersection of the lines of action locating the elastic centre. The directions of R_v and R_h are vertical and horizontal, respectively, only when the pile arrangement is symmetrical.

Now that the elastic centre has been located, the external load applied to the pile cap can be replaced by a force through the elastic centre equal and parallel to the external load and by a couple of moment equal to the moment of the external load about the elastic centre. A triangle of forces is drawn for the force through the elastic centre, known in magnitude and direction, and its components R'_v and R'_h, known in direction only; that is, in the direction of R_v and R_h. Then the forces in the piles due to R'_v and R'_h can be determined from the force polygons by proportion

$$\left(\frac{R'_v F_v}{R_v} \text{ and } \frac{R'_h F_h}{R_h} \right)$$

In order to determinate the loads in the piles due to the moment of the external load about the elastic centre, the distance x from the elastic centre perpendicular to each pile is measured. It is assumed that the deformation in each pile is axial and proportional to the distance x or, if the rotation is θ, the deformation is $\pm x\theta$. Hence the force in any pile is proportional to EAx/L_e and the moment of that force about the elastic centre is proportional to EAx^2/L_e. For equilibrium, the sum of the moments of the forces in the piles about the elastic centre must equal the applied moment. The value of EAx^2/L_e is calculated for each pile and the force in, say, pile No. 1 is given by

$$\frac{EAx_1/L_e}{\sum(EAx^2/L_e)} \times M$$

where M is the moment of the external load about the elastic centre. Where the piles are identical, the force in the pile reduces to $x_1 M/\sum x^2$. The total force on each pile due to vertical and horizontal translation and moment is found by algebraic addition. The calculations are most readily made in tabular form. In order that a reasonable standard of accuracy can be attained in these graphical techniques, it is suggested that the structural scale should be at least $1/50$ and that the deformation, force and other diagrams should be roughly in the proportions relative to the structural diagram shown in Fig. A.2.

Sign convention is important. The deflection diagram is drawn for a horizontal translation in the same direction as the horizontal component of the applied load and, in the corresponding force polygon, compression is taken positive and tension negative. Vertical translation produces compression in all the piles. The force in each pile due to the moment is compressive or tensile respectively, when the pile axis and the applied load are on the same side or opposite sides of the elastic centre.

Certain special cases call for comment. When the piles are in two groups in different directions, with the piles in each group parallel to each other, the elastic centre lies at the intersection of the axes through the centre of gravity of each group parallel with the piles in that group. Calculations are reduced in this case by resolving the load through the elastic centre along the axes of each group and the forces due to moment are determined in the normal way. When a pile group is symmetrical about a vertical axis, R_v lies on the axis of symmetry, and it is necessary to determine R_h only to locate the elastic centre. Where the axes of all piles meet at a point (the elastic centre), an unbalanced moment is developed, unless the resultant passes through the elastic centre also, and this moment is resisted only by the soil at the sides of the piles. It is desirable that the pile layout is such that the elastic centre is on or near the line of thrust of the external load in order that the effect of moment is a minimum.

A.1.2 Fixed-ended piles

Where piles are assumed to be fixed at the ends, it can be shown that the effect of fixity can be replaced by additional external forces perpendicular to the piles.

Table A.1

	End conditions		No. of dummy piles	Position of dummy piles	Length of dummy piles
Case	Top	Bottom			
I	Hinged	Fixed	1	At base of cap	$\frac{1}{3}\frac{A}{I}L_r^3$
II	Fixed	Free (assumed hinged for force polygon)	1	At centroid of lateral reactions (position assessed by designer)	$\frac{1}{3}\frac{A}{I}L_p^3$
III	Fixed	Fixed	2	One pile at base of cap	$\frac{1}{3}\frac{A}{I}L_r^2$
				One pile at point two-thirds of distance between base of cap and point of fixity	$\frac{1}{9}\frac{A}{I}L_r^2$

These forces are assumed to be transmitted through dummy piles, each of which has the same cross-section as the real pile, and both real and dummy piles are treated as hinged at each end. The position and length of the dummy piles for various conditions is given in Table A.1, the values for which are abstracted from CP 2 (ISE, 1951). A is the cross-sectional area and I is the second moment of area of the real pile, L_r is the distance between the underside of the cap and the lower position of fixity (Cases I and III) and L_p is the distance between the underside of the cap and the intersection of real and dummy piles (Case II).

The point of fixity in Cases I and III must be decided by the designer and, where the point of a pile is driven into a stratum of firm material, it is likely to occur between the top of that stratum and a point one-third of the distance down the length embedded in the firm material. Where a friction pile is driven through uniform material which affords lateral resistance to horizontal thrust, in addition to vertical resistance, the effective length L_e may not coincide with the length L_r. Cummings (1935) derived the expression

$$L_r = \sqrt[5]{\left(216\frac{EIN}{\gamma}\right)}$$

where N is a non-dimensional coefficient, determined by the elasticity of the soil, for which a value of 0.005 is suggested for very dense sand, and γ is the density of the soil.

Figure A.2 (f) shows a pile layout similar to that in Fig. A.2 (a) but, in this case, analysis is based on the assumption of pile-ends fixed at the cap and in the soil.

Deflection diagrams and force and link polygons are drawn in the usual way for the entire assembly of real and dummy piles. The calculations for this case are also best made in tabular form. Deformations and forces in both real and dummy piles are included in the calculations, although no direct provision is made in the design for resisting the forces in the dummy piles. Reference to expressions in Table A.1 reveals that the length of the dummy piles is invariably great.

EXAMPLE A.1*

The general arrangement of the group is the same as that shown in Fig. A.2. The spacing of the heads of the piles in each row of four is 1.4, 1.0 and 1.4 m. The piles are of reinforced concrete, 350 mm by 350 mm section, reinforced longitudinally with eight 25 mm diameter mild steel bars and are 12 m long and driven at a rake of 5 in 1. There is a central vertical load on each row of the group of 1200 kN and a lateral load of 100 kN acting horizontally from left to right at a height of 1.8 m above the tops of the piles. Assume

$$E_c = 13.8 \text{ kN mm}^{-2}, \ m = 15 \text{ and } 40 \text{ mm cover to reinforcement.}$$

Static method

The loading can be reduced to a single resultant, the forces in each parallel pair of piles can be treated as a single force acting mid-way between each pair and the resulting triangle of forces can be determined graphically or by calculation.

Alternatively, number the piles 1, 2, 3 and 4 from left to right and let the force in each be F_1, F_2, F_3 and F_4 respectively. Then resolving vertically

$$\frac{5}{\sqrt{26}}(F_1 + F_2 + F_3 + F_4) - 120 = 0$$

and resolving horizontally

$$\frac{1}{\sqrt{26}}(F_1 + F_2 - F_3 - F_4) + 10 = 0$$

Now the forces in the piles constituting one pair are equal and solving these two equations $F_1 = F_2 = 178$ kN and $F_3 = F_4 = 433$ kN.

Elastic method

(a) Hinged-ended piles

The elastic centre lies at the intersection of the axes through the centre of force of each pair of piles and is thus vertically over the centre of the group at a height of 6.0 m above the pile heads. The load through the elastic centre is resolved into two component directions along the axes of each pair of piles. Deformation diagrams are not required for the solution in this particular case but the deformations are included in Table A.2 in order to show how they are treated in computations when required. The forces F_m due to the moment about the elastic centre are given by $x_n M / \Sigma x^2$ where $M = -100 \times 4.0 = -400$ kN.

*The editor is indebted to A.R. McCarthy for reworking Example A.1 in SI units.

Table A.2

	Pile number			
	1	2	3	4
Pile deformation for unit				
horizontal displacement	−0.196	−0.196	+0.196	+0.196
Resultant force F_h (kN)	−127.5	−127.5	+127.5	+127.5
Pile deformation for unit				
vertical displacement	+0.981	+0.981	+0.981	+0.981
Resultant force F_v (kN)	+306.0	+306.0	+306.0	+306.0
Distance of pile axis				
from elastic centre x (m)	−0.686	+0.686	−0.686	+0.686
x^2	0.470	0.470	0.470	0.470
Resultant force F_m (kN)				
($\Sigma x^2 = 1.880$)	+146.0	−146.0	+146.0	−146.0
Total force $F = F_h +$				
$F_v + F_m$ (kN)	+324.5	+32.5	+579.5	+287.5

(b) Fixed-ended piles

Equivalent area of pile section = 1775 cm² (concrete units). Equivalent second moment of area of pile section about axis through centre = 186 928 cm⁴ (concrete units).

For fixity at top and bottom of pile, lengths of dummy piles are (Table A.1, Case III) 54 695 m at the pile head and 18 232 m at a point 4.0 m above the pile foot. For each real pile the head dummy pile is denoted by subscript h and the lower dummy pile by subscript b. Height of elastic centre above head of piles is 6.48 from graphical solution. Table A.3 shows calculations based on graphical determinations.

It will be observed in these elastic methods that the deformations are converted to strains and these are a measure of forces and are plotted as such. Where the forces determined from strains in the dummy piles are considerably smaller than the forces in the real piles, it may be convenient to neglect them when constructing the force polygon.

It will be noted that the difference between the values obtained by assuming hinged-ended and fixed-ended conditions is not of practical significance in this case. The forces in the dummy piles due to vertical and horizontal translation and moment are very small and the maximum total force in any dummy pile is less than 3 kN.

A.2 NUMERICAL ANALYSIS OF PILE GROUPS

The method outlined below was evolved by Sawko (1968) and is based on stiffness: analysis reduces to the solution of three simultaneous equations for any pile system. Bending stiffness of piles can be included or neglected and rake or number of piles in the group does not complicate the solution. Piles are assumed encastré at the cap and fixed or pinned at the feet. Furthermore, it is assumed that no lateral stress is imposed on the sides of the piles. The following symbols are employed in the analyses.

$\delta_1 \delta_2 \delta_3$ Horizontal, vertical and rotational displacements of pile cap.

$\Delta_1 \Delta_2 \Delta_3$ Pile distortions.

Table A.3

	Pile Number											
	1	2	3	4	1_h	1_b	2_h	2_b	3_h	3_b	4_h	4_b
Pile deformation for unit horizontal displacement	-0.196	-0.196	+0.196	+0.196	+0.981	+0.981	+0.981	+0.981	-0.981	-0.981	-0.981	-0.981
Pile length (m)	12	12	12	12	54,695	18,232	54,695	18,232	54,695	18,232	54,695	18,232
Pile strain for unit horizontal displacement $\times 10^2$	-16.3	-16.3	+16.3	+16.3	+0.0179	+0.0538	+0.0179	+0.0538	-0.0179	-0.0538	-0.0179	-0.0538
Resultant force F_h, from diagram (kN)	-127.5	-127.5	+127.5	+127.5	+0.1400	+0.4208	+0.1400	+0.4208	-0.1400	-0.4208	-0.1400	-0.4208
Pile deformation for unit vertical displacement	+0.981	+0.981	+0.981	+0.981	+0.196	+0.196	+0.196	+0.196	+0.196	+0.196	+0.196	+0.196
Pile strain for unit vertical displacement $\times 10^2$	+81.75	+81.75	+81.75	+81.75	+0.00358	+0.01075	+0.00358	+0.01075	+0.00358	+0.01075	+0.00358	+0.01075
Resultant force F_v, from diagram (kN)	+306.0	+306.0	+306.0	+306.0	+0.0134	+0.0402	+0.0134	+0.0402	+0.0134	+0.0402	+0.0134	+0.0402
Distance of pile axis from elastic centre x(m)	-0.59	+0.78	-0.78	+0.59	-6.72	-14.71	-6.46	-14.45	+6.46	+14.45	+6.72	+14.71
x^2	0.348	0.608	0.608	0.348	45.158	216.384	41.732	208.803	41.732	208.803	45.158	216.384
$x/L_e \times 10^4$	-491.67	+650.00	-650.00	+491.67	-1.23	-8.07	-1.18	-7.93	+1.18	+7.93	+1.23	+8.07
$x^2/L_e \times 10^4$ $\left(\sum \dfrac{x^2}{L_e} \times 10^4 = 2092.2 \right)$	290	507	507	290	8.26	118.68	7.63	114.53	7.63	114.53	8.26	118.68
Resultant force $F_m = \dfrac{xM}{L_e} \left/ \sum \dfrac{x^2}{L_e} \right.$ kN ($M = 100 \times 4.68 = 468\,\text{kNm}$)	+110.0	-145.4	+145.4	-110.0	+0.28	+1.81	+0.26	+1.77	-0.26	-1.77	-0.28	-1.81
Total force $F = F_h + F_v + F_m$ (kN)	+288.5	+33.1	+578.9	+323.5	+0.433	+2.271	+0.413	+2.231	-0.387	-2.151	-0.407	-2.191

n	Number of pile in group.
T_n	Transformation matrix for determining pile distortions from pile cap displacements for pile n.
$[a]$	Geometric matrix.
$[s]$	Stiffness matrix of unassembled structure.
W, H, M	External vertical and horizontal loads and moments acting on pile cap.
$[P]$	Matrix of external loads and moments acting on pile cap.
$[S]$	Stiffness matrix of pile group w.r.t. chosen axes.
α_n	Angle of inclination of pile n with vertical.
$A_n L_n$	Cross-sectional area and length of pile n.
e_n	Distance of pile n from origin measured as shown in Fig. A.3.
E	Modulus of elasticity of pile material.

A.2.1 Bending stiffness neglected

A typical pile group is shown in Fig. A.3(a). Under the action of applied in-plane loading (Fig. A.3(b)) the pile cap deflects as a rigid body with three degrees of freedom (Fig. A.3(c)).

Any arbitrary set of carthesian axes can be selected but it is convenient to employ horizontal and vertical axes having an origin on the underside of the pile cap. All the forces and displacements are considered with reference to these axes. In the absence of bending, only axial effects in piles are important and the analysis takes a very simple form.

The relationship between the displacements of the pile cap $(\delta_1 \delta_2 \delta_3)$ and axial distortions Δ_2 in each pile can be determined by applying, in turn, unit values of rigid body displacements to the pile cap and finding the corresponding effect on axial displacements of piles in the pile group (Fig. A.4). For a typical pile n with

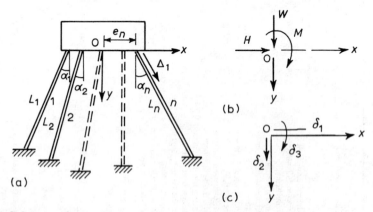

Figure A.3 System of co-ordinates, loads and displacements in a typical pile group. (a) Elevation on pile group; (b) loads acting at o, (c) displacements at o.

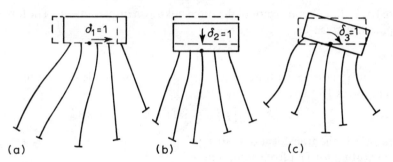

Figure A.4 Unit rigid body displacements of pile group. (a) Horizontal; (b) vertical; (c) rotational.

eccentricity e_n from the origin and inclined at an angle α_n to the y axis (Fig. A.3a), the transformation matrix for determining Δ_2 produced by unit values of pile cap distortions is:

$$\begin{array}{c|c|c} \delta_1 = 1 & \delta_2 = 1 & \delta_3 = 1 \end{array}$$
$$T_n = [\, + \sin \alpha_n \mid + \cos \alpha_n \mid + c_n \cos \alpha_n \,] \Delta_2 \text{ for pile } n.$$

Values thus obtained for each pile are inserted in a geometric matrix.

$$[a] = \begin{bmatrix} T_1 \\ T_2 \\ T_n \end{bmatrix}$$

This matrix has three columns, corresponding to the three cap displacements, and the number of rows equal to the number of piles in the group (when only axial effects are considered). It will be observed that the terms of matrix $[a]$ are inserted by inspecting the geometry of the pile group and the formulation of the matrix for a complete pile group presents no difficulty.

The axial stiffness of each pile $(s_n = A_n E_n / L_n)$ is next inserted into the diagonal stiffness matrix s of the unassembled structure

$$[s] = \begin{bmatrix} s_1 & & \\ & s_2 & \\ & & s_n \end{bmatrix},$$

and the applied loads into the loading matrix

$$[P] = \begin{bmatrix} H \\ W \\ M \end{bmatrix}$$

The three matrices $[a]$, $[s]$ and $[P]$ having been formulated, analysis proceeds by a series of matrix operations. The distortions in the piles are given by

$$[\Delta] = [a][\delta] \tag{A.1}$$

where $[\delta]$ is the column vector of the three unknown displacements. The forces in the piles are given by

$$[F] = [s][\Delta] = [s][a][\delta] \tag{A.2}$$

By contragradience with Equation A.1, the external loads $[P]$ are related to internal forces $[F]$ by

$$[P] = [a]^{\mathrm{T}}[F] \tag{A.3}$$

where $[a]^{\mathrm{T}}$ is the transform of matrix $[a]$.

Substituting for $[F]$ from Equation A.2

$$[P] = [a]^{\mathrm{T}}[s][a][\delta] = [S][\delta] \tag{A.4}$$

where $[S]$ is the required stiffness matrix of the pile group with respect to the chosen axes.

Equation A.4 relates the applied loading $[P]$ to the unknown displacements $[\delta]$ and is the equilibrium equation of the structure. Stiffness matrix $[S]$ comprises known numerical coefficients and is a function of the geometry of the structure. Solution of Equation A.4 thus yields the required displacements $[\delta]$ of the pile cap and forces in individual piles are obtained by matrix multiplication defined in Equation A.2.

In many problems it is convenient to work in numerical terms once matrices $[a]$ and $[s]$ are established. It is, however, of interest to examine matrix $[S]$ in general terms. Matrix multiplication $[s][a]$ (Equation A.2) gives

$$[s][a] = \begin{bmatrix} s_1 & & \\ & s_2 & \\ & & s_n \end{bmatrix} \begin{bmatrix} T_1 \\ T_2 \\ T_n \end{bmatrix} = \begin{bmatrix} s_1\,T_1 \\ s_2\,T_2 \\ s_n\,T_n \end{bmatrix}$$

and stiffness matrix $[S]$ becomes

$$[S] = [T'_1\,T'_2\,T'_n] \begin{bmatrix} s_1\,T_1 \\ s_2\,T_2 \\ s_n\,T_n \end{bmatrix} = \sum_{i=1}^{n} [T_i^{\mathrm{T}} s_i T_i] \tag{A.5}$$

The stiffness matrix is therefore the sum of contributions of each individual pile of the group. For a typical pile n in Fig. A.3(a), this contribution becomes

$$[T^{\mathrm{T}} s T]_n = \begin{bmatrix} \sin \alpha \\ \cos \alpha \\ e_n \cos \alpha \end{bmatrix} \frac{AE}{L} [\sin \alpha \cos \alpha\; e_n \cos \alpha]$$

$$= \frac{AE}{L} \begin{bmatrix} \sin^2 \alpha & \sin \alpha \cos \alpha & e_n \sin \alpha \cos \alpha \\ \sin \alpha \cos \alpha & \cos^2 \alpha & e_n \cos^2 \alpha \\ e_n \sin \alpha \cos \alpha & e_n \cos^2 \alpha & e_n^2 \cos^2 \alpha \end{bmatrix} \tag{A.6}$$

In the special case when the pile is vertical ($\alpha = 0$)

$$[T^T sT]_n = \frac{AE}{L}\begin{bmatrix} 0 & 0 & 0 \\ 0 & 1 & e_n \\ 0 & e_n & e_n^2 \end{bmatrix}$$

A.2.2 Bending stiffness included

When bending stiffness is included in the analysis, the shearing and rotational deformations Δ_1 and Δ_3 of each pile have to be considered in addition to the axial deformation Δ_2. With the notation of Fig. A.5 it can be verified that the stiffness matrix for a typical member m with respect to the three member displacements $(\Delta_1 \Delta_2 \Delta_3)$ is given by

$$[s_m] = \begin{bmatrix} \dfrac{12EI}{L^3} & 0 & -\dfrac{6EI}{L^2} \\ 0 & \dfrac{EI}{L} & 0 \\ -\dfrac{6EI}{L^2} & 0 & \dfrac{4EI}{L} \end{bmatrix}$$

when the foot is rigidly fixed and

$$[s_m] = \begin{bmatrix} \dfrac{3EI}{L^3} & 0 & -\dfrac{3EI}{L^2} \\ 0 & \dfrac{EA}{L} & 0 \\ -\dfrac{3EI}{L^2} & 0 & \dfrac{3EI}{L} \end{bmatrix}$$

when the foot is pinned.

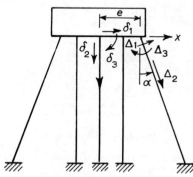

Figure A.5 Definition and signs of pile cap displacements and pile distortions when bending stiffness is included.

The transformation matrix for obtaining pile distortions from cap displacements is

$$[T_n] = \begin{bmatrix} \overset{\delta_1=1}{+\cos\alpha} & \overset{\delta_2=1}{-\sin\alpha} & \overset{\delta_3=1}{-e\sin\alpha} \\ +\sin\alpha & +\cos\alpha & +e\cos\alpha \\ 0 & 0 & +1 \end{bmatrix} \begin{matrix} \text{for } \Delta_1 \\ \Delta_2 \\ \Delta_3 \end{matrix}$$

Here Δ_2 corresponds to axial displacement and the second row of the matrix is identical with the transformation matrix when bending is neglected.

These are the only new features introduced in the analysis when bending stresses are included. The geometric matrix, unassembled stiffness matrix and stiffness matrix for the complete pile group take the same form as before. It is a simple matter of matrix multiplication to verify that for a typical pile rigidly fixed at both ends

$$[s_m][T_n] = \frac{AE}{L} \begin{bmatrix} \left(\frac{12\,I}{L^2\,A}c^4\right) & \left(-\frac{12\,I}{L^2\,A}c^3 s\right) & \left(-\frac{12\,I}{L^2\,A}ec^3 s - \frac{6\,I}{L\,A}c^2\right) \\ (sc) & (c^2) & (ec^2) \\ \left(-\frac{6\,I}{L\,A}c^3\right) & \left(\frac{6\,I}{L\,A}c^2 s\right) & \left(\frac{6\,I}{L\,A}ec^2 s + 4\frac{I}{A}c\right) \end{bmatrix} \quad (A.7)$$

and

$$[S]_n = [T_n^T s_m T_n] = \frac{AE}{L} \begin{bmatrix} \left(\frac{12\,I}{L^2\,A}c^5 + s^2 c\right) & \left(-\frac{12\,I}{L^2\,A}c^4 s + sc^2\right) & \left(-\frac{12\,I}{L^2\,A}ec^4 s - \frac{6\,I}{L\,A}c^3 + esc^2\right) \\ \left(-\frac{12\,I}{L^2\,A}c^4 s + sc^2\right) & \left(\frac{12\,I}{L^2\,A}c^3 s^2 + c^3\right) & \left(\frac{12\,I}{L^2\,A}ec^3 s^2 + \frac{6\,I}{L\,A}c^2 s + ec^3\right) \\ \begin{pmatrix} -\frac{12\,I}{L^2\,A}ec^4 s \\ -\frac{6\,I}{L\,A}c^3 + esc^2 \end{pmatrix} & \begin{pmatrix} \frac{12\,I}{L^2\,A}ec^3 s^2 \\ +\frac{6\,I}{L\,A}c^2 s + ec^3 \end{pmatrix} & \begin{pmatrix} \frac{12\,I}{L^2\,A}e^2 c^3 s^2 + \frac{12\,I}{L\,A}ec^2 s + e^2 c^3 \\ +4\frac{I}{A}c \end{pmatrix} \end{bmatrix}$$

$$(A.8)$$

EXAMPLE A.2*

Bending stiffness ignored

Consider the pile group shown in Fig. A.6(a). The loads indicated are per slice of the pile group equal to the longitudinal spacing of the piles. With the notation, and with $s = \sin\alpha$, $c = \cos\alpha$, the matrices become, since L is measured vertically,

*The editor is indebted to Professor F. Sawko for permission to include the method of analysis in this Appendix and to F.W. Mathews of Brighton Polytechnic for reworking Example A.2 in SI units.

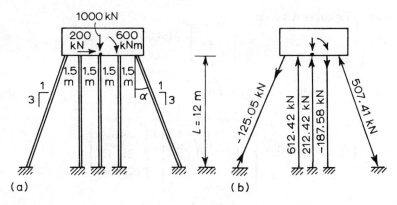

Figure A.6 Analysis of typical pile group. (a) Loading and dimensions; (b) axial force distribution.

$$[a] = \begin{bmatrix} -s+c-2ac \\ 0+1-a \\ 0+1 \quad 0 \\ 0+1+a \\ +s+c+2ac \end{bmatrix} \quad [s] = \frac{AE}{L}\begin{bmatrix} c & & & \\ & 1 & & \\ & & 1 & \\ & & & 1 \\ & & & & c \end{bmatrix}$$

$$[s][a] = \frac{AE}{L}\begin{bmatrix} -sc+c^2-2ac^2 \\ 0+1 \quad -a \\ 0+1 \quad 0 \\ 0+1 \quad +a \\ +sc+c^2+2ac^2 \end{bmatrix}$$

$$[S] = [a]^T[s][a] = \frac{AE}{L}\begin{bmatrix} 2s^2c & 0 & 4asc^2 \\ 0 & 3+2c^3 & 0 \\ 4asc^2 & 0 & 8a^2c^3+2a^2 \end{bmatrix}$$

It will be observed that for a symmetrical piling system the three equations can be separated to two sets of one and two equations each.

With the numerical values from Fig. A.6(a) the stiffness matrix becomes

$$[S] = \frac{AE}{L}\begin{bmatrix} 0.189737 & 0 & 1.70763 \\ 0 & 4.70763 & 0 \\ 1.70763 & 0 & 19.86867 \end{bmatrix}$$

Applied loading vector

$$[P] = \begin{bmatrix} 200 \\ 1000 \\ 600 \end{bmatrix}$$

Solution of Equation A.4 gives

$$[\delta] = \frac{L}{AE} \begin{bmatrix} 3454.066 \\ 212.421 \\ -266.664 \end{bmatrix}$$

and the forces in each pile are derived from

$$[F] = [s][a][\delta] = \begin{bmatrix} -0.3 + 0.9 - 2.7 \\ 0 \; +1.0 - 1.5 \\ 0 \; +1.0 - 0 \\ 0 \; +1.0 + 1.5 \\ +0.3 + 0.9 + 2.7 \end{bmatrix} \begin{bmatrix} 3454.066 \\ 212.421 \\ -266.664 \end{bmatrix}$$

$$= \begin{bmatrix} -125.047 \\ 612.418 \\ 212.421 \\ -187.575 \\ 507.405 \end{bmatrix} \text{kN} \qquad \begin{matrix} \text{pile no. } 1 \\ 2 \\ 3 \\ 4 \\ 5 \end{matrix}$$

The axial force distribution is shown in Fig. A.6(b), positive forces indicating compression in a pile, negative tension.

Bending stiffness included

For the pile group of Fig. A.6(a) the stiffness matrix becomes

$$[S] = \frac{AE}{L} \begin{bmatrix} \left(\dfrac{12\,I}{L^2 A}(3 + 2c^5) + 2s^2 c\right) & (0) & \left(-\dfrac{6\,I}{LA}(3 + 2c^3) + 4asc^2 - \dfrac{48a\,I}{L^2 A}c^4 s\right) \\ (0) & \left((3 + 2c^3) + \dfrac{24\,I}{L^2 A}c^3 s^2\right) & (0) \\ \left(-\dfrac{6\,I}{LA}(3 + 2c^3) + 4asc^2 - \dfrac{48a\,I}{L^2 A}c^4 s\right) & (0) & \left(\dfrac{96a^2\,I}{L^2\,A}c^3 s^2 + 2a^2(1 + 4c^3) + 4\dfrac{I}{A}(3 + 2c) + \dfrac{48a\,I}{LA}c^2 s\right) \end{bmatrix}$$

When $I = 0$, this reduces to the form obtained above.
Substituting numerical values and assuming 0.5 m diameter solid circular piles,

$$\frac{I}{A} = \frac{D^2}{16} = 0.15625$$

$$[S] = \frac{AE}{L} \begin{bmatrix} 0.19564 & 0 & 1.6689 \\ 0 & 4.7078 & 0 \\ 1.6689 & 0 & 20.2034 \end{bmatrix}$$

Comparing this with the set where bending stiffness is neglected, it is clear that bending stiffness plays a negligible part in reducing the deflections and hence the distribution of axial forces in the piles. This applies to many cases in practice.

It follows that bending moments and shearing forces can be estimated by first ignoring the influence of bending stiffness to determine cap displacements. Pile forces and moments are then obtained from Equation A.7, the corresponding matrices including bending. Thus, in spite of neglecting flexure, bending moments, being a function of cap displacements, can be estimated with sufficient accuracy for practical purposes. Simple multiplication gives the values of moments at pile cap as

$$
\begin{bmatrix}
-40.16 \\
-43.64 \\
-43.64 \\
-43.64 \\
-40.16
\end{bmatrix}
\begin{matrix}
\text{for pile 1} \\
\text{for pile 2} \\
\text{for pile 3} \\
\text{for pile 4} \\
\text{for pile 5}
\end{matrix}
$$

Moments at the pile feet can be similarly calculated.

REFERENCES

Åas, G. (1965) A study of the effect of vane shape and rate of strain on the measured values of *in-situ* shear strength of soils. *Proc. 6th Int. Conf. Soil Mech. Found. Eng.*, **1**, 141.

Abbiss, C.P. (1981) Shear wave measurements of the elasticity of the ground. *Géotechnique*, **31**, 91.

Abbiss, C.P. (1983) Calculation of elasticities and settlements for long periods of time and high strains from seismic measurements. *Géotechnique*, **33**, 397.

Abbs, A.F. and Sinclair, T.J.E. (1979) Computer modelling of stability beneath the edge of an oil storage tank. *Proc. Inst. Civil Eng.*, **67** (2), 597.

Abdul-Baki, A. and Beik, L.A. (1970) Bearing capacity of foundations on sand. *J. Soil Mech. Found. Div. (ASCE)*, **96**, 545.

Abelev, Y.M. and Askalonov, V.V. (1957) The stabilization of foundations of structures on loess soils. *Proc. 4th Int. Conf. Soil Mech. Found. Eng.*, **1**, 259.

ACI (1958) Pressures on formwork. Report of ACI Comm. 622, *J. ACI*, **55**, 173.

ACI–ASCE (1962) Report of Comm. on: Shear and diagonal tension. *Proc. ACI*, **59**, 1, 277, 353.

ACI (1966) Committee Report 436: Suggested design procedures for combined footings and mats. *J. ACI* **63**, 1041; Discuss., **64** (1967), 1537.

ACI (1977) *Standard 318–77, Building code requirements for reinforced concrete.*

ACI (1978a) Standard specification for end bearing drilled piers: Proposed ACI Standard. *J. ACI*, **75**, 429.

ACI (1978b) *Standard 347–78, Recommended practice for concrete formwork.*

ACI (1982) *Foundations for Equipment and Machinery*, Special Publication 78.

Adams, D. (1973) Preboring – its effects and usefulness. *Construction News – Piling and Found. Suppl.*, 13 December, 48.

Adams, J.I. (1965) The engineering behaviour of Canadian muskeg. *Proc. 6th Int. Conf. Soil Mech. Found. Eng.*, **1**, 3.

Adams, J.I. and Radharkrishna, H.S. (1971) Uplift resistance of augered footings in fissured clay. *Can. Geotech. J.*, **8**, 452.

Adamson, J.N. and Scott, R.A. (1973) Borehole investigations and logging methods in shaft sinking and tunnelling. *The Min. Eng.*, **132**, 181.

Adler, F. (1964) Jig for the construction of a 17-storey block of flats at Barras Heath, Coventry. *Proc. Inst. Civil Eng.*, **27**, 433.

Agar, M. and Irwin–Childs, F. (1973) Seaforth Dock, Liverpool: planning and design. *Proc. Inst. Civil Eng.*, **54** (1), 255.

Agbezuge, L.K. and Deresiewicz, H. (1974) On the indentation of a consolidating half-space. *Isr. J. Technol.*, **12**, 322.

Agrément Board (1981) *The assessment of torque-expanded anchor bolts when used in dense aggregate concrete.* MOAT 19: 1981, Garston.

Aizlewood, H., and Heathcote, N.B. (1970) Coventry inner ringroad: design and construction. *Proc. Inst. Civil Eng.*, **45**, 199.

990

Ajaz, A. and Parry, R.H.G. (1976) Bending test for compacted clays. *J. Geotech. Eng. Div.* (*ASCE*), **102**, 929.

Akatsuka, Y. (1968) Pressure on forms of prepacked concrete. *J. ACI*, **65**, 390.

Akinmusuru, J.O. and Akinbolade, J.A. (1981) Stability of loaded footings on reinforced soil. *J. Geotech. Eng. Div.* (*ASCE*), **107**, 819.

Akroyd, T.N.W. (1966) Workmanship for a structural concrete code. *The Struct. Eng.*, **44**, 369; **45** (1967), 219.

Aldrich, H.P. (1965) Precompression for support of shallow foundations. *J. Soil Mech. Found. Div.* (*ASCE*), **91** (March), 5.

Al-Jassar, S. and Hawkins, A.B. (1977) Some geotechnical properties of the main carbonate lithologies within the carboniferous limestone formation of the Clifton Gorge, Bristol. *Proc. Conf. Rock Eng.*, Univ. Newcastle-upon-Tyne, 393.

Al-Khafaji, A-A.W.N. and Andersland, O.B. (1981) Ignition test for soil organic content measurement. *J. Geotech. Eng. Div.* (*ASCE*), **107**, 465.

Allan, B.J. (1971) Aesthetics in bridge design. *J. Inst. Highway Eng.*, **XVIII** (May).

Allan, B.J. (1976) Some notes on significance of form in bridge engineering. *Proc. Inst. Civil Eng.*, **60**, 79.

Allaway, R.G. (1962) Restoration of the smelter building at Nkana, Northern Rhodesia. *The Struct. Eng.*, **40**, 229.

Allen, D.N. de G. and Severn, R.T. (1960) The stresses in foundation rafts. *J. Inst. Civil Eng.*, **15**, 35.

Allen, D.N. de G. and Severn, R.T. (1961) Composite action between beams and slabs under transverse load. *The Struct. Eng.*, **39**, 149, 235.

Allen, D.N. de G. and Severn, R.T. (1962) Composite action between beams and slabs under transverse load. *The Struct. Eng.*, **40**, 191.

Allen, D.N. de G. and Severn, R.T. (1963) The stresses in foundation rafts. *J. Inst. Civil Eng.*, **25**, 257.

Allen, D.N. de G. and Severn, R.T. (1964) Composite action between beams and slabs under transverse load. *The Struct. Eng.*, **42**, 429.

Allen, D.N. de G. and Southwell, R.V. (1950) Relaxation methods applied to engineering problems – XIV: Plastic straining in two dimensional stress systems. *Phil. Trans Roy. Soc. Ser. A*, **242**, 379.

Alpan, I. (1957) An apparatus for measuring swelling pressure in expansive soils. *Proc. 4th Int. Conf. Soil Mech. Found. Eng.*, **1**, 3.

Al-Saadi, R. and Brooks, M. (1973) A geophysical study of Pleistocene buried valleys in the Lower Swansea Valley, Vale of Neath and Swansea Bay. *Proc. Geol. Ass.*, **84**, 135.

Amaryan, L.S. *et al.* (1973) Consolidation laws and mechanical-structural properties of peaty soils. *Proc. 8th Int. Conf. Soil Mech. Found. Eng.*, **2** (2), 1.

Amer, A.M. and Awad, A.A. (1974) Permeability of cohesionless soils. *J. Geotech. Eng. Div.* (*ASCE*), **100**, 1309.

Anandakrishnan, M. and Varadarajulu, G.H. (1963) Laminar and turbulent flow of water in sand. *J. Soil Mech. Found. Div.* (*ASCE*), **89** (September), 1.

Andersen, K.H. *et al.* (1980) Cyclic and static laboratory tests on Drammen clay. *J. Geotech. Eng. Div.* (*ASCE*), **106**, 499.

Andersen, K.H. and Stenhamar, P. (1982) Static plate loading tests on overconsolidated clay. *J. Geotech. Eng. Div.* (*ASCE*), **108**, 918.

Anderson, J.G.C. and Blundell, C.R.K. (1965) The sub-drift rock-surface and buried valleys of the Cardiff District. *Proc. Geol. Ass.*, **76**, 367.

Anderson, J.K. and Brown, C.D. (1964) Design and construction of the Kingsferry Lifting Bridge, Isle of Sheppey. *Proc. Inst. Civil Eng.*, **28**, 449.

Anderson, J.K. *et al.* (1965) Forth Road Bridge. *Proc. Inst. Civil Eng.*, **32**, 321.

Anderson, W.F. (1974) Factors influencing the measured properties of Glasgow region till. *Ground Eng.*, **7** (2), 20.

Anderson, W.F. *et al.* (1983) Overall stability of anchored retaining walls. *J. Geotech. Eng.*, **109**, 1416.

Andrawes, K.L. and El-Sohky, M.A. (1973) Factors affecting coefficient of earth pressure K_0. *J. Soil Mech. Found. Div. (ASCE)*, **99**, 527.

Andrews, W.C. (1944) Foundations in combustible material. *The Struct. Eng.*, **22**, 53.

Ang, A.H.S. and Harper, G.N. (1964) Analysis of contained plastic flow in plane solids. *J. Eng. Mech. Div. (ASCE) EMS*, **90**, 397.

Anon, (1937) Sinking deep caissons by the sand-island method. *The Eng.*, **163**, 556.

Anon (1949) New river wall, South Bank, Lambeth, London. *Concrete and Constr. Eng.*, **44**, 349.

Anon (1960) Form follows function in the new lecture theatre, Edinburgh University. *Concrete Quart.*, **47**, 2.

Anon (1961) Simshill Secondary School, Glasgow. *Concrete Quart.*, **51**, 29.

Anon (March, 1962) Horizontal clay seam triggered Wheeler Lock failure. *Civil Eng.*, 89.

Anon (1963) Cathodic protection of marine structures. Sea Action Committee. *Proc. Inst. Civil Eng.*, **25**, 193.

Anon (July, 1968) Major re-development in Paris. *Concrete*, **2**, 280.

Anon (March, 1970a) Bourne Hall, Ewell, Surrey. *Concrete*, 101.

Anon (1970b) The logging of rock cores for engineering purposes. *Quart. J. Eng. Geol.*, **3**, 1.

Anon (1971) *The Production of Soil Engineering Maps for Roads and The Storage of Materials Data*, TRH2, National Institute for Road Reseach, Pretoria.

Anon (1972a) The preparation of maps and plans in term of engineering geology. *Quart. J. Eng. Geol.*, **5**, 295.

Anon (1972b) Moving house after 350 years. *Contract J.*, 3 February.

Anon (1973) *Proc. Symp. Field Instrumentation in Geotech. Eng.*, Butterworth, London.

Anson, M. (1964) An investigation into a hypothetical deformation and failure mechanism for concrete. *Mag. Concrete Res.*, **16**, 47, 73.

Anson, M. and Newman, K. (1966) The effect of mix proportions and method of testing on Poisson's ratio for mortars and concrete. *Mag. Concrete Res.*, **18**, 115; Discuss. (1967) **19**, 118.

Antonakis, C.J. (1972) A problem of designing and building for a structure at sea (Royal Sovereign Lighthouse). *Proc. Inst. Civil Eng.*, **52** (1), 95; Discuss. (1973) **54** (1), 673.

Apelt, C.J. and Isaacs, L.T. (1968) Bridge piers – hydrodynamic force coefficients. *J. Hyd. Div. (ASCE)*, **94**, 17.

API (1972) *American Petroleum Institute, RP 2A*. March.

Appendino, M. and Comastri, A. (1970) *Prove di carico supali di ghiaia eseguiti in terreno coesivo mediante vibroflottazione*, E.N.E.L., Milano, Italy.

Arbhabhirama, A. and Dinoy, A.A. (1973) Friction factor and Reynolds Number in porous media flow. *J. Hyd. Div. (ASCE)*, **99**, 901.

Armitage, J.S. (1981) The anatomy and exercise of engineering judgement. *The Struct. Eng.*, **59A**, 167.

Arnold, M. (1980) Prediction of footing settlements on sand. *Ground Eng.*, **13** (2), 40.

Arulanandan, K. *et al.* (1971) Undrained creep behaviour of a coastal organic silty clay. *Géotechnique*, **21**, 359.

Arutiunian, N.K. (1966) *Some problems in the theory of creep in concrete structures*, Pergamon Press, Oxford.

Arvidsson, K. (1975) Non-uniform shear wall-frame systems with elastic foundations. *Proc. Inst. Civil Eng.*, **59** (2), 139.

ACSE (1976) *Subsurface Investigation for Design and Construction of Foundations of Buildings*, Manuals and Reports in Engineering Practice No. 56, ASCE.

ASCE (1980) Expansive soils. *Proc. Int. Conf. ASCE.*

Astill, A.W. (June, 1974) Foundation design and CP110. *Civil Eng.*, 26.

Astill, A.W. and Al-Sajir, D.K. (1980) Compression bond in column-to-base joints. *The Struct. Eng.*, **58B**, 1.

ASTM (1966a) *Testing Techniques for Rock Mechanics*, Special Technical Publication 402.

ASTM (1966b) *Determination of Stress in Rock*, Special Technical Publication 429.

ASTM (1969) *Determination of Stress in Rock – a State of the Art Report*, Special Technical Publication.

ASTM (1975) *Performance Monitoring for Geotechnical Construction*, STP 584.

Atherton, F.G. and Garrett, W.S. (1959) History of cementation in shaft sinking. *Symp. on Shaft Sinking and Tunnelling*, Inst. Min. Eng., p. 394.

Atkins, W.S. (1950) *New Steelworks, Port Talbot, Planning and Design*, Structural Paper 25, Inst. Civil Eng.

Atkinson, J.H. and Potts, D.M. (1977) Subsidence above shallow tunnels in soft ground. *J. Geotech. Eng. Div. (ASCE)*, **103**, 307.

Attewell, P.B. (1977a) Appraisal of face stability at a limestone quarry in Shropshire. *Proc. Conf. Rock Eng.*, Univ. Newcastle-upon-Tyne, p. 361.

Attewell, P.B. (ed.) (1977b) *Proc. Conf. on Engineering Properties of Rocks of Mid-Jurassic Age and Older in the UK*, Univ. Newcastle-upon-Tyne.

Attewell, P.B. and Farmer, I.W. (1973) Attenuation of ground vibrations from pile driving. *Ground Eng.*, **6** (4).

Au, T. and Baird, D.L. (1960) Bearing capacity of concrete blocks. *J. ACI*, **31**, 869, 1467.

Au, W.-C. and Chae, Y.S. (1980) Dynamic shear modulus of treated expansive soils. *J. Geotech. Eng. Div. Proc. ASCE*, **106**, 255.

Audric, T. and Bouquier, L. (1976) Collapsing behaviour of some loess soils from Normandy. *Quart. J. Eng. Geol.*, **9**, 265.

Aynsley, W.J. and Hewitt, G. (1960) Subsidence observations over shallow workings, including pneumatic stowing and rapidly advancing faces. *Trans. Inst. Min. Eng.*, **120**, 552.

Bagnold, R.A. (1939) Interim report on wave pressure research. *Proc. Inst. Civil. Eng.*, **12**, 202.

Baguelin, *et al.* (1977) The pressuremeter and foundation engineering. *Trans. Tech. Publ.*

Bahmeier, H.F. (May, 1950) Construction of Davis Dam adapted to foundation conditions. *Civil Eng.*, 23.

Baker, A.L.L. (1949) *Reinforced Concrete*, Concrete Publications, London.

Baker, A.L.L. (1957) *Raft Foundations* – Soil-line Method of Design, 3rd Ed, Concrete Publications, London.

Balaam, N.P. (1978) *Load-settlement Behaviour of Granular Piles*, Ph.D. thesis, University of Sydney.

Baldovin, G. and Santovito, D. (1973) Tunnel construction in high swelling clays. *Proc. 8th Int. Conf. Soil Mech. Found. Eng.*, **2** (2), 13.

Baligh, M.M. and Azzouz, A.S. (1975) End effects on stability of cohesive slopes. *J. Geotech. Eng. Div. (ASCE)*, **101**, 1105.

Baligh, M.M. *et al.* (1980) Cone penetration in soil profiling. *J. Geotech. Eng. Div. (ASCE)*, **106**, 447.

Balla, A. (1961) The resistance to breaking out of mushroom foundations for pylons. *Proc. 5th Int. Conf. Soil Mech. Found. Eng.*, **1**, 569.

Balla, A. (October, 1962) Bearing capacity of foundations. *J. Soil Mech. Found. Div. (ASCE)*, **88**, 13.

Ballard, W.E. (1962) The application and inspection of protective coatings on structural steel. *Proc. Inst. Civil Eng.*, **22**, 227; Discuss., **24** (1963), 291.

Bally, *et al.* (1969) On stress–strain in highly compressible triphasic soils. *Proc. 7th Int. Conf. Soil Mech. Found Eng.*, **1**, 1.

Bals, R. (1931) *Beitrag zur der Vorausberechnung berbaulicher Senkungen.* Mitt a.d. Marks.

Bandis, S. *et al.* (1981) Experimental studies of scale effects on the shear behaviour of rock joints. *Int. J. Rock Mech. Mining Sci.*, **18**, 1.

Banerjee, P.K. (1978) Analysis of axially and laterally loaded pile groups, in Scott, C.R. (ed.). *Developments in Soil Mechanics* – 1, 317, Applied Science Publishers, Barking.

Banerjee, P.K. and Driscoll, R.M.C. (1978) *Programme for the analysis of pile groups of any geometry subjected to horizontal and vertical loads and moments*, PGROUP, (2.1) HECB/B/7, Dept. Transport, HECB, London.

Baracos, A. and Bozozuk, K. (1957) Seasonal movements in some Canadian clays. *Proc. 4th Int. Conf. Soil Mech. Found. Eng.*, **1**, 264.

Barden, L. (1962a) Distribution of contact pressure under foundations. *Géotechnique*, **12**, 181; Corresp., **13** (1963), 87.

Barden, L. (1962b) An approximate solution for finite beams resting on elastic soil. *Civil Eng. and Public Works Rev.*, **57**, 1429.

Barden, L. (1963a) The Winkler model and its application to soil. *The Struct. Eng.*, **41**, 279.

Barden, L. (1963b) Stresses and displacements in a cross-anisotropic soil. *Géotechnique*, **13**, 3.

Barden, L. (1965a) Consolidation of clay with non-linear viscosity. *Géotechnique*, **15**, 345.

Barden, L. (1965b) Contact pressures under circular slabs. *The Struct. Eng.*, **43**, 153.

Barden, L. (1968) Primary and secondary consolidation of clay and peat. *Géotechnique*, **18**, 1.

Barden, L. (1972) The influence of structure on deformation and failure in clay soil, *Geotechnique*, **22**, 159.

Bareš, R. and Massonnett, C. (1968) *Analysis of Beam Grids and Orthotropic Plates*, (English translation by J. Vanek), Crosby Lockwood, London.

Barkan, D.D. (1957) Foundation engineering and drilling by the vibration method. *Proc. 4th Int. Conf. Soil Mech. London*, Paper 2, p. 3.

Barkan, D.D. (1962) *Dynamics of Bases and Foundations*, McGraw-Hill, New York.

Barnbrook, G. (1981) *House Foundations for the Builder and Building Designer*, Cement and Concrete Ass. Wexham Springs, Slough.

Bartak, A.J.J. and Shears, M. (1972) The new tower for the Independent Television Authority at Emley Moor, Yorkshire, *The Struct. Eng.*, **50**, 67.

Bartlett, V.F. and Danks, J.Y. (1941) Steel pedestals for heavy columns. *J. Inst. Civil Eng.*, **17**, 153.

Barton, M.E. and Coles, B.J. (1984) The characteristics and rates of the various slope degradation processes in the Barton Clay cliffs of Hampshire. *Quart. J. Eng. Geol.*, **17**, 117.

Barton, N. (1973) Review of a new shear-strength criterion for rock joints. *Eng. Geol.*, **7**, 287.

Barton, N. (1981) Estimation of *in-situ* joint properties, Nasliden mine. *Proc. Conf. on Application of Rock Mechanics to Cut and Fill Mining*, Univ. of Luleå, Sweden, Inst. Mining and Met.

Barton, N. and Kjaernsli (1981) Shear strength of rockfill. *J. Geotech. Eng. Div. (ASCE)*, **107**, 873.

Base, G.D. (September, 1960) Some tests on a particular design of reinforced concrete structural hinge. *Reinforced Concrete Rev.*, 449.

Base, G.D. (1965) *Tests on Four Prototype Reinforced Concrete Hinges.* Cement and Concrete Association Research Report 17.

Bassett, R.H. (1970) Discussion. *Proc. Conf. Ground Eng.*, Inst. Civil Eng., p. 89.

Bast, S.C. (July, 1960) United Engineering Center, 1. An all-welded structure, Civil Eng. (ASCE) p. 62.

Basu, A.K. and Chapman, J.C. (1966) Large deflexion behaviour of transversely loaded rectangular orthotropic plates. *Proc. Inst. Civil Eng.*, **35**, 79.

Bauman, V. and Bauer, G.E.A. (1974) The performance of foundations on various soils stabilized by the vibro-compaction method. *Can. Geotech. J.*, **11**, 509.

Baxter, J.W., Lee, D.J. and Humphries, E.F. (1972) Design of Western Avenue Extension (Westway). *Proc. Inst. Civil Eng.*, **52**, 177.

BCSA (1955) *Welded Details for Single Storey Portal Frames*, British Constructional Steelwork Association, publication No. 9.

Becker, J.C. (July, 1972) Alaska builds highway over muskeg and permafrost. *Civil Eng.*, 75.

Beeby, A.W. (1978) *Concrete in the Oceans: Report No. 1, Cracking and Corrosion*, Cement and Concrete Association, Wexham Springs, Slough.

Beevers, C. and Wardell, K. (1954) Recent research in mining subsidence. *Trans. Inst. Min. Eng.*, **114**, 223.

Bell, A.L. (1915) The lateral pressure and resistance of clay and the supporting power of clay foundations. *Proc. Inst. Civil Eng.*, **199**, 233.

Bell, R.A. and Iwakiri, J. (1980) Settlement comparison used in tank-failure study. *J. Geotech. Eng. Div. (ASCE)*, **106**, 153.

Belshaw, T. (1973) The use of the deep sounding static penetrometer in the Napier District, New Zealand. *New Zealand Eng.*, 15 May.

Bennett, P.B. and Elliott, D.H. (1975) *Physiology and Medicine of Diving and Compressed Air Work*, Baillière, Tindall and Cassell, London.

Berezantsev, V.G. (1961) Load bearing capacity and deformation of piled foundations. *Proc. 5th Int. Conf. Soil Mech. Found. Eng.*, **2**, 11, Dunod, Paris.

Bergdahl, U. and Broms, B.B. (1967) New method of measuring *in-situ* settlements. *J. Soil Mech. Found. Div. (ASCE) SM* 5, **93**, 51.

Berggren, R.A. (June, 1961) Determining vertical stress beneath a footing. *Civil Eng.*, 71.

Bergstrom, S.G. (1946) *Circular Plates with Concentrated Loads on an Elastic Foundation*, Bull. No. 6, Swedish Cement and Concrete Research Institute, Stockholm.

Berry, F.G. (1979) Late Quaternary scour-hollows and related features in central London. *Quart. J. Eng. Geol.*, **12**, 9.

Berry, P.L. (1983) Application of consolidation theory for peat to the design of a reclamation scheme by preloading. *Quart. J. Eng. Geol.*, **16**, 103.

Berry, P.L. and Poskitt, T.J. (1972) The consolidation of peat. *Géotechnique*, **22**, 27.

Berry, P.L. and Vickers, B. (1975) Consolidation of fibrous peat. *J. Geotech. Eng. Div. ASCE*, **101**, 741.

Bessey, G.E. and Lea, F.M. (1953) The distribution of sulphates in clay soils and groundwater. *Proc. Inst. Civil Eng.*, **2** (1), 159.

Bethel, R. (1968) Underpinning a landmark. *Civil Eng.*, **38** (August), 67.

Bhandari, R.K.M. (1977) Countering liquifaction by subsoil densification at a refinery complex in Assam, India. *Proc. 4th World Conf. on Earthquake Eng.*, New Delhi.

Bhushan, K. and Haley, S.C. (1976) Stress distribution for heavy embedded structures. *J. Geotech. Eng. Div. (ASCE)*, **102**, 807.

Biarez, J. and Capelle, J.F. (1961) Contribution a l'étude de la rotation des fondations. *Proc. 5th Int. Conf. Soil Mech. Found. Eng.*, **2**, 367.

Biarez, J. et al. (1965) Equilibre limite d'écrans verticaux soumis á une translation ou une rotation. *Proc. 6th Int. Conf. Soil Mech. Found. Eng.*, **2**, 368.

Bieniawski, Z.T. (1974) Estimating the strength of rock materials. *J. S. Afr. Inst. Mining Met.*, **74**, 312.

Bieniawski, Z.T. and van Heerden, W.L. (1975) The significance of *in-situ* tests on large rock specimens. *Int. J. Rock Mech. and Min. Sci.*, **12**, 101.

Biery, P.F. and Delleur, J.W. (1962) Hydraulics of single span arch bridge constrictions. *J. Hydraul. Div. (ASCE)*, **88** (March), 75.

Bingham, D.M. and Soane, A.J.M. (1970) Unusual foundations over the Mersey Tunnel. *The Struct. Eng.*, **48**, 153.

Bingham, T.G. and Lee, D.J. (1969) The Mancunian Way elevated road structure. *Proc. Inst. Civil Eng.*, **42**, 459.

Binquet, J. and Lee, K.L. (1975a) Bearing capacity tests on reinforced earth slabs. *J. Geotech. Eng. Div. (ASCE)*, **101**, 1241.

Binquet, J. and Lee, K.L. (1975b) Bearing capacity analysis of reinforced earth slabs. *J. Geotech. Eng. Div. (ASCE)*, **101**, 1257.

Biot, M.A. (1935) Effect of certain discontinuities on the pressure distribution in a loaded soil. *J. Appl. Phys.*, **6**, 367.

Biot, M.A. (1937) Bending of an infinite beam on an elastic foundation. *J. Appl. Mech.*, **4**.

Biot, M.A. and Clingan, F.M. (1942) Bending settlement of a slab resting on a consolidating foundation. *J. Appl. Phys.*, **13**. 35.

Bishop, A.W. (1955) The use of the slip circle in the stability analysis of slopes. *Géotechnique*, **5**, 7.

Bishop, A.W. (1962) Correspondence on Potyondy, 1961, *Géotechnique*, **12**, 72.

Bishop, A.W. (1966) The strength of soils as engineering materials. *Géotechnique*, **16**, 91.

Bishop, A.W. (1967) Progressive failure with special reference to the mechanism causing it. *Proc. Geotech. Conf.*, Oslo, **2**, 142.

Bishop, A.W. and Edlin, A.K.G. (1950) Undrained triaxial tests in saturated sands and their significance in the general theory of shear strength. *Géotechnique*, **2**, 13.

Bishop, A.W. and Lovenbury, H.T. (1969) Creep characteristics of two undisturbed clays. *Proc. 7th Int. Conf. Soil Mech. Found. Eng.*, **1**, 29.

Bishop, A.W. and Morgenstern, N. (1960) Stability coefficients for earth slopes. *Géotechnique*, **10**, 129.

Bishop, A.W. *et al.* (1965) Undisturbed samples of London Clay from the Ashford Common shaft: strength–effective stress relationships. *Géotechnique*, **15**, 1.

Bishop, A.W. *et al.* (1973) Strength and deformation measurements in soils. *Proc. 8th Int. Conf. Soil Mech. Found. Eng.*, **1** (1), 57.

Bishop, R.F. *et al.* (1945) The theory of indentation and hardness tests. *Proc. Phys. Soc.*, **57**, 147.

Bjerrum, L. (1957) Norwegian experiences with steel piles to rock. *Géotechnique*, **7**, 73.

Bjerrum, L. (1963) Discussion. *Proc. Europe. Conf. Soil Mech. Found. Eng.*, Wiesbaden, **3**, 135.

Bjerrum, L. (1967) Engineering geology of Norwegian normally consolidated marine clays as related to settlements of buildings. *Géotechnique*, **17**, 81.

Bjerrum, L. (1973) Geotechnical problems involved in foundations of structures in the North Sea. *Géotechnique*, **23**, 319.

Bjerrum, L. and Eggestad, A. (1963) Interpretation of loading tests on sand. *Proc. Europe. Conf. Soil Mech. Found. Eng.*, Wiesbaden.

Bjerrum, L. and Eide, O. (1956) Stability of strutted excavations in clay. *Géotechnique*, **6**, 32.

Bjerrum, L. and Overland, A. (1957) Foundation failure of an oil tank in Fredrikstad, Norway. *Proc. 4th Int. Conf. Soil Mech. Found. Eng.*, **1**, 287.

Bjerrum, L. and Rosenqvist, I. Th. (1956) Some experiments with artificially sedimented clays, *Géotechnique*, **6**, 124.

Bjerrum, L. *et al.* (January, 1965) Measuring instruments for strutted excavations. *J. Soil Mech. Found. Div. (ASCE)*, **91**, 111.

Bjerrum, L. *et al.* (1969) Reduction of negative skin friction on steel piles to rock. *Proc. 7th Int. Conf. Soil Mech. Found. Eng.*, **2**, 27.

Bjerrum, L. *et al.* (1972a) Earth pressures on flexible structures – a state-of-the art report. *Fifth Europe. Conf. Soil Mech. Found. Eng.*, 167.

Bjerrum, L. *et al.* (1972b) Hydraulic fracturing in field permeability testing. *Géotechnique*, **22**, 319.

Black, J.C. *et al.* (1982) Hydrological assessment and grouting at Selby. *Proc. ASCE Spec. Conf. on Grouting* (in *Geotech. Eng.*) New Orleans, 665.

Black, W.P.M. (1961) The calculation of laboratory and *in-situ* values of California Bearing Ratio from bearing capacity data. *Géotechnique*, **11**, 14.

Black, W.P.M. (1962a) A method of estimating the California Bearing Ratio of cohesive soils from plasticity data. *Géotechnique*, **12**, 271.

Black, W.P.M. (1962b) *A Description of the Rubber Pot Bearing*, Road Research Laboratory Note No. LN/174.

Black, W.P.M. (1971) *Notes on Bridge Bearings*, Road Research Laboratory Report LR 382.

Black, W.P.M. and Listner, N.W. (1979) *The Strength of Clay Fill Subgrades: Its Prediction in Relation to Road Performance*, Laboratory Report 889, Transport and Road Research Laboratory.

Bland, J.A. (1983) Fitting failure envelopes by the method of least squares. *Quart. J. Eng. Geol.*, **16**, 143.

Blight, G.E. (1970) *In-situ* strength of rolled and hydraulic fill. *J. Soil Mech. Found. Div. (ASCE)*, **96**, 881.

Bolognesi, A.J.L. and Moretto, O. (1957) Properties and behaviour of silty soils originated from loess formations. *Proc. 4th Int. Conf. Soil Mech. Found. Eng.*, **1**, 9.

Bolton, M.D. (1981) Limit state design in geotechnical engineering. *Ground Eng.*, **14** (6), 39.

Bolton, M.D. and Pang, P.L.R. (1982) Collapse limit states of reinforced earth retaining walls. *Géotechnique*, **32**, 349.

Bolton, R. (1972) Stresses in circular plates on elastic foundations. *J. Eng. Mech. Div. (ASCE) EM* 3, **98**, 629.

Bond, D. (1961) The influence of foundation size on settlement. *Géotechnique*, **11**, 121.

Booker, J.R. and Poulos, H.G. (1976) Analysis of creep settlement of pile foundations. *J. Geotech. Eng. Div. (ASCE)*, **102**. 1.

Booth, G.H. and Wormwell, F. (1961) Corrosion of mild steel by sulphate-reducing bacteria. Effect of different strains of organisms. *First Int. Cong. Metallic Corrosion*, London, **1**, 341.

Borden, R.H. *et al.* (1982) Creep behaviour of grouted sand. *Conf. Grouting and Geotech. Eng.*, New Orleans, 450.

Borowicka, H. (1936) Influence of rigidity of a circular foundation slab on the distribution of pressure over the contact surface. *Proc. 1st Int. Conf. Soil Mech. Found. Eng.*, **2**, 144.

Borowicka, H. (1943) Eccentrically loaded rigid plates on elastic isotropic subsoil. *Ingenieur Archi.*, **1**, 1.

Boscawen, H.E. *et al.* (1963) Some civil engineering aspects of the erection of extra-high-voltage towers and conductors. *Proc. Inst. Civil Eng.*, **26**, 125.

Boulton, G.S. and Paul, M.A. (1976) The influence of genetic processes on some geotechnical properties of glacial tills. *Quart. J. Eng. Geol.*, **9**, 159.

Boulton, N.S. (1951) The flow pattern near a gravity well in a uniform water-bearing medium. *J. Inst. Civil Eng.*, **36**, 534.

Boulton, N.S. (1954) The drawdown of the water-table under non-steady conditions near a pumped well in an unconfined formation. *J. Inst. Civil Eng.*, **3** (3), 564.

Boulton, N.S. (1963) Analysis of data from non-equilibrium pumping tests allowing for delayed yield from storage. *Proc. Inst. Civil Eng.*, **26**, 469.

Boussinesq, J. (1885) *Application des potentiels a l'étude de l'équilibre et du mouvement des solides élastiques*, Gauthier-Villars, Paris.

Bowden, S.R. (1968) *Analysis of Sulphate-Bearing Soils in which Concrete is to be Placed*, Current Paper CP3/68, Building Research Station.

Bower, J.E. (1973) Predicted pullout strength of sheet-piling interlocks. *J. Soil. Mech. Found. Div. (ASCE)*, **99**, 765.

Bowles, D.S. and Ko, H.-Y. (eds) (1984) Probablistic characterization of soil properties: bridge between theory and practice, Proc. Symp. ASCE.

Bowels, J.E. (1974) *Analytical and Computer Methods in Foundation Engineering*, McGraw-Hill, New York.

Bowley, M.J. (1979) *Analysis of Sulphate-bearing Soils in which Concrete is to be Placed*, Current Paper CP2/79, Building Research Establishment.

Boyd, G. (1958) Tall towers; the design of anchor bolts to resist overturning. *Proc. Inst. Civil Eng.*, **10**, 193.

Bozozuk, M. (1963) The modulus of elasticity of Leda Clay from field measurements. *Can. Geotech. J.*, **1**, 43.

Bozozuk, M. (1972) Downdrag measurements on a 160 ft floating pipe test pile in marine clay. *Can. Geotech. J.*, **9**, 127.

Bozozuk, M. *et al.* (1962) Deep bench marks in clay and permafrost areas. In *Field Testing of Soils*, Spec. Tech. Pub. 322, p. 265, ASTM.

Bozozuk, M. and Burn, K.N. (1960) Vertical ground movements near elm trees. *Géotechnique*, **10**, 19.

Brand, E.W. and Shen, J.M. (1984) A note on the principle of the mid-point circle. *Géotechnique*, **34**, 123.

Braswell, A.M. (1958) Settlement of oil storage tanks. *J. Soil Mech. Found. Div. (ASCE)* **84**.

Bratchell, G.E. *et al.* (1974) The performance of two large oil tanks founded on compacted gravel at Fawley, Southampton, Hampshire. *Proc. Conf. on Settlement of Structures*, Pentech Press, p. 3.

Brauns, J. (1978) Initial bearing capacity of stone columns and sand piles. *Symp. on Soil Reinforcing and Stabilizing Techniques*, Sydney.

BRE (1975) *Concrete in sulphate-bearing soils and ground-waters*. Digest 174, Building Research Establishment.

BRE (1979) *Fill and Hardcore*, Digest 222, Building Research Establishment.

BRE (1980a) *Accuracy in Setting-out*, Digest 234, Building Research Establishment.

BRE (1980b) *Low-rise Building on Shrinkable Clay Soils: Parts 1, 2 and 3*, Digests, 240, 241 and 242, Building Research Establishment.

BRE (1981a) *Concrete in Sulphate-bearing Soils and Groundwaters*, Digest 250, Building Research Establishment.

BRE (1981b) *Assessment of Damage in Low-rise Buildings*, Digest 251, Building Research Establishment.

BRE (1983a) *Fill: Part 1 Classification and Load Carrying Characteristics*, Digest 274; *Fill: Part 2 Site Investigation, Ground Improvement and Foundation Design*, Digest 275; *Hardcore*, Digest 276, Building Research Establishment.

BRE (1983b) *Vibrations: Building and Human Response*, Digest 278, Building Research Establishment.

Brebner, A. and Wright, W. (1952) An experimental investigation to determine the variation of sub-grade modulus of sand loaded by plates of different breadths. *Géotechnique*, **3**, 307.

Briggs, H. (1929) *Mining Subsidence*, Edward Arnold, London.

Briggs, H. and Ferguson, W. (1933–34) Investigation of mining subsidence at Barbauchlaw Mine, West Lothian. *Trans. Inst. Min. Eng.*, **85**.

Brink, A.B.A. and Kantey, B.A. (1961) Collapsable grain structure in residual granite soils in Southern Africa. *Proc. 5th Int. Conf. Soil Mech. Found. Eng.*, **1**, 611.

British Geotechnical Society (1970) *Proc. Conf. on in-situ Investigations in Soils and Rocks*.

British Standards Institution (1972) *Recommendations for the Co-ordination of Dimensions in Building: Tolerances and Fits for Building*, DD 22.

British Steel Corporation (1971) *Larssen Steel Sheet Piling*, and *Steel Piling Products*, GPO Box 4, South Teesside Works, Middlesbrough.

Broadbent, C.D. and Ko, K.C. (1972) Rheologic aspects of rock slope failures. Stability of slopes. *Proc. 13th Symp. Rock Mech.*, 573.

Broch, E. and Franklin, J.A. (1972) The point load strength test. *Int. J. Rock Mech. Min. Sci.*, **9**, 669.

Bromhead, E.N. (1979) Factors affecting the transition between various types of mass movement in coastal cliffs consisting of overconsolidated clay with special reference to Southern England. *Quart. J. Eng. Geol.*, **12**, 291.

Bromhead, E.N. *et al.* (1983) Engineering implications of earth surface processes– engineering geomorphology. *Quart. J. Eng. Geol.*, **16**, 259.

Broms, B.B (1964a) Lateral resistance of piles in cohesive soils. *J. Soil Mech. Found. Div. (ASCE)*, **90** (SM2), 27.

Broms, B.B. (1964b) Lateral resistance of piles in cohesionless soils. *J. Soil Mech. Found., Div. (ASCE)*, **90** (SM3), 123.

Broms, B.B. (1971) Lateral earth pressures due to compaction of cohesionless soils. *Proc. 4th Budapest Conf. Soil Mech. Found. Eng.*, Akademiai Kiado, p. 373.

Broms, B.B. (1980) Soil sampling in Europe: state-of-the-art. *J. Geotech. Eng. Div. (ASCE)*, **106**, 65.

Broms, B.B. and Borman, P. (1979) Lime columns – a new foundation method. *J. Geotech Eng., Div. (ASCE)*, **105**, 539.

Broms, B.B. and Ingelson, I. (1972) Lateral earth pressure on a bridge abutment. *Fifth Europe. Conf. Soil Mech. Found. Eng.*, Madrid, p. 117.

Broms, B.B. and Ingelson, I. (1979) Earth pressure against the abutments of a rigid frame bridge. *Géotechnique*, **21**, 15.

Brook, D.H. and Westwater, R. (1956) The use of explosives for demolitions. *Proc. Inst. Civil Eng.*, **4** (3), 862.

Brook, N. (1977) A method of overcoming both shape and size effects in point load testing. *Proc. Conf. on Eng. Properties of Rocks of Mid-Jurassic Age and Older in the UK*, Univ. Newcastle-upon-Type, p. 53.

Brooks, W.T. *et al.* (1960) The 16 inch ERW Tube Plant, Shotton. *The Struct. Eng.*, **38**, 334.

Brown, C.B. and Sheu, M.S. (1975) Effects of deforestation on slopes. *J. Geotech. Eng. Div. (ASCE)*, **101**, 147.

Brown, C.D. and Mead, P.F. (1973) London Bridge. Papers 7595 and 7597. *Proc. Inst. Civil Eng. Part 1*, **54**, 56.

Brown, E.T. and Trollope, D.H. (1970) Strength of a model of jointed rock. *J. Soil Mech. Found. Div. (ASCE)*, **96**, 685.

Brown, J.D. and Meyerhof, G.G. (1969) Experimental study of bearing capacity in layered clays. *Proc. 7th Int. Conf. Soil Mech. Found. Eng.*, **2**, 45.

Brown, J.D. and Peterson, W.G. (1964) Failure of an oil storage tank founded on a sensitive marine clay. *Can. Geotech. J.*, **1**, 205.

Brown, P.D. and Robertshaw, J. (1953) The *in-situ* measurement of Young's modulus for rock by a dynamic method. *Géotechnique*, **3**, 283.

Brown, P.T. (1968) The effect of local bearing failure on behaviour of rigid circular rafts. *Civil Eng. Trans. Inst. Eng. Aust.*, **10**, 190.

Brown, P.T. (1969a) *Raft Foundations, Post Graduate Course in Analysis of the Settlement of Foundations*, School of Civil Engineering, Univ. Sydney, Australia.

Brown, P.T. (1969b) Numerical analyses of uniformly loaded circular rafts on elastic layers of finite depth. *Géotechnique*, **19**, 301

Brown, P.T. and Gibson, R.E. (1972) Surface settlement of a deep elastic stratum whose modulus increases linearly with depth. *Can. Geotech. J.*, **9**, 467.

Brown, P.T. and Lee, I.K. (1972) Structure foundation interaction analyses. *J. Struct. Div. (ASCE) ST11*, **99**, 2413.

Brown, R.E. (1977) Vibroflotation compaction of cohesionless soils. *J. Geotech. Eng. Div. (ASCE)*, **103**, 1437.

Brown, R.E. and Glenn, A.J. (1976) Vibroflotation and Terra-Probe comparison. *J. Soil Mech. Found Div. (ASCE)* **102**, 1059.

Brown, S.F. and Pell, P.S. (1967) Subgrade stress and deformation under dynamic load. *J. Soil Mech. Div. (ASCE) SM1*, **93**, 17.

Bruckshaw, J.M. and Mahanta, P.C. (1961) The variation of the elastic constants of rocks with frequency. *Early Papers, European Association of Exploration Geophysicists.*

Brunsden, D. *et al.* (1975) Large scale geomorphological mapping and highway engineering design. *Quart. J. Eng. Geol.,* **8**, 227.

BSI (1957) *CP 114:1957 The Structural Use of Reinforced Concrete in Buildings.* British Standards Institution, London.

BSI (1969a) *BS 449:1969 The Use of Structural Steel in Building.*

BSI (1969b) *CP 114:1969 The Structural Use of Reinforced Concrete in Buildings, Part 2 (Metric Units).*

BSI (1969c) *CP 2011:1969 Safety Precautions in the Construction of Large Diameter Boreholes for Piling and other Purposes.*

BSI (1970a) *BS 1881:1970 Methods of Testing Concrete, Part 4 (Strength Tests); Part 5 (Other Tests such as Modulus of Elasticity).*

BSI (1970b) *CP 2007:1970 Design and Construction of Reinforced and Prestressed Concrete Structures for the Storage of Water and other Aqueous Liquids.*

BSI (1971) *CP 112:1971 The Structural Use of Timber. Part 2 (Metric Units).*

BSI (1972a) *BS 153:1972 Specification for Steel Girder Bridges.*

BSI (1972b) *BS 4360:1972 Specification for Weldable Structural Steels.*

BSI (1972c) *CP 110:1972 The Structural Use of Concrete.*

BSI (1972d) *CP 2004:1972 Code of Practice For Foundations.*

BSI (1972e) *DD 22:1972 Recommendations for the Coordination of Dimensions in Building. Tolerances and Fits for Building. Calculation of Work Sizes and Joint Clearances for Building Components.*

BSI (1973a) *BS 913:1973 Pressure Creosoting of Timber.*

BSI (1973b) *CP 102:1973 Protection of Buildings Against Water from the Ground.*

BSI (1973c) *CP 1021:1973 Code of Practice for Cathodic Protection.*

BSI (1974a) *BS 4072:1974 Wood Preservation by Means of Water-borne Copper/ Chrome/Arsenic Compositions.*

BSI (1974b) *CP2012:1974 Code of Practice for Foundations for Machinery, Part 1 Foundations for Reciprocating Machines.*

BSI (1975) *BS 1377:1975 Methods of Testing Soils for Civil Engineering Purposes.*

BSI (1977) *BS 5493:1977 Code of Practice for Protective Coating of Iron and Steel Structures Against Corrosion.*

BSI (1978) *BS 5573:1978 Code of Practice for Safety Precautions in the Construction of Large Diameter Boreholes for Piling and Other Purposes.*

BSI (1978/9) *Draft Code of Practice on Maritime Structures.*

BSI (1981a) *BS 5930:1981 Code of Practice for Site Investigations.*

BSI (1981b) *BS 6031:1981 Code of Practice for Earthworks.*

BSI (1982a) *BS 6235:1982 Code of Practice for Fixed Offshore Structures.*

BSI (1982b) *BS 5975:1982 Code of Practice for Falsework.*

BSI (1982c) *BS 6187:1982 Code of Practice for Demolition.*

BSI (1984) *BS 6472:1984 Guide to Evaluation of Human Exposure to Vibration and Shock in Buildings* (1 Hz to 80 Hz).

B.S.P. Pocket Book (1969) *Technical Information on the Design and Construction of Sheet Pile Structures and Bearing Pile Foundations.* The British Steel Piling Co. Ltd, Claydon, Ipswich, Suffolk.

Buchan, S. *et al.* (1972). Relations between the acoustic and geotechnical properties of marine sediments. *Quart. J Eng. Geol.,* **5**, 265.

Buchholz (1930–31) Erdwiderstand auf Ankerplatten Jahrb. *Hatenbautechn. Ges.,* **12**.

Buckton, E.J. and Cuerel, J. (1943) The new Waterloo Bridge. *J. Inst. Civ. Eng.,* **20**, 145.

Building Research Station (1952) *The short bored pile foundation.* Digest 42.

Building Research Station (1957) *Costs of house foundations on shrinkable clays.* Digest 96.

Bulgakov, E.N. (1961) Podpornaya stenka iz zbornogo zhelezobetona. *Promishlennoye Stroitelstvo,* **7**, 54.

Burland, J.B. (1973) Shaft friction of piles in clay – a simple fundamental approach. *Ground Eng.*, **6**, 30.

Burland, J.B. (1977) Settlement of granular materials, *Proc. 9th Int. Conf. Soil Mech. Found. Eng.*, **2**, 517.

Burland, J.B. (1978) Application of the finite element method to prediction of ground movements; in *Developments in Soil Mechanics – 1* (ed. C.R. Scott) Applied Science Publishers, Barking.

Burland, J.B. and Cooke, R.W. (1974) The design of bored piles in stiff clays. *Ground Eng.*, **7**, 28.

Burland, J.B. and Davidson, W. (1976) *A Case Study of Cracking of Columns Supporting a Silo Due to Differential Foundation Settlement*, Building Research Establishment CP 42/76.

Burland, J.B. and Hancock, R.J.R. (1977) Underground car park at the House of Commons, London: geotechnical aspects. *The Struct. Eng.*, **55**, 87.

Burland, J.B. and Lord, J.A. (1969) Discussion on Session A. *Proc. Conf.* In-Situ *Investigations in Soils and Rocks*, Brit. Geotech. Soc., p. 61.

Burland. J.B. and Lord, J.A. (1970) *The Load–Deformation Behaviour of Middle Chalk at Mundford, Norfolk: A Comparison Between Full-scale Performance and* In-Situ *and Laboratory Measurements*, Building Research Station CP 6/70.

Burland, J.B. and Moore, J.F.A. (1973) The measurement of ground displacement around deep excavations; in *Field Instrumentation in Geotechnical Engineering* (British Geotechnical Society) Butterworth, London.

Burland, J.B. and Wroth, C.P. (1975) *Settlement of Structures*, p. 611, British Geotechnical Society, Pentech Press, London.

Burland, J.B. *et al.* (1966) The behaviour and design of large diameter bored piles in stiff clay. *Proc. Symp. Large Bored Piles*, Inst. Civil Eng., p. 51.

Burland, J.B. *et al.* (1973) A field and theoretical study of the influences of non-homogeneity on settlement. *Proc. 8th Int. Conf. Soil Mech Found. Eng.*, **1** (3), 39.

Burland, J.B. *et al.* (1977a) A study of ground movement and progressive failure caused by a deep excavation in Oxford Clay *Géotechnique*, **27**, 557.

Burland, J.B. *et al.* (1977b) Behaviour of foundations and structures, *Proc. 9th Int. Conf. Soil Mech. Found. Eng.*, 495.

Burland, J.B. *et al.* (1981) The overall stability of free and propped embedded cantilever retaining walls. *Ground Eng.*, **14** (5), 28.

Burmister, D.M. (1943) The theory of stresses and displacements in layered systems and applications to the design of airport runways. *Proc. Highway Res. Board*, **23**, 126.

Burmister, D.M. (1962) Prototype load-bearing tests for foundations of structures and pavements. Field Testing of Soils, ASTM Special Pub. **332**, 98.

Burmister, D. (1945) The general theory of stresses and displacements in layered systems. *J. Appl. Phys.*, **16**, 89, 126, 296.

Burnett, A.D. and Epps, R.J. (1979) The engineering geological description of carbonate suite rocks and soils. *Ground. Eng.*, **12** (2), 41.

Burnett, A.D. and Fookes, P.G. (1974) A regional engineering study of the London Clay in the London and Hampshire Basins. *Quart. J. Eng. Geol.*, **7**, 257.

Butterfield, R. and Banerjee, P.K. (1971a) The elastic analysis of compressible piles and pile groups. *Géotechnique*, **21**, 43.

Butterfield, R. and Banerjee, P.K. (1971b) The problem of pile group-pile cap interaction. *Géotechnique*, **21**, 135.

Butterfield, R. and Banerjee, P.K. (1981) *Boundary Element Methods in Engineering Science*, McGraw-Hill, New York.

Butterfield, R. and Douglas, R.A. (1981) *Flexibility coefficients for the design of piles and pile groups*, CIRIA Technical Note 108.

Buttling, S. and Wood, L.A. (1982) A failed raft foundation on soft clays – investigation and analysis. *Ground Eng.*, **15** (July), 40.

Button, S.J. (1953) The bearing capacity of footings on a two-layer cohesive soil. *Proc. 3rd Int. Conf. Soil Mech. Found. Eng.*, **1**, 332.

Button, S.J. (1961) Rapid determination of consolidation settlements *Proc. 5th Int. Conf. Soil Mech. Found. Eng.*, **1**, 615.

Caine, N. (1982) Toppling failures from Alpine Cliffs on Ben Lomond, Tasmania. *Earth Surface Processes and Landforms*, 7, 133.

Cannon, R.W. *et al.* (1981) Guide to the design of anchor bolts and other steel embedments. *Concrete Int.: Design and Construction*, **3** (7), 28.

Cantrell, A.H. (1953) Some major problems in railway civil engineering maintenance. *Proc. Inst. Civil Eng.*, **2** (2), 106.

Capper, P.L., Cassie, W.F. and Geddes, J.D. (1980) *Problems in Engineering Soils*, 3rd edn, E. and F.N. Spon, London.

Capps, J.F. and Hejj, H. (1968) Laboratory and field tests on a collapsing sand in Northern Nigeria. *Géotechnique*, **18**, 506.

Carden, S.G. *et al.* (1983) A problem of surface methane emission. *Mun. Eng.*, **110**, 133.

Carden, D.R. *et al.* (1980) *Earth Pressures Against an Experimental Rtaining Wall Backfilled with Silty Clay*, Report 946, Transport and Road Research Laboratory.

Carder, D.R. *et al.* (1977) *Experimental Retaining Wall Facility–Lateral Stress Measurements with Sand Backfill*, Report 766, Transport and Road Research Laboratory.

Carey, R. and Cumming, C.G. (1961) The design and construction of Erith Jetty. *Proc. Inst. Civil Eng.*, **18**, 15.

Carlson, E.D. and Fricano, S.P. (1961) Tank foundations in Eastern Venezuela. *J. Soil Mech. Found. Div. (ASCE)* **87**, 69; *Discuss.*, **88** (1962), 177, 233.

Carlson, L. (1948) Determination *in-situ* of the shear strength of undisturbed clay. *Proc. 2nd Int. Conf. Soil Mech. Found. Eng.*, **1**, 265.

Carrier, W.D. and Christian, J.T. (1973) Rigid circular plate resting on a non-homogeneous elastic half-space. *Géotechnique*, **23**, 67.

Carroll, D.M. *et al.* (1977) *Air Photo-Interpretation for Soil Mapping*, Technical Monograph No. 8, Soil Survey of England and Wales.

Carstens, M.R. (May, 1966) Similarity laws for localized scour. *J. Hyd. Div. (ASCE)*, **92**, 13.

Carter, P.G. and Sneddon, M. (1977) Comparison of Schmidt hammer, point load and unconfined compression tests in carboniferous strata, *Proc. Conf. Engineering Properties of Rocks of Mid-Jurassic Age and Older in the U.K.* University of Newcastle-upon-Type, p. 197.

Casagrande, A. (1965) Role of the calculated risk in earthwork and foundation engineering. *J. Soil Mech. Found. Div. (ASCE)*, **91**, 1.

Casagrande, L. (1949) Electro-osmosis in soils. *Géotechnique*, **1**, 159.

Casagrande, L. (February, 1973) Comments on conventional design of retaining structures. *J. Soil Mech. Found. Eng. Div. (ASCE)*, 181.

Cass, J.R. (October 1959) Subsurface explorations in permafrost areas. *J. Soil. Mech. Found. Eng. Div. (ASCE)* **85**, 31.

Cement and Concrete Association (1966) Two prestressed concrete bridges. *Concrete Quart.*, **68**, 6.

Cement and Concrete Association (1977) *The Appearance of Concrete Highway Structures*, 46.019, Wexham Springs, Slough.

Cement and Concrete Association (1982) *Guide to Good Practice – Demolition of Reinforced and Prestressed Concrete Structures*, Wexham Springs, Slough.

CCE (1948) Prestressing applied to strengthening a church tower in Staffordshire. *Concrete and Constr. Eng.*, **43**, 281.

Chae, Y.S. *et al.* (1965) Dynamic pressure distribution beneath a vibrating footing. *Proc.*

6th Int. Conf. Soil Mech. Found. Eng., **2**, 22.

Chakravorty, A.K. and Ghosh, A. (1975) Finite difference solution for circular plates on elastic foundations. *Int. J. Numerical Methods Eng.*, **9**, 73.

Chalmers, A. *et al.* (1979) A modified form of aquifer depletion/recovery test for assessing potential water makes into deep excavations. *Proc. 4th Int. Conf. Rock Mech.*, Montreux, **2**, 67.

Chamecki, S. (1956) Structural rigidity in calculating settlements. *J. Soil Mech. and Found. Eng. Div. (ASCE) SM*1, **82**, 1.

Chan, H.T. and Kenney, T.C. (1973) Laboratory investigation of permeability ratio of New Liskeard varved soil. *Can. Geotech. J.*, **10**, 453.

Chandler, R.J. (1974) Lias clay: the long-term stability of cutting slopes. *Géotechnique*, **24**, 21.

Chandler, R.J. (1969) The effect of weathering on the shear strength properties of Keuper Marl. *Géotechnique*, **19**, 321.

Chandler, R.J. (1970) The degradation of Lias clay slopes in an area of the east Midlands. *Quart. J. Eng. Geol.*, **2**, 161.

Chandler, R.J. (ed.) (1978) Report of conference on engineering problems associated with ground conditions in the Middle East. *Quart. J. Eng. Geol.*, **11**, 1.

Chandler, R.J. Martins, J.P. (1982) An experimental study of skin friction around piles in clay. *Géotechnique*, **32**, 119.

Chandler, R.J. and Skempton, A.W. (1973) The design of permanent cutting slopes in stiff fissured clays. *Géotechnique*, **24**, 457; *Discuss.*, **25**, 425.

Chandler, R.J. *et al.* (1981) Engineering geology applied to construction. *Quart. J. Eng. Geol.*, **14**, 149.

Chang, J.C. and Forsyth, R.A. (1977a) Design and field behaviour of reinforced earth wall. *J. Geotech. Eng. Div. (ASCE)*, **103**, 677.

Chang, J.C. and Forsyth, R.A. (1977b) Finite element analysis of reinforced earth wall. *J. Geotech. Eng. Div. (ASCE)*, **103**, 711.

Chaplow, R. *et al.* (1977) The description of rock masses for engineering purposes. *Quart. J. Eng. Geol.*, **10**, 355.

Chapman, D.A. *et al.* (1978) NHBC structural requirements for housing. *The Struct. Eng.*, **56A**, 3.

Chapman, T.G. (1957) Two-dimensional ground-water flow through a bank with vertical faces. *Géotechnique*, **7**, 35.

Chard, B.M. and Symons, I.F. (1982) *Trial Trench Construction in London Clay: A Ground Movement Study at Bracknell*, Laboratory Report 1051, Transport and Road Research Laboratory.

Charge, J. (1972) The raising of the old Wellington Inn and Sinclair's Oyster Bar. *The Struct. Eng.*, **50**, 483.

Charles, J.A. (1976) The use of one-dimensional compression test and elastic theory in predicting deformations of rockfill embankments. *Can. Geotech. J.*, **13**, 189.

Charles, J.A. (1978) Methods of treatment of clay fills. *Conf. on Clay Fills*, Inst. Civil Eng., London, General Report, p. 315.

Charles, J.A. and Soares, M.M. (1984) Stability of compacted rockfill slopes. *Géotechnique*, **34**, 61.

Charles, J.A. and Watts, K.S. (July, 1982) A field study of the use of the dynamic consolidations ground treatment technique on soft alluvial soil. *Ground Eng.*, **15**, 17.

Charles, J.A. and Watts, K.S. (1983) Compressibility of soft clay reinforced with granular columns. *Eighth Europe. Conf. Soil Mech. Found Eng.*, Helsinki, **1**, 347, Balkema, Rotterdam.

Charlton, T.M. (1973) *Energy Principles in Applied Statics*, 2nd edn Blackie, London.

Chattopadhyay, R. and Ghosh, A. (1969) Analysis of circular plates on semi-infinite elastic subgrade. *Indian J. Technol.*, **7**, 312.

Chen, W.F. (1975) *Limit Analysis and Soil Plasticity. Developments in Geotechnical Engineering*, vol. 7, Elsevier, Amsterdam.

Chen, W.F. and Drucker, D.C. (1969) Bearing capacity of concrete blocks or rock. *J. Eng. Mech. Div. (ASCE)*, **95**, 955.

Cheung, Y.K. and Nag, D.K. (1968) Plates and beams on elastic foundations – linear and non-linear behaviour. *Géotechnique*, **18**, 2, 250.

Cheung, Y.K. and Zienkiewicz, O.C. (1965) Plates and tanks on elastic foundations – an application of the finite element method. *Int. J. Solids and Structures*, **1**, 451.

Chiarella, C. and Booker, J.R. (1975) The time-settlement behaviour of a rigid die resting on a deep clay layer. *Quart. J. Mech. Appl. Math.*, **28**, 317.

Christian, J.T. (1968) Undrained stress distribution by numerical methods. *J. Soil Mech. Div. (ASCE)*, SM6, **94**, 1333.

Christiansen, K.P. (1963) The effect of membrane stresses on the ultimate strength of the interior panel in a reinforced concrete slab. *The Struct. Eng.*, **41**, 261.

Christie, I.F. (1965) Secondary compression effects during one-dimensional consolidation tests. *Proc. 6th Int. Conf. Soil Mech. Found. Eng.*, **1**, 198.

Christie, I.F. (1966) The solution of consolidation problems by general-purpose analogue computer. *Géotechnique*, **16**, 131.

Christman, H.E. (February, 1960) Bolts stabilize high rock slopes. *Civil. Eng.*, 62.

Chu, K.-H. and Afandi, O.F. (1966) Analysis of circular and annular slabs for chimney foundations. *J. ACI*, **63**, 1425; *Discuss.*, **64** (1967), 1613.

CIRIA (1965) *The Pressure of Concrete on Formwork*, Research Report 1.

CIRIA (1974) *A Comparison of Quay Wall Design Methods*, Technical Note 54.

CIRIA (1975) *Medical Code of Practice for Work in Compressed Air*, Report 44.

CIRIA (1980) *Pile Load Testing Procedures*, (PD7).

CIRIA/UEG (1984) *Principles of Safe Diving Practice.*

Claesson, A.I.M. and Horvat, E. (1974) Reducing negative friction with bitumen slip layers. *J. Geotech. Eng. Div. (ASCE)*, **100**, 925.

Clark, P.J. *et al.* (1973) Plate friction load control devices – their application and potential. *Proc. Inst. Civil Eng.*, **55** (2), 335.

Clarke, K.B. (1961) A method of failure of peat slopes. *Géotechnique*, **11**, 59.

Clarke, N.W.B. (1963) The wide trench condition and its effect on the loads imposed on rigid underground conduits. *Proc. Inst. Civil Eng.*, **26**, 105; **28**, 599.

Clarke, N.W.B. and Young, O.C. (1962) Loads on underground pipes caused by vehicle wheels. *Proc. Inst. Civil Eng.*, **21**, 91.

Clarke, R.H. (1977) Earthworks in soft chalk: performance and prediction. *The Highway Eng.*, **24** (3), 18.

Clarke, V.L. (1973) *Behaviour and Design of Pile Caps with Four Piles*. Technical Report 4248.9, Cement and Concrete Association, Wexham Springs, Slough.

Clarkson, T.E. and Ropkins, J.W.T. (1977) Pipe-jacking applied to large structures. *Proc. Inst. Civil Eng.*, **62** (1), 539.

Clayton, C.R.I. (1977) Chalk in earthworks – performance and prediction. *The Highway Eng.*, **24** (2), 14.

Clemence, S.P. and Finbarr, A.O. (1981) Design considerations for collapsible soils. *J. Geotech. Eng. Div. (ASCE)*, **107**, 305.

Clevenger, W.A. (1958) Experiences with loess as foundation material. *Trans. ASCE*, **123**, 151.

Coates, D.F. (1970) *Rock Mechanics Principles*, Mines Branch Monograph 874, Ottawa, Canada.

Coates, D.F. and Gyenge, M. (1965) Plate loading tests on rock for deformation and strength properties. *Symp. Testing Tech. for Rock Mech.*, ASTM, S.T.P. **402**, 19.

Coates, F.W. and Taylor, R.S. (1976) Hartlepool Power Station: major civil engineering features. *Proc. Inst. Civil Eng.*, **60** (1), 95.

Coates, R.H. and Slade, L.R. (1958) Construction of circulating-water pump house at Cowes Generating Station, Isle of Wight. *Proc. Inst. Civil Eng.*, **9**, 217.

Cochrane, S.R. and Montgomery, F.R. (1980) Are designers happy with probability-based partial safety factors. *Proc. Inst. Civil Eng.*, **69** (2), 863.

Cole, K.W. and Burland, J.B. (1972) Observations of retaining wall movements associated with a large excavation. *Proc. 5th Europe. Conf. Soil Mech. Found. Eng.*, **1**, 445; reproduced as CP8/72, Building Research Station.

Coleman, J.D. *et al.* (1963) *The Moisture Characteristics, Structure and Composition of a Red Clay Soil from Nyeri, Kenya*, Laboratory Note 383, Road Research Laboratory.

Conard, R.F. (1969) Tests of grouted anchor bolts in tension and shearing. *J. ACI*, **66**, 725.

Constrado (1980) *Holding Down Systems for Steel Stanchions*, Constructional Steel Research and Development Organization, Croydon.

Cooke, R.W. (1975) *The Settlement of Friction Pile Foundations*, Building Research Station CP12/75.

Cooke, R.W. (1978) The design of piled foundations; in *Developments in soil mechanics – 1* (ed. C.R. Scott) Applied Science Publishers, Barking.

Cooke, R.W. and Price, G. (1973a) Horizontal inclinometers for the measurement of vertical displacement in the soil around experimental foundations. *British Geotechnical Society, Symposium on Field Instrumentation*, Butterworths, London, pp. 112–25.

Cooke, R.W. and Price, G. (1973b) Strains and displacements around friction piles. *Proc. 8th Int. Conf. Soil Mech.*, Moscow.

Cooke, R.W. and Whitaker, T. (1961) Experiments on model piles with enlarged bases. *Géotechnique*, **11**, 1.

Cooke, R.W. *et al.* (1979) Jacked piles in London Clay: a study of load transfer and settlement under working conditions. *Géotechnique*, **29**, 113.

Cooke, R.W. *et al.* (1980) Jacked piles in London Clay: interaction and group behaviour under working conditions. *Géotechnique*, **30**, 97.

Cooke, R.W. *et al.* (1981) Some observations of the foundation loading and settlement of a multi-storey building on a piled raft foundation in London Clay.*Proc. Inst. Civil Eng.*, **70** (1), 433.

Cooling, L.F. (1942) Soil mechanics and site exploration. *J. Inst. Civil Eng.*, **18**, 37.

Cooling, L.F. and Gibson, R.E. (1955) Settlement studies on structures in England. *Conf. on Correlation Between Calculated and Observed Stresses and Displacements in Structures*, Inst. Civil Eng. Prelim. p. 295.

Cooling, L.F. and Ward, W.H. (1948) Some examples of foundation movements due to causes other than structural loads. *Proc. 2nd Int. Conf. Soil Mech. Found. Eng.*, **2**, 162.

Cornfield, G.M. (1961) Simplified Hiley formula for reinforced concrete piles. *Eng., London*, **192**, 44.

Cornfield, G.M. (1968) A new empirical formula for base driven cased piles. *Ground Eng.*, **6** (3), 14.

Coull, A. (1974) Stiffening of coupled shear walls against foundation movement. *The Struct. Eng.*, **52**, 23.

Coull, A. and Chantaksinopas, B. (1974) Design curves for coupled shear walls on flexible bases. *Proc. Inst. Civil Eng.*, **57** (2), 595.

Cox, H.L. and Mitchell, S.E. (1952) The measurement of very slow movements in large structures. *Proc. Inst. Civil Eng.*, **1** (1), 682; *Discuss.*, **2** (1) (1953), 458.

Coyle, H.M. and Bartoskewitz, R.E. (1976) Earth pressure on precast panel retaining wall. *J. Geotech. Eng. Div. (ASCE)*, **102**, 441.

Coyle, H.M. and Costello, R.R. (1981) New design correlations for piles in sand. *J. Geotech. Eng. Div. (ASCE)*, **107**, 965.

Cracknell, D.W. (1963) The Runnymede Bridge. *Proc. Inst. Civil Eng.*, **25**, 325; **27** (1964), 648.

Craig, R.N. (1983) *Pipe Jacking: a State-of-the-art Review*, Technical Note 112, CIRIA.

Crawford, C.B. and Burn, K.N. (1969) Building damage from expansive steel slag backfill. *J. Soil Mech. Found. Div. (ASCE)*, **95**, 1325.

Crawford, C.B. and Johnston, G.H. (1971) Construction on permafrost. *Can. Geotech. J.*, **8**, 236.

Crawford, C.B. and Sutherland, J.G. (1971) The Empress Hotel, Victoria, British Columbia. Sixty-five years of foundation settlements. *Can. Geotech. J.*, **8**, 77.

Creasy, L.R. *et al.* (1965a) Radio towers. *The Struct. Eng.*, **43**, 323.

Creasy, L.R. *et al.* (1965b) Museum Radio Tower. *Proc. Inst. Civil Eng.*, **30**, 33.

CRME (1953) Milton Ernest water tower. *Contractor's Record and Municipal Eng.*, 14 January, 20.

Crockett, J.H.A. (1958) Modern forging hammer foundation. *Civil Eng. and Public Works Rev.*, **53**, 657, 798, 909, 1029.

Crockett, J.H.A. and Hammond, R.E.R. (1947) Reduction of ground vibrations into structures. Structural Paper No. 18. Inst. Civil Eng., p. 3.

Crockett, J.H.A. and Hammond, R.E.R. (1948) Natural oscillation of ground and industrial foundations. *Proc. 2nd Int. Conf. Soil Mech. Found. Eng.*, **3**, 88.

Crockett, J.H.A. and O'Neill, D.B. (1959) Research and development of shock-controlled foundation for a heavy-gun laboratory testing hammer. *Proc. Inst. Civil Eng.*, **13**, 133; **15** (1960), 278.

Croney, D. and Jacobs, J.C. (1968) *The Frost Susceptibility of Soils and Road Materials*, Transport and Road Research Laboratory Report 90.

Cronin, H.J. (1980) Post-tensioned foundations as economical alternative. *Civil Eng.*, **14** (10), 12.

Crosby, J.W. *et al.* (1981) Geotechnical applications of borehole geophysics. *J. Geotech. Eng. Div. (ASCE)*, **107**, 1255.

Crowser, J.C. *et al.* (1974) Settlement and contact pressure distribution of a mat supported silo group on an elastic subgrade. *Brit. Geotech. Soc. Conf. on Settlement of Structures*, 344, Pentech Press, London.

Cummings, A.E. (1935) Discussion (p. 1355) of *Lateral Pile-loading Tests*, by L.B. Feagin (1935) *J. ASCE*, **61**, 1335.

Cummings, E.M. (1960) Cellular cofferdams and docks. *Trans. ASCE*, **125**, 13.

Cushing, J.J. and Moline, R.M. (1975) Curved diaphragm cellular cofferdams. *J. Geotech. Eng. Div. (ASCE)*, **101**, 1055.

Cuthbert, L.G. and Poskitt, T.J. (1983) Development of instruments for offshore piles. *Ground Eng.*, **16** (1), 29.

Dallard, N.J. (1971) Design and construction of embankments on an alluvial plain. *Proc. Inst. Civil Eng.*, **49**, 157.

Dalrymple, T. (1964) *Handbook of Applied Hydrology*, Section 25-I (ed. V.T. Chow) McGraw-Hill, New York.

Dalton, J.C.P. and Hawkins, P.G. (1982) Fields of stress – some measurements of the *in-situ* stress in a meadow in the Cambridgeshire countryside. *Ground Eng.*, **15** (4), 15.

Danilevsky, A. (1982) Safety factor of dams and retaining walls. *J. Geotech. Eng. Div. (ASCE)*, **108**, 47.

D'Appolonia, D.J. *et al.* (1968) Settlement of spread footings on sand. *J. Soil Mech. Div. (ASCE) SM3*, **94**, 735.

D'Appolonia, D.J. *et al.* (March, 1970) Discussion of settlement of spread footings in sand. *J. Soil Mech. Found. Div. (ASCE) SM2*, **96**, 754.

D'Appolonia, D.J. *et al.* (1971) Initial settlement of structures on clay. *J. Soil Mech. Found. Div. (ASCE) SM10*, **97**, 1359.

Darragh, R.D. (January, 1965) Controlled water tests to preload tank foundations. *J. Soil Mech. Found. Div. (ASCE)*, **91**, 303.

Dash, B.P. *Discussion, Symposium on Ground Treatment by Deep Compaction*, Inst. Civil Eng., p. 100.

Davidson, R. and Perez, J.-Y. (1982) Properties of chemically grouted sand at locks and dam No. 26. *Conf. Grouting in Geotech. Eng. ASCE*, New Orleans, p. 433.

Davie, J.R. (1973) *Behaviour of Cohesive Soils under Uplift Forces*, Ph.D. Thesis, University of Glasgow, Scotland.

Davie, J.R. and Sutherland, H.B. (1977) Uplift resistance of cohesive soils. *J. Geotech. Eng. Div. (ASCE)*, **103**, 935.

Davies, C. (1962) Structural engineering aspects of the Millbank Tower Block, London, *The Struct. Eng.*, **40**, 3.

Davies, J.D. (1962a) The influence of support conditions on the behaviour of cylindrical concrete tanks. *Proc. Inst. Civil Eng.*, **23**, 379; *Discuss.*, **25** (1963), 412.

Davies, J.D. (1962b) Influence of support conditions on the behaviour of long rectangular tanks. *J.ACI*, **34**, 601.

Davies, J.D. (1963a) Analysis of long rectangular tanks resting on flat rigid supports. *J. ACI*, **60**, 487; *Discuss.*, 1775.

Davies, J.D. (1963b) Bending moments in square concrete tanks resting on flat rigid supports. *The Struct. Eng.*, **41**, 407.

Davies, J.D. (1964) The influence of support conditions on the behaviour of square concrete tanks. *Mag. Concrete Res.*, **16**, 153; *Discuss.*, **17** (1965), 152.

Davies, M.C.R. and Parry, R.H.G. (1983) Shear strength of clay in centrifuge models. *J. Geotech. Eng.*, **109**, 1331.

Davis, A.G. (1968) The structure of the Keuper Marl. *Quart. J. Eng. Geol.*, **1**, 145.

Davis, A.G. and Chandler, R.J. (1973) *Further Work on the Engineering Properties of Keuper Marl*, Report 47, CIRIA.

Davis, D.F. (1970) The contribution of shape factor to pile bearing capacity. *Found. Facts*, **6** (1).

Davis, E.H. and Booker, J.R. (1973) The effect of increasing strength with depth on the bearing capacity of clays. *Géotechnique*, **23**, 551.

Davis, E.H. and Christian, J.T. (1971) Bearing capacity of anisotropic cohesive soil. *J. Soil Mech. Found. Div. (ASCE)*, **97**, 753.

Davis, E.H. and Lee, I.K. (1969) One-dimensional consolidation of layered soils. *Proc. 7th Int. Conf. Soil Mech. Found. Eng.*, **2**, 65.

Davis, E.H. and Poulos, H.G. (1968) The use of elastic theory for settlement prediction under three-dimensional conditions. *Géotechnique*, **18**, 67.

Davis, E.H. and Poulos, H.G. (1972) Rate of settlement under two and three-dimensional conditions. *Géotechnique*, **22**, 1, 95; *Discuss.*, 533.

Davis, P. *et al.* (1981) Mytilus, a soil compaction vessel. *10th Int. Conf. Soil Mech. Found. Eng.*, Stockholm, p. 641.

De Beer, E.E. (1965) The influence of the transverse dimensions of a pile on the point resistance. *De Ingenieur*, **3, 5**; *Bouw en Waterbouwkunde*, **1, 2**.

De Beer, E. (1957) The influence of the width of a foundation raft on the longitudinal distribution of the soil reactions. *Proc. 4th Int. Conf. Soil Mech. Found. Eng.*, **1**, 269.

De Freitas, M.H. and Watters, R.J. (1973) Some field examples of toppling failure. *Géotechnique*, **23**, 495.

De Freitas, M.H. *et al.* (1981) Mudrocks of the United Kingdom. *Quart. J. Eng. Geol.*, **14**, 241.

De Jong, J. and Morgenstern, N.R. (1973) Heave and settlement of two tall building foundations in Edmonton, Alberta. *Can. Geotech. J.*, **10**, 261.

De Josselin de Jong, G. (1957) Application of stress functions to consolidation problems. *Proc. 4th Int. Conf. Soil Mech. Found. Eng.*, **1**, 320.

Delapierre, J. and Dufour, Ch. (1980) Analyses of horizontal loading tests on Franki piles. *Ground Eng.*, **13** (2), 32.

de Mello, V.F.B. (1969) Foundations of buildings in clay. *Proc. 7th Int. Conf. Soil Mech. Found. Eng.*, State of the Art Volume.

de Ruiter, J. and Fox, D.A. (September, 1976) Site investigations for the North Sea Forties Field. *Ground Eng.*, **9**, 25.

de Ruiter, J. (1971) Electric penetrometer for site investigations. *J. Soil Mech. Found. Div. (ASCE)*, **97**, 457.

De Simone, S.V. and Gould, J.P. (1972) Performance of two mat foundations on Boston blue clay. *ASCE Specialty Conference, Purdue University*, **1** (2), 935.

Dearman, W.L. *et al.* (1977) Engineering geological mapping of the Tyne and Wear conurbation, North-East England. *Quart. J. Eng. Geol.*, **10**, 145.

Dearman, W.R. (1976) Weathering classfication in the characterisation of rock: a revision. *Bull. Int. Ass. Eng. Geol.*, **13**, 123.

Dearman, W.R. and Fookes, P.G. (1974) Engineering geological mapping for civil engineering practice in the United Kingdom. *Quart. J. Eng. Geol.*, **7**, 223.

Deere, D.U. (1968) Geological considerations; in *Rock Mechanics in Engineering Practice* (eds. K.G. Stagg and O.C. Zienkiewicz), Wiley.

Deere, D.U. and Davisson, M.T. (1961) Behaviour of grain elevator foundations subjected to cyclic loading. *Proc. 5th Int. Conf. Soil Mech. Found. Eng.*, **1**, 629.

Denny, D.F. (1951) Further experiments on wave pressures. *Proc. Inst. Civil Eng.*, **35**, 330.

Department of the Environment (1969) *Specification for Road and Bridge Works*, HMSO.

Department of the Environment (1977) *Standard Highway Loadings*, Technical Memorandum (Bridges) BE1/77.

Department of the Environment (1976) *Design of Elastomeric Bridge Bearings*, Technical Memorandum (Bridges) BE1/76.

Derbyshire, P.H. (1984) Continuous flight auger piling in the U.K. *Conference on Piling and Ground Treatment*, ICE, London.

Derrington, J.A. (1977) TPl: the construction of gas treatment platform No. 1 for the Frigg Field for Elf-Norge A/S. *The Struct. Eng.*, **55**, 61: *Discuss.*, **56A** (1978), 177.

Desai, C.S. and Christian, J.T. (1977) *Numerical Methods in Geotechnical Engineering*, John Wiley, London.

Descans, L. (1954) Cellular sheet piling structures – a detailed study of the stressing in circular walls. Reprint in English from No. 1 and 2 (January and February), *L Ossature Métallique*, Centre Belgo-Luxembourgeois d'Information de l'Acier, Brussels.

Di Biagio, E. and Kjaernsli, B. (1961) Struts loads and related measurements on contract 63a of the Oslo subway. *Proc. 5th Int. Conf. Soil Mech. Found. Eng.*, **2**, 395.

Di Biagio, E. (1982) Monitoring the foundation performance of offshore gravity base structures. *Ground Eng.*, **15** (8), 24.

Dick, D.R.R. (1959) Berkeley Power Station, with particular reference to the design of the reactor building. *The Struct. Eng.*, **37**, 67.

Dickens, H.B. and Gray, D.M. (1960) Experience with a pier-supported building over permafrost. *J. Soil Mech. Found. Div. (ASCE)*, **86** (October), 1.

Di Gioia, A.M. and Nuzzo, W.L. (June 1972) Fly ash as structural fill. *J. Power Div., (ASCE)*, **98**, 77.

Diver, M. and Paterson, A.C. (1977) Large cooling towers: the present trend. *The Struct. Eng.*, **55**, 431.

Dixon, R.K. (1967) New techniques for studying seepage problems using models. *Géotechnique*, **17**, 236.

Dobson, W.D. *et al.* (1959) The co-ordination of surface and underground development at Peterlee. *Trans. Inst. Mining Eng.*, **119**, 270.

Dodge, A. (1964) Influence functions for beams on elastic foundations. *J. Struct. Div. (ASCE).*, **90**, 63.

Doughty, P.S. (1968) Joint densities and their relation to lithology in the Great Scar Limestone. *Proc. Yorks. Geol. Soc.*, **36**, 479.

Douglas, B. *et al.* (November, 1981) Field measurements of lateral pressures on concrete wall forms. *Concrete International: Design and Construction*, **3**, 56.

Douglas, D.J. and Davis, E.H. (1964) The movement of buried footings due to moment and horizontal load and the movement of anchor plates. *Géotechnique*, **14**, 115.

Douglass, P.M. and Voight, B. (1969) Anisotropy of granites: a reflection of microscopic fabric. *Géotechnique*, **19**, 376.

Dowling, J.W.F. (1968) The classification of terrain for road engineering purposes. *Proc. Conf. Civil Eng. Problems Overseas.* Inst. Civil Eng.

Dowling, J.W.F. and Beaven, P.J. (1969) Terrain evaluation for road engineers in developing countries. *The Highway Eng.*, **16** (6), 5.

Dowling, J.W.F. and Williams, F.H.P (1964) The use of aerial photographs in materials surveys and classification of landforms. *Proc. Conf. Civil Eng. Problems Overseas*, Inst. Civil Eng, p. 209.

Dowrick, D.J. (1970) The use of prestressing in foundation strengthening at York Minster. *6th Int. Congr. Fédération International de la Précontrainte*, The Concrete Soc.

Dowrick, D.J. (1977) *Earthquake Resistant Design*, Wiley, New York.

Dowrick, D.J. (1981) Earthquake risk and design ground motions in the UK offshore area. *Proc. Inst. Civil Eng.*, **71** (2), 305.

Draper, L. (1963) Derivation of a 'design wave' from instrumental records of sea waves. *Proc. Inst. Civil Eng.* 26, 291.

Drapkin, B. (1955) Grillage beams on elastic foundations. *Proc. ASCE Mech. Div.*, **81**, 771.

Drashevska, L. (1962) Review of recent USSR publications in selected fields of engineering soil science. *Rev. Eng. Geol. Soc. Amer.*, **1**, 197.

Drnevish, V.P. and Gray (eds) (1981) *Acoustic Emissions in Geotechnical Engineering Practice.* Special Technical Publication 750, ASTM.

Drucker, D.C and Prager, W. (1952) Soil mechanics and plastic analysis or limit design. *Quart. J. Appl. Math.*, **10**, 157.

Drysdale, R.G. (1973) Variation of concrete strength in existing buildings. *Mag. Concrete Res.*, **25**, 201.

Dudley, J.H. (1970) Review of collapsing soils. *J. Soil Mech. Found. Div. (ASCE).* **96**, 925.

Dumas, F. and Lee, K.L. (1980) Cyclic movements of offshore structures on clay. *J. Geotech. Eng. Div. (ASCE)*, **106**, 877.

Dumbleton, M.J. (1968) *The Classification and Descritpion of Soils for Engineering Purposes: a Suggested Revision of the British System*, Report LR 182, Transport and Road Research Laboratory.

Dumbleton, M.J. (1981) *The British Soil Classification System for Engineering Purposes: Its Development and Relation to Other Comparable Systems*, Laboratory Report 1030, Transport and Road Research Laboratory.

Dumbleton, M.J. and West, G. (1967) *Studies of the Keuper Marl: Stability of Aggregation under Weathering*, Report 85, Transport and Road Research Laboratory.

Dumont-Villares, A. (1956) The underpinning of the 26-storey Companhia Paulista de Seguras building, Sao Paulo, Brazil. *Géotechnique*, **6**, 1.

Duncan, J.M. and Clough, W.G. (1971) Finite element analysis of Port Allen Locks. *J. Soil Mech. Found. Div. (ASCE) SM*8, **97**, 1053.

Duncan, N and Hancock, K.E. (1966) The concept of contact stress in the assessment of the behaviour of rock masses as structural foundations. *Proc. 1st Congr. Int. Soc. Rock Mech.*, **2**, 487.

Dunham, C.W. (1962) *Foundations of Structures*, 2nd edn. McGraw-Hill, New York.

Dunican, P. (ed) (1975) Design and construction of deep basements. *Proc. Symp. Inst. of Struct. Eng. Suppl.*, 1978.

Dunican, P. and Martin, J. (1969) The rebuilding of the Stock Exchange: the first phase. *The Struct. Eng.*, **47**, 431.

Duvivier, J. (1939) Cliff-stabilization works in London Clay. *J. Inst. Civil Eng.*, **14**, 412.

Duviver, S. and Henstock, P.L (1979) Installation of the piled foundations and production modules on Occidental's Piper A platform. *Proc. Inst. Civil Eng.*, **66** (1), 407; *Discuss.*, **68** (1) (1980), 281.

Dvořák, A. (1966) Tests of anisotropic shales for foundations of large bridges. *Proc. 1st Cong. Int. Soc. Rock Mech.*, **2**, 537.

Early, K.R. and Dyer, K.R. (1964) The use of a resistivity survey on a foundation site underlain by karst dolomite. *Géotechnique*, **14**, 341.

East, E.W. (1951) Foundations on shrinkable clays. *J. Inst. Mun. Eng.*, **78**, 273.

Eastwood, W. (1955) The bearing capacity of eccentrically loaded foundations on sandy soils. *The Struct. Eng.*, **33**, 181.

Eckel, E.B. (ed) (1958) *Landslides and Engineering Practice*, Special Report 29, US Highway Research Board.

Eden, W.J. *et al.* (1973) Measured contact pressures below raft supporting a stiff building. *Can. Geotech. J.*, **10**, 180.

Edwards, P.B. (1969) Dimensional co-ordination for metric structures. *The Struct. Eng.*, **47**, 475.

Edwards, P.B. and Rigg, R.B. (1961) The design and construction of extension to British European Airways engineering base at London Airport. *The Struct. Eng.*, **39**, 17.

Edwards, R.J.G. *et al.* (1982). Land surface evaluation for engineering practice. *Quart. J. Eng. Geol.*, **15**, 265.

Egan, P.C. (1973) Cryogenic gas storage above and below ground (Informal discussion). *Proc. Inst. Civil Eng.*, **54** (1), 543.

Eggestad, Aa. (1963) Deformation measurements below a model footing on the surface of dry sand. *Proc. Europe. Conf. Soil Mech. Found. Eng.*, Wiesbaden, p 233.

Eisenstein, Z. and Morrison, N.A. (1973) Prediction of foundation deformation in Edmonton using an *in-situ* pressure probe. *Can. Geotech. J.*, **10**, 193.

Elfgren, I. *et al.* (1980) *Anchor Bolts in Reinforced Concrete Foundation. Short Time Tests*, Research Report 36, University of Luleå.

Elfgren, I. *et al.* (1981) Fatigue of anchor bolts in reinforced concrete foundations. (Paper presented at the IABSE Colloquium on fatigue of steel and concrete structures, Lausanne, 1982.) University of Luleå.

Ellis, B.R. (1979) A study of dynamic soil–structure interaction. *Proc. Inst. Civil Eng.*, **67** (2), 771.

Elson W.K. (1968) Experimental investigation of the stability of slurry trenches. *Géotechnique*, **18**, 37.

Elson, W.K. (1984) *Design of Laterally Loaded Piles*, Report 103, CIRIA, London.

Emerson, M. (1979) *Bridge Temperatures for Setting Bearings and Expansion Joints*, Supplementary Report 479, Transport and Road Research Laboratory.

Endo, M. *et al.* (1969) Negative skin friction acting on steel pipe pile in clay. *Proc. 7th Int. Conf Soil Mech. Found. Eng.* **2**, 85.

Enriquez, R.R. and Fierro, A. (June 1963) A new project for Mexico City. *Civ. Eng.*, 36.

Esrig, M.I. (1970) Stability of cellular cofferdams against vertical shear. *J. Soil Mech Found. Div. (ASCE)* **96**, 1853.

Essenburg, F. (1962) Shear deformations in beams on elastic foundations. *J. Appl. Mech.*, *Ser. E*, **29** (2).

Evans, H.E. (1962) A note on the average coefficients of permeability for a stratified soil mass. *Géotechnique*, **12**, 145.

Evans, H.R. and Shanmugam, N.E. (1979) The elastic analysis of cellular structures containing web openings. *Proc. Inst. Civil Eng.* **67** (2), 1035.

Evans, R.H. (1942) Effect of rate of loading on mechanical properties of some materials. *J. Inst. Civil Eng.*, **18**, 296.

Evans, R.H. and Wood, R.H. (1937a) Transverse elasticity of building materials. *Engineering*, **143**, 161.

Evans, R.H. and Wood, R.H. (1937b) Transverse elasticity of natural stones. *Proc. Leeds Phil. Lit. Soc. (Scientific Section)*, **3**, 340.

Evans, R.S. (1981) An analysis of secondary rock toppling failures – the stress redistribution method. *Quart. J. Eng. Geol.* **14**, 77.

Evison, F.F. (1966) The seismic determination of Young's modulus and Poisson's ratio for rocks *in-situ. Géotechnique*, **16**, 118.

Ewan, V.J. and West, G. (1981) *Reproducibility of Joint Orientation Measurements in Rock*, Supplementary Report 702, Transport and Road Research Laboratory.

Ewan, V.J. *et al.* (1981) *Reproducibility of Joint Spacing Measurements in Rock*, Laboratory Report 1013, Transport and Road Research Laboratory.

Eyles, N. and Sladen, J.A. (1981) Stratigraphy and geotechnical properties of weathered lodgement till in Northumberland, England. *Quart. J. Eng. Geol.*, **14**, 129.

Eyre, W.A. (1973) The revetment of rock slopes in the Clevedon Hills for the M5 motorway. *Quart. J. Eng. Geol.*, **6**, 223.

Faber, O. (1933) Pressure distribution under bases and stability of foundations. *The Struct. Eng.*, **11**, 116.

Fairhurst, C. (1964) On the validity of the Brazilian test for brittle materials. *Int. J. Rock Mech. Min. Sci.*, **1**, 535.

Fancutt, F. and Hudson, J.C. (1960) The choice of protective schemes for structural steelwork. *Proc. Inst. Civil Eng.*, **17**, 405; Discuss. **21** (1962), 558.

Fardis, M.N. and Veneziano, D. (1981) Estimation of SPT-N and relative density. *J. Geotech. Eng. Div. (ASCE)*, **107**, 1345.

Farmer, I.W. *et al.* (1971) The effect of bentonite on the skin friction of cast in place piles. *Conference on Behaviour of Piles*, London, 15–17 September, Inst. Civil Eng.

Farrady, R.V. and Charlton, F.G. (1983) *Hydraulic Factors in Bridge Design*, Hydraulics Research Station, Wallingford.

Farshad, M. and Shahinpoor, M. (1972) Beams on bilinear elastic foundations. *Int. J. Mech. Sci.*, **14**, 441.

Feda, J. (1963) Validity of some settlement computation theories as tested in laboratory conditions on granular soils. *Proc. Europe. Conf. Soil Mech. Found. Eng.*, Wiesbaden, **1**, 61.

Feda, J. (1978) *Stress in Subsoil and Methods of Final Settlement Calculation.* Developments in Geotechnical Engineering, Vol. 18, Elsevier, Amsterdam.

Feld, J. (1943) Discussion of paper by F.M. Masters. *Trans. ASCE*, **108**, 143.

Feld, J. (May, 1965) Tolerance of structures to settlement. *J. Soil Mech. Found. Div. (ASCE)*, **91**, 63.

Fellenius, B.H. (1972) Downdrag on piles in clay due to negative skin friction. *Can. Geotech. J.*, **9**, 323.

Fellenius, B.H. (1974) High quality precast concrete piles. *Ground Eng.*, **7** (2), 28.

Fellenius, B.H. (1980) The analysis of results from routine pile load tests. *Ground Eng.*, **13** (6), 19.

Fellenius, B.H. and Broms, B.B. (1969) Negative skin friction for long piles driven in clay. *Proc. 7th Int. Conf. Soil Mech. Found. Eng.*, **2**, 93.

Fewtrell, A.C. (1949) The New Hawkesbury River Railway Bridge, New South Wales, Australia. *J. Inst. Civil Eng.*, **32**, 419.

Filonenko-Borodich, M.M. (1940) Some approximate theories of the elastic foundation (in Russian). *Uchenyie Zapiski Moskovskogo Gosudarstvennogo Universiteta Mekhanica*, **46**, 3.

Finn, W.D.L. and Shead, D. (1973) Creep and creep rupture of an undisturbed sensitive clay. *Proc. 8th Int. Conf. Soil Mech. Found. Eng.*, **1** (1), 135.

Finn, W.D.L. *et al.* (1978) Liquefaction of thawed layers in frozen soils. *J. Geotech. Eng. Div. (ASCE)*, **104**, 1243.

FIP (1977) *Recommendations for the Design and Construction of Concrete Sea Structures*, Cement and Concrete Ass., Wexham Springs, Slough.

FitzGibbon, M.E. (1976) *Large Pours for Reinforced Concrete Structures*, Current Practice Sheet No. 28; See also (1976) *Large Pours, Heat Generation and Control*, CPS No. 35 and (1977) *Large Pours, Continuous Casting*, CPS No, 36. Cement and Concrete Ass., Wexham Springs, Slough.

FitzHugh, M.M. *et al* (1947) Shipways with cellular walls on a marl foundation. *Trans. ASCE*, **112**, 298.

Fjeld, S. (1963) Settlement damage to a concrete-framed structure. *Proc. Europe. Conf. Soil Mech. Found. Eng.*, Wiesbaden, **1**, 391.

Flaate, K. (1966) Factors influencing the results of vane tests. *Can. Geotech. J.*, **3**, 18.

Flaeschentraeger, H. (1958). Consideration on ground movement phenomena, *Colliery Eng.*, **35**, 342.

Flay, G.F. (1944) Moving a big boiler without dismantling. *Eng. News-Record*, **132**, 178.

Fleming, W.G.K. (1984) *Pile faults and diagnosis*. Proc. Int. Conf. on Structural Faults, Engineering Technics Press, Edinburgh.

Fleming, D.E. and Wolf, W.H. (1963) Testing program for lateral pressure of concrete. *J. ACI.* **60**, 567.

Fleming, W.G.K. and Thorburn, S. (1984) *Recent Piling Advances, Piling and Ground Treatment*, ICE, London.

Flint, A.R. and Edwards, L.S. (1970) Limit state design of highway bridges. *J. Inst. Struct. Eng.*, **48**, 23; Discuss., 371.

Flint, A.R. and Neill, J.A. (1977) Engineering aspects of the National Theatre. The Struct. Eng., **55**, 19.

Flügge, W. (ed.) (1962) *Handbook of Engineering Mechanics*, McGraw-Hill, New York.

Fondedile (1975) Advertisement, *New Civil Eng.*, 19 June.

Fookes, P.G. (1965) Orientation of fissures in stiff overconsolidated clay of the Siwalik System. *Géotechnique*, **15**, 195.

Fookes, P.G. and Best, R. (1969) Consolidation characteristics of some late Pleistocene periglacial metastable soils in east Kent. *Quart. J. Eng. Geol.*, **2**, 103.

Fookes, P.G. and Denness, B. (1969) Observational studies on fissure patterns in Cretaceous sediments of south-east England. *Géotechnique*, **19**, 453.

Fookes, P.G. and Higginbottom, I.E. (1975) The classification and description of near-shore carbonate sediments for engineering purposes. *Géotechnique*, **25**, 406.

Fookes, P.G. and Sweeney, M. (1976) Stabilization and control of local rock falls and degrading rock slopes. *Quart. J. Eng. Geol.*, **9**, 37.

Fookes, P.G. and Wilson, D.D. (1966) The geometry of discontinuities and slope failures in Siwalik clay. *Géotechnique*, **16**, 305.

Fookes, P.G. *et al.* (1971) Some engineering aspects of rock weathering with field examples from Dartmoor and elsewhere. *Quart. J. Eng. Geol.*, **4**, 139.

Foot, R.E. (1962) The design and construction of the new headquarters building for the Daily Mirror Newspapers Ltd at Holborn Circus. *The Struct. Eng.*, **40**, 247.

Foott, R. and Ladd, C.C. (1981) Undrained settlement of plastic and organic clays. *J. Geotech. Eng. Div. (ASCE)*, **107**, 1079.

Forbes, H. (1951) The geochemistry of earthwork. *J. ASCE*, **116**, 637.

Forde, M.C. and Whittington, H.W. (1983) Resistivity Testing of Piles. *Proc. Int. Conf. on Non-destructive testing, London*, Engineering Technics Press, Edinburgh.

Forrest, J.B. and MacFarlane, I.C. (1969) Field studies of response of peat to plate loading. *J. Soil Mech. Found. Div. (ASCE)*, **95**, 949.

Fournier, J. (1977) *Comportement Mécanique des Argiles après Choc*. Conservatoire National des Arts et Métiers, Thesis, Laboratoire de Géologie Appliquée.

Fox, E.N. (1948) The mean elastic settlement of a uniformly loaded area at a depth below the ground surface. *Proc. 2nd Int. Conf. Soil Mech. Found. Eng.*, **1**, 129.

Fox, L. (1948) *Computation of Traffic Stresses in a Simple Road Structure*, Road Research Technical Paper 9, HMSO.

FPS/BR MCA (1977) *Modifications to Specification for Cast-in-place Concrete Diaphragm Walling* (1973) *and Specification for Cast-in-place Piles formed under Bentonite Suspension* (1975), Federation of Piling Specialists and British Ready Mixed Concrete Association.

Fragaszy, R.J. and Cheney, J.A. (1981) Drum centrifuge studies of overconsolidated slopes. *J. Geotech. Eng. Div.* (*ASCE*), **107**, 843.

Francis, A.J. (1964) Analysis of pile groups with flexural resistance. *J. Soil Mech. Found. Div.* (*ASCE*), **90**, (SM3) 1.

Francis, A.J. *et al.* (1963) The behaviour of slender point-bearing piles in soft soil. *Symposium on the Design of High Buildings*, University Press, Hong Kong, p. 25.

Franklin, A.G. *et al.* (1973) Compaction and strength of slightly organic soils. *J. Soil Mech. Found Div.* (*ASCE*), **99**, 541.

Franklin, J.A. and Denton, P.E. (1973) The monitoring of rock slopes. *Quart. J. Eng. Geol.*, **6**, 259.

Fraser, C.K. and Lake, J.R. (1967) *A Laboratory Investigation of the Physical and Chemical Properties of Burnt Colliery Shale*, Report 125, Transport and Road Research Laboratory.

Frederick, D. (1957) Thick rectangular plates on an elastic foundation. *Trans. ASCE*, **122**, 1069.

Freer, R (1968) ELDO SLV launching base in French Guiana: special design factors. *Proc. Inst. Civil Eng.*, **40**, 61.

Frischmann, W.W. *et al.* (1967) Features in the design of Drapers Gardens Development. *The Struct. Eng.*, **45**, 47.

Frischmann, W.W. *et al.* (1983) National Westminster Tower: design. *Proc. Inst. Civil Eng.*, **74** (1), 387.

Frohlich, O.K. (1934) *Druckverteilung in Baugrunde*, Springer-Verlag OHG, Berlin.

Fudo Construction Co. Ltd (1963) Publicity/technical literature.

Furley, A.E. and Curtis, D.C. (1981) *The Instrumentaion of the Foundations of the National Westminster Bank Tower*, Technical Note 103, CIRIA.

Galin, L.A. (1961) *Contact Problems in the Theory of Elasticity* (translation from the Russian, eds H. Moss and I.N. Sneddon), North Carolina State College, School of Physical Sciences and Applied Mathematics.

Galletly, G.D. (1959) Circular plates on a generalized elastic foundation. *J. Appl. Mech., Trans ASME, Ser. E* **81** (26), 297.

Gardner, N.J. and Ho, P.T.-J. (1979) Lateral pressure of fresh concrete. *J. ACI*, **76**, 809.

Garner, J.B. and Heptinstall, S.M. (August/September, 1974) Aerial photo interpretation of engineering and soil surveys. *The Highway Eng.*, **21**, 24.

Gazetas, G. (1981) Machine foundations on deposits of soft clay overlain by a weathered crust. *Géotechnique*, **31**, 387.

Geddes, J.D. (1966) Stresses in foundation soils due to vertical subsurface loading. *Géotechnique*, **16**, 231.

Geddes, J.D. (1969) Boussinesq-based approximations to the vertical stresses caused by pile-type subsurface loadings. *Géotechnique*, **19**, 509.

Geddes, J.D. (ed.) (1981) *Proc. 2nd Int. Conf. Ground Movements and Structures*, Pentech Press, London.

Geddes, J.D. and Cooper, D.W. (1962) Structures in areas of mining subsidence. *The Struct. Eng.*, **40**, 79.

Gedney, D.S. and McKittrick, D.P. (October, 1975) Reinforced earth: a new alternative for earth-retention structures. *Civil Eng.*, p 58.

Gelson, W.E. and Plank, G.A. (1960) The new highway bridges across the Tigris at Amara and Kut in Iraq. *Proc. Inst. Civil Eng.*, **16**, 33; (1961) **18**, 179.

Geological Society of America (1963) *Rock Color Chart*.

George, P.J. and Sladden, P.R. (1980) Certification of the Heather platform. *Ground Eng*, **13 (1)**, 15.

George, P. and Wood, D. (eds) (1976) Offshore soil mechanics. *Symposium, Lloyd's Register of Shipping*, Crawley, West Sussex.

Gerrard, C.M. *et al.* (1971) Instrumentation of raft foundations in Perth. *Proc. 1st Aust. N.Z. Conf. Geomech.*, Melbourne, **1**, 361.

Gerwick, B.C. (1954) Hollow precast concrete units of great size form bridge substructure. *Civil Eng., ASCE*, **24**, 235.

Gerwick, B.C. (1965) Bell-pier construction, recent developments and trends. *J. ACI*, **62**, 1281.

Ghali, A. (1958) Analysis of cylindrical tanks with flat bases by moment distribution methods. *The Struct. Eng.*, **36**, 165.

Gibbs, H.J. and Bara, J.P. (1962) Predicting surface subsidence from basic soil tests; in *Field Testing of Soils*, Special Publication 322, p. 231, ASTM.

Gibbs, H.J. and Holtz, W.G. (1957) Research on determining the density of sands by spoon penetration testing. *Proc. 4th Int. Conf. Soil Mech. Found. Eng.* **1**, 35.

Gibson, A.H. (1952) *Hydraulics and its Applications*, Constable, Edinburgh.

Gibson, D. (1957) Buildings without foundations on moving ground. *RIBA J.*, **65**, 47.

Gibson, R.E. (1950) Discussion, Wilson, G. The bearing capacity of screwed piles and screwcrete cylinders. *J. Inst. Civil Eng*, **34**, 382.

Gibson, R.E. (1963) An analysis of system flexibility and its effect on time-lag in pore-water pressure measurements. *Géotechnique*, **13**, 1.

Gibson, R.E. (1966) A note on the constant head test to measure soil permeability *in situ*. *Géotechnique*, **16**, 256.

Gibson, R.E. (1967) Some results concerning the displacements and stresses in a non-homogeneous elastic half-space. *Géotechnique*, **17**, 58.

Gibson, R.E. (1970) An extension to the theory of the constant head *in situ* permeability test. *Géotechnique*, **20**, 193.

Gibson, R.E. (1974) The analytical method in soil mechanics. *Géotechnique*, **24**, 115.

Gibson, R.E. and Lo, K.Y. (1961) *A Theory of Consolidation for Soils Exhibiting Secondary Compression*. Norwegian Geotechnical Institute Publication 41 (also *Acta Polytechnica Scandinavica*, 296/191, Ci 10).

Gibson, R.E. and McNamee, J. (1957) The consolidation settlement of a load uniformly distributed over a rectangular area. *Proc. 4th Int. Conf. Soil Mech. Found. Eng.*, **6**, 297.

Gibson, R.E. *et al.* (1970) Plane strain and axially symmetric consolidation of a clay layer on a smooth impervious base. *Quart. J. Mech. Appl. Math.*, **23**, 505.

Gifford, E.W.H. and Butler, A.A.W. (1966) Elephant and Castle shopping centre. *Proc. Inst. Civil Eng.*, **33**, 93.

Gifford, E.W.H. and Taylor, P. (1964) The restoration of ancient buildings. *The Struct. Eng.*, **42**, 327.

Gifford, E.W.H. *et al.* (August 1969) The design and construction of the Calder Bridge on the M1 motorway. *Proc. Inst. Civil Eng.*, **43**, 527.

Girault, P. (1965) Discussion on Thon and Coltrin (1958). *J. Soil Mech. Found. Div. (ASCE)*, **91**, 128.

Girijavallabhan, C.V. (1970) Stresses in restrained cylinder under axial compression. *J. Soil Mech. Found. Div. (ASCE)*, **96**, 783.

Gladwell, G.M.L. (1975) Unbonded contact between a circular plate and an elastic foundation; in (eds A.D. de Pater and J.J. Kalker) *The Mechanics of the Contact Between Deformable Bodies*, Proc. IUTAM Symposium, Enschede, Delft University Press, The Netherlands, p. 99.

Gladwell, G.M.L. (1976) On some unbonded contact problems in plane elasticity theory. *J. Appl. Mech. (Trans. ASME)*, **43**, 263.

Gladwell, G.M.L. (1980) *Contact Problems in the Classical Theory of Plasticity*, Sijthoff and Noordhoff, Aphen ann den Rijh, The Netherlands.

Glassman, A. (1972) Behaviour of crossed beams on elastic foundations. *J. Soil Mech. Found. Div. (ASCE)*, **98**, (SMI) 1.

Glossop, R. (1945) Soil mechanics in foundations and excavation; in *The Principles and Application of Soil Mechanics*, Inst. Civil Eng.

Glossop, R. (1954) Discussion on Nowson, W.J.R. The history and construction of the foundations of the Asia Insurance Building, Singapore. *Proc. Inst. Civil Eng.*, **3** (1), 407.

Gnaedinger, J.P. (April, 1961) Grouting to prevent vibration of machinery foundations. *J. Soil Mech. Found. Div. (ASCE)*, **87**, 43.

Godden, R.J. *et al.* (1970) The development of Leith Harbour. *Proc. Inst. Civil Eng.*, **45**, 1.

Goldbeck, A.T. (1924) Friction tests of concrete on various sub-bases. *Public Roads*, **5**, 19.

Golder, C.R. (January, 1970) Presplitting and rockbolts cut costs at Stockton Dam. *Civil Eng*, p. 45.

Golder, H.Q. (1965) State-of-the-art of floating foundations. *J. Soil Mech. Found. Div. (ASCE)*, **91**, 81.

Golder, H.Q. and Gass, A.A. (1962) Field tests for determining permeability of soil strata; in *Field Testing of Soils*, Special Technical Publication 332, p. 29, ASTM.

Golder, H.Q. and Palmer. D.J. (1955) Investigation of a bank failure at Scrapsgate, Isle of Sheppey, Kent. *Géotechnique.*, **5**, 55.

Golder, H.Q. and Sanderson, A.B. (October, 1961) Bridge foundation preloaded to eliminate settlement. *Civil Eng.*, p. 62.

Golder, H.Q. *et al.* (1961) An unusual case of underpinning and strutting for a deep excavation adjacent to existing buildings. *Proc. 5th Int. Conf. Soil Mech. Found. Eng.*, **2**, 413.

Golder, H.Q. *et al.* (1970) Predicted performance of braced excavation. *J. Soil Mech. Found. Div. (ASCE)*, **96**, 801.

Goldfinger, H. (March, 1960) Permanent steel used as subway shoring. *Civil Eng.*, ASCE, **30**, 47.

Golecki, J. and Knops, R.J. (1969) Introduction to a linear elastostatics with variable Poisson's ratio. *Zesz. Nauk. Akad. Gorn. Hutn.*, **30** (204), 81.

Goodier, J.N. and Loutzenheiser, C.B. (1965) Pressure peaks at the ends of plane strain rigid die contacts (elastic). *J. Appl. Mech.*, **32**, 462.

Goodman, R.E. and Bray, J.W. (1976) Toppling of rock slopes. *Proc. Conf. Rock Eng.*, Univ. of Colorado, Vol. 2, ASCE., p. 201.

Goodman, R.E. *et al.* (1968) A model for the mechanics of jointed rock. *J. Soil Mech. Found. Div. (ASCE)*, **94**, 637.

Gorbunov-Posadov, M.I. (1941) *Slabs on Elastic Foundations*, Gosstroiizdat, Moscow.

Gorbunov-Posadov, M.I. (1949) *Beams and Slabs on Elastic Foundations*, Mashstroiizdat, Moscow.

Gorbunov-Posadov, M.I. (1951) *Design of Structures on Elastic Foundations*, Gosstroiizdat, Moscow.

Gorbunov-Posadov, M.I. and Serebrjanyi, R.V. (1961) Design of structures on elastic foundations. *Proc. 5th Int. Conf. Soil Mech. Found. Eng.*, **1**, 643.

Gorbunov-Posadov, M.I. (1965) Calculations for the stability of a sand bed by a solution combining the theories of elasticity and plasticity. *Proc. 6th Int. Conf. Soil Mech. Found. Eng.*, **3**, 51.

Goughnour, R.R. (1983) Settlement of vertically loaded stone columns in soft ground. *Proc. 8th Europe. Conf. Soil Mech. Found. Eng.*, Helsinki, **1**, 235, Balkema, Rotterdam.

Goughnour, R.R. and Bayuk, A.A. (1979a) Analysis of stone columns–soil matrix interaction under vertical load, *C.R. Coll., Int. Reinforcement des Sols*, ENPC-LCPC, Paris, p. 271.

Goughnour, R.R. and Bayuk, A.A. (1979b) A field study of long term settlements of loads supported by stone columns in soft ground. *C.R. Coll. Int. Reinforcement des Sols*, ENPC-LCPC, Paris, p. 279.

Gould, J.P. (1970) Lateral pressures on rigid permanent structures. *1970 Specialty Conf. Lateral Stresses in the Ground and Design of Earth-retaining Structures*, Soil Mech. Found. Eng. Div. ASCE, p. 219.

Gourlay, A.W. and Carson, C.S. (1982) Grouting plant and equipment. *Conf. Grouting in Geotech. Eng.*, ASCE, New Orleans, p. 121.

Graf, E.D. (1969) Compaction grouting technique. *J. Soil Mech. Found. Div. (ASCE)* **95**, 1151.

Grainger, P. *et al.* (1973) The application of the seismic refraction technique to the study of the fracturing of the Middle Chalk at Mundford, Norfolk. *Géotechnique*, **23**, 219.

Grant, K. (1970) *Terrain Classification for Engineering Purposes of the Maree Area, South Australia*, Div. Soil Mech. Tech. Paper 4, CSIRO.

Grant, R. *et al.* (1974) Differential settlement of buildings. *J. Geotech. Eng. Div. (ASCE)*, **100**, 973.

Grasshoff, H. (1957) Influence of flexural rigidity of superstructure on the distribution of contact pressure and bending moments on an elastic combined footing. *Proc. 4th Int. Conf. Soil Mech. Found. Eng.*, **1**, 300.

Gray, H. (1957) Field vane shear tests of sensitive cohesive soils. *Trans. ASCE*, **112**, 844.

Gray, H. and Nair, K. (1967) A note on the stability of soil subject to seepage forces adjacent to sheet pile. *Géotechnique*, **17**, 136.

Gray, R.E. and Meyers, J.F. (1970) Drive subsidence and support methods in Pittsburgh area. *J. Soil Mech. Found. Div. ASCE*, **96**, 1267.

Green, B. (1954) Guildhall: provision of new foundations to the north wall by underpinning and other means. *Proc. Inst. Civil Eng.*, **3** (1), 201.

Green, H. (1961) Long-term loading of short-bored piles. *Géotechnique*, **11**, 47.

Green, N.B. (1957) Design of floating slab foundation. *J. ACI*, **28**, 889; Discuss., 1359.

Green, P.A. (1971) Some aspects of the foundation design for the Commercial Union Building. *Proc. Symp. on Interaction of Structure and Foundation*, p. 118, Midland Soil Mech. Found. Eng. Soc., Birmingham.

Greensmith, J.T. and Tooley, M.J. (eds) (1982) Sea-level movements during the last deglacial hemicycle (about 15,000 years), IGCP Project 61. *Proc. Geol. Ass.*, **93**, 3.

Greenwood, D.A. (1970) Mechanical improvement of soils below ground surface. *Conf. Ground Eng. Inst. Civil Eng.*, p. 11.

Greenwood, D.A. (1974) Differential settlement tolerances of cylindrical steel tanks for bulk liquid storage. *Conf. Settlement of Structures*, Cambridge, p. 361.

Greenwood, D.A. (1982) Discussion. *Conf. Grouting in Geotech. Eng.*, ASCE, New Orleans, Vol. 2, p. 13.

Greenwood, D.A. and Hutchinson, M.T. (1982) Squeeze grouting unstable ground in deep tunnels. *Conf. Grouting in Geotech. Eng.*, ASCE, New Orleans, p. 631.

Greinig, J.F. *et al.* (1957) The design and construction of the foundations and pressure shell of the 8ft × 8ft high-speed wind tunnel at the Royal Aircraft Establishment, Bedford. *Proc. Inst. Civil. Eng.*, **8**, 383.

Griffiths, D.V. (1982) Computation of bearing capacity factors using finite elements. *Géotechnique*, **32**, 195.

Gromko, G.J. (1974) Review of expansive soils. *J. Geotech. Eng. Div. (ASCE)*, **100**, 667.

Grond, G.J.A. (1950) Disturbance of coal measures strata due to mining. *Iron and Coal Trades Rev.*, **160**, **161**. See also NCB Inf. Bull No. 51349.

Gudehus, G. (ed.) (1977) *Finite Elements in Geomechanics*, Wiley, London.

Gutt, W.H. and Harrison, W.H. (1977) *Chemical Resistance of Concrete*, Current Paper 23, Building Research Establishment; also *Concrete*, **11** (5), p. 35.

Habel, A. (1937) Die auf dem elastisch-isotropen Halbraum aufruhende zentral-symmetrisch belastete elastisch Kreisplatte. *Der Bauingenieur*, **18**, 188.

Haddow, T.H. (1967) The roof of the East Kilbride swimming pool. *The Struct. Eng.*, **45**, 435.

Haefeli, R. *et al.* (1953) The behaviour under the influence of soil creep pressure of the concrete bridge built at Klosters by the Rhaetian Railway Co., Switzerland. *Proc. 3rd Int. Conf. Soil Mech. Found. Eng.*, **2**, 175.

Hahn, F.A. (July, 1964) Box type mat solves foundation problem. *Civil Eng.*, p. 62.

Haines, R.P. *et al.* (1968) The Brize Norton hangar. *The Struct. Eng.*, **46**, 45.

Halstead, P.E. (1954) An investigation of the erosive effect on concrete of soft water of low pH value. *Mag. Concrete Res.*, **6**, 93.

Halton, G.R. *et al.* (1965) Vacuum stabilization of subsoil beneath runway extension at Philadelphia International Airport. *Proc. 6th Int. Conf. Soil Mech. Found. Eng.*, **2**, 61.

Hambly, E.C. (1979) *Bridge Foundations and Substructures*, HMSO, London.

Hamilton, J.J. (1963) Volume changes in undisturbed clay profiles in Western Canada. *Can. Geotech. J.*, **1**, 27.

Hamilton, S.B. (1945) The design of independent foundations. *The Struct. Eng.*, **23**, 403.

Hammer, M.J. and Thompson, O.B. (November, 1966) Foundation clay shrinkage caused by large trees. *J. Soil Mech. Found. Div. (ASCE)*, **92**, 1.

Hammond, A.J. *et al.* (1980) Design and construction of driven cast *in situ* piles in stiff fissured clays; in *Recent Developments in the Design and Construction of Piles*, Inst. Civil Eng., p. 157.

Hammond, R.E.R. (March, 1959) Vibration controlled foundation at Saltley. *Iron and Steel.*

Hamrol, A. (1961) A quantitative classification of the weathering and weatherability of rocks. *Proc. 5th Int. Conf. Soil Mech. Found. Eng.*, **2**, 771.

Hancock, P.L. (1968) Joints and faults: the morphological aspects of their origins. *Proc. Geol. Ass.*, **79**, 141.

Hancock, P.L. (1969) Jointing in the Jurassic limestones of the Cotswold Hills. *Proc. Geol. Ass.*, **80**, 219.

Hancock, R.J.R. and Adams, H.C. (1962) The gas turbine facility at Pyestock. *Proc. Inst. Civil. Eng.*, **21**, 503.

Hanna, M.M. and Dawood, R.H. (1965) Stresses in circular tank floors and walls subject to settlement on uniform soils. *Proc. Inst. Civil Eng.*, **30**, 167.

Hanna, T.H. (1963) Model studies of foundation groups in sand. *Géotechnique*, **3**, 334.

Hanna, T.H. (1973) *Foundation Instrumentation*, Rock and Soil Mechanics Series, vol. 1 (3), Trans Tech. Publishers, Germany.

Hanna, T.H. and West, A.S. (November, 1976) Diaphragm walls and anchorages: review of Institution of Civil Engineers seminar. *Ground Eng.*, **9**, 26.

Hanrahan, E.T. *et al.* (1967) Shear strength of peat. *Proc. Geotech. Conf.*, Oslo, **1**, 193.

Hansbö, S. (1977) Dynamic consolidation of rockfill at Uddovalla Shipyard. *Proc. 9th Int. Conf. Soil Mech. Found Eng.*, Tokyo, **2**, 241.

Hansen, F.J. (1974) North Sea structures – a new breed? *The Struct. Eng.*, **52**, 221.

Hansen, J.B. (1953) *The Stabilizing Effect of Piles in Clay*, Christian and Nielson Post, Denmark.

Hansen, J.B. (1961) *A General Formula for Bearing Capacity*, Bull. 11, Danish Geotech. Inst.

Hansen, J.B. (1961) The bearing capacity of sand, tested by loading circular plates. *Proc. 5th Int. Conf. Soil Mech. Found. Eng.*, **1**, 659.

Hansen, J.B. (1966) *Comparison of Methods for Stability Analysis*, Bull. 21, Danish Geotech. Inst.

Hansen, J.B. and Gibson, R.E. (1949) Undrained shear strength tests of anisotropically consolidated clays. *Géotechnique*, **1**, 189.

Hansen, J.B. and Inan, S. (1969) Tests and formulas concerning secondary consolidation. *Proc. 7th Int. Conf. Soil Mech. Found. Eng.*, **1**, 45.

Harding, H.J.B. (1949) Site investigations, including boring and other methods of sub-surface exploration. *J. Inst. Civil Eng.*, **32**, 111.

Harding, H.J.B. (1981) *Tunnelling History and my own Involvement*, Golder Associates, Toronto, p. 136.

Harding, H.J.B. and Glossop, R. (1951) The influence of modern soil studies on the construction of foundations. *Bldg Res. Congr. Proc., Division* 1, 146.

Hardy, R.M. and Ripley, C.F. (1961) Horizontal movements associated with vertical movements. *Proc. 5th Int. Conf. Soil Mech. Found. Eng.* 1, 665.

Harr, M.E. (1966) *Foundations of Theoretical Soil Mechanics*, McGraw-Hill, New York.

Harr, M.E. *et al.* (1969) Euler beams on a two parameter foundation model. *J. Soil Mech. Div. (ASCE)*, 95 (SM4), 933.

Harris, C. (1977) Engineering properties, groundwater conditions, and the nature of soil movement on a solifluction slope in North Norway. *Quart. J. Eng. Geol.*, 10, 27.

Harris, G.M. (1976) Foundations and earthworks for cylindrical steel storage tanks. *Ground Eng.*, 9 (5), 24.

Harrison, J.V. and Falcon, N.L. (1936) Gravity collapse structures and mountain ranges, as exemplified in South-Western Iran. *Quart. J. Geol. Soc.*, 92, 91.

Harrison, T.A. (1983) *Pressure on Formwork when Concrete is Placed in Wide Sections*, Cement and Concrete Ass., Wexham Springs, Slough.

Harrison, W. and Richardson, A.M. (1967) Plate load tests on sandy marine sediments, lower Chesapeake Bay. *Proc. Int. Res. Conf. Marine Geotechnique*, Illinois, USA, p. 274.

Haseltine, B.A. *et al.* (1977) The resistance of brickwork to lateral loading: design of walls to resist lateral loads. *The Struct. Eng.*, 55, 422.

Haswell, C.K. (1969) Thames Cable Tunnel. *Proc. Inst. Civil Eng.*, 44, 323.

Hawkins, A.B. (1979) Case histories of some effects of solution/dissolution in the Keuper rocks of the Severn Estuary region. *Quart. J. Eng. Geol.*, 12, 31.

Hawkins, N.M. (1968a) The bearing strength of concrete loaded through rigid plates. *Mag. Concrete Res.*, 20 (62), 31.

Hawkins, N.M. (1968b) The bearing strength of concrete loaded through flexible plates. *Mag. Concrete Res.*, 20 (63), 95.

Haydon, R.E.V. and Hobbs, N.B. (1977) The effect of uplift pressures on the performance of a heavy foundation on layered rock. *Proc. Conf. Rock Eng.*, University of Newcastle-upon-Tyne.

Healy, P.R. and Weltman, A.J. (1980) *Survey of Problems Associated with the Installation of Displacement Piles*, CIRIA Report PG8, Property Services Agency Civil Engineering Technical Guide No. 26.

Hearne, T.M. *et al.* (1981) Drilled-shaft integrity by wave propagation method. *J. Geotech. Eng. Div. (ASCE)*, 107, 1327.

Heasman, A. (1936) The preservation of ancient monuments. *The Struct. Eng.*, 14, 351.

Heathcote, F.W.L. (1965) Movement of articulated buildings on subsidence sites. *Proc. Inst. Civil Eng.* 30, 347.

Heijnen, W.J. and Lubking, P. (1973) Lateral soil pressure and negative friction on piles. *Proc. 8th Int. Conf. Soil Mech. Found. Eng.*, 2(1), 143.

Heil, H. (1969) Studies on the structural rigidity of reinforced concrete building frames on clay. *Proc. 7th Int. Conf. Soil Mech. Found. Eng.*, 2, 115.

Helenelund, K.V. (1965) Torsional field shear tests. *Proc. 6th Int. Conf. Soil Mech. Found. Eng.*, 1, 240.

Hendron, A.J. *et al.* (1970) Compressibility characteristics of shales measured by laboratory and *in situ* tests. *Determination of the in-situ Modulus of Deformation of Rock*, Special Technical Publication 477, p. 137, ASTM.

Henkel, D.J. (1956) Discussion on Watson, J.D. Earth movement affecting L.T.E. Railway in deep cutting east of Uxbridge. *Proc. Inst. Civil Eng.*, 5(2), 320.

Henkel, D.J. (1961) Slide movements on an inclined clay layer in the Avon Gorge in Bristol. *Proc. 5th Int. Conf. Soil Mech. Found. Eng.*, 2, 619.

Henkel, D.J. (1982) Geology, geomorphology and geotechnics. *Géotechnique*, **32**, 175.

Henkel, D.J. and Skempton, A.W. (1955) A landslide at Jackfield, Shropshire, in heavily over-consolidated clay. *Géotechnique*, **5**, 131.

Henry, F.D.C. (1955) Soil mechanics aspects of the design of foundations for structures. *Engineering*, **180**, 454.

Henry, F.D.C. (1956) *Design and Construction of Engineering Foundations.* E. & F.N. Spon, London, p. 416.

Hertwig, A. and Lorenz, H. (1935) Das Dynamische Bodenuntersuchungsverfahren. *Der Bauingenieur*, **16**, 279.

Hertwig, A. *et al.* (1933) *Die Ermittlung der für das Bauwesen wichtigsten Eigenschaften des Bodens durch erzwungene Schwingun.* DEGEBO No. 1. Springer, Berlin.

Hertz, H. (1884) On the equilibrium of floating elastic plates (in German). *Wiedemanns Ann.*, **22**, 449.

Hetényi, M. (1936) Analysis of bars on elastic foundation. *Final report 2nd Int. Congr. Bridge Struct. Eng. (Berlin–Munich).*

Hetényi, M. (1946) *Beams on Elastic Foundations*, University of Michigan Press, Ann Arbor.

Hetényi, M. (1966) Beams and plates on elastic foundations and related problems. *Appl. Mech. Rev.*, **19** (2), 95.

Hewitt, J. (Rep.) (1980) Waves and wave forces inshore. Report on seminar, with useful list of references. *Proc. Inst. Civil Eng.*, **68**, (1), 323.

Higginbottom, I.E. and Fookes, P.G. (1970) Engineering aspects of periglacial features in Britain. *Quart. J. Eng. Geol.* **3**, 85.

Hill, R. (1950) *Mathematical Theory of Plasticity*, Oxford University Press, London.

Hinch, L.W. and Martin, P.L. (1976) The foundation for an oil storage tank on Quaternary soils near Bridgwater, Somerset. *Quart. J. Eng. Geol.* **9**, 237.

Hird, C.C. *et al.* (1978) The development of centrifugal models to study the influence of uplift pressures on the stability of a flood bank. *Géotechnique*, **28**, 85.

HMSO (1952) *Soil Mechanics for Road Engineers*, London.

HMSO (1957a) *Handbook of Hardwoods*, Ministry of Technology Forest Products Research laboratory.

HMSO (1957b) *Handbook of Softwoods*, Ministry of Technology Forest Products Research Laboratory.

HMSO (1960) *The Durability of Reinforced Concrete in Sea-water*, National Building Studies Research Paper 30.

HMSO (1976) *Specification for Road and Bridge Works; Supplement* (1978), Department of Transport, HMSO, London.

HMSO (1981) *A Guide to the Diving Operations at Work Regulations*, HMSO, London.

Ho, M.M.K. and Lopes, R. (1969) Contact pressure of a rigid circular foundation. *J. Soil Mech. Found. Div. (ASCE)*, **95**, 791.

Hoare, D.J. (March, 1984) Geotextiles as filters. *Ground Eng.*, **17**, 29.

Hobbs, D.W. (1970a) The behaviour of broken rock under triaxial compression. *Int. J. Rock Mech. Min. Sci.* **7**, 125.

Hobbs, D.W. (1970b) Stress–strain–time behaviour of a number of Coal Measures rocks. *Int. J. Rock Mech. Min. Sci.*, **7**, 149.

Hobbs, N.B. (1963) An analytical approach to the bearing capacity and settlement of the large diameter bored pile. *Proc. Inst. Civil Eng.*, **25**, 177.

Hobbs, N.B. and Healy, P.R. (1979) *Piling in Chalk*, CIRIA report PG6, CIRIA London.

Hodgson, F.T. and Bryan, A.W. (1975) Redevelopment of the south side of Victoria Street, London. *The Struct. Eng.*, **53**, 463.

Hodgson, J.M. (ed.) (1974) *Soil Survey Field Handbook*, Soil Survey, Rothamsted Experimental Station, Harpenden.

Höeg, K. and Murarka, R.P. (1974) Probabilistic analysis and design of a retaining wall. *J. Geotech. Eng. Div. (ASCE)*, **100**, 349.

Höeg, K. *et al.* (1968) Settlement of a strip load on elastic-plastic soil. *J. Soil Mech. Found. Div.* (*ASCE*), **94** (SM2), 431.

Hoek, E. (1973) Methods for the rapid assessment of the stability of three-dimensional rock slopes. *Quart J. Eng. Geol.* **6**, 243.

Hoek, E. (1983) Strength of jointed rock masses. *Géotechnique*, **33**, 187.

Hoek, E. and Bray, J.W. (1977) *Rock Slope Engineering*, Inst. Mining and Met., London.

Hoek, E. and Brown, E.T. (1980) Empirical strength criterion for rock masses. *J. Geotech. Eng. Div.* (*ASCE*), **106**, 1013.

Hoek, E. and Londe, P. (1974) Surface workings in rock. *Proc. 3rd Cong. Int. Soc. Rock Mech.*, **1A**, 612.

Hogben, N. *et al.* (1977) Estimation of fluid loading on offshore structures. *Proc. Inst. Civil Eng.*, **63** (2), 515; Discuss., **65** (2), 953.

Hogg, A.H.A. (1938) Equilibrium of a thin plate symmetrically loaded resting on an elastic foundation of infinite depth. *Phil. Mag.*, Ser. 7, **25**, 576.

Hognestad, E. (1953) Shearing strength of reinforced concrete column footings. *J. ACI*, **25**, 189.

Holl, D,L. (1938) Thin plates on elastic foundation. *Proc. 5th Int. Congress Appl. Mech.*, Cambridge, Mass., John Wiley, New York. p. 73.

Hollingshead, G.W. and Raymond, G.P. (1972) Field loading tests on muskeg. *Can. Geotech. J.*, **9**, 278.

Hollingworth, S.E. *et al.* (1944) Large-scale superficial structures in the Northampton ironstone field. *Quart J. Geol. Soc.*, **100**, 1.

Holmes, M. (1961) Steel frames with brickwork and concrete infilling. *Proc. Inst. Civil Eng.* **19**, 473.

Holmes, M. (1963) Combined loading on infilled frames. *Proc. Inst. Civil Eng.*, **25**, 31.

Holtz, W.G. and Gibbs, H.J. (1956) Engineering properties of expansive clays. *Trans. ASCE*, **121**, 641.

Hoole, R. (ed.) (1978) Engineering behaviour of glacial materials. *Proc. Conf. Univ. of Birmingham*, Geobooks, Norwich.

Hooper, J.A. (1973) Observations on the behaviour of a piled-raft foundation on London Clay. *Proc. Inst. Civil Eng.*, **55** (2), 855.

Hooper, J.A. (1974) Analysis of a circular raft in adhesive contact with a thick elastic layer. *Géotechnique*, **24**, 561.

Hooper, J.A. (1975) Elastic settlement of a circular raft in adhesive contact with a transversely isotropic medium. *Géotechnique*, **25**, 691.

Hooper, J.A. (1976) Parabolic adhesive loading of a flexible raft foundation. *Géotechnique*, **26**, 511.

Hooper, J.A. (1978) Foundation interaction analysis; in *Developments in Soil Mechanics – 1* C.R. Scott, Ed, Applied Science Publishers. Barking.

Hooper, J.A. (1983a) Analysis and design of a large raft foundation in Baghdad. *Proc. Inst. Civil Eng.*, **74** (1), 837.

Hooper, J.A. (1983b) Interactive analysis of foundations on horizontally variable strata. *Proc. Inst. Civil Eng.*, **75** (2), 491.

Hooper, J.A. (1983c) Non-linear analysis of a circular raft on clay. *Géotechnique*, **33**, 1.

Hooper, J.A. (1984) Raft analysis and design – some practical examples. *The Struct. Eng.*, **62A**, 233.

Hooper, J.A. and Butler, F.G. (1966) Some numerical results concerning the shear strength of London Clay. *Géotechnique*, **16**, 282.

Hooper, J.A. and West, D.J. (1983) Structural analysis of a circular raft on yielding soil. *Proc. Inst. Civil Eng.* **75** (2), 205.

Horne, M.R. (1964) The consolidation of a stratified soil with vertical and horizontal drainage. *Int. J. Mech. Sci.*, **6**, 187.

Horner, P.C. (1980) *Earthworks*, Inst. Civil Eng.

Hoskin, B.C. and Lee, E.H. (1959) Flexible surfaces on viscoelastic subgrades. *J. Eng. Mech. Found. Div. (ASCE)*, **85** (EM4), 11.

Houghton, D.L. *et al.* (1969) Preplaced aggregate concrete for structural and mass concrete. *J. ACI* **66**, 785.

Houlsby, A.C. (1982a) Cement grouting for dams. *Conf. Grouting in Geotech. Eng.*, ASCE, New Orleans, p. 1.

Houlsby, A.C. (1982b) Optimal water/cement ratio for rock grouting. *Conf. Grouting in Geotech. Eng.*, ASCE, New Orleans, p 317.

Houston, W.N. and Herrmann, H.G. (1980) Undrained cyclic strength of marine soils. *J. Geotech. Eng. Div. (ASCE)*, **106**, 691.

Hovland, H.J. (1977) Three-dimensional slope stability analysis method. *J. Geotech. Eng. Div. (ASCE)*, **103**, 971.

Howorth, G.E. (1937) The construction of the Lower Zambesi Bridge. *J. Inst. Civil Eng.*, **4**, 369; Correspondence, **6**, 365.

Howorth, G.E. and Smith, H.S. (1947) The new Howrah bridge, Calcutta: construction. *J. Inst. Civil Eng.*, **28**, 211.

HRA (1983) *Hydraulic Factors in Bridge design*, Hydraulics Research Association, Wallingford.

Hsieh, T.K. (1960) Discussion on Stresses on foundation rafts. *Proc. Inst. Civil Eng.*, **17**, 339.

Huang, Y.H. (1969) Influence charts for two-layer elastic foundations. *J. Soil Mech. Found. Div. (ASCE)*, **95**, 709.

Huang, Y.H. (1977) Stability coefficients for sidehill benches. *J. Geotech. Eng. Div. (ASCE)*, **103**, 467.

Hubbert, M.K. and Willis, D.G. (1957) Mechanics of hydraulic fracturing. *J. Am. Inst. Min. Met. Eng.*, **210**, 153.

Huder, J. (1969) Deep braced excavation with high ground water level. *Proc. 7th Int. Conf. Soil Mech. Found. Eng.*, **2**, 443.

Hudson, J.A. and Priest, S.D. (1979) Discontinuities and rock mass geometry. *Int. J. Rock Mech. Mining Sci.*, **16**, 339.

Hudson, W.R. and Matlock, M. (1966) Analysis of cracked pavement slabs with non-uniform load and support. *ASCE Structural Engineering Conference*, Miami, Florida, Preprint 280.

Hueckel, S.M. (1957) Model tests on anchoring capacity of vertical and inclined plates. *Proc. 4th Int. Conf. Soil Mech. Found. Eng.*, **2**, 203, Butterworths, London.

Hueckel, S.M. (1965) Distribution of passive earth pressure on the surface of a square vertical plate embedded in soil. *Proc. 6th Int. Conf. Soil Mech. Found. Eng.*, **2**, 381.

Hueckel, S.M. and Kwasniewski, J. (1961) Essais sur modèle réduit de la capacité d'ancrage d'éléments rigides horizontaux, enfouis dans la sable. *Proc. 5th Int. Conf. Soil Mech. Found. Eng.*, **2**, 417.

Hueckel, S.M. *et al.* (1965) Distribution of passive earth pressure on the surface of a square vertical plate embedded in soil. *Proc. 6th Int. Conf. Soil Mech. Found. Eng.*, **2**, 381.

Hughes, J.M.O. and Withers, N.J. (May, 1974) Reinforcing of soft cohesive soils with stone columns. *Ground Eng.*, **7**, 42.

Hughes, J.M.O. *et al.* (1977) Pressuremeter tests in sands. *Géotechnique*, **27**, 455.

Hughes, T.A. and Turner, F.W. (1953) A gas holder foundation tank at Bristol. *Reinf. Concrete Rev.*, 151.

Hulse, R. *et al.* (1980) The strength of reinforced brickwork spanning openings in wall panels. *Proc. Inst. Civil Eng.*, **69** (2), 835.

Hulse, R. *et al.* (1981) *Brickwork beams with reinforced bed joints. Proc. Inst. Civil Eng.*, **71** (2), 921.

Husband, H.C. (1947) Discussion on paper by Crockett and Hammond (1947). Structural Paper No. 18, Inst. Civil Eng., p. 41.

Husband, H.C. and Best, K.H. (1953) The reconstruction of a soaking pit building. *The Struct. Eng.*, **31**, 30.

Hutchinson, J.N. (1972) Field and laboratory studies of a fall in Upper Chalk cliffs at Joss Bay, Isle of Thanet; in R.H.G. Parry (ed.) *Stress–Strain Behaviour of Soils*, Foulis, Cambridge, p. 692.

Hutchinson, J.N. *et al.* (1980) Additional observations on the Folkstone Warren landslides. *Quart. J. Eng. Geol.*, **13**, 1.

Hyde, A.F.L. and Brown, S.F. (1976) The plastic deformation of a silty clay under creep and repeated loading. *Géotechnique*, **26**, 173.

ICE (1966) *Proc. Symp. Large Bored Piles*, London, Inst. Civil Eng.

ICE (1970) *Conf. Behaviour of Piles*, London, Inst. Civil Eng.

ICE (1971) *Behaviour of Piles*, Inst. Civil Eng.

ICE (1974) *Proc. Conf. on Diaphragm Walls and Anchorages*, Inst. Civil Eng.

ICE (1976) *Manual of Applied Geology for Engineers*. Inst. Civil Eng.

ICE (1977a) *Proc. Conf. Design and Construction of Offshore Structures*, Inst. Civil Eng.

ICE (1977b) *Ground Subsidence*, Inst. Civil Eng.

ICE (1978) *Piling: Model Procedures and Specifications*, Inst. Civil Eng.

ICE (1979) *Corrosion in Civil Engineering*, Inst. Civil Eng.

ICE (1979) *Numerical Methods in Offshore Piling*, Inst. Civil Eng.

ICE (1980a) *Recent Developments in the Design and Construction of Piles*, Inst. Civil Eng.

ICE (1980b) *Proc. Conf. on Safe Construction for the Future*, Inst. Civil Eng.

ICE (1983) *Design in Offshore Structures*, Inst. Civil Eng.

ICE (1984) *Piling and Ground Treatment*, Inst. Civil Eng.

Ineson, J. (1959a) The relation between the yield of a discharging well at equilibrium and its diameter, with particular reference to a Chalk well. *J. Inst. Civil Eng.*, **13**, 299.

Ineson, J. (1959b) Yield–depression curves of discharging wells, with particular reference to Chalk wells, and their relationship to variations in transmissibility. *J. Inst. Water Eng.*, **13**, 119.

Ingold, T.S. (1979a) The effects of compaction on retaining walls. *Géotechnique*, **29**, 265.

Ingold, T.S. (1979b) Retaining wall performance during backfilling. *J. Geotech. Eng. Div. (ASCE)*, **105**, 613.

Ingold, T.S. (1980) Lateral earth pressures – a reconsideration. *Ground Eng.*, **13** (4), 39.

Ingold, T.S. (1981) A laboratory simulation of reinforced clay walls. *Géotechnique*, **31**, 399.

Ingold, T.S. (1982) A reconsideration of retaining wall design. *The Struct. Eng.*, **60B**, 83.

Ingold, T.S. (1983) the design of reinforced soil walls by compaction theory. *The Struct. Eng.*, **61A**, 205.

Ingoldby, H.C. and Parsons, A.W. (1977) *The Classification of Chalk for Use as a Fill Material*, Report 806, Transport and Road Research Laboratory.

Ingra, T.S. and Baecher, G.B. (1983) Uncertainty in bearing capacity of sands. *J. Geotech. Eng. Div. (ASCE)*, **109**, 899.

Int. Ass. Eng. Geology (1976) *Engineering Geological Maps: a guide to their preparation*, UNESCO Press.

Ireland, H.O. *et al.* (1970) The dynamic penetration test: a standard that is not standardized. *Géotechnique*, **20**, 185.

ISE (1947) Interim report on mining subsidence and its effect on structures. *The Struct. Eng.*, **25**, 315.

ISE (1951) *CP2 Earth Retaining Structures*. Inst. Struct. Eng.

ISE (1966) *Proc. Conf. Industrialized and the Structural Engineer*, Inst. Struct. Eng.

ISE (1969) *The Shear Strength of Reinforced Concrete Beams*, Inst. Struct. Eng.

ISE (1974) *Report on Structure Soil Interaction in Relation to Buildings*, Special Study Group, Inst. Struct. Eng.

ISE (1978) *Structure–soil Interaction: a state-of-the art report*, Inst. Struct. Eng.

Ismael, N.F. and Klym, T.W. (1979) Uplift and bearing capacity of short piers in sand. *J. Geotech. Eng. Div. (ASCE).*, **105**, 579.

Ismael, N.F. and Vesic, A.S. (1981) Compressibility and bearing capacity. *J. Geotech. Eng. Div. (ASCE)*, **107**, 1677.

ISRM (1974) *Suggested Methods for Rockbolt Testing*, Committee on Field Tests Document No. 2, Int. Soc. for Rock Mech.

ISRM (1974) *Suggested Methods for Determining Shear Strength.*

ISRM (1975) *Recommendations on Site Investigation Techniques.*

ISRM (1975) *Terminology.*

ISRM (1977) *Suggested Methods for Determining Hardness and Abrasiveness of Rocks.*

ISRM (1977) *Suggested Methods for Determining Tensile Strength of Rock Materials.*

ISRM (1977) *Suggested Methods for Determining the Strength of Rock Materials in Triaxial Compression.*

ISRM (1977) *Suggested Methods for Monitoring Rock Movements using Borehole Extensometers.*

ISRM (1977) *Suggested Methods for Petrographic Description of Rocks.*

ISRM (1978) *Suggested Methods for Determining in situ Deformability of Rocks.*

ISRM (1978) *Suggested Methods for Determining the Uniaxial Compressive Strength and Deformability of Rock Materials* (revised).

ISRM (1978) *Suggested Methods for Determining Water Content, Porosity, Density, Absorption and Related Properties and Swelling and Slake-Durability Index Properties.*

ISRM (1979) *Suggested Methods for Pressure Monitoring using Hydraulic Cells.*

ISRM (1980) *Basic Geotechnical Description of Rock Masses.*

ISRM (1981) *Suggested Methods for Geophysical Loading of Boreholes.*

Iwinski, T. (1967) *Theory of Beams*, 2nd edn (Translated from Polish by E.P. Bernat), Pergamon Press, Oxford.

Iyengar, K.T.S. (1965) Matrix analysis of finite beams on elastic foundations. *Inst. Eng. J.*, **45**, 837.

Iyengar, K.T.S.R. and Anantharamu, S. (1963) Finite beam-columns on elastic foundation. *J. Eng. Mech. Div. (ASCE)*, **89** (EM6), 139.

Iyengar, K.T.S.R. and Anantharamu, S. (1965) Influence lines for beams on elastic foundations. *J. Struct. Div. (ASCE)*, **91** (ST3), 45.

Jack, B.J. (September 1971) Anchored diaphragm walls in sand: wall design. *Ground Eng.*, **4**, 15.

Jackson, J.O. and Fookes, P.G. (1974) The relationship of the estimated former burial depth of the lower Oxford Clay to some soil properties. *Quart. J. Eng. Geol.*, **7**, 137.

Jacobs, J.A. (1950) Relaxation methods applied to problems of plastic flow. *Phil. Mag.*, **41**, 349.

Jaeger, J.C. (1969) *Elasticity, Fracture and Flow*, Methuen, London.

Jaeger, J.C. (1972) *Rock Mechanics and Engineering*, Cambridge.

Jaeger, J.C. and Cook, N.G.W. (1969) *Fundamentals of Rock Mechanics*, Chapman and Hall, London.

Jaeger, J.C. and Cook, N.G.W. (1976) *Fundamentals of Rock Mechanics*, 2nd edn, Methuen, London.

Jain, S.C. and Fischer, E.E. (1980) Scour around bridge piers at high flow velocities. *J. Hyd. Div. (ASCE)*, **106**, 1827.

Jain, S.C. et al. (1979) Yield line analysis of square footings. *J. Geotech. Eng. Div. (ASCE)*, **105**, 1355.

James, A.N. (1963) Discussion – Session 3. *Conf. Grouts and Drilling Muds in Eng. Practice*, Inst. Civil Eng., p. 168.

James, A.N. and Kirkpatrick, I.M. (1980) Design of foundations of dams containing soluble rocks and soils. *Quart. J. Eng. Geol.*, **13**, 189.

James, E.L. and Jack, B.J. (1974) A design study of diaphragm walls. *Proc. Conf. Diaphragm Walls and Anchorages*, Inst. Civil Eng., p. 41.

Janbu, N. (1969) The resistance concept applied to deformations of soils. *Proc. 7th Int. Conf. Soil Mech. Found. Eng.*, **1**, 191.

Janbu, N. *et al.* (1956) *Veiledning ved løsning av fundamenteringsoppgaver*, Publication 16, Norwegian Geotech. Inst.

Jarvis, M.G. *et al.* (1979) *Soils of Berkshire*, Soil Survey Bulletin No.8, Soil Survey of England and Wales.

Jeary, A.P. (1974) Damping measurements from the dynamic behaviour of several large multi-flue chimneys. *Proc. Inst. Civil Eng.*, **57** (2), 321.

Jennings, J.E. (1961) A comparison between laboratory prediction and field observation of heave of buildings on desiccated soils. *Proc. 5th Int. Conf. Soil Mech. Found. Eng.*, **1**, 689.

Jennings, J.E. and Knight, K. (1957) The additional settlement of foundations due to a collapse of structure of sandy subsoils on wetting. *Proc. 4th Int. Conf. Soil Mech. Found. Eng.*, **1**, 316.

Jennings, J.E. and Robertson, A. MacG. (1969) The stability of slopes cut into natural rock. *Proc. 7th Int. Conf. Soil Mech. Found. Eng.*, **2**, 585.

Jennings, J.E. *et al.* (1965) Sinkholes and subsidences in the Transvaal Dolomite of South Africa. *Proc. 6th Int. Conf. Soil Mech. Found. Eng.*, **1**, 51.

Jennings, R.C. (1953) West Virginia Highway bridge piers built inside cellular cofferdams. *Civil Eng.*, **23**, 428.

Jessberger, H.L. and Beine, R.A. (1981) Heavy tamping: theoretical and practical aspects. *Proc. 10th Int. Conf. Soil Mech. Found. Eng*, Stockholm. p. 695.

Jirsa, J.O. *et al.* (1966) Tests of a flat slab reinforced with welded wire fabric. *J. Struct. Div. (ASCE)*, **92**, 199.

Johannessen, I.J. and Bjerrum, L. (1965) Measurement of the compression of a steel pile to rock due to settlement of the surrounding clay. *Proc. 6th Int. Conf. Soil Mech. Found. Eng.*, **2**, 261.

Johansen, K.W. (1962) *Yield Line Theory*, Cement and Concrete Ass., Wexham Springs, Slough.

John, K.W. (August, 1962) An approach to rock mechanics. *J. Soil Mech. Found. Div. (ASCE)*, **88**, 1.

John, K.W. (1968) Graphical stability analysis of slopes in jointed rock. *J. Soil Mech. Found. Div. (ASCE)*, **94**, 497; **95** (1969) 685.

Johns, E.A. *et al.* (February, 1963) Oahe Dam: influence of shale on Oahe power structures design. *J. Soil Mech. Found. Div. (ASCE)*, **89**, 95.

Johnson, D. *et al.* (1983) An evaluation of ground improvement at Belawan Port, North Sumatra. *Proc. 8th Europe. Conf. Soil Mech. Found. Eng.*, Helsinki, **1**, 45, Balkema, Rotterdam.

Johnston, G.H. (ed) (1981) *Permafrost: Engineering Design and Construction*, Wiley, New York.

Jonas, E. (September, 1964) Subsurface stabilization of organic silty clay by precompression. *J. Soil Mech. Found. Div. (ASCE)*, **90**, 363.

Jones, C.J.F. (1963) *The performance of a CLASP system school subjected to mining subsidence*, MSc Thesis, University of Newcastle-upon-Tyne.

Jones, C.J.F.P. (1979) Current practice in designing earth retaining structures. *Ground Eng.*, **12** (6), 40.

Jones, C.J.F.P. and Edwards, L.W. (1980) Reinforced earth structures situated on soft foundations. *Géotechnique*, **30**, 207.

Jones, C.J.F.P. and Sims, F.A. (1975) Earth pressures against the abutments and wing walls of standard motorway bridges. *Géotechnique*, **25**, 731.

Jones, O.T. (1944) The compaction of muddy sediments. *Quart. J. Geol. Soc.*, **100**, 137.

Jones, W.A. (1943) Charts and a direct method of cantilever retaining walls. *J. Am. Civ. Eng.* **40**, 5.

Jorden, E.E. (January, 1977) Settlement in sand – methods of calculating and factors affecting. *Ground Eng.*, **10**, 30

Judson, J.C. and Morris, C.J.E. (1974) Drax power station. *Proc. Inst. Civil Eng.*, **56** (1), 559.

Jumikis, A.R. (1971) *Foundation Engineering*, Intext, London.

Jumikis, A.R. (1971) Soil vertical stress influence charts for rectangles. *J. Soil Mech. Found. Div. (ASCE)*, **97**, 521.

Jurgenson, L. (1934) The application of theories of elasticity and plasticity to foundation problems. *J. Boston Soc. Civil Eng.*, **21**, 206.

Just, D.J. *et al.* (1971) Finite element method of analysis of structures resting on elastic foundations. *Symp. Interaction of Structure and Foundation, The Midland Soil Mech. Found. Eng. Soc.*, p. 108.

Kaderabek, T.J. and Reynolds, R.T. (1981) Miami Limestone foundation design and construction. *J. Geotech. Eng. Div. (ASCE)*, **107**, 859.

Kalousek, G.L. and Benton, E.J. (1970) Mechanism of seawater attack on cement pastes. *J. ACI*, **67**, 187.

Kany, M. (1959) *Berechnung von Flächengründungen*, W. Ernst und Sohn, Berlin

Kany, M. (1965) Theory and applicability of best economical dimensioning of foundation groups. *Proc. 6th Int. Conf. Soil Mech. Found. Eng.*, **2**, 93.

Karafiath, L. (1957) Foundation of a blast furnace constructed on loess soil and the computation of settlement. *Proc. 4th Int. Conf. Soil Mech. Found. Eng.*, **1**, 324.

Karal, K. (1977a) Application of energy method. *J. Geotech. Eng. Div. (ASCE)*, **103**, 381.

Karal, K. (1977b) Energy method for soil stability analyses. *J. Geotech. Eng. Div. (ASCE)*, **103**, 431.

Karmalsky, V. and Korner, G. (1956) Design of bridge piers embedded in cohesionless material, taking into account their flexibility. *Proc. Inst. Civil Eng.*, **5**, 535.

Karol, R.H. and Swift, A.M. (1961) Grouting in flowing water and stratified deposits. *Symp. on Grouting. J. Soil Mech. Found. Eng. Div. (ASCE)*, **87**, 125.

Kashef, A-A.I. (July, 1965) Exact free surface of gravity wells. *J. Hyd. Div. (ASCE)*, **91**, 167.

Kassiff, G. (1957) Compaction and shear characteristics of remoulded Negev loess. *Proc. 4th Int. Conf. Soil Mech. Found. Eng.*, **1**, 56, Butterworths, London.

Kassiff, G. and Zeitlen, J.G. (April, 1962) Behaviour of pipes buried in expansive clays. *J. Soil Mech. Found. Div. (ASCE)*, **88**, 133.

Kaufman, R.I. and Sherman, Jr, W.G. (1964) Engineering measurements of Port Allen lock. *J. Soil Mech. Found. Div. (ASCE)*, **90** (SM5) 221.

Kavanagh, T.C. and Johnson, S.M. (July, 1966) Maintenance – the systems approach. *Civil Eng.*, p. 31.

Kavazanjian, E. and Mitchell, J.K. (1980) Time-dependent deformation behaviour of clays. *J. Geotech. Eng. Div. (ASCE)*, **106**, 611.

Kay, J.N. (1975) Interlock tension in steel sheet piling. *J. Struct. Div. (ASCE)*, **101**, 2093.

Kay, J.N. (1976) Sheet pile interlock tension – probabilistic design. *J. Geotech. Eng. Div. (ASCE)*, **102**, 411.

Kay, J.N. and Avalle, D.L. (1982) Application of screw-plate to stiff clays. *J. Geotech. Eng. Div. (ASCE)*, **108**, 145.

Kay, J.N. and Cavagnaro, R.L. (1983) Settlement of raft foundations. *J. Geotech. Eng. Div. (ASCE)*, **109**, 1367.

Kay, J.N. and Parry, R.H.G. (September, 1982) Screw plate tests in a stiff clay. *Ground Eng.*, **15**, 22.

Kazi, A. and Knill, J.L. (1969) The sedimentation and geotechnical properties of the Cromer Till between Happisburgh and Cromer, Norfolk. *Quart. J. Eng. Geol.*, **2**, 63.

Kédzi A. (1972) Stability of rigid structures. *5th Europe. Conf. Soil Mech. Found. Eng.*, **2**, 105.

Keedwell, M.J. (1971) The rheology of clays. *Proc. Symp. on Interaction of Structure and Foundation, Birmingham*, p. 38.

Keller, G.H. (1967) Shear strength and other physical properties of sediments from some ocean basins. *Proc. Conf. Civ. Eng. Oceans*, ASCE.

Keller, G.H. (1969) Engineering properties of some sea-floor deposits. *J. Soil Mech. Found. Div. (ASCE)*, **95**, 1379.

Kenney, T.C. (1963) Permeability ratio of repeatedly layered soils. *Géotechnique*, **13**, 325.

Kenney, T.C. (July, 1964) Pore pressures and bearing capacity of layered clays. *J. Soil Mech. Found. Div. (ASCE)*, **90**, 27.

Kenney, T.C. and Uddin, S. (1974) Critical period for stability of an excavated slope in clay soil. *Can. Geotech. J.*, **11**, 620.

Kerensky, O.A. and Dallard, N.J. (1968) The four-level interchange between M4 and M5 motorways at Almondsbury. *Proc. Inst. Civil Eng.*, **40**, 307.

Kerensky, O.A. and Little, G. (1964) Medway Bridge: design. *Proc. Inst. Civil Eng.*, **29**, 19.

Kerensky, O.A. *et al.* (1972) The Erskine Bridge. *J. Inst. Struct. Eng.*, **50** (4), 147; Discuss., **50** (11), 451.

Kerisel, J. (1975) Old structures in relation to soil conditions. *Géotechnique*, **25**, 433.

Kerisel, J. and Quatre, M. (1968a) Settlement under foundations – calculation using the triaxial apparatus. Part 1. *Civ. Eng. Public Works Rev.*, **63**, 531.

Kerisel, J. and Quatre, M. (1968b) Settlement under foundations – calculation using the triaxial apparatus. Part II. *Civil. Eng. Public Works Rev.* **63**, 661.

Kerr, A.D. (1964) Elastic and viscoelastic foundation models. *J. Appl. Mech., Trans. ASME*, **31**, 491.

Kerr, A.D. (1965) A study of a new foundation model. *Acta Mech.*, **1**, 135.

Kerr, A.D. (1966) *Bending of circular plates confining an incompressible liquid*. US Army Cold Regions Research and Eng. Lab. Research. Report 187.

Kesler, C.E. (Chairman) (1970) *Expansive Cement Concretes – Present State of Knowledge*, Report by ACI Comm. 223. *J. ACI*, **67**, 583.

King, G.J.W. (1967) Correspondence on Gray and Nair (1967), *Géotechnique*, **17**, 434.

King, G.J.W. and Chandrasekaran, V.S. (1974) An assessment of the effects of interaction between a structure and its foundation. *Brit. Geotech. Soc. Conf. on Settlement of Structures*, Pentech Press, London, p. 368.

King. G.J.W. and Cockroft, J.E.M. (1972) The geometric design of long cofferdams. *Géotechnique*, **22**, 619.

King, G.J.W. and Pandey, P.C. The analysis of infilled frames using finite elements. *Proc. Inst. Civil Eng.*, **65** (2), 749.

King, H.J. and Orchard, R.J. (1959) Ground movement in the exploitation of coal seams. *Colliery Guardian*, **198** (5120), 471.

King, H.J. and Smith, H.G. (1954) Surface movements due to mining. *Colliery Eng.*, **31**, 322.

King, H.J. and Whetton, J.T. (1958) Aspects of subsidence and related problems. *Trans. Inst. Min. Eng.*, **118**, 663.

King, J.C. and Bindhoff, E.W. (1982) Lifting and levelling heavy concrete structures. *Conf. Grouting in Geotech. Eng., ASCE*, New Orleans, p. 722.

Kinnear, R.G. (1963) *The Formwork Pressure Balance*. Technical Report TRA/373, Cement and Concrete Ass., Wexham Springs, Slough.

Kirkpatrick, W.M. and Khan, A.J. (1984) The reaction of clays to sampling stress relief. *Géotechnique*, **34**, 29.

Kirsch, K. (1979) Erfahrungen mit der Baugrundverbesserung durch Tiefenruttler, *Geotechnik*, **1**, 21.

Kjaernsli, B. (1958) Test results, Oslo Subway. *Proc. Conf. Earth Pressure Problems*, **2**, 108.

Kleiman, W.F. (September, 1964) Use of surcharges in highway construction. *J. Soil Mech. Found. Div. (ASCE)*, **90**, 331.

Klein, A.M. and Crockett, J.H.A (1953a) Design and construction of a fully vibration-controlled forging hammer foundation. *J. ACI*, **24**, 421.

Klein, A.M. and Crockett, J.H.A. (January, 1953b) Forging hammer foundation built to control destructive vibrations. *Civil Eng.*, p. 30.

Knill, J.L. and Jones, K.S. (1965) The recording and interpretation of geological conditions in the foundation of the Roseires, Kariba and Latigan Dams. *Géotechnique*, 15, 94.

Knodel, P.C. (1981) Construction of large canal on collapsing soils. *J. Geotech. Eng. Div. (ASCE)*, 107, 79.

Ko, H-Y and Davidson, L.W. (1973) Bearing capacity of footings in plane strain. *J. Soil Mech. Found. Div. (ASCE)*, 99, 1.

Koerner, R.M. (1970) Effect of particle characteristics on soil strength. *J. Soil Mech. Found. Div. (ASCE)*, 96, 1221.

Koerner, R.M. et al. (1978) Acoustic emission monitoring of soil stability. *J. Geotech. Eng. Div. (ASCE)*, 104, 571.

Kögler, F. and Scheidig, A. (1938) *Baugrunde unde Bauwerk*, W. Ernst und Sohn, Berlin.

Komernik, A. and David, D. (1969) Prediction of swelling pressure of clays. *J. Soil Mech. Found. Div. (ASCE)*, 95, 209.

Kong, F.K. and Sharp, G.R. (1973) Shear strength of lightweight reinforced concrete deep beams with web openings. *The Struct. Eng.*, 51, 267.

Kong, F.K. and Sharp, G.R. (1977) Structural idealization for deep beams with web openings. *Mag. Conrete Res.*, 29, 81.

Kong, F.K. et al (1970) Web reinforcement effects on deep beams. *J. ACI*, 67, 1010.

Kong, F.K. et al. (1972a) Deep beams with inclined web reinforcement. *J. ACI*, 69, 172.

Kong, F.K. et al (1972b) Shear analysis and design of reinforced concrete deep beams. *The Struct. Eng.*, 50, 405.

Kong, F.K. et al. (1975) Design of reinforced concrete deep beams in current practice. *The Struct. Eng.*, 53, 173.

Kong, F.K. et al. (1978) Structural idealization for deep beams with web openings: further evidence. *Mag. Concrete Res.*, 30, 89.

Korenev, B.G. (1954) *Problems of Calculations of Beams and Slabs on Elastic Foundations* (in Russian), Gosstroiizdat, Moscow.

Korenev, B.G. (1957) A die resting on an elastic half-space the modulus of elasticity of which is an exponential function of depth (in Russian). *Dokl. Akad. Nauk SSSR*, 112, 5.

Korenev, B.G. (1960) *Structural Mechanics in the USSR 1917–1957*, Pergamon Press, London.

Koutsoftas, D. and Fisher, J.A. (1976) *In situ* undrained shear strength of two marine clays. *J. Geotech. Eng. Div. (ASCE)*, 102, 989.

Koutsoftas, D.C. and Fischer, J.A. (1980) Dynamic properties of two marine clays. *J. Geotech. Eng. Div. (ASCE)*, 106, 645.

Kraft, L.M. et al. (1981) Friction capacity of piles driven into clay. *J. Geotech. Eng. Div. (ASCE)*, 107, 1521.

Kramrisch, F. and Rogers, P. (1961) Simplified design of combined footings. *J. Soil Mech. Found. Div. (ASCE)*, 88, 19.

Kravtchenko, J. and Sirieys, P.M. (eds) (1966) *Proc. IUTAM Symp. Rheology and Soil Mechanics*, Grenoble, 1964, Springer-Verlag, Berlin.

Król W. (1964) *Statyka Fundamentów Zelbetowych* (in Polish), Wyndanictwo Arkady, Warsaw.

Krsmanović, D. (1961) Free foundation beams with two supports. *Proc. 5th Int. Conf. Soil Mech. Found. Eng.*, 1, 705.

Krsmanović, D. (1967) Initial and residual shear strength of hard rocks. *Géotechnique*, 17, 145.

Krynine, D.P. (1947) *Soil Mechanics*, McGraw-Hill, New York.

Kubik, L.A. (1980) Predicting the strength of reinforced concrete deep beams with web openings. *Proc. Inst. Civil Eng.*, 69 (2), 939.

Kuhn, S.H. and Williams, A.A.B. (1961) Scour depth and soil profile determinations in river beds. *Proc. 5th Int. Conf. Soil Mech. Found. Eng.*, 1, 487.

Kulhawy, F.H. *et al.* (1979) Uplift testing of model drilled shafts in sand. *J. Geotech. Eng. Div. (ASCE)*, **105**, 31.

Kumar, P. (1977) Design of deep beams with partially predefined reinforcement geometry. *Proc. Inst. Civil Eng.*, **63** (2), 563.

Lacerda, W.A. and Houston, W.N. (1973) Stress relaxation in soils. *Proc. 8th Int. Conf. Soil Mech. Found. Eng.*, **1** (1), 221.

Ladanyi, B. (1983) Shallow foundations on frozen soil: creep settlement. *J. Geotech. Eng. Div. (ASCE)*, **109**, 1434.

Lajtai E.Z. (1969) Strength of discontinuous rocks in direct shear. *Géotechnique*, **19**, 218.

Lake, L.M. and Simons, N.E. (1970) Investigations into the engineering properties of chalk at Welford, Theale, Berkshire, Paper 3, *In situ investigation in Soils and Rocks: B.G.S. Conference*, May 1969, p. 23.

Lama, R.D. and Vutukuri, V.S. (1974, 1978) *Handbook on Mechanical Properties of Rocks* (four volumes dealing with laboratory and *in situ* testing techniques and results). Trans Tech. Publications.

Lambe, T.W. (September, 1964) Methods of estimating settlement. *J. Soil Mech. Found. Div. (ASCE)*, **90**, 43.

Lambe, T.W. (1969) Reclaimed land in Kawasaki City, Japan. *J. Soil Mech. Found. Div. (ASCE)*, **95**, 1181.

Lambe, T.W. (1970) Braced Excavation. *1970 Specialty Conf. Lateral Stresses in the Ground and Design of Earth Retaining Structures*, Cornell University, Soil Mech. Found. Eng. Div. ASCE, p. 149.

Lambe, T.W. (1973) Prediction in soil engineering. *Géotechnique*, **23**, 149.

Lambe, T.W. and Whitman, R.V. (1969) *Soil Mechanics*, Wiley, New York.

Lambe, T.W. *et al.* (1970) Measured performance of braced excavation. *J. Soil Mech. Found. Div. (ASCE)*, **96**, 817.

Lambe, T.W. *et al.* (1981a) Safety of a constructed facility: geotechnical aspects. *J. Geotech. Eng. Div. (ASCE)*, **107**, 339.

Lambe, T.W. *et al.* (1981b) Instability of Amuay cliffside. *J. Geotech. Eng. Div. (ASCE)*, **107**, 1505.

Lane, E.W. and Borland, W.M. (1954) River bed scour during floods. *Trans ASCE.*, **119**, 1072.

Lane, V.P. and Harriman, H. (1975) Minimum cost design using a pre-planned nodal search method. *Proc. Inst. Civil Eng.*, **59** (2), 237.

L'Appolonia, E. *et al.* (1970) *1970 Specialty Conf. Lateral Stresses in the Ground and Design of Earth-Retaining Structures*, Cornell University, Soil Mech. Found. Eng. Div. ASCE.

Larionov, A.K. (1965) Structural characteristics of loess soils for evaluating their constructional properties. *Proc. 6th Int. Conf. Soil Mech. Found. Eng.*, **1**, 64.

Larkin, L.A. (1968) Theoretical bearing capacity of very shallow footings. *J. Soil Mech. Found. Div. (ASCE)*, SM6, 1347.

La Russo, R.S. (1963) Wanapum development – slurry trench and grouted cut-off. *Proc. Symp. Grouts and Drilling Muds in Engineering Practice*, Butterworth, London, p. 22.

Laursen, E.M. (February, 1960) Scour at bridge crossings. *J. Hyd. Div. (ASCE)*, **86**, 39; Discuss. May, August and November (1960), July (1961).

Laursen, E.M. (May, 1963) An analysis of relief bridge scour. *J. Hyd. Div. (ASCE)*, **89**, 93.

Lawton, V.H. (1951) *Examples of Structural Steel Design*, Part 3, BCSA.

Lax, D. and Bunclark, F.T. (1960) The design and construction of the new assembly building for the Ford Motor Company Limited, Dagenham. *The Struct. Eng.*, **38**, 317.

Lazebnik, G.E. (1970) Investigation of soil pressure distribution under foundation mats of buildings. *J. Soil Mech. Found. Eng. Div. (ASCE)*, **6**, 405.

Lee, C.F. and Lo, K.Y. (1976) Rock squeeze study of two deep excavations at Niagara Falls. *Proc. Conf. Rock Engineering*, Univ. of Colorado, **1**, 116, ASCE.

Lee, D.H. (1950) *Sheet Piling, Cofferdams and Caissons*, Concrete Publications, London.
Lee, I.K. (1965) Foundations subject to moment. *Proc. 6th Int. Conf. Soil Mech. Found. Eng.*, **2**, 108.
Lee, I.K. (1968) *Soil Mechanics – Selected Topics*, Butterworths, London.
Lee, I.K. and Harrison, H.B. (1970) Structure and foundation interaction theory. *J. Struct. Div. (ASCE)*, **96**(ST2) 177.
Lee, K.L. and Focht, J.A. (1975) Liquefaction potential at Ekofisk tank in North Sea. *J. Geotech. Eng. Div. (ASCE)*, **101**, 1.
Lee, K.L. and Morrison, R.A. (1970) Strength of anisotropically consolidated compacted clay. *J. Soil Mech. Found. Div. (ASCE)*, **96**, 2025.
Lee, K.L. *et al.* (1973) Reinforced earth retaining walls. *J. Soil Mech. Found. Div. (ASCE)*, **99**, 745.
Lee, S.L. *et al.* (1961) Continuous beam-columns on elastic foundation *J. Eng. Mech. Div. (ASCE)*, **87**, 55.
Leech, T.D.J. and Pender, E.B. (1961) Experience in grouting rock bolts. *Proc. 5th Int. Conf. Soil Mech. Found. Eng.*, **2**, 445.
Leggatt, A.J. and Bratchell, G.E. (1973) Submerged foundations for 100,000 ton oil tanks. *Proc. Inst. Civil Eng.*, **54** (1), 291.
Lekhnitskii, S.G. (1960) *Theory of elasticity of an anisotropic elastic body* (translated from the Russian by P. Fern), Holden-Day, San Francisco.
Lekkerkerker, J.G. (1960) Bending of an infinite beam resting on an elastic half-space. *Proc. K. Ned. Akad. Wet.*, Ser. B, **63**, 484.
Lemcoe, M.M (1960) Stresses in layered elastic solids. *J. Eng. Mech. Div. (ASCE)*, **86**, (EM4) 1.
Leonards, G.A. (1949) Analysis of building frames with unsymmetrical differential settlement of the foundations. *J. ACI*, **45**, 645.
Leonards, G.A. (1962) *Foundation Engineering*, McGraw-Hill, New York.
Leonards, G.A. (1982) Investigation of failures. *J. Geotech. Eng. Div. (ASCE)*, **108**, 187.
Leonards, G.A. and Harr, M.E. (1959) Analysis of concrete slabs on ground. *J. Soil Mech. Found. Div. (ASCE)*, **85** (SM3), 35.
Leonards, G.A. *et al.* (1980) Dynamic compaction of granular soils. *J. Geotech. Eng. Div. (ASCE)*, **106**, 35.
Leonhardt, F. (1970) Modern design of television towers. *Proc. Inst. Civil Eng.*, **46**, 265.
Leussink, H. *et al.* (1966) *Versuche über sohldruckverteilung unter starren gründungs köppern auf Kohärsionless sand*, Veröffentlichungen des Institutes für Bodenmechanik und Felsmechanik der Technischen Hochschule Fridericiana Universität, Karlsruhe.
Levinton, Z. (1947) Elastic foundations analysed by the method of redundant reactions. *Proc. ASCE*, **73**, 1529.
Lewis, W.A. (1967) Full scale studies of the performance of plant in the compaction of soils and granular base materials. *Proc. Inst. Mech. Eng.*, **181** (2A), 79.
Lewis, W.A. and Croney, D. (1966) The properties of chalk in relation to road foundations and pavements. *Proc. Symp. Chalk in Earthworks and Foundations*, Inst. Civil Eng.
L' Herminier, R. *et al.* (1957) Etude et observation concernant le radier et fondation du premier réacteur atomique au centre de Marcoule. *Proc. 4th Int. Conf. Soil Mech. Found. Eng.*, **1**, 307.
L'Herminier, R. *et al.* (1965) Expérimentation en laboratoire de la capacité portante des sols. *Proc. 6th Int. Conf. Soil Mech. Found. Eng.*, **2**, 117.
Li, S-T, and Liu, T.C-Y (1970) Prestressed concrete piling – contemporary design practice and recommendations. *J. ACI* **67**, 201.
Liao, S. and Sangrey, D.A. (1978) Use of piles as isolation barriers. *J. Geotech. Eng. Div. (ASCE)*, **104**, 1139.
Liauw, T.C. and Lee, S.W. (1977) On the behaviour and analysis of multi-storey infilled

frames subject to lateral loading. *Proc. Inst. Civil Eng.*, **63** (2), 641.

Lilwall, R.C. (1976) *Seismicity and Seismic Hazard in Britain*, Seismological Bulletin No. 4, Institute of Geological Sciences, HMSO.

Lin, K.K. *et al.* (1971) Beams on one way elastic foundations. *J. Boston Soc. Civil Eng.*, **58**, 164.

Lindley, P.B. (1962) *Design and use of natural rubber bridge bearings*, Natural Rubber Producers Research Association Technical Bulletin No. 7.

Lindner, E. (1976) Swelling rock: a review. *Proc. Conf. Rock Engineering*, Univ. of Colorado, **1**, 141, ASCE.

Lindsell, P. and El-Dharat, A. (1979) Model analysis of a bridge abutment. *The Struct. Eng.*, **57A**, 183.

Linell, K.A. and Ledrow, J.C.F. (1981) *Soil and Permafrost Surveys in the Arctic*, Oxford University Press.

Little, A.L. (1961) *Foundations*, Edward Arnold, London.

Little, A.L. and Price, V.E. (1958) The use of an electronic computer for slope stability analysis. *Géotechnique*, **8**, 113.

Littlejohn, G.S. (1966) *Soil–Structure Interaction in Mining Areas with Particular Reference to Horizontal Subgrade Restraint*, PhD thesis, University of Newcastle-upon-Tyne.

Littlejohn, G.S. (1970a) Soil anchors. *Proc. Conf. Ground. Eng.*, Inst. Civil Eng., p. 33.

Littlejohn, G.S. (1970b) Discussion. *Proc. Conf. Ground Eng.*, Inst. Civil Eng., p. 115.

Littlejohn, G.S. (1973) Monitoring foundation movements in relation to adjacent ground. *Ground Eng.* **7** (4), 17.

Littlejohn, G.S. (1975a) Observation of brick walls subjected to mining subsidence. *Settlement of Structures* (British Geotechnical Society), Pentech Press, London. p. 384.

Littlejohn, G.S. (1975b) Rock anchors – state of the art. Part 1: Design. *Ground Eng.*, **25** (May); **41** (July).

Littlejohn, G.S. (1979a) Surface stability in areas underlain by old coal workings. *Ground Eng.* **12** (2), 22.

Littlejohn, G.S. (1979b) Consolidation of old coal workings. *Ground Eng.*, **12** (4), 15.

Littlejohn, G.S. (November, 1980) Design estimation of the ultimate load-holding capacity of ground anchors. *Ground Eng.*, **13**, 25.

Littlejohn, G.S. and Bruce, D.A. (1975, 1976) Rock anchors – state of the art. Part 2: Construction. *Ground Eng.*, **8** (5), 34; **8** (6), 36; Part 3: Stressing and testing, *Ground Eng.*, **9** (2), 20; **9** (3), 55; **9** (4), 33.

Littlejohn, G.S. *et al.* (April, 1971) Anchored diaphragm walls in sand – some design and construction considerations. *J. Inst. Highway Eng.*, **18**, 15.

Littlejohn, G.S. *et al.* (January, 1972) Anchored diaphragm walls in sand. *Ground Eng.*, **5**, 12.

Lo, K.Y. (1970) The operational strength of fissured clays. *Géotechnique*, **20**, 57.

Lo, K.Y. (1972) An approach to the problem of progressive failure *Can Geotech. J.* **9**

Lo, K.Y. and Lee, C.F. (1973) Analysis of progressive failure in clay slopes. *Proc. 8th Int. Conf. Soil Mech. Found. Eng.*, **1** (1), 251.

Lo, K.Y. and Milligan, V. (1967) Shear strength properties of two stratified clays. *J. Soil Mech. Found. Div. (ASCE)*, **93**, 1.

Long, A.E. and Bond, D. (1967) Punching failure of reinforced concrete slabs. *Proc. Inst. Civil Eng.*, **37**, 109.

Lord, E.R.F. and Smith, W.E. (1976) The misuse of SPT N value correlations with Upper Chalk grades. *Géotechnique*, **26**, 217.

Lorenz, H. (1934) Neue Ergebnisse der dynamische Baugrunduntersuchung. *ZVDI* **78** (12), 379.

Loudon, A.G. (1952) The computation of permeability from simple soil tests. *Géotechnique*, **3**, 165.

Lowe, J. (1974) New concepts in consolidation and settlement analysis. *J. Geotech. Eng. Div. (ASCE)*, **100**, 574.

Lowe, J.R. and Byrne, B. le C. (1963) Prestressed concrete warehouse for Thos Heiton and Co. Ltd., Dublin. *The Struct. Eng.*, **41**, 281.

Lowery, L. *et al.* (1969) *Pile Driving Analysis State-of-the-Art*, Texas A and M University, Final Research Report 33–13.

Ludvig, B (1981) Direct shear tests of filled and unfilled joints. *Proc. Conf. Application of Rock Mechanics to Cut and Fill Mining*, Univ. of Lulea, Sweden, Inst. Mining and Met.

Lugeon, M. (1937) *Barrages et Géologie*, Rouge et Cie, Lausanne.

Lumb, P. (1962) The properties of decomposed granite. *Géotechnique*, **12**, 226.

Lumb, P. (1965) The residual soils of Hong Kong. *Géotechnique*, **15**, 180; **16** (1966) 78.

Lumb, P. (1970) Safety factors and the probability distribution of soil strength. *Can. Geotech. J.*, **7**, 225.

Lupini, J.F. *et al.* (1981) The drained residual strength of cohesive soils. *Géotechnique*, **31**, 181.

Lur'e, A.I. (1964) *Three-Dimensional Problems of the Theory of Elasticity*, (translated from the Russian by D.B. McVean), Interscience, New York.

Lysmer, J. and Richart, F.E., Jr (January, 1966) Dynamic response of footings to vertical loading. *J. Soil Mech. Found. Div. (ASCE)*, **92** (SMI) 65.

Lytton, R.L. and Meyer, K.T. (1971) Stiffened mats on expansive clay. *J. Soil Mech. Found. Div. (ASCE)*, **97**, 999.

MacDonald, J. (1952) Foundations for Ringsend Power Station, Dublin. *Géotechnique*, **3**, 143.

MacDonald, D.H. and Skempton, A.W. (1955) A survey of comparisons between calculated and observed settlements of structures on clay. *Conf. on Correlation between Calculated and Observed Stresses and Displacements in Structures*, prelim vol., Inst. Civil Eng., p. 318.

MacFarlane, I.C. (ed.) (1969) *Muskeg Engineering Handbook*, University of Toronto.

MacKenzie, T.R. (1955) *Strength of Deadman Anchors in Clay*, Master's thesis, Princeton University.

MacLeod, I.A. and Abu-el-Magd, S.A. (1980) The behaviour of brick walls under conditions of settlement. *The Struct. Eng.*, **58A**, 279.

MacLeod, I.A. and Paul, J.G. (1984) Settlement monitoring of buildings in Central Scotland. *Géotechnique*, **34**, 99.

McRoberts, E.C. (1982) Shallow foundations in cold regions: design. *J. Geotech. Eng. Div. (ASCE)*, **108**, 1338.

Madsen, O.S. (1978) Wave-induced pore pressures and effective stresses in a porous bed. *Géotechnique*, **28**, 377.

Mandel, J. (1965) Interference plastiques de semelles filantes. *Proc. 6th Int. Conf. Soil Mech. Found. Eng.*, **2**, 127.

Magnel, G. (1949). Applications of pre-stressed concrete in Belgium. *J. Inst. Civil. Eng.*, **32**, 161.

Mainstone, R.J. (1972) *On the Stiffnesses and Strengths of Infilled Frames*, CP2/72, Building Research Staion.

Mainstone, R.J. (1974) *Supplementary Note on the Stiffnesses and Strengths of Infilled Frames*, CP13/74, Building Research Station.

Maitland, J.K. and Schroeder, W.L. (1979). Model study of circular sheetpile cells. *J. Geotech. Eng. Div. (ASCE)*, **105**, 805.

Majid, K.I. and Cunnell, M.D. (1976) A theoretical and experimental investigation into soil–structure interaction. *Géotechnique*, **26**, 331.

Majid, K.I. and Rahman, M.A. (1982) Non-linear analysis of structure–soil systems. *Proc. Inst. Civil Eng.*, **73**(2), 53.

Malter, H. (1958) Numerical solutions for beams on elastic foundations. *J. Struct. Div. (ASCE)*, **84** (ST2), Paper 1562.

Mandel, J. (1963) Interférence plastique de fondations superficielles. *Proc. Int. Conf. Soil Mech. Found. Eng.*, Budapest, p. 267.

Mandel, J. (1965) Interférence plastiques de semelles filantes. *Proc. 6th Int. Conf. Soil Mech. Found. Eng.*, **2**, 127.

Manning, G.P. (1933) *Reinforced Concrete Arch Design*, Pitman, London.

Manning, G.P. (1972) *Design and Construction of Foundations*, 2nd ed. Cement and Concrete Association, Wexham Springs, Slough.

Mansfield, E.H. (1964) *The Bending and Stretching of Plates*, Pergamon Press, Oxford.

Mansur, C.I. and Dietrich, R.J. (July, 1965) Pumping test to determine permeability ratio. *J. Soil Mech. Found. Div. (ASCE)*, **91**, 151.

Manuel, R.F. *et al.* (1971) Deep beam behaviour affected by length and shear span variations. *J. ACI*, **68**, 954.

Marchetti, S. (1980) *In-situ* tests by flat dilatometer. *J. Geotech. Eng. Div. (ASCE)*, **106**, 299.

Marcuson, W.F. and Bieganousky, W.A. (1977) Laboratory standard penetration tests on fine sands. *J. Geotech. Eng. Div. (ASCE)*, **103**, 565.

Marcuson, W.F. and Curro, J.R. (1981) Field and laboratory determination of soil moduli. *J. Geotech. Engng. Div. (ASCE)* **107**, 1269.

Margason, G. *et al* (1968) *The Effect of Tides on Subsoil Pore-water Pressures at a Site on the Proposed Shoreham By-pass*, Report 195, Transport and Road Research Laboratory.

Marion, H. and Mahfouz, G. (1974) Design and construction of the Ekofisk artificial island. *Proc. Inst. Civil Eng.*, **56**(1), 497; Discuss. (1975) **58**(1), 631.

Mariupol'skii, L.G. (1965) The bearing capacity of anchor foundations. *Osnovaniya Fundamenty i Mekhanika Gruntov*, **3**(1), 14 (available in English translation from Consultants Bureau, New York, p. 26).

Marr, J.E. (1958) A new approach to the estimation of mining subsidence. *Trans. Inst. Min. Eng.*, **118**, 692.

Marr, W.A. and Christian, J.T. (1981). Permanent displacements due to cyclic wave loading. *J. Geotech. Eng. Div. (ASCE)*, **107**, 1129.

Marr, W.A. *et al.* (1982) Criteria for settlement of tanks. *J. Geotech. Eng. Div. (ASCE)*, **108**, 1017.

Marsh, T.J. and Davies, P.A. (1983) The decline and partial recovery of groundwater levels below London. *Proc. Inst. Civil Eng.* **74**(1), 263.

Marshall, W.T. (1944) Experiments on reinforced column bases. *J. Inst. Civil Eng.*, **22**, 49.

Marsland, A. (1953) Model experiments to study the influence of seepage on the stability of a sheeted excavation in sand. *Géotechnique*, **3**, 223.

Marsland, A. (1971a) Laboratory and *in-situ* measurement of the deformation moduli of London Clay. *Symp. on the Interaction of Structure and Foundation*, The Midland Soil Mech. Found. Eng. Soc., p. 7.

Marsland, A. (1971b) Large *in-situ* tests to measure the properties of stiff fissured clays. *Proc. 1st Aust.–NZ Conf. Geomech.*, **1**, 180.

Marsland, A. (1971c) *The Shear Strength of Stiff Fissured Clays*, CP21/71, Building Research Establishment.

Marsland, A. (1972) Model studies of deep *in-situ* loading tests in clay. *Civil Eng. Public Works Rev.*, **67**, 695.

Marsland, A. (1973a) *Site Investigation for Foundations*, B392/73, Building Research Establishment.

Marsland, A. (1973b) *Large In-situ Tests to Measure the Properties of Stiff Fissured Clays*, CP1/73, Building Research Establishment.

Marsland, A. (1974) *Comparisons of the Results from Static Penetration Tests and Large In-situ Plate Tests in London Clay*, CP87/74, Building Research Station.

Marsland, A. (1977) The evaluation of the engineering design parameters for glacial clays. *Quart. J. Eng. Geol.*, **10**, 1.

Marsland, A. and Butler, M.E. (1968) *Strength Measurements on Stiff Fissured Barton Clay from Fawley (Hampshire)*, CP30/68, Building Research Establishment.

Marsland, A. and Randolph, M.F. (1977) Comparisons of the results from pressuremeter tests and large *in-situ* plate tests in London Clay. *Géotechnique* **27**, 217; reprinted as CP10/78, Building Research Station.

Marsland, A. *et al.* (1980) Soil profile mapping in relation to site evaluation for foundations and earthworks. *Bull. Int. Ass. Eng. Geol.*, **21**, 139.

Marsland, A. *et al.* (1983) The behaviour of a bridge abutment foundations on Keuper marl during and after construction. *Proc. Inst. Civil Eng.*, **74**,(1), 917.

Marston, A. (1930) *The Theory of External Loads on Closed Conduits in the Light of the Latest Experiments*, Iowa State College Bulletin 96. *Proc. 9th Ann. Meeting Highway Res. Board*, Washington DC, p. 138.

Martin, D.J. (1980) *Ground Vibrations from Impact Pile Driving during Road Construction*, Report SR544, Transport and Road Research Laboratory.

Martin and Robinson (1980) Private communication.

Martin, G.R. *et al.* (1975) Fundamentals of liquefaction under cyclic loading. *J. Geotech. Eng. Div. (ASCE)*, **101**, 423.

Martin, G. R. *et al.* (1980) Pore-pressure dissipation during offshore cyclic loading. *J. Geotech. Eng. Div. (ASCE)*, **106**, 981.

Martin, I. and Ruiz, S. (1959) Folded plate raft foundation for a 24 storey building. *J. ACI*, **31**, 121; Discuss., 949.

Martin, L.H. (1979) Methods for the limit state design of triangular steel gusset plates. *Building and Environment*, **14**, 147.

Marvin, E.L. (1972) Viscoelastic plate on poroelastic foundation. *J. Eng. Mech. Div. (ASCE)*, **98**(EM4), 911.

Mascardi, C. (1981) Compattozione dinamica. 10 Ciclo di conferenze dedicate ai problemi di meccanica dei terreni e ingegneria delle fondazioni. Politecnico di Torino Ingegneria.

Mason, J. and Frost, A.D. (1963) Stag Place Development. *The Struct. Eng.*, **41**, 347; (1963) **42**, 169.

Massarsch, K.R. and Broms, B.B. (1976) Lateral earth pressure at rest in soft clay. *J. Geotech. Eng. Div. (ASCE)*, **102**, 1041.

Massarsch, K.R. and Broms, B.B. (1983) Soil compaction by vibro-wing method. *Proc. 8th Europe. Conf. Soil Mech. Found. Eng.*, Helsinki, **1**, 275, Balkema, Rotterdam.

Masters, F.M. (1951). Philadelphia's Penrose Avenue Bridge opened to traffic. *Civil Eng.*, **21**, 578.

Mathes, G.M. (June, 1982) Spilled petroleum recovered from atop water table of Mississippi Aquifer. *Civ. Eng. (ASCE)*, p. 58.

Matheson, G.D. (1983) *Rock Stability Assessment in Preliminary Site Investigations – Graphical Methods*, Laboratory Report 1039, Transport and Road Research Laboratory.

Matlock, H. and Ripperger, E.A. (1956) Procedures and instrumentation for tests on a laterally loaded pile. *Proc. 8th Texan Conf. Soil Mech. Found. Eng.*

Matlock, H. and Wayne, I.B. (1963) Bending and buckling of soil supported structural elements, Paper 32. *Proc. 2nd Pan. Am. Conf. Soil Mech. Found. Eng.*, Brazil.

Matsu, M. (1967) Study on the uplift resistance of footing (1). *Soil and Foundation Tokyo, Japan* **7**(4), 1.

Mattes, N.S. and Poulos, H.G. (1969) Settlement of a single compressible pile. *J. Soil Mech. Found. Div. (ASCE)*.

Mautner, K.W. (1948) Structures in areas subjected to mining subsidence. *The Struct. Eng.*, **26**, 35.

Maxwell-Cook, P.V. (ed.) (1974) Prestressed concrete foundations and ground anchors. *7th Congress, Fédération Internationale de la Précontrainte.*

May, M.A. (1964). Setting out tall buildings. *Proc. Inst. Civil Eng.*, **28**, 405.

Mayer, A. (1963). Recent work in rock mechanics. *Géotechnique*, **13**, 99.

Maynard, D.P. and Davis, S.G. (1974) The strength of *in-situ* concrete. *The Struct. Eng.*, **52**, 369; (1975) **53**, 277.

Mayne, J.R. and Cook, N.J. (1978) *On Design Procedures for Wind Loading*, CP25/78, Building Research Establishment.

Mayne, P.W. and Kulhawy, F.H. (1982) K_0-OCR relationships in soil. *J. Geotech. Eng. Div. (ASCE)*, **108**, 851.

McCann, D.M. and Hobbs, P.R. (1977) Static and dynamic elastic moduli of Jurassic limestones from the Dorset coast. *Proc. Conf. Rock Eng.*, Univ. Newcastle-upon-Tyne, p. 155.

McCarter, W.J. *et al.* (1981) An experimental investigation of the earth-resistance response of a reinforced concrete pile. *Proc. Inst. Civil Eng.*, **70**(2), 1101.

McClelland, B. (1974) Design of deep penetration piles for ocean structures. *J. Geotech. Eng. Div. (ASCE)*, **100**, 709.

McFarlane, I.H. and Tomlinson, M.J. (1974) Site investigations for structural foundations. *The Struct. Eng.*, **52**, 57.

McGown, A. (1971) The classification for engineering purposes of tills from moraines and associated land forms. *Quart. J. Eng. Geol.*, **4**, 115.

McGown, A. and Derbyshire, E. (1977) Genetic influences on the properties of tills. *Quart. J. Eng. Geol.*, **10**, 389.

McGown, A. *et al.* (1974) Fissure patterns and slope failures in till at Hartford, Ayrshire. *Quart. J. Eng. Geol.*, **7**, 1.

McGown, A. *et al.* (1980) Recording and interpreting soil macrofabric data. *Géotechnique*, **30**, 417.

McKinley, D. (January, 1965) Field observation of structures damaged by settlement. *J. Soil Mech. Found. Div. (ASCE)*, **91**, 249.

McKittrick, D.P. (1979) Reinforced earth: application of theory and research to practice. *Ground Eng.*, **12**(1), 19.

McMeekin, R.D. (1964) A review of industrialized building. *The Struct. Eng.*, **42**, 63.

McNamee, J. (1951) Seepage into a sheeted excavation. *Géotechnique*, 1, 229.

McNamee, J. and Gibson, R.E. (1960). Plane strain and axially symmetric problems of the consolidation of a semi-infinite clay stratum. *Quart. J. Mech. Appl. Math.*, **2**, 210.

McNaughton, F.M. (1953) Bridge foundations in difficult ground. *J. Inst. Mun. Eng.*, **79**, 505.

McNulty, J.F. and O'Brien, J.S. (February, 1961) Underpinning of a wind tunnel. *J. Soil Mech. Found. Div. (ASCE)*, **87**, 1.

McRoberts, E.C. and Morgenstern, N.R. (1974) The stability of thawing slopes. *Can. Geotech. J.*, **11**, 447.

McTrusty, J.W. (1959) Control of mining subsidence. *Colliery Eng.*, **36**, 122.

Mears, T.F. and Charman, W.R. (1966) The design and construction of cylindrical television masts in Great Britain. *The Struct. Eng.*, **44**, 5.

Measor, E.O. and Williams, G.M.J. (1962) Features in the design and construction of the Shell Centre, London. *Proc. Inst. Civil Eng.*, **21**, 475; Discuss., **24**, 409.

Meigh, A.C. (1950) *Model Footing Tests on Clay*, MSc (Eng) thesis, University of London Library.

Meigh, A.C. (1968) Foundation characteristics of the Upper Carboniferous rocks. *Quart. J. Eng. Geol.*, **1**, 87.

Meigh, A.C. (1970) Some driving and loading tests on piles in gravel and chalk, paper 2. *Conference on Behaviour of Piles*, Inst. Civil Eng. p. 9.

Meigh, A.C. (1976) The Triassic rocks, with particular reference to predicted and observed performance of some major foundations. *Géotechnique*, **26**, 393.

Meigh, A.C. and Greenland, S.W. (1965) *In-situ* testing of soft rocks. *Proc. 6th Int. Conf. Soil Mech. Found. Eng.*, **1**, 73.

Meigh, A.C. and Nixon, I.K. (1961) Comparison of *in-situ* tests for granular soils. *Proc. 5th Int. Conf. Soil Mech. Found. Eng.*, **1**, 499.

Meigh, A.C. *et al.* (1973) Field and laboratory creep tests on weak rocks. *Proc. 8th Int. Conf. Soil Mech. Found. Eng.*, **1**, 265.

Ménard, L. (1965) Règles pour le calcul de la force portante et du tassement des fondations en fonction des résultats pressiométriques. *Proc. 6th Int. Conf. Soil Mech. Found. Eng.*, **2**, 295.

Mencl, V. (1966) Mechanics of landslides with non-circular slip surfaces with special reference to the Vaiont slide. *Géotechnique*, **16**, 329; (1967) **17**, 170.

Menzies, B.K. and Simons, N.E. (1978) Stability of embankments on soft ground; in *Developments in Soil Mechanics – 1* (ed. C.R. Scott), Applied Science Publishers, Barking.

Mercer, R. (1982) The stabilization of Craig-y-Dref, Tremadog, *Ground Eng.*, **15**, 28.

Meyerhof, G.G. (1947) The settlement analysis of building frames. *The Struct. Eng.*, **25**, 369.

Meyerhof, G.G. (1951) The ultimate bearing capacity of foundations. *Géotechnique*, **2**, 301.

Meyerhof, G.G. (1952) The tilting of a large tank on soft clay. *Proc. South Wales Inst. Eng.*, **67**, 53.

Meyerhof, G.G. (1953a) The bearing capacity of concrete and rock. *Mag. Concrete Res.*, **4**, 107.

Meyerhof, G.G. (1953b) The bearing capacity of foundations under eccentric and inclined loads. *Proc. 3rd Int. Conf. Soil Mech. Found. Eng.*, **1**, 440.

Meyerhof, G.G. (1953c) Some recent foundation research and its application to design. *The Struct. Eng.*, **31**, 151; Discuss. (1954) **32**, 156.

Meyerhof, G.G. (1955) Influence of roughness of base and ground-water conditions on the ultimate bearing capacity of foundations. *Géotechnique*, **5**, 227.

Meyerhof, G.G. (1956) Penetration tests and bearing capacity of cohesionless soils. *J. Soil Mech. Found. Div. (ASCE)*, **82** (**SM1**), Paper 866.

Meyerhof, G.G. (1957) The ultimate bearing capacity of foundations on slopes. *Proc. 4th Int. Conf. Soil Mech. Found. Eng.* **1**, 384.

Meyerhof, G.G. (1959) Compaction of sands and bearing capacity of piles. *J. Soil Mech. Found. Div. (ASCE)* **85** (SM6), 1.

Meyerhof, G.G. (1962) Load-carrying capacity of concrete pavements. *J. Soil Mech. Found. Div. (ASCE)*, **88**, 89.

Meyerhof, G.G. (1963) Some recent research on the bearing capacity of foundations. *Can. Geotech. J.*, **1**, 16.

Meyerhof, G.G. (1965) Shallow foundations. *J. Soil Mech. Found. Div. (ASCE)*, **2**, 21.

Meyerhof, G.G. (1970) Safety factors in soil mechanics. *Can. Geotech. J.*, **7**, 349.

Meyerhof G.G. (1972) Stability of slurry trench cut in saturated clay. *Proc. Conf. Performance of Earth and Earth Support Structures*, Purdue University, **1** (2).

Meyerhof, G.G. (1973) The uplift capacity of foundations under oblique loads. *Can. Geotech. J.*, **10**, 64.

Meyerhof, G.G. (1974) Ultimate bearing capacity of footings on sand layer overlying clay. *Can. Geotech. J.*, **11**, 223.

Meyerhof, G.G. (1976) Bearing capacity and settlement of pile foundations. *J. Geotech. Eng. Div. (ASCE)*, **102**, 197.

Meyerhof, G.G. and Adams, J.I. (1968) The ultimate uplift capacity of foundations. *Can. Geotech. J.*, **5**, 225.

Meyerhof, G.G. and Chaplin, T.K. (1953) The compression and bearing capacity of cohesive layers. *Brit. J. Appl. Phys.*, **4**, 20.

Meyerhof, G.G. and Rao, K.S.S. (1974) Collapse load of reinforced concrete footings. *J. Struct. Div (ASCE)*, **100**, 1001.

Milano, J. (May, 1961) Cathodic protection of marine terminal facilities. *J. Waterways Harbours Div. (ASCE)*, **87**, 27.

Millet, R.A. and Perez, J.-Y. (1981) Current USA practice: slurry wall specifications. *J. Geotech. Eng. Div. (ASCE)*, **107**, 1041.

Milligan, V. (August, 1982) Experience with Canadian varved clays. *J. Soil Mech. Found. Div. (ASCE)*, **88**, 31.

Milligan, V. and Lo, K.Y. (1970) Observations on some basal failures in sheeted excavations. *Can. Geotech. J.*, **7**, 136; (1971) **8**, 346.

Milović, D.M. (1965) Comparison between the calculated and experimental values of the ultimate bearing capacity. *Proc. 6th Int. Conf. Soil Mech. Found. Eng.*, **2**, 142.

Milović, D.M. (1970) Stresses and displacements in an anisotropic medium due to a circular load. *Proc. 2nd Cong. Int. Soc. Rock Mech.*, **3**, 479.

Milović, D.M. (1973) Stresses and displacements produced by a ring foundation. *Proc. 8th Int. Conf. Soil Mech. Found. Eng.*, **1**, 167.

Milović, D.M. *et al.* (1970) Stresses and displacements in an elastic layer due to inclined and eccentric load over a rigid strip. *Géotechnique*, **20**, 231.

Minikin, R.R. (1950) *Winds, Waves and Maritime Structures*, Griffin, London.

Ministry of Transport *et al.* (1966) *Roads in Urban Areas*, HMSO.

Ministry of Transport (1966) *Rules for the Design and Use of Freyssinet Concrete Hinges in Highway Structures*, Memorandum 577/1, HMSO.

Ministry of Transport (1968) *Headroom Standards*, Technical Memorandum (Bridges) BE14, HMSO.

Ministry of Transport (1969) *The Appearance of Highway Bridges*, HMSO.

Ministry of Works (1951) *Mining Subsidence; Effects on Small Houses*, Nat. Build. Studies Spec. Rep. 12, HMSO.

Miranda, C. and Nair, K. (1966) Finite beams on elastic foundations. *J. Struct. Div. (ASCE)*, **92**, 131.

Mirata, T. (1969) A semi-empirical method for determining stresses beneath embankments. *Géotechnique*, **19**, 188.

Mitchell, C.W. (1973) *Terrain Evaluation*, Longman, London.

Mitchell, J.K. (1968) *In place Treatment of Foundation Soils*, Specialty Conference, Cambridge, Mass., ASCE, p. 93.

Mitchell, J.K. and Durgunoglu, H.T. (1973) *In-situ* strength by static cone penetration test. *Proc. 8th Int. Conf. Soil Mech. Found. Eng.*, **1** (2), 279.

Mitchell, J.K. and Gardner, W.S. (1971) Analysis of load bearing fills over soft subsoils. *J. Soil Mech. Found. Div. (ASCE)*, **97** (SM11), 1549.

Mitchell, J.K. and Houston, W.N. (1969) Causes of clay sensitivity. *J. Soil Mech. Found. Div. (ASCE)*, **95**, 845.

Mitchell, J.K. and Katti, R.K. (1981) Soil improvement – state of art report. *Proc. 10th Int. Conf. Soil Mech. Found. Eng.*, Stockholm, Balkema, Rotterdam.

Mitchell, J.K. and Lunne, T.A. (1978). Cone resistance as measure of sand strength. *J. Geotech. Eng. Div. (ASCE)*, **104**, 995.

Mitchell, J.M. (December, 1973) Assessing large diameter piles. *The Consulting Eng.*, p. 37.

Moe, J. (1961) *Shearing Strength of Reinforced Concrete Slabs and Footings under Concentrated Loads*, Development Department Bulletin D47, Portland Cement Ass.

Mohan, A. (1975) Heat transfer in soil–water–ice systems. *J. Geotech. Eng. Div. (ASCE)*, **101**, 97.

Mohan, D. and Chandra, S. (1961) Frictional resistance of bored piles in expansive clays. *Géotechnique*, **11**, 294.

Mohan, D. *et al.* (1978) Remedial underpinning of steel tank foundation. *J. Geotech. Eng. Div. (ASCE)*, **104**, 639.

Moore, J.F.A. (1974a) Mapping major joints in the Lower Oxford Clay using terrestrial photogrammetry. *Quart. J. Eng. Geol.*, **7**, 57.

Moore, J.F.A. (1974b). A long-term plate test on Bunter Sandstone. *Proc. 3rd Int. Cong. Rock Mech.*, **2B**, 724; reprinted as CP23/75, Building Research Establishment.

Moore, J.F.A. and Longworth, T.I. (1979) Hydraulic uplift of the base of a deep excavation in Oxford Clay. *Géotechnique*, **29**, 35.

Moore, J.G. *et al.* (1977) The effect of leaching on engineering behaviour of a marine sediment. *Géotechnique*, **27**, 517.

Moore, W.W. (January, 1964) The Golden Gateway: foundation design. *Civil Eng.*, p. 33.

Moorhouse, D.C. and Baker, G.L. (1969) Sand densification by heavy vibratory compactor. *J. Soil Mech. Found. Div. (ASCE)*, **95**, 985.

Moran, B.J.J. (1949) The use of vegetation in stabilizing artificial slopes. *Proc. Conf. Biol Civ. Eng.*, Inst. Civil Eng.

Morandi, R. (1960) An underground hall in prestressed concrete. *Concrete Quart.*, **47**, 14.

Morgan, A.V. (1971) Engineering problems caused by fossil permafrost features in the English Midlands. *Quart. J. Eng. Geol.*, **4**, 111.

Morgenstern, N. (1963) Stability charts for earth slopes during rapid drawdown. *Géotechnique*, **13**, 121.

Morgenestern, N.R. (1981) Geotechnical engineering and frontier resource development. *Géotechnique*, **31**, 305.

Morgenstern, N.R. and Eigenbrod, K.D. (1974) Classification of argillaceous soils and rocks. *J. Geotech. Eng. Div. (ASCE)*, **100**, 1137.

Morgenstern, N.R. and Eisenstein, Z. (1970) Methods of estimating lateral loads and deformations. 1970 *Specialty Conf. Lateral Stresses in the Ground and Design of Earth-Retaining Structures*, Soil Mech. Found. Eng. Div. ASCE, p. 51.

Morgenstern, N.R. and Price, V.E. (1965) The analysis of the stability of general slip surfaces. *Géotechnique*, **15**, 79.

Morgenstern, N.R. and Proce, V.E. (1967) A numerical method for solving the equations of stability of general slip surfaces. *Computer J.*, **9**, 388.

Morgenstern, N.R. and Vaughan, P.R. (1963) Some observations on allowable grouting pressures. *Symp. Grouts and Drilling Muds in Eng. Practice*, Inst. Civ. Eng., p. 36.

Morris, C.J.E. *et al.* (1971) Reconstruction of Guildhall Precincts – City of London. *The Struct. Eng.*, **49**, 121.

Morris, D. (1966) Interaction of continuous frames and soil media. *J. Struct. Div. (ASCE)*, **92**, 13.

Morse, R.K. and Thornburn, T.H. (1961) Reliability of soil map units. *Proc. 5th Int. Conf. Soil Mech. Found. Eng.*, **1**, 159.

Mortimore, R.N. (1979) *The Relationship of Stratigraphy and Tectofacies to the Physical Properties of the White Chalk of Sussex*, five volumes, PhD thesis, Brighton Polytechnic Library.

Mortimore, R.N. (1982) The stratigraphy and sedimentation of the Twronian to Campanian in the south province of England. *Zitteliana* (in press).

Mosonyi, E.F. (May, 1973) Estimate of the critical flood discharge. *Civil Eng. Public Works Rev.*, p. 420.

Moulton, L.K. *et al.* (1958) Scour at bridges caused by floods. *Civil Eng. Public Works Rev.*, **53**, 669.

Moum, J. and Rosenqvist, I. Th. (1957) On the weathering of young marine clay. *Proc. 4th Int. Conf. Soil Mech. Found. Eng.*, **1**, 77.

Mozingo, R.R. (1967) General method for beams on elastic supports. *J. Struct. Div.* (*ASCE*), **93** (ST2), 177.

Mueller, R.E. (1982) Multiple hole grouting method. *Conf. Grouting in Geotech. Eng.*, ASCE, New Orleans, p. 792.

Muhs, H. (1965) General discussion on shallow foundations and pavements, Session 4, Div. 3. *Proc. 6th Int. Conf. Soil Mech. Found. Eng.*, **3**, 436.

Muhs, H. and Weiss, K. (1973) Inclined load tests on shallow strip footings. *Proc. 8th Int. Conf. Soil Mech. Found. Eng.*, **1**, 173.

Munsell Color Co. Inc., Baltimore (1954) *Munsell Soil Colour Charts.*

Murray, R.T. (1971) *Embankments Constructed on Soft Foundations: Settlement Study at Avonmouth*, Report LR 419, Transport and Road Research Laboratory.

Murray, R.T. (1973) *Embankments Constructed on Soft Foundations: Settlement Studies near Oxford*, Report LR 538, Transport and Road Research Laboratory.

Murray, R.T. (1978) Developments in two- and three-dimensional consolidation theory; in *Development in Soil Mechanics* (ed. C.R. Scott), Applied Science Publishers, Barking.

Murray, R.T. (1980) Fabric reinforced earth walls: development of design equations. *Ground Eng.*, **13**, 29.

Murray, R.T. and Irwin, M.J. (1981) *A Preliminary Study of TRRL Anchored Earth.* Supplementary Report 674, transport and Road Research Laboratory.

Murray, W.A. and Monkmeyer, P.L. (1973). Validity of Dupuit–Forchheimer equation. *J. Hyd. Div. (ASCE)*, **99**, 1573.

Murrell, S.A.F. (1965) The effect of triaxial stress systems on the strength of rock at atmospheric temperatures. *Geophys. J. Roy. Astr. Soc.*, **10**, 231.

Musannif, A.A.B. (1980). Demolition of concrete buildings. *Proc. Inst. Civil Eng.*, **68** (1), 91.

Muspratt, M.A. (1969). Behaviour of footing slabs. *Build. Sci.*, **4**, 109.

Musso, G. (November, 1979) Jacked pipe provides roof for underground construction in busy urban area. *Civ. Eng. (ASCE)*, **49**, 79.

Mynett, A.E. and Mei, C.C. (1982) Wave-induced stresses in a saturated poro-elastic sea bed beneath a rectangular caisson. *Géotechnique*, **32**, 235.

Nadai, A. (1925) *Die elastischen Platten*, Springer, Berlin.

Nagai, S. (1968) The pressure of partial standing waves. *J. Waterways Harbours Div.* (*ASCE*), **94**, 273.

Naghdi, P.M. (1972) Theory of plates and shells. *Encyclopaedia of Physics*, **6a/2**, p. 425 (ed. C. Truesdell), Springer, Berlin.

Naghdi, P.M. and Rowley, J.C. (1953). On the bending of axially symmetric plates on elastic foundations. *Proc. 1st MidWestern Conf. Solid Mech.* p. 119.

Nahrgang, E. (1974) Verformungsverhalten eines weichen bindigen Untergrundes. *Veröffentlichungen des Inst. für Bodenmechanik u. Telsmechanik, Univ. Karlsruhe, H.,* p. 60.

Nash, W.A. and Ho, F.H. (1960) Finite deflections of a clamped circular plate on an elastic foundation. *6th Cong. Int. Ass. Bridge Struct. Eng*, p. 61.

Nathan, S.S. (1978) Ultimate flexural strength of combined footings. *Proc. Inst. Civil Eng.*, **65** (2), 113.

National Coal Board (1975) *The Subsidence Engineer's Handbook*, NCB London.

Naylor, D.J. (1978) Stress–strain laws for soil; in *Developments in Soil Mechanics* (ed. C.R. Scott, Applied Science Publishers, Barking.

Naylor, D.J. and Jones, D.B. (1973) The prediction of settlement within broad layered fills. *Géotechnique*, **23**, 589.

Nelson, J.D. and Thompson, E.G. (1977). A theory of creep failure in overconsolidated clay. *J. Geotech. Eng. Div. (ASCE)*, **103**, 1281.

Neill, C.R. (1965) Measurements of bridge scour and bed changes in a flooding sand-bed river. *Proc. Inst. Civil Eng.*, **30**, 415.

Neville, A.M. (1963) A study of deterioration of structural concrete made with high-alumina cement. *Proc. Inst. Civil Eng.*, **25**, 287; (1964) **28**, 57.

Neville, A.M. (1966) Failure of a prestressed concrete reservoir. *Proc. Inst. Civil Eng.*, **34**, 335.

Neville, A.M. (1971) *Hardened Concrete: Physical and Mechanical Aspects*, ACI Monograph No. 6, Iowa State University Press, Ames, Iowa.

Nevin, C.M. (1949) *Principles of Structural Geology*, Wiley, New York.

Newberry, C.W. and Eaton, K.J. (1974) *Wind Loading Handbook*, HMSO.

Nevill, D. (1961) A laboratory investigation of two red clays from Kenya. *Géotechnique*, **11**, 302.

Newman, K. (1966) Concrete systems; in *Composite Materials* (ed. L. Holliday), Elsevier, Amsterdam., p. 336.

Newman, K. and Lachance, L., (1964) The testing of brittle materials under unifiorm uniaxial stresses. *Proc. ASTM*, **64**, 1044.

Newmark, N.M. (1942) *Influence Charts for Computation of Stresses in Elastic Foundations*, Bulletin 40, Univ. of Illinois, p. 12.

NHBC (1977) *Registered House-builder's Foundations Manual: Preventing Foundation Failures in New Dwellings*, National House Building Council.

Nichols, T.C. (1980) Rebound, its nature and effect on engineering works. *Quart. J. Eng. Geol.*, **13**, 133.

Nicholson, A.J. (1963) Discussion, Session 2. *Symp. on Grouts and Drilling Muds in Eng. Practice*, Inst. Civil Eng., p. 108.

Nixon, I.K. (1949) Correspondence concerning $\phi = 0$ analysis. *Géotechnique*, **1**, 208, 274.

Nixon, I.K. (1954) Some investigations on granular soils with particular reference to compressed-air sand sampler. *Géotechnique*, **4**, 16.

Noorany, I. and Gizienski, S.F. (1970) Engineering properties of submarine soils: state-of-the-art review. *J. Soil Mech. Found. Div. (ASCE)*, **96**, 1735.

Nordlund, R.L. (1963) Bearing capacity of piles in cohesionless soils. *J. Soil Mech. Found. Div. (ASCE)*, **89** (SM3), 1.

Norlund, R.L. and Deere, D.U. (1970) Collapse of Fargo grain elevator. *J. Soil Mech. Found. Div. (ASCE)*, **96**, 585.

Northwood, R.P. and Sangrey, D.A. (1971) The vane test in organic soils. *Can. Geotech. J.*, **8**, 69.

Ockleston, A.J. (1958) Arching action in reinforced concrete slabs. *The Struct. Eng.*, **36**, 197.

O'Connor, K.A. (1975) Civil engineering aspects of Heysham nuclear power station. *Proc. Inst. Civil Eng.*, **58** (1), 377.

Oehlers, D.J. and Johnson, R.P. (1981) The splitting strength of concrete prisms subjected to surface strip or patch loads. *Mag. Concrete Res.*, **33**, 171.

Ogden, H. and Orchard, R.J. (1959) Ground movements in North Staffordshire. *Trans. Inst. Min. Eng.*, **119**, 259.

Ohde, J. (1942) Die berechnung der Sohldruckverteilung unter Gründungskörpern. *Der Bauingenieur*, **23**, 99.

Olander, H.C. (1970) Design of cylindrical concrete water tanks. *J. Struct. Div. (ASCE)*, **96**, (ST5) 947.

Olsen, K.A. (1958) The re-siting of structures. *Proc. 50th Anniversary Conf. Inst. Struct. Eng.*, p. 364.

Olsen, R.E. and Mesri, G. (1970) Mechanisms controlling compressibility of clays. *J. Soil Mech. Found. Div. (ASCE)*, **96**, 1863.

Olszak, W. (ed.) (1959) Non-homogeneity in elasticity and plasticity. *Proc. Int. Union Theor. Appl. Mechanics Symposium*, Warsaw, 2–9 September 1958, Pergamon Press, New York. p. 1959.

O'Neill, M.W. *et al.* (1977) Analysis of three-dimensional pile groups with nonlinear soil response and pile–soil–pile interaction. *Proc. 9th Offshore Technology Conf.*, **2**, 245.

O'Neill, M.W. (1983) Side load transfer in driven and drilled piles. *J. Geotech. Eng. Div. (ASCE)*, **109**, 1259.

O'Neill, M.W. and Ghazzaly, O.I. (1977) Swell potential related to building performance. *J. Geotech. Eng. Div. (ASCE)*, **103**, 1363.

O'Neill, M.W. and Poormoayed, N. (1980). Methodology for foundations on expansive clays. *J. Geotech. Eng. Div. (ASCE)*, **106**, 1345.

O'Neill, M.W. and Reese, L.C. (1972) Behaviour of bored piles in Beaumont Clay. *J. Soil Mech. Found. Div. (ASCE)*, **98**, 195.

O'Neill, M.W. *et al.* (1982a) Installation of pile group in overconsolidated clay. *J. Geotech. Eng. Div. (ASCE)*, **108**, 1369.

O'Neill, M.W. *et al.* (1982b) Load transfer mechanisms in piles and pile groups. *J. Geotech. Eng. Div. (ASCE)*, **108**, 1605.

O'Neill, M.W. (1983) Side load transfer in driven and drilled piles. *J. Geotech. Eng. (ASCE)*, **109**, 1259.

Orchard, R.J. (1953) Recent developments in predicting the amplitude of mining subsidence. *J. Roy. Inst. Chart. Surv.*, **33**, 864.

Orchard, R.J. (1956) Surface effects of mining – main factors. *Colliery Guardian*, **198**, 159.

Orchard, R.J. (1961) Underground stowing. *Colliery Guardian*, **203**, 258.

Orchard, R.J. and King, H.J. (1959) Ground movement in the exploitation of coal seams. *Colliery Guardian*, **198**, 471.

Ore, E.L. and Straughan, J.J. (1968). Effect of cement hydration on concrete form pressure. *J. ACI*, **65**, 111.

O'Riordan, N.J. (April, 1982) The mobilization of shaft adhesion down a bored, cast-in-situ pile in the Woolwich and Reading Beds. *Ground Eng.*, **15**, 17.

Orme, D.H. and Thorburn, S. (1970) System-built flats on a deep buried quarry. *The Struct. Eng.*, **48**, 5.

O'Rourke, T.D. (1981) Ground movements caused by braced excavations. *J. Geotech. Eng. Div. (ASCE)*, **107**, 1159.

Osborne, B.R. (1970) Site investigation and foundation problems in Bristol. *J. Inst. Mun. Eng.*, **97**, 75.

Ovesen, N.K. (1962) *Cellular Cofferdams, Calculation Methods and Model Tests*, Bulletin 14, Danish Geotechnical Institute.

Oweis, I. and Bowman, J. (1981) Geotechnical consideration for construction in Saudi Arabia. *J. Geotech. Eng. Div. (ASCE)*, **107**, 319.

Packshaw, S. (1945) Discussion on Terzaghi, K. (1944) *Proc. ASCE*, **71**, 541.

Packshaw, S. (1946) Earth pressure and earth resistance. *J. Inst. Civil Eng.*, **25**, 233.

Packshaw, S. (1962) Cofferdams. *Proc. Inst. Civil Eng.* **21**, 367; Discuss., **24**, 91.

Padfield, C.J. and Schafield, A.N. (1983) The development of centrifugal models to study the influence of uplift pressures on the stability of a flood bank. *Géotechnique*, **33**, 57.

Pagan, A.R. (July, 1963) Hydraulic design of small bridges. *Civil Eng.*, p. 35.

Page, A.W. (1979) A non-linear analysis of the composite action of masonry walls on beams. *Proc. Inst. Civil Eng.*, **67** (2), 93.

Page, A.W. (1981) The biaxial compressive strength of brick masonry. *Proc. Inst. Civil Eng.*, **71** (2), 893.

Page, F.A. *et al.* (1981) Design and construction of the Northfleet Hope container terminal, Tilbury Docks. *Proc. Inst. Civil Eng.*, **70** (1), 623.

Palmer, D.J. and Stuart, J.G. (1957) Some observations on the standard penetration test and a correlation of the test with a new penetrometer. *Proc. 4th Int. Conf. Soil Mech. Found. Eng.*, **1**, 231.

Panak, J.J. *et al.* (1972) Slab foundation subjected to complex loads. *J. ACI*, **69**, 630.

Pande, G.N. and Zienkiewicz, D.C. (1980) *Proc. Int. Symp. on Soils under Cyclic and Transient Loading*, Balkema, Rotterdam.

Pannett, R.J. (March, 1974) Port Victoria: Seychelles. *Consulting Eng.*, **38**, 35.

Pariset, E. *et al.* (November, 1966) Formation of ice covers and ice jams on rivers. *J. Hyd. Div. (ASCE)*, **92**, 1.

Park, R. (1964) Tensile membrane behaviour of uniformly loaded rectangular reinforced concrete slabs with fully restrained edges. *Mag. Concrete Res.*, **16**, 39.

Park, R. (1965) The lateral stiffness and strength required to ensure membrane action at the ultimate load of a reinforced concrete slab-and-beam floor. *Mag. Concrete Res.*, **17**, 29.

Parkes, E.W. (1956) A comparison of the contact pressures beneath rough and smooth rafts on an elastic medium. *Géotechnique*, **6**, 183.

Parry, R.H.G. (1971) A direct method of estimating settlements in sands from S.P.T. values. *Proc. Symp. Interaction of Structure and Foundation*, Midland Soil Mech. Found. Eng. Soc., Birmingham, p. 29.

Parry, R.H.G. (ed.) (1972) *Proc. Symp. Stress–Strain Behaviour of Soils*, Foulis, Cambridge.

Parry, R.H.G. (1978a) Interpreting the standard penetration test. *Ground Eng.*, **11** (4), 6.

Parry, R.H.G. (1978b) Estimating foundation settlements in sand from plate bearing tests. *Géotechnique*, **28**, 107; Discuss., **28**, 481.

Parry, R.H.G. (1980) A study of pile capacity for the Heather platform. *Ground Eng.*, **13** (2), 26.

Parry, R.H.G. and Swain, C.W. (1977) Effective stress methods of calculating skin friction on driven piles in soft clay. *Ground Eng.*, **10** (3), 24; Discuss. (1978) **11** (8), 47.

Parsons, A.W. (1967) *Earthworks in Soft Chalk: a Study of Some of the Factors affecting Construction*, Report 112, Transport and Road Research Laboratory.

Parsons, J.D. (1976) New York's glacial lake formation of varved silt and clay. *J. Geotech. Eng. Div. (ASCE)*, **102**, 605.

Parthasarathy, A. and Blyth, F.G.H. (1959) The superficial deposits of the buried valley of the River Devon near Alva, Glackmannan, Scotland. *Proc. Geol. Ass.*, **70**, 33.

Pasternak, P.L. (1954) *On a new method of analysis of an elastic foundation by means of two foundation constants* (in Russian), Gosudarstvennoe Izdatelstro Liberaturi po Stroitelstvu i Arkhitekure, Moscow.

Pasternak, P.L. (1956) Principles of a new method of analysis of rigid and flexible foundations on elastic ground. *Sb. Trud. Mosk. Inzh.-Stroit. Inst.*, No. 14.

Paton, J. *et al.* (1968) Special features of Hamilton By-Pass Motorway (M74). *Proc. Inst. Civil Eng.*, **41**, 247.

Patterson, W.S. *et al.* (1982) *Fatigue Strength of Concrete in Seawater*, Cement and Concrete Ass., Wexham Springs, Slough.

Paulding, B.W. (1970) Coefficient of friction of natural rock surfaces. *J. Soil Mech. Found. Div. (ASCE)*, **96**, 385.

Pauw, A. (1953) *A dynamic analogy for foundation–soil systems. Symposium on Dynamic Testing of Soils*, Special Technical Publication 156, p. 90, ASTM.

Pearson, R. and Money, M.S. (1977) Improvements in the Lugeon or packer permeability test. *Quart. J. Eng. Geol.*, **10**, 221.

Peate, T.S. (1956) Mining subsidence. *J. Inst. Mun. Eng.*, **83**, 341.

Peck, R.B. (1962) Art and science in subsurface engineering. *Géotechnique*, **12**, 60.

Peck, R.B. (1969) Advantages and limitations of the observational method in applied soil mechanics. *Géotechnique*, **19**, 171.

Peck, R.B. and Bazarra, A.R.S.S. (1967) Discussion of paper by D'Appolonia *et al. J. Soil Mech. Found. Div. (ASCE)*, SM3, **93**, 305.

Peck, R.B. and Berman, S. (1948) Measurements of pressures against a deep shaft in plastic clay. *Proc. 2nd Int. Conf. Soil Mech. Found. Eng.*, **3**, 300.

Peck, R.B. and Bryant, F.G. (1953) The bearing capacity failure of the Transcona elevator. *Géotechnique*, **3**, 201.

Peck, R.B. *et al.* (1953) *Foundation Engineering*, Wiley, New York.

Pelletier, J. (1953). The construction of Tignes Dam and Malgovert Tunnel. *Proc. Inst. Civil Eng.*, **2** (2), 480.

Pengelly, C.D. *et al.* (1955) Structural model studies of concrete slab foundations. *J. ACI*, **26**, 961.

Penircioglu, H. (1965) Adjustable box foundation as a measure against excessive settlement and tilting. *Proc. 6th Int. Conf. Soil Mech. Found. Eng.*, **2**, 174.

Penman, A.D.M. (1971) *Rockfill*, CP15/71, Building Research Station.

Penman, A.D.M. (1977) Soil–structure interaction and deformation problems with large oil tanks. *Proc. Int. Symp. Soil–Structure Interaction*, Roorkee, **1**, 521; reproduced as CP14/78, Building Research Station.

Penman, A.D.M. and Charles, J.A. (1971) *Measuring Movements of Engineering Structures*, CP32/71, Building Research Station.

Penman, A.D.M. and Charles, J.A. (1975) *The Quality and Suitability of Rockfill used in Dam Construction*, CP87/75, Building Research Station.

Penman, A.D.M. and Godwin, E.W. (1974) Settlement of experimental houses on land left by opencast mining at Corby. *Conf. Settlement of Structures*, British Geotech. Soc., London, p. 53.

Penman, A.D.M. and Watson, G.H. (1965) The improvement of a tank foundation by the weight of its own test load. *Proc. 6th Int. Conf. Soil Mech. Found. Eng.*, **2**, 169.

Penman, A.D.M. and Watson, G.H. (1967) Foundations for storage tanks on reclaimed land at Teesmouth. *Proc. Inst. Civil Eng.*, **37**, 19.

Penner, E. *et al.* (1973) Floor heave due to biochemical weathering of shale. *Proc. 8th Int. Conf. Soil Mech. Found. Eng.*, **2** (2), 151.

Perlow, M. and Richards, A.F. (1977) Influence of shear velocity on vane shear strength. *J. Geotech. Eng. Div. (ASCE)*, **103**, 19.

Perpich, W.M. *et al.* (1965) Desiccation of soil by trees related to foundation settlement. *Can. Geotech. J.*, **2**, 23.

Perrott, W.E. (1965) British practice for grouting granular soils. *J. Soil Mech. Found. Eng. Div. (ASCE)*, **6**, 57.

Peterson, R. and Peters, H. (1963) Heave of spillway structures on clay shales. *Can. Geotech. J.*, **1**, 5.

Peurifoy, R.L. (December, 1965) Lateral pressure of concrete on formwork. *Civ. Eng. (ASCE)*, **35**, 60.

Pfister and Mattock (May, 1963) *High Strength Bars as Concrete Reinforcement: Part 5, Lapped Splices in Concentrically Loaded Columns*, Portland Cement Association.

Philcox, K.T. (1962) Some recent developments in the design of high buildings in Hong Kong. *The Struct. Eng.*, **40**, 303; (1963) **41**, 171.

Pickett, G. and Janes, W.C. (1953) Bending under lateral load of a circular slab on an elastic solid foundation. *Proc. 1st MidWestern Conf. Solid Mech.*, p. 112.

Pickett, G. and McCormick, F.J. (1951) Circular and rectangular plates under lateral load and supported by an elastic solid foundation. *Proc. 1st US Nat. Cong. Appl. Mech.*, p. 331.

Pickett, G., *et al.* (1951) *Deflections, Moments and Reactive Pressures for Concrete Pavements*, Bulletin 65, Kansas state College.

Pihlainen, J.A. (December, 1959) Pile construction in permafrost. *J. Soil Mech. Found. Div. (ASCE)*, **85**, 75.

Pike, C.W. and Saurin, B.F. (1952) Buoyant foundations in soft clay for oil refinery structures at Grangemouth. *J. Inst. Civil Eng.*, **1**, 301.

Pipes, L.A. (1943) *Applications of the Operational Calculus to the Theory of Structures*, Publication 375, Graduate School of Engineering, Harvard University.

Pippard, A.J.S. and Baker, J.F. (1968) *The Analysis of Engineering Structures*, Arnold, London, p. 211.

Pister, K.S. (1961) Viscoelastic plate on a viscoelastic foundation. *J. Eng. Mech. Div. (ASCE)*, **87** (EM1), 43.

Pister, K.S. and Westmann, R.A. (1962) Bending of plates on an elastic foundation. *J. Appl. Mech., Trans. ASME*, Ser. E, **29** (2), 369.

Pister, K.S. and Williams, M.L. (1960) Bending of plates on a viscoelastic foundation. *J. Eng. Mech. Div. (ASCE)*, **86** (EM5) 31.

PJA (1975) *Jacking Concrete Pipes*, Pipe Jacking Association.

PJA (1981) *A Guide to Pipe Jacking Design*, Pipe Jacking Association.

Polshin, D.E. and Tokar, R.A. (1957) Maximum allowable non-uniform settlement of structures. *Proc. 4th Int. Conf. Soil Mech. Found. Eng.*, **1**, 402.

Popov, E.P. (1951) Successive approximations for beams on an elastic foundation. *Proc. ASCE*, **116**, 1083.

Posey, C.J. (1949) Why bridges fail in floods. *Civil Eng.*, **19**, 94.

Posey, C.J. (May, 1963) Scour at bridge piers – protection of threatened piers. *Civil Eng.*, p. 48.

Potts, D.M. and Burland, J.B. (1983) *A Parametric Study of the Stability of Embedded Earth Retaining Structures*, Special Report 813, Transport and Road Research Laboratory.

Potts, D.M. and Martins, J.P. (1982) The shaft resistance of axially loaded piles in clays. *Géotechnique*, **32**, 369.

Potts, E.L.J. (1949) Ground subsidence from mining. *Engineering*, **168**, 321.

Potyondy, J.G. (1961) Skin friction between various soils and construction materials, *Géotechnique*, **11**, 339.

Poulos, H.G. (1967) Stresses and displacement in an elastic layer underlain by a rough rigid base. *Géotechnique*, **17**, 378.

Poulos, H.G. (1971) Behaviour of laterally loaded piles: I, single piles and II, pile groups, *J. Soil Mech. and Foundn. Div. ASCE*, **97** (SM5) 711.

Poulos, H.G. (1977) Estimation of pile group settlements. *Ground Eng.*, **10** (2), 40.

Poulos, H.G. (1979a) Settlement of single piles in nonhomogeneous soil. *J. Geotech. Eng. Div. (ASCE)*, **105**, 627.

Poulos, H.G. (1979b) Group factors for pile-deflection estimation *J. Geotech. Eng. Div. (ASCE)*, **105**, 1489.

Poulos, H.G. (1980) *User's guide to programme DEFPIG—deformation and analysis of pile groups*, School of Civil Engineering, University of Sydney.

Poulos, H.G. (1982) The influence of shaft length on pile load capacity in clays. *Géotechnique*, **32**, 145.

Poulos, H.G. and Davis, E.H. (1974) *Elastic Solutions for Soil and Rock Mechanics*, Wiley, New York.

Poulos, H.G. and Davis, E.H. (1975) Prediction of downdrag forces in end-bearing piles. *J. Geotech. Eng. Div. (ASCE)*, **101**, 189.

Poulos, H.G. and Davis, E.H. (1980) *Pile Foundation Analysis and Design*, John Wiley & Son, New York.

Poulos, H.G. and Mattes, N.S. (1969) The analysis of downdrag in end-bearing piles. *Proc. 7th Int. Conf. Soil Mech. Found. Eng.*, **2**, 203.

Poulos, H.G. and Randolph, M.F. (1983) Pile group analysis: a study of two methods. *J. Geotech. Eng. Div. (ASCE)*, **109**, 355.

Poulos, H.G. *et al.* (1976) Method of calculating long-term creep settlements. *J. Geotech. Eng. Div. (ASCE)*, **102**, 787.

Prakash, S. and Puri, V.K. (1981) Dynamic properties of soils from *in-situ* tests. *J. Geotech. Eng. Div. (ASCE)*, **107**, 943.

Prakash, S. and Saran, S. (1971) Bearing capacity of eccentrically loaded footings. *J. Soil. Mech. Found. Div. (ASCE)*, **97**, 95.

Premchitt, J. and Brand, E.W. (1981) Pore pressure equalization of piezometers in compressible soils. *Géotechnique*, **31**, 105.

Prevost, J.H. *et al.* (1981a) Offshore gravity structures: centrifugal modelling. *J. Geotech. Eng. Div. (ASCE)*, **107**, 125.

Prevost, J.H. *et al.* (1981b) Offshore gravity structures: analysis. *J. Geotech. Eng. Div. (ASCE)*, **107**, 143.

Price, D.G. (1971) Engineering geology in the urban environment. *Quart. J. Eng. Geol.*, **4** (3), 191.

Price, D.G. and Knill, J.L. (1967) The engineering geology of Edinburgh Castle Rock. *Géotechnique*, **17**, 411.

Price, D.G. *et al.* (1969) Foundation of multi-storey blocks on the coal measures with special reference to old mine workings. *Quart. J. Eng. Geol.*, **1**, 271.

Price, N.J. (1958) A study of rock properties in conditions of triaxial stress; in *Mechanical Properties of Non-Metallic Brittle Materials* (ed. W.H. Walton), Butterworth, London.

Price, N.J. (1966) *Fault and Joint Development in Brittle and Semi-brittle Rock*, Pergamon, London.

Priebe, H. (1976) Abschatzung des Setzungsverhaltens eines durch Stopfverdichtung verbesserten Baugrundes. *Die Bautechnik*, **53**, 160.

Priebe, H. (1978) Abschatzung des Scherwiderstandes eines durch Stopfverdichtung verbesserten Baugrundes. *Die Bautechnik.*, **55**, 281.

Pryke, J.F.S. (1960) Underpinning and jacking buildings affected by mining subsidence or other differential movement. *The Chartered Surveyor*, **92**, 560.

Pryke, J.F.S. (September, 1967) Moving structures. *The Consulting Eng.*, p. 85.

Pryke, J.F.S. (1975) Differential foundation movement of domestic buildings in South East England – Distribution investigation, causes and remedies; in *Settlement of Structures*, British Geotechnical Society, Pentech Press, London, p. 403.

Purkayastha, R.D. and Char, R.A.N. (1977) Stability analysis for eccentrically loaded footings. *J. Geotech. Eng. Div. (ASCE)*, **103**, 647.

Purushothamaraj, P. *et al.* (1974) Bearing capacity of strip footings in two layered cohesive-friction soils. *Can. Geotech. J.*, **11**, 32.

Pynford (1977) *Technical Information*, Pynford Ltd.

Qadar, A. (1981) The vortex scour mechanism at bridge piers. *Proc. Inst. Civil Eng.*, **71** (2), 739.

Quigley, R.M. *et al.* (1973) Oxidation and heave of black shale. *J. Soil Mech. Found. Div. (ASCE)*, **99**, 417.

Ractliffe, A.T. (1983) The basis and essentials of marine corrosion in steel structures. *Proc. Inst. Civil Eng.*, **74**, (1), 899.

Radharkrishna, H.S. and Adams, J.I. (1973) Long-term uplift capacity of augered footings in fissured clay. *Can. Geotech. J.*, **10**, 647.

Raffle, J.F. and Greenwood, D.A. (1961) The relation between rheological characteristics of grouts and their capacity to permeate soil. *Proc. 5th Int. Conf. Soil Mech. Found. Eng.*, p. 789.

Ramakrishnan, V. and Ananthanarayana, Y. (1968) Ultimate strength of deep beams in shear. *J. ACI*, **65**, 87.

Ramaswamy, S.D. *et al.* (1979) Treatment of peaty clay by high energy impact. *J. Geotech. Eng. Div. (ASCE)*, **105**, 957.

Ramelot, C. and Vandeperre, L.J. (1950) A study of foundations for electric power transmission pylons. *C. R. Rech., IRSIA*.

Randolph, M.F. (1980) *PIGLET: A computer programme for the analysis and design of pile groups under general loading conditions*, Cambridge University Engineering Research Report, Soils TR91.

Randolph, M.F. (1981) The response of flexible piles to lateral loading. *Géotechnique*, **31**, 247.

Randolph, M.F. (1981) Piles subjected to torsion. *J. Geotech. Eng. Div. (ASCE)*, **107**, 1095.

Randolph, M.F. (May, 1983) Settlement considerations in the design of axially loaded piles. *Ground Eng.*, **16** (4), 28.

Randolph, M.F. and Wroth, C.P. (1978) Analysis of deformation of vertically loaded piles. *J. Geotech. Eng. Div. (ASCE)*, **104**, 1465.

Randolph, M.F. and Wroth C.P. (1978) A simple approach to pile design and the evaluation of pile tests, *Behaviour of deep foundations*, ed. R. Lundgren, ASTM STP670, 484.

Randolph, M.F. and Wroth, C.P. (1979) An analysis of the vertical deformation of pile groups. *Géotechnique*, **29**, 423.

Randolph, M.F. and Wroth, C.P. (1981) Application of the failure state in undrained simple shear to the shaft capacity of driven piles. *Géotechnique*, **31**, 143.

Randolph, M.F. and Wroth, C.P. (October, 1982) Recent developments in understanding the axial capacity of piles in clay. *Ground Eng.*, **15** (7), 17.

Randolph, M.F., *et al.* (1979) Driven piles in clay – the effects of installation and subsequent consolidations. *Géotechnique*, **29**, 361; Discuss., **31**, 291.

Ranganatham, B.V. and Hendry, A.W. (1963) The ultimate flexural strength of reinforced concrete rafts. *Mag. Concrete, Res.*, **15**, 159.

Rao, B.S.R. and Rao, J.V. (1973) Seepage into sheet pile cofferdam. *J. Hyd. Div. (ASCE)*, **99**, 1515.

Rao, H.B.A (1961) Thé design of machine foundations related to the bulb of pressure. *Proc. 5th Int. Conf. Soil Mech. Found. Eng.*, **1**, 563.

Rao, N.S.V.K. *et al.* (1971) Variational approach to beams on elastic foundations. *J. Eng. Mech. Div. (ASCE)*, **97** (EM2), 271.

Raphael, J.M. and Goodman, R.E. (1979) Strength and deformability of highly fractured rock. *J. Geotech. Eng. Div. (ASCE)*, **105**, 1285.

Rawlinson, J. and Stott, P.F. (1962). The Hammersmith Flyover. *Proc. Inst. Civil Eng.*, **23**, 565; Discuss., **27**, 793.

Ray, K.C. (1958) Influence of lines for pressure distribution under a finite beam on elastic foundation. *J. ACI*, **30**, 729; Discuss., 1459.

Raymond, G.P. (1967) The bearing capacity of large footings and embankments on clays. *Géotechnique*, **17**, 1.

Raymond, S. (1961) Pulverized fuel ash as embankment material. *Proc. Inst. Civil Eng.*, **19**, 515.

Reddy, A.S. and Srinivasan, R.J. (1970) Bearing capacity of footings on anisotropic soils. *J. Soil Mech. Found. Div. (ASCE)*, **96**, 1967.

Reese, L.C. (1978) Design and construction of drilled shafts. *J. Geotech. Eng. Div. (ASCE)*, **104**, 95

Reese, L.C. and O'Neill, M.W. (1969) *Field tests of bored piles in Beaumont Clay*. ASCE Annual Meeting, Chicago.

Regan, P.E. (1974). Design for punching shear. *The Struct. Eng.*, **52**, 197.

Rehnman, S.E. and Broms, B.B (1972) Lateral pressure on basement wall. Results from full-scale tests. *Proc. 5th Europe. Conf. Soil. Mech. Found. Eng.*, p. 189.

Reiding, F.J., Middendorp, P. and Van Brederode, P.J. (1984) A Digital Approach to Sonic Pile Testing, *Proc. 2nd Int. Conf. on the Application of Stress Wave Theory on Piles*, Royal Swedish Academy of Engineering Sciences, Stockholm.

Rein, R.G. *et al.* (1975) Creep of sand–ice system. *J. Geotech. Eng. Div. (ASCE)*, **101**, 115.

Reisser, S.M. and Wright, K.M (1958) The design and construction of the Pelham Bridge, Lincoln. *The Struct. Eng.*, **36**, 399.

Reissmann, H. (1954) Bending of circular and ring shaped plates on an elastic foundation. *J. Appl. Mech., ASME*, **76**, 45.

Reissner, E. (1936) Stationäre, axialsymmetrische durche eine Schüttelnde Masse erregte

Schwingungen eines homogenelastichen Halbraumes. *Ingenieur-Archiv.*, **7** (6), 381.

Reissner, E. (1945) The effect of transverse-shear deformation on the bending of elastic plates. *J. Appl. Mech., Trans. ASME*, **67**, A-69.

Reissner, E. (1955) Stresses in elastic plates over flexible foundations. *J. Eng. Mech. Div. (ASCE)*, **81**, Paper 690.

Reissner, E. (1958) Deflection of plates on viscoelastic foundation. *J. Appl. Mech., Trans. ASME*, **80** (3), 144.

Rejman, W. (1955) Stability of reinforced concrete retaining walls and abutments. *J. Am. Civ. Eng.*, **26**, 1013.

Rennie, I.B. and Fried. P. (1979) An account of the piling problems encountered and the innovative solutions devised during the installation of the main 'A' tower in New Zealand. *Offshore Technology Conference*, Houston, Texas, Paper OTC 3442, 30 April, p. 723.

Reynolds, C.E. (1964) *Reinforced Concrete Design*, 6th edn, Concrete Publications Ltd, London.

Reynolds, C.E. (1972) *Reinforced Concrete Designer's Handbook*, 7th edn, Cement and Concrete Ass. Wexham Springs, Slough.

Reynolds, C.E. and Steedman, J.C. (1974) *Reinforced Concrete Designer's Handbook*, 8th edn, Viewpoint Publications Ltd.

Rhines, W.J. (1969) Elastic plastic foundation model for punch-shear failure. *J. Soil Mech. Div. (ASCE)*, **95** (SM3), 819.

Richart, F.E. (1948) Reinforced concrete wall and column footings. *J. ACI*, **45**, 97, 237.

Richart, F.E., Jr. (1962) Foundation vibrations. *Trans. ASCE*, **127** (1), 863.

Richart, F.E. (1975) Some effects of dynamic soil properties on soil–structure interaction. *J. Geotech. Eng. Div. (ASCE)*, **101**, 1197.

Richart, F.E. and Zia, P. (1962) Effects of local loss of support on foundation design. *J. Soil Mech. Found. Div. (ASCE)*, SM1, Paper 3056.

Richart, F.E. *et al.* (1970) *Vibrations of Soils and Foundations*. Prentice-Hall, New Jersey.

Ridehalgh, H. (1958) Shoreham Harbour development. *Proc. Inst. Civil Eng.*, **11**, 285.

Rigby, C.A. and Dekema (1952) Crack resistant housing. *Public Works S. Africa*, **11**, 95.

Ritchie, A.G.B. (1962). The pressures developed by concrete on formwork. *Civ. Eng. Public. Works Rev.*, **57**, 1027.

Ritchie, J.O.C. (1957) Watertowers. *The Struct. Eng.*, **35**, 14.

Roark, R.J. (1965) *Formulas for Stress and Strain*, 3rd edn, McGraw-Hill, New York.

Roberts, D.V. (1961) Foundations for cylindrical storage tanks. *Proc. 5th Int. Conf. Soil Mech. Found. Eng.*, **1**, 785.

Roberts, D.V. and Darragh, R.D. (1962) Areal fill settlements and building foundation behaviour at the San Francisco Airport. *Field Testing of Soils*, ASTM Special Technical Publication 322, p. 211.

Roberts, P.W. (1960) Adverse weather – Arctic and sub-arctic. *Civil Eng.*, **30**, 44.

Robinsky, E.I. and Bespflug (1973) Design of insulated foundations. *J. Soil Mech. Found. Eng. (ASCE)*, **99**, 649.

Robinson, R.J. (1964) *Piers, Abutments and Formwork for Bridges*, Crosby Lockwood & Son, London.

Rocha, M. (1970) A new method for the determination of deformability in rock masses. *Proc. 2nd Cong. Int. Soc. Rock Mech.*, Beograd, **1**, 423.

Rodin, S. (1952) Pressure of concrete on formwork. *Proc. Inst. Civil Eng.*, **1** (1), 709.

Rodin, S. (1961) Experiences with penetrometer with particular reference to the standard penetration test. *Proc. 5th Int. Conf. Soil Mech. Found. Eng.*, **1**, 517.

Rodin, S. *et al.* (1974) Penetration testing in United Kingdom. *Proc. Europe. Symp. on Penetration Testing.*

Romstad, K.M. *et al.* (1976) Integrated study of reinforced earth – I. Theoretical formulation. *J. Geotech. Eng. Div. (ASCE)*, **102**, 457.

Rosenqvist, I. Th. (1953) Considerations on the sensitivity of Norwegian quick clays. *Géotechnique*, **3**, 195.

Rosenqvist, I. Th. (1965) Fundamental properties of some Norwegian magmatic and metamorphic rocks. *Proc. 6th Int. Conf. Soil Mech. Found. Eng.*, **1**, 109.

RoSPA (1976) *Supervisor's Guide to the Construction Regulations*, Royal Society for the Prevention of Accidents, London.

Ross, K. *et al.* (1972) The new dry dock at Belfast. *Proc. Inst. Civil Eng.*, **51**, 269.

Rossyiskii, V.A. (1961) *Sbornye zhelezobetonnye podpornye stenki*, Gilsa, Kiev.

Rowe, P.W. (1951) Cantilever sheet piling in cohesionless soils. *Engineering*, **172**, 316.

Rowe, P.W. (1952) Anchored sheet-pile walls. *Proc. Inst. Civil Eng.*, **1**, 27.

Rowe, P.W. (1957) Sheet-pile walls in clay. *Proc. Inst. Civil Eng.*, **7**, 629.

Rowe, P.W. (1959) Measurement of the coefficient of consolidation of lacustrine clay. *Géotechnique*, **9**, 107.

Rowe, P.W. (1964) The calculation of the consolidation rates of laminated, varved or layered clays, with particular reference to sand drains. *Géotechnique*, **14**, 321.

Rowe, P.W. (1972) The relevance of soil fabric to site investigation practice. *Géotechnique*, **22**, 195.

Rowe, P.W. and Peaker, K. (1965) Passive earth pressure measurements. *Géotechnique*, **15**, 57.

Rowe, P.W. and Shields, D.H. (1965) The measured horizontal coefficient of consolidation of laminated, layered or varved clays. *Proc. 6th Int. Conf. Soil Mech. Found. Eng.*, **1**, 342.

Rowe, R.E. *et al.* (1965) New concepts in the design of structural concrete. *J. Inst. Struct. Eng.*, **43**, 399; Discuss., **44**, 127, 411.

Rowe, R.R. (1957) Rigid culverts under high overfills. *Trans. ASCE*, **122**, 410.

Rumer, R.R. and Drinker, P.A. (September, 1966) Resistance to laminar flow through porous media. *J. Hyd. Div. (ASCE)*, **92**, 155.

Rutledge, P.C. (January, 1970) Utilization of marginal lands for urban developments. *J. Soil Mech. Found. Div. (ASCE)* **96**, 3.

Rvachev, V.L. (1956) Pressure of a strip-like punch on an elastic half-space. *Prinkl. Math. Mekh* (English Trans. *Mech. and Appl. Math.*), **20**, 2.

Rvachev, V.L. (1958) On bending of an infinite beam on elastic half-space. *Prinkl. Math. Mekh* (English Trans. *Mech. and Appl. Math.*), **22**, 984.

Sadowsky, M. (1928) Zweidimensionale probleme der Elastizitätstheorie. *Z.A.M.M.*, **8**, 107.

Sainsbury, R.N. and King, D. (1971) The flow induced oscillation of marine structures. *Proc. Inst. Civil Eng.*, **49**, 269.

Sainsbury, R.N. and Shipp, S.J.I. (1983) National Westminster Tower: some aspects of construction. *Proc. Inst. Civil Eng.*, **74**, (1), 435.

St John, H.D. (1983) The measurement of performance of offshore piled foundations – a review. *Ground Eng.*, **16** (2), 24.

Saito, A. (1977) Characteristics of penetration resistance of a reclaimed sandy deposit and their change through vibratory compaction. *Soil and Found.*, **17**.

Salas, J.A.J. and Belzunce, J.A. (1965) Résolution théorique de la distribution des forces dans les pieux. *Proc. 6th Int. Conf. Soil Mech. Found. Eng.* **2**, 309.

Salehy, M.R. *et al.* (1977) The occurrence and engineering properties of intraformational shears in Carboniferous rocks. *Proc. Conf. Rock Eng.*, Univ. Newcastle-upon-Tyne, p. 311.

Salmon, C.G. *et al.* (1957) Moment–rotation characteristics of column anchorages. *Trans. ASCE*, **122**, 132.

Salmon, C.G. *et al.* (1964) Laboratory investigation of unstiffened triangular bracket plates. *Proc. ASCE Struct. Div.*, **90**, 257.

Samuels, R. (January, 1958). Controlled vibration in blasting at close quarters. *Civil Eng.*, p. 35.

Samuels, S.G. and Cheney, J.E. (1974) Long-term heave of a building on clay due to tree removal. *Conf. Settlement of Structures*, British Geotech. Soc., London, p. 212.

Sandegren, E. *et al.* (1972) Behaviour of anchored sheet-pile wall exposed to frost action. *Proc. 5th Europe. Conf. Soil Mech. Found. Eng.*, p. 285.

Sanglerat, G. (1972) The penetrometer and soil exploration; in *The Interpretation of Penetration Diagrams – Theory and Practice*, Elsevier, Amsterdam.

Saunders, M.K. and Fookes, P.G. (1970) A review of the relationship of rock weathering and climate and its significance to foundation engineering. *Eng. Geol.*, **4**, 289.

Saurin, B.F. (1949) Correspondence concerning $\phi = 0$ analysis. *Géotechnique*, **1**, 272.

Savin, G.N. (1940) On the additional pressure transferred by the foot of a perfectly rigid stamp onto an elastic anisotropic base caused by a neighbouring load (in Russian). *Dokl. Akad. Nauk SSSR*, No. 7.

Savran, M. (1962) Moment–load charts for symmetrical footings subjected to combined bending and axial load. *J. ACI*, **34**, 73, 1263.

Sawko, F. (1972) A note on the computer analysis of foundation rafts resting on plastic soil: a technical note. *The Struct. Eng.*, **50**, 171.

Scherman, K.A. (1969) Discussion on Session A. *Proc. Conf. In-Situ Investigations in Soils and Rocks*. British Geotech. Soc., London, p. 50.

Schiffman, R.L. (1963) The viscoelastic compression of soil–water systems. *Proc. 4th Int. Cong. Rheology.*

Schiffman, R.L. *et al.* (1964) The secondary consolidation of clay. *IUTAM Symp. Rheology and Soil Mech.*, Grenoble (eds J. Kravtchenkoand P.M. Sirieys), Springer-Verlag, Berlin, p. 273.

Schosser, F. (1978) Research on reinforced earth: mechanism, behaviour and design methods. *1st Colombian Geotechnical Seminar*, (*Socjedad Colombiana de Geotecnia*), Bogota, September 1978.

Schmertmann, J.H. (1969) *Dutch Friction-Cone Penetrometer Exploration of Research Area at Field 5, Eglin Air Force Base, Florida*, Contract Rep. S69-4, US Army Eng. Waterways Exp. Station, Vicksburg, Miss.

Schmertmann, J.H. (1970) Static cone to compute static elastic settlement over sand. *J. Soil Mech. Found. Div. (ASCE)*, **96**, 1011.

Schmertmann, J.H. (1979) Statics of SPT. *J. Geotech. Eng. Div. (ASCE)*, **105**, 655.

Schmertmann, J.H. and Crapps, D.K. (1980) Slope effects on house shrink–swell movements. *J. Geotech. Eng. Div. (ASCE)*, **106**, 1327.

Schneebeli, G. (1964) Le stabilité des trenchées profoundes forées en présence de boue. *Houille Blanche*, **19**, 815.

Schroeder, W.L. and Bynington, M. (1972) Experiences with compaction of hydraulic fills. *Proc. 10th Annual Eng. Geol. Soil Eng. Symp.*, Moscow, Idaho.

Schroeder, W.L. and Maitland, J.K. (1979) Cellular bulkheads and cofferdams. *J. Geotech. Eng. Div. (ASCE)*, **105**, 823.

Schroeder, W.L. *et al.* (1977) Performance of a cellular wharf. *J. Geotech. Eng. Div. (ASCE)*, **103**, 153.

Schultze, E. (1961) Distribution of stress beneath a rigid foundation. *Proc. 5th Int. Conf. Soil Mech. Found. Eng.*, **1**, 807.

SCI (1980) *Proc. Conf. on Reclamation of Contaminated Land*, Society of Chemical Industry.

Scott, C.R. (ed.) (1978) *Developments in Soil Mechanics*, Applied Science Publishers, Barking.

Scott, P.A. and Roberts, G. (1958) The Volta Bridge. *Proc. Inst. Civil Eng.*, **9**, 395.

Scott, R.A. (1963) Fundamental considerations governing the penetrability of grouts and their ultimate resistance to displacement. *Symp. on Grouts and Drilling Muds in Eng. Practice*, Inst. Civil Eng., p. 10.

Scott, R.A. and Pearce, R.W. (1976) Soil compaction by impact. *Symp. Ground Treatment by Deep Compaction*, Inst. Civil Eng., p. 19.

Scott, R.F. and Ko, H.-Y. (1969) Stress-deformation and strength characteristics. *Proc. 7th Int. Conf. Soil Mech. Found. Eng.*, State-of-the-Art Volume, p. 1.

Scriven, W.E. and Pilgrim, W.R. (1965) Deflexions and bending moments produced by a load distributed over a small area of an infinite circular slab on an elastic foundation. *The Struct. Eng.*, **43**, 345.

Sedykh, E.K. (1964) On the diagram of reactive pressures, under the base of a rigid foundation. *J. Soil Mech. Found. Eng. (ASCE)*, **3**, 153.

Seed, H.B. (1965) Settlement analyses, a review of theory and testing procedures. *J. Soil Mech. Found. Div. (ASCE)*, **91** (SM2), 39.

Seed, H.B. and Chan, C.K. (October, 1959) Structure and strength characteristics of compacted clays. *J. Soil Mech. Found. Div. (ASCE)*, **85**, 87.

Seed, H.B. and Whitman, R.V. (1970) Design of earth retaining structures for dynamic loads. *1970 Specialty Conf. Lateral Stresses in the Ground and Design of Earth-Retaining Structures*, ASCE, p. 103.

Seed, H.B. *et al.* (July, 1964a) Clay mineralogical aspects of the Atterberg limits. *J. Soil Mech. Found. Div. (ASCE)* **90**, 107.

Seed, H.B. *et al.* (November, 1964b). Fundamental aspects of the Atterberg limits. *J. Soil Mech. Found. Div. (ASCE)*, **90**, 75.

Selvadurai, A.P.S. (1973) Bending of an infinite beam resting on a porous elastic medium. *Géotechnique*, **23**, (3), 407.

Selvadurai, A.P.S. (1976) The response of a rigid circular plate resting on an idealized elastic-plastic foundation. *Int. J. Mech. Sci*, **18**, 463.

Selvadurai, A.P.S. (1977a) Axisymmetric flexure of an infinite plate resting on a finitely deformed imcompressible elastic halfspace. *Int. J. Solids Struct.*, **13**, 357.

Selvadurai, A.P.S. (1977b) The influence of a Mindlin force on the axisymmetric interaction between an infinite plate and an elastic halfspace. *Lett. Appl. Eng. Sci.*, **5**, 379.

Selvadurai, A.P.S. (1978) The time dependent response of a deep rigid anchor in a viscoelastic medium. *Int. J. Rock Mech. Min. Sci. Geomech. Abstr.*, **15**, 11.

Selvadurai, A.P.S. (1979a) The interaction between a uniformly loaded circular plate and an isotropic elastic halfspace: A variational approach. *J. Struct. Mech.*, **7**, 231.

Selvadurai, A.P.S. (1979b) Elastic analysis of soil–foundation interaction. *Developments in Geotechnical Engineering*, vol. **17**, Elsevier, Amsterdam.

Selvadurai, A.P.S. (1980a) Elastic contact between a flexible circular plate and a transversely isotropic elastic halfspace. *Int. J. Solids Struct.*, **16**, 167.

Selvadurai, A.P.S. (1980b) The flexure of a thick circular raft resting on an isotropic elastic medium. *Proc. Int. Conf. Structural Foundations on Rock*, Sydney, **1**, 161.

Selvaduari, A.P.S. and Adjeleian, J. (1977) Axisymmetric interaction of a thick plate and an internally loaded elastic halfspace. *CANCAM 77, 6th Can. Cong. Appl. Mech.*, Vancouver, BC, p. 25.

Selvadurai, A.P.S. and Kempthorne, R.H. (1980) Plane strain contact stress distribution beneath a rigid footing resting on a soft clay. *Can. Geotech. J.*, **17**, 114.

Selvadurai, A.P.S. and Moutafis, N. (1975) Some generalized results for an orthotropic elastic quarter-plane. *Appl. Sci. Res.*, **30**, 295.

Selvadurai, A.P.S. and Nicholas, T.J. (1979) A theoretical assessment of the screw plate test. *Proc. 3rd Int. Conf. Numerical Methods in Geomech.*, Aachen, A.A. Balkema,. **3**, 1245.

Selvadurai, A.P.S. *et al.* (1980) Screw plate testing of a soft clay. *Can. Geotech. J.*, **17**, 465.

1050 *References*

Semple, R.M. (1981) Partial coefficient design in geotechnics. *Ground Eng.*, **14** (6), 47.

Serafim, J.L. and Lopes, J.J.B. (1961) *In-situ* shear tests and triaxial tests of foundation rocks of concrete dams. *Proc. 5th Int. Conf. Soil Mech. Found. Eng.*, **1**, 533., Dunod, Paris.

Serota, S. (May, 1972) Site investigation rigs: design and use. *Ground Eng.*, **5**, 18.

Serota, S. and Jennings, R.A.J. (1959) The elastic heave of the bottom of excavations. *Géotechnique*, **9**, 62, 145.

Severn, R.T. (1962) The deformation of a rectangular slab with free edges under transverse loads. *Mag. Concrete Res.* **14**, 33.

Severn, R.T. (1966) The solution of foundation mat problems by finite-element methods. *The Struct. Eng.*, **44**, 223; Correspondence (1967) **45**, 215.

Shalon, R. and Raphael, R. (1959) Influence of sea water on corrosion of reinforcement. *J. ACI*, **55**, 1251.

Shannon, W.L. and Strazer, R.J. (March, 1970) Tied-back excavation wall for Seattle First National Bank. *Civ. Eng. (ASCE)*, **40**, 62.

Shelson, W. (1957) Bearing capacity of concrete. *J. ACI*, **29**, 405.

Shen, C.K. *et al.* (1976) Integrated study of reinforced earth – II. Behaviour and design. *J. Geotech. Eng. Div. (ASCE)*, **102**. 577.

Shen, C.K. *et al.* (1981a) Ground movement analysis of earth support system. *J. Geotech. Eng. Div. (ASCE)*, **107**, 1609.

Shen, C.K. *et al.* (1981b) Field measurements of an earth support system. *J. Geotech. Eng. Div. (ASCE)*, **107**, 1625.

Shen, H.W. *et al.* (1969) Local scour round bridge piers. *J. Hyd. Div. (ASCE)*, **95**, 1919.

Sherif, M.M and Mackey, R.D. (1977) Pressures on retaining wall with repeated loading. *J. Geotech. Eng. Div. (ASCE)*, **103**, 1341.

Sherwood, P.T. (1975a) *The Use of Waste and Low-Grade Materials in Road Construction:* 2. *Colliery Shale*, Report 649, Transport and Road Research Laboratory.

Sherwood, P.T. (1975b) *The Use of Waste and Low-Grade Materials in Road Construction:* 3. *Pulverised Fuel Ash*, Report 686, Transport and Road Research Laboratory.

Sherwood, P.T. and Ryley, M.D. (1966) *The Use of Stabilized Pulverized Fuel Ash in Road Construction: a Laboratory Investigation*, Report 49, Transport and Road Research Laboratory.

Shevehenko, L.K. (June, 1972) Stroitelstvo Prichalov na Kam A Ze. *Transportnoye Stroitelstvo*, p. 16.

Shibasaki, M. and Ohta, S. (1982) A unique underpinning of soil solidification utilizing super-high pressure liquid jet. *Conf. Grouting in Geotech. Eng.*, ASCE, New Orleans, p. 680.

Shirkov, V.N. *et al.* (1971) A circular plate on a non-linearly deforming base. *Proc. 4th Conf. Soil Mech. Found. Engng.*, Budapest, p. 757.

Shockley, W.G. (1950) Field test determines foundation modulus for Bayou Bodcau Dam. *Civil Eng.*, **20**, 446.

Shtaerman, I.Ia. (1956) Distribution of pressure under the foundation caused by the presence of a plastic zone. (in Russian). *Sb. Trud. Mosk. Inzh.-Stroit. Inst.*, No. 14.

Siegel, C. (1962) *Structure And Form In Modern Architecture*, Crosby Lockwood and Son, London.

Siegel, R.A. *et al.* (1981) Random surface generation in stability analysis. *J. Geotech. Eng. Div. (ASCE)*, **107**, 996.

Siemonsen, F. (1948) *Die Lastaufnahmekräfte im Baugrund und ihre Auswirkung auf die spannungen in einem Fundament. Abhandlungen über Bodenmechanik und Grundbau,* Eirich Schmidt Verlag, Berlin. p. 120.

Sigvaldason, O.T. (1966) The influence of testing machine characteristics upon the cube and cylinder strength of concrete. *Mag. Concrete Res.*, **18**, 197.

Sikso, H.A. and Johnson, C.V. (1964) Pressure cell observations – Garrison Dam project. *J. Soil Mech. Found. Div. (ASCE)*, **90**, (SM5) 157.

Sills, G.C. *et al.* (1977) Behaviour of an anchored diaphragm wall in stiff clay. *Proc. 9th Int. Conf. Soil Mech. Found. Eng.*, **2**, 147.

Simons, N.E. and Menzies, B.K. (1978) The long-term stability of cuttings and natural clay slopes; in *Developments in Soil Mechanics – 1.* (ed. C.R. Scott), Applied Science Publishers, Barking.

Simpson, A.G. (1976) Foundation settlements at London Bridge. *Ground. Eng.*, **9**, (7).

Simpson, B. *et al.* (1979) A computer model for the analysis of ground movements in London clay. *Géotechnique*, **29**, 149.

Simpson, B., *et al.* (1981) An approach to limit state calculations in geotechnics. *Ground Eng.*, **14** (6), 21.

Sims, F.A. and Bridle, R.J. (August, 1964) The design of concrete hinges. *Concrete and Constr. Eng.*, p. 276.

Sims, F.A. and Bridle, R.J. (November, 1966) Bridge design in areas of mining subsidence. *J. Inst. Highway Eng.*, p. 19.

Singh, A. and Mitchell, J.K. (1968) General stress–strain–time function for soils. *J. Soil Mech. Found. Div. (ASCE)*, **94**, 21.

Singh, A. and Mitchell, J.K. (1969) Creep potential and creep rupture of soils. *Proc. 7th Int. Conf. Soil Mech. Found. Eng.*, **1**, 379.

Singh, J.P. *et al.* (1977) Design of machine foundations on piles. *J. Geotech. Eng. Div. (ASCE)*, **103**, 863.

Sinha, S.N. (1963) Large deflections of plates on elastic foundations. *J. Eng. Mech. Div. (ASCE)*, **89**, 1.

Skempton, A.W. (1942) An investigation of the bearing capacity of a soft clay soil. *J. Inst. Civil Eng.*, **18**, 307.

Skempton, A.W. (1944) Notes on the compressibility of clays. *Quart. J. Geol. Soc.*, **100**, 119.

Skempton, A.W. (1945) A slip in the west bank of the Eau Brink Cut. *J. Inst. Civil Eng.*, **24**, 267.

Skempton, A.W. (1946) Earth pressure and the stability of slopes; in *The Principles and Application of Soil Mechanics*, Inst. Civil Eng.

Skempton, A.W. (1948a) The geotechnical properties of a deep stratum of post-Glacial clay at Gosport. *Proc. 2nd Int. Conf. Soil Mech. Found. Eng.*, **2**, 145.

Skempton, A.W. (1948b) Vane tests in the alluvial plain of the River Forth. *Géotechnique*, **1**, 111.

Skempton, A.W. (1951) The bearing capacity of clays. *Proc. Building Res. Cong.*, London, **1**, 180.

Skempton, A.W. (1954a) A foundation failure due to clay shrinkage caused by poplar trees. *Proc. Inst. Civil Eng.*, **3** (1), 66.

Skempton, A.W. (1954b) The pore pressure coefficients *A* and *B*. *Géotechnique*, **4**, 143.

Skempton, A.W. *et al.* (1955a) Settlement analyses of six structures in Chicago and London. *J. Inst. Civil Eng.*, **4**, (1), 525.

Skempton, A.W. (1955b) Foundations for high buildings. *Proc. Inst. Civil Eng.*, **4** (3), 246.

Skempton, A.W. (1957) Discussion: The planning and design of the new Hong Kong airport. *Proc. Inst. Civil Eng.*, **7** (2), 275.

Skempton, A.W. (1959) Cast-*in-situ* bored piles in London clay. *Géotechnique*, **9**, 153.

Skempton, A.W. (1964) Long-term stability of clay slopes. *Géotechnique*, **14**, 77.

Skempton, A.W. (1966) *Symposium on Large Bored Piles*, Inst. Civil Eng.

Skempton, A.W. (1968) Discussion on Mangla. *Proc. Inst. Civil Eng.*, **41**, 133.

Skempton, A.W. (1970a) The consolidation of clays by gravitational compaction. *Quart. J. Geol. Soc.*, **125**, 373.

Skempton, A.W. (1970b) First-time slides in over-consolidated clays. *Géotechnique*, **20**, 320.

Skempton, A.W. and Bishop, A.W. (1950) The measurement of the shear strength of soils. *Géotechnique*, **2**, 90.

Skempton, A.W. and Bjerrum, L. (1957) A contribution to the settlement analysis of foundations on clays. *Géotechnique*, **7**, 168.

Skempton, A.W. and Henkel, D.J. (1953) The post-Glacial clays of the Thames Estuary at Tilbury and Shellhaven. *Proc. 3rd Int. Conf. Soil Mech. Found. Eng.*, **1**, 302.

Skempton, A.W. and Hutchinson, J.N. (1969) Stability of natural slopes and embankment foundations. *Proc. 7th Int. Conf. Soil Mech. Found. Eng.*, State-of-the-Art Volume, p. 291.

Skempton, A.W. and La Rochelle, P. (1965) The Bradwell slip: a short-term failure in London Clay. *Géotechnique*, **15**, 221.

Skempton, A.W. and MacDonald, D.H. (1956) The allowable settlements of buildings. *Proc. Inst. Civil Eng.*, **5** (3), 727.

Skempton, A.W. and Northey, R.D. (1952) The sensitivity of clays *Géotechnique*. **2**, 30.

Skempton, A.W. and Sowa, V.A. (1963) The behaviour of saturated clays during sampling and testing. *Géotechnique*, **13**, 269.

Skempton, A.W. and Ward, W.H. (1953) Investigations concerning a deep cofferdam in the Thames estuary clay at Shellhaven. Géotechnique, **3**, 248, 346.

Skempton, A.W. *et al.* (1955) Settlement analysis of six structures in Chicago and London. J. Inst. Civil Eng. **4** (1), 525.

Skempton, A.W. *et al.* (1969) Joints and fissures in the London Clay at Wraysburg and Edgware. *Géotechnique*, **19**, 205; (1970) **20**, 208.

Skipp, B.O. (September, 1961) Corrosion and site investigation. *Corrosion Technol.*, **8**, 269.

Sliwinski, Z.J. (1973) Bentonite control in diaphragm walling. *Construction News*, Piling and Foundation Supplement, 13 December.

Sliwinski, Z.J. and Fleming, W.G.K. (1984) *The integrity and performance of bored piles.* Piling and Ground Treatment Conference, ICE, London.

Smeardon, R.F.J. *et al.* (1967) Engineering works at Tilbury Docks: 1963–67. *Proc. Inst. Civil Eng.*, **38**, 177.

Smith, B.S. and Rahman, K.M.K. (1972) The variations of stress in vertically loaded brickwork walls. *Proc. Inst. Civil Eng.*, **51**, 689.

Smith, B.S. and Riddington, J.R. (1978) The design of masonry infilled steel frames for bracing structures. *The Struct. Eng.*, **56B**, 1.

Smith, E.A.L. (1962) Pile driving analysis by the wave equation. *Trans. ASCE*, **127** (1), 1145.

Smith, G.N. (1981) Probability theory in geotechnics – an introduction. *Ground Eng.*, **14** (7), 29.

Smith, H. (1954) The welded structure in a pump-house foundation. *Brit. Welding J.*, **1**, 13, 223.

Smith, I.M. and Molenkamp, F. (1980) Dynamic displacements of offshore structures due to low frequency sinusoidal loading. *Géotechnique*, **30**, 179.

Smith, J.W. and Zar, M. (1964) Chimney foundations. *J. ACI*, **36**, 673; Discuss., 1657.

Smith, M.A. *et al.* (1982) Colloquium: reclamation of contaminated land for building. *The Struct. Eng.*, **60A**, 5.

Smith, P.H. (1962) Field trials on fly ash. *Contract J.* 20 September, p. 319.

Smith, P.H. (July, 1968) The use of PFA in grouting and foundation work. *Ground Eng.* **1**, 21.

Smith, P. *et al.* (1970) Guide to joint sealants for concrete structures. *J. ACI*, **67**, 489.

Smolira, M. (1973) Analysis of infilled shear walls. *Proc. Inst. Civil Eng.*, **55** (2), 895.

Smoltczyk, H.U. (1967) Stress computation in soil media. *J. Soil Mech. Found. Div. (ASCE)*, **93** (SM2), 101.

Smoots, V.A. and Benton, P.H. (August, 1961) Compacted earth fill for a power-plant foundation. *Civil Eng.*, p. 54.

Smorodinov, M.I. *et al.* (1967) *Pile Driving Equipment*, Chapter 3: *Vibro-pile Drivers and Hammers*.

Sneddon, I.A. (1946) Boussinesq's problem for a flat-ended cylinder. *Proc. Camb. Phil. Soc.*, **62**, 29.

Sneddon, I.N. *et al.* (1975) Bonded contact of an infinite plate and an elastic foundation. *Lett. Appl. Eng. Sci.*, **3**, 1.

Sogge, R.L. (1981) Laterally loaded pile design. *J. Geotech. Eng. Div. (ASCE)* **107**, 1179.

Sokolovsky, V.V. (1960) *Statics of Granular Media*, Pergamon, Oxford.

Sokolovski, V.V. (1965) *Statics of Soil Media* (3rd edn, translated from the Russian), Butterworths, London.

Somerville, A.L. (November, 1955) The holding power of bolts in concrete. *Chartered Civil Eng.*, p. 43.

Sommer, H. (1965) A method for the calculation of settlements, contact pressures and bending moments in a foundation including the influence of flexural rigidity of the superstructure. *Proc. 6th Int. Conf. Soil Mech. Found. Eng.* **2**, 197.

Sorota, M.D. and Kinner, E.B. (1981) Cellular cofferdam for Trident drydock: design. *J. Geotech Eng. Div. (ASCE)*, **107**, 1643.

Sorota, M.D. *et al.* (1981) Cellular cofferdam for Trident drydock: performance. *J. Geotech. Eng. Div. (ASCE)* **107**, 1657.

Sowa, V.A. (1970) Pulling capacity of concrete cast *in situ* bored piles. *Can. Geotech. J.*, **7**.

Sowers, G.F. (1962) Shallow Foundations, in *Foundation Engineering* (ed. G.A. Leonards, McGraw-Hill, New York.

Sowers, G.F. (September, 1964) Fill settlement despite vertical sand drains. *J. Soil Mech. Found. Div. (ASCE)*, **90**, 289.

Sowers, G.F. (1975) Failures in limestone in humid subtropics. *J. Geotech. Eng. Div. (ASCE)*, **101**, 771.

Sowers, G.F. and Kennedy, C.M. (*c.* 1953) *Effect of Repeated Load Application on Soil Compaction Efficiency*, Georgia Inst. of Technology, Atlanta.

Sowers, G.F. *et al.* (1957) The residual lateral pressures produced by compaction. *Proc. 4th Int. Conf. Soil Mech. Found. Eng.*, **2**, 243.

Sowers, G.F. *et al.* (1965) Compressibility of broken rock and the settlement of rockfills. *Proc. 6th Int. Conf. Soil Mech. Found. Eng.*, **2**, 561.

Sparks, P.R. and Menzies, J.B. (1973) The effect of rate of loading upon the static and fatigue strengths of plain concrete in compression. *Mag, Concrete Res.*, **25**, 73.

Spencer, A.J.M. (1965) The solution of plane elastic/plastic problems by relaxation methods. *Appl. Sci. Res.*, **12A**, 391.

Spencer, C.B. (1941) Steel-frame building moved by new method. *Civ. Eng. (ASCE)*, **11**, 659.

Sreekantiah, H.R. (1982) Rocking vibrations of footings. *J. Geotech. Eng. Div. (ASCE)*, **108**, 905.

Sridharan, A. and Rao, A.S. (1982) Mechanisms controlling the secondary compression of clays. *Géotechnique*, **32**, 249.

Stagg, K.G. and Treharne, G. (1970) Some experiments on the ultimate bearing capacity of rock-type materials. *Proc. 2nd Cong. Int. Soc. Rock Mech.*, **1**, 517.

Stagg, K.G. and Zienkiewicz, O.C. (Eds). (1968) *Rock Mechanics in Engineering Practice*, Wiley, London.

Steedman, C.W. (1940) The structural design of sewers. *J. Inst. Mun. Eng.,* **66**, 701.

Steffen, O.K.H. *et al.* (1975) Recent developments in the interpretation of data from joint surveys in rock masses. *6th Regional Conf. for Africa on Soil Mech. Found. Eng.,* **2**, 17

Steffens, R.J. (1964) *Symp. Application of Prestressed Concrete to Machinery Structures,* Prestressed Concrete Development Group, London.

Stepanov, E.N. (1972) Zhelezobetonnye podpornye stenki. *Transportnoye Stroitelstvo,* **6**, 38.

Stetzler, B.U. (1982) Mechanical behaviour of silicate grouted soils. *Conf. Grouting. in Geotech. Eng.,* ASCE, New Orleans, p. 498.

Stewart, M. and Beaven, P.J. (1980) *Seismic Refraction Surveys for Highway Engineering Purposes,* Laboratory Report 950, Transport and Road Research Laboratory.

Stewart, R. (1885) Discussion on 'Some notes on subsidence and draw' by J.S. Dixon. *Trans. Min. Inst. Scotland,* **7**.

Stott, J.P. (1961) Tests on material for use in sliding layers under concrete road slabs. *Civil Eng. Public Works Rev.,* **56**, 1297.

Stow, G.R.S. (1962) Modern water-well drilling techniques in use in the United Kingdom. *Proc. Inst. Civil Eng.,* **23**, 1; Discuss. (1963) **25**, 218.

Stroud, M.A. (1974) The Standard Penetration Test in Clays and Soft Rocks, *Proc. European Symposium on Penetration Testing,* Stockholm.

Stuart, J.G. (1962) Interference between foundations with special reference to surface footings on sand. *Géotechnique,* **12**, 15.

Stuart, J.G. and Hanna, T.H. (1961) Groups of deep foundations: a theoretical and experimental investigation. *Proc. 5th Int. Conf. Soil Mech. Found. Eng.,* **2**, 149.

Suklje, L. (1969) *Rheological Aspects of Soils Mechanics,* Interscience, London.

Suklje, L. and Vidmar, S. (1973) Critical loads depending on layer thickness. *Proc. 8th Int. Conf. Soil Mech. Found. Eng.* **1** (3), 253.

Suter, G.T. and Hendry, A.W. (1975) Shear strength of reinforced brickwork beams. *The Struct. Eng.,* **53**, 249.

Sutherland, H.B. (1963) The use of *in-situ* tests to estimate the allowable bearing pressure of cohesionless soils. *The Struct. Eng.,* **41**, 85.

Sutherland, H.B. (1965) Model studies for shaft raising through cohesionless soils. *Proc. 6th Int. Conf. Soil Mech. Found. Eng.,* Montreal, **2**, 410.

Sutherland, H.B. (General Reporter) (1974) *Conf. Settlement of Structures,* Review Paper, Session I, University Press, Cambridge.

Sutherland, H.B. (1975) Granular materials: Review Paper. *Proc. Conf. Settlement of Structures,* Cambridge, p. 473.

Sutherland. H.B. and Gaskin, P.N. (1970) Factors affecting the frost susceptibility characteristics of pulverized fuel ash. *Can. Geotech. J.,* **7**, 69.

Sutherland, H.B. and Lindsay, J.A. (1961) The measurement of load distribution under two adjacent column footings. *Proc. 5th Int. Conf. Soil Mech. Found. Eng.,* **1**, 829.

Sutherland, H.B. *et al.* (June, 1968) Engineering and related properties of pulverised fuel ash. *J. Inst. Highway Eng.,* **15**, 19.

Svec, O.J. and Gladwell, G.M.L. (1973) A triangular plate bending element for contact problems. *Int. J. Solids Struct.,* **9** (3), 435.

Swain, R.J. (November, 1960) Bottom of San Francisco Bay evaluated for trans-bay tube. *Civil Eng.,* p. 66.

Swatek, E.P. (1967) Cellular cofferdam design and practice. *J. Waterways Harbours Div. (ASCE),* **93**, 109.

Swiger, W.F. (June, 1957) Prestressed sheetpile underpinning. *Civ. Eng. (ASCE),* **27**, 66.

Swiger, W.F. and Estes, H.M. (October, 1959) Major power station foundation in broken limestone. *J. Soil Mech. Found. Div. (ASCE),* **85**, 77.

Sykes, J.F. *et al.* (1974) Finite element permafrost thaw settlement model. *J. Geotech. Eng. Div. (ASCE),* **100**, 1185.

Symons, I.F. (1968) *The Application of Residual Shear Strength to the Design of Cuttings in*

Overconsolidated Fissured Clays. Report LR227, Transport and Road Research Laboratory.

Symons, I.F. (1983) Assessing the stability of a propped, *in-situ* retaining wall in overconsolidated clay. *Proc. Inst. Civil. Eng.,* **75**, (2)617.

Szava-Kovats, L.J. (1967) Design of combined footings using support reaction and moment influence lines of continuous beams on elastic supports. *J. ACI,* **64**, 312.

Szilard, R. (1974) *Theory and Analysis of Plates; Classical and Numerical Methods,* Prentice-Hall, Englewood Cliffs, New Jersey.

Talbot, A.N. (1913) *Reinforced Concrete Wall and Column Footings,* Bulletin 67, University of Illinois, Engineering Experiment Station.

Talbot, W.J. (Chairman) (1973) Recommendations for design, manufacture and installation of concrete piles, Report by ACI Committee 543. *J. ACI,* **70**, 509.

Tang, W.H. (1981) Probabilistic evaluation of loads. *J. Geotech Eng. Div. (ASCE),* **107**, 287.

Tattersall, F. (1958) Design, construction and maintenance of service reservoirs. *J. Inst. Water Eng.,* **13**, 21.

Tattersall, F. *et al.* (1955) Investigation into the design of pressure tunnels in London Clay. *Proc. Inst. Civil Eng.,* **4**, 400.

Tavenas, F. and La Rochelle, P. (1972) Accuracy of relative density measurements. *Géotechnique,* **22**, 549; **23**, 301.

Taylor, D.W. (1937) Stability of earth slopes. *J. Boston Soc. Civil Eng.,* **24**, 197.

Taylor, D.W. (1948) *Fundamentals of Soil Mechanics,* Wiley, New York.

Taylor, H.P.J. and Clarke, J.L. (1976). Some detailing problems in concrete frame structures. *The Struct. Eng.,* **54**, 19.

Taylor, M.E. (1965) *Rubber Bridge Bearings – a Survey of Present Knowledge,* Laboratory Note LN/822, Road Research Laboratory.

Taylor, M.E. (1972) *PTFE in Highway Bridge Bearings,* Report LR 491, Transport and Road Research Laboratory.

Taylor, R.L. and Brown, C.B. (March, 1967) Darcy flow solutions with a free surface. *J. Hyd. Div. (ASCE),* **93**, 25.

Taylor, R. *et al.* (1966) Effect of the arrangement of reinforcement on the behaviour of reinforced concrete slabs. *Mag. Concrete Res.,* **18**, 85.

Tchalenko, J.S. (1968) The microstructure of London Clay. *Quart. J. Eng. Geol.,* **1**, 155.

Tedd, P. and Charles, J.A. (1981) *In situ* measurement of horizontal stress in overconsolidated clay using push-in spade-shaped pressure cells. *Géotechnique,* **31**, 554.

Teng, C.Y. (1949) Determination of the contact pressure against a large raft foundation. *Géotechnique,* **1**, 222.

Teng, W.C. (1962) *Foundation Design,* Prentice-Hall, Englewood Cliffs, New Jersey.

Terzaghi, K. (1934) Large retaining wall test. *Eng. News Record,* 1 February, 22 February, 8 March, 20 March, 19 April.

Terzaghi, K. (1935) The actual factor of safety in foundations. *The Struct. Eng.,* **13**, 126.

Terzaghi, K. (1943) *Theoretical Soil Mechanics,* Wiley, New York.

Terzaghi, K. (1944) Stability and stiffness of cellular cofferdams. *Proc. ASCE,* **70**, 1015; Discuss., 1625; (1945) **71**, 225, 367, 541, 741, 966.

Terzaghi, K. (1945) *Rock Defects and Loads on Tunnel Supports,* Publication 418, Graduate School of Engineering, Harvard University.

Terzaghi, K. (1946) *Stress Conditions for the Failure of Saturated Concrete and Rock,* Publication 425, Graduate School of Engineering, Harvard University.

Terzaghi, K. (1954) Anchored bulkheads. *Trans. ASCE,* **119**, 1243.

Terzaghi, K. (1955) Evaluation of coefficients of subgrade reaction. *Géotechnique,* **5**, 197; **6**, 94.

Terzaghi, K. (1962a) Measurement of stresses in rock. *Géotechnique.* **12**, 105, 352; (1963) **13**, 96.

Terzaghi, K. (1962b) Stability of steep slopes on hard unweathered rock. *Géotechnique,* **12**, 251.

Terzaghi, K. and Peck, R.B. (1967) *Soil Mechanics in Engineering Practice* (2nd edn), Wiley, New York.

Terzaghi, R.D. (1965) Sources of error in joint surveys. *Géotechnique*, **15**, 287.

Tetior, A.N. and Litvinenko, A.G. (1971). Foundations for tower-like structures. *Proc. 4th Conf. Soil Mech. Found. Eng.*, Budapest, p. 819.

Thoft-Christensen, P. (ed.) (1982) *Probabilistic Characterization of Soil Properties: Bridge between Theory and Practice*, NATO Advanced Study Institute, Series E, Applied Sciences No. 70, Martinus Nijhoff, The Hague.

Thomas, D. (1965) Static penetration tests in London Clay. *Géotechnique*, **15**, 174; **16**, 76.

Thomas, D. (1968) Deep sounding test results and the settlement of spread footings on normally consolidated sands. *Géotechnique*, **18**, 472; **19**, 316.

Thomas, E. (1962) Stabilization of rock by bolting, In *Reviews in Engineering Geology* (eds T. Fluhr and R.F. Legget), **1**, 257, Geological Society of America.

Thomas, F.G. (1953) The strength of brickwork. *The Struct. Eng.*, **31**, 35.

Thompson, T.V. (1966) Building high in a typhoon zone – Hang Seng Bank, Hong Kong. *The Struct. Eng.*, **44**, 171.

Thoms, R.L. (1960) Experimental study of beams on elastic foundations. *J. Eng. Mech. Div. (ASCE)*, **86**, 107.

Thon, J.G. and Coltrin, G.L. (1958) Morro Bay steam electric plant. *Trans. ASCE*, **123**, 207.

Thorburn, S. (1963) Tentative correction chart for the standard penetration test in non-cohesive soils. *Civil Eng. Public Works. Rev.*, **58**, 752.

Thorburn, S. (1970) Discussion on Thorburn, S. and MacVicar, R.S.L. (1970) *Conf. on Behaviour of Piles*, Inst. Civil Eng., p. 54.

Thorburn, S. (1976). The static penetration test and the ultimate resistance of driven piles in fine-grained non-cohesive soils. *The Struct. Eng.*, **54**, 205.

Thorburn, S. and Beevers, C. (1981) Lateral displacement of low embankments on alluvium containing thin peat layers. *Proc. Inst. Civil Eng.*, **70** (1), 277.

Thorburn, S. and Thorburn, J.Q. (1977) *Review of problems associated with the construction of cast in place piles*. Rept PG2, CIRIA, London.

Thorburn, S. *et al.* (1981) The importance of the stress histories of cohesive soils and the cone penetration test. *The Struct. Eng.*, **59A**, 87.

Thorburn, T.H. *et al.* (1970) *Engineering Soil Report: Dewitt County*, Illinois, Bulletin 505, University of Illinois Engineering Experiment Station.

Thornburn, T.H. and Larsen, W.R. (October, 1959). A statistical study of soil sampling. *J. Soil Mech. Found. Div. (ASCE)*, **85**, 1.

Thornton, D.L. (1951) *Mechanics Applied to Vibrations and Balancing*, Chapman and Hall, London.

Thorogood, R.P. (1975) *Accuracy of in-situ concrete*, CP99/75, Building Research Station.

Thrower, E.H. (1968, 1971) *Calculations of Stresses and Displacements in a Layered Elastic Structure: Parts 1 and 2*, Reports LR160 and LR373, Transport and Road Research Laboratory.

Tickell, R.G. (1977) Continuous random wave loading on structural members. *The Struct. Eng.*, **55**, 209.

Tickell, R.G. *et al.* (1976) Long-term wave loading on offshore structures. *Proc. Inst. Civil Eng.*, **61** (2), 145.

Timoshenko, S. (1941) *Strength of Materials*, vol, 2, Van Nostrand, New York.

Timoshenko, S. and Goodier, J.N. (1970) *Theory of Elasticity*, McGraw-Hill, New York.

Timoshenko, S. and Woinowsky-Krieger, S. (1959) *Theory of Plates and Shells* (Engineering Society Monographs, 2nd edn), McGraw-Hill, New York, p. 143.

Timoshenko, S.P. and Young, D.H. (1965) *Theory of Structures*, McGraw-Hill, New York.

Ting, E.C. (1973) Unified formulation of two-parameter foundation model. *Z.A.M.M*, **53**, 636.

Tison, L.J. (1940) Erosion autour de piles de ponts en Rivière. *Annales des Travaux Publics de Belgique*, **41**, 813.

Todd, D.K. (1980) *Ground Water Hydrology*, Wiley, New York.

Tomlinson, K. (1957) The adhesion of piles driven in clay soil. *Proc. 4th Int. Conf. Soil Mech. Found. Eng.*, London, p. 66.

Tomlinson, M.J. (1954) Site investigation for maritime and river works. *Proc. Inst. Civil Eng.*, **3** (2), 225.

Tomlinson, M.J. (1970) Some effects of pile driving on skin friction. *Conf. Behaviour of Piles*, London, Inst. Civil Eng.

Tomlinson, M.J. (1975) *Foundation Design and Construction*, Pitman, London.

Tomlinson, M.J. (1976) Preface: Piles in weak rock. *Géotechnique*, **26**, 1.

Tomlinson, M.J. and Francis, H.W.A. (1963) Site investigation for motorways. *J. Inst. Highway Eng.*, **20** (6), 23.

Tomlinson, M.J. *et al.* (1961) Symposium on large diameter bored piles. *Reinforced Concrete Rev.*, **5**, 673.

Tomlinson, M.J. *et al.* (1978) Foundations for low-rise buildings. *The Struct. Eng.*, **56A**, 161; Discuss. (1982) **60A**, 242; reprinted as CP 61/78, Building Research Establishment.

Toms, A.H. (1949) The effect of vegetation on the stabilization of artificial slopes. *Proc. Conf. Biol. Civ. Eng.*, Inst. Civil Eng., p. 99.

Toms, A.H. (1966) Chalk in cuttings and embankments. *Proc. Symp. Chalk in Earthworks and Foundations*, Inst. Civil Eng., p. 43.

Toolan, F.E. and Fox, D.A. (1977). Geotechnical planning of piled foundations for offshore platforms. *Proc. Inst. Civil Eng.*, **62** (1), 221; Discuss. (1978) **64**, 261.

Toombs, A.F. *et al.* (1982) *Ground Movements Caused by Deep Trench Construction in an Urban Area*, Laboratory Report 1040, Transport and Road Research Laboratory.

Tornaghi, R. (1978a) *Iniezioni, Seminario su Consolidamento di Terreni e Rocce in Posto nell'ingegneria Civile*, Stresa, p. 342.

Tornaghi, R. (1978b) *Iniezioni, Seminario su Consolidamento di Terreni e Rocce in Posto nell'ingegneria Civile*, Stresa, p. 315.

Touma, F.T. and Reese, L.C. (1974) Behaviour of bored piles in sand. *J. Geotech. Eng. Div. (ASCE)*, **100**, 749.

Trow, W. and Bradstock, J. (1972) Instrumented foundations for two 43-storey buildings on till, Metropolitan Toronto. *Can. Geotech. J.*, **9**, 290.

TRRL (1974 to 1978). *The Use of Low-Grade Materials in Road Construction*: Sherwood, P.T. (1974) *Guide to Materials Available*, LR647; Sherwood, P.T. (1975) *Colliery Shale*, LR649; Sherwood, P.T. (1975) *Pulverised Fuel Ash*, LR686; Roe, P.G. (1976) *Incinerated Refuse*, LR728; Tubey, L.W. (1978) *China Clay Sand*, LR817; Burns, J. (1978) *Spent Oil Shale*, LR818; Sherwood, P.T. *et al.* (1977) *Miscellaneous Wastes*, LR819. Transport and Road Research Laboratory.

TRRL (1978a) *Settlement and Stability of Earth Embankments on Soft Foundations*, Supplementary Report 399, Transport and Road Research Laboratory.

TRRL (1978b) *Terrain Evaluation for Highway Engineering and Transport Planning*, Supplementary Report 448, Transport and Road Research Laboratory.

TRRL (1979) *Reinforced Earth and Other Composite Soil Techniques*, Proceedings of symposium, Special Report 457, Transport and Road Research Laboratory.

Tsai, N.C. and Westmann, R.E. (1967) Beams on tensionless foundation. *J. Eng. Mech. Div. (ASCE)*, **93** (EM5), 1.

Tschebotarioff, G.P. (1951) *Soil Mechanics, Foundations and Earth Structures*, McGraw-Hill, New York.

Tschebotarioff, C.P. (1965) Analysis of a high crib wall failure. *Proc. 6th Int. Conf. Soil Mech. Found. Eng.*, **2**, 414.

Tschebotarioff, G.P. and Ward, E.R. (1948) The resonance of machine foundations and the soil coefficients which affect it. *Proc. 2nd Int. Conf. Soil Mech. Found. Eng.*, **1**, 309.

Tsinker, G.P. (1983) Anchored sheet pile bulkheads: design practice. *J. Geotech. Eng.*, **109**, 1021.

Tsytovich, N.A. *et al.* (1970) The problem of determining the modulus of deformation of *in situ* rock masses. *Proc. 2nd Cong. Int. Soc. Rock. Mech.*, **1**, 301.

Tucker, R.L. and Poor, A.R. (1978) Field study of moisture effects on slab movements. *J. Geotech. Eng. Div. (ASCE)*, **104**, 403.

Tunstall, M.J. (December, 1973) Vibration of chimneys by vortex shedding. *Concrete*, p. 41.

Turnbull, W.J. and Mansur, C.I. (1973) Compaction of hydraulically placed fills. *J. Soil Mech. Found. Div. (ASCE)*, **99**, 939.

Turnbull, W.J. *et al.* (1961) Stresses and deflections in homogeneous soil masses. *Proc. 5th Int. Conf. Soil Mech. Found. Eng.*, **2**, 337.

Turner, A.E. (July, 1962) Uplift resistance of transmission tower footings. *J. Power Div. (ASCE)*, **88**, 17.

Twine, M.E. (1980) The diver in the offshore construction industry. *Proc. Inst. Civil Eng.*, **68** (1), 455.

Tyrrell, F.C. (November, 1959) The big dish. *Civil Eng.*, p. 56.

UEG (1975) *The Principles of Safe Diving Practice, 1975, Part 1: General principles and air diving*, Report UR2, Underwater Engineering Group, CIRIA.

Ueshita, K. and Meyerhof, G.G. (1967) Deflection of multi-layer soil systems. *J. Soil Mech. Found. Div. (ASCE)*, **93** (SM5), 257.

Umansky, A.A. (1933) *Analysis of Beams on Elastic Foundation* (in Russian), Central Research Institute of Auto-Transportation, Leningrad.

Underwood, L.B. *et al.* (March, 1964) Rebound in redesign of Oahe Dam hydraulic structures. *J. Soil Mech. Found. Div. (ASCE)*, **90**, 65.

United States Steel Export Co. Design extracts from former catalogs – USS sheet piling.

Uppal, J.Y. and Kemp, K.O. (1973) An instability theory of failure for concrete. *Mag. Concrete Res.*, **25**, 21.

US Bureau of Reclamation (1960) *Earth Manual*, US Government Printing Office.

US Navy (1971) *Soil Mechanics, Foundation and Earth Structures*, NAVFAC OM7 Design Manual.

Valera, J.E. and Donovan, N.C. (1977) Soil liquefaction procedures – a review. *J. Geotech. Eng. Div. (ASCE)*, **103**, 607.

Vallabhan, C.V. and Alikhanlov, F. (1982) Short rigid piers in clays. *J. Geotech. Eng. Div. (ASCE)*, **108**, 1255.

Vallejo, L.E. (1979) An explanation for mudflows. *Géotechnique*, **29**, 351.

Van de Graaf, H.C. and Smits, A.P. (1983). Offshore site investigation by rotary drilling from a diving bell. *Ground Eng.*, **16** (1), 19.

Van der Veen, C. (1953) Loading tests on concrete slabs at Schibol Airport. *Proc. 3rd Int. Conf. Soil Mech. Found. Eng.*, **2**, 133.

Van der Veen, C. and Boersma, L. (1957) The bearing capacity of a pile predetermined by a cone penetration test. *Proc. 4th Int. Conf. Soil Mech. Found. Eng.*, **2**, 72, Butterworth, London.

Van Eekelen, H.A.M. and Potts, D.M. (1978) The behaviour of Drammen Clay under cyclic loading. *Géotechnique*, **28**, 173.

Varga, L. (1965) The compressibility of loess soils. *Proc. 6th Int. Conf. Soil Mech. Found. Eng.*, **1**, 395.

Vasiljev, B.D. (1955) *Osnovaniya i Fundamenty*, Gilsa, Moscow.

Vaughan, P.R. (1969) A note on sealing piezometers in boreholes. *Géotechnique*, **19**, 405.

Vaughan, P.R. (Chairman) (1979) *Proc. Conf. Clay Fills*, Inst. Civil Eng.

Vaughan, P.R. and Walbancke, H.J. (1973) Pore pressure changes and the delayed failure of cutting slopes in overconsolidated clays. *Géotechnique*, **23**, 531.

Vaughan, P.R. *et al.* (1983) Squeeze grouting of stiff-fissured clay after a tunnel collapse. *Proc. 8th Europe. Conf. Soil Mech. Found. Eng.*, Helsinki, **1**, 171, Balkema, Rotterdam.

Vavasour, P. and Wilson, J.S. (1966) Cumberland Basin Bridges Scheme: Part 1, Planning and design. *Proc. Inst. Civil Eng.*, **33**, 261.

Veder, C. (1963) Excavation of trenches in the presence of bentonite suspensions for the construction of impermeable and load bearing diaphragms. *Proc. Symp. Grouts and Drilling Muds in Engineering Practice*, Butterworths, London, p. 181.

Venter, J. *et al.* (1954) *The Protection of Buildings against Earth Movements*, N.C.B Translation A978/DJS – unpublished.

Verdeyen, J. and Nuyens, J. (1965) Calcul des rideaux d'ancrage de palplanches. *Proc. 6th Int. Conf. Soil Mech. Found. Eng.*, **2**, 417.

Vesić, A.B. (1961a) Bending of beams resting on isotropic elastic solid. *J. Eng. Mech. Div. (ASCE)*, **87**, 35.

Vesić, A.B. (1961b) Beams on elastic subgrade and Winkler's hypothesis. *Proc. 5th Int. Conf. Soil Mech. Found. Eng.*, **1**, 845.

Vesić, A.B. and Johnson, W.H. (1963) Model studies of beams resting on a silt subgrade. *J. Soil Mech. Found. Div. (ASCE)*, **89**, 1.

Vesić, A.S. (1969) *Experiments with Instrumented Pile Groups in Sand*, Special Tech. Publication 444, ASTM. p. 177.

Vesić, A.S. (1971) Breakout resistance of objects embedded in ocean bottom. *J. Soil Mech. Found. Div. (ASCE)*, **97**, 1183.

Vesić, A.S. (1972) Expansion of cavities in infinite soil mass. *Proc. ASCE*, **98** (SM3), 265.

Vesić, A.S. and Barksdale, R.D. (1963) *Theoretical Studies and Cratering Mechanisms Affecting the Stability of Cratered Slopes*, Final Report, Project A-655, Engineering Experiment Station, Georgia Institute of Technology, Atlanta, Ga, p. 103.

Vesić, A.S. *et al.* (1965) An experimental study of dynamic bearing capacity of footings on sand. *Proc. 6th Int. Conf. Soil Mech. Found. Eng.*, **2**, 209.

Vesić, A.S. (1973). Analysis of ultimate loads of shallow foundations. *J. Soil Mech. Found. Div. (ASCE)*, **99**, 45.

Vetter, C.P. (1938) Design of pile foundations. *J. ASCE*, **64**, 311.

Vick, E.H. *et al.* (1965) Southern Outfall Works of the London County Council. *Proc. Inst. Civil Eng.*, **30**, 369; (1966) **33**, 711.

Vidal, H. (1966) La terre armée. *Ann. Inst. Tech. Bâtim. Trav. Publics.*

Vijayvergiya, V.N. and Focht, J.A. (1972) A new way to predict the capacity of piles in clay. *4th Annual Offshore Technical Conf.*, Houston, Texas.

Vlazov, V.Z. (1949). *Structural Mechanics of Thin Walled Three-dimensional Systems* (in Russian), Gosstroiizdat, Moscow, Leningrad.

Vlazov, V.Z. and Leontiev, U.N. (1966) *Beams, Plates and Shells on Elastic Foundations* (translated from the Russian), Israel Program for Scientific Translations, Jerusalem.

Voight, B. (1973) Correlation between Atterberg plasticity limits and residual shear strength of natural soils. *Ge{otechnique*, **23**, 265; Discuss., **23**, 600.

Voight, B. (1978, 1980) *Rockslides and Avalanches*, vol. 1: *Natural Phenomena*; vol. 2: *Engineering Sites*, Elsevier, Amsterdam.

Voight, B. and Faust, C. (1982) Frictional heat and strength loss in some rapid landslides. *Géotechnique*, **32**, 43.

Voight, B. and Panseall, W. (1970) State of predictive art in subsidence engineering. *J. Soil Mech. Found. Div. (ASCE)*, **96**, 721.

Volterra, E. (1947) *Sul problema generale della piastra poggiata su suolo elastico*, Publicazioni dell'Instituto per le applicazioni del Calculo No. 201, Consiglio Nazionale delle Ricerche, Rome.

Volterra, E. and Chung, R. (1955) Constrained circular beams on elastic foundations. *Trans. ASCE*, **120**, 301.

Wagner, A.A. (1957) The use of the Unified Soil Classification System by the Bureau of

Reclamation. *Proc. 4th Int. Conf. Soil Mech. Found Eng.*, **1**, 125.

Wakeling, T.R.M. (1970) A comparison of the results of standard site investigation methods against the results of a detailed geotechnical investigation in the Middle Chalk at Mundford, Norfolk. *Conf. In-situ Investigations in Soils and Rocks*, Inst. Civil Eng., p. 17.

Wakeling, T.R.M. and Jennings, R.A. (1976). Some unusual structures in the river gravels of the Thames Basin. *Quart. J. Eng. Geol.*, **9**, 255.

Wakeling, T.R.M. *et al.* (1983) The influence vegetation on the swelling and shrinkage of clays. *Géotechnique*, **33**, 87; Discuss. (1984) **34**, 139.

Waldorf, Q.A. *et al.* (July 1963) Foundation modulus tests for Karadj Arch Dam. *J. Soil Mech. Found. Div. (ASCE)* **89**, 91.

Walker, L.K. and Darvall, P. le P. (1973) Dragdown on coated and uncoated piles. *Proc. 8th Int. Conf. Soil Mech. Found. Eng.*, **2** (1), 257.

Waller, R.A. (1971) The design of tall structures with particular reference to vibration. *Proc. Inst. Civil Eng.*, **48**, 303.

Walmsley, J. (1981) Service reservoir design, with particular reference to economics and the 'simply supported wall' concept. *The Struct. Eng.*, **59A**, 285.

Wang, W.L. and Yen, B.C. (1974) Soil arching in slopes. *J. Geotech. Eng. Div.* (ASCE), **100**, 61.

Ward, A.M. and Bateson, E. (1947) The new Howrah Bridge, Calcutta: design of the structure, foundations and approaches. *J. Inst. Civil Eng.*, **28**, 167.

Ward, J.C. (September, 1964) Turbulent flow in porous media. *J. Soil Mech. Found. Div. (ASCE)*, **90**, 1.

Ward, L.E. (1953) The design and construction of a three bay aluminimum aircraft hangar at London Airport. *The Struct. Eng.*, **31**, 103.

Ward, W.H. (1948) A coastal landslip. *Proc. 2nd Int. Conf. Soil Mech. Found. Eng.*, **2**, 19.

Ward, W.H. (1957) The use of simple relief wells in reducing water pressure beneath a trench excavation. *Géotechnique*, **7**, 134.

Ward, W.H. (1971) *Some Field Techniques for Improving Site Investigation and Engineering Design*, CP30/71, Building Research Station.

Ward, W.H. and Burland, J.B. (1973) *The Use of Ground Strain Measurements in Civil Engineering*, CP13/73, Building Research Establishment.

Ward, W.H. *et al.* (1965) Properties of the London Clay at the Ashford Common shaft: in-*situ* and undrained strength tests. *Géotechnique*, **15**, 321.

Ward, W.H. *et al.* (1968) Geotechnical assessment of a site at Mundford, Norfolk, for a large proton accelerator. *Géotechnique*, **18**, 399.

Wardell, K. (1953) Some observations on the relationship between time and mining subsidence. *Trans. Inst. Min. Eng.*, **113**, 471.

Wardell, K. (1954) Mining subsidence. *J. Roy. Inst. Chart. Surv.*, **33**, 53.

Wardell, K. (1957) Minimisation of surface damage. Colliery Eng., **34**, 361.

Warkentin, B.P. and Bozozuk, M. (1961) Shrinkage and swelling properties of two Canadian clays. *Proc. 5th Int. Conf. Soil Mech. Found. Eng.*, **1**, 851.

Warner, J. (1982) Compaction grouting – the first 30 years. *Conf. Grouting in Geotech. Eng.*, ASCE, New Orleans, p. 694.

Wasilkowski, I.F. (1956) Complete protection of structures against damage due to mining subsidence. *Civ. Struct. Eng. Rev.*, **10**, 81.

Waterhouse, R.W. and Sills, A.N. (1952) Thaw-blast method prepares permafrost foundation for Alaska plant. *Civil Eng.*, **22**, 126.

Watson, G.N. (1944) *A Treatise on the Theory of Bessel Functions*, Cambridge University Press.

Watt, A.J. *et al.* (1967) Loading tests on structures founded on soft cohesive soils strengthened by compacted granular columns. *Proc. 3rd Asian Conf. Soil Mech. Found Eng.*, Haifa, **1**, 248.

Weber, W.G. (1969) Performance of embankments constructed over peat. *J. Soil Mech. Found. Div. (ASCE)*, **95**, 53.

Webster, R. and Beckett, P.H.T. (1970) Terrain classification and evaluation using air photography: a review of recent work at Oxford. *Photogrammetria*, **26**, 51.

Weissman, G.F. (1965) Measuring the modulus of subgrade soil reaction. *Mat. Res. Stand.*, **5**, 71.

Weissman, G.F. (1972) Tilting foundations. *J. Soil Mech. Found. Div. (ASCE)*, **98** (SM1), 59.

Weissman, G.F. and White, S.R. (1961) Small angular deflections of rigid foundations. *Géotechnique*, **11**, 186.

Weitsman, Y. (1970) On foundations that react in compression only. *J. Appl. Mech.*, p. 1019.

Weltman, A.J. (ed.) (1980) *Noise and Vibration from Piling Operations*, CIRIA Report PG9/PSA Civil Engineering Technical Guide 32, CIRIA and Property Services Agency.

Weltman, A.J. and Healy, P.R. (1978) *Piling in Boulder Clay and other glacial fills*, CIRIA Report PG5.

West, G. and Dumbleton, M.J. (1975) *An Assessment of Geophysics in Site Investigation for Roads in Britain*, Report LR680, Transport and Road Research Laboratory.

West, H.W.H. *et al.* (1977) The resistance of brickwork to lateral loading: experimental methods and results of tests on small specimens and full sized walls *The Struct. Eng.*, **55**, 411.

West, J.M. and Stuart, J.G. (1965) Oblique loading resulting from interference between surface footings on sand. *Proc. 6th Int. Conf. Soil Mech. Found. Eng.*, **2**, 214.

West, R.E. (1971) The new Manchester Road Bridge in the Port of London. *Proc. Inst. Civil Eng.*, **48**, 161.

Westergaard, H.M. (1918) The resistance of pile groups. *Eng. Construction*, 22 May.

Westergaard, H.M. (1938) *A Problem of Elasticity Suggested by a Problem in Soil Mechanics*, Timoshenko Anniversary Volume, MacMillan, New York.

Westergaard, H.M. (1943) Stress concentration in plates loaded over small areas. *Trans. ASCE*, **108**, 831.

Westergaard, H.M. (1948) New formulas for stress in concrete pavements of airfields. *Trans. ASCE*, **113**, 425.

Whetton, J.T. (1957) A general survey of the ground movement problem. *The Engineer*, **203**, 562.

Whitaker, T. (1957) Experiments with model piles in groups. *Géotechnique*, **7**, 147.

Whitaker, T. (1970) *The Design of Piled Foundations*, Pergamon Press, Oxford.

Whitaker, T. and Cooke, R.W. (1961) A new approach to pile testing. *Proc. 5th Int.Conf. Soil Mech. Found. Eng.*, **2**, 171.

Whitaker, T. and Cooke, R.W. (1965) Bored piles with enlarged bases in London Clay. *Proc. 6th Int. Conf. Soil Mech. Found. Eng.*, **2**, 342.

Whitaker, T. and Cooke, R.W. (1966) An investigation of the shaft and base resistance of large diameter piles in London Clay, Paper 1. *Symp. on Large Bored Piles*, Inst. Civil Eng.

White, A. *et al.* (1961) Field study of a cellular bulkhead. *J. Soil Mech. Found. Div. (ASCE)*, **87**, 89.

White, B.G. *et al.* (1961) The design and construction of Bombay Marine Oil Terminal. *Proc. Inst. Civil Eng.*, **18**, 121.

White, L.S. (1953) Transcona elevator failure: eye-witness account. *Géotechnique*, **3**, 209.

White, L.S. and Gardner, G.A. (1950) Government Offices, Whitehall Gardens: the special problem of re-siting of an historic building. *J. Inst. Civil Eng.*, **34**, 222.

Whitman, R.V. (1970) Hydraulic fills to support structural loads. *J. Soil Mech. Found Div. (ASCE)*, **96**, 23.

Whitman, R.V. (1984) Evaluating calculated risk in geotechnical engineering. *J. Geotech. Eng.*, **110**, 145.

Whitney, C.S. (1957) Ultimate shear strength of reinforced concrete flat slabs, footings, beams and frame members without shear reinforcement. *J.ACI*, **29**, 265.

Wiegel, R.L. (Ed.) (1970) *Earthquake Engineering*, Prentice-Hall, Englewood Cliffs, New Jersey.

Wijk, G. (1979a) *Some New Aspects of the Brazilian Test*, Report DS 1979: 5, Swedish Detonic Research Foundation.

Wijk, G. (1979b) *The Effect of the Sample Shape on the Point Load Test Strength Index*, Report DS 1979: 16, Swedish Detonic Research Foundation.

Wijk, G. and Hirengen, P. (1979) *Sclerograph Measurements on Rock Materials*, Report DS 1979: 6, Swedish Detonic Research Foundation.

Wild, P.A. and Haslam, E.F. (July 1962) Towers and foundations for project EHV. *J. Power Div. (ASCE)*, **88**, 69.

Wilkinson, W.B. (1967) Correspondence on Gibson (1966). *Géotechnique*, **17**, 68.

Wilkinson, W.B. (1968) Constant head permeability tests in clay strata. *Géotechnique*, **18**, 172.

Williams, G.M.J. (1957) The design of the foundation of the Shell Building, London. *Proc. 4th Int. Conf. Soil Mech. Found. Eng.*, **1**, 457.

Williams, G.M.J. and Rutter, P.A. (1967) The design of two buildings with suspended structures in high yield steel. *The Struct. Eng.*, **45**, 143.

Williams, H. and Stothard, J.N. (1967) Rock excavation and specification trials for the Lancashire–Yorkshire Motorway, Yorkshire (West Riding) Section. *Proc. Inst. Civil Eng.*, **36**, 607.

Williams, I.E. (January, 1960) Exceptional job engineering pays off on bridge construction. *Civil Eng.*, p. 60.

Wilson, E.J. (1975) Engineering geology and site investigation. Part 3: Investigation methods. *Ground Eng.*, **8** (5), 21.

Wilson, G. (1941) The calculation of the bearing capacity of footings on clay. *J. Inst. Civil Eng.*, **17**, 87.

Wilson, N.E. (1962) Stabilization of a tilting foundation. *The Struct. Eng.*, **40**, 113.

Wilson, W.S. and Sully, F.W. (1949) *Compressed-air Caisson Foundations*, Works Construction Div., Paper 13, Inst. Civil Eng.

Wilson, W.S. and Sully, F.W. (1952) The construction of the caisson forming the foundation to the circulating-water pump-house for the Uskmouth Generating Station. *Proc. Inst. Civil Eng.*, **1** (3), 335.

Wiłun, Z. and Starzewski, K. (1972) *Soil Mechanics in Foundation Engineering, Vol. 1: Properties of Soils and Site Investigations*, Intertext, London.

Wiłun, Z. and Starzewski, K. (1975) *Soil Mechanics in Foundation Engineering*, 2nd edn, Vol. 2, Surrey University Press.

Windle, D. and Wroth, C.P. (1977) The use of a self-boring pressuremeter to determine the undrained properties of clays. *Ground Eng.*, **10** (6), 37.

Winkler, E. (1867) *Die lehre von der elastizitat unde festigkeit*, H. Dominicus, Prague.

Wiss, J.F. (1967) Damage effects of pile driving vibrations. *Highway Res. Board*, **155**.

Wiss, J.F. (1981) Construction vibrations: state-of-the art. *J. Geotech. Eng. Div. (ASCE)*, **107**, 167.

Wood, A.M.M. (1956) Folkestone Warren landslips: investigations 1948–50. *Proc. Inst. Civil Eng.*, **4** (2), 410.

Wood, A.M.M. (1959) Correspondence on Skempton and Bjerrum (1957). *Géotechnique*, **9**, 29.

Wood, I.R. (1981) Report of Symposium on vertical drains. *Géotechnique*, **31**, 1.

Wood, L.A. and Larnach, W.J. (1974) The effects of soil–structure interaction on raft foundations. *Proc. Conference on Settlement of Structures*, Cambridge.

Wood, R.H. (1952) *The Composite Action of Brick Panel Walls Supported on Reinforced Concrete Beams*, Research Paper 13, National Building Studies, HMSO.

Wood, R.H. (1958). The stability of tall buildings. *Proc. Inst. Civil Eng.*, **11**, 69.

Wood, R.H. and Simms, L.G. (1969) *A Tentative Design Method for the Composite Action of Heavily Loaded Brick Panel Walls Supported on Reinforced Concrete Beams*, CP26/69, Building Research Station.

Woods, W.D., Barnett, N.E. and Sagesser, R. (1974) Holography – a new tool for soil dynamics. *J. Geotech. Eng. Div. (ASCE)*, **100**, 1231.

Wormwell, F. (1969) Research and advisory work on corrosion prevention at the National Physical Laboratory. *J. Inst. Mun. Eng.*, **96**, 137.

Woźniak, Cz. (1970) *Siakowe dźwigary powierzchniowe*, Państwowe Wydawnictwo Naukowe, Warszawa.

Wright, H.J. (1962) Cathodic protection. *Proc. Inst. Civil Eng.*, **21**, 811; (1963) **24**, 288.

Wright, S.G. (ed.) (1983) *Geotechnical Practice in Offshore Engineering*, ASCE.

Wright, S.J. and Reese, L.C. (1979) Design of large diameter bored piles. *Ground Eng.*, **12** (8), 17.

Wright, W. (1952) Beams on elastic foundations – solution by relaxation methods. *The Struct. Eng.*, **30**, 169.

Wroth, C.P. (1972) General theories of earth pressures and deformations. *Proc. 5th Europe. Conf. Soil Mech. Found.*, **2**, 33.

Wu, T.H. (1974) Uncertainty, safety, and decision in soil engineering. *J. Geotech. Eng. Div. (ASCE)*, **100**, 329.

Wu, T.H. and Wong, K. (1981) Probabilistic soil exploration: case history. *J. Geotech. Eng. Div. (ASCE)*, **107**, 1693.

Wyllie, D.C. (1980) Toppling rock slope failures: examples of analysis and stabilization. *Rock Mech.*, **13**, 89.

Yen, B.C. and Scanlon, B. (1975) Sanitary landfill settlement rates. *J. Geotech. Eng. Div. (ASCE)*, **101**, 475.

Yitzhaki, D. (1966) Punching strength of reinforced concrete slabs. *J. ACI*, **63**, 527.

Zanbak, C. (1983) Design charts for rock slopes susceptible to toppling. *J. Geotech. Eng.*, **109**, 1039.

Zaretsky, Yu K. and Tsytovich, N.A. (1965) Consideration of heterogeneity and non-linear deformation of the base in the design of rigid foundations. *Proc. 6th Int. Conf. Soil Mech. Found. Eng.*, **3**, 222.

Zeevaert, L. (1959) Reduction of point bearing capacity of piles because of negative friction. *Proc. 1st Pan. Am. Conf. Soil Mech.*, **3**, 1145.

Zemochkin, B.N. and Sinitsyn, A.P. (1947) *Practical Method for Calculation of Beams and Plates on Elastic Foundations* (in Russian), Stroiizdat, Moscow.

Zienkiewicz, O.C. (1971) *The Finite Element Method in Engineering Science*, McGraw-Hill, London.

Zienkiewicz, O.C. and Holister, G.S. (1965) *Stress Analysis*, Wiley, London.

Zienkiewicz, O.C. and Stagg, K.G. (1965) *The In-situ Testing of Rock Deformability*. Research Report 2, CERA.

Zwamborn, J.A. (1981) Umfolozi road bridge hydraulic model investigation. *J. Hyd. Div. (ASCE)*, **107**, 1317.

INDEX

Note: Pages on which figures and tables appear are indicated by *italic numbers*. Footnotes are indicated by a suffix 'n'.